The Biodiversity of African Plants

Cover: Begonias from the rain forests of West and Central Africa (all drawings x 0.4).

Front, from left to right: *Begonia quadrialata* Warb. subsp. *nimbaensis* Sosef, *Begonia adpressa* Sosef, *Begonia atroglandulosa* Sosef subsp. *tshelaensis* Sosef, *Begonia scutifolia* Hook.f., *Begonia prismatocarpa* W.J. Hooker subsp. *prismatocarpa*, *Begonia scutulum* Hook.f.

Side: *Begonia pseudoviola* Gilg.

Back, from left to right: *Begonia pulcherrima* Sosef, *Begonia minuta* Sosef, *Begonia potamophila* Gilg., *Begonia atroglandulosa* Sosef subsp. *atroglandulosa*, *Begonia gentilii* De Wild., *Begonia quadrialata* Warb. subsp. *quadrialata* var. *quadrialata*.

Drawings by W. Wessel-Brand; from Sosef 1994a, see page 611.

Cover design by X.M. van der Burgt.

The Biodiversity of African Plants

Proceedings XIVth AETFAT Congress

22 - 27 August 1994, Wageningen, The Netherlands

Edited by

L.J.G. van der Maesen
X.M. van der Burgt
J.M. van Medenbach de Rooy

Department of Plant Taxonomy - Herbarium Vadense
Wageningen Agricultural University
The Netherlands

KLUWER ACADEMIC PUBLISHERS
DORDRECHT / BOSTON / LONDON

A C.I.P. Catalogue record for this book is available from the Library of Congress

ISBN 0-7923-4095-7

Published by Kluwer Academic Publishers,
P.O. Box 17, 3300 AA Dordrecht, The Netherlands.

Kluwer Academic Publishers incorporates
the publishing programmes of
D. Reidel, Martinus Nijhoff, Dr W. Junk and MTP Press.

Sold and distributed in the U.S.A. and Canada
by Kluwer Academic Publishers,
101 Philip Drive, Norwell, MA 02061, U.S.A.

In all other countries, sold and distributed
by Kluwer Academic Publishers Group,
P.O. Box 322, 3300 AH Dordrecht, The Netherlands.

Colophon

Department of Plant Taxonomy - Herbarium Vadense
Wageningen Agricultural University
P.O. Box 8010, 6700 ED Wageningen
The Netherlands
Tel.: +31 317 483160 Fax.: +31 317 484917
E-Mail: office@secr.pt.wau.nl

Printed on acid-free paper

Printed in the Netherlands

Contents

Biodiversity in African lowland rain forests

Biodiversity in African savannas

Ecology and conservation

Generic delimitations in flowering plants

Foreword

We have the pleasure to present to you the proceedings of the XIVth congress of the "*Association pour l'Etude Taxonomique de la Flore d'Afrique Tropicale*" (AETFAT). The XIVth AETFAT congress was held from 22 to 27 August 1994, at the Wageningen International Conference Centre in Wageningen, The Netherlands.

The congress was attended by 215 participants, 104 of which came from 28 African countries. During the congress 46 papers and more than 100 posters were presented, the texts of most of which are presented in these Proceedings.

The expenses of the congress were covered by grants from the Wageningen Agricultural University, the Royal Netherlands Academy of Sciences, and others; see page XVI for a complete list. Funds for the participation of 55 members from Africa came from a number of sponsors, in particular from the Directorate General for International Cooperation, the Technical Centre for Agricultural and Rural Co-operation, the Tropenbos Foundation, and Shell Nederland. We express our sincerest thanks to all contributors.

The convenors who organized the symposia, and edited the papers afterwards, are gratefully acknowledged. The editors also acknowledge the valuable help of J.J. Bos and A.J.M. Leeuwenberg, both with preparations in the Organizing Committee, as well as with final editing. The printing was undertaken by Kluwer Academic Publishers, in collaboration with A.C. Plaizier.

The publication took longer than anticipated because the amount of work involved by far overruled our expectations, and because we had many other things to do. Of all lectures abstracts are given in both English and French; 35 of the latter were benevolently translated by P. Bamps from Meise, Belgium. The posters are only provided with bilingual abstracts if these were available from the author. Some of the English abstracts were drafted by the editors. Punctuation follows either English or French usage, depending on the language of the paper. In many cases the tables have been recomposed to improve comprehension and reduce space. Some of the figures, if not essential for the understanding of the text, are not included in order to reduce space. The authors of botanical names have not been verified or added in all cases. In order to save time the authors have not been sent proofs. The lectures and posters of the symposium "Phytosociology in tropical Africa" are not included in these proceedings.

After a successful Congress we wish to thank all those who participated in the organization thereof, the visitors of the congress, and those who prepared the Proceedings. We hope that our common goal - the furtherance of African botany in its widest sense and the taxonomy of African plants in particular - will be served and that we will meet again in the future.

Wageningen, March 1996
The Editors

Photo: M. Wijnbergh

This volume is dedicated to
emeritus Prof. Dr. H.C.D. de Wit (1909-..),
who founded and stimulated
the study of African plants in Wageningen.

Sponsors

The AETFAT Secretariat acknowledges with sincere gratitude the financial and material assistance provided by the following institutions and companies. Without their generous help the realization of the XIVth Congress would not have been possible. In particular the support for travel and subsistence of 55 African participants is much appreciated.

ABN-AMRO Bank, Wageningen

AMEV Nederland N.V., Wageningen

Anti-Apartheidsfonds, Wageningen

Backhuys Universal Book Services, Leiden

Commission of the European Communities, Directorate-General for Development, Training and Cultural Cooperation, Brussels, Belgium

Directorate-General for International Cooperation, Ministry of Foreign Affairs, the Hague

IBN-DLO, Institute for Forestry and Nature Research, Wageningen

Kluwer Academic Publishers, Dordrecht

Kniphorst International Booksellers, Wageningen

Le Jardin Botanique National de Belgique, Meise, Belgium

RABO Bank, Utrecht

Rotary Club, Ede

Royal Netherlands Academy of Sciences, Amsterdam

Shell Nederland B.V., Rotterdam

Technical Centre for Agricultural and Rural Co-operation, Ede

The Tropenbos Foundation, Wageningen

Wageningen Agricultural University, Wageningen

The assistance of the Wageningen International Congress Centre in Wageningen, and of the Congress Bureau of the Wageningen Agricultural University, is greatly appreciated.

Monographs and databases

22 August 1994

Convenors: I.M.M. Hedberg, O. Hedberg

Monographs, databases and biodiversity conservation

O. HEDBERG
Department of Systematic Botany, Uppsala University, Villavägen 6, S-752 36 Uppsala, Sweden.

Abstract
The primary tasks of systematic botany are to explore, describe and classify the flora of the world, thus providing an efficient information storage and retrieval system for plants, plant products, and plant conservation. Work on regional Floras has been prioritized, but from both scientific and practical points of view it would be better to concentrate on monographic work. Inventories followed by monographic studies are urgently needed also to facilitate rational activities for biodiversity conservation. While the herbaria of the world are per se important databases, the use and networking of computers enormously facilitate stock-taking of the global plant resources for all purposes.

Résumé
La systématique botanique a comme tâches primordiales l'exploration, la description et la classification de la flore du monde, ce qui implique un stockage de l'information et un système de consultation efficaces pour les plantes, les produits végétaux et la conservation des plantes. On a donné la priorité à l'étude des flores régionales, mais tant d'un point de vue scientifique que pratique, il serait plus avisé de se concentrer sur des études monographiques. Des inventaires suivis d'études monographiques sont également d'une urgente nécessité pour promouvoir une action rationnelle dans le domaine de la conservation de la biodiversité. Cependant que les herbiers du monde représentent par essence même d'importantes banques de données, l'utilisation des ordinateurs et leur interconnexion facilitent énormément le stockage des données concernant l'ensemble des ressources végétales, quels qu'en soient les usages.

Key words
Biodiversity, Conservation, Databases, Herbaria, Monographs, Taxonomists.

Introduction

Since the time of Linnaeus the primary task of systematic botany has been to explore, describe and classify the flora of the world. The resulting system of classification provides the framework required for efficient information storage and retrieval concerning plants and plant products. It is consequently of crucial importance not only for systematic and evolutionary botany but also for applied botany in areas such as plant breeding, agriculture, forestry, horticulture, range management, watershed management, pharmacology, ecology and conservation (compare e.g. I. Hedberg 1988). As is well known to every taxonomist working on tropical material those primary tasks are far from completed. It has been estimated that about 50,000 vascular plant species are still undescribed. While up to a decade or two ago this caused relatively little concern, the rapidly increasing depletion of biological diversity discussed below now calls for drastically accelerated fulfilment of our task.

Types of current research

Most contributions to systematic botany today can be sorted into three categories:

L.J.G. van der Maesen et al. (eds.), The Biodiversity of African Plants, 3-9.

detailed evolutionary studies, treatments for local or regional floras, and monographic work. As discussed by O. Hedberg (1988) the prospects for and interest in these three areas vary considerably between different botanists and botany departments - e.g. due to variation in possibilities for funding.

For gaining scientific merits at university departments the first category has often been most popular, not least because by using the newest methods available, the botanist may feel near the "cutting edge of science". But the trust in new methods has often been grossly exaggerated (compare O. Hedberg 1995).

Easiest to finance has earlier been the second category - floras are needed for every country.

The third category, monographic work, has evidently been most difficult to recruit, largely because of the large amounts of work, time and money required, making it awkward in academic competition. And yet there seems to be widespread agreement that the most economic and scientifically satisfactory approach would be to start with world monographs of all genera and families, from which pertinent data for each country can be extracted (compare e.g. Gillett 1958, O. Hedberg 1988, Stace 1980 & 1989, Stuessy 1993).

Biodiversity: importance and depletion

While interest in protecting plant species threatened by extinction has long been evident in many northern countries, it was only after intense inventories and revisions had been started in the tropics that botanists realized the scope and speed of species elimination that is going on. While earlier the species were usually considered one by one or in small groups (compare e.g. Lucas & Synge 1978) the increasing awareness of the mutual interdependence of different species has in recent years focused the interest on ecosystems, not least on rain forests and, to a lesser degree, on grasslands, and on the importance of these ecosystems for man. The conversion of tropical moist forests was summarized by Myers (1980) and examples of the effects of deforestation in the Third World are given in Sutlive et al. (1980 & 1981).

Examples of threatened plant species in the Americas were given by Prance and Elias (1977), and a global survey for all kinds of organisms was published by the World Conservation Monitoring Centre (Groombridge 1992). The rate of extinction was estimated by Wilson (1994) in the following way: "Reduction in the area of tropical rain forest at the current rate can be expected to extinguish or doom to extinction about half a percent of the species (about 27,000) in the forest each year…"

As regards Africa a survey of the remaining natural vegetation, conservation measures taken, and further needs for conservation of vegetation in each of the African countries South of the Sahara was given already in Hedberg & Hedberg (1968).

The concept *Biodiversity*, now very much in vogue, seems to have been little used before 1980. Its break-through was effected at the National Forum on Biodiversity in Washington in 1986, one of the results of which was an impressive volume called Biodiversity (Wilson 1988). The new term has since then often been used both for ecosystems and species - as well as for genetic variation within species.

The biodiversity of the tropics is unquestionably of great global significance, as exemplified by the report on "Underexploited Tropical Plants with Promising

Economic Value" (Anon. 1975). And, as noted by Brill (1979): "We are entering an age in which genetic wealth, especially in tropical areas such as rain forest, until now a relatively inaccessible trust fund, is becoming a currency with high immediate value." An important fact often overlooked in scientific papers is that the tropical forests are even more important to the 40% of the world's population living in the tropics. Their almost boundless usefulness was summarized e.g. by Oldfield (1980: 278). Increasing awareness of the progressive deterioration of the flora and vegetation of the world has triggered a number of conferences, reports and books, like "The Sinking Arc" (Myers 1979), the "World Conservation Strategy" (compare Allen 1980), 'The Conservation of Ecosystems and Species" (Jones 1987), the "Global Biodiversity Strategy" (1992a, b), "Biodiversity - Implications for global food security" (Swaminathan & Jana 1992), but insufficient attention has been paid to the basic scientific work required for efficient conservation.

Conservation

Conservation of diversity in plants can in principle be done in three different ways: *ex-situ* conservation in gene banks, *ex-situ* conservation in botanic gardens or "field gene banks", and *in-situ* conservation in national parks and reserves. Each of these three methods has its advantages and disadvantages.

Ex-situ conservation of frozen seeds as practised in conventional gene banks can hold large amounts of samples in a small space, but it requires considerable inputs of labour and energy, it can only store part of the genetic variability in the parent population, and it does not conserve the coevolutionary environment. From the start the gene banks stored only material of crop plants but in recent years the collecting has also been extended to a number of potentially useful and little studied wild plant species, e.g. by the SADCC Gene Bank (Anon. 1989) and the Kew Seed Bank (Jodrell Newsletter 1991).

A vital prerequisite for meaningful conservation of plant genetic resources in gene banks is adequate naming and full documentation of each sample. This has, however, sometimes been neglected in the past. Thus in the largest gene bank of the world, at Fort Collins in USA, "lack of description makes up to 95% of the collections practically useless for the users and hampers collecting efforts" (Perlas & Rifkin 1986). Some suggestions for improvement of gene bank collecting were recently published by van Hintum (1994).

Ex-situ conservation in field gene banks (botanic gardens) can be applied to plants with recalcitrant seeds but requires large resources of land and labour, and suffers the same limitations as frozen seed storage.

In-situ conservation in national parks and reserves is on the one hand the least expensive method, and maintains both intrapopulation variation and coevolutionary contacts. It therefore provides the best possibilities for long-term conservation of genetic resources in wild plants. On the other hand it implies restrictions in land use for the areas reserved, which has sometimes brought hardships on the resident human population.

Each of the three conservation methods also requires its own agents. *Ex-situ* conservation of seeds must of course be done in proper gene banks. Field gene banks for *ex-situ* conservation should in most cases be attached to national botanic gardens or arboreta. And *in-situ* conservation in national parks and other reserves should as a rule be initiated and supervised by national herbaria, and must always pay due attention to the rights and needs of the local people.

Prerequisites for biodiversity conservation

For conservation purposes attention has up to now been mainly concentrated on collecting and storing material and data for individual species, whereas the scientific infrastructure required for an integrated survey of our plant resources has been largely neglected.

In order to effect efficient conservation of botanical diversity we must first know what we have, which calls for stock-taking, description and classification of the flora of the world. Large parts, perhaps a fifth of its vascular plant species, are still unknown to science (compare e.g. Prance 1984).

Inventories

As mentioned above most of the diversity occurs in the tropics, where competent botanists at present are few and the infrastructure required insufficient or missing (compare e.g. Anon. 1980, House of Lords 1991). Inventorying therefore requires collaboration between botanists from the North, who have far better access to the infrastructure needed, and botanists from the South, who not only live in the actual environment but also command local knowledge about the actual or potential use of the plants concerned. Equally urgent is to drastically increase the number of trained taxonomic botanists employed in tropical countries. Each country should have at least one national herbarium with adequate localities and staff, provided with at least one four-wheel-drive vehicle, with funds for field work, and a computer for documentation of the national flora. Without such basic resources our attempts at conservation cannot be successful.

Description and classification

Parallel to the inventory phase must come a stage of description and classification. This may be done to some extent in the South, but the amount of herbarium material and literature required for it makes it easier to perform in the larger herbaria in the North. Thus intense collaboration is also needed for this phase.

The overwhelming amount of work needed seems, however, up to recently to have prevented serious efforts to present a workable program for this phase. "Systematic agenda 2000", presented earlier this year by a consortium of North American systematists, obviously attempts to rectify the situation. But apart from being too big - the objective is to cover all kinds of organisms in 25 years at a cost of 3 billion dollars per year - it is only a vision, nothing near to a program which could be discussed and modified.

In my opinion it would be impossible to have one programme covering the global diversity in that way. It is absolutely necessary to use a number of independent task forces capable of taking care of different shares of the biodiversity. For higher plants this function might be filled by a *Linnaeus Foundation* of the sort I proposed in 1991 to the Swedish Academy of Sciences. The purpose of this foundation should be:

* To put up a central office to see about coordination of tasks, distribution of grants and supervision of results, and publishing of the monographs and revisions produced.
* To provide reasonable salaries for competent taxonomic botanists doing monographic work on the world's flora, and provide the travel grants required for such work and for the collecting expeditions which may be needed.
* To organize and finance training of taxonomic botanists, particularly from tropical countries.
* To solicit and, when necessary, help to fund creation or strengthening of at least one reasonably staffed and equipped national herbarium in each tropical third world country, and support botanists in those countries.

Further details and rough outlines may be obtained from the author.

One of the most important tasks of the Linnaeus Foundation should be to support monographic studies, since monographs attempt to cover all species known of the group concerned in their entire distribution, and to achieve a stable nomenclature. By starting with monographic work and excerpting the relevant data required for each country and region one can avoid much unnecessary duplication of work and attain a reasonably definite nomenclature, to the benefit of biodiversity stock-taking. Because of the amount of work and resources required it is often difficult, not to say impossible, to mobilize national grants for such monographic work. It should therefore be financed and coordinated by an independent international body.

Monographic work heavily depends on databases provided by herbaria, especially the larger European ones with comprehensive tropical collections. Utilization of herbarium material is greatly facilitated when data from the labels are stored in computers in the way recently started for incoming collections in many herbaria. However, the bulk of the old material has not been treated in this way and is, though equally important, not so comfortably available. This may be compared to literature search by computers: as a rule nothing published before 1965 is available in this way. It must also be remembered that even those herbaria in which computerization has not yet been introduced, must be looked upon as extremely important databases.

One of the greatest difficulties in this connection is that, because of competition with newer branches of biology, taxonomic research has in recent decades lagged behind in funding. The support for important biological collections has declined, and the number of students embarking on taxonomic studies has diminished (House of Lords 1991).

Awaiting improvement in financing and infrastructure it would seem worthwhile to follow the suggestion by Inga Hedberg at the Fourth East African Wildlife Symposium in Arusha in 1978: "Start small, think big, act now!"; in other words encourage as much as possible inventories, description and classification of new material, aiming at monographic work, and networking of computerized documentation.

For the time being we have to do with the local or regional revisions available and use their classification and nomenclature to document all available information on the taxonomy, ecology and distribution of the vascular plants of each country and region. To establish which species are threatened on a regional or global scale one must then pool the information from the countries where it occurs through a database network of the type described by Green (1994). A Conservation Data Centre Network of this kind has already been established in the Americas (Global Biodiversity Strategy 1992a: 154). For Africa a similar network is urgently needed. In its preparation it is important to concentrate efforts of inventorying and conservation to those menaced ecosystems which are particularly rich in biodiversity, such as, e.g., western and southern Cameroon with adjacent parts of Nigeria and Gabon, and the forests of the Eastern Arc mountains in East Africa (compare Anon. 1980, Hedberg & Hedberg 1968, Iversen 1988 & 1991). The final step is to bring home the relevant information to the authorities concerned, nationally and internationally, as briefly summarized in Global Biodiversity Strategy (1992b), and solicit the public and financial support required for conservation. The lists of species and ecosystems in need of conservation and the conservation measures applied must then be continually updated according to new discoveries made.

References

Allen, R.A., 1980. How to save the world. Strategy for world conservation. London.

Anon. 1975. Underexploited Tropical Plants with Promising Economic Value. National Academy of Sciences. Washington, D.C.

Anon. 1980. Research Priorities in Tropical Biology. National Academy of Sciences. Washington, D.C.

Anon. 1989. A Regional Plant Genetic Resources Centre SADCC. Plan of Operation 1989-1992. Stockholm, SIDA.

Brill, W.J., 1979. Nitrogen Fixation: Basic to Applied. American Scientist 67: 458-465.

Gillett, J.B., 1958. Ways and means of promoting the production of monographs of families and genera. Mem. Soc. Brot. 13: 107-111.

Global Biodiversity Strategy. 1992a. Guidelines for Action to Save, Study and Use Earths Biotic Wealth Sustainably and Equitably. WRI, IUCN, and UNEP in collaboration with FAO and UNESCO.

Global Biodiversity Strategy. 1992b. A Policy-makers' Guide. WRI, IUCN, and UNEP in collaboration with FAO and UNESCO. Baltimore, Maryland.

Green, D.G., 1994. Databasing diversity - a distributed, public-domain approach. Taxon 43: 51-62.

Groombridge, B. (ed.), 1992. Global Biodiversity. Status of the Earths Living Resource. A Report compiled by the World Conservation Monitoring Centre. London, Glasgow, New York, Tokyo, Melbourne, Madras.

Hedberg, I. (ed.), 1988. Systematic Botany - a key science in tropical research and documentation. Symb. Bot. Upsal. 28 (3).

Hedberg, I. & O. Hedberg, 1968. Conservation of Vegetation in Africa South of the Sahara. Acta Phytog. Suec. 54.

Hedberg, O., 1988. Classical taxonomy versus modern systematic methods in African botany. Monogr. Syst. Bot. Missouri Bot. Gard. 25: 567-573.

Hedberg, O., 1995. Cladistics in taxonomic botany - master or servant? Taxon 44: 3-11.

Hintum, T.J.L. van, 1994. Drowning in the Genepool. Managing genetic diversity in genebank collections. Alnarp.

House of Lords, 1991. Select Committee on Science and Technology. Systematic Biology Research. Vol. 11. Report.

Iversen, S.T., 1988. Conservation of the Usambara mountains. Botanical values at the species and phytogeographical level. In: Hedberg, I. (ed.), Systematic Botany - a key science for tropical research and documentation. Symb. Bot. Upsal. 28 (3): 218-243.

Iversen, S.T., 1991. The Usambara Mountains, NE Tanzania: Phytogeography of the vascular plant flora. Symb. Bot. Upsal. 29 (3): 1-234.

Jodrell Newsletter 9, October 1991. Kew.

Jones, G.E., 1987. The Conservation of Ecosystems and Species. London, New York, Sydney.

Lucas, G. & H. Synge, 1978. The IUCN Plant Red Data Book. Morges, Switzerland.

Myers, N., 1979. The Sinking Ark. A new look at the problem of disappearing species. Oxford.

Myers, N., 1980. Conversion of Tropical Moist Forests. National Academy of Sciences. Washington, D.C.

Oldfield, M.L., 1980. Tropical deforestation and genetic resource conservation. In: Sutlive, V.H., N. Altshuler & M.D. Zamora (eds.). Blowing in the wind: Deforestation and long-range implications. Studies in Third World Societies, 14: 277-345.

Perlas, N. & J. Rifkin, 1986. Default of the Gene Banks. Garden May/June 1986, Agricultural Research Service, USDA.

Prance, G.T., 1984. Completing the Inventory. In: Heywood, V.H. & D.M. Moore (eds.), Current Concepts in Plant Taxonomy. Systematics Association Special Volume No. 25: 365-396.

Prance, G.T. & T.S. Elias, 1977. Extinction Is Forever. Threatened and Endangered Species of Plants in the Americas and Their Significance in Ecosystems Today and in the Future. New York.

Stace, C.A., 1980. Plant Taxonomy and Biosystematics. Baltimore.

Stace, C.A., 1989. Plant Taxonomy and Biosystematics. Second Edition. London, Melbourne, Auckland.

Stuessy, T.F., 1993. The role of creative monography in the biodiversity crisis. Taxon 42: 313-321.

Sutlive, V.H., N. Altshuler, & M.D. Zamora, 1980. Blowing In The Wind: Deforestation and Long-range Implications. Studies in Third World Societies, 14.

Sutlive, V.H., N. Altshuler, & M.D. Zamora, 1981. Where Have All The Flowers Gone? Deforestation in the Third World. Studies in Third World Societies No. 13. Williamsburg, Virginia.

Swaminathan, M.S. & S. Jana, 1992. Biodiversity. Implications for global food security. Madras.

Systematics Agenda 2000. Charting the Biosphere. A Global Initiative to discover, describe and classify the worlds species. New York 1994. American Museum of Natural History and New York Botanical Garden.

Wilson, E.O. (ed.), 1988. Biodiversity. Washington, D.C.

Wilson, E.O., 1994. In: Systematics Agenda 2000, p. 2.

Conspectus of the vascular plants of Madagascar: a taxonomic and conservation electronic database

G.E. SCHATZ [1], P.P. LOWRY II [1], M. LESCOT [2], A.E. WOLF [2],
S. ANDRIAMBOLOLONERA [1], V. RAHARIMALALA [1] & J. RAHARIMAMPIONONA [1]
[1] Missouri Botanical Garden, P.O. Box 299, St. Louis, MO 63166-0299, U.S.A.
[2] Muséum National d'Histoire Naturelle, 16 rue Buffon, Paris 75005, France.

Abstract
The Conspectus of the Vascular Plants of Madagascar Project is an international multi-institutional effort involving the Missouri Botanical Garden (MO), the Muséum National d'Histoire Naturelle (P), the Parc Botanique et Zoologique de Tsimbazaza (TAN), and the Centre National de la Recherche Appliquée au Développement Rural (TEF), whose goal is to make widely accessible information on all vascular plant names attributed to Madagascar. In addition to names and bibliography, the database will hold a maximum number of exsiccatae to voucher the occurrence of taxa within the protected areas system, and to permit automated mapping and analyses of distribution patterns. By maintaining the database in an updatable electronic form, and by incorporating GIS and image technology, we hope the Conspectus will both facilitate new and revised treatments for the *Flore de Madagascar*, as well as serve the needs of conservation NGO's and the Malagasy people.

Résumé
Le projet de "Conspectus des plantes vasculaires de Madagascar" résulte d'une collaboration internationale et pluriinstitutionnelle impliquant le Missouri Botanical Garden (MO), le Muséum National d'Histoire Naturelle (P), le Parc Botanique et Zoologique de Tsimbazaza (TAN) et le Centre National de la Recherche Appliquée au Développement Rural (TEF). Leur but est de permettre un très large accès à l'information concernant tous les noms de plantes vasculaires connues de Madagascar. En plus des noms et de la bibliographie, la banque de données prendra en compte un maximum d'échantillons d'herbier, attestant la présence des taxons à l'intérieur d'un système d'aires protégées et permettant leur cartographie automatisée et l'analyse de leurs modes de distribution. En actualisant la banque de données par gestion électronique et en y incorporant les systèmes d'information géographique et la technologie de représentation picturale, on peut espérer que ce Conspectus facilitera la tâche de ceux qui entreprennent soit un nouveau traitement, soit une révision d'un ancien traitement de la *Flore de Madagascar*, et qu'il jouera son rôle dans la conservation tant auprès des organismes officiels de Madagascar qu'auprès des organisations non gouvernementales.

Key words
Conservation, Conspectus, Database, Information technology, Madagascar, Taxonomy.

Abbreviations
AID - U.S. Agency for International Development
ESRI - Environmental Systems Research Institute, Inc.
GIS - Geographical Information Systems
GPS - Global Positioning System
NGO - Non-Governmental Organization
NSF - U.S. National Science Foundation

L.J.G. van der Maesen et al. (eds.), The Biodiversity of African Plants, 10-17.

Introduction

During this final decade of the twentieth century, we are witness to a staggering revolution in global communications and information technology. Barely ten years after the first IBM desktop computers, we find ourselves linked in cyberspace, sending e-mail instantaneously half way around the globe in a single keystroke, accessing the Gray Card Index through the Internet, searching for plant names in Index Kewensis on CD-ROM, and viewing images of Australian plants, even listening to Australian bird calls, through the World Wide Web. But at this same point in time, we are also witness to - indeed, agents of - the greatest extinction spasm since the end of the Cretaceous. The rest of the species with whom we share the planet, and whom we now refer to somewhat impersonally as "biodiversity", face an increasingly bleak and finite future. For all of us who are concerned with conserving the very plant resources necessary to sustain life, but who are simultaneously being swept along into the coming new millennium's "Age of Information", the question becomes: What is the connection between the biodiversity crisis and the explosion in information technology? How can we most effectively utilize these new technologies to achieve our conservation goals? The Conspectus of the Vascular Plants of Madagascar Project, an international multi-institutional project involving the Missouri Botanical Garden (MO), the Muséum National d'Histoire Naturelle (P), the Parc Botanique et Zoologique de Tsimbazaza (TAN), and FO.FI.FA. (the Centre National de la Recherche Appliquée au Développement Rural) (TEF) is attempting to answer these questions by assimilating, and making readily accessible, information on all vascular plant names described from or attributed to Madagascar.

Status of the Flora of Madagascar

How many species of vascular plants are there in Madagascar? The truth is that we still have only crude estimates of the total number: 8,500 spp.? (White 1983); 10,000 spp.? (Humbert 1959); or 12,000 spp.? (Dejardin et al. 1973). In contrast to the floras of North and South Africa, and like the tropical African flora (Morat & Lowry II 1993), knowledge of the Malagasy flora must still be considered as only fragmentary. As of 1974, Keraudren-Aymonin (1976) estimated that 5,900 species in 153 families had been published in the *Flore de Madagascar*, the first fascicle of which appeared in 1936.

Since 1974, 23 families [including the first revised family treatment (Iridaceae: Goldblatt 1991), which added 8 species for the family in Madagascar] have been published, accounting for an additional 330 species. All or most of 40 families remain to be treated, of which 14 contain over 100 species each. Among these large, intractable families, four (Euphorbiaceae, Leguminosae, Gramineae, and Rubiaceae) rank among the largest in the Flora, comprising 500-600 species each, and thus collectively at least 20% of the total Flora. As Jacobs (1969) has remarked, treatments of large families are the exception in our business, and might be tackled most effectively by a team of workers, as is currently the case with Malagasy Leguminosae [D. Du Puy (K), J.N. Labat (P), P. Phillipson (GRA), R. Rabevohitra (TEF), J.F. Villiers (P)] and Euphorbiaceae [G. McPherson (MO) and A. Radcliffe-Smith (K)]. At the very minimum, the 40 families yet to receive comprehensive treatments in the *Flore de Madagascar* will yield 3,000 additional species.

But what is the status of the 176 families already treated in full or in part? Polhill (1990) has suggested that the useful lifespan of any flora is only 40 years, which has prompted Morat & Lowry II (1993) to declare that virtually all Madagascar family treatments prior to 1960 will need to be revised. As one example, compare the treatments of several genera of Flacourtiaceae (*Homalium*, *Ludia*, and *Scolopia*) initially published in the *Flore de Madagascar* in 1946 by Perrier de la Bâthie, and revised 26 years later by Sleumer (1972a, 1972b, 1973). Perrier recognized 34 species of *Homalium* based upon 203 collections; Sleumer placed 6 of those species into synonymy, while describing 9 new species, thus recognizing 37 in all based upon 737 collections. Perrier recognized 6 species of *Ludia* based upon only 20 collections; Sleumer placed one species into synonymy, made two new combinations, and described 16 new species, thus recognizing 23 species based upon 151 collections. Perrier recognized 10 species of *Scolopia* based upon only 17 collections; Sleumer combined and moved two of those species to *Ludia*, placed another into synonymy, and described 7 new species, thus recognizing in total 14 species based upon 73 collections.

The critical element in these revisions is the three to seven fold increase in the number of collections available for study. Dorr et al. (1989) has estimated that 260,000 collections from Madagascar exist in the world's herbaria. However, many of these specimens have never been consulted in the course of the preparation of *Flore de Madagascar* treatments. In particular, woody families treated prior to 1955 mostly failed to benefit from the Service Forestier (SF) series of collections, established by R. Capuron in 1948 and now numbering over 34,000. Indeed, virtually all of the novelties described by Sleumer in Flacourtiaceae were based on SF collections. Moreover, despite the additional collections available to Sleumer by 1972, the majority of species he recognized were insufficiently known, if one arbitrarily sets a lower limit of six collections to represent a species. Of the 37 species of *Homalium*, 7 were known only from the type, and 7 more from 5 or fewer collections; of the 23 species of *Ludia*, 6 were known only from the type, and 9 more from 5 or fewer collections; of the 14 species of *Scolopia*, 3 were known only from the type, and 7 more from 5 or fewer collections. Even when the 20,000 collections made by the Missouri Botanical Garden and associates during the past ten years are factored in, the collecting effort is still far from complete in Madagascar.

It is not at all unreasonable to assume that the total number of vascular plant species in Madagascar will exceed the estimate of 10,000 by Humbert, and probably approach the figure of 12,000 proposed by Dejardin et al. However, with 1/3 of the Flora remaining to be treated, and another 1/3 badly in need of revision, we simply lack the necessary data to go beyond mere speculation.

What is the Madagascar Conspectus?

Given the incomplete and outdated status of much of the Flora, and the widely dispersed nature of primary literature pertaining to the as yet untreated species, a conspectus, or a "concise" summary lacking descriptions, can potentially address the immediate need for botanical input into current conservation initiatives, as well as facilitate future "monographic" *Flore de Madagascar* treatments. The Madagascar Conspectus has several antecedents in the African-Madagascar region. From 1895-1898, the *Conspectus*

Florae Africae of Durand & Schinz began to enumerate African plants with their full references, secondary references, and a general statement of distribution, but without specimen citation or keys, and hence, was essentially a nomenclator similar to the current *Enumeration des Plantes à Fleurs d'Afrique Tropicale* by Lebrun & Stork (1991). Since 1937, the *Conspectus Florae Angolensis* has included full references, synonymy as it pertains to Angola, habit and ecology, distribution, multiple specimen citation by district, and keys for the plants of Angola. With the exception of keys, the Madagascar Conspectus follows both the format and the spirit of *Conspectus Florae Angolensis*, whose founder, L.W. Carisso (1937) stated in his introduction:

> "The immediate object of the *Conspectus Florae Angolensis* is to make a critical list of the vascular flora of Angola and to provide a working basis for investigators, students, and workers who need botanical information for the solution of economical and agricultural problems."

The Madagascar Conspectus database is a subset of the larger TROPICOS database housed at the Missouri Botanical Garden, from which the recent *Catalogue of the Flowering Plants and Gymnosperms of Peru* (Brako & Zarucchi 1993) was derived, and upon which other ongoing database projects covering the plants of China, Ecuador, and Mesoamerica are based. The TROPICOS database consists primarily of three interlinked components: data associated with Names, including Types; Bibliographic data; and Exsiccatae, or specimen, data. From the onset of the Madagascar Conspectus project, the goal has been to maximize the accessibility of the data *in Madagascar* as it becomes available. Therefore, rather than publishing a snapshot of the database at some point in time for hard-copy distribution, the data resides in Madagascar in its electronic form on a PC version of TROPICOS, and thus can be continually updated. Data can also be entered in Madagascar (primarily specimen data at TAN and TEF) and then incorporated into the main system in St. Louis. Paris is linked directly to TROPICOS through an Internet connection. Maintaining the database in its electronic form in Madagascar, and therefore also maintaining an unlimited capacity to add both new and historical data, in effect constitutes a repatriation of data back to its source. The Madagascar Conspectus database currently holds over 25,000 Names records, 2,000 Bibliographic references, and nearly 50,000 geo-coded Exsiccatae records.

Exsiccatae

A "living" database especially provides the opportunity to incorporate the maximum amount of specimen data, from which both historical and contemporary distributions, and therefore critical conservation status, must ultimately be derived. All contemporary collections generated through Missouri Botanical Gardens' programs in Madagascar, anticipated to number 10,000 in 1994, are immediately captured, as they have been for the past ten years. With regard to historical collections, several approaches are currently being followed. In Madagascar, all specimens with sufficient label data at both TAN and TEF will be entered into the database, which will, in effect, computerize those herbaria. In Paris, specimen data entry is focused upon Types and vouchers that document the occurrence (or absence, if no such vouchers exist) of taxa in each of the provinces, and more importantly, within the Protected Areas system (Nicoll & Langrand 1989), thus providing a direct indication of conservation status.

As with historical collections from any region, their usefulness as vouchers for distribution is often hindered by the absence of geographical coordinates. Thus, one of the most important aspects of the Madagascar Conspectus is the post facto assigning of latitude/longitude/elevation to as many historical collections as possible. In Madagascar, work on the two important series of collections, the Service Forestier (SF) and Réserves Naturelles (RN) series is progressing. The localities of over 16,000 SF collections have been assigned coordinates, encompassing a total of nearly 2,000 distinct localities. All collections in the RN series originate from the 12 original Réserves Integrales Naturelles (each of which has been assigned a range of coordinates) and are thus of especially important conservation significance. In Paris, all of the master localities from Humbert's ten missions to Madagascar, representing 23,600 collections, have been assigned coordinates, as have the collections of Leandri and Morat. Along with locality information from contemporary collections (coordinates for which are now derived directly from GPS units), the post facto assigning of coordinates to historical exsiccatae will provide the basis for a *Gazetteer to the Botanical Collecting Localities* in Madagascar, to be issued in hard-copy within the next several years. Similar gazetteers for the *Flore d'Afrique Centrale* by Bamps (1982), the *Flora of Tropical East Africa* by Polhill (1988), and the *Flore du Cameroun* by Letouzey (1968) serve as indispensable reference works for both taxonomic and conservation users. Post facto assigning of coordinates to historical collections will also enable us to utilize them in GIS analyses of both past and present distribution, as well as allow us to model potential distribution in relation to remaining habitat parameters. Collaboration between the Missouri Botanical Garden and ESRI, the makers of ARCVIEW2 and other GIS software, will result in automated mapping and distribution analyses during the next phase of the Madagascar Conspectus Project.

Proteaceae: a recently published family

The Proteaceae of Madagascar, recently published in the *Flore de Madagascar* (Bosser & Rabevohitra 1991), serve to illustrate the capacity of the Madagascar Conspectus database to incorporate new data, including differences in taxonomic opinion. Bosser & Rabevohitra recognize three genera native to Madagascar: *Faurea*, a diverse genus throughout Africa, with a single endemic species in Madagascar, for which they recognize two varieties; endemic *Dilobeia* with two species, one of which they describe therein; and endemic monotypic *Malagasia*. As Bosser & Rabevohitra were preparing their treatment of Proteaceae, S.K. Marner (1989) published a new species of *Faurea* (*F. coriacea* S.K. Marner) from Madagascar based on one (Humbert 23669) of the four original syntypes of *F. forficuliflora* Baker var. *elliptica* Humbert (1959). Although *F. coriacea* is validly published, it is unfortunate that Marner failed to cite *F. forficuliflora* var. *elliptica* in synonymy. He does cite two (Humbert 22515 and 23442) of the other three original syntypes of var. *elliptica* as paratypes, but the fourth (Humbert 23780) is curiously omitted. Presumably Bosser & Rabevohitra were unaware of Marner's paper, for, in recognizing *F. forficuliflora* var. *elliptica*, they fail to mention *F. coriacea*, remark that two of the original syntypes (Humbert 22515 and 23669 - the type of *F. coriacea*) could not be refound (undoubtedly, they were on loan to Marner!), and lectotypify var. *elliptica* with Humbert 23780, the one syntype Marner omitted! At this point, it becomes a question of whose taxonomic opinion to accept. Valid names for the *Faurea* on the Marojejy Massif exist at both the ranks of species and variety.

With regard to *Malagasia*, the database records the original basionym description by Capuron (1963) as *Macadamia alticola*, the description of the genus *Malagasia* and resulting new combination by Johnson & Briggs (1975), and acceptance by Bosser & Rabevohitra (1991) in the *Flore de Madagascar*. The Exsiccatae portion of the database also reflects that, since the publication of the *Flore*, the species has been recollected at the type locality (Réserve Spéciale d'Ambohitantely) (Schatz et al. 3586), the first collection since 1967. Moreover, specimen data associated with the Exsiccatae record for Schatz et al. 3586, indicate that leaves were placed in silica gel for DNA preservation, and that a photo accompanies the collection. A sample set of 200 vouchered photos of Malagasy plants, including *Malagasia alticola*, have recently been digitized, and will soon be accessible through the World Wide Web. Both *Malagasia* and *Dilobeia* are pivotal taxa for ongoing phylogenetic and biogeographic studies of Proteaceae, in that they represent two separate relict disjunctions from Australian/Indo-Malesian Proteaceae, which probably date to the late Cretaceous/early Tertiary (Johnson & Briggs 1975).

Training and institutional capacity building

As we contemplate the progress of flora projects in the Africa-Madagascar region and the large gaps that still exist in our knowledge (Morat & Lowry II 1993, Polhill 1990), the issue of resources, particularly human resources, should be considered (Burley et al. 1990). Regardless of whether "concise, monographic, or semi-monographic" formats are adopted (Morin et al. 1989, Workshop on the Future of Flora Malesiana Plenary Discussion 1990), or whether interim solutions such as an electronic conspectus are opted for, floristic projects today must include more than taxonomic treatments. Long-term conservation of plant resources, as well as, in all likelihood, the flora projects themselves, will be achieved only through in-country training and capacity-building. The Madagascar Conspectus Project has been fortunate thus far to have received U.S. governmental support from the National Science Foundation (NSF) and Agency for International Development (AID). With NSF support, the three Malagasy members of Equipe Conspectus have received extensive training in all facets of botanical data management, as well as English language, and were able to spend a two-month "stage" at MO during 1993. AID has provided funds for complete facility renovation and upgrading at TAN and TEF, including herbarium cases (fabricated in Madagascar), freezers and microwaves for insect pest control, air conditioners, microscopes, computers, etc. Increased capacity for Malagasy scientists to conduct research *in Madagascar* is an essential component in any biodiversity conservation strategy.

Conclusion

We believe that the key element in biodiversity conservation is *not* the demonstration of direct human benefits, nor even of ecosystem functioning, but rather consists instead in building a more personal relationship with, and understanding of, the species around us. Familiarity will engender stewardship; whereas, lack of familiarity all too often leads to neglect, and even loss. Herein lies the great potential for information technologies to play a role in removing biodiversity from the realm of generic abstraction, from the exclusive scholarly domain of arcane Latin descriptions that are the very basis of our knowledge of that biodiversity, and thus from the private purview of our professional society. In so doing, these technologies bring about a new synthesis never before possible. Our goal for the next phase of the Madagascar Conspectus is to increase the accessibility of information on the diverse, endemic, and highly threatened plant resources of Madagascar by:

1. incorporating GIS technology for mapping and distribution analyses;
2. adding digitized images of Malagasy plants and their habitats;
3. scanning original illustrations from type descriptions, and possibly the types themselves.

All will be linked to the underlying taxonomic database, and displayed through a graphical user interface, both on PC in Madagascar, and on the Internet. Increased accessibility will serve the dual purpose of facilitating new and revised family treatments for the *Flore de Madagascar*, as well as providing critical conservation data to Malagasy government agencies and numerous conservation NGO's, all of whom are committed to preserving Madagascar's natural heritage for future generations.

References

Bamps, P., 1982. Flore d'Afrique Centrale (Zaire-Rwanda-Burundi): Répertoire des lieux de récolte. Jardin botanique national de Belgique, Meise.

Bosser, J. & R. Rabevohitra, 1991. Proteaceae. Fl. Madagasc. 57: 47-69.

Brako, L. & J.L. Zarucchi, 1993. Catalogue of the Flowering Plants and Gymnosperms of Peru. Monogr. Syst. Bot. Missouri Bot. Gard. 45: 1-1286.

Burley, J.S., P.F. Stevens & E.A. Wood, 1990. Resources for inventorying plants in the southeast Asian tropics: plans for the future. Fl. Malesiana Bull., Special Volume, 1: 37- 40.

Capuron, R., 1963. Contributions à l'étude de la flore de Madagascar. XI. Présence à Madagascar d'un représentant du genre *Macadamia* F.v.M. (Protéacées). Adansonia, n.s., 3: 370-373.

Carrisso, L.W., 1937-1982. Conspectus Florae Angolensis: elaborado pelo instituto botanico de Coimbra com a colaboracao do Museu britanico (British museum), Lisboa.

Dejardin, J., J.-L. Guillaumet & G. Mangenot, 1973. Contribution à la connaissance de l'élément non endémique de la flore malgache (végétaux vasculaires). Candollea 28: 325-391.

Dorr, L.J., L.C. Barnett & A. Rakotozafy, 1989. Madagascar. In: Campbell, D.G. & H.D. Hammond (eds.), Floristic Inventory of Tropical Countries: 236-250, New York Botanical Garden.

Durand, Th. & H. Schinz, 1895-1898. Conspectus Florae Africae. Jardin Botanique de l'Etat, Bruxelles.

Goldblatt, P., 1991. Iridaceae (2me édition). Fl. Madagasc. 45: 3-45.

Humbert, H., 1959. Origines présumées et affinités de la flore de Madagascar. Mém. Inst. Sci. Madagascar, sér. B, Biol. Vég. 9: 149-187.

Jacobs, M., 1969. Large families - not alone! Taxon 18: 252-262.

Johnson, L.A.S. & B.S. Briggs, 1975. On the Proteaceae - the evolution and classification of a southern family. Bot. J. Linn. Soc. 70: 83-182.

Keraudren-Aymonin, M., 1976. Progrès accomplis dans la connaissance de la Flore de Madagascar et des Comores. Boissiera 24: 635-639.

Lebrun, J.P. & A.L. Stork, 1991. Enumération des Plantes à Fleurs d'Afrique Tropicale. Conservatoire et Jardin botaniques de la Ville de Genève.

Letouzey, R., 1968. Les Botanistes au Cameroun. IV. Index Géographique. Flore du Cameroun 7: 66-87.

Marner, S.K., 1989. New and noteworthy species of *Faurea* (Proteaceae) from the evergreen forests of Malawi and elsewhere. Bull. Jard. Bot. Belg. 59: 427-431.

Morat, Ph. & P.P. Lowry II, 1993. Do we really know as much as we think about the flora of the Africa-Madagascar region? XV Intl. Bot. Congr. Abstracts, p. 33.

Morin, N.R., R.D. Whetstone, D. Wilken & K. Tomlinson (eds.), 1989. Floristics for the 21st Century. Monogr. Syst. Bot. Missouri Bot. Gard. 28: 1-163.

Nicoll, M.E. & O. Langrand, 1989. Madagascar: Revue de la conservation et des aires protégées. World Wide Fund for Nature, Gland, Switzerland.

Perrier de la Bâthie, H., 1946. Flacourtiacées. Fl. Madagasc. 140: 1-131.

Polhill, D., 1988. Flora of Tropical East Africa - Index of Collecting Localities. Royal Botanic Gardens, Kew.

Polhill, R.M., 1990. Production rates of major regional Floras. Fl. Malesiana Bull., Special Volume, 1: 1-20.

Sleumer, H., 1972a. Révision du genre *Ludia* Comm. ex Juss. (Flacourtiacées). Adansonia, n.s., 12: 79-102.

Sleumer, H., 1972b. A taxonomic revision of the genus *Scolopia* Schreb. (Flacourtiaceae). Blumea 20: 25-64.

Sleumer, H., 1973. Révision du genre *Homalium* Jacq. (Flacourtiacées) en Afrique (y compris Madagascar et les Mascareignes). Bull. Jard. Bot. Belg. 43: 239-328.

White, F., 1983. The vegetation of Africa. A descriptive memoire to accompany the UNESCO/-AETFAT vegetation map of Africa. Natural Resources Research Series, No. 20. The UNESCO Press, Paris.

Computerisation of the East African Herbarium: development of a regional plant information service

T.R. PEARCE [1], D. FILER [2] & E. VANDEN BERGHE [1]
[1] National Museums of Kenya, P.O. Box 40658, Nairobi, Kenya.
[2] Oxford Forestry Institute, South Parks Road, Oxford, United Kingdom.

Abstract
The increased emphasis on the provision of botanical information to a wider audience of biodiversity researchers has led to increased demands on the already limited staff capacity at the East African Herbarium. Information is held in a variety of formats in a number of locations within the Herbarium and the key to accessing this information in an efficient and effective manner is the gradual compilation of this information into a computerised data management system. A pilot study has been undertaken to address the issues that will need to be reconciled in order to pursue this objective in the herbarium. A strategy for the continued accession of a variety of data types has been worked up and the opportunities that this process presents for the future of the Herbarium in both a regional and international context are discussed.

Résumé
La demande sans cesse croissante d'informations d'ordre botanique de la part d'un éventail de plus en plus large de chercheurs intéressés à la biodiversité a vite saturé la capacité d'y répondre de la part d'un effectif déjà très réduit de l'East African Herbarium. Ces informations sont dispersées sous diverses formes, à différents endroits de l'herbier, et la solution pour accéder de manière efficace et rapide à ces informations est de les grouper dans un système de gestion informatisée. Une étude-pilote a été entreprise pour analyser les problèmes qui se posent en fonction de cet objectif. Une stratégie pour l'accroissement incessant de données variées a été développée et les perspectives que ce procédé laisse entrevoir pour l'avenir de l'herbier, tant du point de vue national qu'international, sont discutées.

Key words
BRAHMS, Databases, East African Herbarium.

Abbreviations
BRAHMS - Botanical Research and Herbarium Management System
EA - The East African Herbarium
GEF - Global Environment Facilities
GIS - Geographical Information Systems
NMK - National Museums of Kenya
ODA - Overseas Development Administration
OFI - Oxford Forestry Institute

Introduction

The modern herbarium

Traditionally designed as depositories for reference material used for taxonomic investigations of floristic regions, herbaria around the world have recently come under increasing pressure to supply a broader range of plant information to a more varied

L.J.G. van der Maesen et al. (eds.), The Biodiversity of African Plants, 18-31.

cadre of users. Whilst the primary role of the herbarium remains rooted in plant systematics, the range of additional services now expected to be provided by the herbarium has sharply increased; distributional data, details of species phenology, indigenous uses and local names, trends in species collections etc. are often needed in making decisions regarding the most appropriate conservation actions or best land use strategy. Most modern herbaria are quite capable of collating such diverse sets of information given the time and resources to sift through the files and specimens secreted away in herbarium cupboards. Having this information "on-line" would however, be a significant improvement to the efficiency of information retrieval. Accessing this stored information, being able to analyze it and thence add to it in a planned, strategic way is the key to developing a healthy botanical institution capable of fully supporting the needs of the biodiversity research community. Once the information is stored electronically an endless range of products and analyses can be generated ranging from routine lists, distribution maps and labels to complex taxonomic treatments or regional phenological summaries.

Currently with around 750,000 accessions and a predicted steady flow of between 2,000 and 5,000 additional specimens per year for the foreseeable future, the East African Herbarium is the second largest herbarium on the African continent after the National Herbarium in Pretoria. It houses the most comprehensive collection of specimens from the East African region and is a vital source of information for local and regional taxonomic treatments and flora production. The common role of the herbarium in the past has been to provide correct identifications to other Government departments and visiting researchers in addition to the loaning of material to other plant taxonomy research institutions. Whilst these continue to be an important service that we provide to a variety of users, we are continually developing our own in-house research programmes. Our four taxonomic staff have made significant contributions to the publication of the Flora of Tropical East Africa as well as publishing a number of papers in the botanical literature. Our library is probably the best equipped in the region where, in addition to the books and journals, we maintain an enormous amount of information on index cards regarding plant uses and indigenous names.

The East African Plant Information Service

The staff of the East African Herbarium are committed to the provision of quality information on the plants of the East African region; to supply this information quickly, and in a way that is immediately useful to the end user. We are aware of the whole variety of the information that is stored on our specimens, in our library, on our thousands of index cards and the many useful reports of field trips to various parts of the continent currently stored away in filing cabinets. We recognise the need to be able to integrate this information with new data that is continually being generated from our own research and from the findings of visiting botanists to the region. We intend to continually update this information to ensure that we remain a centre of excellence in the region that people can rely on for the provision of quality data. In essence we are aiming to provide a comprehensive East African Plant Information Service, where access to the whole range of information we hold is literally at the press of a button.

With the ever increasing demands on our limited staff, the ability to provide the quality and quantity of data required has frequently been limited. This has led to the decision, taken some time back, to investigate the long term transfer of all the specimens and related data to an electronic database. Our approach has been one of caution through the initiation of a pilot project to establish skills that were not yet present in the Herbarium, to determine the actual costs of realising our objectives and to eventually plan a strategy that would be flexible enough to cope with the stochastic nature of the necessary overseas funding. A pilot project was initiated in 1993 to firstly test the operation of an existing database system in the herbarium using "real" herbarium data and, in so doing, understand the issues that would need to be addressed in order to develop this information service.

This paper therefore discusses the approach that is being used to introduce computerised databases to the East African Herbarium and raises the issues that have been addressed in the process. It is not intended to detail the technical side of databases nor to promote the various systems we operate. Rather, it outlines the difficulties encountered and the implications of computerisation on the more traditional methods of routine herbarium management by discussing the ways that these issues have been addressed by the East African Herbarium.

Project implementation

Any database system used to assist in the management of herbarium information needs to perform a variety of tasks and achieve a range of output options. BRAHMS (Botanical Research And Herbarium Management System) is an example of an information system designed for the storage and processing of botanical data. It has been developed as a tool to assist with botanical research, both monographic and floristic, and to provide curatorial support in the herbarium. Whilst it can be regarded as primarily a specimen-based system, the collation and processing of data relating to taxa can be fully integrated with data derived from specimens; the focus on herbarium specimens and specimen-related data reflects the central role of the specimens in plant systematics/taxonomy. Any botanist or herbarium worker who has had to type the same piece of data, such as a collector, species or locality name more than once, stands to gain from using a botanical database. Botanists, who have to prepare lengthy, fastidiously formatted specimen citation lists, using data that has already been punched for labels or other forms of research output, are in a position to make considerable savings in time and effort in these and related tasks. Table 1 summarises the data management services currently provided by BRAHMS.

BRAHMS 3.2 is an IBM compatible, single user application, operating efficiently on 386/486 machines and can be run across a network as a single user system. A full, multi-user version is now being developed. Projects can have single or multiple data entry points. Data can be freely imported/exported to and from the system in a range of formats. Improved links to spreadsheet packages (e.g. Excel and Lotus) have been forged which is ideal for generating tabulated output, that can be imported into WP packages (e.g. Word, WordPerfect and Wordstar). Links with MapInfo have been developed specifically tailored to the needs of the National Museums of Kenya (where this particular GIS is operational) in order to facilitate the production of good quality printed maps. BRAHMS is a fully relational system based on the Microsoft FoxPro database management system, and has been developed at the Oxford Forestry Institute (OFI) under a Forestry Research Programme grant from the ODA.

Table 1. Summary of data management services provided by BRAHMS.

Activity	Example products and services
Collections management	General specimen cataloguing; logging accessions; monitoring incoming and outgoing loans, gifts and exchanges; label/det slip production; addresses and mailing lists; visitor enquiry services.
Revisions/monographic research	General taxonomic descriptions; nomenclatural details; citation lists; distributional data (maps or text); ecological information; publications.
Floristic/geographic research	Annotated species check lists; geographical--based output; Flora and florula preparation.
Species descriptions	A compilation of information held at the species level or derived by scanning specimen records.
Historical studies	Collector itineraries; collection patterns and frequencies; anthropological studies.
General research	Planning field visits; ecological research; phenological studies; monitoring associated collections (e.g. photos, seed collections, and pollinators).
Additional benefits	Data standards established; minimize retyping of data; easier data editing and maintenance; data exchange encouraged.

The pilot phase of the project aimed not to assess the suitability of BRAHMS (which had been done by a number of other institutions around the world) but more at understanding the implications of initiating a computerisation programme in the Herbarium. A suitably large set of specimen data was required to assess these implications and initially the genus *Acacia* was chosen not just because of the convenient size of the collection (around 5,500 specimens in EA) but also because the distributional information held in the collection was required as part of an ongoing project at the OFI on the genus in Eastern and Southern Africa. The collection had recently been curated by Mr. Chris Fagg of OFI which left only a few specimens without good determinations.

Following on from this initial data entry phase, the EA Plant Conservation and Propagation Unit contributed to the database by entering all 5,000 orchid specimens to the system thus giving them access to the wealth of information of conservation importance to their programme. The whole of the family Apocynaceae have also been added by a university student preparing the family for publication in FTEA as part of her

PhD research. The genera *Fuirena* and *Ceropegia* have been added by EA taxonomic scientists and a number of site-based check-listing initiatives have also used the facilities of BRAHMS to keep track of the specimens from particular areas of interest. A total of some 15,000 specimens has been logged and within the next year at current rates we hope to have this figure increased to about 50,000 specimen records.

Initially one BSc graduate was hired for a period of six months and was supported by an overseas volunteer for two of the six months for which funding had been secured. Currently we have three BSc data loggers hired on six month contracts. They have the use of two notebook computers and a number of desktop computers. The master data set is housed in a powerful desktop computer where administration and management of the data set is carried out. The project has been supported throughout by the ODA-funded Plant Conservation and Propagation Unit and the Oxford Forestry Institute.

The various problems that we have thus far encountered has enabled us to develop a future strategy for the compilation of botanical information into electronic data files. This strategy, whilst providing a good framework within which to proceed, will inevitably have to be adapted to the variable nature of outside funding which this type of venture will require. Rather than sticking to a rigid model where data entry begins at the monocots and proceeds to the dicots etc., it will be the research priorities of the institution and its collaborating projects which will dictate the programme of data entry.

Development of a data entry and management strategy for the East African Herbarium

Many issues were raised during the pilot phase and the strategy that we have used to address these issues were tailored to the resources available to us in the Herbarium. Our strategy therefore may not be entirely suitable for other herbaria to be adopted in total but it does provide practical guidelines for other institutions intent on embarking on computerisation.

Hardware requirements

Most herbaria are short of space, maintaining flexibility in terms of moving machines from one area of the collection to another is therefore an important consideration. Generally speaking, it is wise to buy the most powerful machines one can afford. Whilst most *x*-base applications will comfortably run on 286 machines, if the Windows environment is chosen for your application then minimum 386 and preferably above will be needed. We chose two 486 notebook computers as much data that had previously been entered (species lists etc.) in Windows based software such as Excel and Access was to be used. It is important, however, to be able to run simple data entry modules which can utilise "outdated" machines which other users have discarded in favour of their more powerful offspring. In BRAHMS for example, we frequently use a simple flat file entry system called RDE (Rapid Data Entry) which can be run on very low power 286 machines. Consequently we have happily inherited a number of 286 laptop machines which are unable to run the more modern Windows based products effectively.

Software options

As mentioned above there will be two major options open to data entry projects. The two should not be seen as mutually exclusive, the fact that the National Museums of Kenya (NMK) operates both in different parts of the institution shows that it is not so much the system that is important *per se* but the ability of data exchange and the way that the data is split up into the packets or fields of information. BRAHMS is currently available as a FoxPro application running in DOS. Much of the entry of species lists, editing of gazetteer and collector series files has however, been carried out in Windows based applications such as EXCEL or ACCESS. Exchange of data files between such programmes nowadays appears to present few problems although a reasonable level of computer literacy will be required at least somewhere in the institution. Data transfer is the key to the debate on software options, as long as the database application you choose is entirely capable of transferring its data to another system and *vice-versa*, then "any modern system will do", as a rule of thumb most applications will be able to import x-base files and the "save as" option should always include an opportunity to create x-base file formats.

Data considerations

A whole variety of data types will need to be collected which will reflect the variety of data stored in the herbarium (see Table 2). Data (not just specimen data) come in two major sorts; those which are particular to the specimen (Primary and Secondary) and those constant for a number of specimens. Data unique to the specimen are often quite limited; they will nearly always consist of the Collector Series and number, field notes, current determination, information such as plant uses and local names and often some additional comments about the specimen. All other information to be found on the specimen is likely to be found on many other specimens in the herbarium and as such should not have to be entered more than once on your database. It is the development of these framework files or standard lists of data items that can radically improve not only the efficiency of data entry but also the quality by reducing the amount of errors as a result of bad typing. Being able to access, for example, collector series names or species names or locality names is of tremendous importance and should form a major part of the initial work of developing an institutional database.

Specimen data

Primary specimen data entry involves physically removing the specimen from the herbarium cupboard and transferring the written information found on labels, det. slips, notes etc. to the data file. This is the basic form of data entry and is the one completed for the 15,000 specimens mentioned above.

Copying text from written notes is a task which can be carried out by people without a strong botanical training. A typical herbarium specimen may contain more than just the "primary data". We call data, that can be taken from the plant material itself through observation, Secondary or "Value Added" data. Noting if the specimen was flowering at the time of collection (important for planning seed collection trips) or taking a variety of measurements for morphological studies is something which requires a certain amount of understanding of the material. Collection of secondary data should therefore be tailored to the requirements of the particular data entry project.

Table 2. Data categories dealt with during computerisation of botanical information.

Data category	Specific examples of data stored
Taxonomic names	Ranging from family to sub-specific ranks with authorities, citations, synonyms (categorized and sorted), quick reference codes, type details.
Associated species information	Vernacular names with language/tribe, descriptive text (in user-defined text memos), numeric and logical descriptors (in user-defined variables), publications.
Geographic names	Ranging from continent to mappable locality, user-defined check list areas, national or regional gazetteer entries.
Collector series names	Single or joint series with provision to store quick reference codes, alternative printing formats and memo comment.
Index Herbariorum and other names/addresses	Index Herbariorum entries including code, full address and telecom details; names/addresses of botanists and other relevant contacts; address group records.
General background data files	Dictionaries of type, habit, habitat, wild or cultivation status categories; abbreviations dictionary.
Herbarium specimens	Storage for all commonly used specimen descriptors including collector name/number, accession number, full locality details (including altitude and map grids), collection date, general field notes, additional collectors, determination and alternative determinations (including names/dates), full duplicate details (including herbarium, determinations and type status), habit category, habitat category, wild/cult status, flower/fruit status coding, associated collections (spirit material, wood samples, photos, pollinators, *etc.*), common names, host plant name, and a range of user-defined attributes.
Accession records	Accession number, details of sending herbarium or individual, dates and related specimen list.
Mapping data	Latitude/longitude points (and alternative grid formats) at specimen and gazetteer level; coastline and political boundary data (low resoution) for all countries.
System configuration details	A wide range of user-established database control features including access passwords, label and citation list design files, alternative month name formats, screen colour settings, printer driver details, *etc*.

Table 3. Explanation of core specimen data fields established by the National Museums of Kenya, Centre for Biodiversity and additional fields included in the current herbarium project.

Core data field	Explanation
Taxonomic name	Split into two data fields
Author	Using standard abbreviated formats (Brummitt & Powell 1992)
Collector series	An agreed format is decided upon on the basis of the most commonly used format on herbarium labels
Date	Date of collection
Locality	Emphasis placed on the second tier administrative division in East Africa (the District) as these are usually small enough to correspond with a particular set of environmental parameters.
Grid reference	Two sets are collected, one for the specimen (this is often given or, for specific projects, can be deduced), one for the locality name
Precision	A useful addition to the locality reference especially when analyzing species with relatively small areas of distribution.
Habitat	Regionally, we are trying to establish an agreed set of broad habitat categories.

The NMK has established a Centre for Biodiversity which *inter alia* ensures that data gathering and data output from the institution is coordinated. Current funding through the GEF (Global Environment Facility) has allowed the Museum to establish a Museum-wide databasing initiative which aims to facilitate the computerisation of biological information throughout the Museum. This central data gathering and processing system runs on an application written in the Windows based program Access. Whilst most other departments in the Museum were effectively starting from scratch in terms of data management, the Herbarium had been gathering data using BRAHMS (and other database applications) for some time prior to the implementation of this initiative. Ensuring that the two database systems were capable of freely exchanging information has been the major challenge facing both the Museum-wide project and the current Herbarium project.

Establishing an agreed set of "core fields" for specimen data, that is the bare minimum set of data that the institution would want to be able to offer about a particular record, is an important first decision to make (see Table 3). It eventually dictates the speed at which data are added to the system; less core fields means more records logged on per hour but less information and hence less querying power of the database. In addition to these Museum-wide core fields we agreed that there was value to be gained in recording at what phenological state the plant was in at the time of collection. This has been of tremendous use to us when we are planning seed collection trips for some of our rare plants. It uses quick codes such as "B" for buds only and "UR" for both unripe and ripe fruit on the plant etc. This takes but a few moments to determine and enter, and it was felt that this secondary data should be included in our core data fields for plant specimens.

Taxonomic framework file

A Regional initiative was started in 1993 between the NMK and the Makerere University Institute of Environment and Natural Resources in Kampala to transfer all accepted plant names appearing in FTEA to computer. This was complemented by a survey of names appearing on EA specimens from families not yet published. Logging the published names has now been completed and has resulted in a reasonably complete list of names of species occurring in our flora area. Whilst this is an easy task to undertake (although quite time consuming) it must be realised that this will only form the basic list on which to build a more valuable data set with references to synonymy and bibliographic citations. This integrates the work of the taxonomist with the management of the herbarium database as it is they who have the vested interest in ensuring that the database reflects the most recent nomenclature for their particular group(s) of interest along with the location of the relevant type specimens and its correct bibliographic citation. Taxonomic files can be easily updated by workers at remote institutions who may have a particular interest in a set of genera or families, by constructing a simple data file structure into which information on citation, type identification etc. can be incorporated. This has an added advantage in fostering collaborative initiatives between herbaria.

Global references to taxonomic nomenclature (see Brummitt 1992, Mabberley 1990) should be used as often as possible for ensuring that the names entering the database correspond to already published names. We were fortunate enough to receive an electronic copy of the List of Families and Genera (Brummitt 1992) and are now able to electronically check new generic names entering the database from specimen data and from taxonomic framework files against this standard publication. We hope soon to obtain Index Kewensis on CD-Rom which will allow us to keep our taxonomic framework file in line with current nomenclature.

The geographical framework file

Most areas have benefitted from some sort of gazetteer work and these are important sources of information when mapping species distributions needs to be carried out. Developing good geographical information as a framework file is an important priority for any biodiversity data management system. We are fortunate in the FTEA region in having a comprehensive index of collection sites compiled from specimens housed at Kew (Polhill 1988). This information was not available in an electronic format and the text needed to be electronically scanned to produce a text file which could then be translated to a database file. Some 25% of the records required manual editing and some were beyond editing and required manual re-entry to the newly created database file.

The structure of any gazetteer can become extremely complicated. The basic necessity however is to ensure that each place name is assigned a grid reference and that the hierarchy (present and historical) of administrative regions is well understood. As with the taxonomic framework file, the data relating to a geographical point can be improved upon at stations remote from the herbarium as resources allow. Easy data checks can be made by the use of computerised mapping packages by comparing point plots with maps of administrative boundaries etc. Establishing the precision of a data entry is also important; an entry referring to "Nairobi" is obviously less valuable than locality entries such as "Nairobi Arboretum". If the grid reference for the former is taken as the centre of the area (which often seems a reasonable thing to do) then the peculiar situation may arise where two grid references are very similar and yet have widely different precision. The solution chosen here is to include an extra field in the database which reflects the precision or size of the geographical unit. This field can be used to set a filter to the desired precision of the georeference. Notes attached to the label may be able to pinpoint the exact population of plants to quite a high degree of accuracy which may not be reflected in the actual locality name. In addition, the precision of the specimen plot can be improved by referring to a map and using the locality qualifying text in the notes to improve the grid reference provided by the gazetteer entry. This, incidentally, highlights the importance of maintaining two sets of fields associated with the geographical data: one for the specimen reference (either given or approximated from locality and field notes), and one for the reference held in the gazetteer file.

Collector series framework file

As one of the key pieces of information enabling specimen to be identified the Collector series and the series number are extremely important pieces of "core data". An understanding of the variety of collectors who have worked in an area and during what period etc. also provides an interesting historical perspective on floristic studies for a particular region.

In the pilot phase of our programme we adopted the concept of the collector "series" as the basic unit of data for this framework file in our data entry system. Whilst information on the individual collectors can be stored separately (and used in other parts of the database such as Dets. & Confs. etc.) it is the collector series which is most frequently referred to in specimen citations, etc. In an ideal world it would be much simpler if each collector series title had its own numerical series as well, so that "Gillett JB" had a numerical sequence from 1 to say 22,000; when this collector then went on a collecting trip with "Polhill R" then there would also be a new numerical sequence 1-10,000 or whatever. This is probably unrealistic to assume contemporary botanists to do and of course it is impossible to alter the reality of how most collections have been done in the past.

Much confusion exists in the area of identifying collector series titles and what they actually mean. Table 4a shows a number of collector series titles generated from actual specimens of *Acacia.* These were then analysed by senior members of the herbarium (and the collectors themselves) and Table 4b clarifies the apparent confusion into an accurate description of the collectors and their series. Correcting simple syntax errors,

Table 4a. A sample of some series titles as they appear on specimens, generated by entering Acacia specimens from the East African Herbarium.

Original series title	Explanation
Luke	Mr. Quentin Luke from the East African Herbarium
Luke Q	As above
Luke WRQ	As above
Robertson A	Mrs Anne Robertson from the East African Herbarium
Robertson SA	As above
Luke & Robertson	Joint collection series #1 during WWF-Coastal Forest Survey
Luke Q & Robertson A	As above
Luke WRQ & Robertson A	As above
Robertson SA & Luke Q	Joint collection series #2 during WWF-Coastal Forest Survey

Table 4b. The same sample corrected and rationalised following confirmation (in this instance) by the collectors themselves.

Corrected series title	Explanation
Luke WRQ	Active collection series from Quentin Luke at the East African herbarium
Robertson SA	Active collection series from Anne Robertson at the East African Herbarium
Luke WRQ & Robertson SA	Collection series #1 (no longer active) from WWF-Coastal Forest Survey 1987-90
Robertson SA & Luke WRQ	Collection series #2 (no longer active) from WWF-Coastal Forest Survey 1987-90

misinterpretations outlined here, supplying additional anecdotal information about the collector series and computer generated summaries (e.g. first/last collect date) can lead to a powerful and more meaningful dataset for the user. It fosters a better understanding of the history of botanical collection in the region and can be important in clarifying the status of particular botanical collections when their origin or authenticity is in doubt. Table 4a is an example of some of the 2,000 or so different collector series titles which were originally generated through the logging of 5,500 herbarium specimens. Table 4b is an example of how we have been able to reduce this list to around 1,200 collector series titles.

We are still improving the value of this data set by incorporating some factual information about the collectors themselves and about the history of their collecting missions by interviewing a number of botanists at the various herbaria in the region; this will complement the work currently being undertaken at Kew. Our aim is not so much to create a comprehensive history of botanical collection in East Africa, as to create a working data set to facilitate specimen data entry and processing for the Herbarium.

Data management and quality control

The balance between quantity and quality of data that an institution holds is one of utmost importance. There exist many datasets with large numbers of records made up of relatively poor quality data. We do not consider that this provides real information which is of immediate value to the end user. Our emphasis has been on providing good quality data that other users of the data can rely upon. Whilst we recognise that no extensive data set can be 100% error free we stick to the premise that the wrong data can sometimes be more dangerous than having no data.

There are a number of ways that we are able to ensure the quality of our dataset is not polluted by bad-quality data from specimen data entry projects or introduction of new taxonomic lists. Early emphasis on the generation of framework data files provides the "look-up" tables which can be accessed during the operation of specimen data entry thus eliminating the need to always enter the taxon name or the Collector series etc. Loading specimen data sets from flat files to the relational system can be carried out "against" these framework files, which ensures that no pollution of the edited files occurs by alerting the Database Manager of new taxon names, collector series or localities etc. All new entries to these files as a result of data entry can be compiled into reports which can be sent back to the data loggers for re-checking.

Security filters can be installed so that password access is required to actually enter or modify data in the database. We frequently operate a quarantine database where a new set of data is loaded from a flat file and the database is used to run checks on the new data. For example if all specimens occurring above a particular altitude (most of Kenya lies below 1,800 m) frequent typing errors in the altitude field are quickly picked up. Similarly with querying on collections made before 1800 and after 1994, the typing errors as well as the very few pre-1800 specimens we have in the Herbarium are easily picked up.

Inevitably errors will slowly creep into the dataset and one of the key tasks of the database Manager is to continually clean the data as it is being accessed. This demands a diligent approach to the use of the dataset, simply keeping a notepad fixed to the bench next to the computer is a simple way of ensuring that there is always a list of editing tasks to be carried out.

Staffing

Data entry has been carried out in the herbarium wings thus minimising the amount of movement of both specimens and people. Data entry into computers is a thankless task, and no problems arise with data quality when it is only done one day a week. The reality however, is that data entry has to be of top quality, for months at a time. It is also true to say that if the data logger perceives a personal advantage in carrying out the work then the quality of data will be greatly improved. Initially we opted for the contractual employment of recently graduated BSc. and MSc. students and set them what we conceived to be easy targets to meet. It was soon realised that the conditions under which data entry operators work is directly proportional to the quality and quantity of data entered. Specimens inevitably take up a large amount of space on the bench and in order to avoid damage, enough space needs to be allocated at the benches. Seating must be comfortable, colour monitors should preferably supplement the LCD screens of notebook computers and a secure cabinet should be provided for each operator who should function as a fully mobile unit within the herbarium.

We have found that concentration spans vary tremendously and what you can do in one day is not likely to be repeated over the whole week. It is sensible to partition the day into a concerted data entry period, say the morning, reserving the afternoon for other tasks like curation or identification. The numbers of specimens that one can enter over a prolonged period never normally exceeds 50 per day which appears rather low. Whilst this could no doubt increase drastically if there was constant supervision our approach has been one of quality data entry rather than quantity. We accept that although data is gathered rather slowly, the quality is such that it becomes acceptable for distribution after only a very brief period of data checking.

Quality of data however is best assured if the person entering the data feels a sense of ownership for that particular data set. More recently we have tried therefore to treat the data entry process as a part of a bigger project on the particular taxonomic group in which the data loggers play an active part. This we feel will lead to a vested interest for the data logger in the ultimate quality of the data that they are entering.

Every database project, no matter how small, requires careful management. At minimum, this involves checking data integrity and regular backing-up of the data files. With the more complex projects, especially those involving several data entry stations, managerial responsibilities increase. In effect, each project needs a database manager. Managerial duties may include:

* Ensuring that software is running correctly. This includes checking hardware configuration, system operation speed, printer driver selection, etc.
* Ensuring system is correctly configured for the particular data entry project and that the framework data files have been correctly initialized.
* Providing in-house system training as needed.
* Coordinating data input activities.
* Explaining backing-up methods and ensuring that all project datasets are backed up correctly.
* Distributing framework files to satellite stations and other data entry projects.
* Checking the databases for integrity.
* General troubleshooting.
* Coordinating projects regionally, as appropriate.

Management of a dataset demands considerable input of effort and if the data and the system are to be fully integrated with the day-to-day working of the institution then a full time, trained data manager must be considered as a priority investment into the staff structure. Data logging can be done with contract employees as part of an "apprentice" type approach, but data management is a staffing commitment that the institution must make early on.

Opportunities for institutional collaborations

Databasing initiatives have many hidden spin-offs to the institution. At the East African Herbarium we now take every opportunity to capture specimen data when queries are sent to us from acknowledged experts around the world. Visiting researchers (if they provide us with enough warning of their arrival) will now be greeted with a complete list of current names in operation at the Herbarium and if their particular group is not so large, a complete dataset of specimen information. This we will be happy to make available to them in return for an input into data checking and editing of the taxonomic lists.

A common scenario is that visiting researchers will first be presented with the current names found on our sheets. This list is then corrected, reducing some names to synonymy and adding newly published names to the list as appropriate. Determination slips can then be produced from the edited data file and these are stuck on to the specimens before the researchers leave the Herbarium. This leaves us with the opportunity to log the specimens with the correct names at the start or re-edit the names in the database if they have already been entered.

Regional collaboration on checking framework files, such as the gazetteer or collector series list, fosters good working relationships with neighbouring herbaria. The National Museums of Kenya has hosted National workshops on databasing biodiversity information and a regional workshop is planned specifically for computerisation of plant data.

Summary points for consideration

The most important premise for embarking on computerisation as a state to which the herbarium wishes to achieve, is that the process must be seen as a long (very long) term investment to which the institution and all its staff must be committed. The database must integrate with the working herbarium right from the start and not as an adjunct which will someday miraculously take over the routine curatorial tasks and free everybody up for more exciting research and field work. Certain far-reaching decisions have to be taken early on to ensure that the process not only has immediate effect in the day to day operation of the herbarium but also to ensure that the database has a future within the institution. The database must work for the staff and must be adaptable enough to cope with the inevitable stochastic approach to funding so characteristic of overseas aid projects.

References

Brummitt, R.K., 1992. Vascular plant families and genera. Royal Botanic Gardens, Kew.

Brummitt, R.K. & C.E. Powell, 1992. Authors of plant names. Royal Botanic Gardens, Kew.

Mabberley, D.J., 1990. The plant book: a portable dictionary of the higher plants. Cambridge University Press, Cambridge.

Polhill, D., 1988. Flora of Tropical East Africa. Index of collection localities. Royal Botanic Gardens, Kew.

Plants of southern Africa database

T.H. ARNOLD & M. KOEKEMOER
National Botanical Institute, Private Bag X101, 0001 Pretoria, South Africa

Abstract
The "Plants of Southern Africa Database" contains a complete list of all taxa that occur in South Africa, Namibia, and Botswana. This database is the start of a computerized flora for Southern Africa and would, at a later stage, also include descriptive data for all taxa.

Key words
Database, List of taxa, Southern Africa.

The "Plants of Southern Africa Database" is based on a book *Plants of southern Africa: names and distribution* (1993), eds T.H. Arnold & B.C. de Wet. This publication contains a complete list of all taxa that occur in South Africa, Namibia and Botswana, and also includes synonyms, literature references and distributions for each taxon. Information for the book is gathered and continuously updated by staff of the National Herbarium, Pretoria.

The database is menu-driven for easy use and includes 347 family names, 3,351 genus names, 24,372 names in current use, 13,882 synonyms, 863 naturalized taxa, 5,964 literature references, tree numbers, provincial and national distribution of taxa, and important general literature.

The database can be interrogated for taxa satisfying specific search criteria either by direct access or via standard reports. It is therefore possible to obtain a list of all the Compositae in the Transvaal, all the Asclepiadaceae in Namibia and Botswana, taxa described by Linnaeus, Leguminosae which have subspecies, naturalized taxa in Swaziland or simply a list of species of *Protea* that are trees.

The database is developed using Data Ease RDBMS and is distributed with a legal run-time version of Data Ease. It can be run on a computer with a 80286 (preferably 80386) processor with 1Mb RAM, VGA, EGA or monochrome monitor, at least 12 Mb disc space and MS-DOS 3.3 or higher.

Data updates and any programme enhancements will be available annually. Staff of the National Herbarium are at present entering data on life form, life cycle, plant height, altitude, habitat and other biotic effects which will be included in the database. This database is the start of a computerized flora for Southern Africa and would, at a later stage, also include descriptive data for all taxa.

Copies of the database are for sale and can be ordered from the Programme Leader: Data Management, National Botanical Institute, Private Bag X101, 0001 Pretoria, South Africa.

L.J.G. van der Maesen et al. (eds.), The Biodiversity of African Plants, 32.

Uses, chromosome number and distribution of *Solanum* species in Uganda

R. BUKENYA-ZIRABA
Botany Department, Makerere University, P.O. Box 7062, Kampala, Uganda.

Abstract
Information on distribution, uses and chromosome numbers are provided for all species, subspecies and cultivar groups in the genus *Solanum* L. (Solanaceae) that occur in Uganda. The distribution is influenced by a number of factors, including rainfall, altitude and type of vegetation, the weedy nature of some species and whether the species are crops or ornamentals.

Key words
Chromosome numbers, Distribution, *Solanum,* Uganda, Useful plants.

Introduction

The genus *Solanum* L. contains some 2,000 species, about 35 of which occur in Uganda. Several species are important as food crops, medicinals, ornamentals or weeds (Heine 1963). The genus is widely distributed throughout the world. It was described by Linnaeus (1753) and has been reclassified innumerable times. A major revision of *Solanum* in Africa, providing the most detailed treatment available today, was done by Bitter (1913, 1917, 1921, 1923). A systematic treatment of African *Solanum* was provided by Jaeger (1985).

The Flora of Tropical East Africa is being produced in parts and Solanaceae has not yet been done. Hence information on species of *Solanum* occurring in Uganda must be sought in other regional Floras, particularly in Flora of West Tropical Africa, ed. 2 (Heine 1963). However, in this Flora not all species of the genus occurring in Uganda are included. Detailed descriptions and a key to all *Solanum* species occurring there was provided by Bukenya (1993).

In this paper information on the distribution of all the Solanums in Uganda, their uses and chromosome numbers, is provided. Such data will be helpful in research on the chemistry, pharmacology, toxicology etc. of *Solanum* species and aid in future collection of germplasm to be used in genetic resources programmes for improvement of cultivated species.

Materials and Methods

Information on distribution of the species is based mostly on herbarium material kept at the Herbarium, Botany Department, Makerere University, Uganda; the Herbarium, Royal Botanic Gardens, Kew, England; the Forest Department Herbarium, Oxford, England; and the Herbarium, Institute of Systematic Botany, Munich, Germany. Fresh specimens were collected by the author from various parts of Uganda, studied and deposited in the Herbarium, Makerere University.

Results

The results of the study are summarized in Table 1 (Uses and chromosome numbers of

L.J.G. van der Maesen et al. (eds.), The Biodiversity of African Plants, 33-37.

Table 1. Uses and chromosome numbers of *Solanum* species occurring in Uganda.

Species	Uses			Chromosome number
	Part plant	Processing	Use	
Solanum aculeastrum Dunal	R L F		T O *	24
S. aculeatissimum Jacq.	Sh		W SR	24
S. aethiopicum Gilo group (cult.)	F	B	V	24
id.	L		T *	
S. aethiopicum Shum group (cult.)	L	B	V	24
S. americanum Mill.	L	B	V	24
S. anguivi Lam.	F	B	V T *	24
S. coagulans Forssk.			TP *	
S. florulentum Bitter	L	B	V	
S. giganteum Jacq.	L F		T *	24
S. grossedentatum A. Rich.	L (F)	B	V	
S. incanum L.	R L F		T *	24
S. kagehense group				24
S. macrocarpon L. group (cult.)	L	B	V	24
id. (semiwild and wild)	R L S		T *	24
S. mammosum L. (cult.)			O	22,24
S. mauritianum Scop. (weed)			O	24
S. melongena L. (cult.)	F	B	V	24,48
S. nakurense C.H. Wright				24
S. nigrum L. ssp. *nigrum*	L	B	V	72
S. renschii Vatke				24
S. sarrachoides Sendtn.				24
S. scabrum Mill. (cult.)	L	B	V	72
S. seaforthianum Andr.			O	
S. tarderemotum Bitter	L	B	V	
S. terminale Forssk. ssp. *terminale*	R		T	24
S. t. ssp. *inconstans* (Wright) Heine	R		T	24
S. t. ssp. *sanaganum* (Bitter) Heine	R		T	24
S. tuberosum L. (cult.)	Tu	B		48
S. usambarense Bitter & Dammer				24
S. villosum Mill.	L	B	V	48
S. welwitschii C.H. Wright				24
S. wendlandii Hook. (cult.)			O	24
S. wrightii Benth.			O	24

Part of plant: F = Fruits; L = Leaves; R = Roots; S = Seeds; Sh = Shoots; Tu = Tubers. Processing: B = Boiled. Use: O = Ornamental; SR = Spiritual Rituals; T = Traditional Medicine; TP = Technical Products; V = Vegetable; W = Witchcraft; * = further information given in Table 2.

Solanum species occurring in Uganda), Table 2 (Uses of some *Solanum* species in traditional medicine and for technical purposes) and Table 3, in which the occurrence of each taxon in the 4 geographical regions in Uganda (according to Flora of Tropical East Africa, cf. Figure 1) is listed.

Table 2. Uses of a number of *Solanum* species in traditional medicine and for technical purposes (cf. Table 1).

Solanum aculeastrum	Roots, leaves and fruits used for treatment of numerous diseases, int. al. abdominal worms, epilepsy, cellulitis, enlarged spleen, menorrhagia, ringworms, respiratory diseases in cattle.
S. aethiopicum Gilo group	Leaves used for treatment of hypertension.
S. anguivi	Infusion from boiled fruits used to cure hypertension.
S. coagulans	For tanning, and coagulating milk.
S. giganteum	Leaves used for treatment of sleeplessness, leaves and fruits for ulcers.
S. incanum	Roots, leaves and fruits used for treatment of numerous diseases, e.g. hookworm, impotence, diarrhoea, astheniamalaise, snake bites, hiccough, dermatosis, menorrhagia, ringworm, sores on eyes of cattle.
S. macrocarpon (wild and semi-wild)	Roots are used for treatment of abdominal worms, brain fever, epilepsy, urticaria; boiled leaves to ease child delivery; seeds for treatment of inguno-scrotal hernia.

Conclusions

Almost all *Solanum* species occurring in Uganda are used by local people (Table 1). Table 3 shows that *S. americanum, S. incanum, S. macrocarpon* and *S. tarderemotum,* (wild growing), *S. anguivi* and the *S. kagehense* group, occur in all four geographical regions. All these taxa, apart from the last mentioned, have weedy characteristics and are adapted to various habitats. The *S. kagehense* group is a large taxon which is adapted to thickets.

Thirteen taxa are fairly well distributed occurring in three regions, while fourteen taxa are relatively rare, occurring in two regions. This is associated with species preference to certain habitats e.g. forest, mountains, drylands and whether they are weedy, crops or ornamentals. Of the rare species occurring only in one region, *S. scabrum* is a minor crop, *S. albicaule* and *S. renschii* are species of arid areas, *S. benderianum* is a montane species; *S. seaforthianum* and *S. wendlandii* are introduced ornamentals and *S. terminale* ssp. *inconstans* occurs in forest.

Regions U4 and U2 have the largest number of taxa. These two are the wettest regions of the country with a bimodal rainfall regime and contain a variety of habitats including forests, woodland, savanna and montane vegetation — hence offering favourable conditions for a large variety of taxa. Since those areas are also easily accessible

Table 3. Occurrence of *Solanum* taxa in the four geographical divisions of Uganda.

Taxa	Occurrence			
	U1	U2	U3	U4
Solanum aculeastrum	-	+	+	+
S. aculeatissimum	+	+	-	+
S. aethiopicum Gilo group	-	+	+	+
S. aethiopicum Shum group	-	+	+	+
S. albicaule Kotschy ex Dammer	+	-	-	-
S. americanum	+	+	+	+
S. anguivi	+	+	+	+
S. benderianum Schimper ex Dammer	-	+	-	-
S. coagulans	+	+	-	-
S. cyaneo-purpureum De Wild.	-	+	-	+
S. florulentum	-	+	-	+
S. giganteum	-	+	+	+
S. grossedentatum	+	+	+	-
S. hastifolium Hochst. ex Dunal	+	-	-	+
S. incanum	+	+	+	+
S. kagehense group	+	+	+	+
S. macrocarpon (cultivars)	-	-	+	+
S. macrocarpon (semi-wild)	-	-	+	+
S. macrocarpon (wild)	+	+	+	+
S. mammosum	-	-	+	+
S. mauritianum	-	+	+	+
S. melongena	-	+	+	+
S. nakurense	-	+	-	+
S. nigrum ssp. *nigrum*	+	-	+	+
S. renschii	+	-	-	-
S. runsoriense C.H. Wright	-	+	+	-
S. sarrachoides	-	+	+	-
S. scabrum	-	+	-	-
S. seaforthianum	-	-	-	+
S. taitense Vatke	+	-	-	+
S. tarderemotum	+	+	+	+
S. terminale ssp. *inconstans*	-	-	-	+
S. terminale ssp. *sanaganum*	-	+	-	+
S. terminale ssp. *terminale*	-	+	+	+
S. tuberosum	-	+	+	+
S. usambarense	-	+	+	-
S. villosum	+	+	+	-
S. welwitschii	-	+	-	+
S. wendlandii	-	-	-	+
S. wrightii	-	+	+	+
Total	15	29	24	30

+: recorded, voucher specimens available in most cases; -: not recorded.
Author citation is given for species not listed in Table 1.

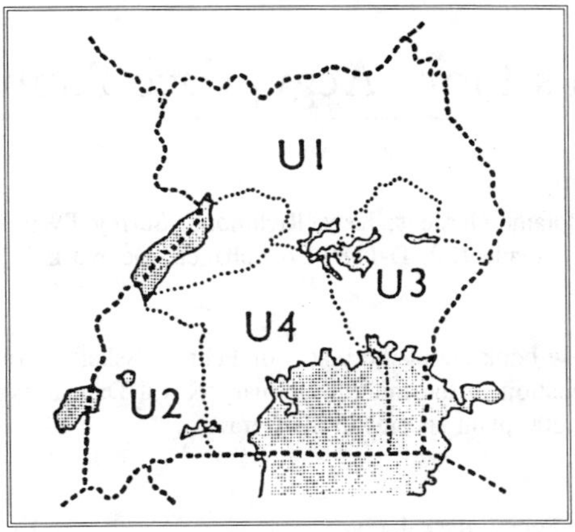

Figure 1. Geographical divisions of Uganda according to the Flora of Tropical East Africa.

they might have been more frequented by plant collectors. U1 has the smallest number of taxa recorded. This region has a monomodal rainfall regime with predominant savanna and arid types of vegetation, hence harbouring mainly taxa adapted to such environments. The region is not very accessible and therefore probably undercollected.

More plant collecting expeditions are required, especially in poorly collected areas, e.g. the northern part of the country, mountainous areas and Lake Victoria islands. Pharmacological studies are needed to authenticate the medicinal value of the species and cytological studies are required especially for those species the chromosome number of which is unknown.

Acknowledgements
I am deeply indebted to Makerere University, UNESCO, DAAD, the Linnean Society of London and the British Council for financial support. I am grateful to Dr. J.F. Carasco for advice and encouragement. Thanks are extended to Dr. J.M. Edmonds for identifying some material belonging to section *Solanum*. I am most grateful to Dr. K. Vollesen and Prof. J. Grau for assistance during herbarium studies at Kew and Munich, respectively.

References

Bitter, G., 1913. Solana Africana 1. Bot. Jahrb. 49: 560-569.
Bitter, G., 1917. Solana Africana 2. Bot. Jahrb. 54: 416-506.
Bitter, G., 1921. Solana Africana 3. Bot. Jahrb. 55: 69-89.
Bitter, G., 1923. Solana Africana 4. Fedde's Repertorium 16: 1-320.
Bukenya, Z.R., 1993. Studies in the taxonomy of genus Solanum in Uganda. Ph.D. Thesis, Makerere University.
Heine, H.H., 1963. Solanaceae. In: Keay, R.W.J. & F.N. Hepper, Flora of West tropical Africa, ed. 2: 325-335. London.
Jaeger, P.M.L., 1985. Systematic studies in the genus Solanum in Africa. Ph.D.Thesis, University of Birmingham.
Linnaeus, C., 1753. Species plantarum, 2 vols. Salvius, Stockholm.

The plants of Pehr Forsskål's Flora Aegyptiaco-Arabica

F.N. HEPPER [1] & I. FRIIS [2]
[1] The Herbarium, Royal Botanic Gardens, Kew, Richmond, Surrey TW9 3AB, United Kingdom.
[2] Botanical Museum, Gothersgade 130, D-1180 Copenhagen, Denmark.

Abstract
Attention is drawn to a new book about the plants of Pehr Forsskål's Flora Aegyptiaco-Arabica. In this new book, his participation in the multidisciplinary Royal Danish expedition is described, and data are given about the 1,200 plant species in his flora.

Key words
Flora Aegyptiaco-Arabica, Pehr Forsskål.

Figure 1. Portrait of Per Forsskål. Idealised engraving published by W. Lagus in "Strödda blad. Nytt och gammalt i historiska och humanistiska ämnen. I. Petter Forsskåls lefverne (Helsingfors, 1877)" from a painting by P. Dahlmann, 1760, now at Salnecke Castle, Uppland, Sweden.

Although the Royal Danish Expedition went through the Mediterranean and the Red Sea reaching Yemen in 1763, hundreds of the plants collected by Forsskål were of tropical African affinity. Most of these were new to science too. Many were described in his "Flora Aegyptiaco-Arabica" published in 1775, but this is a difficult work to use and there has been confusion in some cases as to which type specimen belongs to which description. Some of the descriptions are too short to validate the names, while some of the names are ambiguous. Other names have been considered of doubtful validity due to their apparent non-binary form, and others have been overlooked due to the confusing format of the book. Nevertheless, Forsskål's collections and his posthumous book are extremely important for taxonomists dealing with both Mediterranean and African flora.

Attention is drawn here to a new book by the authors, published by the Royal Botanic Gardens, Kew, in collaboration with the Botanical Museum, Copenhagen, where most of Forsskål's herbarium is preserved. Other specimens are lodged in the

L.J.G. van der Maesen et al. (eds.), The Biodiversity of African Plants, 38-39.

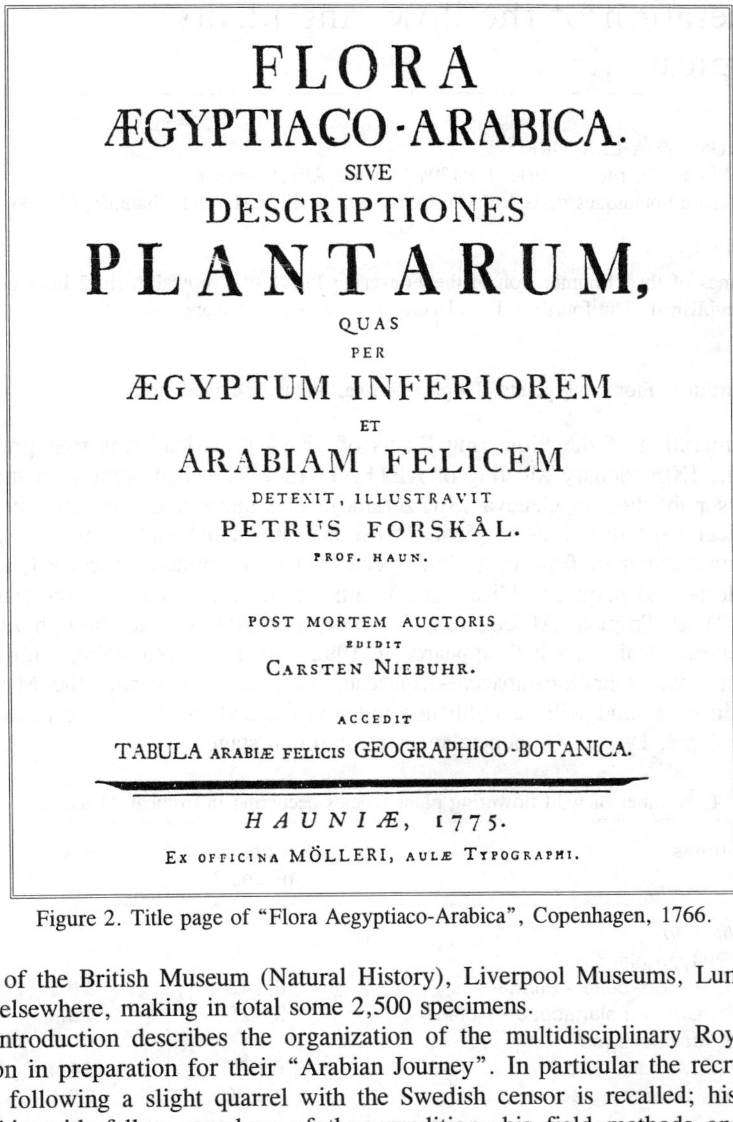

FLORA
ÆGYPTIACO-ARABICA.
SIVE
DESCRIPTIONES
PLANTARUM,
QUAS
PER
ÆGYPTUM INFERIOREM
ET
ARABIAM FELICEM
DETEXIT, ILLUSTRAVIT
PETRUS FORSKÅL.
PROF. HAUN.

POST MORTEM AUCTORIS
EDIDIT
CARSTEN NIEBUHR.

ACCEDIT
TABULA ARABIÆ FELICIS GEOGRAPHICO-BOTANICA.

HAUNIÆ, 1775.
EX OFFICINA MÖLLERI, AULÆ TYPOGRAPHI.

Figure 2. Title page of "Flora Aegyptiaco-Arabica", Copenhagen, 1766.

herbaria of the British Museum (Natural History), Liverpool Museums, Lund University and elsewhere, making in total some 2,500 specimens.

The introduction describes the organization of the multidisciplinary Royal Danish expedition in preparation for their "Arabian Journey". In particular the recruitment of Forsskål following a slight quarrel with the Swedish censor is recalled; his turbulent relationship with fellow members of the expedition, his field methods and his correspondence with Linnaeus, who received some seeds that were grown at Uppsala and described as new species. Notes are provided on the subsequent publication and format of the "Flora Aegyptiaco-Arabica" and the "Icones Rerum Naturalium", and on the nomenclatural interpretation of these works in the light of the decisions made at the Tokyo Botanical congress, 1993, on how to define binary nomenclature.

The Catalogue is preceded by a gazetteer of localities with old and modern names. Each of the 1,200 species is set out with relevant synonymy, vernacular name, locality and herbarium specimens, together with location and IDC microfiche reference.

Enumeration of the flowering plants of tropical Africa

J.-P. LEBRUN [1] & A.L. STORK [2]
[1] CIRAD-EMVT, 10, rue P. Curie, F 94704 Maisons Alfort, France.
[2] Cons. et Jardin botaniques de la Ville de Genève, C.p. 62, CH 1292 Chambésy/Ge, Switzerland.

Abstract
Three volumes of the "Enumeration of the Flowering Plants of Tropical Africa" have been or will shortly be published. The fourth and final volume is at its initial stage.

Key words
Family repartition, Flowering plants, Tropical Africa, Status of enumeration.

The "Enumeration of the Flowering Plants of Tropical Africa" was first presented in 1978, at the IXth Plenary Meeting of AETFAT. In 1991 it had come to maturity, and Vol. I was published in Geneva (Switzerland). The introductory chapter summarizes the botanical exploration of tropical Africa and its actual status. It also gives the approximate number of flowering plant species for each country concerned, as well as a list of floras and plant checklists. The Enumeration itself, following the sequence of "Flora of West Tropical Africa" (ed. 2), comprises Annonaceae through Euphorbiaceae/Pandaceae. Vol. II, which appeared in 1992, covers the remaining families of the Dialypetalae, viz. Chrysobalanaceae-Apiaceae. Vol. III, comprising the Monocotyledonae, is in press and will be published towards the end of 1994. The manuscript of the final volume, IV, the Gamopetalae, is at its initial stage.

Table 1. Number of wild flowering plant species occurring in tropical Africa.

Volumes	Number of species	%
Published		
Dialypetalae:		
I. Annonaceae - Pandaceae	6,145	26.2
II. Chrysobalanaceae - Apiaceae	5,425	23.1
Not yet published		
III. Monocotyledonae	± 5,600	24.1
IV. Gamopetalae	± 6,230	26.6
Provisional total	± 23,400-24,000	

Table 1 summarizes the number of wild angiosperm species of tropical Africa, and the six largest families are listed in Table 2. The Dialypetalae comprise 150 families, 1,314 genera and about 11,570 species. The 20 largest families are listed in Table 3 in decreasing order, together with large genera with more than 50 species. Some large genera belonging to families not listed in Tables 2 and 3 are given in Table 4.

L.J.G. van der Maesen et al. (eds.), The Biodiversity of African Plants, 40-41.

Table 2. The six largest angiosperm (sub-)families in tropical Africa.

Family	Number of species	% of total number of species
1. Papilionoideae	± 2,062	8.8
2. Rubiaceae	± 1,750	7.5
3. Compositae	± 1,674	7.2
4. Gramineae	± 1,482	6.3
5. Euphorbiaceae	± 1,224	5.2
6. Orchidaceae	± 1.155	4.9
Remaining families		60.1

Table 3. List of the number of species in some of the largest angiosperm (sub-)families in Africa that have genera comprising more than about 60 species.

1. Papilionoideae	2,062	5. Euphorbiaceae	1,224	12. Malvaceae	300
Crotalaria	454	*Euphorbia*	374	*Hibiscus*	157
Indigofera	306	*Phyllanthus*	112	13. Melastomataceae	280
Tephrosia	132	*Monadenium*	64	*Dissotis*	86
Millettia	95	6. Orchidaceae	1,155	14. Anacardiaceae	259
Eriosema	89	*Habenaria*	220	*Trichoscypha*	67
Rhynchosia	81	*Polystachya*	168	15. Mimosoideae	239
Aeschynomene	74	*Eulophia*	125	*Acacia*	116
2. Rubiaceae	1,750	7. Cyperaceae	864	16. Sapindaceae	232
Pavetta	178	*Cyperus*	157	*Allophylus*	74
Psychotria	175	8. Caesalpinioideae	398	17. Tiliaceae	230
3. Compositae	1,674	*Cassia*	63	*Grewia*	122
Vernonia	333	9. Annonaceae	318	18. Capparidaceae	222
Helichrysum	131	*Uvaria*	63	19. Combretaceae	195
Senecio	131	10. Sterculiaceae	311	*Combretum*	140
4. Gramineae	1,482	*Cola*	96	20. Loranthaceae	189
Eragrostis	142	11. Vitaceae	300	21. Moraceae	182
Panicum	119	*Cyphostemma*	190	*Ficus*	92
Digitaria	97	*Cissus*	77	22. Cucurbitaceae	179

Table 4. Additional African genera selected because of their species richness.

Genus	Number of species	Family	Total number of species in family
Begonia	114	Begoniaceae	114
Commiphora	111	Burseraceae	136
Impatiens	108	Balsaminaceae	108
Thesium	99	Santalaceae	103
Rinorea	86	Violaceae	98
Dichapetalum	82	Dichapetalaceae	89
Beilschmiedia	77	Lauraceae	85
Kalanchoe	53	Crassulaceae	101

Thunberg, Hoffmann and the *Phytographische Blätter*

R.O. MOFFETT
Department of Botany, University of the North (Qwaqwa Campus),
P.Bag X13, Phuthaditjhaba, 9866, Republic of South Africa.

Abstract
The attention of taxonomists is drawn to Hoffmann's *Phytographische Blätter*, a very rare, early 19th century publication which included the description by Thunberg of 60 species from 19 families of Cape plants. The circumstances surrounding the relative obscurity of the *Phytographische Blätter*, which appeared only once, is also explained.

Key words
Cape flora, Hoffmann, *Phytographische Blätter*, Thunberg.

Thunberg and the *Phytographische Blätter*

I wish to draw the attention of taxonomists working on the Cape flora to an early, little-known article by C.P. Thunberg in Hoffmann's *Phytographische Blätter* of 1803.

My attention to this article was first drawn by a footnote to the treatment of *Rhus glauca* in *Flora Zambesiaca* (Fernandes & Fernandes 1966), where the authors pointed out that *Rhus glauca* Desf. (1809), was predated by *Rhus glauca* Thunb. (1803). According to Stafleu & Cowan (1979), there are copies of the very rare *Phytographische Blätter* in Geneva (G) and New York (NY), and I was privileged to be able to study Thunberg's article in the copy in the library of the Conservatoire et Jardin Botanique in Geneva while researching *Rhus* in 1982.

The article which bears the place and date Uppsala, 26 February 1802, is titled *Novae Species Plantarum Capensium examinatae et descriptae a C.P. Thunberg* and is the first one in the *Blätter*. It is 30 pages long and comprises detailed descriptions of 60 species representing 19 families (Table 1).

A scan of a list of southern African plants and their major synonyms (Arnold & De Wet 1993) and a catalogue of Cape plants (Bond & Goldblatt 1984), revealed some interesting puzzles. What, for example, has become of the following names, to list but a few: *Ixia fenestrata, Phylica hirsuta, Protea reticulata* and *Restio elongatus*? Does *Protea tenuifolia* Thunb. not predate *P. tenuifolia* R. Br. (now *P. scabra* R. Br.) or *Sorocephalus tenuifolia* R. Br., whichever is conspecific, if at all? These intriguing questions need to be answered by the relevant experts in those groups, and as Thunberg's article is so rare, I have provided each of the libraries of the National Botanic Institute at Kirstenbosch and Pretoria with a photocopy.

Hoffmann and the *Phytographische Blätter*

To find out why Thunberg contributed to the *Phytographische Blätter* and why this journal is so rare, I have looked more closely at it and at its editor Hoffmann with the help of Professor Paul Hiepko of Berlin.

Georg Franz Hoffmann was born at Markbreit in Bavaria on 13-1-1760 and died in

L.J.G. van der Maesen et al. (eds.), The Biodiversity of African Plants, 42-46.

Table 1. List of species in *Novae Species Plantarum Capensium, examinatae et descriptae a C.P. Thunberg*.

Iridaceae	*P. verticillata*	Celastraceae
Ixia radians	*P. macrocephala*	*Celastrus acuminatus*
I. bicolor	*P. reticulata*	*C. obtusus*
I. fenestrata	*P. scabrida*	*C. ellipticus*
I. squalida	*P. marginata*	*C. crispus*
I. elliptica	*P. obtusata*	
		Rutaceae
Cucurbitaceae	Rubiaceae	*Diosma oblonga*
Bryonia cordata	*Galium horridum*	*D. cuspidata*
		D. umbellata
Cyperaceae	Penaeaceae	*D. deusta*
Schoenus aggregatus	*Penaea acuta*	*D. ustulata*
Cyperus hirtus		
	Convolvulaceae	Amaranthaceae
Gramineae	*Cuscuta africana*	*Achyranthes*
Holcus setifolius		*verticillata*
	Urticaceae	
Restionaceae	*Parietaria lanceolata*	Apiaceae
Restio nutans		*Anethum capense*
R. elongatus	Bruniaceae	
R. bifidus	*Brunia comosa*	Anacardiaceae
R. cuspidatus	*B. phylicoides*	*Rhus rosmarinifolia*
R. squamosus		*R. glauca*
	Rhamnaceae	*R. mucronatum*
Proteaceae	*Phylica glabrata*	*R. crenatum*
Protea villosa	*P. hirsuta*	*R. dimidiatum*
P. odorata	*P. rosmarinifolia*	*R. dissectum*
P. tenuifolia		
P. virgata	Campanulaceae	Aizoaceae
P. venosa	*Campanula bracteata*	*Pharnaceum*
P. rugosa	*C. stellata*	*mucronatum*
P. pyramidalis	*Lobelia setacea*	*P. distichum*
P. hirsuta	*L. scabra*	

Moscow on 17-3-1826. His portrait appears in the HI-IAPT Portraits of Botanists series as no. 18 (*Taxon* 21: 374), where his date of birth is incorrectly given as 30 or 31-1-1761. He studied at Erlangen, near Nurnberg, where at the age of 29 years he became Associate Professor of Medicine. Three years later he moved to Göttingen and in 1797 became Professor of Medicine and Botany and Director of the Botanic Garden. In 1801 he was visited by J.W. v. Goethe who needed help in unravelling some of the mysteries of the cryptogams, a subject which in Goethe's words was becoming more and more inaccessible to him. In 1804 Hoffmann accepted the post of Professor of Botany at the University of Moscow and there he remained until his death (Wagenitz 1988).

The *Phytographische Blätter* arose as the medium for papers by members of the *Societas Phytographica Gottingensis*, a society which Hoffmann was instrumental in forming at Göttingen in 1803. In the foreword to the first (and only) volume Hoffmann stated (my translation): "The plan to establish a literary society, whose sole purpose is to spread the science of plants, was supported right from the beginning with so much participation that the present phytographical papers could be published and whose continuation (i.e. the society) needs no further evidence than the following list of members". Then followed a list of honorary, full and associate members which is a veritable "who's who" of European botany at that time. Honorary members included Goethe, Jacquin, Schreiber, Smith, Thunberg and Waldstein, and among the Ordinary members were Andrews, Cavanilles, Link, Persoon, Sowerby, Sprengel and Willdenow, to name only a few (Hoffmann 1803).

Hoffmann's move to Moscow was described in a note in the *Botanische Zeitung* (Anon. 1804) as somewhat strange and, in substantiation, the editor published a letter of his which stated (my literal translation): "I am about to travel to Moscow about Whitsun, thereby sidestepping my opponents in a convenient manner. I had already received the call in May 1803 as well as opportunities to go to Wurzburg and Heidelberg. Since ... [words omitted] I had to show that elsewhere, mainly far north, one can find more humanity and feelings for science than here. The terms I was offered in Russia are very favourable and I will also create a Botanic Garden in Moscow. I will move the Phytographical Society to Moscow and continue the Journal there. I will be 300 miles closer to the Director and Secretary now. Delavigne is namely Professor of Natural Science in Charkow, which is 100 miles behind Moscow. From here Professors Grellmann, Cappel and Bohle are going to Moscow. We are already enjoying the new Fatherland".

Hoffmann's intention to continue with the journal in Moscow did not materialise and no further volumes appeared. It is also assumed that the Phytographical Society ceased to be active.

I have been unable to find out what exactly troubled Hoffmann at Göttingen. Perhaps the key lies in the words omitted (by the editor?) in the letter cited in the above note. His personnel records in the archives of the University of Göttingen give no indication of his unhappiness (pers. comm. Prof.Dr em. G. Wagenitz, 15-3-94). In a paper on the contribution of the German academics at the University of Moscow, Stieda (1930) stated that although Hoffmann was respected scientifically and was continuously busy, he nevertheless felt uncomfortable at Göttingen, and after some hesitation accepted the call to Moscow.

According to Stieda (1930), Hoffmann's private life was beyond reproach and he lived only for his science. The beginning of the 19th century was, of course, politically highly unstable with Napoleon Bonaparte about to spread his tentacles eastwards and this might also have played a part in Hoffmann's move (Albrecht-Carrié 1973). Prof. Wagenitz (pers. comm. cited) suggests, however, that campus, rather than national politics was the probable cause of the move and that it could have been the order of precedence in the Faculty of Medicine and his wishes for the Botanic Garden not being sufficiently met.

Although the exodus from Göttingen resulted in the demise of the Phytographical Society, Hoffmann himself remained very active and continued to publish many articles, notably on cryptogams and Umbelliferae (Apiaceae) (Stafleu & Cowan 1979).

Professor Hiepko, who studied a copy of the *Phytographische Blätter* in the Royal Library of Berlin, provided the following list and annotations of the contents of the single volume:

Contents of *Phytographische Blätter* (1803)

(No authorship is given for articles V, VII & VIII and these were obviously written by Hoffmann himself).

Acknowledgements

I wish to thank Professor P. Hiepko, Director of the Herbarium, Botanischer Garten und Botanisches Museum, Berlin-Dahlem and Prof.Dr em. G. Wagenitz of the Systematisch-Geobotanisches Institut, Universität Göttingen who provided valuable written comments and Dr. J. Schröder of the Department of Mathematics at the Qwaqwa Campus of the University of the North, who helped me with the translations. I also wish to thank the Research Committee at the Qwaqwa Campus of the University of the North for their support.

References

Albrecht-Carrié, R., 1973. Europe 1500-1848. Littlefield, Adams & Co., Totowa, New Jersey.

Anon., 1804. Botanische Notizen. Botanische Zeitung, Regensburg, 3 (16): 255-256.

Arnold, T.H. & B.C. De Wet (eds.), 1993. Plants of Southern Africa: names and distribution. Memoirs of the Botanical Survey of South Africa, 62.

Bond, P. & P. Goldblatt, 1984. Plants of the Cape Flora, a descriptive catalogue. Journal of South African Botany, Supplementary Volume 13.

Fernandes, R. & A. Fernandes, 1966. *Rhus* L. In: Exell, A.W., A. Fernandes & H. Wild, (eds.), Flora Zambesiaca, Volume 2,2. Crown Agents for Oversea Governments and Administrations, London.

Hoffmann, G.F., 1803. Vorword und Mitgliederverzeichnisse. In: Hoffmann, D.G.F. (ed.), Phytographische Blätter, 1: iii-x.
Stafleu, F.A. & R.S. Cowan, 1979. (Eds.). Taxonomic Literature, Edition 2, Volume 2: 237-242.
Stieda, W., 1930. Deutsche Gelehrte als Professoren an der Universität Moskau. Abhandlungen der Philologisch-Historischen Klasse der Sächsischen Akademie der Wissenschaften, 40, 5: 42-44.
Thunberg, C.P., 1803. Novae Species Plantarum Capensium examinatae et descriptae a C.P. Thunberg. In: Hoffmann. D.G.F. (ed.), Phytographische Blätter, 1: 1-34. (Although the title page of *Phytographische Blätter* cites Hoffmann's initials as D.G.F., he was always known as Georg Franz and it is not known what the D stands for.)
Wagenitz, G., 1988. Göttinger Biologen 1737-1945. Göttinger Universitätsschriften, Serie C: Kataloge, 2: 82.

Biodiversity of succulents and African arid regions

22 August 1994

Convenor: N. Jürgens

Biodiversity of arid islands in tropical Africa: the succulents of inselbergs

W. BARTHLOTT & S. POREMBSKI
Botanical Institute, University of Bonn, Meckenheimer Allee 170, D-53115 Bonn, Germany.

Abstract
Inselbergs are isolated rock outcrops of considerable geological and geomorphological age. They occur throughout tropical and subtropical Africa and Madagascar. Microclimatically and edaphically they can be considered as arid islands even in perhumid zonobiomes. Thus they bear a flora differing almost totally from the surrounding vegetation. Succulents are amongst the characteristic elements: for the first time a short annotated systematic survey of their occurrence on inselbergs is provided with notes on the geographical distribution and the centres of diversity.

Résumé
Les inselbergs sont des affleurements rocheux isolés d'un âge géologique et géomorphologique considérable. On les rencontre un peu partout en Afrique tropicale et subtropicale et à Madagascar. D'un point de vue microclimatique et édaphique, on peut les considérer comme des îlots arides, même au sein de biomes fortement humides. En conséquence, ils possèdent une flore presque totalement différente de la végétation environnante. Les plantes succulentes y représentent l'un des éléments caractéristiques; c'est la première fois qu'un aperçu systématique et détaillé de leur présence sur des inselbergs est présenté, en prenant en considération leur distribution géographique et leurs centres de diversité.

Key words
Africa, Biodiversity, Inselbergs, Madagascar, Succulents.

Introduction

Inselbergs are solitary, often monolithic structures rising usually abruptly from surrounding plains. They occur throughout the humid and semihumid tropics (map in Porembski & Barthlott 1992). As microclimatically and edaphically arid islands they bear a biotic cover dramatically different from their surroundings. They provide an excellent model for studying aspects of island ecology, geographical differentiation, and mechanisms controlling diversity (survey in Barthlott et al. 1993, Porembski et al. 1994).

In Africa and Madagascar inselbergs occur throughout the tropical and southern subtropical zonobiomes. They are particularly frequent in:

(1) a zone extending from Guinea/Sierra Leone to Nigeria/Cameroon,
(2) South of the Congo river from the highlands of Angola extending into Namibia,
(3) along the rift valley system from Sudan and Somalia into Zimbabwe, and
(4) in SW and Central Madagascar.

The virtually nude rock surface is usually dark due to a crust of cyanobacteria and lichens. Vascular plants usually colonize only smaller portions of the rocks and often form characteristic community types: monocotyledonous mats (*Afrotrilepis pilosa* (Boeck.) J. Raynal (Figure 2) in the West and *Coleochloa setifera* (Ridley) Gilly (Figure 6) in Eastern Africa and Madagascar, locally *Xerophyta* spp.; Hambler 1964)),

L.J.G. van der Maesen et al. (eds.), The Biodiversity of African Plants, 49-57.

wet flush vegetation, communities in seasonal rock-pools and an ephemeral flush vegetation usually adjacent to cyperaceous mats. Additionally, characteristic species in rock fissures occur; larger inselbergs often provide habitats for semi-deciduous forests.

Obviously, the vegetation of inselbergs is derived from very different sources: ephemeral therophytes and water plants; species from savannas and seasonally inundated lateritic crusts ("bowals"), trees from semi-deciduous forests. Additionally, taxa like epilithic orchids, some ferns and *Rhipsalis* in Madagascar are derived from the epiphytic canopy vegetation. True succulents of the genera characteristic for the arid and semi-arid regions extend their area on inselbergs like on arid islands throughout tropical Africa (e.g. *Aloe*, succulent *Euphorbia* spp.). They usually inhabit the cyperaceous mats or occur in rock fissures.

Arid islands: The abiotic conditions on inselbergs

The geology and geomorphology of African inselbergs are well known (survey in Bremer & Jennings 1978). Usually they consist of precambrian granites and gneisses. The typical dome-shaped tropical monoliths ("bornhardts") are often fragmented in smaller portions ("kopjes") in Southern and Eastern Africa. Inselbergs are geomorphologically very stable and old landscape elements: the estimation of their age usually exceeds ten million years.

Due to the lack of a continuous substrate cover retaining water after rainfalls rock outcrops are an edaphically dry habitat. This is also true for the microclimatic conditions: for example on inselbergs and even on smaller rock outcrops in the perhumid Tai rain forest in SW Ivory Coast the air temperature over sun-exposed rock surfaces exceeds 50°C whereas the temperature in the surrounding forest understorey remains under 30°C; in the same localities the relative air humidity drops under 30% in relation to almost 90% in the surrounding forest understorey (unpublished data from Szarzynski et al., in prep.). Thus inselbergs form mosaic-like patchy xeric habitats: "arid islands" providing growing sites for isolated populations of succulents and other xerophytic plants even within rain forest systems.

Succulents of African and Malagasy inselbergs

Despite the vast literature on African and Malagasy succulents no survey is available of the succulents growing on inselbergs and related rock outcrops. The following short annotated alphabetical family list is a result of field studies (in Ivory Coast, Guinea, Senegal, Cameroon, Namibia, South Africa, Malawi, Zimbabwe, Ruanda, Zaire, and Madagascar), herbarium work, and the consideration of published literature.

Agavaceae. The Central American *Furcraea foetida* (L.) Haw., widely cultivated for its fibre, is one of the very few aggressive neophytes invading inselbergs (in particular in Central Madagascar) often almost completely displacing the original vegetation (Figure 6).

Aizoaceae. The large predominantly S African family surprisingly plays no role (unfavourable conditions for the root system?) apart from a few *Delosperma* spp. extending into E Africa and Madagascar.

Aloaceae. *Aloe* extends from its southern centre of diversity on rock outcrops throughout tropical Africa with a few species up to the Upper Guinea forests (*A.*

buettneri A. Berger, *A. schweinfurthii* Bak.). Inselbergs in Madagascar provide a distinct evolutionary centre (Figure 7) for dwarf species of *Aloe* (e.g. *A. bakeri* Scott Elliot, *A. haworthioides* Bak.).

Apocynaceae. Caudiciform *Adenium* in E Africa; in Madagascar (Figs. 5-6) miniature *Pachypodium* spp. (e.g. *P. densiflorum* Bak., *P. rosulatum* Bak.) exclusively on rock outcrops.

Araceae. *Zamioculcas zamiifolia* (Lodd.) Engl. is lithophytic in E Africa.

Asclepiadaceae. The predominantly S and E African genera *Brachystelma* (Meve & Porembski 1993), *Caralluma*, and *Ceropegia* extend with a few and partially widespread species on rock outcrops throughout tropical Africa (Plowes 1989), *Stapelianthus* with c. 9 spp. endemic in Madagascar.

Balsaminaceae. The stem-succulent *Impatiens zenkeri* Warburg in C Africa and *I. tuberosa* Perr. in Madagascar.

Begoniaceae. Two *Begonia* spp. with succulent stems in E Africa and Madagascar.

Cactaceae. The neotropic family occurs with the epiphytic *Rhipsalis baccifera* (J. Miller) Stearn throughout tropical Africa, Madagascar, and Ceylon. This species developed in Madagascar a secondary evolutionary centre with dwarf neotenic taxa (Barthlott 1983) restricted to inselbergs (e.g. *R. baccifera* ssp. *horrida* (Bak.) Barthlott, Figure 8). *Opuntia* spp. occur as neophytic invaders.

Commelinaceae. The leaf succulent *Cyanotis lanata* Benth. is the most characteristic element of virtually all inselbergs in W and C Africa (Figure 3), a few related species occur throughout E Africa and Madagascar.

Compositae. *Kleinia* and *Senecio* extending from S through E Africa into Madagascar with few species; *Othonna* occurs on inselbergs in Namibia.

Crassulaceae. Several spp. of *Adromischus, Cotyledon* and *Tylecodon* in S and E Africa; *Crassula* with a few ephemeral spp. extending throughout the whole region. *Kalanchoe* has its center of diversity on the inselbergs in Madagascar extending into E Africa; a few spp. (e.g. *K. tubiflora* (Harvey) Hamet = *K. delagonesis* Ecklon & Zeyh.) as ruderal neophytes throughout the region even extending to inselbergs in the neotropics.

Dracaenaceae. *Sansevieria* is widespread in E Africa, extending with *S. liberica* Gér. & Labr. into perhumid tropical W Africa.

Euphorbiaceae. Apart from a few *Monadenium* spp. (S and E Africa) succulent *Euphorbia* spp. extend from S Africa (e.g. Kimberley 1989) throughout tropical Africa (Newton 1992). Several species related to *E. unispina* N.E. Br. (Figure 2) occur throughout the savanna regions of West Africa extending on rock outcrops into perhumid rain forests. A large number of small *Euphorbia* spp. occurs in Madagascar exhibiting speciation within neighbouring inselbergs by geographic isolation resulting from their highly unfavourable dispersal biology (in particular the *E. milii* Des Moulin-group (Figure 7). The succulent ornithochorous *Elaeophorbia drupifera* (Thonner) Stapf is common in semideciduous forest of inselbergs in C and W Africa.

Labiatae. The family has evolved several succulent taxa more or less restricted to rock outcrops: *Aeollanthus* spp., *Solenostemon graniticola* A. Chev. in W Africa (Figure 1), *Perrierastrum oreophilum* Guillaumin in Madagascar, and many species of

Plectranthus throughout tropical Africa and Madagascar (Figure 8).

Moraceae. Few *Dorstenia* spp. in S and E Africa; some slightly pachycaulous *Ficus* spp. in rock crevices common throughout the region.

Moringaceae. Only occasionally pachycaulous *Moringa* spp. from S Africa extending into Madagascar.

Orchidaceae. Several epiphytic orchid genera have evolved epilithic species characteristic for inselbergs and rock outcrops in E Africa and Madagascar. Dominating are in particular monopodial "angraecoid" taxa with succulent leathery leaves (e.g. *Angraecum* spp., *Jumellea* spp., *Sobenikoffia robusta* (Schltr.) Schltr.).

Portulacaceae. Few annual *Portulaca* spp. in E Africa; *Calandrinia* spp. occasionally neophytic on inselbergs.

Vitaceae. Caudiciform *Cyphostemma* spp. occasional in S Africa; stem succulent *Cissus quadrangularis* L. and related species are widespread throughout and occasionally invading inselbergs.

Geographical distribution and centres of diversity

A geographical comparison reveals basic differences of the biodiversity between different regions. The species composition of West African inselbergs is rather uniform with low alpha- and beta-diversity (sensu Magurran 1988) probably due to paleoclimatic fluctuations. Eastern and Southeastern Africa are much more diverse and show a considerable degree of endemism. The largest diversity was observed in Madagascar with locally restricted endemics thus resulting in a high beta-diversity. Tentatively some mechanisms regulating biodiversity on inselbergs are discussed.

Basically the systematic composition and the species richness of the succulent flora of inselbergs and related rock outcrops reflect the general diversity (Lebrun 1976) of xerophytes within the region: most diverse are SE Africa (e.g. Malawi: Porembski 1995) and Madagascar (e.g. Rauh 1973) - the Cape is to be excluded because no inselbergs in the strict geomorphological sense occur. However, the area of the genera involved extends through these insular xeric habitats into the semi-deciduous and rain forests of Central and West Africa (Newton 1989). Apart from the monotypic *Perrierastrum* and *Stapelianthus* in Madagascar - the latter not restricted to inselbergs - no succulent genera are exclusively bound to inselbergs nor have they evolved there.

This is in remarkable contrast to the monocotyledons with endemic inselberg genera such as *Afrotrilepis, Coleochloa* and *Microdracoides*. Thus the situation also differs from comparable inselbergs in the Brazilian forests and caatingas (Barthlott et al. 1993) with endemic succulent genera like *Coleocephalocereus* (Cactaceae). Remarkable for African inselbergs is the almost complete absence of Aizoaceae and the presence of succulent Labiatae.

Being geomorphologically old and stable structures, inselbergs probably played an important historical role as refugia for succulents and species of open grasslands in the paleoclimatically ever-changing scenario between forests and savannas. They provide patchy mosaic-like xeric habitats which can function as stepping-stones for the migration ("island-hopping") of succulents and other plant groups of open vegetation types through larger forest zones (e.g. weedy Gramineae and Cyperaceae). Today they often provide the last intact sites within fire-destroyed or agricultural landscapes for these plants (e.g. Figure 6).

Acknowledgements
The inselberg-project is financed by a grant of the Deutsche Forschungsgemeinschaft (Ba 605/4-2). We acknowledge the additional information by our students and cooperators within this project based on field studies in Guinea and Ivory Coast (N. Biedinger, S. Dörrstock, J.P. Mund, J. Szarzynski), Cameroon (E. Fischer), Ruanda and Zaire (N. Biedinger, E. Fischer), Madagascar (E. Fischer , D. Supthut, I. Theisen), and Zimbabwe (R. Seine).

References

Barthlott, W., 1983. Biogeography and evolution in neo- and paleotropical Rhipsalinae. In: K. Kubitzki (ed.), Dispersal and distribution: 214-218, Sonderbd. naturbd. naturwiss. Ver. Hamburg 7. P. Parey, Hamburg.

Barthlott, W., A. Gröger & S. Porembski, 1993. Some remarks on the vegetation of tropical inselbergs: diversity and ecological differentiation. Biogeographica 69: 17-36.

Bremer, H. & J. Jennings (eds.), 1978. Inselbergs/Inselberge. Z. Geomorph. N.F., Suppl. 31.

Hambler, D.J., 1964. The vegetation of granitic outcrops in Western Nigeria. J. Ecol. 52: 573-594.

Kimberley, M.J., 1989. The succulent Euphorbias of Zimbabwe: a checklist. Euphorbia J. 6: 8-27.

Lebrun, J.P., 1976. Richesses spécifiques de la flore vasculaire des divers pays ou régions d'Afrique. Candollea: 11-15.

Magurran, A.E., 1988. Ecological diversity and its measurement. Chapman & Hall.

Meve, U. & S. Porembski, 1993. *Brachystelma* Sims (Asclepiadaceae) in West Tropical Africa. Bot. Jb. Syst. 115: 315-324.

Newton, L., 1989. West Tropical Africa as an outlier in succulent plant distribution. Excelsa 14: 11-13.

Newton, L.E., 1992. An annotated and illustrated checklist of the succulent Euphorbias of West Tropical Africa. Euphorbia J. 8: 112-122.

Plowes, D.C.H., 1989. The succulents of Sudan. Excelsa 14: 57-69.

Porembski, S. & W. Barthlott, 1992. Struktur und Diversität der Vegetation westafrikanischer Inselberge. Geobot. Kolloq. 8: 69-80.

Porembski, S. & W. Barthlott, 1993. Ökogeographische Differenzierung und Diversität der Vegetation von Inselbergen in der Elfenbeinküste. In: W. Barthlott, C.M. Naumann, K. Schmidt-Loske & K.-L. Schuchmann (eds.), Animal-Plant-Interactions in Tropical Environments: 149-158. Results of the Annual Meeting of the German Society for Tropical Ecology, Bonn 1992.

Porembski, S., W. Barthlott, S. Dörrstock & N. Biedinger, 1994. Vegetation of rock outcrops in Guinea: granite inselbergs, sandstone table mountains and ferricretes - remarks on species numbers and endemism. Flora 189: 315-326.

Porembski, S., 1995. Notes on the vegetation of inselbergs in Malawi. Flora.

Rauh, W., 1973. Über die Zonierung und Differenzierung der Vegetation Madagaskars. Tropische und Subtropische Pflanzenwelt, 1. Mainz, Akad. Wiss. Lit., 146 p.

Figure 1. *Solenostemon graniticola* (Labiatae), Mt. Niangbo, Ivory Coast.

Figure 2. *Euphorbia unispina* (Euphorbiaceae) in mat of poikilohydrous *Afrotrilepis pilosa* (Cyperaceae), Northern Ivory Coast.

Figure 3. *Cyanotis lanata* (Commelinaceae) in flower, Comoé National Parc, Ivory Coast.

Figure 4. Rock pool colonized by *Sansevieria liberica* (Dracaenaceae) and various Cyperaceae, Comoé National Parc, Ivory Coast.

Figure 5. *Pachypodium rosulatum* (Apocynaceae) in Central Madagascar.

Figure 6. Inselbergs in Central Madagascar within a landscape destroyed by fire. The rock in the foreground bears the original vegetation with mats of *Coleochloa setifera* (Cyperaceae) and the caudiciform *Pachypodium densiflorum*. The vegetation cover of the inselberg in the background is formed by the neophytic Central American *Furcraea foetida* (Agavaceae) which has displaced the original flora.

Figure 7. Cushion of *Euphorbia milii* (Euphorbiaceae) and *Aloe acutissima* Perr. (Aloaceae) in Central Madagascar.

Figure 8. Cushion of succulents on an inselberg in SW Madagascar with *Stapelianthus decaryi* Choux. (Asclepiadaceae), *Rhipsalis baccifera* ssp. *horrida* (Cactaceae) and several small plants of a succulent *Plectranthus* sp. (Labiatae); associated with poikilohydric monocotyledons: *Coleochloa setifera* (Cyperaceae) and *Xerophyta dasylirioides* Bak. (Velloziaceae).

Patterns and characteristics of the flora of the Succulent Karoo Biome, southern Africa

C. HILTON-TAYLOR
Conservation Biology Research Unit, National Botanical Institute,
Private Bag X7, Claremont 7735, South Africa.

Abstract
The Succulent Karoo situated in the arid south-western portion of southern Africa is recognised as a hot-spot of global significance in terms of its floral diversity and high concentration of endemic taxa. The Succulent Karoo with approximately 5000 taxa, 40% of which are endemic to the region, is probably the richest arid flora in the world. Analysis of the endemic flora shows that it is not a random assemblage with regards to taxonomic affinity and growth form. The endemics are also not randomly distributed, but tend to occur in discrete centres of endemism. The growth form profiles of the endemics in each centre are distinctive indicating different selective forces. Conservation planning and management need to take the characteristics of the flora, particularly the endemic component, into consideration.

Résumé
Le Karoo à plantes succulentes, qui se situe dans la partie sud-occidentale aride de l'Afrique du Sud, est considéré comme un haut-lieu mondial en termes de diversité floristique et en forte concentration de taxons endémiques. Le Karoo à plantes succulentes, avec environ 5000 taxons dont 40 % d'endémiques à la région, possède probablement la flore aride la plus riche du monde. Une analyse de la flore endémique montre qu'il ne s'agit pas d'un assemblage aléatoire si l'on considère les affinités taxonomiques et les formes biologiques. Les endémiques ne sont pas non plus réparties de façon aléatoire, mais ont tendance à occuper des centres d'endémisme discontinus. Les types de formes biologiques dans chaque centre sont bien distincts, démontrant ainsi diverses voies de sélection. Il faut tenir compte, dans le plan de conservation et d'aménagement, des caractéristiques de la flore et plus particulièrement de ses composantes endémiques.

Key words
Diversity, Endemism, Floristics, Semi-desert, Southern Africa, Succulent Karoo.

Introduction

It has long been recognised that southern Africa (Namibia, Botswana, Lesotho, Swaziland and South Africa) has an extremely rich and diverse flora, comprising some 24,500 taxa (Arnold & De Wet 1993). It is also argued that southern Africa has one of the highest species densities of any region in the world (Cowling & Hilton-Taylor 1994). An examination of the distribution of plant diversity in southern Africa shows that the diversity is not evenly spread across the whole subcontinent, but tends to be concentrated in distinct areas which may be termed hot-spots *sensu* Myers (1988, 1990). Hot-spots are areas with high concentrations of endemic species and which are experiencing high rates of habitat loss and/or modification. Eight such hot-spots have been identified in southern Africa - the Wolkberg, Maputaland, Eastern Mountain Region, Pondoland, Albany, Cape, Kaokoveld and Succulent Karoo (Cowling & Hilton-Taylor 1994, Figure 1).

L.J.G. van der Maesen et al. (eds.), The Biodiversity of African Plants, 58-72.

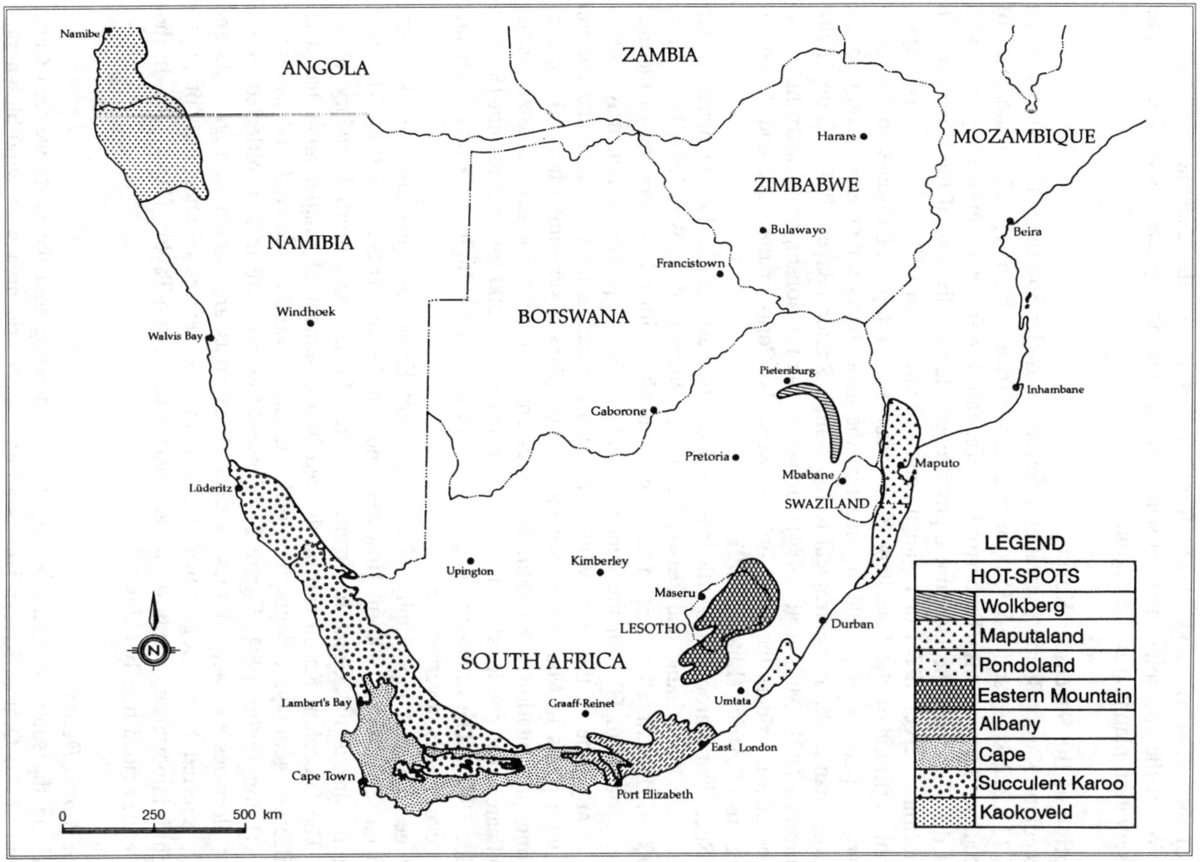

Figure 1. Map of the *Flora of southern Africa* (FSA) region (Botswana, Lesotho, Namibia, South Africa and Swaziland), with the shaded areas indicating the location of hot-spots mentioned in the text, and in particular the boundaries of the Succulent Karoo. Figure obtained from Cowling & Hilton-Taylor (1994).

The Succulent Karoo hot-spot is a semi-desert area which, as its name indicates, is dominated by succulent plant species. In this paper I will be examining (a) the size and taxonomic composition of the Succulent Karoo flora; (b) the levels of endemism in the flora; (c) the taxonomic and growth form composition of the endemic taxa; (d) the distribution of the endemics; and (e) will comment on the implications of these results for conservation planning in the region.

Description of the Succulent Karoo

The Succulent Karoo Biome (hereafter referred to as the Succulent Karoo) covers an area of 111,212 km^2, some 5.35% of southern Africa (Figure 1). The boundaries of the Succulent Karoo (sensu Rutherford & Westfall 1986) are roughly congruent with those of the Western Cape Domain, a phytogeographic subdivision of the Karoo-Namib phytochorion (Werger 1978) or regional centre of endemism (White 1983). The phytogeographic affinity of the Succulent Karoo has been a topic of discussion for many years, with a number of people arguing that the area shows far greater affinity to the Cape Flora than to the Palaeotropical Karoo-Namib Region (Bayer 1984, Hilton-Taylor 1987, Jürgens 1991, Moll et al. 1984). However, the relationships are far from simple when one also considers the close links between the floras of the Cape and Afromontane Regions of Africa (Linder 1990).

The Succulent Karoo can be distinguished on the basis of the low (a mean of 235 mm per year) but reliable and seasonal rainfall (Rutherford & Westfall 1986). More than 40% of the rainfall occurs in the winter months, with a large portion of the area receiving more than 60% in the winter. There is a small portion which may receive rainfall at any time of the year, while the eastern extremes have summer and autumn rainfall with peaks in March and November. Fog plays a very important role in supplementing precipitation from rain. Measurements from the southern coastal section have indicated that precipitation from fog can amount to 300 mm per annum (Schulze & McGee 1978). The minimum temperatures are sufficiently high to prevent frosts and if frost occurs, it is never severe.

Both geology and topography of the Succulent Karoo are quite diverse. Altitudes range from sea level to 1900 m. However, most of the area is below 800 m. The most striking feature is the Great Escarpment, which forms the eastern boundary of the region. The Succulent Karoo can be divided into a number of smaller areas based on their different geology, topography, total amount of rainfall received and season of rainfall (Hilton-Taylor 1994, Figure 2). These differences are in turn reflected by the different floristics of each of these areas. These areas are shown in Figure 2, and although referred to as geographical areas, many are centres of endemism of phytogeographic importance. Each of these areas (see Hilton-Taylor (1994) for further details) is described briefly below.

Southern Namib Desert

Lying along the southern Namibian coast and extending into the north-western Cape, just south of the Orange River. This is an extremely arid area with wind-blown sand dunes and rocky outcrops as dominant features. Rainfall is very low with much of the area receiving less than 50 mm of rain per annum. As a consequence of the harsh climate and shifting dunes, vegetation is sparse. However, concentrations of succulent

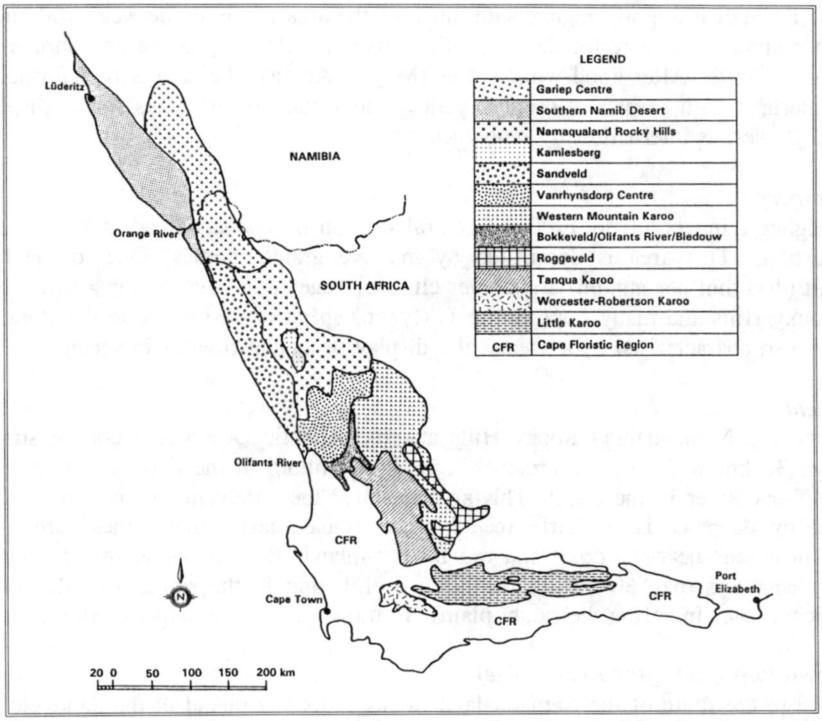

Figure 2. Map of the geographical subdivisions of the Succulent Karoo. The shaded areas indicate the main geographical areas that comprise the Succulent Karoo (Hilton-Taylor 1994). The Bokkeveld, Olifants River valley and Biedouw valley, which are indicated as a single unit here, are in fact a number of distinct areas which are either transitional between the Succulent Karoo and Cape Floristic Region or which are part of the latter region, but because of their small size are difficult to map at this scale.

and other xerophytic species can be found on rocky outcrops which receive some precipitation from the regular coastal fogs.

Gariep Centre
Located in the south-western corner of Namibia and the north-western corner of Namaqualand, straddles the Orange River and extends from Aus in the north to Steinkopf in the south. This is a very mountainous region with a complex geology. Sandy plains are found in the inter-montane valleys. The southern part of the Gariep Centre is known as the Richtersveld. Rainfall is generally very low, usually less than 50 mm per annum, however, the amount received is largely dependent on local physiography. Fog and coastal mists also support the rich xerophytic vegetation.

Namaqualand Rocky Hills
Occurs to the south of the Gariep Centre and extends from Steinkopf in the north to Bitterfontein in the south. The area is dominated by granite hills separated by sandy

plains. Rainfall is slightly higher with much of the area receiving between 100 and 200 mm per annum. As a result the vegetation cover is relatively dense and shrubs are a common feature. After good winter rains (May to August) the area is transformed into a wonderland with carpets of orange, yellow and white flowers. This spring display of annual flowers is internationally renowned.

Kamiesberg
The higher lying (1200 to 1700 m) central portion of Namaqualand is known as the Kamiesberg. This area is dominated by massive granite domes. Due to the higher rainfall (400 mm per annum) and cooler climate, vegetation cover is far greater than in the Rocky Hills and many Cape elements (fynbos species) are present in the flora. This area is also characterised by a spectacular display of annual flowers in spring.

Sandveld
Between the Namaqualand Rocky Hills and the Atlantic Ocean is a coastal strip (on average 30 km wide) which stretches from Port Nolloth in the north to the mouth of the Olifants River in the south. This area is called the Sandveld, because it is characterised by deep sands of fairly recent origin (Quaternary sands), these are usually white in colour near the coast and red further inland. Rainfall along this coastal strip varies from less than 50 mm in the north to 150 mm in the south, fog also plays a significant role. In spring the sandy plains are transformed to a carpet of flowers.

Vanrhynsdorp Centre (or Knersvlakte)
Situated to the south of the Namaqualand Rocky Hills and inland of the Sandveld. This is an area of rolling hills and plains which are covered in small white quartz pebbles, and which have very saline soils. Rainfall ranges between 100 and 200 mm per annum. The dwarf and often sparse vegetation cover is dominated by succulent species.

Western Mountain Karoo and Roggeveld
Found to the east of the Vanrhynsdorp Centre on top of the Great Escarpment. This is a fairly flat plateau-like area which stretches from Niewoudtville and Calvinia in the north to Sutherland in the south. It is characterised by very fertile, red clay soils interspersed with dolerite outcrops and other inselbergs. Due to the higher elevation, the climate is both cooler and wetter than areas to the west. Rainfall is generally more than 400 mm per annum. The vegetation cover is generally dense and the flora is dominated by geophytic species.

Tanqua Karoo
Between the Great Escarpment and the Cedarberg range of the Cape Fold Mountains, lies an interior plain known as the Tanqua Karoo. This area falls in a rain shadow, hence it is extremely arid with rainfall seldom exceeding 50 mm per annum. Occasionally, winter storms penetrate the mountains and bring rain to this area transforming the dry, sparsely vegetated landscape into a floral display of annual species.

Worcester-Robertson Karoo and Little Karoo
These two areas lie between the Cape Fold Mountains in the inter-montane valleys.

Shales and quartzites are the dominant geological formations and together with the higher rainfall, these allow for the establishment of fairly luxuriant shrubland. The flora here shows far greater affinities with the adjoining Cape Flora and sub-tropical thickets found to the east.

Size and Composition of the Flora

The figures published to-date on the size of the Succulent Karoo flora have all been based largely on speculation. Leistner (1979, this volume) estimated that there were between 4000 and 5000 phanerogams in the whole of southern Africa's arid region, an area which includes the entire Karoo-Namib Region and a large part of the Kalahari. White (1983) stated that there were 3500 species in the Karoo-Namib Region, whereas Hilton-Taylor (1987) estimated that the flora of the Karoo-Namib probably exceeded 7000 taxa, with between 35 and 50% endemic to the region. Gibbs Russell (1987), focusing on just a small portion of the Succulent Karoo (an area she termed the core region) and using only information from specimens housed at the National Herbarium in Pretoria, calculated that the flora of this core area comprised 2125 taxa with 29% of them endemic. Le Roux (1988) compiled a checklist of Namaqualand taxa as a basis for the *Flora of Namaqualand* project (Le Roux & Boucher in prep.). This checklist initially comprised 3500 taxa, but the numbers have decreased as work on the *Flora of Namaqualand* has progressed.

It was decided that as the *Flora of Namaqualand* had progressed so far and as we have an inventory of plants in the Cape Flora (Bond & Goldblatt 1984), an area which includes both the Worcester-Robertson Karoo and Little Karoo, it was reasonable to attempt a compilation of a Succulent Karoo checklist. In addition to using the above two publications, all the taxonomic literature published so far was consulted, published and unpublished checklists for all the nature reserves and national parks in the area were also utilised. The information from all these sources was further supplemented by examining specimen material in a number of herbaria (BM, BOL, K, NBG, PRE, STE). Only higher plants, pteridophytes, gymnosperms and angiosperms, were included in the checklist. Using this checklist, the size and composition of the Succulent Karoo flora could be determined.

In compiling a checklist of a flora of this size, taxonomic opinions play an important role. Ideally it would be best to adopt as uniform a treatment of the lower taxa as possible, especially for those of infraspecific rank. This is largely because one author's species may be another author's subspecies or variety. For the purposes of the checklist, infraspecific taxa have generally been used as the lowest common denominator (hence the terms "taxa" or "taxon" have been used rather than "species"). However, in certain genera where there was no recent or adequate revision and where there was some doubt about infraspecific status (e.g. as in *Haworthia* and *Oxalis*) it was decided not to use the infraspecific taxa described. Therefore in some groups the numbers of taxa occurring in the Succulent Karoo are probably a gross under-estimate, but these are probably balanced by very large genera such as *Ruschia* which are also in need of revision and where there will probably be a drastic reduction in the number of taxa. The results obtained here are not definitive, they are merely a first approximation. As taxonomic work and further exploration of the flora proceeds, the total number of taxa recorded in the Succulent Karoo will change.

Table 1. The size of the Succulent Karoo flora in terms of numbers of families, genera, endemic genera, taxa and endemic taxa per major group of plants and in total (numbers in brackets under endemic genera indicate near-endemics which could be added to the endemic totals).

Major group	Families	Genera	Endemic genera	Taxa	Endemic taxa
Pteridophyta	7	10	0 (0)	27	4
Monocotyledonae	25	192	7 (2)	1188	523
Dicotyledonae	86	528	51 (7)	3634	1427
Total	118	730	58 (9)	4849	1954

While compiling the checklist, each taxon was checked to see whether it was confined to the Succulent Karoo or not (i.e. endemic or non-endemic). If endemic, the geographical areas where it occurred were recorded. Each taxon was also assigned a growth form category. The categories used were as follows: trees, tall shrubs (generally taller than 1 m), dwarf shrubs (shorter than 1 m), forbs (short-lived perennials or herbs), annuals, succulent trees, succulent shrubs, dwarf succulents, annual succulents, graminoids (grasses, sedges, etc.), geophytes, vines or creepers, and parasites (root and stem). The succulent categories included both stem and leaf succulents.

The numbers of families, genera, endemic genera, taxa and endemic taxa occurring in the Succulent Karoo are shown in Table 1 according to the major classes, namely Pteridophytes, Monocotyledons and Dicotyledons. In total there are 118 families in the Succulent Karoo, which represent some 33% of the families found in the *Flora of southern Africa* (FSA) region. There are 730 genera, 28% of the genera in the FSA region, 58 of which are endemic to the Succulent Karoo. An additional nine genera could be termed near-endemics as most of their taxa are confined to the Succulent Karoo, but they have either one taxon which occurs mainly outside the region, or a taxon which occurs mainly in the Succulent Karoo but which has disjunct records from outside the area. There are 4849 taxa in the Succulent Karoo, which represents some 20% of the flora of the FSA region. Out of these, 1954 taxa (40.3% of the total flora) are endemic to the Succulent Karoo.

Southern Africa has a very high species to genus ratio of 9.4, a ratio which is normally only found in island floras such as New Zealand, Australia and New Caledonia (Cowling & Hilton-Taylor 1994). Such a high ratio also implies that phytogeographically, the flora of the region is relatively unique and that there has been extensive speciation within the area. The species to genus ratio for the Succulent Karoo is 6.9, which although less than that for the flora of southern Africa, is still relatively high. Comparing the size of the Succulent Karoo flora with other arid regions of the world (Figure 3) it appears that the Succulent Karoo is the richest arid flora in the world. Unfortunately there are no published data available for those arid areas of the world which are environmentally most similar to the Succulent Karoo. Nevertheless it is clear that the Succulent Karoo has a very rich and unique flora for an arid area.

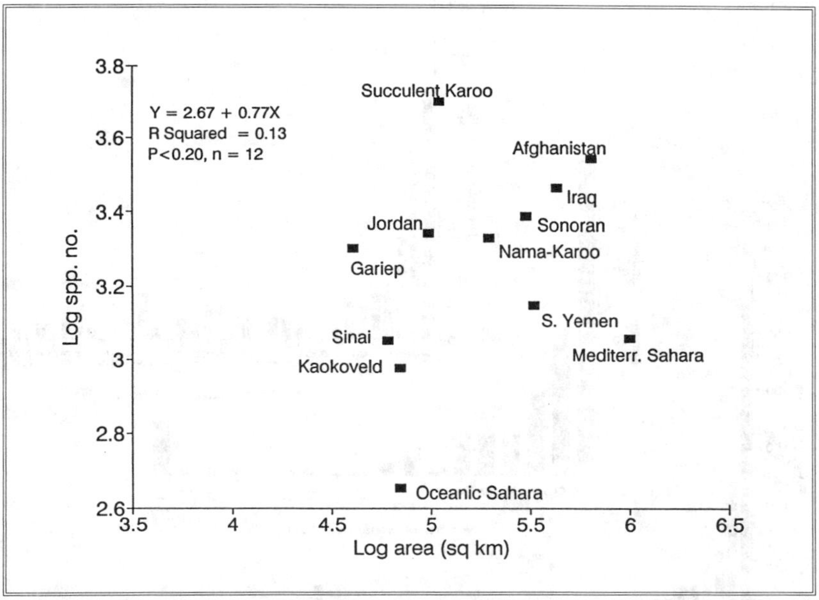

Figure 3. Species-area relationships (vascular plants only) for the arid and semi-arid regions of the world. Figure obtained from Cowling & Hilton-Taylor (1994) and slightly modified according to the latest information available.

A highly skewed negative exponential frequency distribution is observed for taxa per family, genera per family and taxa per genus (Figures 4a, b and c). Monospecific families, monogeneric families and monospecific genera make up 20% (23/118), 39% (46/118) and 38% (275/730) of the total in each case. Figure 4a shows that although most families in the Succulent Karoo are represented by few taxa, there are 28 families which have more than 34 taxa (see below). Similarly with Figures 4b and 4c, most families have few genera and most genera have few taxa, although 10 families have more than 17 genera and 27 genera have more than 34 taxa (see below).

Figures 5a and 5b show the numbers of taxa in the 20 largest families and the 20 largest genera, whilst Figure 5c shows those families which have more than 20 genera. There are ten families with more than 100 taxa, the largest being the Aizoaceae (which includes the former Mesembryanthemaceae) with 861 taxa and 102 genera (this is a conservative estimation of the numbers of taxa in this taxonomically difficult family). This family is also the largest in the whole southern African flora (Gibbs Russell 1985). The second largest family, also the second largest in southern Africa, is the Compositae with 648 taxa in 90 genera. Numbers then drop sharply with Iridaceae the third largest family (5th largest in southern Africa) with 334 taxa, 28 genera; followed by Leguminosae (3rd largest in southern Africa) with 264 taxa, 30 genera; Scrophulariaceae (9th in southern Africa) 208 taxa, 21 genera; Gramineae (7th in southern Africa) 173 taxa, 60 genera; and Crassulaceae (20th in southern Africa) 171 taxa, 4 genera. Other families with more than 100 taxa include Hyacinthaceae (163 taxa, 23 genera), Asclepiadaceae (155 taxa, 27 genera) and Oxalidaceae (114 taxa, 1 genus).

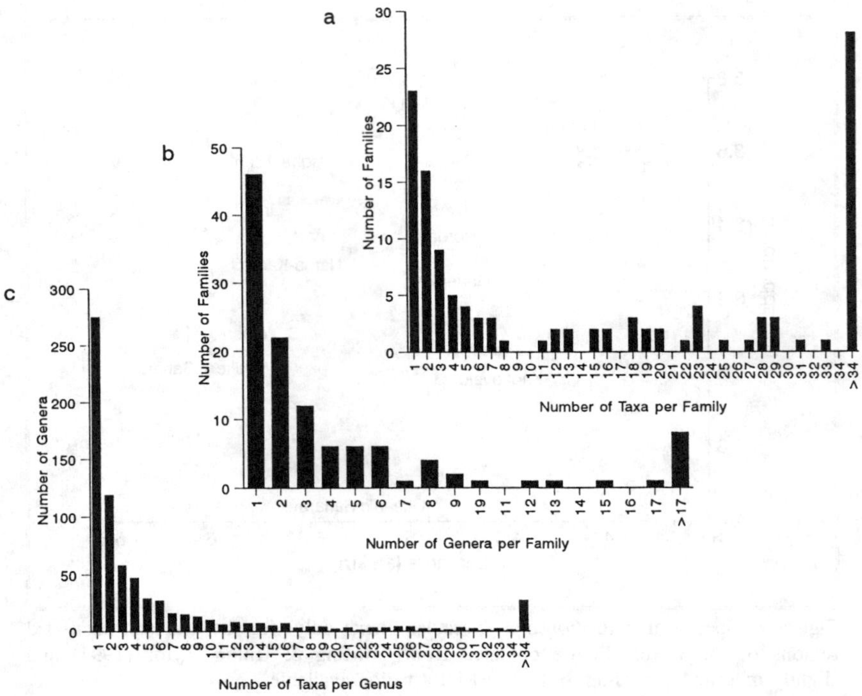

Figure 4. Frequency distribution of (a) families in terms of the numbers of taxa per family; (b) families in terms of the numbers of genera per family; and (c) genera in terms of the numbers of taxa per genus.

The two largest genera, *Ruschia* (136 taxa) and *Conophytum* (116 taxa), both belong to the family Aizoaceae. While the latter has recently been revised (Hammer 1993) the former is in urgent need of revision and may well be separated into a number of genera with taxon numbers greatly reduced. Other large genera include *Oxalis* (114 taxa) and *Crassula* (110), *Euphorbia* (77), *Pelargonium* (72), *Senecio* (72), *Eriospermum* (65), *Othonna* (61), and *Drosanthemum* (55). A number of genera from a number of families therefore contribute substantially to the flora, however the families Aizoaceae, Compositae, Gramineae, Leguminosae, Iridaceae, Asclepiadaceae, Hyacinthaceae and Scrophulariaceae make the largest contribution.

Analysis of the Endemic Flora

Taxonomic affinity

In examining the flora endemic to the Succulent Karoo we can ask the question - are endemics a taxonomically heterogeneous group, representative of the whole flora, or do certain families have more endemics than would be expected? To answer this question, the numbers of endemic taxa and non-endemic taxa were calculated for each family and similarly for the whole flora if that family was removed from the flora. The figures for each family were then compared to the figures for the total flora using the contingency table analysis provided in Statgraphics (STSC 1988). For each analysis

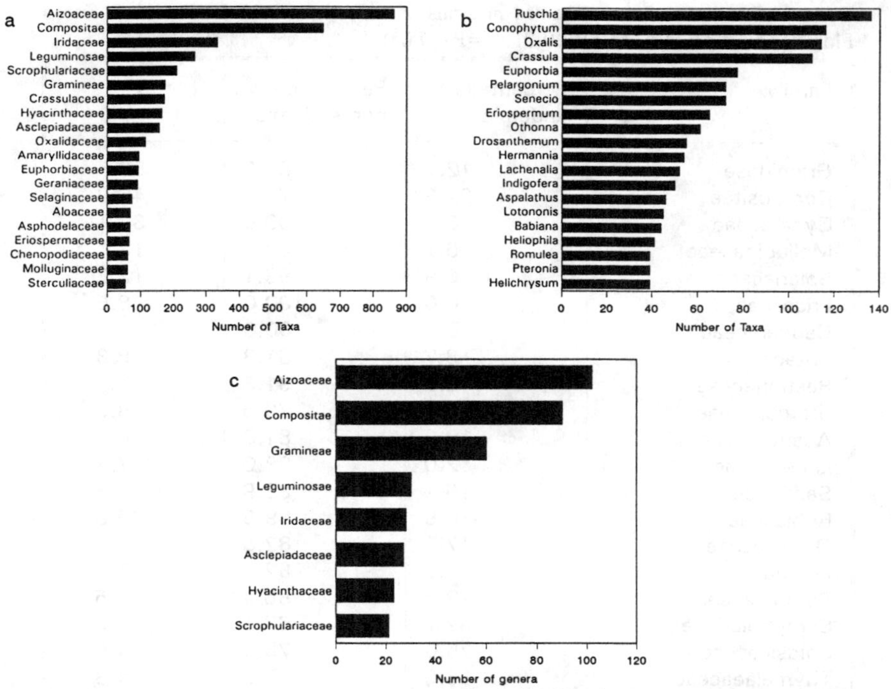

Figure 5. Number of taxa in (a) the twenty largest families in the Succulent Karoo, (b) the twenty largest genera in the Succulent Karoo; (c) shows the number of genera per family for families that contain more than twenty genera.

a chi-squared value, and the probability of getting such a value was obtained. If not significant, i.e. $p > 0.05$, then the family was assumed to have the same levels of endemism as expected for the whole flora. For those families which had significant chi-squared values, the numbers of endemics in the family were compared to the numbers expected for the whole flora, if lower, then the family was considered to be under-represented and if greater, over-represented i.e. the observed number of endemics in that family was greater than expected.

Fifty families had sufficient species to be analyzed, 22 of these had observed levels of endemism the same as expected, 20 families were under-represented (Table 2) and 8 families were over-represented (Table 3).

In general, the under-represented families are families with fairly wide, and in some cases global, distributions and all have their centres of greatest diversity situated elsewhere. Therefore it is not surprising that they should be under-represented in terms of endemics. The chi-squared values obtained for the families over-represented were all highly significant, with the Aizoaceae the most significant of all. Interestingly, the family Aloaceae, which is not that large compared with the other families, contains a large number of endemic taxa. From these results it is concluded that the endemics are not a random heterogeneous group, but that they tend to occur mainly in eight families.

Table 2. Families under-represented in terms of endemic taxa in the Succulent Karoo Flora (***=P<0.001, **=P<0.01, *=P<0.05).

Family	Percentage endemic	Percentage non-endemic	x^2
Gramineae	12.7	87.3	55.5 ***
Compositae	27.9	72.1	46.9 ***
Cyperaceae	0.0	100.0	33.1 ***
Molluginaceae	16.7	83.3	13.1 ***
Solanaceae	6.9	93.1	12.2 ***
Ericaceae	0.0	100.0	8.5 **
Capparaceae	0.0	100.0	8.5 **
Rutaceae	8.7	91.3	8.3 **
Restionaceae	8.7	91.3	8.3 **
Sterculiaceae	20.4	79.6	8.2 **
Acanthaceae	18.2	81.8	8.1 **
Lobeliaceae	12.0	88.0	7.2 **
Santalaceae	16.1	83.9	6.6 *
Rubiaceae	10.5	89.5	5.8 *
Orchidaceae	17.2	82.8	5.5 *
Polygalaceae	17.2	82.8	5.5 *
Boraginaceae	13.6	86.4	5.5 *
Scrophulariaceae	32.7	67.3	4.9 *
Brassicaceae	25.0	75.0	4.5 *
Thymelaeaceae	15.0	85.0	4.3 *

Table 3. Families over-represented in terms of endemic taxa in the Succulent Karoo flora (***=P<0.001).

Family	Percentage endemic	Percentage non-endemic	x^2
Aizoaceae	63.0	37.0	222.2 ***
Iridaceae	58.7	41.3	49.6 ***
Eriospermaceae	78.5	21.5	38.3 ***
Amaryllidaceae	71.6	28.4	38.1 ***
Oxalidaceae	64.9	35.1	28.4 ***
Crassulaceae	53.8	46.2	12.8 ***
Asclepiadaceae	54.2	45.8	12.3 ***
Aloaceae	58.0	42.0	8.4 ***

Growth form composition

As mentioned previously, all the taxa including the endemics were classified according to their dominant growth form. In examining these growth forms it was found that some had so few representatives that it was best for them to be grouped with another category. Succulent trees and non-succulent trees were grouped together, as were vines, climbers and parasites, also tall and dwarf shrubs.

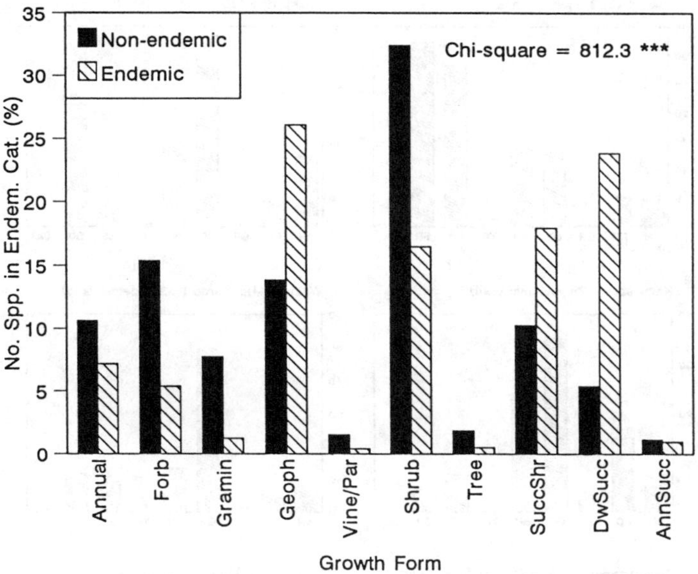

Figure 6. The association between endemism and growth form in the Succulent Karoo flora is illustrated by the percentage of species in each of the endemic categories i.e. endemic or non-endemic, plotted for each growth form. Annual = annuals, Forb = forbs/herbs (short-lived perennials), Gramin = graminoids (grasses & sedges), Geoph = geophytes (bulbs, corms, etc.), Vine/Par = vines, climbers and parasites, Shrub = non-succulent shrubs (tall and dwarf), Tree = trees (non-succulent and succulent), SuccShr = succulent shrubs, DwSucc = dwarf succulents (i.e. stemless, leaf succulents), AnnSucc = annual succulents. There is a very strong association between endemism and growth form as shown by the highly significant chi squared value. There is a very large percentage of non-endemic shrubs, whereas most of the geophytes and succulents are endemic.

Examining the flora we can once again ask whether the endemics are a random assemblage with regard to a biological attribute such as growth form, or do certain growth forms have more endemics than expected? To answer this question I looked at the association between endemism and growth form using the contingency table analysis described above. The high chi-squared value of 812.3 obtained is highly significant and indicates that there is a very strong association between growth form and endemism (Figure 6). While non-endemics tend to be mainly shrubs, forbs and to a lesser extent annuals, the endemics are mainly geophytes, dwarf succulents and succulent shrubs (Figure 6). Trees, vines, parasites, climbers and annual succulents form a very small component of the total flora and have very few endemics. Graminoids, although fairly common in the flora, all tend to be widespread common taxa with very few endemics. The endemics are therefore not a random assemblage with regards to growth form.

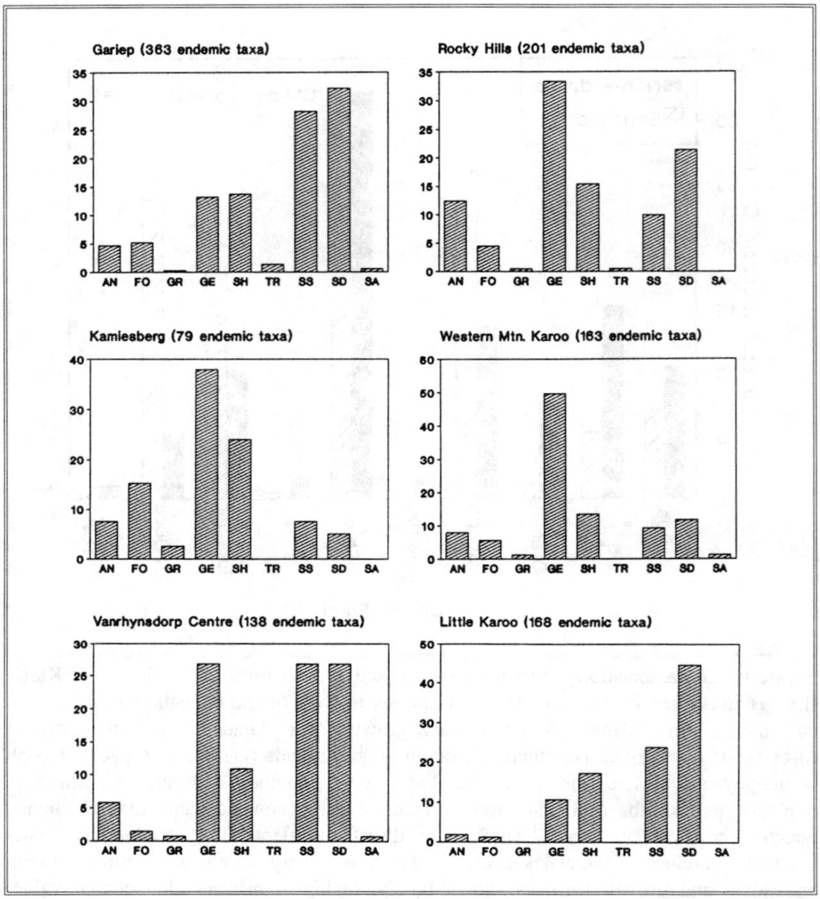

Figure 7. Numbers of endemics in the six major centres of endemism in the Succulent Karoo and the growth form profiles of the endemics in each of those centres (AN = annuals, FO = forbs (short-lived perennials), GR = graminoids, GE = geophytes, SH = non-succulent shrubs (tall and dwarf), TR = trees (both non-succulent and succulent), SS = succulent shrubs, SD = dwarf succulents, SA = annual succulents).

Distribution of endemics and growth form profiles

The Succulent Karoo endemics, like the flora of southern Africa, are not evenly distributed across the landscape. They tend to occur in distinct areas, which are usually termed centres of endemism (Hilton-Taylor 1994). Six centres of endemism can be recognised in the Succulent Karoo, namely the Gariep Centre with 363 endemic taxa, the Namaqualand Rocky Hills with 201 endemics, the Little Karoo with 168 endemics, the Western Mountain Karoo with 163 endemics, the Vanrhynsdorp Centre with 138 endemics and the Kamiesberg with 79 endemic taxa (Figure 7).

Looking at the growth form profiles (Figure 7) of the endemics in each of the centres, it is clear that the profiles change considerably as one moves from one centre to another. In the Gariep Centre the endemics tend to be largely dwarf succulents and succulent shrubs, to the south in the Namaqualand Rocky Hills endemics are predominantly geophytes or dwarf succulents and in the Kamiesberg, the central high-lying portion of the Rocky Hills, it is the geophytes and shrubs which are most represented as endemics. To the south in the Vanrhynsdorp Centre, endemics are equally likely to be geophytes, dwarf succulents or succulent shrubs, whereas in the Little Karoo which has quartz patches reminiscent of the Vanrhynsdorp Centre, it is the dwarf succulents which predominate as endemics.

Conclusions

From the above analysis it is clear that with nearly 5000 taxa the Succulent Karoo has a very rich flora for an arid region and that as shown, compared with other arid regions of the world it is probably the richest. The Succulent Karoo also has a very unique flora with more than 40% of the taxa endemic to the region. There has been considerable speciation within the region possibly reflecting a long period of aridity, but also indicating the heterogeneity of geology, soils and climate which has allowed for the development of very steep ecological gradients and hence the dense packing of many taxa. It is also clear that this speciation has not occurred in all taxonomic groups, as the flora is dominated by certain families and certain genera.

The endemic component of the flora is not a random assemblage with regards to taxonomic affinity and growth form. The endemics are also not randomly distributed, but are clumped in certain areas. The endemics of each area do not contain a mixture of the various growth forms, as each area has a distinct growth form profile. This indicates that different selective forces have been operating in different parts of the Succulent Karoo. This aspect needs to be investigated further.

The results obtained above, have significance for conservation planning and management of the Succulent Karoo. At present, less than two per cent of the Succulent Karoo is conserved and the area has been identified many times as one of the major conservation priorities in southern Africa (see Hilton-Taylor & Le Roux 1989). The distribution of the flora and the endemics need to be taken into account when a Gap Analysis is done to determine the optimal number and location of nature reserves in the Succulent Karoo. Management strategies for the Succulent Karoo are mainly aimed at maintaining the economic potential of the vegetation of the area for grazing purposes. Management also needs to consider the maintenance of the botanical diversity of the area, especially the endemic taxa, which form a substantial portion of the flora. Management strategies therefore need to take into account the distribution of the endemics and especially the growth form profiles of the different areas. Some areas are more suitable to heavy grazing than others and endemics are therefore less likely to be influenced. Mismanagement of the Succulent Karoo, as is currently happening in most parts, will inevitably lead to species loss and the extinction of much of this unique flora (see Hilton-Taylor & Le Roux 1989, Hilton-Taylor 1993, 1994).

Acknowledgements

I am extremely grateful to Annelise le Roux for allowing me to use the information gathered for the *Flora of Namaqualand* project. I thank the Directors of the herbaria visited for permission to examine their specimens. Many thanks to Dave McDonald for his comments on an earlier draft.

References

Arnold, T.H., & B.C. De Wet, 1993. Plants of southern Africa: names and distribution. Memoirs of the National Botanical Survey of South Africa No. 62: 1-825.

Bayer, M.B., 1984. The Cape Flora and the Karoo - a winter rainfall biome versus a fynbos biome. Veld and Flora 70: 17-19.

Bond, P. & P. Goldblatt, 1984. Plants of the Cape Flora: a descriptive catalogue. Journal of South African Botany Supplement 13: 1-455.

Cowling, R.M. & C. Hilton-Taylor, 1994. Patterns of plant diversity and endemism in southern Africa: an overview. In: B.J. Huntley (ed.), Botanical Diversity in Southern Africa: 33-54. Strelitzia 1. National Botanical Institute, Pretoria.

Gibbs Russell, G.E., 1985. Analysis of the size and composition of the southern African flora. Bothalia 17: 213-227.

Gibbs Russell, G.E., 1987. Preliminary floristic analysis of the major biomes in southern Africa. Bothalia 17: 213-227.

Hammer, S.A., 1993. The Genus *Conophytum*. Succulent Plant Publications, Pretoria.

Hilton-Taylor, C., 1987. Phytogeography and origins of the Karoo flora. In: R.M. Cowling & P.W. Roux (eds.), The Karoo Biome: a preliminary synthesis. Part 2 - Vegetation and history: 70-95. South African National Scientific Programmes Report 142. CSIR, Pretoria.

Hilton-Taylor, C., 1993. Karoo succulents threatened with extinction. Our Living World September 1993: 14.

Hilton-Taylor, C., 1994. Western Cape Domain (Succulent Karoo). In: S.D. Davis & V.H. Heywood (eds.), Centres of Plant Diversity: a Guide and Strategy for their Conservation. Vol. 1 Europe, Africa, SW Asia and the Middle East: 204-217. IUCN Publications Unit, Cambridge.

Hilton-Taylor, C. & A. Le Roux, 1989. Conservation status of the Fynbos and Karoo Biomes. In: B.J. Huntley (ed.), Biotic Diversity in Southern Africa: concepts and conservation: 202-223. Oxford University Press, Cape Town.

Jürgens, N., 1991. A new approach to the Namib Region. I: Phytogeographic subdivision. Vegetatio 97: 21-38.

Leistner, O.A., 1979. Southern Africa. In: D.W. Goodall & R.A. Perry (eds.), Arid-land Ecosystems: 109-143. Cambridge University Press, Cambridge.

Le Roux, A., 1988. Namaqualand checklist. Chief Directorate Nature and Environmental Conservation, Cape Province, Cape Town. Unpublished document.

Le Roux, A. & C. Boucher, in prep. The Flora of Namaqualand.

Linder, H.P., 1990. On the relationship between the vegetation and floras of the Afromontane and the Cape Regions of Africa. Mitteilungen aus dem Institut für Botanik, Hamburg 23b: 777-790.

Moll, E.J., C. Hilton-Taylor & M.L. Jarman, 1984. Exactly what is the Cape Floristic Kingdom. In: B. Dell (ed.), Proceedings of the Fourth International Conference on Mediterranean Ecosystems: 113-114. University of Western Australia, Nedlands.

Myers, N., 1988. Threatened biotas: "hotspots" in tropical forests. The Environmentalist 8: 1-20.

Myers, N., 1990. The biodiversity challenge: expanded hot-spots analysis. The Environmentalist 10: 243-255.

Rutherford, M.C. & R.H. Westfall, 1986. Biomes of southern Africa - an objective categorization. Memoirs of the Botanical Survey of South Africa No. 54: 1-98.

Schulze, R.E. & O.S. McGee, 1978. Climatic indices and classifications in relation to the biogeography of southern Africa. In: M.J.A. Werger (ed.), Biogeography and Ecology of Southern Africa: 19-52. Junk, The Hague.

STSC, Inc., 1988. Statgraphics. Statistical Graphics System by Statistical Graphics Corporation version 4.0. Rockville, Maryland, U.S.A.

Werger, M.J.A., 1978. Biogeographical divisions of southern Africa. In: M.J.A. Werger (ed.), Biogeography and Ecology of Southern Africa: 145-170. Junk, The Hague.

White, F., 1983. The Vegetation of Africa. UNESCO, Paris.

Taxonomy and biogeography of the anomalous genus *Wellstedia*

M. THULIN & A.N.B. JOHANSSON
Department of Systematic Botany, Uppsala University, Villavägen 6, S-752 36 Uppsala, Sweden.

Abstract
Five species are here recognized in *Wellstedia*: *W. socotrana* on Socotra, *W. somalensis* sp. nov. and *W. laciniata* sp. nov. in Somalia, *W. filtuensis* in Ethiopia, and *W. dinteri* in Namibia and the adjacent Cape Province. A preliminary phylogenetic analysis suggests that *W. dinteri* is the sister group to the rest of the genus, while *W. filtuensis*, *W. somalensis* and *W. socotrana* are vicarious sister species. *Wellstedia* has usually been treated either as a separate subfamily within the Boraginaceae or as a family of its own close to Boraginaceae. Its most notable feature is the capsular fruits (here shown to be hygrochastic), capsules otherwise being unknown in Boraginaceae. It is suggested that the nearest relative of *Wellstedia* may be *Tiquilia* in Boraginaceae subfamily Ehretioideae.

Résumé
Cinq espèces ont été reconnues au sein du genre *Wellstedia*: *W. socotrana* au Socotra, *W. somalensis* sp. nov. et *W. laciniata* sp. nov. en Somalie, *W. filtuensis* en Ethiopie et *W. dinteri* en Namibie et dans la Province du Cap qui lui est adjacente. Une première analyse phylogénétique laisse apparaître que *W. dinteri* est le groupe "soeur" vis-à-vis du reste du genre, tandis que *W. filtuensis*, *W. somalensis* et *W. socotrana* sont des espèces "soeurs" vicariantes. *Wellstedia* a été généralement traité soit comme sous-famille séparée au sein des Boraginaceae, soit comme famille à part voisine des Boraginaceae. Sa caractéristique la plus notable est la présence de fruits capsulaires (il est démontré ici qu'ils sont hygrochastiques), alors que l'existence de capsules n'est pas connue par ailleurs chez les Boraginaceae. Il semblerait que le genre le plus voisin de *Wellstedia* soit *Tiquilia* dans la sous-famille des Ehretioideae des Boraginaceae.

Key words
Biogeography, Boraginaceae, Phylogeny, Pollen morphology, Taxonomy, *Wellstedia*.

Introduction

Wellstedia Balf. f. is a genus mainly of dwarf shrubs with three previously described species, one on Socotra and in northern Somalia, one in southeastern Ethiopia, and one in Namibia and the Cape Province. The genus is taxonomically isolated and is sometimes regarded as a family on its own, Wellstediaceae, but usually it is treated as a subfamily within Boraginaceae. However, it deviates from all other members of the Boraginaceae by having a 2-locular ovary developing into a 1-2-seeded capsule. The genus is often mentioned in connection with both biogeographical and taxonomic discussions, but the morphology, taxonomy, biology and distribution of the species has remained rather fragmentarily known. This work aims at a revision of the genus at species level, and a discussion of its phylogeny, biogeography and taxonomic position.

Material and methods

The work is based on the study of herbarium material and on field work in Somalia. The following herbaria have been consulted at personal visits or by material obtained

L.J.G. van der Maesen et al. (eds.), The Biodiversity of African Plants, 73-86.

on loan: BM, E, FT, K, S and UPS (abbreviations as in Holmgren et al. 1990).

Pollen preparations were made according to the acetolysis method (Erdtman 1952). One accession from each species was studied except for *W. somalensis*, from which two accessions were examined. Size measurements were made on about 10 pollen grains from each sample in an ordinary light microscope, and surface structure was studied by scanning electron microscopy (SEM).

The phylogenetic analysis was made using the program HENNIG86 version 1.5 (Farris 1988). All specimens cited have been seen, if not otherwise indicated.

Historical review

The first species of the genus *Wellstedia* was discovered by I.B. Balfour and his co-travellers in 1880 during an expedition to Socotra. The name *Wellstedia* was given in honour of the British Lieutenant J.R. Wellsted, who explored the Island for the Indian Government in 1834.

A first description of the genus and its single species, *W. socotrana* Balf.f., was published by Balfour (1884). A few years later he published a more detailed and illustrated account (Balfour 1888).

A second species, *W. dinteri* Pilger from Namibia, was described by Pilger (1912). Hunt (1969), in a review of the genus, subdivided *W. dinteri* into the two varieties var. *dinteri* and var. *gracilior* Hunt, and also reported *W. socotrana* from Somalia. A third species, *W. filtuensis* Hunt & Lebrun, was described some years later (Hunt & Lebrun 1975).

Position of the genus

Balfour (1888), in a discussion of the taxonomic position of *Wellstedia*, stated that the most noteworthy features of the genus are "its depressed Boragineoid habit and alternate leaves, the calyx adnate to the ovary, the gamophyllous tetramerous regular floral envelopes, four equal epipetalous stamens, the two-celled ovary with a solitary pendulous ovule in each loculus, the loculicidally septifragal capsule containing a single exalbuminous seed embraced by the isolated septum, and the large embryo with accumbent cotyledons and superior radicle,... altogether, making a combination with which I can find no parallel". He placed *Wellstedia* at the end of the Gamopetalae as a "genus anomalum", but believed the plant to have its closest affinity with the Boraginaceae. However, Convolvulaceae, Loganiaceae, Scrophulariaceae and Verbenaceae were mentioned as other possibly related families.

Gürke (1897) placed *Wellstedia* in Boraginaceae, but as a "zweifelhafte Gattung". He pointed to its boraginaceous habit, but also mentioned its resemblance in some characters with Verbenaceae and Scrophulariaceae.

Pilger (1912) placed *Wellstedia* in its own subfamily, Wellstedioideae Pilger, within Boraginaceae, and this taxonomy has been followed by most subsequent authors, e.g. Pilger & Krause (1914), Gilg (1924), Lemeé (1935), Erdtman (1952), Emberger (1960), Melchior (1964), Hunt (1969), Hunt & Lebrun (1975), and Cronquist (1981). Pilger (1912) also suggested that there might be some relationship between *Wellstedia* and *Coldenia* L. (Boraginaceae).

Novák (1943), however, treated *Wellstedia* as a family on its own, Wellstediaceae (Pilger) Novák. Merxmüller (1960), unaware of Novák's work, published Wellstediaceae as a new family, thereby creating a later homonym. Other authors treating the group as a distinct family include Dostál (1957), and Takhtajan (1987). Still, all these authors seem to regard Wellstediaceae as more closely related to Boraginaceae than to any other family.

To search for a possible sister group of *Wellstedia* we have surveyed a large number of genera of Boraginaceae. The only possible candidate we have managed to come up with is *Tiquilia* Pers., a genus of some 27 species distributed in south-eastern North America and north-eastern South America (Richardson 1977). This was previously part of *Coldenia* (see above), but *Coldenia* s. str. is now regarded as a monotypic genus in the Old World (Richardson 1977). *Tiquilia* is placed in Boraginaceae subfamily Ehretioideae and consists of dwarf shrubs or herbs confined to arid regions and fairly similar to *Wellstedia* in habit. *Tiquilia*, like *Wellstedia* (and other members of subfamily Ehretioideae), has a terminal style (in some species), while the style is gynobasic throughout in subfamily Boraginoideae. *Tiquilia* also has apical placentation (otherwise very rare in Boraginaceae), and 4-merous flowers are present in the closely related *Coldenia* (otherwise very rare in Boraginaceae). Still, *Tiquilia* differs very markedly from *Wellstedia*, e.g., in its mericarpous fruit with the ovary maturing into four one-seeded nutlets, in its two free styles with capitate stigmas, and in its venation of the corolla with many veins diverging in the lobes.

A relationship between *Wellstedia* and *Tiquilia* would imply that the capsular fruit of *Wellstedia* is derived from a mericarpous ancestor. It would also imply that *Wellstedia* should be included as a highly derived member of Boraginaceae subfamily Ehretioideae. We hope soon to be able to test this still very hypothetical relationship with molecular methods. For a discussion of pollen morphology, see under pollen.

Morphological aspects

Capsule

The shape of the mussel-like capsule is similar in all species of *Wellstedia* (Figures 4F & 5H), but the morphology of the dissepiment varies and is taxonomically useful.

The dissepiment of *W. somalensis* and *W. filtuensis* is crescent-shaped with a small stalk-like placenta at one side, while that of *W. socotrana* is more elongated, but also with a small placenta.

The dissepiment of *W. laciniata* stands out by having a peculiar cap on the top (Figure 4G), while *W. dinteri* has a dissepiment with placentas on both sides, making it resemble a golf-peg.

The ovary is 1- or 2-ovulate, but the capsules are 1-seeded by abortion in all species except *W. dinteri*, where 2 seeds are normally formed.

Indumentum

The hairs in *Wellstedia* are 1-celled, appressed and silvery, and are provided with spinules arranged in longitudinal lines.

In *W. dinteri* and *W. laciniata* some of the hairs have multicellular bases incrustated with calcium. These multicellular bases are particularly prominent in *W. dinteri*.

Hunt (1969) reported the presence of multicellular peltate glands in *Wellstedia* and stated that this type of glands is otherwise unknown in Boraginaceae, but are found in Scrophulariaceae and occasionally in Verbenaceae. However, no multicellular peltate

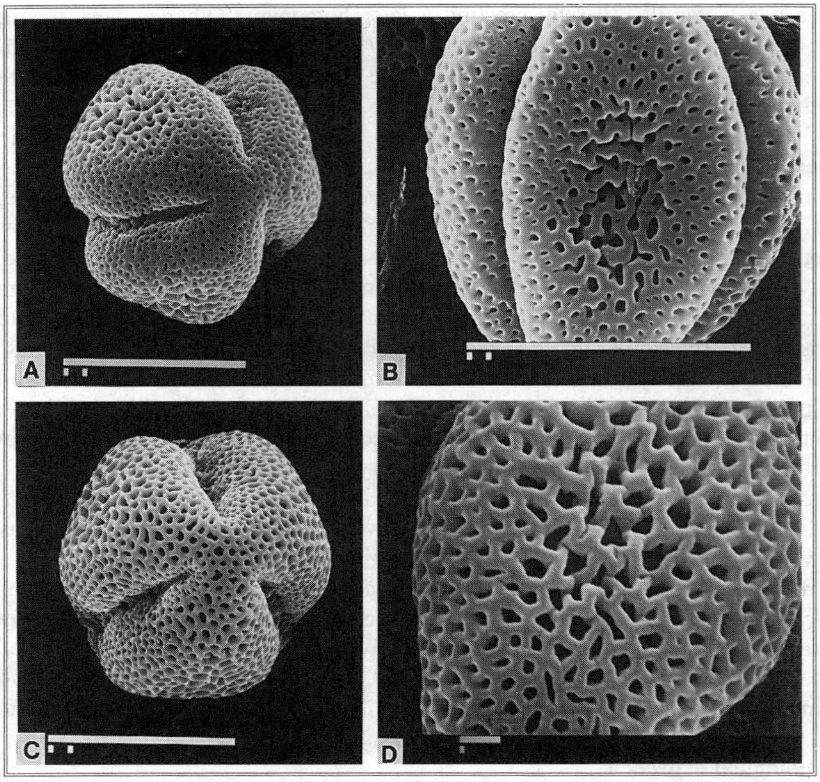

Figure 1. Pollen grains and exine structures in *Wellstedia* (SEM). A, B: *W. filtuensis*, from Gilbert, Sebsebe & Vollesen 8109. C, D: *W. somalensis*, from Thulin & Warfa 5453. Scale bar in A, B and C: 10 μm, in D: 1μm.

glands have been found in *Wellstedia* in this study, and presumably Hunt's report of such glands is based on a misinterpretation of the multicellular hair-bases found particularly in *W. dinteri*. This type of multicellular hair-bases is actually found in numerous taxa within Boraginaceae.

Pollen

The only species of which the pollen morphology has been previously described is *W. dinteri* (Erdtman 1952). The pollen grains of this were described as follows: "3-colporoidate, suboblate (13-15 μ). Amb rounded-hexagonal. Sexine as thick as nexine, reticulate". Our own study of all species except *W. socotrana* has resulted in the following pollen description for the genus:

Pollen grains 3-colporate, oblate to subprolate, 12-25 x 8-15 μm. Amb triangular-hexagonal. Apocolpium diameter 2-3 μm. Colpi 11.3-14 μm long, 1-3 μm wide, colpus membrane granular. Ora rounded, 0.4-0.7 μm in diameter. Exine 1-3 μm thick. Sexine perforate to reticulate, lumina diminishing in size towards apocolpia and colpus margin; reticulum usually coarser at center of mesocolpia (Figure 1).

Table 1. Characters used in the cladistic analysis. The numbers correspond to those in the data matrix (Table 2).

1. Leaves, including petiole, at least 15 mm long (0); leaves up to 8 mm long (1).
2. Indumentum with understorey of short, fine hairs (0); indumentum uniform, without understorey of short, fine hairs (1).
3. Hairs, or some of them, with broad multicellular base (0); hairs without obvious multicellular base (1).
4. Ovary 2-ovulate (0); ovary 1-ovulate (1).
5. Capsule normally 2-seeded (0); capsule 1-seeded (1).

Table 2. Data matrix for *Wellstedia*. Characters numbered as in Table 1.

Species	1	2	3	4	5
W. dinteri	0	0	0	0	0
W. filtuensis	0	1	1	1	1
W. laciniata	0	0	0	0	1
W. socotrana	1	1	1	1	1
W. somalensis	1	1	1	1	1

W. dinteri has the smallest pollen grains (12-14 x 8-10 μm), while the largest are found in *W. laciniata* (22-25 x 12-15 μm). *W. filtuensis* and *W. somalensis* take an intermediate position (pollen grains 16-20 x 10-13 μm).

The 3-colporate, reticulate pollen grains of *Wellstedia* have a general similarity to, e.g., those of *Tiquilia* in Boraginaceae subfamily Ehretioideae (see under Position of the genus). However, with the coarse reticulum at the centre of the mesocolpia, they also have a striking resemblance to the pollen grains of some genera of Hydrophyllaceae (Constance & Chuang 1982). The significance of this resemblance should be studied further.

Dispersal

The species of *Wellstedia* generally grow in semi-desert areas with an open vegetation and a low and erratic rainfall. According to Ulbrich (1928) *W. dinteri* is a "Steppenläufer" ("tumbleweed"), i.e. entire shoots more or less roll up like a ball in dry weather, get loose, and are dispersed by the wind. Whether this applies also to all the other species of the genus is uncertain, but can very well be the case.

Also, by simple experiments on herbarium material we have been able to prove that all species within the genus show hygrochasy, i.e. the seeds are released after wetting (Zohary 1962, van der Pijl 1972). In the dry hardening capsules, which appear to remain on the plant for a long time, the apical valves are close together, but after moistening with water they move outwards and dissemination can take place. This phenomenon, where dissemination of the seeds does not take place immediately after the fruit has ripened, but is delayed until the beginning of the rains, is known particularly in plants of arid regions. Another hygrochastic genus in the Horn of Africa and Socotra is *Xylocalyx* Balf. f. in the Scrophulariaceae (Thulin 1987).

Phylogeny

A preliminary cladistic analysis of *Wellstedia* was performed using the five informative characters listed in Table 1. The corresponding data matrix is given in Table 2.

As the relationships of *Wellstedia* are uncertain *W. dinteri* was used as an outgroup for the rest of the species. *W. dinteri* is the only species in the genus with 2-ovulate ovaries normally developing into 2-seeded capsules, while the other species either have 2-ovulate ovaries developing into 1-seeded capsules or 1-ovulate ovaries. This is assumed to indicate a basal position for *W. dinteri* within the genus.

The single resulting cladogram is entirely without homoplasy and is presented in Figure 2. This shows *W. laciniata* as the closest relative of *W. dinteri*, while *W. filtuensis*, *W. somalensis* and *W. socotrana* form a clade with *W. filtuensis* as sister species to the other two.

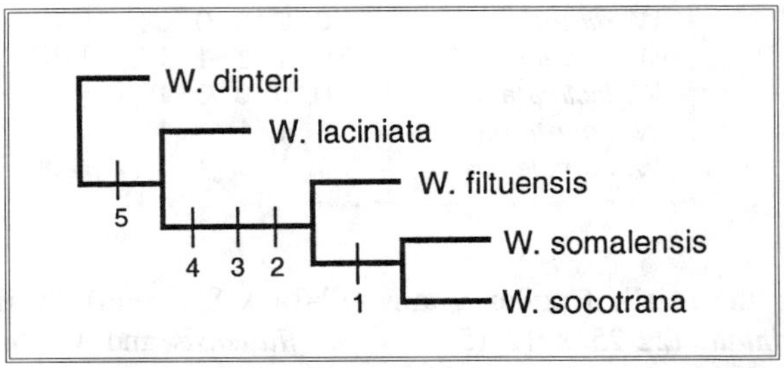

Figure 2. Cladogram of *Wellstedia*. The numbers of the characters correspond to those of the character list (Table 1) and the data matrix (Table 2).

Biogeography

The genus *Wellstedia* is known from Socotra, north-eastern Somalia, south-eastern Ethiopia, southern Namibia and adjacent parts of the Cape Province.

W. socotrana is distributed on Socotra, *W. somalensis* and *W. laciniata* in northern Somalia, *W. filtuensis* in south-eastern Ethiopia, and *W. dinteri* in Namibia and the Cape Province (Figure 3A & B). All species are narrow endemics with sympatric distributions except *W. somalensis* and *W. laciniata*, which have partly overlapping ranges. Expressed in terms of the regional phytochoria recognized in Africa by White (1983), *W. socotrana*, *W. somalensis*, *W. laciniata*, and *W. filtuensis* are all confined to the Somalia-Masai region, while *W. dinteri* belongs to the Karroo-Namib region.

The genus *Wellstedia* is an excellent example of a genus with a disjunction between the arid regions of north-east and south-west Africa (Thulin 1994). Another genus with a similar distribution is *Kissenia* with *K. capensis* Endl. in Namibia and the Cape Province and *K. arabica* R.Br. ex Chiov. in the Horn of Africa and southern Arabia.

The disjunction between the arid regions of north-east and south-west Africa is usually explained by the postulation of an arid corridor during arid phases of the

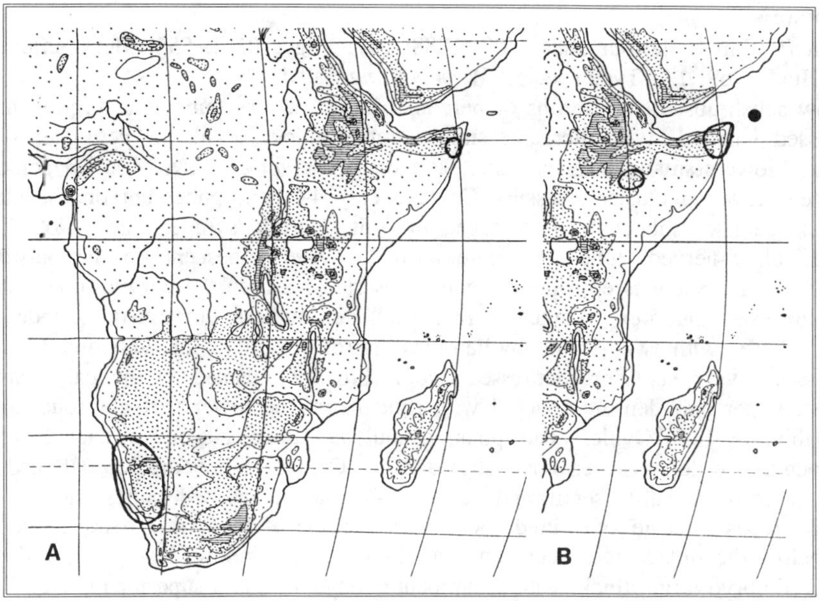

Figure 3. Known distributions of species of *Wellstedia*. A: *W. dinteri* (south-west), *W. laciniata* (north-east). B: from west to east *W. filtuensis*, *W. somalensis* and *W. socotrana* (dot).

Pleistocene, but the possibility of long distance dispersal has also been mentioned. Thulin (1994) agrees that the disjunct distributions of the fairly many *species* with a general north-south disjunction in Africa are Pleistocene relics or results of long distance dispersal. However, he also argues that the more confined disjunctions between south-western Africa and north-east tropical Africa/Arabia (as in *Wellstedia*) practically always concern disjunct genera or genera with vicarious species in the two regions that may well reflect a more ancient link between these floras, perhaps dating back to late Tertiary climatic changes in Africa.

A biogeographical interpretation of the cladogram in Figure 2 would imply that the first vicarious event in the history of *Wellstedia* is a separation between the south-west African and the north-east African populations, where the south-western population became *W. dinteri*. In north-eastern Africa *W. laciniata* in Somalia is the most basal surviving taxon, while vicarious events separating populations in Ethiopia, Somalia and Socotra resulted in the isolation of respectively *W. filtuensis*, *W. somalensis* and *W. socotrana*. Although *W. laciniata* and *W. somalensis* today have partly sympatric distributions in Somalia the biogeography of *Wellstedia* is compatible with an allopatric speciation model as these two species are not sister species.

Recent data on endemism in Somalia (Thulin 1994) shows two major zones particularly rich in endemics, one in the north-east, ranging from 47°E in the west to Cape Guardafui in the east, and one in the central part in the Hobyo area. The two new species of *Wellstedia* from Somalia described here both occur in the first-mentioned area of endemism.

Taxonomy

Wellstedia Balf.f. Balfour 1884: 407, 1888: 247; Pilger 1912: 558; Merxmüller 1960: 619; Hunt 1969: 1. - Type species: *W. socotrana* Balf.f.

Low subshrubs, dwarf shrubs or perennial herbs, silvery pubescent throughout with appressed hairs. Leaves alternate, simple, rather thick, sessile or petiolate; stipules absent. Flowers small, hermaphrodite, actinomorphic, in axillary cymes or sometimes in 1-few-flowered clusters, sessile. Calyx deeply 4-lobed, pubescent on the outside; lobes lanceolate, subequal, ± enlarged in fruit. Corolla sympetalous, 4-lobed; tube cylindrical, 8-nerved; lobes imbricate in bud, ovate, pubescent on the outside, 3-nerved, equal. Stamens epipetalous, equal, inserted between the corolla lobes, slightly bent inwards, glabrous; filaments free, subulate; anthers dorsifixed, rounded with cordate base, with two parallel pollen-sacs, introrsely dehiscent with one fissure on each sac. Ovary superior, depressed ovoid, broadly attached to the calyx-base, 2-locular, upper part densely covered with rigid hairs, the lower part glabrous; dissepiment thin, hyaline. Ovules anatropous, pendulous, subapical, one in each loculus or one loculus sterile; style terminal; stigma 2-fid. Capsule subobcordate, flattened, with rounded corners, hard, septifragal, erect, 2-locular, appressed pubescent, 1- or 2-seeded. Seeds truncate and pitted above with a circle of long rigid hairs in the upper part below the pitted area, acute and shortly pubescent below; testa thin; endosperm absent. Embryo large, thick, with accumbent cotelydons and a superior radicle.

Genus of five species distributed in Socotra, south-eastern Ethiopia, north-eastern Somalia, southern Namibia and the northern Cape Province of South Africa.

Key to the species

1 Some leaves with petiole 6-20 mm long; corolla lobes with fringed margin . . . . 2. *W. laciniata*
- Leaves tapering below into a petiole-like base, but without a distinct petiole; corolla lobes entire 2
2 Leaves 18-55 mm long 3
- Leaves 3-8 mm long 4
3 Plant with indumentum of long rigid appressed hairs and an understorey of much shorter and finer and more spreading hairs; ovary 2-ovulate; capsule usually 2-seeded . . . . 1. *W. dinteri*
- Plant with uniform indumentum of long appressed hairs; ovary 1-ovulate; capsule 1-seeded . . . . 3. *W. filtuensis*
4 Flowers solitary in the leaf axils; corolla-tube pubescent outside, with 4 thin thickenings inside . . . . 4. *W. somalensis*
- Flowers in axillary cincinni; corolla tube glabrous outside, with 4 conspicuous thickenings in the mouth of the tube. . . . . 5. *W. socotrana*

1. *W. dinteri* Pilger

Pilger 1912: 559; Merxmüller 1960: 619; Hunt 1969: 4. - Types: Namibia, Karasberge, Sandverhaar, I.1910, *Dinter 1193* (B, syntype, destroyed); Namib, *Dinter 1379* (B, syntype, destroyed); Auchab, VIII.1907, *Range 691* (B, syntype, destroyed; SAM, isosyntype, not seen); Karibis, I.1910, *Dinter 1250* (B, syntype, destroyed; SAM isosyntype, not seen).

W. dinteri var. *gracilior* Hunt 1969: 6. - Type: Namibia, Buellsport, *Strey 2140* (K, holotype;

BOL, PRE, isotypes, not seen).

Dwarf subshrub or shrublet; branches congested. Leaves oblanceolate, 18-55 x 2-6 mm, narrowing below into a petiole-like base, acute at the apex, densely clothed with rigid appressed hairs up to c. 2 mm long, some of which with broad multicellular base, and an understorey of much shorter and finer often more spreading hairs. Flowers pink to white, mostly arranged in dense cincinni. Calyx lobes 1-2.5 mm long. Corolla tube 1-1.5 mm long, cylindrical, with a rim-like thickening c. 1 mm from the base, near the throat, pubescent outside; lobes 0.5-1 mm long, obtuse, entire. Stamens shorter than the corolla lobes; filaments 0.25-0.5 mm long, slender; anthers 0.25-0.5 mm long. Ovary 2-ovulate; style c. 0.5-1 mm long, glabrous; stigma lobes c. 0.25 mm long. Capsule normally 2-seeded, oblique, 3-4 x 4-5 mm; dissepiment c. 2 mm long. Seeds 1.5-2 mm long.

Distribution and habitat. W. dinteri is distributed in the southern half of Namibia and in the northernmost part of the Cape Province in South Africa (Figure 3A). It occurs in sandy or stony undulating plains, on rocky slopes or in dried-up depressions at altitudes up to c. 1300 m.

Taxonomic remarks. Hunt (1969) indicated *Dinter 1193* in Berlin as the holotype of *W. dinteri*, but this was merely the syntype cited first by Pilger (1912) and Hunt's action can scarcely be regarded as a proper typification of the species. Also, the duplicate of *Dinter 1193* cited as an isotype in SAM apparently was wrongly labelled as the locality and date (April 1913, i.e. after the publication of *W. dinteri*) are not in accordance with the protologue. Possible lectotypes are the isosyntypes *Range 691* and *Dinter 1250* in SAM, but as these have not been seen, no formal lectotypification is proposed here.

W. dinteri is unique in the genus by having normally 2-seeded capsules. It also has a shorter style than the other species.

Two varieties of *W. dinteri* were recognized by Hunt (1969): var. *dinteri* with thicker branches, longer leaves and flowers arranged in dense cincinni, and var. *gracilior* with a more slender habit, smaller leaves and flowers mostly solitary and well separated on lax branchlets. However, the variation seems continuous and we prefer not to subdivide the species. Also Hunt pointed to the existence of intermediate collections and stated that "the significance of the variation described cannot at present be very satisfactorily assessed".

Specimens other than those cited by Hunt (1969). Namibia. Bethanien District: 10 km NE of Sandverhaar, road between Aus and Seeheim, 1.III.1968, *Kers 2239* (S). Keetmanshoop District: 11 km from Keetmanshoop on road to Seeheim, 6.V.1976, *Oliver, Müller & Steenkamp 6322* (K); NE of Rosh Pinah, 7.V.1976, *Oliver & Müller 6397* (K). Warmbad District: Klein Karas, 28.IV.-1928, *Örtendahl 169* (UPS); Klein Karas-Aiais, 29.VII.1931, *Örtendahl 590* (UPS), 7.VIII.1931, *Örtendahl 646* (S, UPS); road between Grünau and Seeheim, 3 km before Klein Karas station, 27.II.1968, *Kers 2328* (S); Otscha Rivier, 26.X.1918, *Örtendahl 260* (UPS).

2. *W. laciniata* Thulin & A.Johanss. sp. nov.

Type: Somalia, Nugaal Region, 95 km E of Sinugif, 40 km NW of Eil airstrip, 1.I.1973, *Bally & Melville 15463* (K, holotype).

Species nova a W. dinteri Pilger foliis externis longe petiolatis, lobis corollae laciniatis, lobis stigmatis obsoletis, et capsulis 1-spermis differt.

Dwarf subshrub or perennial herb, c. 8 cm, forming rosettes on the ground; branches ± decumbent, congested. Outer leaves ovate, with blade 10-15 x 8-14 mm and distinct petiole 6-20 mm long, the inner leaves smaller and ± sessile; indumentum of rigid appressed hairs up to 2 mm long, some of which with broad multicellular bases, and an understorey of much finer shorter and more spreading hairs. Flowers in very dense cincinni, colour unknown. Calyx lobes 2-3.5 mm long. Corolla tube 1.5-2 mm long, widening upwards, with a rim-like thickening c. 1 mm from the base, pubescent in upper part outside, glabrous below; lobes c. 1 mm long, obtuse, with fringed margins. Stamens shorter than the corolla lobes; filaments c. 0.5 mm long; anthers c. 0.5 mm long. Ovary 2-ovulate; style 1.25-1.5 mm long, glabrous; stigma lobes extremely short, obsolete.

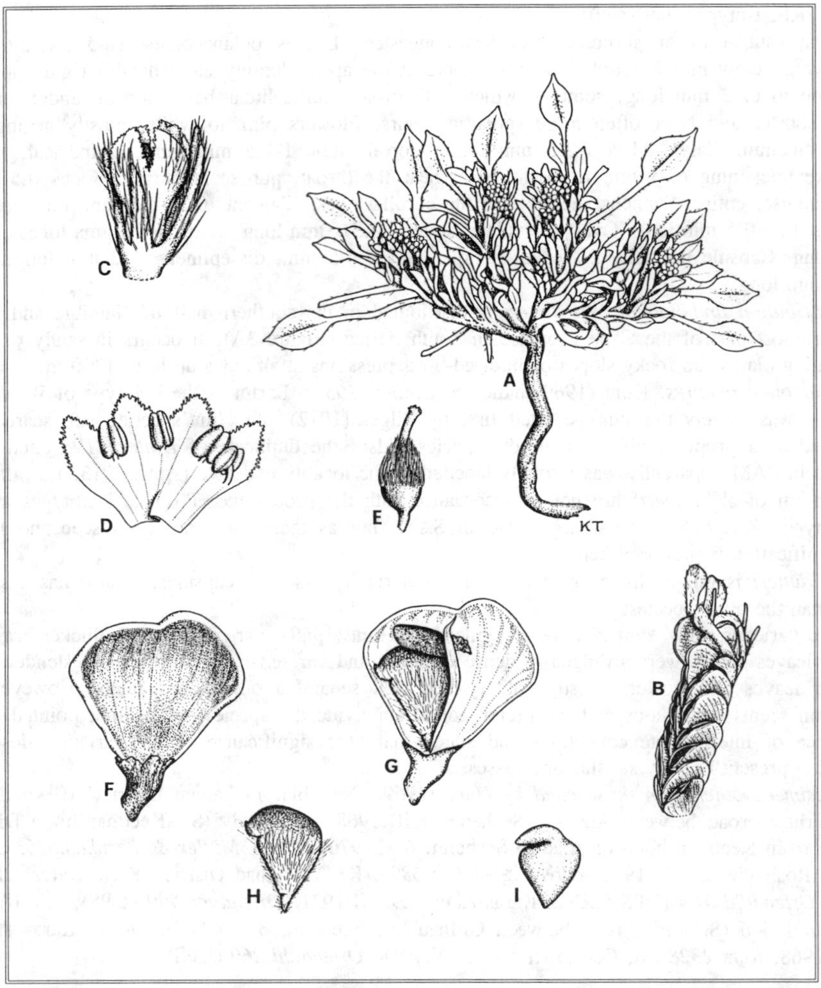

Figure 4. *Wellstedia laciniata*. A: Habit, x0.5. B: Fruiting branchlet, x1.4. C: Flower, x4. D: Corolla opened out, x4. E: Ovary with style, x4. F: Capsule, x4. G: Seed in capsule, x4. H: Seed, x4. I: Embryo, x4. A, G, H from *Bally & Melville 15463*; B, F, I from *Lavranos & Carter 23508*; C, D, E from *Beckett 1078*.

Capsule 1-seeded, oblique, c. 3 x 4 mm; dissepiment c. 2 mm long, with a thick cap-like placenta. Seed 1.5-2 mm long. - Figure 4.

Distribution and habitat. *W. laciniata* is distributed in the Bari and Nugaal Regions of northeastern Somalia, but it is not known from the northernmost mountains (Figure 3A). It grows in rocky limestone slopes with open vegetation of low bushes of *Acacia* and *Commiphora* at altitudes up to 680 m.

Taxonomic remarks. *W. laciniata* is probably most closely related to *W. dinteri*, but differs markedly in its distinctively petiolate leaves (Figure 4A), laciniate corolla-lobes (Figure 4D), obsolete stigma lobes (Figure 4E), cap-like placentae (Figure 4G), and 1-seeded capsules (Figure 4G).

Specimens other than the type. Somalia. Bari Region: 150 km NE of Gardo on road to Iskusciuban, 9°55'N, 50°09'E, 14.XI.1986, *Lavranos & Carter 24670* (K); 9°23'N, 49°21'E, 12.VI.-1981, *Beckett 1078* (K). Nugaal Region: 3 km N of Bedey, E of Eil, 8°03'N, 49°57'E, 4.X.1985, *Lavranos & Carter 23508* (K).

3. *W. filtuensis* Hunt & Lebrun

Hunt & Lebrun 1975: 222. - Type: Ethiopia, Sidamo Region, crest of Mt Filtu, 12.XI.1972, *Rippstein 827* (K, holotype; ALF, M, P, isotypes, not seen).

Shrublet or subshrub to c. 40 cm tall; branches fairly long and laxly arranged. Leaves oblanceolate, 22-45 x 4-6 mm, densely covered with closely appressed hairs up to c. 2 mm long, without obvious multicellular bases. Flowers yellowish-green or white turning reddish, arranged in dense shortly stalked cincinni. Calyx lobes 2-3 mm long. Corolla tube 1.5-2 mm long, cylindrical, with a thin rim-like thickening near the base, pubescent in upper part outside, glabrous below; lobes 0.5-1.5 mm long, obtuse, entire. Stamens shorter than the corolla lobes; filaments c. 0.75 mm long; anthers c. 0.5 mm long. Ovary 1-ovulate; style 1.2-1.8 mm long, glabrous; stigma lobes c. 0.25 mm long. Capsule 1-seeded, oblique, 3-5 x 3.5-5.5 mm; dissepiment c. 2 mm long. Seed 2-2.5 mm long.

Distribution and habitat. W. filtuensis is distributed in the Bale and Sidamo Regions in south-eastern Ethiopia (Figure 3B). It grows in low open bushland on limestone slopes and outcrops, often in disturbed areas, at altitudes between 1300 and 1670 m.

Taxonomic remarks. W. filtuensis differs from *W. dinteri* and *W. laciniata* in its uniform indumentum of long appressed hairs without any understorey of short, fine hairs, in its corolla-tube with the rim-like thickening nearer to the base, and in its 1-ovulate ovary. From *W. somalensis* and *W. socotrana* it differs, e. g. in its much larger leaves.

Specimens other than the type. Ethiopia. Sidamo Region: 60 km from Negele on road to Welensu Ranch, S from road to Melca Guba, 4°59'N, 39°47'E, 23.V.1983, *Gilbert, Ensermu & Vollesen 7739* (K, UPS); Negele, 26-29.IX.1939, *Corradi 6252* (FT); 100 km E of Negele on road to Filtu, 32 km W of Filtu, 5°13'N, 40°26'E, 9.VI.1986, *Gilbert, Sebsebe & Vollesen 8109* (K, UPS). Bale Region: 45 km N of Ganale River on road from Negele and Bitata to Dolo Menna, 11.XI.1986, *Gilbert & Ermias 8469* (K, UPS).

4. *W. somalensis* Thulin & A. Johanss. sp. nov.

Type: Somalia, Bari Region, Galgala, 10°59'N, 49°03'E, 1.XII.1986, *Thulin & Warfa 6202* (UPS, holotype; K, MOG, isotypes).

W. socotrana sensu Hunt 1969: 4, pro parte quoad specim. Bally 10948, non Balf. f. Species nova a W. socotrana Balf.f. floribus axillaribus solitariis, tubo corollae extus pubescenti et intus leviter incrassato, et stylo longiore non dense pubescenti differt.

Low cushion-forming shrublet, up to c. 10 cm, branches ± congested. Leaves ± obovate, 3-7 x 1.5-4 mm, narrowing below into a petiole-like base, completely covered with appressed hairs up to c. 1 mm long hairs, without obvious multicellular bases. Flowers solitary in the leaf axils, yellowish or yellowish white. Calyx lobes 1.2-3 mm long. Corolla tube 1.5-2.5 mm long, cylindrical, pubescent outside in the upper part, glabrous below, with 4 thin thickenings 0.5-1 mm above the base, forming a ± continuous rim; lobes c. 1.2 mm long, subacute, entire. Stamens exserted, about as long as the corolla lobes; filaments c. 1 mm long; anthers c. 0.5 mm long. Ovary 1-ovulate; style 2-3 mm long, sparsely pubescent in lower part or glabrous; stigma lobes 0.25-0.5 mm long. Capsule 1-seeded, oblique, c. 2-3 x 3-4 mm; dissepiment c. 2-2.4 x 0.6-1 mm, with a c. 0.4 mm long placenta. Seed c. 2 mm long. - Figure 5.

Distribution and habitat. W. somalensis is fairly widespread in north-eastern Somalia and is known from the Sanaag, Bari and Nugaal Regions (Figure 3B). It occurs in sparse open vegetation with *Acacia*, *Commiphora*, *Boswellia* and *Mimusops angel* on rocky limestone slopes at altitudes between about 200 and 1300 m.

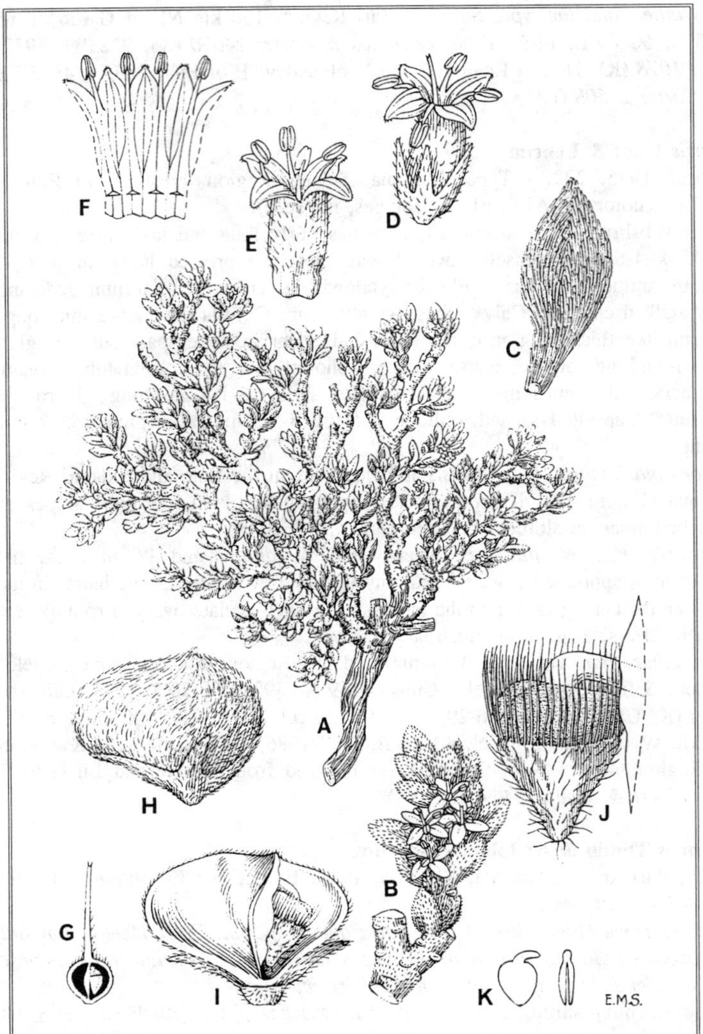

Figure 5. *Wellstedia somalensis* (from Hunt 1969). A: Habit, x0.66. B: Flowering shoot, x2. C: Leaf, x4. D: Flower, x5.3. E: Corolla and stamens, x5.3. F: Corolla opened out, x5.3. G: Vertical section of gynoecium, x5.3. H: Capsule, x5.3. I: Seed in capsule, x5.3. J: Seed, x10.5. K: Embryo, x10.5. All from *Bally 10948*.

Taxonomic remarks. Hunt (1969) treated *Bally 10948* as a first record of *W. socotrana* from the African mainland and several other collections here included in *W. somalensis* have previously also been identified as *W. socotrana*. However, *W. somalensis* (Figure 5) differs markedly from *W. socotrana* in having solitary flowers in the leaf axils (flowers not arranged in distinct cincinni), in its indumentum of longer hairs (hairs up to c. 1 mm long versus up to c. 0.5 mm long), in its corolla tube being pubescent outside and with 4 thin thickenings 0.5-1 mm above the base of the tube inside (corolla tube not glabrous outside and with 4 marked thickenings in the mouth between the lobes), and in its glabrous or slightly hairy and 2-3 mm long style (style densely hairy and 1-1.5 mm long). It should be noted that the stigma lobes of *W. somalensis* are actually more distinct than shown in Figure 5G.

The colour of the flowers in *Bally 10948* was described as brick red, but this probably refers to dried-up flowers only.

Scortecci s.n. (FT), 28.IX.1957, a sterile collection from the Abal area (c. 11°18', 49°33'E) in north-eastern Somalia is similar to *W. somalensis*, but has leaves up to 25 x 6 mm and probably represents an undescribed species. Typical *W. somalensis* is also known from this area, but fertile material of the large-leaved form is needed to show whether it is conspecific or not.

Specimens other than the type. Somalia. Sanaag Region: eastern Al Madu Range, 8.X.1956, *Bally 10948* (K). Bari Region: Galgala, 64 km SSW of Bosaso at eastern end of Al Madu Range, 10.I.1973, *Bally & Melville 15715* (K); Abal, Bocc pass, 27.IX.1957, *Scortecci s.n.* (FT); Abal, 28.IX.1957, *Scortecci s.n.* (FT); Habeno, 14.IX.1957, *Scortecci s.n.* (FT); Bargal--El Gal, 30.XI.-1970, *Bavazzano & Lavranos s.n.* (FT); Tohen--Cape Guardafui, 3.XII.1970, *Bavazzano & Lavranos s.n.* (FT); Karkaar Mts., between Gardo and Iscusciuban, God Adde, 23.XI.1985, *Thulin & Warfa 5453* (K, MOG, UPS); between Sagaleh and the sea (Bender Beila), 7.I.1954, *Merla, Azzaroli & Fois s.n.* (FT). Nugaal Region: 48 km SW of Gardo, 9°18'N, 48°46'E, 17.XI.1980, *Hemming & Watson 3133* (K); 16 km N of Eil and about 2 km N of airstrip, 3.I.1973, *Bally & Melville 15518* (K).

5. *Wellstedia socotrana* Balf. f.

Balfour 1884: 408, 1888: 249; Hunt 1969: 4, pro parte, excl. specim. Bally 10948. Types: Socotra, without precise locality, II-III, 1880, *Balfour, Cockburn & Scott 569* (K, lectotype, selected here).

Low cushion-shaped shrublet; branches strongly congested. Leaves oblanceolate to narrowly obovate, 6-8 x 2-3 mm, densely pubescent with appressed hairs up to c. 0.5 mm long, without obvious multicellular bases. Flowers in axillary dense cincinni, colour unknown. Calyx-lobes c. 1.5 mm long. Corolla tube c. 1.5 mm long, cylindrical, glabrous outside, with 4 conspicuous thickenings in the mouth between the lobes; lobes c. 1.5 mm long, spreading, subacute, entire. Stamens exserted, almost as long as corolla lobes; filaments c. 0.75 mm long; anthers c. 0.5 mm long. Ovary 1-ovulate; style densely appressed hairy all along, 1-1.5 mm long; stigma lobes c. 0.25 mm long. Capsule 1-seeded, oblique, c. 3.6 x 5 mm; dissepiment c. 3 x 1.4 mm, with a c. 0.4 mm long placenta. Seed c. 2.5 mm long.

Distribution and habitat. W. socotrana is only known from Socotra (Figure 3B), where it has not been seen for more than a century. It was searched for in vain by one of us (M. T.) in January-February 1994. The only information on locality and habitat of the species given is "in campis Socotrae Insulae crescens" (Balfour 1884).

Taxonomic remarks. A second syntype of *W. socotrana*, *Hunter s.n.*, was mentioned by Balfour (1884, 1888). This has been searched for in vain at BM, E and K. Balfour's own material at K, that was also the basis for the illustration of the species in Balfour (1888) is here selected as a lectotype, although the number *569* is not indicated on the sheet.

W. socotrana differs from all other species in the genus in its conspicuous thickenings in the mouth of the corolla tube, densely hairy style, and large dissepiment of the capsule.

For the record of *W. socotrana* from Somalia by Hunt (1969), see under *W. somalensis*.

Acknowledgements

We are indebted to Ms Kerstin Thunberg for the drawing in Figure 1, to the Bentham-Moxon Trustees for permission to reproduce the illustration of *W. somalensis*, and to Ms Elisabeth Grafström, Palynological Laboratory, Swedish Museum of Natural History, Stockholm, for the pollen photographs.

References

Balfour, I.B., 1884. Diagnoses plantarum novarum phanerogamarum socotrensium. Proc. Roy. Soc. Edinb. 12: 402-411.

Balfour, I.B., 1888. Botany of Socotra. Trans. Roy. Soc. Edinb. 31.

Constance, L. & T.I. Chuang, 1982. SEM survey of pollen morphology and classification in Hydrophyllaceae (Waterleaf family). Amer. J. Bot. 69: 40-53.

Cronquist, A., 1981. An integrated system of classification of flowering plants. Columbia University Press, New York.

Dostál, J., 1957. Botanicka nomenklatura. Prague.

Emberger, L., 1960. Les vegetaux vasculaires 2. Paris.

Erdtman, G., 1952. Pollen morphology and plant taxonomy. Stockholm.

Farris, J.S., 1988. Hennig86, ver. 1.5. Reference. New York, published by the author.

Gilg, E., 1924. Syllabus der Pflanzenfamilien, ed. 9 & 10. Berlin.

Gürke, M., 1897. Borraginaceae. In A. Engler & K. Prantl (eds.), Die natürlichen Pflanzenfamilien IV 3a. Leipzig.

Holmgren, P.K., N.H. Holmgren & L.C. Barnett, 1990. Index Herbariorum, Part 1, The herbaria of the world, ed. 8. Regn. Veg. 120.

Hunt, D.R., 1969. *Wellstedia socotrana, W. dinteri* var. *dinteri, W. dinteri* var. *gracilior*. Hook. Ic. Pl. 37: tab. 3665-3667.

Hunt, D.R. & J.P. Lebrun, 1975. A new *Wellstedia* (Boraginaceae) from Ethiopia. Kew Bull. 30: 222.

Lemée, A.M.V., 1935. Dictionnaire descriptif de genres de plantes phanérogames 6. Paris.

Melchior, H., 1964. Syllabus der Pflanzenfamilien, ed. 12. Berlin.

Merxmüller, H., 1960. Wellstediaceae. Mitt. Bot. Staatss. München 3: 619-622.

Novák, F.A., 1943. Wellstediaceae. Prát, Rostlinopis 9: 530.

Pijl, L. van der, 1972. Principles of dispersal in higher plants. Berlin.

Pilger, R., 1912. Die Gattung *Wellstedia* in Südwestafrika. Bot. Jahrb. Syst. 46: 558-561.

Pilger, R. & K. Krause, 1914. Nachträge IV zu IV. 3a. In: A. Engler & K. Prantl (eds.), Die natürlichen Pflanzenfamilien. Leipzig.

Takhtajan, A., 1987. Sistema magnoliofitov. Moskva.

Thulin, M., 1987. A new species of *Xylocalyx* (Scrophulariaceae) from Somalia. Nord. J. Bot. 7: 267-269.

Thulin, M., 1994. Aspects of disjunct distribution and endemism in the arid parts of the Horn of Africa, particularly Somalia. Proc. XIIIth AETFAT Congress, Zomba.

Ulbrich, E., 1928. Biologie der Früchte und Samen. Berlin.

White, F., 1983. The vegetation of Africa. Natural Resources Research 20. UNESCO, Paris.

Zohary, M. 1962. Plant life of Palestine. The Ronald Press Company, New York.

Diversification in an arid world: the Mesembryanthemaceae

H.-D. IHLENFELDT
Institut für Allgemeine Botanik und Botanischer Garten, Universität Hamburg, Ohnhorststrasse 18, D-22609 Hamburg, Germany.

Abstract
The family Mesembryanthemaceae (c. 2000 species, 116 genera) is the dominating family in the succulent flora of the Succulent Karoo Region (southern Africa), to which the vast majority of species of this family are restricted. The success of this family is due to an extremely broad spectrum of highly specialized life forms and life strategies. The family must have developed a tempo of evolution which is probably unrivalled within Angiosperms, since the mega-niche Succulent Karoo opened up only 5 million years BCE at most. The remarkable biodiversity of this family is the result of a complex interaction between the availability of numerous diverse niches, often very small in extent, a climate characterized by steep gradients, and a strong genetic drift caused by a very low rate of gene exchange among populations and by frequent breakdown of populations. Evidence is presented that a special genetic potentiality plays a crucial role in the evolutionary processes of this family. This potentiality allows fast and easy rearrangement of high-ranking regulatory "homeotic" genes. This process is strongly canalized by the shortage of resources, i.e. changes in one place of the life cycle usually force changes in other places. By repeated changes in homeotic genes, this family has succeeded in evolving numerous novel and diverse complex life forms and life cycles in spite of a very limited set of subordinate ontogenetic programs. An enlarged version of the presented paper has been published under the same title in Ann. Rev. Ecol. Syst. 25 (1994): 521-546.

Résumé
La famille des Mesembryanthemaceae (environ 2000 espèces et 116 genres) est la famille dominante de la flore des plantes succulentes de la Région à végétation de plantes succulentes du Karoo; la majorité des espèces de cette famille y est d'ailleurs endémique. Cette prédominance s'explique par un spectre extrêmement large des formes biologiques hautement spécialisées et des stratégies de survie. La famille a dû développer un rythme d'évolution probablement inégalé chez les Angiospermes, si l'on tient compte que la niche géante du Karoo avec ses plantes succulentes ne s'est formée au plus tôt qu'il n'y a que 5.000.000 d'années BCE. La biodiversité remarquable de cette famille résulte d'une interaction complexe entre la disponibilité de nombreuses niches diversifiées, souvent de superficie très réduite, un climat caractérisé par de forts gradients, et une forte dérive génétique due à un taux vraiment faible d'échange entre populations et à un appauvrissement fréquent de celles-ci. Il est prouvé qu'une potentialité génétique spéciale joue un rôle crucial dans les processus d'évolution de cette famille. Cette potentialité permet un réarrangement rapide et aisé des gènes "homéotiques" régulateurs de rang élevé. Ce processus est fortement canalisé suite à l'insuffisance des ressources, c'est-à-dire que des changements en un stade du cycle de vie provoque généralement des changements aux autres stades. A la suite de changements répétés dans les gènes homéotiques, cette famille est parvenue à développer de nombreuses formes biologiques complexes nouvelles et diverses et de nouveaux cycles de vie, malgré un ensemble très limité de programmes ontogéniques subordonnés. Une version plus détaillée de cet article a été publiée sous le même titre dans Ann. Rev. Ecol. Syst. 25 (1994): 521-546.

Key words
Adaptation, Arid habitats, Evolution, Homeotic genes, Mesembryanthemaceae.

L.J.G. van der Maesen et al. (eds.), The Biodiversity of African Plants, 87.

Tubers and bulbs as an adaptive strategy in the Kalahari

B.J. HARGREAVES
National Museum, Private Bag 00114, Gaborone, Botswana.

Abstract
Many plant groups which grow as shrubs, vines and trees elsewhere appear as geophytes in Botswana. This is an adaptation to the Kalahari which has deep sands, little surface water and periodic wildfires. This adaptation is found in numerous unrelated plant families, e.g. in the Cucurbitaceae, Passifloraceae, Vitaceae, Pedaliaceae, Leguminosae, Apocynaceae.

Key words
Geophytes, Kalahari, Uses.

Introduction

The Kalahari* is a unique environment. Although the average rainfall is well above what most consider true desert (ranging from over 600 mm per year in the northeast of Botswana to just under 200 mm in the southwest), deep sands (up to 300 m deep) left from a prehistoric desert pose great difficulties for the flora and fauna. In addition, the rainfall deviates considerably from the average in any one season and periodic droughts are common. The vegetation which has been selected by this environment is a dry savanna ranging from savanna woodland in the northeast to a mix of shrub and grassland savanna in the southeast (Weare & Yalala 1971).

This exemplifies the hypothesis of Walter (1979) in which dry woodlands and savannas have grasses which exploit superficial moisture and woody plants utilizing deep moisture, thus avoiding competition. Walker & Noy-Meir (1982) have developed this theme further. Superficial-rooting trees such as some *Combretum* species give rise to thicket vegetation while some, such as *Baikiaea plurijuga* Harms, are very deep-rooted and exploit deep pockets of moisture. Calvert (1986) takes this further to explain the distribution of *Baikiaea* and miombo (*Brachystegia* species). The latter is an adaptable group which exploits both layers. A third factor which has created the Kalahari vegetation is fire. Although wildfires are more controlled today, they played a significant role in selecting trees and grasses which can recover after periodic burning.

In addition to trees and grasses, there are numerous plants which have adapted to the Kalahari by going underground for much of the year. I wish to concentrate on this group in this paper. Geophytes are quite common in this environment as they are protected both from periodic drought and fire by the overlying sand. In general they have water and food storage organs in the form of tubers and bulbs. They are thus classifiable as succulents, although there is considerable debate on this point. Personally, I believe an organ can be water storing (or succulent) at the same time that it is used for food storage.

* Kalahari is more correctly spelled Kgalagadi, but I will bow to English convention for the sake of clarity.

L.J.G. van der Maesen et al. (eds.), The Biodiversity of African Plants, 88-91.

Stem and leaf succulents are not common in the Kalahari, being generally limited to the rocky peripheral areas (Hargreaves 1990). As I shall demonstrate, many groups which have ground succulence elsewhere have become geophytes in the Kalahari. This phenomenon crosses many taxonomic lines and is found in totally unrelated plants.

Leguminosae

One of the most striking geophytes in Botswana is the morama bean, *Tylosema esculentum* (Burchell) A. Schreiber, a member of the Caesalpinioidae of the Leguminosae. Tubers have been weighed at as much as 277 kg (Bergström & Skarpe 1981). From these enormous tubers annual procumbent shoots spread out over the surrounding sand. It is the sand itself that allows such growth, providing a soft, flexible medium to spread into. The tubers are edible, but it is mainly the bean which provides a commercially processed food. The morama bean is closely related to the genus *Bauhinia* and in fact was formerly classified as *Bauhinia esculenta* Burchell. *Bauhinia petersiana* Bolle occurs as a non-succulent woody shrub on the eastern edge of the Kalahari. Elsewhere in southern Africa species of *Bauhinia* reach tree size.

Another example of the geophytic trend in the Leguminosae is the genus *Elephantorrhiza* in the subfamily Mimosoidae. In the northeast of Botswana we find *E. goetzei* (Harms) Harms, the large-bean elephant-root, which is a tree up to 7 m in height. In the southeast we find *E. burkei* Benth., the sumac elephant-root which is usually a shrub and can be as low as 30 cm. In the southwest we find *E. elephantina* (Burch.) Skeels which has hardly anything above ground other than leaves. Both *E. burkei* and *E. elephantina* have large underground rhizomes but only *E. elephantina* dies back completely to ground level every year. They both provide food and medicine and are a good source of tannin as well.

Although the plough-breaker (*Erythrina zeyheri* Harv.) is a geophyte found from the South African Transvaal to Lesotho, no such geophytic species occur in the Kalahari. The only erythrinas in Botswana are *E. lysistemon* Hutch. of the Southeast which is a typical large tree and *E. mendesii* Torre which is a rhizomatous shrub of the North. The only geophytic members of the Papilionoideae in Botswana are three species of small-tubered *Eriosema*.

Apocynaceae

Further examples of geophytic adaptation are found in the Apocynaceae. *Adenium oleifolium* Stapf forms a large underground tuber and only short branches and leaves above ground. Elsewhere species of *Adenium* grow as caudiciform succulents, mainly above ground. Interestingly a plant of *A. oleifolium* which I transplanted to shallow soil has thickened its above-ground parts to look more like adeniums elsewhere. The poisonous tubers are used medicinally.

Asclepiadaceae

Fockea angustifolia K. Schum. has a tuber which may be buried as deep as 1 m underground. Although young tubers are carrot-shaped, the huge old ones are quite irregular. The crisp, juicy tuber is sweet and edible. *Fockea lugardi* N.E. Brown also has an edible tuber. In contrast, *Fockea multiflora* K. Schum. is a succulent vine from the North of Botswana. Other tuberous Asclepiadaceae are found in Botswana, including representatives of the genera *Brachystelma* and *Ceropegia*, but their tubers are small. There are numerous stem succulents including carrion flowers (*Stapelia* spp.) and *Sarcostemma viminale* (L.) R. Br.

Asclepiadaceae (Periplocoideae)
The related subfamily Periplocoideae is represented by tuberous species of *Raphionacme* such as *R. burkei* N.E. Brown, which is an emergency source of water, *R. monteiroae* (Oliv.) N.E. Brown and *R. hirsuta* (E. Mey) R.A. Dyer ex Phill.

Convolvulaceae
The Convolvulaceae are generally vines but may be succulent shrubs such as *Ipomoea adenioides* Schinz. Many are annuals but some are tuberous-rooted perennials. Examples of the latter are *Ipomoea bolusiana* Schinz which is widespread and has an edible tuber, and *I. holubii* Bak.

Pedaliaceae
They have a wide range of forms in Botswana ranging from non-succulent annuals to succulent perennials. *Sesamothamnus lugardii* N.E. Brown is a spectacular succulent caudiciform shrub. *Pterodiscus aurantiacus* Welw. and *P. ngamicus* N.E. Brown are small caudiciform plants while *P. speciosus* Hook. has a completely underground tuber. *Harpagophytum procumbens* (Burch.) DC ex Meisn. and *H. zeyheri* Decne. have non-succulent annual stems from tubers. The central tuber produces side tubers. Medicine is made commercially from these tubers.

Vitaceae
This is another family showing extreme variability. While some are woody shrubs such as *Cissus cornifolia* (Bak.) Planch. and others succulent vines such as *Cissus quadrangularis* L. and *C. rotundifolia* (Forssk.) Vahl, most species in Botswana are tuberous. *Cyphostemma cirrhosum* (Thunb.) Descoings ex Wild & R.B. Drumm. and *C. congestum* (Bak.) Descoings ex Wild & R.B. Drumm. have somewhat succulent leaves on annual stems. The leaves of *C. schlechteri* (Gilg & Brandt) Descoings ex Wild. & R.B. Drumm. are quite succulent. No *Cyphostemma* species in Botswana are caudiciform like those of Namibia. The oxalic acid concentration in some of the fruits is so high that it can cause poisoning.

Passifloraceae
Another variable family is the Passifloraceae. Two species, *Adenia glauca* Schinz and *A. spinosa* Bratt-Davy, found in eastern Botswana, are caudiciform, while two others, *A. digitata* (Horv.) Engl. and *A. repanda* (Burch.) Engl., found further West, are vines with large underground tubers. Reports on the poisonous properties vary, but *A. glauca* is generally regarded as non-toxic whereas *A. digitata* is the most deadly.

Cucurbitaceae
The Cucurbitaceae also vary. Many, such as the well-known tsama melon (a wild and sometimes bitter variety of the common water melon, *Citrullus lanatus* (Thunb.) Matsum. & Nakai) are annual prostrate vines. Others have caudiciform bases or

tubers. *Corallocarpus bainesii* (Hook.f.) A. Meeuse, *C. triangularis* Cogn., *C. welwitschii* (Naud.) & Hook.f. ex Welw. and *C. wildii* C. Jeffrey are generally caudiciform, although the stem may be deep enough in the soil to be called tuberous. The base of *C. bainesii* is not eaten because of its bitter taste, while that of *C. welwitschii* is bitter yet eaten. *Kedrostis hirtella* (Naud.) Cogn. and *K. foetidissima* (Jacq.) Cogn. are similar to the *Corallocarpus* species. *Momordica repens* Bremek., *M. cardiospermoides* Klotzsch and *M. kirkii* (Hook.f.) C. Jeffrey have large tuberous roots. The *M. repens* tuber is used medicinally and *M. cardiospermoides* may have a tuber of 10 m.

Acanthosicyos naudianus (Sond.) C. Jeffrey is related to water melons (and in fact was at one time included in the genus *Citrullus*), but it has a large root which becomes woody but is juicy and edible when young (although bitter). Other species with tubers are: *Cucumis kalahariensis* A. Meeuse, *C. zeyheri* Sond. (although other *Cucumis* species are annuals), *Trochomeria debilis* (Sand.) Hook.f. and *T. macrocarpa* (Sand.) Hook.f. The tuber of *T. debilis* is edible.

Other families in Botswana with tubers are: Portulacaceae with species of *Talinum* (edible tubers), Mesembryanthemaceae with *Nananthus* and *Mestoklema*, and Euphorbiaceae with species of *Jatropha* (although *J. spicata* Pax is a non-tuberous shrub). *Euphorbia davyi* N.E. Brown and *E. duseimata* R.A. Dyer have tuberous stems merging into roots. The former is fatter and is found on the eastern edge of the Kalahari while the latter is thinner with underground branching and grows in deep sand.

Monocotyledonae

A few, such as *Sansevieria aethiopica* Thunb. and *S. pearsonii* N.E. Brown, have succulent rhizomes. Others such as *Cyanotis speciosa* (L.f.) Hassk. and the ground orchids have small tubers. Most, however, have water and storage in leaves or bulbs. The bulbs tend to be closer to the surface than the tubers (although some, such as *Crinum* species can be quite deep) and in many cases have at least the upper part exposed. It is said that more exposed bulbs and tubers are more poisonous, but there are no references available to substantiate this. At any rate *Scilla nervosa* (Burch.) J.P. Jessop, which is quite exposed, was seen to be grazed in the Kalahari. Bulbs are found in numerous lilies and amaryllids. Because of their general shallowness they do not seem as good an adaptation to the Kalahari as are tubers.

References

Bergström, R. & C. Skarpe, 1981. The tuber of morama (Tylosema esculenta). Botswana Notes and Records 13: 156-158.

Calvert, G.M., 1986. The ecology and management of the Kalahari Sand Forest vegetation of south-western Zimbabwe. In: G.D. Pearce (ed.), The Zambezi Teak Forests, Proceedings of the First International Conference on the Teak Forests of Southern Africa, March 1984, Livingstone, Zambia, Forest Department, Ndola, Zambia.

Hargreaves, B.J., 1990. The succulents of Botswana. National Museum, Gaborone, Botswana.

Walker, B.H. & I. Noy-Meir, 1982. Aspects of the stability and resilience of savanna ecosystems. In: B. Huntley & B. Walker (eds), Ecology of tropical savannas, Ecological Studies 42, Springer Verlag, Berlin.

Walter, H., 1979. Vegetation of the earth and ecological systems of the geo-biosphere (2nd ed.). Heidelberg Science Library, Springer-Verlag, New York.

Weare, P.R. & A. Yalala, 1971. Prov. vegetation map. Botswana Notes and Records 3: 131-152.

New collections of Mesembryanthema in South African grasslands

M. MÖSSMER [1], G.F. SMITH [1], H.E.K. HARTMANN [2] & C.M. SMIT [3]
[1] National Botanical Institute, Private Bag X101, Pretoria 0001, South Africa.
[2] Institut für Allgemeine Botanik und Botanischer Garten, Universität Hamburg, Ohnhorststrasse 18, D-22609 Hamburg 52, Germany.
[3] Department of Botany, University of Pretoria, Pretoria 0001, South Africa.

Abstract
Recent surveys of the Mesembryanthema in the southern Transvaal and northern Orange Free State, South Africa, have extended the known geographical distribution ranges of three genera, *Mossia* N.E. Br., *Ebracteola* Dinter & Schwantes and *Chasmatophyllum* Dinter & Schwantes, in both northerly and easterly directions. The lack of records from this summer rainfall area is probably due to collection neglect and not to the absence of plants. The reduction of habitats by agricultural and mining activities may also be an important contributing factor.

Key words
Chasmatophyllum, Conservation, Distribution, *Ebracteola*, Grassland, *Mossia*.

Introduction

Recent surveys in the southern Transvaal and northern Orange Free State (OFS), South Africa, have extended the known geographical distribution ranges of three Mesembryanthema genera, *Mossia* N.E. Br., *Ebracteola* Dinter & Schwantes and *Chasmatophyllum* Dinter & Schwantes. The centre of distribution for the Mesembryanthema (sensu Hartmann 1991) lies in the extreme southwestern part of the African continent (Hartmann 1991). The highest diversity of taxa is found in the arid winter rainfall area of the western Cape Province, South Africa, with a significant secondary centre in the eastern Cape Province with its bimodal rainfall distribution. Compared to these regions, the climatically severe inland escarpment of southern Africa is poor in the number of naturally occurring Mesembryanthema genera (Hartmann 1989). In this summer rainfall area, Mesembryanthema are mostly confined to azonal localities and have consequently been neglected by collectors (Hartmann 1991).

In our study, plants were collected in two separate areas, one on the border between the southwestern Transvaal and the northern OFS, and the other in the southeastern Transvaal. These areas are both subject to mining activities. The first area falls within the gold- and uranium-rich Witwatersrand basin (Reimold 1992), whereas the southeastern Transvaal is an important source of coal (Tyson et al. 1988), supplying electricity to the major urban and industrial centres of the Pretoria-Witwatersrand-Vereeniging (PWV) region. Additionally, both areas fall within the agriculturally significant Grassland Biome (Rutherford & Westfall 1986). Owing to widespread cultivation and livestock production, little natural vegetation has survived on arable soils (Smith & Hartmann 1992). The remaining vegetation is mostly confined to non-arable soils and is dominated by grasses (Gramineae) (Smith & Hartmann 1992).

L.J.G. van der Maesen et al. (eds.), The Biodiversity of African Plants, 92-97.

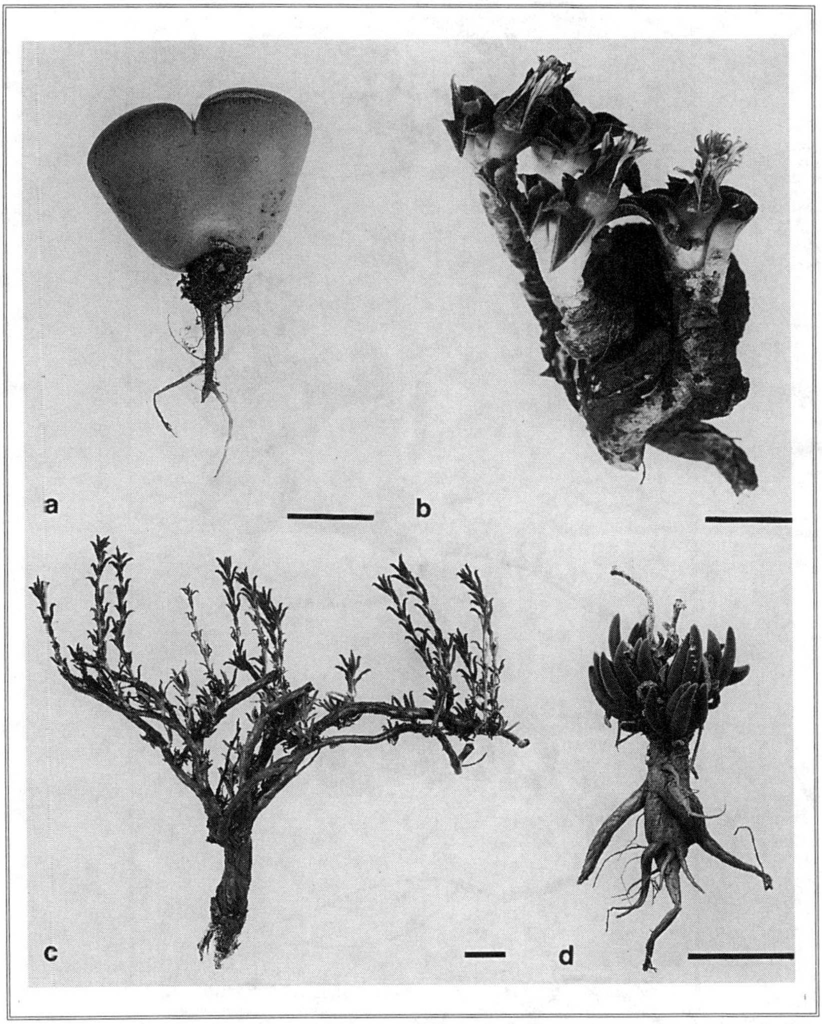

Figure 1. Growth forms. a. *Lithops lesliei*, b. *Nananthus* cf. *transvaalensis*, c. *Ruschia* sp., d. *Delosperma pontii* (scale bar = 20 mm).

Narrow strips and small patches in the grassland formation are occupied by plant communities of marshes and river banks and those of hillocks and rocky ridges (Smith & Hartmann 1992).

Methods

Distribution maps are based on specimens in BOL and PRE, as well as collections made by the authors, lodged in PUC, HBG and PRU. In the case of *Ebracteola wilmaniae* (L. Bolus) H.F. Glen, the localities given by Glen (1986: Figure 10) were incorporated.

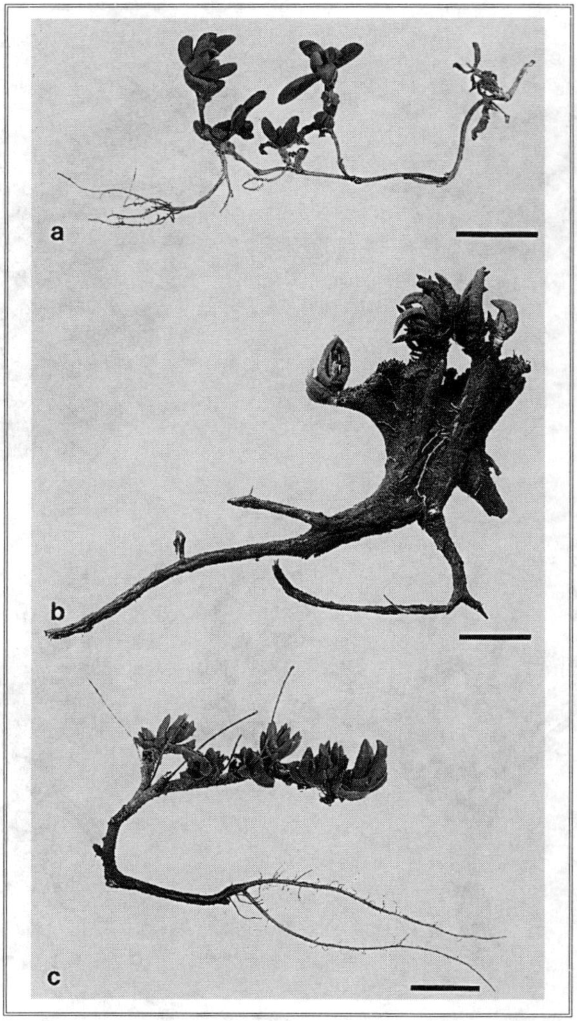

Figure 2. Growth forms. a. *Mossia intervallaris*, b. *Ebracteola wilmaniae*, c. *Chasmatophyllum musculinum*. (scale bar = 20 mm).

Results

Representatives of seven Mesembryanthema genera were collected. Four genera, *Lithops* N.E. Br. (Figure 1a), *Delosperma* N.E. Br. (Figure 1d), *Ruschia* Schwantes (Figure 1c) and *Nananthus* N.E. Br. (Figure 1b), were already known to occur naturally in the areas studied. The geographical distribution ranges of three genera, *Mossia*, *Ebracteola* and *Chasmatophyllum* have not previously included the study areas.

Mossia intervallaris (L. Bolus) N.E. Br. (Figure 2a) was collected exclusively in very shallow depressions in a loamy soil on sandstone outcrops (Smith & Hartmann

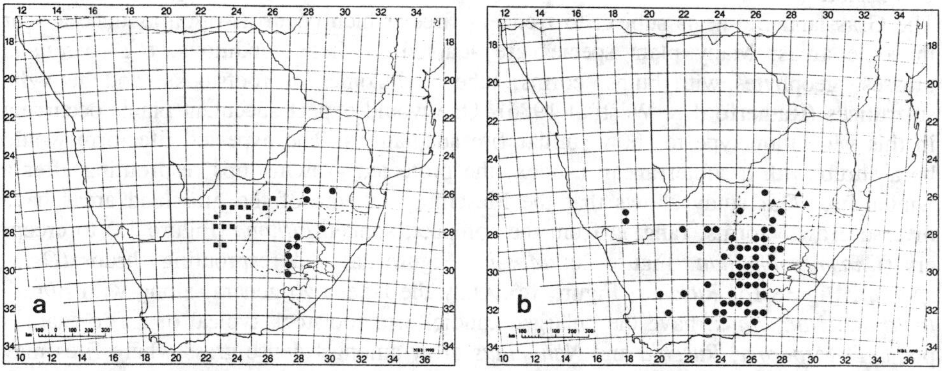

Figure 3. Distribution maps. a. ● known distribution of *Mossia intervallaris*, ■ known distribution of *Ebracteola wilmaniae*, ▲ new locality of both species. b. *Chasmatophyllum musculinum* (● known distribution, ▲ new localities).

1992). A northern population of this monotypic genus is centred around Johannesburg in the Transvaal, whilst a southern population occurs in Lesotho (Figure 3a) (Hargreaves 1989). Specimens have also been collected from Barkly East in the north-eastern Cape Province. Our accession from the northern OFS indicates that these populations might be linked and that further intensive sampling in the central and eastern OFS will confirm a continuous distribution pattern.

Ebracteola wilmaniae (Figure 2b), a dwarf clump-forming succulent with an enlarged caudiciform rootstock (Glen 1986), was also collected in shallow depressions underlain by solid bedrock. Although occurring in a similar habitat at the same locality as *M. intervallaris*, the two species were never found growing in association. The most easterly representative of a predominantly Namibian genus, *E. wilmaniae* has not previously been collected in the Orange Free State. The most north-easterly known locality of the species is Makwassie in the Transvaal, with our collection extending the known distribution by almost 150 km in an easterly direction (Figure 3a). A specimen *F. van der Merwe 2263* (PRE), identified as *E. wilmaniae*, was collected near Middelburg in the Transvaal in the 1930s. There is some doubt as to the validity of this identification, and the locality was not included on the distribution map.

Chasmatophyllum musculinum (Haw.) Dinter & Schwantes (Figure 2c) was found in shallow depressions on rocky outcrops. The species, a creeper, has a compact growth form in very rocky localities but develops extensive ground covers in deeper soils. It spreads with short, fat, horizontal rhizomes, developing thickish adventitious roots very quickly (Smith & Hartmann 1992). Despite occurring widely in the southern Orange Free State, *C. musculinum* has not been recorded from the extreme north of the region. Our collections show that the distribution range of this species includes the northern Orange Free State and the southeastern Transvaal, extending the known range by approximately 200 km (Figure 3b). It is noteworthy that rainfall figures across the east-west distribution of *C. musculinum* range from as little as 150 mm in the arid west to 750 mm in the east (Weather Bureau 1986).

Discussion

The Grassland Biome is subject to frequent fires, which occur as often as once a year in some areas. Many plant species are adapted to these conditions, e.g. perennial grasses, geophytes with bulbs, corms, tubers, rhizomes or rootstocks, and geoxylic suffrutices (Rutherford & Westfall 1986). One would expect succulent plants occurring in this vegetation type to show similar fire adaptations. The expected life form would be a shrub with a subterranean caudex, the protected growing tips resprouting after a winter fire. Our study shows that the Mesembryanthema of the grassland areas have diverse growth forms, ranging from the compact, sunken *Lithops* (Figure 1a) to creepers (*Chasmatophyllum*, Figure 2c; *Mossia*, Figure 2a) and resprouting shrubs (*Delosperma*, Figure 1d; *Ruschia*, Figure 1c). Only the latter two genera, along with *Ebracteola* and *Nananthus*, have the enlarged caudiciform rootstock typical of a fire-adapted plant. *Delosperma*, *Ruschia* and *Nananthus* grow amongst dense grass where fires may occur periodically. The rootstock, and sometimes the base of the stem, must be able to survive high fire temperatures for the plant to be able to resprout in the following season. The microhabitat of the other three genera clearly plays the major role in fire avoidance, rather than biological adaptation to a regular fire regime. They occur in areas marginal to grasslands, such as rocky outcrops, that do not support dense grass and consequently are not subjected to high fire temperatures.

The central-southern African plateau is traditionally seen as impoverished in its mesembryanthemoid flora (Hartmann 1991). Our study shows that this lack of records might well be a reflection of the state of field work in the region, rather than a true absence of plants. This can be ascribed to both a general and a specific collection neglect. According to Gibbs Russell et al. (1984), both areas investigated in this study are undercollected in general: the PRECIS database shows that only 0—10 specimens from these areas (quarter degree grid squares) are housed in the National Herbarium (PRE). A more recent map (Fish 1994), also identifies our study areas as "high priority collecting areas", with less than 100 species recorded for the relevant quarter degree squares. This is surprising as the areas included in our survey lie in close proximity to some of the major southern African herbaria (PRE, PRU, PUC, J, JRAU).

Plant collectors continue to neglect succulent plants in the field. Succulents generally make poor herbarium specimens and although various methods for preparing specimens prior to pressing have been described (Baker et al. 1985; Chudovska 1979; Leuenberger 1982; Logan 1986), these methods are time-consuming and often expensive (Smith 1991). This problem is compounded by the lack of South African experts to identify accessions, especially in the Mesembryanthema. Field workers in the grassland regions also tend to concentrate on the collection of grasses and non-succulent forbs, whilst ignoring the succulent plant component. The *Nananthus* cf. *transvaalensis* (Rolfe) L. Bolus specimen we collected in Potchefstroom is a case in point. The species has been known to occur in this area for more than 40 years (Louw 1951), yet not a single specimen has ever been deposited in a herbarium. It may well be that further investigations in similar undercollected summer rainfall areas will reveal higher genus frequencies, thus expanding the known distribution ranges of Mesembryanthema representatives.

The species recorded in this study were mostly found in pockets of non-arable land, some of them in highly specialized habitats, e.g. shallow depressions in rocky ridges

(Smith & Hartmann 1992). Widespread use of land for rain-fed cropping has in all likelihood eradicated many other Mesembryanthema populations with a preference for deeper loamy soils (Smith & Hartmann 1992). We want to emphasize the need for protecting these remaining sites from destructive human activities like stone quarrying (Smith & Hartmann 1992) and to conserve them as valuable locations of rare highveld succulent taxa.

Acknowledgements

We thank Prof A.E. van Wyk for assistance with the distribution maps, Mrs A. Romanowski for her excellent photography, and Pascale Hartzer for preparing lists of specimens housed in PRE. One of us (HEKH) received financial assistance from the University of Hamburg.

References

Baker, M.A., M.W. Mohlenbrock & D.J. Pinkava, 1985. A comparison of two new methods of preparing cacti and other stem succulents for standard herbarium mounting. Taxon 34: 118-121.

Chudovska, O., 1979. Ein neues Konservierungsverfahren für Sukkulenten. Kakteen und andere Sukkulenten 30: 94-96.

Fish, L., 1994. The National Plant Collecting Programme — why the National Herbarium, Pretoria (PRE) needs amateur collectors. PlantLife 10: 16-17.

Gibbs Russell, G.E., E. Retief & L. Smook, 1984. Intensity of plant collecting in southern Africa. Bothalia 15: 131-138.

Glen, H.F., 1986. Numerical taxonomic studies in the subtribe Ruschiinae (Mesembryanthemaceae) - Astridia, Acrodon and Ebracteola. Bothalia 16(2): 203-226.

Hargreaves, B.J., 1989. Succulents in the snow. Succulents of Lesotho. Part 2. Cactus and Succulent journal (U.S.) 61: 81-86.

Hartmann, H.E.K., 1989. Biological adaptation in Mesembryanthema. Excelsa 14: 39-44.

Hartmann, H.E.K., 1991. Mesembryanthema. Contributions from the Bolus Herbarium 13: 75-157.

Leuenberger, B.E., 1982. Microwaves: a modern aid in preparing herbarium specimens of succulents. The Cactus and Succulent Journal of Great Britain 44: 42-43.

Logan, J., 1986. A pre-pressing treatment for Begonia species and succulents. Taxon 35: 671-684.

Louw, W.J., 1951. An ecological account of the vegetation of the Potchefstroom area. Memoirs of the Botanical Survey of South Africa 24: 1-105.

Reimold, W.U., 1992. Dimension stone quarrying and the environment; a geologist's view of the case of the Vredefort structure, south-west of Johannesburg. Environmental Planning and Management 3: 4-11.

Rutherford, M.C. & R.H. Westfall, 1986. Biomes of southern Africa - an objective classification. Memoirs of the Botanical Survey of South Africa 63: 1-94.

Smith, G.F., 1991. Preparing herbarium specimens of small succulents. PlantLife 5: 12.

Smith, G.F. & H.E.K. Hartmann, 1992. Mesembryanthema in the Potchefstroom-Parys-Vredefort triangle, South Africa. Aloe 29: 70-73.

Tyson, P.D., F.J. Kruger & C.W. Louw, 1988. Atmospheric pollution and its implications in the eastern Transvaal Highveld. S. Afr. Nat. Sci. Program. Rep. 150: 1-114.

Weather Bureau, 1986. Climate of South Africa. Climate statistics up to 1984. Weather Bureau, Department of Environment Affairs, Pretoria.

Biodiversity in African lowland rain forests

23 August 1994

Convenors: F.J. Breteler, J.F. Villiers

Centres of plant diversity in Africa

H.J. BEENTJE
Royal Botanic Gardens, Kew, Richmond, Surrey TW9 3AE, United Kingdom.

Abstract
At the Malawi AETFAT meeting a symposium was held to discuss the sites to be selected for the "Centres of Plant Diversity" project of the IUCN. Thirty sites have now been chosen. Criteria and methodology for their choice will be discussed and some parallels and differences with animal biodiversity in Africa will be highlighted.

The need for qualitative and quantitative measurements is imperative; even though the best sites or "hot spots" are already known, their designation needs to be backed up by hard facts. This leads to thoughts about the need for consensus on measuring or grading biodiversity, on both regional and global scales, and with regards to other disciplines. The value of indicator species seems limited; more promising are rapid and medium-term inventories of interesting areas, and some of the more practical methods of botanical inventory are briefly discussed. Ease of use is one of the main requirements for such methods; but they also need to be globally applicable, so that diverse regions can be compared.

Résumé
Pendant le congrès de l'AETFAT du Malawi, un symposium s'est tenu pour débattre des sites à choisir pour le projet de l'UICN intitulé "Centres of Plant Diversity". Trente sites ont été désignés. Les critères et la méthodologie de leurs choix seront discutés; quelques comparaisons et différences avec la biodiversité animale en Afrique seront mis en évidence.

Le besoin de mesures qualitatives et quantitatives est impératif; même quand les meilleurs sites ou les "points chauds" sont déjà connus; leur désignation doit être renforcé par des faits solides. Cela conduit à des réflexions à propos de la nécessité d'un consensus sur une biodiversité mesurée ou évaluée, sur les échelles régionales ou globales, et en ce qui concerne d'autres disciplines. La valeur des espèces indicatrices semble limitée; les inventaires rapides ou à moyen terme des aires intéressantes semblent plus prometteurs, et quelques'unes des méthodes les plus pratiques d'inventaire botanique sont brièvement discutées. La facilité d'utilisation est une des principales exigences pour de telles méthodes; mais elles ont aussi besoin d'être globalement applicable, de telle sorte que diverses régions puissent être comparées.

Key words
Conservation, Endemism, Indicators, Sampling.

African Centres of Plant Diversity

This is an update of the Malawi AETFAT presentation, at which Steve Davis and I discussed the Centres of Plant Diversity (CPD) Project run by WWF and IUCN, which will result in a series of three publications giving an outline of the main plant diversity centres on a world-wide scale.

There are both weaknesses and strengths in such a project: weaknesses, in that the depth of focus is very limited, due to the scope of the project; and also, that the focus itself is based on debatable premises. The strengths are that the scope gives a chance to compare standardised entries on a global scale; and, also, that it shows the gaps in our knowledge, and which gaps are the widest.

Since the seminar during the Malawi AETFAT congress, eighty-four African CPD

L.J.G. van der Maesen et al. (eds.), The Biodiversity of African Plants, 101-109.

Table 1. African Regional Centres of Endemism.

Guineo-Congolian	
Sapo National Park	Liberia
Taï National Park	Côte d'Ivoire
Mont Nimba	Guinea, Liberia, Côte d'Ivoire
Cross R. National Park	Nigeria
Korup National Park	Cameroon
Mt. Cameroon	Cameroon
Dja R. forests	Cameroon
Cristal Mts	Gabon
Maiko National Park	Zaïre
Salonga National Park	Zaïre
Mayombe	Congo, Cabinda, Zaïre
Bwindi/Impenetrable forest	Uganda
Zambezian	
Kundelungu	Zaïre
Upemba National Park	Zaïre
Mahale-Karobwa Hills	Tanzania
Zambezi Source area	Zambia
Somali-Masai	
Cal Madow	Somalia
Hobyo	Somalia
Sudanian	
Garamba National Park	Zaïre
Karoo-Namib	
Kaokoveld	Angola, Namibia
Succulent Karoo	Namibia, South Africa
Cape	
Cape Floristic Region	South Africa
Indian Ocean Coastal Belt Mosaic	
Rondo Plateau	Tanzania
Maputaland-Pondoland region	Mozambique, South Africa, Swaziland
Afromontane	
Mt Kenya	Kenya
East Usambara Mountains	Tanzania
Mt Mulanje	Malawi
Afro-alpine	
Afro-alpine region	E & NE Africa
Drakensberg alpine region	Lesotho, South Africa
Mediterranean	
High Atlas	Morocco

sites have been determined, as well as the "top-thirty" sites selected for data sheet treatment, in somewhat more depth (Beentje et al. 1994; Table 1). The choice was based on a pre-determined number of sites available for each Regional Centre of Endemism (White 1983) or comparable plant geographical area; within these areas, sites were chosen based on what was perceived to be the highest plant species diversity combined with high numbers of endemic species.

There were several *problems* in the selection of sites:

* Sites have been chosen on their individual merits; we have not used the principle of complementarity (Vane-Wright et al. 1991), which involves the selection of the highest priority site, whereupon the second (and later) choices are based on the fact that you do not wish to replicate taxa which occur in your first choice. This presupposes a level of knowledge of species presence/absence and distribution notably absent from many African sites. Once data are more complete, complementarity increases the efficiency of selection (Pressey et al. 1993). On a continental scale such as CPD, complementarity is employed (albeit intuitively, and in a restricted way only) through the Regional Centres of Endemism/Phytochorion approach.

* The "hit-parade" approach of priorities. Of course, we cannot condone the clear-cutting of the Kakamega forests, just because the Ituri Forest is richer. But it makes sense to give priority to areas containing the richest and most highly endemic biota (Raven 1980), given, on the one hand, the urgent need to protect nature, as well as a rising tide of public interest and, on the other, limited financial resources and political constraints.

Chalker (1990) states that "agreement on a global list of centres of biological diversity for priority conservation" is urgently needed – and this is from a politician who is able to implement funding for such purposes, since she is the British minister for Overseas Development. The CPD publication is the first major step on the plant side, with the selection of truly interesting and worthwhile sites, based on sensible criteria, and agreed upon by many experts.

* Selection based on standardised criteria. The presence of at least 1000 species, of which at least 10% should be endemic (either to the area, or the phytochorion) suggested as a criterion for sites for the African chapter has been taken up as a global criterion for mainland sites (Davis & Heywood 1994); islands have different criteria, "due to their often depauperate nature" – a decision with which I personally disagree, since it detracts from the standardisation.

* Lack of data comparability. The global scale of the project highlights our lack of knowledge of many tropical areas, when compared with more temperate ones. Within Africa, our knowledge on high diversity sites varies enormously; the obvious risk is that the best-known sites get chosen, and maybe even studied, while the lesser-known sites are left out and slide into obscurity. It is amazing (at least to me) how little we know about many famous sites such as the Taï and Ituri forests: how many plant species occur, even in rough numbers? And this is not superficial, "Guinness-book-of-

record" information; these are the questions that lay people will ask, and that includes politicians, when they are interested in, or concerned with, biodiversity and its conservation. And when they turn to biologists for an answer, we had better have one.

* Lack of habitat comparability. There is a risk of inflating species-richness by combining habitats: in many cases, a block of pristine habitat will contain fewer species than a mosaic of habitats, since such an admixture would contain many widespread pioneer and secondary species; whereas a primary habitat would be more valuable.

* Lack of hard facts. There are advantages, as well as disadvantages, to the global selection of sites on set criteria. The one used here, on a global scale, is based on all plant species; alpha-taxonomy for all of these is still incomplete, and distribution areas are not known for many taxa; therefore, the results are inconclusive. The method of selection is unscientific, since it is based on partial data, and the organisation of these data is intuitive rather than quantitative. Weighting methods are ill-defined and intuitive.

A solution would be to use the indicator method: recognition of priority areas based on narrow taxonomic groups for which detailed knowledge is available, and the taxonomy well worked out. The great advantage of this method is that quantitative methods can be used to produce selections based on science rather than conjecture. However, as the basis is limited (due to the small number of taxa used), the applications are also limited. This is also the easiest method for directing the selection of sites: the choice of indicator influences the outcome. The use of larger numbers of taxa reduces this effect, but makes it much more difficult to use quantitative techniques. And as long as we cannot give a rough total of species on Mount Cameroon or in the Bwindi Forest, the use of data-manipulation programs on a world-wide basis is premature.

Naturally these methods are not mutually exclusive. The method used in CPD clearly shows the need for more hard data. There is very little information for many of the CPD sites, and that means that the scientific base of the project is too thin. Nevertheless, I do believe that the selections are more or less correct, and that choices need to be made *now*, for the sake of conservation; but this does not mean we can sit back and relax, the choices need to be confirmed, or corrected, by good sampling world-wide, including Africa.

If any proposal for conservation is to be politically effective it will need a consensus, which can only be produced by integrating results from various methods – and various disciplines. The Manaus Workshop of 1990 identified priority areas for conservation in Amazonia based on the integrated proposals from plant systematicians, plant ecologists, zoologists (from five disciplines), geomorphologists and climatologists, as well as conservationists (Prance 1994); this is urgently needed for Africa as well. In cases where the data do not agree, an integrated balance will have to be worked out.

Pomeroy (1993) compares species-richness for forest birds and forest trees for 12 protected areas in Uganda; he concludes there is a tendency for the numbers in the two groups to be correlated, but the scatter is wide. In a brief comparison of key forests for threatened birds in Africa (Collar & Stuart 1988) with CPD forest sites, a reasonable measure of agreement is apparent (Table 2).

Table 2. Comparison of key forests for threatened birds with CPD forest sites.

Plants: / Birds:	15 high priority forests	30 medium priority forests	Not mentioned
30 high priority forests	6	11	13
30 medium priority forests	5	5	20
Not mentioned	4	14	n/a

The main advantage of the CPD project is that the approach is world-wide rather than piecemeal, which gives a basis (however preliminary) for comparison on a global scale. The project provides a baseline for criticism and fine-tuning, and gives the first world-wide comparative selection of sites of high plant diversity.

Data gathering

Pressey et al. (1993) advocate the use of WORLDMAP on a global level; this method identifies key regions for biodiversity, measured by the number of species and the taxonomic differences between them.

I would suggest that what is first needed, in many cases, is actual field work rather than the manipulation of scanty, and often old, data: a good example is that of the palms of Madagascar. One could have worked with the data in existence in 1990 (Jumelle & Perrier 1945) which enumerates 115 palm species; 18 months of field work and 2 years of herbarium work have resulted in a classification with 186 species (Dransfield & Beentje, in prep.). The WORLDMAP palm hot spot in Madagascar before the field work would have been Analamazaotra with 21 species in 9 genera; six of these genera are now synonyms of some of the remaining three, so generic diversity would have been inaccurate. And so would the hot spot: this would now be the Mananara area with 43 species in 5 genera.

A program like WORLDMAP can only work properly when based on good data; that means good alpha taxonomy: floras and monographs. For the African continent, and especially for "hot-spot" countries such as Cameroon, Gabon, Zaïre or Tanzania, the floras are not yet complete. So for the moment, programs like WORLDMAP can only be based on taxa for which the data are good; these taxa are limited in number, and that means the results of such exercises are limited in scope; they can serve as pointers, but they cannot replace fieldwork which produces hard data for selected sites. Plant distribution data in floras and revisions are based on collections which are not uniform on a continent-wide basis; there are still large (relatively) white spots on our maps where collection intensity has been low (data sheet sites such as Sapo, the Monts de Cristal, Itombwe spring to mind; see also Campbell & Hammond 1989). There is only one solution to these white spots: fieldwork sampling. And I would advocate concentration on tropical forest, since c. 50% of all 250,000 plant species are associated with tropical forest (Soulé 1990).

No shortcuts can replace proper fieldwork. However, there are methods for measuring diversity on the ground which are quicker than others. A total inventory of Taï

Forest is not within sight; but good sampling methods will give a reasonable idea of species numbers, as well as much more information on distribution, age representation, vegetation structure...

Intensity and scale of inventory or sampling methods vary widely. Reitsma (1988) inventoried four one-hectare permanent plots in primary forest in Gabon; this is a prime example of a thorough quantitative inventory with complete vouchering, which provides optimal data. It allows comparison with similar one-hectare sites in other parts of the world, including Korup in Cameroon (Gartlan et al. 1986 for Korup; Ashton 1964 for Brunei; Prance et al. 1976 and Gentry 1988a for Amazonia). But this is a time-consuming method. A relatively rapid method, taking days rather than months, is that of Gentry (Gentry 1982, 1988b). This is a method I would like to recommend as the standard for rapid sampling of alpha-diversity of interesting sites, especially tropical forest. A good field knowledge of the area is essential, but once this is present it can be quickly translated into results; several transects of 50 × 2 metres can be undertaken in a day, and ten of these are required to bring the sampled area to 0.1 hectare. The great advantage of the method is its simplicity, the ease of laying a narrow transect in dense forest (as compared with larger transects), and the comparability with other forest sites: Gentry (1988) lists 85 sites sampled, only six of which are from Africa (Nigeria, Cameroon, Gabon) and one from Madagascar; I am unaware of more recent Gentry-samples for the African continent, apart from three unpublished ones for Kenya (Beentje & Mungai, in sched.; Beentje & Khayota, in sched.). The method provides data such as species lists, size classes and numbers of individuals which can provide the basis for much interpretative work.

In addition to this plea for more sampling, I would like to reiterate the need for "overview" floras utilising the format of the Flora of West Tropical Africa for countries such as Madagascar, Zaire, Tanzania.

Data use

A real risk posed by the lack of consensus about biodiversity measuring systems is cited by Rodda (1993): the U.S. Congress asked a managing agency to maintain the level of diversity in the hardwood forests of part of the U.S.A. Because of the lack of consensus in the scientific community over the best measure of species diversity, Congress left the choice of diversity index to the managing agency; the agency chose to measure diversity in stand age; the forests were replanted as monocultural pine plantations, with maximal age diversity.

Some criteria used recently for the quantification of biodiversity are:

1. Numbers of species, without weighting.
2. Indicator species.
3. Numbers of endemic species.
4. Numbers of species, weighting by rare/vulnerable/endangered species.
5. Taxic diversity.
6. Other values: "flagship" or high profile species.

1. Numbers of species, without weighting

Species-diversity, or species-richness, is commonly the criterion used for identifying

Centres of Plant diversity - this is almost a tautology. Is just adding up the numbers of species the best method for identifying the areas with the highest conservation priority? A lot depends on the homogeneity of the area under study: an area with several vegetation types, with several succession stages in each, will quite possibly be richer in species (with pioneer and other secondary species) than an equal-sized area with a homogeneous and mature vegetation type. The advantage of this method is that it is an easy one, and it can be based on samples taken in standard-size transects, which allows for easy comparison.

The quantitative method of Margules et al. (1988) uses algorithms to define the smallest number of sites in which maximum diversity occurs; their work dealt with 118 plant species in 432 sites. This is an example of the "critical area method" in which all species are treated as taxonomically equal, but presupposes the total number of species, as well as all suitable sites, for a given area is known. Methods like these are not applicable on a continent-wide scale for the foreseeable future.

2. Indicator species

Of course you can massage the figures: if you want grassland conservation, you select herbivore diversity and terrestrial orchids; if you want rain forest, you select leaf litter Arachnoidea and epiphytic orchids. Indicator groups can show what you want them to show; the only objective measure would be the total number of species, or as large a sample as possible.

3. Numbers of endemic species

The number of endemic species should be an important contributing factor - when endemic is taken to imply a limited distribution range. Bibby et al. (1992) and Pomeroy (1993) use 50,000 square kilometre for bird species; for plant species quantitative data like these might be much harder to come by; in the CPD project we have chosen to interpret endemic as endemic to a certain phytochorion.

The critical area method uses quantitative methods to select as the first priority area the one that is richest in endemics; followed by the one richest in endemics *excluding the endemics from the first area*, and so forth, until the lowest number of sites is found in which all species are represented. All taxa are treated as taxonomically equal, but of course endemism determines the sites in the first place, and so is a heavily weighted factor. Ackery & Vane-Wright (1984) came up with 31 critical areas for 157 species of milkweed butterflies; Collins & Morris (1985) found 51 critical areas for 573 species of swallowtail butterflies. Though this is a straightforward method, the mathematics of dealing with 300,000 or so plant species are mind-boggling. Obviously this is a method for monographs rather than overviews.

4. Numbers of species, weighting by rare/vulnerable/endangered species

This is a very appealing method, since it strikes an immediate chord: rarity is what it is all about, so if special habitats start to disappear, their remnants will be full of rare taxa, which will increase the weighting.

The main problem is that rarity listings are lacking for most plant families. Weighting can only be applied when the weighting factor can be linked to the majority of species concerned. This does not preclude the usefulness of rarity as a flagship factor;

this is a very high profile argument, but its use in a quantitative sense is limited by our lack of knowledge.

5. Taxic diversity (Vane-Wright et al. 1991)
This recently proposed method is based on the presumption that not all species are of equal value: *Welwitschia mirabilis* is worth more, to a conservationist, than yet another species of *Grewia*. It seems reasonable to give priority (given two equally threatened taxa) to that taxon not closely related to other living species. The method advocated by Vane-Wright et al. requires a cladistic analysis of all taxa involved in the decision-making, and results in a score for each taxon. This score then gives greater weight to highly distinct taxa. Krajewski (1994) compares seven phylogenetic diversity measures. The obvious problem is one of time, and the lack of data needed to draw up generic- or even species-specific cladograms for all of Africa's plants. The method seems more applicable to small areas and/or restricted groups than to global scale.

6. Other values: "flagship" or high profile species
Is an African Violet worth more than a *Zimmermannia*? It certainly is, in the eyes of the general public, and so this factor has to be taken into consideration. On the other hand we must not get carried away with this, since in the same public eye a single lemur with large mournful eyes probably equals fifty endemic plant species.

It is clear that if we, the scientific community, wish to communicate with the wider public and the politicians to advocate conservation of natural communities, there is a need for priorities to be stated; before priorities can be stated, there is an urgent need for a consensus on a biodiversity index; and to be useful, such an index needs to be fairly simple, straightforward, and based on understandable factors: species numbers, with a certain weighting for highly distinct taxa, very rare taxa, and/or taxa with limited ranges.

The Centres of Plant Diversity project is a start, on which we should build; in the meantime, it will serve as a focus for short-term action plans. The way forward is a refinement of the current selection, based on more facts. And that shows the urgent need for more biogeographical knowledge and rapid inventories in high diversity areas; and both need to be based on monographs and floras. Without continuing work in alpha-taxonomy everything else is based on quicksand.

References

Ackery, P.R. & R.I. Vane-Wright, 1984. Milkweed Butterflies. British Museum (Natural History), London.

Ashton, P., 1964. Ecological studies in the mixed dipterocarp forests of Brunei State. Oxford Forest mem. 25: 1-75.

Beentje, H.J., B. Adams & S.D. Davis, 1994. Regional overview: Africa. In: Davis, S.D. & V.H. Heywood (eds.), 1994. Centres of Plant Diversity; a guide and strategy for their conservation. WWF/IUCN, Gland.

Beentje, H.J. & G.M. Mungai (in sched.). Woody plant species diversity at the 0.1 hectare level in a dry upland forest in Kenya.

Beentje, H.J. & B.N. Khayota (in sched.). Woody plant species diversity at the 0.1 hectare level in two moist lowland forests in Kenya.

Bibby, C.J., N.J. Collar, M.J. Crosby, M.F. Heath, Ch. Imboden, T.H. Johnson, A.J. Long, A.J. Stattersfield & S.J. Thirgood, 1992. Putting biodiversity on the map. International Council for Bird Preservation, Cambridge, England.

Campbell, D.G. & H.D. Hammond (eds.), 1989. Floristic inventory of tropical countries. N.Y. Botanical garden, New York.

Chalker, L., 1990. Extracts from a speech to Friends of the Earth Conference. British Overseas Development 12: 3.

Collar, N.J. & S.N. Stuart, 1988. Key forests for threatened birds in Africa. International Council for Bird Preservation, Cambridge.

Collins, N.M. & M.G. Morris, 1985. Threatened swallowtail butterflies of the world. The IUCN Red Data book. IUCN, Gland & Cambridge.

Davis, S.D. & V.H. Heywood (eds.), 1994. Centres of Plant Diversity; a guide and strategy for their conservation. WWF/IUCN, Gland.

Dransfield, J. & H.J. Beentje (in prep.). The palms of Madagascar. Royal Botanic Gardes, Kew & the International Palm Society.

Gartlan, J.S., P.M. Newbery, D.W. Thomas & P.G. Waterman, 1986. The influence of topography and soil phosphorus on the vegetation of Korup Forest Reserve, Cameroun. Vegetatio 65: 131-148.

Gentry, A.H., 1982. Patterns of neotropical plant species diversity. Evol. Biol. 15: 1-84.

Gentry, A.H., 1988a. Tree species richness of Upper Amazonian forests. Proc. Natl. Acad. U.S.A. 85: 156-159.

Gentry, A.H., 1988b. Changes in plant community diversity and floristic composition on environmental and geographical gradients. Ann. Missouri Bot. Gard. 75: 1-34.

Jumelle, H. & H. Perrier de la Bâthie, 1945. Palmiers. In: Humbert, H., (ed.) - Flore de Madagascar et des Comores 30^{e} famille. Imprimerie officielle, Antananarivo.

Krajewski, C., 1994. Phylogenetic measures of biodiversity: a comparison and a critique. Biological Conservation 69, 1: 33-39.

Margules, C.R., A.O. Nicholls & R.L. Pressey, 1988. Selecting networks of Reserves to Maximise Biological Diversity. Biological Conservation 43: 63-76.

Pomeroy, D., 1993. Centers of high Biodiversity in Africa. Conservation Biology 7, 4: 901-907.

Prance, G.T., W.A. Rodriguez & M.F. da Silva, 1976. Inventario florestal de um hectare de mata de terra firme km 30 de estrada Manaus - Italotiara. Acta Amazonica 6: 9-35.

Prance, G.T., 1994. The use of phytogeographic data for conservation planning. In: Forey P.L., C.J. Humphries & R.I. Vane-Wright (eds.), Systematics and Conservation Evaluation. Systematics Association Special Volume 50: 145-163. Clarendon Press, Oxford.

Pressey, R.L., C.J. Humphries, C.R. Margules, R.I. Vane-Wright & P.H. Williams, 1993. Beyond opportunism: key principles for systematic reserve selection. Trends in Ecology and Evolution 8, 4: 124-128.

Raven, P.H. (ed.), 1980. Research priorities in Tropical Biology. National Academy of Sciences, Washington D.C.

Reitsma, J.M., 1988. Végétation forestière du Gabon/Forest vegetation of Gabon. Tropenbos Foundation, Ede, the Netherlands.

Rodda, G.H., 1993. How to lie with biodiversity. Conservation Biology 7, 4: 959-960.

Soulé, M., 1990. The real work of systematics. Ann. Missouri Bot. Gard. 77: 4-12.

Vane-Wright, R.I., C.J. Humphries & P.H. Williams, 1991. What to protect? – systematics and the agony of choice. Biological Conservation 55: 235-254.

White, F., 1983. The vegetation of Africa: a descriptive memoir to accompany the Unesco/-AETFAT/UNSO vegetation map of Africa. Natural resources research XX. Unesco, Paris.

Mapping plant biodiversity on Mount Cameroon

M. CHEEK [1], S. CABLE [1], F.N. HEPPER [1], N. NDAM [2] & J. WATTS [3]

[1] The Herbarium, Royal Botanic Gardens, Kew, Surrey, TW9 3AE, United Kingdom.
[2] Limbe Botanic Garden, Limbe, P.O. Box 437, SW. Province, Cameroon.
[3] 36 Kingsley Road, Pinner, Middlesex, HA5 5RB, United Kingdom.

Abstract
The Gulf of Guinea is believed to have been one of the two major Pleistocene refugia in Africa. Mount Cameroon lies towards the centre of this area and is one of Africa's centres of plant diversity. At least 50 species and 3 genera are strictly endemic and 50 more species near endemic, to the main massif and the foothills. Large areas of indigenous forest occur on the slopes, and the mountain, West Africa's tallest at 4,000 m, is unusual in having natural vegetation extending from sea-level at the coast to the sub-alpine grassland at the top. An active volcano, the mountain's soils are extremely fertile and the surviving forest is under some pressure from logging and farming. Since 1992, a series of botanical inventories have been carried out of the mountain's forests. Approaching 10,000 specimen numbers have been gathered in the course of these inventories and this paper sets out the progress made in analyzing the data.

Résumé
Il semble que le Golfe de Guinée ait été l'un des deux refuges principaux du Pléistocène en Afrique. Le Mont Cameroun se situe à peu près au centre de cette région et constitue l'un des centres de diversité végétale de l'Afrique. Au moins 50 espèces et 3 genres sont strictement endémiques et 50 autres espèces subendémiques du massif principal et des piedmonts. De grandes étendues de forêt naturelle occupent les pentes; la montagne, la plus haute de l'Afrique occidentale avec ses 4000 m d'altitude, est tout à fait remarquable du fait que sa végétation naturelle s'étale depuis le niveau de la mer jusqu'aux formations herbeuses subalpines du sommet. C'est un volcan actif, les sols de la montagne sont extrêmement fertiles et la forêt qui subsiste est plus ou moins menacée par l'exploitation du bois et l'agriculture. Depuis 1992, une série d'inventaires botaniques des forêts de cette montagne ont été effectués. Environ 10.000 échantillons ont été récoltés au cours de ces inventaires et cet article rend compte des progrès réalisés dans l'analyse des données.

Key words
Endemics, Inventory, Mt. Cameroon, Plant biodiversity, Refugia.

Introduction

One of the richest areas for plant species in Africa is the coastal belt, perhaps 100 km wide, that extends around the Gulf of Guinea from SE. Nigeria, east of the Cross River, through Cameroon to adjoining Gabon. Evidence for this richness is the fact that nearly half of the 7,000-8,000 species of the "Flora of West Tropical Africa" region are found in the area of former British Cameroon (Keay & Hepper 1954-72, Hepper 1972).

One explanation for the exceptional diversity of this area is probably that it corresponds with one of the two main Pleistocene refugia postulated for Africa (Maley 1987 & 1991, Hamilton 1976), the other being E. Kivu in eastern Zaire. Subsidiary refugia are postulated for Guinea-Ivory Coast and Tanzania. It is notable however, that all of the plant families endemic to tropical Africa: Huaceae, Lepidobotryaceae,

L.J.G. van der Maesen et al. (eds.), The Biodiversity of African Plants, 110-120.

Medusandraceae, Octoknemataceae, Hoplestigmataceae, and Dioncophyllaceae (excepting Barbeyaceae, a single species of arid NE. Africa and Arabia) are concentrated in the Gulf of Guinea and usually entirely absent from the Tanzania and Kivu areas, or very poorly represented (Scytopetalaceae with only one or two species in Kivu). These families are either entirely restricted to the forests of the Gulf of Guinea, or largely so, with the Guinee-Ivory Coast forests a secondary centre, though the reverse is true with Dioncophyllaceae.

Mt. Cameroon, in SW. Province, Cameroon, at 4,095m high is the tallest mountain in West Africa and possibly the last where natural vegetation remains unbroken from sea-level to the sub-alpine zone at the summit, with all intermediate altitudinal vegetation types intact. The surrounding lowland forests are, however, richer in terms of biodiversity and under most serious threat of destruction from conversion to plantations and small holder's farms. It was largely in order to conserve these lowland forests that the Limbe Botanic Garden and Rainforest Genetic Conservation Project (hereafter referred to as the Project) was set up on Kew's advice in 1988 (Hepper et al. 1986, Hepper & Bovey 1987).

The Project is supported jointly by the Governments of Cameroon (Forestry Dept.) and Britain (Overseas Development Administration) and has been staffed by about 15 Cameroonian officers and technicians, 60 contract workers and 3 British officers. The headquarters are sited at the southeastern foot of Mt. Cameroon in Limbe (formerly Victoria) at the old botanic garden and herbarium, founded in 1892 during the German colonial period. The botanic garden of about 48 ha, and its buildings have been rehabilitated and can now be used for education and amenity by local people, and as a centre for visiting Cameroonian and international scientists. The herbarium is being developed and there is a small, but excellent botanical library.

Botanical inventories

Towards the end of 1991, the Project's mid-term review advised that botanical inventory of the protected areas should be given priority in order to identify and locate conservation priorities, i.e. rare and threatened species, in order to aid management. It is the purpose of this paper to review progress in that direction.

The Project was originally set up to protect two proposed forest reserves: Mabeta-Moliwe to the east of the main massif, and Etinde, constituting much of the southern part of the mountain proper and the subpeak Mt. Etinde. Following botanical inventories of these areas in 1992, the Project elected to continue with its survey work in the unexplored western foothills during 1993, initially in the Onge forest, then in the already established Mokoko River Forest Reserve. Figure 1 gives the disposition of protected areas around the mountain. Table 1 shows the putative new taxa identified so far and Table 2 shows progress made in specimen collection and determination from these areas.

The Southern Bakundu Forest Reserve, as a result of the Cambridge 1948 expedition, and botanizing by Brenan, Keay and Richards, is comparatively well-known botanically and is famous for providing the new plant family Medusandraceae. In the rain-shadow of the main massif, it is drier than other foothill areas, occurring within the 2 m isohyet (Hawkins & Brunt 1965) and species of drier forest, not found elsewhere on the mountain, such as *Triplochiton scleroxylon* K. Schum. are found here.

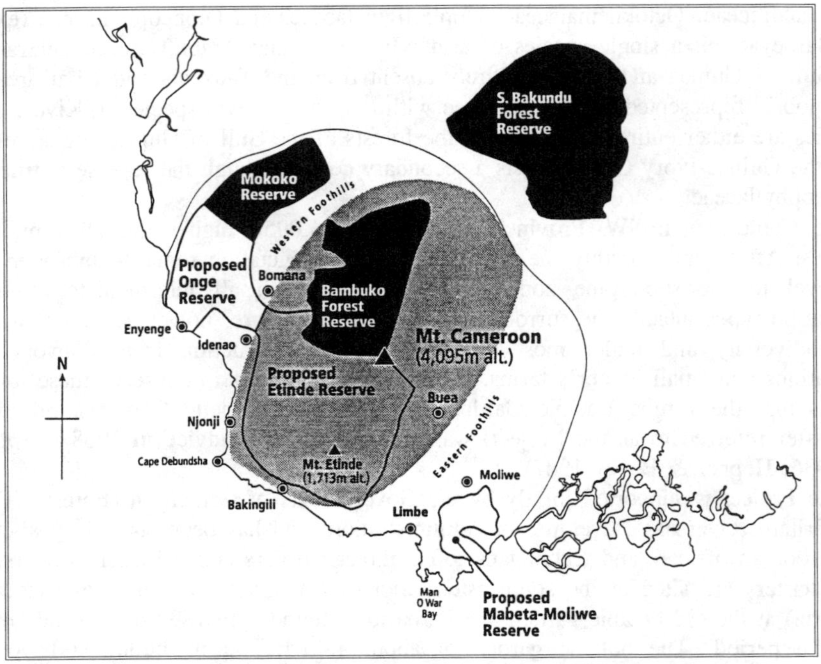

Figure 1. Map of Mt. Cameroon showing massif, foothills and protected areas.

Since the inventory work of 1948 the forest has largely been felled. More seriously, Nigerian yam farmers have brought large areas under cultivation, and an ITTO forestry scheme was begun in 1993 to improve timber yields by removing "secondary" species. The effect of these on the flora of this forest is unknown and needs assessing.

The Bambuko Forest Reserve is not well known botanically. Positioned on the west side of the main massif, it is believed to be dominated by *Hypselodelphys* and *Aframomum* thicket, rather than forest. A botanical inventory of this forest would be worthwhile.

The present paper is an extension and elaboration of a review on the progress on Mt. Cameroon (Cheek & Hepper 1994). The progress made with the botanical inventories and mapping the plants of Mt. Cameroon follows.

Mabeta-Moliwe

Mabeta-Moliwe consists of about 36 km² and is situated a few kilometres to the east of Limbe, between the settlements of Moliwe to the north, and Mabeta to the south. It is situated on old volcanic rocks, and constitutes part of Mt. Cameroon's eastern foothills. It extends from sea-level to about 400 m. The eastern boundary is marked by a steep north-south ridge from which proceed eastward five lesser ridges, giving a steep terrain of ridges and valleys. The valleys are each drained by separate seasonal and permanent streams and rivers giving rise to pockets of freshwater swamp forest where drainage is impeded. At the coast a distinct littoral forest community is present and there is a small area of mangrove.

Table 1. Provisional new taxa from Mount Cameroon.

Family	Species	E	M	O
Anacardiaceae	*Trichoscypha* sp. nov.	*		*
Ancistrocladaceae	*Ancistrocladus* sp. nov.		*	
Anisophylleaceae	*Anisophyllea* sp. nov.		*	*
Annonaceae	*Isolona* sp. nov. 1	*		
	Isolona sp. nov. 2			*
	Monanthotaxis sp. nov.	*		
	Piptostigma sp. nov. 1 & 2			*
	Uvaria sp. nov.	*	*	
Araceae	*Culcasia* sp. nov. 1 (of Ntepe)		*	
	Culcasia sp. nov. 2 (of Ntepe)	*	*	
	Culcasia sp. nov. 3	*		
	Culcasia sp. nov. 4	*		
	Nephthytis sp. nov.	*		
Balsaminaceae	*Impatiens* sp. nov. 1 & 2	*		
Burmanniaceae	cf. *Oxygyne* sp. nov.	*		
Celastraceae	*Salacia* sp. nov. 1		*	*
	Salacia sp. nov. 2		*	
Euphorbiaceae	*Drypetes* sp. nov. 1 & 2		*	
Flacourtiaceae	*Dovyalis* sp. nov.			*
Icacinaceae	*Pyrenacantha* sp. nov.		*	
Melastomataceae	*Warneckea* sp. nov.		*	
Moraceae	*Ficus* sp. nov.	*		
	Dorstenia poinsettifolia Engl. var. nov.	*	*	
	Dorstenia sp. nov.	*		
Myrsinaceae	*Embelia* sp. nov. 1	*		
	Embelia sp. nov. 2	*		
Orchidaceae	*Angraecopsis* sp. nov.	*		
	Cribbia sp. nov.	*		
Rubiaceae	*Coffea* sp. nov.	*		
	Oxyanthus sp. nov.			*
	Psychotria sp. nov. 1 & 2		*	
	Psychotria sp. nov. 3 & 4	*		
	Rutidea sp. nov.		*	
	Sacosperma sp. nov.		*	
Sapindaceae	*Pancovia* sp. nov.			*
	Placodiscus sp. nov.			*
Scytopetalaceae	*Rhaptopetalum* sp. nov. 1	*		
	Rhaptopetalum sp. nov. 2			*
Sterculiaceae	*Cola* sp. nov.			*
	Leptonychia sp. nov. 1		*	
	Leptonychia sp. nov. 2			*
	Scaphopetalum sp. nov.		*	
Verbenaceae	*Vitex* sp. nov.		*	
Vitaceae	*Cissus* sp. nov.			*
Xylariaceae (Fungi)	*Xylaria*		*	
Zingiberaceae	*Aframomum* sp. nov.	*		

E = Etinde; M = Mabeta-Moliwe; O = Onge

Table 2. Fertile specimens collected on Mount Cameroon 1992-1994.

Area	Number	Named
Etinde	2,900	1,700
Mabeta-Moliwe	2,000	1,300
Mokoko	1,500	
Onge	2,600	700
Totals	9,000	3,700

Though far from pristine, the forest was originally proposed for protection since it is the best surviving example of forest between Douala and Mt. Cameroon, the remaining land being taken up by plantations of rubber, oil-palm, bananas, tea, coffee and cocoa, largely established during the colonial era. A timber inventory, performed to establish the forest composition and "opportunity cost" of the proposed reserve, showed that the commonest trees by number (>20 cm DBH) are *Tapura africana* Oliv. (Dichapetalaceae, c. 6%), *Strombosia scheffleri* Engl. (Olacaceae, c. 10%) and *Coelocaryon preussii* Warb. (Myristicaceae, c. 7%). The main emergent tree species are *Desbordesia glaucescens* (Engl.) Van Tiegh. (Irvingiaceae), *Lophira alata* Banks ex Gaertn. f. (Ochnaceae) and *Ceiba pentandra* L. (Bombacaceae).

It was not anticipated that the forest would prove to be of exceptional interest for conservation purposes. However, a botanical inventory carried out by the Project and a team of Cameroonian and British botanists largely in April to June 1992 (Cheek 1992), showed that a total of 67 rare, strictly endemic or new species occur there. Rare species are defined as those known previously from four or less collections in Cameroon. New species are not necessarily known only from Mabeta-Moliwe, several are also known from collections from the Korup (*Salacia*), or S. Bakundu (*Scaphopetalum*) or Douala-Edea (*Psychotria*) but have not yet been published. The number of probable new species currently exceeds twenty (the first ten of which have been written up for publication) and about the same number of strictly endemic species or varieties taking into account those previously recorded from Moliwe, including the near endemic genus *Oxygyne* (Burmanniaceae), known only from Moliwe (in Africa), but with possibly a new species from Etinde, and *Neoschumannia* (Asclepiadaceae) recorded only from Man O'War Bay (just outside the current boundary of the proposed reserve). The largest number of endemic species occurs in Rubiaceae (6), and the remainder are scattered through many different families and life-forms, from saprophytes (see Cheek & Ndam, this congress) to autotrophs, herbs, shrubs, climbers and trees.

The proportion of probable new species (c. 20) to fertile collections made in the 1992 inventory (about 1300) is extremely high for Africa, indeed, it seems to exceed the usual for the Amazon Basin, renowned for being both highly biodiverse and poorly worked. Since publication of the initial report (Cheek 1992) some additional data have come to light, but the overall picture of the high plant biodiversity in this forest remains unchanged.

The explanation for the high number of new species seen in this small area must partly be that most earlier collectors in this part of Cameroon (e.g. Mann, Mildbraed, Swarbrick, Maitland and Thomas) have ignored this small forest, secluded from any major route, in favour of the main massif of Mt. Cameroon. Schlechter and Stammler, two Germans who operated at the beginning of the century were the exceptions. They more or less confined themselves to orchids and orchid-relatives (Burmanniaceae), but made some very important discoveries.

The explanation for the high number of apparent endemics must stem partly from the wide range of habitats and habitat-interfaces present in Mabeta-Moliwe and the fact that it represents both:

1. The last surviving piece of coastal lowland rain forest between Limbe and Douala. No doubt many of the species (and perhaps the vegetation types) now restricted to Mabeta-Moliwe were formerly widespread in this area.
2. Part of the foothills which encircle Mt. Cameroon. While Mt. Cameroon is known as a centre of endemism, it is clear that endemic species are not confined to the main massif. A great deal of the vegetation on the mountain proper must be relatively recent secondary forest given the numerous lava-flows. It may be that areas secure from lava-flows such as Mt. Etinde and foothill areas such as Onge, Mokoko, Bakundu as well as Mabeta-Moliwe are as important as sites for palaeo-endemic species as the remainder of the mountain. Each of the sites may have its cluster of unique species.

Etinde

The second, and most important Project protected area is the proposed Etinde reserve, occupying the southern/south-eastern part of the main massif and occupying about 360 km^2. It takes its name from "Small Mt. Cameroon", or Mt. Etinde, a subsidiary peak about 1710 m tall, slightly isolated from the main massif, far older, volcanically inactive and overlooking Limbe. The proposed boundary extends from 200 m (in some places as low as 60 m) on the seaward side to the peak of the main massif at 4,095 m. An outline botanical inventory of this area was carried out largely in September and October 1992 by the Project and a mixed team of Cameroonian and British botanists. Based on this work a report classifying and enumerating vegetation types and known endemic species was produced the following December (Thomas & Cheek 1992).

During the course of the main inventory about 2500 fertile specimens were collected and identification was nearly completed. Probable new species include at least two new epiphytic *Impatiens* species, a new species of *Thismieae* (possibly *Oxygyne*) and two new species of *Coffea*.

The main massif of Mt. Cameroon has been visited by many botanists. In all, 42 strictly endemic species, (known only from the massif) are recorded, of which 19 are restricted to the high altitude grassland or forest/grassland interface; 50 taxa are ranked as near endemics, that also occur nearby, e.g. Bamenda Highlands, Oku, Kupe, Obudu Plateau (Nigeria) while manyh (29 out of 50 species) are found in Bioko (Fernando Po) too. Of these 50 taxa 21 are restricted to the grassland or grassland/forest interface.

Many more new endemic species are likely to be uncovered as identification proceeds. The richest areas for new species, as indicated by the 1992 inventory work, are Mt. Etinde itself and the lower altitude forest near the coast from Njonji to Bomana.

The area from Mapanja to Buea has proved less exciting. The apparent richness of Mt. Etinde is probably attributable to its freedom from lava flows (the main massif is still active and erupts every 20 years or so). Wet coastal forest is generally rich in species and the stretch from Njonji to Bomana has only been accessible by a good road in the last fifteen years and for this reason may have been neglected by earlier collectors who would have followed the usual route to the peak from Buea.

A high proportion of the endemic species of the main massif of Mt. Cameroon occur not in the forest, but at higher altitudes in the grassland or at the forest/grassland boundary (see above). Because the grassland was suspected of harbouring a large number of endemic species, it has been seen as a conservation priority and was given special attention during the inventory work of 1992. It is feared that fires set in the grasslands every year might pose a threat to these species. Although much analysis remains to be done, field observations and preliminary identifications of some of the specimens at the National Herbarium, Yaoundé, enable us to remark that the majority of these high altitude endemic species were found to be present in abundance during September to October.

On the basis of age, endemic species can be divided into two groups. Neo-endemics are species of limited distribution and of biologically recent origin, at the beginning of their evolutionary life. Their parent species may occur close by, or at least be identified. Palaeo-endemic species, on the other hand, are of biologically ancient origin, further towards the end of their evolutionary life, and often have no obvious parental species. They are evolutionarily isolated and particularly genetically unique.

The majority of the high-altitude endemic species appear to be neo-endemics. Their closest relatives are European or East African montane species which have probably arrived by long-distance dispersal relatively recently and, in isolation from their parent populations, undergone genetic drift. An example is *Succisa trichotocephala* Baksay (Dipsacaceae). Its nearest relatives are European. Formerly more species were believed to fall into this category, but later analysis has shown the difference with the parental species to be insignificant.

At lower altitude, in the forest, neo-endemics again seem to account for much of the endemism. Their origin differs. *Allophylus bullatus* Radlk., for example, is part of a complex undergoing speciation. *Impatiens grandisepala* Grey-Wilson is believed to be of hybrid origin (Grey-Wilson 1981). Both postulated parents occur on the mountain. The same is probably true of at least one of the new species of *Impatiens* and the species of *Begonia*. More work is needed in polarising these endemic species.

Palaeo-endemics are particularly well illustrated by monotypic or near monotypic genera which contain species so isolated genetically that they are accorded generic status. They are of particularly high conservation priority. It is notable that such genera do not occur on the main massif itself, but in the area around Mt. Cameroon, particularly in the foothill areas. Examples are: *Neoschumannia* (Asclepiadaceae: Man O'War Bay), *Oxygyne* (Burmanniaceae: Moliwe), *Ossiculum* (Orchidaceae: Mungo) and (ditypic) *Medusandra* (Medusandraceae: S. Bakundu, Kumba, Kupe, N. Korup to Mamfe). *Batesanthus* (Asclepiadaceae) does occur on Mt. Cameroon, but is comparatively widespread, occurring also in Nigeria, Bamenda and Mabeta-Moliwe.

The explanation for the apparent paucity on the main massif of Mt. Cameroon of palaeo-endemic species, which are more common in the foothills and surrounding forests, is probably that (as discussed under "Mabeta-Moliwe") whilst the Mt. Cameroon area generally is believed to be part of an important Pleistocene refuge, Mt. Cameroon itself is of relatively recent origin, younger than the high mountains of East Africa. It still erupts every 20 years or so, sometimes producing large lava-flows that obliterate everything in their path. It is not surprising that the foothill areas, more secure from lava-flows, should be richer in palaeo-endemic species than the main massif (Cheek 1992).

Onge

The proposed Onge forest reserve is part of the western foothills. It has broadly similar topography and geology to the Mabeta-Moliwe area. There are similarities to Mabeta-Moliwe in the altitudinal range, in the patches of swamp forest, in the numerous small rivers and in the volcanic surface boulders. It is also far from pristine and has a number of old logging trails and abundant *Pycnanthus angolensis* (Welw.) Warb. However, it is less deeply ridged and the soils have a sedimentary element (Hawkins & Brunt 1965). It also differs in that there is no mangrove forest, coastal forest no longer survives, and that there are well-developed riverine communities. A forest corridor connecting with the Etinde forest still remains.

Although rated with a similar rainfall to Mabeta-Moliwe (Hawkins & Brunt 1965), the plant species present indicate a much higher degree of aerial humidity; epiphytic *Impatiens hians* Hook f. var. *bipindensis* (Gilg) Grey-Wilson and *I. niamniamensis* Gilg occur near sea-level, as do several species of terrestrial *Begonia* and epiphytic *Dischistocalyx* (Acanthaceae), none of which occur in Mabeta-Moliwe.

Tree and plot data are not yet available for this forest, but it is remarkable that *Tapura africana* (Dichapetalaceae), one of the commonest trees, and *Heckeldora staudtii* (Harms) Staner (Meliaceae), one of the commonest under shrubs of Mabeta-Moliwe are very much rarer in Onge as are the conspicuous and numerous canopy-emergents of that forest, *Desbordesia glaucescens* (Irvingiaceae). In some areas *Oubangia alata* Bak.f. (Scytopetalaceae) forms large, almost single-species stands. This taxon is unknown from either Mabeta-Moliwe or Etinde though it dominates large areas of the Korup National park to the West.

Intensive collecting in Onge was carried out in March and October-November 1993, using similar techniques to those employed for Etinde and Mabeta-Moliwe, mostly along cutlines made to the northwest from the massif's perimeter road, normal to the Onge River. Specimen determination is still in progress at Kew. Already a very high number of apparently new taxa has come to light, notably a monopodial and hirsute *Oxyanthus* (Rubiaceae) about 30 cm tall, and sapindaceous trees in the genera *Pancovia* and *Placodiscus*. Some of the new taxa first discovered in Mabeta-Moliwe (such as a large-leaved *Psychotria* (Rubiaceae) and Etinde (one of the new *Rhaptopetalum* species) have also been located in the Onge area. There is every indication that, when analysis is complete, this forest will prove as rich in plant biodiversity as Mabeta-Moliwe, though with a considerable difference in composition and species. This is illustrated by the genus *Cola*, where identifications are complete and of which 19 species have been collected in Mabeta-Moliwe and Onge combined, but only 8 species are common to both forests. Several of these species are rather rare and since these were only collected once or twice in one of these forests, the discrepancy in species may be partly explained by inadequate sampling. However, this cannot explain why one of the commonest understorey shrubs, the most frequently collected of all *Cola* species from Mabeta-Moliwe, *Cola flavo-velutina* K. Schum., is totally absent from Onge, where it is replaced in the same niche by *Cola praeacuta* Brenan & Keay, itself absent from Mabeta-Moliwe. The best explanation for discrepancies such as these may be the micro-refugia postulated by Sosef (1994) and Maley (1987). Rather than the Gulf of Guinea constituting a single large continuous refuge, it may have consisted of a collection of smaller ones. Perhaps the eastern foothills of Mt. Cameroon belong to a different microrefuge than that of the western foothills.

As far as is known, the western foothills of Mt. Cameroon have remained completely unexplored botanically until the present work, apart from some collections made by the primatologist, Thomas Struhsaker, in the 1970's. Although much has been done to rectify this deficit recently, there is no doubt that more inventory work is needed during those months not already botanized in.

Mokoko

The Mokoko forest reserve (c. 92 km^2), was originally considered as part of the Project by Hepper et al. (1986). It butts onto the northeastern corner of the Onge forest and contains the headwaters of the Onge River, with many small, fast running streams. It has a similar history of exploitation and a similar vegetation, but with a more closed canopy, more medium-sized trees and a more open shrub layer. Wildlife densities, particularly of bush pigs and chimpanzees, are higher than usual around the mountain. Parts of the Mokoko forest are unusually dynamic, large erosion gullies 30 m deep give rise to disturbance from landslides and patches of pioneers such as *Macaranga hurifolia* Beille, *Alchornea cordifolia* (Schum. & Thonn.) Müll. Arg. and *Musanga cecropioides* R. Br. ex Tedlie develop. In other areas, stands of *Oubangia alata*, as in the Onge Forest, are found. *Tapura africana* is locally common. *Baillonella toxisperma* Pierre, absent from Etinde and Mabeta-Moliwe, but well-known further west, in the Korup, is present. After a reconnaissance in 1993, inventory field-work was carried out in April to June 1994, resulting in about 1,500 specimens which have yet to be named.

Data Processing

Preliminary naming of inventory specimens has been performed by Project staff at the National Herbarium of Cameroon in Yaoundé (YA) with assistance from their botanists. Specimens have then been sent for naming to R.B.G., Kew (K), from where, when necessary, they have been relayed to specialists around the world.

A specimen database of all species likely to occur in SW. Cameroon, proposed in 1990, was set up in 1992 by Karen Sidwell, at Kew, with funding from ODA. All specimens listed in Keay & Hepper (1954-72) from the southern part of "British Cameroon" and adjoining parts of Nigeria and "French Cameroon" were included, together with specimen data extracted from the completed parts of "Flore du Cameroun". BRAHMS (Botanical Research And Herbarium Management System) software, developed at the Dept. of Plant Sciences, Oxford, by Denis Filer was employed. The Kew herbarium was trawled for all specimens from the area in the Rubiaceae by Sally Dawson in 1993 and the plan is to add more families in this manner if labour becomes available. Until mid 1993, all specimen data from recent inventory work was entered into a Microsoft Works database by the Project. This has not yet been transferred to BRAHMS. Since mid 1993, all collections have been entered using the BRAHMS utility RDE (Rapid Data Entry). All recent collections have grid references entered. We plan to use the gazetteer of Letouzey (1968) to convert, with computer assistance, locality names on older specimens in the database into grid references. A mapping function for BRAHMS is being developed at OFI by William Hawthorne by modifying his FROGGIE software from work in Ghana. When this is ready, we hope to be able to map instantly, on screen, all known localities for any plant species around the mountain, for which we have data. We should also have coverage for most of SW. Cameroon and adjoining Cross River State, Nigeria. The database currently encompasses 3-4,000 species and c. 8,000 specimen entries.

The Future

The first five-year term of the Project ran until mid 1993, with an extension until mid 1994. A second term of funding is being approved by ODA (August 1994) and the Govt. of Cameroon, and the new term of three years will coincide with the start of long term GTZ (German Overseas Aid) support for the conservation work on Mt. Cameroon. Support from GEF is also likely to be forthcoming. The prospect is for a larger, combined project which will include the whole mountain and its surrounding foothills. GTZ have already indicated that inventory work in additional forests will be a priority in their scheme for producing an integrated land-use plan (Kirsch-Jung, personal communication). The soils on the main massif are extremely fertile and it is accepted that they must be of proven conservation value before they are withdrawn from cultivation.

Although the main collecting phase is now complete, further work needs to be done. The Bambuko Forest reserve deserves investigation, as does the northern part of the main massif. Supplementary inventory work is needed in most forests in the months not already botanized, and by specialists in critical plant groups. Earthwatch have pledged to continue support for botanical inventory work in this area. Biologists who are interested in working in these forests are urged to contact the authors for more information.

Acknowledgements

We are grateful to our institutions for supporting our work, and to the North and West African desk of the British ODA, particularly John Howarth and John Gilbert, advised by John Hudson, for funding the greater part of the botanical inventory work described above. We are also grateful to Earthwatch, particularly Sally Moyes, for assisting in the inventory of the Onge Forest and for supplementary inventory work in the Etinde and Mabeta-Moliwe forests. We thank Denis Filer of the Department of Plant Sciences, Oxford, for his comments on a draft of this paper.

References

Cheek, M., 1992. A Botanical Inventory of the Mabeta-Moliwe Forest. Government of Cameroon/-RBG, Kew/ODA.

Cheek, M. & F.N. Hepper, 1994. Progress on the Mount Cameroon Rain forest Genetic Resources Project. In: Wildlife Conservation in West Africa. Nigerian Field Society (UK), Occasional Paper No. 1: 15-19.

Grey-Wilson, C., 1981. Balsaminacées. Flore du Cameroun 22.

Hamilton, A.C., 1976. The Significance of Patterns of Distribution. Palaeoecology of Africa 9: 63.

Hawkins, M. & M. Brunt, 1965. Soils and Ecology of West Cameroon. FAO, Rome.

Hepper, F.N., 1972. Numerical Analysis of the "Flora of West Tropical Africa" Ed. II. Kew Bulletin 27 (2): 305-307.

Hepper, F.N., A.J.B. Mitchell & G. Bell, 1986. Botanic Garden and Plant Genetic Resources, SW Cameroon - 1st Mission Report to ODA and Cameroon Govt. ODA/RBG, Kew.

Hepper, F.N. & M. Bovey, 1987. Botanic Garden and Plant Genetic Resources, SW Cameroon - 2nd Mission Report to ODA and Cameroon Govt. ODA/RBG, Kew.

Keay, R.J.W. & F.N. Hepper, 1954-72. Flora of West Tropical Africa. Revised Edition. Crown Agents, London.

Letouzey, R., 1968. Etude phytogéographique du Cameroun. Le Chavalier, Paris.

Maley, J., 1987. Fragmentation de la Forêt Dense Humide Africaine et Extension des Biotopes Montagnards du Quarternaire Récent. Palaeoecology of Africa 18: 307-334.

Maley, J., 1991. The African Rain Forest Vegetation and Palaeoenvironments during Late Quaternary Climatic Change. Climatic Change 19: 79-98.

Sosef, M.S.M., 1994. Refuge Begonia's. Taxonomy, phylogeny and historical biogeography of *Begonia* sect. *Loasibegonia* and sect. *Scutobegonia* in relation to glacial rain forest refuges in Africa. Wageningen Agricultural University Papers, 94.1.

Thomas, D. & M. Cheek, 1992. Vegetation and Plant Species on the South Side of Mount Cameroon in the Proposed Etinde Reserve. Government of Cameroon/RBG, Kew/ODA.

Les Légumineuses Mimosoïdées ombrophiles de basse altitude africaines et malgaches: biodiversité et répartition

J.F. VILLIERS
Laboratoire de Phanérogamie, Muséum National d'Histoire Naturelle,
16 rue Buffon, 75005 Paris, France.

Résumé
Les Légumineuses Mimosoïdées africaines de forêts ombrophiles de basse altitude sont riches de 65 espèces. L'étude de leur répartition est faite en fonction des principales subdivisions chorologiques et des types forestiers. Des centres de biodiversité sont reconnus et l'importance du domaine camerouno-congolais est mise en évidence. Un centre de biodiversité malgache existe dans les formations forestières orientales.

Abstract
The Mimosoideae subfamily of the Leguminosae in African lowland moist forests consists of a number of 65 species. The distribution of these species is studied in the major phytochoria and vegetation types. Main centres of biodiversity are recognized and the cameroonian-congolese domain emerges as very important. The eastern forests of Madagascar also constitute a centre of biodiversity.

Key words
Africa, Biodiversity, Endemism, Madagascar, Mimosoideae.

Introduction

Les Légumineuses Mimosoïdées forment un vaste ensemble essentiellement tropical et subtropical regroupant environ 58 genres et 3.100 espèces.

Une étude de la biodiversité des différents taxons de cette sous-famille peut actuellement être entreprise en Afrique et à Madagascar. La connaissance taxonomique est suffisante même si quelques problèmes restent à résoudre. Les données sur la répartition sont assez bien connues grâce aux nombreuses flores régionales et aux monographies génériques publiées par de nombreux auteurs (Brenan 1959 & 1970, Exell & Mendonça 1956, Gilbert & Boutique 1952, Keay 1956, Villiers 1984a, 1984b, 1990, Villiers & Guinet 1989). Par ailleurs, des collections abondantes, quoique irrégulièrement réparties, sont facilement accessibles.

En Afrique et à Madagascar, les Légumineuses Mimosoïdées se retrouvent dans tous les milieux. Cependant une première observation permet de reconnaître une plus grande biodiversité au niveau des savanes que dans les forêts denses, avec en particulier deux genres largement diversifiés en Afrique: *Acacia* et *Albizia*, et cinq à Madagascar: *Acacia, Albizia, Calliandra, Dichrostachys, Mimosa.*

Nous nous intéresserons tout d'abord aux taxons de l'Afrique continentale puis ensuite à ceux de Madagascar.

L.J.G. van der Maesen et al. (eds.), The Biodiversity of African Plants, 121-130.

Les Legumineuses Mimosoïdées d'Afrique

Le massif forestier dense humide de basse altitude forme une vaste étendue presque continue depuis la côte occidentale de l'Afrique jusqu'à l'océan indien vers l'Est. En fait, nous reconnaîtrons plusieurs zones (d'après White 1983): le massif forestier ouest africain (OA), le massif forestier camerouno-congolais (CC) de loin le plus étendu, la mosaïque du lac Victoria (VI), le massif forestier oriental avec deux parties qui sont la zone de transition de Zanzibar-Inhambane (ZI) et la mosaïque régionale du Tongo-land-Pondoland (TP).

Ce vaste ensemble de formations forestières denses ombrophiles de basse altitude regroupe 16 genres pour 65 espèces; dans le cadre de cette étude nous ne traiterons pas les taxons présents dans les forêts humides mais introduits même s'ils ont tendance à se naturaliser comme par exemple *Adenanthera pavonina* L., *Leucaena leucocephala* (Lam.) De Wit, *Albizia saman* (Jacq.) F. Muell., ou d'autres à très large répartition parfois plus ou moins pantropicaux comme *Mimosa pigra* L., *Mimosa pudica* L., ou *Neptunia oleracea* Lour.

Ils se répartissent de la façon suivante: *Acacia* 5 espèces, *Adenopodia* 2 espèces, *Albizia* 15 espèces, *Aubrevillea* 1 espèce, *Calpocalyx* 11 espèces, *Cylicodiscus* 1 espèce, *Entada* 4 espèces, *Fillaeopsis* 1 espèce, *Newtonia* 10 espèces, *Parkia* 2 espèces, *Pentaclethra* 2 espèces, *Piptadeniastrum* 1 espèce, *Pseudoprosopis* 6 espèces, *Tetrapleura* 2 espèces, *Xylia* 2 espèces.

On remarquera que quelques genres sont strictement forestiers: *Calpocalyx, Cylicodiscus, Fillaeopsis, Pentaclethra, Piptadeniastrum, Tetrapleura*. Exception faite de *Calpocalyx*, ces taxons sont monospécifiques ou paucispécifiques (2 espèces pour *Pentaclethra* et *Tetrapleura*). Ils sont des genres plus ou moins primitifs pour les Légumineuses Mimosoïdées d'Afrique.

Tous les autres genres ont des représentants en savane et en forêt, généralement plus nombreux dans le premier type de formation comme *Acacia*, *Albizia* et *Entada*. Chez *Acacia* et *Entada*, les espèces forestières sont des lianes qui sont habituellement considérées comme évoluées. *Newtonia* et *Pseudoprosopis* sont les seuls genres à avoir plus d'espèces en forêt ombrophile qu'en savane.

Les espèces ont des aires très variables; dans un premier temps nous les envisagerons en fonction des zones de végétation, puis ensuite par rapport aux grands types de forêts.

A. Taxons à large répartition

Un nombre non négligeable de taxons occupe la totalité ou la plus grande partie des formations denses ombrophiles de basse altitude:

Acacia kamerunensis Gand., CC, OA, VI.
Acacia pentagona (Schum.) Hook. f., CC, OA, PT, VI, ZI.
Albizia adianthifolia (Schum.) W.F. Wight, CC, OA, PT, VI, ZI.
Albizia altissima Hook. f., CC, OA, VI.
Albizia ferruginea (Guill. & Perr.) Benth., CC, OA, VI.
Albizia glaberrima (Schumach. & Thonn.) Benth., CC, OA, VI, ZI.
Albizia zygia (DC.) Macbr., CC, AO, VI, ZI.
Entada gigas (L.) Fawc. & Rendle, CC, OA.
Entada mannii (Oliv.) Tiss., CC, OA, VI.

Entada rheedii Spreng., CC, AO, PT, VI, ZI.
Parkia filicoidea Oliv., CC, OA, VI, ZI.
Piptadeniastrum africanum (Hook. f.) Brenan, CC, OA, VI.
Tetrapleura tetraptera (Schumach. & Thonn.) Taub., CC, AO, VI, ZI.

Quelques autres espèces ont aussi des aires vastes mais limitées aux massifs forestiers ouest africain et camerouno-congolais: *Adenopodia scelerata* (A. Chev.) Brenan, *Aubrevillea platycarpa* Pellegr., *Cylicodiscus gabunensis* Harms, *Newtonia duparquetiana* (Baill.) Keay, *Parkia bicolor* A. Chev., *Pentaclethra macrophylla* Benth.

Ces 19 espèces représentent 29,23% des Légumineuses Mimosoïdées des forêts ombrophiles de basse altitude. Certaines espèces, comme *Entada gigas* et *Entada rheedii*, dépassent largement le cadre africain et se retrouvent en Amérique tropicale ou dans les forêts asiatiques. Quelques autres comme *Albizia adianthifolia* et *Acacia pentagona*, seront retrouvées à Madagascar.

Ces taxons, au moins pour le premier groupe, sont des plantes colonisatrices de chablis ou des limites forestières qui peuvent ensuite se maintenir lorsque le couvert forestier est reformé. Elles se rencontrent en général aussi bien dans les forêts denses sempervirentes que dans les forêts semi-caducifoliées. Elles sont souvent pourvues de diaspores transportables par le vent.

B. Les espèces du massif forestier ouest africain

29 espèces sont présentes dans le massif forestier ouest africain, ce qui représente 44,61% de la totalité des espèces. Outre les taxons à large répartition, on trouve plusieurs taxons endémiques à cette région:

Calpocalyx aubrevillei Pellegr.
Calpocalyx brevibracteatus Harms
Albizia dinklagei (Harms) Harms
Albizia rhombifolia Benth.
Newtonia aubrevillei (Pellegr.) Keay
Newtonia elliotii (Harms) Keay
Pseudoprosopis bampsiana Lisowski
Pseudoprosopis sericea (Hutch. & Dalz.) Brenan
Tetrapleura chevalieri (Harms) Bak. f.
Xylia evansii Hutch.

Ces espèces représentent 15,38% de l'ensemble des Légumineuses Mimosoïdées et 34,48% de celles reconnues dans le massif forestier ouest africain. Elles se répartissent en trois ensembles:

* Espèces occupant l'ensemble de la zone (pouvant parfois être observées dans des régions plus sèches): *Albizia dinklagei*, *Xylia evansii*.
* Espèces présentes dans l'extrême ouest (parfois jusqu'en Casamance): *Albizia rhombifolia, Newtonia elliotii, Pseudoprosopis bampsiana, Pseudoprosopis sericea*. Ce sont des sassandriennes (sensu Guillaumet 1967).
* Espèces présentes de la Sierra Leone au Ghana: *Calpocalyx aubrevillei, Calpocalyx brevibracteatus, Newtonia aubrevillei, Tetrapleura chevalieri*.

La plupart de ces taxons endémiques montrent des diaspores relativement lourdes sans dispersion à longue distance.

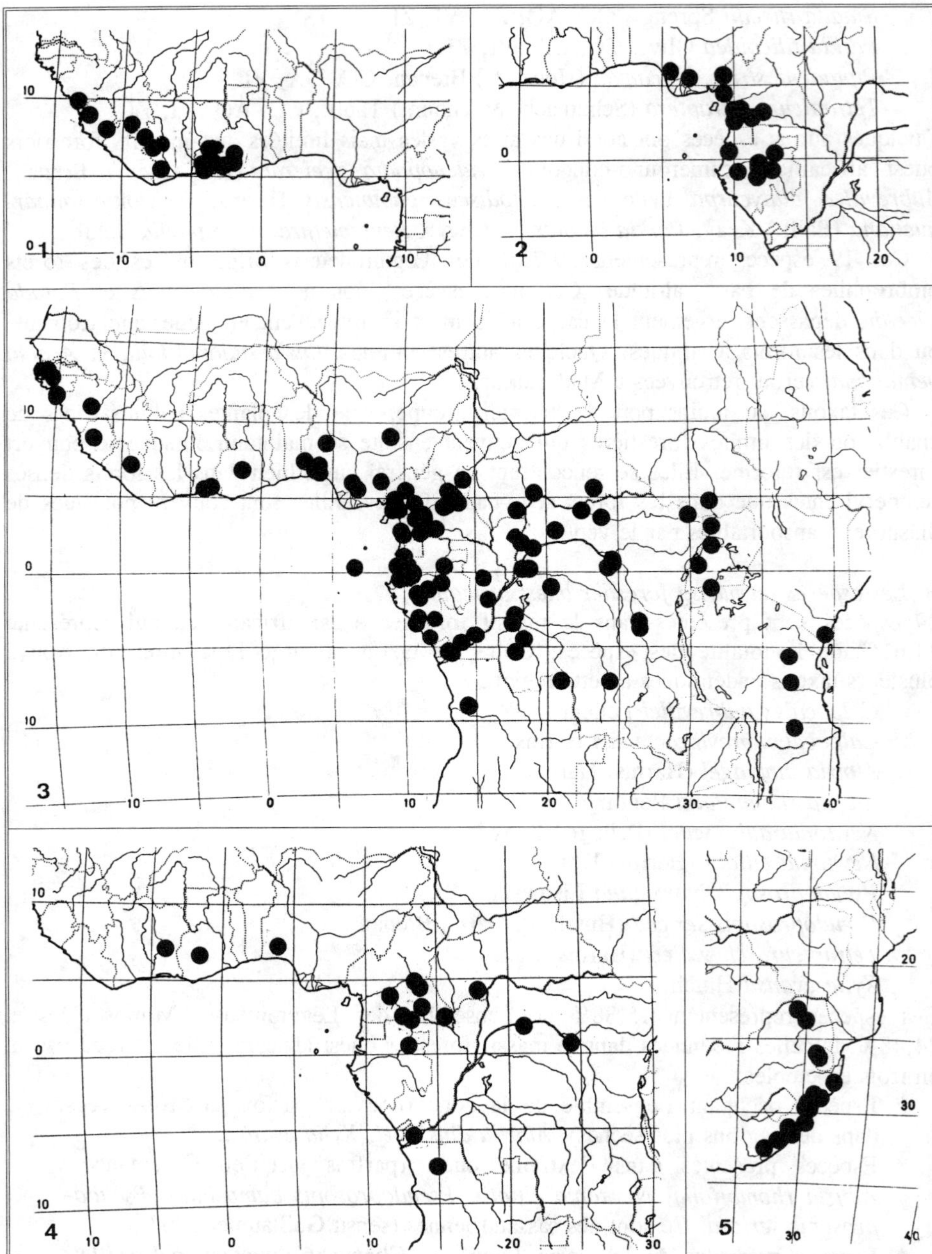

Figure 1. Répartition des Légumineuses Mimosoïdées d'Afrique: 1, *Xylia evansii*; 2, *Newtonia griffoniana*; 3, *Tetrapleura tetraptera*; 4, *Adenopodia scelerata*; 5, *Adenopodia spicata*.

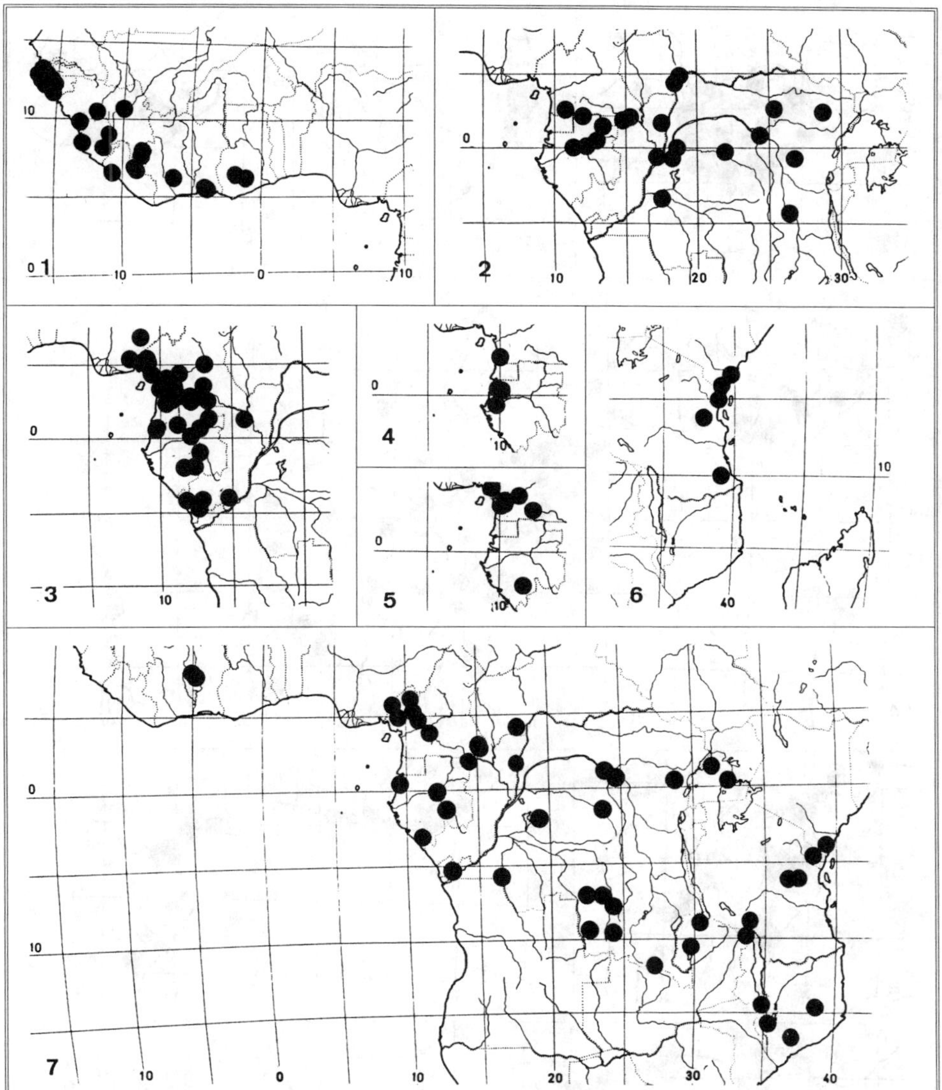

Figure 2. Répartition des Legumineuses Mimosoïdées d'Afrique: 1, *Albizia dinklagei*; 2, *Albizia laurentii*; 3, *Calpocalyx dinklagei*; 4, *Calpocalyx heitzii*; 5, *Calpocalyx ngouniensis*; 6, *Newtonia paucijuga*; 7, *Parkia filicoidea*; 8, *Tetrapleura chevalieri*; 9, *Albizia obliquifoliata*; 10, *Calpocalyx winkleri*.

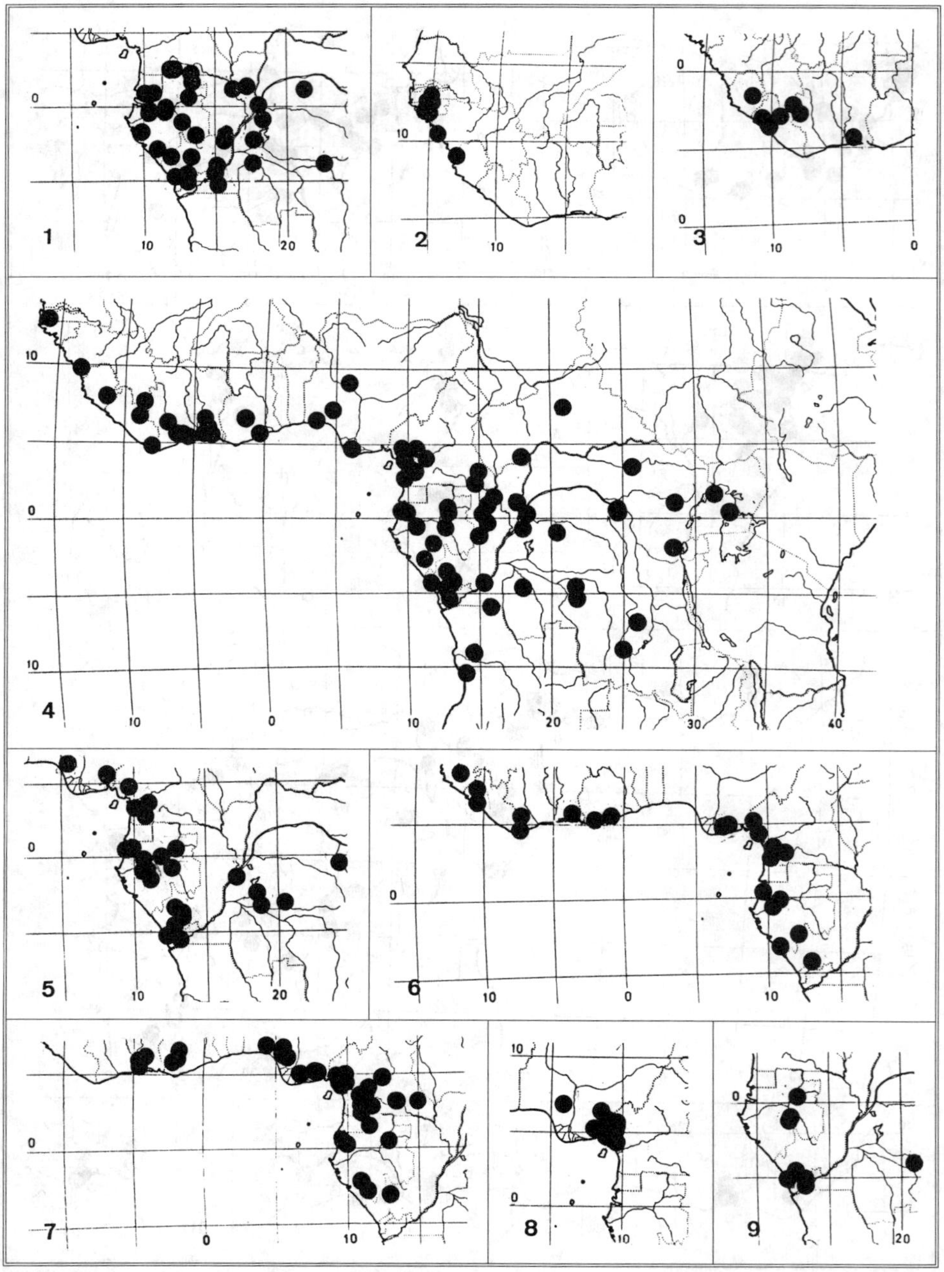

Figure 3. Répartition des Légumineuses Mimosoïdées d'Afrique: 1, *Pentaclethra eetveldeana*; 2, *Albizia rhombifolia*; 3, *Newtonia aubrevillei*; 4, *Piptadeniastrum africanum*; 5, *Fillaeopsis discophora*; 6, *Newtonia duparquetiana*; 7, *Cylicodiscus gabunensis*; 8, *Calpocalyx cauliflorus*; 9, *Newtonia glandulifera*.

Parmi les espèces à large répartition certaines n'occupent pas la totalité de la zone vers l'Ouest: *Adenopodia scelerata* (Ouest de la Côte d'Ivoire), *Aubrevillea platycarpa* (Côte d'Ivoire), *Cylicodiscus gabunensis* (Côte d'Ivoire), *Newtonia duparquetiana* (Sierra Leone). On constate donc qu'il existe, dans le massif forestier ouest africain un centre de biodiversité important aligné le long de la côte atlantique et pratiquement limité par l'isohyète 2.000 mm. La répartition particulière des espèces endémiques est probablement à mettre en relation avec l'existence de refuges forestiers au cours des variations paléoclimatiques de l'ère quaternaire.

C. Les espèces du massif forestier camerouno-congolais

Ce massif forestier, le plus important d'Afrique par sa superficie, regroupe 47 espèces ce qui représente 72,30% des taxons présents dans les formations denses ombrophiles de basse altitude. Outre toutes les espèces à large répartition, un nombre important est limitée à cette zone:

Acacia ciliolata De Wild.
Acacia lujae De Wild.
Albizia laurentii De Wild.
Albizia leptophylla Harms
Albizia obliquifoliata Harms
Albizia welwitschii Oliv.
Calpocalyx atlanticus Villiers
Calpocalyx brevifolius Villiers
Calpocalyx cauliflorus Hoyle
Calpocalyx dinklagei Harms
Calpocalyx heitzii Pellegr.
Calpocalyx klainei Harms
Calpocalyx letestui Pellegr.
Calpocalyx ngouniensis Pellegr.
Calpocalyx winkleri (Harms) Harms
Entada camerunensis Villiers
Fillaeopsis discophora Harms
Newtonia devredii Evrard
Newtonia glandulifera (Pellegr.) Gilb. & Bouti.
Newtonia grandifolia Villiers
Newtonia griffoniana (Baill.) Bak. f.
Newtonia leucocarpa (Harms) Gilb. & Bouti.
Newtonia scandens Villiers
Pentaclethra eetveldeana De Wild. & Th. Dur.
Pseudoprosopis claessensii (De Wild.) Gilb. & Bouti.
Pseudoprosopis gilletii (De Wild.) Villiers
Pseudoprosopis uncinata Evrard
Xylia ghesquieri Robyns

Ces 28 espèces endémiques du massif forestier camerouno-congolais forment 43,07% de la flore des Légumineuses Mimosoïdées et 59,57% de celles de cette zone. La répartition de ces taxons est inégale et on peut reconnaître trois types d'aires:

* Les aires plus ou moins parallèles à l'Equateur: certaines Mimosoïdées se répartissent plus ou moins régulièrement parallèlement à l'Equateur. Outre la plupart des espèces à large répartition, quelques endémiques sont reconnues comme *Acacia ciliolata, Acacia lujae, Albizia laurentii, Albizia leptophylla, Fillaeopsis discophora, Newtonia leucocarpa, Pentaclethra eetveldeana, Pseudoprosopis claessensii*; elles représentent 28,57% des espèces particulières à la zone.
* Les aires "atlantiques": un nombre important de Légumineuses Mimosoïdées a une aire qui se limite plus ou moins à la bordure du littoral atlantique. Outre quelques taxons à large répartition comme *Cylicodiscus gabunensis* ou *Newtonia duparquetiana*, on observe un nombre élevé d'endémiques: *Calpocalyx atlanticus, Calpocalyx brevifolius, Calpocalyx*

cauliflorus, Calpocalyx dinklagei, Calpocalyx heitzii, Calpocalyx klainei, Calpocalyx ngouniensis, Calpocalyx winkleri, Newtonia glandulifera, Newtonia grandifolia, Newtonia griffoniana, Newtonia scandens, Pseudoprosopis gilletii; elles forment 46,42% des endémiques. La plupart d'entre elles a une aire limitée par l'isohyète 2.000 mm. Cinq taxons ne dépassent pas ou guère le secteur atlantique biafréen (sensu Letouzey 1985), dans sa partie la plus humide limitée par l'isohyète 3.000 mm.

* Les aires de la "cuvette congolaise": quelques espèces ont des aires limitées au centre de la cuvette congolaise, dans une zone plus ou moins englobée par l'isohyète 2.000 mm: *Albizia obliquifoliata, Albizia welwitschii, Newtonia devredii, Pseudoprosopis uncinata, Xylia ghesquieri*; elles représentent 17,85% des endémiques.

Seul *Entada camerunensis* a une aire limitée à la périphérie septentrionale du massif forestier camerouno-congolais.

On remarque donc qu'il existe dans cette zone trois centres importants de biodiversité. Celui du littoral atlantique reste le plus riche en taxons. Les aires de la plupart des endémiques de ce centre de biodiversité ont des aires qui correspondent sensiblement aux refuges forestiers quaternaires tels qu'ils ont été reconnus par Maley (dans Sosef 1991).

D. La mosaïque du Lac Victoria

Dans la zone centrée sur le Lac Victoria, la végétation est une mosaïque de formations forestières denses humides et zambéziennes sèches.

Les Légumineuses Mimosoïdées reconnues dans la forêt dense humide de basse altitude sont presque toutes des taxons dont l'aire s'étend jusqu'au massif forestier oriental: *Acacia pentagona, Albizia zygia, Entada rheedii, Parkia filicoidea, Tetrapleura tetraptera*. Quelques autres voient ici se marquer les limites orientales de leurs aires: *Acacia kamerunensis, Albizia altissima, Albizia ferruginea, Albizia glaberrima* var. *glaberrima, Entada gigas, Piptadeniastrum africanum*.

En fait les taxons retrouvés ici font pour la plupart parti du massif forestier camerouno-congolais. *Albizia grandibracteata* Taub. est la seule espèce dont l'aire est limitée à cette zone, débordant sur la végétation montagnarde. *Albizia petersiana* (Bolle) Oliv. var. *petersiana*, caractéristique des formations denses humides de la côte orientale d'Afrique, fait son apparition dans la zone du Lac Victoria.

Les Légumineuses Mimosoïdées trouvées dans cette région représentent 20% du nombre total des espèces.

E. Le massif forestier oriental (la zone de transition de Zanzibar-Inhambane et la mosaïque régionale du Tongoland-Pongoland)

Ces deux ensembles sont composés de forêts denses humides de basse altitude assez réduites et de formations sèches. La flore des Mimosoïdées est limitée; on y retrouve quelques espèces à large répartition comme: *Albizia adianthifolia* (ZI, PT), *Albizia glaberrima* var. *glabrescens* (Oliv.) Brenan, *Albizia zygia* (ZI), *Acacia pentagona* (ZI, PT), *Entada rheedii* (ZI, PT), *Parkia filicoidea* (ZI), *Tetrapleura tetraptera* (ZI).

Quelques taxons ont des aires limitées au massif forestier oriental ou à l'une ou l'autre de ses composantes:

Acacia kraussiana Benth. (PT)
Adenopodia spicata (E. Mey.) C. Presl (PT)
Albizia forbesii Benth. (ZI,PT)
Albizia petersiana (Bolle) Oliv. (ZI)
Albizia zimmermannii Harms (ZI)
Newtonia paucijuga (Harms) Brenan (ZI)
Pseudoprosopis euryphylla Harms (ZI)

Ces espèces représentent 21,53% du domaine forestier dense de basse altitude; celles qui sont particulières à cette zone forment 50% de sa flore.

F. Forêt dense humide sempervirente et forêt dense humide semi-caducifoliée

Deux grands types de formations forestières se partagent le massif forestier dense humide de basse altitude: la forêt dense humide sempervirente et la forêt dense humide semi-caducifoliée.

Il est difficile de mettre en évidence une répartition différente des Légumineuses Mimosoïdées entre ces deux formations. Nombre d'espèces de la forêt semi-caducifoliée pénètrent dans la forêt sempervirente à l'occasion de chablis ou des défrichements. On peut cependant reconnaître quelques taxons plus affines du second type forestier comme *Albizia ferruginea, Albizia glaberrima, Entada camerunensis* (seule espèce à avoir été trouvée exclusivement dans la forêt semi-caducifoliée, en lisière), *Entada mannii, Parkia filicoidea*.

Nombre de ces espèces peuvent se rencontrer dans les savanes périguinéennes où elles peuvent recoloniser les recrus. Plusieurs sont des composantes de la flore de la mosaïque du Lac Victoria, ce qui laisserait penser que les formations forestières de cette région sont de type semi-caducifolié.

Les Legumineuses Mimosoïdées de Madagascar

Les formations forestières denses humides de Madagascar sont limitées à la côte orientale de l'île. Elles renferment un nombre restreint d'espèces (plusieurs taxons n'ont pas encore été décrits). Quinze espèces sont à ce jour reconnues.

Trois taxons sont communs avec les forêts denses humides de basse altitude d'Afrique: *Acacia pentagona, Albizia adianthifolia, Entada rheedii*. Cette dernière espèce se retrouve aussi dans les formations plus sèches de la région occidentale.

Les autres espèces sont toutes endémiques de la forêt ombrophile orientale: *Albizia viridis* Fournier, *Calliandra alternans* Benth., *Calliandra kony* R. Vig., *Calliandra pervillei* Drake, *Calliandra scottiana* R. Vig., *Dichrostachys myriophylla* Bak., *Entada louvelii* (R. Vig.) Brenan, *Lemurodendron capuronii* Villiers & Guinet, *Mimosa myriacantha* Bak., *Mimosa nossibiensis* Benth., *Mimosa tanalarum* R. Vig., *Parkia madagascariensis* R. Vig.

Il est certain, compte tenu de la richesse floristique et de l'endémisme important que la région orientale des forêts denses humides de Madagascar forme un centre de biodiversité.

Conclusion

L'étude de la biodiversité des Légumineuses Mimosoïdées des forêts denses ombrophiles de basse altitude nous a permis de mettre en évidence la remarquable richesse en taxons du massif forestier camerouno-congolais et plus particulièrement de la partie occidentale de cette région avec un haut niveau d'endémisme. Le massif ouest africain est aussi un pôle important de diversité avec nombre de taxons particuliers. De telles répartitions sont difficilement compréhensibles au vu des conditions climatiques actuelles. Elles ne pourront probablement pas être expliquées sans le recours à l'étude des variations paléoclimatiques de l'ère quaternaire (voire plus anciennes) et des refuges forestiers dont l'existence est maintenant admise.

Bibliographie

Brenan, J.P.M., 1959. Flora of Tropical Africa, Leguminosae subfamily Mimosoideae: 1-173.

Brenan, J.P.M., 1970. Flora Zambesiaca 3(1): 7-153.

Exell, A.W. & F.A. Mendonça, 1956. Conspectus Florae Angolensis 2(2): 254-299.

Gilbert, G. & R. Boutique, 1952. Flore du Congo belge et du Ruanda-Urundi 3: 137-233.

Guillaumet, J.L., 1967. Recherches sur la végétation et la flore de la région du Bas Cavally (Côte d'Ivoire). Mém. O.R.S.T.O.M. 20: 1-247.

Keay, R.W.J., 1956. Flora of West Tropical Africa 1(2): 484-504.

Letouzey, R., 1985. Notice de la carte phytogéographique du Cameroun au 1:500000, 4. TV: Domaine de la forêt dense humide toujours verte: 95-142.

Lock, J.M., 1989. Legumes of Africa - a check-list. Royal Botanic Gardens Kew: 62-99.

Ross, J.H., 1975. Flora of Southern Africa 16(1): 1-159.

Ross, J.H., 1979. A conspectus of the African Acacia species. Mem. of the Bot. Surv. of South Africa 44: 1-155.

Sosef, M.S.M., 1991. New species of *Begonia* in Africa and their relevance to the study of glacial rain forest refuges. Wageningen Agricultural University Papers 91(4): 120-151.

Villiers, J.F., 1984a. Le genre *Pseudoprosopis* Harms (Mimosacées) en Afrique. Bull. Jard. Bot. Nat. Belgique 53(3/4): 417-436.

Villiers, J.F., 1984b. Le genre *Calpocalyx* (Leguminosae - Mimosoideae) en Afrique. Bull. Mus. natn. Hist. nat., Paris, 4ème sér., 6, 1984, sect. B, Adansonia, 3.

Villiers, J.F., 1990. Contribution à l'étude du genre *Newtonia* Baillon (Leguminosae Mimosoideae) en Afrique. Bull. Jard. Bot. Nat. Belgique 60(1/2): 119-138.

Villiers, J.F. & Ph. Guinet, 1989. *Lemurodendron* Villiers & Guinet, genre nouveau de Leguminosae Mimosoideae de Madagascar. Bull. Mus. natn. Hist. nat., Paris, 4ème sér., 11, 1989, sect. B, Adansonia, 1: 3-10.

White, F., 1983. The vegetation of Africa. UNESCO, Natural Resources Research 20: 1-356.

Biodiversité morphogénétique, exemples des plantes plagiotropes

A. AMOUGOU
Université de Yaoundé I, Faculté des Sciences, B.P. 812, Yaoundé, Cameroun.

Résumé
Après avoir rappelé quelques notions de l'architecture végétale, l'auteur donne une clé de détermination, une brève définition et quelques exemples pour chaque modèle de croissance. Une nouvelle voie de recherche systématique peut être ouverte: la phylogénie des modèles architecturaux.

Abstract
After reviewing architectural criteria in the first part of this paper, the author presents a dichotomic key of growth models with plagiotropic primary axis, and gives a brief definition with a few examples for each growth model. A new area of research in taxonomy is proposed: the phylogeny of architectural models.

Key words
Biodiversity, Growth model, Plagiotropy, Plant architecture, Taxonomy.

Rappels de quelques notions architecturales

Les marqueurs architecturaux

Après la germination, les premiers stades de développement de l'axe épicotyle d'une plantule sont marqués par l'adoption d'une orientation. L'axe obtenu est alors orthotrope ou plagiotrope. Le type de fonctionnement du méristème apical sera à l'origine de la nature monopodiale ou sympodiale de l'axe en cours d'édification. Plus ou moins précocement suivant les espèces, l'axe primaire amorce la ramification liée soit au dédoublement du méristème apical, soit à l'entrée en activité des méristèmes axillaires. La chronologie de développement entre les différents méristèmes de la plante en cours d'édification est capitale dans la configuration morphologique du jeune plant ramifié.

Ces différentes étapes de la vie de la plante sont désignées sous le de marqueurs architecturaux. Nous résumons dans la figure 1 les différentes alternatives, autant que possible dans l'ordre d'intervention ontogénique et/ou phylétique, et qui s'excluent mutuellement.

La notion de modèle de croissance

La notion de modèle de croissance a été largement discutée par Cremers (1973, 1974), Hallé & Oldeman (1970), Hallé et al. (1978) et Oldeman (1968). Le modèle de croissance répond mieux à une définition qu'à un type systématique. Cette définition donne des limites précises à l'intérieur desquelles un exemple peut s'inclure ou non.

L'abondance ou la rareté d'un modèle de croissance sont également significatives et doivent exprimer, comme dans d'autres disciplines scientifiques, le taux de diversification ou de stagnation, le degré d'évolution ou d'ancienneté, le taux de répartition géographique ou d'endémisme de l'unité considérée...

L.J.G. van der Maesen et al. (eds.), The Biodiversity of African Plants, 131-139.

1. Orientation des axes
    1.1. Orthotropie
    1.2. Plagiotropie

2. Fonctionnement des méristèmes
    2.1. Fonctionnement illimité ou monopodial
        2.1.1. Monopode orthotrope
        2.1.2. Monopode plagiotrope
    2.2. Fonctionnement limité ou sympodial
        2.2.1. Sympode orthotrope
        2.2.2. Sympode plagiotrope
            2.2.2.1. S. p. par apposition d'axes orthotropes à croissance illimitée
            2.2.2.2. S. p. par substitution
                2.2.2.2.1. S. p. s. d'axes orthotropes à croissance limitée
                2.2.2.2.2. S. p. s. d'axes plagiotropes à croissance limitée

3. Ramification
    3.1. Ramification dichotomique
        3.1.1. R. d. isotone
        3.1.2. R. d. anisotone
            3.1.2.1. R. d. a. végétative
            3.1.2.2. R. d. a. florale
    3.2. Ramification latérale ou sympodiale
        3.2.1. R. l. acrotonique
            3.2.1.1. R. l. a. rythmique
            3.2.1.2. R. l. a. continue
        3.2.2. R. l. basitonique
        3.2.3. R. l. épitonique ou centrale

4. Chronologie de développement
    4.1. Syllepsie
    4.2. Prolepsie

Figure 1. Differentes marqueurs architecturaux.

La notion de modèle de croissance n'est ni statistique, ni géographique. Elle est l'expression morphologique d'un génome stable dans la nature. On exclut ainsi les séquences génétiques nouvelles éventuellement observables chez des espèces et hybrides agronomiques et horticoles (Amougou 1991).

Les modèles de croissance sont applicables à tous les groupes de plantes vasculaires chez lesquelles il est aisé de distinguer une ramification vraie, i.e. aussi bien les herbes, les arbres et les lianes, les Monocotylédones et les Dicotylédones, que les Phanérogames, les Préphanérogames et les Cryptogames vasculaires.

On aboutit à une conception globale et unitaire du modèle de croissance qui n'exclut nullement sa simplicité. Par exemple, chaque caractère d'un marqueur architectural donné présente un visage très synthétique, puisqu'il inclut parfois en son sein de très nombreux états de caractères correspondant à autant de séquences génétiques différentes:

* la ramification rythmique rassemble les ramifications: pseudo-verticillées, opposées-décussées, et enfin verticillées, cette dernière pouvant elle-même être à 3-4-5-n éléments par verticille;
* le fonctionnement limité inclut les arrêts de croissance par: nécrose, floraison, parenchymatisation, et différenciation d'organes végétatifs divers (épines, vrilles, crochets, etc.).

La plagiotropie
D'une manière générale, on distingue plusieurs types de plagiotropie qui sont (Koriba 1958):
* la plagiotropie monopodiale (p.e. *Cleistopholis* Pierre ex Engler, *Pterocarpus* L.)
* la plagiotropie sympodiale dans laquelle on retrouve (Figure 2):
  - la plagiotropie par apposition d'axes orthotropes à croissance lente. C'est le type *Terminalia* (axes II de: *Terminalia catappa* L. et de *T. superba* Engl. & Diels, etc.);
  - la plagiotropie par substitution d'axes orthotropes à croissance limitée par nécrose, floraison, parenchymatisation, ou différenciation d'organes végétatifs divers. C'est le type *Rothmannia* (axes II de: *R. hispida* K. Schum.) Fagerlind, *R. megalostigma* (Wernham) Keay, etc.);
  - la plagiotropie par substitution d'axes plagiotropes à croissance limitée, généralement par floraison. C'est le type *Bertiera* (axe I de *Calliandra calothyrsus* Meissn.; axes II de *Bertiera aethiopica* Hiern, *B. ituriensis* K. Krause, ...).

Du point de vue phylétique, tous les auteurs sont d'avis que la plagiotropie est certainement plus évoluée que l'orthotropie (Amougou 1982, Bancilhon 1971, Hallé et al. 1978, Roux 1968). Comme pour confirmer que l'ontogenèse récapitule la phylogenèse, Roux constate qu'il existe toujours "chez l'individu (...) un stade initial orthotrope, plus ou moins long et complexe, avant que n'apparaisse l'architecture plagiotrope". C'est ce qu'Hallé et al. (1978) appellent de l'"orthotropie résiduelle".

On note avec ces derniers auteurs la faible différenciation architecturale de la plagiotropie par rapport à l'orthotropie (Amougou 1982, 1991). Les modèles de croissance à axe I plagiotrope représenteraient au plus le 1/3 numérique de ceux à axe I orthotrope. De plus, ils manifestent une forte tendance au thérophytisme. En effet, plus du tiers (6 sur 15) d'entre eux décrits ci-dessous seraient exclusifs aux thérophytes.

Breves définitions des modèles de croissance a axes I plagiotropes
Généralités

Dès la germination et la levée de la plantule, le marqueur primaire qui est l'orientation des axes permet de distinguer deux lignées évolutives: la lignée orthotrope et la lignée plagiotrope. Chacune de celles-ci présente trois séries évolutives qui comportent respectivement: des axes végétatifs à croissance illimitée, exclusivement, un mélange d'axes végétatifs à croissance illimitée et d'autres à croissance limitée, des axes végétatifs à croissance limitée, exclusivement.

Les configurations morphogénétiques qui en dérivent se répartissent en quatre sous-groupes suivants:

Figure 2. Quelques types de plagiotropie. **a**, plagiotropie par apposition observée chez *Terminalia superba* (Combretaceae). **b**, plagiotropie par substitution d'axes orthotropes à floraison terminale chez *Rothmannia hispida* (Rubiaceae). **c**, plagiotropie par substitution d'axes plagiotropes à floraison terminale chez *Bertiera brevifolia* (Rubiaceae).

* le premier sous-groupe comprend des structures non ramifiées, dites élémentaires;
* le deuxième sous-groupe comprend des structures articulées comportant des modules ou articles non différenciés;

* le troisième comprend des structures différenciées où l'on distingue un tronc monopodial et des axes II;
* et le quatrième comprend des structures articulées différenciées où l'on distingue un tronc sympodial et des axes II distincts.

Un tableau à double entrée illustre les diverses configurations obtenues, réèlles ou dont l'existence est très probable dans la nature (Figure 3).

Les modèles de croissance

Modèle de Takhtajan: Structure élémentaire non ramifiée. Plante construite par l'activité d'un unique méristème apical plagiotrope à fonctionnement illimité. Ce modèle est l'équivalent plagiotrope du modèle de Corner. Exemples: *Elatostema mannii* Wedd., *E. sessite* J. R. & F. Forst., *Procris crenata* C. B. Robinson (Urticaceae) (Tiani 1989). Nous dédions ce modèle à A. Takhtajan pour ses travaux sur l'origine et l'évolution des Angiospermes de 1969 et 1976.

Modèle théorique 5[*]: Structure élémentaire, non ramifiée. Plante construite par l'activité d'un unique méristème apical plagiotrope, à fonctionnement limité. Structure séquentielle monocarpique. Ce modèle est l'équivalent plagiotrope du modèle de Holttum. Son existence est très probable dans la nature, car il constitue le stade d'évolution directement supérieur au modèle de Takhtajan.

Modèle d'Emberger: Structure articulée non différenciée, dichotome. L'axe I se développe par l'activité d'un méristème plagiotrope qui subit une dichotomie. L'un des articles-fils prend plus de vigueur que l'autre, et son méristème subit de la même manière une dichotomie comme le précédent. Ainsi de suite... Ce modèle est l'équivalent plagiotrope du modèle de Fisher. Exemples: *Selaginella myosorus* (Sw.) Alst., *S. versicolor* Spring. Nous dédions ce modèle à L. Emberger pour son importante contribution à la Systématique des végétaux vasculaires de 1960.

Modèle de Troll (Amougou 1982): Structure articulée non différenciée. Chaque article est un axe plagiotrope mixte, avec possibilité d'une orthotropie résiduelle dans sa partie basale. La ramification est centrale ou épitone. Ce modèle est l'équivalent plagiotrope de celui de Champagnat. Exemples: *Euphorbia milii* Des Moul. (Tiani 1989), *Atroxima liberica* Stapf (Cremers 1973), *Carpolobia alba* G. Don, *C. glabrescens* Hutch. & Dalz., *C. lutea* G. Don.

Modèle de Rossignol: Structure articulée non différenciée, à allure scorpioïde, en sympodes linéaires. Chaque article est un sympode plagiotrope à croissance limitée. Ce modèle est l'équivalent plagiotrope du modèle de Chamberlain. Exemple: *Euphorbia hirta* L. (= *E. pilulifera* L. cf. Chev.) (Euphorbiaceae) (Tiani 1989). Nous dédions ce modèle à Line Rossignol née Bancilhon de l'Université Paris-Sud (Orsay) qui a consacré ses travaux à la morphologie et la systématique des Euphorbiaceae du genre *Phyllanthus* (1985).

---

[*] Cette numérotation est à la suite des modèles de croissance comportant un axe I orthotrope.

			Groupe II:
Sous-groupes:		Séries évolutives:	I. Axes à croissance illimitée
Structures élémentaires			Takhtajan 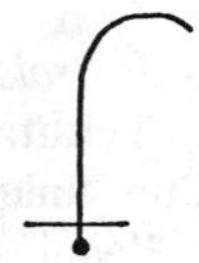
Structures articulées non différenciées	Structures dichotomes		Emberger 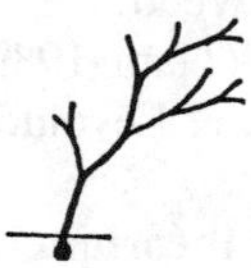
	Structures sympodiales	Epitones ou acrotones	Troll 
		Basitones	Edelin 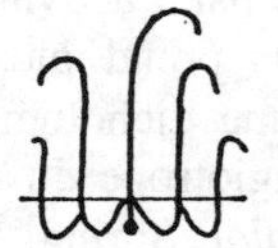
Structures différenciées	Axes II monopodiaux		Hallé  Oldeman 
	Axes II sympodiaux		
Structures articulées différenciées			

Figure 3. Allure physionomique des modèles de croissance comportant un axe I plagiotrope. La lignée évolutive correspondante serait la plus évoluée. Toutes les descriptions dans ce

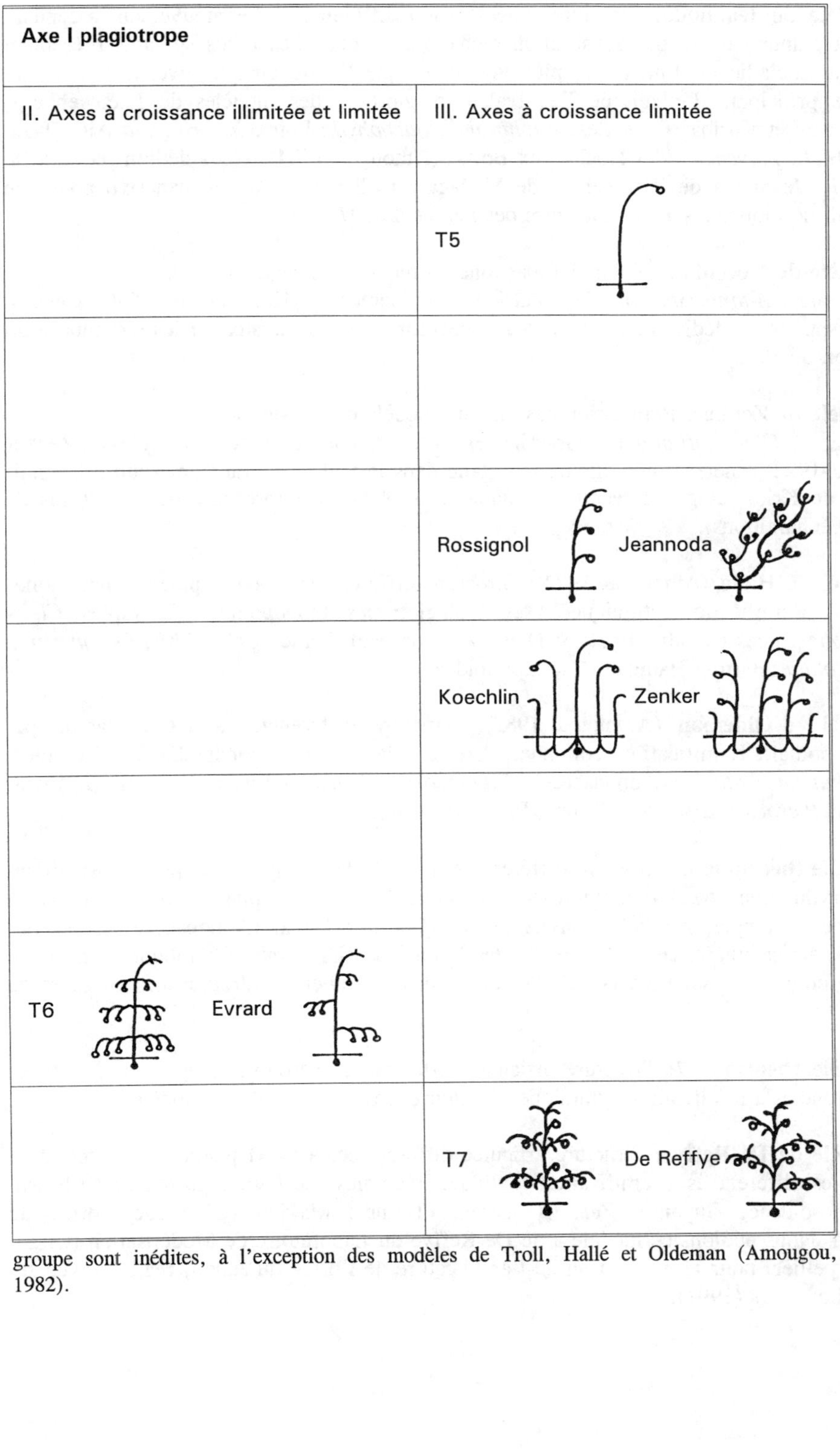

groupe sont inédites, à l'exception des modèles de Troll, Hallé et Oldeman (Amougou, 1982).

Modèle de Jeannoda: Structure articulée non différenciée. Le mouvement ascendant de la plante s'opère par substitution dichasiale isotone. Les relais suivants présentent une anisocladie où l'un d'eux plus vigoureux que l'autre chaque niveau, fera figure d'axe principal. Il s'agit de l'équivalent plagiotrope des modèles de Leeuwenberg, Letouzey et Koriba. Exemples: *Euphorbia glaucophylla* Poir., *E. prostrata* Ait. (Tiani 1989), *E. polycnemoides* Hochst. ex Boiss. (Amougou 1984). Nous dédions ce modèle à V.R. Jeannoda de l'Université de Madagascar, E.E.S.S., à Antananarivo pour son travail de pionnier sur l'architecture des herbes de 1977.

Modèle de Koechlin: Réplication basitone du modèle théorique 5.
Exemples: *Alpinia speciosa* D. Dietrich (Zingiberaceae) (Hallé et al. 1978), *Centhoteca* spp. Nous dédions ce modèle à J. Koechlin pour ses travaux sur les Gramineae du Gabon de 1962.

Modèle de Zenker: Réplication basitone du modèle de Rossignol.
Exemple: *Olyra latifolia* L. (Gramineae). Nous dédions ce modèle à Georges Zenker (1855-1922), ancien colon allemand installé dans le Sud-Ouest du Cameroun à Bipindi, près de Kribi, et grand récolteur botanique dans la zone forestière du pays (plus de 6000 échantillons).

Modèle d'Hallé (Amougou 1982): Structure différenciée à axe I plagiotrope, monopodial. Ramification rythmique. Axes II plagiotropes, monopodiaux. Exemples: *Cleistopholis patens* (Benth.) Engl. & Diels (Annonaceae) (Hallé et al. 1970), *Tetraberlinia bifoliolata* (Harms) Hauman (Caesalpinioideae).

Modèle d'Oldeman (Amougou 1982): Structure différenciée à axe I plagiotrope, monopodial. Ramification continue. Axe II plagiotropes monopodiaux. Exemples: *Annona muricata* L. (Annonaceae), *Klainedoxa gabonensis* Pierre ex Engl. (Irvingiaceae), *Pterocarpus soyauxii* Taub. (Papilionoideae).

Modèle théorique 6: Structure différenciée à axe I plagiotrope, monopodial. Ramification rythmique. Axes II plagiotropes par substitution d'axes plagiotropes à croissance limitée. Exemple: *Monodora myristicifolia* (Gaertn.) Dunal (Annonaceae). Nous dédions, amicalement, ce modèle à Charles Evrard de l'Université Catholique de Louvain (Belgique), pour ses travaux sur les végétations en milieux hydromorphes au Zaïre de 1968.

Modèle théorique 7: Structure articulée différenciée, acrotone, plagiotrope. Articles différenciés à ramification rythmique. Existence très probable de ce modèle.

Modèle de De Reffye: Structure articulée différenciée, à axe I plagiotrope sympodial. Articles différenciés à ramification continue. Exemples: *Calliandra calothyrsus* Meissn. (Mimosoideae, Guyane), *Zelkova serrata* (Thunb.) Makino (Ulmaceae), orme du Japon. Nous dédions ce modèle à de De Reffye du Laboratoire de Modélisation (Cirad, Montpellier) pour la description de l'architecture de l'orme du Japon, réalisée avec ses collaborateurs (1991).

Conclusion

Les plantes du groupe II comportant un axe primaire plagiotrope se répartissent en 15 modèles de croissance. En réalité, il nous paraît que l'extrême diversité du monde végétal des forêts ombrophiles africaines de basse altitude et la grande complexité morphogénétique qui pourrait en résulter ne serait encore que partiellement transcrite dans les 15 modèles de croissance plagiotropes ci-dessus décrits. Les nombreux espaces vacants de la figure 3 traduiraient des potentialités pour de nombreuses autres possibilités.

On note que l'architecture végétale utilise des critères purement morphologiques, phénotypiques. Leur utilisation est donc basée sur une observation empirique s'opposant à la déduction ou à l'interprétation. Au niveau spécifique, l'architecture constitue un caractère stable dans les conditions naturelles, surtout en ce moment où l'on parle de plus en plus de la taxonomie numérique. Il resterait alors à préciser, dans la limite des erreurs habituellement tolérables, la phylogénie de tous les marqueurs architecturaux ci-dessus définis.

Une nouvelle voie de recherche systématique peut être ainsi ouverte.

Bibliographie

Amougou, A., 1982. Deux modèles de croissance nouveaux. Essai sur la révision des arbres plagiotropes. Rev. Sci. et Tech. 2: 55-72.

Amougou, A., 1984. Un mode de croissance particulier chez *Euphorbia polycnemoides* Hochstetter (Euphorbiaceae-Crotonoideae). Ann. Fac. Sci. III (2): 5-12.

Amougou, A., 1991. Revue illustrée des modèles de croissance architecturaux. In: Edelin C. (Ed.). L'arbre. Biologie et développement: 560-561, Nat. Monspeliensia h.s.

Bancilhon, L., 1971. Contribution à l'étude taxonomique du genre *Phyllanthus* (Euphorbiaceae). Boissiera 18.

Cremers, G., 1973, 1974. Architecture de quelques lianes d'Afrique tropicale. Candollea 28: 249-280; 29: 57-110.

De Reffye, Ph., P. Dinouard & D. Barthélémy, 1991. Modélisation et simulation de l'architecture de l'orme du Japon *Zelkova serrata* (Thunb.) Makino (Ulmaceae): la notion d'axe de référence. In: Edelin C. (Ed.). L'arbre. Biologie et développement: 251-266, Nat. Monspeliensia h.s.

Hallé, F., & R.A.A. Oldeman, 1970. Essai sur l'architecture et la dynamique de croissance des arbres tropicaux. Masson & Cie, Paris.

Hallé, F., R.A.A. Oldeman & P.B. Tomlinson, 1978. Tropical trees and forests. An architectural analysis. Springer-Verlag, Berlin Heidelberg New York.

Koriba, K., 1958. On the periodicity of tree growth in the tropics, with reference to the mode of branching, the leaf fall, and the formation of resting buds. Gard. Bull. Singapore 17 (1): 11-81.

Oldeman R.A.A., 1968. Une disjonction entre la sexualité et la dédifférenciation végétative chez le *Mabea piriri* Aubl. (Euphorbiaceae). C. R. Acad. Sc., Paris 267: 1358-1359.

Rossignol, L. & M., 1985. Architecture et tendances évolutives dans le genre *Phyllanthus* (Euphorbiaceae). Bull. Mus. natn. Hist. nat., 4e sér., sect. B, Adansonia 7(1): 67-80.

Roux, J., 1968. Sur le comportement des axes aériens chez quelques plantes à rameaux végétatifs polymorphes: le concept de rameau plagiotrope. Ann. Sc. Nat. Bot., 12 sér., t. IX: 109-256.

Tiani, A.M., 1989. Architecture des Euphorbiacées herbacées et des Urticacées. Thèse 3e cycle, Univ. de Yaoundé.

Analyse de la biodiversité végétale dans la Forêt des Abeilles, Gabon

K. GESNOT, K. VAN ESSCHE & J. LEJOLY
Laboratoire de Botanique systématique et de phytosociologie,
CP 169, Av. F. Roosevelt 50, Université Libre de Bruxelles, 1050 Bruxelles, Belgique.

Abstract
A biodiversity analysis of the vegetation in the Forêt des Abeilles (Gabon) has been undertaken using the transects method. An area of 30.5 ha was studied for the trees with a dbh > 70 cm and 1 ha for dbh > 10 cm. A main characteristic of this tropical rain forest is its high diversity in Caesalpinioideae. A total of 27 species were identified for this subfamily, 20 of which being trees with a dbh > 70 cm and 18 with a dbh > 10 cm. The emergent strate is ruled by the two species: *Paraberlinia bifoliolata* (Caesalpinioideae) and *Aucoumea klaineana* (Burseraceae). Mean species number of trees with dbh > 10 cm is 84 species per 0.5 ha and 125 species per ha.

Key words
Forêt des Abeilles, Gabon, Forest vegetation.

Introduction

Cette étude a été réalisée dans le cadre du projet CEE/STD3 intitulé "Les forêts d'Afrique centrale, essai comparé sur la biodiversité et les relations sol - végétation - faune". La Forêt des Abeilles est située au centre du Gabon, à l'est de la rivière Offoué (Figure 1). La zone étudiée est localisée à la confluence de la rivière Makandé avec l'Offoué (0°40'39"S, 11°54'35"E). Le site d'étude a été choisi en raison de son intérêt biologique exceptionnel et plus particulièrement parce qu'il abrite une espèce de singe endémique, *Cercopithecus solatus* (Gautier et al. 1987).

Matériel et méthodes

Un système de layonnage a été mis en place. Le long de ces layons, les dbh (diameter at breast height ou diamètre à hauteur de poitrine, c'est-à-dire à 1,30 m de hauteur) des arbres ont été mesurés sur une bande de 5 m de large pour les arbres à dbh > 10 cm et sur une bande de 50 m de large pour les arbres de dbh > 70 cm. La forêt étudiée ayant été exploitée il y a trois ans, les relevés pour les essences de dbh > 10 cm ont été choisis dans des zones non endommagées par l'exploitation.

Les résultats discutés ici portent sur un inventaire i) des espèces à dbh > 70 cm effectué sur 6,1 km de long x 50 m de large, couvrant donc une surface de 30,5 ha. (figure 1, layons 8, 12, 32 et 0); ii) des espèces à dbh > 10 cm effectué sur 2 km de long x 5 m de large, couvrant donc une surface de 1 ha.

Résultats

Pour les essences à dbh > 70 cm, 58 taxons appartenant à 22 familles ont été recensés pour 410 pieds (13,4 pieds/ha). Les espèces les plus abondantes sont reprises dans le tableau 1. Les familles les plus diversifiées en nombre d'espèces sont les *Leguminosae-Caesalpinioideae* (20 espèces). Sur un total de 13,4 troncs/ha, les plus abondantes

L.J.G. van der Maesen et al. (eds.), The Biodiversity of African Plants, 140-144.

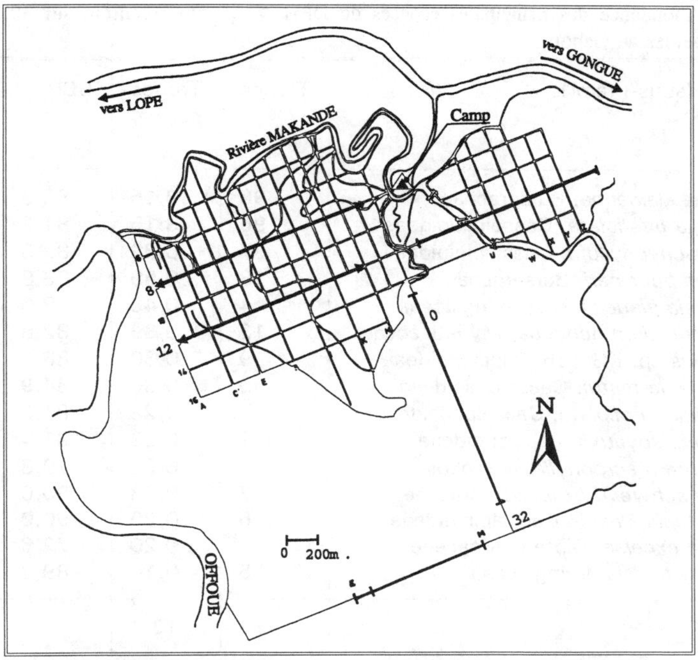

Figure 1. Carte de la station de la Makandé, Gabon.

sont les Caesalpinioideae (5,8 troncs/ha, dont 3,2 pour *Paraberlinia bifoliolata* Pellegr., 0,9 pour *Dialium pachyphyllum* Harms) et les Burseraceae (4,0 troncs/ha, dont 3,2 troncs/ha pour *Aucoumea klaineana* Pierre). Les familles les plus abondantes sont reprises dans le tableau 2.

Pour les essences à dbh > 10 cm, 125 taxons appartenant à 34 familles ont été recensés pour 469 pieds sur une surface de un hectare. Les 26 espèces les plus abondantes, avec 5 pieds ou plus par hectare, sont reprises dans le tableau 3. Les familles les plus diversifiées par le nombre d'espèces (tableau 4) sont les Leguminosae-Caesalpinioideae (18 espèces), les Euphorbiaceae (9 espèces), les Ebenaceae (9 espèces) et les Olacaceae (7 espèces). Les familles les plus représentées par le nombre de troncs sont également les Leguminosae-Caesalpinioideae (112 pieds/ha) dont 39 pieds/ha pour *Paraberlinia bifoliolata*, 14 pieds/ha pour *Pentaclethra eetveldeana* De Wild. & Th. Dur., 13 pieds/ha pour *Dialium pachyphyllum*, 8 pieds/ha pour *Scorodophloeus zenkeri* Harms, 7 pieds/ha pour *Hymenostegia klainei* Pierre ex Pellegr., les Burseraceae (64 pieds/ha) dont 33 pieds/ha pour *Santiria trimera* (Oliv.) Aubr., les Ebenaceae (38 pieds/ha), les Leguminosae-Mimosoideae (23 pieds/ha), les Anacardiaceae et les Sapindaceae (21 pieds/ha); les Rubiaceae et les Guttiferae (19 pieds/ha).

Discussion et conclusion

En conclusion, la forêt étudiée présente une densité très élevée de Caesalpinioideae (24% des essences de dbh > 10 cm et 43% des essences de dbh > 70 cm). Cette famille a également une diversité exceptionnelle, aussi bien en espèces de diamètre

Tableau 1. Abondance des principales espèces de DBH > 70 cm recensées sur 30,5 ha dans la Forêt des Abeilles au Gabon.

Espèces, (sous-) famille	Troncs	Troncs par ha	DBH (cm)	Densité relative (%)
Aucoumea klaineana, Burseraceae	96	3,15	88,3	23,4
Paraberlinia bifoliolata, Caesalpinioideae	96	3,15	81,2	23,4
Dialium pachyphyllum, Caesalpinioideae	27	0,89	82,9	6,6
Dacryodes buettneri, Burseraceae	18	0,59	83,8	4,4
Desbordesia glaucescens, Irvingiaceae	14	0,46	78,0	3,4
Scyphocephalium ochocoa, Myristicaceae	12	0,39	82,8	2,9
Pterocarpus sp. KG 135, Papilionoideae	9	0,30	88,1	2,2
Sindoropsis le-testui, Caesalpinioideae	9	0,30	84,9	2,2
Tetraberlinia bifoliolata, Caesalpinioideae	7	0,23	81,3	1,7
Pterocarpus soyauxii, Papilionoideae	7	0,23	91,4	1,7
Pentaclethra macrophylla, Mimosoideae	7	0,23	79,8	1,7
Canarium schweinfurthii, Burseraceae	7	0,23	90,0	1,7
Hymenostegia klainei, Caesalpinioideae	6	0,20	90,0	1,5
Marquesia excelsa, Dipterocarpaceae	6	0,20	72,6	1,5
Klainedoxa trilesia, Irvingiaceae	5	0,16	89,2	1,2
Total	410	13,4		

Tableau 2. Abondance et diversité des principales (sous-) familles de dbh > 70 cm recensées sur 30,5 ha dans la Forêt des Abeilles au Gabon.

(Sous-) Famille	Troncs	Troncs par ha	Densité relative (%)	Espèces par famille	Diversité relative (%)
Caesalpinioideae	176	5,77	42,9	20	34,5
Burseraceae	121	3,97	29,5	3	5,2
Irvingiaceae	25	0,82	6,1	5	8,6
Myristicaceae	19	0,62	4,6	4	6,9
Papilionoideae	18	0,59	4,4	3	5,2
Mimosoideae	11	0,36	2,7	4	6,9
Dipterocarpaceae	6	0,20	1,5	1	1,7
Annonaceae	5	0,16	1,2	2	3,4
Ochnaceae	4	0,13	1,0	1	1,7
Olacaceae	4	0,13	1,0	2	3,4
Simaroubaceae	4	0,13	1,0	1	1,7
Total	410	13,4		58	

Tableau 3. Abondance des principales espèces de dbh > 10 cm recensées sur 1 ha dans la Forêt des Abeilles au Gabon.

Espèce	Famille ou sous-famille	Troncs par ha	DBH moyen (cm)	Densité relative (%)
Paraberlinia bifoliolata	Caesalpinioideae	39	37,9	8,3
Santiria trimera	Burseraceae	33	21,5	7,1
Grewia coriacea	Tiliaceae	16	16,5	3,4
Eriocoelum macrocarpum	Sapindaceae	14	24,6	3,0
Pentaclethra eetveldeana	Mimosoideae	14	38,8	3,0
Dacryodes sp. KG 115	Burseraceae	13	19,0	2,8
Dialium pachyphyllum	Caesalpinioideae	13	30,2	2,8
Garcinia smeathmannii	Guttiferae	11	14,3	2,4
Napoleonaea sp. KG 122	Lecythidaceae	11	16,8	2,4
Dacryodes buettneri	Burseraceae	10	41,0	2,1
Corynanthe mayumbensis	Rubiaceae	8	14,9	1,7
Scorodophloeus zenkeri	Caesalpinioideae	8	25,7	1,7
Trichoscypha acuminata	Sapindaceae	8	15,0	1,7
Diospyros sp. KG 76	Ebenaceae	7	15,4	1,5
Gilbertiodendron sp. KG 145	Caesalpinioideae	7	19,5	1,5
Hymenostegia klainei	Caesalpinioideae	7	30,4	1,5
Sorindeia nitida	Anacardiaceae	7	16,4	1,5
Aidia micrantha	Rubiaceae	6	13,1	1,3
Desbordesia glaucescens	Irvingiaceae	6	20,2	1,3
Pterygopodium oxyphyllum	Caesalpinioideae	6	18,2	1,3
Treculia obovoidea	Moraceae	6	25,3	1,3
Diogoa zenkeri	Olacaceae	5	25,2	1,1
indét. KG 117	Leguminosae	5	21,7	1,1
Plagiostyles africana	Euphorbiaceae	5	35,6	1,1
Scyphocephalium ochocoa	Myristicaceae	5	65,1	1,1
Xylopia aethiopica	Annonaceae	5	34,5	1,1
Total		469		

supérieur à 10 cm qu'en espèces de diamètre supérieur à 70 cm, avec 27 espèces différentes pour ces deux catégories. Deux essences émergentes à caractère grégaire dominent la forêt, *Paraberlinia bifoliolata* et *Aucoumea klaineana* qui représentent à elles deux près de 50% de la densité relative. *Aucoumea klaineana* n'est représenté que par des individus de grand diamètre, les jeunes individus n'ont pas été observés.

Reitsma (1988) a dénombré 85 espèces d'arbres et lianes à dbh > 10 cm sur un quadrat de 1 ha dans la Forêt d'Ekobakoba au Gabon (près de la station d'étude de Makokou) pour une densité de 438 pieds/ha. Ces chiffres comparés à ceux de la Forêt des Abeilles montrent que les densités à l'hectare de ces deux forêts sont comparables mais que l'on observe un nombre d'espèces supérieur dans la Forêt des Abeilles. Ce nombre supérieur semble être principalement dû au fait que dans la Forêt des Abeilles l'étude a été effectuée par la méthode du transect et que la zone étudiée présente une succession rapprochée de sommets et de vallées (terrain très vallonné).

Tableau 4. Abondance et diversité des principales (sous-) familles de dbh > 10 cm recensées sur 1 ha dans la Forêt des Abeilles au Gabon.

Famille ou sous-famille	Troncs par ha	Densité relative (%)	Nombre d'espèces	Diversité relative (%)
Caesalpinioideae	112	23,9	18	14,4
Burseraceae	64	13,6	6	4,8
Ebenaceae	38	8,1	9	7,2
Mimosoideae	23	4,9	5	4,0
Anacardiaceae	21	4,5	6	4,8
Sapindaceae	21	4,5	5	4,0
Guttiferae	19	4,0	5	4,0
Rubiaceae	19	4,0	5	4,0
Euphorbiaceae	17	3,6	9	7,2
Tiliaceae	16	3,4	1	0,8
Olacaceae	15	3,2	7	5,6
Lecythidaceae	14	3,0	3	2,4
Annonaceae	12	2,6	5	4,0
Irvingiaceae	11	2,4	3	2,4
Total	469		125	

Dans les forêts denses primaires de la réserve du Dja au Cameroun, des relevés linéaires de biodiversité effectués selon la même méthode du transect ont donné de 66 à 105 espèces différentes (n = 8, moyenne 92) par tronçon de 1 km de long et 5 m de large soit par 0,5 ha (Lejoly et al. 1994). Les valeurs obtenues de la même manière à la Forêt des Abeilles sont légèrement inférieures: 84 espèces différentes par tronçon de 1 km avec n = 2.

Remerciements
Nous remercions le Docteur Annie Gautier-Hion de la Station Biologique de Paimpont (Université de Rennes) et Monsieur Paul Posso de l'Institut de Recherche en Ecologie Tropicale (IRET - Libreville) de nous avoir permis d'effectuer nos recherches sur le site de la Makandé, dans la Forêt des Abeilles. Nous remercions Augustin Moungazi, technicien de l'IRET, qui a aidé sur le terrain à la détermination de beaucoup d'arbres.

Bibliographie

Gautier, J.P., J.N. Loireay, F. Moysan & R.W. Cooper, 1987. Une nouvelle espèce gabonaise de cercopithèque *Cercopithecus lhoesti solatus* (Harrisson, 1984): distribution géographique - mesures de protection à prendre. CIRMF, Gabon.

Lejoly J., B. Sonke & K. Van Essche, 1994. Utilisation de la méthode du transect en vue de l'étude de la biodiversité dans la Réserve de Faune du Dja (Cameroun). Communication présentée au 14e Congrès de l'AETFAT à Wageningen, 22-27 août 1994.

Reitsma, J.M., 1988. Végétation forestière du Gabon. - Forest vegetation of Gabon. Tropenbos technical series; 1. Tropenbos Foundation, Ede.

Aperçu sur les groupements secondaires à Marantaceae, Zingiberaceae et Costaceae de la Forêt des Abeilles (Gabon)

M.M. DHETCHUVI
Université Libre de Bruxelles, Laboratoire de Botanique systématique et de Phytosociologie, C.P. 169, 28 avenue Paul Héger, 1050 Bruxelles, Belgique.

Résumé
Les groupements à Marantaceae, Zingiberaceae et Costaceae sont recensés en milieu naturel au Gabon dans la Forêt des Abeilles aux environs de Makandé. Il s'agit de groupements qui se développent sur des terrains dénudés après l'exploitation forestière. Ils constituent un stade herbacé de la dynamique évolutive des zones secondarisées dans la Forêt des Abeilles.

Abstract
The plant communities of Marantaceae, Zingiberaceae and Costaceae are studied in Gabon at the "Forêt des Abeilles" near Makandé. These communities develop after logging of the forest. The three families are the main components of the communities which form an herbaceous stage of the forest regrowth at the "Forêt des Abeilles".

Key words
Costaceae, Gabon, Marantaceae, Secondary vegetation, Zingiberaceae.

Introduction

Les Marantaceae, Zingiberaceae et Costaceae font l'objet d'une attention particulière au cours de ces dernières années lors des études de la régénération de la forêt dense africaine (Bebwa 1993, Lubini 1986, Mosango 1990, Schnell 1952). Les 3 familles regroupent des espèces liées au milieu humide, souvent éclairé ou de sous-bois à couvert peu fermé des forêts secondaires et des forêts remaniées. Dans ces types de végétation, les représentants de ces familles participent souvent à la constitution des groupements végétaux en tant qu'espèces caractéristiques ou compagnes. Notre but est de donner un aperçu général sur les groupements à Marantaceae, Zingiberaceae et Costaceae. Cette étude veut aussi aborder la notion de biodiversité dans les formations secondaires auxquelles sont liés ces groupements. Elle a été réalisée dans le cadre du projet CEE/DG12/STD3 intitulé: "Forêts d'Afrique centrale: essai comparé sur les relations sol-végétation-faune".

Matériel et méthode

L'étude réalisée au Gabon dans la Forêt des Abeilles, durant un séjour continu de 30 jours, a permis d'effectuer, en utilisant la méthode de Braun-Blanquet (1932), 13 relevés dans les groupements localisés dans des zones de chablis, de coupe et de débardage. Treize terrains à végétation homogène ont été localisés et des superficies variant de 25 à 200 m^2 y ont été délimitées où toutes les espèces ont été inventoriées. Les échantillons d'herbier de référence sont déposés à l'herbarium du Jardin Botanique de Meise (BR).

L.J.G. van der Maesen et al. (eds.), The Biodiversity of African Plants, 145-149.

Tableau 1. Inventaire des espèces des groupements à Marantaceae et Zingiberaceae et Costaceae des formations secondaires dans la Forêt des Abeilles (Légende: P = présence; Rec. moy. & RM = recouvrement moyen; AE = autres espèces).

Numéro d'ordre	1	2	3	4	5	6	7	8	9	10	11	12	13		
RELEVES (n°)	13	16	17	12	9	8	15	10	14	6	5	7	19		
Surface (m^2)	64	200	104	65	77	60	40	128	28	105	36	90	64		
Recouvrement total (%)	100	100	100	100	90	90	100	95	100	90	80	80	95		
Strate arbustive: hauteur (m)	7	6	3,5	7	2	3,0	2,5	3,5	5	3	2,8	2,5	2,5		
Strate herbacée: hauteur (m)	2	1,5	2	1,2	1,5	1,8	1,8	1,5	2	2,5	1,5	2	-		
Pente (%)	-	3	3	0	2	0	2	8	2	3	3	0	0		RM
Total espèces	35	93	82	47	42	53	44	41	61	52	49	58	110	P	(%)
Marantaceae															
Haumania liebrechtsiana J. Léon.	4	3	3	3	3	3	2	1	2	2	2	2	2	V	26,34
Megaphrynium macrostachyum M.-Redh.	-	1	-	-	1	+	+	-	1	+	1	-	+	IV	0,83
Hypselodelphys poggeana M.-Redh.	2	1	+	1	-	-	+	+	-	-	-	1	-	III	1,77
Sarcophrynium prionogonium K. Schum.	-	+	1	-	-	-	-	-	-	+	+	+	+	III	1,77
Marantochloa congensis J. Léon. & Mullen.	-	1	-	+	-	+	-	-	-	+	-	+	-	II	0,25
Hypselodelphys hirsuta (Loes.) Koechlin	-	-	+	-	-	+	-	-	-	-	-	-	+	II	0,04
Sarcophrynium schweinfurthianum M.-Redh.	-	+	+	-	-	-	-	-	-	-	-	-	-	I	0,03
Marantochloa sp.	-	2	-	-	-	-	-	-	-	-	-	-	-	I	1,15
Megaphrynium trichogynum Koechlin	-	-	-	-	-	-	-	1	-	-	-	-	-	I	0,19
Rec. moy. Marantaceae (%)	77,5	60,4	40,6	40,2	40	38,1	15,4	5,2	17,5	15,6	17,7	17,9	15,6		32,37
Zingiberaceae															
Aframomum giganteum K. Schum.	-	+	1	+	1	1	3	3	2	4	2	1	1	V	13,87
Aframomum sp. (Dhetchuvi n° 1316)	1	2	1	-	2	1	4	3	4	1	3	-	-	IV	18,46
Aframomum longepetiolatum Koechlin	-	+	-	-	1	+	-	1	1	+	-	-	-	III	0,62
Aframomum citratum (Perr.) K. Schum.	-	+	-	-	-	-	-	-	-	-	-	-	-	I	0,01
Renealmia cincinnata (K. Schum.) Bak.	-	-	-	-	-	-	-	-	-	-	+	+	-	I	0,03
Renealmia congolana De Wild. & Th. Dur.	-	-	-	-	-	-	-	-	-	-	-	-	+	I	0,01
Rec. moy. Zingiberaceae (%)	2,5	15,6	5	0,2	20	5,2	100	77,5	80	65,2	52,7	2,7	2,7		33

Tableau 1 (Suite).

Costaceae															
Costus dubius K. Schum.	+	+	+	1	-	2	-	-	-	2	-	1	+	IV	2,75
Costus ligularis Bak.	-	-	-	-	-	-	-	-	-	-	-	-	2	I	1,15
Costus mamboumiensis Pellegr.	-	-	+	-	-	-	-	-	-	-	-	-	-	I	0,01
Rec. moy. Costaceae (%)	2,7	15,8	5,4	2,7	20	20,2	100	77,5	80	80,2	52,7	5,2	17,9		3,91
Espèces du Caloncobo-Tremion (CT)															
Harungana madagascariensis Lam. ex Poir.	-	-	-	+	-	-	-	-	-	-	-	+	-	I	0,03
Caloncoba welwitschii (Oliv.) Gilg	-	-	-	-	-	+	-	-	-	-	-	-	-	I	0,02
Leptactina leopoldi-II Büttner	-	-	-	-	-	-	-	-	+	-	-	-	-	I	0,02
Espèces du Musango-Terminalietea (MT)															
Xylopia aethiopica (Dun.) A. Rich.	+	1	+	+	2	2	+	2	+	1	+	2	4	V	9,90
Macaranga monandra Müll. Arg.	+	3	1	4	+	+	+	1	-	+	+	1	+	V	8,4
Musanga cecropioides R. Br.	+	-	-	+	-	+	-	-	+	-	+	+	-	III	0,09
Espèces du Strombosio-Parinarietea (SP)															
Aucoumea klaineana Pierre	+	2	-	+	+	+	+	+	+	-	-	-	+	IV	1,28
Dialium pachyphyllum Harms	-	+	+	-	+	-	+	+	-	+	+	-	+	IV	0,12
Strombosia scheffleri Engl.	-	+	+	+	-	-	-	-	-	+	+	+	+	III	0,11
Espèces du Mytragynetea (M)															
Tetracera alnifolia Willd.	-	1	1	+	2	1	-	2	+	2	1	-	1	IV	4,45
Treculia africana Decne.	-	+	-	-	-	-	-	-	+	-	-	-	-	I	0,03
Uapaca paludosa Aubrév. & Léandri	-	-	-	-	-	-	-	-	-	-	+	-	+	I	0,03
Espèce de l'Hyparrhenietea (H)															
Maprounea africana Müll. Arg.	-	+	-	+	-	-	-	-	-	+	+	2	-	II	1,22
Espèces de Lycopodietea cernui (LC)															
Gleichenia linearis (Burn.) Cl.	-	+	-	1	1	-	+	-	+	+	1	-	-	III	0,64
Lycopodiella cernua L.	1	+	-	1	-	1	+	-	-	-	1	+	-	III	0,82
Espèces du Ruderali-Manihotetea (RM)															
Paspalum conjugatum (Schult.) Berg.	-	-	-	1	-	1	-	-	-	-	-	-	-	I	0,38
Pityrogramma calamelanos (L.) Link	-	+	-	-	-	1	-	-	+	-	+	+	+	III	0,27
Rec. moy. CT + MT + SP + M + H + LC + RM + AE	49	146	110	127	122	143	96	22	46	5	123	116	128		28

Résultats

Le dépouillement des 13 relevés donne une liste de 177 espèces, mais dont quelques-uns seulement sont reprises, à titre indicatif, dans le tableau 1. Elles y sont regroupées d'après leur statut phytosociologique. La liste complète des Marantaceae, Zingiberaceae et Costaceae rencontrées dans ces formations secondaires est donnée.

Le coefficient de présence est indiqué pour chaque espèce, ainsi que leur recouvrement moyen. Le recouvrement moyen total est également calculé différemment pour les Marantaceae et les Zingiberaceae. La dernière ligne du tableau reprend le recouvrement moyen total des espèces autres que les Marantaceae, Zingiberaceae et Costaceae.

Discussion et conclusions

Les groupements recensés dans les zones de chablis, de coupe et de débardage sont notamment ceux à *Aframomum giganteum* (Oliv. & Hanb.) K. Schum., à *Aframomum* sp. et à *Haumania liebrechtsiana* (De Wild. & Th. Dur.) J. Léonard. L'ensemble des espèces constitutives de ces groupements donne une indication sur les espèces colonisatrices des zones dénudées et sur la régénération de la forêt dans cette partie du Gabon. Les trois espèces ont en effet le recouvrement moyen le plus élevé de toutes les autres, c'est-à-dire 26,34% pour *Haumania liebrechtsiana*, 18,46% pour *Aframomum* sp. et 13,87% pour *Aframomum giganteum*.

Dans l'ensemble, les Marantaceae, les Zingiberaceae et les Costaceae ont un recouvrement moyen de 69,28% contre 92,27% constitués par le recouvrement moyen total de toutes les autres espèces réunies. C'est une proportion fort élevée, ce qui prouve la prédominance surtout des Marantaceae et des Zingiberaceae dans les formations secondaires étudiées.

Les Marantaceae et les Zingiberaceae rentrent souvent dans la constitution des groupements secondaires des régions forestières. C'est le cas des espèces signalées comme compagnes dans les formations végétales secondaires étudiées au Zaïre (Bebwa 1993, Lubini 1986, Mosango 1990 & 1991), ainsi qu'en Afrique de l'Ouest (Mbarga Bindzi 1992, Schnell 1952).

Dans son travail sur la végétation adventice et postculturale de Kisangani et de la Tshopo (Haut-Zaïre), Lubini (1986) a décrit une association à *Megaphrynium macrostachyum* (Benth.) Milne-Redhead et *Aframomum laurentii* (De Wild. & Th. Dur.) K. Schum. Elle colonise les champs abandonnés, les friches forestières et les sous-bois des forêts secondaires. Notre étude montre l'existence de ces types d'associations; elles seront décrire en d'autres occasions. Il s'agit probablement de l'association à *Haumania liebrechtsiana*, l'association à *Aframomum giganteum* et l'association à *Aframomum* sp.

En ce qui concerne la biodiversité des groupements étudiés, ont peut signaler le nombre important d'espèces retrouvées sur des surfaces si réduites. Ces groupements jouent sans doute le rôle essentiel dans la régénération au sein de la Forêt des Abeilles. La présence de 79 espèces (soit 45% des 177 espèces recensées) de forêt primaire donne une idée sur l'importance de ce type de milieu dans la repousse de la forêt. Notons aussi que le tiers des espèces de Marantaceae de la flore du Gabon se retrouve dans les 13 relevés représentant dans l'ensemble, une superficie de 1061 m^2. Les Zingiberaceae et les Costaceae sont moins représentées en ce qui concerne le nombre d'espèces.

Cette étude montre l'avantage qu'offre ces types de milieu dans la régénération non seulement des Marantaceae, Zingiberaceae et Costaceae, mais également de la plupart des espèces de la Forêt des Abeilles, principalement les espèces ligneuses suivantes: *Anthocleista vogelii, Aucoumea klaineana, Canarium schweinfurthii, Cleistopholis glauca, Dacryodes buettneri, Dialium pachyphyllum, Paraberlinia bifoliolata, Pterocarpus soyauxii, Pycnanthus angolensis, Scorodophloeus zenkeri, Strombosia scheffleri, Xylopia aethiopica, Zanthoxylum gilletii.*

Remerciements

Je remercie le Dr Annie Gautier de l'Université de Rennes et le Prof. J. Lejoly, pour avoir rendu possible la présente étude.

Bibliographie

Bebwa, B., 1993. Écologie quantitative des jeunes stades de la reconstitution forestière en région tropicale (Ile Kongolo, Masako; Zaïre). Thèse doct., Univ. Libre Bruxelles. 324 p.

Braun-Blanquet, J., 1932. Plant sociology. Translation of Fuller, G.D. & H.S. Conard "Pflanzensoziologie", 1st ed. McGraw-Hill book Co. Inc., New York & London, 337 p.

Lubini, A., 1986. Végétation adventice et postculturale de Kisangani et de la Tshopo (Haut-Zaïre). Bull. Jard. Bot. Nat. Belg. 56: 315-348.

Mbarga Bindzi, M.-A., 1992. Processus de reconstitution de la forêt dense mésophile guinéenne: cas du secteur forestier de la région de Yaoundé (Cameroun). Thèse doct., Univ. Paris-Sud, 169 p.

Mosango, M., 1990. Contribution à l'étude botanique et biogéochimique de l'écosystème forêt en région équatoriale (Ile Kongolo; Zaïre). Thèse doct., Univ. Libre Bruxelles. 446 p.

Mosango, M., 1991. La forêt secondaire âgée à *Zanthoxylum gilletii* de l'île Kongolo (Zaïre). Belg. Journ. Bot. 124 (2): 152-166.

Schnell, R., 1952. Végétation et flore de la région montagneuse du Nimba (Afrique occidentale française). Mém. Inst. Fr. Afr. N. 22. 604 p. + planches.

Utilisation de la méthode du transect en vue de l'étude de la biodiversité dans la Réserve de Faune du Dja (Cameroun)

J. LEJOLY [1], B. SONKÉ [2] & K. VAN ESSCHE [1]

[1] Laboratoire de Botanique systématique et de Phytosociologie, Université Libre de Bruxelles, CP 169, Av. F. Roosevelt 50, 1050 Bruxelles, Belgique.

[2] Département des Sciences Biologiques, Ecole Normale Supérieure, Université de Yaoundé I, B.P. 8225, Yaoundé, Cameroun.

Abstract

Biodiversity study by transect method in the "Réserve de Faune du Dja" (Cameroun). On a transect of 29 km, 8.8 km belonging to typical rain forest was analysed. This forest is characterized by 66 to 105 different species of trees and lianas with dbh ≥ 10 cm for each section of 1 km x 5 m (0.5 ha; n = 8, mean 92 species).

Key words

Biodiversity, Cameroun, Rain forest, Transect.

Introduction

La réserve de Faune du Dja est située à une centaine de kilomètres au sud-est de Yaoundé. Avec 500.000 ha, elle constitue une des plus grandes réserves du Cameroun et est couverte en majeure partie par une forêt dense hétérogène mésophile (Letouzey 1968). Cette réserve est délimitée de façon naturelle par la rivière Dja qui forme une boucle autour d'elle. La partie nord de la réserve est soumise à une intense activité humaine du fait des villages situés sur l'axe Somalomo - Ekom. Pour une bonne intégration de ces populations dans la gestion de cette réserve un programme de monitoring et d'aménagement de la réserve est actuellement en cours d'installation et utilise des transects comme base de collecte de données pluridisciplinaires. Les études effectuées sur ces transects sont les inventaires des mammifères, des oiseaux et surtout l'inventaire des arbres à dbh ≥ 10 cm (diameter at breast height, le diamètre à la hauteur de la poitrine c'est-à-dire à 1,3 m de hauteur).

Matériel et méthode

Un premier transect long de 29 km a été mis en place en direction nord-sud. Le long de ce transect, les dbh des arbres et des lianes ont été mesurés sur une bande de 5 m de large pour les arbres à dbh ≥ 10 cm et sur une bande de 50 m de large pour les gros arbres émergents à dbh ≥ 70 cm. Ce transect traverse 37 dépressions marécageuses qui représentent 14,6% du total. Le 1er km est partiellement secondarisé; cette secondarisation résulte de l'activité humaine des villages environnants. Sur les 28 km suivants, à caractère primaire, la terre ferme est occupée par les 4 grands types de forêts primaires suivants: 1: sur rochers granitiques superficiels; 2: dégradées avec de grandes clairières envahies de rotangs; 3: avec grande abondance de *Uapaca* spp.; 4: typique (sur sol profond, avec petits chablis localisés et sans dominance marquée de *Uapaca* spp).

L.J.G. van der Maesen et al. (eds.), The Biodiversity of African Plants, 150-154.

Tableau 1. Abondance des principales essences à dbh ≥ 70 cm recensés sur 44,4 ha de forêts denses primaires sur le transect de Djolimpoum dans la Réserve de Faune du Dja au Cameroun.

Espèce	(Sous-) Famille	A	B	C	D	E
Alstonia boonei	Apocynaceae	44	0,99	101,7	0,87	7,12
Pentaclethra macrophylla	Mimosoideae	40	0,90	82,3	0,49	6,47
Petersianthus macrocarpus	Lecythidaceae	35	0,79	81,9	0,42	5,66
Desbordesia glaucescens	Irvingiaceae	30	0,68	90,4	0,46	4,85
Distemonanthus benthamianus	Caesalpinioideae	30	0,68	88,7	0,43	4,85
Erythrophleum suaveolens	Caesalpinioideae	27	0,61	92,3	0,42	4,37
Piptadeniastrum africanum	Mimosoideae	26	0,59	108,8	0,57	4,21
Cylicodiscus gabunensis	Mimosoideae	21	0,47	112,7	0,50	3,40
Pycnanthus angolensis	Myristicaceae	19	0,43	87,5	0,26	3,07
Uapaca paludosa	Euphorbiaceae	17	0,38	86,2	0,23	2,75
Duboscia macrocarpa	Tiliaceae	14	0,32	101,5	0,28	2,27
Fagara macrophylla	Rutaceae	13	0,29	80,1	0,15	2,10
Musanga cecropioides	Moraceae	13	0,29	89,9	0,19	2,10
Pterocarpus soyauxii	Papilionoideae	13	0,29	92,9	0,21	2,10
Terminalia superba	Combretaceae	12	0,27	86,2	0,16	1,94
Celtis zenkeri	Ulmaceae	11	0,25	85,1	0,14	1,78
Heisteria trillesiana	Olacaceae	11	0,25	79,2	0,12	1,78
Klainedoxa gabonensis	Irvingiaceae	11	0,25	87,1	0,15	1,78
Ongokea gore	Olacaceae	11	0,25	87,1	0,15	1,78
Hylodendron gabunense	Caesalpinioideae	10	0,23	79,5	0,11	1,62
Baillonella toxisperma	Sapotaceae	9	0,20	136,8	0,32	1,46
Divers (72 espèces)		168	3,78		2,66	27,18
Total (94 espèces)		618	13,91		9,75	100

A = Troncs; B = Troncs/ha; C = DBH moyen (cm); D = St (m^2/ha); E = Densité relative (%).

Résultats

Les résultats discutés ici portent sur l'étude de la forêt dense typique dominante et comportent:

i un inventaire des essences à dbh ≥ 70 cm effectué sur 8883 m de long x 50 m de large et couvrant donc une surface de 44,4 ha,

ii un inventaire des essences à dbh ≥ 10 cm effectué également sur 8883 m de long x 5 m de large et couvrant donc une surface de 4,44 ha.

Pour les essences à dbh ≥ 70 cm, 94 taxons appartenant à 32 familles ont été recensés pour un total de 618 pieds soit 13,9 pieds par ha. Le nombre de taxons par tronçon de 1 km (soit 5 ha) varie de 29 à 39 (n = 8, moyenne 33). Les essences les mieux représentées (> 0.2 troncs/ha, ou encore un tronc tous les 200 m en balayant la bande de 50 m de large), sont au nombre de 22 et sont reprises au Tableau 1. Les 7 espèces ayant plus de 0,5 tronc/ha sont: *Alstonia boonei* De Wild., *Pentaclethra macrophylla* Benth., *Petersianthus macrocarpus* (P.B.) Liben, *Desbordesia glaucescens* (Engl.) Van Tiegh., *Distemonanthus benthamianus* Baill., *Erythrophleum suaveolens* (Guill. & Perr.) Brenan et *Piptadeniastrum africanum* (Hook.f.) Brenan; ensemble, elles représentent 5,24 pieds/ha soit 38%).

Tableau 2. Abondance et diversité des principales familles d'arbres à dbh ≥ 70 cm recensés sur 44.4 ha de forêts denses primaires sur le transect nord/sud de Djoloimpoum dans la Réserve de Faune du Dja au Cameroun.

(Sous-) Famille	A	B	C	D	E	F	G
Mimosoideae	93	2,09	96,5	1,63	15,05	7	7,4
Caesalpinioideae	80	1,80	91,5	1,24	12,94	9	9,6
Irvingiaceae	49	1,10	94,1	0,81	7,93	4	4,3
Apocynaceae	44	0,99	101,1	0,84	7,12	1	1,1
Lecythidaceae	35	0,79	81,9	0,42	5,66	1	1,1
Meliaceae	29	0,65	99,0	0,56	4,69	10	10,6
Papilionoideae	26	0,59	92,0	0,41	4,21	4	4,3
Olacaceae	26	0,59	82,4	0,32	4,21	4	4,3
Sapotaceae	24	0,54	108,7	0,55	3,88	5	5,3
Myristicaceae	23	0,52	85,3	0,30	3,72	3	3,2
Euphorbiaceae	22	0,50	84,0	0,28	3,56	6	6,4
Divers (21 familles)	167	1,17		2,40	27,02	40	
Total (32 familles)	618	13,91		9,75	100	94	100

A = Troncs; B = Troncs/ha; C = DBH moyen (cm); D = St (m^2/ha); E = Densité relative (%); F = Nombre de espèces; G = Diversité relative (%).

Les familles les plus diversifiées sont reprises au Tableau 2. Viennent en tête les Meliaceae (10 espèces, 0,7 pied/ha) avec *Entandrophragma angolense* (Welw.) C.DC., *E. candollei* Harms, *E. cylindricum* (Sprague) Sprague, tous avec ≤ 0.1 tronc/ha. Directement après viennent les Caesalpinioideae (9 espèces, 1,8 pieds/ha) avec *Distemonanthus benthamianus* (0,7 tronc/ha), *Erythrophleum suaveolens* (0,6 tronc/ha) et *Hylodendron gabunense* Taub. Viennent ensuite les Mimosoideae (7 espèces; 2,1 pied/ha) avec *Pentaclethra macrophylla* (0,9 tronc/ha), *Piptadeniastrum africanum* (0,6 tronc/ha) et *Cylicodiscus gabunensis* Harms (0,5 tronc/ha).

Pour les essences à dbh ≥ 10 cm, 220 taxons appartenant à 43 familles ont été recensés pour 2677 pieds (soit 603 pieds/ha). Le nombre de taxons par tronçon de 1 km (0,5 ha) varie de 66 à 105 (n = 8, moyenne 92). Les 31 essences les mieux représentées (> 5 troncs/ha) sont reprises au Tableau 3. Les espèces les plus abondantes sont *Tabernaemontana crassa* Benth. (Apocynaceae, 34 pieds/ha), *Petersianthus macrocarpus* (P.B.) Liben (Lecythidaceae, 24) et *Santiria trimera* (Oliv.)Aubr. (Burseraceae, 21).

Les familles les plus diversifiées par le nombre d'espèces sont reprises au Tableau 4. Viennent en tête les Euphorbiaceae (24 espèces), les Annonaceae (15), les Meliaceae (13), les Rubiaceae (13), et les Sapotaceae (12). Les familles représentées par le plus grand nombre de troncs sont les Euphorbiaceae (65 pieds/ha) dont *Dichostemma glaucescens* Pierre (11 pieds/ha) et *Antidesma laciniatum* Muell. Arg. (7), suivies des Annonaceae (52) dominées par *Polyalthia suaveolens* Engl. & Diels (19) et *Anonidium mannii* (Oliv.) Engl. & Diels (14).

Tableau 3. Abondance des principales essences à dbh ≥ 10 cm sur recensés sur 4,44 ha de forêts denses primaires sur le transect de Djolimpoum dans la Réserve de Faune du Dja au Cameroun.

Espèce	(Sous-) Famille	A	B	C	D	E
Tabernaemontana crassa	Apocynaceae	154	34,7	16,3	0,94	5,8
Petersianthus macrocarpus	Lecytidaceae	108	24,3	31,0	2,46	4,0
Santiria trimera	Burseraceae	94	21,2	18,8	0,68	3,5
Polyalthia suaveolens	Annonaceae	84	18,9	18,0	0,61	3,1
Pentaclethra macrophylla	Mimosoideae	73	16,4	35,6	2,21	2,7
Trilepisium madagascariense	Moraceae	66	14,9	18,9	0,49	2,5
Anonidium mannii	Annonaceae	62	14,0	19,4	0,53	2,3
Heisteria trillesiana	Olacaceae	60	13,5	31,4	1,42	2,2
Desbordesia glaucescens	Irvingiaceae	59	13,3	24,4	1,02	2,2
Dichostemma glaucescens	Euphorbiaceae	52	11,7	24,1	0,62	1,9
Carapa procera	Meliaceae	49	11,0	17,9	0,30	1,8
Celtis zenkeri	Ulmaceae	46	10,4	26,6	0,87	1,7
Trichilia rubescens	Meliaceae	42	9,5	18,1	0,29	1,6
Sorindeia grandifolia	Anacardiaceae	38	8,6	15,2	0,17	1,4
Trichoscypha acuminata	Anacardiaceae	37	8,3	16,7	0,20	1,4
Lepidobotrys staudtii	Lepidobotryaceae	36	8,1	18,8	0,28	1,3
Celtis tesmannii	Ulmaceae	33	7,4	29,9	0,66	1,2
Strombosiopsis tetandra	Olacaceae	33	7,4	20,0	0,28	1,2
Antidesma laciniatum	Euphorbiaceae	32	7,2	15,9	0,15	1,2
Coelocaryon preussii	Myristicaceae	32	7,2	32,1	0,70	1,2
Corynanthe pachyceras	Rubiaceae	31	7,0	19,0	0,23	1,2
Drypetes sp	Euphorbiaceae	31	7,0	15,6	0,16	1,2
Strombosia scheffleri	Olacaceae	31	7,0	30,0	0,59	1,2
Strombosia pustulata	Olacaceae	28	6,3	21,2	0,25	1,0
Hylodendron gabunense	Caesalpinioideae	25	5,6	32,0	0,57	0,9
Mareyopsis longifolia	Euphorbiaceae	25	5,6	13,2	0,08	0,9
Uapaca staudtii	Euphorbiaceae	25	5,6	19,6	0,19	0,9
Pterocarpus mildbraedii	Papilionoideae	24	5,4	22,6	0,39	0,9
Staudtia kamerunensis	Myristicaceae	23	5,2	19,7	0,20	0,9
Blighia sapida	Sapindaceae	22	5,0	17,2	0,14	0,8
Garcinia mannii	Guttiferae	22	5,0	15,2	0,10	0,8
Divers (118 espèces)		1173	264,1		20	44
Total (220 espèces)		2677	602,7		37,29	100

A = Troncs; B = Troncs/ha; C = DBH moyen (cm); D = St (m^2/ha); E = Densité relative (%).

Discussion et conclusion

La méthode du transect permet d'obtenir des critères objectifs répétitifs de la biodiversité d'un site forestier. La même méthode étant en cours d'utilisation dans des forêts de 5 autres pays d'Afrique centrale, les données obtenues dans la Réserve du Dja ont pu être comparées: ainsi le nombre de taxons différents à dbh ≥ 10 cm par tronçon de 1 km (soit 0,5 ha) qui est de 92 (n = 8) dans le Dja, est supérieur aux valeurs obtenues en forêt de Ngotto (Lejoly 1994), République centrafricaine (84, n = 2) et en forêt des Abeilles, Gabon (85, n = 2) (Gesnot et al. 1996).

Tableau 4. Abondance et diversité des principales familles d'arbres à dbh ≥ 10 cm recensés sur 4,44 ha de forêts denses primaires sur le transect nord/sud de Djoloimpoum dans la Réserve de Faune du Dja au Cameroun.

(Sous-) Famille	A	B	C	D	E	F	G
Euphorbiaceae	291	65,5	21,8	3,24	10,9	24	10,9
Annonaceae	230	51,8	19,6	2,13	8,6	15	6,8
Olacaceae	178	40,1	27,3	3,18	6,6	9	4,1
Apocynaceae	166	37,7	18,0	1,31	6,2	5	2,3
Meliaceae	141	31,7	22,5	2,29	5,3	13	5,9
Mimosoideae	119	26,8	32,2	3,28	4,4	9	4,1
Burseraceae	110	24,8	18,5	0,77	4,1	5	2,3
Caesalpinioideae	109	24,5	27,9	2,29	4,1	9	4,1
Lecytidaceae	108	24,3	31,0	2,45	4,0	1	0,5
Anacardiaceae	104	23,4	17,6	0,79	3,9	8	3,6
Moraceae	98	22,1	22,0	1,16	3,7	6	2,7
Irvingiaceae	96	21,6	23,2	1,42	3,6	5	2,3
Ulmaceae	79	17,8	28,0	1,53	3,0	2	0,9
Rubiaceae	73	16,4	17,1	0,45	2,7	13	5,9
Myristicaceae	69	15,5	30,5	1,66	2,6	4	1,8
Sapindaceae	53	11,9	17,6	0,34	2,0	5	2,3
Papilionoideae	49	11,0	26,5	1,18	1,8	6	2,7
Guttiferae	47	10,6	25,0	0,75	1,8	7	3,2
Sapotaceae	45	10,1	26,6	0,90	1,7	12	5,5
Divers (25 familles)	512	115,3		6,17	19,1	62	28,1
Total (44 familles)	2677	602,7		37,29	100	220	100

A = Troncs; B = Troncs/ha; C = DBH moyen (cm); D = St (m²/ha); E = Densité relative (%); F = Nombre de espèces; G = Diversité relative (%).

Remerciements
Cette étude a été réalisée dans le cadre du projet Ecofac (CEE) intitulé «Conservation et utilisation rationnelle des écosystèmes forestiers d'Afrique centrale». Nous remercions les responsables camerounais et européens de ce programme pour nous avoir faciliter les recherches sur le terrain. Nous remercions aussi M. Paul Mezili, ancien technicien à l'Herbier National du Cameroun, qui a aidé sur le terrain à l'identification de beaucoup d'arbres.

Bibliographie

Gesnot, K., K. van Essche & J. Lejoly, 1996. Analyse de la biodiversité végétale dans la forêt des Abeilles (Gabon). This volume.

Lejoly, J., 1994. Mise en place des transects en vue des inventaires de biodiversité dans la forêt de Ngotto (République centrafricaine). Projet Ecofac, Agreco-CTFT, Bruxelles: 109 p.

Letouzey, R., 1968. Etude phytogéographique du Cameroun. Ed. Lechevalier: 508 p.

Structure du peuplement des Ptéridophytes d'une forêt ombrophile à Madagascar: réactions à un stress de nature anthropique

F. RAKOTONDRAINIBE [1], G.M. RAHAJASOA [2] & H.L. RANARIJAONA [2]
[1] Ecole Pratique des Hautes Etudes, 16 rue Buffon, 75005 Paris, France.
[2] Service de Botanique, Université d'Antananarivo, B.P. 909, Madagascar.

Résumé
La végétation du Parc National de Ranomafana se présente comme une mosaïque de formations forestières dont chacun des éléments représente un stade de l'évolution régressive ou progressive d'une forêt dense ombrophile de moyenne altitude. La structure et la composition floristique du peuplement des Ptéridophytes sont analysées sur une parcelle de forêt primaire témoin et sur deux parcelles soumises à des pressions anthropiques de nature et d'intensité différentes.

Abstract
The vegetation of Ranomafana National Park appears as a mosaic of forest types, each element of which represents a stage of the regressive or progressive evolution of a mid-altitude tropical rain forest. The structure and floristic composition of Pteridophyte populations were analyzed in a control plot in primary forest and two plots subjected to human influences of different types and intensities.

Key words
Human influence, Madagascar, Pteridophyte, Rain forest.

Introduction

Le Parc National de Ranomafana dans lequel a été menée cette étude, est situé dans la partie Sud Est de Madagascar (21°16' S, 47°28' E), à 265 km au Sud de Tananarive. Il s'étend sur une superficie de 41.600 ha, à une altitude comprise entre 400 et 1374 mètres. La région est soumise à un climat de type équatorial de basse et moyenne altitude (Humbert & Cours Darne 1965) caractérisé par une température moyenne annuelle de 21,7° et des précipitations moyennes annuelles comprises entre 2.400 et 2.600 mm La végétation climacique est une forêt dense ombrophile, sempervirente (Koechlin et al. 1974).

Sur le pourtour de la zone forestière, environ 12.000 habitants regroupés dans une vingtaine de villages vivaient sur les réserves de la forêt avant la création du Parc National. Dans les années 1930-1940, l'exploitation sélective de bois d'oeuvre a été importante et, sur certaines parcelles, la forêt a même été abattue puis brûlée pour permettre l'installation de cultures temporaires. Aussi, la végétation du Parc se présente-t-elle actuellement comme une mosaïque de formations forestières dont chaque élément représente un stade de l'évolution régressive ou progressive d'un même écosystème.

Nous avons essayé d'utiliser le milieu naturel, varié et dynamique, du Parc National de Ranomafana, pour mettre en évidence et caractériser les réactions d'adaptation d'un peuplement de Ptéridophytes (fougères) à un stress d'origine anthropique.

L.J.G. van der Maesen et al. (eds.), The Biodiversity of African Plants, 155-160.

Méthodes d'étude

La végétation du Parc étant en de nombreux points hétérogène en raison du relief ondulé, nous avons opté pour des surfaces échantillons de type transect. Un transect topographique de 8 x 130 m, débutant dans une vallée et remontant sur un versant court, a été délimité dans trois types de forêts: une forêt primaire témoin peu ou pas perturbée, une forêt secondarisée à la suite de l'exploitation sélective de bois d'oeuvre, une forêt reconstituée sur une zone défrichée et cultivée il y a environ 50 ans, aujourd'hui interdite d'accès et bien protégée. L'emplacement des transects a été choisi sur des versants de même orientation (Est-Nord-Est), de même altitude (950 à 1100 m) et de pente approximativement identique (en moyenne 30%).

Les résultats sont exprimés par les paramètres suivants: Richesse floristique (= nombre de taxons, espèces ou variétés), Densité (= nombre d'individus ou de colonies pour les espèces coloniales de petite taille), Indice de diversité de Shannon-Weaver (Shannon & Weaver 1948), Indice de régularité (Daget 1976). La structure de la diversité est représentée par des diagrammes "rangs-fréquences" (Frontier & Pichod-Viale 1991). Le degré de similitude des listes floristiques des relevés pris 2 à 2 est évalué à l'aide de l'indice de Jaccard (Jaccard 1901).

Pour chaque transect (T1, T2, T3), les résultats seront donnés d'abord globalement, puis séparément pour les 2 zones topographiques, le mi-versant et le bas de versant. La limite entre ces 2 zones topographiques se situe généralement au niveau d'une rupture de pente et/ou d'une discontinuité floristique. Sa position a été déterminée en calculant les quantités d'informations apportées par le mode de répartition des taxons le long du transect, selon la méthode exposée par Godron (1966) et Farinas (1982) et déjà appliquée à Madagascar à un peuplement de Ptéridophytes (Rakotondrainibe & Guillerm 1990). La bande de transition entre un bas de versant et un mi-versant, dans laquelle les deux communautés floristiques se mélangent, a été volontairement écartée pour le calcul des différents indices.

Figure 1. Diagramme rangs-fréquences des relevés floristiques effectués sur des transects topographiques, dans 3 types de forêts. Haut: forêt primaire témoin; centre: forêt exploitée, en cours de secondarisation; bas: forêt reconstituée sur un terrain laissé en jachère. S = surface du relevé, D = densité, Rf = Richesse floristique, H = diversité, R = régularité du peuplement des Ptéridophytes.

1. *Lastreopsis currori* (Mett.) Tindale
2. *Asplenium sandersonii* Hook.
3. *Cyathea borbonica* Desv. var. *laevigata* Bonap.
4. *Christella distans* (Hook.) Holtt.
5. *Amauropelta bergiana* (Schl.) Holtt.
6. *Asplenium inaequilaterale* Willd.
7. *Pteris elongatiloba* Bonap.
8. *Asplenium unilaterale* Lam.
9. *Arthropteris monocarpa* (Cord.) C.Chr.
10. *Asplenium mannii* Hook.
11. *Elaphoglossum zakamenense* Tard.
12. *Pleopeltis excavata* (Boory) Moore
13. *Elaphoglossum sieberi* (Hook. & Grev.) Moore
14. *Asplenium normale* Don.
15. *Hymenophyllum sibthorpioides* Mett.
16. *Asplenium adiantoides* (L.) C. Chr.
17. *Cyathea decrescens* Mett.
18. *Blotiella madagascariensis* (Hook.) Tryon
19. *Grammitis gilpinae* (Bak.) Tard.
20. *Pellaea angulosa* (Bory) Bak.
21. *Blechnum microbasis* (Bak.) C. Chr.
22. *Selaginella lyallii* (Hook. & Grev.) Spring

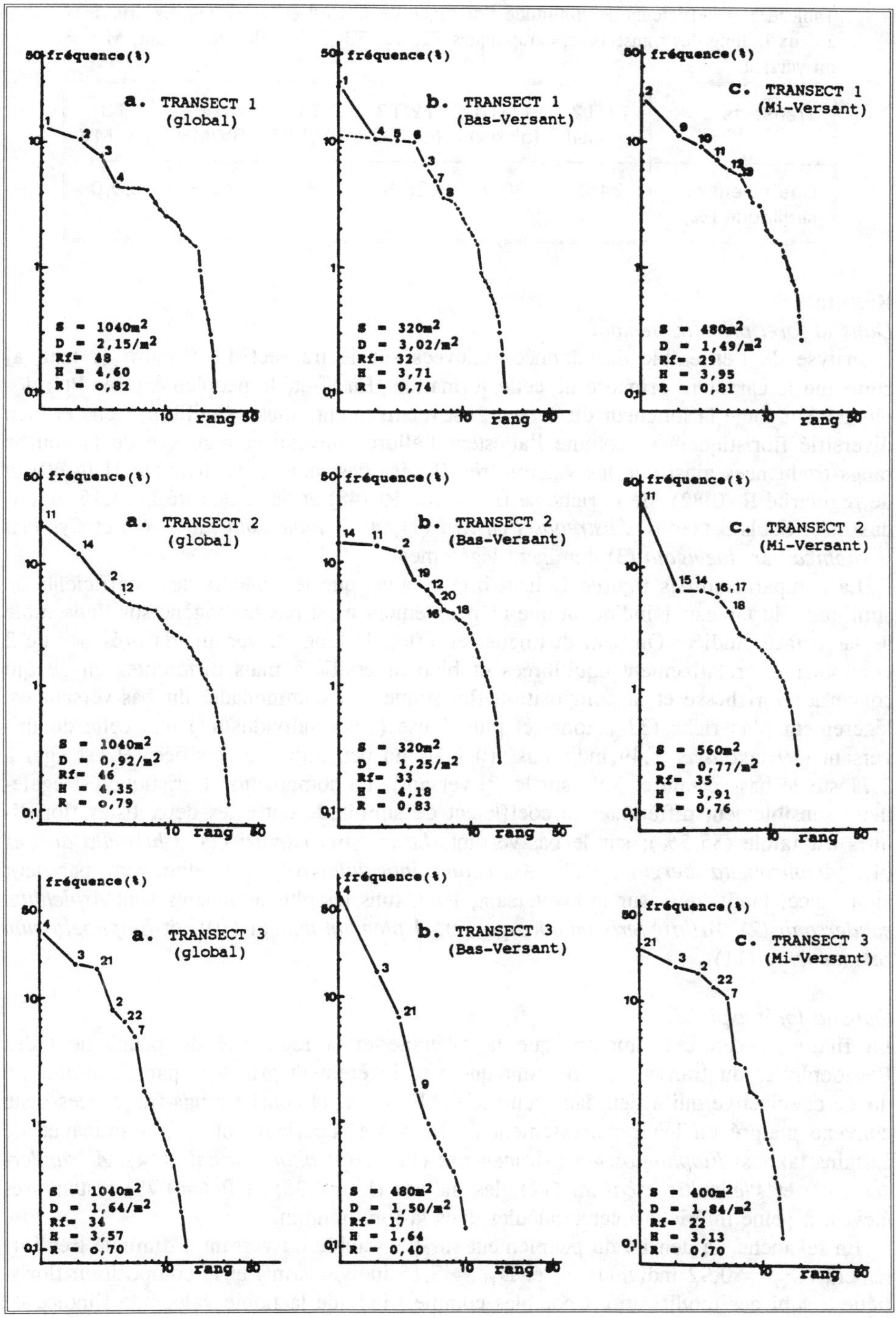
fréquence(%)
rang
a. TRANSECT 1 (global)
S = 1040m²
D = 2,15/m²
Rf= 48
H = 4,60
R = o,82
b. TRANSECT 1 (Bas-Versant)
S = 320m²
D = 3,02/m²
Rf= 32
H = 3,71
R = 0,74
c. TRANSECT 1 (Mi-Versant)
S = 480m²
D = 1,49/m²
Rf= 29
H = 3,95
R = 0,81
a. TRANSECT 2 (global)
S = 1040m²
D = 0,92/m²
Rf= 46
H = 4,35
R = o,79
b. TRANSECT 2 (Bas-Versant)
S = 320m²
D = 1,25/m²
Rf= 33
H = 4,18
R = 0,83
c. TRANSECT 2 (Mi-Versant)
S = 560m²
D = 0,77/m²
Rf= 35
H = 3,91
R = 0,76
a. TRANSECT 3 (global)
S = 1040m²
D = 1,64/m²
Rf= 34
H = 3,57
R = o,70
b. TRANSECT 3 (Bas-Versant)
S = 480m²
D = 1,50/m²
Rf= 17
H = 1,64
R = 0,40
c. TRANSECT 3 (Mi-Versant)
S = 400m²
D = 1,84/m²
Rf= 22
H = 3,13
R = 0,70

Tableau 1. Coefficients de similitude floristique de Jaccard entre les relevés pris deux à deux le long des transects topographiques T1, T2, T3. BV = Bas de versant, MV = mi-versant.

Transects	T1/T2 (global)	T1/T3 (global)	T2/T3 (global)	T1 BV/MV	T2 BV/MV	T3 BV/MV
Coefficient de similitude (%)	34,3	30,1	26,9	35,5	47,8	29,0

Résultats

Dans la forêt primaire témoin

L'analyse de l'ensemble des données relevées sur le transect T1 (Figure 1: haut a) confirme le caractère primaire de cette formation. En effet, le peuplement des Ptéridophytes, sur toute la longueur du versant, est relativement dense, équilibré, riche et bien diversifié floristiquement comme l'attestent l'allure convexe et régulière de la courbe rangs-fréquences ainsi que les valeurs très élevées des indices de diversité H (4,60) et de régularité R (0,82), de la richesse floristique Rf (48) et de la densité D (2,15 individus/m2). Seuls 3 taxons, *Lastreopsis currori* (1), *Asplenium sandersonii* (2) et *Cyathea borbonica* var. *laevigata* (3) dominent légèrement.

La comparaison des figures 1: haut b et c, ainsi que les valeurs des coefficients de similitude du tableau 1 indiquent que ce peuplement n'est pas homogène sur l'ensemble de la surface étudiée. On peut distinguer en effet, le long du versant, la présence de 2 communautés relativement équilibrées et bien diversifiées mais différentes en ce qui concerne la richesse et la composition floristique. La communauté du bas-versant est légèrement plus riche (32 taxons) et plus dense (3,02 individus/m^2) que celle du mi-versant (29 taxons et 1,49 individus/m^2) mais un peu moins diversifiée (H est égal à 3,71 sur le bas-versant et 3,95 sur le mi-versant). La composition floristique est également sensiblement différente: le coefficient de similitude entre les deux listes floristiques est faible (35,5%); sur le bas-versant, *Lastreopsis currori* (1), *Christella distans* (4), *Amauropelta bergiana* (5), *Asplenium inaequilaterale* (6), dominent par leur abondance, tandis que, sur le mi-versant, les taxons les plus abondants sont *Asplenium sandersonii* (2), *Arthropteris monocarpa* (9), *Asplenium mannii* (10), et *Elaphoglossum zakamenense* (11).

Dans la forêt exploitée

La figure 1: centre a, montre que la diversité et la régularité du peuplement des Ptéridophytes du transect T2 ne sont que très légèrement affectées par l'exploitation douce et sélective qui a lieu dans cette forêt. En effet, la courbe rangs-fréquences reste convexe malgré un léger redressement de sa partie supérieure dû à la dominance de certains taxons: *Elaphoglossum zakamenense* (11), *Asplenium normale* (14), *A. sandersonii* (2) et *Pleopeltis excavata* (12); les indices H (=4,35) et R (=0,79) restent très élevés, à peine inférieur à ceux calculés dans la forêt témoin.

En revanche, la densité du peuplement sur l'ensemble du versant a diminué très fortement (D_{T2} = 0,92 individus/m^2 et D_{T1} = 2,15 individus/m^2) et la composition floristique a subi des modifications notables comme l'indique la faible valeur de l'indice de similitude entre les listes floristiques des transects T1 et de T2 (PJ = 34,3%).

La comparaison des figures 1: centre b et c, indique que la composition floristique et la structure du peuplement des Ptéridophytes le long du versant du transect T2 est relativement homogène: la richesse floristique est sensiblement la même sur le bas-versant et le mi-versant (33 et 35 taxons); l'indice de Jaccard (tableau 1) est assez élevé (47,8%), et plusieurs espèces abondantes sur le mi-versant telles qu' *Asplenium normale* (14), *A. adiantoides* (16), *Elaphoglossum zakamenense* (11) et *Blotiella madagascariensis* (18) se retrouvent dans la partie supérieure de la courbe rangs-fréquences du bas-versant. Par contre, la communauté des Ptéridophytes du bas-versant est caractérisée par une diversité et une densité plus grande (H = 4,18; R = 0,83; D = 1,25 individus/m^2) que celles qui sont observées sur le mi-versant (H = 3,91; R = 0,76; D = 0,77 individus/m^2).

Dans la forêt reconstituée sur un terrain laissé en jachère

La structure du peuplement des Ptéridophytes est différente de celle que l'on observe dans les deux transects précédents (Figure 1: bas a; tableau 1). En effet, la courbe rangs-fréquence très redressée montre que le peuplement des Ptéridophytes est peu équilibré, dominé par l'abondance d'un groupe de 6 taxons hémi-sciaphiles: *Christella distans* (4), *Cyathea borbonica* var. *laevigata* (3), *Blechnum microbasis* (21), *Asplenium sandersonii* (2), *Selaginella lyallii* (22) et *Pteris elongatiloba* (7), ce qui se traduit par des indices de diversité H et de régularité R plus faibles. La richesse floristique du versant n'est pas très élevée (34 espèces); par contre, la densité du peuplement est importante (1,64 individus/m^2). L'ensemble de ces observations caractérise un écosystème jeune, en voie de structuration.

La comparaison des figures 1: bas b et c, fait ressortir une nette différence dans la composition floristique et la structure des communautés de Ptéridophytes des 2 zones topographiques du mi-versant et du bas-versant. En effet, l'indice de similitude floristique est faible (29,0%); parmi les taxons abondants, seuls *Cyathea borbonica* var. *laevigata* (3) et *Blechnum microbasis* (21) sont communs aux deux zones; la courbe rangs-fréquences du bas-versant est plus redressée, presque linéaire par rapport à celle du mi-versant ce qui entraîne une diminution très nette de l'indice de diversité (H = 1,64) et de régularité (R = 0,40). La communauté des Ptéridophytes du bas-versant est moins diversifiée, moins équilibrée et plus pauvre floristiquement que celle du mi-versant.

Conclusions

Cette étude permet de dégager des résultats originaux concernant la dynamique du peuplement des Ptéridophytes dans un écosystème tropical déstabilisé à la suite d'une agression externe. Dans la parcelle de forêt primaire témoin, c'est-à-dire dans un écosystème stable et mature, les Ptéridophytes constituent un peuplement très riche, diversifié et dense. Ce peuplement, dans lequel ont pu être mises en évidence deux communautés inféodées, l'une au bas de versant l'autre au mi-versant, apparait donc, également, bien structuré dans l'espace. Une agression permanente mais de faible intensité, exercée sur cet écosystème (cas de la forêt exploitée de la parcelle 2), se traduit par une modification de la composition floristique du sous-bois: les espèces les plus sciaphiles et les moins tolérantes disparaissent, tandis que le milieu s'enrichit en espèce hémi-sciaphiles. Cette modification s'effectue, dans le cas présent, très progressivement et apparemment sans compétition forte entre les différents taxons, de telle sorte que le peuplement reste toujours diversifié et conserve la même richesse floristique. On note cependant une forte diminution de la densité et surtout une simplification de la structure du peuplement le long du versant.

Les observations faites dans la forêt reconstituée sur une jachère ont permis de constater que, malgré la présence d'un environnement forestier favorable, le repeuplement d'un sous-bois en Ptéridophytes est un phénomène très lent. Après une cinquantaine d'années, la communauté des fougères présente encore les caractéristiques floristiques d'un système jeune, en voie de structuration, peu diversifié et dominé par quelques espèces pionnières robustes, à multiplication rapide. Les différences floristiques observées entre les deux zones topographiques ont pu être engendrées soit par une vitesse de restructuration du peuplement variable le long du versant, soit par une intensité plus ou moins forte de la pression anthropique exercée dans le passé sur ce même versant, soit, plus probablement, par les deux phénomènes à la fois.

Tous ces faits démontrent que les Ptéridophytes constituent un matériel de choix pour étudier les réactions internes d'un écosystème tropical à une échelle spatio-temporelle réduite. En effet, ils sont généralement abondants et diversifiés en milieu tropical; leur biologie et leur enracinement superficiel font qu'ils réagissent rapidement à des micro-variations des facteurs environnants; souvent peu tolérants vis-à-vis de l'humidité et de la lumière, leur présence ou leur absence dans un milieu donné est donc toujours significative.

Remerciements
Ces recherches ont été réalisées dans le cadre du Projet 343-9.802/42303 de conservation du Parc National de Ranomafana, mené conjointement par Duke University (U.S.A.) et le Ministère des Eaux & Forêts (Madagascar).

Bibliographie

Daget, J., 1976. Les modèles mathématiques en écologie. Masson, 172 p.

Farinas, M., 1982. Comparaison de plusieurs méthodes d'étude de la structure horizontale et verticale de formations végétales; application au cas de deux anciennes friches languedociennes. Thèse de doctorat de 3ème cycle, U.S.T.L., Montpellier, 430 p.

Frontier, S. & D. Pichod-Viale, 1991. Ecosystème: Structure, Fonctionnement, évolution. Masson, Paris, 392 p.

Godron, M., 1966. Application de la théorie de l'information à l'étude de l'homogénéité et de la structure de la végétation. Oecol. Plant. 1: 187-197.

Humbert, H. & G. Cours-Darne, 1965. Notice de la carte de Madagascar. Trav. Sect. Sci. Tech. Inst. Franc. Pondicherry, h.s. 6: 46-78.

Jaccard, P., 1901. Distribution de la flore alpine dans le bassin des Dranses et dans quelques régions voisines. Bull. Soc. Vaudoise Sc. Nat. 37: 241-272.

Koechlin, J., J.L. Guillaumet & Ph. Morat, 1974. Flore et végétation de Madagascar. J. Cramer, Vaduz, 687 p.

Rakotondrainibe, F. & J.L. Guillerm, 1990. Modèle de répartition horizontale des Ptéridophytes le long d'un gradient d'humidité lié à la topographie: le cas d'une forêt naturelle d'altitude à Madagascar. Candollea 45: 591-598.

Shannon, C.E. & W. Weaver, 1948. The mathematical theory of communication. University of Illinois Press, Urbana, 1963 éd., IV, 125 p.

Biodiversity in African savannas

23 August 1994

Convenor: C. Geerling

The subcanopy flora in the dynamics of the Kalahari Thornveld

O.A. LEISTNER
National Botanical Institute, Private Bag X101, Pretoria, 0001 South Africa.

Abstract
Kalahari Thornveld occupies an area of some 390,000 km² in Namibia, Botswana and South Africa. It constitutes the southern portion of the arid or fine-leaved savanna on Kalahari sand co-dominated by grasses such as species of *Eragrostis* and *Stipagrostis*, and trees and shrubs, especially of the genus *Acacia*. Microclimatic and edaphic conditions in the subcanopy habitat under trees and shrubs often differ widely from those in the grassland. The subcanopy is unstable and subject to much disturbance, mainly due to its rich and varied fauna. A total of 132 plant species was recorded as typical for, or exclusive to the subcanopy habitat. These species tend to be widely to very widely distributed, and endemism was very low. Almost 50% were hemicryptophytes, 28% annuals and 9% succulents. The most common species were almost exclusively dispersed by animals, especially birds. Leguminosae, the family of most of the pioneer canopy formers, was almost absent from the subcanopy.

Résumé
Le "Veld" épineux du Kalahari occupe une superficie de quelque 390.000 km² en Namibie, au Botswana et en Afrique du Sud. Il constitue la partie méridionale des savanes arides ou à feuilles étroites des sables du Kalahari, à dominance de Gramineae comme *Eragrostis* et *Stipagrostis*, ainsi que d'arbres et arbustes appartenant principalement au genre *Acacia*. Les conditions microclimatiques et édaphiques dans le milieu sous couvert d'arbres et d'arbustes diffèrent souvent de façon importante de celles en formation herbeuse. Le milieu sous couvert est instable et sujet à une grande perturbation, principalement en raison de sa faune riche et variée. On a relevé un total de 132 espèces typiques ou exclusives dans ce milieu. Ces espèces ont tendance à être largement à très largement réparties et l'endémisme y est très faible. A peu près 50% d'entre elles sont des hémicryptophytes, 28% des annuelles et 9% des succulentes. Les espèces les plus communes sont à peu près exclusivement disséminées par les animaux, principalement par les oiseaux. Les Leguminosae, famille représentative de la plupart des groupements pionniers de la voûte forestière, sont quasiment absentes de la strate sous couvert.

Key words
Kalahari Thornveld, Savanna dynamics, Seed dispersal, Southern Africa, Subcanopy flora.

Introduction

Savanna is a mosaic of two main habitats: a matrix dominated by hemicryptophytes, mainly grasses, and a subcanopy habitat under trees and shrubs in which young trees, shrubs, perennial herbs or annuals, or a mixture of any of these, may dominate. The flora of the subcanopy habitat has rarely been treated as a separate entity in savanna studies. This paper aims to throw more light on this flora, its contribution towards the biodiversity of the Kalahari Thornveld (KTV) and its role in the dynamics of this vegetation type. Strategies in the competition between subcanopy and matrix plants are discussed mainly in the light of their seed dispersal. It is hypothesized that the subcanopy habitat is a recruitment area for species unable to compete in the matrix, at

L.J.G. van der Maesen et al. (eds.), The Biodiversity of African Plants, 163-179.

least in their juvenile state. It is postulated further that seed dispersal strategies provide one of the most important clues to the dynamics of the savanna system.

Study area, environment and methods

Definition and delimitation of Kalahari Thornveld

The term Kalahari Thornveld (KTV), as used here, essentially agrees with the Kalahari Thornveld Proper (Veld Type 16a) of Acocks (1988). This vegetation type covers an estimated 390,000 km^2 in E and SE Namibia, S and SW Botswana, the northern part of Northern Cape Province and almost the entire Northwest Province of South Africa. In Namibia it coincides fairly well with the Camelthorn Savanna (Central Kalahari) and the Mixed Tree and Shrub Savanna (Southern Kalahari) of Giess (1971) and in Botswana with Arid Shrub Savanna, Southern Kalahari Bush Savanna and Central Kalahari Bush Savanna of Weare & Yalala (1971). On the vegetation map of Africa (White 1983) it falls inside the Kalahari/Karoo Namib Transition and Kalahari *Acacia* Wooded Grassland and Deciduous Bushland. It also includes all of the Southern Kalahari as circumscribed by Leistner (1967). Kalahari Thornveld (KTV) constitutes roughly the southern half of the so-called arid or fine-leaved savanna on Kalahari sands, co-dominated by "sweet" or "white" C_4 grasses, especially *Eragrostis* and *Stipagrostis*, and spinescent, fine-leaved trees and shrubs, mainly *Acacia* (Huntley 1982).

Vegetation

In the southwest, the most arid part of the KTV, with little more than 150 mm rain per year, the region is covered by an extremely open bush savanna. With the increase in annual precipitation towards north and east, the KTV gradually changes into a sparse mixed bush and tree savanna, and eventually, in the moistest parts with an annual rainfall around 550 mm, it becomes a tree savanna usually mixed with shrubs.

Phytogeography

It is widely accepted that phytogeographically most of the KTV falls within the Zambezian Domain of the Sudano-Zambezian Region as defined by White (1983). Opinions, however, differ on the position of the boundary between the KTV and the Karoo Namib Domain which lies south of the KTV (Rutherford & Westfall 1986, 1994, Werger 1986, White 1983). The biodiversity of the KTV increases from the arid southwest to the more mesic north and east as the desert species of the Nama-Karoo are replaced by the Sudano-Zambezian savanna elements.

Environmental conditions

Conditions in the subcanopy habitat tend to differ markedly from those in the grassland matrix. The degree of difference obviously depends on numerous factors such as the density, height and permanence of the canopy and the activity of the subcanopy fauna. In the subcanopy habitat the net radiation is markedly less; the rainfall is reduced but the water regime is more favourable, and the nutrient content, as well as the microbial activity, is higher (Scholes & Walker 1993). Concentrations of N, P, K and Ca, in particular, have frequently been found to be higher under trees than in the surrounding grassland (Belsky et al. 1993, Perkins & Thomas 1993). The high mineral concentrations are at least partly due to the excrements of animals sheltering under the trees.

Conditions under trees and under shrubs may differ significantly. *Trees* typically have an umbrella shape with a dense to fairly dense crown, at least during the warmer months. They provide shade for animals of all sizes but little or no protection from predators for small animals. Their subcanopy is subject to disturbance from large mammals, herbivores as well as carnivores. *Shrubs* mostly have a sparse canopy, even in summer but often have many densely placed stems, often arising from a network of rhizomes with suckers. They provide protection for smaller plants from large herbivores but, as a result of root competition, are not ideal seed beds. Especially the dense, freely suckering species constitute effective traps for flying diaspores and for sand and offer shelter to numerous small animals.

Fauna

The fauna of the KTV is rich and varied and comprises a host of seed dispersers. Areas of considerable size, particularly in Botswana and in the northernmost part of the Northern Cape Province fall within game reserves or are not intensively farmed. Large herbivores include springbok (*Antidorcas marsupialis*), gemsbok (*Oryx gazella*), eland (*Taurotragus oryx*) and blue wildebeest (*Connochaetes taurinus*). Black-backed jackal (*Canis mesomelas*) and bat-eared fox (*Otocyon megalotis*) are among the larger fruit-eaters. The rich rodent fauna of the region includes numerous mice and rats and the porcupine (*Hystrix africaeaustralis*) (Smithers 1983). Livestock kept in the region includes cattle, sheep and goats. The KTV has a varied avifauna ranging from ostrich (*Struthio camelus*) to numerous seed-eaters and several, mostly visiting, fruit-eaters (Maclean 1970). Among the reptiles of the KTV are numerous lizards and the leopard tortoise (*Geochelone pardalis*) which, as Milton (1992) has shown, can play an important role in the dispersal of seeds.

Selection of species

The list of 132 species, one subspecies and one variety (both counted here as species in the statistics) of flowering plants from the subcanopy habitat of the KTV is based very largely on data collected by Acocks (unpublished notes) at 129 survey sites. He listed all species that he observed exclusively or mainly under trees or shrubs, at least in a given survey site, with a T (for tree) or t (for shrub) followed by a downward-pointing arrow. It should be noted that the records by Acocks cover mainly the southeastern parts of the KTV and might therefore not be considered representative of the region as a whole. However, it will be demonstrated that the majority of species of the subcanopy habitat are widely distributed. With minor additions from my own notes from many parts of the KTV, the list is considered an adequate representation of the subcanopy flora of the entire region. Species commonly found both in matrix vegetation and under trees or shrubs in the same sample area have not been listed as part of the subcanopy flora. The naming of plants is according to Arnold & De Wet (1993).

Distribution within southern Africa

The information given on Table 1 was taken directly from Arnold & De Wet (1993). The symbols used are explained in the captions of the Tables. They do not yet reflect the new provinces delimited in South Africa in 1994. Data on endemism within the KTV were gathered mainly in the National Herbarium (PRE), Pretoria.

Table 1. The 40 most common subcanopy species, recorded in 129 survey sites by J.P.H. Acocks (unpublished notes), arranged primarily in order of frequency and secondarily according to the Engler system.

Species	T	t	Distribution: In southern Africa	Distribution: Outside southern Africa	Seed dispersal	Life form	Family
Pollichia campestris	29	3	T O N C I S L B	Africa	B, Vi	H	Illecebraceae
Pavonia burchellii	25	-	T O N C I S B	Africa	B, Vi	H	Malvaceae
Lantana rugosa	16	1	T O N C I S L B	Africa, Asia	B	H	Verbenaceae
*Achyranthes aspera var. aspera**	16	-	T O N C I S L B	Cosmopolitan	Ve	T/H	Amaranthaceae
Protasparagus suaveolens	12	-	T O N C I S L B	Adjoining states	B	H	Asparagaceae
Pupalia lappacea	10	2	T O N C I S B	Old World	Ve	T:1	Amaranthaceae
Eragrostis biflora	12	-	T O C I B	-	R/W	T:1	Gramineae
Setaria verticillata	10	1	T O N C I S B	Old World tropics	Ve	T:1	Gramineae
Protasparagus cooperi	7	3	T O N C I S L B	Adjoining states	B	H	Asparagaceae
Coccinia sessilifolia	7	3	T O C I B	-	Vf, B	H/Gc	Cucurbitaceae
Antizoma angustifolia	8	1	T O C I B	-	B	H/Gc	Menispermaceae
Protasparagus retrofractus	4	3	C I	-	B	H	Asparagaceae
Pentarrhinum insipidum	5	2	T O N C I S B	Africa	W	Hc	Asclepiadaceae
Sporobolus fimbriatus	2	4	T O N C I L B	Africa	R/W	H	Gramineae
*Chenopodium album**	6	-	T O N C I S L B	Cosmopolitan	Vi	T:3	Chenopodiaceae
Amaranthus thunbergii	6	-	T O N C I S L B	Africa	Vi	T:1	Amaranthaceae
Panicum maximum	4	1	T O N C I S B	Africa, Madagascar	R/W, B	H	Gramineae
Commelina benghalenis	4	1	T O N C I S L B	Pantropical	Vi/R	T:1	Commelinaceae
*Chenopodium glaucum**	4	1	T O C	Cosmopolitan	Vi	T:3	Chenopodiaceae
Withania somnifera	5	-	T O N C I L B	Old world	B, Vi, W	H	Solanaceae
Solanum capense	4	1	T O N C I L	-	Vf	H	Solanaceae
Solanum coccineum	4	1	T O N C S B	-	Vf	H	Solanaceae
Kedrostis africana	5	-	T O N C I	-	Vf, B	H/Gc	Cucurbitaceae

Table 1 (Continued).

Species	T	t	Distribution		Seed dispersal	Life form	Family
			In southern Africa	Outside southern Africa			
Aerva leucura	4	-	T O N C I S B	Africa	Vi	H	Amaranthaceae
*Oxalis corniculata**	4	-	T O N C S L B	Cosmopolitan	A, R/W	T/G	Oxalidaceae
Pergularia daemia	3	1	T N C I S B	Old World tropics	W	Hc	Asclepiadaceae
Teucrium trifidum	4	-	T O N C L B	Endemic	R/W	H	Labiatae
Nidorella resedifolia	4	-	T O N C I S L B	Africa	W	T:3	Compositae
*Bidens bipinnata**	4	-	T O N C L	Cosmpolitan	Ve	T:1	Compositae
Panicum coloratum	2	1	T O N C I S L B	Africa	R/W	H	Gramineae
Tricholaena monachne	3	-	T O N C I S B	Africa	R/W	H	Gramineae
Eustachys paspaloides	3	-	T O N C I S L B	Africa +	R/W	H	Gramineae
Triraphis andropogonoides	3	-	T O N C B	-	R/W	H	Gramineae
Sansevieria aethiopica	3	-	T O C I B	Adjoining states	Vi, B	S	Dracaenaceae
*Alternanthera pungens**	3	-	T O N C I S L B	Pantropical	Ve	H	Amaranthaceae
Talinum caffrum	1	2	T O N C I S B	Africa	Vi, B	H/G	Portulacaceae
Kalanchoe rotundifolia	3	-	T O N C I S B	Africa	R/W	S	Crassulaceae
Sida cordifolia	3	-	T N S B	Pantropical	Ve	H	Malvaceae
Corallocarpus schinzii	3	-	C I	-	Vf, B	H/Gc	Cucurbitaceae
*Conyza bonariensis**	3	-	T O N C I S L B	Africa	W	T:1	Compositae

1st column: * = naturalized

2nd column: T = number of records under trees; t = number of records under shrubs

3rd column: T = Transvaal; O = Orange Free State; N = Natal; C = Cape; I = Namibia; S = Swaziland; L = Lesotho; B = Botswana

4th column: adjoining states are Angola, Zambia, Zimbabwe and Mozambique; Africa, in one or more adjoining countries and beyond; Africa + = also in countries near the continent

5th column: A = autochory; B = birds; R = rain; Ve = attached to vertebrates; Vf = fruit-eating vertebrates; Vi = ingested by vertebrates; W = wind (Vertebrates excludes birds)

6th column: H = hemicryptophyte; G = geophyte; S = succulent; T = therophyte; T:1 = flowering mainly in warm months; T:3 = flowering ± throughout the year; c = climber or scrambler.

Distribution outside southern Africa
These data were established mainly by reference to standard works on the African flora and are only approximate.

Life forms
The life forms given on Table 1 are the entities originally recognized by Raunkiaer (1934) as defined by Rutherford & Westfall (1994).

Seed dispersal
Information on seed dispersal is based on own observations and on literature. No experiments were made with transported subcanopy seeds, and the data are generally based on circumstantial evidence. A number of species, including most grasses, with small light seeds lacking features that could obviously aid in their dispersal, have been classified under R/W (rain/wind). The dispersal agents listed on Table 1 are in most cases only those typically associated with the structural features of the propagule in question. On Table 5 only the first dispersal agent mentioned for a species on Table 1 has been taken into consideration.

The following classification of dispersal agents was implemented:
A - autochorous: dispersal is effected by the plant itself.
An - ants: no observations were made on dispersal by ants.
B - birds: plant species with small fruit (usually less than 10 mm in diameter) with a conspicuous, often reddish fleshy covering and a hard core are put into this group even though other, usually larger frugivores may also consider the small fruit worth eating. Fleshy fruit, too large for a bird to swallow whole, are classified under Vf (vertebrates, frugivores) even though birds very often eat them by pecking them to pieces. Certain hard dry seeds may also be voided intact after ingestion by birds (Van der Pijl 1982). Birds also disperse seeds with their nest material (Dean et al. 1990).
R - rain: no observations were made in the KTV on water dispersal of seeds.
V - vertebrates. This category excludes birds.
Ve - vertebrates, epizoochorous: seeds are borne inside structures which can become attached to animals by means of barbs or hooks.
Vf - vertebrates, frugivores: mainly fleshy fruit too large for a bird to swallow whole, and therefore in many cases more effectively distributed by larger frugivores such as antelopes and members of the Canidae.
Vi - vertebrates internal (endozoochorous): seeds with a hard testa or other covering which are usually ingested unintentionally with other edible parts of the plant and often voided intact (Janzen 1984). Seeds ingested by rodents are probably mostly destroyed in the process (Dean & Milton 1991).
W - wind.

Results
The subcanopy flora
With a mere 134 species the subcanopy flora of the KTV, which covers an estimated area of 390,000 km^2, is poorer than the poorest desert flora. However, this total only

Table 2. Representation of largest families. The twelve largest families are ranked according to number of species. Percentages of smaller families are in brackets.

Area	Subcanopy in Kalahari Thornveld		Subcanopy + matrix					
			Southern Kalahari (Leistner 1967)		Kalahari Gemsbok Nat. Park (V. Rooyen et al. 1988)		Vryburg Shrub Bushveld (Werger 1986)	
Total no. of spp.	134		438		397		709	
Ranking of families	%	R	%	R	%	R	%	R
Compositae	15.6	1	11.9	2	14.3	1	14.3	1
Gramineae	14.9	2	14.6	1	10.6	3	12.6	2
Leguminosae	—	—	7.8	3	11.1	2	5.9	4
Liliaceae(+Aspr.)	3 (Aspr.)	9	6.2	4	7.3	4	7.5	3
Aizoaceae	(1.5)	—	4.3	6	6.5	5	5.1	5
Scrophulariaceae	2.2	12	4.8	5	4.0	6	3.7	6
Sterculiaceae	(2.2)	—	2.7	7	3.0	7	3.1	7
Asclepiadaceae	6.7	3	(2.0)	—	2.8	9	2.7	8
Cucurbitaceae	5.2	5	2.5	8	2.5	10	(1.6)	—
Chenopodiaceae	2.2	11	2.3	9	2.8	8	1.8	11
Solanaceae	4.5	7	2.0	11	(1.8)	—	2.0	10
Euphorbiaceae	—	—	2.3	10	2.3	12	2.4	9
Zygophyllaceae	—	—	2.3	12	(1.5)	—	(1.1)	—
Acanthaceae	3.7	8	(1.1)	—	(1.8)	—	1.8	12
Cruciferae	5.2	4	(0.7)	—	(0.5)	—	?	—
Amaranthaceae	4.5	6	(2.3)	—	2.5	11	?	—
Malvaceae	2.2	10	(1.4)	—	(1.2)	—	?	—
Total of largest families	69.9	—	69.7	—	64.0	—	62.9+	—

includes species recorded as typical for, or exclusive to the subcanopy habitat, at least in a given survey site.

As Werger (1973, 1986) has demonstrated, lists of the largest families of a region, ranked according to their number of species, provide a quick and simple means for defining floral regions and for assessing their floristic and phytogeographical affinity. Such lists have been prepared for the subcanopy flora of the KTV, the Southern Kalahari (Leistner 1967) in the southwest section of the KTV, the Kalahari Gemsbok National Park (from data by Van Rooyen et al. 1988) in the extreme southwest corner of the KTV, and the Vryburg Shrub Bushveld (Werger 1986, from unpublished lists by Acocks) which borders on the southeastern KTV (Table 2). The subcanopy flora is here compared to floras comprising both subcanopy and matrix. This was done because many species are not exclusive to either habitat and it was therefore not possible to separate them clearly.

Table 3. Distribution of subcanopy species. Figures are rounded-off percentages.

Subcanopy species: / Distribution:	All species (134)	40 most common species (Table 1)
In all provinces/countries in southern Africa	22	32
Endemic to southern Africa	36	25
Endemic or largely endemic to Kalahari Thornveld	6	2

From Table 2 it can be seen that the lists from the Southern Kalahari, the Kalahari Gemsbok National Park and the Vryburg Shrub Bushveld are very similar and that the subcanopy flora differs markedly from them. The lists are briefly compared:

Families poorly represented in subcanopy include Leguminosae of which only one species was recorded whereas in the other three lists it ranks second to fourth. Liliaceae sens. lat. ranks highly in the three complete floras; in the subcanopy it is only represented by Asparagaceae which ranks ninth. Scrophulariaceae and Sterculiaceae are much less important in the subcanopy than in the other three floras. Aizoaceae, an important family in the three complete listings, is poorly represented in the subcanopy. Euphorbiaceae is not among the 12 largest families of the subcanopy, nor is Zygophyllaceae.

Families well represented in the subcanopy include the following: Asclepiadaceae ranks third in the subcanopy, against eighth or lower in the others. Cucurbitaceae is markedly higher in the subcanopy, and so are Solanaceae, Cruciferae, Amaranthaceae and Malvaceae.

Families with similar representation in all lists are Compositae, Gramineae and Chenopodiaceae.

Distribution and endemism

Some distribution data are summarized in Table 3. It is immediately apparent that the majority of species are very widely distributed and that endemism is exceptionally low.

Only one species, *Tridentea marientalensis*, is considered to be truly endemic to the KTV. Three other Stapeliae (Asclepiadaceae), namely *Huerniopsis decipiens, Orbeopsis knobelii* and *Pachycymbium lugardii* are largely confined to the region. *Helinus spartioides* (Rhamnaceae), and *Corallocarpus schinzii* (Cucurbitaceae) can be described as having their main distribution area in the KTV. *Corallocarpus schinzii* is the only one of the above species listed among the 40 most common subcanopy species (Table 1). Among these 40 species seven are given as naturalized and the majority are widely to very widely distributed.

Life forms

The life forms assigned to all subcanopy species have been condensed into a life form spectrum (Table 4). For the sake of comparison to a few other spectra, the life form

Table 4. Life form spectra (%).

Area	S	M	N	Ch	H	G	T	O
1. Kalahari Thornveld subcanopy	9	2	7	6	47	1	28	-
2. Kalahari Gemsbok N.P. (Van Rooyen et al. 1988)	1	1	7	15	31	13	31	-
3. Southern Kalahari (Leistner 1967)	2	3	8	12	34	8	33	-
4. Ghanzi area (Cole 1986)	2	8	10	16	28	8	28	-
5. Normal spectrum (Raunkiaer 1934)	1	17	20	9	27	3	13	10

S = succulent; M = microphanerophyte; N = nanophanerophyte; Ch = chamaephyte; H = hemicryptophyte; G = geophyte; T = therophyte; O = others.

GH, considered to be intermediate between geophytes and hemicryptophytes (Leistner 1967, Werger 1986) is included under hemicryptophytes in the spectra of the subcanopy and of the Southern Kalahari. The spectrum of the Kalahari Gemsbok National Park (KGNP) (Van Rooyen et al. 1988) has been slightly modified to show the succulents as a distinct life form. The Ghanzi area, for which Cole (1986) drew up a life form spectrum, is in Botswana just north of the KTV.

The subcanopy spectrum is briefly compared to the others on Table 4, dealing with each life form in turn:

* *Succulents*. The high figure for succulents is one of the most striking differences between the subcanopy and the other spectra. The percentage almost reaches levels recorded for the Succulent Karoo (Werger 1986).
* *Microphanerophytes*. The percentage of this life form agrees well with those of the KGNP and the Southern Kalahari, both from the more arid parts of the KTV. In the moister Ghanzi area it almost reaches the level of the normal spectrum.
* *Nanophanerophytes* have a similar representation in all spectra excepting the normal spectrum which is far richer in this life form.
* *Chamaephytes* are markedly less well represented in the subcanopy than in the spectra from the KGNP, the Southern Kalahari and the Ghanzi area. In the normal spectrum this life form has a low percentage.
* *Hemicryptophytes* are the best represented life form in all spectra but the percentage of 47 for the subcanopy is remarkably high.
* *Geophytes*. Only a single typical geophytic species was recorded, namely the orchid *Bonatea speciosa*. This is in striking contrast to the other spectra in which this life form is well represented, with the exception of the normal spectrum in which geophytes also play a minor role.
* *Therophytes*. A percentage of around 30 is comparable to that of many arid regions in different parts of the world. For the KTV as a whole it is, however, considered to be very high, seeing that the more mesic parts of the region receive an annual rainfall around 450 mm.

Table 5. Seed dispersal agents of subcanopy species. Figures are rounded-off percentages.

Dispersal agent: Subcanopy species:	B	Vf	Vi	Ve	W	R/W	A	A/R	An
All 134 spp.	14	12	13	7	19	28	2	5	1
20 most common spp. in Table 1	40	5	20	15	5	15	0	0	0
40 most common spp. in Table 1	20	13	17	15	10	22	3	0	0

A = autochory; An = ants; B = birds; R = rain; Ve = attached to vertebrates; Vf = fruit-eating vertebrates; Vi = ingested by vertebrates; W = wind. (Vertebrates excludes birds).

Tree vs. shrub canopy
The frequency at which Acocks (unpublished notes) recorded a species under trees or shrubs was noted for all species, and is given for the 40 most common species on Table 1. When all species are considered, it is clear that most records were made under trees and that only a few species were seen more often under shrubs than under trees. Species recorded under both trees and shrubs accounted for about 13%, and those under shrubs for some 18% of the total. Families of which more than half the representatives were recorded from underneath shrubs are Santalaceae, Aizoaceae, Mesembryanthemaceae, Portulacaceae, Euphorbiaceae, Celastraceae, Gentianaceae, Asclepiadaceae, Cucurbitaceae and Campanulaceae.

Seed dispersal
The dispersal agents operative in the subcanopy of the KTV have been condensed in Table 5. Concerning dispersal by birds, it is very conspicuous that the percentage of this agent is highest among the most common species and drops off sharply when the rarer species are also taken into consideration. Dispersal by birds can also involve dry, well-protected seeds like the nuts of *Lithospermum* species (Van der Pijl 1982). Nest material collected by birds includes inflorescences of *Panicum maximum* and *Cynodon dactylon*, both prominent in the subcanopy, as well as stems of *Lepidium africanum*, and usually wind-dispersed propagules of certain Asclepiadaceae and of Compositae like *Pentzia incana* (Dean et al. 1990). Dispersal by wind, prominent in Asclepiadaceae and Compositae had its lowest percentage for the 20 commonest species and increased towards the group of all 134 species. The percentages for R/W (rainwater/wind) were strongly influenced by species of Gramineae and were also highest for the all 134 species group.

Autochory, dispersal by the plant itself, scored low throughout; autochory linked to rain, attained a somewhat higher score for the all 134 species group, thanks largely to a few species of Acanthaceae and two rarely recorded members of the Mesembryanthemaceae. Only *Polygala hottentotta* was tentatively listed as dispersed by ants because seeds of this genus have an aril which is eaten by ants without damage to the seed (Bresinsky 1963 quoted by Isaacs et al. 1993). Species of *Polygala* with flattened seed and a flattened, membranous capsule may also be dispersed by wind (Verkerke 1985). In the light of findings by Bews (1917), Van der Pijl (1982) and Bond & Slingsby (1983) it is assumed that ants play a more important role in seed dispersal in the KTV than here reflected.

Totals of seed dispersal effected by animals were as follows: 20 most common species - 80%; 40 most common species - 65%; all 134 species - 47%. These figures are based on the assumption that all grasses not obviously adapted for dispersal by adhesion to animals are transported by rainwater and/or wind.

The percentage of fleshy-fruited species in the total flora of the Southern Kalahari (Leistner 1967) was 10%, for the subcanopy of the KTV 40%.

Discussion

Three major groupings of plants are here distinguished, somewhat simplistically, in order to put the subcanopy flora into perspective within the savanna mosaic. These groupings will be referred to as matrix, canopy and subcanopy.

Matrix

The dominant plants of the matrix are perennial grasses represented by relatively few species but by vast numbers of individuals. These plants are generally long-lived and reproduce mainly vegetatively by rhizomes. In the sparse matrix of the arid southwest KTV sexual reproduction plays a more important role, and it is here that several wind-dispersed species of the genus *Stipagrostis* are locally dominant. In the established communities of perennial grasses, such as they occur particularly in the moister regions to the north and east, the greatest hazard faced by colonizing seedlings is probably the competition from neighbouring plants (Fenner 1985).

Canopy

Species included in this grouping are trees and shrubs adapted to compete with the hemicryptophytes under matrix conditions. They create a new habitat in the matrix which will here be simply referred to as the subcanopy. Among the canopy formers three groups can be distinguished, those in which seeds are dispersed by large herbivores, those with seeds dispersed by fruit eaters and those dispersed by wind.

Herbivore-dispersed canopy formers: this group comprises the trees and shrubs best adapted to compete in the grassland matrix, at least in the moister parts of the KTV. They germinate well in the subcanopy but their survival rate is reduced in that habitat, probably by the low light levels (Smith & Shackleton 1988). Most of the species in this group belong to the Leguminosae, and as will be shown later, this family is very poorly represented in the subcanopy.

The two character trees of the KTV, *Acacia haematoxylon* in the arid parts and *A. erioloba* in the rest of the region, both have indehiscent edible pods. Unless they are eaten, and the seeds ingested by large herbivores, the survival rate of seeds is very low because they are prone to attack by seed-boring bruchid beetles (Coe & Coe 1987, Leistner 1961, 1990). The following species, all of the Leguminosae, are prominent canopy formers in the KTV: *Acacia mellifera* subsp. *detinens*, *A. luederitzii* var.

luederitzii, *A. hebeclada* subsp. *hebeclada* and *Dichrostachys cinerea* subsp. *africana*. Species of *Prosopis* introduced from the Americas for their edible pods, and now aggressive invaders in farming areas in the KTV (Harding & Bate 1991, Henderson 1991), also belong in this group. The ability to develop rhizomes and suckers provides some species in this category with an added advantage in competing in the grass matrix; they include *A. mellifera* subsp. *detinens*, *A. hebeclada* var. *stolonifera* and *Prosopis* species (Johnston 1962).

Fructivore-dispersed canopy formers: as could be expected, species in this group are mostly part of the "second line of attack" on the matrix. Their seeds are carried by birds and other animals, such as antelopes, black-backed jackal and other members of the Canidae, into sites in which the grass sward has been disturbed, where soil conditions have often been changed by grazing and burrowing animals (Dean & Milton 1991), and where the shading by woody species results in less extreme microclimatic conditions. Most of the canopy formers in this group are here considered part of the subcanopy flora. They include *Ehretia rigida*, two species of *Grewia, G. flava* and *G. retinervis*, *Ziziphus mucronata* and two species of *Lycium*, *L. austrinum* and *L. hirsutum*. *Boscia albitrunca*, found throughout the KTV, also belongs here. It is not listed as part of the subcanopy flora because adult specimens are often encountered in the matrix, even though germination, and at least early development, usually, if not always, take place in the subcanopy. This also explains why *Ehretia rigida* and *Grewia retinervis* are recorded only once in the list of subcanopy species.

Wind-dispersed canopy formers: with the exception of *Terminalia sericea* which occurs in patches in the moister parts of the KTV, species in this group are only prominent in the arid parts of the region which merges into the Nama-Karoo (Rutherford & Westfall 1994). The most common representative is *Rhigozum trichotomum* which usually grows in low-lying areas with relatively firm sand, often overlying calcrete. With its dense network of suckering rhizomes it reproduces mainly vegetatively and thus competes very effectively with the sparse grass matrix of the region (Leistner 1967, Moore 1989). In overgrazed or otherwise disturbed matrix it is an aggressive invader (Mostert et al. 1971). It is seldom found in the subcanopy. Another species in this group is *Catophractes alexandri* which can also spread by means of rhizomes (Cole 1986).

Subcanopy

Floristics: the 134 species listed as the subcanopy flora in the KTV obviously do not constitute a flora in a formal sense but are merely an assemblage of species that finds a niche in this special habitat created by a relatively small number of woody species. They are mostly elements of the Sudano-Zambezian Region or they belong to a group of pantropical or cosmopolitan weeds. Numerous species especially in Amaranthaceae, Chenopodiaceae and Compositae are well-known ruderals.

In the following discussion, frequent reference will be made to Bews (1917) and Mostert (1958) who reported on more mesic savanna to the east of the KTV, Bews from Natal and Mostert from the central Orange Free State.

Comparison of the subcanopy flora to others in and near the KTV (Table 2) shows up some striking differences in the representation of families. Leguminosae, family of most of the canopy formers and one of the largest in the matrix, is almost totally absent in the subcanopy. It could be speculated that this is due to the high light requirements of local members of the family, perhaps in combination with factors such as the high nitrogen content of the soil often found in the subcanopy (Belsky et al. 1993, Perkins & Thomas 1993). Bews (1917) found that Leguminosae and Compositae are not common under thorntrees and commented that they seem to prefer full sunlight. Mostert (1958) reported that not a single species of Leguminosae was found in the shade-loving community.

Bulbous plants were very rare in the subcanopy of the KTV. Liliaceae (excluding Asparagaceae), one of the "Big Four" in the Sudano-Zambezian Region (Ross 1973, Werger 1973) was not recorded. By contrast, Bews (1917) reported numerous bulbous plants, all Liliaceae (sensu Dyer 1976) under thorntrees although he found them more characteristic of the open veld. Examples were species of *Ornithogalum, Drimia, Albuca* and *Scilla*. Mostert (1958) found *Dipcadi viride* and species of *Ornithogalum* most frequently in the subcanopy. Bews (1917) also refers to several other genera or species listed from the KTV, such as *Asparagus (Protasparagus)*, succulents and climbers of the Asclepiadaceae and members of Acanthaceae, Labiatae and Verbenaceae. None of the numerous ferns which he recorded were seen in the KTV subcanopy. Mostert (1958) mentions several ruderals also listed for the KTV such as species of *Chenopodium, Lepidium, Achyranthes, Cineraria* and *Atriplex*.

The four most frequently recorded species in the subcanopy of the KTV are all widely distributed in Africa or even beyond. They are all dispersed mainly by animals, three of them by birds: *Pollichia campestris* is almost ubiquitous, *Pavonia burchellii* (part of a species complex extending as far as Malaysia) almost equally common, *Lantana rugosa* with red to purplish berries, and the ruderal *Achyranthes aspera* with its sharply pointed perianth that becomes attached to passing animals.

The two best known species in the subcanopy flora of the KTV, also mentioned by Mostert (1958), are the grasses *Setaria verticillata* and *Panicum maximum*. *Setaria verticillata*, the ruderal bur bristle grass (Gibbs Russell et al. 1990), best known for its nuisance value, does provide some grazing and may leave dense masses of highly inflammable matter which, when set alight, may cause the demise of the host tree (Van der Walt & Le Riche 1984). Davidse (1986) questioned the dispersal value of the bristles in *Setaria* since the spikelets disarticulate above the bristles. However, he may have overlooked the fact that the entire inflorescence usually becomes entangled in the pelage of the dispersal agent. *Panicum maximum* has been described by Acocks (Zacharias 1990) as "one of our very best sweet grasses which should be found in abundance in every patch of bush and under every tree within its range". Scholes & Walker (1993) refer to it as the classic subcanopy species of southern African savannas. They also mention that it has a higher productivity in full sunlight than in shade but that the establishment of seedlings can apparently only occur in a shaded environment.

Vegetation development: Bews (1917) described how the plants of the subcanopy may eventually outcompete the canopy tree under which they grew up. This was not observed in the KTV, at least not in the case of trees. Chamaephytes, on the other hand, a life form apparently not well adapted to conditions on deep sandy soil (Leistner 1967, Rutherford & Westfall 1994) are readily ousted by perennial grasses that have developed in their protection.

Distribution and endemism: Table 3 shows that species of the subcanopy tend to be widely to very widely distributed. Even though the subcanopy habitat has existed as long as the savanna, it appears that few, if any, species have evolved to exploit this habitat. Perhaps *Panicum maximum* comes nearest to this model. The most common species, *Pollichia campestris*, is not confined to the subcanopy throughout its range. Even in the somewhat moister adjoining areas in Botswana it frequently grows in the open (Story 1958). The same can be assumed for the majority of subcanopy species. Just as the mesic species in a riverine forest can follow the course of a river, or even a superficially dry riverbed, into arid regions, so can plants of the subcanopy "island-hop" into the driest savanna.

Life form: Succulents scored very highly in the life form spectrum of the subcanopy (Table 4). These plants generally have a very shallow root system which could be laid bare by a single storm on the loose sandy substrate of the KTV (Leistner 1967). Succulents are therefore found mainly in the subcanopy where soils tend to be firmer and stablized by the woody canopy former. This applies in particular to many-stemmed shrubs under which most of the succulent Stapeliae (Asclepiadaceae) were recorded.

The low percentage of chamaephytes in the subcanopy, compared to those in spectra 2, 3 and 4 (Table 4), can perhaps be connected with the adverse effect of trampling by large animals. Spectra 2 and 3 are furthermore both from areas which merge into the Nama-Karoo Biome in which the substrate tends to be firmer and more favourable for the development of chamaephytes (Rutherford & Westfall 1994). The Ghanzi area (spectrum 4) is also rich in calcrete outcrops with a firm substrate (Cole 1986).

The remarkably low count of geophytes in the subcanopy can perhaps also be partly attributed to the trampling by large animals and perhaps to the rich rodent fauna of the region. Furthermore, most geophytes in the Kalahari region are found on fine compact soil (Leistner 1967) which is not commonly found under trees.

The high percentage of therophytes approaches that of typical desert regions. It can probably be related to the unstable disturbed conditions in the subcanopy which favour short-lived, often weedy species with a high seed production.

Seed dispersal: Bews (1917) remarked that wind played a relatively unimportant role in the dispersal of "Thorn Veld" species and that birds were the chief dispersal agent. This certainly also applies to the 20 most common species of the subcanopy in the KTV (Table 5). The position is already different when 40 most common species are considered, and reversed when all 134 species are taken into account. Bews (1917), however, apparently included all fleshy fruit transported by animals under "birds" and confined his statement to the woody component of the "Thorn Veld".

A major source of uncertainty in the calculations for Table 5 is our poor knowledge of the dispersal of many of the grasses. In the case of dominant perennial matrix species sexual reproduction may be of minor importance, but for species largely confined to the subcanopy it is obviously of major significance. It is known that endozoochorous dispersal of grass takes place (Janzen 1984, Milton 1992), but to what extent rainwater and wind play a role is open to speculation.

Even considering these uncertainties, it can be postulated that animals are the most important dispersal agent for the subcanopy flora as a whole. The most common species of that habitat are almost exclusively dispersed by animals, especially birds.

Conclusions

Mainly large herbivores carry the seed of the pioneer canopy formers into the grassland matrix of the savanna. At the same time they may cause disturbances that reduce the competition by perennial grasses and promote the development of the seedlings. These pioneer canopy formers, which are also the "climax" species, mostly belong to the Leguminosae, a family almost absent from the subcanopy of the KTV.

Once a canopy has been created, a second group of woody species, mostly with fleshy fruit, is introduced by birds and other frugivores. Most of these species, once established, can compete in the matrix, at least in the more mesic parts of the KTV. Microclimatic and edaphic conditions in the subcanopy generally differ significantly from those in the matrix. As a result, many species, not adapted to conditions in the matrix, find a niche in the savanna and enrich its biodiversity. At the same time the fauna changes and destabilizes the subcanopy by activities such as trampling, feeding, excreting and burrowing.

As could be expected, the subcanopy flora of 132 species, one subspecies and one variety is a heterogeneous assemblage of species, not a flora in the strict sense. Floristically it differs markedly from the total flora of the KTV. The majority of its species are widely to very widely distributed, and the rate of endemism is very low. *Panicum maximum* is singled out as a species that may partly have evolved under subcanopy conditions, a habitat that is as old as the savanna itself. Succulents, hemicryptophytes and especially annuals, many of them ruderals, are well represented in the life form spectrum of the subcanopy. Geophytes are almost absent. Animals, particularly birds and other frugivores, were found to be the most important dispersal agents.

Acknowledgements

I dedicate this paper to the memory of John Acocks, mentor and often my co-author. I thank Joan Mulvenna for her very special effort to get the manuscript onto paper and diskette.

References

Acocks, J.P.H., 1988. Veld types of South Africa, 3rd edn. Memoirs of the Botanical Survey of South Africa No. 57.

Arnold, T.H. & B.C. de Wet, 1993. Plants of southern Africa: names and distribution. Memoirs of the Botanical Survey of South Africa No. 62.

Belsky, A.J., S.M. Mwonga, R.G. Amundson, J.M. Duxbury & A.R. Ali, 1993. Comparative effects of isolated trees on their undercanopy environments in high- and low-rainfall savannas. Journal of Applied Ecology 30: 143-155.

Bews, J.W., 1917. The plant succession in the Thorn Veld. South African Journal of Science 14: 163-172.

Bond, W.J. & P. Slingsby, 1983. Seed dispersal by ants in shrublands of the Cape Province and its evolutionary implications. South African Journal of Science 79: 231-233.

Bothma, J. du P., 1966. Notes on the stomach contents of certain Carnivora (Mammalia) from the Kalahari Gemsbok National Park. Koedoe 9: 37-40.

Coe, M. & C. Coe, 1987. Large herbivores, acacia trees and bruchid beetles. South African Journal of Science 83: 624-635.

Cole, M.M., 1986. The savannas. Biogeography and geobotany. Academic Press, London.

Davidse, G., 1986. Fruit dispersal in the Poaceae. In: Soderstrom, T.R. et al. (eds.), Grass systematics and evolution: 143-155, Smithsonian Institution Press, Washington.

Dean, W.R.J. & S.J. Milton, 1991. Patch disturbances in arid grassy dunes: antelope, rodents and annual plants. Journal of Arid Environments 20: 231-237.

Dean, W.R.J., S.J. Milton & W.R. Siegfried, 1990. Dispersal of seeds as nest material by birds in semiarid Karoo shrubland. Ecology 71: 1299-1306.

Dyer, R.A., 1976. The genera of southern African flowering plants vol. 2. Department of Agricultural Technical Services, Pretoria.

Fenner, M., 1985. Seed Ecology. Chapman & Hall, London.

Gibbs Russell, G.E., L. Watson, M. Koekemoer, L. Smook, N.P. Barker, H.M. Anderson & M.J. Dallwitz, 1990. Grasses of southern Africa. Botanical Survey Memoirs, South Africa, No. 58.

Giess, W., 1971. A preliminary vegetation map of South West Africa. Dinteria 4: 1-114.

Harding, G.B. & G.C. Bate, 1991. The occurrence of invasive Prosopis species in the north-western Cape, South Africa. South African Journal of Science 87: 188-192.

Henderson, L., 1991. Invasive alien woody plants of the northern Cape. Bothalia 21: 177-189.

Huntley, B.J., 1982. Southern African savannas. In: Huntley, B.J. & B.H. Walker (eds.), Ecology of tropical savannas, Ecological studies 42: 101-119, Springer-Verlag, Berlin, Heidelberg, New York.

Isaacs, M.J.H., F.M. Weitz & C.T. Johnson, 1993. Seed-coat characteristics of selected southern African species of Polygala L. (Polygalaceae). South African Journal of Botany 59: 592-596.

Janzen, D.H., 1984. Dispersal of small seeds by big herbivores: foliage is the fruit. American Naturalist 123: 338-353.

Johnston, M.C., 1962. The North American mesquites Prosopis Sect. Algarobia (Leguminosae). Brittonia 14: 72-90.

Leistner, O.A., 1961. On the dispersal of Acacia giraffae by game. Koedoe 4: 101-104.

Leistner, O.A., 1967. The plant ecology of the southern Kalahari. Botanical Survey Memoir No. 38. Government Printer, Pretoria.

Leistner, O.A., 1990. Verbreitung der Samen der Giraffen-Akazie. Naturwissenschaftliche Rundschau 43: 18-19.

Maclean, G.L., 1970. An analysis of the avifauna of the southern Kalahari Gemsbok National Park. Zoologia Africana 5: 249-273.

Milton, S.J., 1992. Plants eaten and dispersed by leopard tortoises Geochelone pardalis (Reptilia: Chelonii) in the southern Karoo. South African Journal of Zoology 27: 45-49.

Milton, S.J., W.R. Siegfried & W.R.J. Dean, 1990. The distribution of epizoochoric plant species: a clue to the prehistoric use of arid Karoo rangelands by large herbivores. Journal of Biogeography 17: 25-34.

Moore, A., 1989. Die ekologie en ekofisiologie van Rhigozum trichotomum (Driedoring). Ph.D. thesis, University of Port Elizabeth, South Africa.

Mostert, J.W.C., 1958. Studies of the vegetation of parts of the Bloemfontein and Brandfort Districts. Memoirs of the Botanical Survey of South Africa No. 31.

Mostert, J.W.C., B.R. Roberts, C.F. Heslinga & P.G.F. Coetzee, 1971. Veld management in the Orange Free State Region. Department of Agricultural Technical Services (Pretoria) 391: 1-97.

Perkins, J.S. & D.S.G. Thomas, 1993. Environmental responses and sensitivity to permanent cattle ranching, semi-arid western central Botswana. In: Thomas, D.S. & R.J. Allison (eds.), Landscape sensitivity: 273-286, John Wiley & Sons, London.

Raunkiaer, C., 1934. The life forms of plants and statistical plant geography. Clarendon Press, Oxford.

Ross, J.H., 1973. An analysis of the Flora of Natal. Bothalia 11: 103-106.

Rutherford, M.C. & R.H. Westfall, 1986. Biomes of southern Africa, 2nd edn. Memoirs of the Botanical Survey of South Africa No. 54.

Rutherford, M.C. & R.H. Westfall, 1994. Biomes of southern Africa. Memoirs of the Botanical Survey of South Africa No. 63.

Scholes, R.J. & B.H. Walker, 1993. An African savanna. Cambridge University Press.

Smith, T.M. & S.E. Shackleton, 1988. The effects of shading on the establishment and growth of Acacia tortilis seedlings. South African Journal of Botany 54: 375-379.

Smithers, R.H.N. 1983. The mammals of the Southern African Subregion. University of Pretoria, Pretoria.

Story, R., 1958. Some plants used by the bushmen in obtaining food and water. Memoirs of the Botanical Survey of South Africa No. 30.

Van der Pijl, L., 1982. Principles of dispersal in higher plants, 3rd edn. Springer-Verlag, Berlin, Heidelberg, New York.

Van der Walt, P.T. & E.A.N. le Riche, 1984. The influence of veld fire on an Acacia erioloba community in the Kalahari Gemsbok National Park. Koedoe Supplement 1984: 103-106.

Van Rooyen, N., D.J. van Rensburg, G.K. Theron & J. du P. Bothma, 1988. A check list of flowering plants of the Kalahari Gemsbok National Park. Koedoe 31: 115-135.

Verkerke, W., 1985. Ovules and seeds of the Polygalaceae. Journal of the Arnold Arboretum 66: 353-394.

Weare, P.R. & A.M. Yalala, 1971. Provisional vegetation map of Botswana. Botswana Notes and Records 3: 131-147.

Werger, M.J.A. 1973. Notes on the phytogeographical affinities of the southern Kalahari. Bothalia 11: 177-180.

Werger, M.J.A., 1986. The Karoo and southern Kalahari. In: Evenari, M., I. Noy-Meir & D.W. Goodall (eds.), Hot deserts and arid shrublands, B, Ecosystems of the world 12 B.

White, F., 1983. The vegetation of Africa: a descriptive memoir to accompany the UNESCO/ AETFAT/UNSO vegetation map of Africa by F. White. UNESCO, Paris.

Zacharias, P.J.K. (compiler), 1990. Acocks' Notes: key grasses of South Africa. Grassland Society of Southern Africa.

Plant species diversity of West African inselbergs

S. POREMBSKI & W. BARTHLOTT
Botanical Institute, University of Bonn, Meckenheimer Allee 170, D-53115 Bonn, Germany.

Abstract
A survey of the diversity of the West African inselberg vegetation is given. Inselbergs are geologically old, often dome-shaped monoliths. Their vegetation is in marked contrast with those of the surroundings. Climatic conditions on them are severe with extremes of temperature and insolation. Among the c. 600 vascular plant species on Ivorian inselbergs Gramineae, Cyperaceae and Papilionoideae are dominating. The ephemeral flush vegetation is the most species-rich community whereas *Afrotrilepis pilosa*-mats are the least diverse. All over the Ivory Coast inselberg vegetation is relatively uniform (low beta diversity). Stochastic influences dominate in small-sized habitats like rock pools leading to higher beta diversity. Species diversity of inselbergs decreases from the savanna towards the rain forest zone. In the drier north of the Ivory Coast less favourable growth conditions prevent a state of community equilibrium allowing the coexistence of weak competitors.

Résumé
Une description comparative de la diversité de la végétation des inselbergs de l'Afrique de l'ouest est donnée. Les inselbergs sont très vieux au point de vue géologique et ce sont souvent des dômes crystallins. Leur végétation se dinstingue nettement de la végétation adjacente. Les conditions climatiques sont extrêmes en ce qui concerne la température et l'ensoleillement. Parmi les 600 espèces de plantes vasculaires, les Gramineae, Cyperaceae et Papilionoideae sont dominantes sur les inselbergs. La végétation des sources marécageuses est la plus riche en nombre d'espèces, les tapis d'*Afrotrilepis pilosa*, par contre, constituent l'habitat le plus pauvre en nombre d'espèces. Dans toute la Côte d'Ivoire, la végétation des inselbergs est relativement homogène (beta-diversité basse). Les influences stochastiques sont dominantes dans les types d'habitat de petit taille comme les mares sur dalle rocheuse avec une beta-diversité plus grande. La diversité des espèces des inselbergs diminue des savanes vers la zone des forêts ombrophiles. Dans la région sèche du nord de la Côte d'Ivoire le stade d'équilibre des communautés ne peut être atteint en raison des conditions plus contraignantes pour les plantes, ce qui permet la coexistance de compétiteurs faibles.

Key words
Biodiversity, Inselbergs, Metapopulations, Rock outcrops, Species richness, West Africa.

Introduction

Being isolated rock outcrops which abruptly rise from surrounding plains, inselbergs are the terrestrial counterparts of oceanic islands. Geomorphologically, inselbergs have been studied in detail (Bremer & Jennings 1978), but their vegetation has been largely neglected. Occurring all over the tropics they are concentrated on the old crystalline shields. In West Africa, they mainly consist of Precambrian granites and gneisses (Leneuf 1959, Wilson 1968). Apart from dome-shaped inselbergs with an absolute height of more than 300 m, small rock outcrops occur hardly protruding 5 m above the surrounding landscape (Figure 1).

Due to extreme ecological conditions ("xeric islands") the vegetation of inselbergs differs markedly from their surroundings. Apart from a preliminary survey (Barthlott et al. 1993) there are no comprehensive accounts of inselberg vegetation on global

L.J.G. van der Maesen et al. (eds.), The Biodiversity of African Plants, 180-187.

scale. West African inselberg vegetation has been dealt with in several regional studies (e.g., Adjanohoun 1964, Bonardi 1966, Hambler 1964, Porembski & Barthlott 1992, 1993, Porembski et al. 1994, Reitsma et al. 1992, Richards 1957, Villiers 1981). Many studies dealing with the relationship between species richness and environmental gradients exist. A recent survey on this topic was provided by Tilman & Pacala (1993). In the following we present an account on species richness of the West African inselberg vegetation along a gradient from the savanna towards the rain forest zone. Inselbergs have not been considered in this way apart from a study concerning species richness of shallow depressions on inselbergs (Porembski et al. 1994).

Study area

Fieldwork was carried out in the Ivory Coast and Guinea, thereby covering a wide range of different climatic zones. The south of the Ivory Coast has an humid equatorial climate ("Af" after the Köppen classification) with an annual rainfall of 1800-2300 mm and a dry season of three months. The northern part has a seasonal climate ("Aw") with a dry period of up to seven months. Annual rainfall lies between 900 and 1600 mm with considerable variation from year to year. A detailed account of the climate of the Ivory Coast was provided by Eldin (1971). In northern Guinea rainfall declines from 1600 to 1250 mm/year with a dry season from four to seven months in the north-east. Within the lowland rain forest zone of southern Guinea rainfall varies between 2500 and 3000 mm/year with three months (December to February) below 100 mm precipitation. The highlands of the Fouta Djalon are characterized by a dry season of five months, cooler temperatures, frequent occurrence of mists and precipitation between 1400 and 1800 mm/year (Griffiths 1972).

Edaphically and microclimatologically inselbergs form dry azonal sites even under a rain forest climate. They are characterized by extremes of water availability, temperature (often reaching more than 50° C) and insolation. The soils on inselbergs are rarely more than 15 cm in depth. Typical is their coarse pore volume, leading to high water conductivity and rapid drying out of the soil. The proportion of clay particles is very low, indicating the poor nutrient status of the soils. Inselbergs occur in all parts of the Ivory Coast and Guinea. However, they are concentrated in the central and northern regions. Inselbergs were studied in all major vegetation zones of the Ivory Coast and Guinea (from Sudanian woodland in the north to the rain forests in the south).

Floristic composition and species richness of the West African inselberg vegetation

Floristic composition and habitats

In the following emphasis is laid upon data obtained in the Ivory Coast. Concerning the vegetation of inselbergs in Guinea it is referred to Porembski et al. (1994).

Physiognomically the vegetation of inselbergs can be classified into several community types (after Barthlott et al. 1993): cryptogamic crusts (cyanobacteria, lichens) on exposed rock faces, drainage channels, rock crevices and boulder-falls, rock pools (waterfilled during the rainy season, Figure 2), shallow depressions (filled with debris), *Afrotrilepis pilosa*-mats (Figure 3), ephemeral flush vegetation (Figure 4), wet flush vegetation, belt forests adjacent to inselbergs, and semi-deciduous forests on inselbergs.

Figure 1. Dome-shaped granite inselberg. Ivory Coast, south of Séguéla.

Figure 2. Rock pool on Ivorian inselberg. The Commelinaceae *Cyanotis lanata* is a characteristic species of this community.

Figure 3. *Afrotrilepis pilosa*-mats are the least diverse community on Ivorian inselbergs.

Figure 4. Ephemeral flush communities constitute the most diverse vegetation type on Ivorian inselbergs. Their species richness is reaching a maximum in the savanna zone. Typical are *Utricularia* spp., *Drosera indica*, Gramineae and Cyperaceae.

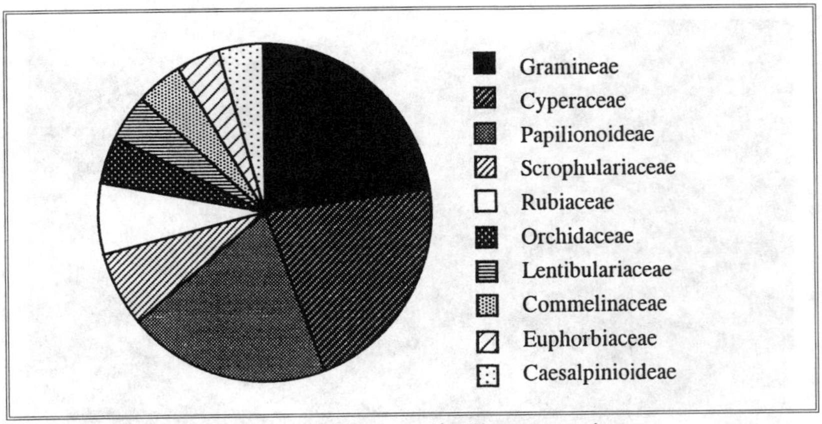

Figure 5. Floristic composition of the Ivorian inselberg vegetations.

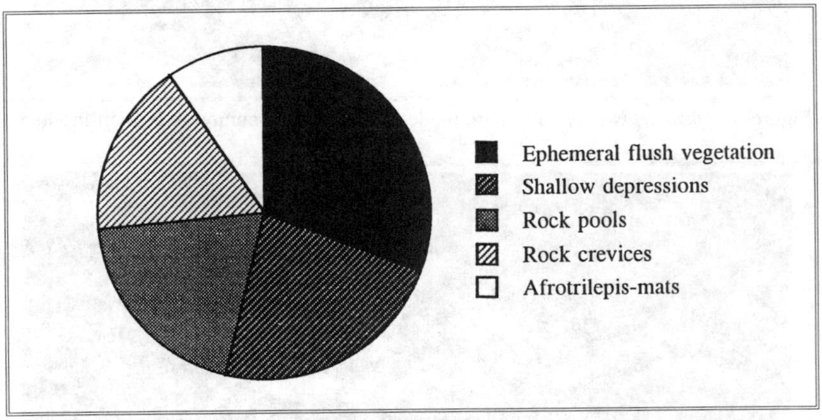

Figure 6. Species number per habitat on Ivorian inselbergs.

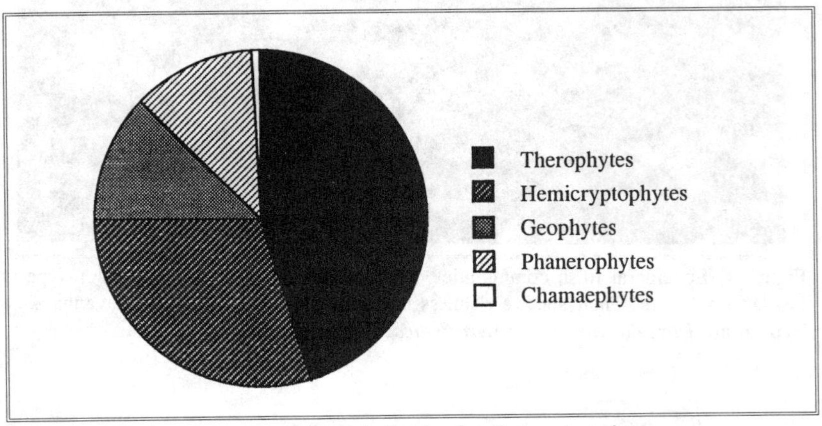

Figure 7. Life forms represented within Ivorian inselberg vegetations.

Since herbaceous communities are the most characteristic components of the vegetation of inselbergs we concentrate on them. Nearly 600 species of vascular plants were recorded on Ivorian inselbergs. Gramineae is the most species-rich family followed by Cyperaceae, Leg.-Papilionoideae and Scrophulariaceae (Figure 5). The ten most species-rich families show the following habitat preferences on inselbergs:

Gramineae (72 species):	ephemeral flush communities, shallow depressions, rock crevices
Cyperaceae (68 species):	rock pools, ephemeral flush communities, *Afrotrilepis pilosa*-mats, rock crevices
Papilionoideae (62 species):	rock crevices, shallow depressions
Scrophulariaceae (23 species):	ephemeral flush communities, rock pools, shallow depressions
Rubiaceae (22 species):	rock crevices, rock pools, shallow depressions
Orchidaceae (15 species):	*Afrotrilepis pilosa*-mats, ephemeral flush communities
Lentibulariaceae (14 species):	ephemeral flush communities, wet flush communities
Commelinaceae (14 species):	shallow depressions, rock crevices, rock pools
Euphorbiaceae (13 species):	rock crevices, *Afrotrilepis pilosa*-mats
Caesalpinioideae (13 species):	shallow depressions, rock crevices

Relatively poor in species but very characteristic for inselbergs are families like Eriocaulaceae, Xyridaceae and Burmanniaceae which mainly occur within ephemeral flush communities. Only a few species (e.g. *Polystachya microbambusa* Kraenzl.) seem to be narrowly restricted to inselbergs as growing sites. About one sixth of all the species found (e.g. *Cyanotis lanata* Benth., *Afrotrilepis pilosa* (Boeck.) J. Raynal, *Lindernia exilis* (Skan) Philcox) occurs mainly on them. Close floristic affinities exist with certain other azonal sites: rock outcrops in general (e.g. ferricretes), poor soils in grassland, seasonally wet ground around ponds. Plants preferably found on fallow land and along roadsides, like *Chamaecrista mimosoides* (L.) Greene are richly represented. For ruderals inselbergs can be considered as natural growing sites.

Due to the harsh ecological conditions on inselbergs, specialists (e.g. poikilohydric species like *Afrotrilepis pilosa*, *Pellaea doniana* Hook.) which cope successfully with extremes of temperature and drought are predominant. Therefore there is only minor variation (low beta-diversity) between the inventories of individual inselbergs. Much more variation (high beta-diversity) can be observed if habitats like rock pools are considered. These are largely influenced by stochastic and sometimes catastrophic events (e.g. unpredictable droughts, chance establishment) leading to an unpredictable inventory.

Habitats and life forms

The ephemeral flush vegetation is the most species-rich community on inselbergs (Figure 6) with Gramineae, Cyperaceae, Lentibulariaceae and Eriocaulaceae as dominating families. Shallow depressions, rock pools and rock crevices follow next, *Afrotrilepis pilosa*-mats are the least diverse community. Ephemeral flush communities characteristically occur at the feet of rocky slopes where water seeps continuously during the rainy season. Sometimes covering not more than a few square meters, more than 30 species may be present with up to half a dozen species of *Utricularia* (e.g. *U. micropetala* Sm., *U. pubescens* Sm., *U. subulata* L.) in a single locality. At the other extreme of species richness *Afrotrilepis pilosa*-mats often occur as island-like patches on precipitous flanks. Obviously highly competitive under these environmental conditions, *Afrotrilepis pilosa* is accompanied by only a few other species (e.g. *Asplenium stuhlmannii* Hieron., *Fimbristylis dichotoma* Vahl).

Regarding life forms therophytic species are most prominent followed by hemicryptophytes and geophytes (Figure 7). In particular in ephemeral flush communities and rock pools therophytes are very abundant, not only in species number but also in terms of individuals.

Species richness along a latitudinal gradient

From the savanna zone in the north towards the rain forests in the south of the Ivory Coast there is a marked change in vegetation types with highest plant diversity in the rain forests (Lebrun 1976).

Apart from floristic differences between savanna and rain forest inselbergs (e.g. "lichen-type" inselbergs in savannas with rock faces covered by lichens versus "cyanobacteria-type" in the rain forest zone (Porembski & Barthlott 1992), there is a latitudinal gradient in species richness. Surprisingly species richness is highest on inselbergs in the savanna zone, and sinking to a minimum in the rain forest region. This inverted diversity gradient is probably due to the fact that species on savanna inselbergs are less isolated because of the existence of ecologically similar azonal habitats (most of all ferricretes). Therefore larger metapopulations (local populations linked by dispersal, Gilpin & Hanski 1991) persist preventing regional extinction which results in a higher species diversity on savanna inselbergs. Moreover the higher probability of unpredictable climatic disturbances in the northern part of the Ivory Coast prevents equilibrium conditions with a few highly competitive species dominating thus allowing less competitive taxa to coexist. It may be supposed that better growth conditions on inselbergs in the rain forest zone lead to equilibrium conditions in most communities resulting in a generally lower species richness. Furthermore it can be expected that species richness of the West African inselberg vegetation as a whole peaks in savanna zones, that is in areas where azonal sites are richly represented. In this regard a promising area for future research would be the Nigerian Jos Plateau, which is fairly well-known for the number of succulents which occur on rock outcrops in this region (Newton 1992).

Acknowledgements

We gratefully acknowledge financial support by the Deutsche Forschungsgemeinschaft (Ba 605/4-2). The authorities and institutions of the Ivory Coast (Direction Général des Eaux et Forêts, IET) and of Guinea (Direction de la Recherche Scientifique, Conakry) are thanked for their permission to conduct research. Prof. L. Aké Assi (Centre National Floristique de l'Université d'Abidjan) kindly provided plant determinations. Furthermore we like to express our thanks to various specialists for the identification of plant material. Field work in the Ivory Coast was supported by Prof. K. E. Linsenmair (Würzburg). For assistance in the field we like to thank N. Biedinger, S. Dörrstock, J.-P. Mund, R. Seine and J. Szarzynski (all Bonn).

References

Adjanohoun, E., 1964. Végétation des savanes et des rochers découverts en Côte d'Ivoire centrale. Mémoires ORSTOM 7.

Barthlott, W., A. Gröger & S. Porembski, 1993. Some Remarks on the Vegetation of tropical Inselbergs: Diversity and Ecological Differentiation. Biogéographica 69: 17-36.

Bonardi, D., 1966. Contribution à l'étude botanique des inselbergs de Côte d'Ivoire forestière. Dipl. Et. sup., Abidjan.

Bremer, H. & J. Jennings (eds.), 1978. Inselbergs/Inselberge. Z. Geomorph. N.F., Suppl. 31.

Eldin, M., 1971. Le Climat. In: Le milieu naturel de la Côte d'Ivoire: 73-108. Mémoires ORSTOM 50.

Gilpin, M.E. & I. Hanski, 1991. Metapopulation dynamics: Empirical and theoretical investigations. Cambridge University Press, Cambridge.

Griffiths, J.F., 1972. Climates of Africa. In: Landsberg, H.E. (ed.): World Survey of Climatology. Vol. 10. Elsevier, Amsterdam.

Hambler, D.J., 1964. The vegetation of granitic outcrops in Western Nigeria. J. Ecol. 52: 573-594.

Lebrun, J.P., 1976. Richesses spécifiques de la flore vasculaire des divers pays ou régions d'Afrique. Candollea: 11-15.

Leneuf, N., 1959. L'altération des granites calco-alcalins et des granodiorites en Côte d'Ivoire forestière et des sols qui en sont dérivés. PhD thesis, Paris.

Newton, L.E., 1992. An annotated and illustrated checklist of the succulent Euphorbias of West Tropical Africa. Euphorbia J. 8: 112-122.

Porembski, S. & W. Barthlott, 1992. Struktur und Diversität der Vegetation westafrikanischer Inselberge. Geobot. Kolloq. 8: 69-80.

Porembski, S. & W. Barthlott, 1993. Ökogeographische Differenzierung und Diversität der Vegetation von Inselbergen in der Elfenbeinküste. In: Barthlott, W., C.M. Naumann, K. Schmidt-Loske & K.-L. Schuchmann (eds.), Animal-Plant-Interactions in Tropical Environments: 149-158. Results of the Annual Meeting of the German Society for Tropical Ecology, Bonn 1992.

Porembski, S., W. Barthlott, S. Dörrstock & N. Biedinger, 1994. Vegetation of rock outcrops in Guinea: granite inselbergs, sandstone table mountains and ferricretes - remarks on species numbers and endemism. Flora 189: 315-326.

Porembski, S., G. Brown & W. Barthlott, 1995. An inverted latitudinal gradient of plant diversity in shallow depressions on Ivorian inselbergs. Vegetatio 117: 151-163.

Reitsma, J.M., A.M. Louis & J.J. Floret, 1992. Flore et végétation des inselbergs et dalles rocheuses: premiere étude au Gabon. Adansonia 14: 73-97.

Richards, P.W., 1957. Ecological notes on West African Vegetation I. The plant communities of the Idanre Hills, Nigeria. J. Ecol. 45: 563-577.

Tilman, D. & S. Pacala, 1993. The maintenance of species richness in plant communities. In: Ricklefs, R.E. & D. Schluter (eds.), Species diversity in ecological communities. University of Chicago Press, Chicago: 13-25.

Villiers, J.F., 1981. Formations climaciques et relictuelles d'un inselberg inclus dans la forêt dense camerounaise. Thèse de Doctorat d'Etat, Université Paris 6.

Wilson, A. F., 1968. Geological report on granites in West Africa. In: Proceedings of the symposium on the granites of West Africa: 123-147. UNESCO, Paris.

Biodiversity in three traditional groves in the Guinea savanna, Ghana

A.A. OTENG-YEBOAH
Department of Botany, University of Ghana, P.O. Box 55, Legon, Ghana.

Abstract
Biodiversity estimates in three traditional groves in the Guinea savanna of northern Ghana have been given. There is an indication of a lower species diversity within than outside the core of the groves. Aspects of life form composition indicate the preponderance of therophytic life forms, particularly outside the core. Thirteen plant (sub-)families, namely: Caesalpinioideae, Combretaceae, Compositae, Euphorbiaceae, Malvaceae, Mimosoideae, Moraceae, Papilionoideae, Rubiaceae and Tiliaceae in the class Dicotyledonae and Araceae, Cyperaceae and Gramineae in the class Monocotyledonae were noted to contribute most species and genera to the flora of the three groves.

Résumé
Une estimation de la biodiversité dans trois bois sacrés de la savane guinéenne dans le nord du Ghana a été réalisée. Il en ressort que la diversité des espèces est moindre au sein de ces bois qu'en dehors. L'analyse de la composition des formes biologiques indique une prépondérance des thérophytes, principalement en dehors du centre. Treize (sous-)familles sont particulièrement représentatives dans la flore de ces trois bois sacrés: ce sont pour les Dicotylédones, les Caesalpinioideae, Combretaceae, Compositae, Euphorbiaceae, Malvaceae, Mimosoideae, Moraceae, Papilionoideae, Rubiaceae et Tiliaceae et pour les Monocotylédones, les Araceae, Cyperaceae et Gramineae.

Key words
Life form categories, Sacred groves, Species diversity.

Introduction

One form of vegetation protection at the indigenous community level in Africa, Asia and South America is the concept of sacred or traditional groves. These are small patches of forest, normally close to settlements, that are managed through local bye-laws, customs, rites and taboos. Several forms of sacred groves exist, and these have been reviewed by Ntiamoa-Baidu (1993) for the Biodiversity Support Program. Altogether, the groves are considered as the abode of either the dead ancestors (*nsamanpow*), or the spiritual guardian(s) (*abosompow*), or in other situations the sacred, totem or tabooed organisms (plant or animal) of the community.

There are an estimated 1453 sacred groves in Ghana (K. Tufour, pers. comm.) scattered in all vegetation zones of the country. A study of these groves in the Guinea savanna ecological zone was undertaken under the framework of UNESCO - MAB. The study formed part of a multidisciplinary project called "Co-operative Integrated Project on Savanna Ecosystems in Ghana (CIPSEG)", and was undertaken in three traditional groves located in three administrative districts of the northern Ghana region (Oteng-Yeboah 1993).

The rationale of the project was to provide relevant data and information upon which an *in-situ* conservation model for the integrated human-oriented sustainable management of the northern Ghana Guinea Savanna (NGGS) ecosystem will be based.

L.J.G. van der Maesen et al. (eds.), The Biodiversity of African Plants, 188-197.

Table 1. Angiosperm class distribution in the sites.

Class	Number of plant families	Number of plant genera	Number of plant species
Dicotyledonae	55	133	169
Monocotyledonae	14	40	51
Total	69	173	220

In-situ conservation refers to the maintenance of biodiversity in the form of individual species, populations of individual species or the entire community at the same place where their evolution has been a continuous process. In the IUCN (1986) categories of management of the worlds legally protected areas, sacred or traditional groves are not listed directly, though they can be referred to category III, under National Monument or Natural Landmark, which is meant to protect and preserve nationally significant features because of their special interest or unique characteristics.

Detailed scientific knowledge and understanding of the composition, functions and structural dynamics of the habitats/ecosystems are lacking. Also lacking is a complete documentation of the knowledge, possessed by the people of these areas, of the uses of species (Oteng-Yeboah 1994). The understanding of the knowledge and uses of biological material for biological resource development and conservation from the genetic through species to ecosystem levels directly enters into the concept of Biodiversity as defined by IUCN in 1990.

The results presented here will provide an insight into the value of the indigenous conservation methods in protecting habitats of special interest to the population.

Materials and methods

The study was conducted in three traditional groves located in the West Dagomba-Tamale (*Malshegu*), Yolon-Kumbungu (*Tale*) and Savelugu-Nantong (*Yiwogu*) districts in the northern region of Ghana. The abbreviations M, T and Y have been used in the tables to refer to Malshegu, Tale and Yiwogu groves respectively. Malshegu is located at longitude 9°20'N and latitude 0°45'W, Tale at 9°25'N and 1°10'W, and Yiwogu at 9°45'N and 0°50'W. Three general methods of observation were adopted, namely:

1 General survey for visual observation, identification and documentation of plant species encountered.
2 Line transect observation at 5 meter intervals along a transect.
3 Quadrat sampling involving the use of 1 x 1 and 25 x 25 m^2 quadrats.

All methods were repeated within and outside the core of each of the groves, the data in the observations refer to this. Nomenclature follows Hutchinson & Dalziel (1954 & 1972) with adaptations where necessary.

Observations

Families and genera

In total 220 species in 173 genera and 69 families were observed in all three traditional groves (Table 1). Thirteen (sub-)families, 3 Monocotyledonae namely Araceae, Cyperaceae and Gramineae, and 10 Dicotyledonae namely Caesalpinioideae, Combretaceae,

Compositae, Euphorbiacae, Malvaceae, Mimosoideae, Moraceae, Papilionoideae, Rubiaceae and Tiliaceae contribute most species and genera to the flora of the groves. The selection of these was based on a minimum number of 4 species in at least 3 genera for the Monocots group, and a minimum number of 6 species in at least 3 genera in Dicots.

The majority of the species are annuals (Therophyte; ThH). It was also obvious that there were more species present in the Tale grove area than in the other two groves. There were four forms of *Indigofera* in all three groves, but it was not possible to identify these to the species level because of incomplete material.

Distribution of species in the different groves

Table 2 shows the presence/absence analysis of species distributions. The distribution analysis is in four categories referring to the manner and area of observation in relation to the total number of species in all groves or specific to each grove and species occurring in all groves or specific to specific groves.

The total number of species per study site showed Tale with the highest diversity (183) followed by Malshegu (121) and Yiwogu (108). Altogether 66 species were common to all sites. Forty three species from the total 220 species were noted to have local distribution only in Tale. The figures of such species for Malshegu and Yiwogu were much lower. Observations made from line transects, and quadrats of 1 x 1 and 25 x 25 m indicate that there were fewer species within the core of the groves than around the core, and that of all cases Tale grove showed the highest species diversity.

Life form composition

Analysis of life form category distribution is shown in Table 3. The herbaceous life form was separated into 6 groups and the shrubby life form into 4 groups. Woody climbers were distinguished into proper lianas and scrambling nanophanerophytes or shrubs. Altogether 15 life form groups were observed, their total numbers in all groves and their representation by grove with typical examples have been provided.

It is obvious that the therophytic life form including the climbers were predominant representing over 50 % of the entire life form spectrum. The least represented life form groups were hydrophytes, epiphytes and woody climbers (lianas).

Discussion

The thirteen most common (sub-)families listed in terms of their species and genera representation recall the usual floristic character of the savanna ecosystems. The (sub)-families Caesalpinioideae, Cyperaceae, Gramineae, Mimosoideae, Papilionoideae and Rubiaceae have always been noted to provide the largest number of species and individuals in the savanna of West Africa (Hopkins & Stanfield 1966, Keay et al. 1960 & 1964, Le Houérou 1980, Oteng-Yeboah 1985).

It was not surprising that there were more dicotyledonous species, genera and families than monocotyledonous ones, considering the observations above. It was also not surprising that there were more herbaceous (forbs and grasses) life forms i.e. therophytes than any life form group. Because of the nature of the environment which experiences a long spell of drought (a dry season of 6 to 9 months), the therophytic life form is an effective strategy for survival.

Table 2. Presence/absence analysis of species distributions.

Survey method	General Survey			Line Transect		1 x 1 m Quadrat		25 x 25 m Quadrat	
	whole area	within grove	outside grove	within grove	outside grove	within grove	outside grove	within grove	outside grove
Total species in all sites	220	96	166	32	43	79	108	37	47
Total species in Malshegu (M)	121	46	85	14	13	42	59	14	22
Total species in Tale (T)	183	87	138	16	27	37	75	23	43
Total species in Yiwogu (Y)	108	32	83	9	23	26	61	6	22
Species common to all sites	66	9	32	1	3	6	25	1	11
Species common to M and T	33	6	18	2	3	4	12	2	8
Species common to M and Y	9	8	11	3	5	7	9	1	1
Species common to T and Y	17	4	14	1	9	2	11	1	8
Species only distributed in M	7	1	6	8	2	22	5	11	1
Species only distributed in T	43	10	33	10	12	20	18	19	15
Species only distributed in Y	8	2	6	5	8	7	5	3	2

Table 3. Life form category distribution in the sites.

Symbol	Definition	No. represented in all sites	No. represented in each site			Example
			M	T	Y	
ThH	Therophyte Herb	112	65	92	79	*Euphorbia hirta*
ThHl	Therophyte Climbing Herb	7	1	6	1	*Momordica charantia*
ChH	Chamaephyte Herb	5	3	5	3	*Gutenbergia macrocephala*
ChH1	Chamaephyte Climbing Herb	6	3	6	0	*Cissus sp.*
ChS	Chamaephyte Shrub	1	0	1	0	*Urena lobata*
HyH	Hydrophyte Herb	1	0	1	0	*Nymphaea lotus*
GrH	Geophyte Herb	14	9	12	3	*Amorphophallus dracontioides*
MPT	Megaphanerophyte Tree	12	5	10	2	*Adansonia digitata*
mPT	Mesophanerophyte Tree	18	12	17	6	*Lannea acida*
MPS	Mesophanerophyte Shrub	2	1	2	1	*Detarium microcarpum*
mpS	Microphanerophyte Shrub	20	14	16	6	*Heeria insignis*
mpE	Phanerophyte Epihyte	1	1	1	1	*Tapinanthus heteromorphus*
mpWl	Microphanerophyte Liana	2	1	2	0	*Saba sp.*
npS	Nanophanerophyte Shrub	17	9	11	6	*Maytenus senegalensis*
NpWl	Nanophanerophyte Liana	1	0	1	0	*Allophylus africanus*

It was obvious that generally there were fewer species occurrences within the core of the groves than there were outside the core of the groves, and that tree species were dominant in the core of the groves. Common tree species included *Acacia hockii*, *Anogeissus leiocarpa*, *Combretum molle*, *Detarium microcarpum* and *Gardenia ternifolia* for Malshegu; *Annona senegalensis*, *Entada africana* and *Terminalia glaucescens*, *Vitellaria paradoxa* for Tale; and *Azadirachta indica*, *Diospyros mespiliformis*, *Pseudocedrela kotschyi* and *Stereospermum kunthianum* for Yiwogu. Also ten well-known forest floor species were predominant within the core of the groves. These species were *Ageratum conyzoides*, *Amorphophallus dracontioides*, *Biophytum petersianum*, *Commelina africana*, *Justicia flava*, *Leucas martinicensis*, *Physalis angulata*, *Synedrella nodiflora*, *Triumfetta rhomboidea* and *Wissadula amplissima*. The presence of these species indicates that the tree cover of the core of the groves was adequate to create typical forest floor conditions for their appearance, and that those species that are not shade-tolerant do not survive within the core of the groves.

Since the outside of the core of the groves does not come under strict management in terms of entry and harvesting, these places have suffered degradation. Consequently, the tree cover is not complete. Ruderals and other opportunistic species therefore easily find establishment, initially as pioneers. This explains why the species diversity is always higher in the areas around the core than within the core of the groves.

Acknowledgements
This contribution is a result of a project on the northern Ghana Savanna Ecosystem under the acronym of CIPSEG, financed through UNESCO from fund in trust (FIT) with the Federal Republic of Germany. The author is grateful to the funding agents as well as to colleagues and students at the Department of Botany, University of Ghana.

References

Hopkins, B. & D.P. Stanfield, 1966. A field key to the Savanna Trees in Nigeria. Ibadan University Press.

Houérou, H.N., 1980. Le rôle des arbres et arbustes dans les pâturages sahéliens. In: Le rôle des arbres au sahel. CRDI Ottawa, Canada.

Hutchinson, J. & J.M. Dalziel, 1954 & 1972. Flora of West Tropical Africa. Crown Agents.

IUCN, 1986. Managing Protected Areas in the Tropics. International Union for Conservation of Natural Resources, Gland, Switzerland.

IUCN, 1990. Biodiversity in Sub-Saharan Africa and its Islands: Conservation, Management and Sustainable Use. Occasional papers IUCN, Species Survival Commission, Gland, Switzerland.

Keay, R.W.J., C.F.A. Onochie & D.P. Standfield, 1960 & 1964. Nigerian Trees, Vol. I and II, Lagos, Nigeria.

Le Houérou, H.N., 1980. Le rôle des arbres et arbustes dans les pâturages sahéliens. In: Le Rôle des arbres au Sahel. CRDI Ottawa, Canada.

Ntiamoa-Baidu, Y., 1993. In Africa Biodiversity: Foundation for the future. Biodiversity Support Program. Washington.

Oteng-Yeboah, A.A., 1985. Ecological Survey of the Savanna vegetation of Sokoto State, Nigeria: Floristic and life form composition. Biologia africana 2 (2): 67-78.

Oteng-Yeboah, A.A., 1993. Ecological and ethnobotanical investigations into three traditional groves in northern Ghana: a base line study report. UNESCO - CIPSEG Project 507/GHA/40.

Oteng-Yeboah, A.A., 1994. DNA Bank depositories in developing countries: the way forward in sub-saharan Africa. Pp. 175-186 in: R.P. Adams et al. (ed.). Conservation of Plant Genes II: Utilization of Ancient and modern DNA. Missouri Botanical Garden, St Louis.

Appendix 1
List of families represented in the groves, and total number of species and genera encountered.

Acanthaceae	D	2	2	Cyperaceae *	M	7	5	Olacaceae	D	1	1
Amaranthaceae	D	4	4	Dioscoreaceae	M	2	1	Onagraceae	D	1	1
Amaryllidaceae	M	1	1	Ebenaceae	D	2	1	Orchidaceae	M	2	2
Anacardiaceae	D	5	3	Euphorbiaceae *	D	6	6	Oxalidaceae	D	1	1
Annonaceae	D	1	1	Flacourtiaceae	D	2	2	Papilionoideae *	P	22	14
Apocynaceae	D	1	1	Gramineae	M	24	17	Passifloraceae	D	1	1
Araceae *	M	4	3	Hypoxidaceae	M	1	1	Pedaliaceae	D	1	1
Balanitaceae	D	1	1	Icacinaceae	D	1	1	Polygalaceae	D	1	1
Bignoniaceae	D	1	1	Iridaceae	M	1	1	Rhamnaceae	D	1	1
Bombacaceae	D	3	3	Labiatae	D	3	3	Rubiaceae *	D	13	11
Boraginaceae	D	1	1	Liliaceae	M	3	3	Sapindaceae	D	1	1
Caesalpinioideae	D	10	7	Loganiaceae	D	1	1	Sapotaceae	D	1	1
Capparaceae	D	3	3	Loranthaceae	D	1	1	Scrophulariaceae	D	4	3
Campanulaceae	D	1	1	Lythraceae	D	1	1	Simaroubaceae	D	1	1
Caryophyllaceae	D	2	1	Malvaceae *	D	9	5	Solanaceae	D	2	2
Celastraceae	D	2	2	Maranthaceae	M	1	1	Sterculiaceae	D	2	2
Cochlospermaceae	D	1	1	Meliaceae	D	3	3	Taccaceae	M	1	1
Combretaceae *	D	8	3	Menispermaceae	D	1	1	Tiliaceae *	D	9	3
Commelinaceae	M	4	2	Mimosoideae *	D	6	5	Urticaceae	D	1	1
Compositae *	D	9	8	Molluginaceae	D	1	1	Verbenaceae	D	3	3
Convolvulaceae	D	1	1	Moraceae *	D	6	1	Vitaceae	D	3	2
Costaceae	M	1	1	Nyctaginaceae	D	2	1	Zingiberaceae	M	1	1
Cucurbitaceae	D	2	2	Nymphaeaceae	D	1	1	Zygophyllaceae	D	1	1

* Families with a high number of species and genera in the flora,
D = Dicotyledonae, M = Monocotyledonae. Leguminosae enumerated as three subfamilies.

Appendix 2
Families with most genera and species, their life form, and their distribution in the three groves.

Family, species	Life form	Distribution M	T	Y
Araceae				
1. *Amorphophallus dracontioides* (Engl.) N.E. Br.	GrH	—	+	—
2. *Anchomanes welwitschii* Rendle	GrH	—	+	—
3. *Stylochiton hypogaeus* Lepr.	GrH	+	+	+
4. *S. lancifolius* Kotschy & Peyr.	GrH	+	+	+
Combretaceae				
1. *Anogeissus leiocarpa* (DC.) Guill. & Perr.	mPT	+	+	+
2. *Combretum ghasalense* Engl. & Diels	mPS	—	+	+
3. *C. glutinosum* Perr. ex DC.	mpS	—	+	—
4. *C. hypopilinum* Diels	mPS	—	+	—
5. *C. molle* R. Br. ex G. Don	mpS	+	+	+
6. *C. nigricans* Lepr. ex Guill. & Perr.	mpS	+	+	+
7. *Terminalia avicennioides* Guill. & Perr.	mpS	+	+	+
8. *T. glaucescens* Planch. ex Benth.	mpS	+	+	—

Compositae		
1. *Acanthospermum hispidum* DC.	ThH	— + +
2. *Ageratum conyzoides* L.	ThH	+ + —
3. *Aspilia africana* (Pers.) C.D. Adams	ThH	— + —
4. *A. helianthoides* (Schum. & Thonn.) Oliv. & Hiern	ThH	— + —
5. *Gutenbergia macrocephala* Oliv. & Hiern	ChH	— + +
6. *Launaea taraxacifolia* (Willd.) Amin ex C. Jeff.	ThH	— + —
7. *Synedrella nodiflora* Gaertn.	ThH	+ + —
8. *Tridax procumbens* L.	ThH	+ + +
9. *Vernonia guineensis* Benth.	ThH	— + —
Cyperaceae		
1. *Cyperus difformis* L.	ThH	— + —
2. *C. rotundus* L.	ThH	+ + +
3. *Fimbristylis bisumbellata* (Forssk.) Bubani	ThH	— + —
4. *F. dichotoma* (L.) Vahl	ThH	+ + +
5. *Fuirena umbellata* Rottb.	ThH	— + —
6. *Kyllinga squamulata* Thonn. ex Vahl	ThH	+ + —
7. *Mariscus squarrosus* (L.) C.B. Clarke	ThH	+ + +
Euphorbiaceae		
1. *Bridelia ferruginea* Benth.	mPT	— + —
2. *Euphorbia convolvuloides* Hochst. ex Benth.	ThH	+ + +
3. *E. hirta* L.	ThH	+ + +
4. *Phyllanthus pentandrus* Schum. & Thonn.	ThH	+ + +
5. *Sapium grahamii* (Stapf) Prain	ChH	— + —
6. *Securinega virosa* (Roxb. ex Willd.) Baill.	mpS	+ + +
Gramineae		
1. *Andropogon gayanus* Kunth	ThH	— + +
2. *Aristida adscensionis* L.	ThH	— — +
3. *A. kerstingii* Pilger	ThH	— — +
4. *A. mutabilis* Trin. & Rupr.	ThH	— + +
5. *A. sieberiana* Trin.	ThH	— — +
6. *Brachiaria stigmatisata* (Mez) Stapf	ThH	— + +
7. *B. jubata* (Fig. & De Not.) Stapf	ThH	— + —
8. *Cenchrus biflorus* Roxb.	ThH	— + —
9. *Chloris barbata* Sw.	ThH	+ — —
10. *Cynodon dactylon* (L.) Pers.	ThH	+ + +
11. *Dactyloctenium aegyptium* (L.) P. Beauv.	ThH	+ + +
12. *Digitaria debilis* (Desf.) Willd.	ThH	+ + +
13. *Echinochloa pyramidalis* (Lam.) Hitch. & Chase	ThH	+ + —
14. *Eleusine indica* (L.) Gaertn.	ThH	+ + +
15. *Eragrostis aspera* (Jacq.) Nees	ThH	+ + +
16. *E. tremula* Hochst. ex Steud.	ThH	+ + +
17. *E. tenella* (L.) P. Beauv. ex Roem. & Schult.	ThH	+ — +
18. *Panicum maximum* Jacq.	ThH	+ + +
19. *Paspalum orbiculare* Forst.	ThH	+ — +
20. *P. polystachyum* R.Br.	ThH	+ — +
21. *Pennisetum pedicellatum* Trin.	ThH	— + +
22. *Setaria pallide-fusca* (Schm.) Stapf & C.E. Hubbard	ThH	+ + +

23. *Sorghum arundinaceum* (Desv.) Stapf	ThH	+	+	+
24. *Sporobolus subglobosus* A. Chev.	ThH	+	+	+
Leguminosae - Caesalpinioideae				
1. *Afzelia africana* Sm.	npS	—	+	—
2. *Burkea africana* Hook.	npS	+	+	—
3. *Chamaecrista absus* (L.) Moench	ThH	—	+	+
4. *Daniellia oliveri* (Rolfe) Hutch. & Dalz.	MPT	—	+	—
5. *Detarium microcarpum* Guill. & Perr.	mpS	+	+	+
6. *Dialium guineense* Willd.	mpS	—	+	—
7. *Piliostigma thonningii* (Schum.) Milne-Redhead	mpS	—	+	—
8. *Senna mimosoides* (L.)Greene	ThH	+	+	+
9. *S. obtusifolia* (L.) Irwin & Barneby	ThH	+	—	—
10. *S. occidentalis* (L.) Link	ThH	—	+	—
Leguminosae - Mimosoideae				
1. *Acacia hockii* De Wild.	mpS	+	+	+
2. *A. macrostachya* Reichb. ex Benth.	mpWL	+	+	—
3. *Dichrostachys glomerata* (Forssk.) Chiov.	mpS	—	+	+
4. *Entada africana* Guill. & Perr.	MPT	+	+	—
5. *Parkia biglobosa* (Jacq.) R.Br. ex G. Don	MPT	+	+	—
6. *Prosopis africana* (Guill. & Perr.) Taub.	MPT	—	+	—
Leguminosae - Papilionoideae				
1. *Abrus precatorius* L.	ThHl	—	+	—
2. *Aeschynomene afraspera* J. Léonard	ThH	—	+	—
3. *Alysicarpus ovalifolius* (Schum. & Thonn.) J. Léonard	ThH	+	+	+
4. *A. rugosus* (Willd.) DC.	ThH	—	+	+
5. *Centrosema pubescens* Benth.	ThHl	+	+	—
6. *Crotalaria confusa* Hepper	ThH	+	+	+
7. *C. goreensis* Guill. & Perr.	ThH	+	+	+
8. *C. retusa* L.	ThH	+	+	+
9. *Desmodium gangeticum* (L.) DC.	ThH	+	+	+
10. *D. ramosissimum* G. Don	npS	+	+	+
11. *Erythrina senegalensis* DC.	mpS	—	+	—
12. *Indigofera sp.* a	ThH	+	+	+
13. *Indigofera sp.* b	ThH	+	+	+
14. *Indigofera sp.* c	ThH	+	+	+
15. *Indigofera sp.* d	ThH	+	+	+
16. *Ostryoderris stuhlmannii* (Taub.) Dunn ex Harms	npS	—	+	—
17. *Sesbania sesban* (L.) Merrill	ThH	—	+	—
18. *Stylosanthes mucronata* Willd.	ThH	+	+	+
19. *Tephrosia purpurea* (L.) Pers.	ThH	+	+	+
20. *Uraria picta* (Jacq.) DC.	ThH	+	+	—
21. *Zornia glochidiata* DC.	ThH	+	+	+
22. *Z. latifolia* Sm.	ThH	—	+	—
Malvaceae				
1. *Abutilon mauritianum* (Jacq.) Medic.	chH	+	—	+
2. *Hibiscus sabdariffa* L.	ThH	+	+	+
3. *H. surattensis* L.	ThH	—	+	+

4. *Sida acuta* Burm. f.	ThH	+	+	+
5. *S. cordifolia* L.	ThH	+	+	—
6. *S. rhombifolia* L.	ThH	+	—	+
7. *S. stipulata* Cav.	ThH	—	+	—
8. *Urena lobata* L.	ChS	—	+	—
9. *Wissadula amplissima* (L.) R.E. Fries	ChH	+	+	+
Moraceae				
1. *Ficus capensis* Thunb.	mPT	+	+	+
2. *F. exasperata* Vahl	mPT	+	+	+
3. *F. gnaphalocarpa* (Miq.) Steud. ex A. Rich.	mPT	—	+	—
4. *F. natalensis* Hochst.	mPT	—	+	—
5. *F. polita* Vahl	mPT	—	+	—
6. *F. thonningii* Blume	mPT	—	+	—
Rubiaceae				
1. *Borreria octodon* Hepper	ThH	—	+	+
2. *B. radiata* DC.	ThH	+	+	+
3. *Crossopteryx febrifuga* (Afzel.) Benth.	ThH	—	+	—
4. *Fadogia agrestis* Schweinf.	ChH	+	+	+
5. *Feretia apodanthera* Del.	npS	+	+	+
6. *Gardenia erubescens* Stapf & Hutch.	mpS	+	+	+
7. *G. ternifolia* Schum. & Thonn.	mpS	+	+	+
8. *Hoslundia opposita* Vahl	mpS	+	+	—
9. *Mitracarpus scaber* Zucc.	ThH	+	+	+
10. *Mitragyna inermis* (Willd.) O. Ktze.	mpT	—	+	—
11. *Nauclea latifolia* Sm.	mpT	—	+	—
12. *Oldenlandia corymbosa* L.	ThH	+	+	+
13. *Pentodon pentandrus* (Schum. & Thonn.) Vatke	mpT	+	+	+
Tiliaceae				
1. *Corchorus aestuans* L.	ThH	+	+	—
2. *C. fascicularis* Lam.	ThH	—	+	+
3. *Grewia barteri* Burret	npS	—	+	—
4. *G. carpinifolia* Juss.	npS/T	—	+	—
5. *G. lasiodiscus* K. Schum.	npS	—	+	—
6. *G. mollis* Juss.	npS	+	+	—
7. *Triumfetta dubia* De Wild.	ThH	—	+	—
8. *T. lepidota* K. Schum.	ThH	+	+	+
9. *T. rhomboidea* Jacq.	ThH	—	—	+

M = Malshegu; T = Tale; Y = Yiwogu; life forms: see Table 3.

Biodiversity of the Maputaland Centre

A.E. VAN WYK
Department of Botany, University of Pretoria, Pretoria 0002, South Africa.

Abstract
The Maputaland Centre [MC] of endemism comprises the northernmost part of the Tongaland-Pondoland Regional Mosaic. With the exception of the narrow Lebombo Range which rises to an elevation of some 600 m, the area is a nearly flat, low-level, sandy coastal plain with a maximum elevation of about 150 m. The vegetation of the MC is exceptionally diverse, consisting of a mosaic of forest, savanna, grassland and swamps, largely determined by local edaphic conditions. About 2,500 species (but probably more) of vascular plants occur in the MC. Of these at least 203 species/infraspecific taxa and 3 genera (*Brachychloa*, *Galpinia*, *Helichrysopsis*) are endemic/near-endemic to the region, and are listed for the first time. The area is also outstanding for its faunistic diversity, which includes many endemics spread over the whole taxonomic spectrum. Biogeographic links of the MC are clearly with the more tropical parts of Africa, with many Afrotropical species reaching the southernmost limit of their range here.

Résumé
Le Centre d'endémisme du Maputaland comprend la partie la plus septentrionale de la Mosaïque régionale du Tongaland-Pondoland. A l'exception de l'étroite chaîne des Lebombo, qui s'élève à quelque 600 m d'altitude, la région se présente sous la forme d'une plaine côtière sablonneuse de basse altitude et pratiquement plate, dont l'altitude maximale plafonne à peu près à 150 m. La végétation y est exceptionnellement diversifiée, constituée d'une mosaïque de forêts, de savanes, de formations herbeuses et de marais, conditionnés en grande partie par l'édaphisme local. On y relève environ 2500 espèces (probablement davantage) de plantes vasculaires. Sur ce total, au moins 203 taxons spécifiques ou infraspécifiques, ainsi que 3 genres (*Brachychloa, Galpinia, Helichrysopsis*), sont endémiques ou subendémiques à la région et c'est la première fois qu'on en dresse l'inventaire. La région est remarquable également par sa diversité faunistique, avec de nombreuses espèces endémiques s'étalant dans l'entièreté du spectre taxonomique. Les liens biogéographiques du Centre d'endémisme du Maputaland avec les parties davantage tropicales de l'Afrique sont bien nets, nombre d'espèces afrotropicales atteignent ici la limite la plus méridionale de leur aire de distribution.

Key words
Biodiversity, Biogeography, Endemism, Flora, Maputaland, Savanna.

Introduction

The Tongaland-Pondoland Regional Mosaic [TPRM] of White (1983) is floristically very diverse and complex. Phytogeograpically this region is part of the Indian Ocean Coastal Belt which is both a Regional Transition Zone and a Regional Mosaic (Moll & White 1978, White 1983). There are, however, at least three clear foci of high floristic endemism in the region, namely the Maputaland Centre in the north, Pondoland Centre in the middle and the Albany Centre in the south (Van Wyk 1990, 1994).

Maputaland, previously known as Tongaland, comprises the northernmost sector of the TPRM. It is at the southern end of the tropics in Africa and many tropical plant and animal species reach the southernmost limit of their range here. Hence, biological, and in particular botanical, interest in Maputaland has hitherto centred mainly on the

L.J.G. van der Maesen et al. (eds.), The Biodiversity of African Plants, 198-207.

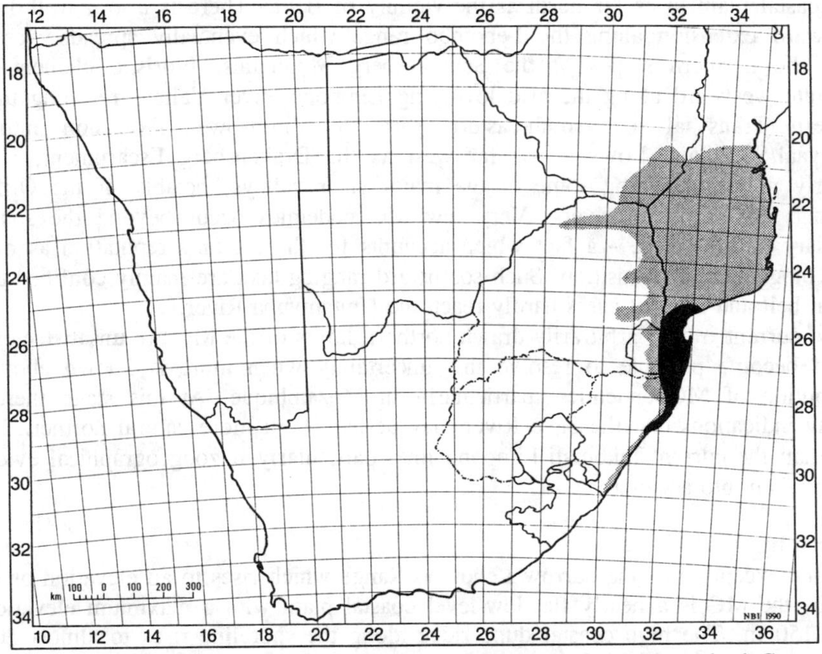

Figure 1. Map of southern Africa showing the location of the Maputaland Centre (solid black; details on boundaries in text). Shading denotes main area in which outliers of certain Maputaland Centre endemics are found.

remarkably sharp biogeographic transformation in the region (Bruton & Cooper 1980, Poynton 1961). Emphasis on Maputaland as a region of biotic transition has largely obscured its status as a regional centre of endemism in own right. Recent formal recognition of the Maputaland Centre (Van Wyk 1994) acknowledges the high levels of endemism and remarkable biodiversity of this region. The present paper supplements the latter contribution. Its principal aim is to present in more detail the floristic evidence on which the region's proposed status as a centre of plant endemism is being based. Brief reference will also be made to the faunal diversity and biogeography of the region.

Delimitation of the Maputaland Centre [MC]

The core area of the MC (c. 26,734 km^2) has been defined by Van Wyk (1994) as that part of southern Mozambique and north-eastern KwaZulu/Natal bounded in the north by the Inkomati-Limpopo River, in the east by the Indian Ocean, in the west by the western foothills of the Lebombo Mountains and in the south by the St. Lucia estuary, from where it extends further along the coast down to near Mtunzini (including the transitional Ngoye Range). Biogeographically the boundaries of the centre are clearly defined, except in the north, where the line is arbitrary. Limits of the MC and its principal outliers, as currently perceived, are shown in Figure 1.

From the core area, many MC elements extend further along the Mozambican coast and coastal plain to as far north as the vicinity of Beira. There is also a well-defined northward extension along the Lebombo range which eventually spreads westward along the northern slopes of the Soutpansberg Mountains. Floristic elements also penetrate westward along the arid low-lying Limpopo River Valley, reaching the far northern Transvaal and south-eastern parts of Zimbabwe. Extension into the Transvaal/Swaziland Lowveld, as far west as the Drakensberg Escarpment, is particularly noticeable along some of the major river valleys, notably of the Olifants, Inkomati and Krokodil rivers. Very few MC endemics occur beyond the southern boundary of the centre - a fact which accounts for the region's reputation as one of sharp biogeographic transition. Such southward ranging taxa are mainly confined to the coastal belt and in most cases hardly reach the Umtamvuna River.

The current, rather arbitrarily drawn northern limits of the MC are unsatisfactory. It should become possible to resolve this uncertainty when more is known about the distribution of MC endemics, particularly in Mozambique. At this stage there are already indications that the Save River may perhaps be a more natural northern boundary than the current Inkomati-Limpopo line, particularly if zoogeographical evidence is also taken into account.

Geography

With the exception of the narrow Lebombo Range which rises to an elevation of some 600 m, the MC is a nearly flat low-level coastal plain with a maximum elevation of about 150 m. The high coastal dune ridge along the shoreline rises to almost 200 m above sea-level in places in KwaZulu/Natal, and is said to be amongst the tallest vegetated dunes in the world.

The Lebombo Range is formed by resistant volcanic rhyolite lavas, whereas the coastal plains consist mainly of Cretaceous to Recent marine sediments. Over most of the MC the soils are infertile, recent aeolian sands. Fertile clayey alluvium occurs in the floodplains of some of the larger rivers (Bruton & Cooper 1980).

A tropical/subtropical climate prevails with no frost in winter. Annual rainfall averages about 1,100 mm along the coast, but declines progressively inland to as low as 500-600 mm on the western plains. The crest of the Lebombo Range receives 800-1,000 mm per year (Bruton & Cooper 1980, Schulze 1982). Relative humidity is generally high, particularly along the coast.

The general character of the region is one of high temperatures, high humidity and relative aridity. There is a pronounced dry season (winter). Mist frequently occurs in the Lebombo Mountains and, through inversion, locally on the plains (particularly in winter).

The MC contains some extensive wetlands, particularly marshes, lakes and estuaries. Most notable are Lake St. Lucia (c. 350 km^2), the largest estuarine system of its kind in Africa, Lake Sibaya (60-70 km^2), the largest freshwater lake in southern Africa, and the Kosi Lake System (c. 37 km^2).

Vegetation

The vegetation of the MC is exceptionally diverse. It consists of a mosaic of mainly extensive savanna communities, arranged in complex patterns, as well as forest (mainly on recent and ancient coastal dunes and Lebombo Mountains), grassland and swamps, largely determined by local edaphic conditions. The vegetation of the KwaZulu/Natal portion of the region has been classified into at least fifteen major types. These are: Lebombo Forest, Lebombo Range Vegetation, Mixed Bushveld, Thicket, Red-sand Bushveld, Floodplain Vegetation, Sand Forest, Pallid-sand Bushveld, Mosi Swamp Vegetation, Papyrus Swamp, Palm Veld, Coast Grassveld, Swamp Forest, Mangroves and Dune Forest (Moll 1977, 1980). On a more refined level, several other aquatic and dune forest communities have been described (Rogers 1980, Weisser 1980). Tinley (1971) recognises no less than 21 different ecosystems in the MC.

Flora

The total number of vascular plant species in the Maputaland Centre is at least 2500, but could well be as high as 3,000. This is considerably more than my previous estimate of 1500 (Van Wyk 1994). Of these, at least 203 species/infraspecific taxa are endemic/near-endemic to the centre (listed in Appendix 1). Note, however, that this is a rather conservative figure. The number of endemics is likely to increase with a better knowledge of the taxonomy and distribution of the region's flora.

Three genera are endemic/near-endemic to the MC (*Brachychloa, Galpinia, Helichrysopsis*). In the light of current taxonomic opinion, *Ephippiocarpa* (= *Callichilia*) and *Inhambanella* (Van Wyk 1994) are no longer considered endemic. There are no families endemic to the MC. The six families with the largest number of endemics are (number of species/infraspecifix taxa in brackets): Asclepiadaceae (13), Rubiaceae (12), Liliaceae *s.l.* (12), Acanthaceae (11), and Compositae (10).

Although many of the endemics are widespread throughout the MC, quite a number occur more locally. Taxa restricted to the southern part of the region include *Dierama sertum, Didymoplexis verrucosa, Helixanthera woodii* (Ngoye only), *Kniphofia leucocephala, Monopsis belliflora, Salpinctium natalense* and *Streptocarpus wendlandii* (Ngoye only). Taxa known at present only from the more northern parts include *Adenopodia schlechteri, Aloe suffulta, Bergia mossambicensis, Chrysocoma mozambicensis, Cyphostemma barbosae, Dolichandrone alba, Krauseola mosambicina* and *Turbina longifolia*. Among some of the apparently very local endemics are species such as *Hemizygia ramosa* and *Pelargonium tongaense*.

Nearly all life-forms and life history types (including saprophytes, hemiparasites and monocarpic plants) are represented among the endemics, with the majority being perennial herbs, shrubs and trees. The presence of quite a number of annuals or short-lived perennials is noteworthy, e.g. *Aneilema arenicola, A. bruneospermum, Brachychloa fragilis, Cineraria pinnata, Cleome bororensis, C. schlechteri, Crassula maputensis, Helichrysum tongense, Lepidium suluense, Nesaea alata, N. wardii* and *Vernonia centaurioides*. Lianas, creepers and scandent shrubs are also particularly prevalent, e.g. *Acacia kraussiana, Albertisia delagoensis, Ancylanthos monteiroi, Basananthe polygaloides, Ceropegia arenaria, Cissampelos hirta, Combretum mkuzense, Commicarpus chinensis* subsp. *natalensis, Entada schlechterii, Hippocratea delagoensis, Ipomoea venosa* subsp. *stellaris, Maerua brevipetiolata, Rhoicissus sessilifolia, Secamone delagoensis, Schotia capitata, Stictocardia laxiflora* var. *woodii, Strophanthus gerrardii, S. luteolus, Tarenna junodii* and *Turbina longifolia*.

A requirement for rocky substrates appears to be one of the factors limiting the distribution of some of those MC endemics largely confined to the Lebombo Mountains (marked with an asterisk in Appendix 1). Most of the endemics, however, favour the sandy substrates that prevail over most of the region but endemics are represented in very many different plant communities, spread over almost the whole spectrum of habitat types.

A significant number of endemics are associated with Sand Forest, perhaps the most remarkable plant community in the MC (Moll 1977, 1980, Moll & White 1978, Wild & Barbosa 1968, mapping unit 5 p.p.). Sand Forest is a distinctive type of very dense, dry, semi-deciduous to deciduous forest that is mainly associated with the generally north-south trending belts of ancient dunes, supposedly representing former shorelines. Particularly rich in woody taxa, these forests are characterized by species such as *Afzelia quanzensis, Artabotrys monteiroae, Balanites maughamii, Cleistanthus schlechteri, Cola greenwayi, Cordyla africana, Craibia zimmermannii, Dialium schlechteri, Ficus tremula, Haplocoelum gallense, Hymenocardia ulmoides, Microcoelia exilis, Monodora junodii, Newtonia hildebrandtii, Pseudobersama mossambicensis, Ptaeroxylon obliquum, Suregada zanzibarensis* and *Zamioculcas zamiifolia*. Endemics of more general occurrence in Sand Forest are *Acacia kraussiana, Albertisia delagoensis, Ancylanthos monteiroi, Brachychloa fragilis, Cassipourea mossambicensis, Cleome bororensis, Combretum mkuzense, Corchorus junodii, Croton steenkampianus, Cussonia arenicola, Erythrophleum lasianthum, Erythroxylum delagoense, Grewia microthyrsa, Hyperacanthus microphyllus, Memecylon sousae, Ochna barbosae, Oxyanthus latifolius, Psydrax locuples, Strophanthus gerrardii, S. luteolus, Sclerochiton apiculatus, Toddaliopsis bremekampii, Warburgia salutaris, Wrightia natalensis,* and many more.

Fauna

For its size the MC is one of the most remarkable areas of biodiversity in the world. Not only are the numbers of endemics high, but they are spread over virtually the whole taxonomic spectrum. There must be few other places in the world with so many rare animals and plants concentrated in such a small area. As with its flora, the MC is the southernmost part of the range of many components of the East African fauna. As in the case of plants, the centre is of exceptional zoogeographical interest because of the sharp biogeographic transformation in the region (Poynton 1961).

Of the more than 472 species of birds in the MC (almost 60% of South Africa's total), 4 species and about 43 subspecies are endemic/near-endemic to the centre. Other endemic/near-endemic species and infraspecific taxa (total indigenous to the centre in brackets) include 14 mammals, mainly of subspecific rank (102 species, about 4 locally extinct), 23 reptiles (about 112 species/subspecies), 3 frogs (45 species/subspecies) and 8 freshwater fishes (67 species). The unspoiled beaches along the Maputaland coast are a major nesting ground for marine turtles. The area also abounds in insect life.

Biogeography

The bulk of the MC flora and fauna is clearly of Afrotropical derivation. Many of the endemic taxa appear to be of fairly recent diversification, an impression supported by the fact that many are differentiated at the infraspecific level only, with nearest relatives still extant. This would also be consistent with the postulated geologically relative recent (Quaternary) origin of the sandy, low-lying coastal plain comprising the greater proportion of Maputaland (Maud 1980).

Although non-endemic or linking plant species occurring in the MC provide connections with several other major African floristic regions, detailed comparisons have not yet been made. Nevertheless, the obvious links with the Somalia-Masai/Zanzibar-Inhambane Regions and areas of deep Kalahari sand (western Kalahari-Highveld Regional Transition Zone and western Zambezian Region) are of particular interest. Compare for example the apparent range disjunction of about 2400 km between northern Tanzania and Maputaland recorded for *Aneilema zebrinum* (Faden 1991). It is clear that any future analyses of the postulated ancient arid corridor(s) linking the Somalia-Masai Region with arid parts of southern Africa, should also involve the hitherto neglected Maputaland Centre (e.g. De Winter 1971, Verdcourt 1969). There are also unexpected links with West Tropical Africa, e.g. *Sorindeia juglandifolia*. In contrast to the Pondoland Centre, Afromontane and Cape linking species are particularly few and of sporadic occurrence.

A link with NE Africa is also signified by some of the MC endemics. For example, in Africa *Wrightia* is represented by the MC *W. natalensis* and a semi-desert Somalia-Masai species, *W. demartiniana* (section *Balfouria*). The remaining species occur in Asia and Australia (section *Balfouria* with one species). *Toddaliopsis* has one species in the MC and two or three others in NE Africa. An avian example is Woodwards' Barbet, *Cryptolybia woodwardii*, which is represented by the small population of the nominate race in the Ngoye Forest at the southern tip of the MC. The second population of the species (*C. w. hylophona*) is also confined to a single forest on the Rondo plateau, south-west of Lindi, in SE Tanzania (Clancey 1985). A somewhat comparable case is that of the Transkei-KwaZulu/Natal population of the Spotted Ground Thrush, *Zoothera guttata guttata*, the closest spatial population being found in certain mountains in Malawi (*Z. g. belcheri*), with yet other detached fragments of the species known from SE Zaire, coastal Kenya and the Imatong Mountains of southern Sudan (Prigogine 1985). Similar Maputaland-NE African connections among mammals include the subspecies of the Red Squirrel (*Paraxerus palliatus*), Meller's Mongoose (*Rhynchogale melleri*), Suni (*Neotragus moschatus*) and the Red Duiker (*Cephalophus natalensis*) (Meester et al. 1986, Skinner & Smithers 1990).

Extra-African links are mainly with Indian Ocean Island and other parts of the Old World. These include a number of littoral plants probably acquired through ocean currents (e.g. *Guettarda speciosa* and *Scaevola sericea*). The rare MC endemic, *Dolichandrone alba*, is the only member of the genus in Africa, with the other species (approximately seven) occurring in tropical Asia and Australia (compare *Wrightia*). A similarly wide disjunction is displayed by the mainly African genus *Commicarpus*. The typical subspecies of the littoral MC endemic, *Commicarpus chinensis* subsp. *natalensis*, also occurs in SE Asia.

Acknowledgements

The Director (Research) of the National Botanical Institute is thanked for the use of plant collecting and distributional data produced by the Pretoria National Herbarium Computerised Information System (PRECIS). Mr. J.C. Hollmann critically read and improved the manuscript. This study was financed in part by the Foundation for Research development and the University of Pretoria.

References

Bruton, M.N. & K.H. Cooper (eds), 1980. Studies on the Ecology of Maputaland. Rhodes University & Wildlife Society of S.A., Grahamstown & Durban.

Clancey, P.A., 1985. The Rare Birds of Southern Africa. Winchester Press, Johannesburg.

De Winter, B., 1971. Floristic relationship between the northern and southern arid areas in Africa. Mitt. Bot. Staatssamml. München. 10: 424-437.

Faden, R.B., 1991. The morphology and taxonomy of *Aneilema* R. Brown (Commelinaceae). Smithsonian Contributions to Botany 76, Smithsonian Institution Press, Washington D.C.

Maud, R.R., 1980. The climate and geology of Maputaland. In: Bruton, M.N. & K.H. Cooper (eds), Studies on the Ecology of Maputaland: 1-7. Rhodes University & Wildlife Society of S.A., Grahamstown & Durban.

Meester, J.A.J., I.L. Rautenbach, N.J. Dippenaar & C.M. Baker, 1986. Classification of Southern African Mammals. Transvaal Museum Monograph 5: 1—359, Transvaal Museum, Pretoria.

Moll, E.J., 1977. The vegetation of Maputaland - a preliminary report on the plant communities and their present and future conservation status. Trees in S. Afr. 29: 31-58.

Moll, E.J., 1980. Terrestrial plant ecology. In: Bruton, M.N. & K.H. Cooper (eds), Studies on the Ecology of Maputaland: 52-68. Rhodes Univ. & Wildlife Soc. of S.A., Grahamstown & Durban.

Moll, E.J. & F. White, 1978. The Indian Ocean Coastal Belt. In: Werger, M.J.A. (ed.), Biogeography and Ecology of Southern Africa. Monographiae Biologicae 31: 563-598. Junk, The Hague.

Poynton, J.C., 1961. Biogeography of south-east Africa. Nature 189: 801-803.

Prigogine, A., 1985. Révision des espèces africaines appartenant au genre *Zoothera*. Gerfaut 75: 258-319.

Rogers, K.H., 1980. The vegetation of the Pongola floodplain: distribution and utilization. In: Bruton, M.N. & K.H. Cooper (eds), Studies on the Ecology of Maputaland: 69-77. Rhodes University & Wildlife Society of S.A., Grahamstown & Durban.

Schulze, R.E., 1982. Agrohydrology and Climatology of Natal. Agricultural Catchments Research Unit Report 14, University of Natal, Pietermaritzburg.

Skinner, J.D. & R.H.N. Smithers, 1990. The Mammals of the Southern African Subregion (2nd ed.). University of Pretoria, Pretoria.

Tinley, K.L., 1971. Lake St. Lucia and its Peripheral Sand Catchment. Wildlife Society of Southern Africa, Durban.

Van Wyk, A.E., 1990. The sandstone regions of Natal and Pondoland: remarkable centres of endemism. In: Heine, K. (ed.), Palaeoecology of Africa 21: 243-257. A.A. Balkema, Rotterdam.

Van Wyk, A.E., 1994. Maputaland-Pondoland Region. In: Davis, S.D. & V.H. Heywood (eds.), Centres of Plant Diversity: A Guide and Strategy for their Conservation. Oxford University Press, Oxford.

Verdcourt, B., 1969. The arid corridor between the north-east and south-west areas of Africa. Palaeoecology of Africa 4: 140-144.

Weisser, P.J., 1980. The dune forest of Maputaland. In: Bruton, M.N. & K.H. Cooper (eds), Studies on the Ecology of Maputaland: 78-90. Rhodes University & Wildlife Society of S.A., Grahamstown & Durban.

White, F., 1983. The Vegetation of Africa. UNESCO, Paris.

Wild, H. & L.A.G. Barbosa, 1968. Vegetation Map of the Flora Zambesiaca Area, with Descriptive Memoir [Supplement to Flora Zambesiaca]. Collins, Salisbury.

Appendix 1
Provisional checklist of vascular plant species endemic or near-endemic to the Maputaland Centre. An asterisk denotes taxa restricted mainly to the Lebombo Mountains.

Gymnosperms
Podocarpaceae
Podocarpus falcatus (Thunb.) R.Br. ex Mirb. *p.p.* [sand/swamp forest form]
Zamiaceae
Encephalartos ferox Bertol.f.
E. lebomboensis I. Verd.*
E. ngoyanus I. Verd.*
E. umbeluziensis R.A. Dyer
E. woodii Sonder

Monocotyledons
Amaryllidaceae
Crinum acaule Baker
Araceae
Gonatopus rhizomatosus Oberm.
Commelinaceae
Aneilema arenicola Faden
A. indehiscens Faden subsp. *lilacinum* Faden
A. bruneospermum Faden
Cyperaceae
Mariscus lebomboensis Vorster* m.s.
Schoenoplectus confusus (N.E. Br.) Lye subsp. *natalitius* Browning
Gramineae
Brachychloa schiemanniana (Schweick.) S.M. Phillips
B. fragilis S.M. Phillips
Eragrostis moggii De Winter var. *moggii*
E. moggii De Winter var. *contracta* De Winter
Iridaceae
Anomatheca laxa (Thunberg) Goldblatt subsp. *azurea* Goldblatt & Hutchings
Dierama sertum Hilliard
Lemnaceae
Wolffiella denticulata (Hegelm.) Hegelm.
Liliaceae *s.l.*
Agapanthus caulescens Spreng. subsp. *gracilis* (Leight.) Leight.*
Aloe komatiensis Reynolds
A. parvibracteata Schönland
A. suffulta Reynolds
A. barberiae Dyer *p.p.* [coastal form with orange flowers]
A. keithii Reynolds*
A. vanbalenii Pillans*
A. spicata L. f.*
A. marlothii A. Berger subsp. *orientalis* Glen & Hardy
Haworthia limifolia Marloth var. *ubomboensis* (I. Verd.) G.G. Sm.*
Kniphofia leucocephala Baijnath
K. tysonii Baker subsp. *lebomboensis* Codd*
Orchidaceae
Bolusiella maudiae (Bolus) Schltr.
Bonatea lamprophylla J. Stewart
Cyrtorchis praetermissa Summerh. var. *zuluensis* (E.R. Harrison) H.P. Linder
Didymoplexis verrucosa J. Steward & E.F. Henn.
Polystachya modesta Rchb.f.*
P. zuluensis L. Bolus*
Palmae
Raphia australis Oberm. & Strey
Restionaceae
Restio zuluensis H.P. Linder

Dicotyledons
Acanthaceae
Asystasia pinguifolia T.J. Edwards
Barleria prionitis L. subsp. *delagoensis* (Oberm.) Brummitt & J.R.I. Wood
B. oxyphylla Lindau
Crossandra fruticulosa Lindau
C. zuluensis W.T. Vos & T.J. Edwards
Dicliptera divaricata Compton
Duvernoia aconitiflora A. Meeuse*
Ecbolium glabratum Vollesen
Metarungia longistrobus (C.B. Clarke) Baden*
Salpinctium natalense (C.B. Clarke) T.J. Edwards
Sclerochiton apiculatus Vollesen
Anacardiaceae
Ozoroa engleri R. & A. Fern.
Rhus kwazuluana Moffett
Apocynaceae
Adenium swazicum Stapf
Callichilia orientalis S. Moore
Strophanthus gerrardii Stapf
S. luteolus Codd
Wrightia natalensis Stapf
Araliaceae
Cussonia arenicola Strey
Asclepiadaceae
Asclepias gordon-grayae Nicholas
Aspidoglossum delagoense (Schltr.) Kupicha
Brachystelma tenue R.A. Dyer
B. chlorozonum E.A. Bruce
B. vahrmeyeri R.A. Dyer
Ceropegia arenaria R.A. Dyer
Orbea paradoxa (I. Verd.) L.C. Leach

Orbea longidens (N.E. Br.) L.C. Leach
Orbeopsis gerstneri (Letty) L.C. Leach subsp. *gerstneri**
Pachycarpus lebomboensis D.M.N. Sm.*
Pachycymbium ubomboense (I. Verd.) M.G. Gilbert
Secamone delagoensis Schltr.
Stapelia unicornis C.A. Lückh.*

Bignoniaceae
Dolichandrone alba (Sim) Sprague

Brassicaceae
Lepidium suluense Marais

Caesalpinioideae
Caesalpinia rostrata N.E.Br.
Dialium schlechteri Harms
Erythrophleum lasianthum Corbishley
Schotia capitata Bolle

Canellaceae
Warburgia salutaris (Bertol.f.) Chiov.

Capparaceae
Cleome bororensis (Klotzsch) Oliv.
C. schlechteri Briq.
Maerua brevipetiolata Killick*
M. nervosa (Hochst.) Oliv.

Caryophyllaceae
Krauseola mosambicina (Moss) Pax & K. Hoffm.

Celastraceae
Hippocratea delagoensis Loes.
Maytenus heterophylla (Eckl. & Zeyh.) N. Robson subsp. *arenaria* N. Robson *p.p.* [Abbott 2685]
M. sp. nov. A [Ward 1499, 2009 & 2561]
Salacia krausii (Harv.) Harv.
S. sp. nov. A [Ward 1136 & 2158]

Chrysobalanaceae
Parinari capensis Harv. subsp. *incohata* F. White

Combretaceae
Combretum mkuzense Carr & Retief
C. woodii Dummer*

Compositae
Chrysocoma mozambicensis Ehr.Bayer
Cineraria pinnata O. Hoffm.
Helichrysopsis septentrionale (Vatke) Hilliard
Helichrysum adenocarpum DC. subsp. *ammophilum* Hilliard
H. silvaticum Hilliard
H. tongense Hilliard
Nidorella tongensis Hilliard
Senecio ngoyanus Hilliard
Vernonia centaurioides Klatt
V. inhacensis G.V. Pope

Convolvulaceae
Ipomoea venosa (Desr.) Roem. & Schult. subsp. *stellaris* (Baker) Verdc. var. *obtusifolia* Verdc.
Stictocardia laxiflora (Baker) Hallier f. var. *woodii* (N.E. Br.) Verdc.
Turbina longifolia Verdc.

Crassulaceae
Crassula maputensis R Fernandes
Kalanchoe neglecta Toelken

Ebenaceae
Diospyros inhacaensis F. White
D. rotundifolia Hiern
Euclea natalensis A. DC. subsp. *rotundifolia* F. White

Elatinaceae
Bergia mossambicensis Wild

Erythroxylaceae
Erythroxylum delagoense Schinz

Euphorbiaceae
Croton steenkampianus Gerstner
C. madandensis S. Moore
Euphorbia clavigera N.E. Br.*
E. confinalis R.A. Dyer subsp. *confinalis**
E. grandicornis Goebel ex N.E. Br. subsp. *grandicornus**
E. keithii R.A. Dyer
E. knuthii Pax subsp. *knuthii**
E. persistens R.A. Dyer
Phyllanthus sp. nov. A [Moll & Nel 5578]
P. sp. nov. B [Compton 31053]
P. sp. nov. C [Ward 1927]

Flacourtiaceae
Dovyalis longispina (Harv.) Warb.
D. revoluta Thom

Geraniaceae
Pelargonium tongaense Vorster

Gesneriaceae
Streptocarpus confusus Hilliard subsp. *lebomboensis* Hilliard & B.L. Burtt*
S. wendlandii Spreng.

Labiatae
Hemizygia ramosa Codd*
Ocimum natalense Ayobangira ex Paton
Plectranthus psammophilus Codd

Lobeliaceae
Monopsis belliflora E. Wimm.

Loranthaceae
Helixanthera woodii (Schltr. & K. Krause) Danser

Lythraceae
Galpinia transvaalica N.E. Br.
Nesaea alata Immelman*
N. wardii Immelman

Malpighiaceae
Acridocarpus natalitius Juss. var. *linearifolius* Launert

Malvaceae
Hibiscus barbosae Exell

Hibiscus meyeri Harv. subsp. *meyeri*
Thespesia acutiloba (Baker f.) Exell & Mendonça
Melastomataceae
Memecylon sousae A. & R. Fernandes
Menispermaceae
Albertisia delagoensis (N.E. Br.) Forman
Cissampelos hirta Klotzsch
Mimosoideae
Acacia kraussiana Meisn. ex Benth.
A. swazica Burtt Davy*
Adenopodia schlechteri (Harms) Brenan
Myrtaceae
Eugenia mossambicensis Engl.
Syzygium cordatum Hochst. *p.p.* [suffruticose form]
Nyctaginaceae
Commicarpus chinensis (L.) Heimerl subsp. *natalensis* Meikle
Ochnaceae
Ochna barbosae N. Robson
O. natalitia (Meisn.) Walp. *p.p.* [suffruticose form]
Papilionoideae
Aspalathus gerrardii Bolus *p.p.* [form with minute flowers]
Baphia kirkii Baker subsp. *ovata* (Sim) Soladoye
Crotalaria dura Wood & Evans subsp. *mozambica* Polhill
C. monteiroi Taub. ex Baker f. var. *monteiroi*
C. schlechteri Baker f.
Indigofera delagoaensis Baker f. ex J.B. Gillett
Tephrosia brummittii Schrire
Passifloraceae
Basananthe polygaloides (Hutch. & K. Pearce) W.J. de Wilde
Periplocaceae
Raphionacme elsana Venter & Verhoeven
R. lucens Venter & Verhoeven
Polygalaceae
Polygala francisci Exell
P. marensis Burtt davy
P. torrei Exell
Polygonaceae
Oxygonum dregeanum Meisn. subsp. *streyi* Germish.
O. robustum Germish.
Rhamnaceae
Ziziphus rivularis Codd*
Z. pubescens Oliv. subsp. *glabra* R.B. Drumm.
Rhizophoraceae
Cassipourea mossambicensis (Brehmer) Alston
C. swaziensis Compton*
Rubiaceae
Ancylanthos monteiroi Oliv.
Coffea racemosa Lour.
Gardenia cornuta Hemsl.
G. volkensii K. Schum. subsp. *volkensii* var. *saundersiae* (N.E. Br.) Verdc.
Hyperacanthus microphyllus (K. Schum.) Bridson
Oxyanthus latifolius Sond.
Pavetta gerstneri Bremek.
Psydrax fragrantissima (K. Schum.) Bridson
P. locuples (K. Schum.) Bridson
Tarenna junodii (Schinz) Bremek.
Tricalysia delagoensis Schinz
T. junodii (Schinz) Brenan var. *junodii*
Rutaceae
Toddaliopsis bremekampii I. Verd.
Vepris carringtoniana Mendonça
Zanthoxylum delagoense Waterm.
Sapindaceae
Allophylus mossambicensis Exell
Atalaya alata (Sim) H.M.L. Forbes*
Scrophulariacae
Alectra dunensis Hilliard & B.L. Burtt
Striga junodii Schinz
Solanaceae
Lycium sp. nov. A [Van Wyk 12291]
Solanum sp. nov. A [Vahrmeijer 739]
S. sp. nov. B [Ward 3840 & 7053]
Sterculiaceae
Hermannia micropetala Harv.
Thymelaeaceae
Gnidia sp. nov. A [Nicholas & MacDevette 2153]
Tiliaceae
Corchorus junodii (Schinz) N.E. Br.
Grewia occidentalis L. var. *litoralis* Wild
G. microthyrsa K. Schum. ex Burrett
G. hornbyi Wild
Turneraceae
Tricliceras laceratum (Oberm.) Oberm.
T. mossambicensis (A. & R. Fern.) R. Fern.
Vahliaceae
Vahlia capensis (L.f.) Thunb. subsp. *vulgaris* Bridson var. *longifolia* (Gand.) Bridson
Verbenaceae
Plexipus schlechteri (Gürke) R. Fern. var. *schlechteri*
Vitaceae
Cyphostemma barbosae Wild & R.B. Drumm.*
Rhoicissus sessilifolia Retief

Les Eriocaulaceae et leurs habitats dans les savanes africaines

V. KIMPOUNI
Cellule de coordination ECOFAC-Congo, B.P. 62, Brazzaville, Congo.

Résumé
Dans les savanes à basse herbe, les Eriocaulaceae colonisent les habitats marécageux ou à humidité très élevée sur sable oligotrophe, cuirasse ferralitique, dalles et dômes rocheux, sol tourbeux et fond de vallée. Quoique omniprésentes, depuis les savanes atlantiques jusqu'aux savanes d'altitudes de l'Est, la diversité spécifique de ces habitats est sous la dépendance du substrat et de l'altitude. Leur aire de distribution étant généralement restreinte, les Eriocaulaceae peuvent être des indicateurs pour l'étude phytosociologique de ces milieux.

Abstract
The Eriocaulaceae occupy the swampy habitat in low herb savanna on sand, laterite, flagstone and rocky outcrops and in riverbeds. Although generally present from the Atlantic savanna to the Eastern hills, the specific diversity of Eriocaulaceae depends on substratum and altitude level. The restricted area of Eriocaulaceae taxa may render them suitable as indicators for phytosociological studies in these habitats.

Key words
Altitude, Eriocaulaceae, Habitat, Species diversity, Substrate.

Introduction

La végétation d'Afrique tropicale est formée de deux grandes zones de savanes séparées en partie par la forêt équatoriale (Kimpouni 1993, White 1983).

Les savanes soudaniennes au Nord sont sur un relief exceptionnellement bas et non escarpé, avec des pics ne dépassant guère les 750 m d'altitude sauf dans le Fouta Djalon (Guinéen supérieur) où les pics atteignent 1900 m. Le climat se caractérise par une pluviosité très faible (moyenne annuelle de 753 mm de pluies à certains endroits).

Dans la zone centrale, domaine de la forêt équatoriale, les trouées caractérisent la présence des savanes incluses. Dans cette partie centrale de la Région Guinéo-Congolaise, les altitudes ne dépassent pas les 1000 m, le relief est en général très bas et plat avant de s'élever brutalement vers la dorsale du Kivu, avec des sommets de 4507 m d'altitude. Cette partie orientale est occupée par la végétation montagnarde. Le climat est du type équatorial avec des précipitations moyennes de 1600 à 2000 mm/an.

La partie Sud, la végétation savanicole des hauts plateaux zambéziens est sur un relief supérieur à 900 m d'altitude, et atteignant 2500 m. Le climat est du type continental avec des températures moyennes annuelles plus élevées que dans la Région Guinéo-Congolaise. Les précipitations moyennes annuelles sont de l'ordre de 500 à 1400 mm de pluie.

Cette étude est basée sur des indications d'herbiers, des notes de terrain et des cartes de distribution phytogéographique des Eriocaulaceae africaines et notamment celles d'Afrique centrale.

L.J.G. van der Maesen et al. (eds.), The Biodiversity of African Plants, 208-213.

Ecologie des Eriocaulaceae

Les Eriocaulaceae africaines a montré qu'elles sont des plantes des formations herbeuses basses, des milieux plus ou moins humides à marécageux ou mêmes découverts, généralement de basse et moyenne altitudes. Dans ces formations, elles croissent dans des milieux très particuliers qui sont les mares saisonnières ou permanentes sur sable ou sur dalles rocheuses et cuirasses ferralitiques recouvertes d'une mince couche de sol regorgée d'eau, les sols tourbeux humides, les endroits humides des dômes rocheux (Inselbergs) et les fonds de vallées (Adjanohoun 1965, Dahlgren et al. 1985, Guillaumet & Adjanohoun 1971, Heywood 1978, Hutchinson 1959 et 1973, Jaeger & Adam 1971, Lisowski et al. 1971, Lisowski & Malaisse 1989, Meikle & Baldwin 1952, Miège 1955, Prance & Johnson 1992, Schmitz 1963, Schnell 1952a, b et 1977). Ces différents habitats sont répartis dans toute la zone tropicale africaine.

Les Eriocaulaceae sont en majeure partie aquatiques ou semi-aquatiques (Dahlgren et al. 1985, Giulietti et al. 1987).

Les espèces aquatiques sont habituellement liées aux mares permanentes (*Eriocaulon melanocephalum* Kunth, *E. setaceum* L., *E. stoloniferum* Welw. ex Rendle, *E. submersum* Welw. ex Rendle et *Mesanthemum reductum* Hess).

Les espèces semi-aquatiques ou mésohygrophytes croissent généralement au bord des mares permanentes ou saisonnières et dans les milieux à humidité très élevée (fond de vallée, tourbières et endroits arrosés en permanence par les eaux arrachées aux chutes par le vent).

Les espèces terrestres à tendance xérophytique sont le plus souvent celles du genre *Paepalanthus*. Leur développement a lieu sur les plages sablonneuses non regorgées d'eau.

Dans les formations herbeuses on reconnait deux types: la savane à haute herbe qui ne se prête pas au développement des Eriocaulaceae, malgré la présence des mares (Koechlin 1961). Cependant, les mares et les marécages des savanes à basse herbe, sur substrat sablonneux et sol tourbeux, constituent les principaux sites en Eriocaulaceae.

Les sables du Kalahari occupent des grandes étendues aussi bien dans les zones de hautes que de basses altitudes. Ils recouvrent les hauts Plateaux zambéziens, et en partie les savanes Guinéo-Congolaises (White 1983). Les autres types de sables oligotrophes couvrent les savanes côtières. Ces sites ont un pH acide compris entre 3,9 et 6,4 (Lisowski & Malaisse 1989, Makany 1976, Malaisse 1969 et 1976, Schnell 1949, Symoens 1968, Symoens & Ohoto 1973).

Les dépressions périodiquement regorgées d'eau et couvertes d'une formation herbeuse en perpétuelle évolution sont aussi des sites à Eriocaulaceae. Ces formations herbeuses sont assez courantes sur les plateaux de la Région Zambézienne.

Les affleurements de cuirasse ferralitique, les dalles rocheuses et les dômes rocheux sont des sites présentant une diversité en espèce aussi grande que celle des mares d'altitude sur sable du Kalahari ou oligotrophe. La végétation est souvent discontinue et s'installe là où il y a une mince couche de sol de 1 à 10 cm et périodiquement regorgée d'eau. Le pH est acide et oscille entre 5,1-5,4 (Lisowski & Malaisse 1989, Malaisse 1969 et 1976, Schnell 1949, Symoens 1968, Symoens & Ohoto 1973).

La flore de ces formations herbeuses, notamment celle des cuirasses ferralitiques, a fait l'objet de plusieurs études en Afrique et est principalement graminéenne (Adjanohoun 1962 et 1965, Ahn 1959, Bellier et al. 1969, Bouillenne et al. 1955, Descoings

1975, Deuse 1960, Duvigneaud 1949 et 1953, Evrard 1968, Germain 1965, Jacques-Félix 1947, Jaeger & Adam 1971, Koechlin 1961, Lebrun 1947, Lisowski et al. 1970 et 1971, Lisowski & Malaisse 1989, Makany 1972 et 1976, Malaisse 1975, Meikle & Baldwin 1952, Reitsma et al. 1992, Schmitz 1971, Schnell 1949 et 1977, Symoens 1955, White 1983). Toutefois, notons que l'association générique, spécifique et variétale est fonction du substrat et varie d'un milieu à l'autre ou d'une région à une autre.

L'effet de l'altitude se manifeste par l'établissement d'une augmentation concomitante du nombre d'espèces avec l'altitude jusqu'à l'étage submontagnard (900-1800 m). Les habitats de très hautes altitudes (montagnards et afro-alpins), caractérisés par un climat beaucoup plus tempéré sont moins propices au développement des Eriocaulaceae, tandis que les habitats de faible altitude sont très pauvres en diversité spécifique. Les habitats offrant les meilleures conditions au développement des Eriocaulaceae sont compris entre 900 et 1800 m d'altitude (Kimpouni 1993, Lisowski & Malaisse 1989, Schnell 1949 et 1977, White 1983). Cette même constatation a été faite dans le Fouta Djalon (Guinéen supérieur) en Angola et en Afrique du sud où les habitats situés sur les hauts plateaux sont des sites très riches en Eriocaulaceae (Hess 1955, Obermeyer 1985, White 1983).

Tableau 1. Répartition verticale des espèces d'Afrique centrale et du Congo. Les chiffres denotent la quantité d'espèces.

Genre	Classes d'altitude			
	I	II	III	IV
Eriocaulon	12	39	5	1
Mesanthemum	1	6	0	0
Paepalanthus	1	0	0	0
Syngonanthus	4	13	4	0

La répartition altitudinale des Eriocaulaceae d'Afrique centrale suivant les étages de végétation définis par Lebrun (1960a, b et c), Ross (1955) et Schnell (1977):

I 0-900 m: espèces planitiaires
II 900-1800 m: espèces des hauts plateaux ou submontagnardes
III 1800-2600 m: espèces montagnardes
IV > 2600 m: espèces afro-alpines

est illustrée par le Tableau 1 montrant l'évolution du nombre d'espèces en fonction de l'altitude (Kimpouni 1993). Notons que les habitats de l'étage submontagnard sont les plus colonisés. Le genre *Eriocaulon* est présent dans les 4 étages de la subdivision altitudinale de la flore africaine. Il se développe mieux dans les stations submontagnardes que dans les stations à basse ou à très haute altitude. *Eriocaulon schimperi* Koern. ex Engl. est la seule espèce qui a un comportement montagnard - subalpin et ne descend pas plus bas.

Le genre *Mesanthemum* Koern. est submontagnard. *M. radicans* (Benth.) Koern. est la seule espèce qui descend en dessous de cet étage pour coloniser les stations de basse altitude et avoir également un comportement planitiaire. Le genre *Paepalanthus* est

uniquement planitiaire et de très basse altitude (480-500 m). Les taxons du genre *Syngonanthus* Ruhl. sont plus présents dans l'étage submontagnard que dans les étages planitiaire et afro-montagnard. Ils n'atteignent pas l'étage afro-alpin.
Certaines espèces ont une aire de répartition altitudinale s'étendant sur 2 étages, tel est le cas d'*E. melanocephalum, E. schimperi, Mesanthemum radicans, Syngonanthus ngoweensis* Lecomte et *S. poggeanus* Ruhl.

Les habitats sur cuirasses ferralitiques, dalles et dômes rocheux sont plus riches en diversité spécifique que les habitats sur sable et sol tourbeux. Elles est encore plus importante lorsque le site est à une altitude comprise entre 900 et 1800 m.

La Région Zambézienne avec son relief assez surélevé, ses hauts plateaux et son sol très métallifère présente un macro-climat favorable au développement des Eriocaulaceae. Par contre, les savanes du nord, excepté Fouta Djalon, sont moins favorables aux développement des Eriocaulaceae.

Le mode de dissémination des diaspores, non efficient, n'est pas un facteur de moindre importance (Kimpouni 1993 et sous presse).

Des travaux de phytosociologie, il ressort que les espèces des Eriocaulaceae sont présentes dans plusieurs Associations (*Desmodio-Echinochloetum equitantis* Schmitz, *Anagallo-Mayacetum baumii* Malaisse, *Utricularieto-Eriocauletum pumili* Schnell), divers groupements végétaux des milieux marécageux et sont à la base de l'Alliance d'*Eriocaulon setacei* Lisowski & Malaisse.

Bibliographie

Adjanohoun, E., 1962. Etude phytosociologique des savanes de la basse Côte d'Ivoire (savanes lagunaires). Vegetatio 11: 1-38.

Adjanohoun, E., 1965. Végétation des savanes et des rochers découverts en Côte-d'Ivoire centrale. Mém. ORSTOM 7: 178 pp, avec annexes.

Ahn, P.M., 1958. The savanna patches of Nzima, South-Western Ghana. J. W. Afr. Sci. Ass. 5: 10-25.

Bellier, L., D. Gillon, Y. Gillon, J.L. Guillaumet & A. Perraud, 1969. Recherches sur l'origine d'une savane incluse dans le bloc forestier du Bas-Cavally (Côte d'Ivoire) par l'étude des sols et de la biocoenose. Cahiers ORSTOM, sér. Biol. 10: 65-94.

Bouillenne, R., J. Moureau & P. Deuse, 1955. Esquisse écologique des faciès forestiers et marécageux des bords du lac Tumba. Mém. Acad. Roy. Sci. Colon., Cl. Sci. Nat. Méd. 8, 3(1): 1-144.

Dahlgren, R.M.T., H.T. Clifford & P.F. Yeo, 1985. The Families of the Monocotyledons: Structure, Evolution and Taxonomy. Springer-Verlag, Berlin: 520 pp.

Descoings, B., 1975. Les grandes régions naturelles du Congo. Candollea 30: 91-120.

Deuse, P., 1960. Etude écologique et Phytosociologique de la végétation des Esobé de la région Est du Lac Tumba (Congo-Belge). Acad. Roy. Sci. Outre Mer, Sci. Nat. Méd., Mem. 8, nouvelle sér. 11 (3): 115 pp.

Duvigneaud, P., 1949. Les savanes du Bas-Congo, Essai de phytosociologie topographique. Lejeunia 10: 192 pp.

Duvigneaud, P., 1953. La Flore et la végétation du Congo Méridional. Lejeunia 16: 95-124.

Duvigneaud, P., 1958. La Végétation du Katanga et des sols métallifères. Bull. Soc. Roy. Belg. Belg. XC, 2: 111-134.

Evrard, C., 1968. Recherches écologiques sur le peuplement forestier des sols hydromorphes de la Cuvette centrale congolaise. Inst. Nat. Et. Agr. Congo (INEAC), Sér. Sci. 110: 295 pp.

Germain, R., 1965. Les biotopes alluvionnaires herbeux et des savanes intercalaires du Congo équatorial. Mém. Acad. Roy. Sci. Outre-Mer, Cl. Sci. Nat. Méd. 8, 15(4): 1-399.

Giulietti, A.M., W.R. Monteiro, S.J. Mayo & J. Stephens, 1987. A preliminary survey of testa sculpture in Eriocaulaceae. Beitr. Biol. Pflanzen. 62(2): 189-209.

Guillaumet, J.L. & E. Adjanohoun, 1971. La végétation de la Côte-d'Ivoire. In: Le milieu naturel de la Côte-d'Ivoire, Mém. ORSTOM 50: 157-232.

Hess, H., 1955. Zur Kenntnis der Eriocaulaceae von Angola und dem unteren Belgischen Kongo. Ber. Schweiz. Bot. Ges. 65: 115-204.

Heywood, V.H., 1978. Flowering plants of the World. Oxford University Press, Oxford: 281-282.

Hutchinson, J., 1959. The families of Flowering plants. Monocotyledons, ed. 2. 2, Clarendon Press, Oxford: 574-576.

Hutchinson, J., 1973. The families of Flowering plants. Ed. 3, Oxford University Press, Oxford: 968 pp.

Jacques-Felix, H., 1947. Le genre Mesanthemum (Eriocaulaceae). Bull. Soc. Bot. Fr. 94: 143-151.

Jaeger, P. & J.-G. Adam, 1971. Aperçu succinct sur la flore et la végétation de l'étage culminal des Monts Loma (Sierra Leone). Mitt. Bot. Staatssamml. München 10: 478-483.

Kimpouni, V., J. Lejoly & S. Lisowski, 1992. Les Eriocaulaceae du Congo. Fragm. Flor. Geobot. 37(1): 127-145.

Kimpouni, V., 1993. Les Eriocaulaceae d'Afrique centrale (Congo, Zaïre, Rwanda, Burundi); Etude Taxonomique et Phyto-géographique. Thèse Doc. Univ. Libre Bruxelles (Inédit): 589 pp.

Kimpouni, V. Analyse phytogéographique des Eriocaulaceae en Afrique et à Madagascar. Mém. Soc. Biogéogr. 3e Sér., 4 (sous presse).

Koechlin, J., 1961. La végétation des savanes dans le Sud de la République du Congo. ORSTOM, Paris: 310 pp.

Lebrun, J., 1947. La végétation de la plaine alluviale au Sud du lac Edouard: Exploration du parc national Albert, mission J. Lebrun, 1937-1938). Publ. Inst. Parcs Nat. Congo belge, Bruxelles, 2 vol.: 800 pp.

Lebrun, J., 1960 (a). Sur les horizons et étages de végétation de divers volcans des Virunga (Kivu-Congo). Bull. Jard. Bot. Etat 30: 255-277.

Lebrun, J., 1960 (b). Sur une méthode de délimitation des horizons et étages de végétation des montagnes du Congo Oriental. Bull. Jard. Bot. Etat 30: 75-94.

Lebrun, J., 1960 (c). Sur la richesse de la flore de divers territoires africains. Bull. Acad. Roy. Sci. Outre-mer, Bruxelles, N.S., 6: 669-690.

Lisowski, S., F. Malaisse & J.J. Symoens, 1970. Plantes rares ou nouvelles pour la flore du Katanga. Bol. Soc. Broter. 44 (2. sér.): 225-244.

Lisowski, S., F. Malaisse & J.J. Symoens, 1971. Une flore des hauts plateaux du Katanga. Mitt. Bot. Staatssamml. München, 10: 51-56.

Lisowski, S. & F. Malaisse, 1989. Groupements végétaux des mares et des anses calmes des rivières du Plateau des Kundelungu. Dans: Symoens, J.J. (éd.), Exploration hydrobiologique du bassin du lac Bangweolo et du Luapula. 18(1): 41 pp.

Makany, L., 1972. Une tourbière tropicale à sphaignes sur les contreforts des Plateaux Batéké (Congo). Ann. Univ. Brazzaville, 8: 93-104.

Makany, L., 1976. Végétation des Plateaux Téké (Congo). Collection: Travaux de l'Université de Brazzaville: 301 pp.

Malaisse, F., 1969. Note à propos de la Flore du plateau des Kundelungu (Haut-Katanga, Rép. dém. Congo). Trav. Serv. Sylv. Pisc. Univ. Off. Congo (Lubumbashi), 2: 10 pp.

Malaisse, F., 1975. Carte de la végétation du bassin de la Luanza. Exploration Hydrologique du bassin du lac Bangweolo et du Luapula. Bruxelles 18(2): 41 pp.

Malaisse, F., 1976. Ecologie de la rivière Luanza. Dans: Symoens, J.J. (ed.), Exploration du bassin du lac Bangweolo et du Luapula. vol. 17(2): 151 pp.

Meikle, R.D. & J.T. Baldin, 1952. Eriocaulaceae and Xyridaceae in Liberia. Dans: Tippo, Bonner, Camp, Cleland, Creighton & Drew (eds.), Amer. J. Bot. 39(1): 44-51.

Miège, J., 1955. Les savanes et les forêts claires de Côte-d'Ivoire. Etudes Eburnéennes, Abidjan, 4: 62-81.

Obermeyer, A.A., 1985. Eriocaulaceae. Dans: Leistner, O.E. (ed.), Flora of Southern Africa which deals with the territories of South Africa, Ciskei, Transkei, Lesotho, Swaziland, Bophuthatswana, South West Africa/Namibia, Botswana and Venda, 4(2): 19-21. Botanical Research Institute, Pretoria.

Ozenda, P., 1982. Les végétaux dans la biosphère. Doin, Paris, 1 vol., 8 + 341 pp.

Prance, G.T. & D.M. Johnson, 1992. Plant collections from the plateau of Serra do Araca (Amazonas, Brazil) and their phytogeographic affinities. Kew Bull. 47(1): 1-24.

Reitsma, J.M., A.M. Louis & J.-J. Floret, 1992. Flore et végétation des inselbergs et dalles rocheuses: première étude au Gabon. Bull. Mus. natl. Hist. nat. Paris, 4e sér., 14, sect. B. Adansonia 1: 73-97.

Ross, R., 1955. Some aspects of the vegetation of Ruwenzori. Webbia 10: 451-457.

Schmitz, A., 1963. Aperçu sur les groupements végétaux du Katanga. Bull. Soc. Roy. Bot. Belg. 96: 223-447.

Schmitz, A., 1971. La végétation de la plaine de Lubumbashi (Haut-Katanga). Publ. INEAC, Sér. Sci., 113: 388 pp.

Schmitz, A., 1988. Révision des groupements végétaux décrits du Zaïre du Rwanda et du Burundi. Ann. Sci. Eco., Musée Royal de l'Afrique centrale Tervuren 17, Bruxelles: 315 pp.

Schnell, R., 1949. Végétation et flore des Monts Nimba (Afrique occidentale française). Ann. Biol., t. 25(7-8): 187-198.

Schnell, R., 1952 (a). Contribution à une étude phyto-sociologique et phytogéographique de l'Afrique occidentale: les groupements et les unités géobotaniques de la Région guinéenne. Melanges Bot., Mém. Inst. Fr. Afr. Noire 18: 43-234.

Schnell, R., 1952 (b). Végétation et flore de la région montagneuse du Nimba. Mém. Inst. Fr. Afr. Noire 22: 604 pp.

Schnell, R., 1977. La flore et la végétation d'Afrique tropicale. Gauthier-Villars, Paris, vol. 1: 459 pp.; vol. 2: 378 pp.

Symoens, J.J., 1955. Les Monts Marungu se distinguent nettement, par leur flore d'affinité afro-orientale, des autres hauts plateaux Katangais. Thèse annexe. Univ. Libre Bruxelles.

Symoens, J.J., 1968. La minéralisation des eaux naturelles. Dans: Exploration hydrobiologique du bassin de lac Bangweolo et du Luapula. Vol. 2 (1): 199 pp.

Symoens, J.J. & E. Ohoto, 1973. Les éléments phytogéographiques de la flore macrophytique aquatique et semi-aquatique du Haut-Katanga. Verh. Internat. Verein. Limnol. 18: 1385-1394.

White, F., 1983. The vegetation of Africa. A descriptive memoir to accompany the UNESCO AETFAT/UNSO/ Vegetation map of Africa. UNESCO, Paris, 1 vol.: 356 pp. Traduction française par P. Bamps, 1986, La végétation de l'Afrique. Mémoire accompagnant la carte de végétation de l'Afrique UNESCO/AETFAT/UNSO. Recherches sur les ressources naturelles 20, ORSTOM-UNESCO, Paris, 1 vol.: 384 pp.

Flora and vegetation of Niokolo-Koba National Park, Senegal

J.E. MADSEN [1,2], D. DIONE [3], A.S. TRAORÉ [3] & B. SAMBOU [1]
[1] Inst. of Environmental Sciences, University Cheikh Anta Diop, Dakar, Senegal.
[2] Inst. of Biological Sciences, Aarhus University, 68 Nordlandsvej, DK-8240 Risskov, Denmark.
[3] Department of Plant Biology, University Cheikh Anta Diop, Dakar, Senegal.

Abstract
The flora of Niokolo-Koba National Park (913,000 ha) is estimated at approximately 1,000 species. Nine 1 ha plots had 3,134 trees in 99 tree species with a dbh ≥ 5 cm. Three plots of gallery forest contained, per hectare, 236-273 trees, 26-44 species, 15-25 families, and a basal area of 16.4-36.2 m^2. Four plots of woodland included, per hectare, 338-1048 trees, 19-30 species, 12-17 families, and a basal area of 7.6-11.3 m^2. The leading tree families in the study plots were: Leguminosae, Combretaceae, and Anacardiaceae.

Résumé
La flore du Parc National du Niokolo Koba (913.000 ha) est estimée à environ 1000 espèces. Neuf parcelles de 1 ha recèlent 3134 arbres de dbh ≥ 5 cm répartis en 99 espèces. Trois parcelles de forêts galeries abritent par ha: 236-273 arbres, 26-44 espèces, 15-25 familles, 16,4-36,2 m^2 de section de base. Quatre parcelles de forêts claires ont par ha: 338-1048 arbres, 19-30 espèces, 12-17 familles, 7,6-11,3 m^2 de section de base. Trois grandes familles se dégagent de cette étude: Leguminosae, Combretaceae et Anacardiaceae.

Key words
Forest structure, Gallery forest, National park, Plant diversity, Savanna vegetation, Senegal.

Introduction

Niokolo-Koba National Park is the largest national park in West Africa and is classified as a World Heritage site by UNESCO (Sayer et al. 1992). It was established in 1925 as a reserve and obtained its present status as a National Park in 1954.

Existing floristic literature dealing with Niokolo-Koba National Park (Adam 1963, 1968 & 1971, Schneider & Sambou 1982) contains information on useful plants, descriptions of habitats, and checklists of plant species. A forthcoming paper (Lawesson in press) is a study on the woody species in Senegal based on multivariate analyses and it also contains data gathered in the Niokolo-Koba National Park. Finally, a recent publication (Lykke 1994) analysed the woody vegetation of yet another National Park in Senegal.

The present study describes structural variation in the principal forest formations in the Niokolo-Koba National Park based on data from nine permanent 1 ha plots. The woody plant families are ranked according to their ecological importance, and plant diversity in the region is discussed as well.

Study area

The study area is located in south-eastern Senegal (12°38'-13°22'N; 12°22'-13°43'W) and covers an area of approximately 913,000 ha. The climate is characterised by a

L.J.G. van der Maesen et al. (eds.), The Biodiversity of African Plants, 214-219.

pronounced rainy season which runs from July to October with an annual precipitation of approximately 800-1100 mm. Average monthly temperature is 25°C in November-January and 33°C in April and May.

The relief is relatively flat except in the area around Mount Assirik which is situated 311 m above sea level. The park is traversed by the large Gambia River and its seasonally dry tributaries, and there are also numerous seasonal swamps (Dupuy 1971).

Vegetation cover

Originally, most of Niokolo-Koba National Park was covered with Sudanian dry forest (Adam 1971). Since ancient times man has used precocious fire in the region, and abandoned villages inside the park testify former intervention of man in the forest ecosystem. Today, the vegetation is still burnt annually and the park service is also using fire as a tool for improving the tourists' ability to view of the animals.

Much of the savanna vegetation at present is made up of woodland dominated by *Pterocarpus erinaceus* Poir. and bush savanna with *Combretum glutinosum* Perr. Exposed areas with crust-like horizons may be almost devoid of woody vegetation. The rivers in the study area are bordered by riparian vegetation and the valleys contain narrow gallery forest with Guinean elements such as *Erythrophleum guineense* G. Don and *Pseudospondias microcarpa* A. Rich. intermixed with patches of bamboo (*Oxytenanthera abyssinica* Munro) and palms (*Borassus aethiopum* Mart.). Numerous temporarily inundated depressions harbour a rich aquatic flora.

Methods and criteria

Nine permanent plots of 1 ha each were established during 1993 and represent woodland, grassland, and gallery forest. The study plots are quadrates (100 X 100 m) or rectangles (500 X 20 m) in the case of narrow gallery forest.

All woody plants including lianas over 5 cm dbh were permanently marked with numbered aluminium tags. In the case of multiple stemmed trees each trunk was measured separately. The tree height was measured with a clinometer or estimated for small trees.

The criteria used for selecting the study plots were the following:

* Four plots of woodland were set up in areas where the forest has a tall stature and appears to be rich in species. The term "woodland" is used in a broad sense and plot 3 could also be regarded as tree savanna. The soils are usually ferruginous with a low litter content.
* Two plots of grassland were selected in order to demonstrate that large areas are but weakly covered with woody vegetation.
* Finally, three plots of gallery forest were established. One site (plot 7) is a quadrat located in the bottom of a shallow valley where the soils dry up shortly after the end of the rainy season. The other sites (plots 8-9) are situated along a narrow gallery forest in the bottom of a 15 m deep ravine. In the distal section (plot 8) the stream-bed dries up periodically. The proximal section (plot 9) is permanently watered by a stream. The clayey soils are rich in litter.

Table 1. Selected data for all woody plants with a dbh ≥ 5 cm in nine 1 ha plots.

Characteristic	Woodland				Grassland		Gallery forest		
	1	2	3	4	5	6	7	8	9
No. of trees	1048	456	418	338	0	66	299	236	273
No. of species	30	29	19	20	0	7	26	44	32
No. of genera	24	26	16	17	0	6	23	42	27
No. of families	14	17	12	14	0	6	15	25	17
Simpson's Index	83.4	87.6	87.1	82.8	-	62.1	84.5	94	93.3
Brillouin's Index	3.0	3.5	3.4	2.8	-	1.7	3.2	4.2	4
No. of stems	1397	550	563	664	0	101	437	432	369
max dbh (cm)	85	75.5	48.8	79	-	39	222.3	230	263
mean dbh (cm)	9.1	13.9	12.3	12.8	-	9.8	20.3	17.8	26.6
s.d. (±)	5	10.2	7.5	10.4	-	6.7	19.8	22.1	30.4
max height (m)	22	19	10	16	-	12	30	33	40
average height (m)	6.6	6.7	4.9	5.8	-	4	8.5	8.5	12.9
s.d. (±)	2	3.3	1.6	2.9	-	2.2	4.4	5.1	8.2
basal area (m^2)	10.5	11.3	7.6	8.8	-	0.9	20.4	16.4	36.2

Table 2. Family Importance Value (FIV) for ecologically important plant families in nine 1 ha plots. The treeless plot 5 has been omitted.

(Sub-) Family	Woodland				Grassland	Gallery forest		
	1	2	3	4	6	7	8	9
Anacardiaceae	16.9	15.1	28.2	17.2	26.0	8.7	22.7	39.6
Annonaceae	15.5	38.9	5.5	10.1	39.0	28.6	6.0	3.9
Bombacaceae	10.1	18.8	22.4	6.6		6.8	9.8	32.7
Caesalpinioideae	59.5	17.5	40.6	19.9		47.0	20.4	18.3
Combretaceae	95.6	66.7	64.7	73.1		67.3	30.5	16.0
Papilionoideae	46.5	46.3	74.5	76.0	122.0	71.7	5.9	
Mimosoideae	15.0			10.0	79.0	10.6	12.3	30.9
Moraceae					17.0		21.0	13.0
Rubiaceae	18.0	5.4	20.5	19.4		20.0	8.4	15.4
Sapotaceae							30.8	42.3
Sterculiaceae		7.4					56.3	
Tiliaceae	3.8	12.9	5.5	4.4		5.3	10.4	3.5
Verbenaceae		44.2	6.7	7.1		6.3	8.8	9.5
"Legumes"	121.0	63.8	115.1	106.6	96.4	129.3	38.6	49.2

Results and discussion

The nine study plots had a total of 3,134 individual trees with 4,513 stems over 5 cm dbh. Quantitative data from the study plots are presented in Table 1 and ecologically important plant families are summarised in Table 2.

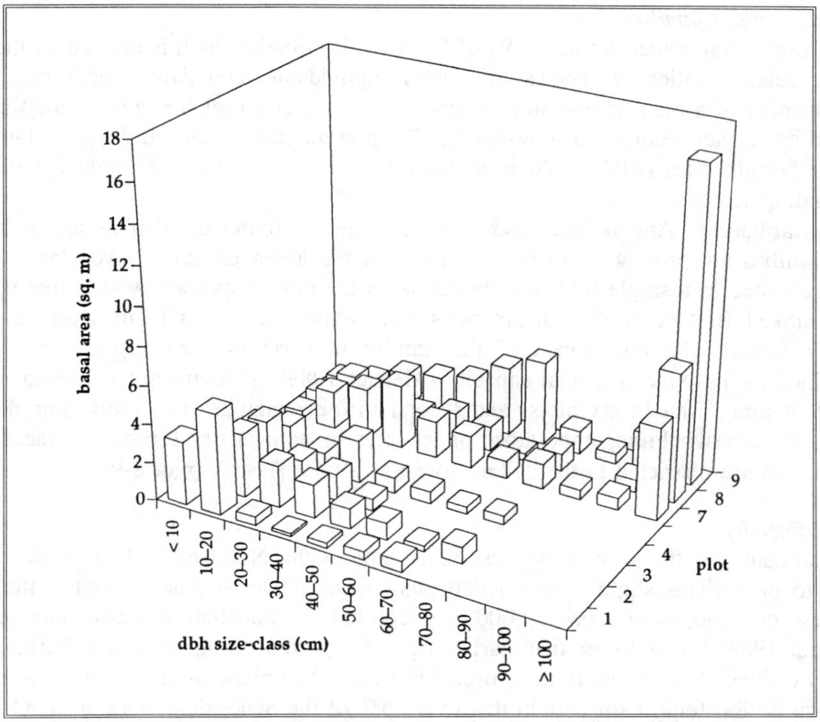

Figure 1. Size-class distribution of basal area in dbh intervals of 10 cm. The plots represent woodland (1-4) and gallery forest (7-9).

Structural differences in woody vegetation

The density of trees in plots of woodland is greater than in plots of gallery forest. However, the trees are smaller. Thus, the average dbh in plots of woodland is much lower than in plots of gallery forest and the thickest trunks in the four one-ha plots of woodland are thinner than the thickest trunks encountered in the three plots of gallery forest.

Likewise, average tree height in plots of woodland have a lower range than in plots of gallery forest and moreover the maximum height of a tree in plots of woodland is also less than in plots of gallery forest. The structural difference between plots of woodland and gallery forest stands out very clearly if basal area is compared. Thus, total basal area for woodland plots is considerably lower than for plots of gallery forest.

Figure 1 illustrates the distribution of basal area in dbh size-classes of 10 cm. It is demonstrated that basal area per ha for small trees (dbh $<$ 20 cm) is greater in plots of woodland than in plots of gallery forest and most of the total basal area per ha in gallery forest comes from the larger size-classes.

With the exception of large trees, which generally have unbranched trunks, a general characteristic of the woody vegetation studied, is the abundance of multi-stemmed trees.

Important plant families

The Family Importance Value (FIV) of Mori et al. (1988), which is defined as the sum of the relative ratios of species (diversity), individuals (abundance), and basal area (dominance) of a plant family in a community, is a useful tool for ranking the families according to their ecological importance. The present study included 31 plant families. The important ones (FIV > 20 in at least one plot) are shown in Table 2 with corresponding scores.

Anacardiaceae, Annonaceae, and Combretaceae are found in all plots and a further five families are present in all but one plot. At the lower extreme, seven families are only recorded in a single plot. Combretaceae is the most important woody tree family. It is ranked first or second in six plots and obtain the 3rd and 7th score in moist gallery forest. The importance of this family in xeric savanna vegetation is often explained by its wind dispersal capacity (Menaut 1983). If legumes are treated as one family it ranks first in six plots, and second in the remainder two plots: this demonstrates the overwhelming importance of legumes in tropical dry forests. Moraceae and Sapotaceae are restricted to the gallery forest at Mount Assirik (plot 8-9).

Plant diversity

We estimate that the species richness of Niokolo-Koba National Park (913,000 ha) is close to or perhaps slightly over 1,000 plant species. Our estimate matches the areal richness of 1060 species per 10,000 km reported for Sahelian and Sudanian regions (Menaut 1983) but is lower than earlier figures reported for the National Park. Adam (1971) estimated the parks flora to probably over 1,500 plant species, and Schneider & Sambou (1982) took it for certain that over 75% of the Senegalese flora is found in the National Park; Senegal has 2,086 species according to Lebrun (1973).

The nine study plots comprise approximately 99 species, 75 genera, and 31 families of woody plants. The number of woody species over 5 cm dbh in each study plot lies in the range of 19-30 species per ha for plots of woodland, 26-44 species per ha for plots of gallery forest, and only 0-7 species per ha for plots of grassland. Indices of diversity for finite populations according to Simpson and Brillouin (formulas and explanation of these indices in Krebs 1989) gave highest scores for the two plots of gallery forest at Mount Assirik (Table 1). This is in line with our expectations although our data are not really comparable because the two plots at Mount Assirik are elongated transects.

Conclusion

The present study carried out in the Sudanian phytochoria demonstrated important structural differences between samples of woodland and gallery forest. In general, the woody vegetation is characterised by a few preponderant plant families.

The gallery forests of Niokolo-Koba National Park constitute extrazonal habitats for numerous Guinean plant species occurring at the northern limit of their natural area of distribution.

Acknowledgements

We would like to thank A.M. Lykke for making useful comments to the manuscript. The study was financed by a grant from DANIDA (StS.4.104.DAN 8.L./203).

References

Adam, G., 1963. Les plantes utiles du Parc National du Niokolo Koba (Sénégal). Notes Africaines 97: 5-21.

Adam, G., 1968. La flore et la végétation du Parc National du Niokolo Koba (Sénégal). Adansonia, sér 2, 8 (4): 439-459.

Adam, G., 1971. Le Milieu Biologique. Flore et Végétation. In: Dupuy, A. R. (ed.), Le Niokolo Koba premier grand Parc national de la République du Sénégal: 43-62, G.I.A., Dakar.

Dupuy, A.R. (ed.)., 1971. Le Niokolo Koba premier grand Parc national de la République du Sénégal. G.I.A., Dakar.

Krebs, C.J., 1989. Ecological methodology. Donnelley & Sons, New York.

Lawesson, J.E., in press. Studies of woody vegetation and flora in Senegal. Opera Bot. 1994.

Lebrun, J.P., 1973. Enumération des Plantes Vasculaires du Sénégal. Etude Botanique No. 2. Institut d'Elevage et de Médecine Vétérinaire des Pays Tropicaux, Maisons-Alfort.

Lykke, A.M., 1994. The Vegetation of Delta du Saloum National Park, Senegal. AAU Reports 33.

Menaut, J.-C., 1983. The Vegetation of African Savannas. In: Bourlière, F. (ed.). Ecosystems of the World 13. Tropical Savannas., pp. 109-149. Elsevier, Amsterdam.

Mori, S.A., B.M. Boom, A.M. de Carvalino & T.S. dos Santos, 1983. Ecological Importance of Myrtaceae in an Eastern Brazilian Wet Forest. Biotropica 15(1): 68-70.

Sayer, J.A., C.S. Harcourt & N.M. Collins (eds.), 1992. The Conservation atlas of tropical forests. Africa. Simon & Schuster, New York etc.: 175-182.

Schneider, A. & K. Sambou, 1982. VI. Prospection botanique dans les parcs nationaux du Niokolo Koba et de Basse Casamance. Mémoires de l'Institut Fondamental d'Afrique Noire 92: 101-122.

Inventory and conservation of economic plant genetic resources in Kenya rangeland: a case study of Turkana district

F. MSAFIRI
Department of Resource Surveys and Remote Sensing, P.O. Box 47146, Nairobi, Kenya.

Abstract
The importance of genetic resources for pastoralists in ASAL areas of Kenya, calls for optimization of plant-use data banks. In Turkana reservation and ownership of grazing areas, **epaka/amire** and **ekwar** respectively, are two of the particular methods also employed by most pastoralists in Kenya's ASALs, including the Masai, and are therefore indirect methods of conservation of biodiversity, particularly of plant species.

Key words
Conservation, Genetic resources, Kenya.

Introduction

Kenya's Arid and Semi-Arid Lands (ASALs), cover about two thirds of the country's land surface. Approximately 20% of the total human population (mostly pastoralists) and about 60% of livestock and wildlife live in these areas (Weiss 1986). Scanty and erratic rainfall hinders crop production in these areas. Therefore, the people depend on renewable natural resources as a source of fodder for both grazers and browsers, fuelwood, building, food, medicine for both livestock and man, making of household utensils etc. The importance of genetic resources for pastoralists in ASAL areas of Kenya, therefore, calls for optimization of plant-use data banks.

Plants are the only significant resources in much of Kenya's ASALs and there is a need for preparing an inventory to determine the status of these resources before recommending their utilization and conservation measures. This paper examines, as an example, the inventory, utilization and conservation of plant genetic resources in Kenya rangeland with particular reference to Turkana District.

Study area

Turkana District covers an area of 72,000 km², in northern Kenya. It has a population of about 225,000 people (Ecosystem Ltd. 1985). The area receives varying amounts of rainfall ranging from 150-200 mm in the centre area to over 400 mm in south. It is bordered by Lake Turkana to the east, Samburu District to the south east, W. Pokot and Baringo districts to the south, Uganda escarpment to the west and Sudan and Ethiopia to the north.

Objectives

* Compile a plant-use data base;
* Form a data base for resource monitoring;
* Maximize plant benefits for livestock feed and their products for domestic consumption.

L.J.G. van der Maesen et al. (eds.), The Biodiversity of African Plants, 220-223.

Methodology

Since the Turkana people are nomadic, an attempt was made to visit their settlements (bomas) and watering points and a very brief on-the-spot interview was carried out on how they use different plants. Information on local names and plant uses was collected. Plant species were collected and identified later in both National Museums of Kenya (NMK) and the herbarium of the Department of Resource Surveys and Remote Sensing (DRSRS) in Nairobi.

Economic plant genetic resources of Turkana district

Virtually all the dominant and common plant species in Turkana are utilized by pastoralists for various uses (Appendix 1). It is important to note that most Turkana pastoralists, both young and old, have an admirable grasp of the names and basic material properties of their local flora: the good stock feeds, the poisonous species, fibre yields, fuelwood, food source, etc. (see also Kuchar 1981). The urgency is therefore in collection and compilation of this bank of knowledge into a form that can be retrieved and used for future conservation strategies

Plant inventory and documentation in Kenya

We cannot talk of conservation of plant genetic resources without knowing our stocks and their status. The Department of Resource Surveys and Remote Sensing (DRSRS) has been developing an inventory of plant genetic resources of Kenya's ASALs for the last fifteen years for the purpose of monitoring range condition and trend. Vegetation is one of the most important parameters in determining the range condition and trend, hence periodical collection of these resources is paramount (Kuchar 1979). In the course of this inventory, collection and identification of plant genetic resources of economic importance have been entered in the data base.

DRSRS and the NMK herbaria have been documenting plants found in Kenya, particularly the following: plant species, habitat, collection date, collector, uses and other important ecological parameters. This is done with a view to set up appropriate research strategies for those plants that have been identified as being rare, based on their last date of collection.

Plants conservation strategies used by Turkana people

As nomads, Turkana people depend almost entirely on plants for their daily requirements. They exploit different economic niches (their cattle including grazers and browsers) and diversify food procurement (Brainard 1981). To be able to meet these important requirements the Turkana people employ silvo-pastoral systems which makes use of the vegetation during both dry and wet seasons. This system entails the setting aside of dry season grazing reserves known as **epaka** or **amire** in the Turkana language (Barrow 1991).

Riverine forest strips and hilly areas are mostly reserved for dry season grazing. These areas are normally owned by clans or individuals (Ekwar-ownership), who have got the right to use the plant resources within them. Some of the most important tree species Turkana people have high regard of are those of food source, e.g. *Dobera glabra, Cordia sinensis, Salvadora persica, Hyphaene compressa* and *Boscia coriacea.* Others include *Acacia tortilis* which is an important fodder source.

Conclusion

In Turkana reservation and ownership of grazing areas, **epaka/amire** and **ekwar** respectively, are two of the particular methods also employed by most pastoralists in Kenya's ASALs, including the Masai, and are therefore indirect methods of conservation of biodiversity, particularly of plant species.

References

Barrow, E.G.C., 1991. Usufruct right to trees: the role of ekwa in dryland central Turkana District, Kenya. Human Ecology 18, 2.

Brainard, J.M., 1981. Herders to farmers: The effects of settlement on the demography of the Turkana population of Kenya. Ph.D. Diss., State of New York, Binghampton.

Ecosystem Ltd., 1985. Turkana District Resources Survey, 1982-1984. Report for the Republic of Kenya, Ministry of Energy and Regional Development, Turkana Rehabilitation project.

Kuchar, P., 1979. A methodology for sampling vegetation in order to monitor condition, trend and plant biomass in Kenya rangeland. Ministry of Planning and National Development, Department of Resource Surveys and Remote Sensing, Nairobi.

Kuchar, P., 1982. Some Kamba plant names and uses: Study on water-point impact on range condition in Kitui District, Kenya. Louis Berger International, Nairobi, Kenya.

Weiss, E., 1986. Plants of arid and semi-arid conditions. Ministry of Agriculture and Livestock Development, German Agricultural Team, Nairobi.

Appendix 1

Some common plants used by Turkana people.

Scientific name	(Sub)Family	Local name	Uses M	Uses F	Uses B
Acacia elatior Brenan	Mimosoideae	Esanyanait	-	x	x
A. mellifera (Vahl) Benth.	Mimosoideae	Ebonyo	-	x	x
A. oerfota (Forsk.) Schweinf. (*A. nubica* Benth.)	Mimosoideae	Epetet	x	-	x
A. paolii Chiov.	Mimosoideae	Eiyelel	x	-	x
A. senegal (L.) Wild	Mimosoideae	Ekunoit	-	x	x
A. tortilis (Forsk.) Hayne	Mimosoideae	Ewoi	-	x	x
Adenium obesum (Forsk.) Roem. & Schult.	Apocynaceae	Egetes	x	-	-
Amaranthus angustifolius Lam. (*A. graecizans* L.)	Amaranthaceae	Louyong'orok	-	x	x
Balanites rotundifolia (Thieg.) Blatt. (*B. orbicularis* Sprague)	Balanitaceae	Ebei	x	x	x
Boscia coriacea Pax	Capparaceae	Erdung	-	x	x
Boswellia neglecta S. Moore (*B. hildebrandtii* Engl.)	Burseraceae	Ekinyate	x	-	x
Cadaba farinosa Forsk.	Capparaceae	Eiva	x	-	x
Calotropis procera (L.) Aiton	Asclepiadaceae	Etesuro	x	-	x
Caralluma somalica N.E. Br.	Asclepiadaceae	Lokoturio	x	-	
Cissus nieriensis T.C.E. Fries (*Cyphostemma nieriense* Descoings)	Vitaceae	Erodo	-	x	x
Cissus quadrangularis L.	Vitaceae	Egis	x	-	-
Cleome parvipetala R.A. Graham	Capparaceae	Enomokere	x	-	-
Commicarpus plumbagineus (Cav.) Standl.	Nyctaginaceae	Lokucin	x	-	x

Cordia gharaf (Forsk.) Asch. (*C. sinensis* Lam.)	Boraginaceae	Edome	-	x	x
Courbonia decumbens Brongn.	Capparaceae	Erut	x	x	x
C. subcordata Gilg	Capparaceae	Erut	x	x	x
Cucumis dipsaceus Ehrenb.	Cucurbitaceae	Ekaleruk	-	x	x
Dactyloctenium bogdanii S.M. Phillips	Gramineae	Ekanduhudu	-	x	x
Dobera glabra (Forsk.) Juss. ex Poiret	Salvadoraceae	Edapal	x	-	x
Eleusine aegyptiaca Desf.	Gramineae	Ekanduhudu	-	x	x
Euphorbia cuneata Vahl	Euphorbiaceae	Echokokile	x	-	-
E. uhligiana (Pax) Engl.	Euphorbiaceae	Emusi	x	-	-
Jasminum fluminense Vell.	Oleaceae	Epelech	x	-	-
Jatropha fissispina Pax	Euphorbiaceae	Lotou	x	-	-
Justicia odora (Forsk.) Lam.	Acanthaceae	Lopara	x	-	x
Kohautia caespitosa Schnizl.	Rubiaceae	Enomokere	x	-	-
Launaea cornuta (O. & H.) C. Jeffrey	Compositae	Echokokile	x	-	-
Lawsonia inermis L.	Lythraceae	Esajait	x	-	x
Leucas pododiskos Bullock	Labiatae	Locikiria	x	-	x
Momordica trifoliolata Hook. f.	Cucurbitaceae	Erikoi	-	x	x
Ricinus communis L.	Euphorbiaceae	Ebune	x	-	-
Salvadora persica L.	Salvadoraceae	Esekon	x	x	x
Sarcostemma viminale (L.) R. Br.	Asclepiadaceae	Eligoi	x	-	-
Seddera hirsuta Hall.f.	Convolvulaceae	Lomanang	x	-	x
Senecio discifolius Oliv. (*Emilia discifolia* (Oliv.) C. Jeffrey)	Compositae	Eididi	x	-	x
Sericocomopsis hildebrandtii Schinz	Amaranthaceae	Ekabonyo	x	-	x
Solanum dubium Fresen.	Solanaceae	Esikilele	-	x	x
S. incanum L. (*S. coagulans* Forsk.)	Solanaceae	Elulelo	x	-	x
Terminalia spinosa Engl.	Combretaceae	Epetait	x	-	-
Vahlia viscosa Roxb.	Saxifragaceae	Lonyang	x	-	x
Ziziphus jujuba Lam. (*Z. mauritiana* Lam.)	Rhamnaceae	Ekalale	-	x	x

M = source of medicine; F = source of food; B = source of browse/building.

The floristic status of grasses of the South Luangwa national park and the Lupande area

P.S.M. PHIRI
Biology Department, University of Zambia, P.O. Box 32379, Lusaka, Zambia.

Abstract
This paper gives an account of the floristic status of grasses which occur in the South Luangwa National Park and the Lupande Game Management Area. In the study area a total of 189 species of grasses belonging to 12 tribes has been recorded. In terms of species number the tribes Andropogoneae, Eragrostideae and Paniceae are the largest.

Key words
Afro-rift valleyane flora, Grasses, Luangwa National Park, Lupande Game Management Area.

Introduction

During the last three decades much has been written on the subjects of the afro-alpine and afro-montane flora of tropical Africa (Killick 1978, Lind & Morrison 1974, White 1978). However, another African geomorphological feature whose flora has not received adequate phytogeographical attention is the African rift valley system for which the term "afro-rift valleyane flora" is proposed. The Luangwa Valley can serve as a type locality in biogeographical studies of the tropical rift valley flora. The aim of this paper is to highlight the floristic status of the grasses which occur in the South Luangwa National Park and the Lupande Game Management Area. This information will be of advantage to the management authorities of the Park and the peoples of Chiefs Nsefu, Jumbe, Kakumbi, Malama, Munkhania and Msoro who live in the valley sector of Chipata District.

The Luangwa Valley is located in the eastern part of Zambia and a greater part of the National Park is situated on the western side of the Luangwa River covering an area of 9,050 km^2. The Lupande area lies on the eastern flank of the Luangwa, encompassing half the size of the Park area. The relief of the valley floor varies from 500 m to 600 m in altitude. This valley experiences mean daily maximum temperatures and minimum temperatures in the range of 32°-36° C and 16°-23° C respectively. The mean annual rainfall, usually received from October to April, amounts to 889 mm.

Material and methods

Plant collections, and floristic and ecological observations, were made during the southern hemispheric summer of 1987-88. Herbarium specimens were examined at the University of Zambia in Lusaka, Chinzombo Research Station near Mfuwe, the British Museum (Natural History), the Royal Botanic Gardens at Kew, and a loan was obtained from Mondlane University in Maputo.

Results and discussion

In the Luangwa Valley 145 families of vascular plants occur, among which the Gramineae ranks as the largest in terms of species number. A total of 65 graminaceous

L.J.G. van der Maesen et al. (eds.), The Biodiversity of African Plants, 224-230.

Table 1. Number of graminoid taxa, habits and species distribution in the South Luangwa National Park and Lupande Game Management Area.

Tribes	Number of taxa				Species distribution		
	G	S	A	P	SL	L	SL+L
Bambuseae	1	1	-	1	-	1	-
Oryzeae	2	4	1	3	3	-	1
Phyllorhachideae	1	1	-	1	1	-	-
Arundineae	1	1	-	1	-	1	-
Aristideae	1	8	4	4	4	1	3
Pappophoreae	1	1	-	1	1	-	-
Eragrostideae	10	43	23	20	22	4	17
Cynodonteae	6	9	5	4	3	3	3
Paniceae	17	76	44	32	38	16	22
Isachneae	1	1	-	1	1	-	-
Arundinelleae	4	8	-	8	6	2	-
Andropogoneae	20	36	14	22	23	3	10
Total	65	189	91	98	102	31	56

G: genera; S: species; A: annual species; P: perennial species; SL: South Luangwa National Park; L: Lupande Game Management Area.

genera including 189 species have been recorded in the South Luangwa National Park and the Lupande Game Management Area. A species list indicating their habitat preferences is given in Appendix 1. The land systems of some habitats that are recognised in the Park, also known to occur in the Game Management Area, were earlier described by Astle et al. (1969). Other habitats have been identified on the basis of the dominant species characterising a given vegetation type and the nature of the topographic unit.

The enumerated grasses belong to 12 tribes. The tribe Paniceae scores the largest number of species, trailed by Eragrostideae and Andropogoneae as indicated in Table 1. A parallel occurrence of some megathermic grass taxa exhibiting a more or less similar tribal composition has been reported in tropical South America (Burkart 1975). The flora of the Luangwa Valley is broadly characterised by the presence of Mopane woodland in which *Colophospermum mopane* (J. Kirk ex Benth.) J. Léonard is the dominant taxon. Prominent grass genera with species that are well adapted to this valleyane environment include *Andropogon*, *Aristida*, *Chloris*, *Dichanthium*, *Digitaria*, *Dinebra*, *Echinochloa*, *Elionurus*, *Enteropogon*, *Eragrostis*, *Hemarthria*, *Heteropogon*, *Hylebates*, *Hyparrhenia*, *Isachne*, *Leptochloa*, *Oryza*, *Oxytenanthera*, *Panicum*, *Pennisetum*, *Phacelurus*, *Rhytachne*, *Sehima*, *Setaria*, *Sporobolus*, *Tragus*, *Tripogon*, *Tristachya* and *Urochloa*.

A number of grasses can be used to provide beneficial pointers to the edaphic conditions of some terrestrial environments. For instance, a grassland community in which *Setaria incrassata* (Hochst.) Hack. occurs in association with *Dinebra retroflexa* (Vahl.) Panza var. *condensata* S.M. Phillips, *Echinochloa frumentacea* Link., *E.*

stagnina (Retz.) P. Beauv. and *Sacciolepis interrupta* (Willd.) Stapf can serve as a clue to the existence of black clay soils in the area. In other aspects of land use management programmes the extent of Mopane woodland can be employed to delimit an area that is underlain by the basic Karroo lithology on which saline soils develop. Therefore, in areas where this woodland has been degraded by elephants, and thus transformed into secondary grassland, the presence of grasses that are known to be restricted to the Mopane woodland would be symptomatic for the existence of the non-arable saline soils. In such locations the occurrence of such grasses as *Brachiaria xantholeuca* (Schinz.) Stapf, *Echinochloa colona* (L.) Link, *Eragrostis setulifera* Pilg., *Panicum phragmitoides* Stapf, *Sporobolus consimilis* Fresen, *S. panicoides* A. Rich. and *Tripogon minimus* (A. Rich.) Steud. constitute a credible index of what could have been the original Mopane woodland.

Many plant species that exhibit an exceptionally wide ecological amplitude are prone to invade agricultural land. In this study area notable taxa with potential weedy characteristics encompass *Aristida adscensionis* L., *A. rhinochloe* Hochst., *Brachiaria deflexa* (Schum.) Robyns, *Chloris virgata* Sw., *Digitaria asthenes* W.D. Clayton, *D. nuda* Schum., *Eleusine indica* (L.) Gaertn. subsp. *africana* (Kennedy-O'Byrne) S.M. Phillips, *E. indica* (L.) Gaertn. subsp. *indica*, *Eragrostis aethiopica* Chiov., *E. arenicola* C.E. Hubb., *E. aspera* (Jacq.) Nees, *E. cylindriflora* Hochst., *E. namaquensis* Schrad. var. *namaquensis*, *E. pilosa* (L.) Beauv., *E. rigidior* Pilg., *E. tenella* (L.) Roem. & Schult. var. *tenella*, *E. viscosa* (Retz.) Trin., *Eriochloa macclounii* Stapf, *Heteropogon contortus* (L.) Roem. & Schult., *H. melanocarpus* (Ell.) Benth., *Leptocarydion vulpiastrum* (De Not.) Stapf, *Leptochloa panicea* (Retz.) Ohwi, *Panicum maximum* Jacq., *Perotis patens* Gand., *Sehima ischaemoides* Forssk., *Setaria pumila* (Poir) Roem. & Schult., *Sporobolus molleri* Hack. and *S. pyramidalis* Beauv. It is interesting to note that about 87% of this category of grasses are annual plants.

A reasonable degree in the biodiversity of the grass taxa is quite evident in the Luangwa. The species composition of grasses that occur in this valleyane flora should elucidate the type of graminoid crop plants whose physiological and biochemical structure can withstand the effects of high temperatures and low rainfall regimes.

Acknowledgements
The author would like to thank the University of Zambia for facilitating travel arrangements to conduct fieldwork, and Prof. D.M. Moore for his guidance and valuable comments on the original draft. Data on botanical inventory of the study area was made possible through herbarium facilities provided by the Royal Botanic Gardens at Kew and the British Museum (Natural History).

References

Astle, W.L., R. Webster & C.J. Lawrence, 1969. Land classification for management planning in the Luangwa Valley of Zambia. J. Appl. Ecol. 6: 143-169.

Burkart, A., 1975. Evolution of grasslands in South America. Taxon 24 (1): 53-66.

Killick, D.J.B., 1978. The afro-alpine regions. In: Werger, M.J.A. (ed.), Biogeography and Ecology of Southern Africa. Junk, The Hague.

Lind, E.M. & M.E.S. Morrison, 1974. East African Vegetation. Longman, London.

White, F., 1978. The afromontane region. In: Werger, M.J.A. (ed.), Biogeography and Ecology in Southern Africa. Junk, The Hague.

Appendix 1

List of grasses recorded in the South Luangwa National Park and the Lupande Game Management Area. For abbreviations see page 230.

	Taxa	Habit	Habitat types
Bambuseae	*Oxytenanthera abyssinica*	Pe	. . . . I . . L . . . .
Oryzeae	*Leersia hexandra*	Pe	. . . . J . . . . . .
	Oryza barthii	An	. . . . J . . . . . .
	O. longistaminata	Pe	. . . . J . . . . . .
	O. punctata	Pe	. . . . J . . . . . .
Phyllorhachideae	*Phyllorhachis sagittata*	Pe	A . . . . L . . . .
Arundineae	*Phragmites mauritianus*	Pe	. . . . I . . L . . . .
Aristideae	*Aristida adscensionis*	An	. B C D . . G . I . . L . . . P
	A. aequiglumis	Pe	. . . . I . . . .
	A. canescens ssp. *canescens*	Pe	. . . . K . . . . .
	A. hordacea	An	. B . . . F G . I . K . . . . .
	A. junciformis ssp. *welwitschii*	Pe	. B C . E . . . I . . . .
	A. recta	Pe	. . . . H I . . . .
	A. rhinochloe	An	. B . . . F . . . . P
	A. scabrivalvis ssp. *scabrivalvis*	An	. . . . . F G . I . . . .
Pappophoreae	*Schmidtia pappophoroides*	Pe	. . C D . F . . I . K . . . . .
Eragrostideae	*Bewsia biflora*	Pe	. . . . . F G . I . . . .
	Dactyloctenium aegyptium	An	. B . . E . . . . . K . . . . .
	D. giganteum	An	. B . . E . . . . . K . . . . .
	Dinebra retroflexa var. *condensata*	An	. . C . . . . . . J K . . . . .
	Eleusine indica ssp. *africana*	An	. . . . H . . K . . . . P
	E. indica ssp. *indica*	An	. . . . H . . K . . . . P
	Eragrostis aethiopica	An	. B . . . . G . . . K L . . . P
	E. arenicola	An	. . . . . F . . . . . L . . . P
	E. aspera	Pe	. . . . L . . . P
	E. chapelieri	Pe	. B . . . F G . . . K . . . . .
	E. cilianensis	An	. . . . K L . . O .
	E. ciliaris	An	. . . . K L . . O .
	E. cylindriflora	An	. . . . L . . . P
	E. gangetica	An	A B C . . . . K L . . . .
	E. hispida	Pe	. B . . . . G . I . . . . . O .
	E. lappula ssp. *divaricata*	Pe	. B . . . F . . . . K . . . . .
	E. namaquensis var. *namaquensis*	An	A . . . . L . . . P
	E. pilosa	An	. B C D . . . . . . K . . . . P
	E. porosa	An	. . . D . . G . . . .
	E. racemosa	Pe	. . . . . F . . . .
	E. rigidior	Pe	A . . . . K L . . . P
	E. rotifer	Pe	. . . . K L . . . .
	E. setulifera	An	. B . . . .
	E. tenella var. *tenella*	An	. . . . I . . . . . . P
	E. trichophora	Pe	. . . . L . . . .
	E. viscosa	An	. B . . E . . . . . K . . . O P
	Leptocarydion vulpiastrum	An	. . . . . F . . I . K L . . . P
	Leptochloa caerulescens	Pe	. . . . J K L . . . .
	L. fusca	Pe	. . . . L . . . .

	L. panicea	An	A . . . . J . . . . . P
	L. uniflora	An	A . . . . . . H . . . .
	Pogonarthria squarrosa	Pe	. . . . . F G . I . K L . . . .
	Sporobolus congoensis	Pe	. . . . . F . . I . . . .
	S. consimilis	Pe	. B . . . .
	S. cordofanus	An	. B C D . . . .
	S. iocladus	Pe	. B C D . . G . . . .
	S. festivus	Pe	. B C D . . G . I . K L . . . .
	S. kentrophyllus	Pe	. B . . . .
	S. molleri	An	. . . . K . . . . P
	S. myrianthus	Pe	. . . . . F G . . . .
	S. panicoides	An	. B C D . . . .
	S. pyramidalis	Pe	. . . D . F . . I . K . . . . P
	Tripogon minimus	Pe	. B C D . . . .
Cynodonteae	*Chloris pycnothrix*	An	A B . . . F . . I . K . . . . .
	C. virgata	An	. B . . . F . . I . K . . . . P
	Cynodon dactylon	Pe	. . . . . F . . . . P
	Enteropogon macrostachys	Pe	. B C . . . .
	Microchloa caffra	Pe	. B C D . F G . I . . L . . . .
	M. indica	An	. B C D . F G . I . . . .
	M. kunthii	Pe	. . C D . . . .
	Perotis patens	An	. B . . . F G . . . K . . . . P
	Tragus berteronianus	An	. B C D . . . .
Paniceae	*Alloteropsis cimicina*	An	. . . . I . . . .
	A. semialata	Pe	. . . . I . . . .
	Brachiaria brizantha	Pe	. . . . . . G . I . K . . N . .
	B. deflexa	An	. B . . . . . . I . K L . . O P
	B. eminii	An	A . . . . K . . . . .
	B. eruciformis	An	A . C . . . . K . . . . .
	B. grossa	An	. B . D . . . H . . . .
	B. jubata	Pe	. . . . . F . . I . . . .
	B. xantholeuca	An	. B . D . . . .
	Cenchrus biflorus	An	. . . . . . G . . . . O .
	Digitaria acuminatissima	An	. . . . K L . . . .
	D. angolensis	An	. . . . . . G . . . K . . . . .
	D. asthenes	An	. . . . P
	D. brazzae	Pe	. . . . I . . . .
	D. compressa	Pe	. . . . . F G . I . . . .
	D. debilis	An	. . . . L . . . .
	D. gayana	An	. . . . . . G . . . K . . . . .
	D. gazensis	Pe	. . . . . F . . . .
	D. leptorhachis	An	. B . . . .
	D. milanjiana	Pe	. B . . . . G H . J K . . . . .
	D. nuda	An	. . . . P
	D. perrottetii	An	. B . D . . . H . . . .
	D. ternata	An	. B C D . . . .
	Echinochloa colona	An	. B C . . . . . . J K . . . . .
	E. frumentacea	An	. . C . . . . . . J . . . . . .
	E. haploclada	Pe	. . . . L . . . .
	E. pyramidalis	An	. . . . J . . . . . .
	E. stagnina	An	. . . D . . . . . J . . . . . .

	Eriochloa macclounii	An	.	.	C	D	.	.	.	.	.	.	K	.	.	.	.	P
	E. meyerana	Pe	.	B	C	.	.	.	.	.	.	.	.	.	M	.	.	.
	Hylebates cordatus	An	A	.	.	.	.	.	.	H	.	.	.	L	.	.	.	.
	Oplismenus burmannii	An	A	.	.	.	.	F	.	H	I	.	K	L	.	.	.	.
	Panicum coloratum var. *coloratum*	Pe	.	.	.	.	.	.	.	.	.	.	K	.	.	.	.	.
	P. comorense	An	.	.	.	.	.	.	.	.	I	.	.	.	.	.	.	.
	P. dregeanum	Pe	A	.	.	.	.	.	.	.	.	.	.	.	.	.	O	.
	P. hanningtonii	An	.	.	.	.	.	.	.	.	.	.	.	.	.	.	O	.
	P. heterostachyum	An	.	.	.	.	.	.	.	.	I	.	.	L	.	.	.	.
	P. hippothrix	An	.	.	.	.	.	.	.	.	.	.	K	.	.	.	.	.
	P. madiperense	An	A	B	C	D	E	.	.	.	.	.	.	.	.	.	O	.
	P. massaiense	An	.	B	C	.	.	.	.	.	.	.	K	L	.	.	.	.
	P. maximum	Pe	.	.	.	.	E	.	.	.	.	.	K	L	.	.	.	P
	P. novemnerve	An	.	.	.	.	.	F	.	.	.	.	.	.	.	.	.	.
	P. phragmitoides	Pe	.	.	C	.	.	.	.	.	.	.	.	.	.	.	.	.
	P. porphyrrhizos	Pe	.	.	.	.	.	.	.	.	.	.	.	.	.	.	O	.
	P. repens	Pe	.	.	.	.	.	.	.	.	.	.	.	.	M	.	O	.
	P. repentellum	Pe	.	.	.	.	.	.	.	.	.	.	K	.	.	.	.	.
	P. schinzii	An	.	.	C	D	.	.	G	.	I	.	.	.	.	.	.	.
	P. subalbidum	An	.	.	.	.	.	.	.	.	.	J	.	L	.	.	.	.
	Paspalum glumaceum	Pe	.	.	.	.	.	.	.	.	.	.	.	L	.	N	.	.
	P. scrobiculatum	Pe	A	.	.	.	.	.	.	.	.	.	K	L	.	N	O	.
	Pennisetum macrourum	Pe	.	.	.	.	.	.	.	.	.	J	.	.	.	.	.	.
	P. polystachion ssp. *atrichum*	An	.	.	.	.	.	.	.	.	.	.	K	L	.	.	.	.
	P. purpureum	Pe	.	.	.	.	.	.	.	.	.	J	.	L	.	.	.	.
	P. unisetum	Pe	.	.	.	.	.	.	.	.	.	.	K	.	.	.	.	.
	Rhynchelytrum longisetum	Pe	.	B	.	.	.	.	G	.	I	.	.	L	.	.	O	.
	R. repens	An	.	.	.	.	.	.	.	.	I	.	.	L	.	.	.	.
	R. subglabrum	Pe	.	.	.	.	.	F	G	.	I	.	K	.	.	.	.	.
	Sacciolepis africana	Pe	.	.	.	.	.	.	.	.	.	.	K	.	.	.	O	.
	S. indica	An	A	.	.	.	.	.	.	.	.	.	.	.	.	.	O	.
	S. interrupta	Pe	.	.	C	.	.	.	.	.	.	J	.	.	.	.	O	.
	S. micrococca	An	.	.	.	.	.	.	.	.	.	.	.	.	.	.	O	.
	S. typhura	Pe	A	.	.	.	.	.	.	.	.	.	K	.	.	.	O	.
	Setaria homonyma	An	A	.	.	.	.	.	.	H	.	.	K	.	.	.	.	.
	S. incrassata	Pe	.	B	C	D	.	.	.	.	.	J	K	.	.	N	O	.
	S. longiseta	Pe	.	.	.	.	.	F	.	H	.	.	K	.	.	.	.	.
	S. petiolata	An	.	.	.	.	.	.	.	.	I	.	.	.	.	.	.	.
	S. pumila	An	.	B	C	.	.	.	.	.	I	J	K	.	.	.	.	P
	S. sagittifolia	An	A	.	.	.	.	.	.	H	.	.	K	.	.	.	.	.
	S. sphacelata	Pe	A	.	.	.	.	F	.	.	.	J	K	.	.	.	O	.
	S. verticillata	An	.	.	.	.	.	.	.	.	.	.	K	.	.	.	.	P
	Stenotaphrum dimidiatum	Pe	.	.	.	.	.	.	.	.	.	.	.	.	.	.	.	P
	Urochloa echinolaenoides	An	.	B	.	.	E	.	.	.	.	.	.	.	.	.	.	.
	U. mosambicensis	An	.	B	.	.	E	.	.	.	.	.	.	.	.	.	.	.
	U. oligotricha	Pe	.	.	.	.	E	.	.	.	.	.	K	.	.	.	.	.
	U. trichopus	An	.	B	.	.	E	.	.	.	.	.	.	.	.	.	.	.
Isachneae	*Isachne angolensis*	Pe	.	.	.	.	.	.	.	.	I	.	.	.	.	.	.	.
Arundinelleae	*Loudetia flavida*	Pe	.	.	.	.	.	F	G	.	.	.	.	.	.	.	.	.
	L. phragmitoides	Pe	.	.	.	.	.	.	.	.	I	.	.	L	.	.	.	.
	Trichopteryx fruticulosa	Pe	.	.	.	.	.	.	.	.	I	.	.	L	.	.	O	.

	T. marungensis	Pe	.	.	.	.	.	.	.	.	.	.	.	.	.	.	O	.
	Tristachya hubbardiana	Pe	.	.	.	.	.	F	.	.	I	.	K	.	.	.	O	.
	T. rehmannii	Pe	.	.	.	.	.	F	G	.	I	.	.	.	.	.	O	.
	T. superba	Pe	.	.	.	.	.	F	.	.	.	.	K	.	.	.	O	.
	Zonotriche inamoena	Pe	.	.	.	.	.	F	G	.	I	.	.	.	.	.	.	.
Andropogoneae	*Andropogon chinensis*	Pe	.	.	.	.	.	F	G	.	I	.	.	.	.	.	O	.
	A. eucomus	Pe	.	.	.	.	.	.	.	.	.	J	.	L	.	.	.	.
	A. gayanus var. *polycladus*	Pe	.	.	.	.	.	F	G	.	I	.	K	.	.	.	.	.
	A. schirensis	Pe	.	.	.	.	.	F	G	.	I	.	K	.	.	.	.	.
	Bothriochloa bladhii	Pe	A	.	.	.	E	F	.	.	.	.	.	L	.	.	.	.
	B. radicans	Pe	.	B	.	.	.	.	.	.	I	.	.	.	.	.	.	.
	Cleistachne sorghoides	An	.	B	C	.	.	.	.	.	.	.	K	.	.	N	.	.
	Cymbopogon caesius	Pe	.	.	.	.	.	.	.	.	.	.	.	.	.	.	O	.
	C. giganteus	Pe	.	.	.	.	.	.	.	.	.	.	K	L	.	.	O	.
	Dichanthium aristatum	Pe	.	.	.	.	.	.	.	.	.	.	K	.	.	.	.	.
	Elionurus tripsacoides	Pe	.	.	.	.	.	F	G	.	.	.	K	.	.	.	.	.
	Elymandra grallata	Pe	.	.	.	.	.	F	G	.	.	.	K	.	.	.	.	.
	Hackelochloa granularis	An	.	.	.	D	.	.	.	.	.	.	K	.	.	.	.	.
	Hemarthria altissima	Pe	.	.	.	.	.	.	.	.	.	J	K	L	.	.	.	.
	Heteropogon contortus	Pe	.	B	.	.	.	F	G	.	I	.	K	.	.	.	O	P
	H. melanocarpus	An	.	B	.	.	.	.	.	.	.	.	K	.	.	.	.	P
	Hyparrhenia anemopaegma	An	.	.	.	.	.	.	G	.	.	.	.	.	.	.	.	.
	H. barteri	An	.	.	.	.	.	.	.	.	I	.	K	.	.	.	.	.
	H. cymbaria	Pe	.	.	.	.	.	.	.	.	.	.	.	L	.	.	.	.
	H. dichroa	Pe	.	.	.	.	.	.	.	.	.	.	K	.	.	.	.	.
	H. filipendula	Pe	.	B	.	.	.	F	G	.	I	.	K	.	.	.	O	.
	H. nyassae	Pe	.	.	.	.	.	.	.	.	.	.	K	.	.	.	.	.
	H. rufa	Pe	.	.	.	.	.	F	.	.	.	.	K	.	.	.	.	.
	H. welwitschii	An	.	B	.	.	.	F	G	.	.	.	.	.	.	.	O	.
	Hyperthelia dissoluta	Pe	.	.	.	.	.	.	.	.	.	.	K	.	.	.	O	.
	Imperata cylindrica	Pe	.	.	.	.	.	.	.	.	.	.	.	L	.	.	.	.
	Rhytachne latifolia	Pe	.	.	C	.	.	.	.	.	.	.	K	.	.	.	.	.
	Rottboellia cochinchinensis	An	.	.	C	.	.	.	.	.	.	.	K	.	.	.	.	P
	Schizachyrium brevifolium	An	.	.	.	.	.	.	G	.	I	.	K	.	.	.	.	.
	S. exile	An	.	.	.	.	.	F	G	.	I	.	K	.	.	.	.	P
	S. jeffreysii	Pe	.	.	.	.	.	F	G	.	I	.	K	.	.	.	.	P
	Sehima ischaemoides	An	.	.	.	.	.	.	.	.	I	.	.	.	.	.	.	P
	Sorghastrum bipennatum	An	A	.	.	.	.	.	G	.	.	.	.	.	.	.	.	.
	Sorghum arundinaceum	An	.	.	.	.	.	.	.	.	.	.	K	L	.	.	.	.
	S. versicolor	An	.	.	.	.	.	.	G	.	I	.	K	.	.	.	.	.
	Thelepogon elegans	An	.	.	C	.	.	.	.	.	I	.	.	.	.	.	.	.

An annual
Pe perennial

A riparian woodland
B tall Mopane woodland
C low Mopane woodland
D scrub (coppiced) Mopane
E *Acacia tortilis* woodland
F Miombo woodland
G low interfluviae woodlands
H thicket on sandy soil
I escarpment and hill slopes
J lagoons, flood channel and water holes
K levees and point bars
L river and stream banks
M salt pan near hot spring
N *Setaria* grassland on black clay
O stream valley heads
P cultivated land, waste ground and roadsides

Les faciès à *Andropogon pseudapricus* des groupements post-culturaux et des savanes arbustives du Nord-Bénin: dissemblance floristique et caractères communs

B. SINSIN, M. OUMOROU & V. OGOUBYI
Faculté des Sciences Agronomiques, Université Nationale du Bénin, B.P. 526, Cotonou, Bénin.

Résumé
L'alliance des *Andropogonion pseudaprici* Sinsin 1993 a été définie pour regrouper les associations soudaniennes post-culturales situées dans les dépressions. D'une manière générale, c'est *Andropogon pseudapricus* qui est l'espèce dominante de ces associations végétales et qui définit par là-même la physionomie des différentes variantes ou faciès. D'autres relevés phytosociologiques effectués dans les savanes arbustives situées à la limite supérieure de la zone soudanienne du nord-Bénin à l'intérieur des aires protégées, ont révélé des faciès à dominance d'*Andropogon pseudapricus* sur des terres non perturbées par les activités agricoles. La comparaison des cortèges floristiques de ces phytocénoses plus sèches et des associations post-culturales des dépressions a montré des différences tant sur le plan physionomique que floristique. La confrontation des degrés de dissemblance et de ressemblance des deux cortèges floristiques a permis de discuter du statut écosociologique des espèces caractéristiques de l'*Andropogonion pseudaprici*.

Key words
Andropogon pseudapricus, Association, Benin, *Desmodietum hirti*, Floristic dissimilarity.

Introduction

Dans une étude antérieure portant sur la typologie et la hiérarchisation des associations végétales de la Zone soudanienne du nord-Bénin (Sinsin 1994), l'alliance de l'*Andropogonion pseudaprici* a été définie comme celle regroupant les associations post-culturales des dépressions humides. D'autres relevés effectués dans les savanes arbustives exploitées essentiellement par des troupeaux de bovins, ont permis de constater l'existence de faciès à *Andropogon pseudapricus* sur des plateaux et dans des dépressions humides exempts de toute trace récente d'activités agricoles.

Il s'est avéré nécessaire de comparer les cortèges floristiques de ces faciès avec ceux des individus d'association (Guinochet 1973) de l'*Andropogonion pseudaprici*.

L'approche sociologique basée sur l'affinité des espèces constituantes des phytocénoses est bien indiquée pour la classification des communautés végétales même si selon Doing (1969) la signification écologique des groupes identifiés paraît difficile à interpréter dans certains cas. Cependant, le traitement des données floristiques obtenues selon la méthode des groupes écologiques peut s'intégrer dans une synthèse phytosociologique dans la mesure où les hiérarchies supérieures tiennent compte des conditions mésologiques (Duvigneaud 1946, Gounot 1969).

Selon certains auteurs (Mason 1977) l'usage des indices de diversité de communautés végétales rencontrées dans des sites différents ou d'une même communauté rencontrée sur des sites variés ou étudiée à des périodes différentes permet de tirer le maximum d'informations écologiques. Enfin les analyses multivariées permettent aussi

L.J.G. van der Maesen et al. (eds.), The Biodiversity of African Plants, 231-238.

de rendre compte du degré d'affinité entre diverses communautés végétales (Djitèye 1988, Lacoste 1976).

L'objet de la présente étude est de faire usage de différentes méthodes d'analyse des phytocénoses afin de statuer sur la variabilité floristique et la signification écologique de l'alliance de l'*Andropogonion pseudaprici*.

Milieu d'étude

L'étude des communautés végétales a été conduite dans deux périmètres-échantillons dans le département du Borgou au nord-Bénin. Les associations post-culturales de l'*Andropogonion pseudapricus* ont été définies à la latitude des 10° N où la pluviométrie moyenne annuelle est de 1150 mm avec six mois de période humide. Les groupements à dominance d'*Andropogon pseudapricus* des savanes arbustives ont été observés et étudiés à la latitude de 11°45 N où la moyenne pluviométrique annuelle est de 1000 mm avec 5 mois de période humide. Pour l'ensemble des périmètres-échantillons les températures moyennes annuelles varient de 25° à 27° C.

D'une manière générale, la végétation est une mosaïque de savanes et de jachères que sillonnent quelques galeries forestières rencontrées le long des cours d'eau. La strate herbacée est dominée par les Gramineae et elle brûle chaque année entre novembre et décembre pendant la saison sèche. Les sols sont ferrugineux tropicaux de texture particulaire et souvent lessivés (Viennot 1978). Les activités qui prédominent dans la région étudiée sont l'agriculture extensive et l'élevage de bovins selon un système transhumant (Sinsin 1991).

Matériel et méthode

L'étude de la composition floristique des phytocénoses a été réalisée selon la méthode des relevés phytosociologiques (Guinochet 1973). Dans les jachères de dépression dominées par *Andropogon pseudapricus*, l'association à *Desmodium hirtum* a été définie à l'issue du traitement d'une matrice phytosociologique de 135 relevés (Sinsin 1993). Dans les nouveaux groupements de savanes dominés par *Andropogon pseudapricus* six relevés phytosociologiques ont été réalisés auxquels nous avons joint six autres relevés de l'association à *Desmodium hirtum* à des fins d'analyses.

Pour le traitement des données, la comparaison des deux faciès de phytocénoses a été faite à l'aide:

* du coefficient de similitude de Sorensen, K, défini par le rapport:
  K = 2c x 100 / (a+b); avec c = nombre d'espèces communes aux phytocénoses A et B; a = nombre d'espèces de la phytocénose A; b = nombre d'espèces de la phytocénose B.
  Pour apprécier un seuil objectif de signification du coefficient de Sorensen nous avons utilisé le coefficient de corrélation de Kendall qui donne le degré de communauté entre deux phytocénoses. Le seuil de probabilité retenu pour le test du KHI 2 est P = 0,001 jugé suffisant pour ce genre d'analyse (Troupin 1966);

* du coefficient de similitude de Jaccard, défini par le rapport:
  P_J = 100 x c / (a+b-c); avec a, b et c comme précédemment décrits.

* de l'analyse factorielle des correspondances (AFC). Dans ce type de traitement nous avons analysé la répartition des relevés de savanes par rapport

au relevé-type de l'association à *Desmodium hirtum* (Sinsin 1994). La superposition du plan contenant les espèces à celui contenant les relevés permet d'apprécier les espèces les plus discriminantes.

Les types biologiques utilisés sont ceux définis par Raunkiaer (cité par Jacques-Félix 1962). Les spectres biologiques présentés sont définis comme étant le rapport de chaque type biologique sur l'ensemble des espèces qui constituent une phytocénose donnée. Le spectre est pondéré lorsqu'il est calculé à partir du recouvrement moyen des espèces (Sinsin 1993).

Résultats

Structure des faciès à Andropogon pseudapricus

Dans les groupements post-culturaux de dépression, les faciès d'association définis par *Andropogon pseudapricus* sont unistrates. Une telle physionomie est due à l'impact des activités agricoles. Les agriculteurs ont l'habitude de couper tous les arbres n'épargnant que quelques pieds de Karité (*Vitellaria paradoxa*) et de néré (*Parkia biglobosa*) à cause de leur valeur socio-économique. L'association qui a été définie dans ce type de milieu, le *Desmodietum hirti* (Sinsin 1993) compte deux ou trois strates selon les sites. On distingue ainsi:

* Une strate inférieure à 72 cm de haut dominée par *Schizachyrium brevifolium* var. *brevifolium*. Cette strate est présente dans l'axe de la dépression quand il y a stagnation d'eau.
* Une strate florifère moyenne à 160 cm de haut dominée par *Sorghastrum bipennatum* en compagnie d'*Andropogon pseudapricus*.
* Une strate florifère supérieure qui culmine entre 240 et 300 cm dominée par *Hyparrhenia involucrata* et/ou *H. rufa* selon l'âge du groupement.

Le tapis herbacé est en général très fermé et le sol est souvent couvert à 100%.

Dans les savanes des latitudes supérieures, les faciès dominés par *Andropogon pseudapricus* sont notés sur les plateaux et sur les talus de certaines rivières. Ici la strate arborescente est présente et marque la différence physionomique au niveau des deux groupements en comparaison. Les espèces ligneuses fréquemment rencontrées sont: *Lannea acida, Crossopteryx febrifuga, Combretum glutinosum, Anogeissus leiocarpa, Sterculia setigera, Vitellaria paradoxa, Detarium microcarpum, Burkea africana* et *Combretum nigricans*.

Au niveau des herbacées des savanes arbustives, on distingue une seule strate dominée par *Andropogon pseudapricus* qui fleurit à 160 cm de hauteur.

Types biologiques

Le Tableau 1 montre les spectres bruts et pondérés des types biologiques de la strate herbacée au niveau des deux phytocénoses. Dans l'ensemble les deux spectres sont similaires et ne présentent pas de différences appréciables. Notons cependant, qu'il s'agit dans les jachères, essentiellement de recrûs des souches de la végétation initialement défrichée alors que dans les savanes, le spectre brut des phanérophytes rend compte des jeunes plants de ligneux.

Le paramètre qui rend le mieux compte de la physionomie du groupement est le spectre biologique pondéré. Ce sont les thérophytes et plus particulièrement *Andropogon pseudapricus* qui définissent le faciès de la strate herbacée.

Tableau 1. Spectres biologiques des deux faciès à *Andropogon pseudapricus* des jachères de dépression et des savanes arbustives (strate herbacée).

Types biologiques	Faciès des savanes arbustives		Faciès des jachères de dépression	
	Spectres bruts (%)	Spectres pondérés (%)	Spectres bruts (%)	Spectres pondérés (%)
Thérophytes	61,9	71,6	59,1	77,6
Hémicryptophytes	7,9	2,0	8,0	9,6
Géophytes	5,8	7,2	10,1	2,1
Chaméphytes	6,4	15,2	8,0	7,9
Phanérophytes	18,0	4,0	14,8	2,8

Analyses quantitatives des compositions floristiques

Coefficients de similitude. Les valeurs obtenues sont 53,4 pour le coefficient de similitude de Sorensen et 36,4 pour le coefficient de similitude de Jaccard.

Le coefficient de corrélation de Kendall calculé a pour valeur -0,58. Son signe négatif est révélateur du poids des espèces exclusives de chaque phytocénose. Le seuil de signification de ce coefficient de corrélation a été calculé au moyen du test de KHI 2 au seuil de probabilité de 0,001. La différence entre les valeurs des coefficients de corrélation observé et théorique n'est pas significative ou en d'autre terme, il y a bien communauté entre les deux phytocénoses comparées floristiquement.

Application de l'analyse factorielle des correspondances (AFC). La comparaison des deux faciès à l'aide de l'AFC en présence-absence a été faite sur l'ensemble des six relevés effectués dans les savanes arbustives auquel nous avons joint le relevé-type du *Desmodietum hirti*. La matrice traitée croise donc un ensemble de 7 relevés phytosociologiques avec celui de 236 espèces recensées.

Le plan factoriel constitué des axes 1 et 3 (Figure 1) qui donne la meilleure représentation montre la position relative du relevé-type du *Desmodietum hirti* (relevé R7) par rapport à l'ensemble des relevés de savane arbustive (relevés R1, R2, R3, R4, R5 et R6). L'axe 1 explique 34,47% de l'inertie totale et sépare nettement le relevé-type du *Desmodietum hirti* de l'ensemble des relevés de savane arbustive. Il existe donc une différence significative dans la composition floristique des deux faciès lorsque l'on se limite au rang de l'association végétale.

Cette différence dans la composition floristique est illustrée à la Figure 2 qui représente le dendrogramme de la classification ascendante hiérarchique des relevés. Ici aussi, le relevé-type du *Desmodietum hirti* est séparé de l'ensemble des relevés de savane arbustive.

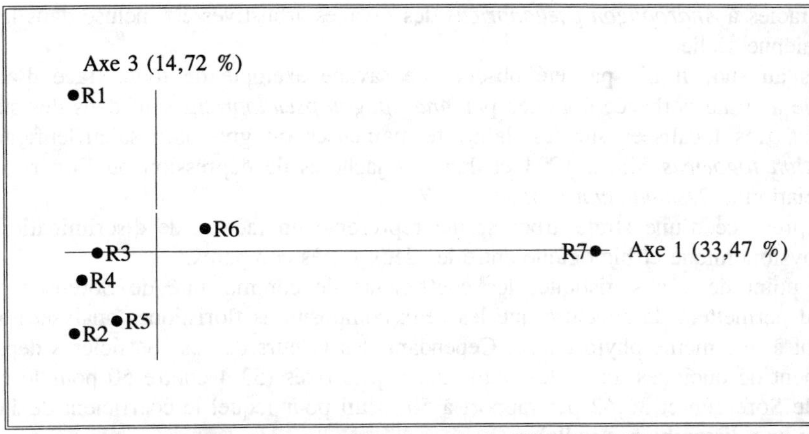

Figure 1. Représentation de la répartition des relevés (R1, R2,…R7) des deux faciès à *Andropogon pseudapricus* dans le plan des axes 1 et 3 de l'AFC.

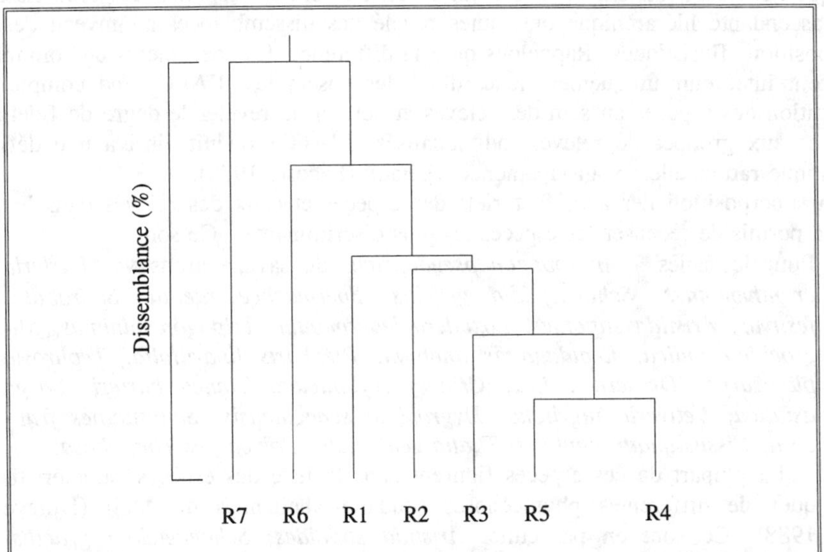

Figure 2. Dendogramme des relevés (R1, R2,… R7) des deux faciès à *Andropogon pseudapricus*.

Discussion

Caractères distinctifs des deux phytocénoses

Du point de vue phytogéographique, les deux faciès en comparaison appartiennent au domaine soudanais (Troupin 1966), qui a scindé en sous-domaine méridional et en sous-domaine septentrional. Ces différences correspondent au passage des groupements véritablement soudaniens aux groupements soudano-sahéliens.

Le faciès à *Andropogon pseudapricus* des savanes arbustives est incluse dans la zone soudanienne sèche.

Plus au sud, il n'a pas été observé de savane exempte de toute trace d'activité agricole à strate herbacée dominée par *Andropogon pseudapricus* sauf dans des stations xériques très localisées sur les dalles ferrugineuses ou grèseuses subaffleurantes du *Loudetion togoensis* Sinsin 1994 et dans les jachères de dépression où l'on rencontre l'association à *Desmodietum hirti*.

La présence d'une strate arborescente représente un facteur de discrimination à la fois physionomique et biologique entre les deux faciès comparés.

Du point de vue statistique, les coefficients de communauté de Sorensen ou de Jaccard permettent de conclure que les deux compositions floristiques analysées appartiennent à une même phytocénose. Cependant, les valeurs de ces coefficients dépassent seulement de quelques unités les seuils théoriques fixés (53,4 contre 50 pour le coefficient de Sorensen et 36,42 par rapport à 50, seuil pour lequel le coefficient de Jaccard indique une dissemblance nulle).

L'application du test de KHI 2 au coefficient de corrélation de Kendall pour apprécier le degré de communauté des deux cortèges floristiques a révélé la présence d'espèces qui s'excluent totalement. L'analyse factorielle des correspondances et la classification ascendante hiérarchique ont toutes révélé des dissemblances au niveau des deux compositions floristiques. Rappelons qu'à la différence des coefficients de communauté qui font intervenir uniquement le cardinal des ensembles, l'AFC rend compte de la répartition des espèces au sein des relevés et permet de révéler le degré de fidelité des espèces aux groupes de relevés individualisés. L'AFC conduit ainsi à une définition floristique rationnelle des groupements végétaux (Lacoste 1976).

La superposition des axes factoriels des espèces et ceux des relevés dans le même plan a permis de recenser les espèces les plus discriminantes. Ce sont:

* Pour le faciès à *Andropogon pseudapricus* de savane arbustive: *Barleria eranthemoides, Schoenefeldia gracilis, Spermacoce ruellia, Sporobolus festivus, Aristida stipoides, Loxodera ledermannii, Tripogon minimus, Microchloa indica, Lepidagathis anobrya, Blepharis linariifolia, Tephrosia platycarpa, Drosera indica, Crinum zeylanicum, Tinnea barteri, Striga asiatica, Vetiveria nigritana, Hygrophila senegalensis, Stylosanthes fruticosa, Cissus quadrangularis, Thalia geniculata, Polycarpea corymbosa.*

La plupart de ces espèces figurent dans la liste des espèces caractéristiques de différentes phytocénoses soudano-sahéliennes du Mali (Djitèye 1988). Ce sont en particulier *Aristida stipoides*, *Schoenefeldia gracilis*, *Tripogon minimus*, *Loudetia togoensis*, *Polycarpea corymbosa*, *Blepharis linariifolia*, *Ctenium elegans*, *Piliostigma reticulatum* et *Guiera senegalensis*. D'autres encore font partie des espèces caractéristiques d'association décrite dans les savanes soudaniennes du sous-domaine méridional et qui sont très rares dans les groupements post-culturaux. C'est le cas en particulier de *Loxodera ledermannii*, espèce caractéristique du *Loxoderetum ledermannii* (Sinsin 1993).

* Pour le *Desmodietum hirti* dont le faciès est aussi dominé par *Andropogon pseudapricus* dans les jachères de dépression, les espèces les plus discriminantes sont: *Nelsonia canescens, Schizachyrium brevifolium* var. *brevi-*

folium, Desmodium hirtum, Sapium grahamii, Spermacoce filifolia, Sacciolepis indica, Scleria pergracilis, Nesaea cordata, Uraria picta, Fuirena ciliaris et Spermacoce paludosa. La plupart de ces espèces sont hygrophiles et leur présence est beaucoup plus liée à l'état hygrométrique du sol. Ce sont aussi des espèces à large distribution phytogéographique (Raynal 1964, Sinsin 1993).

Niveaux de communauté et de dissemblance entre les deux faciès
Les résultats issus du traitement de la matrice phytosociologique par ces deux méthodes d'analyse multivariée qui respectent la dualité entre deux ensembles de relevés avec leurs compositions floristiques spécifiques, permettent de conclure que les deux faciès comparés appartiennent à deux associations différentes. L'association à *Desmodium hirtum* décrite dans les jachères de dépression est donc floristiquement différente du groupement dominé par *Andropogon pseudapricus* des savanes arbustives du sous-domaine soudanais septentrional. L'espèce caractéristique de ce groupement est *Barleria eranthemoides*. Son absence dans les savanes soudaniennes méridionales et son abondance dans sa station permettent de faire du *Barleria eranthemoides* une espèce caractéristique d'association.

Cependant, les deux listes floristiques comparées présentent également des similitudes confirmées sur le plan quantitatif par l'utilisation des coefficients de communauté de Sorensen et de Jaccard. Les espèces communes aux deux faciès représentent respectivement 70% du nombre total des espèces du *Desmodietum hirti* et 76% du nombre total des espèces du faciès à *Andropogon pseudapricus* des savanes arbustives. Nous pouvons donc regrouper ces deux phytocénoses au sein d'une même alliance, l'alliance de l'*Andropogonion pseudaprici* (Sinsin 1994). Cette alliance pourra être morcelée en "variante anthropogène" (Duvigneaud 1946) directement reconnaissable sur le terrain à travers les traces d'activité agricole; tel est le cas du *Desmodietum hirti* des jachères de dépression.

Conclusion

L'application de méthode floristique statistique simple (Gounot 1969) et l'analyse factorielle des correspondances au traitement de listes floristiques présentant un même faciès défini par *Andropogon pseudapricus* a permis d'apprécier la variabilité des phytocénoses pouvant être incluses dans l'*Andropogonion pseudaprici*. La comparaison des faciès herbacés à dominance d'*Andropogon pseudapricus* des savanes arbustives du sous-domaine soudanais septentrional (Troupin 1966) avec l'association à *Desmodium hirtum* des jachères de dépression à l'aide des coefficients de communauté de Sorensen et de Jaccard a révélé une similitude au niveau des deux groupements. Cette similitude a été estimée suffisante pour regrouper les deux phytocénoses au sein de l'*Andropogonion pseudaprici* dont l'une des associations anthropogènes est le *Desmodietum hirti* des jachères de dépression.

L'analyse des facteurs édaphiques des faciès dominés par *Andropogon pseudapricus* dans les savanes arbustives du sous-domaine soudanais septentrional et la synthèse phytosociologique des divers relevés effectués dans des faciès similaires dans les pays soudano-sahéliens (Djitèye 1988) permettront de mieux préciser la définition éco-floristique de l'*Andropogonion pseudaprici*.

Bibliographie

Djitèye, M.A., 1988. Composition, structure et production des communautés végétales sahéliennes: application à la zone de Niono (Mali). Thèse doct. Univers. Paris-Sud Orsay, France.

Doing, H., 1969. Sociological species groups. Acta Bot. Neerl. 18: 398-400.

Duvigneaud, P., 1946. La variabilité des associations végétales. Bull. Soc. Roy. Bot. Belg., 78: 107-134.

Gounot, M., 1969. Méthode d'étude quantitative de la végétation. Ed. Masson & Cie. Paris.

Guinochet, M., 1973. Phytosociologie. Ed. Masson & Cie, Paris.

Jacques-Félix, H., 1962. Les graminées d'Afrique tropicale. I. Généralités, classification, description des fleurs. IRAT, Paris.

Lacoste, A., 1976. Relations floristiques entre les groupements prairiaux du *Triseto-Polygonion* et les mégaphorbiaies (*Adenostylion*) dans les Alpes Occidentales. Vegetatio 31 (1): 161-176.

Mason, C.F., 1977. The performance of a diversity index in describing the zoobenthos of two lakes. J. Appl. Ecol. 14: 363-367.

Raynal, J., 1964. Etude botanique des pâturages du centre de recherches zootechniques de Dahra-Djoloff, Sénégal. ORSTOM, Section Botanique, Paris.

Sinsin, B., 1991. Exploitation des pâturages naturels dans un système traditionnel d'élevage bovin dans le périmètre Nikki-Kalalé (nord-Bénin). IV Conférence Internationale des Terres de Parcours. Montpellier, France: 660-662.

Sinsin, B., 1993. Phytosociologie, écologie, valeur pastorale, productivité et capacité de charge des pâturages naturels du périmètre Nikki-Kalalé au nord-Bénin. Thèse doct. Univers. Libre de Bruxelles, Belgique.

Sinsin, B., 1994. Individualisation et hiérarchisation des phytocénoses soudaniennes du nord-Bénin. Belgian Journ. Bot. 127(1): 87-103.

Troupin, G., 1966. Etude phytocénologique du Parc National de l'Akagera et du Rwanda Oriental. Recherche d'une méthode d'analyse appropiée à la végétation d'Afrique Inter-tropicale. Thèse d'agrégation, Université de Liège, Belgique.

Viennot, M., 1978. Notice explicative de la carte pédologique de reconnaissance de la R.P. Bénin. Feuille de Bembéréké. ORSTOM, Paris.

Ecology and conservation

24 August 1994

Convenor: V.H. Heywood

A floristic classification of the indigenous forests of Transkei, South Africa

S.G. CAWE
Botany Department, University of Transkei, Umtata, South Africa.

Abstract
Comprehensive tree lists from 86 indigenous forests in the subtropical and Afrotemperate regions of Transkei, South Africa were analyzed using a multivariate classification technique (TWINSPAN). Eight forest types were distinguished, four from each region. Altitude and forest size were the main determinants of forest composition and species-richness in both the subtropical and Afrotemperate forests. Large forest tracts especially in the coastal lowlands contain many rare species whereas smaller forest patches are poorer in species. This study does not support previous subjective classifications of the forests.

Résumé
Les listes détaillées de 86 forêts indigènes des régions subtropicales et afrotempérées du Transkei, en Afrique du Sud, ont été analysées en utilisant la méthode d'analyse multivariée TWINSPAN. Huit types de forêts ont été distingués, à raison de 4 par région. L'altitude et les dimensions de la forêt ont été les principaux facteurs déterminants de la composition floristique et de la richesse spécifique tant dans les forêts subtropicales que afrotempérées. Les grandes étendues de forêt, principalement dans les plaines côtières, renferment de nombreuses espèces rares, tandis que les peuplements plus restreints sont spécifiquement plus pauvres. Cette étude ne corrobore pas les classifications subjectives antérieures des forêts.

Key words
Classification, Floristics, Forests, South Africa.

Introduction

The indigenous forests of Transkei, one of the former homelands of South Africa (Figure 1), have been the subject of several classifications for more than four decades. Sim (1907) gave one of the earliest accounts of these forests, but Acocks' (1953) classification is the most widely known. The latter classification is agro-ecological and has often been used as a basis for land-use planning (McKenzie 1984). In keeping with the Clementsian ideas which were then prevalent, Acocks regarded the forests as the climatic climax vegetation over most of the Transkei and based his vegetation units (called "Veld Types") mainly on the floristic composition of the forests. In terms of Acocks' classification, the indigenous forests of Transkei comprise five tropical forest types: (i) Typical Coast-Belt Forest, (ii) Dune Forest (iii) Pondoland Coastal Plateau Sourveld, (iv) Ngongoni Veld and (v) Eastern Province Thornveld and two temperate forest types: (i) Dohne Sourveld and (ii) Highland Sourveld. According to Acocks, these forest types (or their remnants) occur in discrete, mappable geographical units.

Based on quantitative floristic data obtained from quadrats, Cawe (1986) recognised four types of Afrotemperate forest which he presumed to be differentiated on the basis of soil-moisture status. Significantly, Cawe's (1986) forest types were not spatially differentiated into discrete geographical units. Later, Cawe (1990) classified the

L.J.G. van der Maesen et al. (eds.), The Biodiversity of African Plants, 241-249.

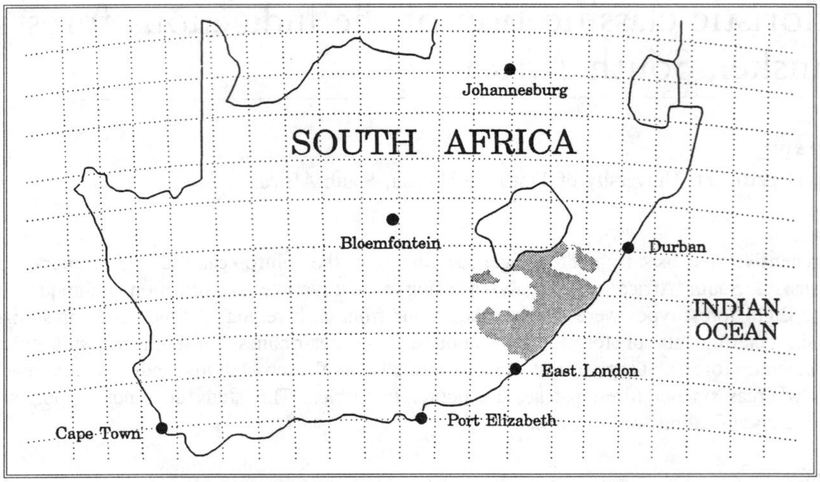

Figure 1. Location map of Transkei (shaded area).

subtropical forests into five main categories: four types of typical coastal forest and dune forest. In this classification the role of humans in changing the floristic composition of the forests was highlighted. More recently, Cooper & Swart (1992), using the same data employed in Cawe's study, adopted Acocks' (1953) classification with minor modifications. Von Breitenbach & Von Breitenbach (1983) recognised the basic dichotomy between the subtropical and Afrotemperate forests, but differed considerably from Acocks (1953), Cawe (1986 & 1990) and Cooper & Swart (1992) in their classification of each of these main forest types, recognising two subtropical and three Afrotemperate forest types. The forest types recognised by Von Breitenbach & Von Breitenbach (1983) also occupy geographically distinct areas. Whereas Acocks (1953) based his classification mainly on lists of all species of higher plants, Cooper & Swart (1992), in common with Von Breitenbach & Von Breitenbach (1983), based their classifications exclusively on lists of the ligneous flora in the various forests. Cooper & Swart (1992) deliberately adopted Acocks' (1953) classification but Von Breitenbach & Von Breitenbach (1983) based their classification on a subjective phytosociological manipulation of their data.

Acocks' (1953) classification is recognised as idiosyncratic (De Winter 1988) and has been the subject of much criticism in recent years. Cowling (1984) points out that:

> "Acocks' criteria for distinguishing veld types are never fully defined and often incorporate vague and untestable statements on history, utilization and dynamics. He completed his major work at a time when the assumptions of Clementsian dynamics were largely accepted. This allowed for the groupings of structurally and floristically unrelated types into a single veld type, based on assumed successional relationships."

Classifications based on this standard work, such as that by Cooper & Swart (1992) share, to a greater or lesser extent, the criticisms that have been made of it. Original classifications such as that by Von Breitenbach & Von Breitenbach (1983) are of particular interest, especially when they differ from the standard work of reference (i.e. Acocks 1953). Given the highly subjective nature of Acocks' (1953) classification, the prominence it enjoys in South African ecological thought, and the mounting criticisms levelled at it in recent times, it is pertinent to re-evaluate the soundness of the whole of his classification of the vegetation of Transkei. In this paper I attempt to do this with respect to the indigenous forests.

Methods

Comprehensive lists of the ligneous flora of 86 forests in both the Afrotemperate and subtropical regions of Transkei, collected by Cooper & Swart (1992), were analyzed using TWINSPAN (Two-Way-Indicator-Species-Analysis), a multivariate classification technique (Hill 1979). The resultant forest classes were then mapped and correlated with some environmental variables such as altitude and rainfall.

Results and discussion

Salient features of the present classification are summarised in Table 1. The basic dichotomy between subtropical forests in the coastal lowlands and the Afrotemperate forests in the hinterland is confirmed. The subtropical forests are much richer in species and tend to be larger than the Afrotemperate forests. Rainfall does not play a significant part in discriminating between these two main forest types.

Subtropical Forest Types

Tree species characteristic of subtropical forest are listed in Table 2. The species appearing on this list are not necessarily exclusive to subtropical forest, but are very common in subtropical forest and very rare in Afrotemperate forest. Relative frequencies of zero are also not absolute; thus *Ocotea bullata* was not encountered on this particular survey in subtropical forest, but is definitely known to occur in some subtropical forests, albeit rarely.

1. Moist Subtropical Forest (Type A)

Moist Subtropical Forest is confined to the high rainfall areas along the coast from Port St Johns northwards. The altitudinal range for these forests is 20-600 m. They fall largely within Acocks' (1953) Typical Coast-Belt Forest (veld type 1) and also include his Pondoland Sourveld Plateau (veld type 3). Forest patches comprising Moist Subtropical Forest are considerably larger than other subtropical forest patches. TWINSPAN indicator species for this forest type are *Cola natalensis* and *Sapium ellipticum*. Large trees typical of Moist Subtropical Forest include *Albizia adianthifolia*, *Chrysophyllum viridifolium*, *Millettia sutherlandii* and *Macaranga capensis*. The last two species tend to occur in dense, almost pure stands. Common shrubs and small trees typical of Moist Subtropical Forest include *Antidesma venosum*, *Bachmannia woodii*, *Colubrina nicholsonii*, *Mackaya bella*, *Mitriostigma axillare* and *Pachystigma macrocalyx*.

An important feature of these large moist forests is the occurrence of rare species such as *Cephalanthus natalensis*, *Cunonia capensis*, *Pseudosalacia polyantha*, *Rhynchocalyx lawsonioides* and *Rinorea domatiosa*, among others. Van Wyk (1989) describes some of these species as Pondoland endemics and further notes that numerous subtropical species reach their southernmost extension within these forests.

Table 1. TWINSPAN classification of the indigenous forests of Transkei and salient features of the various forest types.

	Subtropical forest				Afrotemperate forest			
Forest type	A	B	C	D	E	F	G	H
Number of forests	9	13	7	3	13	14	15	12
Mean forest size (ha)	876	336	370	177	240	219	221	34
Mean annual rainfall (mm)	1222	1048	1008	1018	994	1083	998	951
Mean altitude (m)	263	250	88	53	1256	1131	1316	1346
Mean species richness	120	97	111	90	67	57	53	41

Table 2. List of trees characteristic of subtropical forest. Only those species with at least 50% frequency in subtropical forest and not more than 5% frequency in Afrotemperate forest have been selected.

Name of species	Relative frequency in subtropical forest (%)	Relative frequency in Afrotemperate forest (%)
Acalypha glabrata	50.0	0.0
Bequaertiodendron natalense	100.0	0.0
Brachylaena uniflora	68.8	1.9
Buxus natalensis	53.1	0.0
Casearia gladiiformis	53.1	1.9
Cassipourea gummiflua	53.1	0.0
Celtis gomphophylla	53.1	0.0
Chaetachme aristata	78.1	0.0
Cordia caffra	56.3	0.0
Croton sylvaticus	81.3	0.0
Cryptocarya latifolia	87.5	1.9
Diospyros natalensis	65.6	0.0
Drypetes gerrardii	78.1	0.0
Duvernoia adhatodoides	81.3	0.0
Erythrina caffra	71.9	0.0
Heywoodia lucens	50.0	0.0
Indigofera natalensis	56.3	0.0
Maerua racemulosa	59.4	3.7
Margaritaria discoidea	62.5	0.0
Maytenus cordata	50.0	0.0
Micrococca capensis	78.1	0.0
Millettia grandis	78.1	0.0
Monanthotaxis caffra	90.6	3.7
Oncinotis tenuiloba	65.6	0.0
Pavetta natalensis	65.6	0.0
Protorhus longifolia	81.3	1.9
Rauvolfia caffra	59.4	0.0
Rawsonia lucida	78.1	1.9
Rothmannia globosa	65.6	0.0
Strychnos henningsii	71.9	0.0
Teclea gerrardii	87.5	0.0
Tricalysia capensis	81.3	0.0
Trichilia dregeana	93.8	3.7
Trichocladus crinitus	68.8	0.0
Turraea floribunda	68.8	0.0

Whereas Van Wyk (1989), noting the relatively high degree of endemism and the convergence of three floras in the area underlain by sandstones, suggests that the remarkable diversity of these forests derives mainly from lithological factors, the quite high rainfall in this area (Dent et al. 1989) possibly plays an even more significant role. In this regard it is important to note that a substantial portion of the flora which is endemic to this area is not strictly confined to the sandstones and that the forest patches comprising Moist Subtropical Forest are not underlain exclusively by sandstones, but by a variety of rock types. In the light of the foregoing, Van Wyk's (1989) "edaphic trapping" hypothesis is difficult to maintain.

2. Medium Moist Subtropical Forest (Type B)

Medium Moist Subtropical Forest is more widespread, appears drier, consists of smaller forest patches, and contains fewer species than Moist Subtropical Forest. The altitudinal range for these forests is from 20-600 m. They fall mainly within Acocks' (1953) Typical Coast-Belt Forest (veld type 1), but also include Ngongoni Veld (veld type 5) and Eastern Province Thorn Veld (veld type 7). It should be pointed out that Acocks (1953) distinguished between veld types 5 and 7 on the basis of floristic composition of the grasslands which he considered derivative from forest. There are no TWINSPAN indicator species for this forest type, prominent taxa within such forests having wide ecological amplitudes. Whereas Moist Subtropical Forest is notable for the presence of Pondoland endemics and the high admixture of Afrotemperate species (Van Wyk 1989), these elements are notable for their absence in Medium Moist Subtropical Forest which shares many species with the moister and the drier types of forest.

3. Dry Subtropical Forest (Type C)

Dry Subtropical Forest is generally confined to lower altitudes than either Moist or Medium Moist Subtropical Forest. The altitudinal range for this forest type is from sea level to 252 m. It is shorter in stature and has a scrub-like physiognomy. Forests comprising this forest type all fall within Acocks' (1953) Typical Coast-Belt Forest (veld type 1). There are no TWINSPAN indicator species for this forest type but *Euphorbia grandidens* and *Buxus macowanii* are fairly characteristic of it. Typical shrubs and small trees include: *Capparis tomentosa*, *Coddia rudis*, *Commiphora harveyi*, *Deinbollia oblongifolia*, *Diospyros scabrida*, *D. simii*, *D. villosa*, *Dombeya tiliacea*, *Homalium dentatum*, *Nuxia congesta*, *Ochna arborea*, *Olea woodiana*, *Schotia latifolia*, *Sclerochiton harveyanus*, and *Suregada africana*.

4. Subtropical Dune Forest (Type D)

Forested coastal dunes are common but not extensive in the study area and the three patches of dune forest included in this survey are the largest. The altitudinal range is from sea level to 60 m. The forests are depauperate in species compared to other types of high forest. The fore dunes are typically dominated by *Carissa macrocarpa* and *Mimusops caffra* whereas further inland *Sideroxylon inerme* is the dominant species. The TWINSPAN indicator species for this forest type is *Allophylus natalensis*, a small tree usually found at the forest fringe. Other characteristic Subtropical Dune Forest species include: *Brachylaena discolor*, *Chrysanthemoides monilifera*, *Eugenia capensis* and *Phoenix reclinata*. The dune forests in the study area are, as in the case of other plant communities, notably poorer in species than their counterparts in Natal. For example, only 130 tree species were recorded in the dune forests compared to 242 tree species recorded by Weisser & Drews (1980) in the forested dunes of Maputaland.

Afrotemperate Forest Types

Von Breitenbach & Von Breitenbach's (1983) classification of the Afrotemperate forests, according to sections of the minor escarpment running from east to west, is not supported by the present classification. Altitude is the main discriminating factor among these forests and not geographic location, with respect to sections of the minor escarpment. Mean forest patch-size and species-richness decrease with increasing altitude (Table 1). Although forest patches within a particular section of the minor escarpment tend to be similar, different forest types within the same section of the escarpment are sometimes within close proximity of one another. For example, in the Umzimkulu district where altitudinal gradients are very steep, all four Afrotemperate forest types (Table 1) are found in close proximity to one another. Patches of forest of the same type occurring on different sections of the minor escarpment are sometimes separated by a different type of forest on an intervening section of the minor escarpment. This complexity in the spatial distribution of the forests belies the simple classifications proposed by either Acocks (1953) or Von Breitenbach & Von Breitenbach (1983).

Although rainfall does not appear to play a significant role in differentiating between the various Afrotemperate forest types, this does not necessarily render Cawe's (1986) classification untenable since rainfall is not a true measure of soil-moisture status.

White (1983) notes that most tree taxa in the Afrotemperate region are widespread. This is certainly true of the species occurring in Transkei. Whereas 36 species are listed as characteristic of subtropical forest (Table 2) only seven characterise Afrotemperate forest (Table 3).

Table 3. Trees characteristic of Afrotemperate forest. Only those species with at least 50% frequency in Afrotemperate forest and not more than 5% frequency in subtropical forest have been selected.

Name of species	Relative frequency in subtropical forest (%)	Relative frequency in Afrotemperate forest (%)
Buddleja auriculata	0.0	53.7
B. salviifolia	0.0	75.9
Euclea crispa	0.0	63.0
Heteromorpha trifoliata	3.1	57.4
Leucosidea sericea	0.0	57.4
Ocotea bullata	0.0	55.5
Podocarpus henkelii	0.0	61.1

1. Mistbelt Afrotemperate Forest (Type E)

Mistbelt Afrotemperate Forest falls mainly within Acocks'(1953) Dohne and Highland Sourveld (veld types 44b and 44a respectively), but also occurs within Ngongoni Veld (veld type 5). The altitudinal range of this forest type is 800-1,490 m with an average altitude of 1,250 m. The TWINSPAN indicator species for this forest type is *Peddiea africana*. Mistbelt Afrotemperate Forest differs from other types of Afrotemperate forest in that it has a higher admixture of subtropical species which include: *Brachylaena uniflora*, *Casearia gladiiformis*, *Cassipourea gerrardii*, *Cryptocarya latifolia*,

Dracaena hookeriana, *Keetia guenzii*, *Maesa lanceolata*, *Monanthotaxis caffra*, *Protorhus longifolia*, *Rawsonia lucida*, *Rhus rehmanniana*, *Strychnos henningsii*, and *Trichilia dregeana*.

2. Moist Afrotemperate Forest (Type F)
This forest type is confined mainly to the Matiwane range with some outliers in the Umzimkulu block. It corresponds to Von Breitenbach & Von Breitenbach's (1983) Group A forests and falls within Acocks' (1953) Dohne and Highland Sourveld (veld types 44b and 44a respectively). The altitudinal range of these forests is 800-1,650 m with mean altitude of 1,130 m. The admixture of subtropical species is less than in Mistbelt Forest. Most of the subtropical species which in Moist Afrotemperate Forest also occur in Mistbelt Forest and include: *Bersama swinnyi*, *Cassipourea gerrardii*, *Clerodendrum glabrum*, *Dalbergia multijuga*, *Dracaena hookeriana*, *Ficus sur*, *Maerua racemulosa*, *Oricia bachmannii*, *Pachystigma macrocalyx*, *Psychotria capensis*, *Strychnos mitis*, *Tricalysia lanceolata*. There are no TWINSPAN indicator species for this forest type. Apart from sharing most of the species listed above, Moist Afrotemperate Forest resembles Mistbelt Forest by having the following important species in common: *Curtisia dentata*, *Ocotea bullata*, *Podocarpus falcatus*, *P. henkelii*, *P. latifolius* and *Prunus africana*. However, some of these species are not equally abundant in both forest types. For example, *P. henkelii* and *Prunus africana* are not common in Moist Afrotemperate Forest, but very common in Mistbelt Afrotemperate forest, and *C. dentata*, *Podocarpus falcatus* and *P. latifolius* are very common in Moist Afrotemperate Forest, but are encountered occasionally in Mistbelt Forest.

3. Middle Altitude Afrotemperate Forest (Type G)
The altitudinal range of forests constituting Middle Altitude Afrotemperate Forest is 1,100-1,650 m with mean altitude of 1,315 m. The forests occur on the same chain of mountains as the Mistbelt Afrotemperate Forest but at a higher altitude. They are poorer in species and hardly contain any subtropical species. The TWINSPAN indicator species for this forest type is *Pteroxylon obliquum*. As the number of species thins out the following gain more prominence: *Buddleja auriculata*, *Englerodaphne pilosa*, *Heteromorpha trifoliata*, *Leucosidea sericea* and *Olinia emarginata* all of which are small to medium sized trees or climbers. Large trees such as *Podocarpus henkelii* and *Prunus africana* are also more common than in Moist Subtropical Forest.

4. High Altitude Afrotemperate Forest (Type H)
This forest type generally occurs above Middle Altitude Afrotemperate Forest at elevations ranging between 1,100-1,770 m. Clusters of High Altitude Afrotemperate Forest are concentrated in Western Tembuland and in Mount Frere district. It is even poorer in species than Middle Altitude Forest and comprises smaller patches of forest. There do not appear to be any species that are exclusive to this forest type, species occurring in these forests being generalists.

Conclusion
There is no support for Acocks' (1953) classification of the forests of Transkei. This implies that the classification of the grasslands is also suspect because it was based

essentially on the composition of the forests to which they were considered to be secondary. Lack of support for Acocks' classification has important implications for land-use plans which are based on Acocks' vegetation map. A new vegetation map of South Africa based on less nebulous criteria than Acocks' is overdue.

Acknowledgements
I wish to thank Messrs K.H. Cooper and W. Swart for making their data available to me, Mrs A.J.N. Bosa for commenting on the manuscript, the University of Transkei and the AETFAT Organising Committee for arranging financial support.

References
Acocks, J.P.H., 1953. Veld Types of South Africa. Mem. Bot. Surv. S. Afr. 28. Botanical Research Institute, Pretoria.
Cawe, S.G., 1986. A quantitative and qualitative survey of the inland forests of Transkei. M.Sc. Thesis. University of Transkei, Umtata.
Cawe, S.G., 1990. Coastal forest survey: a classification of the coastal forests of Transkei and an assessment of their timber potential. Unpubl. Report, University of Transkei, Umtata.
Cooper, K.H. & W. Swart, 1992. Transkei Forest Survey. Wildlife Society of Southern Africa, Durban.
Cowling, R.M., 1984. A syntaxonomic and synecological study in the Humansdorp region of the Fynbos Biome. Bothalia 15: 175-227.
De Winter B., 1988. Foreword to third edition. In: Acocks, J.P.H., Veld Types of South Africa. Mem. Bot. Surv. S. Afr. 57: iv. Botanical Research Institute, Pretoria.
Dent, M.C., S.D. Lynch & R.E. Schulze, 1989. Mapping mean annual and other rainfall statistics over southern Africa. Agricultural Catchments Research Unit Report No. 27. University of Natal, Pietermaritzburg.
Hill, M.O., 1979. TWINSPAN - A FORTRAN program for arranging multivariate data in an ordered two-way table by classification of the individual and attributes. Cornell University, Ithaca, New York.
McKenzie, B., 1984. Ecological considerations of some past and present land-use practices in Transkei. Unpubl. Ph.D. Thesis, University of Cape Town, Cape Town.
Sim, T.R., 1907. The forests and forest flora of the Colony of the Cape of Good Hope. Taylor & Henderson, Aberdeen.
Van Wyk, A.E., 1989. The sandstone regions of Natal and Pondoland: Remarkable centres of endemism. In: Heine, K. (ed.), Palaeoecology of Africa and the surrounding islands: 243-257, A.A. Balkema, Rotterdam.
Von Breitenbach, F., & J. Von Breitenbach, 1983. Notes on the natural forests of Transkei. J. Dendrol. 3: 17-53.
Weisser, P.J. & B.K. Drews, 1980. List of vascular plants of the forested dunes of Maputaland. In: Bruton, M.N. & K.H. Cooper (eds.), Studies on the ecology of Maputaland: 91-101. Rhodes University and The Natal Branch of the Wildlife Society of Southern Africa. Rhodes and Durban.
White, F., 1983. The vegetation of Africa: A descriptive memoir to accompany the Unesco/-AETFAT/UNSO vegetation map of Africa. Unesco, Switzerland.

Conservation status of mopane woodlands in Malawi: a case study of Mua-Tsanya Forest Reserve

A.C. CHIKUNI
National Herbarium and Botanic Gardens of Malawi, P.O. Box 528, Zomba, Malawi.

Abstract
A study to determine the current status of mopane woodland conservation in Malawi showed that it is endangered. Increased human population and overdependence on natural resources for firewood, building materials, tool handles, wood for charcoal making and carvings, and other non-woody products such as medicine, food, fodder, etc, are the major causes of deforestation of mopane woodlands. Although mopane woodlands are protected by the forestry legislation, this has not been successfully executed due to lack of a general environmental legislation on biodiversity conservation. For maximum protection of the total biodiversity of mopane woodlands, a general environmental framework should be established. All legislation which impinges on the environment should be reviewed to reflect new environmental needs. A brief overview of bottlenecks concerning the conservation of mopane woodlands such as demographic, poverty and legislative problems is given.

Résumé
Une étude ayant pour but de déterminer le statut actuel de la conservation de la forêt claire à mopane au Malawi a indiqué que celle-ci est en danger. L'augmentation croissante de la démographie et son impact démesuré sur les ressources naturelles, notamment en ce qui concerne le bois de chauffage, les matériaux de construction, les manches d'outils, le charbon de bois et le bois à sculpter, ainsi que d'autres produits non ligneux tels les médicaments, les aliments, les fourrages, etc., sont les causes principales de la déforestation de la forêt claire à Mopane. Bien que les forêts claires à mopane soient protégées par la législation forestière, celle-ci n'a pas été bénéfique étant donné qu'il n'existe pas de loi sur l'environnement en général impliquant la conservation de la biodiversité. Pour obtenir une protection maximale de toute la biodiversité des forêts claires à Mopane, il faudrait établir un système législatif concernant tout l'environnement. Il faudrait revoir toute législation qui est conflictuelle avec l'environnement, en vue de rencontrer les nouvelles exigences qu'il implique. Un bref aperçu des impasses pouvant contrarier la conservation des forêts claires à mopane est donné, notamment les problèmes de démographie, de pauvreté et de législation.

Key words
Colophospermum mopane, Conservation, Malawi, Mopane woodlands, Socio-economic factors.

Introduction

Colophospermum mopane (Kirk ex Benth.) Kirk ex J. Léonard (*Copaifera mopane* Kirk ex Benth.), locally known as tsanya (Chichewa language) or mopane (Tumbuka language), grows up to 10-15 m. Mopane is one of the most distinctive vegetation groups and forms pure stands, hence the term mopane woodland or *Colophospermum* woodland is an acceptable term (Palgrave 1977, Pullinger & Kitchin 1982, Rattray 1961). In Malawi the mopane woodland covers only 2360 km^2 of the country. It covers over 60 km^2 of Liwonde National Park, part of Lengwe National Park, Majete,

L.J.G. van der Maesen et al. (eds.), The Biodiversity of African Plants, 250-258.

Mwabvi and Vwaza Marsh Wildlife Reserves (Shorter 1989). The vegetation community is also common in the middle Shire valley i.e. Chingale (Namatunu Forest) in Zomba and Mua-Tsanya Forest Reserve in Dedza.

Mopane woodlands in protected areas such as national parks, wildlife and forest reserves are fairly well protected, though they suffer from heavy wood poaching for building poles and fuelwood. However, those outside these areas are endangered.

The study area Mua-Tsanya Forest Reserve, is found in Dedza District in the Central Region of Malawi. It was gazetted in 1932 as a forest reserve and is managed by the Forest Department. The population of surrounding villages is estimated at 12,000 and is increasing at a rate of 3.2% per annum (Malawi Government 1991). The main economic activity is subsistence agriculture and crops grown include maize, cotton, beans and cassava. A few estates such as Diena and Bwanje, grow barley tobacco and also keep livestock. The study was undertaken to determine the socio-economic importance of the mopane woodland and its resources, conservation status and possible problems concerned with conservation of mopane woodlands in particular of Mua-Tsanya Forest Reserve.

Methodology

A structured questionnaire was administered to the villagers around the Mua-Tsanya Forest Reserve to determine the socio-economic importance of the reserve in June 1994. Additional information obtained from the methodology of participatory rural appraisal (PRA). PRA techniques used included semi-structured interviews with individuals, key informants and group discussions. A general observation to establish the selling of fuelwood was made in the surrounding villages, roadsides and both Mtakataka and Golomoti markets.

A floristic survey was conducted in Mua-Tsanya Forest Reserve for five days in January 1994, to get an objective picture of the conservation status and plant diversity of the woodland. In total nine (10 x 10 m) quadrats were spaced at an interval of 1 km along three transects. In each quadrat the following were recorded: botanical names of all plant species, diameter at breast height (dbh) for trees and shrubs, tree height and number of seedlings. Anthropogenic pressure in the forest was assessed by noting tree felling, ring barking, bark stripping, whole plant uprooting, pruning and encroachment in form of settlement and cultivation. Herbarium specimens collected in these areas over the past years were studied.

Results

In total 70 plant species were recorded during the floristic survey of which 18% are trees, 47% shrubs and 35% herbs. Mopane covers the largest area and it is the most dominant species. *C. mopane* and *Acacia hockii* have frequencies above 50% while the rest have frequencies of less than 50%. *C. mopane* has the highest number of seedlings per hectare followed by *Commiphora africana*, *Albizia anthelmintica* and *Dioscorea cochleari-apiculata*. 30% of mopane trees have dbh of up to 10 cm, 28% of dbh between 10-20 cm, 24% had dbh of 20-30 cm while only 4.3% had dbh of more than 30 cm. Of the woody species *Acacia hockii*, *Acacia goetzei*, *C. mopane*, *Sclerocarya birrea* subsp. *caffra*, and *Grewia bicolor* measured more than 10 m. On average *C. mopane* measured 10 m high. However, 43.8% of the mopane trees were smaller than

Figure 1. Mopane trees are the largest component of building poles in the study area. Here the wood is used in a church.

Figure 2. Fuelwood of mopane trees for sale along the roadside.

5 m, 20.8% accounted for mopane trees of 5-10 m while 35% of the trees measured more than 10 m.

The PRA results showed that Mua-Tsanya Forest Reserve is fairly well protected though there was an estimated wood poaching of 389 trees/ha. All other pockets of mopane outside the boundaries of Mua-Tsanya Forest Reserve show signs of severe exploitation. The biggest threat to the mopane woodland in the area is deforestation which is due to clearance for subsistence farming, settlement, building poles and fuel-wood. Mopane trees are preferred for building poles (Figure 1), and also for firewood and charcoal (Figure 2); agricultural residues such as maize, pigeonpeas and tobacco stalks are not used for cooking as is the case in many parts of Southern region of Malawi. The woodland is also habitat for many useful plants which are commonly used for thatch, fodder, medicines, food, as well as building poles and household tools (Table 1).

Discussion

The relatively high seedling and sapling density indicate the high regeneration capacity of the reserve. However, a number of seedlings and saplings may be destroyed by livestock grazing in the reserve. Considering the standing density of 556.0 trees/ha, an estimated density of 389 trees/ha destroyed is alarming; but there is need to establish the destruction rate.

Mopane stands on customary land suffer full anthropogenic impact compared to the mopane trees in the reserve. With increasing population and depletion of these satellite mopane stands, the reserve is threatened. Although the reserve suffers from extensive poaching of wood, the original size has reduced by 2% over the past 60 years (Malawi Government 1993).

Although the Malawi Government is making enormous efforts in biodiversity conservation and sustainable utilization, there are still problems that impinge upon the conservation of plant resources as, in case of Mua-Tsanya Forest Reserve, increased human population and overdependence on natural resources for firewood, building materials, tool handles, wood for charcoal making and carvings, and other non-wood products such as medicine, food, and fodder. The following is a brief account of the problems of conservation of mopane woodlands.

Population Growth

The population growth rate for the study area is 3.2% per annum. Migration of settlers from south to central regions and floods of 1986 which forced such villages as Kamwendo, Mdulambale and Mbendela to resettle near the reserve have contributed to this population growth. The vegetation is estimated to be reducing at the rate of 1.2% per annum. With the vegetation resources decreasing every year, the population exerts enormous pressure on the mopane woodlands. As long as the population growth rate remains as high as 3.2% per annum extensive developments in agriculture, forestry and industry will be on the increase. This has created isolated patches or islands of mopane woodlands which are facing danger as their boundaries are under human pressure.

Poverty

Fifty-five percent of the total population and ninety percent of the rural population is living below the poverty line (Africa, South of the Sahara, 1993). Poverty therefore leads the poor to trade off long-term sustainable resource use for short-term consumption of stocks, since they depend entirely on the existing natural resource endowment for a living and other needs (World Bank 1991).

Table 1. Some useful plant species of Golomoti savanna woodland, especially those recorded in mopane woodland.

Botanical name	Local name	Local use
Acacia polyacantha subsp. *campylacantha*		Treating snake bites and gonorrhoea.
Adansonia digitata	Malambe	The fruit is edible.
Anthericum pterocaulon		Leaves are eaten as a vegetable.
Azanza garckeana	Matowo	The fruit is edible.
Cissus integrifolia	Mtambe	Boiled leaves are eaten as a vegetable.
Colophospermum mopane	Tsanya	House, bridge and tobacco barn construction, wood carvings, farm tools, pestles and mortars, firewood, charcoal. A powder from leaves is a cure to sores and wounds. The leaves are fed to livestock.
Dalbergia melanoxylon	Phingo	House construction and wood carvings.
Dregea cinerea subsp. *africana*		An antidote to scorpion and snake bites, and an aphrodisiac.
Euphorbia indica & *E. inaequilatera*		Fish poison, especially in stagnant water.
Friesodielsia obovata	Mjinkha	The stem is used for weaving food stores. The fruits are edible by humans; snakes also like them (black mamba).
Panicum monticola, *P. maximum*		Fed to livestock.
Setaria palustris	Wanje	Grass species that is commonly used as a thatch for houses, stables and tobacco barns.
Sterculia africana	Ngoza	The exocarp is burnt and its ashes filtered, the filtrate is used as potash for cooking vegetables. The seed are roasted, pounded to fine powder and added to vegetables as substitute for groundnuts. The bark is stripped for string fibre. Oil is also extracted from roasted seeds.
Tamarindus indica	Bwemba	The fruit is eaten and is liked by children.

Plant resources are largely used in the construction of the traditional houses (ngomi, thokoti and libanda) in Malawi. Most fuelwood for domestic uses and tobacco industries are harvested from indigenous forests, because other sources of energy such as electricity, gas and paraffin are expensive. Medicinal plants such as *Dichrostachys cinerea*, *Acacia polyacantha* subsp. *campylacantha*, *D. nitidula*, are a few that are harvested from most forest reserves. Unsustainable methods of harvesting the medicinal plants such as ring-barking and uprooting of whole plant are the commonest. Fruit trees such as *Cleistochlamys kirkii*, *Tamarindus indica*, *Flueggea virosa*, and edible vegetables such as *Commiphora africana*, *Anthericum pterocaulon* are collected by the local people. Very often these food plant species are overexploited so that their original distribution has been reduced and/or the actual species have become rare or threatened. Handles for cultivation equipment such as hoes and axes, crop storage facilities, and food processing equipment (mortars and pestles) are made from wood. Wooden cooking sticks and serving spoons are commonly used when preparing the traditional dish nsima. It is apparent from the above that low income is forcing Malawians to use plant resources for a living.

Overdependence on subsistence agriculture

About 91% of Malawi's population lives in the rural areas and depends on subsistence agriculture for their livelihood (Sakanda 1994). The subsistence agriculture being practised is apparently extensive, resulting in shifting from one area to another in search of fertile land. It is common practice that after tree felling to open up a new field, the stumps are set on fire. Fields are abandoned after one or two growth seasons due to infertility of the soils. Regeneration of the mopane is limited in such cases because of destruction of propagules (seeds, seedlings or rootstock-suckers) by fire and the loss of soil fertility (Figure 3). In heavily grazed areas regeneration capacity of most herbaceous species was also affected. With the increasing population more areas of mopane woodland are likely to be turned into farmland.

Overdependence on fuelwood for energy

Biomass and fuelwood are the main sources of energy (charcoal, firewood). Urban dwellers despite being close to other sources of energy (hydro- and solar electricity, coal, gas) also use fuelwood for heating because they cannot afford these other sources of energy (Sakanda 1994). Fuelwood harvesting for rural industries (for tobacco curing, brick making, fish-smoking) and urban household use has a greater impact on the conservation of mopane in Golomoti because most of the wood used is stemwood derived from clear felling. Since the rural population needs income to purchase other necessities of life, sale of fuelwood becomes the easiest source of income. Rural people supply fuelwood to suburban centres such as Salima, Mtakataka, and Golomoti.

Legislative problems

Biodiversity conservation in Malawi is not governed by any act of Parliament. Malawi puts a lot of emphasis on the protection of crop plants under the plant protection act which is concerned with giving phytosanitary certificates to exported and imported plants (Malawi Government 1970). The only instrument for the protection of wild plants seems to be the forest legislation (Malawi Government 1964). However the

Figure 3. Tree felling followed by burning of the stumps are the common procedures for land clearing for cultivation and settlement. This does not allow regeneration of mopane trees after the land is abandoned.

Forest Act is a subsidiary Act, because when activities that affect the forest resources and biodiversity as a whole are carried out under other Acts, such as Land Act, Public Roads Act, Electricity Act and Telecommunications Act, the Forestry Department is not contacted. It would appear therefore that conservation of plant resources is not given any priority. For instance the railway line from Lilongwe to Blantyre cut through the mopane woodland of Namatunu Forest (Zomba). The road from Golomoti which joins the Mua-Monkey-Bay Road at Kapiri destroyed a large portion of mopane woodland. Mua-Monkey-Bay Road divided the Mua-Tsanya Forest Reserve into two. Estate expansion (Diena, Bwanje Spearhead Holdings) has consumed a larger portion of the woodland. Under the Electricity Act, construction is taking place to supply electricity to the estates and huge areas of mopane woodland are being destroyed in the process (Figure 4).

Although the Forestry Department controls plant collection, a plant collection permit issued by the Forestry Research Institute of Malawi (FRIM) to plant collectors is not part of the legislation. There is no instrument regulating the exportation and importation of wild plants but the collection permit has been used by some botanists as an export licence. The non-availability of trade restrictions in Malawi has resulted in rare, protected and endangered plants being collected and exported illegally.

Possible solutions to the problems of conservation in Malawi

Multidisciplinary and multisectoral approaches are needed if conservation and sustainable use of forests and other natural resources is to be achieved. The government and non-governmental organisations should work together so that alternative sources of energy (gas, electricity, solar) should be made available to the people at affordable

Figure 4. Mopane woodland being cleared along electricity powerline corridor. An example of conflicting legal acts.

prices; income-generating activities should be introduced for the people. It is difficult at the moment to move away from subsistence farming. However the estate sector which presently emphasizes the production of tobacco, tea, coffee and rubber should embark on food crop production so as to supplement the little that the rural farmers produce. The remaining fragmented pockets of mopane woodlands on customary land should be afforded conservation status and hence it is recommended that the government should gazette them.

The lack of a general environmental framework legislation makes the whole processes of plant conservation complicated. The fact that the Forest Department was established to exploit timber resources and that their policing powers have been mainly to protect these resources against encroachment from outside are enough evidence to support the establishment of legislation on protection and conservation of wild plants. There is need therefore, to deal with such matters at the ministerial level. It is therefore recommended that an Environment Act should be established and that where necessary the Forestry Act, Game Act, Plant Protection Act, Roads Act, Telecommunications Act and Wildlife Act should be reviewed to reflect the requirements of a new Ministry of the Environment. The National Herbarium Act should also be reviewed so that threatened, rare and endemic plants are listed. Export and import procedures should be laid down clearly. There is great need therefore to establish a plant conservation unit within the National Herbarium Department which will, among other things, organize and coordinate the making of inventories, determine conservation priorities, promote education and public information, issue plant collection and trade permits or licences.

Conclusion

The contribution by the plant resources of the mopane woodlands towards socio-economic development is evidence enough to warrant maximum protection of its botanical diversity. However, despite efforts by government departments and other non-governmental organisations, the loss of plant diversity in mopane woodlands mainly from habitat destruction, overharvesting of plant resources, developmental projects, agricultural expansion, increased population continue. The lack of a general environmental framework legislation on biodiversity conservation causes problems.

The efforts of the different institutions need to be coordinated and it is recommended that a Ministry of Environment should be created to establish a national conservation strategy and provide guidelines. Introduction of the environmental act may take time but as a matter of urgency the government should gazette the remaining mopane forest on the customary land so that the resources receive maximum protection.

Acknowledgements

I would like to thank Messrs M. Mwanyambo, D.A. Kamundi and Z.L.K. Magombo for their guidance, assistance, suggestions and for their critical reading of the manuscript. Great appreciations are due to the herbarium staff in Zomba who assisted the author in collecting data during the survey. I am also thankful to Mr. Ndawala of The National Statistical Office in Zomba for providing recent figures on population of the study area. Finally, I am thankful to Ms Chadaine Mangasasa for typing this paper.

References

Africa, South of the Sahara, 1993. 22nd Edition, Malawi, Europa Publishers Limited, England. pp. 515-529.

Malawi Government, 1991. Malawi Population and Housing Census, 1987, Summary of Final Results Volume 1 National Statistical Office, Government Printer, Zomba.

Malawi Government, 1993. Forest Resources Mapping and Biomass Assessment for Malawi. Satellitbild, Kiruma, Sweden, in conjunction with the Department of Forestry.

Malawi Government, 1964. Laws of Malawi: Forestry Act Chapter 63: 01. Government Printer, Zomba.

Malawi Government, 1970. Laws of Malawi: Plant Protection Act Chapter 64: 01. Government Printer, Zomba.

Palgrave, K.C., 1977. Trees of Southern Africa. Cape Town, C. Struik.

Pullinger, J.S., Kitchin, A.M. 1982. Trees of Malawi, with some herbs and climbers. Blantyre Printers.

Rattray, J.M., 1961. Vegetation types of Southern Rhodesia. Kirkia 2: 660.093.

Sakanda, G.P., 1994. Deforestation, a future challenge to the Malawi Forestry Sector. A paper presented at the First Meeting on Measures to Increase Sustainable Utilization of Natural Resources in Malawi.

Seyani, J.H., 1990. The potential value of Malawi Flora in development. Paper presented at a Conference on Science and Technology for Development. Ryalls Hotel, Blantyre.

Seyani, J.H., 1987. The conservation of plant genetic resources in Malawi. A paper presented at a training workshop on Genetic Resources Conservation in Africa, Lusaka, Zambia.

Shorter, C., 1989. An introduction to the common trees of Malawi. Wildlife Society of Malawi.

World Bank, 1991. Economic Report on Environmental Policy.

Deforestation in southern Côte d'Ivoire: a high-resolution remote sensing approach

C. CHATELAIN, L. GAUTIER & R. SPICHIGER
Conservatoire et Jardin botaniques de la Ville de Genève,
Case Postale 60, CH-1292 Chambésy/GE, Switzerland.

Abstract
The assessment of deforestation and the understanding of the processes involved at the local scale, need the use of high-resolution satellite images, together with field observations. The results are compared with the ones obtained with low resolution images. Vegetation and land use maps of 3 sites in the forest zone of Côte d'Ivoire (Yapo, Divo and Zagné) show various intensities of deforestation. The Zagné study reconstructs the history of forest fragmentation with an old map made with aerial photographs and three satellite images. Management of national forestry heritage is not adapted to the processes involved; more than half of the "forêts classées" can be considered as non-forest. Outside protected areas, primary and secondary forests have completely disappeared, except in land managed under traditional land ownership systems. Conservation needs to be in harmony with short and long term interests of local communities.

Résumé
L'estimation détaillée de la couverture forestière réelle et la compréhension des processus de la déforestation implique l'utilisation d'images satellites à haute résolution, accompagnées de données de terrain. Les résultats sont comparés avec ceux issus d'images à faible résolution. La réalisation de cartes de la végétation de trois sites en zone forestière de Côte d'Ivoire (Yapo, Divo et Zagné) montre des intensités de déforestation variées. Le site de Zagné traite de l'historique de la fragmentation forestière à l'aide d'une carte faite à partir d'anciennes images photographiques et trois images satellitaires. La gestion du patrimoine forestier national ne semble pas adaptée à l'ampleur des processus impliqués. En effet, plus de la moitié des forêts classées ont disparu. De plus, en dehors de ces forêts classées, on ne trouve presque plus de forêts primaires ni même de forêts secondaires, mis à part dans les zones dont l'utilisation du sol est de type traditionelle. La conservation ne peut s'envisager sans l'implication et l'intrégration des communautés locales dans la gestion du patrimoine.

Key words
Conservation, Côte d'Ivoire, Deforestation, Remote sensing.

Introduction

Remote sensing is widely used today to assess deforestation at continental or regional scales, mainly with low resolution (1 km) satellite images, like those of AVHRR/-NOAA. These data show that deforestation in Côte d'Ivoire (Ivory Coast) has reached alarming proportions. At the beginning of the 1980s, FAO-UNEP (1981) estimated that the forest cover in the moist forest zone had decreased from a total area of 13.2 million ha to 4.5 million ha. Relying on AVHRR/NOAA satellite images from the end of the 1980s, Païvinen et al. (1992) showed that, at that time, the forest cover was only 2.7 million ha, of which the National Park of Taï constituted the main continuous block. The annual deforestation rate, reaching 6.5%, would be the highest in the world (Schmidt 1990).

L.J.G. van der Maesen et al. (eds.), The Biodiversity of African Plants, 259-266.

To protect the forest against abusive timber exploitation and further deforestation by cultivators, the French colonial administration established from the 1930s a network of so-called "Forêts Classées". Today their total area, together with National Parks, reaches c. 3.18 million ha in the forest zone. However, overlaying an outline of their limits on an AVHRR/NOAA image clearly shows that more than 50% of this area can be considered as deforested. Outside these relatively protected areas, sizeable primary forest patches have almost totally disappeared.

At the local scale, the size of the pixels of AVHRR/NOAA imagery brings confusion between forest/non-forest mosaic inside one single pixel (mixels) and depleted or young secondary forests. Therefore, it is difficult to obtain results with a higher resolution than a simple forest/non-forest dichotomy, with a very restricted forest class taking into account only the pure forest pixels. In this respect, the non-forest class is very broad and includes, apart from pure non-forest pixels: the forest/non-forest mixels, the secondary an depleted forests, and also all the intermediate situations. Consequently, very different landscapes are confused into the non-forest class, ignoring the small-size forest fragments as well as the secondary vegetation.

In order to get a better understanding of the processes involved in deforestation, we need to have a detailed assessment of the residual forest resources at the local scale. We propose the use of high-resolution (20-30 m pixel size) satellite images together with field results. At this scale, the pixel size approaches the true quantum of heterogeneity in the field, being close in size to the area of the crown of a big forest tree.

Method

A thematic vegetation mapping has been undertaken by making an automatic supervised classification of these high-resolution images. To achieve this goal, we first identify both on the image and in the field a set of samples which appear homogenous and which represent as broadly as possible the range of colours (composition of spectral values) encountered on the image, and all the vegetation types of the area in the field. The principle of the classification is then to group all the pixels of the considered area in classes according to the similarity of their spectral response with the set of samples.

In all the major vegetation types, vegetation surveys are made following a method that we presented at the 1991 AETFAT congress (Gautier et al. 1994) This method consists of a linear 200 m survey, together with four or five 400 m^2 samples for quantitative biometric parameters. Numerous rapid physiognomic surveys are added to provide additional information or to verify the classification. All these surveys are precisely located with a GPS system, if the canopy is not too dense to prevent its use. Following each field trip, a new classification is made taking into account the field information acquired. Following each classification, new areas where ground information is needed are selected.

In each detailed survey, we acquired data concerning floristic composition (important species with an estimate of their cover) and vegetation structure (vertical distribution of the vegetation; number of stems < 2 m high; basal area; biomass estimate). Apart from enabling us to refine our classifications, these data are also very important to show which parameters of the vegetation have an influence on the radiance measured by the satellite. We hope to get a more precise knowledge of what information a satellite image is able to give, and what cannot reasonably be expected.

In this field, our first results confirm that apart from monospecific plantations, floristic composition does not generally show on the satellite images. On the other hand, the total amount of vegetation and its vertical distribution seem to be very sensible parameters. Data processing is still under way, but until now results of simple linear correlations between quantitative parameters and the radiance of the pixels in the different channels are not very meaningful. In the better cases, they can only explain 58% of variance. This exemplifies the complexity of the problem and highlights the overlapping of the quantitative parameters, and thus the need to experiment with multiple regression models. However, we still continue our surveys with this method, which has shown to be very practical in describing very different types of vegetation, especially including depleted forests and secondary formations.

Results

We have been working in 3 sites in the forest area, each one illustrating different intensities of deforestation. For the two first sites, where most of deforestation occurred long ago, we tried to get a better understanding of the processes involved by making a vegetation and land use map and comparing the results with other kinds of information, such as geographical and economic data. In the third site, where deforestation is only recent, we had the opportunity to reconstruct the history of forest fragmentation by comparing an old map and three satellite images over a 35 years period.

1. Divo (5°42'N; 5°15'W)

This c. 40 x 60 km study site is located in the semi-evergreen forest. Deforestation has been extreme (Figure 1), resulting in a landscape with up to 44% of coffee and cocoa plantations, interspersed with rice fields in the depressions. Semi-evergreen forests, which only account for 5% of the area, are restricted to small granitic hills. The traditional shifting cultivation system has been heavily disrupted by the important amount of land used for the perennial cash crops. Consequently, the fallow periods had to be severely shortened and secondary forest fragments are very rare.

Two "Forêts Classées" were originally included in this area, but their forest cover has completely vanished. In the case of the "Forêt classée" of Divo and its surroundings, the AVHRR/NOAA image only shows the disappearance of the forest. On the Landsat high-resolution image, the different vegetation types can be quantified and it appears that forest fragments are even more numerous outside than inside the previously protected area, where their surface accounts for less than 1%. Unshaded cocoa plantations represent 31%; herbaceous vegetation (young fallows) 25%; secondary thickets 12%; cocoa under shade trees 5%.

Compared with the south of the study area, near the FC of the Gô, where human settlement is ancient and many more forest fragments persist in a kind of mosaic, the previous FC of Divo shows much more intense human settlement which is recent and due to immigration of outside populations.

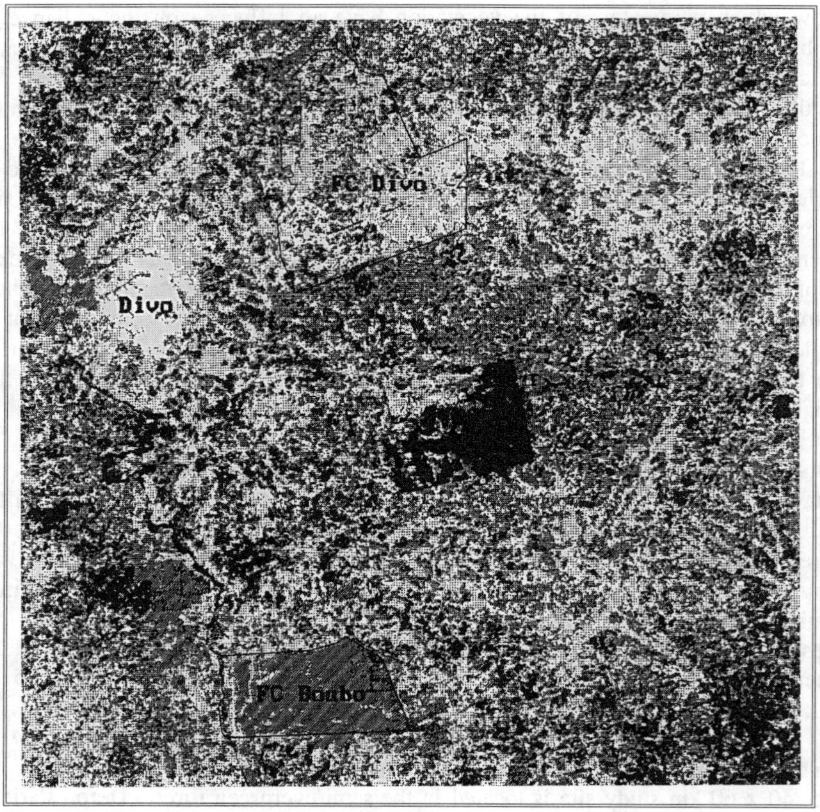

Figure 1. Vegetation map of the area of Divo (1 cm = 3.8 km). Black: forest or cacao plantation beneath forest; grey: coffee, cacao, oil palm plantation, or water; light grey: shrub vegetation or arable land; white: bare soil or built-up area. From Landsat TM 1990.

2. Yapo (5°45'N; 4°05'W)

The study area of c. 60 x 25 km is located in the evergreen forest. This site has been chosen to include the "Forêt Classée de Yapo" (Figure 2). The variation of its limits give interesting insights to the status of protection of this area.

The 1970 map of the "Forêts Classées" shows a forest block of c. 39,600 ha. Various parts have been withdrawn from the protected area, and today there are only 25,800 ha still under protection. Each of the various excluded areas exemplify various degrees of deforestation. For instance, the "Aké Béfiat" zone has been withdrawn from protection in 1986 to create a buffer zone. The disappearance of any forest element, secondary or depleted, is due to a massive immigration from outside. A further penetration into the restricted protected area is to be feared, despite the presence of forest service employees. In contrast, the western fringe of the FC which was excluded much earlier, still shows a high proportion of forest. It is worth mentioning that this area is still run by traditional land use rules, with few immigrants.

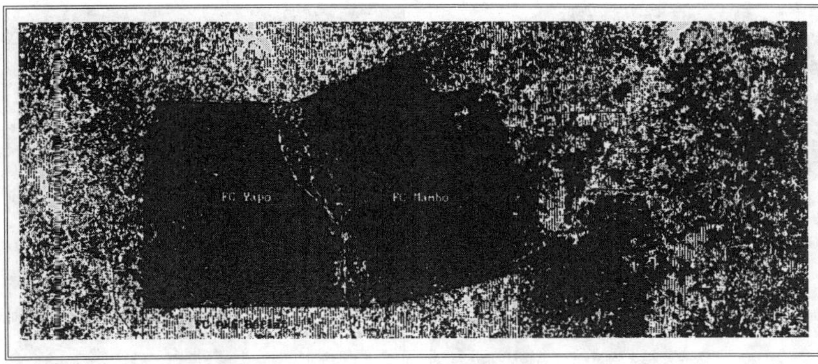

Figure 2. Vegetation map of the area of Yapo (1 cm = 3.8 km). Black: primary or secondary rain forest; grey: oil palm, cacao, cola plantations, shrub vegetation, arable land, or water; white: bare soil or built-up area. From Landsat TM 1990.

Because the city of Abidjan, with its demand for wood and charcoal, is located only 80 km to the south and with such a large forest block occupying 45% of the study site, it might be expected that there would be almost no forest left outside the protected area. However, this is far from being the case and land use is much less intensive than in the case of Divo. Again, this can partly been explained by socio-economic considerations. This region and especially its western part has long been devoted to a typical forest product: the cola nut. This has maintained a field/forest mosaic landscape for a long time. The financial needs of the local population are partially ensured by the sale of the nuts and the need to plant coffee or cocoa has not been raised to the same level as in other parts of the country. Besides, according to local farmers, the soils would be unsuitable for these cash-crops. Plantation of other perennial crops (banana, pineapple) is recent and located mainly south of the FC. In other places, secondary forest fragments are more numerous and constitute a network suitable for conservation.

3. Zagné (6°10'N; 7°35'W)

The study area is situated west of the National Park of Taï, also in the evergreen forest zone. The remoteness of the area has allowed forest to be maintained until recently, and it was thus possible to undertake monitoring using a topographic map made with 1956 aerial photographs and three satellite images (1974; 1984; 1990) (Figure 3). The detailed results of this study will be presented elsewhere (Chatelain et al. in press). This monitoring was conducted on an area of c. 100 x 100 km. Two of the satellite images were Landsat MSS, with a spatial resolution of 80 m. At this scale, only four vegetation classes could be taken into consideration. As exemplified by the results, the limits of the protected areas do not show on the map of 1974. Deforestation only reaches the borders of the protected areas in the 1984-1990 years. Removal from protection of several blocks of the FC of Cavally has allowed the deforestation front to proceed further. In some places it even begins to encroach on the protected areas.

A detailed classification of the latest 1990 Landsat TM image was conducted at full (30 m) resolution on a window of c. 36 x 27 km. In this area we were able to identify the smaller forest fragments which did not show up at the MSS resolution. In that way

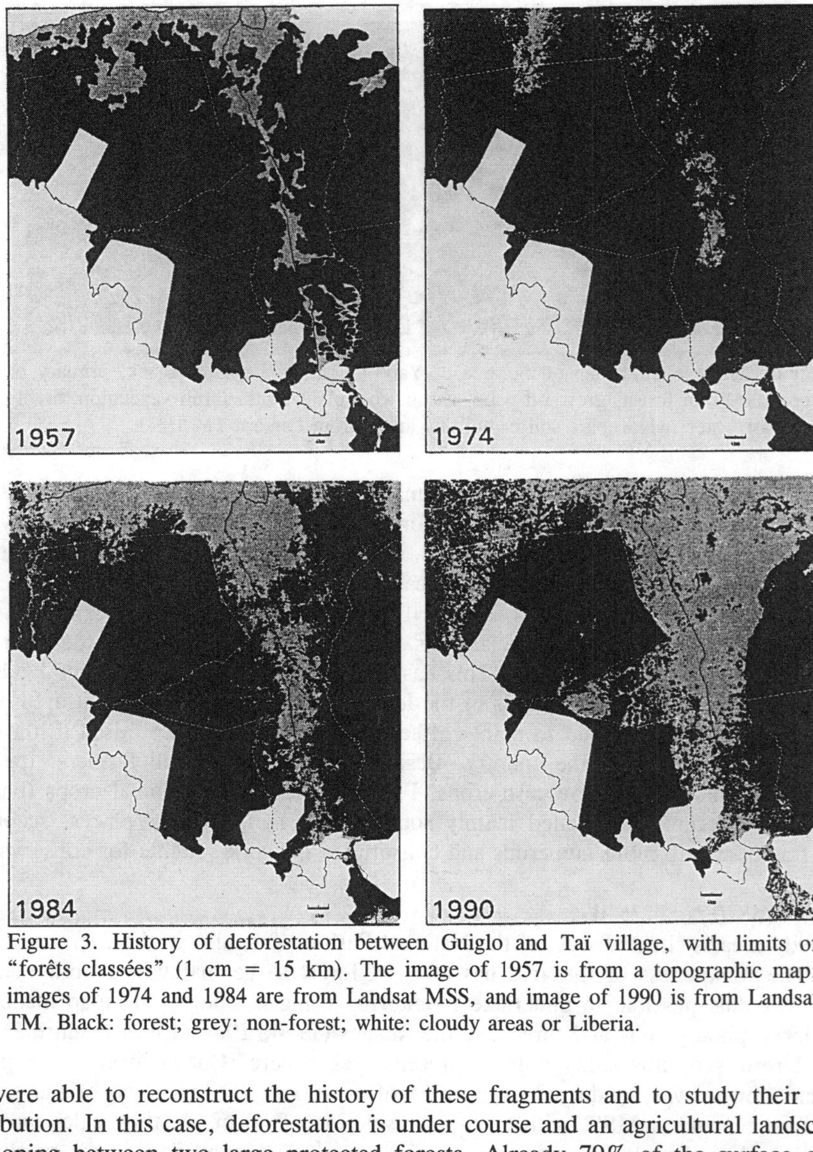

Figure 3. History of deforestation between Guiglo and Taï village, with limits of "forêts classées" (1 cm = 15 km). The image of 1957 is from a topographic map; images of 1974 and 1984 are from Landsat MSS, and image of 1990 is from Landsat TM. Black: forest; grey: non-forest; white: cloudy areas or Liberia.

we were able to reconstruct the history of these fragments and to study their spatial distribution. In this case, deforestation is under course and an agricultural landscape is developing between two large protected forests. Already 79% of the surface outside protected areas is deforested. Numerous forest fragments are still present but they will all soon be destroyed to give way to cocoa plantations. Many of the forest fragments present in 1984 had disappeared 6 years later. In the central and northern part of the area, the processes involved are similar to the ones which took place some decades ago in Divo. In the south area, where outside populations are very few, forest fragments and young secondary forest are not cleared, preserving the mosaic.

Discussion

AVHRR/NOAA data have clearly shown that outside the protected areas in Côte d'Ivoire, important forest blocks have disappeared and that most of the forest zone of the country can be considered as non-forest. Nevertheless, viewed through high resolution satellite images as well as in the field, it appears that behind this very broad non-forest class, different landscapes are hidden. The three study sites briefly illustrated here highlight this point.

The landscapes encountered in the two first examples differ markedly in the amount of forest preserved. In the case of Divo the protected areas have been completely deforested, whereas in Yapo an important part has been preserved until now. Furthermore, in Divo forest fragments are very rare on the whole study site, whereas in Yapo, they still represent a sizeable surface outside the "Forêt Classée", forming a network of secondary or depleted forest fragments. In the third case, Zagné, deforestation processes are rather recent and still proceeding at a high rate, leaving forest fragments behind the front, which will very probably disappear soon.

These examples stress the need of a global forest resource assessment at high resolution. An attempt has been made in 1993 by the Direction Générale des Grands Travaux on behalf of the SODEFOR. Unfortunately, it focused mainly on the protected areas. Outside, only forest fragments more than 1000 ha were taken into consideration. Yet, for some regions like Divo, they represent the last refuge for forest species. With the increasing remoteness of any forest element in space and time, they constitute gene banks of increasing importance. Together with secondary and depleted forests, they should be given serious consideration in any development initiatives. In many cases, although primary forest has already disappeared, there is still hope to save the forest regeneration potential.

One interesting feature is shared by our three study sites. Deforestation is the most intense where outside populations have settled. Either Baoulé coming from the centre of the country or Mossi from the Sahelian countries are strangers to the moist forest zone. They are either small-scale planters or employees in industrial plantations. They have to cope with a preexisting land ownership and are forced to colonize remote areas, as it is the case in Zagné, or areas that were recently removed from protection as in the other cases. Often being young persons new in this environment, the link between them and their new habitat is weak. They have little knowledge about local species and their uses, and generally do not feel the need for forest conservation.

This is certainly not the case among the local populations who are established in their villages for a very long time. On their territories forest has also receded in the last decades simply to give enough cultivation land to feed an increasing population and to establish cash-crops. However, their knowledge of the environment as well as the presence among them of elderly people who remember the time when an important part of the daily needs were insured by hunting and gathering, tends to put a brake on the destruction of the forest. In a sense, a certain degree of protection is achieved by local communities (Bernard 1991).

Conflicts between these two kinds of communities are not rare and if external populations overwhelm the natives, as it is the case around Zagné, it can be feared that the general feeling of responsibility towards the environment might be totally lost.

At the national level, conservation management policies are presently restricted to the retreat of the protected area's boundaries to coincide with what is left of forest inside. The idea behind this type of management which initiated around 1967 (BDPA-CTFT 1967), was to leave to local populations the forest they had already started to cultivate, but to hold firmly on to the remaining forest. However, the excluded areas become invaded by foreigners and are soon totally deforested. The forest microenvironment is definitely lost. Furthermore, deforestation is facilitated to get one step further in the remaining forest (if any). It is now clear that the forest management authorities did not have the power to ensure the protection of the core of the reserve. This policy clearly shows today its limitations and its disastrous consequences.

The trend that appeared since 1986 was to make buffer zones with the outer fringes of the protected areas that had already been encroached upon by human activities. As long as these buffer zones are respected, with no permanent or temporary human settlement in them, this can lead to concrete results. But since the authorities have no means to exert any proper control on these buffer zones, the colonisers invariably invade them which makes this good idea no more efficient than the preceding one.

As elsewhere in the tropics, ignoring the basic needs of the local populations in conservation initiatives has led to disastrous results. The need for a global conservation strategy in harmony with short-and long-term interests of local communities is urgently needed if the species of the western block of the Guineo-Congolian centre of endemism are to be considered worth protection by the international community.

Acknowledgements

This study has been financed by the Swiss National Fund for Scientific Research (grants 31-26294.89 and 5001-035120) The authors wish to thank the Centre Suisse de Recherches Scientifiques in Côte d'Ivoire for field facilities, Mr Hy Dao of the University of Geneva for digital maps of "forêts classées" and UNEP-GRID Geneva for consultation of NOAA-AVHRR images.

References

Bernard, 1991. Les problèmes fonciers des forêts tropicales africaines. Bois et Forêts des Tropiques 227: 13-16.

BDPA-CTFT, 1967. Etude de reboisement et de protection des sols. Etude des zones vulnérables. Fascicule 11, CTFT, Abidjan. 193 pp.

FAO-UNEP, 1981. Tropical Forest Resources Assessment Project. Forest resources of tropical africa. Part II. Country Briefs. UNEP-FAO, Rome. 587 pp.

Gautier, L.; C. Chatelain & R. Spichiger, 1994. Presentation of a relevé method for vegetation studies based on high resolution satellite imagery. Proceedings XIIIth AETFAT Congress, Zomba, Malawi: 1339-1350.

Päivinen, R.; J. Pitkänen & R. Witt, 1992. Mapping closed tropical forest cover in West Africa using NOOA AVHRR-LAC data. Silva Carelica 21: 27-51.

Schmidt, R. 1990. The management of tropical forest for sustainable development. Asean Seminar. Jakarta, 24-25 January 1990. 21 pp.

Plant biodiversity conservation in Kenya: the plant conservation and propagation unit of the East African Herbarium - a case study

S.W. SIMIYU
Plant Conservation and Propagation Unit, The East African Herbarium,
P.O. Box 45166, Nairobi, Kenya.

Abstract
Biodiversity is under threat in Kenya for the same underlying reason as anywhere else in the world. Three quarters of the population and 80% of the forests reside in one fifth of the total land area. Biological resource wealth and alternative human land use are strongly correlated creating potential conflict in interest. The East African Herbarium Plant Conservation and Propagation Unit was established in 1992 with the objective of ensuring the survival and proliferation of Kenya's endemic and threatened flora. The Unit has continued to focus on the conservation of the rare, endangered and endemic species through preliminary data collection and validation, field surveys and assessments, propagation of target species with subsequent conservation action and carrying out associated research. This initiative is a complement to those of other national institutions and NGO's dealing with plant biodiversity issues in Kenya, and will provide a basis for sustainable development to enhance the future security of the nation.

Résumé
La biodiversité est en danger au Kenya pour les mêmes raisons fondamentales que partout ailleurs dans le monde. Trois-quarts de la population et 80% des forêts sont concentrées dans un cinquième de l'ensemble du territoire. La richesse en ressources biologiques d'une part et l'utilisation des terres par l'homme d'autre part sont intimement liées, provoquant ainsi un conflit d'intérêt potentiel. Le "East African Herbarium Plant Conservation and Propagation Unit" a été créé en 1992 avec pour but d'assurer la sauvegarde et la prolifération de la flore endémique et menacée du Kenya. Cette "Unité" a continué à se concentrer sur la conservation des espèces rares, endémiques et menacées en rassemblant et vérifiant les données préliminaires, en effectuant des prospections de terrain et des évaluations, en reproduisant des espèces bien ciblées dont un suivi est assuré tant au point de vue conservation que recherche. Cette initiative, complémentaire à celles prises par d'autres institutions nationales et non gouvernementales en rapport avec les problèmes de la biodiversité au Kenya, devra fournir une base de développement durable qui soit à même de renforcer la sauvegarde du patrimoine national.

Key words
Biodiversity, Conservation, East African Herbarium, Kenya, Plant Conservation and Propagation Unit.

Background

Over geological time all species have a finite span of existence. Species extinction is therefore a natural process which occurs without the intervention of man. However, it is beyond question that extinctions caused directly or indirectly by man are occurring at a rate which by far exceeds any reasonable estimates of background extinction rates (WCMC 1992). Biodiversity is under threat in Africa for the same underlying reason as elsewhere in the world; the conflict between supply and demand, in terms of there

L.J.G. van der Maesen et al. (eds.), The Biodiversity of African Plants, 267-276.

being limited supply of the earth's resources and an increasing demand on them to meet the needs and growing aspirations of the population (Stuart et al. 1990).

In Kenya, the broad causal factors leading to biodiversity loss are demographic pressures and failure in policy (Anon. 1992).

Kenya covers an area of 587,900 km² of which 576,700 km² is land surface while the rest is inland waters. About 80% of the land surface is arid and semi-arid, though carrying 25% of the total human population and slightly more than half of the animal population (Republic of Kenya 1994). Currently, with a population estimated at 24 million, most of the population and 80% of the forests reside in one fifth of the total land area (Anon. 1992). Being a country with a predominantly agro-based economy (Kang'ethe 1989), biological resource wealth and alternative human use of land are strongly correlated creating potential conflict in interest.

Plant Biodiversity Status in Kenya

Kenya sits at the intersection of four major phytogeographical regions (see White 1983) presenting a wide variety of plant biodiversity zones. The predominant region is of Somalia-Masai type characterised by dry *Acacia* and/or *Commiphora* dominated scrubland; while the Afromontane unit containing afro-alpine elements on the highest peaks, the floristically rich Zanzibar-Inhambane regional mosaic along the coastal belt, the Lake Victoria regional mosaic in the West, and the Guineo-Congolian element in the Kakamega area form the smaller phytochoria (Stuart et al. 1990).

A recent checklist based on the records held at the East African Herbarium listed a total of 6817 plant species for Kenya (excluding algae) of which 86% are angiosperms (Anon. 1992). About 7000 species may seem a low number for a country of Kenya's size, but this is remarkably high given the fact that semi-arid and arid lands, which support low but interesting diversity, form the major component of the terrestrial ecosystem. Of this total, 392 species were determined as national endemics, 336 as regional endemics with 6 known extinctions and 258 species as threatened.

Beentje (1987) identified a total of 1800 woody plants of which 270, that is 15%, were rare. He further listed 92 species of woody plants as globally rare, vulnerable and endangered (Beentje 1988a, b). One third of these occur in the coastal forests of Kenya, while the Taita Forests and dry forests of Central Kenya were habitats also found to contain a large number of rare woody species. A further 69 taxa of endemic shrubs and lianas were also noted for Kenya. A breakdown of distribution of endemic and/or rare taxa is given in Table 1. From his survey, Beentje (1987) noted that areas not well known floristically included the Boni forests in Lamu, Mandera District and Mau forest. However, the Mau forests have been recently surveyed in part by the just concluded Kenya Indigenous Forest Conservation Programme (KIFCON) which was a joint initiative of the Government of Kenya Forest Department and the British Government through ODA.

Plant endemism in Kenya is high. Though not restricted only to one or a few sites, it is also rather widely scattered over many sites and in increasingly fragmented forest patches. The UNEP/NBU Kenya country study (Anon. 1992) identified 64 sites of plant endemism, some protected, some proposed for protection and others which definitely will be lost in the future.

Table 1. Distribution of endemic and/or rare woody plants in Kenya (Adapted from Beentje 1988b).

Botanically important areas	Distribution of rare species	Vulnerable or endangered species	Strict endemic species
Coastal forests	32	25	12
Taita Hills forests	10	6	7
Central Kenya - Dry forests	9	6	6
N.E. Kenya - Dry forests	10	2	3
Central Kenya - Riverine Forests	5	4	3
Central Kenya - Moist Forests	3	2	3
North/ Central Hills	7	-	2
Coast, not forest	10	1	-
Coast, littoral thicket	2	2	-

For forest vegetation, Sayer (1992) has defined the East African coastal rain forests and smaller patches of highland forests as centres of endemism. Most of these habitats are found suitable for rainfed agriculture and have become small and fragmented due to extreme encroachment pressure. The Taita Hills are a classic example. Being part of the disjunct arc of ancient crystalline mountains of East Africa and having been isolated from the main tropical African forests blocks in the West and Central Africa for some 5-10 million years (Lovett 1989), these forest relics present high endemism not only of plant but also of animal biodiversity. The once extensive forests have been heavily utilised, leaving only small remnants of indigenous vegetation cover on the hilltops and ridges of Ngangao, Chawia, Vuria, Mbololo and Sagalla, while formerly denuded patches have been reforested with exotic tree species. These hills are home to 10 endemic tree and herb species, of which 6 are endangered including the cycad, *Encephalartos kisambo* Faden & Beentje (Cycadaceae) and the African Violet, *Saintpaulia teitensis* B.L. Burtt (Gesneriaceae) which are only known from single localities.

Further work on coastal forests undertaken by the Coast Forest Survey (CFS) Project (Robertson & Luke 1993) recommended that the extremely species-rich relict patches of indigenous coastal forests with several endemic species which have been protected for sacred purposes, locally known as Kayas, be gazetted as National Monuments as they are too small to qualify for the National Protected Network criteria. Proposals to formalise legal protection and draft management plans for these sites are now being designed by the Coast Forest Conservation Unit of the National Museums of Kenya. The need to establish priority sites for protection based on a cross section of representative vegetation types influenced by species richness and the occurrence of rare species (Beentje & Khayota 1987) cannot be overemphasized. It is encouraging that the Kenya Wildlife Services, with the national mandate for *in-situ* conservation, together with the Forest Department have identified some of these sites in their current development plans.

The East African Herbarium Plant Conservation and Propagation Unit

Historical data on plant extinction in general are very limited. For example, all of 45 Africa's recorded plant extinctions in the WCMC database come from South Africa alone (Fraser et al. 1993). This does not imply that the other countries in Africa are not facing rapid biodiversity loss with actual field extinctions, but rather illustrates the lack of detailed field work to ascertain plant biodiversity status and trends on the other parts of the continent. Conservation in Kenya, for example, has been in the past directed to resources that are of immediate and direct economic value such as wildlife, natural forests, water catchment areas and a few plant species such as Aloes (Kang'ethe 1989).

During a workshop on "Endangered Resources for Development" held at the National Museums of Kenya in 1984 (Anon. 1984), the need to conserve Kenya's plant genetic resources both at species and community levels was highlighted. This was in light of the noted increasing risks especially as many of the habitats are becoming threatened through encroachment and over-exploitation of areas that are suitable for rainfed agriculture besides the overstocking and unsustainable development within the fragile ecosystems of the arid and semi-arid lands.

The National Environmental Secretariat (Ministry of the Environment and Natural Resources), noting the alarming rate of plant genetic erosion in Kenya, identified the National Museums of Kenya as the focal point for matters related to genetic resources. During the National Workshop on Plant Genetic Resources (Beentje & Khayota 1987), it was recommended that the National Herbarium be the focal institution for plant germplasm collection and documentation. Following this recommendation, the National Herbarium was given the mandate to seek funds internally and externally in support of vegetation inventories, germplasm collection and the documentation of wild species (Beentje & Khayota 1987). It was resolved that collaboration was essential between the Herbarium and related National institutions in the field of plant genetic resources conservation namely; the Kenya Agricultural Research Institute (KARI), Kenya Forestry Research Institute (KEFRI) and the Universities. This was to ensure that both domesticated and wild plant genetic resources were sufficiently surveyed, collected and documented. Recommendation no.10 from the *in-situ* working group during this workshop stated that "to meet the conservation objectives, endangered species should be identified along with their habitats and the identification of these endangered species should be done by the National Museums."

The East African Herbarium, Plant Conservation and Propagation Unit (PCPU) was consequently established in 1992 by a collaborative effort between the British Government and the Government of Kenya on a bilateral aid agreement. The broad objective for the PCPU was "To ensure the survival and proliferation of Kenya's threatened and endemic flora", by:

* Compiling a database of all the rare/threatened/endemic plants.
* Assessing conservation priorities through extensive field investigations.
* Securing germplasm for long-term storage and active propagation programmes.
* Implementing appropriate conservation measures through site-based protection and re-introduction programmes.

The option to use a "species-based" approach to conservation was necessitated by the fact that two other National institutions, namely Kenya Wildlife Services (KWS) and the Forest Department (FD) had the mandate for implementing and facilitating "site-based" conservation through the National Protected Areas Network. KWS is responsible for all the game reserves and national parks while the FD and local government authorities are in charge of national and nature reserves, all of which are governed by different national policies.

Though often said to have a narrow focus and often an apparent "emergency room" strategy that is energy-intensive, expensive and sometimes prone to failure (Losos 1993), this "fine-filter" approach leads to rescue of species on the brink of extinction hence retaining valuable genetic diversity and has the advantage of avoiding the practical and political pitfalls inherent in the "coarse filter" approach. In Kenya's case, the species conservation approach is being used to complement as well as fill gaps of other existing conservation programmes.

Methods

The PCPU has continued to focus on rare, endangered and endemic species only as the indigenous food, crop, and medicinal plants are at present within the remits of programmes under other National institutions i.e KARI which houses the National Genebank, KEFRI with the Forestry Seed Centre and the Kenya Medical Research Institute (KEMRI) which co-ordinates the Traditional Medicine Research and Conservation Programme. However, indigenous species of horticultural and local commercial importance are also being investigated by the PCPU due to threat of over-exploitation.

The approach used by the PCPU currently includes preliminary data collection and validation, field surveys and assessments, propagation of target species and associated research with subsequent conservation action being implemented.

Data collection and validation

With over 600,000 accessions of plant specimens from the East African Region, the East African Herbarium poses as the best primary source of data on distribution of plant species in Kenya. Hence, a Herbarium survey by families, followed by computer input of specimen passport data facilitates generation of information on species distributions and related pertinent data such as habitat specificity and range, flowering and fruiting seasons as well as associated species. The geographical distribution of the species is used as an indicator of its endemicity and frequency. Often, no further information is available on the species and this is used as the rough guideline in determining its conservation status. Library searches are done to complement and update information so far gathered. Key reference works consulted include the Flora of Tropical East Africa, field guides, monographs and journal reprints as well popular natural history magazines, newsletters and bulletins. The list of species identified from the above screening exercise as having restricted distribution, forms the preliminary "Candidate list" that is sent out to various contacts and authorities for further comments.

Due to the high number of species on our lists, which are generated by the family series, priority is established using the geographical matrix approach developed by Rabinowitz (1981), and its subsequent revisions when sufficient data are available. This model necessitates knowledge of local population size, habitat specificity and geographic range of species. The final document forms our "Rare List" of target species for conservation action. Some of the families currently targeted are included in

Table 2. Target (sub-)families for Screening by PCPU.

Family	Rationale	Preliminary list	Updated list	Final list
Aloeaceae	Succulent species	*	—	—
Annonaceae	Food plants	*	*	—
Apocynaceae	Medicinal plants	*	*	—
Asclepiadaceae	Succulent species	*	*	—
Caesalpinioideae	Dominant species, agroforestry species	*	—	—
Cycadaceae	Horticultural value	*	—	—
Gesneriaceae	Horticultural value	*	*	—
Melastomataceae	General interest	*	—	—
Mimosoideae	Agroforestry species	*	—	—
Orchidaceae	Horticultural value	*	*	*
Oleaceae	Medicinal value	*	—	—
Papilionoideae	Food plants	*	—	—
Podocarpaceae	Forestry species	*	—	—
Rubiaceae	Food/Medicinal value	*	—	—
Rutaceae	Food/medicinal value	*	—	—

* Done, — Not done.

Table 2. Though not all families have final lists, species identified in Beentje (1988b) are also being targeted as their status has already been determined as "rare" by one author having followed a similar process as the one mentioned above.

Field survey and assessments

Documented populations as outlined from the distribution of rare species on the lists are physically verified and surveys for new populations made. Variables such as population size, structure and range are assessed while pertinent notes on the reproductive biology of the species are taken. Specific observations are made on the populations' regeneration capacity in terms of seed set and eventual recruitment of juveniles. Presence or absence of pathogens, predators, pollinators etc. are noted. Land use patterns, anthropogenic effects on populations and habitats as well as use by animals are assessed to determine operative threats and often local people are interviewed to generate added information.

This information is summarised in standard formats which are used for subsequent design of species conservation action plans. For documentation, further research and analysis, herbarium specimens and soil samples are collected. Germplasm collections are made at appropriate times (i.e. cuttings in the absence of ripe seeds and seeds upon maturity). As a general principle, to conserve a broad spectrum of genes from the target species, the entire geographic range of the species is targeted. Where possible, a stratified random sampling design is used and a minimum size of 200 individuals sampled. This is done for 3-5 localities if the taxon distribution is more widespread, but all populations are visited if it is found in less than three localities. In general, sampling and collecting is done following guidelines of Falk & Holsinger (1991).

Table 3. A list of some *ex-situ* accessions held by PCPU.

Species, (sub-)family	No. of pops.	Veg. stock	Gene bank stock
1 *Baphia keniensis* Brummitt, Papilionoideae	5	*	*
2 *Millettia leucantha* Vatke, Papilionoideae	10	*	*
3 *Millettia oblata* Dunn ssp. *teitensis* Gillett, Papilionoideae	2	*	—
4 *Albizia tanganyicensis* Bak. f. ssp. *adamsoniorum* Brenan, Mimosoideae	8	*	*
5 *Coffea fadenii* Bridson, Rubiaceae	1	*	I
6 *Psychotria petitii* Verdc., Rubiaceae	3	*	I
7 *Psychotria crassipetala* Petit, Rubiaceae	1	*	I
8 *Saintpaulia rupicola* B. L. Burtt, Gesneriaceae	3	*	*
9 *Euphorbia wakefieldii* N. E. Br., Euphorbiaceae	1	*	—
10 *Angraecum eburneum* Bory ssp. *giryamae* (Rendle) Cribb & Senghas, Orchidaceae	1	*	P
11 *Ansellia africana* Lindl., Orchidaceae	5	*	P
12 *Saintpaulia teitensis* B. L. Burtt., Gesneriaceae	1	*	*
13 *Croton alienus* Pax, Euphorbiaceae	6	*	*
14 *Oxystigma msoo* Harms, Caesalpinioideae	1	*	—
15 *Cynometra lukei* Beentje, Caesalpinioideae	1	*	—
16 *Gigasiphon macrosiphon* (Harms) Brenan, Caesalpinioideae	1	*	—

* Present, — None, I Intermediate category, P Propagules only.

Propagation

Both macro- (e.g. use of leafy stem cuttings) and micropropagation techniques are currently utilised within the programme to bulk up species that exist in low numbers with the aim of generating *ex-situ* conservation collections. Depending on the species physiology and reproductive biology/success, trials of different propagation techniques are investigated and the most effective method adopted. The most frequently used techniques at the moment are vegetative propagation using leafy stem cuttings and offsets as in most of the leguminous tree species, seed germination in Legumes, Gesneriaceae and Rubiaceae, whilst germination using micropropagation techniques is being utilised for orchid species. Table 3 presents a summary of species in the conservation collection.

Vegetative stock is at the moment maintained at the PCPU conservation garden for research and educational purposes as well as (re-)introduction programmes. This will be developed as a display garden and will focus on promoting plant species conservation awareness in Kenya.

Conservation action

Since PCPU does not have a mandate to carry out *in-situ* conservation action at present, proposals and action plans for management of species and "sites of special scientific interest" are designed and these are passed on to relevant implementing agencies. The major conservation effort at the moment is therefore the rescue of genes (seeds) for *ex-situ* storage. Germplasm collected using appropriate sampling guidelines to capture a broad range of genotypes is stored in hermetically sealed containers according to IBPGR (now IPGRI) standards with a moisture content of 5-6% in deep freezer at -20 degrees Centigrade. At the moment, the orthodox seeds are stored in sealed plastic bags (though aluminium cans will be used in the future), while species with intermediate and/or recalcitrant seeds (which form a large majority of moist forest species) are maintained as vegetative stock. Lack of extensive planting sites as field genebanks is a major constraint though collaborative links are being utilised to overcome this problem. The PCPU seed bank is for short-term storage whereas germplasm under long-term storage as well as duplicates is currently maintained at the Kenya Forestry Seed Centre. The East African Herbarium and related national institutions together with the Kenya Horticultural Society are drafting a proposal for the development of regional Botanic Gardens which will in the near future form an integral part of PCPU.

Associated Research
Complementary research work is carried out on the target species to facilitate conservation work. The study of symbiotic relationships especially mycorrhizae is essential to any conservation work in Africa as most of the indigenous species display these relationships. For any successful introduction or re-introduction work to be carried out, symbiotic relationships have to be well understood unless other options are available. At present, mycorrhizal relationships in the Caesalpinioideae are being particularly studied, especially as members of this legume subfamily form dominant species in plant communities in East Africa.

Taxonomic research is being undertaken to establish delimitation of target taxa as correct taxonomic concepts form the basis for proper survey and conservation of threatened plants. This is especially so as sometimes plants are reported as abundant that are actually rare and vice versa, due to wrong identifications. Current work includes that on the genus *Saintpaulia* Wendl. (Gesneriaceae), endemic to East Africa, of which the taxonomic status of the Kenyan species is not well understood.

Current Constraints

Scope
Attempting to determine conservation status as well as implement conservation action is a daunting task when one is confronted with over 6,500 species. Prioritization of families as has been done presents a conservation dilemma as some species become at risk while "awaiting their turn" whereas those not in immediate danger of extinction are targeted. This is more so especially for species in threatened habitats, not naturally occurring within the protected areas network or those that belong to families not known to contain taxa immediate of economic potential. Manpower requirements and logistics to carry out this task are in themselves limiting factors. With a staff comprising of the project manager, three senior research scientists, two technicians and two casual employees, the capacity of the PCPU is in itself limited.

Approach

It is often argued that species conservation does not cater for the conservation of the ecological services of the ecosystem. As a result, this approach can be seen as expensive with minimal returns especially when the species targeted have no known economic potential. Collaborative and complementary approaches therefore become more beneficial in enhancing plant biodiversity conservation in Kenya. However, since the institutions carrying out this work in Kenya fall within different Government Ministries each with different policies, priorities and regulations, collaboration efforts sometimes are not easily implemented, hence affecting the success of the targeted work.

Facilities

At the moment the PCPU has limited physical facilities. Only one conservation garden in one climate zone, and a lack of field genebanks, limit the amount of *ex-situ* conservation work that can be done.

Data

Lack of sufficient information on species biology, population dynamics, seed biology and conservation options are major constraints. This is compounded with the historical and current lack of capacity to carry out monitoring work to ascertain trends. This presents a setback in efforts to determine the conservation status of taxa especially in a global sense as most models require knowledge of these variables. Systems such as those proposed by IUCN (Mace & Lande 1991) can thus not be easily utilised due to data limitations.

Concluding Remarks

The need for genetic resource conservation in Kenya cannot be overemphasized. The country's expanding population estimated to reach 37 million by the end of the century will necessitate higher and sustainable agricultural productivity to ensure total food sufficiency. Due to the current economic crisis, further loss of habitats to agricultural and commercial production systems will become inevitable. However, it is encouraging to note that Kenya is making concrete proposals for the achievement of sustainable development as embodied in "Agenda 21". With the Seventh National Development Plan, for the period 1994-1996, whose theme is "Resource Mobilisation for Sustainable Development," the government of Kenya has expressed a commitment to a balanced approach to development that does not compromise the environment. This has been reflected in the recently published Forestry Master Plan, the National Environmental Action Plan and the introduction of the Environmental Impact Assessment legislation. All these will soon be compiled into a National Conservation Strategy.

In the meantime, the "fine filter" approach remains a sensible and relevant approach as a contribution to the conservation of plant genetic resources in Kenya. Initiatives such as those outlined for the East African Herbarium Plant Conservation and Propagation Unit as well as others in related institutions will go a long way in ensuring sustainable resource utilisation as well as providing a basis for sustainable development for the secure future of the nation.

Acknowledgements
The establishment of the Plant Conservation and Propagation Unit of the East African Herbarium has been possible through a bilateral arrangement between the British Government through the Overseas Development Agency and the Government of Kenya.

References

Anon. 1984. Endangered resources for development: proceedings of a workshop on the status and options for management of plant communities in Kenya. NES/ETMA, Nairobi.

Anon. 1992. The costs, benefits and unmet needs of biological diversity conservation in Kenya. A study prepared for the Government of Kenya and The United Nations Environment Programme. National Biodiversity Unit - Macroeconomica Ltd.

Beentje, H.J. 1987. Baseline knowledge of the wild plants and priorities for action. In: Beentje, H.J. & B.N. Khayota (eds.), Proceedings of the Workshop on Plant Genetic Resources: Utafiti 2 (2): 17-69.

Beentje, H.J. (ed.) 1988a. An ecological and floristic study of the forests of the Taita Hills in Kenya. Utafiti 1 (2): 23-66.

Beentje, H.J. 1988b. Atlas of the rare trees of Kenya. Utafiti 1 (3): 71-123.

Beentje, H.J. & B.N. Khayota (eds.) 1987. Proceedings of the Workshop on Plant Genetic Resources: Utafiti 2 (2): 17-69.

Falk, D.A. & K.E. Holsinger (eds.) 1991. Genetics and conservation of rare plants. Oxford University Press, New York.

Fraser, D.M., D.M. Smith, R.M. May, R. Pellew, T.H. Johnson & K.R. Walter 1993. How much do we know about the current extinction rates. Tree 8 (10): 375-378.

Kang'ethe L. (ed.) 1989. National expedition on genetic resources and habitats, June 1988. KENGO - Kenya Energy Non-Governmental Organisations, Nairobi.

Losos, Elizabeth, 1993. The Future of the U.S. Endangered Species Act. Tree 8 (9): 332-335.

Lovett, J. 1989. Feeling Forests. Swara Magazine 12 (6): 28-30.

Mace M.G. & R. Lande 1991. Assessing the extinction threats: Towards a re-evaluation of IUCN threatened species categories. Conservation Biology 5 (2): 148-157.

Rabinowitz, D. 1981. Seven forms of rarity. In: Synge, H: (ed.), The Aspects of Rare Plant Conservation. John Wiley & Sons Ltd, New York.

Republic of Kenya, 1994. National Development Plan: 1994-1996. Printed by the Government Printer, Nairobi.

Robertson, S.A. & W.R.Q. Luke, 1993. Kenya coastal forests: the report of the NMK/WWF Coast Forest Survey. WWF - World Wide Fund for Nature, Project 3256.

Sayer, J.A. 1992. The conservation atlas of tropical forests: Africa. IUCN - World Conservation Union. Macmillan Publishers Ltd, U.K.

Stuart, S.N., R.J. Adams & M.D. Jenkins 1990. Biodiversity in sub-Saharan Africa and its islands. IUCN - The World Conservation Union, Gland, Switzerland.

WCMC 1992. Global Biodiversity. Status of the Earth's Living Resources. Chapman & Hall, London.

White, F. 1983. The Vegetation of Africa. A descriptive memoir to accompany the UNESCO/-AETFAT/UNSO vegetation map of Africa. UNESCO.

An ecological and ethnobotanical study of wild or spontaneous coffee, *Coffea arabica* in Ethiopia

MESFIN TADESSE [1] & LISANEWORK NIGATU [2]
[1] National Herbarium, P.O. Box 3434, Addis Abeba, Ethiopia.
Current address and correspondence: Ohio State University, Department of Plant Biology, 1735 Neil Avenue, Columbus, OH 43210-1293, U.S.A.
[2] Jimma Junior College of Agriculture, P.O. Box 307, Jimma, Ethiopia.

Abstract
The ecology of "wild" or spontaneous *Coffea arabica* L. found in the Harenna forest, Bale, Ethiopia, is studied. In total, 61 species of trees and shrubs, 48 species of herbs and 19 species of climbers and stragglers were found in association with arabica coffee in the 45 sampled plots. Species richness was noted at altitudes between 1550 m and 1650 m. Of the exchangeable cations Ca ions are the most prevalent in the coffee stands and some relationships between availability of cations and coffee plants were noted. The coffee trees are slender, attaining a maximum circumference of only 40 cm. The population structure is suggestive of a young community also indicating that it represents "recent" introduction in the Harenna forest. Maximum development of the coffee plants was noted between 1390 m and 1450 m, and 1550 m to 1560 m. From ethnobotanical studies a number of issues emerged with the most thought-provoking being the dwindling number in the indigenous selectionists and conservationists in the coffee growing districts in Ethiopia.

Résumé
Il s'agit de l'étude écologique de *Coffea arabica* L. à l'état sauvage ou spontané, dans la forêt d'Harenna (Bale) en Ethiopie. Quarante-cinq parcelles ont été échantillonnées et on relève, en association avec *Coffea arabica*, 61 espèces d'arbres et arbustes, 48 espèces herbacées et 19 espèces de plantes grimpantes ou rampantes. La plus grande richesse spécifique est observée entre 1550 et 1650 m d'altitude. Parmi les cations échangeables, ce sont les ions de Ca qui prévalent dans les peuplements de caféiers et on observe une certaine relation entre la disponibilité en cations et les plantes de caféiers. Les pieds de caféiers sont élancés, le maximum de leur circonférence ne dépassant pas 40 cm. La structure du peuplement semble indiquer le caractère de jeunesse de l'association, signifiant également qu'il s'agit d'une introduction "récente" dans la forêt d'Harenna. Le développement maximal des plants de caféiers a été observé entre 1350 et 1450 m et entre 1550 et 1560 m. Des études ethnobotaniques, il ressort un certain nombre de questions, dont l'une mérite particulièrement l'attention, à savoir le nombre de plus en plus décroissant de sélectionneurs et de conservateurs indigènes dans les régions producteurs de café en Ethiopie.

Key words
Coffea arabica, Coffee, Ecology, Ethnobotany, Ethiopia.

Introduction

Although the evidence linking the spread of arabica coffee from the Ethiopian highlands to Yemen, from where it spread first into Asia, then to Europe and South America (La Roque 1732) is still scanty (Mesfin, in prep.), Ethiopia is widely believed to be the centre of proliferation and probably also the centre of origin of the plant (cf. Meyer 1965). Coffee, produced exclusively from *Coffea arabica* L., is the most important plant in the Ethiopian economy as a foreign exchange earner. In the 1960's it accounted for about 55-60% of the total export earnings of the country (Huffnagel

L.J.G. van der Maesen et al. (eds.), The Biodiversity of African Plants, 277-294.

1961) and in the 1990's it accounts for about 50% (Anon. 1993).

For a long time, *Coffea arabica* has been known as a cultivated plant and, as such, it has received and continues to receive much attention from many government workers, scientists, and plantation owners. The wild or spontaneous arabica coffee, which is known only from the Ethiopian highlands, has not been given similar attention.

An FAO Coffee Mission made major field studies on wild or spontaneous coffee in Ethiopia between 1964-65 (FAO 1968). The Mission studied the botany and distribution of coffee and also collected germplasm, sample specimens, insects and pathogens. Prior to this study, there were few sporadic reports on the extent of coffee cultivation (e.g. Branzanti 1942, Strenge 1956, Spaletta, in Huffnagel 1961), production (e.g. various reports by the former Coffee Board of Ethiopia and the Ministry of Coffee and Tea Development), and germplasm collection (e.g. Sylvain 1955; Strenge 1956). This lack of information on the wild or spontaneous coffee in Ethiopia could partly be attributed to the inaccessibility of the coffee growing areas until relatively recently. Logan (1946), who produced an account of the forests of central and southern Ethiopia, wrote that "prior to the Italian occupation little or nothing was done to assess their value as a national asset or to conserve and develop them", probably due to limited access to the forests. With regards to the size of the forests in these regions, he wrote that these have been quite extensive but receding or declining due to development of towns and cities.

In the 1960's and in the early 1970's, the Coffee Board within the Ethiopian Ministry of Agriculture, and in the late 1970's and up to the present, the Ministry of Coffee and Tea Development, have been responsible for all activities in connection with the production, improvement, and commerce of coffee. The Institute of Agricultural Research (IAR) in collaboration with the Plant Genetic Resources Centre (PGRC/E), has the mandate to carry out research on coffee. In the early 1970's, the Addis Abeba University, upon receiving requests from the Coffee Board, also conducted research on spontaneous coffee (cf. Tewolde B.G. Egziabher 1978). Noting the inadequacy in both field and laboratory-oriented work on wild or spontaneous coffee, studies were conducted between 1986 and 1990 in a previously botanically little known area in south-eastern Ethiopia, i.e. the Harenna forest in Bale region. Some of the important results obtained from the studies are presented in this paper but prior to providing the details in connection with this work, descriptions of coffee growing areas in other parts of Ethiopia where the senior author has also observed or studied coffee plants since 1973 will be provided.

The FAO Coffee Mission recorded wild or spontaneous (wof-zerash in Amharic, meaning spread by birds), naturalized and cultivated arabica coffee from about 78 sites (i.e. villages, towns, research stations, forests, etc.) from eight administrative regions in Ethiopia (Meyer 1968). Many of these records are from Keffa (c. 37) and Illubabor (c. 14) regions and are either in the midst of forests or in partially cleared areas in forests. Some of the cultivations have recently been state-run, e.g. the Bebeka plantation in Keffa region.

The vegetation and flora of coffee growing sites

Starting from early times coffee is known to grow naturally in Keffa region (cf.

Richards 1847) in southern Ethiopia, and probably also in Wellega, Illubabor, Sidamo, Bale and Arssi where currently large- to small-scale cultivations are found. Coffee is cultivated in Harerge and to a small extent also in Gonder, Wello, Shewa, and Gojam. In Gojam, the peninsula of Zeghie in Lake Tana and the surrounding fields were known to yield quite a substantial amount for export to the northern regions of the country. The coffee in this region is believed to have been introduced from Keffa (Richards 1847, Huffnagel 1961). It is also believed that the coffee from Keffa might have found its way to neighbouring Yemen in early times (Mesfin, in prep.).

The Western and South-Western Highlands
The vegetation in western and south-western Ethiopia (Wellega, Illubabor and Keffa) has been studied by Logan (1946), Tewolde B.G. Egziabher & Mesfin Tadesse in 1973 (unpublished data, specimens deposited in Herb. ETH), Mesfin Tadesse in 1978, 1981 and 1990 (unpublished data, specimens deposited in Herb. ETH), Chaffey (1979), Abdu Mohammed (1974), Tewolde B.G. Egziabher (1978, 1986), Friis et al. (1982), Friis & Mesfin (1990), and Mesfin (1992). These regions are the centres of prolifera-tion and probably also of origin of arabica coffee (Richard 1847, Strenge 1956, Meyer 1965, Fernie 1966). Richard (1847) considered Ennarea (now Suntu) in Keffa as the home of arabica coffee. He also stated that the coffee in northern Ethiopia and elsewhere is introduced.

The components of the humid upland forest where also spontaneous coffee is found are more or less the same as those in the Harenna forest with particular tree species dominating in different localities. The major species are: *Aningeria adolfi-friederici*, *Albizia gummifera* (J.F. Gmel.) C.A.Sm., *A. schimperiana* Oliv., *A. grandibracteata* Taub., *Blighia unijugata* Bak., *Cassipourea malosana*, *Ekebergia capensis* Sparrm., *Euphorbia ampliphylla* Pax, *Ficus sur*, *F. vasta* Forssk., *F. thonningii* Bl., *Hallea rubrostipulata* (K. Schum.) K. Leroy, *Ilex mitis* (L.) Radlk., *Macaranga capensis*, *Ocotea kenyensis*, *Olea welwitschii*, *Polyscias fulva*, *Schefflera abyssinica* (Hochst. ex A. Rich.) Harms, *Prunus africana* (Hook. f.) Kalkm., *Sapium ellipticum* (Krauss) Pax, and *Syzygium guineense* subsp. *afromontanum*. Only few detailed ecological studies have been made in these forests. There are more species of *Ficus*, a genus of rather wet or moist habitats, in the western highlands than in Harenna forest. *Erythrina brucei* Schweinf. and *Euphorbia ampliphylla*, species that are characteristic of the moist montane forests of the western highlands are not present in the community in which coffee is found in Harenna forest. *Erythrina brucei* occurs as an intermediate storey tree in the wetter and more humid portions of Harenna forest between 1980 m and 2400 m. Relationships between the coffee trees and environmental factors, i.e. both edaphic and biotic, have been studied for only one region in Toba, Illubabor (Tewolde B.G. Egziabher 1978, 1986).

The South and South-Eastern Highlands
The forests on the Ethiopian plateau have been depleted as they have been used for generations as sources of fuel, timber, construction material, utensils, etc. For this reason secondary forests and other serial stages in the regrowth of forests are common on the plateau. Like in other regions of the country, the forests in the south and south-eastern highlands have not been adequately documented. The vegetation of

Harerge was first studied by Gillett (1941) and later on by Chaffey (1979) who recorded *Juniperus procera* Hochst. ex A. Rich., *Olea europaea* L. subsp. *cuspidata* (Wall. ex DC.) Cifferri (syn. *O. africana* Mill.) and *Podocarpus falcatus* (syn. *P. gracilior* Pilg.) as the major tree species in the forests within these highlands. Other large tree species include *Allophylus abyssinicus*, *Celtis africana*, *Croton macrostachyus*, *Dombeya torrida* (J.F. Gmel.) Bamps, *Ekebergia capensis*, *Olea hochstetteri* and *Prunus africana*.

There are no naturally occurring forest coffee trees in Harerge but there are large tracts of cultivated fields throughout the highland. In this region, the coffee seedlings are often planted in the shade of *Cordia africana*, *Podocarpus falcatus*, *Croton macrostachyus*, and *Juniperus procera*. Coffee is also cultivated in the open and is interplanted with other economically useful plants, e.g. *Nicotiana tabacum* L., *Catha edulis* Forssk., *Capsicum annuum* L., and *Manihot esculenta* L.

The vegetation on the highlands of Sidamo have been studied by Friis et al. (1982) and Hailu Sharew (1982). The forests in which naturalized coffee is found is composed of *Croton macrostachyus*, *Millettia ferruginea*, *Cordia africana*, *Ilex mitis*, *Olea europaea* subsp. *cuspidata*, *Albizia schimperiana* and *Schefflera abyssinica*. Small-scale cultivation (inter-mixed with *Ensete ventricosum* and *Catha edulis*) is also found in the shade of *Podocarpus falcatus*. In the Dilla-Wondo-Yirga Chefe area, the major tree species are *Cordia africana*, *Allophylus abyssinicus*, *Albizia schimperiana*, *Millettia ferruginea*, *Erythrina brucei*, *Croton macrostachyus*, *Podocarpus falcatus*, *Polyscias fulva*, *Aningeria adolfi-friederici*, *Prunus africana* and *Schefflera abyssinica* - all typical elements of the mixed wet evergreen forests of the plateau.

The vegetation on the highlands in Bale have only recently been studied. One of the earliest reports about the forests in this region was that of Russ in 1945, who stated that there is "...no estimate of this forest" and that "...it may exceed 10,000 km^2 in area". Within this vast area, spontaneous or wild coffee was located in the Harenna forest in 1986 (Mesfin 1986).

The Harenna Forest

The Harenna forest lies between 6° and 7° N and 39° and 40° E. It is one of the few remaining natural forests in Ethiopia and currently constitutes the largest sub-section of the Bale Mountains National Park. The earliest botanical study on the Harenna forest was made by H. F. Mooney between 1958 and 1962 (Mooney 1963). As a result of the completion of road-building through the forest, it was, later on, also studied by Friis, Gilbert and Vollesen in 1984 (Friis 1986), Mesfin Tadesse (1986), Hillman (1986), Lisanework Nigatu (1987), and Lisanework Nigatu and Mesfin Tadesse (1989). The species composition of the forest is similar to the south-west Ethiopian humid upland forests but it also contains the following species which are not known elsewhere in the country: *Filicium decipiens* and *Alangium chinense*. Recent ecological studies in this forest (Lisanework Nigatu 1987, Lisanework Nigatu & Mesfin Tadesse 1989) revealed that there are "... five homogeneous clusters of vegetation..." and that "the highest number of significant contrasts among the clusters were observed to result from variations in altitude." Spontaneous coffee is found in the lower end of the forest. A more thorough ecological investigation of this area was made between 1989 and 1991.

Coffee in the Harenna Forest
"Wild" or spontaneous coffee is found in the lowermost portion of the Harenna forest extending between 1320 m and 1700 m. There is a narrow transition zone between the forest and the *Combretum-Terminalia-Croton*-woodland between 1500 m and 1450 m along the road starting about 10 km W of Dello Menna town. The transition zone as well as the lowermost end of the forest is occupied by mixed *Warburgia salutaris* (Bertol. f.) Chiov., *Filicium decipiens* and the emergent *Podocarpus falcatus*.

Materials and methods

Vegetation Sampling
The "wild" or spontaneous coffee trees, shrubs and seedlings found in the lower end of the Harenna forest were counted from 40 quadrats along increasing altitudinal gradients in April 1990. In November 1990, five more quadrats were also sampled from the upper and lower altitudinal limits of the coffee plants in order to observe their performance at these altitudes. Sampling was done within quadrats each measuring 20 x 20 m and laid at random at 200 m intervals. Sampling was started at 1500 m, the lower limit of the dry *Podocarpus-Warburgia-Filicium* forest along the Dello Menna-Goba road through the forest. To the east of the road the altitude falls to 1320 m and here the forest is relatively moist and is dominated by *Warburgia*, *Syzygium* and *Podocarpus*. The trees and shrubs found in each quadrat and their seedlings were counted. For the herbaceous layer and climbers/stragglers, presence of species was recorded and for the most ubiquitous ground cover, the Braun-Blanquet cover/abundance scale (Mueller-Dombois & Ellenberg 1974, as modified by Van der Maarel 1979) was used. For plants that could not be identified on the spot voucher specimens were collected and these are deposited in the National Herbarium (ETH). For the coffee plants, in addition to the density, circumference, crown diameter and height were also measured.

Environmental Data
Altitude, slope and aspect were recorded for each quadrat. Soil samples at depths of 0-10 cm, 40-60 cm, and 80-100 cm were collected from eleven representative quadrats by digging a 1 m^3 pit in each quadrat. The samples were air-dried and passed through a 2 mm sieve prior to chemical and physical analysis which was performed at the Soil Laboratory of the Ministry of Agriculture, Addis Abeba, following standard procedures (Allen 1989). Soil colour was determined using Munsell's soil colour charts; bulk density by drying the samples at 105°C; soil texture using the hydrometer method; pH in 1:2 (v/v) soil-water suspension; electric conductivity in 1:2 (v/v) soil-water; organic carbon by the Walkely-Black wet oxidation method; available P by the Bray II method and total N by the Kjeldahl method. Cation exchange capacity (CEC) and exchangeable cations were analyzed after extraction with 1N ammonium acetate at pH 7. Ca and Mg ions were measured using atomic absorption spectrometry and Na and K ions by flame photometry. Extractable micronutrients were determined by the EDTA titration method. Light intensity (in 1000 lux) was measured using light meter both within and outside (in full sunlight) the quadrats.

Table 1. Percentage co-occurrence of *Coffea arabica* with forest species in sample quadrats of Harenna forest, Bale, Ethiopia.

Species	Family	%
Celtis africana	Ulmaceae	95.6
Podocarpus falcatus	Podocarpaceae	88.9
Strychnos mitis	Loganiaceae	82.2
Vepris daniellii	Rutaceae	77.8
Bersama abyssinica	Meliaceae	75.6
Diospyros abyssinica	Ebenaceae	75.6
Oxyanthus speciosus	Rubiaceae	73.3
Croton macrostachyus	Euphorbiaceae	73.3
Warburgia salutaris	Canellaceae	71.1
Polyscias fulva	Araliaceae	68.9
Olea hochstetteri	Oleaceae	66.7
Ehretia cymosa	Boraginaceae	64.4
Lepidotrichilia volkensii	Meliaceae	62.2
Allophylus macrobotrys	Sapindaceae	60.0
Syzygium guineense	Myrtaceae	57.8
Ocotea kenyensis	Lauraceae	51.1
Rhamnus prinoides	Rhamnaceae	51.1

Questionnaires

In an attempt to contribute to the question of the most likely place of origin of *Coffea arabica*, a survey of coffee growing districts in Keffa, Wellega and Illubabor in Ethiopia was made in 1990 and questionnaires were distributed to farmers in these areas as well as to those in Sidamo and Harerge through the regional offices of the Ministry of Coffee and Tea Development. From the responses obtained and from personal observations (by M.T.), a number of issues emerged and these are addressed in this paper.

Results and discussion

Vegetation analysis

Sixty-one species of trees and shrubs, 48 species of herbs and 19 species of climbers and stragglers were encountered in the sample quadrats (Appendix 1). The species with which coffee was found to co-occur at 50% or greater than 50% are given in Table 1. Of the woody plants, 43 species (2 introduced) were observed to regenerate in the forest community (Table 2). In terms of germination and establishment of seedlings, *Celtis africana* and *Podocarpus falcatus*, two shade-tolerant species, were found to display greater capabilities for producing progenies in the forest community than the other species with *Celtis africana* producing seedlings in 39 out of the 45 sampled quadrats (86.7%), and *Podocarpus falcatus* in 29 quadrats (i.e., 64.4%). The other species (Table 3) were observed to display somewhat lesser frequency of germination and establishment of seedlings: *Croton macrostachyus* (46.7%), a species known to perform better in clearings or in gaps in forests or at forest margins, *Warburgia salutaris* (44.4%), *Strychnos mitis* (35.6%), *Diospyros abyssinica* (35.6%), *Bersama abyssinica* (35.6%), *Filicium decipiens* (31.1%), *Lepidotrichilia volkensii* (28.9%), and

Table 2. Percentage frequency of seedlings of 20 of the most widespread indigenous forest species in coffee stands in Harenna forest, Bale, Ethiopia.

Species	%	Species	%
Celtis africana	86.7	*Rhamnus prinoides*	24.4
Podocarpus falcatus	64.4	*Millettia ferruginea*	24.4
Croton macrostachyus	46.7	*Oxyanthus speciosus*	22.2
Warburgia ugandensis	44.4	*Cassipourea malosana*	22.2
Strychnos mitis	35.6	*Canthium oligocarpum*	22.2
Diospyros abyssinica	35.6	*Vepris daniellii*	20.0
Bersama abyssinica	35.5	*Maytenus arbutifolia*	20.0
Filicium decipiens	31.1	*Galiniera saxifraga*	20.0
Lepidotrichilia volkensii	28.9	*Teclea nobilis*	17.8
Allophylus abyssinicus	28.9	*Suregada procera*	17.8

Table 3. Minimum and maximum number of seedlings of arabica coffee and indigenous forest species in coffee stands with calculated mean (M) and standard deviation (SD). Area 400 m^2.

Species	Min	Max	M	SD
Coffea arabica	6	1264	167.4	108.6
Celtis africana	8	139	25.8	20.3
Strychnos mitis	1	136	20.8	15.9
Diospyros abyssinica	3	68	18.7	7.2
Suregada procera	1	52	16.3	12.0
Filicium decipiens	1	19	11.0	6.7
Galiniera saxifraga	1	60	9.2	7.1
Croton macrostachyus	1	102	8.9	5.7
Podocarpus falcatus	1	57	8.1	6.0
Allophylus macrobotrys	1	35	6.4	3.6
Canthium oligocarpum	2	9	6.0	2.3
Bersama abyssinica	1	11	5.9	2.6
Lepidotrichilia volkensii	2	10	4.9	2.6
Solanum giganteum	1	19	4.3	2.4
Millettia ferruginea	2	9	4.2	2.5
Calpurnea aurea	1	11	4.1	3.0
Rhamnus prinoides	1	7	4.0	1.8
Cassipourea malosana	1	7	3.9	2.0
Warburgia salutaris	2	8	3.8	1.7
Oxyanthus speciosus	1	8	3.6	2.1
Vepris daniellii	1	6	2.8	1.9
Teclea nobilis	1	5	2.6	1.2
Maytenus arbutifolia	1	4	2.1	1.1

Table 4. Distribution of number of sample quadrats with the most ubiquitous herbaceous species in estimated cover/abundance classes.

Species	3-7 %	7-13 %	13-25 %	25-50 %	>50 %
Acanthus eminens	-	1	-	-	-
Achyranthes aspera	2	-	-	-	-
Asystasia gangetica	-	1	-	-	1
Desmodium repandum	8	5	-	-	-
Hypoestes aristata	4	10	8	1	-
Hypoestes forskahlii	8	10	10	2	-
Hypoestes verticillaris	-	-	1	1	-
Isoglossa punctata	9	1	-	-	-
Justicia forskahlii	2	3	2	1	1
Oplismenus compositus	7	6	14	11	2
Pentas lanceolata	4	-	-	-	-
Phaylopsis imbricata	4	1	-	-	-
Thalictrum rhynchocarpum	2	-	-	-	-

Allophylus abyssinicus (28.9%). The following species displayed high fecundity (measured in terms of number of seedlings per quadrat): *Celtis africana* (8-139), *Croton macrostachyus* (1-102), *Diospyros abyssinica* (3-68), *Filicium decipiens* (1-19), *Podocarpus falcatus* (1-57), *Strychnos mitis* (1-136) and *Warburgia salutaris* (2-8). Of the herbaceous species, *Desmodium repandum*, *Hypoestes aristata*, *Justicia forskahlii*, *Oplismenus compositus*, and *Achyranthes aspera* occur in over 50% of the quadrats with the highest concentration being at 1550 m. *Oplismenus compositus* followed by *Justicia forskahlii*, *Hypoestes aristata* and *H. forskahlii* make up over 50% of the ground cover in that part of the forest in which arabica coffee was found (Table 4). Of the climbing plants only *Jasminum abyssinicum* and *Landolphia buchananii* occurred in over 50% of the quadrats with the highest concentration being at 1550 m. A full list of species is given in Appendix 1.

The number of species associated with the coffee plants increased with increasing altitude and up to 1650 m; species richness was noted between 1550 m and 1650 m (Table 5). Species diversity for the tree and shrubby species was highest at 1550 m and for the herbaceous and climbing plants at 1600 - 1650 m. Seedling development was also noted as the highest at 1400 m and 1550 m. As indicated in Tables 6 and 7, there is co-variation between species diversity and the concentration of exchangeable cations, phosphorous, conductivity, and altitude (see below).

The coffee trees are slender, attaining a maximum circumference of only 40 cm with the range being (0.5-) 3-37 (-40) cm. As indicated in Table 8, the coffee population exhibit skewed distribution patterns. At particularly lower elevations, i.e. 1300 - 1450 m, the dominant members are plants with low stem-diameters. Trees with relatively higher d.b.h. (diameter at breast height) classes and more evenly spread patterns were obtained between 1500 m and 1550 m. This population structure is thus suggestive of a young community and, compared to forest coffee in, for instance, Illubabor, where trees with d.b.h. of 140 cm have been observed, the coffee trees in the Harenna

Table 5. Species richness along an altitudinal gradient in coffee stands in Harenna forest, Bale, Ethiopia.
A: number of mature tree and shrub species.
B: number of mature tree and shrub species and their seedlings.
C: number of species of trees, shrubs and herbs.
D: number of species of trees, shrubs, herbs and climbers.

Altitude (m)	A	B	C	D
1320	9	19	25	26
1400	8	16	23	25
1450	12	20	24	26
1500	17	17	24	25
1550	20	22	31	35
1600	15	23	33	36
1650	18	23	35	38
1700	13	17	26	28

Table 6. Light measurements in 1000 lux at various altitudinal levels. The measurements for sample plots of similar altitudes were averaged.

Altitude (m)	In sample plot	In full sun (outside plot)
1320	6.8	54.1
1400	12.7	107.4
1450	11.3	82.3
1500	17.2	92.0
1550	24.5	99.4
1600	28.4	106.2
1650	18.8	105.1
1700	14.6	86.5

forest appear to be young. From interviews obtained from the inhabitants at the lower end of the forest near Dello Mena town who also cultivate coffee, the coffee in the Harenna forest is probably introduced from Sidamo. Practically all of the farmers interviewed stated that it was brought by the legendary "Shaa-de-Liya". Thus the coffee in this forest is most probably not truly wild. But after this initial introduction the coffee plant seems to have spread within the lowermost portion of the forest. The crown-diameter of the coffee trees ranges between 0.2 m and 5.7 m and the height between (1.0-) 1.5 m and 6.3 m. Measured in terms of biomass, the maximum development of the coffee plants was noted to be between 1390 m and 1450 m and 1550 m to 1600 m. The upper altitudinal limit for the coffee plant in the Harenna forest is 1700 m while the lower limit appears to be defined by changes in the vegetational composition and the moisture content of the forest. The number of mature coffee plants

Table 7. Measured and mean values (at 1400 m and 1600 m) for environmental factors along increasing altitudinal gradients and soil depth in coffee stands in Harenna forest, Bale, Ethiopia.

Alt.	Soil d.	Na	K	Ca	Mg	CEC	P
1320	0-10	0.62	0.53	36.62	3.01	50.60	29.20
	40-60	0.61	0.12	19.92	1.56	17.52	23.20
	80-100	0.44	0.08	13.88	1.72	19.48	3.00
1400	0-10	0.73	1.55	33.13	2.64	41.76	9.50
	40-60	0.51	0.27	10.69	1.69	25.29	2.37
	80-100	0.55	0.30	8.46	2.38	21.38	4.37
1500	0-10	0.61	0.40	30.09	1.93	37.94	24.40
	40-60	0.50	0.08	13.23	1.21	24.50	22.80
	80-100	0.55	0.08	10.06	1.03	14.24	23.00
1550	0-10	0.57	0.49	27.42	4.97	47.86	4.14
	40-60	0.66	0.08	9.92	1.20	22.75	1.54
	80-100	0.60	0.08	7.80	0.60	38.88	0.94
1600	0-10	0.62	0.94	36.84	5.66	49.03	30.40
	40-60	0.60	0.49	13.29	3.08	31.02	24.20
	80-100	0.52	0.10	12.86	2.55	23.55	26.45
1650	0-10	0.74	0.65	32.54	6.44	49.83	5.34
	40-60	0.58	0.08	9.90	2.83	-	4.54
	80-100	0.58	0.08	8.10	2.33	-	2.77
1700	0-10	0.81	1.03	35.43	8.87	58.26	5.54
	40-60	-	-	-	-	-	-
	80-100	0.68	0.12	2.13	0.53	33.23	6.94

encountered in the dry *Podocarpus-Warburgia-Filicium* forest at 1500 m is less than that in the rather moist *Warburgia-Syzygium-Podocarpus* forest at 1320 m. Light measurement, which is an indirect measure of the photosynthetic efficiency of green plants, was also highest for samples located between 1550 m and 1600 m (Table 6).

Soil Analysis

The soil in the Harenna forest has been characterized and described in earlier papers (Lisanework Nigatu 1987, Lisanework Nigatu & Mesfin Tadesse 1989). It was reported to be rich in the basic exchangeable cations, i.e. Ca. Mg, K and Na ions with Ca ions being the most prevalent. In those studies, only a few samples were analyzed from the stands in which arabica coffee is found. Further proof to the earlier results has been obtained from the present study in which soil samples from eleven quadrats representing different altitudes and seemingly different associations of species within the lower stratum of the Harenna forest consisting of *Coffea arabica* were analyzed.

The soil is dark reddish-brown, silty-clay throughout between altitudes 1490 m and 1510 m and also between 1540 m and 1610 m. Between 1510 m and 1540 m the soil is reddish brown for some distance below the surface and deep or dark red further inwards. Rocks were encountered at depths of 40, 50, 70 and 80 cm only in quadrats 37 (alt. 1390 m), 45 (alt. 1650 m), 4 (alt. 1500 m) and 39 (alt. 1390 m), respectively, and termites only in quadrat 11 (alt 1540 m). In about 50% of the quadrats, roots were

Table 7. (Continued). Alt. = altitude (m), Soil d. = soil depth (cm). Na, K, Ca, Mg, P, Si, Cl: elements, CEC = cation exchange cap., Cond. = conductivity, pH = acidity, Sl = slope angle (°), Sd = slope direction, T.N. = total N, O.C. = organic C.

Cond.	pH	Si	Cl	Sl	Sd	T.N.	O.C.
0.604	6.2	64	35	12	S	0.052	0.733
0.352	6.3	41	57			0.161	3.338
0.222	6.2	46	53			0.058	1.072
0.483	6.2	65	32	7	SE	0.097	1.593
0.158	5.7	17	82			0.094	1.585
0.146	5.4	21	79			0.048	0.808
0.404	6.4	56	43	6	ENE	0.302	4.406
0.195	5.9	30	69			0.094	1.931
0.172	6.4	14	85			0.079	2.065
0.381	5.4	75	20	13	S	0.239	6.881
0.157	5.3	20	78			0.080	1.923
0.121	5.6	20	78			-	0.571
0.510	6.6	70	29	11	E	0.335	5.171
0.230	5.8	33	66			0.119	2.257
0.180	5.5	26	73			0.127	2.455
0.584	5.8	84	14	2	SSE	0.161	2.760
0.141	4.9	30	68			0.112	2.241
0.130	5.0	30	68			0.210	2.276
0.854	5.5	81	16	12	W	0.131	-
-	-	23	76			-	-
0.030	3.8	23	76			0.070	3.086

observed up to a depth of 80 cm and in quadrats 42 (alt. 1400 m), 43 (alt. 1560 m), and 44 (alt. 1700 m) up to 100 cm.

The data on soil texture (Table 7) indicate that at the 0-10 cm soil depth, siltation generally increases with increasing altitude while clay particles are more prevalent at lower altitudes. For instance, the ratio between the silt and clay particles at 1650 m is 6 to 1, while at 1320 m it is 1.8 to 1. At the 40-60 cm and at the 80-100 cm soil depths, clay particles become more prevalent and show clear trends of increment with increasing altitude.

The concentration of K and Na ions peaked at 1400 m but, like Ca and Mg ions, they showed a decrease up to 1500 m and 1550 m, respectively. At an altitude of 1550 m and above, the concentration of the cations increased with increasing altitude. With increasing soil depth, the concentration of the cations decreased. Conductivity, which is indicative of the total amount of water soluble salts (i.e. plant nutrients) and phosphorous, likewise decreased between 1300 m and 1550 m but gradually increased at higher elevations. The lowest value for electrical conductivity, like those of Na, Ca and total nitrogen, was noted at 1550 m. The surface soil pH, which is a measure of the saturation of exchangeable cations, was also noted to be at its lowest at 1550 m. The pH of the soil in which coffee is found ranged between 5.3 and 6.6. The content of soil nitrogen and organic carbon increased from 1320 m to 1500 m, respectively, but both decreased with increasing altitude. Thus, the drop in the concentration of the

Table 8. Distribution of coffee trees and shrubs of various d.b.h. (diameter at breast height) classes along increasing altitudinal gradient in Harenna forest, Bale, Ethiopia.

d.b.h. (cm)	Altitude (m) 1320	1400	1450	1500	1550	1600	1650	1700	Total
0.3-5	10	6	5	1	4	15	-	-	41
6-10	39	55	46	5	32	5	1	5	188
11-15	13	11	7	11	11	4	1	3	61
16-20	3	6	5	6	7	3	1	-	31
21-25	-	2	2	8	3	1	-	-	16
26-30	-	2	1	1	1	1	-	-	6
31-35	-	1	-	4	-	-	-	-	5
36-40	-	-	-	1	-	-	-	-	1

Table 9. Comparison of responses, as percentage of total, offered by farmers to questions related to the presence of forest coffee and its utilization.

Question no. and answer	Region				
	Keffa	Illubabor	Harerge	Sidamo	Wellega
11 Coffee	68.3	58.3	53.9	57.4	70.7
Coffee + others	92.7	80.8	73.5	77.9	78.0
Others	7.3	19.2	26.5	22.1	22.0
18 Yes	53.7	45.9	0.0	32.2	17.5
No	46.0	47.3	96.8	64.7	82.5
I do not know	0.0	6.8	3.2	3.0	0.0
19 Yes	37.5	24.4	0.0	19.4	7.7
No	62.5	75.6	100.0	80.6	92.3
24 Yes	82.1	71.1	5.6	44.2	35.1
No	17.9	28.9	94.4	55.8	64.9

basic soil nutrients at 1500 m to 1550 m, where also a higher species richness and high seedling development was noted, could be attributed to root uptake and translocation into vegetative tissues. Although relatively sloping sample plots (the range being 2° - 15°) were encountered between 1500 m and 1600 m and again at 1700 m, leaching of nutrients due to slope could not have been the major cause for the reduced nutrient status of the sample plots.

Ethnobotanical studies

Comparison of responses offered by farmers to questions related to forest coffee and its utilization (Table 9) revealed that a higher percentage of farmers in Keffa and Illubabor know about the existence of forest coffee than those in other regions. Higher

Table 10. Comparison of responses, as percentage of total, offered by farmers to questions related to the existence of old coffee trees.

Answer	Region				
	Keffa	Illubabor	Harerge	Sidamo	Wellega
Yes	82.9	71.3	86.0	81.5	80.5
No	4.8	13.3	10.0	13.3	17.1
I do not know	12.2	15.3	4.0	5.2	2.4

Table 11. Comparison of responses offered by farmers to questions related to differences between forest and cultivated coffee trees in yield, aroma, and age (A) and other atributes (B) expressed as percentages of total. C refers to those who stated as having no differences and D to those who did not know the difference.

Answer	Region				
	Keffa	Illubabor	Harerge	Sidamo	Wellega
A	96.8	42.2	2.9	28.7	4.9
B	0.0	61.7	20.4	16.2	21.9
C	0.0	1.9	3.9	0.0	4.9
D	3.2	1.9	73.8	60.3	68.3

Table 12. Comparison of responses, as percentage of total, offered by farmers concerning knowledge of any locality noted for the quality of its coffee.

Answer	Region				
	Keffa	Illubabor	Harerge	Sidamo	Wellega
Yes	59.5	36.4	70.4	74.3	50.0
No	10.8	28.1	14.8	7.1	5.0
I do not know	29.7	35.5	14.8	18.6	45.0

share percentages of forest coffee in relation to source of income, presence in district and in cultivated fields are recorded from Keffa and Illubabor.

As indicated in Table 10, a significant proportion of the farmers in each region seem to know the existence of large coffee trees. The figure for Harerge region, where, as indicated in Table 9, no forest coffee is known, is the highest (i.e. 86%). In response to the query concerning differences between forest and cultivated coffee trees in berry yield, aroma, tree age, etc. (Table 11), a large proportion of the farmers in Harerge (c. 74%) had responded negatively thus indirectly indicating that they are not able to differentiate forest from cultivated coffee trees. From Table 11, it also appears that farmers in the Keffa and Illubabor regions have better knowledge concerning those

Table 13. Comparison between the number of children of the farmers interviewed and those engaged in coffee farming in the major coffee growing districts in Ethiopia.

Region	District	Number of families	Number of children	Number and percentage of children engaged in farming	
Illubabor	Limuseka	11	47	45	95.7
	Didessa	10	71	54	76.1
	Gambella	24	135	74	54.8
	Geba	27	203	101	49.8
	Gore	82	509	210	41.3
Keffa	Gimira	19	133	69	51.9
	Decha	5	20	10	50.0
	Keffa	5	29	7	24.1
	Yeki	12	64	12	18.8
Wellega	Kelem	14	109	64	58.7
	Airaguliso	11	93	54	58.1
	Boji	16	97	44	45.4
Sidamo	Arbeguna	14	124	92	74.2
	Gelaya	15	112	75	67.0
	Shebedino	18	173	105	60.7
	Sidama	36	232	133	57.3
	Dalle	17	140	65	46.4
	Borena	34	301	107	35.5
Harerge	Darolebu	19	140	123	87.9
	Messela	13	124	108	87.1
	Bokekuni	16	149	127	85.2
	Habbobo	8	75	62	82.7
	Bedeno	26	249	176	70.7
	Melka-Belo	21	235	94	40.0

characteristics than farmers in the other regions. With regards to the existence of a locality noted for its good quality coffee, over 50% (Table 12) of the farmers in all regions except in Illubabor responded in the affirmative. From Table 12, it is also observed that more affirmative responses (i.e. 70%) were provided by farmers in the Harerge and Sidamo regions, where there are large cultivations of coffee, and less affirmative answers were supplied by farmers in Illubabor.

The conservation of the genetic variability of both wild or spontaneous and cultivated coffee in Ethiopia has been in focus for many years, e.g. Fernie (1966), Amare Getahun and Krikorian (1973), Melaku Werede (1984), Bayetta Belachew and Mesfin Amhea (1986). Various methods and strategies of conservation of the arabica coffee genetic resource base have been proposed. Some experimental as well as field studies have been made, e.g. Tewolde B.G. Egziabher (1978, 1986), Owuor & van der Vossen (1981), Berthaud and Charrier (1988). *In-situ* conservation of coffee through forest preservation has also been proposed, e.g. Melaku Werede (1984), Tewolde B.G. Egziabher (1990), Charrier & Berthaud (1990). In Ethiopia, the small farmers are the *in-situ* conservationists since a long time ago and for a long time to come, but no major

Table 14. Comparison of responses, as percentage of total, offered by farmers about the life span and duration of berry yield of coffee trees. 1 = equivalent to the human life span; 2 = indefinite.

Question and answer		Region				
		Keffa	Illubabor	Harerge	Sidamo	Wellega
Life span of coffee trees	1	94.9	51.8	84.8	84.6	100
	2	-	22.7	14.1	16.2	-
Duration of berry yield	1	55.0	53.8	35.5	54.3	33.3
	2	27.5	56.7	61.3	37.8	61.5

socio-economic studies have been made in and around coffee growing regions, e.g. changes in the family structure, and in the growth of towns and institutions. Having this in mind, a question was included in the survey questionnaire to find if there are any tendencies within the coffee farming families interviewed.

Proportionally more farmers in Sidamo and Harerge are engaged with their children in the coffee fields than those in other regions except perhaps in Limuseka in Illubabor (Table 13). Farmers in the latter region stated that a large number of their children are either government employees (e.g. civil servants, soldiers, etc.) and traders, or are attending school, working in coffee farms only during spare time. Obviously, upon finishing school these children will look for employment in nearby towns and cities rather than going back to the coffee fields. The dwindling number of indigenous selectionists and conservationists is alarming and causes concern for all engaged in the conservation of the coffee gene pool.

In response to the query on the life of a coffee tree (Table 14) over 84% of the farmers in all regions except in Illubabor responded, as anticipated, by saying that its life span is comparable to that of human beings. Only 52% of the farmers in Illubabor responded in a similar manner and about 23% stated that the coffee tree lives indefinitely. In response to the query on the duration a coffee tree continues to yield berries, slightly over 56% of the farmers in Illubabor, Wellega and Harerge said that this goes on indefinitely whereas farmers in Keffa and Sidamo equated this to the human life span. These subjective answers provided by farmers have indicative values, i.e. they are indicative of the most likely areas where one might find long-producing coffee trees.

Acknowledgements
Grants from UNESCO (MAB Programme) and the Ministry of Coffee and Tea Development, and logistical support from the Ethiopian Flora Project are gratefully acknowledged. We also thank the Director and staff of the Soil Laboratory Department of the Ministry of Agriculture, Ethiopia. M. T. thanks Dr. Birru Abebe, Berhanu Antoine, Zewdi Berhe, Mesfin Alemayehu, Meleskachew Amha, all in the Ministry of Coffee and Tea Development and Prof. D. J. Crawford, Ohio State University, for the word processing facilities in his laboratory.

References

Abdu Mohammed, 1974. An ecological study of the forest in Toba Wereda, Illubabor, Ethiopia. BSc. Thesis, Department of Biology, Addis Abeba University, Ethiopia.

Allen, S.E. 1989. Chemical Analysis of Ecological Materials. 2nd ed., Blackwell Scientific Publ., Oxford.

Amare Getahun & A.D. Krikorian, 1973. Chat: Coffee's rival from Harar, Ethiopia. I. Botany, cultivation and use. Econ. Bot. 27 (4): 357-389.

Anon., 1993. Whitaker's Almanack. J. Whitaker & Sons Ltd., London.

Bayetta Belachew & Mesfin Ameha, 1986. Ethiopian National Coffee Collections. Institute of Agricultural research, IAR Progress Report (mimeographed), Department of Coffee, 1983-84: 6-18. IAR, Addis Abeba, Ethiopia.

Berthaud, J. & Charrier, A., 1988. Genetic resources of Coffee. In: Clark, R.J. & R. Macra (eds.), Coffee Agronomy 4: 1-42, Elsevier, London.

Branzanti, C.E., 1942. La coltura del caffe nel Gimma. Agric. Col. 36: 1-8.

Chaffey, D.R. 1979. South-East Ethiopia forest inventory project. A reconnaissance inventory of forest in south-east Ethiopia. Land Res. Dev. Center, Surbiton, Project Report 31: 1-316.

Charrier, A. & J. Berthaud, 1990. Use and value of genetic resources of *Coffea* for breeding and their long-term conservation. Mitt. Inst. Allg. Bot. Hamburg, Band 23a S. 53-64.

Food and Agricultural Organization, 1968. FAO coffee Mission to Ethiopia, 1964-65. Rome.

Fernie, L.M., 1966. Some impressions of Coffee in Ethiopia. Tanganykia Coffee News, Jan./-March, 1966.

Friis, I., F.N. Rasmussen & K. Vollesen, 1982. Studies in the flora and vegetation of South-West Ethiopia. Opera Bot. 63: 1-70.

Friis, I., 1986. The Forest Vegetation of Ethiopia. Acta Univ. Ups. Symb. Bot. Ups., 26 (2): 31-47.

Friis, I. & Mesfin Tadesse, 1990. The Evergreen Forests of Tropical NE Africa. Mitt. Inst. Allg. Bot. Hamburg, Band 23a, S. 249-263.

Gillett, J.B. 1941. The plant formation of western British Somaliland and the Harar province of Abyssinia. Bull. Misc. Inf. Kew, No. 2: 37-75.

Hailu Sharew, 1982. An ecological study of a forest in Jemjem, Sidamo. MSc. Theisis, Department of Biology, Addis Abeba University, Addis Abeba, Ethiopia.

Hillman, J.C. 1986. Bale Mountains National Park: Management Plan. Wildlife Conservation Organization, Addis Abeba, Ethiopia.

Huffnagel, H.P. 1961. Agriculture in Ethiopia. FAO, Rome.

La Roque, G., 1732. Voyage to Arabia Felix, 1708-1710. London.

Lisanework Nigatu, 1987. An ecological study of the vegetation of the Harenna forest. MSc. Thesis, Department of Biology, Addis Abeba University, Addis Abeba, Ethiopia.

Lisanework Nigatu & Mesfin Tadesse, 1989. An ecological study of the vegetation of the Harenna forest, Bale, Ethiopia. SINET: Ethiop. J. Sci. 12 (1): 63-93.

Logan, W.E.M., 1946. An introduction to the forests of central and southern Ethiopia. Imperial Forestry Institue, London, Paper 24.

Melaku Werede, 1984. Coffee genetic resources in Ethiopia: conservation and utilization with particular reference to CBD resistence. In: Abdel-Maguid, A.A. (ed.), Proceedings of the First Regional Workshop on Coffee Berry Disease, Caire-Zaytoon, Egypt.

Mesfin Tadesse, 1986. Botanical Survey. In: Hillman, J.C. (ed.), Harenna Forest Expedition - Preliminary Report (mimeographed), Wildlife Conservation Org., Addis Abeba, Ethiopia.

Mesfin Tadesse, 1992. A survey of the evergreen forests of Ethiopia. In: Sue Edwards & Zemede Asfaw (eds.), The Status of Some plant Resources in Parts of Tropical Africa. Botany 2000: East and Central Africa. NAPRECA Monograph Series No. 2. pp. 1-18.

Meyer, F.G., 1965. Notes on wild *Coffea arabica* from southwestern Ethiopia, with some historical considerations. Econ. Bot. 19 (2): 136-151.

Meyer, F.G., 1968. Alphabetical list of FAO Coffee Mission localities in Ethiopia. In: FAO (ed.), FAO Coffee Mission to Ethiopia, 1964-65: 110-115, Rome.

Mooney, H.F., 1963. An account of two journeys to Araenna mountains in Bale province (SE Ethiopia), 1958 and 1959-60. Proc. Linn. Soc., 174: 127-151.

Mueller-Dombois, D. & H. Ellenberg, 1974. Aims and Methods of Vegetation Ecology. Wiley and Sons, New York.

Owuor, J.B.O. & H.A.M. Van der Vossen, 1981. Interspecific hybridization between *Coffea arabica* L. and tetraploid C. canephora P. ex Fr. I. Fertility in F1 hybrids and backcrosses to *C. arabica*. Euphytica 30: 861-866.

Richard, A. 1847-1951. Tentamen Florae Abyssinicae, 2 vols. Paris.

Strenge, H. von, 1956. Wild Coffee in Keffa Province of Ethiopia. Trop. Agriculture (Trin.), 33 (4): 297-301.

Sylvain, P.G., 1955. Some observations on *Coffea arabica* L. in Ethiopia. Turrialba 5 (1-2): 37-53.

Tewolde B.G. Egziabher, 1978. Some vegetative parameters of coffee, *Coffea arabica* L., proportional to yield. SINET: Ethiop. J. Sci. 1 (1): 51-57.

Tewolde B.G. Egziabher, 1986. Preliminary studies on the ecology of a natural coffee (*Coffea arabica*) forest with emphasis on coffee. Acta Univ. Ups. Symb. Bot. Ups. 26 (2): 146-156.

Tewolde B.G. Egziabher, 1990. The importance of Ethiopian forests in the conservation of arabica coffee gene-pools. Mitt. Inst. Allg. Bot. Hamburg Band 23a S. 65-72.

Van der Maarel, E. 1979. Transformation of cover-abundance values in phytosociology and its effects on community similarity. Vegetatio, 39: 97-114.

Appendix 1

A list of tree, shrub, herbaceous and climbing species found together with *Coffea arabica* in the Harenna forest, Bale, Ethiopia. The numbers in brackets refer to frequency of occurrence, i.e. number of quadrats in which the species was found. Voucher specimens are: four digit numbers of Mesfin Tadesse; two digit numbers of Lisanework Nigatu.

Trees and shrubs

Alangium chinense (lour.) Harms (2)
Allophylus abyssinicus (Hochst.) Radlk. (7)
A. macrobotrys Gilg. (27)
Aningeria altissima (A. Chev.) Aubrev. & Pellegr. (1)
A. adolfi-friederici Robyns & Gilbert (4)
Bersama abyssinica Fresen. (34)
Calpurnea aurea (Lam.) Benth. (20)
Canthium oligocarpum Hiern (22)
Carissa edulis (Forssk.) Vahl (14)
Cassipourea malosana (Bak.) Alston (13)
Celtis africana Burm. f. (43)
C. gomphophylla Bak. (3)
Chionanthus mildbraedii (6)
Citrus aurantium L. (25)
Cordia africana Lam. (3)
Croton macrostachyus Del. (33)
Dalbergia lactea Vatke (3)
Diospyros abyssinica (Hiern) F. White (34)
Dracaena steudneri Engl. (5)
Ehretia cymosa Thonn. (29)
Erythrococca trichogyne Prain (10)
Erythroxylum fischeri Engl. (1)
Euclea schimperi (A. DC.) Dandoy (4)
Fagaropsis angolensis (Engl.) Del.(1)
Ficus sur Forssk. (1)
Filicium decipiens (Wight.& Arn.) Thun. (20)
Galiniera saxifraga (19)
Justicia schimperiana (Hochst. ex A. Rich.) T. Anders. (9)
Lepidotrichilia volkensii (Guerke) Leroy (28)
Macaranga capensis (Baill.) Sim. (9)
Maesa lanceolata Forssk. (1)
Maytenus arbutifolia (Hochst. ex A. Rich.) Wilczek (12)
M. gracilipes (Welw. ex Oliv.) Exell. (2)
Millettia ferruginea (Hochst.) Bak. (20)
Ocotea kenyensis Robyns & Wilczek (23)
Olea hochstetteri Bak. (30)
O. welwitschii (Knobl.) Gilg. & Schell (1)
Oxyanthus speciosus DC. (33)

Phyllanthus sepialis Muell. Arg. (11)
Podocarpus falcatus (Thunb.) R.Br. ex Mirb. (40)
Polyscias fulva (Harms) Harms (31)
Psydrax schimperiana (A. Rich.) Bridson (3)
Pterolobium stellatum (Forssk.) Chiov.. (12)
Rhamnus prinoides L'Herit. (23)
Rhus natalensis Krauss (13)
Rubus steudneri Schweinf. (19)
Rytigyna neglecta Robyns (1)
Senna petersiana (1)
Solanum giganteum Jacq. (13)
Strychnos mitis S. Moore (37)
Suregada procera (Prain) Croizat (15)
Syzygium guineense (Willd) DC. subsp. *afromontana* F. White (26)
Teclea nobilis Del. (22)
T. simplicifolia (Engl.) Verdoon (10)
Trema orientalis (L.) Blume (6)
Turraea holstii Guerke(1)
Vepris daniellii (Pichi-Sermolli) Kokwaro (35)
Vernonia hochstetteri Sch. Bip. ex Walp. (3)
Warburgia salutaris (Bertol. f.) Chiov. (32)

Herbs
Abutilon sp. (14), 8982, 8996
Acalypha psilostachya Hochst. ex A. Rich. (8)
Acanthus eminens C. B. Clark (20)
Achyranthes aspera L. (23)
Adenostemma sp. (1)
Aerangis luteoalba (Kranzl.) Schltr. (1)
Ageratum conyzoides L. (4)
Asplenium aethiopicum (Burm. f.) Kaulf. (1)
A. sandersonii (3)
Asplenium sp. (10), 8997, 9051
Asystasia gangetica T. Anders. (3)
Capsicum frutescens L. (1)
Celosia schweinfurthiana Schinz (1)
Cheilanthus farinosa (2)
Chlorophytum sp. (1)
Conyza sp. (1)
Cuscuta sp. (1)
Cynoglossum coeruleum Hochst ex DC. (2)
Cyperus sp. (11), 9023
Desmodium repandum Vahl (26)
Geophila repens (L.) Johnston (10)
Geranium arabicum Forssk. (6)
Girardinia sp. (1)
Guizotia abyssinica Cass. (1)
Hypoestes aristata Roem. & Schult. (30)
H. verticillaris (L.f.) Roem. & Schult. (2)
Impatiens rothii Hook. f. (4)
Isoglossa punctata (Vahl) Brummitt (17)
Justicia flava Vahl (2)
J. forskahlii Vahl (40)
Lagenaria siceraria (Molina) Standl. (11)
Oplismenus compositus (L.) P. Beauv. (44)
Panicum monticolum Hook. f. (4)
Pentas lanceolata K. Schum. (13)
Peperomia abyssinica Miq. (7)
P. tetraphylla Hook. & Arn. (1)
Phaylopsis longifolia Sims. (7)
Physalis peruviana L. (1)
Pilea sp. (2)
Plectranthus sp. (4), 9082
Polypodium sp. (2)
Sanicula europaea L. (3)
Scadoxus multiflorus Rafin. (1)
Scleria sp. (1)
Tectaria gemmifera (Fee) Alston (2)
Thalictrum rhynchocarpum Q. Dillon & A. Rich (11)
Triumfetta rhomboidea Jacq. (2)
Vernonia urticifolia A. Rich. (1)
V. wollastonii S. Moore (1)

Climbers/Stranglers
Adenia rumicifolia Engl. & Harms (3)
Caesalpinia volkensii Harms (2)
Cissus quadrangularis L. (1)
Glycine sp. (1), 9022, 26
Gouania longispicata Engl.
Helinus mystacinus E. Mey. ex Steud. (12)
Hippocratea africana (Willd.) Loes. (7)
Jasminum abyssinicum Hochst ex DC. (30)
Landolphia buchananii (Hall.f.) Stapf (26)
Paullinia pinnata L. (12)
Pupalia lappacea Juss. (1)
Solanum benderianum Sch. Bip. ex Engl. (2)
Stephania abyssinica Walp. (10)
Stellaria mannii Hook. f. (1)
Toddalia asiatica Lam. (4)
Zehneria sp. (1)
Unidentified sp. (1), 9027

Implications of the Convention on Biological Diversity for cooperation in botanical research

TEWOLDE B.G. EGZIABHER [1] & I. FRIIS [2]
[1] The National Herbarium, Addis Ababa University,
PO Box 3434, Addis Ababa, Ethiopia.
[2] The Botanical Museum, University of Copenhagen,
Gothersgade 130, DK-1123 Copenhagen, Denmark.

Abstract
The Convention is summarized. Benefits: North-South cooperation on research and conservation, funding and political support for this; the role of taxonomy is re-invigorated by the Convention. Problems: Exclusion of existing *ex-situ* collections from the Convention; regulations of transfers of biological material will increase bureaucracy, further complicating collecting. Ethical problems arise: risks of use of genetic material in violation of the Convention; risks of biotechnology from Southern collections may harm Southern exports; biotechnology could endanger safety of Southern environments as well as Northern; and scientists could infringe upon intellectual rights regarding knowledge and technology of traditional Southern communities. AETFAT and the Convention: with common interests, Northern and Southern botanists can openly discuss these problems; AETFAT can develop a code of conduct; AETFAT can disseminate information on these issues. For traditional taxonomic research, entanglement with the Convention may be avoided by denaturing of the DNA of herbarium specimens sent across boundaries; procedures to dispatch viable plant material and DNA samples for basic research have to be developed to comply with the Convention.

Résumé
Condensé de la Convention. Avantages: coopération Nord-Sud dans le domaine de la recherche et de la conservation, ainsi que support financier et politique; l'importance de la taxonomie est revivifiée par la Convention. Problèmes: exclusion de la Convention des collections *ex-situ* existantes; les règlements concernant le transport du matériel biologique vont accroître la bureaucratie et par conséquent compliquer la récolte du matériel. Des problèmes éthiques se posent: risques d'utiliser du matériel génétique en violation de la Convention; risques de voir la biotechnologie développée à partir des collections du Sud appauvrir les exportations du Sud; la biotechnologie pourrait mettre en danger la sécurité tant de l'environnement du Sud que de celui du Nord; les scientifiques pourraient empiéter sur les droits intellectuels concernant le savoir et la technologie des communautés traditionnelles du Sud. L'AETFAT et la Convention: en raison d'intérêts communs, les botanistes du Nord et du Sud peuvent discuter librement de ces problèmes; l'AETFAT peut définir un code de bonne conduite; l'AETFAT peut assurer la diffusion de ces débats. En ce qui concerne la recherche taxonomique traditionnelle, les problèmes avec la Convention peuvent être évités en dénaturant le DNA des spécimens d'herbier envoyés hors frontières; il faudra définir les règles pour que l'expédition du matériel végétal viable et d'échantillons de DNA soit conforme aux prescriptions de la Convention.

Key words
Conservation, Convention of Biodiversity, Ethical Problems, Intellectual Rights, Plant Genetic Resources, Transfer of Resources.

Note
This lecture was scheduled after the symposium "Biodiversity of African savannas", but the lecture fits better in the symposium "Ecology and conservation".

L.J.G. van der Maesen et al. (eds.), The Biodiversity of African Plants, 295-305.

Introduction

The AETFAT meeting in Wageningen is the first meeting after the United Nations Conference on Environment and Development (UNCED) in Rio de Janeiro in 1992. An integral part of that conference was the Convention on Biological Diversity.

For many biologists the acceptance of the Convention on Biological Diversity represented the official recognition of the importance of biological diversity and its conservation. Much information on biodiversity had been compiled prior to the conference (World Conservation Monitoring Centre 1992), and "biodiversity" is now a household word. We assume for example that the Convention on Biological Diversity and the preparations for it has determined the choice of "Biodiversity of African plants" as the theme for this AETFAT meeting.

Much has changed with the signing of the Convention of Biological Diversity, also in relation to what has happened and what will happen within the framework of AETFAT. Of the three main objectives of the Convention, the conservation of biological diversity, the sustainable use of its components and the fair and equitable sharing of its benefits, the members of AETFAT are probably most familiar with the conservation aspect, but previous conferences have also addressed themes relating to the aspect of sustainable use. The third aspect, that of a fair and equitable sharing of the benefits of biodiversity, is the most controversial and the one which turns views held at previous AETFAT meetings upside-down. Until recently it was accepted, e.g. by international organization like FAO, that genetic resources are open access resources to be collected and utilized freely by anybody. They were the common heritage of mankind.

The genetic resources of Africa and their utilization in this spirit was one of the themes of the AETFAT meeting in Hamburg in 1988. The session reported on a number of missions to collect plant genetic resources in Africa; most comprehensive was that described by Rajanaidu (1990), reporting on collecting of genetic resources of oil palms by Malaysian and Nigerian institutions from natural stands in Ivory Coast, Cameroon, Zaire, Tanzania and Madagascar.

In the negotiations for the Convention of Biological Diversity the developing countries queried the principle of common heritage to genetic resources, and questioned the fairness of gene-poor developed countries obtaining genetic resources freely from the gene-rich developing countries, manipulate the genes, patent the products and selling these patented products at exorbitant prices to the country where the genetic resources were collected (Hendrickx et al. 1994).

This principle is radically changed with the signing of the Convention of Biological Diversity, which explicitly states that the countries of origin have sovereign rights over their biological diversity. That change will have strong and lasting implications on botanical research. The authors of this paper have therefore found it relevant to examine the Convention itself and some of its implications to botanical research in Africa and to botanical collaboration between North and South.

A General Evaluation of the Convention on Biological Diversity

The United Nations Conference on Environment and Development was born out of the optimism which ensued when the East-West polarization in geopolitics had virtually disappeared, and when it thus looked as if the world would be more united than in the past. However, the Conference also highlighted the fact that the world was still deeply and fundamentally divided into a North and a South with vastly different economic capacity, as had indeed been shown previously by many (e.g. Bedjaoui 1979, Crosby 1986, Independent Commission on International Development Issues under the Chairmanship of Willy Brandt 1980, Payer 1974). This gap became very important during the negotiations at UNCED, and the conflicting interests of the North and the South have left marks on almost all the Articles of the Convention of Biological Diversity, and it is necessary to examine the Convention in the light of the North-South division.

In the following, we will first look at the general flow of resources envisaged in the Convention, and then at a number of more specified points.

UNCED noted the South-North flow of resources, and, in Agenda 21 and in the Convention on Biological Diversity proposed a countering North-South flow. However, the benefits to accrue in exchange with biodiversity and the funding on which the countering flow was to be based were to be on mutually agreed terms. Since the North is much more powerful than the South, this means that the formulation of these terms depends on the good will of the North (Articles 15.7, 16, 20). According to the Convention on Biological Diversity, the financial expression of this collective good will of the North is to be channelled through the Global Environment Facility (Articles 20.2, 21, 39) which functions under the World Bank, UNEP and UNDP.

Together with the International Monetary Fund, the World Bank is the institution responsible for the modernization of the old South-North system of flow of resources (Hancock 1991, Khor 1994, Payer 1974, Rich 1994), but yet there are no signs that the situation will change, hence the eagerness of the South to find other mechanisms for the provision of financial resources to the South, as for example expressed in the need for a new institutional structure for the provision of financial resources to be negotiated by the Conference of the Parties (Article 39). The first of these Conferences will take place in November 1994.

The Convention covers 42 Articles (Table 1), which we will try to summarize. States which have biological diversity in their territories, which are mostly in the South, are recognized by the Convention as sovereign over that biological diversity (Articles 3 & 15.1), in agreement with the fundamental principle of international law on sovereignty over national natural resources. The North is recognized as the owner of biotechnology (meaning genetic engineering) (Article 16) and finance (Article 20). Biological diversity is seen as essential for making biotechnology useful (Articles 15.6, 15.7, 19.2).

Financial resources are seen as coming from North to South in way of a fair and equitable sharing of the commercialisation of biological diversity from the South (Article 15.7), and to help Southern countries fulfil the obligation put on them by the Convention (Article 20.2).

Biotechnology is seen as coming from North to South in exchange for biological diversity (Articles 15.6, 15.7, 16.1, 16.3, 18.2, 19.1, 19.2) and to help in the conservation and sustainable use of biological diversity (Article 16.1), "under fair and most favourable terms, including on concessional and preferential terms ..." (Article 16.2). Some indigenous and traditional technologies are also visualized as flowing from South to North (Article 18.4). As the current quest for information all over the

Table 1. The Articles of the Convention on Biological Diversity

Art. 1. Objectives	Art. 22. Relationship with Other International Conventions
Art. 2. Use of Terms	Art. 23. Conference of the Parties
Art. 3. Principle	Art. 24. Secretariat
Art. 4. Jurisdictional Scope	Art. 25. Subsidiary Body on Scientific, Technical and Technological Advice
Art. 5. Cooperation	Art. 26. Reports
Art. 6. General Measures for Conservation and Sustainable Use	Art. 27. Settlement of Disputes
Art. 7. Identification and Monitoring	Art. 28. Adoption of Protocols
Art. 8. *In-situ* Conservation	Art. 29. Amendment of the Convention or Protocols
Art. 9. *Ex-situ* Conservation	Art. 30. Adoption and Amendment of Annexes
Art. 10. Sustainable Use of Components of Biological Diversity	Art. 31. Right to Vote
Art. 11. Incentive Measures	Art. 32. Relationship between this Convention and Its Protocols
Art. 12. Research and Training	Art. 33. Signature
Art. 13. Public Education and Awareness	Art. 34. Ratification, Acceptance or Approval
Art. 14. Impact Assessment and Minimizing Adverse Impacts	Art. 35. Accession
Art. 15. Access to Genetic Resources	Art. 36. Entry Into Force
Art. 16. Access to and Transfer of Technology	Art. 37. Reservations
Art. 17. Exchange of Information	Art. 38. Withdrawals
Art. 18. Technical and Scientific Cooperation	Art. 39. Financial Interim Arrangements
Art. 19. Handling of Biotechnology and Distribution of its Benefits	Art. 40. Secretariat Interim Arrangements
Art. 20. Financial Resources	Art. 41. Depositary
Art. 21. Financial Mechanism	Art. 42. Authentic Texts

South by representatives of Northern institutions shows (GRAIN 1993b, World Resources Institute 1993) the North knows that its is not only biodiversity but also traditional biotechnologies that it needs from the South. The South, impressed by the glamour of genetic engineering, does not think much of this wealth which it has, and it is often giving it away by default of protective action.

Biological diversity is seen as coming from South to North in exchange for the promise of sharing in the benefits arising from its commercialization by the North (Article 15.7) and through bilateral agreements (Articles 15.2 & 15.4). The bilateral agreements will presumably be based on exchanges of biodiversity for financial resources, genetic engineering, and human resources development (Articles 15.6 & 15.7). Though the North does not have much biological diversity to make available, the South can have access to whatever it has on the same terms as the North will take from the South (Article 15).

Since biological diversity is amenable to pilfering, a prior informed consent (PIC) of the country of origin is required if export and possession of it is to be legal (Article 15.5). In practice, a weak Southern country has no way of checking on such legality outside its borders. The impact of this stipulation, therefore, depends on the good will of each Northern country. The existence of this good will in the countries of the North is so far untested. In Denmark, for example, it has been debated if a legislation should be adopted that makes any import not complying with the PIC provisions of the exporting country unlawful or criminal, and that illegal exports of genetic resources should be confiscated and returned to the country of origin, but no regulations have yet been enforced (Hendrickx et al. 1994).

Information is seen as flowing from South to North associated with biological diversity (Articles 8.j & 17) and from North to South associated with biotechnology (Articles 15.6, 15.7, 16.3, 17, 19.3, 19.4). But while Northern technology is to be protected as intellectual property (Article 16.2), Southern technology is to be respected, preserved and maintained (Article 8.j), and to be given away free (Article 17). This legalization of a free South-North flow but a priced North-South flow of technology and knowledge thereof, obviously continues the current South-North imbalance.

Existing *ex-situ* collections are exempt from being governed by the Convention. This requires some explanation: The most valuable biodiversity is that of crops; the North has now in its gene banks extensive collections of biodiversity of virtually all the crops and many related species from the South (Crucible Group 1994, p. 75, Fowler & Mooney 1990, pp. 147-200, GRAIN 1993a). Many institutions have been searching the South intensively for genetic material before the Convention came into force so as to augment their *ex-situ* collections (GRAIN 1993b, World Resources Institute 1993).

The only article which might be workable without invocation of Northern good will is that on the possibility of developing a protocol on biosafety (Article 19.3) so that the South could be protected from becoming an experimental ground for the North's transgenic organisms, which could be dangerous to human and environmental well-being (Fry & Day 1992, Levy & Miller 1989, Regal 1987, 1989, 1994). The United States says that transgenic organisms are as safe as those generated through normal sexual reproduction. Europe states that such a protocol would be useless until the South has enough people trained in genetic engineering. Therefore, it seems that there will be no biosafety protocol since Southerners trained in genetic engineering keep being lost to the North in the "brain drain".

The view of the South is therefore that the convention is partly unsatisfactory. However, had the funding of the Convention been more clearly defined, had *ex-situ* collections in the North not been excluded, and had there not been intellectual property rights protection of the North, but not protection of the indigenous and traditional knowledge of the South, then the equation of the Convention matching the rights and obligations of South and North would probably have been generally accepted by the South as fair.

Benefits

Assuming that Northern goodwill will be forthcoming, the Convention affords some important benefits to botanical and zoological teaching and research, as well as for *in-situ* conservation of biodiversity. This is the obvious reason for the generally rather positive reception which the convention has received in biological circles.

Botanists, mostly from the North, have so far had to campaign hard to raise support for teaching and research in the classical botanical disciplines. In the last few decades, the fascination of molecular biology had largely relegated other fields of botany into being seen as somewhat archaic or even irrelevant. Now, the Convention recognizes biological diversity, defined as "the variability among living organisms from all sources, including, inter alia, terrestrial, marine and other aquatic ecosystems and the ecological complexes of which they are part; this includes diversity within species, between species and of ecosystems" (Article 2). Biological diversity thus broadly defined is not only to be conserved under *in-situ* (Article 8) and *ex-situ* (Article 9) conditions, but the necessary research to do that has to be made, and the necessary human resources to be trained (Article 12); there are even to be public education and awareness drives (Article 13). To this end, there is to be cooperation among countries (Article 5), and financial resources are to be made available (Article 20). According to the text of the Convention, these financial resources will only be made available as agreed between the country concerned and the Global Environment Facility (Article 21.2), and the experience of Southern countries on coming to fair agreements with the World Bank is rife with controversies. Nevertheless, there are clearly also possibilities for bilateral arrangements, and initiatives involving such arrangements are being taken by several countries, including Sweden and Denmark. Moreover, the Convention does give biology the emphasis needed to counter decades of a reductionist view which had taken focus away from the whole, to what should be only component parts, for example molecular biology. As such, the Convention provides potential support and rationale for funding for botanical teaching and research.

In particular, it is to be noted that plant taxonomy and, in general, systematics as a whole had come to be considered unfashionable. The recognition by the Convention of all species and infra-specific taxa as worthy of conservation has pushed scientists to evaluate both the proportion of organisms we know, now conservatively estimated to about 16% of the assumed total number (World Conservation Monitoring Centre 1992, p. 37), and the rate of extinction, now estimated to range from 0.1% to 1.1% of all species per year (World Conservation Monitoring Centre 1992, p. 203).

The Convention expects each country to "develop national strategies, plans or programmes for conservation and sustainable use of biodiversity…." (Article 6(a)). When we relate this undertaking to the meaning of biological diversity (definition in Article 2), the need for taxonomists and ecologists becomes self-evident. This will in turn re-invigorate teaching and research in systematics and ecology, reversing the previous trend of closures of departments of systematics for lack of support.

The likely increase in support applies also for the work of Northern scientists in the South and for involving Southern scientists in Northern research. Article 16.6 requires "each Contracting Party … [to] endeavour to develop and carry out scientific research based on genetic resources provided by other Contracting Parties with the full participation of, and where possible in, such Contracting Parties". Grants for botanical research have so far very often been too skimpy to enable the grant holders to involve Southern scientists, let alone enabling them to do their research entirely in the South. The Convention will thus help to modify or even to correct this weakness.

Problems specific to research

In spite of the above benefits, the Convention also has implications which may considerably complicate botanical research in the future.

Article 15.3 states that the biological material governed by the Convention "are only those that are provided by Contracting Parties...". This means that the germplasm collected before a country from which it was collected became a "Contracting Party" is not governed by the Convention. Since only those countries that had already ratified the Convention became "Contracting Parties" in December 1993, it means that all biological materials collected from those countries before then and those collected even later from the other countries, have undetermined ownership. This means that, by default if nothing else, they belong to the country in which they are stored.

This ambiguity of ownership has already caused problems. For example, one of the authors of this paper (TBGE) was involved in a Panel of Experts convened by FAO and CGIAR to suggest ways of handling the extensive CGIAR germplasm collections. The Panel recommended that as an intergovernmental body, the FAO Commission would be an appropriate body to have the collections under its auspices. Following that, in the meeting on the Convention held in Nairobi in June 1994, the World Bank made a bid - which did not succeed - to have the collections entrusted to it. Given this atmosphere of uncertainty, researchers and conservators are bound to find it ethically difficult to deal with these materials, and may want to avoid them for this reason. This is neither good for science, nor for the basis of our food.

The implementation of any legislation brings about new bureaucrats and bureaucracy. Bureaucracy is irksome in any country, and researchers are, on the whole, intolerant of it. The bureaucracy which is being brought in by the Convention on Biological Diversity is international in origin and is thus likely to be even less sensitive to the problems of researchers localized in small corners of the world.

Because the legislation is international, a scientist as an individual is going to find it increasingly difficult to have access to biodiversity in the future, probably even for basic research in plants without economic importance. The legal owner of biodiversity is a state (Articles 3 & 15.1). The individual scientist has, therefore, to deal with the state of the country of origin whenever he is requiring biological material for research outside that country. These dealings have, in order to be successful, to result in an agreement of Prior Informed Consent (a PIC certificate) from the country of origin; Article 15, paragraph 4 and 5 provides for a PIC procedure. This cannot be but a daunting prospect for individual scientist. According to Article 15.4, "access, [to biodiversity] where granted, shall be on mutually agreed terms...". This means that the state authorities of the country of origin will want to negotiate to arrive at "mutually agreed terms". It is unlikely that a state will negotiate with scientists individually as, when it comes to enforcing the terms of the agreement, it will have no jurisdiction over the individual, who will be living in another country. For this reason, a country of origin will most likely want the state of the country of the individual collector to stand guarantor for her/him, but a fixed procedure is not yet established, and yet no country in the North is to our knowledge requiring PIC certificates from the country of origin to allow import of genetic resources.

The PIC principle was applied for the first time as an instrument in environmental issues in the UNEP London Guidelines for the Exchange of Information on Chemicals in International Trade. It was then incorporated in the 1989 Basel Convention on the Control of Transboundary movements of Wastes and Their Disposal and in the FAO International Code of Conduct on the Distribution and Use of Pesticides (Hendrickx et al. 1994). In all these cases, the PIC procedure emphasizes the shared responsibilities of exporting and importing countries. Will this also be the case when PIC procedures are to be applied to transfer of genetic resources from the South to the North?

All countries that sign and ratify the Convention will have to establish a PIC procedure, or explicitly state that such a procedure is not required. Minimum requirements to a PIC procedure are (Hendrickx et al. 1994): (1) a national authority to issue the PIC certificates has to be designated and will have to be responsible for all applications for access and the follow-up procedures. (2) It will be necessary to specify the range of organisms for which PIC certificates are required; states of origin will naturally require PIC certificates for genetic resources of their most valuable crop plants, but how far is the requirement to be extended to other organisms? Will the country require PICs for all kinds of biodiversity? (3) It is necessary to state fees for processing application and issuing PIC certificates. (4) Issue general regulations, and (5) Try to follow future development, once the genetic resources are out of the country of origin.

A procedure could therefore be created in analogy with the procedures for research clearance already established in most African countries. A PIC certificate for export of genetic resources could include a range of provisions or conditions making access dependent on specified conditions for the subsequent sharing of benefits and knowhow. This is bound to frustrate both foreign and local researchers and conservators though, as time goes by and states gear themselves up to implement the Convention, this frustration should reduce.

It has been suggested that although the conditions for issuing a PIC certificate under the Convention can be imposed unilaterally by the country of origin, the agreement will be the result of a negotiated agreement where the laws of give-and-take will rule in a bargaining market; it has been assumed that a risk of loosing a good deal remains if the country of origin insists on unacceptable conditions (Hendrickx et al. 1994). This may be true for material of biodiversity with a very high economic potential, but the risks of complication for students of taxonomic groups with limited economic potential seems considerable. The complications with obtaining the PIC certificate are likely to try the patience of researchers and, in those among them with the sense of responsibility not so highly developed, increase the risk of irresponsible collecting and clandestine exchange of material.

Apart from the bureaucratic problems, there are also a number of ethical problems: Not all the members of the private sector can relish the idea of sharing research results and the benefits arising from the commercializing of biodiversity with the country of origin (Article 15.7). It can, therefore, be expected that at least some private sector enterprises will want to obtain biological materials in an illicit manner. This is made easier by the fact that the material to control comprises very small components of plants and animals extremely hard to identify as "genetic resource" and very easily smuggled out of its country of origin. There already is a history of private sector firms which "acquired and patented" material developed by international institutes of agricultural research (Fowler & Mooney 1990, pp. 186-187). It can thus be foreseen that researchers will continue to inadvertently pass on biological materials to private sector third parties and, unknowingly, become instrumental in defrauding the country of origin. Some of them may even be financially tempted to be involved in the illicit transfer of the materials.

Some of the biological materials can be used to produce transgenic organisms that are amenable to factory production, or extension of production into previously unsuitable regions, of commodities hitherto imported from a particular country or countries of origin. Alternatively, cells could be kept in tissue culture yielding the desired commodity in a Northern factory (Panel IV 1992, p. 18 & Annex V, Perlas 1994, pp. 79-81, RAFI 1989, 1990, 1992). If such commodity substitution were based on biological material obtained under the terms of the Convention, the application of Article 15.7 would, at least partially, rectify the damage to the economy of the country of origin; but most probably not so if the material is illicitly obtained in spite of the fact that Article 15.5 makes it illegal, since a poor Southern country will have very little capability to ensure the enforcement of this article. This puts a heavy moral burden on a researcher working on biological materials, especially on crops.

A scientist handling biological materials can, with or without her/his consent, have that material used to produce a transgenic organism which can damage human or environmental well-being (Fry & Day 1992, Levy & Miller 1989, Regal 1987, 1989, 1994). The scientist may even become an agent of dispersal into new territory of dangerous transgenic organisms.

An aspect of an ethical problem of which the majority of taxonomists and ecologists are probably unaware is that of intellectual property rights. As already pointed out in Section 2, the international scene is biased in favour of the protection of Northern intellectual achievement as intellectual property rights, allowing free access to Southern "indigenous and traditional" intellectual achievements. This issue has recently been reviewed in some detail by a group of researchers from formal and informal institutions, from commercial interests, and in trade and policy (Crucible Group 1994). A botanist could easily unwittingly assist the existing free South-North flow of traditional technology and biological knowledge with the result that it is commercialized in the rich North without anything of value accruing to the Southern communities who obviously should own, but cannot benefit from, the commercial advantages. This is a heavy responsibility on the researcher made even heavier by the usual lack of awareness on the issue.

AETFAT and the Convention on Biological Diversity

Many of the problems described here require thoughtful trained minds with actual field experience to help resolve them. Since the issues arise because of the North-South divide, the trained minds have to include both Northerners and Southerners if those issue are to be evaluated comprehensively and their resolutions to be formulated successfully.

AETFAT's members include both Northern and Southern scientists, who have staked their careers on the future smooth relationships between North and South for continued access to field work in Africa for the Northerners, and to herbaria and libraries in Europe and America for the Southerners. Therefore, both sides have a vested interest to bridge the gap. They also have an intimate knowledge of both North and South to examine the issue comprehensively so long as they work together in a genuine participatory spirit.

In this setting, AETFAT can develop for its scientists a code of conduct which might pre-empt the inevitable international bureaucracy. This code of conduct would minimize tension between researchers and countries from which biological materials will be acquired. Equally importantly, such a code of conduct would be seen to be working and thus greatly influence the development of international bureaucracies and thwart their development of excesses in the name of rigour.

To help make appropriate revisions of the code of conduct and to act as reference for the development by states of international procedures of biological material transfer, it would be advisable for AETFAT to start a data bank or bibliography on the various implications of the Convention and disseminate the information to help its members and also governments who must grapple with the problem.

Recommendations

From the above considerations emerge three possible recommendations which may be further discussed by AETFAT.

6.1. For conventional taxonomic work, entanglement with the Convention could be eliminated by merely denaturing the DNA of plant specimens so that they are no longer of any significance as source of genes, let alone as source of live seeds or embryos. The denaturing is needed because, otherwise, DNA from any scrap of biological material could, at least theoretically, be transferred into another organism to form a transgenic organism, or a specimen of the same species with a new combination of genes. Whether such a procedure is realistic or not in any given case is not to the point, the mere possibility may create an atmosphere of unnecessary suspicion between authorities and researchers which inhibits the transportation of herbarium material between the country of origin and other countries, be it on loan, or as part of an exchange programme. A microwave oven large enough to take herbarium sheets could be used to denature the specimens. It is already often used for killing potential herbarium pests.

6.2. For research that requires live material or intact DNA, AETFAT should develop well thought out procedures which ensure compliance with the Biodiversity Convention without being a hindrance to ordinary taxonomic research. Previous procedures do not apply, as the Convention of Biological Diversity does not work in the same way as the CITES Convention. In its present Congress, AETFAT could establish a committee or task force to draft such a set of procedures in a comparatively short time, say 6 months. This set of procedures may be discussed again or approved by the next Congress, but could in any case be brought to the notice of the membership as soon as it is ready since, even if not generally approved, it might still help members and their institutions to develop a good local code of conduct.

6.3. The present AETFAT Congress should also initiate the establishment of a "data bank" on information of relevance to the implications of the Convention, perhaps both in relation to taxonomic work and to gathering of information on traditional uses of biodiversity. Another committee or task force could be authorised to study and launch this; a column for such information in the AETFAT Bulletin might be adequate. There is probably no need for the approval of a committee's work by a later AETFAT Congress; the general idea could be approved now, and the committee or task force may be able to implement what is decided straight away.

References

Bedjaoui, M., 1979. Towards a new international economic order. UNESCO, Paris.

Crosby, A.W., 1986. Ecological imperialism. Cambridge University Press, Cambridge.

Crucible Group, The, 1994. People, plants and patents. International Development Research Centre, Ottawa.

Fowler, C. & P. Mooney, 1990. Shattering: food, politics and the loss of genetic diversity. The University of Arizona Press, Tucson, AZ.

Fry, J.C. & M.J. Day (ed.), 1992. Release of genetically engineered and other micro-organisms. Cambridge University Press, Cambridge.

GRAIN (Genetic Resources Action International), 1993a. A Decade in Review. Seedling 10-1: 3-12.

GRAIN, 1993b. Bargaining over the benefits of biodiversity. Seedling 10-3: 2-10.

Hancock, G., 1991. Lords of poverty. Mandarin Paperbacks, London.

Hendrickx, F., V. Koester & C. Prip, 1994. Access to genetical resources: a legal analysis. In: Sanchez, V. & C. Juma (eds.), Biodiplomacy - Genetic Resources and International Relations: 139-153.

Independent Commission on International Development Issues under the Chairmanship of Willy Brandt, 1980. North-South: a programme for survival. Pan Books, London and Sydney.

Khor, M., 1994. South-North resource flows and their implication for sustainable development. Resurgence, No. 46: 14-25.

Levy, S.B. & R.V. Miller (eds.), 1989. Gene transfer in the environment. Mc Graw-Hill, New York.

Panel IV of Expert Panels Established to Follow-Up on the Convention on Biological Diversity, 1992. Report of Panel IV: Consideration of the need and modalities of a protocol setting out appropriate procedures including, in particular, advanced informed agreement in the field of the safe transfer, handling and use of any living modified organism resulting from biotechnology that may have adverse effect on the conservation and sustainable use of biological diversity. UNEP, Nairobi.

Payer, C., 1974. The debt trap. Monthly Review Press, New York.

Perlas, N., 1994. Overcoming illusions about biotechnology. Third world Network, Penang.

RAFI (Rural Advancement Fund International), 1989. Vanilla and biotechnology. RAFI Communique, Updated June, 1989. RAFI, Pittsburgh, N.C.

RAFI, 1990. Biotechnology and castor oil. RAFI Communique, January/February, 1990. RAFI, Pittsburgh, N.C.

RAFI, 1992. Genetic engineering of pyrethrines, early warning for East African pyrethrum farmers. RAFI Communique, June, 1992. RAFI, Pittsburgh, N.C.

Rajanaidu, N., 1990. Major developments in oil palm (Elaeis guineensis) breeding. Mitt. Inst. Allg. Bot. Hamburg 23a: 39-52.

Regal, P.J., 1987. Safe and effective biotechnology: mobilizing scientific expertise. In: Fowle III, John R. (ed.), Application of biotechnology: environmental and policy issues. Westview Press, Colorado: 145-164.

Regal, P.J., 1989. Biotechnology jitters: will they blow over? Biotechnology Education Vol. 1, no. 2: 51-55.

Regal, P.J., 1994. Scientific principles for ecologically based risk assessment of transgenic organisms. Molecular Ecology Vol. 3: 5-13.

Rich, B., 1994. Managing the Earth: the World Bank, environmental impoverishment and the crisis of development. Earthscan, London.

World Conservation Monitoring Centre, 1992. Global biodiversity. Chapman & Hall, London.

World Resources Institute, 1993. Biodiversity prospecting. World Resources Institute, Washington, D.C.

Preliminary study of threatened plants of Mozambique

S.O. BANDEIRA, L. MARCONI & F. BARBOSA
Department of Biological Sciences, Eduardo Mondlane University,
P.O. Box 257, Maputo, Mozambique.

Abstract
A preliminary list of threatened plants following the IUCN categories of threat is presented for Mozambique. The species are grouped into *Extinct, Endangered, Vulnerable, Rare, Indeterminate* and *Insufficiently Known* categories. Mozambique is unevenly and little collected; this is exacerbated by the recent war which prevented most collecting efforts throughout the country for many years. Within the signing of Peace Accord it will now be possible to determine the conservation status of plants throughout the country. The list presented here is therefore preliminary, and will be updated as new information comes to light.

Key words
Conservation, Endangered, Rare, Threats, Vulnerable.

Introduction

Mozambique covers an area of 784,755 square kilometres extending from Tanzania in the north (10°30') to Natal, South Africa, in the south (26°52'). It also shares boundaries with Malawi, Zambia, Zimbabwe, South Africa and Swaziland in the west, while the Indian Ocean washes the eastern shore.

In terms of biogeographic affinities approximately 80% of Mozambique is included in the Zambezian Regional Centre of Endemism. A broad band of the Zanzibar - Inhambane Regional Mosaic extends northwards along the coast from Xai-Xai (25°03'S) to the Tanzanian border (10°26'S). A small area of the Tongoland - Pondoland Regional Mosaic occurs in the southern coastal region whilst small isolated areas of Afro-montane elements occur in the highlands of central Mozambique. The country harbours 5500 species of higher plants (World Conservation Monitoring Centre 1992), and seven broad vegetation communities can be identified: miombo woodland, mopane woodland, undifferentiated woodland, afromontane elements, halophytic vegetation, swamp vegetation and coastal mosaics (Figure 1).

Eleven percent of the area of the country has a conservation status and includes national parks, reserves, wildlife utilization units (game reserves) and forest reserves. In addition to this area under protection, a major concern for other conservation measures has to do with the conservation of botanical diversity wherever it is very localized or threatened. In view of this, this paper has the following objectives:

1. To identify threatened species.
2. To identify major pressures and threats to those species.
3. To determine the IUCN threatened categories for those species.
4. To suggest conservation measures for them.

Methodology

Species names were obtained from literature (Grandvaux Barbosa 1968, Schouten

L.J.G. van der Maesen et al. (eds.), The Biodiversity of African Plants, 306-309.

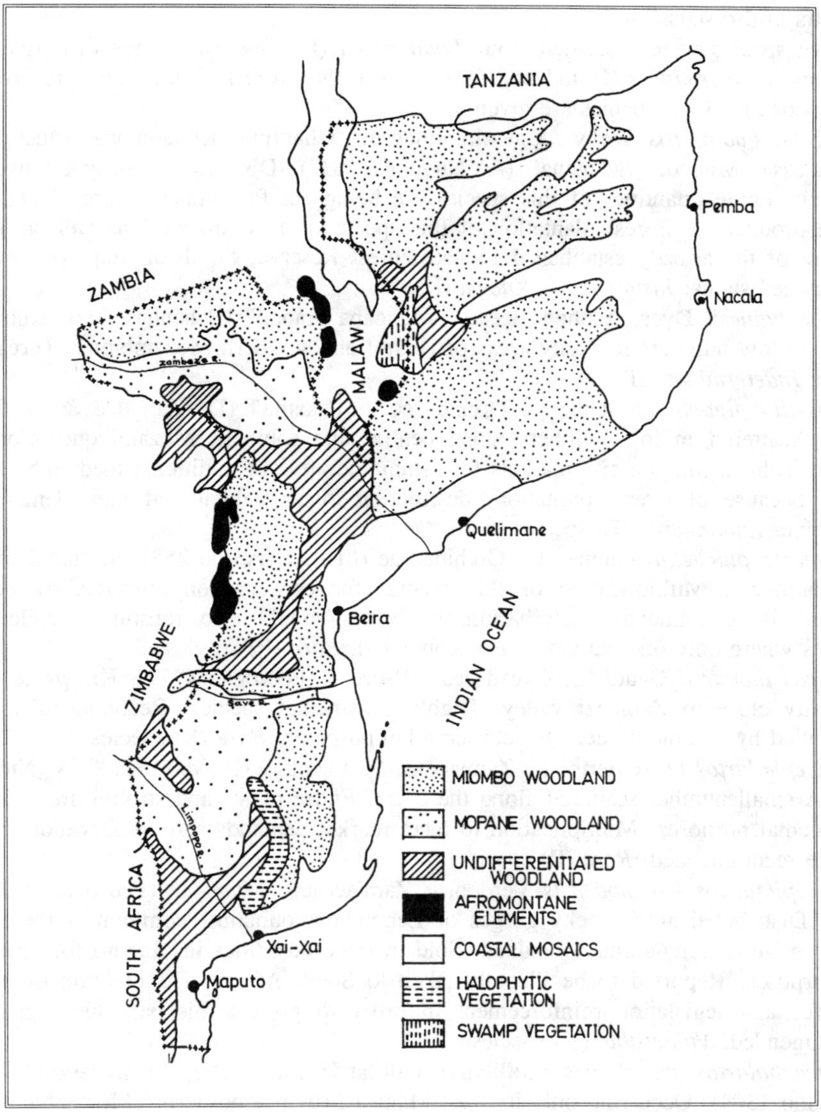

Figure 1. Broad vegetation types in Mozambique.

1992, World Conservation Monitoring Centre 1993, Bandeira et al. in press) and unpublished information at LMU (Eduardo Mondlane University) and LMA (National) Herbaria. IUCN (International Union for Conservation of Nature) threat categories (World Conservation Monitoring Centre 1992) were used to group the different species.

Major pressures or threats to these species were identified using herbarium voucher specimens, field observations and informal questions to local people.

Results and discussion

Twelve species were identified: four *Endangered* (E), one *Vulnerable* (V), five *Rare* (R), one *Indeterminate* (I) and one *Insufficiently Known* (K). The species are arranged alphabetically. Local names are given.

Afzelia quanzensis Welw., Caesalpinioideae, "Chanfuta, Kongonuwa, Mucopócoa, Muçacossa, Muoco, Ngeugena" (A. Nuvunga 1032). Distributed nearly all over the country. Large quantities of this species are being cut for timber, charcoal and fuelwood production. Forest plantation of this species is recommended as well as management of the already established Liquati Forest Reserve, South of Maputo Province. Threatened status: *Insufficiently Known* (K).

Aloe bainesii Dyer, Asphodelaceae, "Mangana grande" (Balsinhas 707). Known to occur in low numbers in bushy areas of central and southern Mozambique. Threatened status: *Indeterminate* (I).

Ansellia gigantea Reichb. f., Orchidaceae, "Pakama" (D. Zunguze & C. Boana 128). Scattered in low numbers along central and southern Mozambique. Common species sold in Maputo city markets for asthma curative in traditional medicine. Threatened because of over-exploitation, destruction of its habitat and high demand for medicine. *Endangered* (E) species.

Bonatea pulchella Summerch., Orchidaceae (M.F. Carvalho 258). Scattered in very low numbers. Multiplication of this species for conservation purposes by botanic gardens is recommended. Herbarium collectors should also refrain to collect the species where only one individual is encountered. *Rare* (R) species.

Cycas thouarsii Gaudich., Cycadaceae (Pedro & Pedrogão 4777). Known to occur naturally close to Zambezi valley. Highly ornamental species. Recommended to be multiplied by botanic gardens for conservation purposes. *Rare* (R) species.

Encephalartos ferox Bertol f., Zamiaceae, "Chihanga" (F. Mapanga & A. Nuvunga 97). A small number scattered along the coast. Entire plants and strobili are taken for ornamental purposes. Multiplication to meet market demands and conservation measures are recommended. *Rare* (R) species.

Encephalartos lebomboensis Verdoorn, Zamiaceae, "Muanga" (Pedro & Pedrogão 438). Distributed along rocky gorges of Lebombo mountains; common at the border between Mozambique and Swaziland. Sold in large quantities in Maputo for ornamental purposes. Reported to be illegally taken to South Africa. The creation of natural reserves and legislation reinforcement in order to protect this particular species is recommended. *Vulnerable* (V) species.

Encephalartos manikensis (Gilliland) Gilliland, Zamiaceae, "Umguarara" (M.F. Carvalho 1304). Occurring only in West Manica Province bordering Mozambique with Zimbabwe on mountain sides and granitic hills (Coates Palgrave 1988). The establishment of a special reserve for this species is recommended. *Rare* (R) species.

Eulophia petersii Reichb. f., Orchidaceae, "Eulophia" (Pedro & Pedrogão 6701). Orchid occurring mainly in sandy areas. *Rare* (R) species.

Podocarpus falcatus (Thunb.) R.Br. ex Mirb., Podocarpaceae, "Msongo, Mulatchen, Um-koba" (Graunoux Barbosa & Lemos, 7502). Occurring in southern Mozambique mainly along Maputo River. Threatened because of increasing settlement and land use in the area. Forestation and propagation recommended. *Endangered* (E) species.

Raphia australis Oberm. & Strey, Palmae, "Rafia" (Gomes e Sousa & N. Rosa 5036). Distributed in few parts of the country along valleys of small rivers; in Maputo occurring only at Bobole Special Reserve: the southernmost colony of raffia palms in Mozambique located at the valley of Bobole River, a tributary to Limpopo River. This species is being threatened because of habitat destruction such as subsistence agriculture inside the Reserve area and settlement of people in the vicinity. This reserve is extended over 12 ha and has approximately 150 specimens. Reinforcement of conservation status of this species is highly recommended. *Endangered* (E) species.

Warburgia salutaris (Bertol. f.) Chiov., Canellaceae, "Chibaha" (J. de Koning, P. Jansen & D. Zunguze 8638). Distributed in southern Mozambique. Being threatened because of over-exploitation for use in traditional medicine to treat throat complications (Jansen & Mendes 1990). High demands also in neighbouring countries (Coates Palgrave 1988, Department of National Parks and Wild Life Management 1991). Plantation for medicinal purposes is recommended. *Endangered* (E) species.

Acknowledgements
We wish to thank the staff of LMU Herbarium (Eduardo Mondlane University) and LMA (National Herbarium) for their contribution to this paper.

References

Bandeira, S.O, J.C. Hatton, P. Munisse & S. Izidine, in press. The ecology and conservation status of plant resources in Mozambique. Symposium "The Conservation and Utilization of Southern African Botanical Diversity", Cape Town, South Africa, 6-11 September 1993.

Coates Palgrave, K., 1988. Trees of Southern Africa. Struik Publishers, Cape Town, 959 pp.

Department of National Parks and Wild Life Management, 1991. Protected species of animals and plants in Zimbabwe. Harare, 69 pp.

Grandvaux Barbosa, L.A., 1968. Moçambique. In: Inga and Olov Hedberg (eds.), Conservation of Vegetation in Africa South of Sahara. Proceedings 6th plenary meeting of AETFAT, Uppsala, 12-16 September 1966. Acta Phytogeographica Sueca, 54: 224-233.

Jansen, P.C.M. & O. Mendes, 1990. Plantas Medicinais - Seu Uso Tradicional em Moçambique. Tomo 3. Instituto Nacional do Livro e do Disco, Maputo, 302 pp.

Schouten, K., 1992. Checklist of CITES Fauna and Flora. Secretariat of the Convention on International Trade in Endangered Species of Wild Fauna and Flora, Lausanne. 238 pp.

World Conservation Monitoring Centre, 1992. Global Biodiversity: Status of the Earth's living resources. Chapman & Hall, London.

World Conservation Monitoring Centre, 1993. Mozambique: conservation status listing of plants. Compiled from the World Conservation Monitoring Centre plants database.

Conservation and *in vitro* propagation of rare Kenyan Orchidaceae, a common goal for conservation and horticulture

B. BYTEBIER, S.W. SIMIYU & T.R. PEARCE
Plant Conservation and Propagation Unit, East African Herbarium, National Museums of Kenya, P.O. Box 40658, Nairobi, Kenya.

Abstract
For a variety of reasons a number of Kenyan orchids are under severe threat of extinction. The Plant Conservation and Propagation Unit (PCPU) of the East African Herbarium (EAH) at the National Museums of Kenya has compiled a list of rare Kenyan orchids. This list, which was raised with the help of both professional and amateur orchid biologists, provides a guideline for orchid conservation priorities. Germplasm, mostly in the form of seeds, is collected from the field and from amateur collections. Through an *in vitro* seed germination approach, we are building up an *ex-situ* collection of endangered orchids as an insurance against genetic erosion or extinction altogether. This collection could in future also be used in re-introduction programmes.

Key words
Conservation, *In vitro* seed germination, Orchidaceae.

Abbreviations
KC - Knudson C medium
MS - Murashige and Skoog medium
PM - Phytamax™ medium

Introduction

Many wild orchids in Kenya are threatened with extinction due to destruction of their natural habitat. There has been a swift decline in Kenya's forests and woodlands and epiphytic orchids have been eliminated along with their host trees. Terrestrial orchids have lost their habitats through agricultural development of marginal lands. Overcollecting of those species of particular horticultural interest have also driven some near to extinction.

The Plant Conservation and Propagation Unit is using both *in-situ* and *ex-situ* conservation efforts to arrest this trend. A list of rare Kenyan orchids has been compiled, using the matrix approach of Rabinowitz (Rabinowitz 1981). Information was gathered from the East African Herbarium, the Royal Botanic Gardens Kew and scientific literature, as well as from amateur orchid enthusiasts. The rare orchids list is a dynamic entity which will continuously be updated through our own surveys and through discussions with enthusiasts who regularly go out in the field.

We are currently trying to locate remaining wild populations of orchids under threat using East African Herbarium records supplemented by knowledge of wild populations from members of the Kenyan Orchid Society. Germplasm in the form of "rescued" plants but more important in the form of seeds is collected and is supplemented, where necessary, by the use of private collections. An active programme of hand-pollinations

L.J.G. van der Maesen et al. (eds.), The Biodiversity of African Plants, 310-312.

of plants from similar localities is being undertaken by using the private collections of various members of the Kenyan Orchid Society.

The relationship between symbiotic fungi and the germination of orchid seed is well known. Equally so is the fact that orchid seeds are particularly amenable to modern *in vitro* propagation techniques, where natural conditions for germination can be well mimicked. Asymbiotic, as well as symbiotic germination techniques, are now preferred methods where the genetic diversity of the species needs to be conserved.

Material and methods

Preferably nearly ripe, but still closed, seed pods are brought to the lab and are sterilized with 20% commercial bleach for 20 minutes. These pods are then dried in a desiccator until dehiscence. Seeds are then separated from the pods. Seed lots are kept for a couple of days longer in the desiccator and are then stored at 4°C for future use. Prior to sowing, seeds are sterilised with 10 % commercial bleach for 10 minutes. A variety of media, including Murashige and Skoog (Murashige & Skoog 1962), Knudson C (Knudson 1946), Vaccin and Went (Vaccin & Went 1949) and Phytamax™ (Sigma), with or without charcoal and/or banana powder, have been used.

Results

Species currently cultivated *in vitro* include *Aerangis confusa* J. Stewart, *A. coriacea* Summerh., *A. luteoalba* (Kraenzl.) Schltr. var. *rhodosticta* (Kraenzl.) J. Stewart, *Angraecum dives* Rolfe, *Ansellia africana* Lindl., *Polystachya bella* Summerh., *P. confusa* Rolfe and *Solenangis wakefieldii* (Rolfe) Cribb & J. Stewart.

Ansellia africana Lindl. has responded particularly well to the asymbiotic germination technique, although large differences in germination percentage and growth rate between different media, different sucrose concentrations and the different provenances used, have been observed. Overall, PM with 30 g/l sucrose has given the best results for germination. Plants can be transferred from petri dishes to flasks, containing PM with 20 g/l sucrose, after about three and a half months. One year after sowing, plants are ready for weaning. The weaning and the initial care of the orchid plantlets is a critical step in the whole process.

Angraecum dives Rolfe germinated well on KC with 30 g/l sucrose. After 4 months plantlets were transferred from petri dishes to jars with KC medium containing 20 g/l sucrose. Plants continued to grow well and are now, after 1 year, ready for weaning.

Solenangis wakefieldii (Rolfe) Cribb & J. Stewart seeds germinated on KC, MS and PM medium, all with 30 g/l sucrose. Growth, however was faster and better on MS. After 4 months plantlets were transferred onto fresh medium, now with 20 g/l sucrose. Growth stopped, and eventually plants died, on PM and KC, but plants on MS were thriving and are now, after one year, ready for weaning.

Polystachya bella Summerh. seeds germinated on both PM and MS with 30 g/l sucrose. Plants were initially thriving on PM, but once transferred to fresh PM with 20 g/l sucrose, stopped growing and turned towards unorganised callus growth. Although the plants on MS were developing slower, they continued to grow normally after transfer to fresh MS with 20 g/l sucrose, four months after sowing.

Recent experiments with *Aerangis confusa* J. Stewart indicate that it only germinates on MS medium.

Conclusion

To date only asymbiotic germination techniques on epiphytic species have been attempted although it is our intention to begin trials on symbiotic *in vitro* germination of terrestrial species.

It is hoped that by offering plantlets for sale at reasonable prices, collection from the wild will be reduced. *Ex-situ* populations are also maintained for future re-introduction programmes. It is realised, however, that unless substantial efforts will be done to protect orchid habitats, many species will become extinct in a near future.

Acknowledgements

The PCPU of the EAH is supported through a bilateral agreement between the ODA (Overseas Development Administration) and the Kenyan Government. B.B. is supported by the VVOB (Flemish Association for Development Cooperation and technical Assistance).

References

Knudson, L., 1946. A new nutrient solution for the germination of orchid seeds. Amer. Orch. Soc. Bull. 15: 214-217.

Murashige, T. & F. Skoog, 1962. A revised media for rapid growth bio-assays with tobacco tissue cultures. Physiol. Plant. 15: 473-497.

Rabinowitz, D., 1981. Seven forms of rarity. In: H. Synge (ed.), Biological aspects of rare plants conservation: 205-217, John Wiley and Sons Ltd, New York.

Vaccin, E.F. & E.W. Went, 1949. Some pH changes in nutrient solutions. Bot. Gaz. 110: 605-613.

Conservation priorities in the arid and semi-arid lands: the case of the hilltop forests of Kenya

F.N.M. GACHATHI
Kenya Forestry Research Institute, P.O. Box 20412, Nairobi, Kenya.

Abstract
The arid and semi-arid lands (ASALs) in Kenya are facing a serious environmental deterioration, largely due to the rapidly increasing population, overgrazing and changing lifestyles. This deterioration is being extended to the hilltop forests within the ASALs as people seek for livelihood in this "saviour sites". There is therefore urgent need for research into the array of issues in this crisis in order to be able to formulate viable conservation and development strategies. Effective conservation has to be broader based and firmly grounded and calls for the involvement of the local people as partners in conservation efforts since the hilltop forests are central to the rural culture.

Key words
Arid and semi-arid lands, Conservation, Hilltop forests, Kenya.

Introduction

There is little documented knowledge on endangered plant communities of Kenya, their geographical distribution, their abundance and their influence on the ecology (Waliaula 1988). This is exacerbated by the fact that literature on the forests of Kenya is rather scattered and piecemeal, and details about forest vegetation are absent in most cases (Beentje 1990). This situation is threatened further by the unprecedented rate at which the plants are being destroyed before their potential use to society and their ecological effects are discovered.

When forests decline or are removed, many forms of plants and animal life that depend on their habitat for survival are invariably lost. This is more serious in the Arid and Semi-Arid Lands (ASALs) where the plant genetic resources are threatened by recurrent droughts, migrating populations, overgrazing and changing lifestyles. These changes disregard traditional knowledge and practice of land use, natural resource management etc. developed by the local people over centuries resulting in serious environmental deterioration in the ASALs. This deterioration is being extended to the hilltop forests as people seek for livelihood in these "saviour" sites.

Physical environment

The ASALs of Kenya are generally characterized by a hot and dry climate. The skies are normally clear, and daytime temperatures may rise to 38°C. The annual rainfall is not only low but also brief and erratic, and ranges from less than 150 mm to 750 mm rendering this vast expanse of land agriculturally marginal. The most suitable land use is livestock husbandry and wildlife conservation, and almost 90% of the 40 gazetted national parks and game reserves are located in the ASALs. Extensive areas of the ASALs are low-lying plains covered by thorn bushes and scattered trees, mainly *Acacia* and *Commiphora* species (Greenway 1993, White 1983). These plains are however characterized by isolated hills and few mountains (Figure 1).

L.J.G. van der Maesen et al. (eds.), The Biodiversity of African Plants, 313-316.

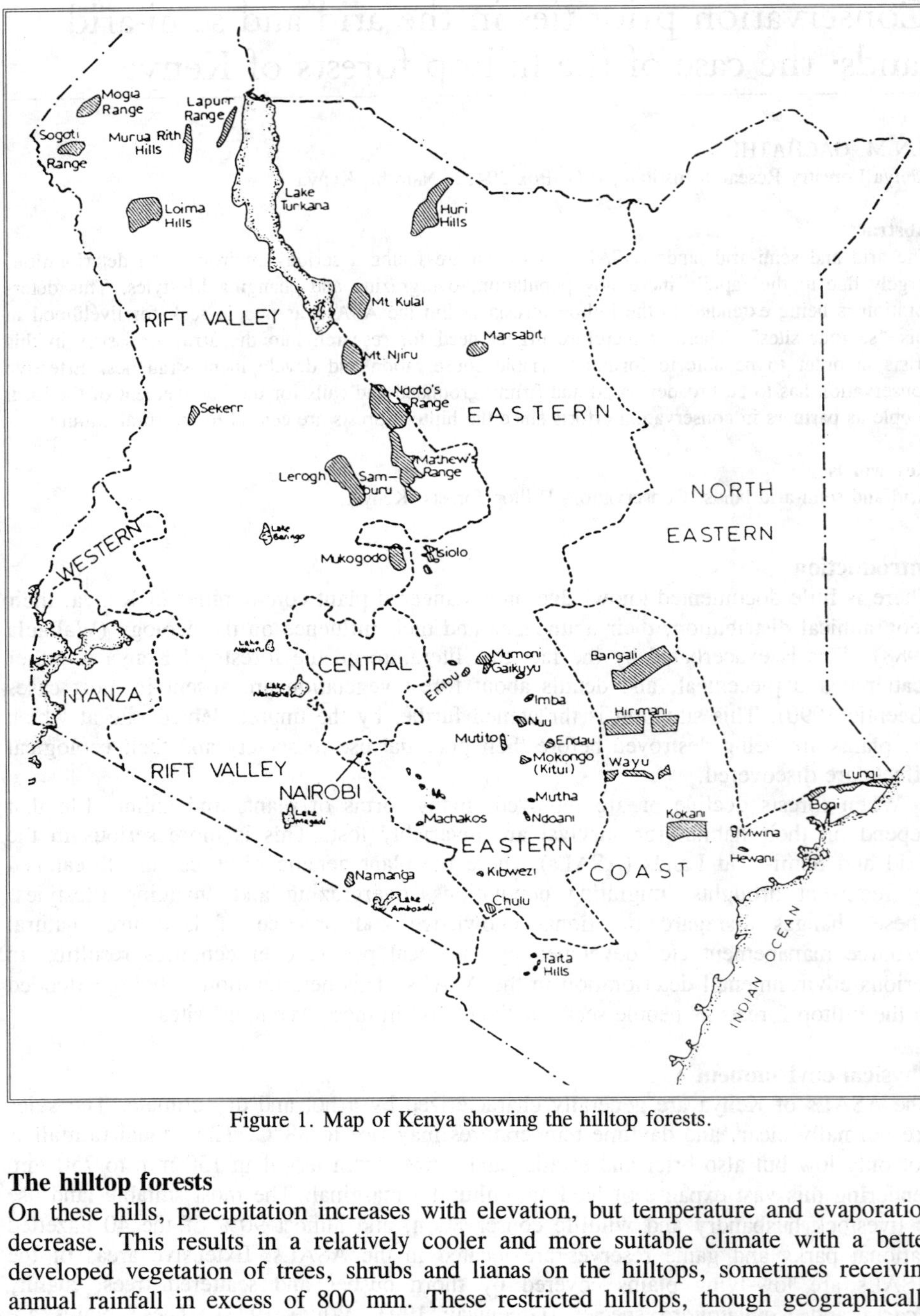

Figure 1. Map of Kenya showing the hilltop forests.

The hilltop forests

On these hills, precipitation increases with elevation, but temperature and evaporation decrease. This results in a relatively cooler and more suitable climate with a better developed vegetation of trees, shrubs and lianas on the hilltops, sometimes receiving annual rainfall in excess of 800 mm. These restricted hilltops, though geographically within the ASALs, are therefore climatically not arid or semi-arid and have mid-altitude forests with tree species of the lowlands and those of the uplands in addition to those found in the ASALs. These are unique sites which are vital habitats to the long term maintenance of biodiversity and other natural processes in the ASALs.

These hilltop forests are of immense antiquity and the species of plants and animals in them have evolved mutual adaptation. On the other hand, the nomads have developed ecological management strategies in harmony with this fragile environment by exploiting different ecological niches. During normal seasons, grazing occurs on the lowland plains but with the onset of drought, animals gradually move to the hills. In this way the nomads have managed both wildlife and their habitats for decades.

The importance of the hilltop forests

During ethnobotanical field studies carried out by Kenya Forestry Research Institute (KEFRI 1991) in Kajiado District, it came out clearly that:

a) Plant medicine use is still common and very popular, and held with a lot of esteem by the pastoralist communities in the ASALs, and in most cases people visit modern hospitals only when the traditional medicine fails.
b) Most plants of use or value to the local community are now found only in the hilltop forests.
c) The hilltop forests contain unique relic stands of plants and animals which have become endemic to specific hills as a result of isolation.
d) The higher hills are water catchments, feeding springs vital for the very existence of life in the ASALs.

Threats to the Hilltop Forests

The ASALs are facing escalating environmental degradation and this degradation is extending to the hilltop forests within them, largely due to the rapidly increasing population and recurrent droughts. Due to the increasing human and livestock population and the effects of drought, the available soil, forests, water and grazing resources are under stress resulting in a wave of migration to the hills. In the past, pastoralists relied on livestock products but they are now changing to the consumption of grains and desire to grow their own crops. The change of land ownership from communal to individual accompanied by change from pastoral to sedentary life style poses an increasing threat to the fragile ecosystems of the ASALs (Herlocker 1979, Trapnell in prep.).

Research Priorities

There is an urgent need for concerted research to investigate the many issues involved in the environmental deterioration of the hilltop forests within the ASALs:

a) Ethnobotanical surveys to identify plant species in the hilltop forests, their uses to the local community, and their abundance;
b) Traditional extraction practices of herbal medicines, opportunities for domesticating species and development of harvesting protocols and sustainable production systems;
c) Forest regeneration, wildlife and plant species, and provenance selection for various zones for better growth, survival and adaptation;
d) Conservation and management of the existing hilltop forests, building on the valuable traditional management strategies.

Conclusion

All conservation initiatives for the hilltop forests must be people-focused and community-driven guided by socio-cultural studies undertaken within the frame work of strengthening and supporting the pastoral economy for which ASALs have an undisputed comparative advantage.

The local communities should be involved in the planning and management of the hilltop forests as opposed to the notion that forests in the ASALs, should be brought under the control of foresters through gazettement.

The pastoralists must be educated in the rationale of conserving and where necessary, preserving the hilltop forests by deferring destructive exploitation today to meet the short-term needs in order to have long-term benefits tomorrow.

References

Beentje, H.J., 1990. The Forests of Kenya. Proceedings XIIth Plenary Meeting of AETFAT, Hamburg.

Greenway, P.J., 1993. A classification of the vegetation of East Africa. Kirkia 9: 1-68.

KEFRI, 1991. A Dryland Forestry Handbook for Kenya. KEFRI, Nairobi.

Herlocker, D., 1979. Vegetation of S.W. Marsabit District. IPAL Techn. Rep. D1 UNEP/MAB. Nairobi.

Trapnell, C.G., in prep. Ecological Considerations in the Management of Indigenous Forests in South West Kenya.

White, F., 1983. The Vegetation of Africa. Natural Resources Research 20. UNESCO, Paris.

Waliaula, S.F.M., 1988. Plants of Baringo District. Technical report 178, Department of Resource Surveys and Remote Sensing (KREMU), Ministry of Planning and National Development, Nairobi.

Forest management systems to sustain resource use and biodiversity: examples from the southern Cape, South Africa

C.J. GELDENHUYS
CSIR Division of Forest Science and Technology, P.O. Box 395, Pretoria, 0001, South Africa.

Abstract
It is demonstrated how suitable forest inventories could be used to assess the scale of the ecological processes of disturbance and recruitment, the resource status of all species, and to indicate the most appropriate management system to sustain resource use. The mountain forests need a higher intensity of disturbance for maintenance of their composition, such as a group selection system for tree harvesting. The platform forests require a low-intensity disturbance for their maintenance, such as a single-tree selection harvesting system.

Key words
Biodiversity, Diameter distribution, Forest, Grain, Management system, Sustainability.

Introduction

Concern is often expressed that the uncontrolled use of forest resources adversely affect the biodiversity of the forests (FAO 1993). Forests in southern Africa provide many products for household needs, and for development of rural and regional economies (Geldenhuys & Van der Merwe 1988, McKenzie 1988, Geldenhuys 1990, Geldenhuys & Lübbe 1991). Some of the species are relatively rare, are over-utilized, and the resource-use management systems do not favour their regeneration or survival.

Resource utilization and biotic diversity of forests can be sustained if the products and values are utilised in relation to the growth potential of the site, and the essential ecological processes of disturbance and recruitment to which the component species are adapted. A first approximation for development of a sustainable resource-use forest management system is obtained through an appropriate forest inventory.

Two general approaches are in use: a) Most often only the volume of utilisable trees of commercial timber species is assessed, i.e. the focus is on immediate, short-term, economic gain. No sustainable utilization is possible. b) In very few cases the resource status of economically and ecologically important species is assessed, i.e. the focus is on long-term, sustained-use management with economic stability and ecological balance. At least three components should be assessed: i) Forest communities and the importance of different species in the different communities (e.g. Geldenhuys 1993a, b, Geldenhuys & Murray 1993); ii) Grain, i.e. the relationship between composition of canopy species in the regeneration and in the canopy of the same stand to gain some understanding of the scale of forest disturbance processes (Midgley et al. 1990, Everard 1992); iii) Shape of diameter distributions within species over different communities, which provide useful insights into the resource status and recruitment requirements or constraints for the specific species (Geldenhuys 1993a, b, Geldenhuys & Murray 1993).

L.J.G. van der Maesen et al. (eds.), The Biodiversity of African Plants, 317-322.

The objective of this paper is to assess and demonstrate the use of forest inventory data in the development of appropriate management systems for sustainable forest resource harvesting and maintenance of biodiversity in forests in South Africa.

Study area

The southern Cape forests (60,500 ha; Geldenhuys 1991) are managed for timber, minor forest products, recreation, and the indirect values of species and habitat conservation (Seydack et al. 1990). The forests occur between the mountains (north; altitude 1000-1500 m) and the coast (12-34 km to the south) at about 34°S, between 22°E and 25°E. The all-year, mean annual rainfall ranges from 400 mm in the west to about 1200 mm in some localities. The proportion of winter rainfall (April to September) varies between 42 and 54%. The area is transitional between the tropical-subtropical and the temperate regions.

The tree communities differ distinctly in patch size and composition between three major landscape units: i) the coastal scarp and steep, open river valleys near the coast, between almost sea level and 240 m altitude; ii) the coastal platform and lower parts of the foothills (altitude 200 to 500 m); and iii) the mountain slopes and gullies, usually above 500 m (Geldenhuys 1991, 1993b). The disturbance regimes in the area are as follows: a) Trees dying standing (most frequent), small windfalls and infrequent lightning strikes (no fire) cause gaps of 0.001-0.2 ha which develop gradually but the gap is not cleared (Geldenhuys & Maliepaard 1983, Midgley & Cameron 1992, Geldenhuys et al. 1994). This is the typical disturbance in the forests of the coastal platform and scarp, and the river valley slopes. b) Infrequent lightning fires, landslides on steep slopes and large wind-blowouts cause larger gaps of 0.05-2.0 ha, often leaving a totally cleared forest floor (Geldenhuys 1990, Geldenhuys et al. 1994). This type of disturbance occurs more frequently on steep slopes in forests of the river valleys. c) Shrubland fires during the dry gusty foehn-like bergwinds destroy forest ecotones, limit forest to wind-shadow areas, and occasionally destroy forest stands (Geldenhuys 1994). This is the typical disturbance in the mountain and foothill forests.

Methods

The forests were sampled with 250 plots along six zones, each covering forests from the coast to the mountains. In each site a circular plot of 11.3 m radius, i.e. 400 m^2, was used to sample a stand representative of the particular area. The following variables were recorded for this study: i) All stems $\geq$5 cm diameter at breast height (DBH), by species and DBH; ii) All stems of 1 to 4.9 cm DBH, counted by species over the entire plot. Frequency of stems $\geq$5 cm DBH were used to classify the tree communities (using TWINSPAN of Hill 1979a) (Geldenhuys 1993b). Forest grain was assessed from a number of plots selected at random from each community. For each plot the data for all tree species that may enter the canopy, were separated into two data sets: the number of stems of 1 to 4.9 cm DBH (data set 1); and the number of stems $\geq$10 cm DBH (data set 2). An ordination diagram was produced (using DECORANA of Hill 1979b) of the mean scores for dataset 1 and 2 for each tree community along axes 1 and 2 (Figure 1). The diameter distribution of important species in each main community (using 5 cm DBH class of 5-9.9; 10-14.9; ... 65-69.9; and 70+ cm) was calculated to show the demographic status of species over different communities

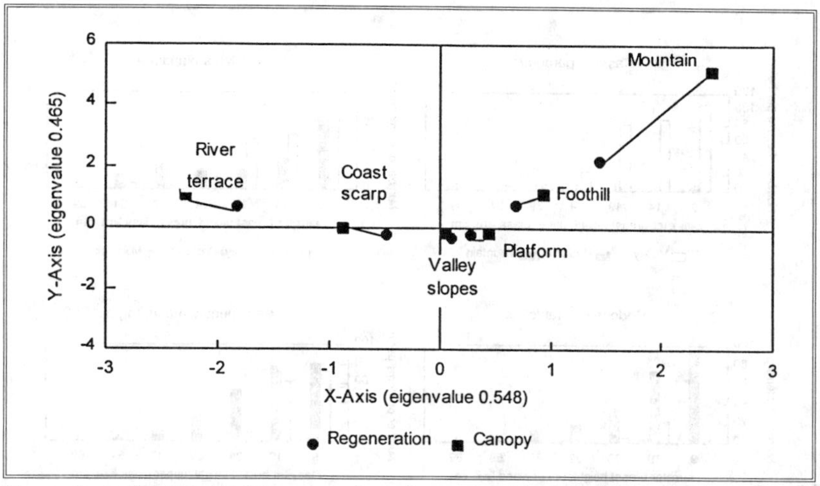

Figure 1. Ordination of the mean scores for composition of the canopy and regeneration of the same stands in different forest communities in the southern Cape forests. The separation or not between the location of the regeneration and canopy scores of the same stands is caused by the disturbance regimes.

(Figure 2). Finally, mean species richness per 400 m^2 plot in each main community was compared, using all stems $\geq$5 cm DBH (Table 1).

Results

Three distinct forest types were identified with very different species: river valley & scarp forest; platform forest; and mountain forest. These were subdivided into eight communities and 18 sub-communities (Geldenhuys 1993b).

The grain of communities is shown for six of the eight communities (Figure 1; all coastal platform communities were combined). Platform forest and river valley-slope forest are fine-grained, i.e. composition of the canopy and regeneration is very similar. The mountain forest is particularly coarse-grained, with foothill, coast scarp and river terrace forests of intermediate grain. The composition of the regeneration of the coarse- and intermediate-grained communities tends towards the coastal platform and river valley-slope forests, which suggest successional trends towards mature forest. An above average number of the mountain, foothill and coast-scarp plots had charcoal present in the litter layer, which suggest relatively recent disturbance.

The diameter distributions of selected species indicate prominence of a species in only one or two of the types (Figure 2), and in a particular type it may be prominent in only some of the communities or sub-communities. Most of the species show the typical inverse J-shaped diameter distributions of shade-tolerant species with expanding populations in closed forest. Both *Ocotea bullata* (Burch.) Baill. and *Cunonia capensis* L., the two dominant canopy trees of mountain forest, have bell-shaped diameter distributions, indicative of a pioneering population in maturing regrowth forest.

Species richness is highest in the river valley and scarp forests and the coastal platform forests, i.e. forest with the small-gap disturbance regime and inverse J-shaped

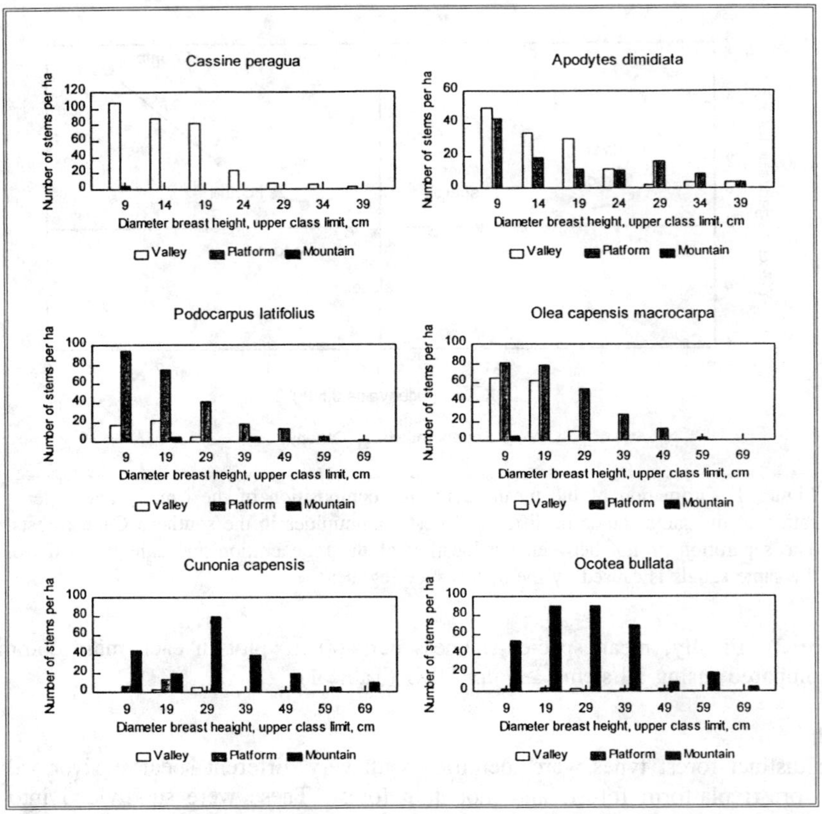

Figure 2. Diameter class distributions of selected species from the three forest types in the southern Cape forests. Note the different diameter class widths for the different species.

diameter distributions (Table 1). It is lowest in the frequently-disturbed mountain forests. The herbaceous growth forms show distinct richness patterns between these different zones (Geldenhuys 1993b), such as an abundance of fern species in the mountain forests, vine species in the coastal forests, and epiphyte species in the fine-grained coastal platform and river valley-slope forests. It is however important to note that the combination of the different forest communities and their serial stages gives the highest species richness.

Discussion

The combined information of community composition, grain and DBH distribution gives the best understanding of the status and dynamics of the forest stands. Table 2 summarises the relevant data for the southern Cape forests which would dictate the preferred forest management system. The proposed forest management systems would provide for a balance of shade-tolerant and shade-intolerant species. This would maximize species richness of the forest by allowing species of all serial stages to persist.

Table 1. Species richness for different forest communities in the southern Cape forests. It increases from the mountain forests (which are relatively frequently disturbed by fire) to the platform, river valley and coastal scarp forests with small-scale disturbance.

Forest zone	Number of species (woody and herbaceous)	
	Mean per 400 m^2 plot	All species in zone
Scarp forest	39	52
River valley	46	61
Platform	34	77
Moist platform	22	38
Foothill	23	26
Mountain	13	20

Table 2. Summary of ecological characteristics and preferred management systems for mountain and coastal platform forests of the southern Cape.

Mountain forests	Coastal platform forests
Characteristics:	Characteristics:
• Frequently disturbed by fire	• Regularly disturbed by small canopy breaks but rarely by fire or large canopy gaps
• Regrowth stage	• Mature forest
• Low species richness	• High species richness
• Few growth forms	• Wide range of growth forms
• Seeds small and dry; or hard and long-living	• Fruits fleshy and seeds short-living
• Regeneration different from canopy	• Regeneration similar to canopy
• Bell-shaped diameter distributions	• Inverse J-shaped diameter distributions
Preferred management system: Group selection system, large gaps	Preferred management system: Single-tree selection system, small gaps

Systems for harvesting forest species should consider the ecological processes of disturbance and recruitment which favour the particular species in order to maintain the biodiversity of the system. If the need arises to harvest other species from the forest, each species should be considered on the basis of its distribution and abundance in the forest, and the site conditions (including disturbance regime) under which it regenerates. Only then can its use be sustained.

Acknowledgements

This study formed part of the activities of the Programme for Ecosystem Management Services, CSIR Division of Forest Science and Technology, Pretoria.

References

Everard, D.A., 1992. On the feasibility of developing a functional classification as a decision support system for the management of the indigenous forests in Natal. Report FOR-DEA 458, Division of Forest Science and Technology, CSIR, Pretoria. 32 pp.

FAO, 1993. The challenge of sustainable forest management: What future for the world's forests? FAO, Rome. 128 pp.

Geldenhuys, C.J., 1990. The present status and future of the southern African evergreen forests. Proceedings XIIth plenary meeting of AETFAT. Mitt. Inst. All. Bot. Hamburg Band 23a, p. 301-320.

Geldenhuys, C.J., 1991. Distribution, size and ownership of forests in the southern Cape. S. Afr. For. J. 158: 51-66.

Geldenhuys, C.J., 1993a. The use of diameter distributions in sustained-use management of forests: examples from southern Africa. In: Piearce, G.D. & D.J. Gumbo, (eds). The ecology and management of indigenous forests in southern Africa. Proceedings of an international symposium, Victoria Falls, Zimbabwe. Zimbabwe Forestry Commission & SAREC, Harare. pp 154-167.

Geldenhuys, C.J., 1993b. Composition and dynamics of plant communities in the southern Cape forests. Report FOR-DEA 612, Division of Forest Science and Technology, CSIR, Pretoria.

Geldenhuys, C.J., 1994. Bergwind fires and the location pattern of forest patches in the southern Cape landscape, South Africa. J. Biogeogr. 21: 49-62.

Geldenhuys, C.J. & W.A. Lübbe, 1991. Sustainability of bark production from the southern Cape forests: possible alternatives. In: Van Warmelo, W. (ed). Proceedings of a workshop: Traditional medicine and plant management in the western Cape, Kirstenbosch, Flora Conservation Committee Report No 91/1. pp 12-16.

Geldenhuys, C.J. & W. Maliepaard, 1983. The causes and sizes of canopy gaps in the southern Cape forests. S. Afr. For. J. 124: 50-55.

Geldenhuys, C.J. & B. Murray, 1993. Floristic and structural composition of Hanglip forest in the Soutpansberg, northern Transvaal. S. Afr. For. J. 165: 9-20.

Geldenhuys, C.J. & C.J. Van der Merwe, 1988. Population structure and growth of the fern *Rumohra adiantiformis* in relation to frond harvesting in the southern Cape forests. S. Afr. J. Bot. 54: 351-362.

Geldenhuys, C.J., C.J. Van der Merwe, & C. Jacobs, 1994. Lightning: a disturbance factor in the mixed evergreen forests of the southern Cape. Report FOR-DEA, Division of Forest Science and Technology, CSIR, Pretoria.

Hill, M.O., 1979a. TWINSPAN - a fortran program for arranging multivariate data in an ordered two-way table by classification of the individuals and attributes. Cornell University, Ithaca.

Hill, M.O., 1979b. DECORANA - a fortran program for detrended correspondence analysis and reciprocal averaging. Cornell University, Ithaca.

McKenzie, B. (ed), 1988. Guidelines for the sustained use of indigenous forests and forest products. Occasional Report No 35, Ecosystem Programmes, Foundation for Research Development, Pretoria. 69 pp.

Midgley, J.J. & M.J. Cameron, 1993. Gap dynamics in the Knysna Forest. Report FOR-DEA 606, Division of Forest Science and Technology, CSIR, Pretoria.

Midgley, J.J., A. Seydack, D. Reynell & D. McKelly, 1990. Fine-grain pattern in southern Cape plateau forests. J. Veget. Sci. 1: 539-546.

Seydack, A.H.W., J.C. Van Daalen, D. Van Dijk, D. Reynell, H. Heyns, A. Jooste, M. Ferguson, W. Pitchford, G. Durrheim & D. Willems, 1990. Yield regulation in the Knysna forests. S. Afr. For. J. 152: 50-61.

How gallery forest turns into savanna: an example from Senegal

A.M. LYKKE
Department of Systematic Botany, Institute of Biology, Aarhus University,
Nordlandsvej 68, DK-8240 Risskov, Denmark.

Abstract
The study was conducted on the vegetation of Delta du Saloum National Park, Senegal. Probably as a consequence of degradation of gallery forests, the density of woody plants was significantly lower in the valley vegetation than in the surrounding savanna. The regeneration density was not significantly different in the two areas. In the valleys 76% of the woody plants were forest species and 51% of the regeneration were forest species, the remainder were savanna species. When the gallery forests degrade, forest species become sparse, leaving the vegetation open for the savanna species to invade. A change seems to take place from gallery forest to transition woodland and finally to savanna.

Résumé
L'étude a été menée sur la végétation du Parc National du Delta du Saloum, Sénégal. La densité des plantes ligneuses dans les bas-fonds, significativement plus faible qu'au niveau des savanes environnantes, apparait comme une conséquence de la dégradation de forêts galeries. La densité de la régénération n'est pas significativement différent dans les deux types de milieu. Dans les bas-fonds, 76% des plantes ligneuses et 51% de la régénération sont des espèces forestières, le reste étant des espèces de savane. Quand les forêts galeries se dégradent, les espèces forestières deviennent clairsemées laissant une végétation ouverte aux espèces de savane qui envahissent le milieu. Un processus de transformation des forêts galeries en forêts claires de transition et finalement en savanes semble être entrain de s'opérer.

Key words
Degradation, Regeneration, Vegetation, Woody plants.

Introduction

In West Africa the woody vegetation of forest and savanna is generally described as being structurally and floristically different, with the boundary between the two being remarkably abrupt and with few plants growing in both areas (Keay 1959, Aubréville 1966, Swaine et al. 1976). The presence of forest species in savanna is therefore an interesting feature (Keay 1959). In the study area the boundary between forest and savanna is blurred structurally as well as floristically; in the valleys savanna species grow among single individuals of forest species and small patches of forest.

This type of vegetation, termed transition woodland and characterized by a mixture of forest and savanna species, has often replaced gallery forest in the study area. Today various stages of degraded gallery forest exist. The spatial differences in the vegetation were used in this study to reveal how the vegetation has changed over time; gallery forest seems to degrade to transition woodland and later to savanna. From knowledge about the former vegetation changes and about the present composition of regeneration, it is possible to indicate how the future vegetation will be. Against this

L.J.G. van der Maesen et al. (eds.), The Biodiversity of African Plants, 323-328.

Figure 1. Transition woodland in a valley of Delta du Saloum National Park. The emergent trees are *Erythrophleum suaveolens* Guill. & Perr., the palm at the right is an *Elaeis guineensis* Jacq.

background it seems that gallery forests and transition woodlands are degrading continuously and that savanna species will take over.

Study area

The study area (13°40'N, 16°30'W) includes about 70 km^2 of Delta du Saloum National Park situated on the west coast of Senegal. Savanna vegetation covers about 95% of the study area, valley vegetation the remaining 5%. The valley vegetation can be described as mainly transition woodland with single standing individuals of forest species, 15-30 m tall, mixed with lower individuals of savanna species and with a ground layer of savanna grasses (Figure 1). Small patches of forest with a closed canopy at about 15 m and with emergent trees, 20-30 m, are found in a few protected areas.

A part of the study area was declared National Park in 1976 the rest in 1986, and it was declared Biosphere Reserve in 1981. The area is subjected to restricted human impact. Precocious fires burn the area every year, and also dry season fires have spread frequently. The average precipitation (1886-1991) was 1109 mm/year with declining tendencies, measured at the nearest meteorological station in Banjul, The Gambia.

Methods

The results were based on data from 128 10×10 m random plots positioned in valley vegetation. All trees, shrubs, and lianas within the plots were identified and counted. Those that were 5 mm in diameter at breast height (dbh) or over were referred to as woody plants (n=340), the smaller ones as regeneration (n=1970). Woody plants with multiple trunks, which were common, were regarded as one individual as long as a connection between trunks was visible. These data were compared to data from 576 similar plots positioned in savanna vegetation.

The term savanna was defined according to Frost et al. (1986), as an area with continuous herbaceous cover dominated by grasses and with a non-continuous woody cover. The term gallery forest was used for forest growing along valleys. The terms forest and transition woodland were defined according to White (1983), in short, areas with trees at least 10 m tall with interlocking crowns, and open areas with a mixture of forest and savanna species, respectively. The term valley vegetation was used for vegetation along valleys ranging from forest to open transition woodland.

Classification of species into forest and savanna species was based on the accounts of Aubréville (1950, 1959), Berhaut (1971-88), and Dugerdil (1970).

For statistics, parametric tests were used as woody plants and regeneration were expected to be normally distributed.

Results

The density of woody plants in valley vegetation (266 individuals/ha) and in savanna vegetation (410 individuals/ha) was significantly different ($t=2.59$, $df=702$, $p=0.01$). Eight of the analyzed plots were situated in a patch of well-preserved gallery forest. The density of woody plants in these plots was 2125 individuals/ha which is significantly different from the other plots in valley vegetation ($t=4.98$, $df=7$, $p=0.001$).

The density of regeneration was not significantly different between valley vegetation (1539 individuals/ha) and savanna vegetation (2177 individuals/ha) ($t=1.74$, $df=163$, $p=0.08$). In the valley vegetation 76% of the woody plants were forest species, the rest was savanna species, whereas regeneration consisted of 51% forest species.

Discussion

It is unknown if the area of the present savanna of Delta du Saloum National Park has recently been covered by forest as described from similar areas (Aubréville 1949, Keay 1959, White 1983). The gallery forests, however, have been wider in extent. That the valleys have earlier supported continuous strips of gallery forest is evident from the remaining forest species in the valley vegetation today. Many of the forest species have tall unbranched trunks indicating that they have been growing in forest formations, and some of the forest species, e.g., *Elaeis guineensis* Jacq., hardly regenerate in open vegetation and therefore also reflect earlier forest formations (Keay 1959). That gallery forests have been wider and denser before is confirmed by local people and by aerial photographs from 1969. Today only small patches of gallery forest are left, and many dead trunks lying in the valleys reveal that degradation is still going on. The degradation is probably caused by frequent fires in co-occurrence with human impact and declining precipitation. Exploitation and fire are often thought to be main causes of forest degradation (Keay 1959, Aubréville 1966, Lawton 1967).

Today the valley vegetation of Delta du Saloum is dominated by transition woodland and the woody plant density is lower in valley vegetation than in savanna vegetation. This confirms that a degradation takes place as the only well-preserved area of gallery forest has a significantly higher density than the rest of the valley vegetation. Strips of transition woodland have often been described where forest and savanna adjoin (Keay 1959, Asare 1962, Hopkins 1965, Swaine et al. 1976). In the study area, however, the transition woodland is enclosed by savanna in most places. Transition woodland is an unstable vegetation type caused either by savanna reverting to forest or by forest degrading to savanna (White 1983). The process of forest replacing savanna has been described from the border between the two in more humid places in West Africa (Miège 1966, Dugerdil 1970, Gautier 1990). In the Delta du Saloum National Park, however, transition woodland reflects degraded gallery forest because of the presence of many old forest trees, as described, and because often no adjoining forest is found. Savanna encroaching upon the forest in spite of forest reserves is also described from Ghana (Asare 1962).

It is of interest to look at the regeneration to preview what will replace the degrading gallery forests. In the valleys the density of the regeneration, contrary to the one of woody plants, is similar to the density in the savanna. Anticipating that the survival of regeneration of forest and savanna species is similar in the valley vegetation, which is open in most places, the density of woody plants in the valleys is in time likely to approach that of the savanna.

The transition woodlands seem to become more similar to savannas as regards both density, as described, and species composition. Savanna species are expected to increase in the valleys as 24% of the woody plants are savanna species and 49% of the regeneration are savanna species. It is anticipated that regeneration of savanna species does not have a higher mortality than that of forest species. In an area with high fire impact the opposite might well be the case. Finally, more savanna species are therefore expected in the areas of former gallery forest.

The process of degradation can be described on basis of the present vegetation formations. The gallery forests degrade to transition woodlands (Figure 2a, b) with a low woody plant density. As the woody plants most exposed to or most sensitive to fire and other forms of impact die, the former forest becomes open exposing the remaining individuals to increased impact. A positive feedback loop occurs. Concerning the water budget another positive feedback loop takes place. As the gallery forests conserve water (Lawton 1967), the soil gets drier when they degrade, and with only little forest left to conserve humidity, the pressure on the remaining forest species is increased. A continuation of the process of depletion of forest species and deterioration of the conditions for forest is expected to finally lead to the development of a savanna vegetation (Figure 2c).

Figure 2. A sequence of vegetation stages visualised by three vegetation types that today are found in the valleys of Delta du Saloum National Park: gallery forest, transition woodland and savanna. a, savanna on the plateau and gallery forest in the valley, b, savanna on the plateau and transition woodland in the valley, c, savanna extended into the valley.

This process is unlikely to be merely a fluctuation. Under the described conditions, with high fire impact, declining precipitation, and so extensive degradation that only small patches of gallery forest are left, it is unlikely that the gallery forests will restore naturally to their earlier extension even under protection.

Conclusion

When gallery forests degrade, transition woodlands are formed, they have a mixture of forest and savanna species and a relatively low density of woody plants. This open vegetation, which is exposed to fire impact, facilitates invasion of savanna species which increase the density again. In this way savanna species seem to invade areas of former gallery forest. As the tendency is stronger for regeneration than for woody plants, this invasion is likely to continue in the future.

Acknowledgements
Thanks to Lis Ellemann, Jens Madsen, and Bienvenu Sambou for useful comments. The study was financed by Danida (grant no. 104.Dan.8/576).

References

Asare, E.O., 1962. A note on the vegetation of the transition zone of the Tain Basin in Ghana. Ghana J. Sci. 2: 60-73.

Aubréville, A., 1949. Climats, forêts et désertifications de l'Afrique tropicale. Société d'Editions Géographiques, Maritimes et Coloniales, Paris.

Aubréville, A., 1950. Flore forestière soudano-guinéenne. Société d'Editions Géographiques, Maritimes et Coloniales, Paris.

Aubréville, A., 1959. La flore forestière de la Côte d'Ivoire. Vol. 1-3. Centre Technique Forestière Tropical, Nogent-sur-Marne.

Aubréville, A., 1966. Les lisières forêt-savane des régions tropicales. Adansonia 6: 175-187.

Berhaut, J., 1971-88. Flore illustrée du Sénégal. Vol. 1-6 & 9. Clairafrique, Dakar.

Dugerdil, M., 1970. Recherches sur le contact forêt-savane en Côte d'Ivoire. I. Quelques aspects de la végétation et de son évolution en savane préforestière. Candollea 25: 11-19.

Frost, P., E. Medina, J.-C. Menaut, Solbrig, M. Swift, & B. Walker, 1986. Responses of savannas to stress and disturbance, a proposal for a collaborative programme of research. Biol. Int. special issue 10.

Gautier, L., 1990. Contact forêt-savane en Côte d'Ivoire centrale: évolution du recouvrement ligneux des savanes de la Réserve de Lamto (sud du V-Baoulé). Candollea 45: 627-641.

Hopkins, B., 1965. Forest and savanna. Heinemann, London.

Keay, R.W.J., 1959. Derived savanna — derived from what? Bull. Inst. Fondam. Afrique Noire sér. A. Sci. Nat. 21: 427-438.

Lawton, R.M., 1967. The conservation and management of the riparian evergreen forests of Zambia. Commonw. Forest Rev. 46: 226-232.

Miège, J., 1966. Observations sur les fluctuations des limites savanes-forêts en basse Cote d'Ivoire. Ann. Fac. Sci. Univ. Dakar 19: 149-166.

Swaine, M.D., J.B. Hall, & J.M. Lock, 1976. The forest-savanna boundary in west-central Ghana. Ghana J. Sci. 16: 35-52.

White, F., 1983. The vegetation of Africa, a descriptive memoir to accompany the Unesco/AETFAT/UNSO vegetation map of Africa. Unesco, Paris.

Molecular taxonomy of *Saintpaulia rupicola* and the genus *Saintpaulia*

S. MACHARIA [1], T.R. PEARCE [2], B. BYTEBIER [2] & R. AMAN [3]
[1] Botany Department, University of Nairobi, P.O. Box 30197, Nairobi, Kenya.
[2] Plant Conservation and Propagation Unit, East African Herbarium, Nat. Museums of Kenya.
[3] Department of Molecular Genetics, Nat. Museums of Kenya, P.O. Box 40658, Nairobi, Kenya.

Abstract
African Violets, genus *Saintpaulia*, (Gesneriaceae), are endemic to coastal Kenya and northern Tanzania. The taxonomic status of the wild species of the African Violet remains unresolved. In addition, there is an almost complete lack of knowledge about the genetic variation among natural populations of the African Violets. We have embarked on a project to assess the chloroplast DNA variation among these African Violets, with special emphasis on the only four remaining natural populations of *Saintpaulia rupicola* B.L. Burtt and *Saintpaulia* sp. nov. from Coastal Kenya.

Key words
Chloroplast DNA, Phylogeny, Restriction site analysis, *Saintpaulia*, *Saintpaulia rupicola*.

Introduction

African Violets are highly endangered. The continued removal of forest cover results in a harsher microclimate than the plants are able to withstand. Quarrying and over-collection from their natural habitats by professional and amateur horticulturalists has contributed to the demise of the natural populations of the species *Saintpaulia rupicola* B.L. Burtt. It occurs in four populations on small limestone outcrops around the Mombasa area in Kenya. Certain morphological characters in this species vary so much between populations that one of the populations has recently been described as a distinct species new to science. It is likely that the only sustainable form of conservation will involve considerable *ex-situ* collection of germplasm and the subsequent crossing of plants in the nursery to bulk population numbers for eventual re-introduction to safe sites either in protected original sites or completely new compatible but protected areas.

In order to prevent gross loss of variation specific to particular populations of *S. rupicola* or other *Saintpaulia* species, a thorough analysis to assess the degree of molecular variation among these populations of *S. rupicola*, as compared to other *Saintpaulia* species, will allow us to make a more considered approach to the degree to which the between population crossing should be carried out.

As an initial approach, we have embarked on a project to assess chloroplast DNA restriction-site variation among these populations of *Saintpaulia rupicola,* including *Saintpaulia* sp. nov., together with a selection of other *Saintpaulia* species, using *Streptocarpus glandulosissimus* Engl. (Gesneriaceae), and *Nicotiana tabacum* L. (Solanaceae) as outgroup genomes for comparison. We hope to contribute to the understanding of the taxonomic delimitation of *S. rupicola* and the genus *Saintpaulia*, in order to contribute to the development of a sound species conservation plan for these species.

L.J.G. van der Maesen et al. (eds.), The Biodiversity of African Plants, 329-330.

Chloroplast DNA offers the best tool for between species phylogenetic studies. Although rare, within species chloroplast DNA (cpDNA) differences, when found, have been quite informative, particularly regarding the origins of several crop plants (Palmer et al. 1988).

Since chloroplast DNA occurs in several thousand copies in the typical leaf cell (Bendich 1987), it is easy to visualize and map cpDNA fragments by hybridizing cloned cpDNA to filter blots containing restriction enzyme digests of total DNA.

Materials and methods

Total DNA extractions, total DNA restriction digests, separation of these digests by agarose gel electrophoresis, double transfer to DNA hybridization membranes, hybridization of tobacco chloroplast DNA to these Southern blots and autoradiography will be done as described by Palmer et al. (1988).

The radioactive labelling of the tobacco chloroplast DNA fragments with ^{32}P will be done by the random priming method described by Maniatis et al. (1989). This chloroplast DNA has been cloned in forty fragments in several plasmids.

References

Bendich, A.J., 1987. Why do chloroplasts and mitochondria contain so many copies of their genome? Bioessays 6: 279-282.

Maniatis T., E.F. Fritsch & J. Sambrook, 1989. Molecular Cloning, a laboratory manual. Cold Spring Harbour Laboratory Press, pp 10.13-10.18.

Palmer, J.D., R.K. Jansen, H.J. Michaels, M.W. Chase & J.R. Manhart, 1988. Chloroplast DNA variation and plant phylogeny. Ann. Missouri Bot. Gard. 75: 1180-1206.

Similarity of nickel distribution in leaf tissue of two distantly related hyperaccumulating species

J. MESJASZ-PRZYBYLOWICZ [1], K. BALKWILL [1], W.J. PRZYBYLOWICZ [2,3], H.J. ANNEGARN [2] & D.B.K. RAMA [4]

[1] C.E. Moss Herbarium, Department of Botany, University of the Witwatersrand, Private Bag 3, WITS 2050, South Africa.

[2] Schonland Research Centre for Nuclear Sciences, University of the Witwatersrand, 2050 Johannesburg, South Africa.

[3] Van de Graaff Group, National Accelerator Centre, P.O. Box 72, Faure 7131, South Africa. On leave from the Academy of Mining and Metallurgy, 30-059 Cracow, Poland.

[4] National Centre for Occupational Health, P.O. Box 4788, Johannesburg 2000, South Africa.

Abstract

Senecio coronatus (Thunb.) Harv. and *Berkheya zeyheri* (Sond. & Harv.) Oliv. & Hiern subsp. *rehmannii* (Thell.) Roessler var. *rogersiana* (Thell.) Roessler (Compositae), two hyperaccumulators from South Africa, are investigated. ACP-AES results show that the level of nickel in both species is 5-6 times higher than the total amount of metal in the associated soil and 50-150 times higher than the plant available Ni in the soil. An elemental, spatial map showed, that the highest nickel concentration was in the outer layer of the leaf in both species.

Key words

Berkheya zeyheri, Hyperaccumulation, Nickel, *Senecio coronatus*, Serpentine.

Introduction

Hyperaccumulation is an unusual plant response to soils enriched with heavy metals. Nickel accumulation from serpentine soil is the most studied and most spectacular example, characterised by uptake to a level exceeding 1000 μg/g (0.1%) of dry mass (Brooks et al. 1977). Despite this the process of hyperaccumulation is poorly understood. The morphological characters of serpentine flora have been defined by Novak (Novak 1928) as serpentinomorphoses. These include xeromorphic foliage, reduction in size and increased root system (Kruckerberg 1984, Proctor & Woodell 1975). Two of the nickel hyperaccumulators from South Africa *Berkheya zeyheri* (Sond. & Harv.) Oliv. & Hiern subsp. *rehmannii* (Thell.) Roessler var. *rogersiana* (Thell.) Roessler and *Senecio coronatus* (Thunb.) Harv., both Compositae, (Mesjasz-Przybylowicz et al. 1994, Morrey et al. 1992) demonstrate xeromorphic adaptation.

The aim of this study was to compare nickel distribution in leaf tissue of two species with histological examination of the investigated leaves. Baker (1987) suggested the use of "more sophisticated analytical technology, such as X-ray microprobe analysis", for better understanding of the mechanisms of metal tolerance. In this study the scanning proton microprobe was used.

Material and methods

Plants and associated soil were collected from serpentine outcrops in the Barberton area (south-eastern Transvaal, South Africa), Kaapsehoop Asbestos Mine (*S. coronatus*) and Diepgezet in Songimvelo Game Reserve (*B. zeyheri*). Ni-sensitive dimethyl-

L.J.G. van der Maesen et al. (eds.), The Biodiversity of African Plants, 331-335.

Table 1. Maximum nickel concentrations in leaf material and total and available nickel in associated soil (μg/g dry weight).

Species	*S. coronatus*	*B. zeyheri*
Ni in leaf *	17252	5849
Total Ni in soil **	3400	936
Available Ni in soil	117	101
% of Ni in soil that is available	3.44	10.79
Ratio of Ni in leaf to total Ni in soil	5.07	6.25
Ratio of Ni in leaf to available Ni in soil	147.45	57.91

* leaf analysis done by ICP-AES

** soil analysis done by Institute for Soil, Climate and Water, Agricultural Research Council, Pretoria.

glyoxime indicator paper was used to identify plants in the field, prior to collection. Leaves were carefully and quickly washed in deionized, distilled water to remove soil and dust from the surface. Leaf specimens were excised with a sharp razor blade and part was immediately freeze-dried. The rest was fixed in formalin acetoalcohol (FAA), embedded in wax, stained (safranin and fast green) using conventional procedure (Johansen 1940) and examined by light microscopy.

Soil samples were collected from the rooting zone of the plants investigated. All soil analyses were done by the Institute for Soil, Climate and Water, Agriculture Research Council in Pretoria using methods described by The Non-affiliated Soil Analysis Work Committee (1990).

Rapid qualitative energy dispersive X-ray fluorescence analyses were performed to select suitable material for further scanning proton microprobe investigations. About 0.1 g of dried plant material was freeze-dried and acid digested, using nitric acid. The solutions were analyzed by ICP-AES. All details of measurements followed those of Yousefi & Rama (1992). Studies of Ni distributions were performed on bulk freeze-dried samples by scanning proton microprobe at the Schonland Research Centre for Nuclear Sciences at the University of the Witwatersrand. Cross-sections of leaves were mounted on aluminium target frames and coated with a thin layer of carbon. Because of their larger size the cross-sections of *S. coronatus* could be glued directly to the target frames at their margins. The smaller cross-sections of *B. zeyheri* were mounted on 1.5% formvar coated frames. Scanning areas were 2.5 x 2.5 mm for *S. coronatus* and 1 x 1 mm for *B. zeyheri*. More detailed information on the experimental conditions during microprobe radiation is reported elsewhere (Mesjasz-Przybylowicz et al. 1994).

Results and Discussion

The nickel concentrations in leaves (Table 1) confirm the results of previous investigations, that both species are nickel hyperaccumulators. Jaffré & Schmid (1974) used the term "hypernickéliphor" for plants containing more than 10,000 μg/g (1.0%) of dry mass, which is true for *S. coronatus*. The amount of Ni in the leaves exceeds the concentrations of plant available metal in the soil by 50-150x and total Ni in soil by 5-6x. Plant available nickel is 3.4% and 10.8% of the total nickel in the soil. The ratios of metal concentrations in leaf to total amount of Ni in soil is 5.07 and 6.25, which is

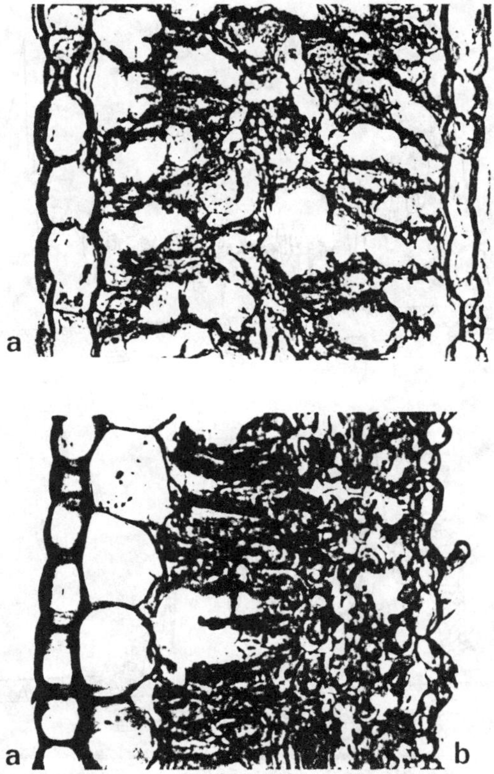

Figure 1. Light micrographs of *Senecio coronatus* (55x) and *Berkheya zeyheri* (67.5x). a. the upper epidermis, b. the lower epidermis.

within the range obtained by Farago & Mahmoud (1984) for *Hybanthus floribundus* (Lindl.) F. Muell (8.4 and 1.9). This means, that Ni is transported to the leaves and accumulated there after uptake from the soil.

Both species demonstrate xeromorphic adaptations, but they are different. *S. coronatus* has glaucous, succulent leaves with a thick, single layer of epidermis on both sides and uniform mesophyll (Figure 1A). *B. zeyheri* has dense hairs on the lower surface and a hypodermis on the upper (Cutler 1978). The mesophyll is differentiated into palisade and spongy parenchyma (Figure 1B) (Cutter 1971, Esau 1963).

Nickel distribution in leaves is not homogeneous (Figure 2): metal is more concentrated close to the surface on both sides of the leaf in both species. The highest amount of Ni is found in the epidermis. In *B. zeyheri*, where the lower epidermis is thin, there is a high level of Ni in the spongy parenchyma close to the epidermis. Besides the two reports on the distribution of Ni in leaves of *Hybanthus floribundus* (Farago & Mahmoud 1984, Severne 1974) there is no previous information about the distribution of Ni in plant tissue. Electron microprobe studies of *H. floribundus* species showed no evidence for the accumulation of nickel in any specific plant tissue (Kelly et al. 1975).

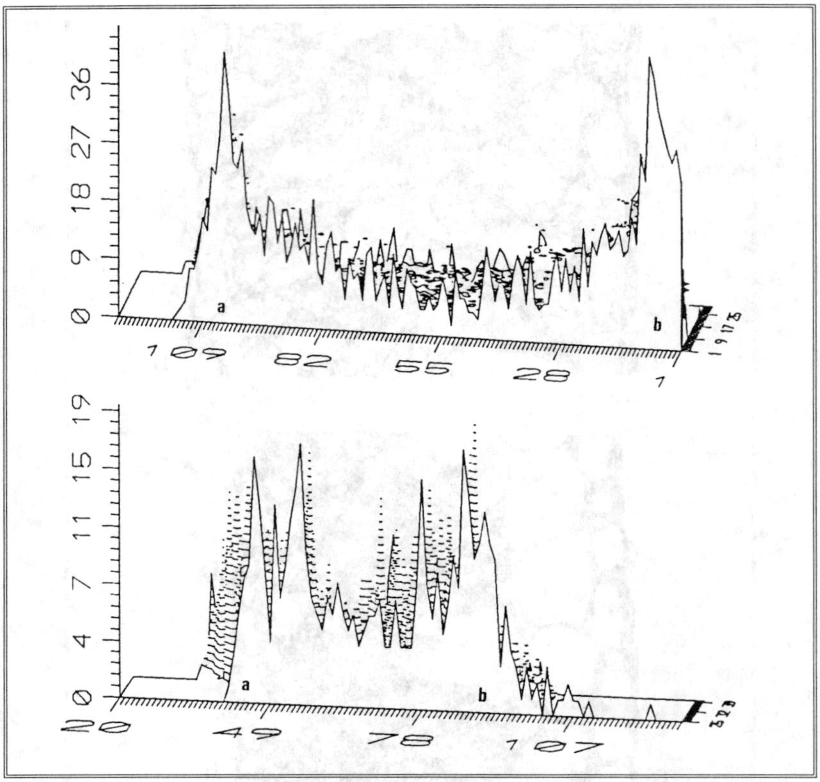

Figure 2. Leaves cross-section with corresponding Ni distribution spatial maps for *Senecio coronatus* (above) and *Berkheya zeyheri* (below). Axis Y $Ni_{K\alpha}$ X-ray counts; X, Z position axes - 2.5 mm (X) transverse section for *S. coronatus* (above) and 1.0 mm (X) for *B. zeyheri* (below); a. the upper epidermis, b. the lower epidermis.

Conclusion

The results obtained clearly show that nickel taken up from soil is accumulated in the leaf with the highest concentrations in the outer layers. Although the two hyperaccumulators are only distantly related, they behave in a similar manner despite differences in the type of tissue at the outer surface of the leaves. It is possible that this metal distribution could help towards xeromorphic adaptation by increasing the water potential in the leaves. It has been hypothesised that nickel hyperaccumulation functions in this way (Baker & Walker 1990, Boyd & Martens 1992, Severne 1974). Further investigations at the cellular level and focused on other hyperaccumulators will give more detail about Ni distributions and thus aid better comprehension of the detoxification mechanism occurring in plants.

Hyperaccumulators should not be seen as mere curiosities. Not only do they illustrate an unusual plant response to elevated soil concentration of metals, but the ability to tolerate high Ni levels may have useful environmental and economic implications and practical uses in the future.

Acknowledgements
The authors are grateful to the University of the Witwatersrand for the provision of a Visiting Research Fellowship and research funds. We also thank E. Storbeck for preparing the spatial maps of Ni distribution.

References

Baker, A.J.M., 1987. Metal tolerance. New Phytol. 106 (Suppl.): 93-111.

Baker, A.J.M. & P.L. Walker, 1990. Ecophysiology of metal uptake by tolerant plants. In: A.J. Shaw (ed.), Heavy Metal Tolerance in Plants: Evolutionary Aspects: 155-157, CRC Press Inc., Boca Raton, Florida.

Boyd, R.S. & S.N. Martens, 1992. The Raison d'Etre for Metal Hyperaccumulation by Plants. In: A.J.M. Baker, J. Proctor & R.D. Reeves (eds.), The Vegetation of Ultramafic (Serpentine) Soils: 279-289, Intercept Andover, Hampshire.

Brooks, R.R., J. Lee, R.D. Reeves & T. Jaffré, 1977. Detection of nickeliferous rocks by analysis of herbarium specimens of indicator plants. J. Geochem. Explor. 7: 49-57.

Cutler, D.F., 1978. Applied plant anatomy. Longman, London, New York.

Cutter, E.G., 1971. Plant Anatomy, Experiment and Interpretation, Part 2, Organs. Edward Arnold, London.

Esau, K., 1963. Plant Anatomy. John Wiley & Sons, Inc., New York, London.

Farago, M.E. & I.E.D.A. Mahmoud, 1984. Plants that accumulate metals (Part vi): Further studies of an Australian nickel accumulating plant. Min. and Environ. 5: 113-121.

Jaffré, T. & M. Schmid, 1974. Accumulation du nickel par une Rubiaceé du Nouvelle Calédonie: *Psychotria douarrei* (G. Beauvisage) Däniker. C.R. Acad. Sci. Paris. Sér. D. 278: 1727-1730.

Johansen, D.A., 1940. Plant Microtechnique. McCraw-Hill, New York, London.

Kelly, P.C., R.R. Brooks & S. Dilli, 1975. Preliminary observations on the ecology and plant chemistry of some nickel-accumulating plants from New Caledonia. Proc. R. Soc. Lond. B. 189: 69-80.

Kruckerberg, A.R., 1984. California serpentines. University of California Press, Berkeley.

Mesjasz-Przybylowicz, J., K. Balkwill, W.J. Przybylowicz & H. Annegarn, 1994. Proton microprobe and X-ray investigations of nickel distribution in serpentine flora from South Africa. Nucl. Instr. and Meth. B 89: 208-212.

Morrey, D.R., K. Balkwill, M-J. Balkwill & S. Wiliamson, 1992. A review of some studies of the serpentine flora of southern Africa. In: A.J.M Baker, J. Proctor & R.D. Reeves (eds.), The vegetation of ultramafic (serpentine) soil: 147-157, Intercept, Andover, Hampshire.

Novak, F.A., 1928. Quelques remarques relatives au problème de la végétation sur les terrains serpentiniques. Preslia 6: 42-71.

Proctor, J., & S.R.J. Woodell, 1975. The ecology of serpentine soils. Adv. Ecol. Res. 9: 255-366.

Severne, B.C., 1974. Nickel accumulation by *Hybanthus floribundus*. Nature 248: 807-808.

The Non-affiliated Soil Analysis Work Committee, 1990. Handbook of Standard Soil Testing Methods for Advisory Purposes. Soil Society of South Africa, Pretoria.

Yousefi, V. & D.B.K. Rama, 1992. Monitoring of air for microbial and metal contamination at selected sites in the vicinity of Johannesburg, South Africa. The Science of the Total Environment 166: 159-167.

Implications of changes in vegetation in Pandero, Murchison Falls Park, Uganda

M. ONEKA
Wageningen Agricultural University, Department of Terrestrial Ecology and Nature Conservation, Postbus 8080, 6700 ER Wageningen, The Netherlands.

Abstract
This report examines changes in the vegetation cover of Pandero over a period of nearly 30 years. The vegetation of the area in 1956 has been observed from aerial photographs. Previous reports provided relevant basic information on the past vegetation physiognomy and floristic composition. These have been compared with 1985 field records. Although the woody vegetation has declined by almost 80%, density of trees in remaining woodland has not changed significantly. Persistence of woody plants of only two size categories (shrubs and trees) in the remnant woodland have helped to retain the open orchard-like structure. This vegetation structure, dominated by *Terminalia glaucescens* Planch. ex Benth. and *Prosopis africana* Taub., is an important aspect of the aesthetic and ecological integrity of the park landscape. Reduction in numbers of the larger herbivores has improved prospects for tree recruitment. In the presence of floristically more diverse shrubs this will inevitably lead to changes in the woodland structure. These changes in the vegetation present an appropriate test case for the discussion of what should be specified in the park vegetation management policies and objectives.

Key words
Conservation strategy, Murchison Falls Park, *Terminalia*, Uganda.

General background

The creation of Murchison Falls Park in 1952 in north-western Uganda (1°50'-2°35'N and 31°22'-31°16'E) provided sanctuary to the wildlife displaced and harassed by the growing human population. The influx, confinement and natural increase in herbivore numbers lead to more trees being killed when fire prevailed. Decline of woody vegetation subsequent to this process has resulted in significant habitat and landscape changes in the park (Buechner & Dawkins 1961, Laws et al. 1975). Since the 1970s poaching increased and has drastically reduced the herbivore populations (Eltringham & Malpas 1980). This has enabled increased regeneration of woody plants in the park. The park management attaches great value to preserving indigenous vegetation. Looking at the rather drastic changes that have occurred in the park vegetation since the areas were gazetted provide an appropriate opportunity to ponder about future directions for the park.

Vegetation is considered here as more than the green mantle over the park landscape. The spatial and temporal variations in the vegetation help determine the availability of food, shelter and general context for life forms besides plants. Together with the abiotic and the other biotic elements, vegetation defines the aesthetic and ecological characteristics which symbolise the integrity of the park's landscapes. In which way the park's integrity is evaluated depends on which of its state is used as reference standard. Such a reference standard could be a state in some specific time in the past, the present state or some imagined state. Through appropriate ecological,

L.J.G. van der Maesen et al. (eds.), The Biodiversity of African Plants, 336-340.

technical and logistical capabilities park management should attempt to reach this standard. In other words, unless these standards are defined evaluation of the management programmes is likely to remain arbitrary. The lack of such standards can be detected in the details of relevant aspects of the park development programmes. The loss of woody vegetation was a negative development for Murchison Falls Park where woody vegetation has not been extensive. The most extensive element of the woody vegetation was *Terminalia* woodland, a patch of which remains in Pandero.

Terminalia woodland

Terminalia woodland is a deciduous tree stand up to 18 m high with an open or closed canopy cover (20-50%) and dominated by *Terminalia* almost in pure stand (Langdale-Brown et al. 1964, Pratt et al. 1966). The term "woodland" as used here equates the "forêt claire" of the Yangambi classification. The ground cover is dominated by herbaceous plants which explains the regular incidence of fire in the vegetation. The woodlands in Murchison Falls Park is dominated by *Terminalia glaucescens* Planch. ex. Benth. (synonyms *T. baimanii* Engl. & Diels., *T. longipes* Engl., *T. passargei* Engl., *T. togoensis* Engl. & Diels.).

Terminalia is a pan-tropical genus in the Combretaceae, with up to 200 tree and shrub species. It is found in rain forests, woodlands, grasslands and shrublands. *T. glaucescens* is found in Africa north of the equator on well-drained soils, between 800-1800 m altitude, in areas with mean annual rainfall of 760-1500 mm and with 2-4 dry months (White 1983). It has been recorded in western Africa from French Guinea to Cameroon and Zaire (Geerling 1982, Hutchinson & Dalziel 1954, Keay 1959), in eastern Africa from Sudan and Ethiopia to Uganda and Tanzania (White 1983). In East Africa it was suggested by Lind & Morrison (1974) that the species was only common in Uganda. According to Eggeling & Dale (1951) *T. glaucescens* is found over about half the country.

The woodland is a plant association endemic to Africa and is very sparingly spread in Uganda (Langdale-Brown et al. 1964, Osmaston 1968) and elsewhere (Keay 1959, Lind & Morrison 1974, White 1983). The woodlands in Murchison Falls Park were mainly plant associations dominated by *T. glaucescens* and *Prosopis africana* Taub. (Buechner & Dawkins 1961). These communities were regarded by the park authorities as important components of the park landscape and habitat diversity (Buechner & Dawkins 1961, Laws et al. 1975). The restoration and the perpetuation of this community is thus an important objective of the park management.

Although *T. glaucescens* occurs over a wide geographical range, *Terminalia* woodland was sparingly dispersed, often as patches in grasslands or at forest fringes (Langdale-Brown et al. 1964, Lind & Morrison 1974, Osmaston 1968). Dawkins (1949) suggested that *Terminalia* woodland was a stage in succession to deciduous forest communities. In accord with Eggeling (1947), he proposed that this succession was from grassland, through *Terminalia* woodland, *Phyllanthus-Terminalia-Albizia* and then forest. In successive stages fire is less severe resulting in simultaneous decrease in fire-adapted species and increase in fire-prone species. Langdale-Brown et al. (1964) have shown that *Terminalia* woodland could be a climax in its own right or a climax induced by fire. All the above mentioned *Terminalia* successional stages have been recorded in parts of Uganda but *Terminalia* woodland is only present in the north where it occurs also in the Murchison Falls Park (Langdale-Brown et al. 1964, Osmaston 1968).

Outside the park, remaining woodlands have dwindled under pressure for farm land. Also, the growing need for charcoal, firewood and building material have resulted in rapid degeneration. The tobacco (*Nicotiana tabacum* L.) growers in northern Uganda relied on such natural resources for up to 60% of their firewood requirements in flue-curing tobacco leaves (pers. obs.). Tree cutting was selective. In the Lango districts, up to 50% of the firewood used consisted of *T. glaucescens*. The species is highly valued for charcoal in northern Uganda. *Terminalia* woodland has thus declined both inside and outside protected areas and it is essential that examples of it can and should be restored and preserved within the park.

Pandero woodland

Pandero is located in the Murchison Falls Park north of the Murchison Falls. It undulates at about 800 m altitude. The soils are shallow, sandy-loam, acidic and low in nutrients especially phosphates (Ollier 1959). Most of the mean annual rainfall of c. 1000 mm falls during April-May and August-November. The mean daily temperatures range from 18°C (minimum) to 32°C (maximum). Some of the herbivores in Pandero are the elephant (*Loxodonta africana* Blumenbach), the buffalo *Syncerus cafer* Sparrman and the hartebeest *Alcelaphus buselaphus* Pallas (Brooks & Buss 1962, Edroma 1986). Pandero has a woodland (tree cover about 20%) surrounded by a tall grassland (tree cover <10%). The woodland in Pandero is remnant of a more extensive *Terminalia* woodland vegetation.

Past and present vegetation structure and floristic composition of the woodland in Pandero indicated that despite about 80% decline in surface area, its tree density has not changed. The floristic composition has, however, changed with very significant increases in *T. glaucescens*. Although the density and diversity of shrubs have increased very significantly, the tree population structure indicates that the above-ground development of the shrubs was being suppressed. Of the two main factors responsible for the decline in woody vegetation in the area, fire has been and remains the least understood (Buechner & Dawkins 1961, Laws et al. 1975). Studies elsewhere (Hopkins 1965, Sabiiti & Wein 1988) showed the value of such insight for better fire management in ecosystems. Study plots initiated in the park have shown that the management of fire and the larger herbivores are vital means to influence the vegetation structure and floristic composition (Lock 1977, Oneka 1991, Smart et al. 1985, Spence & Angus 1971, Wheater 1972).

Discussion and recommendations

The mentioned landscape changes are considered harmful to the park management goals and so needed to be understood, brought under control and monitored. Where should priority of vegetation management be put at this stage: in preserving diversity of the plant species or the diversity of life forms? Or should the emphasis be put on the relative stability of the structures? These issues point to the need for reference standards to evaluate the effectiveness and appropriateness of the conservation programmes.

The aspects physiognomic structure and floristic composition are linked to each other. Giving priority to one of the aspects should not necessary be at the cost of the other. The floristic composition of a plant community influences its physiognomic structure. On the other hand prevailing physiognomic structure also influences the floristic composition of the community. In the Pandero case, the predominance of two species (*T. glaucescens* and *P. africana*) have resulted in the peculiar open structure. The floristically more diverse shrub layer, however, warns for possible changes in the composition of a future woodland in the area. The woodland with a more diverse tree composition is unlikely to remain an open structure, a salient feature of the former *Terminalia* landscape.

Plant species tend to be widespread but plant associations which they characterise are less common. As noted above, *Terminalia* species are widespread but associations in which they are prominent are not common. Through selective destructive pressures, significant changes occur in the floristic composition and physiognomic structure of the vegetation. Under such pressure plant species could be locally exterminated. In the Pandero case the preservation of *P. africana* could be severely affected in this way. The species is not widespread and its regeneration in the park is subdued. Outside the park both selective and indiscriminate cutting are eliminating most specimens. This issue highlights the need to identify as early as possible plant species that may need to be accorded special attention in developing monitoring, research and management programmes.

In emphasizing the preservation of plant associations the likelihood of succession highlights the need to decide *à priori* whether to aim at preserving the association even though it might be a succession stage. The *Terminalia* woodland in Pandero is probably a stage in succession to a more closed woodland. It is likely however that prevailing fire and herbivore factors could maintain the present open structure. If there are significant changes in the main environmental factors, fire and the herbivores, the vegetation would change. Analysis and the monitoring of the implications (management economics, aesthetic and ecological concerns) of vegetation management alternatives could help update the development policies, objectives and strategies.

These issues highlight the need for sound understanding of the biogeography, autecology and the synecology of the plants as a basis for their conservation. In the absence of such an understanding it is, therefore, advisable that management-oriented research and monitoring are developed as integral parts of the plant conservation programmes as early as possible. The research and monitoring programmes could help develop appropriate conservation policies and strategies to meet conservation needs.

Acknowledgements

Thanks to P. Ketner for commenting on an earlier draft. Costs of preparing the manuscript and poster were paid for by the Wageningen Agricultural University (Department of Terrestrial Ecology and Nature Management).

References

Brooks, A.C. & I.O. Buss, 1962. Past and present status of the elephant in Uganda. J. Wildl. Mgmt. 26: 38-50.

Buechner, H.K. & H.C. Dawkins, 1961. Vegetation changes induced by elephant and fire in Murchison Falls National Park, Uganda. Ecology 42: 752-766.

Dawkins, H.C., 1949. Timber planting in the *Terminalia* woodland of northern Uganda. Emp. For. Rev. 28: 226-247.

Edroma, E.L., 1986. Drastic decline in numbers of animals in Uganda National Parks. Rangelands: a resource under siege. Proceedings 2nd Intern. Rangelands Congress, Adelaide, Australia, 13-18 May 1984. Australian Academy of Sciences, 1986: 546-548.

Eggeling, W.J., 1947. Observations on the ecology of the Budongo rain forest. Uganda J. Ecol. 34: 20-87.

Eggeling, W.J. & I.R. Dale, 1951. The indigenous trees of the Uganda Protectorate. University Press, Glasgow.

Eltringham, S.K. & R.C. Malpas, 1980. The decline in elephant numbers in Rwenzori and Kabalega Falls National Parks, Uganda. Afr. J. Ecol. 18: 73-86.

Geerling, C., 1982. Guide de terrain des ligneux sahéliens et soudano-guinéens. Meded. Landbouwhogeschool Wageningen 82-3.

Hopkins, B., 1965. Observation on savanna burning in the Olokomeji Forest Reserve, Nigeria. J. Appl. Ecol. 2: 367-381.

Hutchinson, J. & J.M. Dalziel, 1954. The flora of West Tropical Africa (2nd ed.) vol. 1, part 1: 227-229. Crown Agents, London.

Keay, R.W.J., 1959. The vegetation map of Africa south of the Tropic of Capricorn. Oxford University Press, London.

Langdale-Brown, I., H.A. Osmaston & J.G. Wilson, 1964. The vegetation of Uganda and its bearing on land-use. Government Printer, Entebbe.

Laws, R.M., I.S.C. Parker & R.C.B. Johnstone, 1975. Elephants and their habitats: the ecology of elephants in North Bunyoro, Uganda. Clarendon Press, Oxford.

Lind, E.M. & M.E.S. Morrison, 1974. East African Vegetation. Longman, London.

Lock, J.M., 1977. Preliminary results from fire and elephant exclusion plots in Kabalega Falls National Park, Uganda. East Afr. Wildl. J. 15: 229-232.

Ollier, C.D., 1959. The soils of Northern Province of Uganda excluding Karamoja. Memoirs of Research Division of Department of Agriculture, Uganda. Series 1, number 3.

Oneka, M., 1991. Vegetation changes induced by exclusion of fire and large herbivores in Murchison Falls National Park, Uganda. In: F.I.B. Kayanja & E.L. Edroma (eds). African Wildlife: Research & Management, UNESCO\ICSU Press, Paris.

Osmaston, H.A., 1968. Conservation of vegetation in Africa south of the Sahara: Uganda. Acta Phytogeogr. Suec. 54: 148-151.

Pratt, D.J., P.J. Greenway & M.D. Gwynne, 1966. A classification of East African Rangelands, with an appendix on terminology. J. Appl. Ecol. 3: 369-382.

Sabiiti, E.N. & R.W. Wein, 1988. Fire behaviour and the invasion of *Acacia sieberiana* into savanna grassland openings. Afr. J. Ecol. 26: 301-313.

Smart, N.O.E., J.C. Hatton & D.H.N. Spence, 1985. The long-term exclusion of large herbivores on vegetation in Murchison Falls National Park, Uganda. Biol. Conserv. 33: 229-245.

Spence, D.H.N. & A. Angus, 1971. African grassland management, burning and grazing in Murchison Falls National Park, Uganda. pp 319-331. In: E. Duffey & A.S. Watt (eds). The scientific management of animal and plant communities for conservation. The 11th Symposium of British Ecological Society, Blackwell, London.

Wheater, R.J., 1972. Problems of controlling fires in Uganda national parks. Tall Timbers Fire Ecology Conference 11: 259-275.

White, F., 1983. The vegetation of Africa. Natural Resources Research 20. UNESCO, Paris.

The conservation status of the genus *Saintpaulia* in Kenya

S.W. SIMIYU, P. MUTHOKA, J. JEFWA, B. BYTEBIER & T.R. PEARCE
Plant Conservation and Propagation Unit, East African Herbarium, National Museums of Kenya, P.O. Box 45166, Nairobi, Kenya.

Abstract
With nearly 22 taxa recognised in the genus *Saintpaulia* Wendl., restricted to the coastal lowlands and Eastern Arc Mountains of Kenya and Tanzania, the two recognised taxa and one new morphologically distinct type of African violets in Kenya represent the northernmost unit of the genus. *Saintpaulia rupicola* B.L. Burtt and *S. sp. ?nov.* are restricted to the limestone outcrops on the coastal strip whereas *S. teitensis* B.L. Burtt occurs in one locality in Taita Hills. Found in low numbers in fragmented populations, the Kenyan taxa are threatened by habitat loss through encroachment by cultivation and extractive utilization of the forest resources as well as quarrying activities. Efforts to conserve these species are being made by the Plant Conservation and Propagation Unit of the East African Herbarium, National Museums of Kenya. Both *in-situ* and *ex-situ* strategies are being utilised to enhance conservation of the genetic diversity of the genus in Kenya.

Key words
Conservation, *Ex-situ*, Gesneriaceae, *In-situ*, *Saintpaulia*.

Taxonomy

The generic name for the genus *Saintpaulia* commemorates St. Paul von Illare, the German District Commissioner of Tanga, in the then Tanganyika, at the end of the 19th Century (Johansson 1978). He sent seeds of *Saintpaulia ionantha* H. Wendland to his father, who later introduced the species into cultivation. Saintpaulias have now attained amazing horticultural popularity, due to their ease of cultivation and propagation, creating an industry generating at least $30 million in retail sales a year (Lovett 1985). To date, 20 species have been published within the genus, some rather closely allied and a number probably new to science. The genus has been particularly studied by Mr. B.L. Burtt of the Royal Botanic Garden, Edinburgh alongwith his other studies in the family Gesneriaceae. However, the taxonomy has proven difficult over time owing to a variety of reasons including paucity of collections (Sheil 1993). Lovett (1993) postulates that *Saintpaulia* is a neo-endemic genus and in general neo-endemic species are represented by taxonomically difficult complexes. Hybridization may play a role in the complexity and origin of some of these taxa. For Saintpaulias, this is evident in the variety of the progeny following hybridisation trials so important in the horticultural industry. The lack of clear-cut taxonomic delimitations presents problems in designing appropriate conservation strategies. The genus *Saintpaulia* is endemic to East Africa being restricted to the coastal lowlands and the Eastern Arc Mountains of Kenya and Tanzania (Lovett 1989) which display high levels of endemism. Two species occur in Kenya, namely *Saintpaulia rupicola* B. L. Burtt and *S. teitensis* B. L. Burtt, while a third unnamed species, *S. sp. ?nov.* (Robertson & Luke 1993) is found along the Kenyan coast.

L.J.G. van der Maesen et al. (eds.), The Biodiversity of African Plants, 341-344.

In-situ status

Of the Kenyan species, *Saintpaulia rupicola* and *S. sp. ?nov.* are confined to the limestone outcrops found in relic patches of forest along the coastal lowlands. Localities for *Saintpaulia rupicola* include limestone outcrops to the North and East of Mombasa at Chasimba, Mwarakaya and Mwachi whereas *S. sp. ?nov.* is found at Kachororoni gorge, some 60 km north of these populations. Populations at these sites are represented by small numbers often less than 100 plants per site. Saintpaulias are inherently habitat-specific and the specialised environment that they require is in itself a limiting factor in their distribution. They are at the same time very sensitive to drought and competition especially from ferns which tend to inhabit a similar niche in these areas.

Populations at Mwachi and Mwarakaya, which must have in the recent past occupied a wider range, are now confined to a few sites on the cliff faces of the limestone outcrops in areas that still have a forest cover. The healthier populations however were found in pockets at Chasimba along the cliff faces under overhanging rocks. Here it was observed that recruitment in terms of seedlings was good. However, the establishment was poor and thence only a few juvenile and mature plants could be found. Scarcity of satisfactory niches appeared to be the main factor affecting population regeneration here. Due to agricultural encroachment and continued felling of trees for fuelwood and building poles on these unprotected sites, the remnant vegetation cover is rapidly being removed. As a result, though the Saintpaulias are not directly being exploited, the loss of canopy cover has put the species at risk of extinction following habitat degradation.

At Kachororoni gorge, a similar trend was observed in the population structure that contains very few juveniles. Recently cut tree stumps and footpaths were evident indicating a lot of human use of the forest resources. The canopy is more open than previously and the mature plants are clearly under stress, reducing the chances of survival for the few mature plants and juvenile recruitment.

The Chasimba and Mwachi populations are threatened not only by habitat loss and opening of the forest canopy due to selective extractive utilisation of the forest but also due to quarrying activities. The Kenya coastal limestone outcrops form part of the Jurassic Kambe limestone series that is the moment being exploited for cement manufacture. There have already been proposals submitted for establishment of factories to utilise these resources hence making the survival of the populations here a chance event.

In the Taita hills, *Saintpaulia teitensis* B. L. Burtt is apparently safe owing to the fact that it occurs on steep cliff faces that are not easily accessible within a forest reserve. The occurrence of *Saintpaulia* species here is not surprising as the Taita Hills are the northern-most outliers of the disjunct arc of forests of Eastern Tanzania and are similar floristically (Lovett 1985). The Taita hills, though formerly covering a more extensive area, are represented by a few forest patches covering no more than 3 km² in total (Beentje et al. 1987). The remnant forest patches are surrounded by cultivation and forest plantations. Further excision of the forests is likely to have adverse effects on the micro-climate of the region hence threatening the survival of the *Saintpaulia* populations.

All the localities of *Saintpaulia* species at the Kenyan coast occur outside the Protected Areas Network. The coastal lowland forest fragments are too small to be included within the Forest Department. At present, sites of scientific interest may be gazetted under the Sites and Monuments Act, though guidelines on how to include sites of conservation interest are not clear. It is recommended that the sites be accorded some form of legal protection to enhance their future survival. The importance of integrating the local community in any conservation strategy is appreciated by both the Plant Conservation and Propagation Unit of the East African Herbarium and the WWF-funded Coastal Forests Conservation Unit, both groups being actively involved in species conservation in Kenya. Apart from looking into the possibilities of gazetting these sites, proposals are in the pipeline for developing them for eco-tourism with benefits accruing to the local community as well as ensuring that the conservation model is sustainable.

The *in-situ* conservation of these sites will bear added benefits as these are the localities for other endemic and rare species and it is hoped that a viable sample of representative habitats will contain and sustain a broader complement of species.

Ex-situ status

Ex-situ conservation is being carried out on the genus primarily as an insurance against the total loss of its genetic diversity should the current threats be operative culminating in the extirpation of the populations in the wild.

Germplasm collections of *Saintpaulia* have been made at all the sites documented in Kenya, following herbarium survey and correspondence with field botanists. The efforts have targeted propagules, both seed and whole plants where possible to ensure sufficient germplasm for propagation as well as short-term storage in local seed banks. The seeds are targeted to provide a wide genetic base for the *ex-situ* collection as the species is not self-pollinating hence field material is not clonal. Further research is being carried out within the unit to assess seed viability and longevity before long-term storage can be attempted. The dust-like nature of the seed has meant that standard methods for seed testing and storage are not entirely suitable for this species.

For the coastal species, seed collections have been made both for germination and short-term storage. Due to low population numbers, whole plants were not collected unless this was done as a rescue effort of the plants that were found suffering from severe exposure following the loss of canopy cover. Mature plants have been collected for *S. teitensis* as the populations were in a healthier state. The seed obtained from these accessions will be used in part for propagation to bulk the accessions for research work and eventual re-introduction trials. Carefully planned cross-pollination will be carried out on plants from populations with very few individuals to broaden the genetic base for re-introduction and translocations.

For short-term storage, collected seed pods have been allowed to dehisce in vials and the seed dried to a moisture content of 5-6%. The seeds were then sealed in plastic bags and stored in freezers at -20°C. Should a horticultural demand for the indigenous species arise, we will be in a position to provide the necessary germplasm.

Constraints

* Taxonomic delimitations within the genus and especially between the Kenyan taxa are still not clear.
* The general biology of the genus is not well understood such as pollination biology and breeding systems.
* Modalities for management of small sites of scientific interest for conservation purposes are not catered for in the national legal framework nor in the conservation policies.

Acknowledgements

This work is being carried out by the Plant Conservation and Propagation Unit, East African Herbarium, National Museums of Kenya. The project acknowledges the generous funding of the Kenya Government and the UK Overseas Development Administration.

References

Beentje, H.J., N. Ndegwa & J. Mutanga, 1987. Forest Islands in the mist. Swara Magazine 10 (1): 20-23.

Johansson, D.R., 1978. Saintpaulias in their natural environment with notes on their present status in Kenya and Tanzania. Biol. Conserv. 14: 45-62.

Lovett, J., 1985. Moist forests of Eastern Tanzania. Swara Magazine 8 (5): 8-9.

Lovett, J., 1989. Feeling forests. Swara Magazine 12 (6): 28-30.

Lovett, J., 1993. Eastern Arc forest flora. In: J.C. Lovett & S.K. Wasser (eds.), Biogeography and Ecology of the East African Coastal Forests: 33-55, Cambridge University Press, Cambridge.

Robertson, S.A. & W.R.Q. Luke, 1993. Kenya Coastal Forests. The report of the NMK/WWF Coast Forest Survey. WWF Project 3256. WWF, World Wide Fund For Nature.

Sheil, D., 1993. What future for the African Violets. Swara Magazine 16 (5): 18-21.

Floristic composition of Gara Muleta and Kundudo mountains, south-eastern Ethiopia: implications for the conservation of biodiversity

D. TEKETAY
Swedish University of Agricultural Sciences, Department of Forest Vegetation Ecology, S-901 83 Umeå, Sweden.
Permanent Address: Alemaya University of Agriculture, Faculty of Forestry, P.O. Box 138, Dire Dawa, Ethiopia.

Abstract
The vegetation of the Gara Muleta and Kundudo mountains in Ethiopia conserves soil, water and both plant and animal diversity. It is under severe pressure of destruction caused by deforestation, fire, agriculture and overgrazing. The results of floristic studies revealed that there are more than 361 species of vascular plants representing 82 families at Gara Muleta and 96 species representing 43 families at Kundudo mountains. Out of the species recorded from Gara Muleta 45 (13%) are endemic to Ethiopia while 38 (11%) are nearly endemic. There are 6 (6%) endemic species at Kundudo mountain. The floristic composition of the two mountains was compared with the floristic composition of five mountain areas (Bale, Gara Ades, Menagesha, Semien and Wof-Washa mountains) in the country and with Afro-alpine areas in East Africa (Kenya, Rwanda-Burundi, Tanzania, Uganda and Zaire) and Usambara mountains. The highest similarity in floristic composition was obtained with Wof-Washa natural forest in both Gara Muleta (39%) and Kundudo (41%) suggesting that the flora in these mountains has some distinctive features. If the present trend of destruction continues the consequence will be many-fold, among which loss of biodiversity is one. Therefore, it is strongly recommended that these areas be afforded all possible protection before it is too late.

Key words
Biodiversity, Conservation, Endemic species, Floristic composition.

Introduction

Mountain ecosystems are important storehouses of biological wealth. As lowlands have been altered by commercial agriculture, industry and urban settlement, the last stronghold of nature is often in the mountains (Anon. 1992). From these storehouses we have taken some of our most current food staples, for instance potatoes from the Andes and sorghum from Ethiopia. Mountains also house wild relatives of cultivated crops on which our future may depend. A lot of medicinal material can be obtained from mountainous areas and these provide actual and potential benefit to all humanity. The montane tropical rain forests, the mountain cloud forests, and indeed mountain wild lands in general are refuge for a great wealth of endemic and endangered species, and plant communities. This great wealth of biological diversity exists because of the great variety of environmental conditions in mountain areas. This variation is caused by changes in elevation over short horizontal distances which results in rapid changes in temperature, light intensity, rainfall and soil. Compounding this altitudinal heterogeneity are the changes caused by exposure to different points of the compass, north-facing slopes being very different from south-facing slopes.

L.J.G. van der Maesen et al. (eds.), The Biodiversity of African Plants, 345-350.

Mountains are also the water towers to modern civilization (Anon. 1992). Mountains intercept the global circulation of air. They force it upwards where it cools and condenses its water vapour into clouds. These clouds provide rain and snow which, if it falls on vegetated slopes, runs off cleaned and slowly into the rivers which cross the plains of the sea. The rivers generate hydropower, irrigation, fishing industries and water transport. They provide water for textiles, mining and light industries. They cool steel plants, smelters, power stations, and oil refineries. In short, they make civilised life possible and always did, even before the industrial revolution. The Ethiopian Highlands, the Atlas, the Himalayas, the mountains of China are only some of the ranges currently in the public eye in their role as water towers (Anon. 1992). Therefore, we need to know more about these banks of biological wealth and water towers, and about their potential for a sustainable flow of benefits.

Gara Muleta and Kundudo massifs are among the highest mountains found in the south-eastern part of Ethiopia. Gara Muleta is the highest summit of the Chercher Highlands. It houses an evergreen cloud forest with many epiphytes. These mountains are unique areas which may be considered as islands of forest region surrounded by semi-arid or arid climatic areas. Such natural restrictions and barriers make them ideal for the study of certain natural phenomena. In effect, they are controlled experimental areas convenient for the study of natural history as well as for the investigation of future trends. Because of their mountainous landscape nature, they are not suitable for cultivating crops and the forest areas are not of a commercial type. However, they are sources of wood for fuel, construction, house-hold and farm implements. They are used for grazing livestock and are important in regulating the watershed in the surrounding areas. Moreover, if protected and developed, the areas would in the future be ideal places for naturalists, challenges for mountaineers, areas of great pleasure for tourists and outdoor laboratories for practical training of students and for research. In spite of the multipurpose services rendered by these areas, they are under severe pressure of destruction caused by deforestation, fire, encroaching agriculture and overgrazing. To underpin the need for the protection and conservation of these areas, the present work was undertaken with the following objectives: (a) to study the floristic composition of the areas which provides foundations for the conservation of biodiversity, (b) to compare the floristic composition of the two mountains with similar areas both inside and outside of the country to show the uniqueness of their flora and (c) to generate relevant information about the mountains for scientists, tourists, teachers, students, mountaineers, etc.

Study Area

Gara Muleta and Kundudo mountains are located in the Harerge Administrative Region at about 9°18' North and 41°44' East and 9°27' North and 42°20' East respectively. Gara Muleta has a horizontal extension of about 16 km and an altitudinal range of 2400-3400 m while Kundudo has an altitudinal range of 2100-2860 m. The Geology, soils and climate of Gara Muleta are described by Murphy (1968) and Uhlig & Uhlig (1990b). There is no information concerning the geology, soils and climate of Kundudo mountain. Gara Muleta is an important catchment for tributaries of the Wabe Shebele river such as the Ramis, Mojo and Gobele rivers. These rivers feed several irrigation schemes in the adjacent dry lands to the south where coffee, banana, sugarcane, etc. are grown. The common feature of these mountains is that the increase in height and slope steepness, together with gravity, cause accelerated runoff and, hence, erosion. Both mountains are situated in the chain of mountains which were described by Hurni (1988, 1992) to have serious erosion.

Different sorts of wild animals inhabit the forests. Among this are Bush-buck, Leopard, Ibex, Oryx, Hyena, Baboons (three kinds), Grivets, Colobus monkey, Jackal, Antelope, Bush pig, Genet, Civet, Crested porcupine, Wart dog, Francoline, Guinea fowl and a variety of birds and reptiles.

Materials and Methods

Gara Muleta mountain was visited several times between 1983-1994 and floristic studies were made. The study of Kundudo, however, is mainly based on herbarium specimens and from literature. At Gara Muleta, species which were readily identifiable were simply recorded while herbarium specimens were collected for some of the species that were difficult to identify in the field. The specimens were then processed and identified at the Alemaya University of Agriculture Herbarium (ACD). In addition, specimens which were collected by different collectors at different times and deposited at ACD as well as relevant references (Gillett 1941, Cufodontis 1953-72, Thulin 1975) were studied. Based on these studies, a checklist of all plant species recorded from Gara Muleta with corresponding information on altitude and aspect of collection area was prepared. Endemic species were identified based on the works of Cufodontis (1953-72) and Hedberg & Edwards (1989). This list was compared (at the species level) with similar works on the Afroalpine Vascular Plants (Kenya, Tanzania, Uganda, Zaire and Rwanda Burundi) (Hedberg 1957), Bale Mountains National Park (Mooney 1963, Hillman 1986), Semien Mountains National Park (Scott 1955, Hurni 1986), Menagesha National Park (Sebsebe Demissew 1988), Gara Ades Mountain (Uhlig & Uhlig 1990a), Usambara Mountains (Iversen 1991) and Wof-Washa Mountains (Demel Teketay and Tamrat Bekele, unpublished). Nomenclature follows Cufodontis (1953-72) and Hedberg & Edwards (1989). Information was also gathered from the local people about the wild animals living in the Gara Muleta mountain.

Results

More than 361 and 96 species of plants representing 82 and 43 families were recorded from Gara Muleta and Kundudo mountains respectively*. Of these 45 (13%) species from Gara Muleta and 6 (6%) species from Kundudo are endemic to Ethiopia while other 38 (11%) species recorded from Gara Muleta were nearly endemic to Ethiopia, found only in one or two countries other than Ethiopia. A comparison of the floristic composition of Gara Muleta and Kundudo mountains with other mountains is given in Table 1. The vegetation at Gara Muleta is quite similar to the Wof-Washa mountains (39%) within Ethiopia and to Usambara mountains (35%) from East African mountains. Similarly, the vegetation at Kundudo is also similar to Wof-Washa (41%) and

* A comparative list of plants recorded from Gara Muleta and Kundudo mountains may be obtained from the author.

Table 1. Comparison of the floristic composition of Gara Muleta and Kundudo mountains with other mountains.

Mountain(s)	Gara Muleta		Kundudo	
	Species in common	%	Species in common	%
Bale	122	35	31	35
Gara Ades	92	26	28	32
Menagesha	57	16	11	13
Semien	67	19	19	22
Wof-Washa	135	39	36	41
Usambara	121	35	27	31
Afro-alpine	52	15	9	10

Usambara mountains (31%). However, these two mountains have their own floristic uniqueness in that more than 60% of the vegetation is different from even the more similar vegetation. On top of this the number of endemic species, especially at Gara Muleta, is relatively high.

At the bottom of the southern face of Gara Muleta mountain, the area is covered with multi-storey mixed deciduous forest dominated by emergent trees of *Ekebergia capensis, Olea hochstetteri* and *Prunus africana* associated with the following species at the intermediate and lower storeys: *Bersama abyssinica, Croton macrostachyus, Dombeya bruceana, Dombeya schimperiana, Erythrina brucei, Euphorbia obovalifolia, Galiniera coffioides, Hagenia abyssinica, Lepidotrichilia volkensii, Maytenus* spp., *Nuxia congesta, Olea africana, Pittosporum* spp., *Rhus* spp., *Schefflera abyssinica, Vernonia* spp. and many other shrubs. The lower storey and the ground vegetation is very rich in species composition and consists of shrubs, climbers, ferns and other herbs. There are also several epiphytes: *Diaphananthe schimperana, Peperomia abyssinica* and *Usnea* spp. However, the vegetation cover, species richness and stature of trees/shrubs decrease with increase in altitude.

The vegetation of the northern side of the mountain is different from the southern side. Out of 206 species for which aspect of the collection areas was available at Gara Muleta, 158 (77%) were collected from the southern face of the mountain. This is mainly caused by the difference in moisture regime on each side of the mountain. This in turn is brought about by the movement of the inter-tropical convergent zone which gives rise to desiccating north-easterly trade winds during the dry season and carrying clouds and rain from south-west direction during the rainy season (Uhlig & Uhlig 1990b). The cloud formation which is frequently observed, especially at the southern face of the mountain, influences the amount of radiation as well as humidity considerably. The result is a luff and lee effect which is reflected in the vegetation.

The vegetation at the bottom of the northern side of Gara Muleta mountain is dominated by the two conifers, *Juniperus procera* and *Podocarpus gracilior* associated with *Ekebergia capensis, Erythrina abyssinica, Gnidia glauca, Hagenia abyssinica, Maesa lanceolata, Olea africana* and *Schefflera abyssinica.* This side of the mountain is poor in its species composition, both in its upper- and under-storey as well as ground vegetation compared to the southern side. Moreover, it has been heavily affected by deforestation. As a result, only scattered trees can be observed in the area which are indicators of the previous forest cover. Zonation and description of the structure of the vegetation as well as constituents of the different strata at the northern side of Gara Muleta mountain are reported by Uhlig & Uhlig (1990b). According to them, four zones of the vegetation can be recognized: the *Juniperus-Podocarpus, Juniperus, Hagenia* and *Erica* belts.

Compared to Gara Muleta, Kundudo mountain is poor in its species composition. On the lower part of the north-eastern slope, some scattered trees of *Juniperus procera* and *Podocarpus gracilior* which are representatives of the previous forest cover are found. These are associated with *Brucea antidysenterica, Dombeya schimperiana, Pittosporum abyssinica, Pavetta oliveriana, Teclea nobilis, Toddalia asiatica* and several other shrubs. The ground vegetation contains many herbaceous plants including several grasses and sedges.

Discussion and Conclusion

The total number of species and the number of endemic species recorded from Gara Muleta mountains are higher than those areas to which it is compared in the country. Moreover, many of the species recorded from Gara Muleta and Kundudo mountains in the present study are not reported from the Ethiopian mountains used for comparison indicating that they have their own unique components. These facts coupled with their actual and potential multipurpose services justify the protection and conservation of the two mountain areas.

Logging for fuelwood and leaf fodder collection, slash and burn subsistence agriculture, land clearing for commercial agriculture, clearing for reservoirs, bulldozing for new mountain roads, felling for transmission lines or pipe lines, harvesting forest game, grazing by livestock and burning for many of the foregoing purposes have a negative impact on the remaining mountain forests in almost all parts of the world (Anon. 1992). Ethiopia is no exception. These activities disrupt the protective as well as the productive functions of the mountain vegetation. They create soil erosion and flooding which causes physical destruction and loss of life at the populated lowlands. There will also be accelerated loss of biodiversity and in some instances extermination/extinction of species. Rivers will dry up sooner after the rains, since less water can be retained by mountains denuded of vegetation cover. This in turn will reduce drinking water supplies and disrupt irrigation systems so vital in lowland areas such as Harerge. In general, the long-term environmental health of mountains matter much to dwellers in the lowlands as to inhabitants of the mountains themselves as well as to the conservation of biodiversity. To improve the existing situation in mountain areas of the world, it requires more knowledge about the mountains and inhabitants, political will, good economic incentives and skilful communication (Anon. 1992).

Ethiopia is facing severe degradation in its mountains as a result of the current population increase of around 3% per year. Land resource scenarios indicate that if present trends continue, Ethiopia will have an overall crisis involving livestock grazing land in less than 15 years, and an arable land crisis in about 25 years (Hurni 1992).

Since Gara Muleta and Kundudo mountains are legacies of generations which are of paramount importance as genetic reservoirs, water catchment areas, sources of forest products, potential areas of education, research, recreation and tourist attraction, it is highly recommended that they be given all possible protection and conservation as a matter of urgency.

Acknowledgements
I would like to thank Alemaya University of Agriculture for logistic support and the Swedish University of Agricultural Sciences, Department of Forest Vegetation Ecology for the facilities used during the write-up of this report. I thank Ato Tamerat Abebe and Ato Abdurazak Abdulahi for their assistance during the study and Ato Getachew Eshete for translating a paper from the German language. I am very grateful to Prof. dr. L.J.G. van der Maesen (AETFAT Secretariat) and Dr. Ir. J.J. Bos for helping me to obtain the financial support to participate in the XIVth AETFAT Congress.

References

Anon., 1992. An Appeal for the Mountains. Mountain Agenda, Switzerland.

Cufodontis, G., 1953-1972. Enumeratio Plantarum Aethiopiae Spermatophyta. Jardin Botanique de l'Etat Bruxelles.

Gillett, J.B., 1941. The plant formations of western British Somaliland and the Harar Province of Abyssinia. Bull. Misc. Inf. Kew 1941: 37-75.

Hedberg, I. & S. Edwards, 1989 (eds.). Flora of Ethiopia. The National Herbarium, Addis Ababa.

Hedberg, O., 1957. Afroalpine Vascular Plants, a taxonomic revision. Symbolae Botanicae Upsalienses XV: 1.

Hillman, J.C., 1986. Bale Mountains National Park Management Plan. Wildlife Conservation Organization, Addis Ababa.

Hurni, H., 1986. Management Plan, Semien Mountains National Park and Surrounding Area. Wildlife Conservation Organization, Addis Ababa.

Hurni, H., 1988. Degradation and conservation of soil resources in the Ethiopian Highlands. Mountain Research and Development 8 (2/3): 123-130.

Hurni, H., 1992. African Mountain and Highland Environment, suitability and susceptibility. In: P.B. Stone (ed), The State of the World's Mountains: 11-44, Zed Books Ltd., London & New Jersey.

Iversen, S.T., 1991. The Usambara Mountains, NE Tanzania: Phytogeography of Vascular Plant Flora. Symbolae Botanicae Upsalienses XXIX: 3.

Mooney, H., 1963. An account of two journeys to the Araenna Mountains in Bale province (Southeast Ethiopia), 1958 and 1959-1960. Proc. Linn. Soc. 172: 127-174.

Murphy, H.F., 1968. A Report on the Fertility Status and Other Data on Some Soils of Ethiopia. College of Agriculture, Haile Sellassie I University and Oklahoma State University, Experimental Station Bulletin No 44.

Scott, H., 1955. Journey to the Semien District, Northern Ethiopia, 1952-53. Webbia XI: 425-450.

Sebsebe Demissew, 1988. The floristic composition of Menagesha State Forest and the need to conserve such forests in Ethiopia. Mountain Research and Development 8 (2/3): 243-247.

Thulin, M., 1975. The genus *Wahlenbergia* s.l. (Campanulaceae) in Tropical Africa and Madagascar. Symbolae Botanicae Upsalienses XXI: 1.

Uhlig, S.K. & K. Uhlig, 1990a. The floristic composition of a natural montane forest in southeastern Ethiopia. Feddes Repertorium 101 (1-2): 85-88.

Uhlig, S.K. & K. Uhlig, 1990b. Die Höhenstufen der Vegetation am Gara Mulatta, Äthiopien. Feddes Repertorium 101 (11-12): 651-664.

Generic delimitations in flowering plants

24 & 25 August 1994

Convenor: A.J. Paton

Monophyly: a criterion for generic delimitation, with special reference to Leguminosae

B.D. SCHRIRE & G.P. LEWIS
Royal Botanic Gardens, Kew, Richmond, Surrey TW9 3AE, United Kingdom.

Abstract
The concepts of systematics, similarity, homology, monophyletic, paraphyletic and polyphyletic groups, generic delimitation and classification are discussed with reference to three recent phylogenetic analyses, above the genus level, in the family Leguminosae. The aim is to suggest explicit methods for recognising genera in an attempt to achieve relative stability of names within a natural classification.

Résumé
Les concepts de systématique, de similarité, d'homologie, de groupes monophylétiques, paraphylétiques et polyphylétiques, de délimitation générique et de classification sont discutés par rapport aux trois analyses phylogénétiques récentes de la famille des Leguminosae au niveau supragénérique. Le but en est de proposer des méthodes précises permettant de reconnaître les genres, en essayant ainsi d'assurer une relative stabilité de la nomenclature au sein d'une classification naturelle.

Key words
Classification, Generic Delimitation, Genus, Leguminosae, Monophyly.

Introduction

The concept of the genus - a group to which a particular species belongs - has its roots in folk science (Clayton 1983) and has persisted because of its value in facilitating ordering the living world (Jeffrey 1987). The recognition of genera (and other taxa) is founded on the ability to distinguish like from unlike; separate constant from variable; extract form from manifestation, i.e. homology from analogy; detect the hierarchically structured order of life and construct classifications based on group-defining features and interrelationships with other taxa.

Algorithms of comparison are used by biologists to understand how organisms are related to one another (Brooks & McLennon 1991). The cornerstone of comparative biology is that order does exist in the biotic world and that nature is arranged hierarchically. The goal is to study *patterns* of relationship among organisms and the distribution of characters in them (Wiley 1981). Other fields of general biology deal with the *processes* generating diversity.

Comparative biology aims to analyze and identify biotic patterns so theories of process may be developed to explain those patterns (Brady 1985, Eldredge & Cracraft 1980). The type of orderliness exhibited by intrinsic features (characters) of organisms is expressed in terms of relative degrees of similarity. Many more properties tend to be shared between some organisms and relatively fewer between others. Orderliness in nature is thus apparent in the hierarchy of similarities among organisms comprising the biota. To study this orderliness - from the perception of a pattern to a conscious understanding of its logical structure - the problem arises at the outset about what is to

L.J.G. van der Maesen et al. (eds.), The Biodiversity of African Plants, 353-370.

be compared or even what "similar" means (Eldredge & Cracraft 1980).

Systematics is basic to comparative biology, dealing with the theory and practice of reflecting the orderliness in nature that has resulted from patterns of phylogenetic ancestry and descent. The primary task of systematics is the recognition and classification of taxa. The realm of taxonomy is the description and naming of sets of taxa within a hierarchical arrangement. All approaches to understanding this hierarchy depend on the pattern of similarity of characters revealing nested sets of taxa (Hennig 1966). Hypotheses about pattern (phylogenetic history) are necessary starting points for virtually all comparative evolutionary studies (Eldredge & Cracraft 1980). Initially the study of pattern should be divorced as much as possible from the study of process, to provide an unbiased base-line for the evaluation of alternative hypotheses about process. This is implicit in recognising the distinction between the systematic method and evolutionary theory (Brady 1985, Humphries & Funk 1984). The usefulness of phylogenetic hypotheses depends on how well they describe and explain observations of natural processes.

Evolutionary systematists (Mayr 1969, 1974, Simpson 1965, 1975) give much weight to the process of evolution in producing classifications that are based on the degree of divergence between ancestral and descendent groups (Janvier 1984). Any classification is consistent - in Simpson's (1961) sense - with what is known about phylogeny, if all the classes in the classification are convex (see Duncan 1980, Estabrook 1978, Meacham & Duncan 1987 for definitions). Many convex groups are paraphyletic, however, and as such (see next section), are artificial classes of taxa (Hennig 1966). The delimitation of taxa and their rank is estimated by criteria which include the evolutionary systematist's personal opinion and "ingenuity" (Simpson 1961). Existing taxonomies were simply reinterpreted in evolutionary terms, i.e. the reality of previously recognised taxa was taken for granted and evolutionary concepts and mechanisms were formulated to account for their existence (De Queiroz 1988, Stevens 1984). Phenetic systematists (Gilmour 1961, McNeill 1979, Sneath & Sokal 1973) emphasize the criterion of overall similarity to determine relationships.

Phylogenetic systematists or cladists (Eldredge & Cracraft 1980, Hennig 1966, Wiley 1981) differ from both the above in employing empirical methods to construct hypotheses of phylogeny and infer classifications, rather than relying on intuition or authority (Wiley et al. 1991).

Phylogenetic systematics is primarily concerned with erecting hypotheses about the pattern of life's history. The aim is to discover the genealogical relationships among the taxa being studied rather than the phenetic relationships, i.e. overall similarity (Hennig 1966, Phillips 1984). Organisms are ranked entirely on the basis of common descent, i.e. the sequence of dichotomies in the inferred phylogeny.

Similarity stands at a basic level in propositions of homology (De Pinna 1991, Patterson 1982, Rieppel 1988, Stevens 1984). Homology means equivalence of parts and is essentially a comparative concept serving as a sorting procedure to investigate the validity of information, i.e. comparisons between homologues are meaningful within a hierarchical context, while comparisons between non-homologues are not (De Pinna 1991). All similarities may be deemed homologous initially (primary homologies of De Pinna 1991), reflecting the expectation that there is a correspondence of parts that can be detected by an observed match of similarities. Similarity or dissimilarity alone, however, does not support or refute homology propositions. A secondary level of homology is the outcome of a pattern-detecting analysis (e.g. parsimony) which is a congruence test of homology. Secondary homology propositions, therefore, are primary homologies that have been evaluated against the framework of a general pattern (cladogram) which accommodates all available evidence under the most parsimonious configuration.

Homologies are the result of evolution by descent, and can arise either as novelties, uniquely derived in a particular group of taxa, or as retained (primitive) characters shared with ancestors of that group (Hennig 1966). "False" homologies, or cases of mistaken similarity, are the result of evolutionary convergence and parallelism. The basic premise of phylogenetic systematics is that resemblances between organisms do not always have the same bearing on relationship. The core concept is the use of derived (apomorphic) characters to reconstruct relationships and group taxa based on common ancestry.

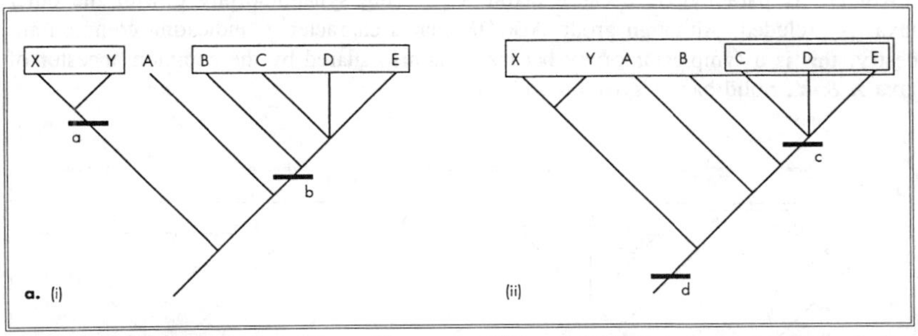

Figure 1a. Monophyletic groups: (i) groups XY and BCDE are monophyletic, being diagnosed by synapomorphies a and b respectively. Each group is a natural product of evolutionary descent comprising all descendants of a common ancestor. (ii) Similarly, groups CDE and XYABCDE are diagnosed by synapomorphies c and d respectively. Solid bar = synapomorphies.

Monophyly

Only shared derived characters (synapomorphies) are evidence of exclusive common ancestry (Hennig 1966). Synapomorphies are hypotheses of homology, i.e. unique group membership characters inherited from a common ancestor (Farris 1974). A monophyletic group (Figure 1a) is diagnosed by the presence of a synapomorphy; it is a natural product of evolutionary descent from a common ancestor and has its own history (De Queiroz 1988, Donoghue & Cantino 1988, Wiley 1981). Such groups are complete systems of common ancestry comprising all the descendants of a common ancestor (Patterson 1982, Platnick 1979), whether those organisms are living or dead, known or unknown.

Homologies based on shared primitive characters (symplesiomorphies) indicate common ancestry, but not immediate and exclusive common ancestry (Janvier 1984). Members of paraphyletic groups (Figure 1b) share synapomorphies with taxa outside the group, therefore they are not natural products of evolution because some descendants of a common ancestor are excluded. Their own history is also part of that of the

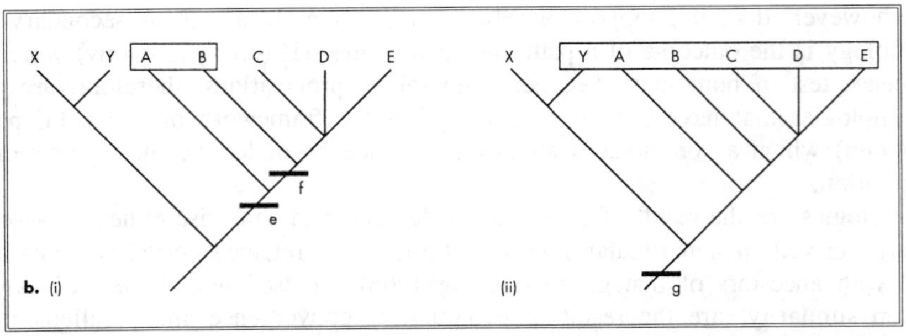

Figure 1b. Paraphyletic groups: (i) group AB is paraphyletic because although taxa A & B share a uniquely derived character (synapomorphy) e, the other descendants of the most recent common ancestor possessing e, i.e. taxa C, D & E are excluded. Taxon B also shares synapomorphy f with taxon CDE, therefore AB is not a natural product of evolutionary descent from a common ancestor. (ii) Similarly, group YABCDE is paraphyletic because taxon X, sharing synapomorphy g with the other taxa, is excluded. Although group ABCDE shares character g, indicating common ancestry, this is a symplesiomorphy because g is also shared by the common ancestor of taxa X & Y. Solid bar = synapomorphies.

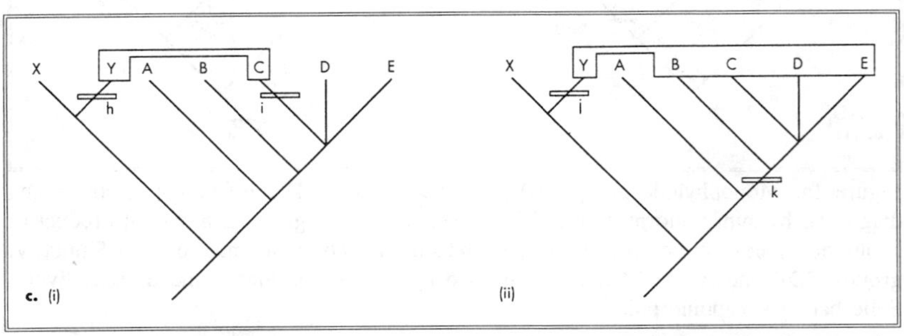

Figure 1c. Polyphyletic groups: (i) taxa Y & C are a polyphyletic group because the membership characters h & i are not uniquely derived. They were found to be different characters, *a posteriori* by cladistic analysis. (ii) Similarly, groups BCDE and taxon Y are based on convergent (non-unique) characters k & j. Such groups represent products of evolutionary descent from different ancestors. Open bar = non-uniquely derived convergent characters.

monophyletic groups which are rooted in them, i.e. Dicotyledons are a paraphyletic group that also includes the early history of the monocotyledons. Paraphyletic groups are incomplete systems of common ancestry, i.e. they are artificial classes of taxa created by taxonomists emphasising the importance of particular traits, as well as phenetic or overall similarity (Donoghue & Cantino 1988, Hennig 1966). The justification for recognising such grade groups in traditional classifications has been that one or more terminal taxa could be removed [from a monophyletic group] on the basis that they have diverged significantly from the rest, thereby introducing arguments about process (anagenesis) into the interpretation of pattern. Paraphyletic groups should not be recognised formally because they are likely to confound the study of evolution (Donoghue & Cantino 1988). Polyphyletic groups (Figure 1c) are also artificial classes of taxa whose membership characters are different, i.e. they are not uniquely derived, being based on convergent or "false" homology. They represent products of evolutionary descent from different most recent common ancestors.

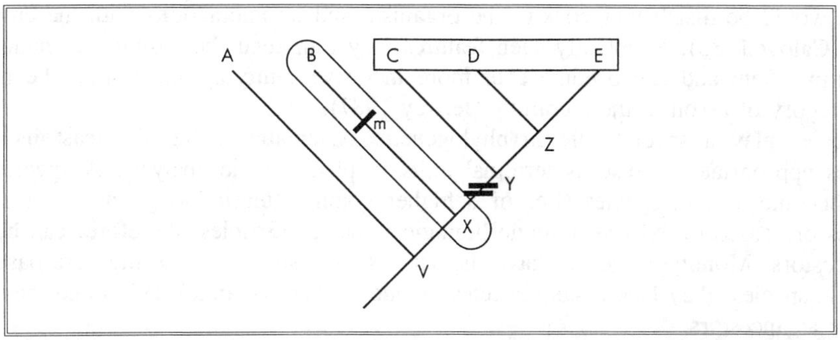

Figure 2. Problem of definitions of monophyly based on the concept of common ancestry. Since X is the most recent common ancestor of taxon BCDE, all descendants are not included if taxon B is defined as a monophyletic group based on X. Taxon B, therefore, can only be diagnosed as a monophyletic group by the presence of synapomorphy m. The most recent common ancestor of taxon CDE is Y and this clade is diagnosed as a monophyletic group by two synapomorphies.

Some phylogeneticists (Farris 1974, Hennig 1966, Wiley 1981) defined monophyletic, paraphyletic and polyphyletic groups in terms of their relation to a common ancestor, i.e. a monophyletic group is a complete system of common ancestry including the common ancestor and all its descendants (Farris 1974). Platnick (1979) and Patterson (1982), however, consider that it is not possible to define such groups based on common ancestry because, for example in Figure 2, X is not just the most recent ancestor of taxon B, it is the common ancestor of taxon BCDE. They reason that theories of characters (homologies) provide evidence of groups which can then be interpreted in terms of common ancestry. No empirical method exists for detecting ancestors from their descendants since there is only one criterion for group recognition, i.e. taxon B can only be diagnosed as a monophyletic group by the presence of synapomorphies.

Monophyletic, paraphyletic and polyphyletic groups can only be inferred by cladistic analysis. A proposed hypothesis of phylogenetic relationships (cladogram) is based on the most parsimonious (congruent) arrangement of the largest number of synapomorphies, until it is refuted following the discovery of new information (Hennig 1966). The true history of a group is unknowable and can only be approached more and more closely by a succession of hypotheses and refutations (Eldredge & Cracraft 1980).

Generic concept

The genus is a mandatory category in the binomial nomenclature of organisms. Taxonomists recognise genera as representing a level of unifying regularities above the species; a level where our perceptions make it advantageous to assign a distinct name

(Jeffrey 1987). Also important is how extensive are the unifying regularities, and the relative advantages of applying a name to a given manifestation of regularities at generic as opposed to any other rank. Clearly defined species groups, for example, may be recognised as infrageneric taxa in a classification.

Such regularities arose as features became fixed in organisms (Alberch 1982, Waddington 1975), i.e. through operating to resist certain kinds of genetic change which would be disadvantageous to the organism and its interactions with the environment (Calow 1983). Eventually such features may transcend the individual phenotypes exhibiting them and can become ever more inclusive, unifying and stable, the higher the category of taxon in the hierarchy (Jeffrey 1987).

Species play a special role in phylogenetic systematics being the least inclusive entities appropriate for use as terminal units in phylogenetic analysis. A species can only become a stem species (i.e. of a higher taxon) after it has given rise to other species or ceased to exist as a single biological species. Species, therefore, can behave as ancestors. Monotypic genera have the same status, since because they are based on a single species, they lack a stem species unique to them (Platnick 1977) and they may behave as ancestors.

Many monotypic genera have been described because a species possesses a number of distinct autapomorphies, i.e. characters unique to that species, making it easily distinguishable from other related species. Divergence is thus used to justify recognising a separate genus, but it is only synapomorphies with other species that are informative about relationships. Distinctive taxa not fitting into recognisable monophyletic groups should always be coded separately in cladistic analyses, so an *a posteriori* assessment can reveal relationships with other taxa. Monotypic genera, therefore, which nest well within the ingroup (i.e. supported by numbers of synapomorphies) should be re-evaluated in terms of the overall relationships within the clade. If, however, after various analyses with different outgroups the divergent species constantly remains unresolved near the base of the cladogram (i.e. with few synapomorphies to hold them in the ingroup), then they may be relictual or isolated taxa which could validly represent monotypic genera.

Higher taxa (which excepting the above contain two or more species and hence a unique stem species) only exist as a consequence of evolutionary descent, i.e. they are monophyletic groups (whole systems) which cannot serve as ancestors of other taxa (Donoghue & Cantino 1988, De Queiroz & Donoghue 1988). Paraphyletic groups are not whole systems and do not accurately convey information about common ancestry and character evolution. There are thus only two categories of grouping with which systematists should be concerned, the species and monophyletic groups of species.

Problems in recognising genera

One of the major difficulties of working at the generic level (compared with species) is availability of data (West 1987). To analyze several genera, terminal taxa must be adequately sampled and in many cases this may involve several hundred species. Characters have to be carefully interpreted and scored for each taxon and special consideration is needed when terminal taxa are polymorphic for certain characters (Nixon & Davis 1991).

Many traditionally circumscribed genera are artificial (paraphyletic or polyphyletic) groups which are only defined by characters not shared by monophyletic groups. They are defined by plesiomorphies and often by characters indicating what they are not, rather than what they are (Eldredge & Cracraft 1980). Such genera were often based on historically euro-centric taxonomies which became inadequate to handle the large influx of new plants coming in from elsewhere over the last 200-300 years (Jeffrey 1987). In consequence very large, unwieldy, core genera, or numerous small segregate or monotypic genera were delimited on the basis of easily observed morphological features, often giving rise to 'single organ' genera, i.e. groups defined on the basis of a particular pod type regardless of affinities indicated by other characters. Circumscription of genera must be decided on a world-wide basis and not in floristic treatments with their artificial boundaries.

Problems also occur predominantly in woody groups of many families, in which it may be difficult to sample taxa and characters for evaluating monophyly. The reasons may be because of:

1) ill-defined, diffuse, lineages widely scattered across the tropics and often with large disjunctions,
2) extreme plasticity of features in any actively evolving group,
3) different suites of characters having evolved in different directions in less derived groups because of relatively less integrated levels of overall organisation compared with more derived groups (Stebbins 1984, Stevens 1986).

Phylogenetic systematics is, nevertheless, a rigorous methodological discipline, providing rules for determining group membership in classifications. It is a stepwise process based on successive approximation and deductive reasoning, allowing the evaluation of alternative phylogenetic hypotheses in the light of the knowledge of the taxa and characters concerned (West 1987). It is the technique of cladistics that is objective, not the hypothesis or classification itself. All the evidence should be logically presented and open to critical re-analysis, and hypotheses must be testable. In contrast there are no rules for group membership in classifications which allow paraphyletic taxa. Ultimately, group membership is subjective without strict monophyly.

Classification

Traditional (alpha) taxonomy and the methods of phylogenetic systematics may be considered as different tools for producing classifications. Different classifications serve different purposes, and the choice of which one depends on what the users of classifications want, how versatile they are, the level of data available and the resources necessary to produce them.

Classification involves naming taxa, and since there are no binding rules for ascribing rank, we must ask what that name communicates about those taxa. Should a name simply be an aid to identify plants and place them within an existing classification, or should that name communicate a level of information about evolutionary relationships between taxa?

There are two fundamentally different ways of looking at plants for the purposes of classification. In traditional taxonomy characters are sought which will help to recognise plants, i.e. unique characters or combinations of characters that distinguish species or genera and indicate how far one taxon has diverged from another. This may be a perfectly acceptable approach for the purpose of simply recognising and naming taxa.

If we wish rather to know how one taxon is related to another, then it is necessary to look at characters differently, i.e. to look for synapomorphies which are the only characters informative about relationships. Taxa are then designated a level in a nested hierarchy of relationships derived from the total evidence of all characters available.

A phylogeny, comprising nested sets of monophyletic groups expressed by hierarchies of derived characters, summarises the historical characteristics of those groups and is the only non-arbitrary basis for classification (Hennig 1966). Regardless of criticisms that cladograms do not represent "true" phylogenies, and therefore classifications based on them are no more objective than those constructed by any other method, phylogenetic classifications have the advantage of being refutable on the basis of the phylogenetic hypotheses from which they were derived (Funk & Stuessy 1978, Janvier 1984).

A system of classification which reflects evolutionary history as accurately as possible seems preferable because:

1) it is presumed there is one true phylogeny of life which provides justification for choosing among alternative taxonomic methods, and is also a standard against which alternative classifications can be evaluated (Donoghue & Cantino 1988). There is no true hierarchy of phenetic similarities and hence no objective criterion by which to choose among various measures of similarity,
2) it facilitates studies of the evolutionary process, i.e. the study of other characters, and for investigations of biogeography, ecology, coevolution, developmental biology and speciation (Funk & Brooks 1990). Artificial taxa that do not reflect the genealogy of the taxa in question cannot be used in such investigations,
3) it has predictive value, i.e. is a source of new information from characters as yet unstudied and an efficient means of storage and retrieval of information. The recognition of paraphyletic and polyphyletic groups decreases the predictive value and information content of a classification.

The theme of this meeting is the study of African biodiversity. It is necessary not only to name plants accurately, but that the name should communicate as much as possible about that plant.

The fundamental difference between phylogenetic systematics and other methods is captured in the distinction between systematisation and classification as the goal of taxonomy (De Queiroz 1988, Donoghue & Cantino 1988, Griffiths 1974). In traditional classifications the taxonomist recognises abstract classes of organisms that are defined by possessing particular characters or sets of characters (Donoghue & Cantino 1988). Taxonomic boundaries are drawn through what are seen to be phenetic gaps.

In contrast, the phylogeneticist's goal is systematisation, i.e. the recognition of individual systems that are the natural outcome of evolutionary descent. Monophyletic groups are identified and diagnosed by synapomorphies, but they are not defined by characters of any sort. They owe their existence to descent from a common ancestor. Hennig (1966) stated that relations between categories in a phylogenetic system exist whether they are recognised or not. Characters are not themselves ingredients of the definition of higher categories, but aids used to apprehend the genetic criteria that lie behind them. The relationship between taxa and characters is "intentional" in phenetic and "evolutionary" classifications because characters determine the make-up of taxa (Donoghue & Cantino 1988).

Since all higher categories are simply monophyletic groups of groups, it is largely irrelevant at what level the groups of species are recognised (West 1987). However, in constructing a phylogenetic classification - a process quite independent of the cladistic analysis - arbitrary choices often following precedent must be made to determine the rank at which to recognise a particular group. There are specific conventions (see below) for translating a cladogram into a phylogenetic classification (Wiley 1981):

1) the Linnean hierarchy should be used with the following conventions to classify organisms,
2) classifications should be maximally useful and informative, i.e. the cladistic hypothesis must be retrievable from the classification,
3) two taxa that are more clearly related to each other than to any others are called sister groups (Hennig 1966) and are given equal taxonomic rank, however much they differ in divergence in number of species (Janvier 1984). Thus when one or a few species are the sister of a very large group, they have to be placed in a supraspecific taxon whose rank is equal to that of the large group,
4) taxa of considerable importance in the group being classified will be retained at their traditional ranks whenever possible, consistent with phylogenetic relationships and the taxonomy of the group as a whole (thereby maintaining nomenclatural stability),
5) no empty or redundant categories will be produced; for example when a taxon containing only one species is given high rank because it is the sister of a taxon of equally high rank, it is not necessary to create a succession of supraspecific taxa for this species, the highest ranked taxon being enough,
6) phyletic sequencing (Humphries & Funk 1984, Nelson 1974) may be used so that terminal taxa of an asymmetrical cladogram can be given equal rank and be listed in order of their branching sequence. Such a classification would reflect the cladogram, yet avoids the proliferation of categories resulting from having to name each dichotomy. This answers a common criticism that phylogenetic classifications would be impossible to use because of the exceedingly large number of ranks that can be generated by completely subordinating the cladogram (Bremer & Wanntorp 1981, Cronquist 1987, Funk & Stuessy 1978). Classifications may thus use a combination of "ranking" and "sequencing" although a statement as to which convention was used, and where, must accompany the classification,
7) when it is not possible to demonstrate that terminal taxa are monophyletic i.e. that they are diagnosed by synapomorphies, and yet no data are available indicating they are either paraphyletic or polyphyletic, the resultant polytomies have been called metaphyletic taxa (Donoghue 1985, Mishler & Brandon 1987). The convention is that terminal taxa in such polytomies (e.g. taxa A, B & C in Figure 3 should be given equivalent ranks and placed (with *sed mutabilis* after the name) at the level of the hierarchy where their interrelationships to other taxa are known,

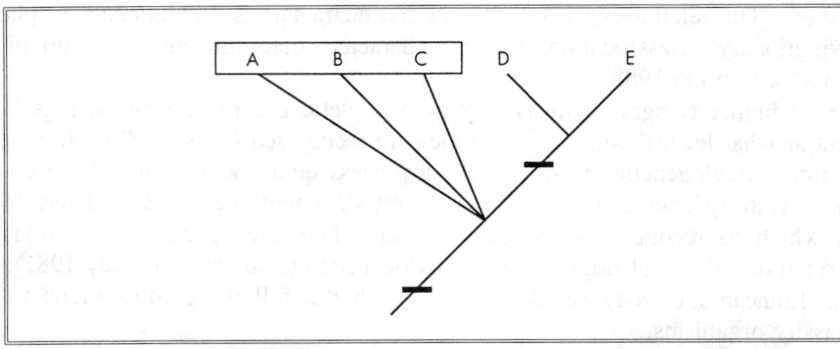

Figure 3. Taxon ABC is metaphyletic (i.e. a metataxon) since it is not possible to demonstrate that the group is either monophyletic, paraphyletic or even (rarely) polyphyletic.

8) a paraphyletic or polyphyletic assemblage, or a group of unknown status may be included in a phylogenetic classification if it is put in quotes and denoted afterwards by *incertae sedis* at the level in the hierarchy that the group is placed. No formal rank will be given to such groups.

Other conventions (Wiley 1981) have been described applying to fossil and hybrid taxa. In recognising genera, a balance will have to be achieved between the disadvantages of obscuring underlying organisation and patterns of similarity by "lumping" terminal taxa, or disrupting natural groups into artificial subfields and losing information by "splitting". A decision will have to be made at which level of universality in the nested set of monophyletic groups the genus should be recognised.

More explicit methods for recognising genera and to help choose between alternative classifications are suggested below:

1) genera must be inferred *a posteriori*, as a result of a robust phylogenetic analysis of an ingroup - and its nearest relatives - on a worldwide basis,
2) apply the appropriate level of unifying regularities consistently across the ingroup, implementing the sister-group criterion of Hennig (1966) and/or phyletic sequencing. Levels of unifying regularities characterising related genera in the same tribe or subtribe as the ingroup should also be a guide to equivalent levels within the ingroup (Vavilov 1922),
3) the level of monophyly ultimately recognised as a genus should be from the node most strongly supported by diagnosing characters. Such a group should also be highly predictive, with members expected to have similar combinations of characters,
4) above all, confusion with other groups must be avoided and genera should be easily recognised,
5) divisions of taxa, and rank, could also be guided by estimates of branch lengths (the measure of distance between taxa and nodes) generated from inferring a cladistic hypothesis (Carpenter 1993, Swofford & Olsen 1990),
6) paraphyletic groups inclusive in the ingroup should be resolved where possible, as these are likely to influence overall generic delimitation in that clade,

7) a broad consensus of views should be sought to find which name changes would be most widely accepted by the botanical community; perhaps by committee for particular families.

The conflict between the strong desire of achieving a natural classification and the need for stability of names, is likely to escalate as the number of cladistic analyses emerging of major plant groups increases. Many users of classifications express concern about nomenclatural stability, especially when changes to existing classifications are proposed, based on poorly supported cladistic studies. Traditional paraphyletic genera, for example, should be abandoned or re-arranged only after a broad consensus of views has been achieved.

While recognition of paraphyletic genera will remain a favoured and less disruptive option in those groups where phylogeny is poorly understood, this should not be the long-term goal of taxonomists. Such genera are far less predictive and informative than those based on monophyly, because evidence from both known and unknown shared characters, as well as historical ecology and phytogeography are lost when monophyletic groups are dismembered. Resistence to the reinterpretation of existing taxonomies in evolutionary terms is a major obstacle to the elimination of paraphyletic groups, and to producing classifications based on evolutionary relationships.

Despite the apparent subjectivity involved in designating which level in any phylogeny is called a genus, the criteria suggested here should help to reconcile widely differing views of classifying the same plant taxa, and go much further towards matching the aims of both the makers and users of classifications, i.e. to achieve relative stability of names within a natural classification.

Generic delimitation in the Leguminosae

Example 1: Tribe Indigofereae (Papilionoideae).

In a recent cladistic analysis of the tribe Indigofereae (Schrire 1994), the genus *Indigofera* L. was found to be polyphyletic as traditionally circumscribed. *Microcharis* Benth. and *Indigastrum* Jaub. & Spach were included as subgenera (Gillett 1958, 1971) and *Vaughania* S. Moore was confused with *Indigofera* (Polhill 1981a).

Of thirty two tribes recognized by Polhill (1981b) in the subfamily Papilionoideae, the Indigofereae form part of an alliance with a diffuse woody group, the Millettieae. The Indigofereae is one of the most advanced, herbaceous, predominantly Old World tropical tribes with only *Indigofera,* the third largest genus in Leguminosae, being pantropical. In the last major review of the tribe (Polhill 1981a) 4 genera were distinguished. *Indigofera* was broadly circumscribed to include some 780 species, while *Phylloxylon* Baill. (7 species), *Rhynchotropis* Harms (2 species) and *Cyamopsis* DC. (4 species) comprised the rest of the tribe.

As a result of this analysis (see Figure 4), *Vaughania* - which was originally treated as a monotypic genus - is upheld as a well substantiated monophyletic group comprising 11 species restricted to Madagascar. *Microcharis* (c. 35 species) and *Indigastrum* (9 species) are also distinct monophyletic groups well supported as derived elements in the sister clade to *Indigofera*. As presently circumscribed, therefore, (i.e. with the exclusion of the above taxa) *Indigofera sens. strict.* becomes a monophyletic group. Since 4 terminal taxa in the cladogram (Figure 4) are genera, by applying phylogenetic sequencing, the remaining 3 taxa *Phylloxylon, Cyamopsis* and *Rhyncho-*

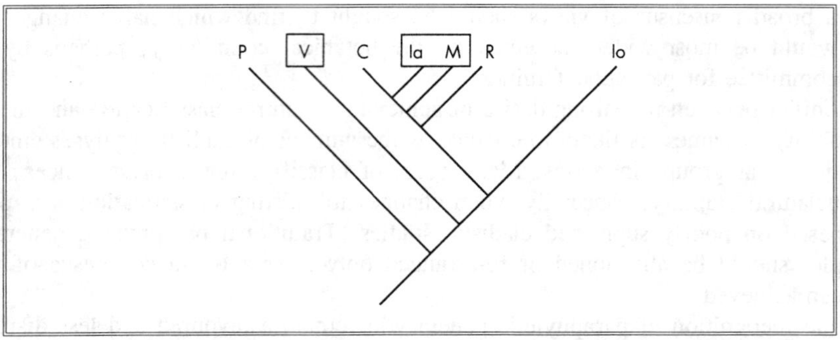

Figure 4. Cladogram of tribe Indigofereae. P = *Phylloxylon*; C = *Cyamopsis*; R = *Rhynchotropis* and Io = *Indigofera*. The taxa in boxes were previously included in a polyphyletic *Indigofera sens. lat*. (Polhill 1981). V = *Vaughania*; Ia = *Indigastrum* and M = *Microcharis*

tropis can be treated at generic level. These 3 genera have been reinstated by Schrire (1994).

The following classification of the tribe Indigofereae is proposed, indicating the branching sequence of the cladogram.

Tribe Indigofereae: Genus *Phylloxylon* Baill.,
Genus *Vaughania* S. Moore
Genus *Indigofera* L.
Genus *Rhynchotropis* Harms
Genus *Microcharis* Benth.
Genus *Indigastrum* Jaub. & Spach
Genus *Cyamopsis* DC.

Another alternative is to recognize *Phylloxylon* at subtribal level, i.e. as a monotypic subtribe. There is no obligation to do so, however, and ranking has been minimised to avoid a proliferation of categories arising by naming each dichotomy.

Example 2: *Caesalpinia* group (Caesalpinieae: Caesalpinioideae).

The genera of the *Caesalpinia* group are not yet as resolved as those in the tribe Indigofereae. This is partly because many features that have become fixed in more advanced groups of legumes are transitional and unstable in the Caesalpinieae (Polhill & Vidal 1981), and there is more controversy as to the delimitation of genera in this basal tribe.

Polhill & Vidal (1981) placed 16 genera containing 176 species in their *Caesalpinia* group of tribe Caesalpinieae. The *Caesalpinia* group cannot be understood without first unravelling the taxonomy of the genus *Caesalpinia* L. *sens. lat*. and this is a complex task given that 25 generic names are currently placed in synonymy. Of these, several were monotypic genera described by Britton & Rose (1930), but in the light of new data some of these have now been subsumed elsewhere.

A morphological cladistic analysis of the *Caesalpinia* group (Lewis & Schrire 1994) demonstrates that the traditional circumscription of the genus *Caesalpinia sens. lat*. is polyphyletic; some species are more closely related to other genera in the group than

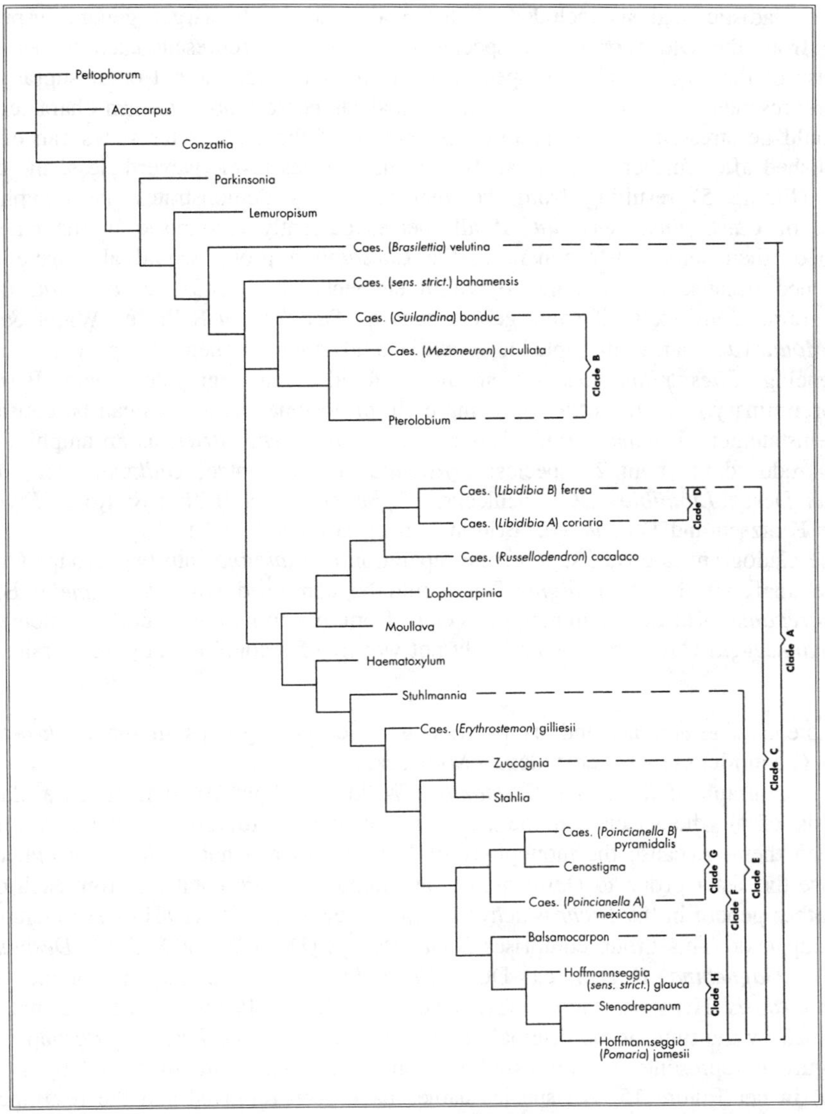

Figure 5. Preliminary cladogram of the *Caesalpinia* group (Leguminosae - Caesalpinioideae - Caesalpinieae).

to each other (see Figure 5). It is too early to reinstate genera presently in synonymy under *Caesalpinia sens. lat.* because all the species have not yet been analyzed, and monophyletic groups, therefore, have not yet been delimited. Some species intermediate between higher taxonomic groups have not been resolved systematically and this is necessary before the exact number of genera in the *Caesalpinia* group can be determined.

The cladistic analysis included 27 terminal taxa. In the larger genera, especially those from the Old World, one species was chosen to represent each higher taxon because of the large number of species involved. Some genera and other supraspecific taxa represented by a single species in the analysis possess non-uniform characters and it should be stressed that the precise distribution of these character states can only be established after further cladistic analysis at the species level. Nevertheless, the cladogram (Figure 5) resulting from the analysis clearly demonstrates the polyphyletic nature of *Caesalpinia sens. lat.* If all species currently recognised in the latter are retained, then some other genera in the *Caesalpinia* group would also have to be subsumed because they are nested within elements of *Caesalpinia sens. lat.* on the cladogram. Conversely, if some genera such as *Pterolobium* R.Br. ex Wight & Arn. and *Moullava* Adans. are upheld, as we would suggest, then by applying phyletic sequencing, *Caesalpinia sens. lat.* should be divided into segregate genera. Based on our preliminary cladogram the following eight provisional candidates can be considered for reinstatement at generic rank, leaving *Caesalpinia sens. strict.* as an amphi-atlantic genus reduced to about 25 species: *Brasilettia* (DC.) Kuntze, *Guilandina* L., *Mezoneuron* Desf., *Libidibia* (DC.) Schlecht., *Russellodendron* Britton & Rose, *Erythrostemon* Klotzsch and *Poincianella* Britton & Rose Groups A and B.

The cladogram does not support the separation of *Libidibia* into two groups (A & B of the analysis), but *Cenostigma* Tul. could be combined with *Poincianella* B, and *Stenodrepanum* Harms (a monotypic genus from Argentina) is a derived element of *Hoffmannseggia* Cav. which is probably not worthy of recognition at generic rank.

Example 3: *Desmanthus* and its putative generic sister groups in the *Dichrostachys* group (Leguminosae - Mimosoideae - Mimoseae).

A monograph of the genus *Desmanthus* Willd. (Luckow 1993) included a cladistic analysis of all the species in the genus in an attempt to discover the relationships between them. Because the monophyly of *Desmanthus* had not been demonstrated and because the sister group to *Desmanthus* was unknown, representatives from each of the four other genera in the *Dichrostachys* group of Lewis & Elias (1981) were included as outgroup taxa. This group comprises *Dichrostachys* (DC.) Wight & Arn., *Desmanthus* Willd., *Gagnebina* Necker ex DC., *Neptunia* Lour. and the monotypic genus *Calliandropsis* Hern. & Guinet (Hernandez & Guinet 1990), which had not been described by the time of the original circumscription of the *Dichrostachys* group.

Figure 6 represents a compressed version of the cladogram presented by Luckow (1993) in her Figure 15. All species names have been removed and the resulting tree collapsed to show the relationships between the genera currently recognised in the *Dichrostachys* group. As in the first two examples, this analysis demonstrates that the traditional circumscription of *Desmanthus* is not monophyletic. This is because *Calliandropsis nervosus*, originally placed in *Desmanthus*, is resolved within *Dichrostachys* (see Figure 6), forming a clade with its sister group *Gagnebina*. Also, *Neptunia* is a derived monophyletic group nested within *Desmanthus*.

Initially it would seem sensible to recognise *Neptunia* as distinct from *Desmanthus* because of the broader stipules and differently shaped pods with a different dehiscence mechanism and seed position.

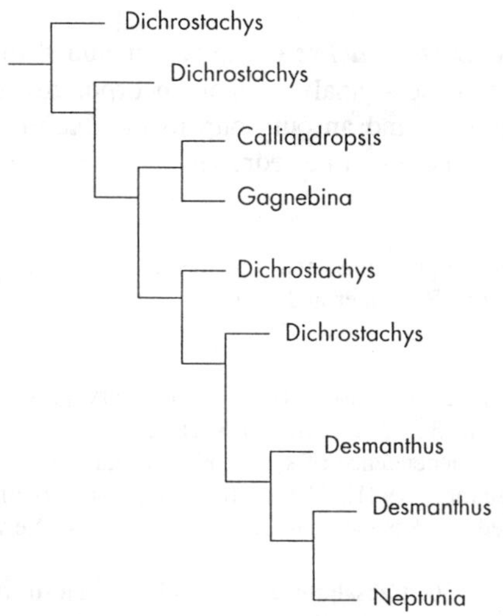

Figure 6. Compressed cladogram of *Desmanthus* and its putative generic sister groups in the *Dichrostachys* group (Leguminosae - Mimosoideae - Mimoseae).

All other characters used to separate *Neptunia* break down because they occur in one or more species of *Desmanthus*.

In the cladistic analysis, *Neptunia* is defined by two reversals to plesiomorphic characters, the fruit and stipule types also occurring in other genera closely related to *Desmanthus*. This weak support for *Neptunia* is obscured when autapomorphies are used as the only justification for distinguishing genera.

If *Neptunia* is recognised as a genus, this would leave *Desmanthus* as a paraphyletic group. This restrictive definition of *Desmanthus* loses information about relationships, shared characters and ecological and phytogeographical connections with *Neptunia*, which is part of the same monophyletic group as *Desmanthus*. The most pragmatic solution in any future analyses, is to recognise all species of *Neptunia* as part of *Desmanthus*, but not to change the nomenclature of the former until more detailed work on the group is available.

Luckow's analysis has not looked at all species in the *Dichrostachys* group apart from *Desmanthus*, and she comments that it would be premature to redraw generic boundaries based on her cladogram because:

1) The characters used to construct the cladogram were selected to resolve relationships among species of *Desmanthus* and *Neptunia*, not among genera of the *Dichrostachys* group. Several characters relevant to an evaluation of the monophyly of the genera were not included, because they were absent or invariant in *Desmanthus*. As an extension of this, a different data set that incorporated the appropriate characters to resolve generic relationships might lead to different conclusions about the relationships among the genera,

2) The sampling of species of *Dichrostachys*, *Gagnebina* and *Neptunia* is incomplete,

3) No outgroup to the *Dichrostachys* group was included in the analysis.

Luckow concluded that a new analysis that incorporates the appropriate data, all species in each of the genera and an outgroup to the *Dichrostachys* group, is required before the lines between genera can be redrawn.

Acknowledgements

We gratefully acknowledge suggestions and comments on the manuscript by M. Chase, A. Bruneau, C. Jeffrey, A. Paton, P. Linder and E. Knox.

References

Alberch, P., 1982. Developmental constraints in evolutionary processes. In: Bonner, J.T. (ed.), Evolution and development: 313-332, Springer-Verlag.

Brady, R.H., 1985. On the independence of systematics. Cladistics 1: 113-126.

Bremer, K. & H.-E. Wanntorp, 1981. The cladistic approach to plant classification. In: Funk, V.A. & D.R. Brooks (eds.), Advances in Cladistics 1: 87-94, New York Botanic Garden, New York.

Britton, N.L. & J.N. Rose, 1930. Caesalpiniaceae (conclusio). North American Flora 23: 301-342.

Brooks, D.R. & D.A. McLennan, 1991. Phylogeny, ecology and behaviour: a research program in comparative biology, University of Chicago Press, Chicago and London.

Calow, P., 1983. Evolutionary principles. Blackie, Glasgow and London.

Carpenter, K.E., 1993. Optimal cladistic and quantitative evolutionary classifications as illustrated by fusilier fishes Teleostei Caesionidae. Syst. Biol. 42: 142-154.

Clayton, W.D., 1983. The genus in practice. Kew Bull. 38: 149-153.

Cronquist, A., 1987. A botanical critique of cladism. Bot. Rev. 53: 1-52.

De Pinna, M.C.C., 1991. Concepts and tests of homology in the cladistic paradigm. Cladistics 7: 367-394.

De Queiroz, K., 1988. Systematics and the Darwinian revolution. Phil. Sci. 55: 238-259.

De Queiroz, K. & M.J. Donoghue, 1988. Phylogenetic systematics and the species problem. Cladistics 4: 317-338.

Donoghue, M.J., 1985. A critique of the biological species concept and recommendations for a phylogenetic alternative. Bryologist 88: 172-181.

Donoghue, M.J. & P.D. Cantino, 1988. Paraphyly, ancestors, and the goals of taxonomy: a botanical defense of cladism. Bot. Rev. 54: 107-128.

Duncan, T., 1980. Cladistics for the practising taxonomist - an eclectic view. Syst. Bot. 5: 136-148.

Eldredge, N. & J. Cracraft, 1980. Phylogenetic patterns and the evolutionary process, Columbia University Press, New York.

Estabrook, G.F., 1978. Some concepts for the estimation of evolutionary relationships in systematic botany. Syst. Bot. 3: 146-150.

Farris, J., 1974. Formal definitions of paraphyly and polyphyly. Syst. Zool. 23: 548-554.

Funk, V.A. & T.F. Stuessy, 1978. Cladistics for the practicing plant taxonomist. Syst. Bot. 3: 159-178.

Funk, V.A. & D.R. Brooks, 1990. Phylogenetic systematics as the basis of comparative biology. Smithsonian Contributions to Botany 73: 1-45.

Gillett, J.B., 1958. *Indigofera (Microcharis)* in tropical Africa with the related genera *Cyamopsis* and *Rhynchotropis*. Kew Bull., Add. Ser. 1: 1-166.

Gillett, J.B., 1971. Indigofereae. In: Milne-Redhead, E. & R.M. Polhill (eds.), Flora of Tropical East Africa 3, Legum.-Papil.: 212-330. Crown Agents for the Colonies, London.

Gilmour, J.S.L., 1961. Taxonomy. In: MacLeod, A.M. & L.S. Cobley (eds.), Contemporary botanical thought: 27-45, Oliver & Boyd, Edinburgh.
Griffiths, G.C.D., 1974. On the foundations of biological systematics. Acta Biotheoretica 13: 85-131.
Hennig, W., 1966. Phylogenetic systematics. University of Illinois Press, Urbana, Illinois.
Hernandez, H. & Ph. Guinet, 1990. *Calliandropsis*, a new genus of Leguminosae, Mimosoideae, from Mexico. Kew Bull. 45: 609.
Humphries, C.J. & V.A. Funk, 1984. Cladistic methodology. In: Heywood, V.H. & D.M. Moore (eds.), Current concepts in plant taxonomy, Systematics Association special volume 25: 323-362, Academic Press, London.
Janvier P., 1984. Cladistics: theory, purpose and evolutionary implications. In: Pollard, J.W. (ed.), Evolutionary theory paths into the future: 39-75. Wiley Interscience, New York.
Jeffrey, C., 1987. The concept of the genus. Austral. Syst. Bot. Soc. Newsletter 53: 27-31.
Lewis, G.P. & T. Elias, 1981. Mimoseae. In: Polhill, R.M. & P.H. Raven (eds.), Advances in legume systematics 2: 155-168. Royal Botanic Gardens, Kew.
Lewis, G.P. & B.D. Schrire, 1994. A reappraisal of the *Caesalpinia* group (Caesalpinioideae: Caesalpinieae) using phylogenetic analysis. In: Crisp, M.D. & J.J. Doyle (eds.), Advances in legume systematics, Part 7, Phylogeny (in press). Royal Botanic Gardens, Kew.
Luckow, M., 1993. Monograph of *Desmanthus* (Leguminosae - Mimosoideae), Systematic Botany Monographs 38: 1-166.
Mayr, E., 1969. Principles of systematic zoology, McGraw Hill, New York.
Mayr, E., 1974. Cladistic analysis or cladistic classification? Zeitsch. Zool. Syst. Evol. 12: 94-128.
McNeill, J., 1979. Purposeful phenetics. Syst. Zool. 28: 465-482.
Meacham, C.A. & T. Duncan, 1987. The necessity of convex groups in biological classification. Syst. Bot. 12: 78-90.
Mishler, B.D. & R.N. Brandon, 1987. Individuality, pluralism, and the phylogenetic species concept. Biology & Philosophy 2: 397-414.
Nelson, G., 1974. Classification as an expression of phylogenetic relationship. Syst. Zool. 22: 344-359.
Nixon, K.C. & J.I. Davis, 1991. Polymorphic taxa, missing valves and cladistic analysis. Cladistics 7: 233-241
Patterson, C., 1982. Morphological characters and homology. In: Joysey, K.A. & A.E. Friday (eds.), Problems of phylogenetic reconstruction: 21-24, Academic Press, New York.
Phillips, R.B., 1984. Considerations in formalising a classification. In: Duncan, T. & T.F. Stuessy (eds.), Cladistics: perspectives on the reconstruction of evolutionary history, Columbia University Press, New York.
Platnick, N.I., 1977. Paraphyletic and polyphyletic groups. Syst. Zool. 26: 195-200.
Platnick, N.I., 1979. Philosophy and the transformation of cladistics. Syst. Zool. 28: 537-546.
Polhill, R.M., 1981a. Indigofereae. In: Polhill, R.M. & P.H. Raven (eds.), Advances in legume systematics 1: 289-291. Royal Botanic Gardens, Kew.
Polhill, R.M., 1981b. Papilionoideae. In: Polhill, R.M. & P.H. Raven (eds.), Advances in legume systematics 1: 191-208. Royal Botanic Gardens, Kew.
Polhill, R.M. & J.E. Vidal, 1981. Caesalpinieae. In: Polhill, R.M. & P.H. Raven (eds.), Advances in legume systematics 1: 81-95. Royal Botanic Gardens, Kew.
Rieppel, O.C., 1988. Fundamentals of comparative biology. Birkhauser Verlag, Basel.
Schrire, B.D., 1994. Cladistic analysis of the tribe Indigofereae (Leguminosae - Papilionoideae). In: Crisp, M.D. & J.J. Doyle (eds.), Advances in legume systematics, Part 7, Phylogeny (in press). Royal Botanic Gardens, Kew.
Simpson, G.G., 1961. Principles of animal taxonomy. Columbia University Press, New York.
Simpson, G.G., 1965. The geography of evolution. Chilton Brooks, Philadelphia.

Simpson, G.G., 1975. Recent advances in methods of phylogenetic inference. In: Luckett, W.P. & F.S. Szalay (eds.), Phylogeny of the primates, a multidisciplinary approach: 3-19, Plenum, New York.

Sneath, P.H.A. & R.R. Sokal, 1973. Numerical taxonomy. W.H. Freeman, San Francisco.

Stebbins, G.L., 1984. Mosaic evolution, mosaic selection and angiosperm phylogeny. Bot. Jour. Linn. Soc. 88: 149-164.

Stevens, P.F., 1984. Homology and phylogeny: morphology and systematics. Syst. Bot. 9: 395-409.

Stevens, P.F., 1986. Evolutionary classification in botany, 1960-1985. Jour. Arn. Arb. 67: 313-339.

Swofford, D.L. & G.J. Olsen, 1990. Phylogeny reconstruction. In: Hillis, D.M. & C. Moritz (eds.), Molecular systematics: 411-501. Sinauer, Sunderland, Massachusetts.

Vavilov, N.I., 1922. The law of homologous series in variation. Journal of Genetics 12: 47-89.

Waddington, C.H., 1975. The evolution of the evolutionist. Edinburgh University Press, Edinburgh.

West, J.G., 1987. The role of cladistics in generic delimitation. Austral. Syst. Bot. Soc. Newsletter 53: 38-47.

Wiley, E.O., 1981. Phylogenetics: the theory and practice of phylogenetic systematics. John Wiley & Sons, New York.

Wiley, E.O., D. Siegel-Causey, D.R. Brooks & V.A. Funk, 1991. The compleat cladist: a primer of phylogenetic procedures. The University of Kansas Museum of Natural History special publication 19, Lawrence, Kansas.

In defence of paraphyletic taxa

R.K. BRUMMITT
The Herbarium, Royal Botanic Gardens, Kew, Richmond, Surrey, TW9 3AE, United Kingdom.

Abstract
It is maintained that a classification based solely on phylogeny* is different from and incompatible with the Linnaean system, and the theory of Hennigian classification is flawed. A purely phylogenetic* system has no class concepts, and, if paraphyletic taxa are not allowed, because every descendant taxon has to be of a lower rank than its ancestor and sunk into it, infinite regression occurs and the whole classification telescopes back into its original ancestral taxon. Every genus since the original one must make another genus paraphyletic, and the latter may or may not be extinct. A cladogram, which purports to represent phylogenetic relationships of taxa, is based on a hierarchy different from that of a Linnaean classification, which represents anagenetic relationships. Neither a cladogram nor a phylogenetic tree is a classification, and subjective decisions must always be taken to impose the limits of taxa and their rank. Existing cladistic classifications became possible probably because either the taxa were incompletely known (many ancestral taxa having become extinct), or the data were inadequate (leaving cladograms unresolved), or because the classification was incomplete (not enough ranks being included). Lowering the rank of one taxon to avoid another being paraphyletic generally merely shifts paraphyly to a different rank. A phylogeny and a classification are both desirable but have different functions, and should be allowed to exist side by side, inter-related but not interdependent, to give maximum predictivity. The widespread and increasing disruptiveness of the application of cladistic theory of classification is an additional reason for rejecting it. The criterion of convexity, which is also based on evolutionary history but allows paraphyletic taxa, is a logical possibility and should be adopted in classification. Taxa should be recognisable by characters, not by position in a putative phylogeny. Examples given of the difficulties of Hennigian classification are the cases of *Ptycholobium*, which makes the currently recognised *Tephrosia* paraphyletic, and *Neuracanthus* sect. *Leucobarleria* which would make sect. *Neuracanthus* paraphyletic, and other cases mentioned include Lemnaceae which makes Araceae paraphyletic, Cruciferae which makes Capparaceae paraphyletic, and the monocots which make the dicots paraphyletic.

* In this paper the words "phylogeny" and "phylogenetic" are used to refer to relationships of ancestry and descent (cladogenesis) only. Linder (1988, p. 208) notes that they are "correctly" used in a broader sense including both cladogenesis and anagenesis.

Résumé
On soutient qu'une classification basée uniquement sur la phylogénie est différente du système linnéen et incompatible avec celui-ci, et que la théorie de la classification selon le système de Hennig n'est pas absente de défauts. Un système purement phylogénétique n'a pas de concept de classes et, si l'on n'accepte pas les taxons paraphylétiques pour la raison que chaque taxon descendant doit avoir un rang inférieur à son ancêtre et y être inclus, une régression sans borne s'opère et toute la classification se confond finalement avec le taxon ancestral d'origine. Chaque genre, à partir du genre d'origine, doit produire un autre genre paraphylétique et ce dernier peut ou non avoir disparu. Un cladogramme, qui suppose représenter les relations phylogénétiques des taxons, a pour base une hiérarchie différente de celle de la classification linnéenne, qui représente des relations anagénétiques. Pas plus un cladogramme qu'un arbre phylogénétique ne constitue une classification, et il est toujours nécessaire de prendre des décisions subjectives pour fixer les limites des taxons et leur rang. Des classifications cladistiques existantes sont devenues possibles, probablement parce que soit les taxons étaient incomplètement connus (de nombreux taxons

L.J.G. van der Maesen et al. (eds.), The Biodiversity of African Plants, 371-384.

ancestraux ayant disparu), soit les données étaient inadéquates (laissant des cladogrammes non résolus), soit parce que la classification était incomplète (pas assez de rangs n'étant inclus). L'abaissement de rang d'un taxon pour éviter qu'un autre ne devienne paraphylétique ne fait généralement que déplacer la paraphylie à un autre rang. Une phylogénie et une classification sont toutes deux désirables mais ont des fonctions différentes et on peut concevoir qu'elles existent côte à côte, interconnectées mais non interdépendantes, afin d'obtenir un maximum d'information prévisible. Le bouleversement croissant et étendu de l'application de la théorie cladistique de la classification est une raison supplémentaire pour la rejeter. Le critère de convexité, qui se base également sur l'historique de l'évolution mais accepte des taxons paraphylétiques, est une possibilité logique et pourrait être adopté dans une classification. Les taxons devraient pouvoir se reconnaître par des caractères et non par leur emplacement dans une phylogénie hypothétique. Comme exemples des difficultés de la classification selon Hennig, on notera le cas de *Ptycholobium*, qui rend paraphylétique *Tephrosia*, genre couramment accepté, tout comme celui de la section *Leucobarleria* de *Neuracanthus*, la section *Neuracanthus* devenant paraphylétique; d'autres cas évoqués se rapportent aux Lemnaceae qui donnent aux Araceae un statut paraphylétique, aux Cruciferae qui donnent aux Capparaceae un statut paraphylétique, ainsi qu'aux Monocotylédones qui donnent aux Dicotylédones un statut paraphylétique.

Key words
Cladistics, Classification, Convexity, Paraphyletic, Phylogeny, Regression.

Introduction

At our last AETFAT Congress, in Malawi in 1991, we held a symposium on specific and infraspecific concepts in the African flora (Seyani & Chikuni 1994, pp. 295-509). In 21 papers presented in that symposium the subject of cladistics was raised only once.

This time we are holding a symposium on generic concepts, and from the start the conveners have found it necessary to raise cladistic classification as a major point of discussion. This change of emphasis is partly due, I think, to the fact that cladistic methods are more commonly used at generic rank than at infraspecific ranks, but it is also a reflection of the very rapidly growing influence of cladistic theory in taxonomic botany. In the last year I have attended three international congresses and have heard many speakers present their cladograms and argue their new taxonomic treatments derived from them. Quite commonly this includes a statement that some previously recognised taxon is now found to be paraphyletic and therefore has to be abandoned.

These statements always cause me concern, because when I was first introduced to the cladistic theory of classification, probably about 20 years ago, it seemed to me that the whole idea of classification without paraphyletic taxa was a logical impossibility. In subsequent years I have often tried to convince my cladistic friends of my arguments over coffee-tables, sadly without success. So I am not sure whether, as a consequence, I have been "invited" or "challenged" to put my coffee-table views forward to an international audience today. I accepted hesitantly, and over the last year I must have decided a dozen times that I am not the person to do this and that I should back out. But the question of paraphyletic taxa is absolutely fundamental to the job of any taxonomist, and none of us can duck the issue. We have to face this question repeatedly on a day to day basis, for these days one cannot even refer to the dicots without a huge question mark arising in one's mind.

The controversy

In proposing his theory of phylogenetic systematics Willi Hennig argued that no paraphyletic group should be accepted as a formal taxon in a classification (Hennig 1966, pp. 146-148). In practical terms for this symposium on generic concepts, this means that according to Hennigian (or cladistic) taxonomy, a classification in which any genus is ancestral to another genus cannot be accepted, and either the ancestral genus has to be split up or the descendant genus has to be sunk into it. There is no doubt that this is a central plank of the cladistic theory of classification (see, for example, Donoghue & Cantino 1988, p. 113). However, many of Hennig's original premises have been cast aside by current cladists (see Hull 1984, p. 15; Linder 1988, p. 209) and now appear naive, and his views on paraphyletic groups are, I believe, now being increasingly challenged in private discussion. Few botanists, however, other than Buck (1986) and Cronquist (1987), have troubled to publicise their opinions. Jeffrey (1987) concluded his paper by noting that "The realization of the objective existence of inevitably paraphyletic taxa might save a good deal of unprofitable searching for non-existent synapomorphies". Notable among those who have stated their opinion in the context of actual examples are Orchard (1992) in opposition to Anderberg (1990, 1992), and Sosef (1994, pp. 81-82). In *Taxon* earlier this year Rieseberg & Brouillet (1994) have pointed out that many plant species are likely to be paraphyletic taxa, because they have given rise to other species. Many botanists will take little convincing that this is an obvious truth, and will feel that exactly the same arguments apply at generic and other ranks as well.

Example 1.
Tephrosia and *Ptycholobium* (Leguminosae)

About 25 years ago I completed an account of some 140 taxa of the genus *Tephrosia* (Leguminosae) for Flora Zambesiaca (unpublished). In doing so I had to take decisions on the status of four peripheral groups which had sometimes been segregated as distinct genera. One of these, *Ptycholobium* Harms, I had already recognised as a genus (Brummitt 1965), and I have often used it as an example in my coffee-table arguments with cladists. In order to present all four cases as I saw them, i.e. from the traditional intuitive taxonomic point of view, I published a review of them all (Brummitt 1980). Most of the 3-400 species of *Tephrosia* have pinnately compound leaves and flowers in a terminal pseudoraceme (several flowers at each node). One apparently rare species, *T. sylitroides* Baker, from Angola, differs from the usual facies of the genus in having leaves trifoliate or unifoliate and flowers all in axillary fascicles, both characters which I instinctively view as derived or apomorphic in the genus, and both of which occur sporadically in other species in other parts of the genus. Its fruits, nonetheless, are the flat, stiff-walled, dehiscent pods characteristic of the genus *Tephrosia*. The five taxa (3 species plus 2 subspecies) which I put in *Ptycholobium* are very similar to *T. sylitroides* in facies (the epithet is derived from the name *Sylitra* which is a synonym of *Ptycholobium*) and have the same trifoliate or unifoliate leaves and axillary flowers. They occur in the drier areas of southern tropical Africa, including Angola, with one species *P. plicatum* (Oliv.) Harms also extending to sub-Saharan Africa and southern Arabian peninsula. In particular *P. biflorum* (E. Mey.) Brummitt from Namibia shows a very close resemblance to

T. sylitroides. *Ptycholobium* differs from *Tephrosia*, however, in having strongly contorted, thin-walled, indehiscent pods, and also in the terminal leaflet having lost its petiolule.

It seems very likely indeed to me that *Ptycholobium* has descended from an ancestor which would be placed within *Tephrosia*. The ancestor might even be conspecific with the present day relict species *T. sylitroides*, but that is pure conjecture. In a cladogram *Ptycholobium* would surely nest right within the genus *Tephrosia*, and I think *T. sylitroides* would prove to be its sister group. So if I recognise *Ptycholobium* as a genus, I make *Tephrosia* as currently recognised a paraphyletic group, and according to cladistic theory this is not a permissible taxonomy.

When I have put this case to my cladistic friends the usual first reaction is that I should make *Ptycholobium* a subgenus. But this would be just as bad, because then subgenus *Tephrosia* would be a paraphyletic taxon. (It does not matter whether the two subgenera I proposed in *Tephrosia* in 1980 are accepted or not). Perhaps a mere section then? But unless I take the unlikely course of making *T. sylitroides* a section on its own, the same arguments apply, and one of the sections of *Tephrosia* would then be paraphyletic, and for all I know we could go right through the hierarchy of ranks and still not recognise *Ptycholobium* at any of them. If I am right in thinking that the species *T. sylitroides* is the sister group, Hennig's argument that the sister group should always be the same rank would perhaps lead me either to treat *T. sylitroides* as a subgenus or section or to treat the whole of *Ptycholobium* as one species. Of course this would be nonsense, but few cladists these days still accept this piece of Hennigian dogma (Linder 1988, p. 209). The alternative cladist position is that I should raise various monophyletic groups within *Tephrosia* to generic rank alongside *Ptycholobium*, even though they have very weak characters which no traditional taxonomist would ever consider to be of generic significance. Such a classification would receive no support in practical taxonomy.

My position on this case is that paraphyletic taxa must be allowed. I reserve the right to recognise *Ptycholobium* as a genus if I want to, but if others want to sink it into *Tephrosia* they also have the right to do that. The chances of generic recognition of such a group probably varies with the number of species it includes. Very probably if *Ptycholobium* had evolved further and included, say, 20 or 30 species, then its chances of recognition would be much higher. As it is, three out of the five taxa concerned have never had a name published under *Tephrosia*, and I believe many taxonomists (but not all) would, like myself, opt to keep up the genus *Ptycholobium*. It is a subjective decision, and there is no obligation for any one course to be followed.

But the important question is not whether *Ptycholobium* should be recognised as a genus or as a subseries. It is whether a group which shows a clear morphological discontinuity from another group, and which is immediately recognisable by a distinctive character which is conspicuously absent from the other group, and whose distinguishing feature (pod structure) is of a similar nature to the characters which are used to define other major groups (usually genera) in the same family, has to be given the same rank as another group (here one or more species of *Tephrosia*) which does not possess any of these characteristics. In other words, are we classifying by characters or by position in a putative phylogeny? This is really the dilemma which faces us all.

Example 2.
***Neuracanthus* sect. *Neuracanthus* and sect. *Leucobarleria* (Acanthaceae)**
My colleague Sally Bidgood and I have recently produced a revision of the genus *Neuracanthus*, recognising 31 species plus 8 subspecies (Bidgood & Brummitt, in press). Four species plus 1 subspecies are in India and Indochina, while the remaining taxa are in tropical Africa (including 8 species in Madagascar) and the southern Arabian peninsula. The genus is very natural and apparently rather isolated in the family, with very uniform and unique pollen and little variation in floral characters, most of the useful taxonomic characters being found in vegetative parts, inflorescences and fruits.

We have recognised three sections. The most easily recognisable of these, sect. *Leucobarleria* (Lindau) Chiov., includes 12 species almost entirely confined to Somalia and adjacent countries. It is easily recognisable by the presence of very reduced sterile lateral inflorescences which form a basket of dense spines at the base of the fertile inflorescences. It also shows a characteristic dense white indumentum on the stem and leaf under-surface, but one or two other species also have this. This section has, with one exception, four seeds per fruit. Nearly all other species in the rest of tropical Africa, including all the Madagascan species, lack the spines and white indumentum, but have only two seeds per fruit, and we have placed these in a sect. *Didymocarpa* Bidgood & Brummitt. This leaves two species in sub-Saharan Africa plus all tropical Asian taxa which have four seeds but lack the characters of sect. *Leucobarleria*. These form our sect. *Neuracanthus*.

Viewing this in evolutionary terms we see sect. *Leucobarleria* as a monophyletic highly derived clade, and since it is usual in Acanthaceae to regard 2-seededness as a more advanced character than 4-seededness we regard sect. *Didymocarpa* as also defined by an apomorphic character state. This leaves sect. *Neuracanthus* as defined by the plesiomorphic character states and no unique synapomorphy, and so not recognisable as a taxon under cladistic theory. What is to be done? Even if we were to allow that the argument for sect. *Didymocarpa* being monophyletic might be a little weak, and it might be better not to give formal status to the group, surely the highly distinctive sect. *Leucobarleria* deserves taxonomic recognition. It was originally recognised as a separate genus *Leucobarleria* by Lindau, but we have accepted Chiovenda's reduction of this to a section of *Neuracanthus*. But as soon as we give it any infrageneric status, we leave the rest of *Neuracanthus* as a paraphyletic group once the relationships have been resolved. Apparently the only way to avoid a paraphyletic taxon is not to recognise any infrageneric taxa.

Two cladist friends to whom I put this case advised me that I should not give any formal status to sect. *Leucobarleria* but merely recognise it as an informal grouping. Another cladist friend went further. Immediately recognising the situation as having parallels in his own work (it is in fact a very common situation), he said that he was thinking of putting forward a proposal to amend the International Code of Botanical Nomenclature to rule that it is not necessary to recognise a nomenclaturally typical infrageneric taxon, i.e. we could have our sections *Leucobarleria* and *Didymocarpa* and not recognise any other section. This is ingenious, but of course it is pure chance that the type species of *Neuracanthus* falls into the paraphyletic group. He had no answer when I asked what we should have done if the type species had been one of those in sect. *Leucobarleria*.

Theoretical considerations

I have given these two examples just to show the sort of practical problems which can arise if we do not allow paraphyletic taxa. But my objections to cladistic classification are not merely practical. I believe that the practical problems arise from a basic theoretical failure.

Cladists are inclined to argue that their classification is given reality because it is a reflection of the pattern of actual descent in evolution. In their review of cladistic methodology, Humphries & Funk (1984, p. 347) have said that "the cladogram and the classification are one and the same". I disagree fundamentally with this statement. Neither a cladogram nor a phylogenetic tree will tell you where the limits of genera are to be drawn. It takes subjective decisions by a taxonomist to construct a classification into sections, genera, tribes, families etc. A cladogram is not a classification. How, then, do we proceed from a phylogenetic tree or a cladogram to a classification? Traditionally taxonomists have considered characters and character combinations and have subjectively defined how genera, families etc. should be delimited. Despite the subjectivity, there is a very wide agreement among taxonomists, though it is certainly not always unanimous. A cladogram does not tell you where to draw taxonomic limits, but cladistic theory does impose a restricting factor on this operation by saying that paraphyletic groups cannot be recognised as taxa. In saying this cladists require that any descendant group must have a lower rank than the group from which it is descended so that it can be included within that ancestral group in a classification. (Thus if *Tephrosia* is a genus then *Ptycholobium*, which is descended from within it, must have a rank lower than genus so that it can be included in the genus *Tephrosia*). So, if we have complete evidence of ancestry, in a phylogenetic tree we have to sink the monophyletic terminal taxa into their paraphyletic ancestors below them, then we have to sink these into their ancestors, these again into the previous generation, and we end up in infinite regression telescoping the whole tree into its ancestral taxon. This telescope effect is why I have argued for 20 years that a purely phylogenetic classification, which does not accept paraphyletic taxa, is a logical absurdity, or at least is completely incompatible with traditional Linnaean taxonomy. (I find the analogy of a telescope useful, but of course telescopes are linear structures and not dichotomously branched, so the analogy should not be pushed too far. Perhaps a better analogy would be that of a whirl-pool which sucks a wider and wider range of taxa in at the top and drags them all down to the base.)

A truly phylogenetic classification would have a completely different structure and the ancestral taxon would be included in every other taxon recognised. The difference is illustrated in Figure 1, where hierarchies and corresponding Venn diagrams depict the hierarchical relationships within (a) a phylogenetic tree, and (b) a Linnaean classification. The significant thing to note is that in the Venn diagrams in (a) the thing which is subdividing is the ancestral taxon and its descendants are the small inner circles, while in (b) what is subdividing is the entire group and the stem taxon is merely one of the inner circles. In (b), unlike (a), the ancestral taxon does not have a direct relationship to all other taxa. The relationships between (a) and (b) are reversed,

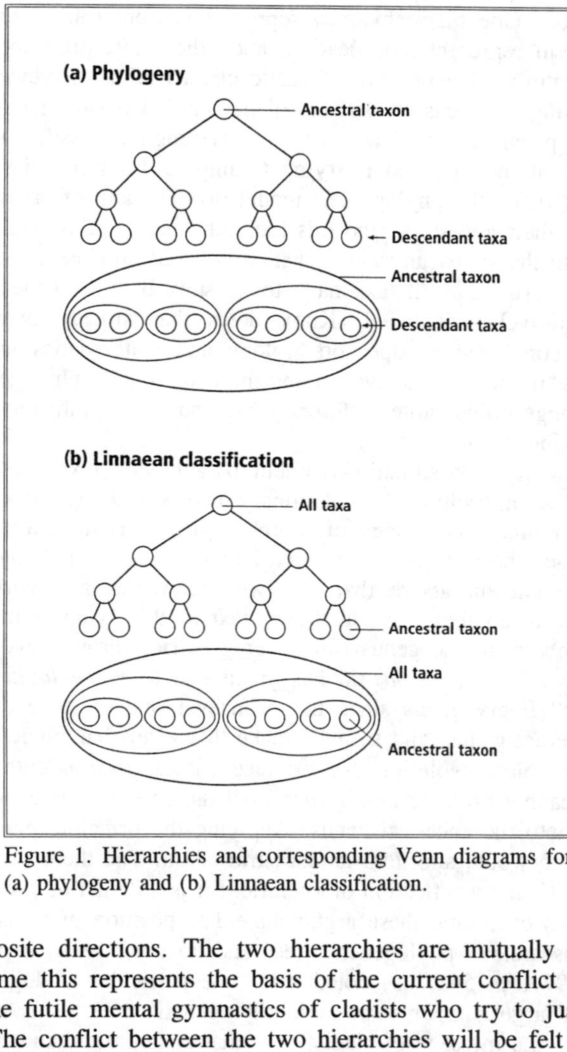

Figure 1. Hierarchies and corresponding Venn diagrams for (a) phylogeny and (b) Linnaean classification.

working in opposite directions. The two hierarchies are mutually incompatible and conflicting. To me this represents the basis of the current conflict in taxonomy and what I see as the futile mental gymnastics of cladists who try to justify equating the two diagrams. The conflict between the two hierarchies will be felt at every level in the taxonomic system if we can elucidate actual phylogeny.

After I had circulated the first draft of this paper to numerous friends recently, I was delighted to receive from Eric Knox, who is speaking later in this symposium, the typescript of an extensive paper which he has already in press in the Botanical Review. In a review of the use of hierarchies as organizational models in systematics he has provided a broad perspective to the philosophical background to this debate (Knox, in press). As I read this I was repeatedly struck by the concordance between my own intuitive views as a taxonomist and his appreciation of the problems from a broader standpoint, particularly on the question of paraphyletic taxa. He has emphasized the dualistic nature of evolution, or, as Darwin put it, "descent *with* modification". He points out that there is no one model which is *the* hierarchy which represents the total course of evolution. One hierarchy may represent descent (phylogeny), while a different hierarchy will represent modification, and other quite different hierarchies may represent other features of evolution. Cladistic classification, however, is "a monistic approach with a singular focus on patterns of descent". Considering my own Figure 1 in relation to his paper, I would now say that Hennigian classification confuses two different hierarchical models, and is trying to impose Linnaean classification, which should be model (b), on the phylogenetic model (a). The aim of (a) is simply to relate all clades back to their ancestors, and this hierarchy has no class concepts. When this model is applied to the Linnaean system, which is based on class concepts, it therefore has the inevitable result that all taxa have to be sunk back into the stem taxon, as I have already intuitively concluded. Knox notes that already over 40 years ago Woodger (1952) concluded a paper on biology and mathematics in a philosophical journal by saying "the taxonomic system and the evolutionary phylogenetic scheme are quite different things doing quite different jobs, and only confusion will result from identifying or mixing them".

The whole idea of a classification without paraphyletic taxa runs contrary to my intuitive feelings as a traditional evolutionary taxonomist. Evolution is continually producing a wider and wider range of morphological variation, and yet cladists say that as it progresses the descendant taxa have to have lower and lower rank. This is reflected in my comment above that the two hierarchies are working in opposite directions. Looked at another way, in the context of this symposium on generic concepts, if the members of one genus can never give rise through evolution to another genus in the way I believe *Tephrosia* has given rise to *Ptycholobium*, how can new genera ever arise? Every genus since the primaeval one must be descended from a member of a different genus, and so must make that genus paraphyletic. But according to Hennigian principles, evolution can produce endless *descendants* showing endless variation, but it cannot produce new *genera*. All the genera in one family would have to be telescoped into the ancestral genus. Applying the principle more broadly, every monophyletic group would sink into its ancestral taxon. The theory is nonsensical, and is seems to me to be not a reflection of evolution; it is a denial of evolution.

How do cladists overcome these arguments? The position of transformed cladistics is that cladograms are not phylogenetic trees and do not purport to indicate ancestry. Thus Platnick (1979, p. 541) has noted that "one particular cladogram might be the result of any of a large number of possible phylogenetic trees", but has still concluded (l.c. p. 542) that "the message of cladistics is that cladograms, and not [phylogenetic] trees, are the necessary and sufficient basis of classification". Further, Donoghue & Cantino (1988, p. 129) have said "Although all groups in a cladogram appear at the tips of branches (species are not assigned to internal nodes), this is not meant to imply that ancestors do not exist or that none of the species in the analysis are ancestral to others". Cladograms do not, in fact, tell us what is ancestral to what. So if evolutionary history is the underlying justification for Hennigian classification, and even that is called into question (my arguments here), why should I base my classification on a cladogram which is at least one stage removed from being a representation of actual evolution? If one defines paraphyly in terms of actual ancestry rather than position in a cladogram, as many do and as for example Donoghue & Cantino suggest (l.c. p. 113), the case for basing one's classification on a cladogram seems to be very suspect.

In the literature on cladistics which I have consulted, I have found virtually no consideration of the infinite regression referred to above. In his well-known botanical critique of cladism Cronquist (1987) on p. 33 got half way to making the point, saying that "The descendant must necessarily be included in the same taxon as its ancestors", but failed to draw the obvious conclusion that all descendants must therefore be successively sunk into the original ancestral taxon. Unfortunately Cronquist then descended to arguments of convenience and tradition, which were rightly attacked by his opponents (Humphries & Chappill 1988, Donoghue & Cantino 1988). Elsewhere I have noted a brief reference by Phillips (1984, p. 264) who wrote "Unfortunately this solution would continue through the entire phylogenetic tree of life on earth, requiring all things to be members of the same genus", but he apparently considered this to be so ridiculous that it was not worth further discussion. Comments similar to some of mine here have been made in a textbook of vertebrate palaeontology by Carroll (1988) who noted that "Approximately half the recognised taxonomic groups, ranging from species to classes, are paraphyletic" (p. 14), and that taxa could be sunk "ad infinitum" (p. 13). However, what seems to me to be the logical conclusion, that a classification without paraphyletic taxa is an impossibility, is nowhere clearly stated.

Practical considerations

You may well ask how it is, if the whole idea is based on a logical absurdity, that many botanists are producing classifications based on cladistic theory which apparently work satisfactorily. I think there are three major factors in this, all related to incompleteness of data. Firstly, and probably most important, is that the complete evolutionary record is seldom available because of extinction. The actual ancestors of our present day taxa usually do not need to be accommodated in our cladograms. The widely divergent views of some cladists on how one should cope with fossils and ancestors were reviewed by Wiley (1981a, pp. 214-223), and I think his conclusion that a "a general system of classification must be capable of handling all organisms and not just descendants" was inescapable. Further discussion is given by Donoghue & Cantino (l.c. pp. 120-124). Where extinction has operated, the telescope effect cannot operate, because there is nothing for the descendant taxa to sink into. But not *all* the ancestral taxa do become extinct. The evolution of *Ptycholobium* did not require *Tephrosia* to become extinct. Where the ancestral taxon has persisted, then we have problems. Secondly cladograms are often unresolved due to inadequacy of data, so we end up with so-called metaphyletic taxa, which is just an admission of ignorance allowing us to skirt around the real problems. And thirdly I think many important problems are avoided by taxonomists dealing with only major taxonomic ranks - family, genus and species. When one applies cladistic principles to subfamilies, subgenera and sections, subspecies etc., then the problems really hit you.

An example of the latter point arose at the Monocot Conference held at Kew in 1993. A paper by Mayo, Bogner & Boyce (in press) presented cladograms of the Arales which showed that the Lemnaceae nest neatly within the heart of the Araceae. The speakers said that they reluctantly felt obliged in theory to sink the Lemnaceae into the Araceae, which would otherwise be paraphyletic. There is nothing new in sinking the Lemnaceae into the Araceae; Engler did it about a century ago. But there are at least two subfamilies in the Araceae and, except in the very unlikely event of a trichotomy, the Lemnaceae must have arisen from one subfamily or the other - so that would then be paraphyletic. And then the Lemnaceae must have arisen within one of the many tribes of Araceae recognised, and, if it is monophyletic, from within one genus of the Araceae (possibly, but not necessarily, now extinct). So if we had the full data available we might end up having to sink the whole of the Lemnaceae (4 genera, c. 30 species) into one genus of Araceae. You will recognise my telescope effect operating again.

This leads me naturally on to a very interesting and challenging paper published earlier this year by Judd, Sanders & Donoghue (1994) in which they report on their phylogenetic analyses of a number of closely related groups of angiosperm families. Among their recommendations, based on their perceived need to avoid paraphyletic families, are the sinking of Araliaceae into Umbelliferae, Asclepiadaceae into Apocynaceae, Capparaceae into Cruciferae, both Dipsacaceae and Valerianaceae into Caprifoliaceae, and all of Moraceae, Cecropiaceae and Cannabaceae into Urticaceae, to quote only those important in our African context. My first objection is not that it is inconvenient or disruptive to lose these well-known families, but that no indication is given of the resulting infrafamilial classifications. The authors themselves do comment (p.42) that "care must be taken not to inadvertently create paraphyletic infrafamilial taxa", but in all cases they completely fail to address the question. They accept that Asclepiadaceae is monophyletic (though acknowledging that this is controversial). If they sink Asclepiadaceae, which is commonly treated as having three subfamilies (Liede & Albers 1994), into Apocynaceae, they must surely make one of the two existing subfamilies of Apocynaceae (see Leeuwenberg 1994) paraphyletic. Then if the Asclepiadaceae is monophyletic it will sink down through tribe, subtribe and presumably even genus (possibly now extinct) within Apocynaceae to find where it eventually nests. This is the logical conclusion of the theory. Every time, sinking one taxon into another only shifts the level of paraphyly down one rank. It is not enough merely to sink Capparaceae into Cruciferae; we need to know how we can cope with this in the intrafamilial classification of Cruciferae. We cannot make the present family Capparaceae into a subfamily Capparoideae, nor a tribe Cappareae, nor a subtribe Capparineae, since all these would be just as paraphyletic as the family. Do Cruciferae *sensu stricto* nest within Cleomoideae, and, if so, where? Are the present subfamily Capparoideae paraphyletic in relation to both subfamily Cleomoideae and the present family Cruciferae? But then is it really necessary to obfuscate the non-botanist's concept of the Cruciferae by including in it the Caper, *Capparis spinosa* L., with a conspicuous mass of stamens in every flower? Who benefits from this? And there may be more trouble ahead to judge by the postscript (p. 51) which notes the Caesalpinioideae, Simaroubaceae, Scrophulariaceae and Ericaceae are also paraphyletic.

These cases show that a cladistic classification may often be very different from a Linnaean one. This is not surprising when the criteria for recognising taxa are so different. In a cladistic classification taxa are defined principally on the putative phylogenetic position, whereas in Linnaean classification taxa are defined by discontinuity of variation. Taxonomists must be aware of the fact that users of their work need to be able to use their classifications. Taxa must be recognisable by characters and identifiable in keys, not by position in a putative phylogeny. What we have seen so far of the effect of the introduction of cladistic classification is only the tip of the ice-berg. If you take any taxon which has been traditionally divided into two subtaxa, it is inherently unlikely that *both* these subtaxa arose simultaneously from their common ancestor, and one will prove to be ancestral to the other and so paraphyletic. Thus the flowering plants were traditionally divided into two parts, monocots and dicots, but now we understand that monocots were derived from early dicots. The dicots are therefore paraphyletic and we are not allowed to recognise them as a taxon. My *Neuracanthus* example is similar in principle, as is almost any genus which is divided into two sections or subgenera, or family divided into two subfamilies. I predict that very many currently recognised taxa, at all levels, will prove to be paraphyletic and susceptible to rejection by cladistic theory.

Alternatives to Hennigian classification

Is it necessary to abandon the principles of Linnaean classification and insist on a purely phylogenetic alternative? Some cladists argue their case very strongly; at a seminar at Kew recently a visitor pronounced, apparently seriously, that "to recognise a paraphyletic genus would be a diabolical sin". There is a strong emotive argument in claiming that cladistic classification is based on evolution, which sounds "scientific". But for others it is not necessary to abandon the evolutionary history of a group as a factor in their classification. Duncan (1980), Estabrook (1986) and Meacham & Duncan (1987) have promoted the adoption of convex groups in a cladogram as taxa. The term "convex", which is adapted from mathematics and which I do not particularly like, was originally proposed by Estabrook (1978). It corresponds essentially with "monophyletic" in the loose pre-cladistic sense, including both "monophyletic" and "paraphyletic" of cladistic terminology, but excluding "polyphyletic". This seems to me to be a sensible approach to classification. It differs essentially from Hennigian classification in allowing paraphyletic taxa, so imposing non-regressive class concepts on phylogeny which otherwise has no class concepts. It approximates with Mayr's "evolutionary classification" (see Mayr & Ashlock 1991, pp. 242-273 and earlier references given there), and since most taxonomists these days would reject clearly polyphyletic taxa, it corresponds well with current "traditional" taxonomy.

Some of the conflict in the literature seems to be due to a failure to allow that different criteria for classification are possible. In the first of the two immediate responses to Cronquist's 1987 paper, Humphries & Chappill (1988, p. 136) have opened their comments on paraphyletic taxa by quoting their fellow responders, saying "Donoghue & Cantino (1988) have shown that paraphyletic groups are unacceptable in classification because they give a distorted and misleading impression of relationships". But this applies only if they insist on putting their own interpretation on somebody else's classification which was based on a quite different understanding. A similar argument was used by Wiley (1981b) in his attempt to show that a classification allowing paraphyletic taxa is inconsistent with its phylogeny, which, as noted by Meacham & Duncan (1987, p. 84), is erroneous if one accepts convexity as one's criterion. If we knowingly accept paraphyletic taxa in our classification, it is a false argument to say that this gives a distorted and misleading impression because the taxa are not all monophyletic in a cladistic sense, or to claim that our classification is inconsistent with its phylogeny.

Some zoologists have already concluded that the system of Linnaean classification and nomenclature which we have operated for over 200 years is incompatible with cladistic classification (De Queiroz & Gauthier 1992, 1994, Griffiths 1974). I think this is becoming steadily more obvious, and of course I agree. But what they want to do is to hi-jack the Linnaean nomenclatural system and apply it, with great difficulty, to a phylogenetic hierarchy, abandoning Linnaean classification. Support for this principle in botany in future is given by Judd et al. (1994: 42). However, to me this seems impractical and unacceptable. An extensive review reaching similar conclusions has been provided by Charig (1982). We should recognise that a phylogeny, represented by a cladogram or phylogenetic tree, is not the same as a classification. A phylogeny shows relationships between phyletic lines and may be valuable in making predictive statements about them. Linnaean classification incorporates the anagenetic relationships between plants, defines taxa and gives them ranks and names. This may also have predictive value. But each must be assessed for what it is and not confused with the other. If one wishes to make deductive statements from the phylogenetic relationships of taxa (say, e.g., in cladistic biogeography) it is illogical to say that you must start with a cladogram, then construct a classification, then make your deductions from the classification. One can make deductions from a cladogram irrespective of the delimitation of taxa and the names applied to them. Abandonment of Linnaean classification in favour of an attempt at a phylogenetic hierarchy must result in *loss* of anagenetic information and some degree of predictivity, while causing very undesirable taxonomic and nomenclatural upsets. A phylogeny and a classification should exist side by side, inter-related but not interdependent, each based on different criteria and serving different functions. In this way maximum utility is preserved.

Conclusion

While all studies which give insight into phylogenetic relationships are to be welcomed, their significance for formal taxonomy needs urgent appraisal. I believe that the distinction between a phylogeny and a classification should be emphasized strongly by those who are concerned about the destabilising effect that Hennigian principles can have on the established Linnaean system. Ideally taxonomists should be able to present both a classification, which may allow paraphyletic taxa, and a putative phylogeny. The theoretical case against paraphyletic taxa is in my opinion negative, and, if I am right, the nearer we get to elucidating actual relationships, the greater will be the conflict between phylogeny and classification. The practical effects of applying cladistic principles to classification and nomenclature are potentially disastrous. In the present decade when molecular data are shedding increasing light on probable phylogenetic relationships, it seems to me more important than ever that we do not allow ourselves to be seduced into wholesale changes to our taxonomic system. Unfortunately there is a danger at the moment that the bigger the upset to existing classifications that a new cladistic study causes, the greater is its presumed scientific value*. I hope that those concerned at this trend will speak up and state their arguments. We should not be ashamed of paraphyletic taxa. They have been acknowledged to exist ever since Darwin, and are an unavoidable and essential feature of our taxonomy.

* No criticism is intended here of any authors cited in this paper.

Acknowledgements
I am very grateful to many colleagues and friends of various taxonomic opinions, many of whom would probably not wish to be associated by name with this paper, for very amicable discussion of these problems. In particular I must thank my colleagues Paul Wilkin, Alan Paton and Brian Schrire, and Neil Brummitt, for referring me to a lot of relevant literature, and Eric Knox for providing me with an advance copy of his important and instructive paper and moral support in times of doubt.

References

Anderberg, A.A., 1990. *Nablonium* is a congener of *Ammobium* (Asteraceae-Gnaphalieae). Telopea 4: 129-135.

Anderberg, A.A., 1992. In defense of the transfer of *Nablonium* to *Ammobium* (Asteraceae-Gnaphalieae), a reply to Orchard. Telopea 5: 13-19.

Bidgood, S. & R.K. Brummitt (in press). A revision of the genus *Neuracanthus* (Acanthaceae). Kew Bull.

Brummitt, R.K., 1965. Nomina conservanda proposita, 2. Regnum Vegetabile 40: 24.

Brummitt, R.K., 1980. Reconsideration of the genera *Ptycholobium*, *Caulocarpus*, *Lupinophyllum* and *Requienia* in relation to *Tephrosia* (Leguminosae-Papilionoideae). Kew Bull. 35: 459-473.

Buck, W.R., 1986. Traditional methods in taxonomy: a personal approbation. Taxon 35: 306-309.

Carroll, R.L., 1988. Vertebrate Palaeontology and Evolution. 698 pp. W.H. Freeman & Co, Oxford.

Charig, A.J., 1982. Systematics in biology: a fundamental comparison of some major schools of thought. In: K.A. Joysey & A.E. Friday (eds.). Problems of Phylogenetic Reconstruction. Systematics Association Special Volume 21. Academic Press, London etc.

Cronquist, A., 1987. A botanical critique of cladism. Botanical Review 53: 1-52.

De Queiroz, K. & J. Gauthier, 1992. Phylogenetic taxonomy. Ann. Rev. Ecol. Syst. 23: 449-480.

De Queiroz, K. & J. Gauthier, 1994. Toward a phylogenetic system of biological nomenclature. TREE 9: 27-31.

Donoghue, M.J. & P.D. Cantino, 1988. Paraphyly, ancestors and the goals of taxonomy: a botanical defense of cladism. Botanical Review 54: 107-128.

Duncan, T., 1980. Cladistics for the practising taxonomist - an eclectic view. Systematic Botany 5: 136-148.

Estabrook, G.F., 1978. Some concepts for the estimation of evolutionary relationships in systematic botany. Systematic Botany 3: 146-150.

Estabrook, G.F., 1986. Evolutionary classification using convex phenetics. Syst. Zool. 35: 560-570.

Griffiths, G.C.D., 1974. On the foundations of biological systematics. Acta Biotheor. 23: 85-131.

Hennig, W., 1966. Phylogenetic Systematics [translated by D.D. Davis & R. Zangerl]. 263 pp. University of Illinois Press, Urbana, Chicago & London.

Hull, D.L., 1984. Cladistic theory: hypotheses that blur and grow. In: T. Duncan & T.F. Stuessy (eds.). Cladistics: perspectives on the reconstruction of evolutionary history: 5-23. Colombia University Press, New York.

Humphries, C.J. & J.A. Chappill, 1988. Systematics as a science: a response to Cronquist. Botanical Review 54: 129-144.

Humphries, C.J. & V.A. Funk, 1984. Cladistic methodology. In: V.H. Heywood & D.M. Moore (eds.). Current Concepts in Plant Taxonomy: 323-362. Systematics Association Special Volume, no. 25. Academic Press, London and Orlando.

Jeffrey, C., 1987. The concept of the genus. Austral. Syst. Bot. Newsl. 53: 27-30.

Judd, W.S., R.W. Sanders & M.J. Donoghue, 1994. Angiosperm family pairs: preliminary phylogenetic analysis. Harvard Papers in Botany 5: 1-51.

Knox, E.B. (in press). The use of hierarchies as organizational models in systematics. Botanical Review.

Leeuwenberg, A.J.M., 1994. Taxa of the Apocynaceae above the genus level. In: J.H. Seyani & A.C. Chikuni (eds.). Proc. XIIIth Plenary Meeting AETFAT, Malawi, 2: 1213-1227.

Liede, S. & F. Albers, 1994. Tribal disposition of genera in the Asclepiadaceae. Taxon 43: 201-231.

Linder, H.P., 1988. A review of cladistics for botanists. S. Afr. Tydskr. Plantk. 54: 208-220.

Mayo, S.J., J. Bogner & P.C. Boyce (in press). What is the Ariflorae? In: P. Rudall et al. (eds.). Monocotyledons: Classification and Phylogeny. Royal Botanic Gardens, Kew.

Mayr, E. & P.D. Ashlock, 1991. Principles of Systematic Zoology, ed. 2. 475 pp. McGraw Hill, New York.

Meacham, C.A. & T. Duncan, 1987. The necessity of convex groups in biological classification. Systematic Botany 12: 78-90.

Orchard, A.E., 1992. *Ammobium* and *Nablonium* (Asteraceae - Gnaphalieae) - an alternative view. Telopea 5: 1-12.

Phillips, R.B., 1984. Considerations in formalizing a classification. In: T. Duncan & T.F. Stuessy (eds.). Cladistics: perspectives on the reconstruction of evolutionary history: 257-272. Colombia University Press, New York.

Platnick, N.I., 1979. Philosophy and the transformation of cladistics. Syst. Zool. 28: 537-546.

Rieseberg, L.H. & L. Brouillet, 1994. Are many plant species paraphyletic? Taxon 43: 21-32.

Seyani, J.H. & A.C. Chikuni (eds.), 1994. Proceedings of the XIIIth Plenary Meeting of AETFAT, Zomba, Malawi, 2-11 April 1991. 1514 pp. in 2 vols. National Herbarium and Botanic Gardens of Malawi, Zomba.

Sosef, M.S.M., 1994. Studies in Begoniaceae, V. Refuge begonias. Taxonomy, phylogeny and historical biogeography of *Begonia* sect. *Loasibegonia* and sect. *Scutobegonia* in relation to glacial rain forest refuges in Africa. Wageningen Agric. Univ. Papers, 94 (1): 1-306.

Wiley, E.O., 1981a. Phylogenetics. 439pp. John Wiley & Sons, New York, Chichester, Brisbane, Toronto, Singapore.

Wiley, E.O., 1981b. Convex groups and consistent classifications. Systematic Botany 6: 346-358.

Woodger, J.H., 1952. From biology to mathematics. Brit. J. Phil. Sci. 3: 1-21.

Generic delimitations in African groups which also have wide distributions outside Africa

F.J. BRETELER
Herbarium Vadense, Postbus 8010, 6700 ED Wageningen, The Netherlands.

Abstract
At least at the generic level, the African continent is not isolated, i.e. many genera are not restricted to that continent. Which genera are endemic and which are not can be established only after extensive and thorough research in most cases. This research may reveal connections not known before or lead to the rupture of entities in two or more parts along continental boundaries. At least some knowledge of the family on a worldwide scale is needed to know where to look for possible connections. These connections with other continents may be to the East or to the West, quite often dependent on the family or subfamily in question. In Africa, the taxonomic exploration of the tropical lowland rain forest started much later than in the other continents, which may have important consequences of priority in nomenclature. Before the colonial era, botanists worked on a mondial or at least on a continental scale. In colonial times taxonomists restricted themselves to a colony, a national possession, frequently, in competition with colleagues from other colonial powers working for their colonies. The more genera and species they published the better, often with severe consequences for the quality of their work. At present taxonomists are gradually returning to the continental or cosmopolitical scale and do more monographic work. A rapid production of a good generic monograph is no longer possible. Examples of several African groups will be given to illustrate the different aspects of the problems encountered.

Résumé
Tout au moins au niveau générique, le continent africain n'est pas isolé, ce qui signifie que de nombreux genres n'y sont pas confinés. La question de savoir quel genre est endémique et quel genre ne l'est pas ne peut être résolue dans la plupart des cas qu'après des recherches étendues et approfondies. Ces recherches peuvent mettre en évidence des connexions qui n'apparaissaient pas auparavant ou conduire au morcellement de certaines entités en deux ou plusieurs fractions correspondant à des frontières continentales. Une certaine connaissance de la famille à l'échelle mondiale est à tout le moins nécessaire pour situer le champ de recherches. Ces connexions avec d'autres continents peuvent s'étendre vers l'est ou vers l'ouest, le plus souvent en fonction de la famille ou de la sous-famille étudiée. En Afrique, l'exploration taxonomique des forêts tropicales ombrophiles planitiaires a commencé bien plus tard que sur les autres continents, ce qui entraîne des conséquences importantes au niveau de la priorité nomenclaturale. Avant l'aire coloniale, les botanistes travaillaient à une échelle mondiale ou au moins continentale. Au temps des colonies, les taxonomistes se sont attachés davantage à une colonie, possession nationale, souvent en concurrence avec des collègues appartenant à d'autres puissances coloniales et travaillant pour leurs colonies. C'était une course au nombre de genres et d'espèces décrits, avec souvent des conséquences néfastes sur la qualité du travail réalisé. A l'heure actuelle, les taxonomistes reviennent progressivement à un travail à l'échelle continentale ou mondiale et s'intéressent davantage à des études monographiques. La production rapide d'une bonne monographie générique n'est plus envisageable. Des exemples de plusieurs groupes africains sont présentés afin d'illustrer les différents aspects des problèmes rencontrés.

Key words
Classification, Continental scale, Floristic poverty, Relationships.

L.J.G. van der Maesen et al. (eds.), The Biodiversity of African Plants, 385-392.

Introduction

We, taxonomists of tropical African plants, are working on a continental scale, at least we usually pretend to do so, although flora treatments for countries or regions are still being prepared. Often this is done in absence of a previous thorough revision of the taxa concerned. This may cause many name changes afterwards, sometimes well before a flora is completed. As for some of these floras it may take many decades or even a century to see the end, intermediate name changes are practically unavoidable. Prance (1981) estimated that at the present speed 300 years would be needed to finish Flora Neotropica.

But can we content ourselves to work on a continental scale? I quote from what the late Gentry (1993) said when he had first visited tropical Africa: "When I first visited tropical Africa, I found that I could recognize to family and even to genus most of the plants I encountered, using Neotropical taxonomic concepts, although many of the plants subsequently turned out to masquerade under different generic names in Africa". In the early days of the taxonomy of tropical plants, let's say before 1880, before the great botanical exploration of tropical Africa got well under way, botanists like De Candolle, Baillon, Bentham & Hooker and others practised their profession on a worldwide scale. Their herbaria were rather small and novelties from Africa were compared with plants from other continents and in doing so many connections between continents at the generic level were established. Later on these connections had to be evaluated and not all of them were maintained. A few examples.

In the genus *Rourea* of Aublet (Connaraceae) described from tropical America, many African species were described, but the genus entirely disappeared from our continent due to excessive splitting, on generic level later on. Recently (Jongkind 1989) it was established that the segregate genera could not be maintained. Only a few new combinations had to be made in order to restore the old situation. An opposite case is seen in the Chrysobalanaceae. For a long time Welwitsch's genus *Dactyladenia* was considered to be a synonym of Aublet's genus *Acioa* from South America and consequently all species were described or combined in that genus until Prance and White (1988) demonstrated that the African species belong to a different genus for which *Dactyladenia* is the oldest name. Many new combinations had to be made.

A case which might not yet be settled is seen in the Ochnaceae. The South American genus *Ouratea*, again of Aublet, was for a long time considered to be present in Africa as well (e.g. Flora of West Tropical Africa, ed.1 (1927) & ed.2 (1958) but Farron (1963) classified the African representatives in 3 segregate genera: *Campylospermum* van Tiegh., *Idertia* Farron, and *Rhabdophyllum* van Tiegh. At this moment it is not at all sure if this classification will stand.

The only species in Africa of the Asiatic genus *Sindora* Miquel (*S. klaineana* Pierre ex Pellegr.) of the Caesalpinioideae was recognized by Pierre as a *Sindora* (see Pellegrin 1919), most likely because Pierre was also familiar with the plants of Indochina from where he knew that genus. Otherwise almost surely a new genus would have been described.

The present generic delimitation in Africa is the result of many events in the past, notably the colonial period when the tropical part of Africa was divided between Belgium, Britain, France, Germany, Italy, Portugal and Spain. In this era of nationalism the taxonomists followed suit in their almost exclusive attention for the colonial possessions. With the botanical exploration of the colonies the great production of new taxa, genera and species, often started in competition between the colonial powers. Taxonomists did not restrict themselves to the tropics or to a continent, but to a colony. Harms (1897) for instance described in a single publication two new genera of the Caesalpinioideae, *Kingiodendron* from Asia and *Oxystigma* from Cameroon. Both genera are so closely related that one has to look very carefully in order to avoid referring specimens from different continents to the same specific name. It is very likely that these two genera will be united, but also that they may prove to be congeneric with *Prioria* Griseb. from the American tropics as well. Harms' classification into 3 different genera is even more amazing when his key to these genera is taken into consideration, it only mentions minor details in order to differentiate between them.

The generic delimitation in African groups which also have wide distributions outside Africa is in fact that of groups with distributions that are already known and those that are still hidden to us. The problems that may be encountered are basically the same for both, but how to unveil what is hidden to us?

The most simple answer is make a world wide revision of your family, subfamily or tribe and you will see how the genera from one continent are related to those in other continents and also if genera have to be split or should be united. It is also the only answer that is taxonomically sound. But if your work, for convenience, remains restricted to a continent like Africa and you still want to do your work properly, without the burden of all the species that occur or may occur elsewhere, what to do? How to proceed?

There are taxonomic publications that can be consulted to see if the continents are connected at the generic level. These publications treat families worldwide, usually at the generic level, like Bentham & Hooker's Genera Plantarum (1862-1883) or Engler & Prantl's Die natürlichen Pflanzenfamilien (1889-1909). But they are of rather limited use, either for being more or less outdated or for not being of a standard that is needed for these purposes. The treatment of the Caesalpinioideae in Die natürlichen Pflanzenfamilien is from 1894 and most new genera for Africa have been published more recently. Hutchinson's Genera of Flowering Plants (1964-1967) has never been completed and is, at least in my opinion, not very critical. For the Leguminosae we are lucky to have at our disposal the comprehensive taxonomic treatment of Polhill & Raven (1981). This treatment may not be a proper revision but it is extremely useful, especially its classification of genera into groups, despite the few errors that may obscure the true relationships.

The problems we face when dealing with groups that have also wide distributions outside Africa are in fact of two kinds:

* *nomenclatural problems*, there may be an older name elsewhere which must be taken into account, and
* *taxonomic problems*, how does the variation elsewhere interfere with a classification for Africa.

Nomenclatural problems

When Cowan was revising the American species of *Macrolobium* Schreb. of the Caesalpinioideae in the early fifties he only considered material from America because the genus was described from that continent.

At the other side of the Atlantic, in Brussels, Léonard was studying the African species of *Macrolobium*. Not knowing the American species too well, Léonard contacted Cowan in order to become more familiar with the variation in the American species and to hear his opinion on the idea of segregating the African species and placing them in one or more separate genera. The discussion between the two taxonomists, as summarized by Cowan in his revision (1953), terminates when he, Cowan (Léonard was apparently not so far) comes to the conclusion that the African species are sufficiently different to justify generic separation. As a result *Macrolobium* is considered to be an exclusively American genus with its closest relatives in Africa. The classification of the former African species into one or more segregate genera is a problem of the taxonomists dealing with Africa. That problem was left to Léonard.

At present *Macrolobium* is indeed considered to be endemic to America and the former African species are currently classified into 4-6 segregates (Léonard 1957, Aubréville 1968) in order to accommodate the large variation that exists in the once African *Macrolobium*.

This example illustrates that the historical advantage of having the oldest name, the advantage of priority, may lead a botanist to restrict his revision to one continent only. If not, the "burden" of the other continent(s) has to be taken on in order to establish whether the same genus is present elsewhere.

African taxonomy, especially that of the heart of Africa, the richest tropical part, started much later than in the other tropical regions of the world. For instance most genera of the Caesalpinioideae, as stipulated above, date from the late 19th century or the 20th century, while those from America (e.g. Aublet 1775) and Asia are from a much earlier date.

Taxonomic problems

Tropical Africa is generally considered to be poor as regards the number of species of flowering plants. Also in terms of endemic families, and usually genera as well, the African tropics have a low biodiversity (Brenan 1978).

This floristic poverty is, as we all know, mainly due to gigantic and destructive climatic changes in the past. These are, very likely, also the cause for the disjunction in distributions, and with it, probably also for the gaps in the morphological variation as experienced by the small number of species per genus in Africa as compared with America.

This poverty is not always evident, at least not in every sense. Quite often the relatively low number of taxa represented in our continent show a magnitude of variation not seen elsewhere in the tropics.

The genus *Begonia* L. of the Begoniaceae is circumtropical in distribution, with many species in America and Asia, but with comparatively few species in Africa. These few African representatives, however, show a variation in morphology and ecology that is much larger than is seen elsewhere (J.J.F.E. de Wilde, personal communication).

Table 1. Distribution of Connaraceae (Breteler 1989).

Genus	Africa	Asia	America
Agelaea	4	2	0
Burttia	1	0	0
Cnestidium	0	0	2
Cnestis	12	1	0
Connarus	7	c. 19	51
Ellipanthus	2	c. 4	0
Hemandradenia	2	0	0
Jollydora	3	0	0
Manotes	5	0	0
Pseudoconnarus	0	0	5
Rourea	12	c. 16	43
Vismianthus	1	1	0
Total species	49	c. 43	101
Total genera	10	6	4

The genus *Rinorea* Aubl. of the Violaceae has more or less the same distribution and again is not very rich in number of species in Africa. But as regards its variation in the flower (from actinomorphic to zygomorphic) the position of the inflorescence (ramiflorous to cauliflorous) and the fruit variation, African *Rinorea* shows a great diversity (J.J. Bos, personal communication).

The Connaraceae, a small circumtropical family, numbers 11 genera. Of these 10 are present in Africa with a total of c. 50 species. Asia has almost the same number of species in 6 genera and America many more species in 4 genera only (Lemmens 1989, Table 1). This means that, in some instances at least, Africa holds a key position as regards the morphological variation. This is certainly true for the tribe *Macrolobieae* Bret. of Caesalpinioideae (Breteler 1995) which, at present counts c. 170 species in c. 20 genera, less than 10 species per genus in Africa. In America there are 75 species in 3 genera only, 25 species per genus. In the same tribe Africa counts 7 monotypic genera and America only 1. Brenan (1965) stated that in the Caesalpinioideae 66,6% or 54 out of 81 genera are endemic to Africa whereas these figures for the more advanced subfamilies Mimosoideae and Papilionoideae respectively, are 36,4% and 26,9%.

Africa's poverty has its exceptions as is shown above. Especially as regards the morphological variation, it can be quite rich. The large gaps in this morphological variation we often experience in Africa might be filled, at least partly, when the close relatives from the other continents are taken into account. Therefore we should not restrict ourselves too exclusively to Africa. This may be illustrated, as I see it, by the following example.

Cynometra L. of the Caesalpinioideae is a rather variable tropical genus of c. 70 species worldwide. It is based on a species from Malesia. The closely related Asiatic genus *Maniltoa* occurs mainly in that area. In my opinion, the two genera are rather obscurely distinguished from one another (Knaap-van Meeuwen 1970). But to decide whether the latter genus can be maintained I think that the African species of *Cynometra* will have to be studied. On the other hand a worldwide revision of *Cynometra*, including *Maniltoa*, would be needed to decide whether the small African satellites of *Cynometra* like *Lebruniodendron* and *Scorodophloeus* really deserve their present status.

The possible relationships of the African *Detarieae* and *Macrolobieae* of the Leguminosae - Caesalpinioideae are being investigated and the preliminary result is summarized in Table 2. Some relationships are very close, e.g. between *Oxystigma* and *Prioria*, as stipulated above, or between *Eurypetalum* and *Eperua*. Others need more investigation e.g. between *Berlinia* and *Dicymbe*. Some genera like the latter two are found in the same group in Cowan & Polhill's (1981) classification, others like *Eperua* and *Eurypetalum* rather far apart.

The strong relationships between Africa and America which are apparent in the Caesalpinioideae, the more primitive member of the Leguminosae, corresponds with the distribution of the primitive character states in the Annonaceae (Schatz & Le Thomas 1993). A similar contribution as regards the more advanced Compositae (Bremer 1993) does not point to such relationships and seems to be more in conformity with the situation in the more advanced subfamilies Mimosoideae and Papilionoideae.

Much emphasis has been laid on Caesalpinioideae, not only because it is my most important subject of research at present, but because the relationships of this subfamily are worldwide and therefore serve as a good example for my conclusion that revising the African taxa as an isolated group should not be done.

But we must be realistic as well. In the meantime we have to produce our floras, for a country (often a former colony) or for a region of our continent, and rarely do our financiers or sponsors have the patience to wait till a basic monograph has been completed first.

When treating families for a flora without a preceding thorough revision one should, so to say, be conservative in attitude, which means not too quickly decide that a new taxon is at hand and that it has to be published.

Of the 18 new taxa, 8 species and 10 varieties, described in the genus *Dichapetalum* (Dichapetalaceae) (Hauman 1958, Léonard 1994) in the course of the treatment for the Flore d'Afrique centrale (Hauman 1958) only one of each could be maintained in the revision for Africa that followed (Breteler 1973-1986). For the treatment of the same genus for the second edition of the Flora of West Tropical Africa (Keay 1958) not a single new taxon was described, but four unnamed taxa, indicated by the letters A-D, were mentioned. Although the latter treatment may seem to be very unsatisfactory, it does not publish a series of new names that have to be abandoned later on.

References

Aublet, J.B.C.F., 1775. Histoire des plantes de la Guiane françoise. 4 vols.

Aubréville, A., 1968. Légumineuses - Caesalpinioidées. Flore du Gabon, 15.

Bentham, G. & J.D. Hooker, 1862-1883. Genera Plantarum. London, 3 vols.

Bremer, K., 1993. Intercontinental relationships of African and South American Asteraceae: A Cladistic Biogeographic Analysis. in Goldblatt, P. (ed.) Biological Relationships between Africa and South America: 105-135. Yale Univ. Press, New Haven & London.

Table 2. Genera of the Detarieae (D) and Macrolobieae[1] (M) (Leguminosae - Caesalpinioideae) in Africa with strong relationships outside Africa.

African genera				Related genera outside Africa			
				America		Asia	
Genus	Tribe	Area	Number of species	Genus	Number of species	Genus	Number of species
Anthonotha	M	Mainland	c. 30	*Macrolobium*	c. 60		
Apaloxylon	D	Madagascar	2			*Hardwickia*	1
Berlinia	M	Mainland	c. 15	*Dicymbe*	13		
Cynometra	D	Circumtropical	c. 70			*Maniltoa*	20-25
Detarium	D	Mainland	3	*Copaifera*	c. 25		
Eurypetalum	D	Mainland	2-3	*Eperua*	14		
Gilbertiodendron	M	Mainland	c. 26	*Macrolobium*	c. 60		
Gossweilerodendron	D	Mainland	2			*Hardwickia*	1
Guibourtia	D	Mainland, America	16-17			*Hardwickia*	1
Hylodendron	D	Mainland	1	*Copaifera*	c. 25		
Isomacrolobium	M	Mainland	4	*Macrolobium*	c. 60		
Oxystigma	D	Mainland	5	*Prioria*	1	*Hardwickia*	1
Pellegriniodendron	M	Mainland	1	*Macrolobium*	c. 60		
Sindoropsis	D	Mainland	1			*Sindora*	18-20[2]
Tessmannia	D	Mainland	11			*Sindora*	18-20[2]

[1] The Macrolobieae only show relationships across the Atlantic because they do not occur in Asia. The Detarieae are circumtropical.
[2] And 1 species in Africa.

Brenan, J.P.M., 1978. Some aspects of the phytogeography of tropical Africa. Ann. Missouri Bot. Gard. 65: 437-478.

Brenan, J.P.M., 1965. The geographical relationships of the genera of Leguminosae in Tropical Africa. Webbia, 19: 545-559.

Breteler, F.J., 1973-1986. The African Dichapetalaceae. Meded. Landbouwhogeschool Wageningen 73-13, 78-10, 79-16, 81-10, 82-8 and Agric. Univ. Wageningen Papers 86-3.

Breteler, F.J. (ed.), 1989. The Connaraceae, a taxonomic study with emphasis on Africa. Wageningen Agric. Univ. Papers 89-6.

Breteler, F.J., 1995. The boundary between *Amherstieae* and *Detarieae* (Caesalpinioideae). In: M.D. Crisp & J.J. Doyle (eds.), Advances in Legume Systematics 7: 53-61. Royal Botanic Gardens, Kew.

Cowan, R.S., 1953. A taxonomic revision of the genus *Macrolobium* (Leguminosae - Caesalpinioideae). Mem. New York Bot. Garden, 8: 257-342.

Cowan, R.S. & R.M. Polhill, 1981. *Amherstieae & Detarieae*. In: Polhill, R.M. & P.M. Raven, (eds), Advances in Legume Systematics, Part 1: 117-142. Royal Botanic Gardens, Kew.

Engler, A. & K. Prantl, 1889-1909. Die natürlichen Pflanzenfamilien. W. Engelmann, Leipzig.

Farron, C., 1963. Contribution à la taxonomie des *Ourateae* Engl. (Ochnacées) Bull. Soc. Bot. Suisse 73: 196-217.

Gentry, A.H., 1993. Diversity and Floristic Composition of Lowland Tropical Forest in Africa and South America. In: Goldblatt, P. (ed.), Biological Relationships between Africa and South America: 500-547. Yale Univ. Press, New Haven & London.

Harms, H.A.Th., 1897. Nachträge zu III. 3. Caesalpinioideae. In: Engler, A. & K. Prantl (eds), Die natürlichen Pflanzenfamilien: 192-198.

Hauman, 1958. Dichapetalaceae. In: Flore du Congo Belge et du Ruanda-Urundi 7: 287-348.

Hutchinson, J., 1964-1967. The genera of flowering plants. 2 vols. Clarendon Press, Oxford.

Hutchinson, J. & J.M. Dalziel, 1927. Flora of West Tropical Africa. Ochnaceae vol. 1. part 1: 189-195. Crown Agents, London.

Jongkind, C.C.H., 1989. In: F.J. Breteler (ed.), The Connaraceae, a taxonomic study with emphasis on Africa. Wageningen Agric. Univ. Papers 89-6: 310-368.

Keay, R.W.J., 1954. Flora of West Tropical Africa. 2nd edition. *Ochnaceae*. vol. 1. part 1: 221-231. Crown Agents, London.

Keay, R.W.J., 1958. Flora of West Tropical Africa. 2nd edition. Chailletiaceae. vol. 1. part 2: 433-439. Crown Agents, London.

Knaap-van Meeuwen, M.S., 1970. A revision of four genera of the tribe Leguminosae - Caesalpinioideae - Cynometreae in Indomalesia and the Pacific. Blumea 18: 1-52.

Lemmens, R.H.M.J., 1989. In: F.J. Breteler (ed.), The Connaraceae, a taxonomic study with emphasis on Africa. Wageningen Agric. Univ. Papers 89-6: 11-13.

Léonard, J., 1957. Genera des *Cynometreae* et des *Amherstieae* africaines (Leguminosae - Caesalpinioideae). Mém. Acad. Roy. Belg. Cl. Sci. 30(2): 1-314.

Léonard, J., 1994. Statistiques des Spermatophytes de la Flore d'Afrique centrale de 1940 à 1990. Bull. Jard. Bot. Nat. Belg. 63: 181-194.

Pellegrin, F., 1919. Bull. Mus. Paris: 653.

Polhill, R.M. & P.H. Raven (eds.), 1981. Advances in legume systematics 1 & 2. Royal Botanic Gardens, Kew.

Prance, G.T., 1981. Flora Neotropica: Where do we stand? Taxon 30: 81-87.

Prance, G.T. & F. White, 1988. The genera of Chrysobalanaceae: a study in practical and theoretical taxonomy and its relevance to evolutionary biology. Philos. Trans. Roy. Soc. London, B. Biological Sciences 320: 1-184.

Schatz, G.E. & A. Le Thomas, 1993. Annonaceae: A primitive Dicot Family with an Ancient Center in Africa - South America. In: P. Goldblatt (ed.), Biological Relationships between Africa and South America: 86-104. Yale Univ. Press, New Haven & London.

Problems with generic delimitation and subdivision in a large genus, *Barleria* (Acanthaceae)

M-J. BALKWILL & K. BALKWILL
C.E. Moss Herbarium, Department of Botany, University of the Witwatersrand,
Private Bag 3, WITS 2050, South Africa.

Abstract
Barleria L. is a large, polymorphic, pantropical genus, the subdivision of which into natural groups has long been subject to question. The publication of seven accounts of the genus has not solved this problem due to the regional nature of the classifications. We undertook a re-assessment of the generic limits and subdivision of *Barleria* on a monographic scale, using a phylogenetic approach. A major thrust of this research was the careful definition of characters and character-states and tracing their distribution within the genus. Many of the problems encountered during this work are typical of large, widespread genera in this and other families. These include: difficulty in obtaining adequate material of all taxa; inability to conduct adequate field work in all areas; restrictions on the kinds of characters that can be effectively utilised; incomplete data sets and difficulties associated specifically with cladistic methodology. A revised infra-generic classification of *Barleria* is proposed in which 2 subgenera and 7 Sections are recognised. This classification, with a full account of the taxonomy, nomenclature and phylogeny, will be published separately.

Résumé
Barleria L. est un grand genre pantropical et polymorphe, dont la subdivision en groupes naturels a depuis longtemps fait l'objet de controverses. La publication de sept traitements concernant le genre n'a pas résolu ce problème, étant donné le caractère régional des classifications adoptées. Nous avons entrepris de réexaminer les limites génériques et la subdivision de *Barleria* dans un cadre monographique et suivant une méthode phylogénétique. Un des objectifs majeurs de cette recherche était de définir avec précision les caractères ainsi que leur hiérarchie, et d'établir leur répartition à l'intérieur du genre. Les nombreux problèmes rencontrés durant ce travail sont ceux que l'on rencontre dans l'étude de grands genres à large distribution, dans cette famille comme dans d'autres. Il s'agit entre autres de la difficulté d'obtenir un matériel adéquat de tous les taxons, de l'impossibilité d'effectuer un travail sur le terrain dans toutes les régions concernées, du nombre réduit des caractères auxquels on peut utilement avoir recours, des groupements de données incomplets et des difficultés inhérentes à la méthodologie cladistique. Une révision de la classification infragénérique de *Barleria* est proposée avec la reconnaissance de 2 sous-genres et 7 sections. Cette classification, accompagnée d'un traitement taxonomique, nomenclatural et phylogénique, sera publiée séparément.

Key words
Barleria, Generic Delimitation, Infra-generic Classification, Phylogeny.

Introduction

Studies of large, variable, widespread genera such as *Justicia* L. (Acanthaceae), *Senecio* (Compositae) and *Erica* (Ericaceae) have tended to follow two divergent approaches: those that recognise numerous, small segregate genera and those that accept a broad generic circumscription. The former situation often arises when divergent treatments of a genus in local floras lead to the description of segregate genera in different regions, because the authors are unaware of the full range of

L.J.G. van der Maesen et al. (eds.), The Biodiversity of African Plants, 393-408.

variation (Clayton 1972 & 1983, Graham 1988, Jeffrey 1987, Stevens 1985 & 1987). This has certainly been the case in large acanthaceous genera of Linnaean origin, such as *Justicia* (Graham 1988). We believe that the only way to adequately address problems of generic delimitation and subdivision in these large genera is to work on a monographic scale, a view supported by several authors such as Clayton (1972 & 1983), Jeffrey (1987) and Stevens (1987). However, large genera, in which the numbers of species run into hundreds, are difficult to deal with adequately on a monographic scale, for numerous reasons. Few taxonomists can be afforded extended periods of time to work on a single genus. Furthermore, taxonomists often tend to work on problems at the species level or may be constrained by having to produce flora accounts for a particular region in a reasonably short period. Other problems include: difficulty in obtaining adequate material of all taxa; inability to conduct fieldwork in all areas in which the genus is represented; limitations on the sources from which characters can be drawn (e.g. it may not be possible to use chromosome numbers or other features that require fresh material, as only herbarium material may be available for certain or all of the taxa) and incomplete data sets (e.g. the only material that was available for study for some species may not have had seeds present, although features of the seeds may be important in the genus). It is often also difficult to find characters that have an unambiguous distribution within the genus and to obtain good correlation between characters. For those wishing to adopt a phylogenetic approach (and particularly cladistic methodology) other difficulties (discussed below) present themselves including that numerous characters arise convergently or undergo reversal or that they behave as non-unique apomorphies or display apomorphic tendencies. This may limit the kinds of characters that can be used and lack of availability of data may be problematical (West 1987).

Obviously, not all of these problems will be equally manifested in all large genera - for example, problems associated with availability of material and fieldwork are not that great in genera such as *Erica*. Although this genus contains many hundreds of species, distributed from northern Europe to the southern tip of Africa (Oliver 1993), 98% of them occur in the Cape Province of southern Africa (Oliver 1988). On the other hand, availability of suitable material is highly problematical in very widespread genera such as *Justicia* and *Barleria,* in which large numbers of species occur in a number of geographically far-flung areas.

At the outset, we wish to emphasise that this paper is not intended as an in-depth theoretical treatise on the use of cladistic methodology or the possible merits or demerits of cladistics over other means of devising classifications. Cladistic methodology and principles are controversial and different authors may argue for different approaches, as is evident in other papers in this symposium. We believe that the classification we have proposed for *Barleria* is the best we could have produced, regardless of whether or not cladistic methodology was used. We do believe, though, that the strict rigour imposed by cladistic principles on the definition of characters and the kinds of characters that can be used, was very helpful in devising a new classification of the genus. Our brief from the symposium organisers was to address specifically the practical problems associated with working on large genera through the use of a particular case study, and it is this that we hope to achieve in this paper. Reference to methodological details will only be made where they relate specifically to the matter of dealing with large genera. A full account of the taxonomy, nomenclature and phylogeny of the taxa referred to throughout this paper will be published separately. The main thrust of the following discussion, relates to the difficulties in dealing with large genera, particularly sampling of taxa and the search for reliable characters in very variable groups.

In this paper, *Barleria* (Acanthaceae) is used as a case study for the problems associated with generic circumscription and subdivision in large, widespread genera. It is hoped that the problems discussed and the ways in which they were addressed will be of general application to other large genera in which similar problems arise. The approach we adopted was, so far as possible, a multidisciplinary one, combining herbarium work with field and laboratory studies wherever possible. Because much of the study was of necessity herbarium-based, emphasis was placed on morphology and various aspects of anatomy. Other aspects of anatomy, karyology and phytochemistry were investigated, but these data sets are incomplete as these studies had to be restricted to those taxa of which fresh material was available. The emphasis in our study was placed on the accurate definition of characters and character-states and on tracing their distribution within the genus and related genera where possible. Taxonomic problems and nomenclatural difficulties at the species level were noted, but were not investigated in any detail.

Before turning to the problems of dealing with large genera in detail, a brief introduction to *Barleria*, to which reference will be made fairly often, is in order. *Barleria* L. (Acanthaceae) is a large, pantropical genus of herbs and shrubs comprising, at a conservative estimate, some 300 species world-wide. Its greatest representation is in Africa (particularly the eastern parts) and Asia, with the greatest centre of diversity (approximately 80) species in east tropical Africa. The only accounts of *Barleria* that have been undertaken on a world-wide basis are those of Nees (1847) and Lindau (1895). All more recent accounts have been undertaken as components of regional floras. The problem with *Barleria* has not so much been one of generic circumscription as it has been one of the satisfactory sub-divisions of this variable assemblage of species into satisfactory natural groups. This problem was first formally identified by Bentham (1876) and, despite the publication of no less than 7 regional revisions in which infra-generic classifications have been proposed (Benoist 1967, Clarke 1885, 1899 & 1901, Hutchinson & Dalziel 1931, Obermeijer 1933, Ridley 1923), the problem of satisfactory subdivision has, to date, not been adequately solved. Almost every character in *Barleria* is highly variable, the variation in some characters being as great within *Barleria* as it is across the entire Acanthaceae. Furthermore, not all combinations of characters are represented in any one geographic area. Thus the previous classifications of the genus, as may be expected, do not fully accommodate taxa from other geographic areas. We felt that if *Barleria* is a single genus, as indeed we believe it is, then it must be possible (and preferable, both philosophically and practically) to produce a single classification for it.

Generic circumscription

As this is a symposium on generic concepts in the African flora, some general comments on how we circumscribe *Barleria* seem appropriate. Those interested in theoretical discussions of generic concepts should consult the appropriate literature

(Anderson 1940, Bartlett 1940, Camp 1950, Clayton 1972 & 1983, Funk 1985, Jeffrey 1987, Sherff 1940, Stevens 1984, 1985 & 1987, Walters 1961, West 1987). Whilst the literature provides us with numerous guidelines for the delimitation of taxa, it does not provide clear criteria for the assignment of rank to these taxa (or, at least, there is considerable disagreement about which of these criteria is acceptable). Previous authors working on *Barleria* have not specifically addressed the question of generic delimitation and it is not surprising, therefore, that there are few statements to be found on the generic "concept" employed by these authors. Furthermore, most taxonomists who have worked on *Barleria* have regarded it as polymorphic but clearly defined, although they have not always said how it should be defined. The concept adopted in our research was not deliberately broad or narrow; the taxon we call *Barleria* is clearly defined, easily recognisable and we believe that all of its members are more closely related to each other than they are to species in other taxa. As generic limits within the Acanthaceae are notoriously messy and in the interests of maintaining nomenclatural stability, we thought it best to retain our taxon at the generic rank and formally recognise infra-generic groups within it. Our concept of this genus is thus a broad one, based on the consistent co-occurrence of three derived character-states: the presence of a distinctly 4-lobed calyx with two inner and two outer lobes; spheroidal, honeycomb pollen and a predominance of double cystoliths in the leaves. Although each of these characters is known to occur individually elsewhere in the Acanthaceae (sometimes in rather distantly related groups where they have probably arisen independently), their consistent co-occurrence is restricted to *Barleria*. We thus take the co-occurrence of these three characters states to support the concept of *Barleria* as a monophyletic assemblage.

Sampling of taxa

Adequate sampling of taxa is one of the fundamental problems a taxonomist faces in dealing with large genera, especially if a cladistic approach is adopted (West 1987). As the number of species in *Barleria* is rather high (approximately 300), it would have been impossible to study every one in detail in a limited period of time. Careful sampling of a wide range of species was necessary to cover the morphological variability and geographic distribution of the genus. As we are based in southern Africa, and this region represents the second highest centre of diversity of the genus (approximately 70 species), we thought it would be satisfactory to initiate the study using mainly southern African species as it was easy to conduct extensive fieldwork and obtain herbarium material. Based on this material, initial working hypotheses were formulated regarding character variability and distribution and possible infrageneric groupings. These working hypotheses were then tested and modified by investigation of as many of the other species as possible. Material was obtained on loan from MO, FT and EA and the entire holdings of *Barleria* at K, E and BM were studied. Surveys of the material concentrated on scoring taxa for the presence or absence of certain characters, which facilitated their placement in broad infrageneric groups. A selection of taxa was then studied in detail across their full geographic range in order to take in as much of the variability as possible. Approximately 240 of the 300 described species of *Barleria* were ultimately studied in this way in a period of 8 weeks, as well as material of numerous undescribed taxa, especially from tropical East Africa. It was not possible to trace material of certain names and the validity of some of the described taxa is not clear. Also, the full extent of synonymy in the genus is not known as *Barleria* is in sore need of a full taxonomic revision. Despite these difficulties, we feel that our sampling of the genus was sufficiently representative to provide a basis for the formulation of a revised infra-generic classification of *Barleria*.

Characters and character-states

Characters and character-states are the crucial pivots around which a taxonomic study revolves. An extensive analysis of characters across the entire genus was necessary in order to provide suitable criteria for delimiting natural infra-generic groups. Wherever possible, the presence or absence of these characters was also established for related genera as this gave a broader indication of their reliability. Cladistics force workers to study characters and taxa in great detail (just as non-cladists do), but with very careful interpretation of evolutionary polarity and homology, and it is this that often leads to discrepancies with previous interpretations (West 1987). This is why it is dangerous to rely on information from the literature. For this reason, great care was taken to establish the homology of the characters used in our study. Although previous classifications have contributed greatly to our understanding of character distributions within the genus, some of the characters used were not homologous, thus leading to the formulation of hypotheses of false similarity. A case in point is that of the presence or absence of spines. This has been used in almost all previous classifications of *Barleria* to divide the genus into two major groups (either series, subgenera or sections). In the classification of Clarke (1899), which has been the most widely used classification of the genus, the presence or absence of spines was used to divide the genus into two Series, each of which contained two sections. Our work has shown clearly that the spines in different parts of the genus are produced in different ways and so are not homologous at all. In *Barleria* it is common for two serial buds to form in the axil of each leaf. Both buds may give rise to a vegetative lateral shoot or the upper bud may give rise to an inflorescence unit instead. In one part of the genus (Section *Prionitis*) the characteristic 2- or 4-rayed spines form from the lower serial bud and the upper bud may give rise either to a vegetative shoot or an inflorescence unit (Figure 1a). These spine systems are thus taken to represent modified vegetative shoots. In Section *Acanthoidea sensu* Clarke, however, the spine systems form from the upper serial bud in the axil and represent sterile, aborted or old (persistent) inflorescences (Figure 1b). The spinous nature of the bracteoles in the fertile inflorescences also adds to the overall spinescence of the plants. Thus any resemblance between the spines in Sections *Prionitis* and *Acanthoidea* is superficial and placing them in the same group on the basis of "spines" is incorrect.

This study was largely herbarium-based, restricting the sources from which data could be drawn. We thus considered it more practical and useful to concentrate on re-assessing characters that have been used before and on those that have been neglected in previous works, rather than to attempt covering the full spectrum of potentially useful characters. In addition, some new characters (such as the presence of particular glands) were discovered and used in the formulation of a revised classification. Although a large number of characters incorporating a wide range of vegetative and floral features was surveyed, some of these had to be eliminated from the data sets

Figure 1. Axillary spines systems in *Barleria*; a) spine systems formed form the lower serial axillary bud in *B. rotundifolia* Oberm. (Section *Prionitis*); b) spine systems formed from upper serial bud in leaf axils in *B. elegans* Nees (Section *Barleria*).

used in the cladistic analyses. We considered it preferable to include a smaller number of reliable characters rather than a larger number of characters whose distribution was not fully known in the genus, or for which the homology and polarity could not be determined with some degree of confidence. In accordance with the theory of phylogenetic systematics, only those character-states believed to represent the advanced condition were used to delimit groups and absence characters (such as the lack of a beak on the capsule) were not used as they were thought to be poor indicators of phylogenetic relationship. In the case of characters that behaved as non-unique apomorphies or showed apomorphic tendencies (Linder 1988), such as the composition of the androecium, the predominance of a derived character state in a group was taken to indicate phylogenetic relationship, particularly if it coincided with the occurrence of another derived character state. Although the use of these kinds of characters is hotly debated in the cladistic literature (see Linder 1988), we felt that they indicate at least some degree of underlying relationship and so used them, with caution, in our study. Several other characters had to be eliminated from the data sets for sound methodological reasons. *Barleria* is a frustrating genus in that there are a number of excellent characters (eg. characteristic hair types) that are highly characteristic of particular groups and so have an unambiguous distribution within the genus. However, many of these characters are also autapomorphies for the major infrageneric taxa and so are not useful in resolving relationships between them.

In our study of *Barleria,* the most useful characters included: basic configuration of the corolla lobes; structure of the stigma; the composition of the androecium; morphology of the capsules; number of seeds in the mature capsule; anatomy of the septum; the nature of the fruit wall and the nature of the hygroscopic hairs on the seeds. Aspects of vegetative morphology, especially the nature of the pubescence, and the structure of inflorescence units were also useful characters for defining smaller groups within several of the larger ones. Some of the results of our work are discussed below to illustrate how reliable characters were found despite the immense variation displayed

Figure 2. Diagrammatic representation of the variation in corolla morphology in *Barleria*. a) *B. monticola* Oberm.: Balkwill & Balkwill 5418 (J); b) *B. irritans* Nees: Mauve 5238 (J); c) *B. repens* Nees: Henrico s.n. (NBG); d) *B. obtusa* Nees: Balkwill 668 (NU); e) *B. elegans* S. Moore: Rushworth 4 (SRGH); f) *B. heterotricha* Lindau: Balkwill & Balkwill 6321 (J); g) *B. bremekampii* Oberm.: Balkwill & Balkwill 5830 (J); h) *B. lancifolia* T. Anders.: Balkwill & Balkwill 4302; i) *B. lugardii* C.B. Cl.: Leach 10 595 (SRGH); j) *B. popovii* Verdc.: Popov GP/SO/315 (BM); k) *B. pretoriensis* C.B. Cl.: Obermeyer 943 (SAM); l) *B. lukwangulensis* Mildbr.: Congdon 209 (K); m) *B. eranthemoides* R.Br. ex C.B. Cl.: Ash 880 (MO); n) *B. albostellata* C.B. Cl.: Balkwill et al. 5811 (J); o) *B. randii* S. Moore: Eyles 88 (SRGH); p) *B. argentea* Balf.f.: Thulin & Warfa 4663 (K); q) *B. rotundifolia* Oberm.: Buitendag 440 (PRE).

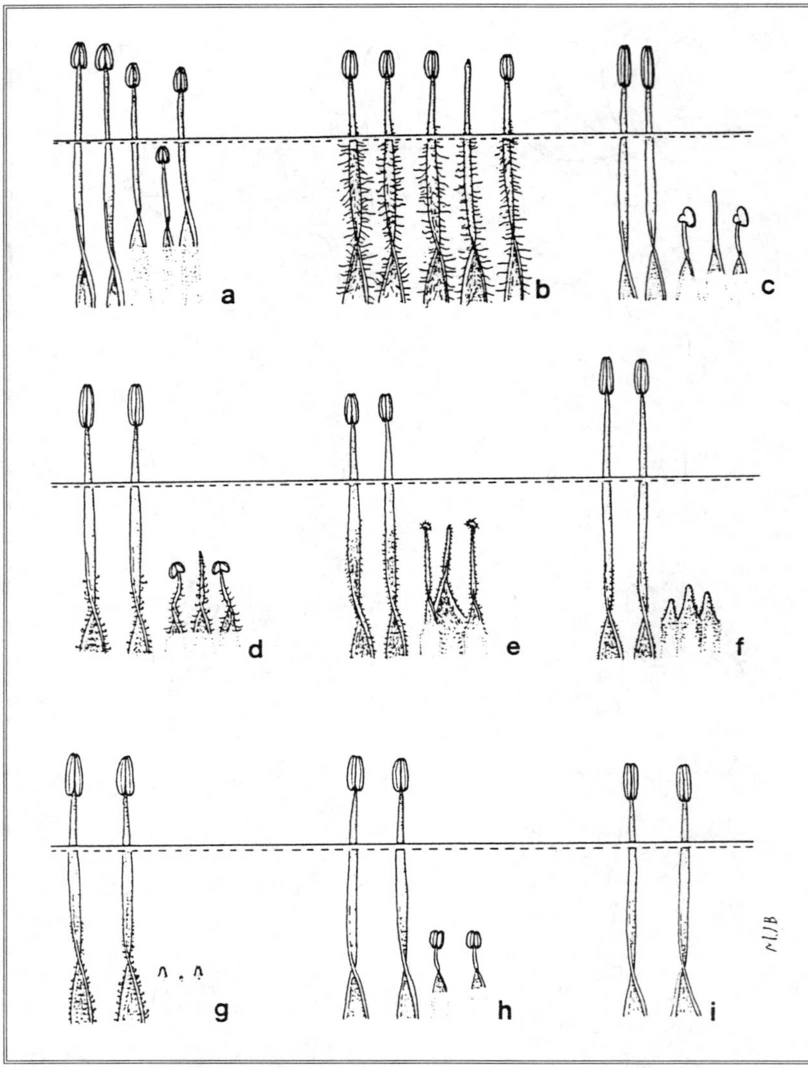

Figure 3. Variation in the composition of the androecium in *Barleria*. a) *B. morrisiana* E. Barnes & C.E.C. Fischer: Barnes 2122 (K); b) *B. ovata* E. Mey. ex Nees: Balkwill, Balkwill & Green 6261 (K); c) *B. rigida* Nees: Herman 561 (PRE); d) *B. jubata* S. Moore: Pearson 9895 (BOL); e) *B. lukwangulensis* Adelyser 163 (K); f) *B. dolomiticola* M. & K. Balkwill: Moffett 1084 (PRE); g) *B. wilmsiana* Lindau: Balkwill, Balkwill & Kidger 8086 (J); h) *B. crossandriformis* C.B. Cl.: Balkwill 1849 (NU); i) *B. hildebrandtii* S. Moore: Gilbert, Ensermu & Vollesen 7723 (K).

in this genus. The names used for the infrageneric taxa are provisional; names for the sections will be formally published in a separate publication after certain nomenclatural technicalities have been clarified.

Corolla morphology (Figure 2)
The morphology of the corolla is highly variable in *Barleria,* but if the basic form of the corolla is studied a number of trends are discernible, including: reduction in the number of corolla lobes from 5 to 4; changes in the relative width and length of the upper and lower portions of the corolla tubes and changes in the arrangement of the corolla lobes leading to increased zygomorphy (Figure 2). An important feature is the basic configuration of the corolla lobes: in the least advanced forms the 5 lobes of the corolla are more or less equal in shape, with two slightly smaller lobes in the upper position and three slightly larger lobes in the lower position. This is termed the "2+3" configuration of the corolla and is predominant in Subgenus *Barleria* (Figure 2a - g). In Subgenus *Prionitis* four of the corolla lobes are in the upper position and only one, which is separated from the corolla tube for a greater distance than the other four, in the lower position. This is called the "4+1" configuration (Figure 2h - q).

Composition of the androecium (Figure 3)
In the Acanthaceae the androecium usually comprises of either four or two stamens and may include a number of staminodes. It is unusual, though, for there to be five parts present, as occurs in some members of *Barleria.* Variation in the composition of the androecium in this genus can be quite bewildering. The androecium may comprise of five, four or two parts (Figure 3). There are always two fertile stamens and the balance of the androecial complement is made up of either fertile stamens (Figure 3a & b) or, more commonly, variously reduced staminodes (Figure 3c - h). A basic distinction can be made between those species with five parts in the androecium (Figure 3a - g) and those with four or less (Subgenus *Prionitis,* excepting Section *Barlerites* and parts of Section *Somalia;* Figure 3h & i). Within the latter groups there is a trend towards increased reduction of the adaxial staminal elements until only two fertile stamens are present and staminodes are absent. Although other aspects of the androecium may initially seem useful for delimiting smaller groups, these characters are labile within certain species, thus rendering them somewhat unreliable.

Morphology of the stigma (Figure 4)
The stigma may be two- or one-lobed in *Barleria,* the two-lobed condition being the primitive state. In two-lobed taxa, the two stigma branches are seldom equally well-developed. They may be sub-confluent and reduced so that the stigma appears capitate (Figure 4a & f) or the stigma branches may be flattened out forming a discoid structure (Figure 4b). In some two-lobed species one of the stigma branches is very much longer than the other so that these stigmas may superficially seem to comprise only one lobe (Figure 4c). In all other parts of the genus the stigma is characteristically reduced to one filiform lobe (Figure 4d & e) although vestiges of the second, ventral lobe may be visible in some specimens. Observations made on fresh stigmas of southern African species indicate that two-lobed stigmas are conspicuously papillate, whilst the one-lobed stigmas are smooth. These observations indicate that further investigations of stigma morphology are warranted.

Capsule morphology (Figure 5)
Capsules in *Barleria* may possess a solid apical beak, a hollow apical beak or no

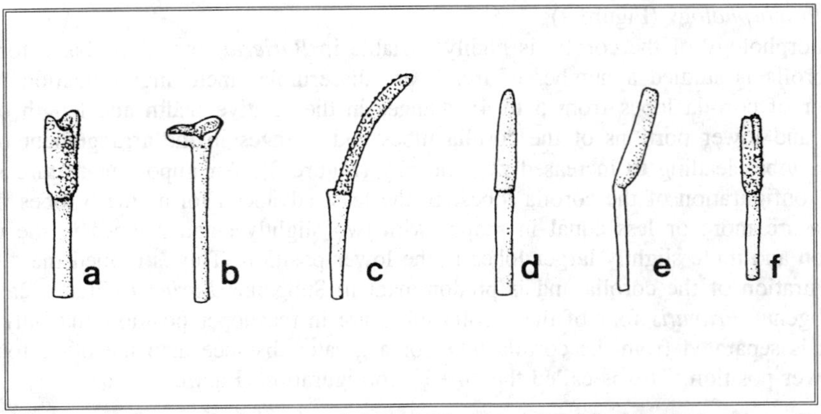

Figure 4. Variation in stigma morphology in *Barleria*. a) 2-lobed, capitate stigma, *B. obtusa*: Balkwill 668 (NU); b) 2-lobed, discoid stigma, *B. monticola*: Balkwill & Balkwill 5418 (J); c) 1-lobed, filiform stigma showing vestiges of second stigma lobe, *B. strigosa* Willd.: Grierson and Long 3382 (E); d) 1-lobed stigma, *B. albostellata*: Balkwill et al. 5811 (J); 1-lobed stigma, *B. crossandriformis*: Balkwill 4308 (J).

distinct beak (Figure 5). In Sections *Prionitis* and *Somalia,* the capsule is turgid at the base and the apex is produced into a stout, cylindrical woody beak (Figure 5a & b). Capsules of this kind contain only two seeds at maturity, as the two lower ovules abort early during the development of the fruit. In Section *Barlerianthus* the capsule possesses a prominent apical beak, but this beak is hollow (Figure 5e & f). These capsules contain four seeds at maturity and are basically spindle-shaped but become markedly narrowed below the point of attachment of the upper seeds. In the rest of the genus the capsules are not beaked, although a small point is usually present at the apex of the capsule (Figure 5 h & i and k & l). In some cases these capsules are laterally compressed and obtrulloid, containing four seeds (the primitive condition) at maturity (Figure 5h & i). In others (Figure 5k & l), the capsules are broadly elliptical to rounded and contain only two seeds at maturity, by abortion of the lower two ovules. The lateral walls of these capsules may be extremely thin, but the narrow flanks of the capsule are woody forming a frame over which the thin walls are "stretched". When these capsules dehisce, the thin walls tear away from the woody flanks for about half of their length, superficially resembling the dehiscence of capsules with elastic placental bases.

Composition of the septum (Figure 5)
The composition of the septum, a previously overlooked character, provides very useful characters for delimiting major infra-generic taxa. The septum may be entirely sclerenchymatous (e.g. *B. prionitis* L.; Figure 5 c) or may comprise woody and membranous (parenchymatous) sections (Figure 5d, g, j, m). In some cases there is a small membranous portion at the tip of the capsule (Figure 5j), or there may be a broad membranous portion above the upper seeds (Figure 5d & g) or almost the entire septum may be membranous (Figure 5m). It is interesting that the two-seeded, unbeaked capsules described before have septa which are almost entirely membranous.

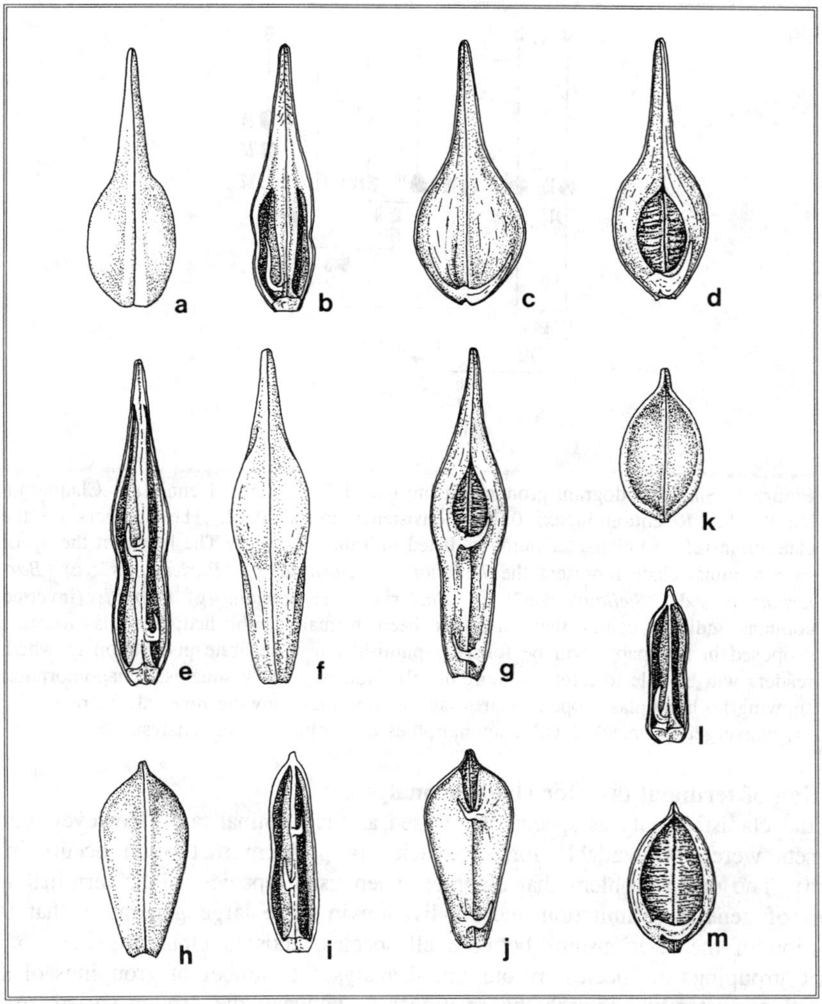

Figure 5. Capsule morphology and the composition of the septum in *Barleria*; all capsules magnified x 3.5. a) closed capsule with solid apical beak, b) opened capsule showing locules and solid apical beak, c) opened capsule showing woody septum, all *B. senensis* Klotzsch: Balkwill & Balkwill 6331 (J); d) opened capsule with woody beak and septum with a broad membranous region, *B. lancifolia*: Balkwill & Balkwill 4302 (J); e) opened capsule showing locules and hollow beak, f) closed capsule with hollow beak, g) opened capsule with hollow beak showing partially membranous septum, all *B. montana*: Ritchie s.n. (E); h) closed 4-seeded capsule without an apical beak, i) opened 4-seeded capsule without an apical beak, j) opened 4-seeded capsule without an apical beak showing small membranous portion at tip of septum, all *B. elegans*: Balkwill et al. 5811 (J); k) unopened 2-seeded capsule with prominent woody flanges, l) opened 2-seeded capsule showing absence of a beak, m) opened 2-seeded capsule showing almost entirely membranous septum, all *B. albostellata*: Balkwill et al. 5811 (J).

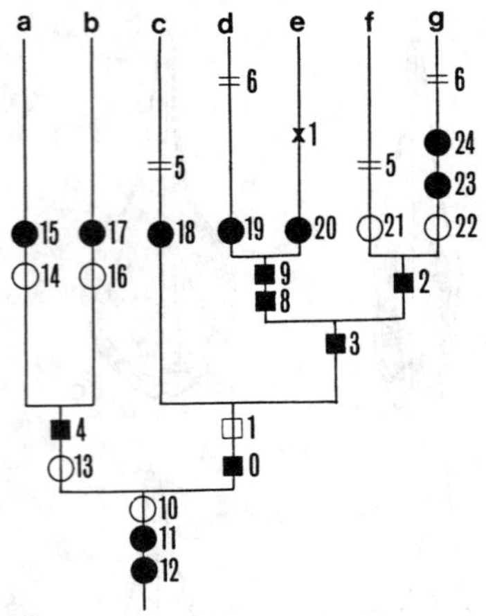

Figure 6. Single cladogram produced using the "ie" routine in Hennig'86. Cladogram length: 11; Retention index: 0.75; Consistency index: 0.72. The numbers on the cladogram refer to character numbers listed in Tables 1 and 2. The letters at the tip of each terminal clade represent the 7 sections: a) *Barleria*; b) "*Barleriopsis*"; c) "*Barlerianthus*"; d) "*Stellatohirtae*"; e) "*Barlerites*"; f) *Somalia*; g) *Prionitis* (inverted commas indicate names that have not been formally published; the classification proposed in this paper will be formally published in a separate publication to which readers will be able to refer for more detail). Notation: black squares: synapomorphies showing no homoplasy; open squares: synapomorphies showing reversal; X: reversals; =: convergences; black dots: autapomorphies not included in the analysis.

Selection of terminal taxa for cladistic analyses

In initial cladistic analyses species were used as the terminal taxa. However, complete data sets were not available for all species (a problem that often occurs in large genera). The other problem that emerges when using species as the terminal taxa in studies of generic delimitation and subdivision in these large genera, is that lack of resolution of the relationships between all species tends to cloud the issue of infrageneric groupings of species. In our initial analyses a number of groupings of species emerged consistently although the relationships between the groups varied. As these groups of species were defined by clear synapomorphies it was decided that these represented the most likely monophyletic units and that they should be used as the terminal taxa in all later analyses. This resulted in the recognition of 7 major infrageneric taxa; smaller groups of species could be identified within four of these seven groups and initially these smaller groups were retained as separate entities in the analyses but not all of them could be fully resolved. The value of a classification is diminished if data from different hierarchical levels is combined (Stevens 1987). We thus thought it preferable for the smaller groups to be included in the larger ones as they represented a different level of organisation within the genus. It is likely that with more information for some of the more poorly known species these smaller groups will be fully resolved and found to represent true clades that could be afforded formal status at a later stage, but at some lower hierarchical level.

Table 1. Character set and polarisations used for cladistic analyses.

No.	Plesiomorphic state (0)	Apomorphic state (1)
0	corolla 2+3 or 4+1	corolla invariably 4+1
1	stigma of 1 or 2 lobes	stigma 1-lobed
2	capsules unbeaked	capsules with a solid terminal beak
3	seeds mostly 4	seeds mostly 2
3a	seeds always 4	seeds always 2
4	septum without narrow membranous portion at tip	septum with narrow membranous portion at tip
5	septum without broad membranous portion at centre	septum with broad membranous portion at centre
6	androecium mostly of 5 parts	androecium mostly 4 or 2 parts
7	hygroscopic hairs variable	hygroscopic hairs gold, silky
8	septum not entirely membranous	septum almost entirely membranous
9	capsule without woody flanges	capsules with woody flanges

A new classification of *Barleria*

For many cladists a cladogram is a classification but others, such as Jeffrey (1987) have argued strongly against this, a major issue being the assignment of rank to different clades and the recognition of paraphyletic groups. The question that needs to be answered is how best to translate the cladogram into a hierarchy of taxonomic ranks without causing too much disruption to the existing nomenclature. Generally only a few nodes of a cladogram should be formally named; if one can establish which nodes have the highest confidence rating, these can be selected for formal naming with the result that the classification established from them will be maximally stable (Linder 1991). There are different ways of establishing confidence in the nodes of a cladogram including statistical methods such as "bootstrapping" (Linder 1991). However, the most commonly used indicator of confidence, and the one used in our study, is the nature of the data supporting the cladogram.

The basal dichotomy shown in the cladogram (Figure 6) was obtained in almost every analysis conducted and is supported by reasonably convincing apomorphies. As it seems unlikely that any future analyses will lead to a change in this basic topology, we propose that the two major clades be assigned formal status as subgenera. We have elected to assign all the terminal taxa in the cladogram equal status as sections. This

Table 2. Additional characters not used in cladistic analyses.

Character	Plesiomorphic state (0)	Apomorphic state (1)
10	cystoliths predominantly single	cystoliths predominantly double
11	pollen variable	pollen globose, honey-combed
12	calyx of 5 lobes, or if of 4 then not in 2 pairs	calyx of 4 lobes in 2 pairs
13	spine systems (from upper bud in leaf axil) absent	spine systems (from upper bud in leaf axils) invariably present
14	calyx herbaceous	calyx scarious
15	inflorescence units variable	inflorescence units invariably based on scorpioid cymes
16	inflorescence units invariably solitary, axillary	inflorescence units invariably aggregated into a terminal synflorescence
17	plants glabrous or pubescent, but pubescence not usually of conspicuous yellow hairs	plants usually with conspicuous pubescence of yellow hairs
18	capsules without prominent hollow beak	capsules with prominent hollow beak
19	plants with pubescence of simple hairs only	plants with pubescence of complex hairs (stellate and dendroid)
20	walls of capsule remaining intact upon dehiscence	walls of capsule tearing upon dehiscence
21	hygroscopic hairs on seeds variable	hygroscopic hairs on seeds white-woolly
22	leaves usually without sunken glands on lower surfaces	leaves always with sunken glands on lower surfaces
23	plants without axillary spine systems formed from lower bud in leaf axils	plants invariably with spines formed from lower bud in leaf axils
24	flowers usually blue, white or pink	flowers usually yellow or apricot yellow

approach was adopted by van Wyk (1991) in his classification of *Lotononis* (Leguminosae). Most importantly, this kind of system ensures that any future rearrangements within the sections (e.g. the recognition of subsections) can be made without necessitating any serious changes to the infra-generic nomenclature.

We thus propose that the species of *Barleria* be placed in 7 sections within 2 subgenera. Of the 7 sections proposed, two (Section *Prionitis* and Section *Somalia)* are retained as previously circumscribed by Clarke (1899), although additional characters have been found to delimit them; one section (*Barleria*) is retained partially as previously circumscribed, but with two of the groups previously included by Clarke (1899) removed and elevated to Sectional status (the affinities of these taxa are also thought to lie more closely with Section *Somalia*, than with Section *Barleria*); two taxa are recognised for the first time at the sectional level and the remaining two sections are recognised here for the first time. Details of the classification and the characteristics of each of the sections proposed will be published separately.

Conclusions

This classification of *Barleria* is the first since the time of Lindau (1895) that deals with *Barleria* on a global scale. Whilst the affinities of the sections may be open to other interpretations, most of the sections are very clearly defined and the classification should provide a stable framework in the context of which future, finer-scale taxonomic research at the species level can be conducted.

The monographic scale at which this work was conducted had a number of important implications and consequences. In families such as the Acanthaceae in which there are numerous large, widespread and variable genera such as *Barleria,* it is absolutely essential that workers obtain a thorough knowledge of the full range of variation before making decisions regarding circumscription of genera and infrageneric groupings. For example, many of the Indian species of *Barleria* are not represented in other geographic areas or are represented by only one or a few species in other regions. Many of the unusual character-states, character-state reversals or combinations of character-states occur in these Indian species. Had the work been limited geographically, these important Indian taxa would not have been seen and the decisions made regarding character state distributions and polarities would have been considerably different and the weaker for it.

A classification based on a limited but reliable data set may not be without flaws but at least represents an advancement over numerous disparate regional classifications, as it provides a consistent treatment for the entire genus. A stable generic classification is highly desirable, not only from a theoretical standpoint, but also in terms of producing a consistent treatment that can be adopted in the various floristic works that are being produced in different areas.

Acknowledgements

We are grateful to the University of the Witwatersrand, South Africa and to the Foundation for Research Development (South Africa) for providing funding for this research. Thanks are due to the Curators of the herbaria from which herbarium material was borrowed or whose facilities were used during this study. In the latter case, special thanks are due to the Keeper and staff of the Herbarium, Royal Botanic Gardens, Kew.

References

Bartlett, H.H., 1940. The concept of the genus I: history of the generic concept in Botany. Bull. Torrey Bot. Club 67 (5): 349-362.

Benoist, R., 1967. Acanthacées. In: Humbert, H., (ed.), Flore de Madagascar et des Comores 1: 134-166.

Bentham, G., 1876. Acanthaceae. In: Bentham, G. & J.D. Hooker, (eds.), Gen. Pl. Volume III, part II, Lovell Reeve & Co., London.

Camp, W.H., 1940. Our changing generic concepts. Bull. Torrey Bot. Club 67 (5): 381-389.

Clarke, C.B., 1885. Acanthaceae. In: Hooker, J.D. (ed.), Fl. Brit. Ind., Lovell Reeve & Co., London.

Clarke, C.B., 1899. See Burkill, I.H. & C.B. Clarke, 1899.

Clarke, C.B., 1901. Acanthaceae. In: Thiselton-Dyer, W.T., (ed.), Fl. Cap. Volume 5, part 1, Lovell Reeve & Co., London.

Clayton, W.D., 1972. Some aspects of the genus concept. Kew Bull. 27 (2): 281-287.

Clayton, W.D., 1983. The genus concept in practice. Kew Bull. 38 (1): 149-153.

Funk, V.A., 1985. Cladistics and generic concepts in the Compositae. Taxon 34 (1): 72-80.

Graham, V.A.W., 1988. Delimitation and infrageneric classification of *Justicia* (Acanthaceae). Kew Bull. 43 (4): 551-624.

Hutchinson, J. & J.M. Dalziel, 1931. Acanthaceae. In: Flora of West Tropical Africa 2 (1): 261-262.

Jeffrey, C., 1987. The concept of the genus. Austral. Syst. Bot. Soc. Newsletter 53: 27-30.

Lindau, G., 1895. Acanthaceae. In: Engler, A. & K. Prantl (eds.), Pflanzenfam. 4 (3a & b): 313-315.

Linder, H.P., 1988. A review of cladistics for botanists. S. Afr. J. Bot. 54 (3): 208-220.

Linder, H.P., 1991. Confidence limits in phylogenies: an example from African Restionaceae. Taxon 40: 253-266.

Nees von Esenbeck, C.G.D., 1847. Acanthaceae. In: De Candolle, A. (ed.), Prod. 11: 223-247, Truettel & Wu'urtz, Paris.

Obermeijer, W., 1933. A revision of South African species of *Barleria*. Ann. Transvaal Mus. 16-2: 123-180.

Oliver, E.G.H., 1988. Studies in the Ericoideae (Ericaceae) VI. The generic relationship between *Erica* and *Philippia* in southern Africa. Bothalia 18-1: 1-10.

Oliver, E.G.H., 1993. Studies in the Ericoideae (Ericaceae) XII. The placing of the genus *Blaeria* into synonymy under *Erica*; nomenclatural and taxonomic changes for the southern African region. Bothalia 23-1: 1-7.

Ridley, R.N., 1923. The flora of the Malay Peninsula 2: 587-588.

Sherff, E.E., 1940. The delimitation of genera from the conservative point of view. Bull. Torrey Bot. Club 67 (5): 375-380.

Stevens, P.F., 1984. Homology and phylogeny: taxonomy and systematics. Syst. Bot. 9: 395-409.

Stevens, P.F., 1985. The genus concept in practice - but for what practice? Kew Bull. 43 (3): 457-464.

Stevens, P.F., 1987. Genera, what and why - some thoughts. Austral. Syst. Bot. Soc. Newsletter 53: 31-38.

Van Wyk, B-E., 1991. A synopsis of the genus Lotononis (Fabaceae: Crotalarieae). Contrib. Bol. Herb. 14: 1-292.

Walters, S.M., 1961. The shaping of angiosperm taxonomy. New. Phytol. 60: 74-84.

West, J.G., 1987. The role of cladistics in generic delimitation. Austral. Syst. Bot. Soc. Newsletter 53: 38-47.

Generic delimitation and relationships in the tribe *Hildebrandtieae* (Convolvulaceae)

SEBSEBE DEMISSEW [1] & D.F. AUSTIN [2]
[1] The National Herbarium, Science Faculty, Addis Abeba University, P.O. Box 3434, Addis Abeba, Ethiopia.
[2] Florida Atlantic University, Department of Biological Sciences, P.O. Box 3091, Boca Raton, FL 33431-0991, U.S.A.

Abstract
In Ethiopia, the tribe *Hildebrandtieae* Peter (1891: 19) includes three genera: *Cladostigma* Radlkofer with three species, *Hildebrandtia* Vatke with nine species, *Sabaudiella* Chiovenda with one species. Morphological investigations of all and palynological studies of representative species have been performed. As a result, the status of the three genera and the relationships between them have been reconsidered and additional characters provided to delimit the genera. The members are restricted mainly to the drier parts of eastern Africa, Madagascar and extending to Arabia. A hypothesis is offered on relationships of the tribe *Hildebrandtieae* with other related tribes within the family Convolvulaceae.

Résumé
En Ethiopie, la tribu des *Hildebrandtieae* Peter (1891: 19) comprend trois genres: *Cladostigma* Radlkofer avec trois espèces, *Hildebrandtia* Vatke avec neuf espèces, et *Sabaudiella* Chiovenda avec une espèce. Un examen morphologique de toutes les espèces et une étude palynologique d'espèces représentatives ont été réalisées. Il en résulte que le statut des trois genres et les relations entre eux ont été reconsidérés, et que des caractères additionnels ont été proposées pour la délimitation des genres. Les membres de cette tribu sont principalement confinés aux régions plus sèches de l'Afrique orientale, de Madagascar et s'étendent jusqu'en Arabie. Une hypothèse concernant les relations de la tribu des *Hildebrandtieae* avec les tribus voisines au sein de la famille a été émise.

Key words
Cladostigma, Convolvulaceae, *Hildebrandtia*, Hildebrandtieae, *Sabaudiella*, Taxonomy.

Introduction

The tribe *Hildebrandtieae* Peter is characterized by the combination of the following characters. Leaves mainly fascicled on short shoots, rarely solitary; flowers unisexual, 4-5-merous; styles two, free with branched stigma lobes; sepals accrescent in the female flowers; fruits utricular to capsular, 1-4 locular, 1-4-seeded, and seeds rugose. Five genera, *Cladostigma* Radlk., *Dactylostigma* Austin, *Hildebrandtia* Vatke, *Pterochlamys* Roberty and *Sabaudiella* Chiov. are known within the tribe.

Characteristics

Unisexual flowers. This is a unique character trait of the tribe, otherwise unknown in the family.

Branched stigma lobes. Branched stigma lobes in the dioecious *Hildebrandtieae* are presumably to increase the chances of the pollen grains from the staminate flowers falling on the stigmas of the pistillate flowers, although the advantage of having

L.J.G. van der Maesen et al. (eds.), The Biodiversity of African Plants, 409-420.

branched stigma lobes in the staminate flowers as well is far from clear. Whatever the advantages of having branched stigma lobes, this character trait also occurs independently in other distantly related genera in the family. The Australian endemic genus *Polymeria* R. Br., in the tribe *Convolvuleae* (Choisy) Choisy, has a 2-locular ovary with one ovule in each locule and a simple style with 4-8 linear stigmatic lobes. The genus *Neuropeltis* Wall. in the tribe *Cresseae* which occur in west tropical Africa and tropical Asia, has 2 free short styles with peltate, kidney-shaped to lobed stigmas approaching to the *Hildebrandtia*.

Accrescent sepals. Accrescent sepals occur also in other tribes to varying degrees. The tribe *Poraneae* Hall. f. for example includes genera (*Calycobolus* Willd., *Cardiochlamys* Oliver, *Cordisepalum* Verdc., *Porana* Burman f., *Poranopsis* Roberty and *Rapona* Baillon) that have conspicuously accrescent sepals similar to those in the *Hildebrandtieae*. The *Poraneae* are characterized by an accrescent, wing-like calyx enclosing a one-seeded indehiscent fruit with a thin to thick chartaceous pericarp, and the style with simple or branched stigma lobes. The possession of accrescent fruiting sepals by the *Hildebrandtieae* and other members of the Convolvulaceae which occur in the drier parts of Africa and Arabia is probably an adaptive mechanism to help with wind dispersal. A similar pattern occurs in the female flowers in *Lippia carviodora* Meikle (Verbenaceae) from the drier parts of East Africa (Sebsebe Demissew & Puff 1988). The members of the tribe *Poraneae* which occur in the wetter parts of tropical America, West Africa and Asia also possess accrescent sepals, probably an adaptive mechanism to help with water dispersal by floating rather than wind dispersal. A member of the genus *Seddera* Hochst., *S. bracteata* Verdc. in the tribe *Cresseae* Benth. & Hook. also has accrescent sepals.

Relationships with other tribes

Hallier (1893), in his monumental scheme showing the relationships of the then known genera in the family Convolvulaceae, placed *Cladostigma* Radlk. and *Hildebrandtia* Vatke within a branch of the *Dicranostyleae* (Meisn.) Hall. f. (Figure 1) together with *Bonamia* Thouars, *Cressa* L., *Evolvulus* L., *Seddera* and *Stylisma* Raf. These taxa are now placed within the tribe *Cresseae* Bentham & Hooker (1873: 880). The tribe *Cresseae* includes the genera *Bonamia*, *Cressa*, *Evolvulus*, *Itzaea* Standley & Steyerm., *Metaporana* N.E. Br., *Neuropeltis* Wall., *Seddera* and *Stylisma*. The *Cresseae* are characterized by the branched style or two free styles, the globose peltate, kidney-shaped or slightly lobed stigmas, dehiscent capsules and non-accrescent sepals (except in *Seddera bracteata*). The genus *Neuropeltis* has accrescent fruiting bracts adnate to the pedicel which show strong parallelism with the accrescent fruiting sepals of the *Hildebrandtieae*.

The genus *Hildebrandtia* was described by Vatke (1876a: 7, 1876b: 864) from Somalia with *H. africana* Vatke as the type species. The genus was then indicated to be related to the genera *Cressa* L. and *Seddera* Hochst. The second genus *Cladostigma* was later described by Radlkofer (1883: 412) from Ethiopia with *C. dioicum* Radlk. as the type species. This genus was indicated to be related to the previous genus *Hildebrandtia*. The third genus *Sabaudiella* was described by Chiovenda (1929: 366) from eastern Ethiopia with *S. aloysii* Chiov. as the type species and its affinity was

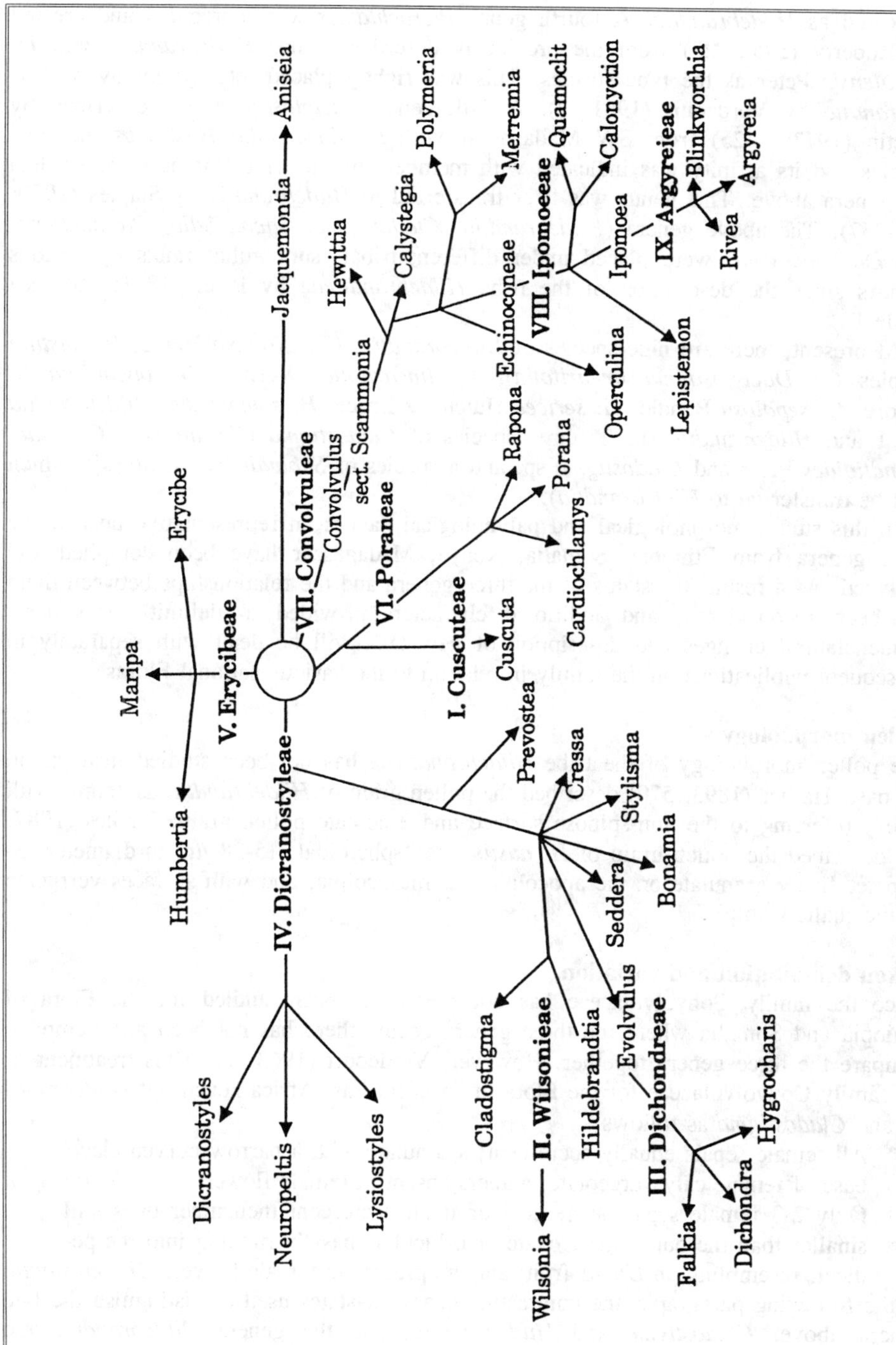

Figure 1. Hallier's (1893) phylogenetic scheme for the Convolvulaceae (redrawn).

indicated as *Hildebrandtia*. A fourth genus *Pterochlamys* was segregated and erected by Roberty (1952: 23) from the previously described genus *Hildebrandtia* with *H. somalensis* Peter as the type species. This was rightly placed into synonymy of *Hildebrandtia* by Verdcourt (1961: 2). A fifth genus *Dactylostigma* was described by Austin (1973a: 425) from SW Madagascar with *D. linearifolia* Austin as the type species and its affinity was indicated with members of the tribe Poraneae rather than the genera above. This genus was later transferred to *Hildebrandtia* by Staples (1990: 254-257). The above genera (*Hildebrandtia, Cladostigma, Sabaudiella, Pterochlamys* and *Dactylostigma*) were placed under different tribal, subfamilial ranks by various authors since the description of the tribe *Hildebrandtieae* by Peter (1891: 19), see Table 1.

At present, there are nine species of *Hildebrandtia*, (*H. africana* Vatke, *H. austinii* Staples (= *Dactylostigma linearifolia*), *H. linearifolia*, Verdc., *H. obcordata* S. Moore, *H. sepalosa* Rendle, *H. sericea* Hutch. & Bruce, *H. somalensis*, *Hildebrandtia* sp. 1 and *Hildebrandtia* sp. 2, three species of *Cladostigma* (*C. dioicum*, *C. hildebrandtioides* Peter and *Cladostigma* sp. and a species of *Sabaudiella*, *S. aloysii* (which will be transferred to *Hildebrandtia*).

In this study, morphological and palynological data from representative taxa of the three genera from Ethiopia, Somalia, Kenya, Madagascar have been compiled and analyzed. As a result, the status of the three genera and the relationships between them has been reconsidered, and additional characters provided to delimit the genera. Nomenclatural changes and description of new taxa will be dealt with separately in subsequent publications on the family in relation to the various national Floras.

Pollen morphology

The pollen morphology of the tribe *Hildebrandtieae* has not been studied in depth in the past. Hallier (1893: 570) described the pollen grain of *Hildebrandtia* as "convolvuli type", referring to the non-spinose surface and 3-colpate pollen grain. Staples (1987: 56) described the pollen grain of *H. austinii* as "spheroidal, 15-28 μm in diameter, 3-colpate, finely granulate on the apocolpia and mesocolpia, and with surfaces verrucate on the shallow colpi."

Taxon delimitation and variation

Since the family Convolvulaceae has not previously been studied for the Flora of Ethiopia and Somalia where the three genera occur, there has not been an attempt to compare the three genera together. However, Verdcourt (1963: 1) in his treatment of the family Convolvulaceae for the Flora of Tropical East Africa keyed out *Hildebrandtia* and *Cladostigma* as follows:

* All female sepals equally accrescent, spathulate with a narrow curved claw; base of female calyx urceolate; anthers absent in female flower: . . . *Cladostigma*

* Only 2-3 female sepals accrescent or if all accrescent then inner ones still smaller than the outer ones, ovate or orbicular, basally running into the pedicel, resembling an *Ulmus* fruit; anthers present in female flower: *Hildebrandtia*

In the following paragraphs the contrasting character-states used to distinguish the two genera above, *Cladostigma* and *Hildebrandtia*, and the genera *Hildebrandtia* and *Sabaudiella* will be reviewed and additional ones will be given.

Table 1. Historical development of the tribal classification of the Convolvulaceae in relation to the Hildebrandtieae.

Baillon (1891: 305-320)
- Série 1. Convolvuleae (*Hildebrandtia, Cladostigma*)
- Série 2. Cresseae
- Série 3. Cuscuteae
- Série 4. Dichondreae

Peter (1891: 13-37)
- Subfam. I. Convolvuloideae
  - Tribus 1. Dichondreae
  - Tribus 2. Dicranostyleae (*Cladostigma*)
  - Tribus 3. Hildebrandtieae **Peter (1891: 19)** (*Hildebrandtia*)
  - Tribus 4. Convolvuleae
  - Tribus 5. Erycibeae
- Subfam. II. Cuscutoideae

Hallier (1893: 567-586); Peter (1897: 376-377)
- Subfam. I. Psiloconiae
  - Tribus 1. Cuscuteae
  - Tribus 2. Wilsonieae
  - Tribus 3. Dichondreae
  - Tribus 4. Dicranostyleae (*Cladostigma* & *Hildebrandtia*)
  - Tribus 5. Poraneae
  - Tribus 6. Erycibeae
  - Tribus 7. Convolvuleae
- Subfam. II. Echinoconiae
  - Tribus 8. Ipomoeae
  - Tribus 9. Argyreieae

Roberty (1952: 22-45)
- Subfam. I. Humbertoideae
- Subfam. I . Dichondroideae
- Subfam. I I. Wilsonioideae
- Subfam. IV. Hildebrandtioideae (*Hildebrandtia, Cladostigma, Pterochlamys* (= *Hildebrandtia*))
- Subfam. V. Eryciboideae
- Subfam. VI. Poranoideae
  - Tribe 1. Neuropeltideae
  - Tribe 2. Dipteropeltideae
  - Tribe 3. Prevosteae (= *Calycobolus*)
  - Tribe 4. Poraneae
  - Tribe 5. Lepistemonopseae
- Subfam. VII. Convolvuloideae
  - Tribe 1. Evolvuleae
  - Tribe 2. Nephrophylleae
  - Tribe 3. Cresseae
  - Tribe 4. Convolvuleae
  - Tribe 5. Ipomoeae
- Subfam. VIII. Argyreioideae
  - Tribe 1. Lysiostyleae
  - Tribe 2. Dicranostyleae
  - Tribe 3. Argyreieae
  - Tribe 4. Blinkworthieae

Roberty (1964: 138-140)
- Subfam. I. Argyreioideae
  - Tribe 1. Blinkworthieae
  - Tribe 2. Argyreieae
  - Tribe 3. Dicranostyleae
  - Tribe 4. Lysiostyleae
  - Tribe 5. Erycibeae
- Subfam. II. Convolvuloideae
  - Tribe 6. Ipomoeeae
  - Tribe 7. Convolvuleae
  - Tribe 8. Cresseae
  - Tribe 9. Nephrophylleae
  - Tribe 10. Evolvuleae
- Subfam. III. Poranoideae
  - Tribe 11. Lepistemonopsideae
  - Tribe 12. Poraneae
  - Tribe 13. Prevosteae (= *Calycobolus*)
  - Tribe 14. Hildebrandtieae (same as before)
  - Tribe 15. Neuropeltideae

Melchior (1964: 427-429)
- Subfam. I. Humbertioideae
- Subfam. II. Dichondreae
- Subfam. III. Convolvuloideae
  - Tribe 1. Wilsonieae
  - Tribe 2. Hildebrandtieae (*Hildebrandtia*)
  - Tribe 3. Erycibeae
  - Tribe 4. Poraneae
  - Tribe 5. Convolvuleae (incl. Ipomoeeae)
  - Tribe 6. Argyreieae
- Subfam. IV. Cuscutoideae

Willis, J.C. (1973: 288)
- Tribe 1. Dichondreae
- Tribe 2. Dicranostyleae
- Tribe 3. Hildebrandtieae (*Cladostigma, Hildebrandtia* & *Sabaudiella*)
- Tribe 4. Convolvuleae
- Tribe 5. Poraneae
- Tribe 6. Ipomoeeae
- Tribe 7. Argyreieae
- Tribe 8. Erycibeae

Austin (1973b: 346-350)

Tribes with "non-spinose pollen" excl. Tribes Ipomoeeae & Argyreieae
- Tribe 1. Erycibeae
- Tribe 2. Cresseae (*Hildebrandtia* & *Cladostigma*)
- Tribe 3. Poraneae
- Tribe 4. Convolvuleae
- Tribe 5. Merremioides
- Tribe 6. Dichondreae

Number and size of accrescent sepals. All sepals in female flowers equally accrescent (*Cladostigma*) versus only 2-3 female sepals accrescent, if all accrescent then inner ones still smaller than the outer ones (*Hildebrandtia*). These contrasting character-states were useful to distinguish the two genera when Verdcourt (1963) used them based on the specimens then available. However, with the accumulation of new specimens since the 1980's in southern Ethiopia, among which is an undescribed species of *Cladostigma* (= *Gilbert* et al. 7653) having only 3 accrescent sepals, the other two remaining smaller, these contrasting character-states could not be used.

Shape of sepals. Sepals spathulate (*Cladostigma*) versus ovate or orbicular (*Hildebrandtia*). This character is also variable in the two genera and could not be used to distinguish them.

Anthers in female flowers. Anthers in female flowers absent (*Cladostigma*) versus present (*Hildebrandtia*). In the specimens so far investigated in the two genera anthers are present only in few species of *Hildebrandtia*. Therefore this character cannot be used to distinguish the two genera.

Base of sepals. Base of sepals clawed (*Cladostigma*) versus basally running into the pedicel (*Hildebrandtia*). These contrasting character-states are still useful to distinguish the two genera.

Fusion of outer fruiting sepals. The genus *Sabaudiella* was distinguished from the genus *Hildebrandtia* mainly by the fusion of the outer fruiting sepals and the accrescent sepal not adnate to the pedicel. However, fruiting sepals from the same specimen show varying degrees of fusion, from more or less open to completely fused outer sepals which are ± non-adnate to partly adnate to the pedicel. Thus it is difficult to distinguish *Sabaudiella* from other members of the genus *Hildebrandtia*.

Female inflorescence type. There are mainly three types of inflorescences: cymose, solitary, or solitary and fasciculate. Although there are no clear cut differences in the inflorescence types, species of the genus *Cladostigma* have cymose and/or solitary inflorescences, while those of the *Hildebrandtia* have solitary or fasciculate inflorescences save one species from Madagascar.

Locule number. There are two types of locules in the tribe *Hildebrandtieae*. All *Cladostigma* species and all species of the *Hildebrandtia* species from mainland Africa and Arabia have two-locular ovaries, while the two species of *Hildebrandtia* from Madagascar have a 1-locular ovary.

Fruit types. There are two types of fruits in the tribe *Hildebrandtieae*. All *Cladostigma* species and all species of the *Hildebrandtia* species from mainland Africa and Arabia have capsular fruits, while the two species of *Hildebrandtia* from Madagascar have a utricular fruit.

Materials and methods

Morphological studies

Herbarium specimens of all species in the tribe were examined. These were obtained on loan from AUA, BM, ETH, K, MO and UPS (abbreviations according to Holmgren et al. 1981).

Light microscope (LM) studies on vegetative and floral parts were made using a Bausch & Lomb dissecting microscope.

Pollen studies

Preparation for Scanning Electron Microscopy (SEM) of pollen was carried out at the Department of Biological Sciences, and SEM micrographs of pollen grains were made at the Faculty of Ocean Engineering, Florida Atlantic University, Boca Raton Florida.

Pollen material was prepared as follows. For each specimen, 2-3 flower buds at the point of opening were placed in a staining dish containing distilled water, covered and left for one hour to soften. The anthers were then dissected out of the flowers and retained in the solution. Other flower parts were re-dried and returned to the herbarium sheets. Using a dissecting needle, the anthers were gently broken apart to release the pollen. The entire contents from the staining dish were then washed into a tube, and centrifuged. The pollen grains were then acetolysed for 3-5 minutes using a standard 9:1 acetolysis solution, followed by 3-4 washes in distilled water. These were then transferred to 70% and to 95% ethanol. The pollen grains are then mounted to a stub with a double adhesive. They were then coated using sputter. The pollen micrographs were taken using JEOL JSM-6100 Scanning Electron Microscope. Data on the morphological and palynological characters are given in Table 2.

Table 2. Data matrix of species of Hildebrandtieae and the outgroup *Seddera*. The characters are numbered in accordance with the text and Table 3. Question marks indicate missing data. Polymorphisms are noted by more than one number.

Species	1	2	3	4	5	6	7	8	9	10	11	12	13	14	15	16	17	18	19	20
Cladostigma dioicum	1	0	01	0	1	0	0	0	1	0	0	0	1	0	1	0	1	1	0	1
C. hildebrandtioides	1	0	01	0	1	0	0	0	1	0	0	0	1	0	0	0	1	1	0	1
C. sp.	1	01	01	0	1	0	1	0	0	0	0	0	1	0	0	0	1	1	0	1
Hildebrandtia africana	1	1	2	1	0	1	0	0	0	1	1	1	1	0	1	0	1	1	0	1
H. austinii	1	0	1	0	0	0	0	0	1	1	0	1	2	1	1	1	1	1	1	0
H. linearifolia	1	1	2	1	0	1	0	0	0	1	0	1	?	?	1	0	1	1	0	1
H. obcordata	1	1	2	1	0	1	0	0	0	1	1	1	?	?	1	0	1	1	0	1
H. sepalosa	1	1	2	1	0	0	0	1	0	1	0	1	1	0	1	0	1	1	0	1
H. sericea	1	1	2	0	0	0	1	0	0	1	0	1	1	0	1	0	1	1	0	1
H. somalensis	1	1	2	0	0	0	1	0	0	1	1	1	01	1	1	0	1	1	0	1
H. sp. 1 (ETH)	1	1	2	1	0	1	0	0	0	1	1	1	?	?	1	0	1	1	0	1
H. sp. 2 (MAD)	1	0	0	0	0	1	0	0	0	1	0	1	0	1	1	1	1	1	1	1
Sabaudiella aloysii	1	1	2	1	0	1	0	0	0	1	1	1	0	0	1	0	1	1	0	1
Seddera arabica	0	0	0	0	0	0	0	0	0	0	0	0	0	0	0	0	0	0	0	0
S. bracteata	0	0	1	0	0	0	1	0	0	0	0	0	0	0	0	0	0	0	0	0

Cladistic analysis
A cladistic analysis was made using data from Tables 2 and 3. The data were analyzed using MacClade programme (Maddison & Maddison 1992) to build the data matrix and review character-states, and PAUP programme (Swofford & Begle 1993) for branch and bound cladistic analysis. The authors are working at present on the phylogenetic relationships within the Convolvulaceae and from preliminary results the tribes *Hildebrandtieae* and *Cresseae* are closely related and probably originated from a common ancestor. Of the other members in the *Cresseae*, *Seddera* was found to be closely related. Thus two species, *S. arabica* Hochst. and *S. bracteata* were used together as an outgroup (Watson & Wheeler 1981) for assessing character polarity and for rooting the cladogram. The analysis included all taxa listed in Table 2.

Table 3. List of the character states and the coding for different states as well as what has been interpreted as the ancestral state.

Characters	States (0: ancestral state)
1. Flower sex	0: bisexual, 1: unisexual
2. Inflorescence location	0: axillary, 1: on short shoots
3. Female/bisexual inflorescence type	0: cymose, 1: solitary, 2: solitary + fasciculate
4. Floral parts	0: five, 1: four
5. Female/bisexual sepals clawed	0: no, 1: yes
6. Accrescent sepals two	0: absent, 1: present
7. Accrescent sepals three	0: absent, 1: present
8. Accrescent sepals four	0: absent, 1: present
9. Accrescent sepals five	0: absent, 1: present
10. Accrescent sepal & pedicel adnation	0: no, 1: yes
11. Corolla colour	0: white, 1: pale yellow
12. Filament lengths	0: ± equal, 1: unequal
13. Pollen size (equatorial view)	0: 40-45 μ, 1: 30-35 μ, 2: 15-20 μ
14. Pollen shape	0: prolate, 1: spheroidal
15. Ovary pubescence	0: pubescent, 1: glabrous
16. Number of styles in male/bisexual flowers	0: two, 1: one
17. Stigma lobes	0: peltate to obovate, 1: branched
18. Fruit pedicel position	0: straight, 1: recurved
19. Fruit locules	0: two, 1: one
20. Seed surface	0: smooth, 1: rugose

Results

Pollen studies

Studies of the pollen grains of 10 species in the tribe Hildebrandtieae showed that they are usually 3-colpate, occasionally 4-colpate in *H. somalensis*, the shape varying from prolate to spheroidal and the surface granulate with or without punctae (Figure 2). The results did not show clear distinctions between the genera.

Cladistic analysis

The analysis resulted in a cladogram of 38 steps long with a consistency index (CI) of 0.71 and retention index (RI) of 0.79. A strict consensus tree of six cladograms is shown in Figure 3. In this analysis, all the cladograms indicate that *Cladostigma* and *Hildebrandtia* are sister groups and *Sabaudiella* is nested within *Hildebrandtia*. The analysis also shows that the genus *Hildebrandtia* has two main groups, the group of species from mainland Africa and another species group restricted to Madagascar.

Generic relationship within the tribe

The genus *Cladostigma* possesses such ancestral characters as the pistillate flowers in cymose inflorescences, sepals clawed at the base, not adnate to the pedicels and filaments ± equal. Thus it is possibly the most primitive genus in the tribe.

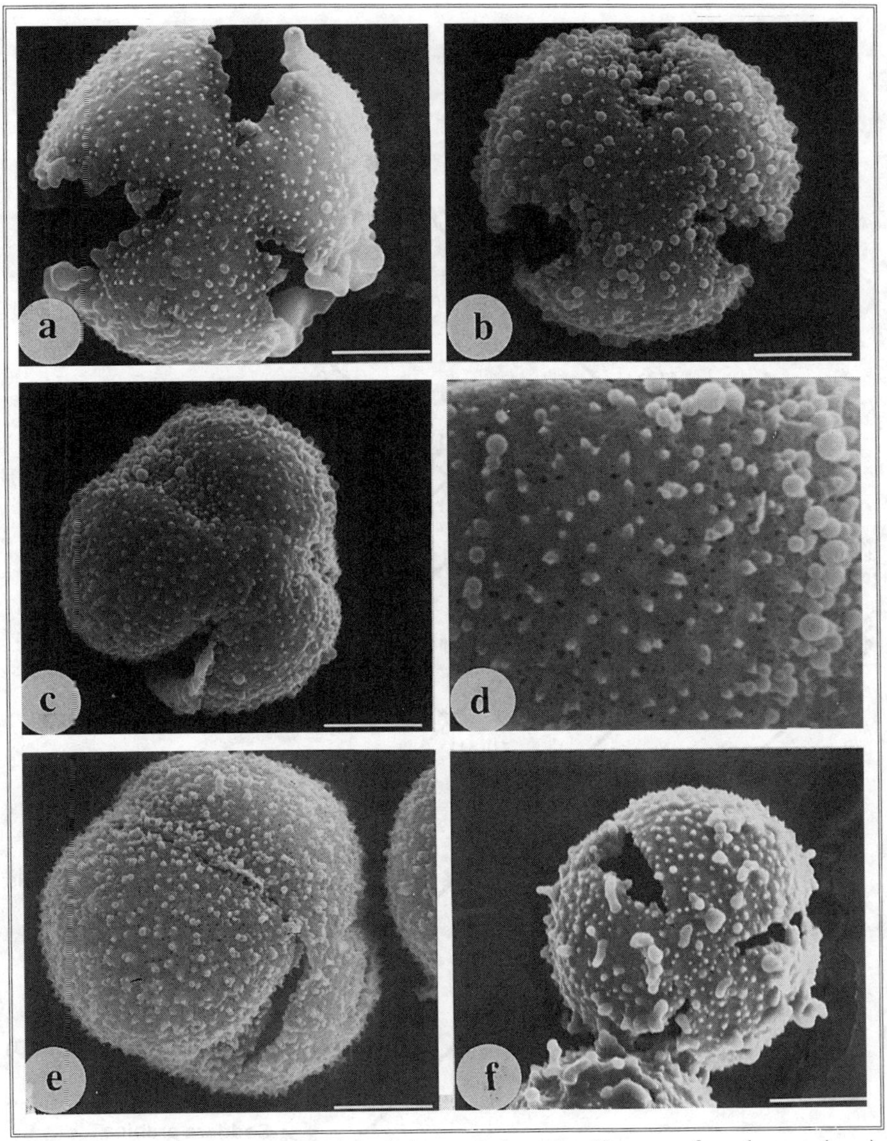

Figure 2. Pollen morphology, SEM (scale bars: whole grains 10 μm; surface close-up 1 μm).
a *Cladostigma dioicum*, 3-colpate grain in polar view, surface granulate.
b *Cladostigma* sp., 3-colpate grain in polar view, surface granulate.
c *C. hildebrandtioides*, 3-colpate grain in semi-polar view, surface granulate.
d *C. hildebrandtioides*, granulate surface close-up with punctae.
e *Hildebrandtia africana*, 3-colpate grain in semi-polar view, surface granulate with punctae.
f *H. somalensis*, 3-colpate grain in polar view, surface granulate.
Pollen samples taken from a: Gilbert & Thulin 139; b: Gilbert et al. 7692; c,d: Hemming 248/1; e: Puff et al. 870507-2/4B; f: Ellis 348.

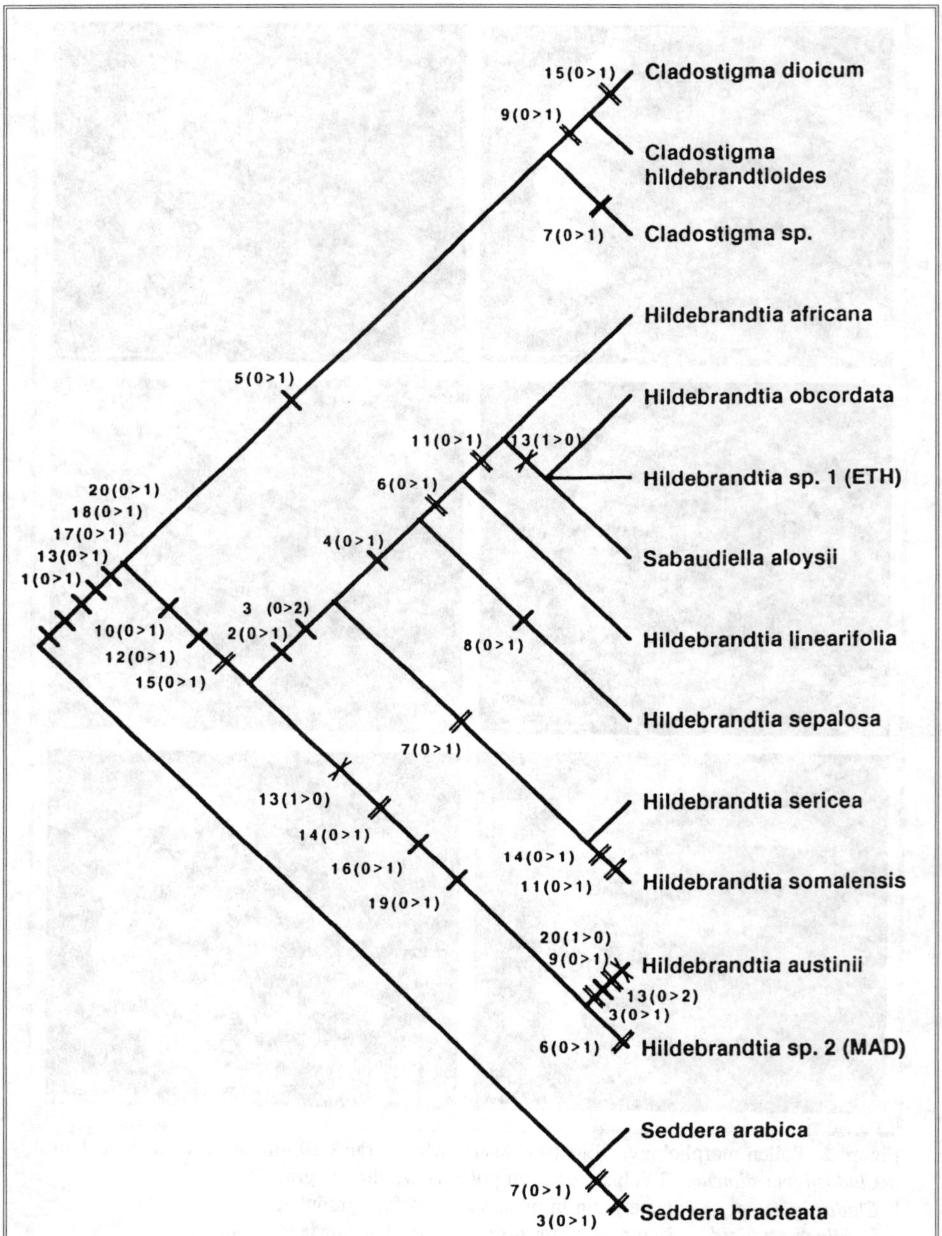

Figure 3. One of the six equally parsimonious cladograms of the species in the *Hildebrandtieae*. *Seddera arabica* and *S. bracteata* are outgroup taxa. Characters are numbered in accordance with the text, Tables 2 and 3. Dark lines = synapomorphies, light lines = parallelisms and cross lines = reversals.

The genus *Hildebrandtia* is distinguished by the partial adnation of the fruiting sepals to the pedicel and filaments unequal in length. *Hildebrandtia* clearly comprises two distinct groups: a group of species mainly in the eastern part of mainland Africa with one species extending to Arabia (*H. africana*) and the other group restricted to Madagascar. The two species known from Madagascar (*H. austinii* and *Hildebrandtia* sp.) are clearly distinguished from the other group by having a 1-locular ovary, one style in the staminate flower and a utricular fruit, both characters not seen in other members of the tribe *Hildebrandtieae*. This separation is also clearly supported by the cladogram. Thus, the resurrection of the genus *Dactylostigma* which was described from Madagascar or at least its recognition at subgeneric level should not be ruled out.

The genus *Sabaudiella* has been distinguished from its closest relative *Hildebrandtia*, by its umbrella-like fused outer calyx at the fruiting stage. However, our studies of the genus have shown that all outer sepals do not always certainly fuse in some specimens to form the umbrella-like outer calyx. In addition, although usually the outer sepals are not adnate to the pedicels, some specimens have shown partial adnation. The genus *Sabaudiella* thus has a number of shared characters with *Hildebrandtia*, and its inclusion in the *Hildebrandtia* is suggested. In the cladogram it was clearly seen together with a group of related species in *Hildebrandtia*.

Conclusion

The tribe *Hildebrandtieae* is related to the tribe *Cresseae* within the family Convolvulaceae. It is characterized by dioecious habit, two free styles with branched stigma lobes, accrescent sepals in the female flowers, fruits utricular to capsular, 1-4 locular, 1-4-seeded, and seeds rugose. It includes two genera *Cladostigma* mainly characterized by the clawed sepals in the female flowers and *Hildebrandtia* mainly characterized by the adnation of the fruiting accrescent sepals with the pedicels. Members of the genus *Hildebrandtia* from Madagascar share unique features: 1-locular ovary, one style in the female flower and indehiscent fruit. Thus, for these species the previously described taxon *Dactylostigma* could be resurrected or recognized as a distinct subgenus within *Hildebrandtia*. The third genus *Sabaudiella* usually included in the tribe is regarded as congeneric with *Hildebrandtia* as the characters delimiting this taxon from *Hildebrandtia* are variable within the single species.

Acknowledgements

S.D. would like to thank the Swedish Agency for Research and Cooperation with Developing countries (SAREC) which enabled field work to be carried out in Ethiopia, and the Fulbright Program that awarded him a grant to carry out research on the generic delimitation of some African taxa in the family Convolvulaceae for nine months at Florida Atlantic University. We wish to thank the keepers, directors and curators of AUA, BM, ETH, K, MO and UPS for making the material for the study available on loan. We also thank Dr. S. Lipka and Mr. A.N. Poulassichidis for their help with the SEM, and Dr. B. Verdcourt (Kew) for kindly reading the manuscript.

References

Austin, D.F., 1973a. *Dactylostigma*, a new genus of Convolvulaceae. Phytologia 25: 425-429.

Austin D.F., 1973b. The American Erycibeae (Convolvulaceae): *Maripa, Dicranostyles* and *Lysiostyles* I. Systematics. Ann. Missouri Bot. Gard. 60 (2): 306-412.

Baillon, H., 1891. Convolvulacées. Histoire des Plantes 10: 305-331.

Bentham, G. & J.D. Hooker. 1973. Convolvulaceae. Genera Plantarum 2: 865-881.
Chiovenda, E., 1929. La Collezzione Botanica di S.A.R. il Duca degli Abruzzi alla sorgente dell'Vebi Scebeli. Nuovo Giorn. Bot. Ital. 36: 360-371.
Hallier, H., 1893. Versuch einer natürlichen Gliederung der Convolvulaceen auf morphologischer und anatomischer Grundlage. Bot. Jahrb. 16: 453-591.
Holmgren, P.K., W. Keuken & E.K. Schofield, 1981. Index Herbariorum 1. The Herbaria of the world. 7th Edn. 452 pp. Bohn, Scheltema & Holkema, Utrecht.
Maddison, W.P. & D.R. Maddison, 1992. MacClade Version 3. Analysis of Phylogeny and Character Evolution. Sinauer Associates Inc. Publishers. Sundeland, Massachusetts, U.S.A.
Melchior, H., 1964. Engler's Syllabus der Pflanzenfamilien, ed. 12(2): 427-429. Berlin.
Peter, A., 1891. Convolvulaceae. In: Engler, A. & K. Prantl (eds.), Die Natürlichen Pflanzenfamilien 4(3a): 1-40. Leipzig.
Peter, A., 1897. Convolvulaceae. In: Engler, A. & K. Prantl (eds.), Die Natürlichen Pflanzenfamilien 4(3a): 375-377.
Radlkofer, L., 1883. Ein Beitrag zur afrikanischen Flora, Convolvulaceae. In: Abhandl. Naturw. Ver. Bremen 8: 412-416.
Roberty, G., 1952. Genera Convolvulacearum. Candollea 14: 11-60.
Roberty, G., 1964. Les genres des Convolvulacées (esquisse). Boissiera 10: 129-156.
Sebsebe Demissew & C. Puff, 1989. Observations on *Lippia carviodora*, a dioecious Verbenaceae. Beitr. Biol. Pflanzen 64: 221-230.
Staples, G.W., 1987. A revision of Porana Burman f. (Convolvulaceae) and an evaluation of the Poraneae. Harvard University, Department of Evolutionary Biology. Ph.D. thesis: 1-328.
Staples, G.W., 1990. Preliminary taxonomic consideration of the Poraneae (Convolvulaceae). Journ. Arn. Arb. 71: 251-258.
Swofford, D.L. & D.P. Begle, 1993. Phylogenetic Analysis Using Parsimony (PAUP) Version 3.1. Center for Biodiversity, University of Illinois Natural History Survey, Champaign, Illinois, U.S.A.
Vatke, W., 1876a. In: Sitzungsber. Ges. natürf. Freunde Berlin 1876: 7, 18 Jan. 1876.
Vatke, W., 1876b. In: Monatsb. Akad. Berlin 14 Dec. 1876.
Verdcourt, B., 1961. Notes on African Convolvulaceae V. Kew Bull. 15(1): 1-18.
Verdcourt, B., 1963. Convolvulaceae. In: Hubbard, C.E. & E. Milne-Redhead (eds.), Flora of tropical East Africa. London, Whitefriars Press, Ltd.
Watrous, L.E. & Q.L. Wheeler, 1981. The outgroup comparison method of character analysis. Systematic Zoology 30: 1-11.
Willis, J.C., 1973. H.K. Airy Shaw, editor. A dictionary of the flowering plants and ferns. 8th ed. London, Cambridge Univ. Press.

The problem of generic delimitation when conflict exists between molecular and morphological data

E.B. KNOX
East African Herbarium, National Museums of Kenya, P.O. Box 45166, Nairobi, Kenya.

Abstract
Congruent molecular and morphological data present few problems for generic delimitation; the diagnostic morphological synapomorphies can be used in keys and generic descriptions, and these characters provide reliable means for field identification and generic assignment of new species. Problems arise when evolution results in drastic modification of the floral and fruit characters in one species (or its resulting clade) such that the diagnostic features of the group are no longer discernible due to the pattern of subsequent modification. If the pattern of subsequent modification is sufficiently extensive, the historical information needed to reconstruct the true phylogeny may not be represented in the morphological features, and a morphologically-based analysis may yield a false reconstruction. In some cases, molecular data may retain the historical information needed to reconstruct the true phylogeny, and from this, the pattern of morphological modification can be inferred, but such conflicts between data sets present problems for morphologically-based generic delimitation in a monophyletic system.

Résumé
Des données moléculaires et morphologiques congruentes ne présentent guère de problèmes quant à la délimitation générique; les synapomorphies morphologiques diagnosiques peuvent être utilisées dans les clés et dans les descriptions génériques, et ces caractères s'avèrent utiles pour une identification de terrain et pour la désignation générique d'une nouvelle espèce. Des problèmes surgissent lorsque l'évolution donne lieu à des modifications radicales des caractères de la fleur et du fruit chez une espèce (ou chez le clade qui en résulte), de sorte que les caractéristiques diagnosiques ne sont plus discernables en raison de la modification subséquente. Si la structure de celle-ci est suffisamment ample, l'information historique nécessaire pour reconstruire la phylogénie réelle peut ne pas être traduite par des caractères morphologiques, et une analyse basée sur la morphologie peut donner lieu à une fausse reconstruction. Dans certains cas, les données moléculaires peuvent conserver l'information historique nécessaire pour reconstruire la phylogénie réelle, et à partir de là, la structure de la modification morphologique peut en être déduite, mais de tels antagonismes entre des groupements de données posent problème quant à la délimitation générique basée sur la morphologie dans un système monophylétique.

Key words
Generic delimitation, Molecular systematics, Monophyly, Paraphyly, Phylogenetic systematics.

Introduction

The sexual system of Linnaean classification used the combinatorial diversity of floral parts to delimit genera and higher groups of plants. The presence or absence, number, position and fusion of floral parts provided a convenient system for identifying and classifying plants. Sexual parts have generally been considered more evolutionarily conservative than vegetative features, and so, have been considered a more reliable guide to phylogenetic relationships. Within certain groups, vegetative features may also show patterns that correspond with sexually-based generic delimitations (see Léonard, this volume). Because of convergent evolution, some groups of plants show a mix-and-

L.J.G. van der Maesen et al. (eds.), The Biodiversity of African Plants, 421-432.

match combination of floral parts, and these groups have been notoriously difficult to place in higher level classification. But even within well-delimited families, generic delimitation frequently relies on a combination of primitive (plesiomorphic) and derived (apomorphic) characteristics. For many herbarium and field-based taxonomists and ecologists, such traditional generic delimitation poses no problem as long as the genera constitute evolutionarily related groups (convex groups sensu Estabrook 1978) that may be readily identified by their key characteristics.

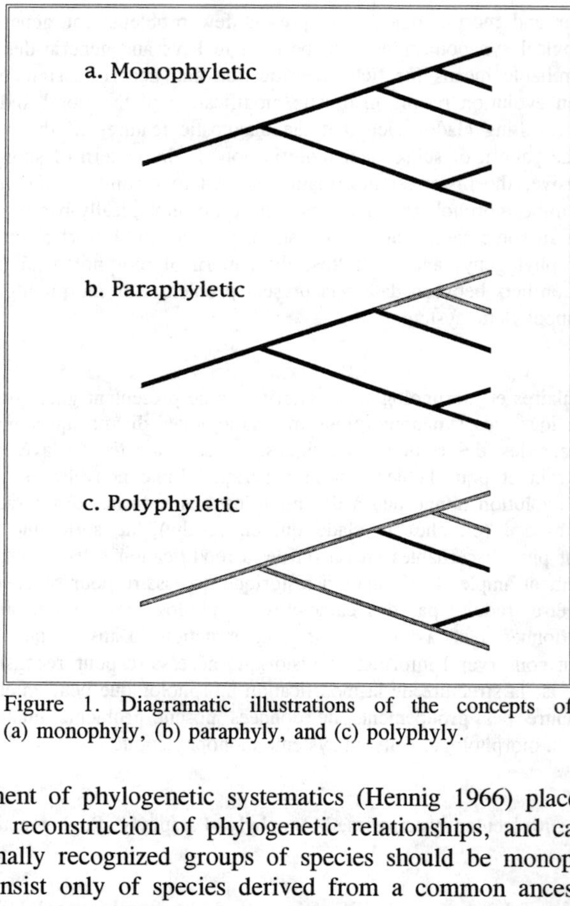

Figure 1. Diagramatic illustrations of the concepts of (a) monophyly, (b) paraphyly, and (c) polyphyly.

The development of phylogenetic systematics (Hennig 1966) placed new emphasis on the historical reconstruction of phylogenetic relationships, and carried with it the dictum that formally recognized groups of species should be monophyletic (i.e., the group should consist only of species derived from a common ancestral species, and furthermore, the group should consist of all species derived from that common ancestor; Figure 1a). This approach disallows groups that are paraphyletic (i.e., a chunk of the genealogical nexus that does not include all of the descendants of a common ancestor; e.g., the Gymnosperms or the Bryophytes; Figure 1b) or polyphyletic (i.e., two or more subgroups that have separate evolutionary histories, and, so, do not constitute a chunk of the genealogical nexus; e.g., Engler and Prantl's Amentiferae or Hutchinson's Herbaceae; Figure 1c). There is general consensus that polyphyletic groups do not constitute "natural groupings" and should be rejected, but there remains considerable disagreement whether formally recognized genera (and higher groups) should be strictly monophyletic or whether morphologically well-delimited paraphyletic groups should be retained. In this volume, B.D. Schrire & G.P. Lewis and R.K. Brummitt presented arguments supporting these two positions, respectively. My intention is to discuss the use of monophyletic and paraphyletic generic delimitation when conflict exists between molecular and morphological data.

The use of molecular data in phylogenetic reconstruction

The advent of molecular systematics provided new lines of evidence for addressing questions of phylogenetic relationship. The ability to investigate changes in DNA sequence and structural organization offers new sources of characters for inferring evolutionary relationships. Chloroplast DNA is commonly used in plant molecular systematics because:

1 it is uniparentally inherited in most cases (usually maternally),
2 the molecule has many copies in photosynthetic tissue,
3 the molecule is evolutionarily conservative and
4 character homology can be assessed across morphologically divergent groups.

By using molecular characters to reconstruct phylogeny, one can then interpret the morphological characters of interest, thereby avoiding the potentially circular reasoning of using morphological characters to infer a phylogeny, which is, in turn, used to interpret the evolution of those same morphological characters.

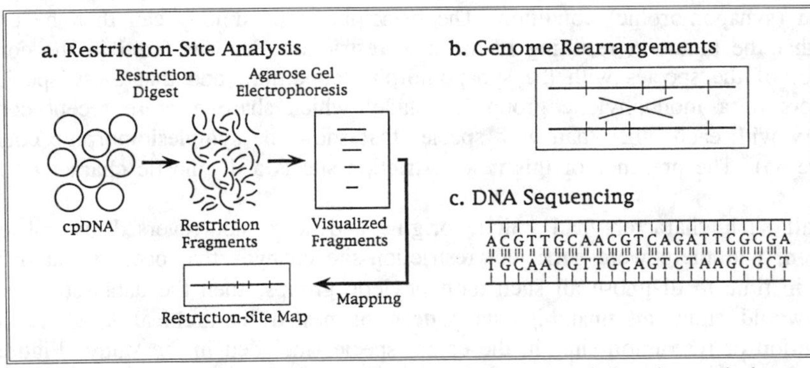

Figure 2. Sources of molecular systematic data: a. Restriction-site analysis uses restriction enzymes to digest DNA samples, and the resulting fragments are separated by size using agarose gel electrophoresis and the visualized fragments are mapped to determine the position of the restriction sites. b. Genome rearrangements, such as inversions (shown here), insertions, and deletions are relatively rare events that make them very useful characters in phylogenetic analysis. c. DNA sequencing uses individual nucleotides in genes or spacer regions as characters.

Molecular systematics investigates DNA evolution by sampling changes in DNA sequence with restriction enzymes (Figure 2a), by mapping rearrangements (Figure 2b), or by determining the sequence of nucleotides for a given segment of DNA (Figure 2c). In restriction-site analysis, different restriction enzymes are used to cut DNA at specific recognition sites (restriction sites) where the matching between the

DNA sequence and the active site of the enzyme allows the cleavage reaction to occur. Various restriction enzymes are commercially available, that recognize different DNA sequences and cut with varying frequency. A molecular systematic study must be designed so that the molecule to be studied, the questions to be addressed, and the restriction enzymes to be used are well coordinated. After digesting a DNA sample with a given enzyme, the resulting fragments are separated by size using agarose gel electrophoresis, and depending on the specific laboratory protocols used, the fragments may be visualized by staining with a fluorescent dye (ethidium bromide) or transferred to a membrane (Southern blotting) and probed using autoradiography (Figure 2a).

The characters of interest in this approach are the gain and loss of restriction sites caused by mutations in the DNA. For example, in a given study, the DNA samples from half the species may have two homologous restriction sites that are 10 kilobases (kb) apart (that result in a 10 kb fragment after digestion), whereas the homologous section of DNA in the other half of the species may have the same two restriction sites, and also a third, resulting in two fragments, one 6 kb, the other 4 kb. Whether this difference in the DNA fragments generated by digestion with a restriction enzyme is due to the gain of a restriction site (resulting in the 6 kb and 4 kb fragments) or the loss of a restriction site (resulting in the 10 kb fragment) is determined by comparison with DNA from one or more closely related species (the outgroup species) that are not members of the monophyletic group (the ingroup) under study. If such comparisons with the outgroup species shows that the 10 kb condition is the shared, primitive (symplesiomorphic) state, then the 6 kb and 4 kb condition is interpreted as the shared, derived (synapomorphic) condition. The principle of parsimony can then be used to infer that the mutation resulting in the new restriction site was fixed in the common ancestor of the species with the synapomorphic condition, and that these species are members of a monophyletic group (or clade) which share a more recent common ancestry with each other than with species that show the symplesiomorphic condition (Figure 3a). The presence of this new restriction site is a diagnostic character for this clade.

If all such characters had unique origins and were not reversed by subsequent mutation, and there were sufficient restriction-site changes that occurred at different points in time to diagnose all such monophyletic groups, then the data set for such a study would show an unambiguous pattern of nested hierarchical structure in the distribution of restriction sites in the extant species included in the study (Figure 3b). Because convergence can occur in molecular evolution, resulting in some characters that are at variance with the nested hierarchical pattern that reflects the true phylogenetic relationships, computer packages for phylogenetic analysis use the principle of parsimony to detect the general pattern of nested hierarchical structure and to identify those characters that are interpreted as resulting from convergent evolution (i.e., are homoplasious).

The same principles apply when working with data on structural organization (in which the characters are rearrangements) or DNA sequence data (in which the characters are individual nucleotides). The purpose in using molecular data for phylogenetic reconstruction is that the generally conservative nature of DNA offers potential characters that can be compared across a diverse group of organisms in order to infer the true phylogenetic relationships.

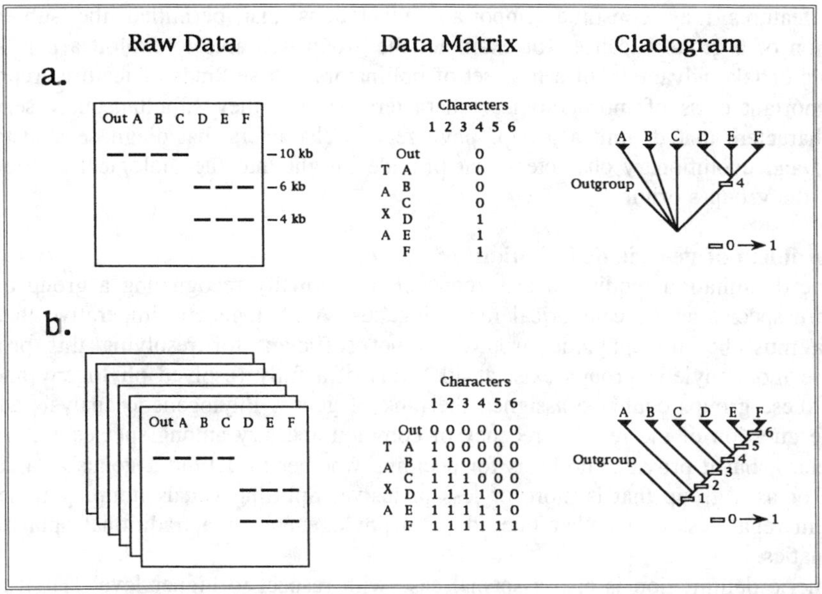

Figure 3. The use of restriction-site data in phylogenetic reconstruction: a. Outgroup comparison allows the evolutionary directionality (polarity) to be determined, and the shared, derived (synapomorphic) character-state (in this case, the 6 kb and 4 kb fragments) is used to diagnose a monophyletic group. b. Many such mutation fixed at different points in time result in a nested hierarchical structure in the data matrix and provide a fully resolved phylogenetic reconstruction.

The use of morphological data in phylogenetic reconstruction

Morphological data can be used in the same manner as molecular data to infer phylogenetic relationships. A shared, derived character (synapomorphy) diagnoses a monophyletic group that evolved from a common ancestral species in which that character was fixed, and a nested hierarchical pattern in character distribution can be used to infer phylogenetic relationships. Hybridization is recognized as an important aspect of plant evolution in which the reticulating lineages confound the nested hierarchical pattern that would otherwise be laid down in a strictly dichotomous process of cladogenic events with complete fixation of traits between speciation events. Phylogenetic analysis not only provides good reconstructions in those cases where evolution has proceeded through simple, dichotomous speciation, but it also provides a good starting point for investigating cases resulting from more complex evolutionary processes. A comparison of molecular and morphological data provides a useful approach to studying complex evolutionary events, particularly because the apparent lack of genetic recombination in the generally uniparentally inherited chloroplast DNA means that the biological features of this molecule closely approximate the assumptions of phylogenetic analysis.

Morphological and molecular sources of data have different strengths and weaknesses. Morphological features are phenotypic expressions of the underlying genotypes and are exposed, to varying degrees, to the effects of natural selection. In some cases, these features may constitute important adaptations that permitted the subsequent radiation of a group because, for example, the group was able to exploit a new set of habitats or take advantage of a new set of pollinators. These kinds of features represent an important class of morphological characters because they simultaneously serve as key characters that delimit a group, phylogenetic characters that diagnose a historical entity, and evolutionary characters that provide insight into the biological circumstances of the group's origin.

The problem of generic delimitation

Generic delimitation requires a class concept for formally recognizing a group of one or more species at the categorical rank of genus. A phylogenetic imperative that "all genera must be monophyletic groups" is not sufficient for resolving this problem because monophyletic groups exist at all levels of a fully resolved phylogeny and not all of these groups could be assigned the rank of genus. Phylogenetic analysis addresses the question of the relative recency of common ancestry among species and groups of species, but it provides no basis for deciding whether to delimit a genus as a certain group or as a group that is more or less inclusive. Splitting versus lumping remains a problem regardless of whether one adopts a phylogenetic or a traditional approach to systematics.

Generic delimitation is also a special case with respect to higher level classification because of the tradition of binomial nomenclature. Specific epithets alone do not uniquely identify plants (e.g., abyssinica); unique species names are derived only through binomial nomenclature (e.g., *Hagenia abyssinica* or *Erythrina abyssinica*). This means that changes in generic delimitation can, in some cases, result in changes in both the generic and specific names of affected species. Changes in familial and higher level classification do not have corresponding effects on species names.

It has been suggested that phylogenetic approaches to classification should minimize disruption of existing nomenclature. This pragmatic approach to conservation of traditional taxonomy is commendable, but it nonetheless fails to provide a solid theoretical basis for applying class concepts to the problem of generic delimitation. The nature of the problem can be profitably examined by contrasting cases in which conflict between molecular and morphological data does and does not exist.

Generic delimitation when no conflict exists between molecular and morphological data

When no conflict exists between molecular and morphological data, the same monophyletic groups are diagnosed by both sources of data. Generic delimitation in these cases faces no new complications other than the long-standing problem of lumping versus splitting. Although phylogenetic systematics carries the dictum that only monophyletic groups may be formally recognized, the exclusive focus on the relative recency of common ancestry (propinquity of descent) provides no class concepts for determining which monophyletic groups should be formally recognized in a system of classification, nor the rank at which these monophyletic groups should be placed (Knox 1995). The source of difficulty in converting a phylogeny into a classification is due to the structural differences between a semi-nested monophyletic hierarchy and a fully nested classification hierarchy (Knox 1995).

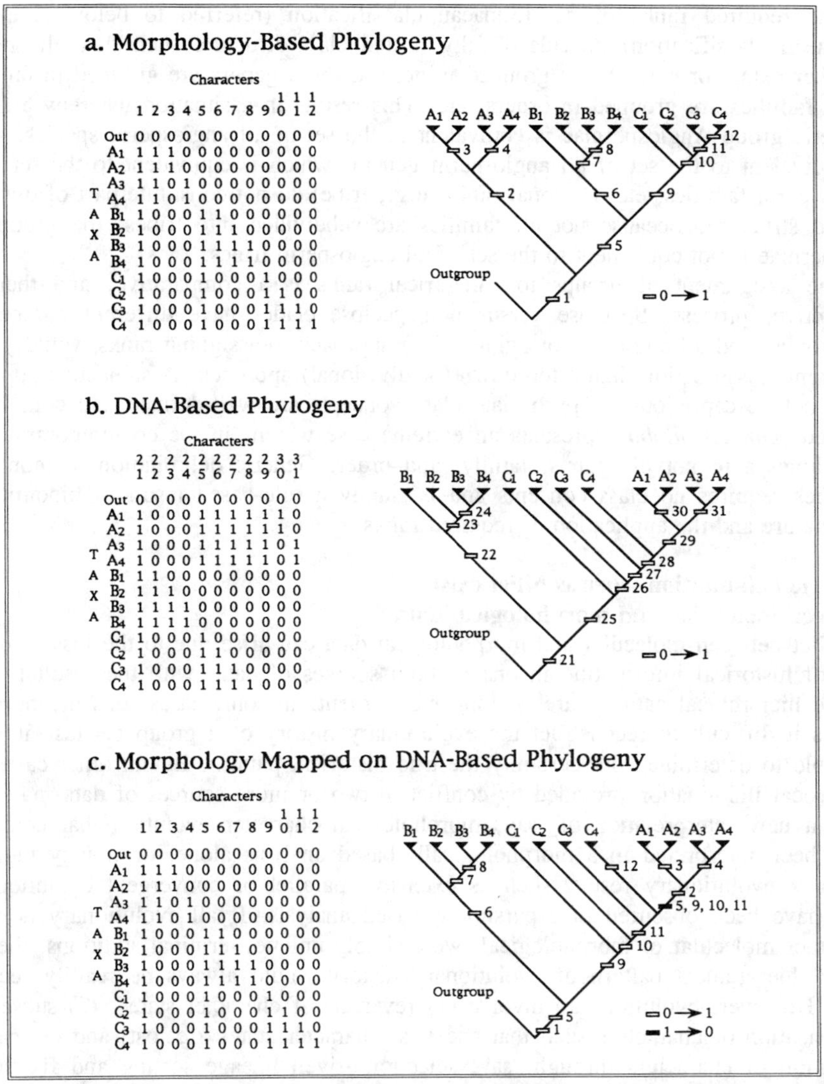

Figure 4. The problem of generic delimitation when conflict exists between molecular and morphological data: a. The morphology-based phylogenetic analysis shows a data matrix with nested hierarchical structure and a phylogeny in which Genera A, B, and C are all diagnosed as monophyletic entities. b. The DNA-based phylogenetic analysis also shows a data matrix with nested hierarchical structure and a phylogeny that agrees with the morphology-based phylogeny in all respects except for the placement of Genus A. c. When morphology is mapped on the DNA-based phylogeny, Characters 5, 9, and 10 are interpreted as having undergone reversal in the evolution of the common ancestral species of Genus A, and Character 11 could either be interpreted as a reversal (as shown) or as parallel gains in Taxa C3 and C4. Possible solutions to the problem of generic delimitation are discussed in the text.

The required ranks of neo-Linnaean classification (referred to below simply as Linnaean classification) provide a fully nested class structure such that all species, whether extant or extinct, are grouped in genera, these genera are grouped in families, these families are grouped in orders, etc. This results in a situation whereby a monophyletic group Angiospermae is equivalent to the set of all angiosperm species, which is equivalent to the set of all angiosperm genera, which is equivalent to the set of all angiosperm families, etc. Optional ranks (e.g., tribe) do not constitute part of this fully nested structure. Because not all families are subdivided into tribes, the group Angiospermae is not equivalent to the set of all angiosperm tribes.

The assignment of groups to categorical ranks is a comparative, and therefore subjective, process. Speciose versus non-speciose clades have different numbers of subgroups, and a bottom-up (or aggregative) approach to assigning ranks would yield a different classification than a top-down (or divisional) approach. A subjective approach need not be capricious if appropriate class concepts are well-defined and consistently applied. *Ginkgo biloba* represents an extreme case which, in the contemporary flora, constitutes a monotypic genus, family, and order. Generic delimitation in monotypic families requires no class concepts and is simply a rule-based result of binomial nomenclature and the application of required ranks.

Generic delimitation when conflict exists between molecular and morphological data

Conflict between molecular and morphological data can arise due to the loss or distortion of historical information in one or both sources of data, with the result that the nested hierarchical patterns are no longer congruent. In some cases, this incongruence makes it difficult to reconstruct the evolutionary history of a group because it is not possible to determine with certainty the true historical pattern. But in other cases, the reciprocal illumination provided by conflict in two or more sources of data may highlight a new interpretation of, say, morphological character evolution that could not have been interpreted in a morphologically-based analysis alone, or it may highlight complex evolutionary results such as extensive parallel or convergent evolution that may have been obscured in a parsimony-based analysis. If all evolutionary novelties (whether molecular or morphological) were simply unique, terminal additions, then the nested hierarchical pattern of evolutionary history would always be readily recoverable. However, evolution can involve (1) reversals of character states, (2) subsequent modification of characters such that ancestral character states are lost, and (3) parallel evolution of characters through, say, selection-driven lineage sorting and fixation of recessive traits such that morphological features displaying homoplasious pattern are nonetheless homologous. These aspects of evolution present problems for phylogenetic reconstruction, and when these problems affect features regarded as generic traits, difficulties can arise with respect to generic delimitation.

As an example, consider the effect of character-state reversal on phylogenetic reconstruction. In a hypothetical case of a morphological analysis (Figure 4a), Character 1 (say, resupinate flowers) diagnoses a family that consists of three genera (Genera A, B, and C), and Character 1 would also serve as a key character for identifying the family. Character 2 (say, adnate filaments) diagnoses (and is a key character for) Genus A, which may consist of two subgenera that are diagnosed by Characters 3 and

4. Character 5 (say, inferior ovaries) is present in Genera B and C but not present in Genus A or members of the closely related families that might be used as outgroups to root the phylogeny. In this example, Character 5 serves to unite Genera B and C as sister-groups and to place Genus A in a basal position within the family because it has superior ovaries. Character 6 (say, exserted stamens) diagnoses Genus B, with Characters 7 and 8 (say, vegetative characters) diagnosing subgeneric groups. Character 9 (say, fleshy fruits) diagnoses Genus C, with characters 10 and 11 (say, growth-form characters) diagnosing subgeneric groups. For the purposes of this example, it does not matter whether the terminal taxa (e.g., A1) are individual species or groups of species. Each terminal taxon would have its own unique features (such as Character 12 for Taxon C4), but these have been otherwise omitted, both for the purpose of illustration and because such unique (autapomorphic) characters are not used in phylogenetic reconstruction. Also, only a single character has been shown for each node in the cladogram. Additional characters that are congruent with the nested hierarchical pattern shown in this example would be represented in the data set as multiple instances of the character-state distribution shown by one of the characters presented. This simplification is strictly for the purpose of illustration and should not be misconstrued as support for single-character taxonomy. In this example, there is no conflict within the morphological data-set, the family and genera are all monophyletic entities for which the diagnostic characters could be used as key characters for purposes of identification, and the common ancestor of the family can be inferred to have had superior ovaries and free stamens.

Now consider the implications in this hypothetical case of possible conflicting results from a molecular analysis. The DNA-based phylogeny (Figure 4b) differs from the morphology-based phylogeny in one important respect: Genus A is no longer the basal member of the family, but, instead, is nested within Genus C, specifically as the sister group to Taxon C4. This placement of Genus A, if correct, makes Genus C paraphyletic, which presents a problem for monophyletic approaches to generic delimitation. In this hypothetical example, there is no basis for deciding whether the morphological or molecular reconstruction best represents the true phylogeny, but in real studies, one could re-examine the original evidence and also seek new evidence to gain insight into the nature of the conflict. In this hypothetical case, suppose that several lines of molecular evidence (e.g., chloroplast DNA restriction-site analysis, nuclear gene sequencing, and a rearrangement shared by Genus A and Taxon C4) all support the DNA-based phylogeny. Alternatively, suppose that it is not clear which of the phylogenetic reconstructions most closely represented the true phylogeny, and the DNA-based phylogeny is only provisionally accepted as part of a process of reciprocal illumination between the molecular and morphological data sets. What, then, are the implications for the morphological evolution and generic delimitation within this group?

If the morphological characters are mapped onto the DNA-based phylogeny (Figure 4c), then four characters can be interpreted as having undergone character-state reversal in the evolution of Genus A. The superior ovaries of Genus A (Character 5) would now be interpreted as secondarily derived instead of primitive within the family. Similarly, the fleshy fruits of Genus C (Character 9) underwent reversal to the dry, capsular state during the evolution of the common ancestor of Genus A. Character 10 (a growth-form character) would also be most parsimoniously interpreted as a reversal, and Character 11 could be interpreted either as a reversal (as shown) or as parallel gains in Taxa C3 and C4. Although the pattern of descent may now be firmly established, and the morphological evolution may be inferred with varying degrees of certainty for each of the characters, the problem of generic delimitation remains. Genera A and B are diagnosed as monophyletic groups by both the morphological and molecular data, so they are not problematic. Genus C is the paraphyletic entity, and it is here that the classification problem lies. At least six possible solutions exist:

1 Retain Genus A and split Genus C into coordinate monophyletic genera (C1, C2, C3, and C4).
2 Lump Genus A with C4 into a single monophyletic genus and recognize C1, C2, and C3 separately.
3 Lump Genus A with C3 and C4 into a single monophyletic genus and recognize C1 and C2 separately.
4 Lump Genus A with C2, C3, and C4 into a single monophyletic genus and recognize C1 separately.
5 Lump Genera A and C into a single monophyletic genus.
6 Recognize Genus C as a paraphyletic genus delimited by Character 9.

Solutions 1 to 5 constitute a continuum of possible solutions, all of which satisfy the cladistic dictum that only monophyletic groups are recognized. Each of these solutions recognizes groups that may be diagnosed using molecular data, but none of these solutions provides groups that have diagnostic morphological characters that could also serve as key characters for identification. There may be a preference within the systematic community for, say, Solution 1 or Solution 5, depending on attitudes toward splitting versus lumping, but this does not invalidate Solutions 2, 3, or 4. The source of the difficulty is that cladistics provides a methodology for inferring patterns of descent, but cladistic theory lacks the class concepts needed to handle the issues of subordination and coordination that are essential for constructing Linnaean classification (Knox 1995).

The recognition of a paraphyletic Genus C (Solution 6) provides a classification in which all entities are chunks of the genealogical nexus (or convex groups). All three genera are delimited by shared homologies, although the fleshy fruits of Genus C (Character 9) are interpreted as plesiomorphic with respect to the capsular fruits of Genus A, and these homologies can serve as key characters for purposes of identification. The knowledge that Genus C is paraphyletic requires that phylogenetically-based studies must work with monophyletic entities such as the Genus A-Genus C clade, or one of the less inclusive clades contained within it, but such work is not hampered by the formal recognition of Genus C because other un-named clades are also present in this phylogeny. The fact that a morphologically coherent, but paraphyletic, Genus C is demonstrated by the molecular data is a great advantage for interpreting ancestor-descendant relationships. These data offer strong support that the common ancestor of Genus A evolved from a species that had the generic characters of Genus C.

Evolutionary reversal is but one example of a more general pattern that results when characters deemed worthy of generic recognition during the evolution of the ancestor of a clade undergo subsequent modification during the evolution of some lineage within

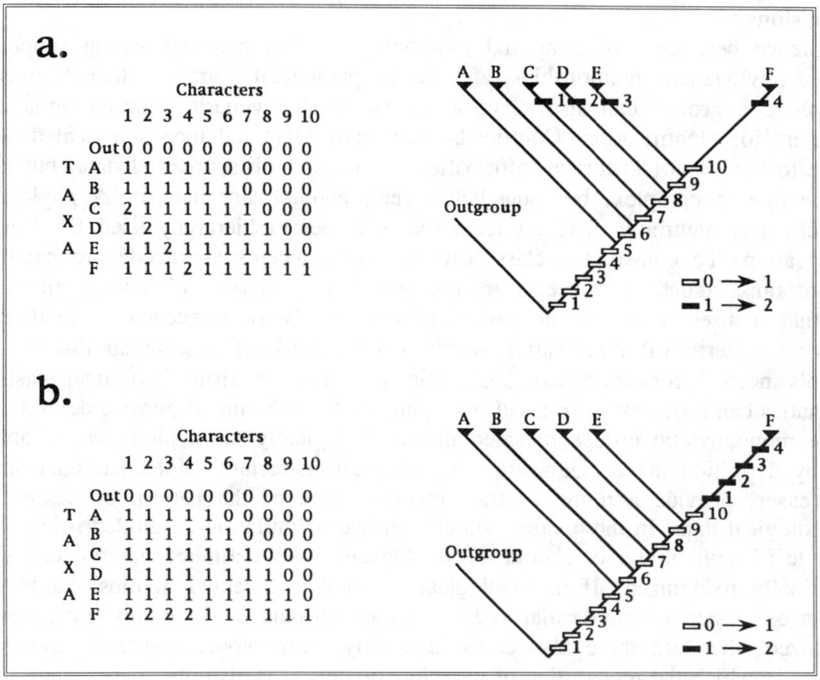

Figure 5. Patterns of subsequent modification: a. Four of the five morphological characters that diagnose the clade A-F undergo subsequent modification, but there is no strong paraphyletic pattern that would support the recognition of a paraphyletic taxon. b. The same four morphological characters as above all undergo subsequent modification in the common ancestral species of Taxon F, which results in four, coordinated, plesiomorphic characters that might justify the recognition of the paraphyletic Taxon A-E.

that clade. Darwin's characterization of evolution as descent with modification underscores the dualistic nature of evolution (Knox 1995). We now have good methods for analyzing the pattern of descent. This is necessary, but not sufficient. What we still lack in systematic theory is a good methodology for analyzing the pattern of modification within that framework of descent. But even with such a methodology in place, there will still be a need for systematists to make decisions based on their knowledge of a group of plants. For example, if all of the generic characters for a group undergo subsequent modification, but in different lineages (Figure 5a), there may be several equally satisfactory classifications that represent the evolutionary pattern seen in a group of plants. But in other cases (Figure 5b), there may be a strong pattern of subsequent modification in one lineage that favours the formal recognition of a paraphyletic group that is delimited by multiple, shared, plesiomorphic homologies. The principal issue in considering this next step of development in systematic theory, regardless of whether one is working from a traditional or phylogenetic perspective, is the need to move away from ad hoc solutions so that problems of generic delimitation can be treated with greater conceptual rigor.

Conclusions

Congruence between molecular and morphological data provides strong support for inferred phylogenetic relationships, and the morphological characters that diagnose the monophyletic groups can also serve as characters for generic delimitation and key characters for identification. Conflict between molecular and morphological data may be due to the loss of historical information in one or both sources of data, but it may also be due to complex, but nonetheless real, evolutionary pattern. A phylogenetic hierarchy is structurally different from the fully nested hierarchy used for Linnaean classification. To construct a classification, class concepts are needed to handle the issue of rank, and these concepts are not provided by theory of phylogenetic systematics that focuses solely on the pattern of descent. Work is needed on methods for analyzing patterns of modification within a framework of descent so that the class concepts needed for constructing classifications have a firm theoretical basis and systematics can move away from ad hoc solutions to problems of generic delimitation.

The monophyletic groups revealed in morphologically-based phylogenetic analyses will, by definition, have diagnostic morphological characters. Molecular data may, in some cases, provide a more accurate reconstruction of the pattern of descent than morphological data. In these cases, should generic delimitation be guided solely by the principle of monophyly, or should morphologically well-delimited paraphyletic genera be formally recognized? If morphological primacy is accepted in those cases where conflict exists between molecular and morphological data, and paraphyletic genera are recognized, then are there also cases in strictly morphologically-based phylogenetic analyses in which the recognition of paraphyletic genera is also justified?

References

Estabrook, G.F., 1978. Some concepts for the estimation of evolutionary relationships in systematic botany. Syst. Bot. 3: 146-158.

Hennig, W., 1966. Phylogenetic Systematics. University of Illinois Press, Urbana, Il., USA.

Knox, E.B., 1995. The use of hierarchies as organizational models in systematics. Botanical Review (in press).

Generic delimitation in the palms of Madagascar

H.J. BEENTJE & J. DRANSFIELD
Royal Botanic Gardens, Kew, Richmond, Surrey TW9 3AE, United Kingdom.

Abstract
Until recently, collections of palms from Madagascar have been few in number, and in many cases fragmentary. This has resulted in a fairly confused taxonomy both at specific and at generic level. A period of palm-dedicated fieldwork has resulted in a better understanding of the family in Madagascar. Before 1987, 14 genera with some 115 species were known; since 1987, three new genera have since been described, while nine genera will be put into synonymy. Some seventy new species will be described, bringing the total number to 190 (of which 185 are endemic).

Résumé
Jusqu'à ces derniers temps, la collection des palmiers de Madagascar n'était connue que par un petit nombre d'espèces et dans bien des cas de façon fragmentaire. Il en est résulté une certaine confusion taxonomique tant au rang générique que spécifique. Une prospection systématique des palmiers sur le terrain a permis d'avoir une meilleure compréhension de la famille à Madagascar. Avant 1987, on connaissait 14 genres et quelque 115 espèces; depuis cette date, trois nouveaux genres ont été décrits, tandis que neuf autres devront être mis en synonymie. Quelque septante nouvelles espèces seront décrites, ce qui portera leur nombre total à 190, dont 185 endémiques.

Key words
Endemism, Genera, Madagascar, Palmae.

Introduction

Madagascar is without doubt one of the most interesting sites for palms. The taxonomic *status quo* in 1985 was 19 genera (in six subtribes) of which eleven were considered endemic, with 106 species: three introduced, two occurring in mainland Africa as well, and 101 endemic species.

The first palms collected in Madagascar which still survive, are Philibert Commerson's specimens (*Neophloga commersoniana* Baill., presumably collected near Tolanaro, then Fort Dauphin; and *Dypsis forficifolia* Mart., probably from the bay of Antongil) dating back to 1770 - 1771 (dates from Lanjouw & Stafleu 1954) and *Phloga nodifera* (Mart.) Salomon, collected by Francisco de Noroña in 1775 (date from Vegter 1983). Collections by Pierre Poivre are more doubtful as to date, but might be older than either of these (?1740 - 1757?); he collected *Dypsis hirtula* Mart., *Chrysalidocarpus lutescens* Wendl. and *Neophloga poivreana* (Baill.) Becc., probably all in the same area between Toamasina (Tamatave) and the Bay of Antongil, we suggest from Ile Ste. Marie, now Nosy Boraha.

The first descriptions of palms from Madagascar were by Martius (1838, 1845) who established the genus *Dypsis* (etymology still unclear) for the four species he described. Since that time, twenty-three genera have been described from Madagascar (table 1).

Due to the poor and fragmentary state of much of the material in their possession, authors up to the middle of this century were often unable to compare like with like, and much of the confusion hitherto has been caused by this. Jumelle and Perrier de la Bâthie, for instance (Jumelle 1927, 1928, 1933a, 1933b; Jumelle & Perrier de la

L.J.G. van der Maesen et al. (eds.), The Biodiversity of African Plants, 433-442.

Table 1. Palm genera accepted for Madagascar.

Date	Genus with author(s)	Status 1987	1994/5
1838	*Dypsis* Mart.	recognized	recognized
1878	*Chrysalidocarpus* Wendland	recognized	=*Dypsis*
1878	*Ravenea* Wendland	recognized	recognized
1881	*Bismarckia* Hildebrandt & Wendland	recognized	recognized
1883	*Phloga* J.D. Hooker	recognized	=*Dypsis*
1894	*Dypsidium* Baillon	=*Neophloga*	=*Dypsis*
1894	*Haplodypsis* Baillon	=*Neophloga*	=*Dypsis*
1894	*Haplophloga* Baillon	=*Neophloga*	=*Dypsis*
1894	*Neodypsis* Baillon	recognized	=*Dypsis*
1894	*Neophloga* Baillon	recognized	=*Dypsis*
1894	*Phlogella* Baillon	=*Neophloga*	=*Dypsis*
1906	*Adelodypsis* Beccari	=*Dypsis*	=*Dypsis*
1906	*Trichodypsis* Beccari	=*Dypsis*	=*Dypsis*
1906	*Vonitra* Beccari	recognized	=*Dypsis*
1912	*Louvelia* Jum. & H. Perrier	recognized	=*Ravenea*
1914	*Macrophloga* Beccari	=*Chrysalidocarpus*	=*Dypsis*
1915	*Beccariophoenix* Jum. & H. Perrier	recognized	recognized
1928	*Antongilia* Jumelle	recognized	=*Dypsis*
1933	*Masoala* Jumelle	recognized	recognized
1933	*Sindroa* Jumelle	=*Orania*	recognized
1955	*Marojejya* Humbert	recognized	=*Masoala*
1984	*Halmoorea* J. Dransfield & Uhl	recognized	=*Orania*
1989	*Voanioala* J. Dransfield	n/a	recognized
1991	*Lemurophoenix* J. Dransfield	n/a	recognized
1994	*Satranala* J. Dransfield & Beentje	n/a	recognized

Bâthie 1912, 1913, 1945) relied heavily on characters such as the "oppositipetiolar ligule", the keel of the leaf rachis, and the way the peduncular bract opens on the inflorescence; we feel that all of these may (occasionally) be specific characters, but not generic ones.

In 1987 Genera Palmarum appeared (Uhl & Dransfield 1987). This publication cleared up some of the confusion, and emphasized the relatively poor knowledge of Madagascar palm genera; at about the same time the Madagascar palm project started at Kew. Until the publication of Genera Palmarum, some 650 palm specimens had been collected in Madagascar; many of these were partial specimens, and did not really contribute to an understanding of relationships. Since that date another 600 accessions have been acquired, of which 500 by the authors, and much has become clearer (Table 1) - though not all!

We hope to publish a monograph on the palms of Madagascar in 1995 (Dransfield & Beentje, in prep.). The following article is an introduction to some of the changes, with the emphasis on the problems we experienced in generic delimitation.

A Conspectus of genera of Madagascar palms (Table 2)

* *Subfamily Coryphoideae: tribe Phoeniceae*

One genus, *Phoenix*, with a single species, *P. reclinata*, also widespread in Africa; it is uncertain whether this is native or introduced.

* *Subfamily Coryphoideae: tribe Borasseae*

This tribe used to be represented in Madagascar by three genera: *Borassus* (two endemic species; the genus is widespread in the Old World) in the subtribe Lataniinae, and in the subtribe Hyphaeninae: *Hyphaene* (*H. coriacea*, also known from East Africa) and the endemic *Bismarckia* with one species: *B. nobilis*. A new monotypic genus, *Satranala* (Dransfield & Beentje, in press), has come to light during recent fieldwork.

* *Subfamily Calamoideae: tribe Calameae*

One genus, *Raphia*, with a single widespread, probably introduced species, *R. farinifera*.

* *Subfamily Ceroxyloideae: tribe Ceroxyleae*

Two genera were hitherto accepted for Madagascar: *Ravenea* (eight spp., also with two spp. in the Comoro Islands); and *Louvelia* (with three spp.); the latter is now a synonym of *Ravenea* (now with 15 species). The other genera of this tribe are from South America (*Ceroxylon*), Juan Fernandez Island in the eastern Pacific (*Juania*) and Queensland, Australia (*Oraniopsis*).

* *Subfamily Arecoideae: tribe Areceae*

\- *Subtribe Oraniinae*

Two genera were accepted for Madagascar: *Orania*, otherwise known from the Malesian region, and the endemic and monotypic *Halmoorea*. Fieldwork has shown that *Halmoorea* is synonymous with *Orania*, now the only member of the subtribe.

\- *Subtribe Dypsidinae* (near-endemic)

This is the largest group of species (82, plus some 60 new spp.), spread (until now) over seven genera, viz.: *Dypsis* (22 spp.), *Chrysalidocarpus* (15 spp. + four spp. from elsewhere), *Neodypsis* (14 spp.), *Phloga* (two spp.), *Neophloga* (25 spp.), *Vonitra* (four spp.), *Antongilia* (one sp.). All are endemic except for *Chrysalidocarpus*, which has a single species on Pemba Island (Tanzania), two on the Comoro Islands, and two species only known from cultivation. All these genera will merge, resulting in a large, variable genus *Dypsis* with some 137 species. Relationships with other subtribes are as yet unclear, but more clarity might develop from a cladistic treatment of the family (Dransfield & Uhl, in sched.).

\- *Subtribe Masoalinae* ined. (endemic): as yet unclear subtribe of uncertain affinities.

In Genera Palmarum the endemic genus *Marojejya* (with two species) was included in the subtribe Sclerospermatinae, together with the West African *Sclerosperma*, but with more material it has become clear these genera belong in different subtribes. *Marojejya* turns out to be related to the monotypic and endemic *Masoala*; recent collections have shown that the genera need to be merged, which will result in the genus *Masoala* with four or five species.

\- *Lemurophoenix*, as yet unclear subtribe of uncertain affinities. A recently described monotypic genus endemic to the Masoala Peninsula; this certainly deserves subtribal status, though its relationships are still unclear.

Table 2. Synopsis of palm genera in Madagascar.

Subfamily Tribe, Subtribe	Genus (spp. on Madag.)	Genus distribution
Coryphoideae		
Phoeniceae	*Phoenix* (1)	widespread in old world
Borasseae		
Lataniinae	*Borassus* (2)	widespread in old world
Hyphaninae	*Hyphaene* (1)	Africa & India
	Satranala (1)	endemic
	Bismarckia (1)	endemic
Calamoideae		
Calameae		
Raphiinae	*Raphia* (1)	also in Africa. Non-native?
Ceroxyloideae		
Ceroxyleae	*Ravenea* (15)	near-endemic
Arecoideae		
Areceae		
Oraniinae	*Orania* (4)	also in Malesia
Dypsidinae	*Dypsis* (137)	near-endemic
(uncertain)	*Masoala* (4 or 5)	endemic
(uncertain)	*Lemurophoenix* (1)	endemic
Cocoeae		
Beccariophoe-nicinae	*Beccariophoenix* (1)	endemic
Elaeidinae	*Elaeis* (1)	also in Africa. Non-native?
Butiinae	*Voanioala* (1)	endemic

* *Subfamily Arecoideae: tribe Cocoeae*

- *Subtribe Beccariophoenicinae* (endemic)

The monospecific *Beccariophoenix* is a most singular genus, and probably an ancient remnant of the line of evolution which has given rise to the Cocoeae (Uhl & Dransfield 1987).

- *Subtribe Elaedinae*

Only *Elaeis* occurs (possibly wild) in the far North, with a single species, the oil-palm (*E. guineensis*).

- *Subtribe Butiinae*

Voanioala is a recently described, monotypic genus; the nearest relative within the subtribe is *Jubaeopsis* in South Africa; other genera in the subtribe are from South America, except for *Cocos* (probably an early introduction to Madagascar) which probably originates from the western Pacific, and *Paschalococos* from Easter Island. *Voanioala* has an extremely high chromosome number: 2n = 596+ (Johnson 1989); the only other high chromosome number in the family is found in *Jubaeopsis* (2n = 160 – 200; Robertson 1976).

Discussion

The dilemma which faced us time and again was when to merge genera, and when to describe a third genus "between" two adjacent genera. Obviously, this is a question of the number and quality of characters distinguishing the taxa, but it is never clear-cut. In Madagascar it is probably easier than elsewhere. In many cases distinction between genera has been very feeble, based on the differences between the two type species, which were clearly distinct; afterwards, authors felt bound by precedents to squeeze new taxa into old genera. Added to this was the fact that the fragmentary material left a lot of scope to the imagination and conjecture. Quite a lot of specimens collected before 1940 by Perrier and other French collectors were not mentioned in the Flore de Madagascar (Jumelle & Perrier 1945), presumably because the authors felt uncertain about their identification.

In several cases our decisions have been made for us: the distinctions between genera became ever vaguer with the increase in good collections, and often disappeared altogether (*Chrysalidocarpus* and *Neophloga*; *Orania* and *Halmoorea*): this might be called "automatic merging".

In other cases, intermediates have been found, and the distinctions which were generic before were whittled down to specific ones: this might be called the "chain-forming" process (*Ravenea* and *Louvelia*; *Masoala* and *Marojejya*).

Then, there were taxa which were so distinct that they could not be accommodated in other, existing, taxa, and these were described as new genera (*Voanioala*, *Lemurophoenix*).

The last and in many ways the most difficult case was that of the new species which had characters of *Bismarckia*, but was distinct from *B. nobilis*, and in several ways intermediate between that taxon and *Medemia*.

Bismarckia is endemic to the grasslands of western Madagascar, while *Medemia* is only known from wadis in southern Egypt and northern Sudan. The fact that *Medemia* and *Bismarckia* had been lumped for a while (Jumelle & Perrier de la Bathie 1945) probably contributed to our agonizing; for some reason, even "normal" taxonomists feel the weight of historical precedents ... But we have decided upon a new genus. The fact that both the new taxon, *Satranala*, and the single known species of *Medemia* are incompletely known makes this a preliminary decision, though it is supported by a so far unpublished cladistic analysis by Dransfield and Uhl. This is in a way a case of pragmatism versus rigorous phylogeny: if we include both *Bismarckia* and the new *Satranala* within *Medemia*, by the same reasoning we would have to sink *Medemia* s.l. into *Hyphaene*. Such large-scale lumping would hide the morphological variability within the subtribe, which we feel is generic rather than specific.

"Automatic merging": the case of Orania and Halmoorea

The latter is distinguished by its two large peduncular bracts, the free sepals, and the much higher number of stamens. Material collected in 1988 and in 1992 showed that *Orania longisquama* can have either one or two peduncular bracts; the number of stamens in a new species (again with either one or two peduncular bracts within a single population) is intermediate between the aforementioned taxa. Therefore, we conclude that *Halmoorea* cannot be upheld as a separate, and with sadness we see it go into synonymy.

"Chain forming": the case of Ravenea and Louvelia

This is a typical case of the first member of a new genus, in this case *Louvelia*, being very different from the existing genus (*Ravenea*), and easily distinguished by a suite of characters - petals connate or free; stigmatic remains on the fruit apical or sub-basal; fruit endocarp sclerified or soft; seed per fruit one or one-several. Then, with the passage of time, new taxa are discovered, gradually blurring the once clear distinctions: two *Louvelia* species described within twenty years of the first already blurred the distinctions, causing the petals character to disappear, though the material was so fragmentary (the genera are dioecious) that this did not become evident immediately. More collections showed that the stigmatic remains character as well as the seed per fruit character could be variable in both genera. During the last eight years five more new species have come to light, and provided evidence that the fruit endocarp character is specific rather than generic. An almost seamless series of species now links the former extremes, and they will be officially united (Beentje, in press).

"Chain forming": the case of Marojejya and Masoala

Marojejya was described based on *M. insignis*, very distinct from the equally monospecific *Masoala*. More recent taxa of *Marojejya*, including the spectacular *M. darianii* as well as two undescribed taxa, enlarged the generic description but did not bring the genera any closer. During the current project we found the fruits of *Masoala madagascariensis*, hitherto unknown; and discovered a new taxon, bridging the gap between the two genera, which was based on the size and branching of the inflorescence, the sexuality of the inflorescence, and some minor flower characters. The new taxon (*Masoala kona*) is neatly intermediate in inflorescence structure, and with regards to the minor flower characters sometimes displays the characters of one genus, and sometimes those of the other. The most unexpected was its being intermediate between a taxon with very short monoecious, unisexual inflorescences (*Marojejya*) and a taxon with, within a single species, long-exserted inflorescences with either triads, showing a central pistillate flower flanked by two staminate ones, or rachillae with proximally purely pistillate flowers and distally purely staminate ones (*Masoala*). *M. kona* has inflorescences proximally male and distally female; this same mixed inflorescence has been found (with the sexes reversed, and much shorter) in a new *Marojejya* taxon. Because of these finds we have decided to unite the genera, and *Masoala* has precedence.

A much more complicated case is present in the subtribe Dypsidinae, currently with seven accepted genera. As is evident from Table 1, there used to be another seven genera in the subtribe. This reflects the uncertainty about characters, which, again, came from very incomplete material; this resulted in the few original taxa receiving generic status, with taxonomists in later years contorting themselves to wrestle new taxa within narrowly circumscribed genera. A good example is the distinction between *Chrysalidocarpus* and *Neophloga*; the most solid character was size and number of pinnae ("Palms large with many pinnae: *Chrysalidocarpus* / Palms small with few pinnae: *Neophloga*") while the type species of each of these genera were of almost exactly the same size, with roughly the same number of pinnae. These two genera form an almost seamless chain of species.

Neodypsis and *Chrysalidocarpus* formed a different problem: they had always been very close. The only consistent difference between them was a single character (endosperm ruminate or homogeneous), which in sister-groups within their tribe could be variable in otherwise homogeneous genera. The ruminations can vary enormously, not only in size but in character; there are species in which the seed coat is sculptured, and the ruminations are caused by the undulations in the outer seed coat; in other species the seed coat is smooth on the outside but the intrusions appear to derive from the inner coat. In a taxon like *Neodypsis baronii* (which bears a strong resemblance in gross morphology to *Chrysalidocarpus lutescens*) the ruminations can be less than half a millimetre deep and difficult to spot; in other taxa the ruminations take up as much space as the endosperm itself. This character might not be a single one, since the ruminations might arise in two (or even three) different ways. In thirteen other genera (out of 58) in the tribe (e.g. from the New World: *Oenocarpus/Jessenia, Euterpe, Prestoea;* Old World: *Iguanura, Hydriastele, Veitchia, Gronophyllum*) this character may vary within a genus; and we would like to suggest this is the case in Dypsidinae, supported by a host of other characters which are constant within the tribe. What had divided the two genera was not so much a real difference, as the weight of history.

The monotypic genus *Antongilia* was based on a single species, distinct from *Neodypsis* by what surely were wrongly observed characters; specimens of the same taxon were also described in two different species in *Chrysalidocarpus*!

Both *Phloga* and *Vonitra* seem to form good genera within the subtribe. *Phloga* has didymous anthers and very distinct ovate and acuminate leaflets, grouped along the rachis; *Vonitra* has branched trunks, many inflorescences in various stages of development, and the curious "piassava", or leaf sheath fibres hanging all around the upper trunk, as well as didymous anthers, but strictly regular pinnae. But *Antongilia*, as well as several taxa found by us in recent years bridge the gap between *Vonitra* and the main group: we can present a series of species with strictly didymous anthers, slowly grading into the normal main group type of anther; the piassava character is variable, e.g. in the *V. nossibensis*, refound in 1992 after eighty years of obscurity; several of the new taxa have unbranched trunks (as has *V. nossibensis*) or trunks branched underground; and some of the new taxa have only a single inflorescence per plant. This line of argument also causes the *Phloga* character of didymous anthers to lose its generic importance, which leaves only the distinct leaflets; and these occur also in *Neophloga mananarensis*, refound in 1992, with characters perfectly matching the bulk of *Neophloga* in all other respects.

Dypsis with three stamens in the staminate flower seemed unambiguously defined - a species has either six stamens, or three, and if it has three it must be a *Dypsis*. So defined, the genus consists of mostly undergrowth palmlets, with inflorescences with (usually) many rachillae - though in the absence of staminate flowers it was very difficult to distinguish *Dypsis* from many *Neophloga* or even *Phloga*. However, there is more to the staminate flower of *Dypsis* than the number of stamens (Table 3). The position and shape of the stamens and the presence or absence of staminodes and pistillode are of significance and variations in these parameters provided the basis for the recognition of further genera: *Adelodypsis* and *Trichodypsis* (Beccari 1906). Later these genera were subsumed once again into *Dypsis*, now as subgenera (Beccari 1912). *Dypsis* in this enlarged sense has members with some of the smallest of all palm flowers, rarely exceeding 1 millimetre in diameter, a factor that makes the crucial examination of the flowers very difficult.

Table 3. Stamen shape and position in *Dypsis* s.l.

	anthers sagittate	anthers didymous	anthers pendulous
stamens 6	+	+	+
stamens 3 antesepalous	+	+	+
stamens 3 antepetalous		+	
staminodes antesepalous		+	
staminodes antepetalous		+	

Given that the base number of stamens is probably six, the three-stamen condition could have arisen by the loss of one of the whorls of stamens. In the subgenus *Dypsis* there are three antesepalous, short stamens, no staminodes and a minute pistillode; in *Adelodypsis* the stamens are also antesepalous but the anthers are sagittate and there is a columnar pistillode in the flower centre; in *Trichodypsis* three antepetalous fertile stamens alternate with three antesepalous staminodes, the anthers are didymous and there is a very small pistillode. The structure and positioning of the stamens gives quite a different appearance to the flower at anthesis. In members of *Trichodypsis* the fertile anthers are closely appressed to each other at the tip of the flower, while the fact that the staminodes are minute and antesepalous means that there is a minute aperture into the flower just above the staminodes next to the gape in the corolla. The natural history of the flower form is not understood and there have been no observations of the pollination of these minute flowers, but we are impressed by the differences between the flowers and suggest that there is probably some functional significance that relates to access by pollinators. That the three stamens can be either antepetalous or antesepalous suggests to us that reduction from six to three has taken place at least twice in evolutionary history. If this is true then in a cladistic sense the genus *Dypsis* (sensu stricto) is polyphyletic and unnatural.

If there are at least two evolutionary lines, as we suggest here, then there are some really remarkable parallels in overall form between taxa with antesepalous stamens and taxa with antepetalous stamens. The most extraordinary example is displayed by *Dypsis zahamenae* ined. and the closely related *D. angustifolia* ined. that have antesepalous stamens, while *D. pachyramea* ined. with almost identical, unusually condensed hairy rachillae has antepetalous stamens. Yet, despite all this variation in flower form among the dypsids with three stamens the homogeneity of the endosperm remains consistent (as it does with the other group of small understorey palms within the dypsids, the former *Neophloga*). However, we need to stress that fruits are unknown for many of these palms. Further new combinations of floral characters have been discovered which cloud the rather neat distinctions between the three subgenera (or genera) of *Dypsis*. An extraordinary unnamed taxon from the Marojejy Massif has three antesepalous stamens and three antepetalous staminodes that are adnate to the central columnar pistillode. One accession of *Dypsis lantzeana* from the Masoala Peninsula (a species that normally has three antesepalous stamens) has consistently a single stamen (still antesepalous), while another collection of the same species has some flowers with two stamens, and some with three. Collections of *Dypsis ambilaensis* ined. display some staminate flowers with four, five or even six stamens next to the normal flower with three.

There is a further development of the anthers, as yet unrecorded. In a species from the Marojejy massif the stamens are of a bizarre form. There is a terete filament topped by a long reflexed bifid connective, the two arms of which are divergent and bear, at their tips, elongate anthers. This form of stamen is somewhat reminiscent of stamen form in the tribe Geonomeae (Arecoideae), an entirely New World tribe, and we have dubbed the stamen form "geonomoid". Two other collections of dypsids with three stamens (*D. thouarsiana*) also show this geonomoid stamen, though in a slightly different form; they are from lowland rain forest much further south.

All this suggests to us that the three-staminate condition has arisen several times by reduction from six. Clearly our hypothesis demands a more rigorous testing using cladistic methods, but as yet our data are incomplete. It would seem that within the Dypsidinae there has been extraordinary radiation with little of the extinction that allows the recognition of disjunctions in variation and the demarcation of genera. It might be noted that the same condition, with either three or six stamens occurring within a single genus, is known to occur in *Geonoma*, *Orania* and *Areca*. In practical terms, the recognition of a single genus *Dypsis* has the great advantage of making identification of incomplete material much easier, at least to generic level!

Are there going to be more major changes in the generic delimitation of the palms of Madagascar? Probably not. The planned cladistic analysis and future molecular studies will, we hope, clear up some of the affinities at the subtribal and higher levels, and it is still possible that more exciting finds will be made in Madagascar. But we feel that the generic *status quo* will be stabilized after our monograph. And that stability will be a nice change.

Acknowledgements

A four-year project entitled "the palms of Madagascar" has been funded by McDonald's Restaurants (UK) and we would like to express our thanks to this organization and to Mr Paul Prescott, who has taken a keen personal interest. In Madagascar, we are grateful for much help from the Director and the staff of the Parc Botanique et Zoologique de Tsimbazaza, Antananarivo.

References

Beccari, O., 1906. Palmarum Madagascariensium Synopsis. Engl. Bot. Jahrb. 38, Beibl. 87: 1-41.

Beccari, O., 1912. Palme del Madagascar. Istituto Micrografica Italiano, Firenze.

Beentje, H.J., (in press). A monograph of *Ravenea* (Palmae: Ceroxyleae). Kew Bull.

Dransfield, J. & H.J. Beentje (in press). *Satranala* (Coryphoideae: Borasseae: Hyphaeninae), a new palm genus from Madagascar. Kew Bull.

Dransfield, J. & H.J. Beentje (in prep.). Palms of Madagascar. Royal Botanic Gardens, Kew.

Johnson, M.A.T., 1989. An unusually high chromosome number in *Voanioala gerrardii* (Palmae: Arecoideae: Cocoeae: Butiinae) from Madagascar. Kew Bull. 44, 2: 207–210.

Jumelle, H., 1927. *Ravenea* et *Louvelia*, palmiers de Madagascar. Ann. Inst. Bot.-Geol. Colon. Marseille 4, 5: 21-51.

Jumelle, H., 1928. Un nouveau genre Malgache de palmiers. Ann. Inst. Bot.-Geol. Colon. Marseille 4, 6, 2: 17-19.

Jumelle, H., 1933a. Nouveaux palmiers de Madagascar. Ann. Inst. Bot.-Geol. Colon. Marseille 5, 1, 1: 5-29.

Jumelle, H., 1933b. Nouvelles observations sur quelques palmiers Malgaches. Ann. Inst. Bot.-Geol. Colon. Marseille 5, 1, 3: 5-19.

Jumelle, H. & H. Perrier de la Bâthie, 1912. Un nouveau genre de palmiers de Madagascar. Comptes Rendus de l'Académie de Paris 155: 410-411.

Jumelle, H. & H. Perrier de la Bâthie, 1913. Palmiers de Madagascar. Ann. Inst. Bot.-Geol. Colon. Marseille 3, 1: 1-91.

Jumelle, H. & H. Perrier de la Bâthie, 1945. Palmiers. In: Humbert, H., (ed.), Flore de Madagascar et des Comores, 30e famille. Imprimerie officielle, Antananarivo.

Lanjouw, J. & F.A. Stafleu, 1954. Index Herbariorum, part II, Collectors A-H. IAPT, Utrecht.

Martius, C.F.P., 1838 & 1845. Historia naturalis palmarum. Weigel, Leipzig.

Robertson, B.L., 1976. Embryology of *Jubaeopsis caffra* Becc. I. J. South African Botany 42: 97–108.

Uhl, N.W. & J. Dransfield, 1987. Genera Palmarum. A classification of palms based on the work of H.E. Moore Jr. L.H. Bailey Hortorium and the International Palm Society, Allen Press, Kansas.

Vegter, I.H., 1983. Index Herbariorum, part II, Collectors N-R. Bohn, Scheltema & Holkema, Utrecht.

Les délimitations des genres chez les Caesalpinioideae africaines (Detarieae et Amherstieae) (1957-1994)

J. LÉONARD
Jardin botanique national, Meise, Belgique.

Résumé
En 1957, l'auteur a essayé de définir les genres africains des Detarieae et des Amherstieae, en utilisant des corrélations aussi larges que possible de caractères de valeur générique bien définis au préalable. Cette classification telle qu'elle a été proposée a été utilisée dans toutes les Flores africaines et a été confirmée par la découverte de diverses nouvelles espèces et par l'étude des plantules de nombreux genres. Une analyse détaillée de quelques modifications proposées ultérieurement par Aubréville montre que ces modifications résultent d'une autre conception de la définition des genres et qu'elles sont basées sur des critères subjectifs qui n'ont pas de valeur générique. Il semble dès lors préférable, en accord avec de nombreux auteurs, de ne pas accepter ces modifications.

Abstract
In 1957, the author tried to define the African genera of the Detarieae and Amherstieae, using as wide as possible correlations of characters of previously well-defined generic value. The proposed classification has been used in all the African Floras and has been confirmed by the discovery of various new species and by the study of the seedlings of many genera. A detailed analysis of the few modifications proposed later by Aubréville shows that these modifications are the consequence of another idea of the definition of the genera, and that they are based on subjective criteria of non-generic value. It seems preferable therefore, in concordance with many authors, not to accept these modifications.

Key words
Africa, Amherstieae, Caesalpinioideae, Delimitation, Detarieae, Genera.

Introduction

La délimitation aussi naturelle que possible des genres est une opération **fondamentale** en systématique. Cette délimitation naturelle des genres non seulement contribue à la *stabilité* des noms des genres, mais aussi permet de tirer des conclusions intéressantes en phytochorologie, en phytogéographie et en phytosociologie.

Pour délimiter les genres, il convient de se baser sur une **corrélation de caractères de valeur générique**. Un seul caractère distinctif ne suffit pas pour séparer deux genres. *Character non facit genus* disaient les Anciens. Il faut, au contraire, rechercher une **corrélation** aussi large que possible de données, les unes reposant sur la morphologie de la fleur et du fruit ainsi que sur des caractères végétatifs, les autres tirées de disciplines connexes telles l'anatomie du bois, la blastogénie (étude des plantules), la palynologie et d'autres encore.

Il convient ensuite, et ceci est fondamental, que les caractères distinctifs utilisés (ou la plupart d'entre eux) aient une **valeur générique**. Pour déterminer la valeur d'un caractère, il faut, *au préalable*, étudier les espèces de **tous** les genres de la famille ou

L.J.G. van der Maesen et al. (eds.), The Biodiversity of African Plants, 443-455.

d'une tribu donnée. Lors de cette étude, il convient de se rappeler que la valeur d'un caractère dépend aussi de la famille étudiée. Tel caractère qui a une valeur générique dans une famille peut ne pas avoir de valeur générique dans une autre famille.

Pour établir une classification générique aussi naturelle que possible, il convient de plus de *se baser sur des critères objectifs* (c'est-à-dire sur des critères résultant des faits eux-mêmes) *et non sur des critères subjectifs*, arbitraires (c'est-à-dire dépendant de l'opinion du botaniste qui les étudie).

Situation en 1957

Sur base de ces principes, nous avons tenté, en 1957 (Léonard 1957), de délimiter, aussi naturellement que possible, les genres africains des *Cynometreae* et des *Amherstieae* en faisant largement appel à des données de blastogénie (le nom *Cynometreae* Benth. est à remplacer par *Detarieae* DC., qui lui est antérieur).

Nous avons ainsi retenu 50 genres africains regroupant un peu plus de 300 espèces africaines, mis 8 genres en synonymie, rétabli 4 genres (*Anthonotha* P. Beauv., *Gilletiodendron* Vermoesen, *Guibourtia* Benn. et *Zenkerella* Taub.) et proposé 7 genres nouveaux (*Colophospermum*, *Gilbertiodendron*, *Lebruniodendron*, *Neochevalierodendron*, *Paramacrolobium*, *Pellegriniodendron* et *Sindoropsis*). De ce fait, plus de 100 espèces connues ont dû être transférées dans d'autres genres.

Quel sort a été réservé à ces conclusions au cours des 37 années qui ont suivi leur formulation?

Les approbations

Les 50 genres retenus, y compris donc les 4 genres rétablis et les 7 nouveaux proposés, ont été acceptés dans toutes les grandes Flores africaines, du moins ceux d'entre eux existant dans les régions couvertes par ces Flores.

Tel a été le cas dans les travaux suivants: Torre & Hillcoat (1956), Keay (1958), Aubréville (1959), White (1962), Hutchinson (1964), Voorhoeve (1965), Brenan (1967), Aubréville (1968a, 1970), Normand & Paquis (1976), Ross (1977), Cowan & Polhill (1981), Watson & Dallwitz (1983), Keay (1989), Lock (1989) et Brummitt (1992).

Deux exceptions sont cependant à signaler sur lesquelles nous reviendrons: d'une part le genre *Trachylobium* Hayne et, d'autre part, les délimitations génériques de certains genres telles que proposées par Aubréville.

Les confirmations

a. *Les espèces nouvelles*

Entre 1957 et 1993, 26 espèces africaines nouvelles ont été décrites appartenant à 15 genres des deux tribus envisagées ici (voir les Index Kewensis, Lock 1989 ainsi que Lebrun & Stork 1992). L'attribution générique de ces 26 espèces nouvelles n'a pas soulevé de problème particulier au niveau de la conception des genres. Ceci constitue une confirmation, indirecte il est vrai, de la délimitation naturelle de ces genres.

b. *Les plantules nouvelles*

Depuis la parution de notre Genera en 1957, les plantules de 55 espèces, appartenant à 25 genres, ont été décrites (Léonard 1994). La structure de ces plantules est *par-*

faitement semblable à celle précédemment décrite (Léonard 1957) des plantules des espèces des genres déjà connus. Cette parfaite identité de structure confirme non seulement les délimitations génériques proposées en 1957 et l'homogénéité des genres retenus, mais aussi la *valeur générique* de la structure des plantules.

Les modifications apportées

a. *Les nouvelles délimitations génériques proposées par Aubréville*

Lors de l'étude des Caesalpinioideae pour la seconde édition de la Flore forestière de la Côte d'Ivoire (1959), pour la Flore du Gabon (1968a) ainsi que pour la Flore du Cameroun (1970), Aubréville a rétabli 3 genres que nous avions mis en synonymie (*Paraberlinia* Pellegr., *Pterygopodium* Harms et *Zingania* A. Chev.), mis en synonymie le genre *Michelsonia* Hauman que nous avions conservé et créé 5 genres nouveaux: *Isomacrolobium*, *Leonardendron* et *Triplisomeris* pour des espèces précédemment placées dans le genre *Anthonotha*, *Leonardoxa* pour 3 espèces précédemment incluses dans le genre *Schotia* Jacq. ainsi que le genre *Toubaouate* pour une espèce rangée dans le genre *Didelotia* Baill.

Ces 5 genres nouveaux ne sont donc pas des genres basés sur des espèces nouvelles, mais bien des genres nouveaux résultant d'une autre classification d'espèces déjà existantes, classification qui est le reflet d'*une autre conception de la délimitation des genres*.

Pour délimiter les 9 genres précités, Aubréville ne s'est pas basé sur une corrélation de caractères, mais il a utilisé des caractères qui, nous l'avions montré en 1957, n'avaient pas une valeur générique au sein des deux tribus envisagées ici. Aubréville n'admettait guère les subdivisions au sein d'un genre (sous-genres, sections), mais bien souvent a simplement élevé les subdivisions existantes au rang de genres.

En d'autres mots, pour délimiter les 9 genres précités, Aubréville s'est basé sur des caractères subjectifs, arbitraires et non sur des caractères objectifs, c'est-à-dire qu'il s'est basé sur ses intuitions plus que sur les faits eux-mêmes. Cette façon de procéder amène à des conceptions génériques variables selon les auteurs et à l'établissement d'autant de systèmes qu'il y a de botanistes à avoir étudié ces genres.

Passons en revue les 9 genres créés, rétablis ou mis en synonymie par Aubréville.

1. *Isomacrolobium* Aubrév. & Pellegr., Bull. Soc. Bot. Fr. 104: 497 (1958).

Les espèces africaines jadis rangées dans le genre *Macrolobium* Schreb. (devenu uniquement américain) ont été distribuées entre 4 genres bien distincts, dont le genre *Anthonotha* (Léonard 1957: 201-215, tableau face p. 208).

Les espèces du genre *Anthonotha* ont été réparties en 5 sections sur la base du nombre et des dimensions des pétales (Léonard 1957: 217 et tableau face p. 208): section *Anthonotha* (1 grand pétale + 4 petits), section *Vouapina* (Harms) J. Léonard (5 pétales de même longueur, parfois 4 ou 6), section *Triplisomeria* (Baill.) J. Léonard (3 grands de même longueur + 2 petits), section *Tetranisomeria* J. Léonard (1 grand + 3 petits) et section *Variabilia* J. Léonard (2-4 de même longueur).

Nous avons montré (Léonard 1957: 17 et 171) qu'au sein de la tribu des *Amherstieae*, dont fait partie le genre *Anthonotha*, le nombre et les dimensions des pétales représentaient des caractères variables qui n'étaient que rarement en corrélation avec des caractères de valeur générique. **A eux seuls** *ces caractères n'ont donc pas une* *valeur générique*, mais peuvent être utilisés, par contre, pour grouper les espèces d'un même genre en *sections*. Tel est bien le cas du genre *Anthonotha*.

En 1958, Aubréville & Pellegrin (loc.cit.: 496-498) ont estimé que le genre *Anthonotha* ainsi défini demeurait hétérogène et qu'il convenait d'en extraire les espèces à pétales tous de même longueur, c'est-à-dire les espèces des sections *Vouapina* (9 espèces) et *Variabilia* (1 espèce). Ils ont proposé de classer les espèces de ces 2 sections dans le nouveau genre *Isomacrolobium*.

En procédant de la sorte, Aubréville & Pellegrin ont accordé, selon leur seule intuition, une valeur générique aux *seuls* caractères du nombre et des dimensions des pétales, sans avoir mis en évidence d'autres caractères corrélatifs qui eux auraient une valeur générique.

En 1959, dans la seconde édition de la Flore forestière de la Côte d'Ivoire, Aubréville (1959: 280-281) précise que le genre *Isomacrolobium* a été proposé pour les espèces d'*Anthonotha* à 5 pétales subégaux et il publie à nouveau la liste des transferts d'espèces d'*Anthonotha* dans le genre *Isomacrolobium*. Dans cette liste figure, pour la première fois, le transfert en *Isomacrolobium* d'*Anthonotha lebrunii* (J. Léonard) J. Léonard, espèce-type de la section *Tetranisomeria* précitée, pourtant caractérisée par la présence d'un grand pétale et de 3 petits pétales.

Les données de blastogénie ne confirment pas le bien-fondé de l'établissement du genre *Isomacrolobium*. Les plantules d'*Isomacrolobium vignei* (Hoyle) Aubrév. & Pellegr. sont, en effet, *parfaitement semblables* aux plantules *si caractéristiques* d'*Anthonotha crassifolia* (Baill.) J. Léonard, d'*A. fragrans* (Baker f.) Exell & Hillc. et d'*A. macrophylla* P. Beauv. (voir les figures des plantules de ces 4 espèces dans de la Mensbruge 1966: 147, 159, 165 et p. 175-176 et dans de Koning 1983: 190, 191, 192, 194 et fig.: 193).

En conclusion, l'établissement du genre *Isomacrolobium*, basé sur un *seul caractère*, d'ailleurs variable et de surcroît *de valeur non générique*, n'est pas justifié. Il convient dès lors de le considérer comme synonyme du genre *Anthonotha* et de replacer dans ce dernier les 11 espèces des sections *Vouapina*, *Variabilia* et *Tetranisomeria*. Cette façon de procéder avait d'ailleurs déjà été adoptée par Cowan & Polhill (1981: 141), par Keay (1989: 224), par Lock (1989) et par Brummitt (1992). *Isomacrolobium hallei*, décrit par Aubréville en 1968, doit, de ce fait, être transféré dans le genre *Anthonotha*.

Anthonotha P. Beauv., Fl. Ow. Benin 1: 70, tab. 42 (1806).

Isomacrolobium Aubrév., Fl. For. Côte Iv., éd. 1, 1: 224 (1936) nomen.

Isomacrolobium Aubrév. & Pellegr., Bull. Soc. Bot. Fr. 104: 497 (1958); Aubrév., Fl. For. Côte Iv., éd. 2, 1: 280, 281, 293 (1959); Fl. Gabon 15: 188 (1968); Fl. Camer. 9: 181 (1970). Espèce-type: *I. leptorrhachis* (Harms) Aubrév. & Pellegr.

Anthonotha hallei (Aubrév.) J. Léonard, comb. nov.— Type: Gabon, N. Hallé 2195 (P holo-).

Isomacrolobium hallei Aubrév., Fl. Gabon 15: 362 (1968).

2. *Leonardendron* Aubrév., Adansonia, sér. 2, 8: 167, pl. 5 (1968).

Nous venons de voir ci-dessus que les espèces de la section *Vouapina* du genre *Anthonotha* ont été placées en 1958 par Aubréville & Pellegrin dans le nouveau genre *Isomacrolobium*. Tel a été notamment le cas d'*Anthonotha gabunensis* J. Léonard, à fleurs avec 5 pétales de même longueur, devenu donc *Isomacrolobium gabunense* (J. Léonard) Aubrév. & Pellegr.

En 1968 cependant, Aubréville (1968b: 163, 167) estime que cette espèce n'est pas à sa place dans le genre *Isomacrolobium*. En effet, dit-il, l'androcée de cette espèce ne se compose que de 5 étamines fertiles et d'aucun staminode, alors que chez les autres espèces il est composé d'un ensemble d'étamines et de staminodes égalant 9. Cette structure de l'androcée du type 5 impose de classer *I. gabunense* dans un genre nouveau, *Leonardendron*. Ce raisonnement n'est guère défendable pour les raisons suivantes:

* ce caractère différentiel "5 étamines" est un caractère qui n'est en corrélation avec *aucun* autre caractère qui aurait une valeur générique
* ce caractère à lui seul n'a pas de valeur générique et manque d'ailleurs de constance: dans la description latine du genre *Leonardendron*, Aubréville lui-même n'indique-t-il pas "stamina 5, interdum solum 3-4"?
* la dissection des fleurs d'*Anthonotha gabunensis* nous a montré la présence fréquente d'un staminode en plus des 5 étamines
* chez certaines espèces, comme *Isomacrolobium vignei*, les 3 étamines fertiles sont accompagnées de 3-6 staminodes selon les fleurs, ce qui montre que l'androcée n'est pas toujours du type 9.

On voit ainsi que le caractère "androcée du type 5" opposé à "androcée du type 9" manque de rigueur. Nous marquons dès lors notre accord (horresco referens!) de placer le genre *Leonardendron* en synonymie du genre *Anthonotha*, comme l'ont d'ailleurs déjà fait Cowan & Polhill (1981: 141), Lock (1989) et Brummitt (1992).

Anthonotha P. Beauv., Fl. Ow. Benin 1: 70, tab. 42 (1806).

Leonardendron Aubrév., Adansonia, sér. 2, 8: 163, 167, pl. 5 (1968); Fl. Gabon 15: 244 (1968); Fl. Cameroun 9: 224 (1970).— Espèce-type: *L. gabunense* (J. Léonard) Aubrév.

Anthonotha gabunensis J. Léonard, Acad. Roy. Belgique, Cl. Sc., Mém. in 8°, 30, 2: 224 (1957).— Type: Gabon, Le Testu 7808 (BR holo-).

Isomacrolobium gabunense (J. Léonard) Aubrév. & Pellegr., Bull. Soc. Bot. Fr. 104: 498 (1958).

Leonardendron gabunense (J. Léonard) Aubrév., Adansonia, sér. 2, 8: 163, 167, pl. 5 (1968); Fl. Gabon 15: 244, pl. LIX (1968); Fl. Cameroun 9: 224 (1970).

Il serait particulièrement intéressant de chercher à obtenir des plantules de cette espèce pour voir si leur structure est bien identique à celle, si particulière, des plantules des autres espèces du genre *Anthonotha*.

3. *Leonardoxa* Aubrév., Adansonia, sér. 2, 8: 178 (1968).

Le cas du genre *Leonardoxa* a fait l'objet d'une note détaillée récente (Léonard 1993).

4. *Michelsonia* Hauman, Bull. Séances Inst. Roy. Col. Belge 23: 478 (1952).

Le genre *Michelsonia* a été établi par Hauman pour 2 espèces, *M. microphylla* (Troupin) Hauman, espèce-type, et *M. polyphylla* (Harms) Hauman, la première ayant été décrite dans le genre *Julbernardia* Pellegr., la seconde en *Berlinia* Sol. ex Hook.f.

En 1957, nous avons admis le genre *Michelsonia* (Léonard 1957: 187). Dans un travail ultérieur (Léonard 1965: 98, 104, 106, 107), nous avons indiqué, d'une part, que *M. polyphylla* était mieux placé en *Tetraberlinia* (Harms) Hauman, ce que confirmait l'anatomie de son bois (Normand & Paquis 1976: 100), et, d'autre part, que les affinités du genre *Michelsonia*, réduit dès lors à son espèce-type, le situaient auprès du genre *Tetraberlinia*. Ces deux genres "paraissent très voisins et leurs affinités demanderont à être davantage précisées" écrivions-nous (1965: 104). Voorhoeve (1965: 215) s'est même demandé si ces deux genres n'étaient pas synonymes.

En 1968, dans la Flore du Gabon (15: 295) et en 1970 dans la Flore du Cameroun (9: 272), Aubréville, sans la moindre justification, place le genre *Michelsonia* en synonymie de *Tetraberlinia* et établit la combinaison *Tetraberlinia microphylla*, sans faire remarquer que la corolle de cette dernière espèce se compose de 5 pétales subégaux, alors que dans sa clef des genres (p. 10) et dans la description du genre (p. 295 et 272), Aubréville indique que la corolle des *Tetraberlinia* est composée d'un grand pétale et de 4 linéaires courts.

Il nous semble préférable, pour le moment et en l'absence d'argument décisif, de conserver le genre *Michelsonia* qui se distingue du genre *Tetraberlinia* notamment par son réceptacle floral subplan avec le périanthe et l'ovaire insérés pratiquement au même niveau et non concave-obconique avec le périanthe et l'ovaire insérés à des niveaux nettement différents (Léonard 1965: 104). Le genre *Michelsonia* est maintenu également par Normand & Paquis (1976: 100), par Cowan & Polhill (1981: 138), par Lock (1989) et par Brummitt (1992).

L'obtention des plantules de *M. microphylla* est vivement souhaitée afin de pouvoir les comparer à celles des *Tetraberlinia*.

Michelsonia microphylla (Troupin) Hauman, Bull. Séances Inst. Roy. Col. Belge 23: 478 (1952); Fl. Congo Belge Ruanda-Urundi 3: 406, fig. 33 (1952); J. Léonard, Bull. Jard. Bot. Etat Brux. 35: 107 (1965).— Type: Zaïre, Michelson 882 (BR holo-).

Julbernardia microphylla Troupin, Bull. Jard. Bot. Etat Brux. 20: 320, fig. 30 (1950).

Tetraberlinia microphylla (Troupin) Aubrév., Fl. Gabon 15: 295 (1968); Fl. Cameroun 9: 272 (1970).

5. *Paraberlinia* Pellegr., Bull. Soc. Bot. Fr. 90: 79 (1943).

Le genre *Paraberlinia* ne comprend qu'une espèce *P. bifoliolata* Pellegr. En 1950, Troupin (1950: 311) a mis ce genre en synonymie avec le genre *Julbernardia*, dont il ne se distinguait que par sa corolle composée de 5 pétales subégaux et très réduits, l'espèce devant dans ce cas porter le nom de *J. pellegriniana* Troupin.

Nous avons partagé l'opinion de Troupin en précisant que ce *seul* caractère différentiel n'avait pas de valeur générique au sein de la tribu des *Amherstieae* et que de plus il était variable au sein même du genre *Julbernardia* (Léonard 1957: 189-190). Le genre *Paraberlinia* n'est pas admis non plus par Hutchinson (1964: 273).

Dans la Flore du Gabon en 1968 et dans la Flore du Cameroun en 1970, Aubréville, sans apporter d'arguments nouveaux, rétablit le genre *Paraberlinia*, estimant que ce seul caractère de dimension des pétales peut suffire à distinguer le genre *Paraberlinia* du genre *Julbernardia*. L'opinion d'Aubréville n'a pas été suivie, ni par Breyne & Evrard (1975: 309-310), ni par Cowan & Polhill (1981: 140), ni par

Keay (1989: 200), ni par Lock (1989: 15), ni par Brummitt (1992), ni par Lebrun & Stork (1992: 52).

La parfaite similitude des plantules de *Paraberlinia bifoliolata* (Breyne & Evrard 1975: 309-310, fig. 1) avec celles de 7 espèces différentes de *Julbernardia* (Léonard 1957: 195-198, pl. 18 et 1995) confirme le bien-fondé de la mise en synonymie du genre *Paraberlinia*.

Julbernardia pellegriniana Troupin, Bull. Jard. Bot. Etat Brux. 20: 311 (1950) nom. nov.— Type: Gabon, Le Testu 7339 (P lecto-).

Paraberlinia bifoliolata Pellegr., Bull. Soc. Bot. Fr. 90: 79 (1943); Aubrév., Fl. Gabon 15: 321 (1968); Fl. Cameroun 9: 292 (1970).— Type: Gabon, Le Testu 7339 (P lecto-).

6. *Pterygopodium* Harms, Engl. Bot. Jahrb. 49: 439 (1913).

En 1950, nous avons montré (Léonard 1950b) que les *seules* différences existant entre les genres *Pterygopodium* et *Oxystigma* Harms (consistance et longueur de l'aile du fruit et nature de la surface de la graine) ne possédaient pas une valeur générique. Le genre monospécifique *Pterygopodium* ne méritait dès lors de n'être considéré tout au plus que comme une section du genre *Oxystigma* (Léonard 1950b). L'anatomie du bois et la parfaite similitude de structure des plantules de ces 2 genres ont confirmé le bien-fondé de leur mise en synonymie (Léonard 1957: 128-130, pl. 12).

Par contre, dans la Flore du Gabon, en 1968, et dans la Flore du Cameroun, en 1970, Aubréville rétablit le genre *Pterygopodium* sans apporter d'arguments nouveaux et tout en reconnaissant qu'il ne se distingue du genre *Oxystigma* que par ses fruits et qu'en l'absence de ces derniers ces deux genres ne peuvent être distingués. Selon Aubréville, les fruits des *Pterygopodium* sont samaroïdes et dispersés par le vent, alors que ceux des *Oxystigma* sont épais et ligneux et dispersés au gré des courants des rivières. C'est ne pas tenir compte de l'existence d'un *Oxystigma gilbertii* J. Léonard, espèce de terre ferme à fruits épais et ligneux, mais munis d'une aile de 0-2 (-3,5) cm de long.

La proposition de rétablissement du genre *Pterygopodium* n'a été suivie ni par Normand & Paquis (1976: 96), ni par Cowan & Polhill (1981: 131), ni par Keay (1989: 200), ni par Lock (1989: 58), ni par Brummitt (1992).

Oxystigma oxyphyllum (Harms) J. Léonard, Bull. Séances Inst. Roy. Col. Belge 21: 753, fig. 1B et 2G-H (1950).— Type: Cameroun, Mildbraed 4538 (P iso-).

Pterygopodium oxyphyllum Harms, Engl. Bot. Jahrb. 49: 439 (1913); Aubrév., Fl. Gabon 15: 144, pl. 32 (1968); Fl. Cameroun 9: 143, pl. 29 (1970).

7. *Toubaouate* Aubrév. & Pellegr., Bull. Soc. Bot. Fr. 104: 491 (1958).

Ce genre ne comprend qu'une seule espèce, *T. brevipaniculata* (J. Léonard) Aubrév. & Pellegr., précédemment décrite dans le genre *Didelotia*.

Dans sa révision du genre *Didelotia*, Oldeman (1964: 212-214, 220-225) a clairement expliqué pourquoi il convenait de placer à nouveau cette espèce dans le genre *Didelotia*. Cette opinion, exprimée également par Hutchinson (1964: 272), a été suivie par Kunkel (1965: 92), Normand & Paquis (1976: 108), Letouzey (1977: 127), Cowan

& Polhill (1981: 142), Watson & Dallwitz (1983: 24), Lock (1989: 10), Brummitt (1992) ainsi que par Lebrun & Stork (1992: 48).

La parfaite similitude de la structure des plantules de cette espèce (Oldeman 1964: 225; Kunkel 1965, fig.: 93c; de la Mensbruge 1966: 174, fig.: 155) et celle des plantules de 3 autres espèces de *Didelotia* (Léonard 1957: 269 et 1995) confirme le bien-fondé de la mise en synonymie du genre *Toubaouate*. Cette dernière est confirmée également par la structure homogène du bois des espèces de *Didelotia*, y compris celui de *D. brevipaniculata* (Normand 1958; Gottwald et al. 1968; Normand & Paquis 1976: 108-109).

Didelotia brevipaniculata J. Léonard, Acad. Roy. Belgique, Cl. Sc., Mém. in 8°, 30, 2: 267 (1957).— Type: Cameroun, Medou 1826 (BR holo-).

Toubaouate brevipaniculata (J. Léonard) Aubrév. & Pellegr., Bull. Soc. Bot. Fr. 104: 492 (1958); Aubrév., Fl. For. Côte Iv., éd. 2, 1: 296, pl. 98 (1959); Fl. Gabon 15: 258, pl. 62 (1968); Fl. Cameroun 9: 233, pl. 53 (1970).

8. *Triplisomeris* (Baill.) Aubrév. & Pellegr., Bull. Soc. Bot. Fr. 104: 497 (1958).

Nous avons vu (Léonard 1957: 217) que les espèces du genre *Anthonotha* ont été réparties en 5 sections sur la base du nombre et des dimensions des pétales et qu'Aubréville & Pellegrin (1958b) ont groupé 3 de ces sections dans le nouveau genre *Isomacrolobium*. Ils ont, d'autre part, élevé la section *Triplisomeria* au niveau de genre, sous le nom de *Triplisomeris*.

Le genre *Triplisomeris* n'est pas basé, lui non plus, sur une *corrélation* de caractères de *valeur générique*, mais essentiellement sur le nombre et les dimensions des pétales, caractères que nous avons montré ne pas avoir à eux seuls une valeur générique au sein de la tribu des *Amherstieae* (Léonard 1957: 17 et 171). Nous proposons dès lors de ne pas retenir le genre *Triplisomeris*, comme l'ont d'ailleurs déjà fait Hutchinson (1964: 276), Cowan & Polhill (1981: 141), Lock (1989: 1 et 3) et Brummitt (1992), mais bien de le considérer comme une section du genre *Anthonotha*.

Il serait intéressant de rechercher les plantules des 3 espèces de cette section [*A. ernae* (Dinkl.) J. Léonard, *A. explicans* (Baill.) J. Léonard et *A. triplisomeris* (Pellegr.) J. Léonard] pour voir si elles sont semblables ou non à celles des espèces des autres sections du genre *Anthonotha*.

Anthonotha P. Beauv. 1806 Sectio Triplisomeria (Baill.) J. Léonard, Acad. Roy. Belgique, Cl. Sc., Mém. in 8°, 30, 2: 225 (1957).— Espèce-type: *A. explicans* (Baill.) J. Léonard.

Vouapa Aubl. Sectio *Triplisomeria* Baill., Adansonia 6: 181 (1865).

Triplisomeris (Baill.) Aubrév. & Pellegr., Bull. Soc. Bot. Fr. 104: 497 (1958); Aubrév., Fl. For. Côte Iv., éd. 2, 1: 286 (1959); Fl. Gabon 15: 192 (1968); Fl. Cameroun 9: 184 (1970).— Espèce-type: *T. explicans* (Baill.) Aubrév. & Pellegr.

9. *Zingania* A. Chev., Rev. Bot. Appl. 26: 601, pl. 22 (1946).

Nous avons mis (Léonard 1952b: 206, 208 et 1957: 265) le genre monospécifique *Zingania* en synonymie du genre *Didelotia* dont, selon nous, il ne se distinguait par aucun caractère de valeur générique.

En 1958 cependant, Aubréville & Pellegrin ont remis le genre *Zingania* à l'honneur en trouvant de petites différences avec le genre *Didelotia*. Cette opinion n'a pas été partagée par Hutchinson (1964: 272).

Dans sa révision du genre *Didelotia*, Oldeman (1964: 210-212, 233) a expliqué en détail pourquoi *Zingania* devait bien être considéré comme synonyme de *Didelotia*. En 1968, dans la Flore du Gabon, Aubréville, sans explication, a, lui aussi, replacé *Zingania* en synonymie de *Didelotia*.

Didelotia minutiflora (A. Chev.) J. Léonard, Bull. Jard. Bot. Etat Brux. 22: 206 (1952); Aubrév., Fl. Gabon 15: 254, pl. 62 (1968).— Type: Gabon, Walker 18 (P holo-).

Zingania minutiflora A. Chev., Rev. Bot. Appl. 26: 601-603, pl. 22 (1946); Aubrév. & Pellegr., Bull. Soc. Bot. Fr. 104: 492 (1958).

Conclusion

Il ressort de l'examen précité des 9 modifications génériques proposées par Aubréville qu'aucune de celles-ci ne peut être acceptée (elles ne l'ont d'ailleurs pas été par de nombreux auteurs) parce qu'elles reposent la plupart sur des caractères subjectifs, arbitraires *et non sur une corrélation de caractères de valeur générique*.

b. *Les genres Trachylobium et Hymenaea*

Le genre *Trachylobium* est un genre tropical africain ne comprenant qu'une seule espèce, *T. verrucosum* (Gaertn.) Oliv., dénommée antérieurement *Hymenaea verrucosa* Gaertn. et répandue sur la côte orientale de l'Afrique, du Kenya au Mozambique, ainsi qu'à Madagascar, à l'île Maurice et aux Seychelles. Les plantules de cette espèce sont *particulièrement caractéristiques* (Léonard 1957: 116, pl. 11).

Il y a une vingtaine d'années déjà, Langenheim et ses collaborateurs ont montré, d'une manière lumineuse et remarquablement documentée, que *T. verrucosum* ne se distinguait pas, au niveau générique, des espèces du genre *Hymenaea* L., genre pluri-spécifique largement répandu en Amérique tropicale. Il n'y a, ont-ils démontré, aucune différence significative de valeur générique entre les inflorescences, les fleurs et les fruits des espèces de ces deux genres. La composition chimique de la résine sécrétée par leurs troncs est identique ainsi que leurs pollens et leurs nombres chromosomiques (Langenheim et al. 1972, 1973, 1974 et Lee & Langenheim 1975). Langenheim & Lee (1974) ont dès lors proposé de replacer *T. verrucosum* dans le genre *Hymenaea*.

Cette mise en synonymie a été adoptée par Cowan & Polhill (1981: 130), Lock (1989: 54) et Brummitt (1992: 432), mais pas par Watson & Dallwitz (1983: 34, 58).

Il est tout à fait remarquable de constater que la structure des plantules de 11 espèces américaines du genre *Hymenaea* (Lee & Langenheim 1975: 15, 17-20, fig. 4-5) est *parfaitement semblable* à celle, *si particulière*, de *Trachylobium verrucosum* (comparer la planche 11 de Léonard 1957 et les fig. 4-5 de Lee & Langenheim 1975). Cette similitude de structure confirme remarquablement l'homogénéité du genre *Hymenaea* et le bien-fondé de la mise en synonymie du genre *Trachylobium*.

c. *Les genres Schotia, Leonardoxa et Normandiodendron*

Nous avons indiqué (Léonard 1957: 102) que les 3 espèces afro-tropicales du genre *Schotia* méritaient de constituer un genre distinct. Nous attendions cependant pour ce faire l'obtention des plantules de *S. africana* (Baill.) Bequaert et des renseignements complémentaires sur l'anatomie du bois de ces 3 espèces. En 1968, Aubréville a créé d'office le nouveau genre *Leonardoxa* pour ces 3 espèces (Aubréville 1968c).

Grâce à l'obtention des plantules de *S. africana* et à l'étude anatomique du bois de ces 3 espèces (Normand 1993), il a été montré (Léonard 1993) que le genre *Leonardoxa* était un bon genre, mais qu'il convenait de le limiter à sa seule espèce-type *L. africana* (Baill.) Aubrév., alors que les 2 autres espèces devaient être classées dans le genre nouveau *Normandiodendron* J. Léonard.

Les plantules de *Schotia*, de *Leonardoxa* et de *Normandiodendron* sont *nettement* distinctes les unes des autres (Léonard 1993: 439, fig. 2) confirmant ainsi le bien-fondé de la séparation de ces 3 genres.

Les problèmes non encore résolus

Indépendamment du cas du genre *Copaifera* L. déjà évoqué (Léonard 1957: 88-89), reste à résoudre le problème des *Cynometra* L. et des *Guibourtia*.

a. *Le genre Cynometra*

Le genre *Cynometra* comprend environ 70 espèces répandues dans toutes les régions tropicales du globe, dont une trentaine en Afrique. L'espèce-type, *C. cauliflora* L., est originaire de l'Inde. Ce genre était extrêmement hétérogène. La révision de ses espèces africaines (Léonard 1951), ayant entraîné la création du genre *Lebruniodendron* J. Léonard, le rétablissement des genres *Gilletiodendron* Vermoesen et *Zenkerella* Taub. ainsi que le transfert de diverses de ses espèces dans les genres *Hymenostegia* (Benth.) Harms, *Plagiosiphon* Harms et *Scorodophloeus* Harms, a rendu le genre *Cynometra* un peu moins hétérogène en Afrique.

Les espèces africaines restantes ont été classées en 3 groupes (Léonard 1951, tableau p. 381), mais font toujours partie d'un ensemble pantropical génériquement hétérogène. Cette hétérogénéité a été constatée également par Aubréville (1968a: 66 et 1970: 77) ainsi que par Normand & Paquis (1976: 88-90) au point de vue de l'anatomie du bois.

L'étude comparée de *toutes* les espèces tropicales de ce genre devrait être entreprise avec l'aide de données blastogéniques, palynologiques et xylologiques. Indépendamment des révisions effectuées par Dwyer (1959) et par Knaap-Van Meeuwen (1970), on trouvera déjà quelques renseignements sur la structure des plantules de diverses espèces africaines dans Léonard (1952a: 317, 319 et 321), Taylor (1960: pl. 30), Voorhoeve (1965: fig. 25-26), de la Mensbruge (1966: 147, 153 et 166) et de Koning (1983: 201).

Cette révision comparée, que nous avions projeté d'entreprendre, n'a pu avoir lieu, en partie par suite de la difficulté de nous procurer des graines fraîches authentifiées d'un nombre suffisant d'espèces américaines et asiatiques afin de pouvoir en étudier les plantules. Cette difficulté a été rencontrée également par Knaap-Van Meeuwen (1970: 2, 4 et 7) qui n'est pas parvenue, elle non plus, à se procurer des graines des *Cynometra* indo-malais (20 espèces), qui, semble-t-il, ne fructifieraient que rarement. La seule plantule obtenue appartient à *C. cauliflora* (l'espèce-type), mais l'auteur, fort malheureusement, dit seulement qu'elle est épigée. Une description et une figure des plantules de l'espèce asiatique *C. ramiflora* L. ont été publiées par de Vogel (1980:

310, fig. 108).

L'hétérogénéité du genre *Cynometra* se reflète également dans la diversité de la structure des plantules déjà connues de ses espèces.

Cette révision comparée de toutes les espèces de *Cynometra* est cependant urgente à entreprendre parce que des espèces africaines aussi importantes que *C. alexandri* C.H. Wright et *C. hankei* Harms, grands arbres communs dans certains types de forêts au Zaïre, n'appartiennent plus que vraisemblablement pas au genre *Cynometra*.

b. *Le genre Guibourtia*

Le genre *Guibourtia* a été rétabli pour des espèces précédemment placées dans le genre *Copaifera* dont il est bien distinct (Léonard 1949: 391-406). Ses espèces ont été classées en 3 sous-genres (Léonard 1950a: 278-281 et 1957: 138-139): les sous-genres africains *Guibourtia* et *Gorskia* (Bolle) J. Léonard ainsi que le sous-genre américain et africain *Pseudocopaiva* (Britton & Wilson) J. Léonard. Dans ce dernier figurent les 4 espèces africaines suivantes: *G. coleosperma* (Benth.) J. Léonard, *G. leonensis* J. Léonard, *G. pellegriniana* J. Léonard et *G. tessmannii* (Harms) J. Léonard.

Ces 4 espèces africaines se distinguent des espèces des deux autres sous-genres par une corrélation de caractères importants basés sur les fruits, les graines et les plantules. L'absence de caractères distinctifs dans les fleurs et dans les organes végétatifs nous avait fait hésiter à placer ces 4 espèces dans un autre genre que *Guibourtia* (Léonard 1950a: 278-281). Par après, nous avons proposé de rétablir le genre *Pseudocopaiva* Britton & Wilson, dont l'espèce-type est originaire de Cuba, et nous nous sommes demandé (Léonard 1957: 153-156) si ces 4 espèces africaines ne devaient pas être exclues du genre *Guibourtia* et même si elles appartenaient bien au genre *Pseudocopaiva*.

Les 4 espèces africaines précitées devraient être extraites du genre *Guibourtia*, mais leur attribution générique future n'est pas encore fixée. La connaissance complète des fruits, des graines et surtout des plantules des espèces américaines de *Pseudocopaiva* devrait permettre de résoudre cette question.

Conclusions

En conclusion de tout ce qui précède, il apparaît clairement que pour délimiter les genres de façon aussi naturelle que possible afin d'établir un système cohérent, il convient de se baser sur une *corrélation de caractères dont la valeur générique aura été préalablement définie au sein de la famille ou de la tribu étudiée*.

Toute délimitation générique reposant sur un ou même sur plusieurs caractères *subjectifs*, dont la valeur n'aura pas été *préalablement* définie au sein de la famille ou de la tribu étudiée, risque d'aboutir à l'établissement de genres mal définis faisant partie d'un système arbitraire que son auteur sera le seul à défendre.

Bibliographie

Aubréville, A., 1959. La flore forestière de la Côte d'Ivoire, éd. 2. Publ. Centre Techn. For. Trop. 15, 1: 241-331.

Aubréville, A., 1968a. Flore du Gabon 15: 1-362.

Aubréville, A., 1968b. Les Césalpinioidées de la flore camerouno-congolaise. Considérations taxinomiques, chorologiques, écologiques, historiques et évolutives. Adansonia, sér. 2, 8: 147-175.

Aubréville, A., 1968c. *Leonardoxa* Aubréville, genre nouveau de Césalpinioidées guinéo-congolaises. Adansonia, sér. 2, 8: 177-179.

Aubréville, A., 1970. Flore du Cameroun 9: 1-339.

Aubréville, A. & F. Pellegrin, 1958a. *Didelotia* Baill. et genres affinés, *Zingania* A. Chev. et *Toubaouate* Aubr. & Pellegr., genre nouveau. Bull. Soc. Bot. Fr. 104: 490-492.

Aubréville, A. & F. Pellegrin, 1958b. De quelques Césalpiniées africaines. Bull. Soc. Bot. Fr. 104: 495-498.

Brenan, J., 1967. Flora of Tropical East Africa, *Leguminosae*, Subfamily *Caesalpinioideae*: 1-230.

Breyne, H. & C. Evrard, 1975. *Julbernardia pellegriniana* Troupin, Césalpiniacée nouvelle pour la flore du Zaïre. Bull. Jard. Bot. Nat. Belg. 45: 307-312.

Brummitt, R., 1992. Vascular plant families and genera. Royal Bot. Gard. Kew.

Cowan, R. & R. Polhill, 1981. *Detarieae*. In: Polhill, R. & P. Raven (eds), Advances in Legume Systematics 1: 117-134; *Amherstieae*, ibid.: 135-142, Royal Bot. Gard. Kew.

De Koning, J., 1983. La forêt du Banco. Med. Landb. Hogesch. Wageningen 83, 1: 1-156 et 1-921.

De la Mensbruge, G., 1966. La germination et les plantules des essences arborées de la forêt dense humide de la Côte d'Ivoire. Publ. Centre Techn. For. Trop. 26: 1-389.

De Vogel, E., 1980. Seedlings of dicotyledons. Structure, development, types. Description of 150 woody Malesian taxa. Centre Agr. Publ. Doc. Wageningen.

Dwyer, J., 1959. The new world species of *Cynometra*. Ann. Missouri Bot. Gard. 45: 313-345.

Gottwald, H., W. Knigge, D. Noack & M. Sachtler, 1968. Anatomische und physikalisch-technologische Untersuchungen an vier liberianischen Holzarten. Mitt. Bundesforsch. Forst. Holzw. Reinbeck 67: 1-48.

Hutchinson, J., 1964. The genera of flowering plants, Oxford Clarendon Press 1: 221-276.

Keay, R., 1958. Flora of West Tropical Africa, ed. 2, 1: 439-484.

Keay, R., 1989. Trees of Nigeria, Oxford Science Publ.: 191-231.

Knaap-Van Meeuwen, M., 1970. A revision of four genera of the tribe *Leguminosae-Caesalpinioideae-Cynometreae* in Indomalesia and the Pacific. Blumea 18: 1-52.

Kunkel, G., 1965. The trees of Liberia, German Forestry Mission to Liberia, Report 3.

Langenheim, J., 1973. Leguminous resin-producing trees in Africa and South America. In: Meggers, B., E. Ayensu & W. Duckworth (eds.), Tropical Forest Ecosystems in Africa and South America, a comparative review: 89-104, Smithsonian Inst. Press, Washington.

Langenheim, J. & Y. Lee, 1974. Reinstatement of the genus *Hymenaea* (*Leguminosae-Caesalpinioideae*) in Africa. Brittonia 26: 3-21.

Langenheim, J., S. Martin & Y. Lee, 1972. Affinity of the African Leguminous resin-producing genus *Trachylobium* with the neotropical genus *Hymenaea*. Brittonia 24: 122-123.

Lebrun, J.-P. & A. Stork, 1992. Enumération des plantes à fleurs d'Afrique tropicale 2, Conservatoire et Jardin botaniques, Genève: 37-57.

Lee, Y. & J. Langenheim, 1975. Systematics of the genus *Hymenaea* (*Leguminosae*, *Caesalpinioideae*, *Detarieae*). Univ. Calif. Publ. Bot. 69: 1-109.

Léonard, J., 1949. *Caesalpiniaceae-Amherstieae* africanae americanaeque. Bull. Jard. Bot. Etat Brux. 19: 383-408.

Léonard, J., 1950a. Nouvelles observations sur le genre *Guibourtia*. Bull. Jard. Bot. Etat Brux. 20: 269-284.

Léonard, J., 1950b. Observations sur les genres africains *Oxystigma* et *Pterygopodium* (*Caesalpiniaceae*). Bull. Séances Inst. Roy. Col. Belge 21: 744-753.

Léonard, J., 1951. Les *Cynometra* et les genres voisins en Afrique tropicale. Bull. Jard. Bot. Etat Brux. 21: 373-450.

Léonard, J., 1952a. *Cynometra*, Flore Congo Belge et Ruanda-Urundi 3: 309-327.

Léonard, J., 1952b. *Caesalpiniaceae africanae* (*Cynometra*, *Didelotia*, *Zingania*, *Cryptosepalum*, *Pynaertiodendron*). Bull. Jard. Bot. Etat Brux. 22: 201-210.

Léonard, J., 1957. Genera des *Cynometreae* et des *Amherstieae* africaines. Essai de blastogénie appliquée à la systématique. Acad. Roy. Belgique, Cl. Sc., Mém. in 8°, 30, 2: 1-314.
Léonard, J., 1965. Description de *Tetraberlinia tubmaniana* sp. nov. Considérations sur *Tetraberlinia* et les genres affins (*Caesalpiniaceae* africaines). Bull. Jard. Bot. Etat Brux. 35: 97-107.
Léonard, J., 1993. Note sur les genres *Schotia* Jacq. et *Leonardoxa* Aubrév. et sur le nouveau genre *Normandiodendron* J. Léonard (*Caesalpiniaceae* africaines). Bull. Jard. Bot. Nat. Belg. 62: 433-451.
Léonard, J., 1994. Nouveaux apports de la blastogénie à la délimitation générique des *Caesalpiniaceae* africaines (*Detarieae* et *Amherstieae*). Bull. Jard. Bot. Nat. Belg. 63: 357-395.
Letouzey, R., 1977. *Didelotia pauli-sitai* R. Letouzey, Césalpiniacée nouvelle du Congo. Adansonia, sér. 2, 17: 125-127.
Lock, J.M., 1989. Legumes of Africa. A check-list. ILDIS, Royal Bot. Garden Kew.
Normand, D., 1958. Anatomie du bois et taxinomie II. Journ. Agr. Trop. Bot. Appl. 5: 297-301.
Normand, D., 1993. Sur la structure anatomique du bois des espèces de *Leonardoxa* Aubrév. (Caesalpiniacées africaines). Bull. Jard. Bot. Nat. Belg. 62: 429-432.
Normand, D. & J. Paquis, 1976. Manuel d'identification des bois commerciaux 2, Afrique guinéo-congolaise. Centre Techn. For. Trop., Nogent-sur-Marne: 80-110.
Oldeman, R., 1964. Revision of *Didelotia* Baill. (*Caesalpiniaceae*). Blumea 12: 209-239.
Ross, J., 1977. Flora of Southern Africa 16, 2: 1-142.
Taylor, C., 1960. Synecology and silviculture in Ghana. Univ. Coll. Ghana.
Torre, A. & D. Hillcoat, 1956. Conspectus Florae Angolensis 2: 162-253.
Troupin, G., 1950. Contribution à l'étude systématique de *Berlinia* et genres voisins (*Caesalpiniaceae-Amherstieae*). Bull. Jard. Bot. Etat Brux. 20: 285-324.
Voorhoeve, A., 1965. Liberian high forest trees. Centre Agr. Publ. Doc., Wageningen.
Watson, L. & M. Dallwitz, 1983. The genera of *Leguminosae-Caesalpinioideae*. Anatomy, morphology, classification and keys. Austral. Nat. Univ. Res. School Biol. Sc., Canberra.
White, F., 1962. Forest Flora of Northern Rhodesia: 94-128.

A new circumscription of the genus *Dolichos* (Leguminosae)

M.P. MUNYENYEMBE
Department of Biology, Chancellor College, University of Malawi,
P.O. Box 280, Zomba, Malawi.

Abstract
A new classification of *Dolichos* L. has been prepared, based on numerical phenetic and cladistic analyses. The details of the methods used and the resultant classification are to be published elsewhere. A summary of the presentation on the AETFAT Congress is presented below.

Résumé
Une nouvelle classification des *Dolichos* L. a été préparée, en se basant sur des analyses numériques phénétiques et cladistiques. Le détail des méthodes utilisées et la classification qui en résulte sont publiées ailleurs. Cette communication n'en représente que l'introduction.

Key words
Circumscription, *Dolichos*, Intra-generic classification.

The genus *Dolichos*

The genus *Dolichos* L. *nom. conserv.* belongs in the subtribe Phaseolinae of the tribe Phaseoleae, subfamily Papilionoideae, family Leguminosae. Until recently the genus has been considered to comprise about 70 species distributed mainly in Africa, with less than 10 species in Asia. With the exclusion of *Macrotyloma* (Wight & Arn.) Verdc. and other smaller segregate genera from *Dolichos* (Verdcourt 1970a, 1970b), the genus now seems to be more or less natural in its circumscription, but questions remain as to its exact size, and circumscription, and of the patterns of evolution within the genus.

A detailed taxonomic study of the genus *Dolichos* L. (Leguminosae) was undertaken with the following objectives:

1 to study the biological diversity of *Dolichos* in order to determine the pattern of resemblances, geographical distribution and evolution among the species. Since the last detailed classification of *Dolichos* species by Baker (1929) more than 20 new species (over a third of all *Dolichos* species) have been described, but very little is known about the relationships between the species.

2 to determine how many species there are and catalogue them. Currently, there is some fragmented data on *Dolichos* species from existing regional and national floras, checklists and published names (e.g. Lebrun & Stork 1992, Lock 1989, Lock & Heald 1994, Lock & Simpson 1991, Tateishi & Ohashi 1977).

The first account of *Dolichos* was that of Linnaeus (1753) who included 12 species in the genus, of which only one, *Dolichos trilobus*, belongs in the modern-day genus *Dolichos*. His description of the genus was based on the type *Dolichos lablab*, which now belongs to the monospecific genus *Lablab* Adans. The current nomenclatural type

L.J.G. van der Maesen et al. (eds.), The Biodiversity of African Plants, 456-458.

for the genus is *Dolichos trilobus* L. (Verdcourt 1968). Since Linnaeus' time there have been many classifications of *Dolichos* which are only of historical interest now, since concepts of the genus itself have changed considerably with time.

Traditionally, *Dolichos* had been the depository for odds and ends in the Phaseoleae (Lackey 1977). Verdcourt (1970a) studied the genus world-wide and succeeded in fragmenting the *Dolichos* complex into distinct units. He gave a new circumscription of the genus *Dolichos* with only about 60 species in Africa and Asia, instead of the 120 or so (Wilczek 1954). From *Dolichos* he segregated the genera *Pseudovigna* (Harms) Verdc., *Pseudeminia* Verdc., *Sinodolichos* Verdc., *Austrodolichos* Verc., *Macrotyloma* (Wight & Arn.) Verdc., *Dipogon* Liebm., *Alistilus* N. E. Br., *Lablab* Adans. and *Decorsea* R. Viguier. Verdcourt's (1970a) study of the genus *Dolichos* for the Flora of Tropical East Africa is the most recent attempt at elucidating intra-generic relationships of *Dolichos*. On the basis of the presence or absence of a tooth on the vexillary stamen and also style thickening and pubescence, he divided *Dolichos* into 3 subgenera: *Chloryllis*, *Dolichos* and *Odontichos*. He recognized the "*Dolichos trilobus* Group" and a group of closely related West African species within the subgenus *Dolichos*, which he called the "*Dolichos subcapitatus* Group".

Although Verdcourt more accurately delimited the genus and recognized some patterns within the genus, his scheme of it is still sketchy and incomplete. He did not assign all the species of *Dolichos* known at the time to the subgenera that he recognized. He did not propose a formal classification of the genus.

This study was herbarium-based. Numerical phenetic and cladistic techniques were used to analyze data from gross morphology and palynology. The analyses revealed new intra-generic patterns of resemblances. These patterns formed the basis for proposing a new classification and delimitation of *Dolichos* with 57 species recognized within *Dolichos* (including 3 new species). Three subgenera, including one new subgenus, were recognized, namely, subgenus *Dolichos*, subgenus *Chloryllis* and the new subgenus *Tenuistylis*. The subgenus *Dolichos* was further divided into 4 sections, which were in turn divided into series. *Dolichos fragrans* Kerr, *D. junghuhnianus* Benth. and *D. subcarnosus* sensu Prain (all belonging in *Dolichos* subgenus *Odontichos* Verdc.) were excluded from *Dolichos* and it was proposed that these species be included in a new genus sharing affinities with both *Dolichos* L. and *Nesphostylis* Verdc. On the basis of gross morphological and palynological evidence, *Dolichos junodii* (Harms) Verdc. was found not to belong in *Dolichos* and it was proposed that the species be transferred to the genus *Nesphostylis* Verdc. with which it shares many characters.

From the results of the cladistic analysis a hypothesis about the possible phylogeny of the genus *Dolichos* was put forward.

Full versions of materials and methods, results and discussion are reported elsewhere (Munyenyembe 1993, Munyenyembe & Bisby in press).

References

Baker, E.G., 1929. The Leguminosae of Tropical Africa. Part II Suborder Papilionaceae. Unitas Press, Ostend.

Lackey, J.A., 1977. A Synopsis of the Phaseoleae (Leguminosae, Papilionoideae). Ph.D. dissertation, Iowa State University, Ames.

Lebrun, J-P. & A. Stork, 1992. Enumération des Plantes à Fleurs d'Afrique Tropicale, Vol. II: 83-84. Conservatoire et Jardin Botaniques de la Ville de Genève, Genève 1992.

Linnaeus, C., 1753. Species Plantarum, Vol. 2: 725-727. Salvius, Stockholm.

Lock, J.M., 1989. Legumes of Africa, a Check-list: 386-448. Royal Botanic Gardens, Kew.

Lock, J.M. & K. Simpson, 1991. Legumes of West Asia, a Check-list. Royal Botanic Gardens, Kew.

Lock, J.M. & J. Heald, 1994. Legumes of Indo-China, A Check-list: 103-104. Royal Botanic Gardens, Kew.

Munyenyembe, M.P., 1993. Biological Diversity of Species in the Leguminosae Genus *Dolichos* L. Ph.D. thesis, University of Southampton.

Munyenyembe, M.P. & F.A. Bisby (in press). Systematic studies of Dolichos L. (Leguminosae).

Tateishi, Y. & H. Ohashi, 1977. The Genus *Dolichos* (Leguminosae) in the Himalayas. Bull. Natn. Sci. Mus., Ser. B (Bot.), 3(2): 71-84.

Verdcourt, B., 1968. The identities of *Dolichos trilobus* L. and *Dolichos trilobatus* L. Taxon 17(2): 170.

Verdcourt, B., 1970a. Studies in the Leguminosae-Papilionoideae for the 'Flora of Tropical East Africa': III. Kew Bull. 24(3): 379-447.

Verdcourt, B., 1970b. A new genus of Leguminosae-Phaseoleae. Kew Bull. 24(2): 322.

Wilczek, R., 1954. Phaseolinae. In: Flore du Congo Belge et du Ruanda-Urundi 6: 206-409. Bruxelles.

Generic delimitation of subfamily Cassinoideae (Celastraceae) in Africa

R.H. ARCHER [1] & A.E. VAN WYK [2]
[1] National Botanical Institute, Private Bag X101, Pretoria, 0001 South Africa.
[2] Department of Botany, University of Pretoria, Pretoria, 0002 South Africa.

Abstract
It has long been a matter of debate whether to treat *Cassine* in a wide sense, or to recognize several segregate genera. The accumulation of new taxonomic evidence from palynology, macromorphology and anatomy, as well as other sources, leads to a better understanding of generic limits in the Celastraceae. Available evidence suggests that *Cassine s.l.* as currently circumscribed, is an extremely artificial taxon and best subdivided into smaller, more homogeneous units. Most of these groups correspond very closely with previously described segregate genera. Eight genera (*Cassine s. str., Elaeodendron, Mystroxylon, Pleurostylia, Allocassine, Lauridia, Maurocenia* and a still to be described new genus) are proposed for the African species of Cassinoideae. *Cassine s. str.* comprises three species endemic to southern Africa.

Key words
Cassine, Cassinoideae, Celastraceae, Genus, Taxonomy.

Introduction

The genus *Cassine* was established by Linnaeus (1753, 1754), based on two southern African species, *C. peragua* and *C. maurocenia* (= *Maurocenia frangula*). The former species was selected as the lectotype of the genus *Cassine* (Loesener 1901), therefore predating subsequent attempts at lectotypification (Jarvis 1992). Subsequently at least ten new genera (*Maurocenia* Mill., *Elaeodendron* Jacq., *Hartogia* L.f. [a new name *Hartogiella* Codd was proposed for this illegitimate generic name (Codd 1983)], *Pleurostylia* Wight & Arn., *Crocoxylon* Eckl. & Zeyh., *Lauridia* Eckl. & Zeyh., *Mystroxylon* Eckl. & Zeyh. and *Pseudocassine* Bredell) were described for the African members of the Cassinoideae. With the exception of *Pseudocassine* [placed in Hippocrateaceae (Loesener 1942b)] and *Crocoxylon*, Loesener recognized these genera as belonging in subfamily Cassinoideae (Loesener 1942a). For the Flora Zambesiaca area, Robson (1965, 1966) described a new genus, *Allocassine*, while maintaining six genera, all considered by him to be natural taxa.

The differentiation of these segregate genera was based mainly on obvious morphological characters such as leaf disposition, floral structure and seed shape. In recent years most authors (particularly in southern Africa) applied the name *Cassine* in a broad sense, comprising all members of Celastraceae *s. str.* with indehiscent fruit except *Allocassine*, *Hartogiella*, *Maurocenia* and *Pleurostylia* (Dyer 1975, Arnold & De Wet 1993). This inclusive generic concept follows the taxonomic opinions of Loesener (1892), Davison (1927), Hou (1962), Codd (1966), Lobreau-Callen (1975), Jessup (1984) and Kostermans (1986). *Cassine s.l.* comprises about 80 species with a pantropical distribution (Mabberley 1987).

This paper emanates from a multidisciplinary approach towards a taxonomic revision of the southern African species of Cassinoideae (Archer 1990, Archer & Van

L.J.G. van der Maesen et al. (eds.), The Biodiversity of African Plants, 459-463.

Wyk 1992, 1993a, 1993b, 1994). Its principal aim is to advocate the recognition of various segregate genera based on a summary of available evidence considered to be of taxonomic significance.

Results and discussions

A synopsis of taxonomically noteworthy results is presented here. Palynology (Archer & Van Wyk 1992), leaf, wood and bark anatomy (Archer 1990, Archer & Van Wyk 1993a, 1993b, 1994, Hartog & Baas 1978) present a wealth of information, especially useful for generic delimitation. Several distinctive pollen, wood and bark anatomical types were recognized each separately or together showing remarkable congruence with the generic delimitation of Loesener (1942a) and Robson (1965).

Characters of fruit morphology and leaf anatomy strongly support the species groups as initially proposed on the basis of wood and bark anatomical, and palynological characters. The fact that this evidence correlates with the findings of macromorphology, is seen as strong support for the validity of the proposed segregate genera. The comprehensive generic key below largely summarizes the diagnostic characters of the genera in easily comprehensible form. It is mainly based on morphological characters, supplemented by particularly significant anatomical evidence. The proposed genera with their estimated number of species, as well as the type species of each genus, are listed in Table 1.

Comprehensive key to the African genera of the Cassinoideae (Celastraceae)

A Leaves alternate, opposite or fasciculate, if opposite then sepals small, c. 1 mm long, c. subcircular with margin entire; ovary 2-4-locular; ovules 2 per locule; fruit drupaceous; pollen tectum reticulate to rugulose reticulate; wood with fibres septate or axial parenchyma in apotrachial bands, vessel elements simple or with perforation plates scalariform; bark relatively thick (3-15 mm; inner bark only), with abundance of large fibresclereids or brachysclereids throughout the secondary phloem, rhytidome often well developed; stomatal complexes laterocytic, latero-cyclocytic or anisocytic:

B Leaves alternate, pubescent or glabrous, elastic threads (trans-1,4 polyisoprene) often present when broken; sepals small, ± deltoid with margin laciniate; endocarp thin and crustaceous; pollen tectum reticulate; wood with axial parenchyma in apotrachial bands, vessel elements simple; bark with phelloderm parenchymatous, rarely sclerified or crystalliferous; leaves with crystals absent in epidermis, stomatal complexes laterocytic or latero-cyclocytic:

C Leaves, young stems and inflorescences pubescent; elastic threads absent; sepals of equal size; ripe fruit red or tinged with red; pollen tectum reticulate with wide (1-3.5 μm) lumina, sexine much thicker than nexine; bark thick (4-10 mm; inner bark only), with large brachysclereids present throughout the secondary phloem; styloid crystals absent; stomatal complexes laterocytic . . . . . . ***Mystroxylon***

C All parts glabrous; elastic threads present in leaf and bark (conspicuously when pieces of leaf are pulled apart); sepals unequal, outermost sepals larger than inner sepals; ripe fruit white or yellow; pollen tectum reticulate, lumina narrow (0.5-1 μm), sexine and nexine of equal thickness; bark relatively thin (2-4 mm; inner bark only), sclerenchymatous elements absent in secondary phloem; styloid crystals of calcium oxalate abundant; stomatal complexes latero-cyclocytic **Genus A**

B Leaves usually opposite or fasciculate, glabrous; elastic threads absent; sepals small, ± subcircular with margin entire; endocarp thick and woody (putamen); pollen tectum reticulate to rugulose-reticulate; wood with fibres septate in apotrachial bands, vessel elements nearly always with scalariform perforation plates; phelloderm crystalliferous and sclerified; leaf epidermal cells with prismatic crystals of calcium oxalate, stomatal complexes anisocytic . . . . ***Elaeodendron***

A Leaves usually opposite; ovary 2-3-locular; ovules 2 per locule; fruit baccate; pollen tectum reticulate or smooth; wood with septate fibres in apotrachial bands, vessel elements always simple; bark relatively thin (1-3 mm; inner bark only), sclerenchymatous elements consists of fibres or groups of isodiametric sclereids in dilated rays, or broad continuous bands of densely packed chambered crystalliferous strands, phelloderm mostly crystalliferous; stomatal complexes cyclocytic:

D Leaves coriaceous with margin revolute; sepals vestigial; ovules pendulous; pollen tectum reticulate; wood rays nearly homocellular with cells predominantly square; bark with broad continuous bands of densely packed lignified chambered crystalliferous strands; rhytidome well developed with periderm ramified; leaves with adaxial hypodermis well developed, 3-5 layers thick . . . . ***Maurocenia***

D Leaves variously textured with margin plane or occasionally revolute; sepals well-developed; ovules erect; pollen tectum reticulate or nearly smooth; wood rays heterocellular or nearly homogeneous; bark with sclereids consisting of fibres and/or groups of small isodiametric sclereids present in dilated rays adjacent to fibres, or bands of lignified chambered crystalliferous strands; rhytidome absent or weakly developed, occasionally well developed; leaves with adaxial hypodermis occasionally present, 1-2 layers thick:

E Shrubs or trees; inflorescence a compound dichasium; flowers pedicellate; ovary 2-3-locular; ripe fruit light to dark brown or purplish, succulent or ± fleshy; pollen tectum reticulate; wood rays heterogenous with square and procumbent cells; bark with sclereids consisting of fibres and small isodiametric sclereids in dilated rays adjacent to fibres, rhytidome weakly developed; raphide clusters of magnesium oxalate or small spheroidal calcium oxalate crystals occasionally present; leaf epidermis with straight anticlinal cell walls; adaxial hypodermis occasionally present . . . . ***Cassine***

E Lianes, scrambling shrubs, or small trees; inflorescence thyrsoid with simple or compound cymules arranged along the rachis; flowers sessile; ovary 2-locular; ripe fruit red or purple, fleshy or leathery; pollen tectum foveolate or smooth and irregularly foveolate; wood rays nearly homocellular with upright or square cells; bark sclereids consisting of fibres an/or bands of lignified chambered strands, rhytidome absent or well developed; leaf epidermis with sinuous anticlinal cell walls; hypodermis absent:

F Lianes or scrambling shrubs; leaves opposite to alternate; sepals triangular, coriaceous; flowers 5-merous; berries elliptic; wood rays consisting of predominantly upright cells; bark with fibres, periderm persistent with no sequential periderm formation, hence phellem well developed . . . . . ***Allocassine***

F Lianes, scrambling shrubs or small trees; leaves opposite; sepals ± suborbicular, fleshy; flowers 4- or 5-merous; berries spheroid; wood rays consisting of predominantly square cells; bark with continuous broad bands of densely packed lignified chambered crystalliferous strands, rhytidome well developed with periderm ramified . . . . ***Lauridia***

A Leaves opposite; ovary with 1 fertile locule; ovules 2-4 (-8?); fruit small, nutlike and lopsided with a persistent lateral style or stylar scar, pericarp thin and dry; pollen tectum smooth and irregularly foveolate; wood with axial parenchyma in apotrachial bands, vessel elements simple; bark thick (3-10 mm; inner bark only), with broad oval to irregular groups of densely packed lignified chambered crystalliferous strands, rhytidome well developed with periderm ramified; stomatal complexes cyclocytic ***Pleurostylia***

Table 1. African genera of the Cassinoideae; type species, estimated number of taxa and geographical distribution.

Genus	Type species	Number of sp./subsp.; Geographical distribution
Allocassine	*A. laurifolia*	1; southern Africa
Cassine	*C. peragua*	3/3; southern Africa
Elaeodendron	*E. orientale* (Mauritius)	± 30 (9 in Africa); S and E Africa, Mauritius, India, Australasia and Neotropics
Genus A	*Cassine eucleiformis* (new comb. required)	2; southern Africa
Lauridia	*L. reticulata*	2; southern Africa
Maurocenia	*M. frangula*	1; Western Cape
Mystroxylon	*M. aethiopica*	1/4; Africa & Madagascar
Pleurostylia	*P. opposita* (India, Australasia and Africa)	4 (3 in Africa); India, Australasia and Africa

References

Arnold, T.H. & B.C. De Wet (eds.), 1993. Plants of southern Africa: names and distribution. Mem. Bot. Surv. S. Afr. 62: 1-825.

Archer, R.H., 1990. The taxonomic status of *Cassine s.l.* (Celastraceae) in southern Africa. M.Sc. Thesis, Univ. of Pretoria.

Archer, R.H. & A.E. van Wyk, 1992. Palynology and intergeneric relationships in some southern African species of subfamily Cassinoideae (Celastraceae). Grana 31: 241-252.

Archer, R.H. & A.E. van Wyk, 1993a. Bark structure and intergeneric relationships in some southern African Cassinoideae (Celastraceae). IAWA Journal 14: 35-53.

Archer, R.H. & A.E. van Wyk, 1993b. Wood structure and generic status of some southern African Cassinoideae (Celastraceae). IAWA Journal 14: 373-389.

Archer, R.H. & A.E. van Wyk, 1994. Bark structure of *Cassine s.l.* (Celastraceae) in southern Africa. In: J.H. Seyani & A.C. Chikuni (eds.), Proceedings XIIIth Plenary Meeting AETFAT, Malawi, 2: 1269-1278.

Codd, L.E., 1966. Celastraceae. The *Cassine* complex. In: New and interesting records of African flowering plants. Bothalia 9: 123-161.

Codd, L.E., 1983. Celastraceae. A new name for *Hartogia* Thunb. ex L.f. In: New and interesting records of African flowering plants. Bothalia 14 (2): 219-222.

Davison, J.D., 1927. Celastraceae. Bothalia 2: 289-346.

Dyer, R.A., 1975. The genera of southern African flowering plants. Vol. 1. Department of Agricultural Technical Services, Pretoria.

Hartog-Van Ter Tholen, R.M. den & P. Baas, 1978. Epidermal characters of the Celastraceae *sensu lato*. Acta Bot. Neerl. 27: 355-388.

Hou, Ding, 1962. Celastraceae I, II. In: C.G.G.J. Van Steenis (ed.), Flora Malesiana, Series I, Vol. 6: 227-421. Wolters-Noordhoff Publishing, Groningen.

Jarvis, C., 1992. Seventy-two proposals for the conservation of types of selected Linnaean generic names, the report of Subcommittee 3C on the lectotypification of Linnaean generic names. Taxon 41: 552-583.

Jessup, L.W., 1984. Celastraceae. In: A.S. George (ed.), Flora of Australia. Vol. 22: 150-180. Australian Government Publishing Services, Canberra.

Kostermans, A.J.G.H., 1986. Notes on Asiatic *Cassine* L. (Celastraceae). Gardens' Bulletin, Singapore 39: 177-191.

Linnaeus, C., 1753. Species Plantarum. Impensis Laurentii Salvii, Stockholm.

Linnaeus, C., 1754. Genera Plantarum. Impensis Laurentii Salvii, Stockholm.

Lobreau-Callen, D., 1975. Deux genres de Celastraceae *Cassine* L. et *Maytenus* Mol. revus à la lumière de la palynologie. Adansonia Sér. 2, 15: 215-223.

Loesener, T., 1892. Celastraceae. In: A. Engler & K. Prantl (eds.), Die natürlichen Pflanzenfamilien Ed. 1, 3, 5: 189-222. W. Engelmann, Leipzig.

Loesener, T., 1901. Celastraceae africanae III. Bot. Jahrb. 28: 150-161.

Loesener, T., 1942a. Celastraceae. In: A. Engler, H. Harms & J. Mattfeld (eds.), Die natürlichen Pflanzenfamilien Ed. 2, 20b: 87-197. Duncker & Humblot, Berlin.

Loesener, T., 1942b Hippocrateaceae. In: A. Engler, H. Harms & J. Mattfeld (eds.), Die natürlichen Pflanzenfamilien Ed. 2, 20b: 198-231. Duncker & Humblot, Berlin.

Mabberley, D.J., 1987. The plant-book. A portable dictionary of the higher plants. Cambridge University Press, Cambridge.

Robson, N.K.B., 1965. New and little known species from the Flora Zambesiaca area XVI, taxonomic and nomenclatoric notes on Celastraceae. Bolm Soc. Brot. 39 (ser. 2): 5-55.

Robson, N.K.B., 1966. Celastraceae (incl. Hippocrateaceae). In: A.W. Exell, A. Fernandes & H. Wild (eds.), Flora Zambesiaca II/2: 355-418. Crown Agents for Overseas Governments and Administrations, London.

Leaf anatomy of the African genera of Commelinaceae: *Anthericopsis* and *Murdannia*

R.B. FADEN [1] & K.E. INMAN [2]
[1] Department of Botany, National Museum of Natural History, Smithsonian Institution, Washington, DC, 20560, USA.
[2] 9 Park Vale Avenue, #10, Allston, MA 02134, USA.

Abstract
Leaf anatomy of *Anthericopsis sepalosa* and nine species of *Murdannia* was studied using scanning electron microscopy, paraffin embedded sections, whole leaf clearings, and epidermal scrapes. *Anthericopsis* was distinctive because of its attenuate margin with basal tannin cells and a broad, discontinuous, adaxial hypodermis. *Murdannia* was characterized by a thin epidermis, patterned cuticle, usually continuous hypodermis, and generally marginal sclerenchyma. Both genera showed several apparently synapomorphic characters, such as lack of a thick midrib, presence of a broad adaxial hypodermis, and a tapered margin, that separated them from other genera of tribe Commelineae. The anatomical support morphological data that show a phylogenetic relationship between the two genera.

Key words
Anatomy, *Anthericopsis*, Commelinaceae, Leaf, *Murdannia*, Phylogeny.

Introduction

The 18 genera of African Commelinaceae belong to three of the four tribes recently recognized in the family (Faden & Hunt 1991). Although these genera are readily distinguishable by morphological characters, their phylogenetic relationships have not been determined. The tribe Commelineae is of particular interest because 12 of its 13 genera occur in Africa and Madagascar (*Aneilema, Anthericopsis**, *Buforrestia, Commelina, Floscopa, Murdannia, Pollia, Polyspatha**, *Pseudoparis**, *Rhopalephora, Stanfieldiella**, and *Tricarpelema* (African endemics marked with an asterisk) and comprise more than 82% of the 275 species of Commelinaceae in the flora. Faden and Hunt (1991) were unable to divide the tribe into subtribes, in contrast to their treatment of tribe Tradescantieae.

Tomlinson (1966, 1969), in his survey of the genera of Commelinaceae, showed that anatomical characters could be used to separate genera that were once taxonomically confused, e.g. *Aneilema* and *Murdannia*, or wrongly grouped together (Brenan 1966, Faden & Hunt 1991). Faden (1993) commented that Tomlinson had not examined a wide enough array of genera and species of Commelineae to shed much light on the relationships among the genera of the tribe. The present paper is part of an ongoing study of the anatomy of Commelineae that is attempting to determine whether all genera of the tribe can be distinguished by their leaf anatomy and whether anatomy can provide phylogenetically useful characters. The two genera studied were *Anthericopsis*, a monotypic African genus of nonforest habitats, and *Murdannia*, a pantropical genus of about 50 species, of which 11 occur in Africa and Madagascar in open habitats (Faden 1983, 1988).

L.J.G. van der Maesen et al. (eds.), The Biodiversity of African Plants, 464-471.

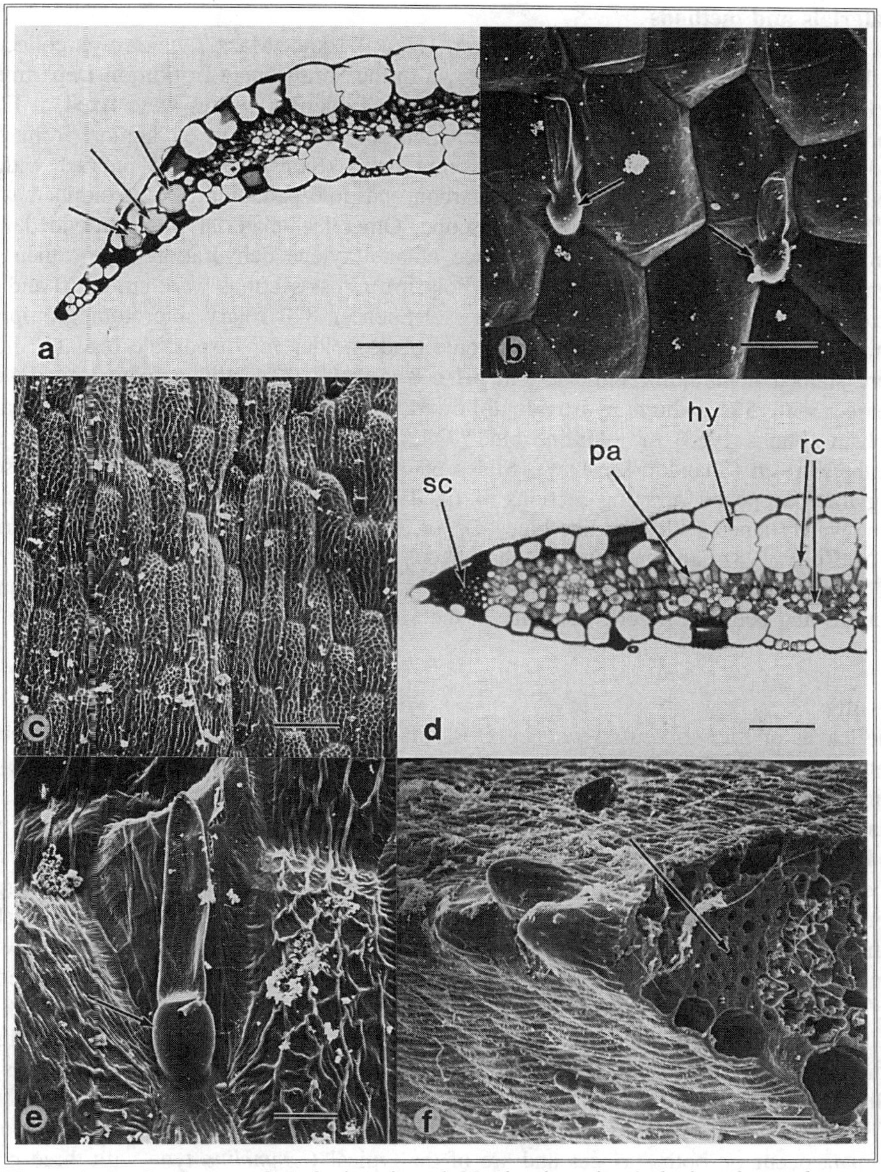

Figure 1. Leaf anatomy of *Anthericopsis* and *Murdannia*. a-b, *Anthericopsis sepalosa*: a, Leaf margin, cross-section: tannin cells (arrows); b. Adaxial surface showing hexagonal epidermal cells and glandular microhairs: middle cells (arrows). c-f, *Murdannia simplex*: c, Adaxial surface showing rectangular epidermal cells with reticulate cuticle; d, Leaf margin, cross-section: hy = hypodermis, pa = palisade, rc = raphide canal, sc = sclerenchyma; e, Glandular microhair: middle cell (arrow) (note reticulate cuticle on epidermal cells); f, Leaf margin, cross-section: marginal sclerenchyma (arrow). Bars = 107 μm in a, 40 μm in b, 100 μm in c, 115 μm in d, 22 μm in e, and 30 μm in f.

Materials and methods

The plants used, except *Murdannia keisak* (Hassk.) Hand.-Mazz., which was collected in the field in northern Virginia, were grown in the Smithsonian Institution Department of Botany greenhouse in Washington, D.C. (see Table 3). Leaves were fixed in FPA (formaldehyde, propionic acid, 50% ethanol, 1:1:18). Pieces of lamina from the midvein, margin and tip were critical point dried (Samsri-PVT-3B), coated with a Cook Model 301T sputter coater with carbon and gold/palladium, and examined with an Hitachi 570 scanning electron microscope. Other leaf material was subdivided into 5 mm strips, processed through a graded ethanol/xylene dehydration series, then infiltrated and embedded in paraffin wax (Polyfin). Cross-sections were cut at 10 and 12 μm and paradermals at 5 μm using an AO Spencer 820 rotary microtome equipped with a Tissue-Tek II Accu-Edge microtome blade holder for disposable blades. Slides were stained in toluidine blue "O" at pH 6.8 (Sakai 1973). Whole fixed leaves were cleared with 5% sodium hydroxide followed by chloral hydrate, stained with basic fuchsin (Fuchs 1963) or toluidine blue "O", and mounted on slides in a permanent synthetic resin (Shandon/Lipshaw). Slides of epidermis were prepared under a dissecting microscope by scraping sections of fixed leaves using fine forceps. The preparations were stained with toluidine blue "O" or Periodic Acid Schiff's Reaction (O'Brien & McCully 1981) and mounted in an aqueous mounting medium (Aqua-Poly Mount). *M. keisak* and *M. japonica* (Thunb.) Faden were studied from unstained freehand sections and peels. All work was done at the Smithsonian Institution.

Results

The leaves of *Anthericopsis sepalosa* (C.B. Clarke) Engl. show a number of distinctive features. They lack a strongly thickened midrib region composed of several close-set veins embedded in parenchyma, having instead a single midvein. The epidermis is relatively thick, comprising more than 50% of the leaf thickness halfway between the midvein and the margin. A one-celled layer adaxial hypodermis is present throughout most of the lamina although it does not reach the margin. It is interrupted by the larger veins, including the midvein. A one-celled palisade layer is readily distinguishable although its cells are not very elongate. The spongy mesophyll is thin and relatively dense. Raphide canals are present in the mesophyll below the hypodermis and above the abaxial epidermis.

The leaf margin is attenuate (Figure 1a). At the base of the taper are one to three tannin cells. In the cleared leaf these can be seen to be elongate and to form chains of cells that run the length of the lamina.

The epidermal cells are arranged in a hexagonal pattern. The cuticle is smooth. Stomata occur on both surfaces and are of the typical *Commelina*-type with three pairs of subsidiary cells (Tomlinson 1966, 1969). On the adaxial surface they occur only along the major veins, but not submarginally. Only microhairs are present. These are short and have a short, ellipsoid middle cell (Figure 1b).

The nine *Murdannia* species studied are diverse in their anatomy but several generalizations can be made. The leaves lack a thickened midrib area but sometimes have a distinct midvein. The epidermis is thin, usually constituting less than 25% of the leaf thickness, but occasionally 25-50%. An extensive adaxial hypodermis is always present. It consists of one to many cell layers and commonly nearly reaches the leaf margin (Figure 1d). Sometimes it is less extensive, and in *M. keisak* it only reaches only halfway to the margin. Usually the adaxial hypodermis is continuous although it may be thinner over the veins. In *M. zeylanica* (C.B. Clarke) G.Brückn. and *M. keisak* it is interrupted over the veins, including, in the latter and sometimes in the former, over the midvein. Abaxial hypodermis may be present but, except in *M. clarkeana* Brenan, it is always interrupted and rarely extends far out into the lamina from the midvein, often being confined to small patches under the main veins.

The palisade is a one-cell layer and not well differentiated from the rest of the mesophyll in *M. acutifolia* (Lauterb. & K. Schum.) Faden, *M. clarkeana*, *M. cryptantha* Faden, and *M. japonica*. The single palisade layer is composed of more elongate and distinct cells in *M. bracteata* (C.B. Clarke) Hong, *M. edulis* (Stokes) Faden, and *M. simplex* (Vahl) Brenan (Figure 1d). In *M. zeylanica* the palisade may be partly two-layered, and in *M. keisak* it is consistently two-layered. In both these species the adaxial cell layer is continuous and well-differentiated whereas the abaxial layer is not very distinct from the underlying mesophyll and may be discontinuous. Raphide canals are abundant in the mesophyll below the hypodermis and above the abaxial epidermis in all species except *M. clarkeana* where they are confined within the very thick adaxial hypodermis.

The lamina always tapers at the margin, sometimes gradually, sometimes abruptly (Figure 1d). The actual margin is commonly acute in cross-section, but in *M. acutifolia* and *M. bracteata* the tip is rounded. The subterete leaves of *M. clarkeana* essentially lack a margin. A distinctive feature of *Murdannia* is the presence of a marginal band of sclerenchyma (Figure 1d, f). Marginal sclerenchyma is lacking, however, in *M. clarkeana* and *M. keisak*. Tannin cells are absent marginally, but subspherical to ellipsoid ones are scattered in the mesophyll in *M. edulis*.

The epidermal cells, particularly on the adaxial surface, are usually more or less rectangular and arranged in columns (Figure 1c). *Murdannia keisak* and *M. zeylanica* are unusual is having hexagonal epidermal cells. The cuticle is always patterned, ranging from striate to reticulate (Figure 1c, e). *Murdannia clarkeana* is the only species in which it is strictly striate. The cuticular pattern is usually conspicuous, but in *M. acutifolia* it is rather faint although clearly visible in the epidermal scapes.

Stomata are usually present on both surfaces, but they are lacking adaxially in *M. clarkeana* and *M. edulis*. In *M. bracteata*, *M. japonica* and *M. simplex* adaxial stomata are present only near the margin. They usually have only four subsidiary cells of the *Tradescantia* type (Tomlinson 1966: Figure 79, Tomlinson 1969), but the arrangement may be irregular. Similar stomata also occur submarginally in *M. acutifolia* and *M. cryptantha*, but these species also have stomata over the veins that usually have six subsidiary cells (irregular arrangement in *M. acutifolia, Commelina* type in *M. cryptantha*). In *M. keisak* and *M. zeylanica* the adaxial stomata all have six subsidiary cells of the *Commelina* type and are more widely distributed, but situated mainly (*M. keisak*) or exclusively (*M. zeylanica*) over the veins and near the margins. Abaxial stomata are much more abundant than adaxial and, except in *M. clarkeana*, they always have six subsidiary cells of the *Commelina* type. In *M. clarkeana* the stomata all have four subsidiary cells of the *Tradescantia* type.

Table 1. Comparison of *Anthericopsis, Murdannia* and the other genera of the tribe Commelineae*.

Anthericopsis	*Murdannia*	Other genera of Commelineae
Antesepalous stamens fertile; antepetalous staminodial	Antesepalous stamens fertile, sometimes one staminodial; antepetalous staminodial, rarely lacking	Arrangements various but never alternating stamens and staminodes
Leaves in a basal rosette, inflorescence scapose	A few species like this	None exactly like this
Leaf vernation supervolute	All species supervolute	All species involute except one *Aneilema*, many *Commelina*, and single African *Tricarpelema*
Lamina sessile	Lamina sessile	Lamina commonly petiolate, sometimes sessile
Thick midrib absent	Thick midrib absent	Thick midrib almost always present
Adaxial hypodermis well developed	Adaxial hypodermis well developed	Adaxial hypodermis almost always confined to the midrib area, rarely well developed or absent
Leaf margin tapered	Leaf margin tapered	Leaf margin usually not tapered, sometimes enlarged
Microhairs with elliptic middle cell	Microhairs with elliptic middle cell	Microhairs usually with a cylindrical, but sometimes elliptic, middle cell

* *Aneilema, Buforrestia, Commelina, Dictyospermum, Floscopa, Pollia, Polyspatha, Pseudoparis, Rhopalephora, Stanfieldiella, Tricarpelema* (after Faden & Hunt 1991).

The glandular microhairs in all species have an ellipsoid middle cell (Figure 1e). Macrohairs present may include papillae, prickle hairs and uniseriate hairs. All three may occur on the margins but commonly only papillae and prickle hairs are present there. Uniseriate hairs and prickle hairs may be scattered on the surfaces, as in some populations of *M. zeylanica*, but often they are confined to an adaxial submarginal band, as in *M. acutifolia* and *M. edulis*.

Table 2. Anatomical differences in the leaves of *Anthericopsis* and *Murdannia*.

Anthericopsis	*Murdannia*
Margin attenuate, only 2 cells thick	Margin less tapered, thicker
Margin with tannin cells, lacking sclerenchyma	Margin lacking tannin cells, usually with sclerenchyma (except *M. clarkeana* and *M. keisak*)
Adaxial hypodermis interrupted over major veins	Adaxial hypodermis usually continuous (except *M. keisak* and *M. zeylanica*)
Macrohairs lacking	Macrohairs usually present (except *M. clarkeana*)
Epidermis midway between midvein and margin >50% of the leaf thickness	Epidermis midway between midvein and margin usually <25%, occasionally 25-50%, of the leaf thickness
Epidermal cells hexagonal	Adaxial epidermal cells usually ± rectangular (except *M. keisak* and *M. zeylanica*)
Cuticle smooth	Cuticle patterned
Adaxial stomata not submarginal	Adaxial stomata, when present, always submarginal and sometimes more widespread
All stomata with 6 subsidiary cells	Adaxial submarginal stomata often with 4 subsidiary cells (except *M. keisak* and *M. zeylanica*)
Bundle sheath indistinct	Bundle sheath distinct at least around smaller bundles

Discussion

Both *Anthericopsis* and *Murdannia* have proven to have a very distinctive leaf anatomy. The former can be recognized by its lack of a thick midrib, presence of adaxial hypodermis throughout the lamina, and attenuate margins with basal tannin cells. The most consistent characters of *Murdannia* are the tapered margins, lack of thick midrib, broad adaxial hypodermis, and patterned cuticle. Marginal sclerenchyma and rectangular epidermal cells also characterize most *Murdannia* species.

It is noteworthy that the combination of the absence of a thick midrib and presence of a tapered margin and broad adaxial hypodermis characterize both these genera and serve to separate them from all other genera of tribe Commelineae, insofar as they have been studied (Table 1). *Anthericopsis* and *Murdannia* also share the character of having small glandular microhairs with an ellipsoid middle cell, a feature not exclusive to them, but otherwise uncommon in the genera of Commelineae.

These apparently synapomorphic characters provide evidence that *Anthericopsis* and *Murdannia* represent an evolutionary line within tribe Commelineae. This conclusion is further supported by other evidence, particularly morphology (Table 1). The anatomical differences between *Anthericopsis* and *Murdannia* are summarized in Table 2.

Both *Murdannia* and *Anthericopsis* were studied previously by Tomlinson (1966, 1969). The new features found in the present study include the presence of marginal tannin cells in *Anthericopsis* and a much greater range of variation within *Murdannia*, including the discontinuous adaxial hypodermis and absence of marginal sclerenchyma in some species, the occurrence of four-celled adaxial stomata near the margin, and exclusively four-celled stomata in *M. clarkeana*, the last a feature completely unknown previously in tribe Commelineae. It is noteworthy that *M. clarkeana* is unusual in many additional features, including subterete leaf shape, very thick adaxial hypodermis with included raphide canals, continuous abaxial hypodermis, and absence of marginal sclerenchyma and macrohairs. *Murdannia keisak* and *M. zeylanica* uniquely share several features that not only separate them from the other species of *Murdannia* studied but also make them approach species of other genera of Commelineae, including, in some features, *Anthericopsis*: two or one- to two-layered palisade, relatively thick epidermis, adaxial hypodermis stopping well short of the margin and interrupted over the major veins, numerous adaxial stomata all of the *Commelina* type, and hexagonal epidermal cells. There are no morphological or ecological explanations for these similarities, and one can only speculate whether they might be phylogenetically significant. Additional species of *Murdannia* need to be studied to resolve this matter.

Two cautions about our results may be noted. Tomlinson (1966) referred to the pattern on *Murdannia* epidermal cells as a character of the wall, not the cuticle. We have neither confirmed nor refuted this but can only state that the pattern is present in the cuticle, with the possible exception of *M. acutifolia*. Tomlinson (1966) noted that anatomical features are often modified by growing conditions. All but one of the species we studied came from greenhouse cultivated plants. Therefore some differences might be found in field-collected material.

Acknowledgements

We wish to thank Stanley Yankowski for technical assistance in all aspects of this research; the National Museum of Natural History Research Training Program and National Science Foundation Research Experience for Undergraduates Grant BIO-9200203 for providing one of us (KEI) the opportunity to participate in this research; Ching-I Peng for supplying plants and seeds; and Audrey J. Faden and other volunteers for caring for the research plants for many years.

References

Brenan, J.P.M., 1966. The classification of Commelinaceae. J. Linn. Soc. Bot. 59: 349-370.

Faden, R.B., 1983. Phytogeography of African Commelinaceae. Bothalia 14: 553-557.

Faden, R.B., 1987. Vegetative and reproductive features of forest and nonforest genera of African Commelinaceae. Monogr. Syst. Bot. Missouri Bot. Gard. 25: 521-526.

Faden, R.B., 1993. Leaf anatomy and generic characterization in the Commelinaceae. Amer. J. Bot. 80 (6), Suppl.: 148.

Table 3. Specimens used.

Species	Collection Number	Place of Origin
Anthericopsis sepalosa (C.B. Clarke) Engl.	Faden & Faden 74/504	Kenya
Murdannia acutifolia (Lauterb. & K. Schum.) Faden	Manierl 65 ex Kew	New Guinea
Murdannia bracteata (C.B. Clarke) Hong	Peng 5042	Taiwan
Murdannia clarkeana Brenan	Faden & Ng'weno 87/58 & 94/5	Kenya
Murdannia cryptantha Faden	A. Faden 2/91	Australia
Murdannia edulis (Stokes) Faden	Peng 11177	Taiwan
Murdannia japonica (Thunb.) Faden	Peng 15249	China
Murdannia keisak (Hassk.) Hand.-Mazz.	Faden & Faden 94/7	Virginia, USA
Murdannia simplex (Vahl) Brenan	Faden & Faden 77/315	Kenya
Murdannia zeylanica (C.B. Clarke) G. Brückn.	Faden & Faden 76/548	Sri Lanka

Faden, R.B & D.R. Hunt, 1991. The classification of the Commelinaceae. Taxon 40: 19-31.

Fuchs, C., 1963. Fuchsin staining with NaOH clearing for lignified elements of whole plants or plant organs. Stain Technol. 38: 141-144.

O'Brien, T P. & M.E. McCully, 1981. The Study of Plant Structure: Principles and Selected Methods, 6.83, Termacarphi Pty. Ltd, Melbourne.

Sakai, W.S., 1973. Simple method for differential staining of paraffin embedded plant material using toluidine blue O. Stain Technol. 48: 247-249

Tomlinson, P.B., 1966. Anatomical data in the classification of Commelinaceae. J. Linn. Soc. Bot. 59: 371-395.

Tomlinson, P.B., 1969. Commelinaceae. In: C.R. Metcalfe (ed.), Anatomy of the Monocotyledons. III. Commelinales-Zingiberales: 12-63, Clarendon Press, Oxford.

Generic delimitation of *Lasiodiscus* (Rhamnaceae)

E.P.N. FIGUEIREDO
Botany Department, University of Reading, Whiteknights,
P.O. Box 221, Reading RG6 2AS, United Kingdom.

Abstract
A cladistic analysis of *Lasiodiscus* Hook. f. and *Colubrina* Rich. ex Brongn. was undertaken, using characters from gross morphology and with 10 of the 13 species of *Lasiodiscus* and 19 of the 32 species of *Colubrina* as terminal taxa. The cladograms indicate that, with the exclusion of *L. articulatus* Capuron several synapomorphies characterize *Lasiodiscus*. The inclusion of this species in the genus obscured the boundary between *Lasiodiscus* and *Colubrina*. *L. articulatus* should be transferred to *Colubrina*.

Key words
Colubrina, *Lasiodiscus*, Rhamnaceae.

Introduction

A taxonomic revision of the genus *Lasiodiscus* Hook. f., which will be published separately, has evidenced a weak distinction between this genus and *Colubrina* Rich ex Brongn., suggesting the possibility of a congeneric status. *Lasiodiscus* consists of 13 species occurring in tropical Africa and *Colubrina* consists of 32 species distributed in Central and South America, Hawaii, SE. Asia, Madagascar and S. Africa (Pondoland).

The two genera have been considered closely allied (Capuron 1965, Johnston 1971) and a feasible congeneric status has previously been postulated (Johnston 1971).

According to Capuron (1965) *Lasiodiscus* and *Colubrina* can be distinguished by seed characters. In *Lasiodiscus* the seed lacks endosperm and an aril, the hilum is circular or elliptic, the coat is coriaceous and the radicle is not protruding, while in *Colubrina* endosperm and an aril are present, the hilum is reduced to a slit, the coat is crustaceous and the radicle is protruding.

In a taxonomic revision of *Colubrina*, Johnston (1971) referred also to the absence of endosperm and to the occurrence of interpetiolar united stipules in *Lasiodiscus*, as diagnostic features for this genus. Interpetiolar, united stipules occur also in one species of *Colubrina* (*C. oppositifolia* Brongn. ex H. Mann) but were interpreted as a consequence of independent evolution. The hilum, seed coat and radicle were not considered by Johnston and the presence of an aril was said to be occasional.

Although *Colubrina* was revised recently, the lack of a taxonomic revision of *Lasiodiscus* made difficult the evaluation of these characters in the delimitation of the two genera.

Material and methods

For the revision of *Lasiodiscus* herbarium material obtained on loan from the following institutions was examined: B, BM, BR, COI, EA, G, GOET, HBG, K, L, LISC, LISJC, LISU, M, MA, MO, P, PO, PRE, S, UPS, WAG. The information on *Colubrina* was retrieved from literature (Johnston 1971) and from the collections at K.

A list of characters was prepared and scored for the 32 species of *Colubrina* and 13

L.J.G. van der Maesen et al. (eds.), The Biodiversity of African Plants, 472-476.

Table 1. Matrix for the cladistic analysis.

No.	Species	1	2	3	4	5	6	7	8	9	10	11	12	13	14	15
1.	*C. arborescens* (Mill.) Sarg.	0	0	0	0	1	0	0	0	1	1	0	0	0	2	0
2.	*C. obscura* (Schrank) M.C. Johnst.	0	0	0	0	?	?	0	0	?	?	0	0	0	2	0
3.	*C. glandulosa* Perkins	?	0	0	0	1	0	0	?	1	1	0	0	0	2	0
4.	*C. cordifolia* Reissek	?	0	0	0	?	0	0	?	1	1	0	0	1	2	0
5.	*C. oppositifolia* Brongn. ex H. Mann	1	1	?	?	1	0	0	?	?	?	0	0	0	2	0
6.	*C. elliptica* (Sw.) Brizicky & Stern	0	0	0	0	1	0	0	?	1	0	0	0	0	1	0
7.	*C. viridis* (M.E. Jones) M.C. Johnst.	0	0	0	0	1	0	0	?	1	?	0	1	0	1	0
8.	*C. heteroneura* (Griseb.) Standl.	0	0	0	0	?	0	0	?	1	0	1	0	0	1	0
9.	*C. retusa* (Pittier) Cowan	?	0	0	0	?	0	0	?	1	0	1	0	0	1	0
10.	*C. triflora* Brongn.	0	0	0	0	1	0	0	?	1	1	0	0	1	0	1
11.	*C. decipiens* (Baill.) Capuron	1	?	0	0	1	0	0	?	?	?	0	0	1	0	0
12.	*C. faraloatra* (H. Perrier) Capuron	0	0	0	0	1	0	0	?	1	1	0	0	1	2	0
13.	*C. alluaudii* (H. Perrier) Capuron	0	0	0	0	?	?	0	?	1	?	0	0	0	0	0
14.	*C. humbertii* (H. Perrier) Capuron	0	0	0	0	?	0	0	1	0	?	0	1	0	0	0
15.	*C. celtidifolia* (Cham. & Schlt.) Schlt.	0	0	0	0	1	0	0	1	1	1	0	0	1	0	1
16.	*C. macrocarpa* (Cav.) G. Don	0	0	0	0	1	0	0	1	1	1	0	0	1	0	1
17.	*C. texensis* (Torr. & Gray) A. Gray	0	0	0	0	1	0	0	1	1	?	0	0	1	0	1
18.	*C. californica* I.M. Johnston	0	0	0	0	?	0	0	1	1	?	0	1	0	0	0
19.	*C. asiatica* (L.) Brongn.	0	0	0	0	1	0	0	0	1	0	0	0	1	0	1
20.	*L. chevalieri* Hutch.	1	1	0	1	0	1	1	1	1	1	0	0	0	0	0
21.	*L. fasciculiflorus* Engl.	1	1	0	0	0	1	1	1	1	1	0	0	0	0	0
22.	*L. usambarensis* Engl.	1	1	0	0	0	1	1	1	1	1	0	0	0	0	0
23.	*L. lebrunii* ined.	1	1	0	1	0	1	1	1	1	1	0	0	0	0	0
24.	*L. gillardinii* Staner	1	1	0	1	0	1	1	1	1	1	0	0	1	0	0
25.	*L. mannii* Hook. f.	1	1	0	1	0	1	1	1	1	1	0	0	0	0	0
26.	*L. marmoratus* C.H. Wright	1	1	1	1	0	1	1	1	1	1	0	0	0	0	0
27.	*L. mildbraedii* Engl.	1	1	0	1	0	1	1	1	1	1	0	0	0	0	0
28.	*L. pervillei* Baill.	1	1	0	1	0	1	1	1	1	1	0	0	0	0	0
29.	*L. palustris* ined.	1	1	0	1	0	1	1	1	1	1	0	0	0	0	0
30.	*L. articulatus* Capuron	0	0	0	0	0	0	0	0	0	0	0	0	1	0	0

species of *Lasiodiscus*. Two species of *Lasiodiscus* and thirteen species of *Colubrina* were subsequently eliminated due to insufficient data.

The following characters were scored for 11 species of *Lasiodiscus* and 19 species of *Colubrina* and used in the construction of a matrix (Table 1): 1. Leaves alternate (0), opposite (1); 2. Stipules lateral (0), interpetiolar (1); 3. Adaxial stipules free (0), united (1); 4. Abaxial stipules free (0), united (1); 5. Endosperm absent (0), present (1); 6. Receptacle margin at the base of the fruit (0), above or at the middle (1); 7. Fruit glabrous (0), velvety (1); 8. Ovary protruding from the middle of the disc (0), sunk (1); 9. Inflorescence terminal (0), axillary (1); 10. Radial walls of fruit persistent (0), caducous (1); 11. Axillary thorns absent (0); present (1); 12. Branchlets not thorn-tipped (0), thorn-tipped (1); 13. One primary nerve (0), several (1); 14. Glands in the lower surface of leaf absent (0), basal and paired (1), scattered (2); 15. Marginal glands on leaf teeth absent (0), present (1).

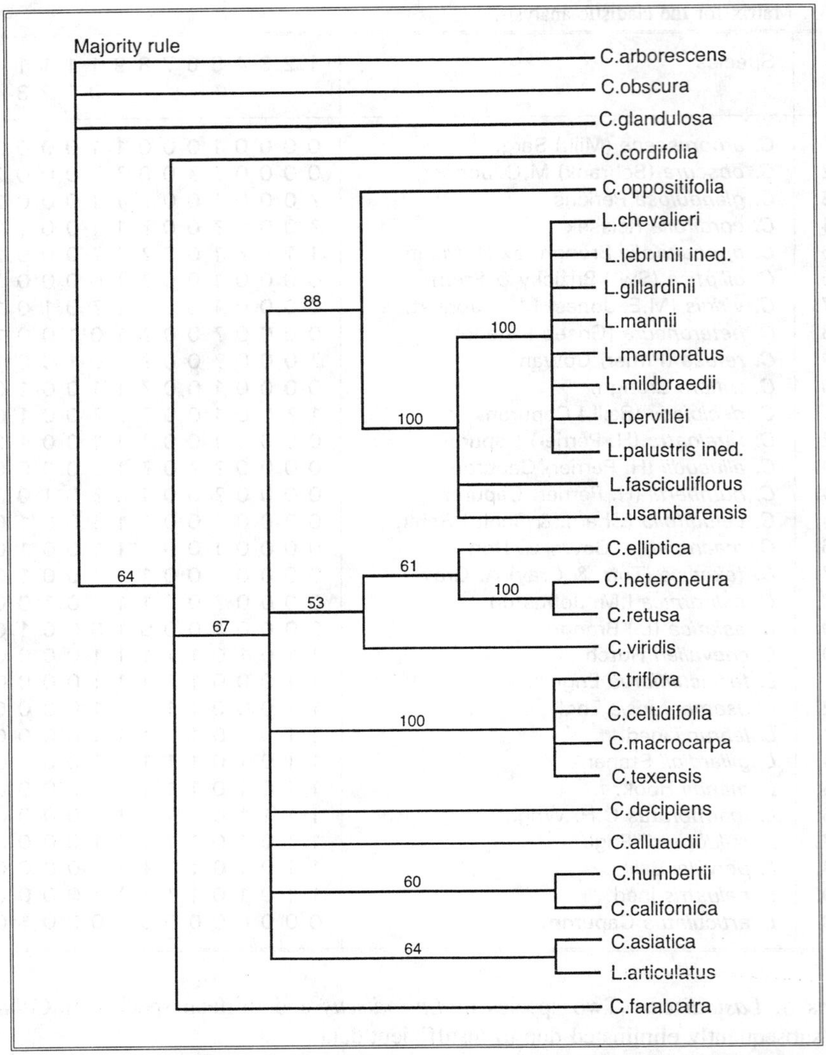

Figure 1. Fifty percent majority rule consensus tree (unrooted) of the 956 most parsimonious trees (C.I. =0.615; R.I= 0.886) obtained from heuristic search.

A cladistic analysis was performed with PAUP (Phylogenetic Analysis Using Parsimony) version 3.1.1 for Macintosh (Swofford 1993), using the Heuristic (branch swapping) option.

Results

The unrooted majority rule consensus tree (Figure 1) of the 956 most parsimonious trees shows that 10 species of *Lasiodiscus* are grouped in a well supported clade, which occurs in all the trees. This clade is characterised by the occurrence of velvety fruits with the receptacle margin at or above the middle. Opposite leaves and interpetiolar stipules also characterise this group, occurring in *C. oppositifolia* as a consequence of parallel evolution. *L. articulatus* Capuron is nested in a group of *Colubrina* species.

Discussion

The examination of the characters used by Johnston (1971) to separate the genera revealed that these are not constant within the two taxa.

In *Lasiodiscus* the stipules are not invariably interpetiolar and united. The majority of species has dorsiventral branchlets with interpetiolar stipules which are free adaxially and united abaxially. Interpetiolar stipules free on both sides of the branchlets occur in two species (*L. fasciculiflorus* Engl. and *L. usambarensis* Engl.) and united stipules were found in one species (*L. marmoratus* C. H. Wright). In *L. articulatus* the stipules are not interpetiolar.

Endosperm is absent in all the species of *Lasiodiscus* examined. In *Colubrina* endosperm varies from abundant to scanty (Johnston 1971) but information is unavailable for several species which prevents the use of this character as a diagnostic feature of the genus.

Lasiodiscus articulatus differs from the other species in the genus by several characters. It has glabrous fruits with a distinct and prominent margin of the receptacle at the base like the species of *Colubrina*. In the other species of *Lasiodiscus* fruits are velvety with the margin of the receptacle at or above the middle. *L. articulatus* also differs from the rest of the genus in having persistent radial walls of the fruit, subterminal inflorescences and the top of the ovary protruding from the middle of the disc. These characters are present in some species of *Colubrina* and absent in the other species of *Lasiodiscus*.

The possibility of a close affinity of *L. articulatus* to *Colubrina,* suggested by the occurrence of these characters, was supported by the cladistic analysis. In the consensus tree *L. articulatus* appears nested in a group *Colubrina* species, while the other species of *Lasiodiscus* form a distinct clade. The classification of *L. articulatus* in the genus based on the absence of endosperm and regardless of other characters shared with *Colubrina*, obscured the real boundary between the two genera. *L. articulatus* will be transferred to *Colubrina* in the revision of *Lasiodiscus*, published separately.

The genus *Lasiodiscus* is a monophyletic group characterised by the opposite leaves and interpetiolar stipules, velvety fruits, and the margin of the hypanthium not prominent, at or above the middle of the fruit. The consensus tree indicates a close relationship of the Hawaiian *Colubrina oppositifolia* to *Lasiodiscus*, which, in accordance with Johnston's analysis (1971), can be considered a consequence of parallel evolution. In fact, although opposite leaves and interpetiolar stipules are also found in this species, most of its features are typical of *Colubrina*.

The lack of herbarium material or data in literature for many species of *Colubrina* prevented the use of more characters in the analysis. The presence of an aril, for instance, which apparently defines the section *Capuronia* M.C. Johnst., containing the Madagascan species, is unclear and has been omitted in this analysis. In the consensus tree the four species from Madagascar, *C. decipiens* (Baill.) Capuron, *C. alluaudii* (H. Perrier) Capuron, *C. humbertii* (H. Perrier) Capuron, and *C. faraloatra* (H. Perrier)

Capuron, are not grouped in a clade, appearing unresolved.

The infrageneric classification of *Colubrina* (Johnston 1971) is not supported by this analysis. The division of the genus in two subgenera is not supported by the consensus tree and species from different sections group together in the same clades. It is not known if *Colubrina* is a paraphyletic group or if it is monophyletic and characterised by a set of unique characters. A detailed survey of this genus for a cladistic analysis with the inclusion of other genera as outgroups might clarify its status and infrageneric classification.

Acknowledgements

This work was funded by a grant from Junta Nacional de Investigacao Cientifica e Tecnologica (Portugal) and was carried out whilst being a postgraduate student at the University of Reading.

References

Capuron, R., 1965. Les Rhamnacées arbustives ou arborescentes de Madagascar. Centre Technique Forestier Tropical. Sect. Botanique.

Johnston, M.C., 1971. Revision of Colubrina (Rhamnaceae). Brittonia 23: 2-53.

Swofford, D.L., 1993. PAUP: Phylogenetic Analysis Using Parsimony (version 3.1.1), Computer program distributed by Illinois Natural History Survey, Champaign, Il.

A study on microsporogenesis and ovule morphology in tropical African Anthericaceae and Asphodelaceae

S. KATIVU
Dept. of Biol. Sciences, Univ. of Zimbabwe, P.O. Box MP 167, Mt. Pleasant, Harare, Zimbabwe.

Abstract
A study on tropical African species of Anthericaceae and Asphodelaceae showed that microsporogenesis is successive in Anthericaceae and simultaneous in Asphodelaceae. Both families have anatropous ovules and frequently have nectaries and raphids. Type of microsporogenesis appears to be an important taxonomic character at the family level.

Key Words
Anthericaceae, Asphodelaceae, Microsporogenesis, Ovule morphology, Tropical Africa.

Introduction

Anthericaceae J.G. Agardh and Asphodelaceae Juss. (order Asparagales) are segregate families of the large heterogeneous group Liliaceae s.l. which were reinstated by Dahlgren et al. (1985). The two families were separated mainly on the basis of chemical and embryological characters. Seed morphology and position of ovules provided further evidence. Asphodelaceae were said to be characterized by its production of anthraquinone compounds and by having anatropous, hemianatropous or orthotropous ovules, simultaneous microsporogenesis and generally arillate seeds. Anthericaceae, on the other hand, were said to be characterized by its lack of anthraquinones and production of steroidal saponins instead, and by having successive microsporogenesis and often campylotropous ovules.

This delimitation placed genera formerly regarded as closely allied in two separate families. *Anthericum sensu* Marais and Reilly, 1978, in Anthericaceae, and *Trachyandra* Kunth, in Asphodelaceae, are of special significance in tropical Africa. The two genera were, until recently (Obermeyer 1962), regarded as congeneric (Baker 1878, 1896, 1898).

Stedje and Nordal (personal communication), working on Ethiopian material of *Anthericum* and *Trachyandra*, pointed out that the morphological characters used by Dahlgren et al. (1985) were not always consistent with the proposed family delimitation. Considering inflorescence characters for instance, although nodes with two or more flowers (or at least with more than one bract) are never found in Asphodelaceae, their presence does not characterize the whole of Anthericaceae as the genus *Anthericum sensu* Kativu & Nordal (1993) has simple nodes. Also, although articulated pedicels largely characterize Anthericaceae, they are generally absent in *Anthericum*, yet they frequently occur in *Asphodelus* (Asphodelaceae).

Stedje and Nordal (personal communication) also observed similarities in karyotype morphology among some taxa of the two families, e.g. between *Jodrellia macrocarpa* Baijnath (Asphodelaceae) and *Chlorophytum tetraphyllum* (L.f.) Bak. (Anthericaceae).

L.J.G. van der Maesen et al. (eds.), The Biodiversity of African Plants, 477-480.

Table 1. Type of microsporogenesis in species of Anthericaceae and Asphodelaceae from tropical Africa (A.B. = Anders Bjornstad, I.N. = Inger Nordal, S.K. = Shakkie Kativu); all voucher specimens deposited at O.

Taxon	Voucher number	Origin	Type of microsporogenesis
Aloe cryptopoda Baker	S.K. 720	Zimbabwe	simultaneous
A. excelsa A. Berger	S.K. 721	Zimbabwe	simultaneous
Bulbine frutescens Willd. (in cult.)	S.K. 705	Zimbabwe	simultaneous
B. latifolia Baker	S.K. 676	Zimbabwe	simultaneous
Chlorophytum andongense Baker	S.K. 311	Zimbabwe	successive
C. bowkeri Baker	S.K. 306	Zimbabwe	successive
C. calyptrocarpum (Baker) Kativu	S.K. 341	Zimbabwe	successive
C. comosum (Thunb.) Jacq.	A.B. 1706	Tanzania	successive
C. galpinii (Baker) Kativu var. *norlindii* (H. Weim.) Kativu	S.K. 262	Zimbabwe	successive
C. ducis-aprutii Chiov.	I.N. 1045	Tanzania	successive
C. graniticola Kativu	S.K. 285	Zimbabwe	successive
C. kirkii Baker	A.B. 2809	Tanzania	successive
C. longifolium Schweinf. ex Baker	I.N. 1507	Zimbabwe	successive
C. macrophyllum (A. Rich.) Aschers	I.N. 1033	Tanzania	successive
C. macrosporum Baker	S.K. 255	Zimbabwe	successive
C. pauper Poelln.	S.K. 254	Zimbabwe	successive
C. pusillum Schweinf. ex Baker	S.K. 247	Zimbabwe	successive
C. subpetiolatum (Baker) Kativu	S.K. 251	Zimbabwe	successive
C. suffruticosum (Milne-Redh.)	A.B. 2809	Tanzania	successive
C. zavattarii (Cufod.) Nordal	I.N. 2284	Ethiopia	successive
Trachyandra saltii (Baker) Oberm.	I.N. 2259	Ethiopia	simultaneous
T. reflexipilosa (O. Kuntze) Oberm.	Mavi 825	Zimbabwe	simultaneous

They concluded that the available cytological evidence did not give strong support to the family delimitation of Dahlgren et al. (1985). Nordal (personal communication) also observed that ovule characters were of no diagnostic value between the two families.

The present study re-assesses the diagnostic value of the type of microsporogenesis and ovule characters (ovule position especially) in the separation of Anthericaceae and Asphodelaceae by extending the investigation to tropical African taxa not treated by Dahlgren et al. (1982, 1985).

Materials and Methods

Microsporogenesis was studied from freshly picked flower buds of 16 species of Anthericaceae and six of Asphodelaceae which were at different stages of development. These were fixed in Carnoy's fluid (Johansen 1940) for 24 hours at 4° C and washed in two series of 70% alcohol. The young anthers were stained overnight in

Table 2. Ovule characters in species of Anthericaceae and Asphodelaceae: Ovular shape anatropous in all species mentioned; all voucher specimens deposited at O.

Taxon	Collected	Voucher	S.N.	R.
Anthericum corymbosum Baker	Ethiopia	I.N. 2243	-	+
Chlorophytum cameronii (Baker) Kativu	Tanzania	A.B. 551	-	-
C. colubrinum (Baker) Engl.	Tanzania	A.B. 1362	+	-
C. elgonense Bullock	Tanzania	A.B. 547	+	-
C. tetraphyllum (L.f.) Baker	Ethiopia	I.N. 2267	+	-
C. tuberosum (Roxb.) Baker	Ethiopia	I.N. 2243	+	-
C. viridescens Engl.	Tanzania	I.B. 265	+	-
Jodrellia macrocarpa Baijnath	Ethiopia	I.N. 2287	-	-
Trachyandra saltii (Baker) Oberm.	Zimbabwe	I.N. 2214	+	+

S.N. = Septal Nectaries, R. = Raphides, + indicates presence of character, - absence; A.B. = Anders Bjørnstad, I.N. = Inger Nordal (formerly I.B. = Inger Bjørnstad).

alcoholic HCl-Carmine stain (Snow 1963) at 60° C. Stained tissues were washed in 70% ethyl alcohol and studied under a Zeiss Universal Research Microscope.

The study on ovules was treated as a pilot investigation of the nature of variation in type, shape and orientation of ovules, and of the occurrence of septal nectaries and raphids. It was based on seven species of Anthericaceae and two of Asphodelaceae. Freshly picked flower buds at various stages of development, were fixed in Karpechenko's modified Navashin's fluid (Johansen 1940), washed and dehydrated in concentration series of alcohol (70%, 90%, 95% and absolute alcohol), embedded in paraffin and sectioned on a microtome at 10-14 micrometres. Sectioned tissues were stained in safranin and fast green, made into permanent mounts and studied under a Zeiss photomicroscope.

Results

Data on type of microsporogenesis is presented in Table 1. Microsporogenesis was simultaneous in Asphodelaceae and successive in Anthericaceae. Observed ovule characters are presented in Table 2. All examined species (both Anthericaceae and Asphodelaceae) had anatropous ovules. The nucellus was slightly curved in *Anthericum corymbosum* and *Chlorophytum tuberosum* (both Anthericaceae). In two species, *Jodrellia macrocarpa* and *Trachyandra saltii* (both Asphodelaceae), the funiculus was more laterally positioned than in the rest, giving the ovules a somewhat transitional shape towards the hemianatropous ovule. Septal nectaries were observed in *Chlorophytum colubrinum, C. elgonense, C. tetraphyllum, C. tuberosum, C. viridescens* and *Trachyandra saltii*. Raphides were observed in cells of the placenta in *Anthericum corymbosum, Chlorophytum elgonense, C. tetraphyllum* and *Trachyandra saltii*.

Discussion and conclusions

Within order Asparagales, most taxa have successive microsporogenesis. Exceptions include Phormiaceae, Doryanthaceae, Asphodelaceae, Dianellaceae and Tecophilaeaceae. These five families, however, are not closely linked morphologically. Instead, Asphodelaceae appear to be related to Anthericaceae.

Microsporogenesis is reported to be successive in Anthericaceae and simultaneous in Asphodelaceae, and is said to be of diagnostic significance at family level (Dahlgren et al. 1982, 1985). The taxonomic relevance of this character in family delimitation is confirmed by the present study.

The family delimitation of Dalgren et al. (1985) presents Anthericaceae as characterized by anatropous, hemianatropous or orthotropous ovules and Asphodelaceae by largely campylotropous ovules. Campylotropous ovules are not found in any of the seven Anthericaceae species examined, even though two of them have slightly curved nucelli. Rather, the anatropous ovule is common to all examined taxa, irrespective of their family. The septal nectaries and raphides observed in some of the taxa appear not to be of taxonomic significance at the family level as they are found in both families. With the possible exception of the arillate seeds of Asphodelaceae (Stedje and Nordal, personal communication), a character not examined in the present study, the use of ovule characters in family delimitation is not supported.

Although it is of no practical use to the herbarium taxonomist, the type of microsporogenesis is synapomorphic and cannot be ignored in the taxonomy of Asphodelaceae and Anthericaceae. Seed characters and anther attachment may provide the basis for the initial separation of the two families.

Acknowledgements

Thanks are due to the following: Inger Nordal supervised the study which is part of a D. Phil. project; Brita Stedje commented on the observations; Robert Bailey Drummond identified some of the specimens; the study was carried out in the Department of Biology, University of Oslo and was funded by the Norwegian Ministry of Development Cooperation.

References

Baker, J.G., 1878. Report on the Liliaceae, Iridaceae, Hypoxidaceae and Haemodoraceae of the Welwitsch's Angolan Herbarium. Trans. Linn. Soc., Bot., Ser. 2, 1: 245-273.

Baker, J.G., 1896. Liliaceae. In: W.T. Thiselton-Dyer (ed.), Flora Capensis 6: 253-423. Lovell Reeve, London.

Baker, J.G., 1898. Liliaceae. In: W.T. Thiselton-Dyer (ed.), Flora of Tropical Africa 7: 385-568. Lovell Reeve, London.

Dahlgren, R.M.T. & H.T. Clifford, 1982. The monocotyledons: a comparative study. Academic Press, London.

Dahlgren, R.M.T., H.T. Cliford & P.F. Yeo, 1985. The families of the monocotyledons: structure, evolution and taxonomy. Springer-Verlag, Berlin.

Johansen, D.A., 1940. Plant microtechnique. McGraw-Hill, New York.

Kativu, S. & I. Nordal, 1993. New combinations of African species of in the genus *Chlorophytum* (Anthericaceae). Nord. J. Bot. 13 (1): 59-65.

Obermeyer, A.A., 1962. A revision of the South African species of *Anthericum, Chlorophytum* and *Trachyandra*. Bothalia 7: 669-767.

Snow, R., 1963. Alcoholic hydrochloric acid-carmine as a stain for chromosomes in squash preparations. Stain Technology 38: 9-13.

On the position of the genus *Pergularia* (Asclepiadaceae)

S. LIEDE
Abteilung Spezielle Botanik, Universität Ulm, 89075 Ulm, Germany.

Abstract
The position of the genus *Pergularia* (Asclepiadaceae) is reconsidered. The genus is transferred from the subtribe Cynanchinae to the subtribe Asclepiadinae. Best argument for the transfer is the unusual 7ß,8ß-Epoxycardenolide glycoside with 4,6-dideoxyhexosulose as sugar component, a structure found only in *Asclepias, Calotropis, Gomphocarpus*, and *Pergularia*. This transfer is supported by the results of an analysis of the ITS region of the nuclear genome. Morphological characters are found to be consistent with the new placement.

Résumé
La position du genre *Pergularia* (Asclepiadaceae) est reconsidérée dans ce travail. Ce genre est transféré de la subtribu Cynanchinae à celle des Asclépiadinae. La meilleure raison pour le transfert est constitué par les 7ß,8ß-epoxycardenolide glycosides, qui se trouvent seulement dans les genres *Asclepias, Calotropis, Gomphocarpus* et *Pergularia*. Ce transfert est confirmé par les résultats de l'analyse du région ITS d'un gène nucléaire. De plus, les données morphologiques sont en accord avec ce nouveau placement.

Key words
Asclepiadaceae, Asclepiadeae, Asclepiadinae, Cynanchinae, *Pergularia*, Systematics.

Abbreviations
Ci - interstaminal corona (parts)
Cs - staminal corona (parts)
Cis - corona of staminal and interstaminal parts, not fused
C(is) - corona of fused staminal and interstaminal parts
(see Liede & Kunze 1993)

Introduction

Currently, seven tribes are recognized in the Asclepiadaceae (Bruyns & Forster 1991, Kunze et al. 1994): Periploceae, Secamoneae, Fockeae, Marsdenieae, Stapelieae, Gonolobeae and Asclepiadeae. Of these, the Asclepiadeae, identified by the synapomorphic character state of hanging pollinia (Kunze 1993, Liede in press) is by far the largest (Liede & Albers 1994). A current long-term project aims at clarifying the relationships between the genera of the tribe. In the course of this project, the genus *Pergularia* L. turned out ill-placed as a genus without obvious affinities.

Within the Asclepiadeae, *Pergularia* L. is well characterized by possessing a rather long, narrow corolla tube, ciliate corolla lobes, a translucent germination pore in exodistolateral position (Rao & Kumari 1979) on the pollinia, and pairwise fruits that are almost always covered with soft spines (Kuriachen & Dave 1989). An early confusion in nomenclature started by R. Brown (1811) has been resolved by N.E. Brown (1907), so that *Daemia* R.Br. is now regarded as a superfluous later name for *Pergularia* L. The marsdeniad species (with erect pollinia) described under *Pergularia*

L.J.G. van der Maesen et al. (eds.), The Biodiversity of African Plants, 481-488.

between 1810 and early this century are now (mostly) placed in *Telosma* Coville. A list of all *Pergularia* names and their affinities is given by Liede (1990). The number of species is not yet established with certainty; it ranges between three and five. The genus is widespread from southern Africa throughout all more arid parts of the continent to Arabia, India and Sri Lanka. Its fairly large, attractive flowers and long flowering period recommend it as ornamental in warmer countries.

The position of the genus has so far not been studied in detail; the general opinion, reflected in placements in numerous floras, was to place it close to *Sarcostemma* R.Br. Holm (1950), in his revision of *Sarcostemma*, after listing the differences between *Pergularia* and *Sarcostemma*, goes as far as to state "Decidedly it does not seem advisable to complicate further the synonymy of the sarcostemmas by changing them all to *Pergularia*". If a subtribal concept for the Asclepiadeae is employed, the genus is thus placed in the subtribe Cynanchinae K. Schum. Doubts regarding this idea have been raised in a recent cladistic analysis of morphological characters of Holm's *Sarcostemma* (Liede unpubl.), in which *Pergularia* and an unusual, but unambiguous American "*Sarcostemma*" species were invariably identified as belonging to the same clade - most likely an effect of long-branch attraction (Felsenstein 1978).

While morphological characters are not sufficient to clarify the position of the genus, published chemical data suggest its affinities. To create an additional independent dataset, a molecular analysis of the ITS-region of the nuclear genome, which has been shown to yield valuable phylogenetic markers on the genus and species level (Baldwin 1992) has been employed.

Materials and methods

Specimens from BOL, MO, and NBG form the bulk of material examined. Living material was available from Kenya, courtesy of Mrs. A. Robertson, and Pakistan (Ali s.n., courtesy of Dr. R.E. Wyatt, Univ. of Georgia, Athens, USA). The chromosome count was undertaken by Dr. U. Meve, Münster. Chemical data have been extracted during a recent survey of published chemical compounds in the Asclepiadaceae (Liede unpubl.). Analysis of the ITS region of the nuclear genome was undertaken at UGA in the laboratories of Dr. M.L. Arnold and Dr. R.A. Price. The ITS dataset was manually aligned and subjected to analysis with PAUP version 3.1. (Swofford 1991). The outgroup was chosen from the *Tylophora*/*Tylophoropsis* complex (Liede & Meve 1994). In this complex the transition in pollinia orientation from erect to hanging is exemplified, so that, according to present knowledge, it represents a basal taxon in the tribe Asclepiadeae.

Results and discussion

The genus *Pergularia* displays a very unusual combination of characters, some of which are not matched in other Asclepiadeae genera. The corolla tube, certainly an apomorphic feature, is the most obvious of these characters. Also most likely apomorphic is the presence of exodistolateral germination pores (Rao & Kumari 1979). The paired follicles, in contrast, remind strongly of the paired follicles in the Periploceae and Secamoneae, the basal tribes in the family (Liede in press, Kunze 1993), and so do the extraordinarily large chromosomes (Figure 2, Meve pers. comm.). The genus thus displays a heterobathmic character pattern.

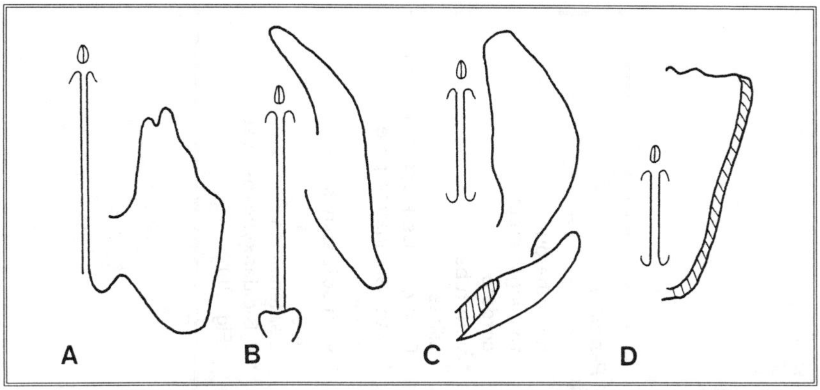

Figure 1. Corona types (schematic). A. *Gomphocarpus* (Asclepiadinae); B. *Pergularia*; C. *Sarcostemma* (Cynanchinae); D. *Cynanchum* (Cynanchinae).

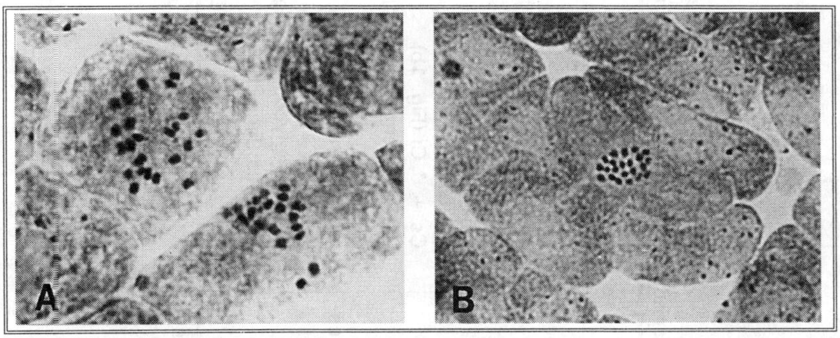

Figure 2. Chromosomes of *Pergularia* and *Sarcostemma*. A. *Pergularia daemia* (Robertson 6913); total karyotype length 26.1 μm. B. *Sarcostemma stocksii* (Bally 11739); total karyotype length 17.6 μm. Photos by U. Meve, c. 1500x.

On first sight, the genus seems to fit well into the subtribe Cynanchinae with its twining habit, heart-shaped leaves, and traces of an interstaminal corona, that is, however, not fused to the staminal corona parts (Table 1, Figure 1). However, a twining habit is most likely a plesiomorphic feature in the Asclepiadeae (Liede 1992), and whether heart-shaped leaves are plesio- or apomorphic, cannot at present be determined. Thus, only the possession of interstaminal corona parts joins *Pergularia* to the Cynanchinae.

As early as in 1962, Mittal et al. established that *Pergularia* possesses cardenolides. This, again, is untypical for Cynanchinae, in which, of all genera tested, only one other genus, *Oxystelma* R.Br., also of uncertain position, was found to contain cardenolides (Srivasta et al. 1993). *Oxystelma* possesses pregnane glycosides in addition to cardenolides (e.g. Trivedi et al. 1988), and pregnane glycosides are typical for all other Cynanchinae genera studied (e.g. Hegnauer 1964, 1989). No pregnane glycosides, however, have yet been reported from *Pergularia*. Cardenolides are ubiquitous

Table 1: Morphological and chemical characteristics of Asclepiadinae, Cynanchinae, *Pergularia*.

	Asclepiadinae	Cynanchinae	*Pergularia*
Roots:	* tuberous	rarely tuberous	not tuberous
Habit:	* erect	mostly twining	twining/* erect
Leaf bases:	cuneate	mostly cordate	cordate
Corolla:	without tube	without tube	* with tube
	occasionally ciliate	not ciliate	ciliate
Corona:	Cs (Fig. 1a)	Cs + * Ci (Fig. 1b)	* C(is) or * Cs + C(is) (Fig. 1c,d)
Pollinia:	without germination pore	without germination pore	* with germination pore
Follicles:	* single	* single	paired
	occasionally * with soft spines	mostly smooth	* with soft spines
Chromosomes:	* small	* small	large
Aglycones:	Cardenolides (Fig. 2a)	* Pregnaneglycosides (Fig. 2b)	Cardenolides (Fig. 2a)
Carbohydrates:	* 4,6-dideoxyhexosulose (Fig. 2a) or hexoses	Hexoses (Fig. 2b)	* 4,6-dideoxyhexosulose (Fig. 2a)

Most likely apomorphic character states are provided with an asterisk (*).

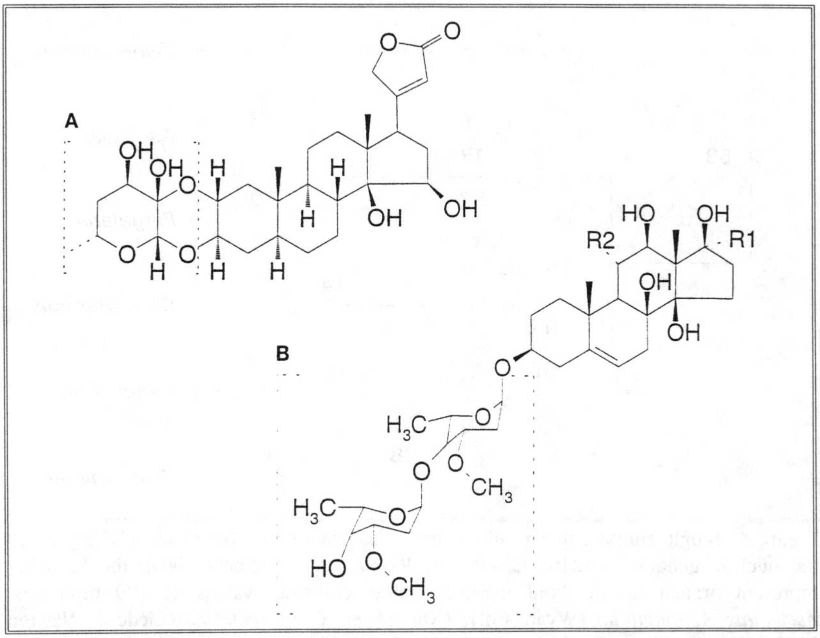

Figure 3. Glycosides in Asclepiadinae and Cynanchinae. A: 7ß,8ß-Epoxycardenolide glycoside with 4,6-dideoxyhexosulose as carbohydrate and cardenolide as aglycone as found in *Pergularia, Asclepias, Calotropis,* and *Gomphocarpus*; B. Pregnanglycoside with a disaccharide as sugar component and a pregnane derivate as aglycon.

in the subtribe Asclepiadinae, where they normally occur alone, but sometimes also together with pregnane glycosides (Hegnauer 1964, 1989). It is known that cardenolides also occur in the Periploceae and Secamoneae (e.g. Hegnauer 1964, 1989). Biochemically, cardenolides and pregnane glycosides are derived via the same pathway (Reichstein 1967, Hegnauer 1989). For these reasons, their presence cannot be rated as an apomorphic character state; rather, the possession of pregnane glycosides has apomorphic character. Cheung and Watson (1980), however, report a very unusual sugar structure, 4,6-dideoxyhexosulose, which has so far only been observed in four genera: *Asclepias* L., *Calotropis* R.Br., *Gomphocarpus* R.Br. (all three members of the subtribe Asclepiadinae) and *Pergularia*. This structure (Figure 3a) is extremely stable against acid hydrolysis, and also highly poisonous. With this clue in mind, one cannot fail to notice that, apart from the interstaminal corona scales (Figure 1b), the corona in *Pergularia* closely resembles a typical Asclepiadinae corona (Figure 1a). More important, all four genera mentioned possess fruits with soft spines at least in some species. In the Cynanchinae, fruits are almost always glabrous; though there are exceptions (e.g., *Cynanchum corymbosum* Wight, *Pentarrhinum* E.Mey.). Also, ciliate corolla lobes are known from *Gomphocarpus* and *Kanahia* R.Br., the latter a genus for which chemical data are lacking (Table 1). As these two characters are certainly apomorphic (compared to other Asclepiadaceae tribes), the hypothesis is proposed that *Pergularia* constitutes a member of the Asclepiadinae; most likely a basal one.

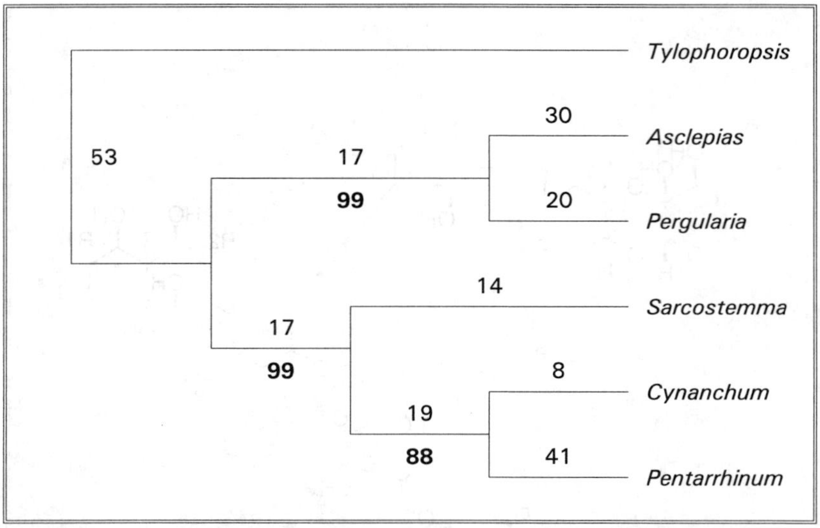

Figure 4. Single shortest tree resulting from analysis of 582 BP of the ITS region of the nuclear genome (l=219, CI=0.936, RI=0.702). Numbers above the branches represent branch length; bold numbers below bootstrap values of 100 replicates. *Asclepias: A. longifolia* (Wyatt s.n.), *Cynanchum: C. altiscandens* (Liede & Newton 2873), *Pentarrhinum: P. insipidum* (Liede 2940), *Pergularia: P. daemia* (Ali s.n.), *Tylophoropsis: T. heterophylla* (Liede & Newton 3155).

To test this hypothesis, an analysis of the genomic ITS sequences of some of the genera involved was performed. For the Asclepiadinae, *Asclepias longifolia* Michx, was chosen as representative, for the Cynanchinae, *Sarcostemma viminale* (L.) R.Br. and *Cynanchum altiscandens* K. Schum. In addition, *Pentarrhinum insipidum* E. Mey. was included into the analysis, because *Pentarrhinum* exhibits striking similarities with *Pergularia* in general habit and some fruit characters. *Pentarrhinum* has been transferred from the Asclepiadinae to the Cynanchinae only recently (Liede & Nicholas 1992). No chemical data are available for *Pentarrhinum*. However, the plants are neither used medicinally nor known as poisonous throughout their large distribution area. Therefore, it can be deduced with some certainty that cardenolides of the *Pergularia*-type are absent from *Pentarrhinum*.

Analysis of 582 BP (90 BP were excluded because data for *Asclepias* were lacking) results in one single tree (Figure 4; l=219, CI=0.936, RI=0.702). *Pergularia* was placed unequivocally into a clade with *Asclepias*, while *Pentarrhinum* and *Cynanchum* were equally unequivocally placed into a clade with *Sarcostemma*. An analysis coding gaps as a single event adds 10 steps to the tree, but does not change its topology.

Conclusions

While a larger molecular dataset including more species of both Asclepiadinae and Cynanchinae would be highly desirable in order to finally resolve the position of *Pergularia*, the evidence presented here seems sufficient to exclude the genus from the

Cynanchinae and to tentatively assign it to the Asclepiadinae, where its closest relative will have to be sought close to *Gomphocarpus*. Consequently, Holm's (1950) earlier statement that *Pergularia* is so closely related to *Sarcostemma* that merging the two genera is an option is not tenable. To resolve relationships within *Sarcostemma*, our studies indicate that *Cynanchum* is more closely related to *Sarcostemma* and thus a more suitable outgroup to study the relationships within the latter than *Pergularia*. As a corollary it can be noted that the earlier judgement to place *Pentarrhinum* into the Cynanchinae close to *Cynanchum* (Liede & Nicholas 1992) is supported by the ITS data.

Acknowledgements
I am deeply indebted to my friends Dr. M.L. Arnold, Dr. S.A. Hodges, Dr. R.A. Price and Dr. R.A. Wyatt at the University of Georgia in Athens, under whose guidance, financial and moral support I was introduced to the use of molecular methods for systematics. For financial support I acknowledge the Deutsche Forschungsgemeinschaft, grants LI496/1-4. For constant support of my Asclepiad work I wish to thank Prof. emer. F. Weberling, Prof. G.K. Gottsberger (both Ulm), Prof. F. Albers and Dr. U. Meve (both Münster).

References

Baldwin, B.G., 1992. Phylogenetic utility of th Internal Transcribed Spacers of nuclear ribosomal DNA in plants: an example from the Compositae. Mol. Phyl. Evol. 1 (1): 3-16.

Brown, N.E., 1907. The genus *Pergularia*. Bull. Misc. Inform. 1907: 323-325.

Brown, R., 1811. On the Asclepiadaceae, a natural order of plants separated from the Apocinae of Jussieu. Mem. Wern. Nat. Hist. Soc. 1: 12-78.

Bruyns, P.V. & P.I. Forster, 1991. Recircumscription of the Stapelieae (Asclepiadaceae). Taxon 40: 381-391.

Cheung, H.T.A. & T.R. Watson, 1980. Stereochemistry of Hexusulose in Cardenolide glycoside of the Asclepiadaceae. J.C.S., Perkin I: 2163-2168.

Felsenstein, J., 1978. Cases in which parsimony or compatibility methods will be positively misleading. Syst. Zool. 27: 401-410.

Hegnauer, R., 1964. 19. Asclepiadaceae. In: Die Chemotaxonomie der Pflanzen 3: 199-223; 640-641. Stuttgart, Birkhäuser.

Hegnauer, R., 1989. Asclepiadaceae. In: Die Chemotaxonomie der Pflanzen 8: 84-95. Stuttgart, Birkhäuser.

Holm, R.W., 1950. The American species of *Sarcostemma* R.BR. (Asclepiadaceae). Ann. Missouri Bot. Gard. 37: 477-560.

Kunze, H., 1993. Evolution of the translator in Periplocaceae and Asclepiadaceae. Pl. Syst. Evol. 185: 99-122.

Kunze, H., U. Meve & S. Liede, 1994. *Cibirhiza albersiana*, a new species of Asclepiadaceae, and establishment of the tribe Fockeeae. Taxon 43: 367-376.

Kuriachen, P.M. & Y. Dave, 1989. Structure and development of fruit wall ornamentations in *Pergularia daemia* (Forsk.) Chiov. (Asclepiadaceae). Proc. Indian Acad. Sci., Pl. Sci. 99 (1): 15-20.

Liede, S., 1990. *Daemia - Pergularia - Telosma*: A nomenclatural confusion. Asklepios 51: 63-68.

Liede, S., 1992. Growth Forms of Cynanchinae in Madagascar. National Geographic Research and Exploration 8 (1): 108-115.

Liede, S., (in press). Anther differentiation in the Asclepiadaceae: Form and Function. In: W.G. d'Arcy & R.C. Keating (eds.): The Anther: Form, Function and Phylogeny. Cambridge University Press.

Liede, S. & F. Albers, 1994. Tribal disposition of Asclepiadaceae genera. Taxon 43 (2): 201-231.

Liede, S. & H. Kunze, 1993. A descriptive system for corona analysis in the Asclepiadaceae. Pl. Syst. Evol. 185: 275-284.
Liede, S. & U. Meve, 1994. A new species of *Tylophoropsis* (Asclepiadaceae), and notes on the genus. Kew Bull. 49 (4): 749-756.
Liede, S. & A. Nicholas, 1992. A revision of the genus *Pentarrhinum* E. Meyer (Asclepiadaceae). Kew. Bull. 47 (3): 475-490.
Mittal, O.P., C. Tamm & T. Reichstein, 1962. Die Glykoside von *Pergularia extensa* (Jacq.) N.E. Br. Helv. Chim. Acta 45: 907-924.
Rao, O.M. & O.L. Kumari, 1979. Germination loci of pollinia and their taxonomical significance. Geobios 6: 163-165.
Reichstein, T. 1967. Cardenolid- und Pregnanglycoside. Die Naturwissenschaften 54: 53-67.
Srivasta, S., M.P. Khare & A. Khare, 1993. Cardenolide Diglycosides from *Oxystelma esculentum*. Phytochemistry 32 (4): 1019-1921.
Swofford, D.L. 1991. PAUP: Phylogenetic Analysis Using Parsimony, Version 3.1. Champaign, Illinois.
Trivedi, R., A. Khare & M.P. Khare, 1988. A pregnane ester tetraglycoside from *Oxystelma esculentum*. Phytochemistry 27 (7): 2297-2300.

Taxonomic studies of the genus *Sansevieria* (Dracaenaceae)

P.K. MBUGUA & D.M. MOORE
Department of Botany, University of Reading, Reading RG6 2AS, England.

Abstract
The importance and the urgency of the study of *Sansevieria* Thunb. with the focus on Kenya is highlighted. A brief history, uses, literature review; world and African distribution are illustrated. Important vegetative and reproductive morphological characters are listed as they are used in construction of a cladogram indicating interrelationships between the species.

Key words
Dracaenaceae, Rhizomes, *Sansevieria*, Succulents, Xerophytes.

Introduction

Taxonomists are faced with the enormous task of recording the vegetation by identifying, naming and classifying the various plant species as a pre-requisite to all research and major uses to which plants are put by man, eg. foods, medicines, fibres etc. In Eastern Africa there is no complete taxonomic account of the family Dracaenaceae in which the genus *Sansevieria* Thunb. is found. According to collection records at B, BM, EA, HBG, K, MO, NAI and P, there are about 22 *Sansevieria* species in Kenya, of which 13 have their type localities there. This shows that Kenya has a very rich flora of this genus. Some species are very widely spread, eg. *S. robusta* N.E.Br., while others have been recorded from very few localities eg. *S. pinguicula* Bally and *S. dawei* Stapf. each having only one specimen representing the species. The types of *S. caulescens* N.E.Br., *S. dawei* Stapf. and *S. dooneri* N.E.Br., for example, are from Kenya, yet none of these species has more than two specimens at Kew. There is therefore an urgent need to study, work out the distribution of many of these species if we are to save some of them before extinction, which is a definite possibility.

History and literature review

The genus *Sansevieria* was established by Thunberg in 1794 with the description of two species, *S. thyrsiflora* and *S. aethiopica*. The name *Sansevieria* is said to have been derived from the name of Raimond de Sangro, Prince of San Severo (born in Naples, Italy 1710 and died 1771). In 1753 Carl Linnaeus (Species Plantarum) included the genus *Sansevieria* in *Aloe*. Adanson (1763) recognized them as a distinct genus, *Cordyline* (Pfennig 1979). Many of the early descriptions are not well detailed. The most recent revision of the group is that of Brown (1915). The name *Sansevieria* however, was proposed for conservation over Adanson's earlier name, *Cordyline*, in 1973 (Marais 1973). This genus has now been moved to Dracaenaceae Salisb. under the recent classification of the monocotyledons by Dahlgren & Clifford (1982) and adopted at Kew, the classification also followed in this paper.

Sansevieria is a genus of xerophytic plants and is known variously as Bow-string Hemp, Snake Plant, Zebra Lily, Mother-in-laws Tongue, Cow Tongue etc. (Agnew

L.J.G. van der Maesen et al. (eds.), The Biodiversity of African Plants, 489-492.

1974). The genus comprises perennial herbs with erect, stiff leaves and short, thick, stoloniferous rhizomes (Brown 1915).

In 1895, Holtze stated the possibility to cultivate *S. zeylanica* and *S. cylindrica* as fibre plants (Holtze 1895). Gerome and Labroy in 1903 described 20 of the more important fibre species and classified them according to leaf shape and certain other vegetative characteristics. In 1893, Dodge reported on the propagation of *Sansevieria* and its potentiality as a fibre crop in Florida (Dodge 1893).

Methodology

All available literature and a large number of herbarium specimens were studied to determine the distribution of the various species in the world, and Kenya, where the field work was undertaken between 1991 and 1993. Various herbaria were requested to loan *Sansevieria* specimens for study (see page 489), from which the localities were recorded. The field studies and lab work involved the study and listing of macro and microscopic morphological characters of all available species. Some of the major characters were leaf size, shape (eg. flat vs. cylindrical), numbers, inflorescence shapes, number of branches, sizes, flower colours etc. A cladogram was constructed using the PULP 3.1 programme. This will be published in a full taxonomic revision of *Sansevieria* in Kenya.

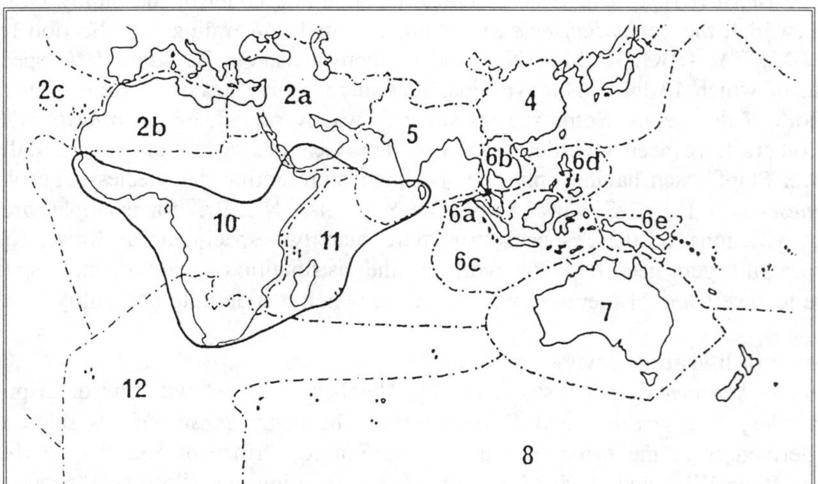

Figure 1. *Sansevieria* Thunb. distribution in the world. Numbers and broken lines indicate the geographical divisions of the world in use at the Kew Herbarium.

Sansevieria distribution

The genus *Sansevieria* Thunb. comprises 50 to 70 species, or even over 80 as stated by Byrom (1950-51). It is chiefly confined to the African continent and the islands near its coast and northwards to Arabia (see Figure 1). Nearly all the species are congregated in the tropical area of the continent with about 3-5 occurring in South Africa and none found in the northern extratropical region. About 2 species are known to inhabit other continents and these are natives of Sri Lanka (Ceylon), India, and Burma (Figure 1). In the West Indies, Central and South America, *Sansevieria* species have been introduced at least some 60 years ago (Biswas 1935).

Uses of *Sansevieria*

Besides being useful ornamental plants (Chahinian 1985), many *Sansevieria* species are considered important for their leaf fibre which is used for fine matting, rope or cordage, hats, bowstrings, articles of clothing, and for medicinal purposes. The parts used medicinally are usually the sap (or juices), and ground or powdered roots and rhizomes (Byrom 1950-51).

Study sites

The field study area was in Kenya. The localities were those from which *Sansevieria* specimens have been collected before and deposited in the herbaria or portrayed the vegetation type amongst which *Sansevieria* thrives. Some of the species found in Kenya include: *S. caulescens* N.E.Br., *S. conspicua* N.E.Br., *S. dooneri* N.E.Br., *S. fischeri* (N.E.Br.) Marais, *S. gracilis* N.E.Br., *S. parva* N.E.Br., *S. pinguicula*, *S. powellii* N.E.Br., *S. raffillii* N.E.Br., *S. robusta* N.E.Br., *S. suffruticosa* N.E.Br., *S. suffruticosa* var. *longituba* Pfennig and *S. volkensii* Gurke.

The habitats of most of the species are mainly bush-land, the sides of anthills amongst bushes, dry river-beds, rocky areas and thick forests eg. along the Rift Valley Escarpment bordering the Kedong Valley.

Floral and vegetative morphology

In the genus *Sansevieria*, the morphology of leaves, eg. size, shape, provide the most important characters in their classification. Other structures of great importance include the stem (where present above the ground) and the inflorescence, e.g. type and size.

S.E.M. investigations have been focused on stomatal shapes and leaf surface structures (Dilcher 1974). The results show that some surfaces are moderately rough with stomata openings which are rectangular with distinct transverse ledge structures e.g. *S. pinguicula* Bally, while others have very rough surfaces, elliptical stomata without any visible internal ledge e.g. *S. sulcata* Boj.

Conclusion

Sansevieria is a tropical genus. Its leaf characters are the most important in its taxonomic study at all levels. Inflorescence structure is important at the sectional level. The delimitation of the genus appears to depend on the succulence, fibrous and non-lustrous character of the leaf. These characters are not found in most *Dracaena* Salisb., the putative sister group of *Sansevieria*.

A classification of the genus based on the phenetic and cladistic analysis will be published in the near future.

Acknowledgements

My participation in the 14th AETFAT Congress was made possible through a sponsorship arranged by the AETFAT Secretariat to which I am grateful.

References

Adanson, M., 1763. Familles des Plantes 2: 54, 543.

Agnew, A.D.Q., 1974. Upland Kenya wild flowers, A flora of ferns and herbaceous flowering plants of upland Kenya. Oxford University Press, London.

Biswas, K., 1935. Notes on the systematic position of *Sansevieria* growing in India with special reference to *S. Laurentii* Willd. J. Bombay Nat. Hist. Soc. 38: 153-157.

Brown, N.E., 1915. *Sansevieria* - a monograph of all known species. Bull. Misc. Inf. (Kew) 1915 (5): 185-261.

Byrom, M.H., 1950-51. Progress with long vegetable fibres crops in peace and war. The yearbook of Agriculture. U.S. Dept. of Agric. Washington DC. U.S. Govt. Printing Office.

Chahinian, B.J., 1985. The *Sansevieria trifasciata* dwarf cultivars: a beginning with no end. Cact. & Succ. J. (U.S.). 57: 199 -203.

Dahlgren, R.M.T. & H.T. Clifford, 1982. The monocotyledons. A comparative study. Academic Press, London.

Dodge, C.R., 1893. Fibre Investigations Report No. 5. U.S.D.A.

Dilcher, D.L., 1974. Approaches to the identification of Angiosperm leaf remains. The Botanical Review 40 (1): 1-157.

Holtze, N., 1895. Experimental cultivation at Port Darwin. Kew Bull. Misc. Inf.: 99-102.

Linnaeus, C., 1753. Species plantarum. Ed. 1. Salvius Stockholm.

Marais, W., 1973. Nomina conservanda proposita. Taxon 22 (1): 158-159.

Pfennig, H., 1979. Grass-like to tree-like: Sansevierias. Cact. Succ. J. (Gr. Britain) 41 (3): 56-60.

Thunberg, C.P., 1794. Prod. Flora Capensis 65.

Generic concepts in *Bidens* and *Coreopsis* (Compositae): an overview

MESFIN TADESSE [1,2], D.J. CRAWFORD [2] & E.B. SMITH [3]
[1] Addis Abeba University, National Herbarium, P.O. Box 3434, Addis Abeba, Ethiopia.
[2] Ohio State University, Department of Plant Biology,
1735 Neil Avenue, Columbus, OH 43210-1293, USA (address for correspondence).
[3] University of Arkansas, Department of Biological Sciences, Fayetteville, Arkansas 72701, USA.

Abstract
Generic concepts in *Bidens* and *Coreopsis* are briefly discussed. A few of the remarks on the historical difficulties in differentiating the two genera are highlighted. Important results from studies concerning these genera especially for species found in Africa, North America and Hawaii are briefly summarized. Previous and current attempts at resolving the generic boundary problem are indicated. Other related genera requiring close study are *Cosmos, Coreocarpus, Oparanthus* and *Petrobium*.

Key words
Bidens, Compositae, *Coreopsis*, Generic concepts, Limits.

Introduction

Bidens L. and *Coreopsis* L., two genera of the Heliantheae (Compositae) with some members known mostly for their weedy habits and ornamental value, have been the subject of detailed systematic, biosystematic and evolutionary studies during the last 25 years. Delimitation of the genera has remained problematical and a variety of views have been proposed as solutions. However, to date no satisfactory differential characters have been found. The search has continued and a large amount of data is available for especially the North American species of *Coreopsis* and the Hawaiian *Bidens*. Currently some molecular data on representative species of African and south American *Bidens* and *Coreopsis* are being assembled at the Ohio State University with financial support from the National Science Foundation. The purpose is to elucidate inter-continental relationships and character evolution in *Bidens* and *Coreopsis*.

The problem of differentiating *Bidens* from *Coreopsis* was probably first recognized by Gray (1862) who, while working on the North American *Bidens*, found that the achenes of some species, while exhibiting all the characters of *Bidens*, included some characters of *Coreopsis*, e.g. antrorse direction of the aristal barbs. Later on other workers also wrote at length about similar findings. A detailed discussion on this is found in Sherff (1937) and Mesfin Tadesse (1984, 1986). In his monographs on *Coreopsis* and *Bidens*, Sherff (1936, 1937) stated that the awn character to separate *Coreopsis* from *Bidens* is worthless and should be abandoned. He then proposed that the two genera be maintained by the following differences of the type materials: *Bidens* - peculiar habit, strongly (i.e. retrorsely) barbed awns and wingless achenes of the Linnaean species; and, *Coreopsis* - peculiar habit and winged achenes of the Linnaean type species. But the wing character is also stated by Sherff (1937: 24) to be unreliable in Hawaiian *Bidens*. The typological concept of genera adopted by Sherff, as indeed by

L.J.G. van der Maesen et al. (eds.), The Biodiversity of African Plants, 493-498.

other later workers also, is difficult to apply and the assignment of taxa to either *Bidens* or *Coreopsis* has thus remained problematic.

Some instances of in-depth studies on *Bidens* and *Coreopsis*

African taxa

Comparative gross morphological, palynological, and anatomical studies were performed on African *Bidens* (incl. *Coreopsis*) by Mesfin Tadesse (1984, 1986, 1993) and Rayner (1992). Both authors, as others before them, concluded that the African taxa belong to one taxon, i.e. *Bidens*. Mesfin (1984: 7), however, took no definite standpoint on the extra-African species of *Coreopsis*. Sixty-three species of *Bidens* are recorded for Africa with sixty of the species being native. The achenes provided the main inter-specific differences between many of the species. Other differences were found in certain features of the capitulum and also the habit of the plants. No major differences in exo- and endo-morphology of pollen grains, anatomy of leaves and petioles and ecology of species of *Bidens* and those formerly kept in *Coreopsis* were found. The following haploid and diploid chromosome numbers are known for African *Bidens* (incl. *Coreopsis*): n = 23; 2n = 32, 34, 36, 42, 48, and 72.

North American Coreopsis

A few selected results of two major contributors to studies on North American *Coreopsis* are given below:

E.B. Smith (1971 - present). Smith (1976) provided the details of a biosystematic study of *Coreopsis* in the eastern United States and Canada. He also gave a key and illustrations to the species and brief descriptions for five sections. Twenty-five out of the 44 species currently recognized for North America (cf. Jansen et al. 1987) were accounted for in the above cited work. The basic chromosome numbers for *Coreopsis* are: x = 12 (5 sections), x = 13 (5 sections), and x = 14 (1 section). Aneuploid series of n = 6, 7, 9 and 10 are known in *Coreopsis* sect. *Coreopsis*.

D.J. Crawford (1969 - present). Starting with systematic studies on Mexican *Coreopsis*, Crawford studied flavonoid chemistry, chromosome numbers, morphology, hybridization, phylogeny, isozymes and cpDNA in North American *Coreopsis* (e.g. Crawford 1969, Crawford & Smith 1983, Crawford & Whitkus 1988, Crawford et al. 1990a, 1990b, 1991, 1992). Besides the phylogenetic studies, Crawford et al. (e.g. 1990b, 1991) also studied inter- and intra-specific as well as inter-sectional character evolution and relationships within sections of *Coreopsis*. Two examples of the latter type of study are given below.

In an attempt to reveal the phylogenetic relationships and character evolution among sections of North American *Coreopsis*, Crawford et al. (1991) found that all species of *Coreopsis* except those in sections *Electra* (DC.) Blake and *Anathysana* Blake, which consist of woody species from Mexico, have two isozymes for plastid phosphoglucose isomerase (PGI). This number was initially thought to represent the ancestral condition but the presence of two plastid PGI isozymes in other diploid Coreopsidinae precludes this assumption and argues for the former to rather represent the derived condition. Moreover, sections *Electra* and *Anathysana* form a monophyletic group based on cpDNA whereas this was not the case for the cladistic analysis of morphological data (cf. Jansen et al. 1987).

Smith (1982) postulated phyletic trends within *Coreopsis* sect. *Coreopsis* by computing relative evolutionary advancement indices. He indicated the annual habit to be a derived condition within sect. *Coreopsis*. In the phylogenetic hypothesis of Jansen et al. (1987), the annual species are supposed to be a monophyletic assemblage. In both phylogenies the likely ancestor of the annual species could not be determined. In the phylogeny derived from cpDNA analysis (Crawford et al. 1991), seven of the nine species of sect. *Coreopsis* (3 perennial and 4 annual species) share five derived restriction site mutations and one length mutation and are grouped together in the resulting cladogram. By contrast, the remaining two perennial species (*Coreopsis auriculata* L. and *C. pubescens* Elliott) showed no mutations. This indicates that, with regards to cpDNA, the three perennial species (*C. grandiflora* Hogg ex Sw., *C. intermedia* Sherff and *C. lanceolata* L.) are closer to the annual species than the other perennial species within sect. *Coreopsis*. Thus the cpDNA data, besides confirming that the perennial habit is primitive in sect. *Coreopsis* also removes *C. auriculata* and *C. pubescens* from being the likely ancestors of the annual species.

Hawaiian Bidens

Sherff (1937) recorded 43 species of *Bidens* with some 20 varieties in Hawaii. Gillett & Lim (1970), Ganders & Nagata (1983, 1990), and Helenrum & Ganders (1985) among others, worked extensively on the Hawaiian *Bidens*. Ganders & Nagata (1983, 1990) recognized 19 species and 8 subspecies as being native to the islands. Helenrum & Ganders (1985) reported that all of the species of *Bidens* are polymorphic but genetically undifferentiated. All Hawaiian *Bidens* have the same chromosome number of $2n = 72$.

Central American and Mexican Bidens and Coreopsis

While many of the studies on Central American and Mexican *Bidens* and *Coreopsis* are based on morphological characters (e.g. McVaugh 1984, Melchert 1976, Melchert & Turner 1990), a few of them also employed hybridization, cytotaxonomy and chemotaxonomy (e.g. Ballard 1986, Hart 1979, Roseman 1986, Smith & Parker 1971). A number of studies also involved the transfer of taxa from *Cosmos* Cav. and *Coreocarpus* Benth. to *Bidens* thus indicating the need to investigate the relationships at the molecular level not only between *Bidens* and *Coreopsis* but also between *Bidens* and the other related genera particularly those found in Mexico.

South American Coreopsis and Bidens, and Bidens from the islands in the South Pacific and South Atlantic Oceans

We have not come across any major work on particularly South American *Bidens* apart from that of Sherff (1937). There are isolated reports on new species and chromosome numbers, e.g. Turner et al. (1967), Sagastegui (1969), Sagastegui & Sanchez Vega (1971). Currently 37 species of *Coreopsis* are known from South America and all of them belong to *Coreopsis* sect. *Pseudoagarista* A. Gray. Many of the species have been differentiated on foliar morphological features. Apart from morphology and few chromosome counts, e.g. Smith et al. (unpublished data), not much is known about these species.

Stuessy (1988) considered *Oparanthus* Sherff and *Petrobium* R. Br. from the Tubuai

Islands of the South Pacific and from St. Helena in the South Atlantic, respectively, as congeneric with *Bidens* but, for reasons outlined below, Karis & Ryding (1994) prefer to keep these apart.

Attempts at resolving the generic boundary problem

Although there is quite a substantial amount of data on *Bidens* and *Coreopsis*, so far no diagnostic markers between the two taxa have been found. The search has, however, continued and it is likely that a suitable solution will be found. The attempts made so far are summarized below.

Previous attempts

1. Treat all materials with intermediate features as products of hybridization between *Bidens* and *Coreopsis*, e.g. Gray 1884: 296, Wiegand 1899: 399 (cf. also Mesfin Tadesse 1986: 190-191). This action was quite common between 1860 and 1890.
2. Keep *Bidens* and *Coreopsis* apart and consider those taxa with similar morphology as other species of *Bidens* but with antrorsely barbed achenial aristae as minor variants of *Bidens*, e.g. Fernald (1903, 1913), Sherff (1937). This was taken as a solution during the years 1890 to 1937.
3. Keep all African *Coreopsis* in *Bidens*. Initially advanced by Wild (1967), this view has been favoured by many workers and to-date almost all species of *Coreopsis* in Africa have been transferred to *Bidens*, e.g. Mesfin Tadesse (1984, 1986, 1993).
4. Treat all Hawaiian species in *Bidens* in spite of the presence of narrow wings on the achenes of some taxa, e.g. Gillett & Lim (1970), Ganders & Nagata (1983, 1990).
5. Keep *Coreopsis* (incl. African species) and *Bidens* as separate genera but subdivide the latter into smaller monophyletic units, e.g. Karis & Ryding (1994).

In a cladistic analysis of the *Coreopsidinae* (as tribe *Coreopsideae*) Ryding & Bremer (1992), subdivided the taxon into three groups. *Coreopsis* (including the African species) was included in group 1 together with mainly American genera. *Bidens* was left unclassified. From their paper, it is not clear which species of *Bidens* and *Coreopsis* were included in the analysis. Moreover, the characters used to define *Coreopsis* are not clearly spelt out. Since the aim of their study was on the entire tribe, Ryding & Bremer (1992) could not examine the two large and most complex genera in any detail. The work of Karis & Ryding (1994) appears to base its conclusion on that of Ryding & Bremer (1992). *Oparanthus* and *Petrobium* were, likewise, kept separate from *Bidens* by Karis & Ryding (1994).

Current attempts

1. Continue the search for diagnostic morphological and anatomical features in *Bidens* and *Coreopsis* (Mesfin & Crawford, in press; Mesfin, Crawford & Smith, in press).
2. Ascertain those characters that could be utilized in a cladistic study of *Coreopsis* and *Bidens* (Mesfin & Crawford, in prep.)

3. Make preliminary studies of restriction site mutations in cpDNA and sequencing of the ITS regions of the nuclear rbDNA to produce phylogenetic hypotheses for comparison with the phylogeny generated from morphological data (Crawford, Mesfin & Smith, in prep.)
4. Continue work on those taxa (species, genera, species complexes, species groups, etc.) with little or no clear-cut demarcation (Crawford, Mesfin & Smith, in prep.)

References

Ballard, R.E., 1986. *Bidens pilosa* complex (Asteraceae) in North and Central America. Amer. J. Bot. 73 (10): 1452-1465.

Crawford, D.J., 1969. Systematic studies on Mexican *Coreopsis* (sect. *Anathysana*), with special reference to the relationships between *C. cyclocarpa* and *C. pinnatisecta*. Bull. Torr. Bot. Club 97: 161-167.

Crawford, D.J., & E.B. Smith, 1983. The distribution of anthoclor floral pigments in North American *Coreopsis* (Compositae): taxonomic and phyletic interpretations. Amer. J. Bot. 70 (3): 355-362.

Crawford, D.J., & R. Whitkus, 1988. Allozyme divergence and the mode of speciation for *Coreopsis gigantea* and *C. maritima* (Compositae). Syst. Bot. 13 (2): 256-264.

Crawford, D.J., J.F. Palmer & M. Kobayashi, 1990a. Chloroplast DNA restriction site variation and the phylogeny of *Coreopsis* sect. *Coreopsis* (Asteraceae). Amer. J. Bot. 77 (4): 552-558.

Crawford, D.J., J.F. Palmer, M. Kobayashi, M.L. Roberts, M. Benkowski & M. Hoffman, 1990b. Phylogenetic implications of differences in number of plastid phosphoglucose isomerase isozymes in North American *Coreopsis* (Asteraceae: Heliantheae, Coreopsidinae). Amer. J. Bot. 77 (1): 54-63.

Crawford, D.J., J.F. Palmer & M. Kobayashi, 1991. Chloroplast DNA restriction site variation, phylogenetic relationships, and character evolution among sections of North American *Coreopsis* (Asteraceae). Syst. Bot. 16 (2): 211-224.

Crawford, D.J., J.F. Palmer & M. Kobayashi, 1992. Chloroplast DNA restriction site variation and the evolution of the annual habit in North American *Coreopsis* (Asteraceae). In: P.S. Soltis, D.E. Soltis & J.J. Doyle (eds.), Molecular Systematics of Plants: 280-294, Chapman and Hall, New York and London.

Fernald, M.L., 1903. A new *Bidens* from the Merrimac valley. Rhodora 5: 90-92.

Fernald, M.L., 1913. Some noteworthy varieties of *Bidens*. Rhodora 15: 74-78.

Ganders, F.R. & K.M. Nagata, 1983. New taxa and new combinations in Hawaiian *Bidens* (Asteraceae). Lyonia 2 (1): 1-16.

Ganders, F.R. & K.M. Nagata, 1990. *Bidens*. In: W.L. Wagner, D.R. Herbst & S.H. Sohmer (eds.), A Manual of the Flowering Plants of the Hawaiian Islands: 267-283.

Gillett, G.W. & E.K.S. Lim, 1970. An experimental study of the genus *Bidens* in the Hawaiian Islands. Univ. Calif. Publ. Bot. 56: 1-63.

Gray, A., 1862. Characters of some Compositae in the collection of the United States South Pacific Exploring Expedition under Captain Wilkens, etc., Proc. Amer. Acad. Arts, 5: 114-146.

Gray, A., 1884. Synoptical Flora of North America, Vol. 1, pt. 2. New York.

Hart, C.R., 1979. The Systematics of the *Bidens ferulaefolia* complex (Compositae). Syst. Bot. 4 (2): 130-147.

Helenrum, K. & F.R. Ganders, 1985. Adaptive radiation and genetic differentiation in Hawaiian *Bidens*. Evol. 39 (4): 753-765.

Jansen, R.K., E.B. Smith & D.J. Crawford, 1987. A cladistic study of North American *Coreopsis* (Asteraceae: Heliantheae). Pl. Syst. Evol. 157: 73-84.

Karis, P.O. & O. Ryding, 1994. Tribe Heliantheae. In: K. Bremer, Asteraceae: Cladistics & Classification: 559-624, Timber Press Inc., Oregon.

McVaugh, R., 1984. *Compositae*. In: W.R. Anderson (ed.), Flora Novo-Galiciana, Vol. 12. The University of Michigan Press, Ann Arbor.

Melchert, T.E., 1976. *Bidens*. In: D.L. Nash & L.O. Williams (eds.), Flora of Guatemala, Vol. 24 (12): 193-214.

Melchert, T.E., & B.L. Turner, 1990. New species, names, and combinations in Mexican *Bidens* (Asteraceae: Coreopsideae). Phytologia 68 (6): 20-31.

Mesfin Tadesse, 1984. The genus *Bidens* in NE tropical Africa. Acta Univ. Ups. Symb. Bot. Upsal. 24 (1).

Mesfin Tadesse, 1986. The morphological basis for inclusion of African species of *Coreopsis* L. in *Bidens* L. (Compositae - Heliantheae). Acta Univ. Ups. Symb. Bot. Upsa. 26 (2): 189-203.

Mesfin Tadesse, 1993. An account of *Bidens* (Compositae: Heliantheae) for Africa. Kew Bull. 48 (3): 437-516.

Rayner, T.M., 1992. A revision of *Bidens* in Africa. PhD. dissertation, University of Reading, England.

Roseman, R.R., 1986. A systematic study of *Bidens* section *Greenmania* in Mexico, Central America, and Jamaica: chemotaxonomy, cytotaxonomy, and phenetics. Ph.D. dissertation, University of Iowa, Iowa City, Iowa.

Ryding, O. & K. Bremer, 1992. Phylogeny, distribution, and classification of the *Coreopsideae* (Asteraceae). Syst. Bot. 17: 649-659.

Sagastegui, A.A., 1969. Cuatro especies nuevas de Compuestas Peruanas. Bol. soc. Arg., Bot. 11 (4): 240-250.

Sagastegui, A.A., & I. Sanchez Vega, 1971. Una nueva especie de *Coreopsis* (Compositae) del Peru. Bol. Soc. Argentina Bot. 13: 337-340.

Sherff, E.E., 1936. Revision of the genus *Coreopsis* L. Field Mus. Nat. Hist. Bot. 11 (6): 279-475.

Sherff, E.E., 1937. The genus *Bidens*, 2 vol. Field Mus. Nat. Hist. Bot. 16: 1-709.

Smith, E.B., 1976. A biosystematic survey of *Coreopsis* in eastern United States and Canada. Sida 6: 123-215.

Smith, E.B., 1982. Phyletic trends in section *Coreopsis* of the genus *Coreopsis* (Compositae). Bot. Gaz. 143 (1): 121-124.

Smith, E.B., & H.M. Parker, 1971. A biosystematic study of *Coreopsis tinctoria* and *C. cardaminefolia* (Compositae). Brittonia 23: 161-170.

Stuessy, T.F., 1988. Generic relationships of *Oparanthus* and *Petrobium*, especially with reference to *Bidens* (Compositae, Heliantheae, Coreopsidinae). Brittonia 40 (2): 195-199.

Turner, B.L., A.M. Powell & J. Cuatrecasas, 1967. Chromosome numbers in Compositae, XI. Peruvian species. Ann. Miss. Bot. Gard. 54: 172-177.

Wiegand, K.M., 1899. Some species of *Bidens* found in the United States and Canada. Torrey Bot. Club 26 (8): 399-422.

Wild, H., 1967. The Compositae of the Flora of Zambesiaca area, I. Kirkia 6 (1): 1-62.

The taxonomy of the genus *Fuirena* (Cyperaceae) in East Africa

A.M. MUASYA
East African Herbarium, National Museums of Kenya, P.O. Box 45166, Nairobi, Kenya.

Abstract
Work on the taxonomy of the genus *Fuirena* Rottb. is in progress. The genus *Fuirena* (Cyperaceae, Scirpeae) comprises about 30 species distributed throughout wet habitats of tropical countries. In East Africa, *Fuirena* is represented by 14 species, five of which are recorded in Uganda, 12 are known from Kenya, and all species grow in Tanzania. The genus *Fuirena* is currently classified into three subgenera, two of which were previously treated as distinct genera, and the three subgenera are delimited on the basis of the morphology and anatomy. Additional data obtained from general morphology and nutlet microstructural features, anatomy, palynology, and cytology, support the current classification.

Key words
Cyperaceae, East Africa, *Fuirena*, Taxonomy.

Introduction

The genus *Fuirena* Rottb. (Cyperaceae, tribe Scirpeae) comprises about 30 species distributed throughout wet habitats of tropical countries (Mabberley 1987). In East Africa, *Fuirena* is represented by 14 species all of which are good fodder plants but seldom occur in large quantities (Napper 1965, Haines & Lye 1983).

Fuirena was described by Rottboell (1773) after the discovery of *F. umbellata* by Rolander in Surinam. Two allied genera were later described by Persoon (*Vanginaria*; 1805) and Turczaninow (*Pentasticha*; 1862), but Oteng-Yeboah (1974) treated these as subgenera of *Fuirena* because of the shared nodose stems, paniculate inflorescences and hairy glumes. Chermezon (1936) classified taxa in the subgenus *Pentasticha* into two sections, *Pseudoscirpus* and *Pseudoisolepis*.

The three subgenera of *Fuirena* are currently delimited on the basis of the morphology and anatomy of vegetative and reproductive structures. Subgenus *Pentasticha* (Turcz.) Oteng-Yeb. is characterised by a trigonous culm, V-shaped leaves in transverse section, and cuticular papillae that overarch the stomata. The two sections of subgenus *Pentasticha* are distinguished by the presence or absence of six needle-like perianth segments, section *Pseudoscirpus* (C.B.Clarke) Chermezon and section *Pseudoisolepis* (C.B.Clarke) Chermezon, respectively. Subgenera *Fuirena* and *Vanginaria* (Pers.) Oteng-Yeb. have six perianth segments. The three inner perianth segments are petal-like structures of various shapes, while the outer three are needle-like. The culm is terete or nearly so, the leaves are thinly crescentiform in transverse section, and cuticular papillae are absent. Subgenus *Fuirena* has leaves with well-developed blades and generally paniculate inflorescences, while subgenus *Vanginaria* has leaf blades absent or greatly reduced to sheaths, and the inflorescences are single terminal clusters of few spikelets, or sometimes solitary.

Fuirena, like the rest of the Cyperaceae, has highly reduced vegetative and

L.J.G. van der Maesen et al. (eds.), The Biodiversity of African Plants, 499-503.

reproductive structures. The genus poses problems of classification, on the basis of morphology alone, as the flowers are very small, the exact morphology of their parts obscure, and the morphology of the inflorescences is difficult to interpret. This has led to different interpretations of the morphology of the taxa and consequently the raising and lowering of the rank of various taxa. To attain a natural system of classification, there is need for taxonomic data from other sources to be incorporated. Sources of taxonomic information included in this study, are macro- and micromorphology, nutlet microstructural features, anatomy, palynology, and cytology. It is of interest whether data from this study support the current classification. The data obtained from this study will be useful in reconstructing the phylogeny and understanding the taxonomy of the genus, and contribute to the Flora of Tropical East Africa (FTEA) account of the Cyperaceae.

Materials and methods

Morphological, anatomical, and palynological studies were based on herbarium specimens located at the East African Herbarium. For anatomical work, herbarium material was revived by placing in cold water and boiling until the original shape and texture was resumed. Material was then fixed in formalin acetic alcohol (FAA) for at least 48 hours. Material for chromosomal studies was collected from various sites in Kenya during and after the April 1992 rainy season.

Various morphological, anatomical, palynological, and cytological characters of the taxa in the genus *Fuirena* were investigated. *Fuirena* material was examined for the morphology of the culms, leaves and inflorescences. The size of macro- and micro-characters were measured using a transparent metric ruler and a calibrated ocular, respectively. Pollen shape and size was also determined. Pollen mother cells were squashed in aceto-orcein for cytological studies. The chromosome number, size and meiotic behaviour were observed.

Results

Observations on the morphology, anatomy, palynology and cytology of the East African *Fuirena* are recorded.

Morphology

Fourteen species of *Fuirena* are recorded in East Africa. Five are found in Uganda, 12 in Kenya, and all 14 grow in Tanzania (Table 1). Two species, *F. abnormalis* and *F. claviseta*, were recently recorded in Kenya. *F. pachyrrhiza* and *F. welwitschii*, previously regarded as varieties of *F. pubescens* by Haines & Lye (1983), have been reinstated as species (Muasya, in press).

The genus has both annual and perennial species. Six of the 14 East African *Fuirena* species are perennials while the other eight are annuals (Table 1). The annuals have a minute root system. The perennial species have either a short tufted rhizome as in *F. stricta*, or a long creeping rhizome with culms spaced at intervals greater than 5 mm. The culm height ranges between 5-140 cm among the East African species (Table 1). Annuals are generally small, in most cases below 50 cm, while perennial species are taller.

Table 1. The species of *Fuirena* in East Africa. A: Distribution (Uganda, Kenya, Tanzania), B: Habit (Annual/Perennial), C: Culm length (cm), D: Chromosome number.

Taxon	A	B	C	D
Genus *Fuirena* Rottb.				
I. Subgenus *Fuirena*				
1. *F. angolensis* (C.B. Clarke) Lye	KT	A	11-39	-
2. *F. ciliaris* (L.) Roxb.				
2a. forma *ciliaris*	KT	A	13-52	-
2b. forma *apetala* (Wingf.) Lye	T	A	up to 24	-
3. *F. claviseta* A. Peter	KT	A	31-47	-
4. *F. leptostachya* Oliv.				
4a. forma *leptostachya*	UKT	A	7-38	-
4b. forma *nudiflora* Lye	UKT	A	21-28	-
5. *F. microcarpa* Lye	T	A	6-27	-
6. *F. ochreata* Kunth	KT	P	24-68	-
7. *F. sagittata* Lye	T	A	6-19	-
8. *F. umbellata* Rottb.	UKT	P	24-116	-
9. *F. zambesica* Lye	KT	A	14-33	-
II. Subgenus *Pentasticha* (Turcz.) Oteng-Yeb.				
A. Sect. *Pseudoscirpus* (C.B. Clarke) Cherm.				
10. *F. stricta* Steudel				
10a. subspecies *stricta*	UKT	P	30-60	-
10b. subspecies *chlorocarpa* (Ridl.) Lye	UKT	P	22-104	19
B. Sect. *Pseudoisolepis* (C.B. Clarke) Cherm.				
11. *F. abnormalis* C.B. Clarke.	KT	A	11-65	12
12. *F. pachyrrhiza* Ridl.	UKT	P	70-139	24
13. *F. pubescens* (Poir.) Kunth				
13a. variety *pubescens*	UKT	P	35-60	21
13b. variety *abbreviata* Lye	U	P	-	-
14. *F. welwitschii* Ridl.	KT	P	13-76	40

Among the Cyperaceae, the genus *Fuirena* differs from related genera in its nodose stem. The internode length ranges between 2.1-33.5 cm. Leaves are attached along the length of the culm. The lower leaves have very short leaf blades or none, while the upper leaves have blades up to 30 cm long. The leaves are hairy, at least on the margins and the apex. *Fuirena* has a typical Cyperaceae leaf with a closed, tubular sheath that terminates in a throat and a flat blade. Among the perennial taxa in the genus, *F. umbellata* has a peculiar phyllotaxy with a 2/5 interval on the spiral. Other species have regular 3-ranked leaves.

The genus has paniculate inflorescences. The spikelets have numerous imbricate glumes which are spirally arranged on a knotted rachis. Sometimes the glumes are arranged in 5 ranks, giving the spikelet an angular morphology, a character used to separate varieties in *F. stricta*. Each floret is enclosed in a glume, which is mostly hairy. A glabrous glume character is used to separate *F. abnormalis* from the other annual species. The midrib of the glumes is excurrent and forms a mucro. The length of the mucro varies among related taxa, and is used to delimit two perennial species, *F. pubescens* and *F. pachyrrhiza*.

Perianth-segment morphology is important in the taxonomy of the genus. The absence or presence, and shape of the perianth-segments has been extensively used in classification of the genus at the subgenus, species subspecies and form ranks (see introduction above).

The spikelets have numerous flowers which mature acropetally. There are three stamens, and the style is deeply 3-cleft at the apex, with the base often persistent on the nutlet as a cylindrical knob. The ovary is trigonous, ripening into a single-seeded nutlet with a sculptured surface of various colours. The colour and sculpturing of the nutlet surface is of taxonomic importance in the genus. Three allied species, previously treated as varieties of *Fuirena pubescens*, can be separated on the basis of the nutlet characters as *F. pubescens* has white nutlets with a smooth surface while *F. pachyrrhiza* and *F. welwitschii* have coloured nutlets with tuberculate surface.

Nutlet surface micromorphology is an additional taxonomic character. Taxa in the subgenus *Pentasticha* can be separated to the respective sections based on the shape of the anticlinal wall of the nutlet surface cells. The shape is sinuous for section *Pseudoscirpus* and straight for *Pseudoisolepis*.

Anatomy

Detailed culm and leaf anatomy of *Fuirena stricta* subsp. *chlorocarpa*, *F. umbellata*, *F. pubescens*, *F. pachyrrhiza*, *F. welwitschii*, *F. claviseta*, and *F. ochreata* has been studied. Leaf anatomy shows no marked structural variation among the species of *Fuirena*. A dorsiventral leaf-type is present, and the V-shaped outline can be used to separate subgenus *Pentasticha* from the other two, whose leaves are flattened to slightly crescentiform. *Fuirena leptostachya* differs diagnostically from other species in the genus in that the median vascular bundles of the leaf are supported by two lateral adaxial sclerenchyma strands instead of a single median strand (Metcalfe 1971).

Observations of the cross-sectional culm shape can be used in the taxonomy of the genus. Among the perennial taxa, the culm is triangular in subgenus *Pentasticha* (*F. stricta*, *F. pachyrrhiza*, *F. welwitschii* and *F. pubescens*), while in subgenus *Fuirena* the culm is pentangular (*F. umbellata*) or spherical (*F. ochreata*). All annual species have a spherical culm cross-section.

Fuirena shows interspecific variation in the structure of the central ground tissue. The tissue is "net type" in *F. umbellata* and *F. ochreata*, but breaks down to form irregular air-cavities, or may even become hollow in other species. Peripheral air-cavities are well-developed in *F. ciliaris* and *F. leptostachya* (Metcalfe 1971).

Palynology

Pollen characters of taxonomic value in Cyperaceae include apertural morphoforms, size, exine ornamentation and stratification. Two patterns of pollen, the *Cyperus* (or *Mapania* type sensu Haines & Lye 1983) and *Carex* type, are seen in Cyperaceae (Padhye & Makde 1980). The pollen of *Fuirena* species examined is of the *Cyperus* type, and has granular exine ornamentation. Interspecific pollen size variation is of taxonomic value.

Cytology

Extensive aneuploidy and polyploidy has been reported in the Cyperaceae and is thought to play a major role in speciation (Rath & Patnaik 1974, Sanyal 1972, Sanyal & Sharma 1972, Tanaka 1948). In *Fuirena*, cytological studies on East African material revealed normal meiosis, small dot-chromosomes, and meiotic numbers n=12, 19, 21, 24, and 40 (Table 1). Previous work on non-East African *Fuirena* material have reported chromosome numbers n=18, 19 and 26 in India (Nijalingappa 1975, Rath & Patnaik 1974), and 2n=48 in Nigeria (Baquar 1978).

Conclusion

The taxonomy of *Fuirena* in East Africa is stable. The current treatment of species, on the basis of morphology, is supported by the current study. Nutlet microstructure, anatomy of culm and leaves, and chromosome numbers provide additional taxonomic characters. Perianth-segment shape and anatomy of leaf and culm are used to delimit the subgenera, sections are separated by use of nutlet microstructure, and species, subspecies, varieties and forms are distinguishable by their gross morphology in addition to their chromosome number and anatomy.

Acknowledgements

Part of this work was submitted for the award of a Masters degree, funded by Moi University. Work done while at EA Herbarium was supported by a grant obtained from the Regional Global Environment Facility (GEF) office. E.B. Knox and S.R. Baquar offered valuable advice.

References

Baquar, S.R., 1978. Cytological studies of some Southern Nigerian Cyperaceae. La Kromosomo II-9: 263-270.

Chermezon, H., 1936. In: Humbert (ed.), Flore de Madagascar, 29e Famille: 158.

Haines, R.W. & K.A. Lye, 1983. The sedges and rushes of East Africa. East African Natural History Society Nairobi.

Mabberley, D.J., 1987. The plant hand-book, A portable dictionary of the higher plants. Cambridge University Press, Cambridge.

Metcalfe, C., 1971. Anatomy of monocotyledons V: Cyperaceae. Oxford Univ. Press, London.

Napper, D.M., 1965. Cyperaceae of East Africa III. Journal of the East African Natural History Society and National Museum.

Nijalingappa, B.H.M., 1975. In: IOPB chromosome number report XLVIII. Taxon 24: 370.

Oteng-Yeboah, A.A., 1974. Taxonomic studies in Cyperaceae-Cyperoideae. Notes Roy. Bot Gard. Edinb. 33: 311-316.

Padhye, M.D. & K.K. Makde, 1980. Pollen morphology of Cyperaceae. J. Palynol. 16: 71-78.

Persoon, C.H., 1805. Synops. Pl. 1: 70.

Rath, S.P. & S.N. Patnaik, 1974. A note on the cytotaxonomy of East Indian species of the genus *Fuirena* Rottb. Bot. Mag. Tokyo 87: 333-336.

Rottboell, C.F., 1773. Desc. Ic. rar. pl.: 70.

Sanyal, B., 1972. Cytological studies on Indian Cyperaceae II. Tribe Cypereae. Cytologia 37: 33-42.

Sanyal, B. & A. Sharma, 1972. Cytological studies in Indian Cyperaceae I. Tribe Scirpeae. Cytologia 37: 13-32.

Tanaka, N., 1948. The problems of aneuploidy (Chromosome studies in Cyperaceae with special reference to the problem of aneuploidy). Biological Contribution in Japan 4: 1-327.

Turczoninow, N., 1862. Bull. Soc. Nat. Moscow 2: 330.

The plant family Apocynaceae in East Africa

E.A. OMINO
Jomo Kenyatta University of Agriculture and Technology, Department of Biological Sciences, P.O. Box 62000, Nairobi, Kenya.

Abstract
A brief summary of *Apocynaceae* species in East Africa is given. The number of genera and species is provided including those in cultivation and of medicinal use. Taxonomy, anatomy, cytology and palynology will be treated in larger revisions.

Key words
Anatomy, Apocynaceae, Cytology, Palynology.

Introduction

The plant family *Apocynaceae* is commonly known as the Dogbane or Oleander family. For a very long time, it has been of considerable interest to botanists, biochemists, phytochemists and many others (Van Beek et al. 1984). According to Connor (1977), the family is renowned for its poisonous plants as well as for its yield of interesting and valuable constituents including alkaloids in *Rauvolfia* and *Catharanthus* and arrow poisons of *Acokanthera*. The family is chiefly found in the tropics and 1/3 of all species occur in Africa.

The family is usually divided into two subfamilies, the *Plumerioideae* with the type genus *Plumeria* from tropical America and the *Apocynoideae* with the type genus *Apocynum* from North America (Leeuwenberg 1983). The subfamilies are further subdivided into a number of tribes and subtribes. Comparative studies have shown that the subdivision of *Plumerioideae* is much more natural than that of *Apocynoideae* (Leewenberg 1988). Although *Apocynoideae* is a coherent grouping, it has so far proved difficult to devise a satisfactory and practical subdivision for it.

Number of species in East Africa

According to the present survey, there are about 75 species in 29 genera in East Africa, of which 10 genera and 30 species are lianas. The cultivated species constitute 22 species in 15 genera and one of these species, *Catharanthus roseus* (L.) G. Don has become naturalised. The most commonly cultivated species are the ornamentals *Nerium oleander* L., *Plumeria rubra* L. and *Thevetia peruviana* (Pers.) Mers.

Medicinal species

For a detailed survey of ethnobotany of *Apocynaceae* species in Kenya see Omino & Kokwaro (1993), whereby 50% of *Apocynaceae* species is reported to be medicinal. The survey of medicinal use of species of *Apocynaceae* in Tanzania and Uganda is in progress.

Palynology

According to the current study and literature from Nilsson (1986) and Erdtman (1952),

L.J.G. van der Maesen et al. (eds.), The Biodiversity of African Plants, 504-506.

Apocynaceae pollen grains are usually solitary (monads) although tetrads have been reported in some genera, that are not represented in East Africa. The outer exine surface is usually smooth and perforate, while the inner exine surface is variously patterned. The pollen morphological diversity is most pronounced in the family *Plumerioideae* which usually has 3- rarely 4-colporate grains. The *Apocynoideae* have 2-4 porate grains.

Anatomy

From the present survey and literature from Metcalfe & Chalk (1950), the most outstanding anatomical characters in the East African *Apocynaceae* include:

1. The occurrence of laticiferous canal with various contents.
2. Dorsiventral leaves.
3. Typically bicolateral vascular bundle of the midrib and petiole.
4. Presence of crystals of various types.
5. Stomata which are either paracytic or anomocytic although other types may also occur.
6. The presence of intraxylary phloem in the young stem.

Cytology

For a detailed study of the cytotaxonomy of *Apocynaceae* in general see Van der Laan & Arends (1985), whereby 90% of the African *Apocynaceae* species are reported to have a chromosome number of x = 11. The present literature indicates that most *Apocynaceae* species have a chromosome number of 2n = 22 as in *Funtumia africana* (Benth.) Stapf and *F. elastica* (Preuss) Stapf (Arends & Van der Laan 1981); *Holarrhena pubescens* (Buch.-Ham.) Wall ex G. Don (De Kruif 1981); *Carissa edulis* (Forssk.) Vahl and *Tabernaemontana ventricosa* Hochst. ex A.DC. (Gadella 1978). *Strophanthus hispidus* DC. and *S. sarmentosus* DC. yielded 2n = 18 (Arends & van der Laan 1982) and *Catharanthus roseus* yielded 2n = 16 (Plaizier 1981).

Acknowledgements

The present study which is still in progress is currently being supported by the Department of Plant Taxonomy, Agricultural University, Wageningen, The Netherlands in conjunction with the Anatomy section of the Jodrell laboratory of The Royal Botanic Gardens, Kew, England, as part of the "Series of Revisions of Apocynaceae" edited by Dr. A.J.M. Leeuwenberg. The author would like to thank Prof. Dr. L.J.G. van der Maesen and Dr. A.J.M. Leeuwenberg of the Department of Plant Taxonomy, in Wageningen and Dr. David Cutler of the Jodrell Laboratory in Kew.

References

Arends, J.C. & F.M. Van der Laan, 1981. Somatic chromosome numbers in *Funtumia*. In: H.J.C. Zwetsloot, A revision of *Farquharia* Stapf and *Funtumia* Stapf (Apocynaceae). Meded. Landb. Wageningen 81, 16: 32-33.

Arends, J.C. & F.M. Van der Laan, 1982. Chromosome numbers. In: H.J.A. Beentje, Monograph of *Strophanthus* DC. (Apocynaceae). Meded. Landb. Wageningen 82, 4.

Connor, H.E., 1977. The poisonous plants of New Zealand. E.C. Keating, Government printer, Wellington, New Zealand.

De Kruif, A.P.M., 1981. A revision of *Holarrhena* R.Br. (Apocynaceae). Meded. Landb. Wageningen 81, 2: 1-40.

Erdtman, G., 1952. Pollen Morphology and Plant Taxonomy of Angiosperms. Stockholm, Almquist and Wiksell.

Gadella, T.W.J., 1978. In: A. Love (ed.), IOPB chromosome number reports IXI. Taxon 27, 4: 375-392.

Leeuwenberg, A.J.M., 1983. Some remarks on the taxonomy of the Plumerioideae (Apocynaceae). Bothalia 14 (3 & 4): 799-801.

Leeuwenberg, A.J.M., 1988. The Nerieae (Apocynaceae-Apocynoideae). Monogr. Syst. Bot. Missouri Bot. Gard. 25: 157-160.

Metcalfe, C.R. & L. Chalk, 1950. Anatomy of the Dicotyledons. Clarendon Press, Oxford.

Nilsson, S., 1986. The significance of pollen morphology in the Apocynaceae. In: S. Blackmore & I.K. Ferguson (eds.), Pollen and spores: Form and function: 359-374. Academic Press, London, etc.

Omino, E.A. & J.O. Kokwaro, 1993. Ethnobotany of Apocynaceae species in Kenya. Journ. of Ethnopharm. 40: 167-180.

Plaizier, A.C., 1981. A revision of *Catharanthus roseus* (L.) G. Don. (Apocynaceae). Meded. Landb. Wageningen 80, 12: 1-40.

Van Beek, T.A., A.J.M. Leeuwenberg, N.G. Bisset, R. Verpoorte & A. Baerheim-Svendsen, 1984. *Tabernaemontana* L. (Apocynaceae): A review of its taxonomy, phytochemistry, ethnobotany and pharmacology. Journ. Ethnopharm. 10: 1-156.

Van der Laan, F.M. & J.C. Arends, 1985. Cytotaxonomy of the Apocynaceae. Genetica 68: 3-35.

Provisional generic delimitation in the *Haumaniastrum/Acrocephalus* complex (Labiatae)

A. PATON
Herbarium, Royal Botanic Gardens, Kew, Richmond, Surrey, TW9 3AE, United Kingdom.

Abstract
Generic delimitation within the "*Acrocephalus* group" of genera is discussed using a preliminary cladistic analysis. The African *Haumaniastrum* and an enlarged Asiatic, Madagascan and African *Acrocephalus* are recognized at this preliminary stage. *Geniosporum* as currently delimited is polyphyletic and it is suggested that the genus should not be recognized, its species being placed either in *Acrocephalus* or *Limniboza*. It is also proposed to merge *Ceratanthus* into *Platostoma*, which is the earlier name. It should be noted that these are preliminary conclusions and that no formal changes are made here.

Résumé
Délimitation générique dans le "*Acrocephalus* group" de genres est discuté en utilisant une analyse cladistique et préliminaire. Le *Haumaniastrum* Africain et un *Acrocephalus* agrandi de l'Asie, Madagascar et l'Afrique sont reconnus. *Geniosporum* comme couramment délimité est polyphyletique et il est suggéré que le genre ne doit pas être reconnu, les espèces étant placées dans *Acrocephalus* ou *Limniboza*. C'est aussi proposé à placer *Ceratanthus* dans *Platostoma* qui est le nom matinal. On doit noter que celles sont des conclusions préliminaires et qu'on ne fait pas de changements formels ici.

Key words
Acrocephalus, Generic delimitation, *Geniosporum, Haumaniastrum*.

Introduction

Acrocephalus was described from Indian material by Bentham (1829) and was soon enlarged to include African and Asian species. Duvigneaud & Plancke (1959) separated the African species into the new genus *Haumaniastrum*, suggesting that as *Acrocephalus*, in the broad sense, was a heterogeneous mixture of forms with similar, dense, head-like inflorescences, the African and Asian species should be separated on account of their different calyx morphology and vegetative structure. However, Robyns (1966) argued that as there were differences in calyx morphology within *Haumaniastrum* itself, there was no reason to give the calyx difference between the African and Asian species more weight than the differences within the African species alone. He also noted that there was no clear discontinuity in vegetative structure and he concluded that *Haumaniastrum*, as circumscribed by Duvigneaud & Plancke, could not be upheld. Robyns suggested the name *Acrocephalus* should be used for both the African and Asian species until a thorough revision of *Acrocephalus* has been completed.

Ryding (1992) divided tribe *Ocimeae* subtribe *Ociminae* into three informal groups using nutlet anatomy. *Acrocephalus* and *Haumaniastrum* belong in the "*Acrocephalus* group". The questions of whether the inflorescence type has arisen independently or not and the relative importance of the difference in calyx type can only be resolved when viewed in a phylogenetic context. Therefore, a preliminary cladistic analysis of

L.J.G. van der Maesen et al. (eds.), The Biodiversity of African Plants, 507-512.

the "*Acrocephalus* group" of genera was carried out using PAUP (Swofford 1991) and MacClade (Maddison & Maddison 1992). The African species of *Acrocephalus/Haumaniastrum* have been revised and a morphologically representative sample were included (Paton in prep.). The other genera in the complex are listed below. The numbers following the generic names represent the number of species used in the analysis, followed by the approximate total number of species in that genus: *Mesona* Bl.(2/3), *Nosema* Prain (3/6), *Ceratanthus* G. Tayl. (5/7), *Platostoma* P. Beauv. (2/3), *Octomeron* Robyns (1/1), *Geniosporum* Benth. (6/12), *Limniboza* R.E.Fries (2/3), *Benguellia* G. Tayl. (1/1), Asiatic *Acrocephalus* (2/3). The herbarium collections of these genera at BM, BR, BRLU, E, K and P were examined to ensure the sample of species used was representative of their morphological variation and geographical distribution. Representative samples of the "*Ocimum* group" and the "*Orthosiphon* group" (Ryding 1992) were used as outgroups for the analysis.

Cladistic study

Details of the 22 characters that were used in the analysis, and the data matrix used to run the analysis, are presented in appendix 1. PAUP found 1300 equally parsimonious trees of length 112, consistency index = 0.7, retention index = 0.84. A strict consensus tree of these is presented in Figure 1. It must be stressed that this analysis is preliminary. In particular, the results of a palynological survey of the group carried out by Madeline Harley at Kew have not yet been included in the analysis.

Preliminary conclusions

The consensus tree largely supports the delimitation of the "*Acrocephalus* group" of Ryding (1992) (Figure 1, Clade A), the exception being *Benguellia*, which is not included in Clade A. All genera in the "*Acrocephalus* group" except *Benguellia* have striate or regularly reticulate nutlets and posterior staminal filaments which are basally swollen or appendiculate and pubescent. *Ocimum* also has appendiculate stamens, but of a different form to the appendages seen in this group. *Benguellia* has similar nutlets to the rest of the "*Acrocephalus* group" but the remaining characters ally the genus to the "*Orthosiphon* group". Data from the analysis of pollen may clarify the position of *Benguellia*.

The preliminary analysis supports the separation of Asiatic *Acrocephalus* and African *Haumaniastrum*. *Acrocephalus* is closely related to some Asiatic members of *Geniosporum* (Clade B), whereas *Haumaniastrum* is more closely related to the African genus *Limniboza, Geniosporum rotundifolium* Briq. from Africa and *G. thymiflorum* Benth. from Madagascar (Clade C). The condensed head-like inflorescence has probably arisen independently in these two groups. Vegetative structure is a very homoplasious character with similar habits and structures evolving independently.

Clade D comprising *Platostoma, Ceratanthus, Mesona, Nosema, Geniosporum rotundifolium, G. thymiflorum* and *Haumaniastrum* is characterized by the median lobes of the calyx lower lip being fused together and pressed up towards the upper lip when the nutlets are immature, so closing the calyx throat. *Haumaniastrum* is a monophyletic group supported by the synapomorphies of an upwardly curved calyx which is flattened in cross-section and which can easily be split in two at maturity, the lateral lobes of the calyx are fused to, and in the same plane as, the upper lobe and the

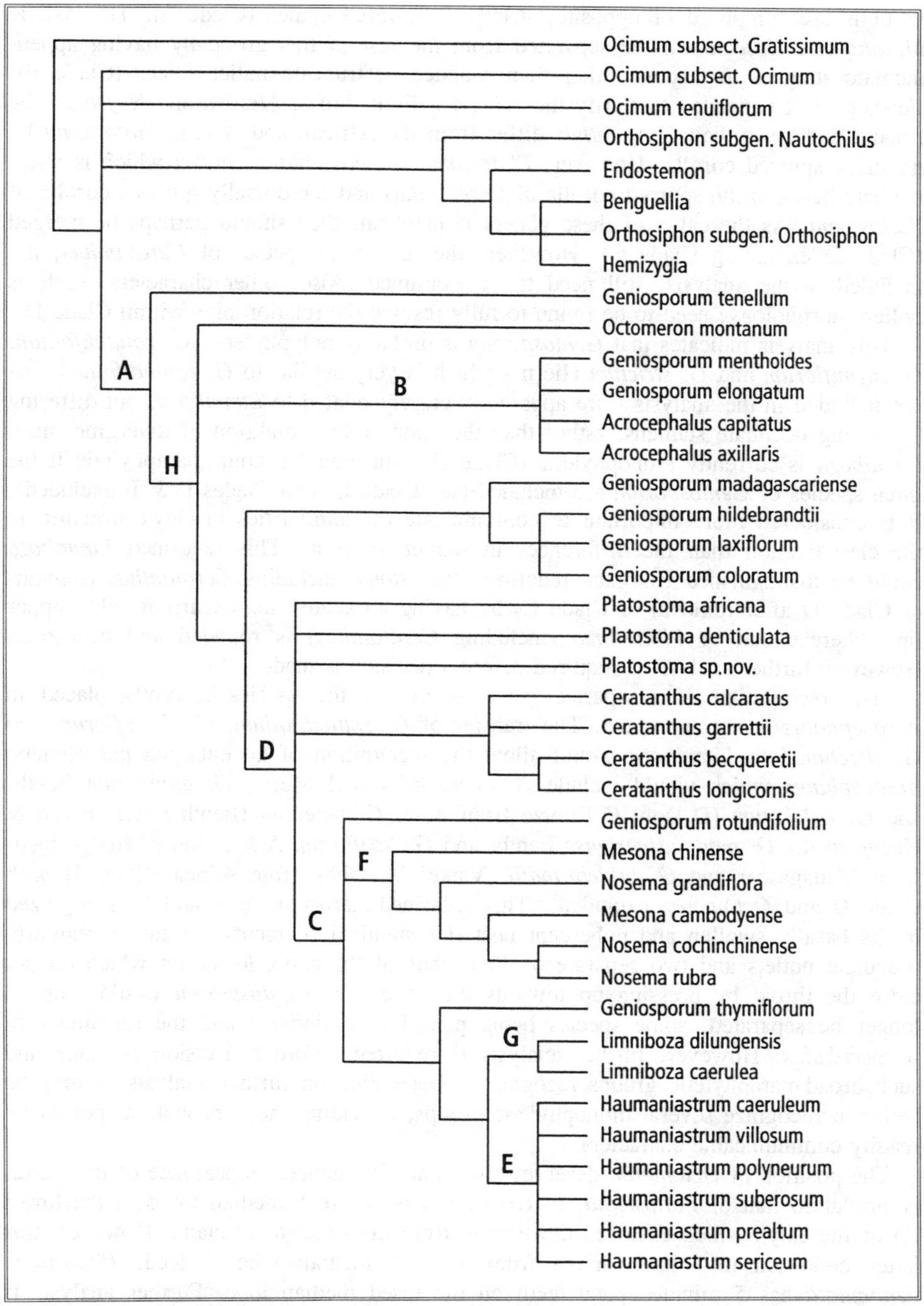

Figure 1. Strict consensus tree generated from 1300 equally parsimonious trees generated by PAUP.

verticils are composed of opposite, strictly 3-flowered cymes (Clade E). The Asiatic *Mesona* and *Nosema* can be separated from the rest of this group by having appendiculate stamens and acute rather than rounded or truncate nutlet apices (Clade F). Further work is needed to clarify the generic delimitation of *Mesona* and *Nosema*. The Asiatic and Australian *Ceratanthus* differ from the African and Asiatic *Platostoma* by having a spurred corolla. However, *Platostoma* sp. nov. has a corolla which is intermediate between the spurred corolla of *Ceratanthus* and the dorsally gibbous corolla of *Platostoma*. As the calyx of these genera is identical, they should perhaps be merged (Clade D excluding Clade C). However, the remaining species of *Ceratanthus*, not included in the analysis, still need to be examined. Also, other characters, such as pollen morphology, need to be found to fully resolve the relationships within Clade D.

This analysis indicates that *Geniosporum* is probably polyphyletic. *G. rotundifolium, G. thymiflorum* and *G. strictum* Hiern - which is very similar to *G. rotundifolium*, but not included in the analysis - are apparently closely related to *Limniboza*, but differing by having declinate stamens, rather than the apomorphic condition of diverging ones. *Limniboza* is currently monophyletic (Clade G), but would become paraphyletic if the three species of *Geniosporum* are included (see Clade C with Clades E & F excluded). It is considered more important to communicate the similarities in calyx structure in the classification than the differences in stamen position. This redefined *Limniboza* could be distinguished from the redefined *Platostoma* including *Ceratanthus* (remains of Clade D after removal of Clade C) by having an acute, non-decurrent calyx upper lip, whereas that of *Platostoma* (including *Ceratanthus*) is rounded and decurrent. However, further analysis is required before a decision is made.

The recognition of a separate genus solely for the species currently placed in *Acrocephalus* is not supported. The transfer of *G. rotundifolium, G. thymiflorum* and *G. strictum* into *Limniboza* would allow the recognition of an enlarged paraphyletic *Acrocephalus* which would include *G. menthoides* (L.) Merr., *G. elongatum* Benth. and *G. coloratum* (D.Don) O.Kuntze from Asia, *G. tenellum* (Benth.) A.J. Paton & Hedge ined., *G. madagascariense* Benth. and *G. laxiflorum* A.J. Paton & Hedge ined. from Madagascar and *G. hildebrandtii* (Vatke) M.Ashby from Africa (Clade H with Clade D and *Octomeron* excluded). This redefined *Acrocephalus* could be recognized by its basally swollen and pubescent posterior staminal filaments, striate or regularly reticulate nutlets and two separate median lobes of the calyx lower lip which do not close the throat by pressing up towards the upper lip. *Geniosporum* would thus no longer be separated, some species being placed in *Limniboza* and the remainder in *Acrocephalus*. However, further analysis is required before a decision is made and such broad paraphyletic groups recognized. Depending on further analysis, it may be better to recognize several monophyletic groups, providing they are well supported by readily communicable characters.

The position of *Octomeron* deserves comment. The general appearance of this genus is similar to that of *Platostoma*. In *Octomeron* there are 5 median lobes to the lower lip of the calyx which excludes *Octomeron* from the *Platostoma* clade. However, this latter condition may have arisen from a fused median lobe. Indeed, *Platostoma denticulata* has 5 minute apical teeth on the fused median lobe. Further analysis is needed to determine the position of *Octomeron*.

Acknowledgements
I am grateful to the directors and staff of the following herbaria for the provision of research material: B, BM, BR, BRLU, COI, E, K, LISC, P and W, and to Brian Schrire for commenting on the manuscript.

References

Bentham, G., 1829. Labiatae. Edward's Bot. Reg. 15: 1282, 1300.

Duvigneaud, P. & J. Plancke, 1959. Les "*Acrocephalus*" arborescents des plateaux Katangais. Biol. Jaarb. 27: 214-257.

Maddison, W.P. & D.R. Maddison, 1992. MacClade version 3.1. Sinauer Assoc. Inc., Sunderland, Massachusetts.

Paton, A., in prep. Revision of the genus *Haumaniastrum*.

Robyns, W., 1966. On the status of *Acrocephalus* Benth. Bot. Not. 119: 185-195.

Ryding, O., 1992. Pericarp structure and phylogeny within the Lamiaceae subfamily Nepetoideae, tribe Ocimeae. Nord. J. Bot. 12: 273-298.

Swofford, D.L., 1991. Phylogenetic analysis using parsimony (PAUP), version 3.0r. Illinois Natural History Survey, Champaign.

Appendix 1

The characters used in the cladistic analysis are enumerated below. The data matrix lists the characters observed in each species. If a species is polymorphic for any character, the states observed for that character are linked by an ampersand (&).

1. *Inflorescence*: verticils laxly disposed (0); verticils condensed into head (1). **2.** *Verticils*: always composed of opposite 3-flowered cymes (0); composed of opposite >3-flowered cymes, showing variation in number of flowers (1); composed of two opposite flowers (2); composed of opposite or ternate irregularly 1-3 flowered cymes (3) **3.** *Calyx tube cross-section*: ± circular, not easily split into upper & lower halves (0); dorso-ventrally flattened with lines of weakness between upper and lower halves (1). **4.** *Fruiting calyx shape*: straight (0); curved upward (1); curved downward (2). **5.** *Fruiting calyx with dorsal swellings*: absent (0); present (1). **6.** *Hairy annulus in calyx throat*: absent (0); present (1). **7.** *Upper calyx lip shape*: apex rounded (0); acute (1); emarginate (2); indistinct from lateral lobes (3). **8.** *Upper lip decurrent*: decurrent (0); not decurrent (1). **9.** *Position of lateral lobe of calyx lower lip*: midway between upper & lower lobes (0); nearer upper lip but still separate (1); entirely fused to upper, but not in same plane as upper (2); entirely fused and in same plane as upper (3); in same plane as upper but distinct at apex only (4). **10.** *Median lobes of calyx lower lip*: 2-lobed (0); 5-lobed (1); fused, apex entire (2); fused, apex denticulate (3); fused, apex emarginate (4). **11.** *Calyx throat*: open (0); closed by median lobes of lower lip pressing up against upper lip (1); lateral & median lobes pressed against upper (2). **12.** *Nutlet sculpturing*: minutely irregularly reticulate (0); with minute irregular raised areas (1); minutely and regularly reticulate (2); minutely striate (3). **13.** *Nutlet apex shape*: rounded (0); truncate (1); acute (2). **14.** *Nutlet apex indumentum*: glabrous (0); pubescent (1); tuberculate (2). **15.** *Corolla tube dorsal surface*: straight (0); gibbous (1); spurred (2). **16.** *Corolla tube curvature*: straight (0); curved downwards (1). **17.** *Corolla upper lip*: 4-lobed (0); 3-lobed (1). **18.** *Posterior staminal filament form*: Tapering towards apex (0); "*Ocimum*" type appendage (1); basally swollen (2); "*Mesona*"-type appendage (3); very small bump, like very small "*Mesona*"-type (4). **19.** *Stamen positioning*: declinate (0); spreading (1). **20.** *Style apex*: bifid, lobes diverging (0); clavate, lobes ± parallel (1). **21.** *Basal style appendage*: absent (0); present on one side of style (1); forming a basal disc (2). **22.** *Disk lobing*: lobes equal (0); anterior lobe larger than others (1).

Data matrix

Mesona chinensis Benth.	010 000 111 2&313 201 113 0001
Nosema cochinchinense (Lour.) Merr.	110 200 0&202 213 201 113 0001
Mesona cambodyensis Doan	010 200 202 213 201 113 0001
N. rubra (Doan) C.Y.Wu	110 200 013 213 201 114 0001
N. grandiflora (Doan) Mukerjee	010 000 202 213 201 113 0001
Platostoma africana P. Beauv.	010 000 001 212 001 102 0001
P. denticulata Robyns	010 000 001 312 001 102 0001
P. sp. nov.	010 000 001 212 001 102 0001
Octomeron montanum Robyns	010 000 000 102 001 102 0001
Geniosporum rotundifolium Briq.	0&110 000 111 2&412 00&11 102 0001
G. hildebrandtii (Vatke) Ashby	010 110 0&201 00&23 111 102 0001
G. thymiflorum Benth.	010 000 111 412 001 002 0001
G. madagascariense Benth.	010 000 011 002 011 102 0001
G. laxiflorum sp. nov. ined.	010 100 011 003 111 102 0001
G. tenellum (Benth.) ined.	000 000 000 0&102 001 102 0001
G. coloratum (D.Don) O.Kuntze	010 100 001 003 011 102 0001
G. menthoides (L.) Druce	010 001 00&10 0&103 001 102 0001
G. elongatum Benth.	0&110 000 010 00&23 001 102&3 0011
Limniboza dilungensis Lisowski & Mielcarek	110 000 111 40&13 001 012 1001
L. coerulea R.E. Fries	110 000 111 413 001 002 1001
Acrocephalus indicus (Burm. f.) O.Kuntze	110 000 010 00&23 001 102 0001
A. axillaris Benth.	010 000 010 00&23 001 102 0001
Ceratanthus calcaratus (Hemsl.) G.Tayl.	010 000 001 212 002 102 0001
C. garrettii (Craib) G.Tayl.	010 000 001 2&312 002 102 0001
C. becqueretii Doan	000 000 001 312 002 102 0001
C. longicornis (F.Muell.) G.Tayl.	000 000 001 212 002 102 0001
Benguellia lanceolata (Gürke) G.Tayl.	000 000 110 003 010 000 0101
Haumaniastrum villosum (Benth.) ined.	101 100 113 413 001 012 0001
H. caeruleum (Oliv.) P.A.Duvign. & Plancke	101 100 113&4 413 001 002 0001
H. polyneurum (S.Moore) P.A.Duvign. & Plancke	101 100 1&313&4 413 001 002 0001
H. suberosum (Robyns & Lebrun) P.A.Duvign. & Plancke	101 100 113&4 413 001 002 0001
H. prealtum (Briq.) P.A.Duvign. & Plancke	101 100 313 213 001 002 0001
H. sericeum (Briq.) ined.	101 100 1&313&4 2&413 001 002 0001
Ocimum subsect. *Gratissimum*	0&100 000 000 410 00&11 101 0000
O. subsect. *Ocimum*	000 001 000 001 001 001 0000
O. tenuiflorum L.	000 000 000 001 001 001 0000
Orthosiphon subgen. *Orthosiphon*	00&1&20 000 000 001 000 000 0101
Orthosiphon subgen. *Nautochilus* (Bremek.) Codd	01&30 000 000 000&1 000 100 0000
Endostemon N.E.Br.	00&20 000&1 000 00&1&20&1 0&10&10 000 0120
Hemizygia (Benth.) Briq.	00&2&30 000 0&10&10 000 000&1 0&100 0101

Leaf anatomy and generic delimitation in South African Amaryllidaceae

P. VORSTER & A.D. SPREETH
Botany Department, University of Stellenbosch,
Private Bag X5018, 7599 Stellenbosch, South Africa.

Abstract
In the South African species of Amaryllidaceae the external morphology of the leaf varies considerably in respect of dimensions, width, shape in transverse section, and texture. It seems as if at least several genera can be recognized on leaf anatomical characteristics; if not by single features at least by combinations of characteristics. The widely different conditions under which the different species and genera occur, are not reflected in the leaf anatomy. The result is an anatomically surprisingly mesophytic leaf structure rather uniform throughout the family.

Key words
Amaryllidaceae, Leaf anatomy, South Africa.

Introduction

The family Amaryllidaceae is well represented in the South African flora, with about 20 genera and more than 200 species. Some genera are almost or completely restricted to the summer rainfall area or the winter rainfall area, and a minority occur in both areas. It would seem as if adaptation to a specific rainfall season is a deeply-engrained characteristic, and that species do not readily migrate from one rainfall area to another.

The South African species are morphologically diverse, but also demonstrate considerable morphological convergence, so that generic delimitation has always been problematic. Because most species produce horticulturally attractive flowers, a vast amount of literature on various aspects has amassed over the years. However, the anatomy of the plants has almost totally been ignored up to now. A study of the leaf anatomy has therefore been initiated with the aim of:

1. To determine the anatomical structure and its diversity in South African species, as part of our data bank on our plants.
2. To determine whether leaf-anatomical characteristics can be used in generic delimitation.

Material and methods

To date not all the genera have been studied, because the species are scattered over a wide geographical area and we were restricted by the availability of documented living material. The species studied can be found in Table 1. Material was fixed in F.A.A., embedded in wax, stained differentially, and sectioned with a sliding microtome.

Results

In the South African species of Amaryllidaceae the external morphology of the leaf varies considerably in respect of dimensions, width, shape in transverse section, and texture. Nevertheless they are conservative in their anatomy.

L.J.G. van der Maesen et al. (eds.), The Biodiversity of African Plants, 513-516.

Table 1. Genera and species studied.

Amaryllis belladonna L.	*Cyrtanthus brachyscyphus* Bak.
Apodolirion lanceolatum (L.f.) Benth.	*C. breviflorus* Harv.
Boophane disticha (L. f.) Herb.	*C. elatus* (Jacq.) Traub
B. flava W.F. Barker ex Snijman	*C. herrei* (Leighton) R.A. Dyer
B. guttata (L.) Herb.	*Gethyllis ciliaris* (Thunb.) Thunb.
B. haemanthoides Leighton	*Haemanthus amarylloides* Jacq.
Boophane Herb. sp.	*H. canaliculatus* Levyns
Brunsvigia josephinae (Red.) Ker-Gawl.	*H. deformis* Hook. f.
B. marginata (Jacq.) Ait.	*H. nortieri* Isaac
B. radula Ait.	*Hessea* Herb. sp.
Brunsvigia Heist. sp. 1	*Nerine sarniensis* (L.) Herb.
Brunsvigia Heist. sp. 2	*N. undulata* (L.) Herb.
Clivia caulescens R.A. Dyer	*Scadoxus membranaceus* (Bak.) Friis & Nordal
C. gardenii Hook.	*S. multiflorus* (Martyn) Raf. subsp. *katharinae* (Bak.) Friis & Nordal
C. miniata (Lindl.) Regel	*S. puniceus* (L.) Friis & Nordal
C. nobilis Lindl.	*Strumaria salteri* W.F. Barker
Crinum campanulatum Herb.	*S. truncata* Jacq.
C. moorei Hook. f.	
Cybistetes longifolia (L.) Milne-Redh. & Schweick.	

In all the species examined, the leaf consists of ab- and adaxially equal or almost equal epidermis with anomocytic stomata, two zones of assimilation tissue inwards of each epidermis, and a central zone of hyaline parenchyma in which the vascular bundles are embedded in a single row over the width of the leaf. Each vascular bundle is surrounded by a sheath of one to three layers of mostly hyaline parenchyma. In many genera large lisogenous cavities (canals) are present in the tissue between the vascular bundles.

Strumaria: many, but not all, species have an indumentum. The stomata are more or less equally distributed over both surfaces, and lisogenous cavities are absent.

Hessea: the leaf anatomy does not exhibit any striking characteristics except that some species have an indumentum, which is uncommon in the family.

Nerine: contains slime canals in the chlorenchyma. Most of the stomata are restricted to the abaxial surface.

Brunsvigia: contains slime canals in the chlorenchyma, and the stomata are equally distributed on both surfaces. In some species there are sclerenchyma strands between the vascular bundles and each epidermis. In some species wing-like extensions of thick-walled cells are present at the leaf margins. In *B. radula* there is a slight enlargement of the leaf axis forming a false midrib. Some species carry an indumentum.

Crinum: contains slime canals in the chlorenchyma, but these are not very conspicuous in transverse section. The leaf axis is usually differentiated to form a false midrib. In some species the stomata are equally distributed over both surfaces, while in others most occur on the abaxial surface.

Table 2. South African Amaryllidaceae: distribution amongst genera of characteristics deviating from the general pattern. Dark grey: in all species which were examined in genus. Light grey: in some species of the genus.

GENERA	CHARACTERISTICS: Sclerenchyma present	Slime canals present	Wing-like extensions to leaf margins present	False midrib differentiated	Indumentum present	Stomata mostly abaxially	Stomata equally on both surfaces	Lisogenous air cavities absent
Strumaria					dark grey		dark grey	dark grey
Hessea					dark grey			
Nerine		dark grey				dark grey		
Brunsvigia	light grey	dark grey	light grey	light grey	light grey		dark grey	
Crinum				dark grey		light grey	light grey	
Ammocharis		dark grey					dark grey	
Cybistetes		dark grey	dark grey				dark grey	
Amaryllis		dark grey					dark grey	
Boophane	dark grey	dark grey	dark grey				dark grey	
Gethyllis		dark grey			light grey			
Apodolirion		dark grey					dark grey	dark grey
Haemanthus		dark grey			light grey	light grey	light grey	dark grey
Scadoxus		?		dark grey		dark grey		dark grey
Cyrtanthus		light grey		light grey			dark grey	light grey
Clivia		dark grey					dark grey	

Ammocharis: contains slime canals in the chlorenchyma, and wing-like extensions of thick-walled cells at the leaf margins, and the stomata are more or less equally distributed over both surfaces.

Cybistetes: contains slime canals in the chlorenchyma, and wing-like extensions of thick-walled cells at the leaf margins, and the stomata are more or less equally distributed over both surfaces.

Amaryllis: contains slime canals in the chlorenchyma, and the stomata are more or less equally distributed over both surfaces.

Boophane: like some species of *Brunsvigia*, it is noteworthy for containing sclerenchyma strands between the vascular bundles and epidermis, on both sides of the leaf. It contains slime canals in the chlorenchyma, wing-like extensions of thick-walled cells at the leaf stomata are more or less equally distributed over both surfaces, and the stomata are more or less equally distributed over both surfaces.

Gethyllis: anatomically this genus is diverse, but generally it contains slime canals in the chlorenchyma, and in several species an indumentum is present.

Apodolirion: contains slime canals in the chlorenchyma, most of the stomata occur on the abaxial surface, and lisogenous air cavities are absent.

Haemanthus: contains slime canals in the chlorenchyma, the stomata are mostly restricted to the abaxial leaf surface, and lisogenous air cavities are absent. Several species carry an indumentum.

Scadoxus: contains slime canals in the chlorenchyma, but these are not very clear in transverse section. The enlargement of the leaf axis to form a false midrib is very conspicuous. The stomata are mostly restricted to the abaxial surface, and lisogenous air cavities are absent.

Cyrtanthus: in all the species, except the geographical anomaly *C. herrei*, slime canals are present in the chlorenchyma. In some species the leaf axis is expanded into a false midrib. The stomata are more or less equally distributed over both surfaces. Lisogenous air cavities are generally present, but absent in *C. herrei*.

Clivia: slime canals are present in the chlorenchyma, but in transverse section these are unclear. Practically all the stomata are restricted to the abaxial surface.

We wish to stress that these results are based on the relatively small sample of species which we examined. It is possible that the picture may change as more species of each genus are studied. The results are summarized in Table 2.

Conclusions

There is considerable interspecific variation within genera. Nevertheless, even though relatively few species within each genus have been examined, it seems as if at least several genera can be recognized on leaf anatomical characteristics; if not by single features at least by combinations of characteristics.

The widely different conditions under which the different species and genera occur, are not reflected in the leaf anatomy, probably because being geophytes the plants can afford to produce leaves only when conditions are not too severe. The result is an anatomically surprisingly mesophytic leaf structure throughout the family.

Glacial forest refuges

25 August 1994

Convenors: M.S.M. Sosef, J.J.F.E. de Wilde

Le cadre paléoenvironnemental des refuges forestiers africains: quelques données et hypothèses

J. MALEY
CNRS/ISEM & ORSTOM, Université de Montpellier II, 34095 - Montpellier, France.

Résumé

Depuis au moins 800.000 ans, durant les phases les plus froides à l'échelle globale, en particulier lors des extensions glaciaires maximum des hautes latitudes, la forêt tropicale africaine a été très fragmentée et elle n'a subsisté que dans une série de refuges éparpillés à travers l'ensemble du massif forestier. La fin du dernier cycle climatique, depuis environ 25 à 30.000 ans, a été marquée par les dernières grandes fluctuations du massif forestier africain. Cette période peut servir de modèle aux fluctuations cycliques qui ont affecté la forêt tropicale africaine depuis 800.000 ans. La dernière phase de mise en place des refuges est intervenue entre environ 20.000 et 15.000 ans BP, soit durant environ 5000 ans. La phase d'extension forestière maximum, synchrone de la dernière période la plus chaude sur le plan global, s'est située entre environ 9000 et 4000 ans BP, avec une durée similaire d'environ 5000 ans. Depuis environ 800.000 ans, on pourrait donc déduire que chacune de ces phases d'extension minimum et maximum n'aurait concerné que 5 à 10% du temps total écoulé. Ce seraient finalement des situations d'extension intermédiaire, plus ou moins en retrait par rapport à l'étendue forestière actuelle, qui auraient couvert entre 80 et 90% du temps total et qui donc auraient été la norme de la majeure partie du Quaternaire.

Abstract

Since at least 800,000 years, during globally cold periods, and especially when glaciers at high latitudes reached their maximum sizes, the African tropical forest was very fragmented, and only survived in a series of relatively small refuges, spread across the original forest area. The end of the last climatic cycle, that started about 25 to 30,000 years ago, is characterized by the latest large fluctuation of the African tropical forest. This last period may help in modelling cyclic changes undergone by the African forest during the last 800,000 years. The last period when the forest had retreated into refuges was between 20,000 and 15,000 BC, and thus lasted about 5,000 years. A maximum extension of the forest, which coincided with the last warm period, occurred between 9,000 and 4,000 BC, and therefore also lasted 5,000 years. It might be deduced that, during the last 800,000 years, each minimum and maximum extension phase lasted no more than 5 to 10% of the total time. This means that during about 80 to 90% of the total time, the African forest exhibited an average extension, that is about the same extension as it has nowadays. The present extension therefore is typical for the greater part of the Quaternary.

Key words
Africa, Paleoenvironment, Rain forest refuges.

1 Introduction: le cadre paléoenvironnemental des variations forestières

Les grandes variations des paléoenvironnements et des végétations survenus sur la zone tropicale africaine depuis la fin du Néogène, peuvent s'interpréter dans un contexte global de variation des températures et en particulier des phases de refroidissement qui ont été synchrones de l'extension des calottes glaciaires sur les zones polaires antarctique et arctique (Maley 1980, et à paraître, a), suivant des rythmes liés aux variations orbitales de la terre autour du soleil (Broecker & Denton 1989). L'évolution en phase des hautes et basses latitudes, particulièrement durant les périodes froides

L.J.G. van der Maesen et al. (eds.), The Biodiversity of African Plants, 519-535.

(McIntyre et al. 1989), est due à l'activité des anticyclones subtropicaux et aux alizés qu'ils génèrent et qui convergent vers l'équateur. L'accroissement du gradient thermique Pôle-Equateur durant les périodes froides, à comparer au phénomène qui survient annuellement durant les hivers arctique ou antarctique, entraîne une accélération des alizés en surface des océans tropicaux. Lorsque la configuration des océans et des continents est favorable, comme c'est le cas pour la majeure partie du Golfe de Guinée, l'accélération des alizés déplace les eaux de surface, ce qui provoque des remontées d'eaux profondes relativement froides (upwellings), y compris au niveau de l'équateur du fait de l'annulation de la force de Coriolis (Merle 1980, Servain et al. 1985, etc.). Ce sont ces eaux froides de surface qui, en modulant l'activité de la mousson dans le sens surtout d'une forte diminution de la pression de vapeur d'eau et, en conséquence, de la convection, ont un rôle climatique aridifiant sur le continent voisin, provoquant une réduction ou une suppression des pluies durant les périodes de remontées d'eaux froides (Mahé & Citeau 1993, Maley 1987, 1989). Corrélativement, du fait de la persistance des nuages, une baisse des températures intervient aussi sur le continent (Maley & Elenga 1993).

Pour la fin du Néogène, les variations révélées par la palynologie dans les dépôts du delta du Niger (Morley & Richards 1993, Poumot 1989) se situent dans ce contexte. Plusieurs fortes extensions des milieux savanicoles, associées à des réductions de l'aire de la forêt dense, sont intervenues vers la fin du Miocène, entre 8 et 5 Ma, y compris en Afrique orientale (Harris 1993, d'après les paléo-faunes), période marquée par de larges et rapides fluctuations eustatiques, en réponse directe aux fortes extensions des glaces polaires (Haq et al. 1987). Toutefois, près du Golfe de Guinée (Morley & Richards 1993, Poumot 1989), ainsi qu'en Afrique orientale (Bonnefille 1983, 1984), le changement majeur est intervenu entre 2,6 et 2,4 Ma, synchrone de la première grande glaciation globale qui a été marquée par la première extension des calottes glaciaires sur l'Hémisphère Nord et par un développement synchrone de l'Antarctique (Broecker & Denton 1989, Robin 1988).

Au cours du Pliocène puis du Pléistocène, l'étude du $\delta^{18}O$ de carottes marines montre un accroissement progressif de l'amplitude des variations glaciaires, avec deux étapes principales, la première survenue vers 2,5 Ma, avec l'apparition de cycles de 41.000 ans et la seconde il y a environ 800.000 ans avec des cycles de 100.000 ans (Start & Prell 1984) (Figure 1). Corrélativement, la variabilité climatique s'est beaucoup accrue depuis 800.000 ans, associée à des phases de refroidissement plus intenses suivant une périodicité dominante de 100.000 ans. Lorsqu'on examine des résultats polliniques couvrant la majeure partie des dépôts du Cénozoïque supérieur du delta du Niger, on constate que c'est vers la base du Pléistocène (1,8 Ma) que les plus forts pourcentages de pollens de Gramineae ont été observés (Morley & Richards 1993), indice d'une forte aridité et probablement d'une importante fragmentation de la forêt dans cette région. Pour la fin du Quaternaire, des données polliniques obtenues sur des enregistrements sédimentaires marins dans le Golfe de Guinée (depuis environ 250.000 ans) et surtout lacustres dans le domaine forestier (depuis environ 30.000 ans) illustrent bien l'impact de ces variations sur la végétation, avec une réduction maximum de la forêt dense entre 20 et 15.000 ans BP.

La répartition des climats étant largement conditionnée par les latitudes, la disposition des continents et ensuite par des facteurs géographiques plus régionaux (montagnes,

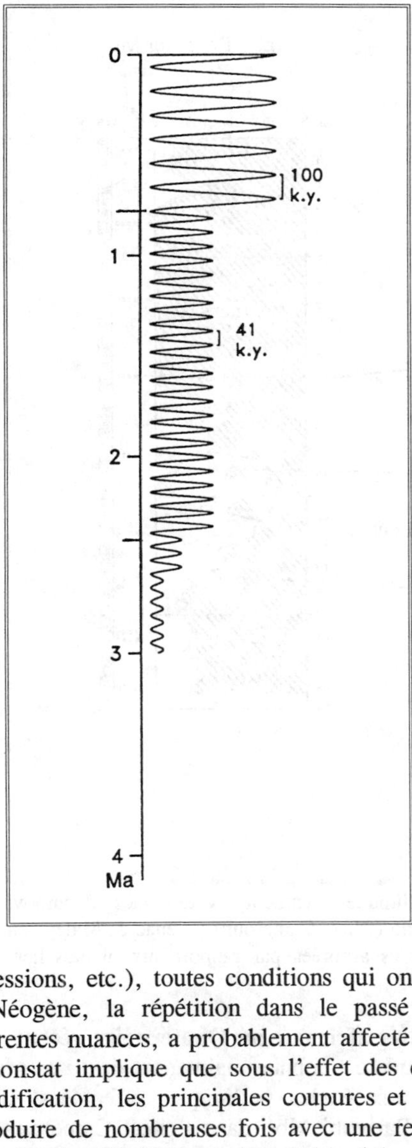

Figure 1. Schéma des grands cycles de variation des calottes glaciaires sur l'Hémisphère nord, avec indication des principales périodicités. Les deux étapes majeures sont l'installation des conditions glaciaires vers 2,5 Ma et l'accroissement de l'amplitude des cycles vers 800.000 ans (extrait de Ruddiman et al. 1989, Figure 8-B).

dépressions, etc.), toutes conditions qui ont été quasiment stables à l'échelle de la fin du Néogène, la répétition dans le passé des changements climatiques, avec leurs différentes nuances, a probablement affecté de manière comparable les mêmes régions. Ce constat implique que sous l'effet des changements climatiques, et ici des phases d'aridification, les principales coupures et fragmentations du bloc forestier ont dû se reproduire de nombreuses fois avec une relative constance géographique, particulièrement depuis 800.000 ans. Etant donné la cyclicité des grands événements climatiques mondiaux, dont on sait par divers enregistrements sédimentaires et micropaléontologiques obtenus dans le Golfe de Guinée (Bonifay & Giresse 1992, Jansen et al. 1984, McIntyre et al. 1989, Prell et al. 1976, Ruddiman et al. 1989) qu'ils ont affecté régulièrement et d'une manière intense la zone équatoriale africaine, la fin du dernier cycle climatique, de 25.000 ans à l'Actuel, période pour laquelle les données sont les plus nombreuses, pourrait donc fournir un modèle des principales fluctuations qui ont affecté la forêt dense africaine.

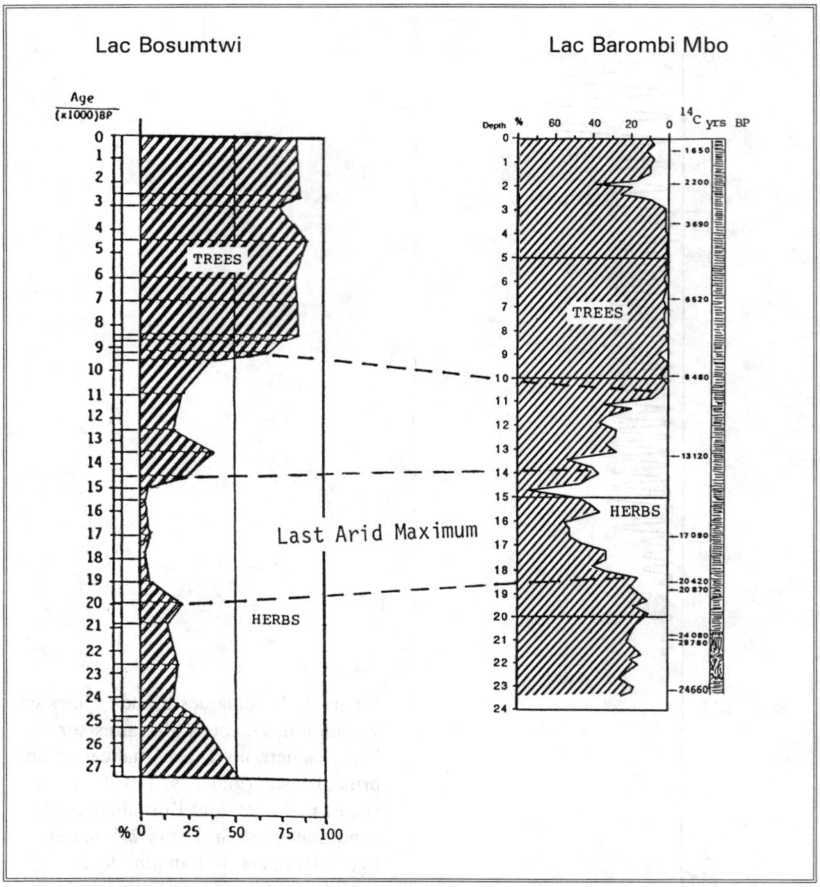

Figure 2. Comparaison des spectres polliniques synthétiques des lacs Bosumtwi (Ghana) (Maley 1989, 1991) et Barombi Mbo (Ouest Cameroun) (Brenac & Maley, en préparation). Courbes des pollens des plantes arborées par rapport aux plantes herbacées: Rapport AP/NAP.

2 Les grandes lignes de l'histoire de la forêt africaine au Quaternaire récent

L'histoire de la forêt africaine de basse altitude est connue maintenant dans ses grandes lignes pour le Quaternaire récent, grâce aux études polliniques effectuées sur les sédiments des lacs Bosumtwi au Ghana et Barombi Mbo dans l'ouest Cameroun. Les sédiments étudiés au Bosumtwi remontent à environ 28.000 ans BP (Maley & Livingstone 1983, Maley 1987, 1989, 1991, en préparation) et au Barombi Mbo à une date voisine (Brenac 1988, Brenac & Maley en préparation, Giresse et al. 1994, Maley & Brenac 1987, Maley 1989, 1991). Au Congo des données palynologiques ont été aussi obtenues sur la végétation montagnarde du Pléistocène terminal (Elenga 1992, Elenga et al. 1991). On a ainsi des informations assez précises pour la fin du stade 3, le stade 2 qui a comporté le dernier maximum glaciaire et enfin le stade 1 qui comporte essentiellement l'Holocène.

En résumé, d'après les publications citées ci-dessus, les données principales suivantes peuvent être présentées, tout d'abord pour la région du *lac Bosumtwi* (Figure 2):

* entre 28.000 et 24.000 ans BP (fin stade 3) existait une forêt avec un élément montagnard (*Olea capensis*, syn. *O. hochstetteri*), sous un climat relativement humide et frais,
* de 24.000 à 19 - 20.000 ans cette forêt régresse progressivement,
* de 19.000 à 15.000 ans, sous l'effet d'un climat très aride, la forêt disparaît (les pollens arborés étaient alors inférieurs à 5%): elle est remplacée par une formation très ouverte qu'on pourrait comparer, non à une savane de basse altitude, mais à une prairie de moyenne altitude comportant des bouquets d'arbres épars, constitués de taxons montagnards et de quelques taxons forestiers dont l'aire s'étend aussi en plaine (cf. Letouzey 1968, 1985, Schnell 1977).
* Ensuite, après 15.000 ans, le climat s'améliorant, la forêt recolonise la région avec une première pulsation positive vers 13.500 ans BP, toujours associée avec un élément montagnard. Un retour à des conditions plus sèches se note entre 12.000 et 10.000 ans BP: cette phase pourrait correspondre en partie au "Dryas récent" des plus hautes latitudes (Talbot & Johannessen 1992).
* Entre 9500 et 8500 ans BP, en même temps que l'élément montagnard disparaît et que donc le climat se réchauffe, la forêt se réinstalle complètement puis se maintient jusqu'à l'Actuel. Les hauts niveaux lacustres de l'Holocène inférieur et surtout de l'Holocène moyen sont aussi la preuve d'un climat nettement plus humide (Maley 1989, 1991, Talbot et al. 1984). Il est ainsi probable que la forêt a largement transgressé au delà de ses limites actuelles, particulièrement à l'Holocène moyen, et qu'ainsi "la coupure actuelle du Dahomey", qui voit les savanes atteindre la mer au niveau du Togo et du Bénin, ait été complètement submergée et recolonisée par la transgression forestière (Maley 1989, 1991). Une preuve de cette transgression existe aussi sur le littoral du Congo avec la présence de souches d'arbres fossiles, typiques d'une forêt dense, dans un secteur actuellement couvert de savanes, et datées de l'Holocène moyen (Schwartz et al. 1990).
* A l'Holocène récent, alors que le lac régressait brutalement vers 3700 ans BP, la forêt s'est maintenue près du lac. On pourrait rattacher à cette régression lacustre un événement plus lointain avec une première phase de régression forestière dans le secteur de la "coupure du Dahomey" (Maley 1989, 1991). L'extension maximale des savanes dans ce secteur serait intervenue probablement un peu plus tard, entre 2500 et 2000 ans BP, lors d'une péjoration climatique généralisée (cf. infra).

Dans l'ouest du Cameroun, pour la région du *lac Barombi Mbo*, les faits et interprétations principales sont les suivants (Figure 2 et 3):

* Entre 25 et 29.000 ans (la sédimentation est perturbée vers la base de la carotte, cf. Giresse et al. 1991) et 20.000 ans BP existait une forêt riche en Caesalpinioideae (Leguminosae) qui incluait aussi un élément montagnard (essentiellement *Olea capensis*, syn. *O. hochstetteri*).

* Une phase sèche est intervenue brutalement vers 20.000 ans BP, faisant régresser la forêt avec irruption de végétations plus ouvertes. Cette situation a perduré jusque vers 14 - 15.000 ans. Les pourcentages des pollens arborés oscillant autour de 50 à 60%, on en déduit que le climat était cependant beaucoup moins aride qu'au Ghana, où les pollens arborés étaient inférieurs à 5%: des îlots forestiers importants ont donc subsisté dans cette région de l'Ouest Cameroun. Cette interprétation est confirmée par des études isotopiques effectuées sur la matière organique détritique, venant du bassin versant, et conservée dans les sédiments de cette époque: les mesures du $\delta^{13}C$ ont donné des valeurs situées en moyenne vers la limite inférieure des valeurs typiquement forestières (Giresse et al. 1994).
* Entre 14.000 et 9500 ans BP on assiste à une phase de recolonisation forestière, avec, avant 10.000 ans, une oscillation négative qu'on pourrait, comme au Bosumtwi, rapporter aussi au "Dryas récent" (Giresse et al. 1994).
* A partir de cette date et jusque vers 3000 ans BP, la forêt a présenté son extension maximum: les pollens d'herbacées, essentiellement les Gramineae, oscillent entre 0 et 3%: la quasi totalité du spectre est alors composée de pollens d'arbres.
* Après 3000 ans on note un accroissement progressif des pollens de Gramineae qui, entre 2500 et 2000 ans BP, atteignent des valeurs de 30 à 40%, indication très claire d'une phase brutale d'ouvertures et de reculs de la forêt.

Figure 3. Evolution des pourcentages relatifs des pollens de 3 taxons caractéristiques venant des dépôts quaternaires (carotte BM-6) du lac Barombi Mbo (Ouest Cameroun) (d'après Brenac & Maley, en préparation). Ages interpolées à partir de 12 radiocarbone dates; pour toutes les questions concernant ce site et les données botaniques, géologiques, stratigraphiques, etc, cf. Brenac (1988), Giresse et al. (1991, 1994), Letouzey (1985), Maley (1989, 1991, 1992).

Gramineae. Ce taxon constitue la quasi totalité des pollens d'herbacées non-aquatiques et il est caractéristique des milieux ouverts de type savane. Son extension a été la plus forte entre 20.000 et 10.000 ans BP. Par contre le fait que les pollens d'arbres ont continué à largement dominer entre 20 et 15.000 ans avec une moyenne de 50 à 60%, montrent que les milieux forestiers ont persisté dans cette région en formant un "refuge" (cf. Giresse et al. 1994).

Alchornea cordifolia. Ce taxon correspond à des arbustes pionniers caractéristiques des forêts "secondaires": ils se développent en abondance dans tous les types d'ouverture du milieu forestier où ils participent activement à la reconstitution de la forêt, comme par exemple lors de la phase de péjoration climatique du début de l'Holocène récent (ca. 3000 - 2000 ans BP).

Caesalpinioideae (total des pollens de Caesalpinioideae). Cette famille est surtout caractéristique des milieux forestiers "primaires" de type sempervirent (Letouzey 1985). Les Caesalpinioideae produisent peu de pollens, de ce fait même 1 ou 2% sont significatifs d'une certaine importance locale (Brenac 1988, Reynaud en préparation). Il faut noter leurs pourcentages élevés avant 20.000 ans indiquant leur forte domination du milieu forestier. Ensuite, bien qu'ils aient baissé fortement ils se maintiennent dans le secteur entre 20 et 15.000 ans. Ils sont ensuite relativement abondants à partir de 14.000 ans jusque vers 3000 ans lorsqu'ils régressent très fortement durant la phase de péjoration ayant culminé entre 2500 et 2000 ans BP. Depuis 2000 ans leur importance est très variable.

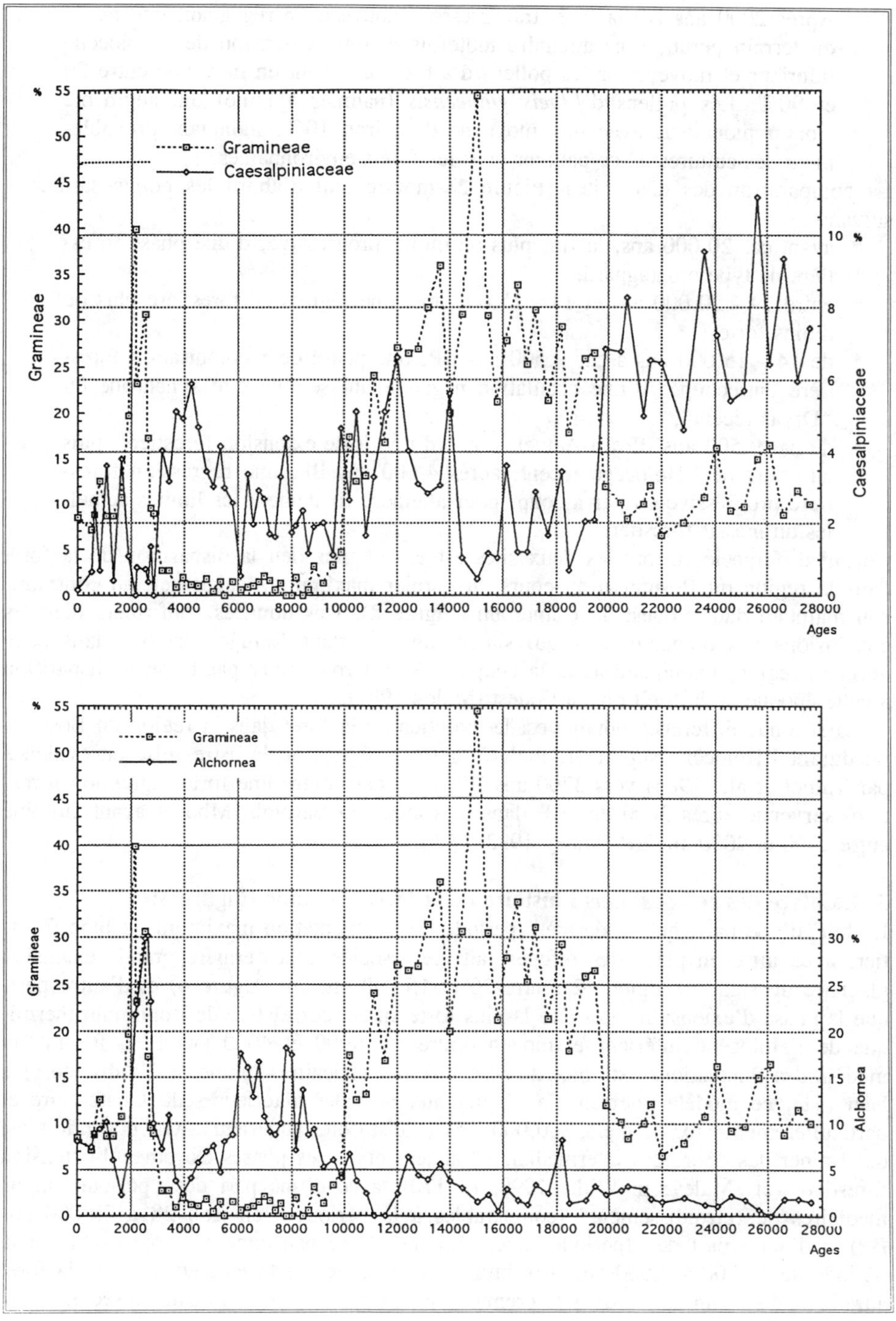
Gramineae
Caesalpiniaceae
%
Gramineae
Caesalpiniaceae
Ages
%
%
Gramineae
Alchornea
Gramineae
Alchornea
Ages
%

* Après 2000 ans BP la forêt transgresse à nouveau en regagnant une partie du terrain perdu, sans atteindre toutefois la forte extension de l'Holocène inférieur et moyen, car les pollens d'arbres se situent en moyenne entre 80 et 90%. Les pollens d'*Elaeis guineensis* (palmier à huile) ont augmenté alors rapidement, avec une moyenne d'environ 10%, indiquant probablement des cultures de ce palmier dans les forêts environnantes.

La comparaison des deux sites (Figure 2) montre tout d'abord les points similaires suivants:

* avant ca. 20.000 ans, la fin, plus ou moins progressive, d'une phase forestière de type montagnard,
* entre 19 - 20.000 ans et 14 - 15.000 ans, une régression forestière plus ou moins forte,
* de 14 - 15.000 ans à ca. 9.500 ans BP, une phase de recolonisation forestière entrecoupée d'une oscillation négative qui semble être synchrone du "Dryas récent",
* de ca. 9.500 ans BP à l'Actuel, d'abord une forte extension forestière, puis au cours de l'Holocène récent, après 4.000 ans BP, une régression forestière progressive ou par à coup, pour atteindre finalement les limites actuelles du massif forestier.

Ce qui différencie surtout les deux sites est en premier lieu la disparition de la forêt dans la région du Bosumtwi au cours du dernier maximum glaciaire et, au contraire, son maintien dans l'ouest du Cameroun (Figure 2). Ces données confirment bien les conclusions des biogéographes qui situent un important Refuge forestier dans cette dernière région, expliquant aussi la coupure de la Cross River par la quasi disparition à cette époque de la forêt plus à l'ouest (Maley 1987).

La seconde différence notable est le maintien de la forêt dans la région du Bosumtwi durant l'Holocène supérieur, malgré la forte régression lacustre mise en évidence par Talbot et al. (1984) vers 3700 ans BP, avec par contre une forte régression forestière survenue après 3000 ans BP dans le secteur du Barombi Mbo et ayant culminé entre 2500 et 2000 ans BP (Maley 1992) (§ 5).

3 La place des refuges dans l'histoire de la forêt africaine (Figure 4)

Le fait, d'une part, que la dernière période de fragmentation maximum du bloc forestier, avec mise en place des refuges, ait été associée à la dernière grande extension glaciaire du stade isotopique 2 (entre 20 et 15.000 ans BP, figure 4) et d'autre part, que la phase d'extension forestière la plus forte ait été corrélative de l'optimum thermique de l'Holocène inférieur et moyen (entre ca. 9000 et 4000 ans BP), fournit un modèle des fluctuations extrêmes du bloc forestier africain. Comme on l'a discuté plus haut (§1), ce modèle pourrait s'appliquer aux périodes antérieures du Quaternaire et particulièrement aux derniers 800.000 ans (Pléistocène supérieur). On pourrait ainsi déterminer des "périodes intermédiaires" situées entre ces phases extrêmes d'extension forestière (cf. Maley, in White 1990, p. 170), avec d'une part des "périodes intermédiaires de refroidissement" (comparables à la période d'environ 28 à 20.000 ans BP) et d'autre part des "périodes intermédiaires de réchauffement" (comparables à la période de 15.000 à 9.500 ans BP) durant lesquelles les surfaces couvertes par la forêt auraient été en diminution dans le premier cas, ou en augmentation dans le second.

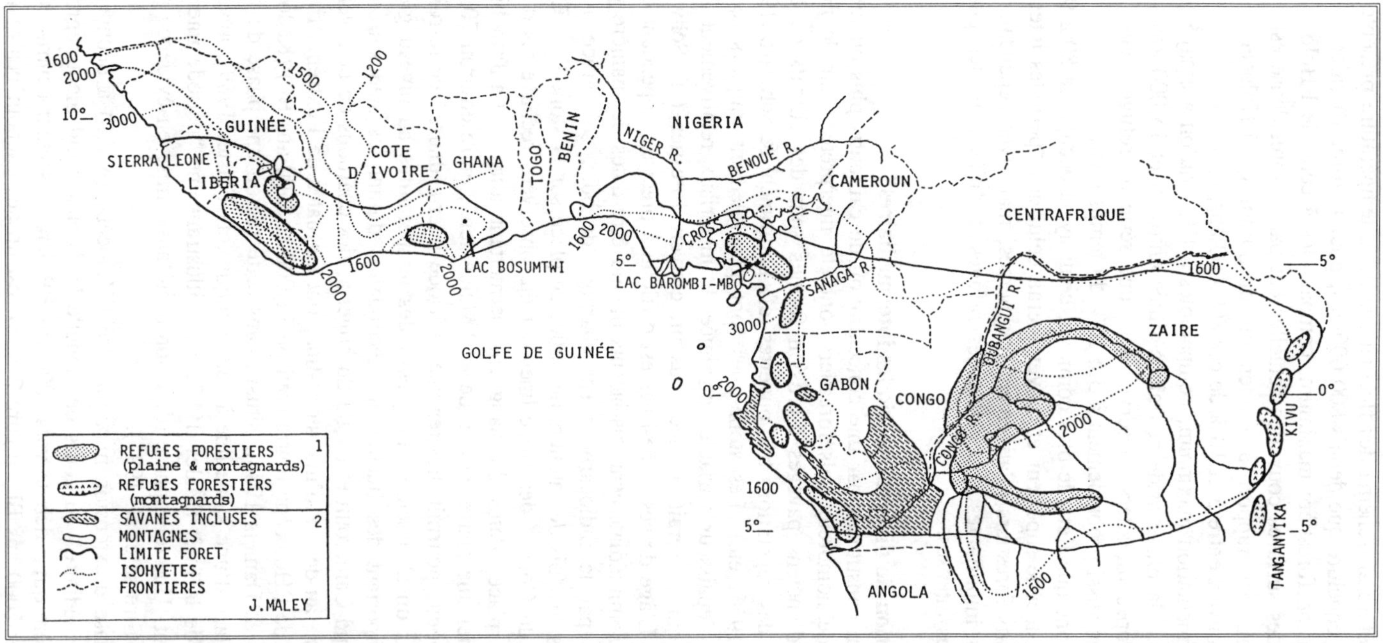

Figure 4. Schéma des refuges forestiers durant la dernière grande phase aride (ca. 18.000 ans BP), adapté et complété de Maley (1987), Figure 1, d'après Van Rompaey (1993, 1994) pour le Liberia et la Côte d'Ivoire, Sosef (1994) pour le Gabon et Congo et Colyn (1987, 1991) pour la cuvette du Congo - Zaïre. Dans l'encadré: 1, 18.000 ans BP; 2, Actuel.

Remarques: Le refuge du Libéria pouvait être relié à celui autour du Mont Nimba. Van Rompaey (1994) et Robbrecht (1994) considèrent que le refuge secondaire du SW Ghana/SE Côte d'Ivoire devait être relié au refuge principal du Libéria. Pour le Mont Cameroun, Cheek & Cable (1994) ont montré que ce secteur était très riche en plantes endémiques: on pourrait donc conclure à un refuge secondaire à rattacher au refuge principal de l'Ouest Cameroun/Oban. Pour le Zaïre oriental, vu en particulier la grande richesse faunique, un refuge forestier de plaine serait aussi possible (T. & J. Hart, commun. pers.)

Ainsi, lorsqu'on se réfère au schéma des fluctuations de la température présenté sur la Figure 1, on pourrait conclure que depuis 800.000 ans, pour la forêt africaine:

* les phases d'extension forestière maximum, comparables à celle de l'Holocène (l'ensemble des surfaces couvertes actuellement par la forêt dense est déjà en net retrait par rapport à l'extension maximum de l'Holocène moyen) auraient couvert entre 5 et 10% de ces 800.000 ans,
* les phases de fragmentation maximum, comparables à l'état qu'on a essayé de reconstituer pour la phase aride et fraîche située entre 20 et 15.000 ans BP, avec les biotopes forestiers conservés dans une série de refuges, auraient correspondus aussi probablement à 5 à 10% du temps total.

Finalement, il semble donc que l'état le plus commun, celui ayant occupé de 80 à 90% du temps durant le Pléistocène supérieur et moyen, serait celui des "périodes intermédiaires", avec des massifs forestiers relativement fragmentés, mais avec cependant des surfaces nettement plus importantes que celles des refuges postulés pour les phases arides et fraîches les plus intenses.

4 Essai de caractérisation de diverses "périodes intermédiaires"

Les stone-lines, qui sont constituées par une concentration de cailloutis plus ou moins grossiers vers la base de nombreux sols forestiers, ont été interprétées par de nombreux auteurs comme d'anciens pavages d'érosion formés sous des climats de type semi-aride et associés à des végétations très ouvertes (cf. Lanfranchi & Schwartz 1991, Schwartz 1990, Vogt 1966, etc). Les stone-lines sont généralement toujours recouvertes par une couche d'épaisseur variable constituée de matériaux relativement fins, appelée colluvium, et qui se serait formée sous un couvert forestier (cf. Schwartz 1990, Vogt 1966, etc). L'âge de ces stone-lines est difficile à préciser, cependant dans certains cas des niveaux immédiatement sus-jacents ont pu être datés de manière plus ou moins précise, soit par le radiocarbone sur charbons de bois, soit grâce à des industries préhistoriques, parfois des ateliers de taille, en place sur la stone-line. C'est ainsi que dans le nord du Congo, des stone-lines ont été observées recouvertes d'ateliers préhistoriques en place, datés du Sangoen, permettant ainsi à Lanfranchi & Schwartz (1991) de situer une phase semi-aride au "Maluékien", entre environ 70.000 et 40.000 ans. Cette époque pourrait illustrer une "période intermédiaire" de refroidissement, caractérisée par un abaissement progressif des températures au niveau global (tendance générale comportant des fluctuations secondaires durant les stades isotopiques 4 & 3). Dans le sud Cameroun et le sud du Nigéria, des charbons de bois venant de la base du colluvium, au-dessus d'une stone-line, ont été datés de l'extrême fin du Pléistocène et du début de l'Holocène (entre environ 11.000 et 8500 ans BP) (Kadomura & Hori 1990, Shaw & Daniels 1984), situant dans cette région une phase de forte érosion avec mise en place d'une stone-line à cette époque (cf. Fölster 1969), avant la réinstallation complète de la forêt (Maley 1993; § 3), illustrant une "période intermédiaire" de réchauffement. Une telle phase d'érosion a été aussi mise en évidence pour le fleuve Congo-Zaïre (Giresse et al. 1991).

Aussitôt après la phase d'extension maximum de l'Holocène (§ 2,3) a débuté une nouvelle "période intermédiaire" au cours de laquelle la forêt a rapidement régressé. Cette régression forestière qui a débuté vers 3000 ans BP dans le secteur central et qui a culminé entre 2500 et 2000 ans BP (Figure 2 & 3), est un des phénomènes majeurs de l'Holocène récent en Afrique tropicale. Elle a affecté la majeure partie du bloc forestier car des témoignages palynologiques et paléobotaniques synchrones ont été obtenus en divers points (Cameroun, Congo, Zaïre, Uganda) (Brenac & Maley en préparation, Hart et al. commun. pers. et cette volume, Maley 1992, Richards 1986, Taylor 1990, Vincens et al. 1994). Une forte réduction des couvertures arborées a été notée aussi en divers points des savanes de l'Afrique nord tropicale, comme au Sénégal (Lézine 1989) et au Tchad (Maley 1981). Cette période a été aussi directement associée à une phase de forte érosion avec des dépôts de sédiments grossiers généralisés, soit sur les interfluves sous forme de stone-lines (Fölster 1969), soit dans les thalwegs sous forme de graviers ou de conglomérats qui constituent la base d'une "Basse Terrasse" (Maley 1981). De tels sédiments grossiers, datés souvent sur bois fossiles ou charbons, ont été retrouvés dans le bloc forestier central (Caratini & Giresse 1979, Giresse commun. pers., Kuete 1990, etc) mais surtout dans les zones des savanes soudano-guinéenne, soudanienne et sahélienne (Cameroun, Congo, Nigéria, bassin du Tchad, nord Ethiopie, Sahel, etc) (Fölster 1969, Maley 1981, 1992, Mawson & Williams 1984, Van Noten & De Ploey 1977). Ces différents phénomènes (régression forestière et érosion), du fait de leur caractère généralisé à la majeure partie de l'Afrique tropicale, sont la conséquence d'un changement climatique abrupt, dans le sens d'une péjoration climatique due probablement à une saisonnalité plus marquée des pluies jusque dans la zone équatoriale (Maley 1981, 1992, à paraître, b). Un autre aspect de cette phase de régression forestière a été le rôle important joué par les feux (Maley, à paraître, b).

5 L'extension des formations montagnardes durant les "périodes intermédiaires" et la question de leur migration éventuelle

La "période intermédiaire" de refroidissement située entre 27.000 & 20.000 ans et la "période intermédiaire" de réchauffement située entre 15.000 et 9000 ans, toutes deux avec d'importantes fluctuations, ont comporté au Bosumtwi et au Barombi Mbo des pourcentages plus ou moins importants des taxons montagnards (§ 2). Ce phénomène a eu une grande ampleur dans le site du Bois de Bilanko au Congo central, situé vers 600 m d'altitude, où, avant 10.000 ans, a été mise en évidence une phase forestière largement dominée par une association typiquement afro-montagnarde, avec environ 60% des spectres polliniques comportant en premier *Podocarpus* avec environ 50% des pollens (déterminés seulement au niveau générique, mais appartenant très probablement à l'espèce *P. latifolius*, cf. Maley et al. (1990), et ensuite *Ilex mitis* et *Olea capensis* avec 10% (Elenga 1992, Elenga et al. 1991). L'abaissement de la température nécessaire à l'extension de cette formation montagnarde à basse altitude, aurait pu être déterminé par un fort développement des couvertures nuageuses et des brouillards dans des milieux de type forêt de nuage (Maley 1989, Maley et al. 1990, Maley & Elenga 1993) (§ 1).

Durant ces deux types de "période intermédiaire" des formations montagnardes ont pu entrer en contact avec des formations forestières de plaine, au gré des accidents topographiques et des conditions climatiques locales, favorisant peut-être des échanges géniques et des phénomènes de spéciation (cf. les exemples actuels donnés par Gentry (1989), pour des forêts de nuage situées dans les Andes Equatoriennes). C'est probablement aussi durant ces périodes que des migrations de taxons montagnards ont pu avoir lieu (Maley 1987, 1989, 1991, Maley et al. 1990), expliquant peut-être pourquoi les montagnes du Cameroun et de l'Est Africain ont en commun de nombreuses espèces ou des espèces très proches, tant dans la flore - 53% des espèces de la forêt montagnarde et 49% de celles des prairies montagnardes sont communes entre le Mont Cameroun et l'Est Africain (Hall 1973) - que dans la faune (Eisentraut 1970). Des études taxonomiques et génétiques pourraient peut-être apporter des éléments de chronologie pour situer ces phénomènes dans le temps.

Concernant la répartition des *Podocarpus* africains, Gymnospermes typiquement montagnardes, deux espèces sont présentes actuellement en Afrique Centrale Atlantique: *Podocarpus latifolius* qui en constitue la quasi totalité et qui se rencontre dans de nombreuses stations continentales (Zaïre, Angola, Congo, Cameroun et NE Nigéria) (Maley et al. 1990, White 1981, 1983) et *Podocarpus mannii* présente seulement sur l'île de São Tomé qui est située presque sur l'équateur et isolée par de grands fonds marins à 300 km du continent. White (1993) émet l'hypothèse que cette espèce aurait pu migrer vers cette île au cours du Quaternaire récent et qu'ainsi il serait possible que les pollens de *Podocarpus* trouvés en abondance avant 10.000 ans au Congo, au Bois de Bilanko (cf. supra), ne soient pas *P. latifolius* mais éventuellement *P. mannii*. Cette hypothèse est hautement improbable, vu d'une part le grand isolement de l'île de São Tomé et surtout le caractère très particulier de *P. mannii* qui est dioïque et qui possède des fruits relativement lourds et volumineux (Woltz 1971, 1986), rendant un transport à longue distance tout à fait improbable. Il faudrait plutôt estimer que ce *Podocarpus* serait rélictuel d'une ancienne lignée Gondwanienne présente déjà en milieu montagnard vers la fin du Crétacé, comme l'atteste de faibles pourcentages de pollens de Podocarpaceae dans certains niveaux Crétacé du fond du Golfe de Guinée (Doyle et al. 1982). Ce caractère rélictuel serait d'autant plus envisageable que le soubassement de São Tomé est constitué d'une formation sédimentaire gréseuse d'âge Crétacé dont on trouve des équivalents sur le continent voisin, comme près de Douala et au Nigéria, ce qui permet de conclure que cette île serait constituée à l'origine par un fragment micro-continental abandonné lors de l'ouverture de l'Atlantique et sur lequel un volcanisme basaltique se serait développé beaucoup plus tard, débutant au Pliocène pour se poursuivre jusqu'à l'époque récente (Gouhier et al. 1974). Une partie de la flore de l'île de São Tomé pourrait donc avoir un caractère rélictuel qu'il serait important d'étudier. Etant donné ces conlusions, on pourrait aussi se poser des questions à propos de *P. latifolius* qui aurait pu exister aussi en Afrique Centrale Atlantique, en particulier sur la Dorsale Camerounaise, depuis le Crétacé et y subsister dans des milieux montagnards très réduits jusque vers la fin du Néogène, au moment où des changements climatiques (§ 1) lui auraient permis de nouvelles phases d'expansion (Maley 1980, à paraître, b).

6 Conclusions générales

Depuis au moins 800.000 ans la forêt africaine a présenté de grandes fluctuations en phase avec les fluctuations des températures et en particulier des extensions glaciaires aux hautes latitudes. Les phases de fragmentation maximum de la forêt, avec mise en place des refuges, n'ont concerné que les phases les plus froides - les deux dernières sont intervenues au cours des stades isotopiques 6 et 2, la dernière n'ayant pas duré plus de 5000 ans - et n'ont affecté que 5 à 10% du temps écoulé. Les phases d'extension forestière maximum, apparemment synchrones des phases les plus chaudes - les deux dernières sont intervenues au cours des stades isotopiques 1 et 5e (Eémien) (Bengo & Maley 1991) - auraient eu des durées similaires d'environ 5000 ans et n'auraient concerné aussi qu'environ 5 à 10% du temps total écoulé. Les 80 à 90% du temps restant ont correspondu à des "périodes intermédiaires" caractérisées par des étendues forestières plus ou moins réduites par rapport à l'état actuel, qui est lui-même déjà en retrait d'un état d'extension maximum. Ces "périodes intermédiaires" ont correspondu à des situations climatiques très variables suivant les époques et les tendances, c'est à dire qu'elles sont intervenues soit durant des phases de réchauffement ou de refroidissement à l'intérieur d'une période glaciaire, soit en période interglaciaire, comme durant l'Holocène récent et dans une ambiance climatique générale plutôt chaude, avec intervention brutale des incendies en milieu forestier. Certaines de ces "périodes intermédiaires" ont été aussi accompagnées par des érosions intenses, de type semi-aride, avec formation de "stone-line", et d'autres périodes par des extensions montagnardes caractérisées floristiquement par *Podocarpus latifolius* et *Olea capensis*.

Remerciements
Certaines des données et conclusions utilisées dans cet article ont été obtenues dans le cadre de recherches financées par l'ORSTOM (Institut Français de Recherche Scientifique pour le Développement en Coopération) et depuis 1992, aussi par le CNRS, dans le cadre du programme international ECOFIT (Ecosystèmes et Paléoécosystèmes des Forêts Intertropicales en rapport avec les Changements Globaux). L'auteur a bénéficié d'informations inédites aimablement communiquées par P. Brenac, R. Dechamps, P. Giresse, T. & J. Hart et R.S.A.R. van Rompaey. Cet article est la contribution no 94-085 de l'Institut des Sciences de l'Evolution de Montpellier (ISEM).

Bibliographie

Bengo, M.D. & J. Maley, 1991. Analyses des flux polliniques sur la marge sud du Golfe de Guinée depuis 135.000 ans. C.R. Acad. Sc., Paris, II, 313: 843-849.

Bonifay, D. & P. Giresse, 1992. Middle to late Quaternary sediment flux and post-depositional processes between the continental slope off Gabon and the mid-Guinean margin. Marine Geology 106: 107-129.

Bonnefille, R., 1983. Evidence for a cooler and drier climate in the Ethiopian uplands towards 2.5 Myr ago. Nature 303: 487-491.

Bonnefille, R., 1984. Cenozoic vegetation and environments of early hominids in East Africa. In: R.O. Whyte (ed.). The evolution of the East Asian environment, Palaeobotany, Palaeozoology and Palaeoanthropology 3: 579-612. Centre Asian St. Publ., Univ. Hong Kong.

Brenac, P., 1988. Evolution de la végétation et du climat dans l'Ouest Cameroun entre 25.000 et 11.000 ans BP. Actes Xème Symposium Ass. Palynologues Langue Française, Trav. Sect. Sci. & Tech. Inst. Français Pondichéry 25: 91-103.

Broecker, W.S. & G.H. Denton, 1989. The role of ocean-atmosphere reorganization in glacial cycles. Geochimica et Cosmochimica Acta 53: 2465-2501.

Caratini, C. & P. Giresse, 1979. Contribution palynologique à la connaissance des environnements continentaux et marins du Congo à la fin du Quaternaire. Comptes Rend. Acad. Sc., Paris, D, 288: 379-382.

Cheek, M. & S. Cable, 1994. Mapping plant biodiversity on Mount Cameroon. In: Programme and abstracts, 14th AETFAT Congress. Dept. of Plant Taxonomy, Wageningen Agricultural University. p. 33 & 138.

Colyn, M.M., 1987. Les Primates des forêts ombrophiles de la cuvette du Zaïre: interprétations zoogéographiques des modèles de distribution. Revue Zool. Afr. 101: 183-196.

Colyn, M.M., 1991. L'importance zoogéographique du bassin du fleuve Zaïre pour la spéciation: le cas des Primates Simiens. Annales Musée R. Afr. Cent., Sc. Zool., Tervuren 264. 250 p.

Doyle, J.A., Jardiné, S. & A. Doerenkamp, 1982. *Afropollis*, a new genus of early Angiosperm pollen, with notes on the Cretaceous palynostratigraphy and paleoenvironments of northern Gondwana. Bull. Centr. Rech. Explor. Prod. Elf-Aquitaine 6: 39-117.

Eisentraut, M., 1970. Eiszeitklima und heutige Tierverbreitung im tropischen Westafrika. Umschau Wiss. & Techn. Frankfurt a. M. 3: 70-75.

Elenga, H., 1992. Végétation et climat du Congo depuis 24.000 ans BP. Analyse palynologique de séquences sédimentaires du pays Batéké et du littoral. Thèse Sc., Univ. Aix-Marseille.

Elenga, H., A. Vincens & D. Schwartz, 1991. Présence d'éléments forestiers montagnards sur les Plateaux Batéké (Congo) au Pléistocène supérieur: nouvelles données palynologiques. Palaeoecology of Africa 22: 239-252.

Fölster, H., 1969. Late Quaternary erosion phases in SW-Nigeria. Bull. Ass. Sénég. ét. Quatern. Ouest Afr., Dakar 21: 29-35.

Gautier, L., 1990. Contact forêt-savane en Côte d'Ivoire centrale: évolution du recouvrement ligneux des savanes de la Réserve de Lamto (sud du V-Baoulé). Candollea 45: 627-641.

Gentry, A.H., 1989. Speciation in tropical forests. In: L.B. Holm-Nielsen, I.C. Nielsen & H. Balslev (eds.). Tropical forests, botanical dynamics, speciation and diversity: 113-134. Academic Press, London.

Giresse, P., J. Maley & K. Kelts, 1991. Sedimentation and palaeoenvironment in crater lake Barombi Mbo, Cameroon, during the last 25,000 years. Sedimentary Geology 71: 151-175.

Giresse, P. Kinga-Mouzeo & D. Schwartz, 1991b. Breaks in the sedimentary and environmental equilibrium in the Congo Basin and incidences on the Oceanic sedimentation during Quaternary. J. Afr. Earth Sc. 12: 229-236.

Giresse, P., J. Maley & P. Brenac, 1994. Late Quaternary palaeoenvironments in the lake Barombi Mbo (Cameroon) deduced from pollen and carbon isotopes of organic matter. Palaeogeogr., Palaeoclimatol., Palaeoecol. 107: 65-78.

Gouhier, J., J. Nougier & D. Nougier, 1974. Contribution à l'étude volcanique du Cameroun: Ligne du Cameroun - Adamaoua. Ann. Fac. Sc. Cameroun 17: 3-48.

Hall, J.B., 1973. Vegetational zones on the southern slopes of Mount Cameroon. Vegetatio 27: 49-69.

Haq, B.U., J. Hardenbol & R.P. Vail, 1987. Chronology of fluctuating sea levels since the Triassic. Science 235: 1156-1167.

Harris, J., 1993. Ecosystem structure and growth of the African savanna. Global & Planet. Change 8: 231-248.

Jansen, J.H.F., T.C. Van Weering, R. Gieles & J. Van Iperen, 1984. Middle and late Quaternary oceanography and climatology of the Zaire-Congo fan and the adjacent eastern Angola basin. Netherlands J. Sea Res. 17: 201-248.

Kadomura, H. & N. Hori, 1990. Environmental implications of slope deposits in humid tropical Africa: evidence from southern Cameroon and western Kenya. Geographical Report, Tokyo Metro. Univ. 25: 213-236.

Kuete, M., 1990. Géomorphologie du Plateau sud-Camerounais à l'ouest du 13°E. Thèse Doctorat Etat, Univ. Bordeaux III.

Lanfranchi, R. & D. Schwartz, 1991. Les remaniements de sols pendant le Quaternaire supérieur au Congo. Evolution des paysages dans la région de la Sangha. Cah. ORSTOM, série Pédologie 26: 11-24.

Letouzey, R., 1968. Etude phytogéographique du Cameroun. Encyclopédie Biologique, Paris, 49, 508 p.

Letouzey, R., 1985. Notice de la carte phytogéographique du Cameroun au 1:500.000. Inst. Carte Intern. Végétation, Toulouse & Inst. Rech. Agron., Yaoundé.

Lezine, A.M., 1989. Late Quaternary vegetation and climate of the Sahel. Quaternary Res. 32: 317-334.

McIntyre, A., W.F. Ruddiman, K. Karlin & A.C. Mix, 1989. Surface water response of the equatorial Atlantic ocean to orbital forcing. Paleoceanography 4: 19-55.

Mahé, G. & J. Citeau, 1993. Relations Ocean - Atmosphère - Continent dans l'espace Africain de la mousson Atlantique. Veille Clim. Satel. 44: 34-54.

Maley, J., 1980. Les changements climatiques de la fin du Tertiaire en Afrique: leur conséquence sur l'apparition du Sahara et de sa végétation. In: M.A.J. Williams & H. Faure (eds.). The Sahara and the Nile: 63-86. Balkema Publ., Rotterdam.

Maley, J., 1981. Etudes palynologiques dans le bassin du Tchad et paléoclimatologie de l'Afrique nord-tropicale de 30.000 ans à l'époque actuelle. Travaux & Documents ORSTOM 129. 586 p.

Maley, J., 1987. Fragmentation de la forêt dense humide africaine et extension des biotopes montagnards au Quaternaire récent: nouvelles données polliniques et chronologiques. Implications paléoclimatiques et biogéographiques. Palaeoecology of Africa 18: 307-334.

Maley, J., 1989. Late Quaternary climatic changes in the African rain forest: the question of forest refuges and the major role of sea surface temperature variations. In: M. Leinen & M. Sarnthein (eds.). Paleoclimatology and Paleometeorology: modern and past patterns of global atmospheric transport. NATO Adv. Sc. Inst., Ser. C, Math. & Phys. Sc. 282: 585-616. Kluwer Acad. Publ., Dordrecht.

Maley, J., 1990. L'histoire récente de la forêt dense humide africaine: essai sur le dynamisme de quelques formations forestières. In: R. Lanfranchi & D. Schwartz (eds.). Paysages Quaternaires de l'Afrique centrale Atlantique: 383-389. Didactiques, ORSTOM, Paris.

Maley, J., 1991. The African rain forest vegetation and palaeoenvironments during late Quaternary. Climatic Change 19: 79-98.

Maley, J., 1992. Mise en évidence d'une péjoration climatique entre ca. 2500 et 2000 ans BP en Afrique tropicale humide. Bull. Soc. Géol. France 163: 363-365.

Maley, J., 1993. The climatic and vegetational history of the equatorial regions of Africa during the Upper Quaternary. In: Th. Shaw, P. Sinclair, B. Andah & A. Okpoko (eds.). The Archaeology of Africa: food, metals and towns: 43-52. Routledge Publ., London & New-York.

Maley, J. (à paraître, a). The African rain forest: principal patterns of vegetation and climate changes from Upper Cretaceous to Quaternary. In: R. Watling, M.D. Swaine & I.J. Alexander (eds.). West African forests: studies in the Guineo-Congolian Domain. The Roy. Soc. Edinburgh Publ.

Maley, J. (à paraître, b). The impact of arid phases on the African rain forest during his geological history. In: W. Weber, A. Vedder & H.S. Morland (eds.). African rain forest ecology and conservation. New York Wildlife Conserv. Soc. Publ., Yale Univ. Press.

Maley, J. & P. Brenac, 1987. Analyses polliniques préliminaires du Quaternaire récent de l'Ouest Cameroun: mise en évidence de refuges forestiers et discussion des problèmes paléoclimatiques. Mém. Trav. Ecole Pratiq. Htes Etudes Montpellier 17: 129-142.

Maley, J. & D.A. Livingstone, 1983. Extension d'un élément montagnard dans le sud du Ghana (Afrique de l'Ouest) au Pléistocène supérieur et à l'Holocène inférieur: premières données polliniques. Comptes Rend. Acad. Sc., Paris, II, 296: 1287-1292.

Maley, J., G. Caballé & P. Sita, 1990. Etude d'un peuplement résiduel à basse altitude de Podocarpus latifolius sur le flanc Congolais du Massif du Chaillu. Implications paléoclimatiques et biogéographiques. Etude de la pluie pollinique actuelle. In: R. Lanfranchi & D. Schwartz (eds.). Paysages Quaternaires de l'Afrique centrale Atlantique: 336-352. Didactiques, ORSTOM, Paris.

Maley, J. & H. Elenga, 1993. Le rôle des nuages dans l'évolution des paléoenvironnements montagnards de l'Afrique tropicale. Veille Climatique Satellitaire, Lannion 46: 51-63.

Mawson, R. & M.A.J. Williams, 1984. A wetter climate in eastern Sudan 2,000 years ago. Nature 309: 49-51.

Merle, J., 1980. Variabilité thermique annuelle et interannuelle de l'Océan Atlantique équatorial est. L'hypothèse d'un "El Nino" Atlantique. Oceanologica Acta 3: 209-220.

Morley, R.J. & K. Richards, 1993. Gramineae cuticle: a key indicator of Late Cenozoic climatic change in the Niger delta. Rev. Palaeobotan. Palynol. 77: 119-127.

Poumot, C., 1989. Palynological evidence for eustatic events in the tropical Neogene. Bull. Centres Rech. Explor. Prod. Elf-Aquitaine 13: 437-453.

Prell, W.L., J.V. Gardner, A.W.H. Be & J.D. Hays, 1976. Equatorial Atlantic and Caribbean foraminiferal assemblages, temperatures and circulation: interglacial and glacial comparisons. Geological Soc. America Memoir 145: 247-266.

Richards, K., 1986. Preliminary results of pollen analysis of a 6,000 year core from Mboandong, a crater lake in Cameroun. Hull Univ. Geography Dept. Misc. Ser. 32: 14-28.

Robin, G. de Q., 1988. The Antarctic ice sheet, its history and response to sea level and climatic changes over the past 100 million years. Palaeogeogr., Palaeoclimatol., Palaeoecol. 67: 31-50.

Robbrecht, E., 1994. Geography of African Rubiaceae with reference to glacial rain forest refuges. In: Programme and abstracts, 14th AETFAT Congress, Dept. of Plant Taxonomy, Wageningen Agricultural University: 104.

Ruddiman, W.F., M. Sarnthein, J. Backman, J.G. Baldauf, W. Curry, L.M. Dupont, T. Janececk, E.M. Pokras, M.E. Raymo, B. Stabell, R. Stein & R. Tiedemann, 1989. Late Miocene to Pleistocene evolution of climate in Africa and the low-latitude Atlantic: overview of Leg 108 results. Proceed. Ocean Drill. Prog., Sc. Res. 108: 463-484.

Schnell, R., 1977. Introduction à la phytogéographie des pays tropicaux. 4: La flore et la végétation de l'Afrique tropicale. Gauthier-Villars Publ., Paris. 378 p.

Schwartz, D., 1990. Relations sols-reliefs-variations paléoclimatiques en Afrique Centrale. In: R. Lanfranchi & D. Schwartz (eds.). Paysages Quaternaires de l'Afrique centrale Atlantique: 186-192, Didactiques, ORSTOM, Paris.

Schwartz, D., 1992. Assèchement climatique vers 3000 B.P. et expansion Bantu en Afrique centrale atlantique: quelques réflexions. Bull. Soc. Géol. France 163: 353-361.

Schwartz, D., B. Guillet & R. Dechamps, 1990. Etude de deux flores forestières mi-Holocène (6000-3000 BP) et subactuelle (500 BP) conservées in situ sur le littoral Pontenegrin (Congo). In: R. Lanfranchi & D. Schwartz (eds.). Paysages Quaternaires de l'Afrique centrale Atlantique: 283-297. Didactiques, ORSTOM, Paris.

Servain, J., J. Picaut & A.J. Busalacchi, 1985. Interannual and seasonal variability of the tropical Atlantic Ocean depicted by sixteen years of sea-surface temperature and wind stress. In: J.C.J. Nihoul (ed.). Coupled Ocean - Atmosphere models: 211-237. Elsevier Sc. Publ.

Shaw, T. & S.G. Daniels, 1984. Excavations at Iwo Eleru, Ondo State, Nigeria. West Afr. J. Archaeology (Univ. Ibadan) 14. 269 p.

Sosef, M.S.M., 1994. Refuge Begonias: taxonomy, phylogeny and historical biogeography of Begonia sect. *Loasibegonia* and sect. *Scutobegonia* in relation to glacial rain forest refuges in Africa. Wageningen Agric. Univ. Papers 94-1: 1-306.

Start, G.G. & W.L. Prell, 1984. Evidence for two Pleistocene climatic modes: data from DSDP site 502. In: A.L. Berger & C. Nicolis (eds.). New perspectives in climate modelling. Developments in Atmospheric Science 16: 3-22. Elsevier Sc. Publ.

Talbot, M.R., D.A. Livingstone, P.G. Palmer, J. Maley, J.M. Melack, G. Delibrias & S. Gulliksen, 1984. Preliminary results from sediment cores from Lake Bosumtwi, Ghana. Palaeoecology of Africa 16: 173-192.

Talbot, M.R. & T. Johannessen, 1992. A high resolution palaeoclimatic record for the last 27,500 years in tropical West Africa from the carbon and nitrogen isotopic composition of lacustrine organic matter. Earth Planet. Sci. Lett. 100: 23-37.

Taylor, D.M., 1990. Late Quaternary pollen records from two Ugandan mires: evidence for environmental change in the Rukiga Highlands of southwest Uganda. Palaeogeogr., Palaeoclimatol., Palaeoecol. 80: 283-300.

Van Noten, F. & J. De Ploey, 1977. Quaternary research in north-eastern Nigeria. Ann. Musée Royal Afrique Centrale, Tervuren, Sc. Hum. 92. 61 p. & pl.

Van Rompaey, R.S.A.R., 1993. Forest gradients in West Africa: a spatial gradient analysis. Doctoral thesis, Wageningen Agric. Univ. 142 p.

Van Rompaey, R.S.A.R., 1994. Locating rain forest refugia in Liberia and western Côte d'Ivoire using biogeographic evidence from mainly endemic Caesalpinioideae species: coastal versus submontane position. In: Programme and abstracts, 14th AETFAT Congress, Dept. of Plant Taxonomy, Wageningen Agricultural University: 108. (see also this volume).

Vincens, A., G. Buchet, H. Elenga, M. Fournier, L. Martin, C. De Namur, D. Schwartz, M. Servant, & D. Wirrmann, 1994. Changement majeur de la végétation du lac Sinnda (vallée du Niari, Sud-Congo) consécutif à l'assèchement climatique holocène supérieur: apport de la palynologie. C.R. Acad. Sc., Sc. Vie, Paris, II, 318: 1521-1526.

Vogt, J., 1966. Le complexe de la stone-line. Mise au point. Bull. Bur. Rech. Géol. & Min. 4: 3-51.

White, F., 1981. The history of the Afromontane archipelago and the scientific need for its conservation. Afr. J. Ecology 19: 33-54.

White, F., 1983. Long-distance dispersal and the origins of the Afromontane flora. Sonderbd. Naturwiss. Ver. Hamburg 7: 87-116.

White, F., 1990. *Ptaeroxylon obliquum* (Ptaeroxylaceae), some other disjuncts, and the Quaternary history of African vegetation. Bull. Mus. Natn. Hist. Nat., B, Adansonia 12: 139-185.

White, F., 1993. Refuge theory, ice-age aridity and the history of tropical biotas: an essay in plant geography. Fragm. Flor. Geobot. Suppl. 2: 385-409.

Woltz, P., 1971. L'évolution vasculaire remarquable des plantules de *Podocarpus mannii* Hooker f. Trav. Lab. For. Toulouse 1 (8): 1-6.

Woltz, P., 1986. Les *Podocarpus*, origines et évolution. Etude des plantules. Comparaison avec les familles de Conifères de l'Hémisphère Sud. Thèse Sc., Univ. Aix-Marseille.

Les *Rinorea* comme indicateurs des grands types forestiers du Cameroun

G. ACHOUNDONG
Herbier National du Cameroun, B.P. 1601, Yaoundé, Cameroun.

Résumé
Les *Rinorea* (Violaceae) sont des bio-indicateurs très sensibles des types forestiers. En sous-bois, les aires de distribution épousent les contours des faciès forestiers, reflets des zones biogéographiques. Au Cameroun, on distingue en particulier les *Rinorea* de la plaine littorale et les *Rinorea* du plateau. A l'intérieur de chaque division, les *Rinorea* s'associent en unités dont les aires coïncident parfaitement aux zones de refuge reconnue par de nombreux auteurs dont Amiet (1987), Maley (1987) et récemment Sosef (1994).

Abstract
Rinorea species (Violaceae) present sensible bioindicators for forest typification. In understorey forest, their distribution often coincides with the forest facies contour. In Cameroon in particular, there are two groups of *Rinorea* species: species of the coastal plain and species of the plateau. In each of these two groups, there are species-associations from which the distributions coincide exactly with the refuge zones well known in Cameroon, as outlined by many authors e.g. Amiet (1987), Maley (1987) and recently Sosef (1994).

Key words
Bioindicator, Cameroun, Distribution, Forest types, *Rinorea*

Introduction

Au terme de la révision des *Rinorea* que nous avons entreprise, la répartition de ces Violaceae sur le territoire camerounais apparaît plus clairement qu'auparavant. La présente étude a pour objectif de comparer les aires de distribution des espèces de *Rinorea* et des principaux types forestiers classiquement reconnus au Cameroun. Nous adoptons ici les subdivisions phytogéographiques reconnues par Letouzey (1985) dans la notice de la carte phytogéographique du Cameroun.

1 La forêt atlantique (anciennement biafréenne Letouzey 1968) ou la forêt à Caesalpinioideae; c'est une forêt sempervirente développée dans la région côtière ou subcôtière recevant d'abondantes précipitations annuelles. Elle comprend deux districts: Le district atlantique littoral caractérisé d'une part par l'abondance relative de *Lophira alata* et *Sacoglottis gabonensis,* d'autre part par la rareté des Caesalpinioideae, et le district atlantique biafréen, zone de grande richesse en Caesalpinioideae.

2 La forêt congolaise; elle aussi sempervirente, forme dans le sud et le centre une apophyse de la forêt répandue dans la cuvette zaïroise.

3 La forêt semi-caducifoliée, ou forêt mésophile à Sterculiaceae et à Ulmaceae. Elle s'étend en situation plus continentale et moins arrosée que les types précédents. Elle atteint au moins sous forme de galerie ou de massifs isolés le pied du plateau de l'Adamaoua. Vers le sud, elle est actuellement en voie d'expansion au dépens des types sempervirents.

L.J.G. van der Maesen et al. (eds.), The Biodiversity of African Plants, 536-544.

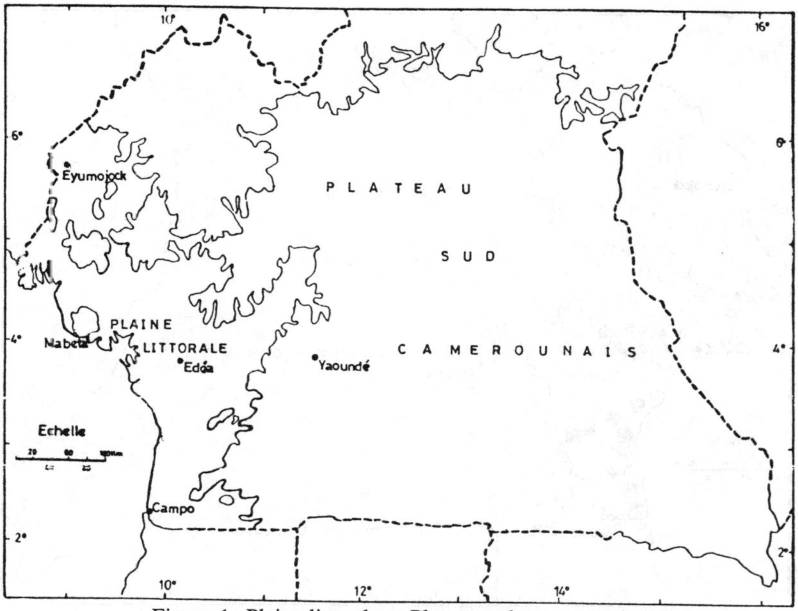

Figure 1. Plaine littorale et Plateau sud-camerounais.
Sur la carte est reportée la courbe 200 m.

Dans la pratique et en simplifiant, on peut considérer que la forêt atlantique correspond à la plaine littorale et que la forêt congolaise et semi-décidue correspondent au plateau sud-camerounais. Ce sont donc ces deux grands ensembles géomorphologiques (Figure 1) qui nous serviront de premier cadre pour analyser la répartition des *Rinorea*.

Les espèces de *Rinorea* endémiques de la plaine littorale

La plaine littorale, une étroite bande côtière qui ne dépasse guère 150 km de large (Morin 1979), est une zone de grande richesse en *Rinorea*. En effet, parmi les 55 espèces répertoriées au Cameroun, 42 soit plus de 76% se rencontrent dans cette zone. Suivant leur dépendance de la plaine, ces taxons littoraux se répartissent en deux groupes: les espèces endémiques de la plaine littorale et les espèces communes à la plaine côtière et au plateau.

Parmi les 42 espèces rencontrées dans la plaine 28, soit plus de 50% des espèces camerounaises, sont exclusives de cette zone et ne se retrouvent pas dans les autres types forestiers. Cependant, leur répartition dans la plaine n'est pas homogène. En effet suivant l'importance et les contours des aires de distribution des espèces on distingue trois groupes: les espèces de *Rinorea* du district littoral, les espèces de *Rinorea* du district biafréen et les espèces de *Rinorea* communes aux deux districts.

Les espèces du district atlantique littoral (Figure 2)

Quelques *Rinorea* de la plaine se rencontrent surtout en forêt littorale à *Lophira alata* et à *Sacoglottis gabonensis*. Certaines espèces sont des taxons à aire extrêmement réduite et parfois limitée à quelques points de récolte entièrement inclus dans la forêt

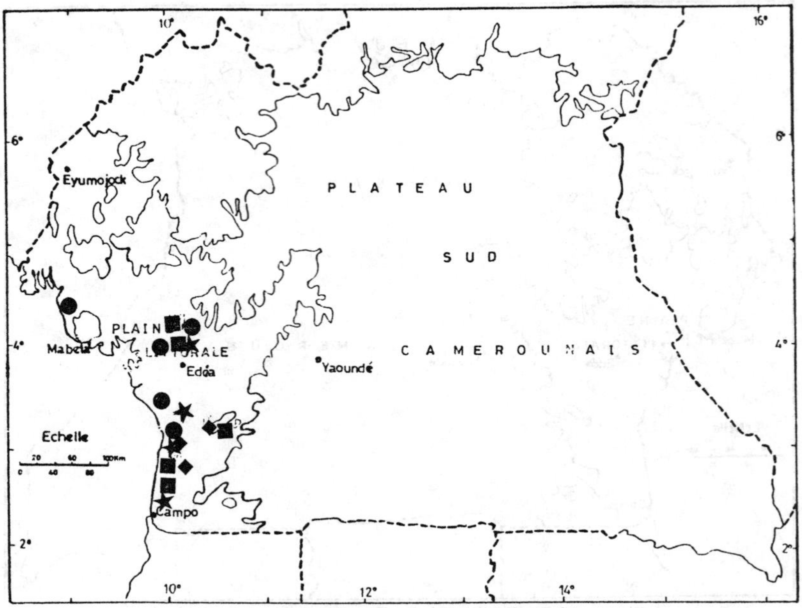

Figure 2. *Rinorea* du district littoral.
♦ *Rinorea microglossa*, ■ *Rinorea caudata*, ★ *Rinorea* sp. 3, ● *Rinorea* sp. 12.

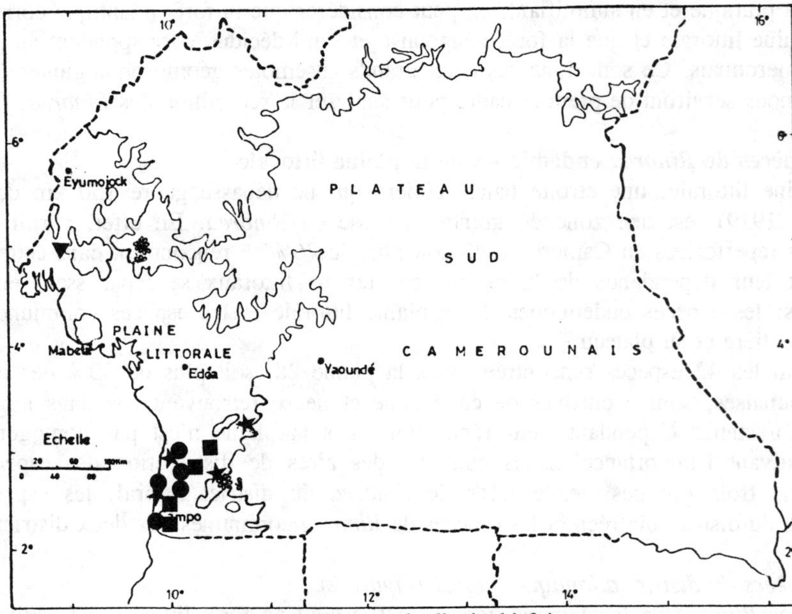

Figure 3. *Rinorea* du district biafréen.
★ *Rinorea albidiflora*, ■ *Rinorea ovata*, ● *Rinorea umbricola*,
♦ *Rinorea soyauxii*, * *Rinorea verrucosa* Chipp., ▾ *Rinorea crassifolia* (Bak.f.) De Wild.

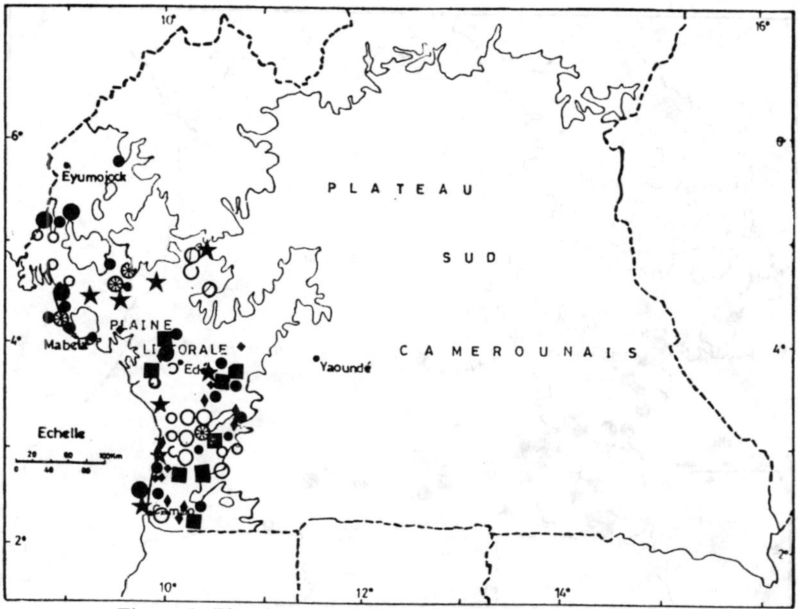

Figure 4. *Rinorea* communs aux districts littoral et biafréen.
• *Rinorea kamerunensis*, ○ *R. longisepala* Engl., ♦ *R. ledermanii* Engl., ● *R. breviracemosa*, ○ *R. exappendiculata*, • *R. parviflora*, ★ *R. gabunensis*, ❁ *R. afzelii*, ■ *R.* sp. 5, ♦ *R.* sp. 11.

littorale. C'est le cas de *Rinorea microglossa* Engl., *Rinorea* sp. 3*, *Rinorea* sp. 12. D'autres sont des espèces à aire nettement centrée sur la forêt littorale, mais débordant vers l'intérieur pour s'irradier plus ou moins loin dans la forêt biafréenne. Les espèces de *Rinorea* les plus significatives à cet égard sont: *Rinorea caudata* (Oliv.) O. Ktze, *Rinorea breviracemosa* Chipp., *Rinorea* sp. 11. Dans la partie élargie du croissant littoral, ces espèces n'atteignent pas le pied du plateau sud-camerounais et de la dorsale camerounaise.

Les espèces du district atlantique biafréen (Figure 3)
Quelques espèces de *Rinorea* généralement répandues dans les forêts biafréennes autour de Campo se raréfient vers le nord et disparaissent avant Edea. Il s'agit de: *Rinorea albidiflora* Engl., *R. ovata* Chipp., *R. umbricola* Engl. et *R. soyauxii* M. Brandt.

Les espèces communes aux deux districts (Figure 4)
La plupart des *Rinorea* de la plaine se rencontrent dans les deux districts, mais l'extension des aires des différentes espèces est inégale. *Rinorea kamerunensis* Engl. est

* Notre communication est extraite d'une révision du genre Rinorea au Cameroun où les espèces nouvelles décrites ne sont pas encore publiées. Elles sont repérées par des numéros. Cependant, à titre indicatif nous donnons en annexe la liste des noms correspondants au numéros.

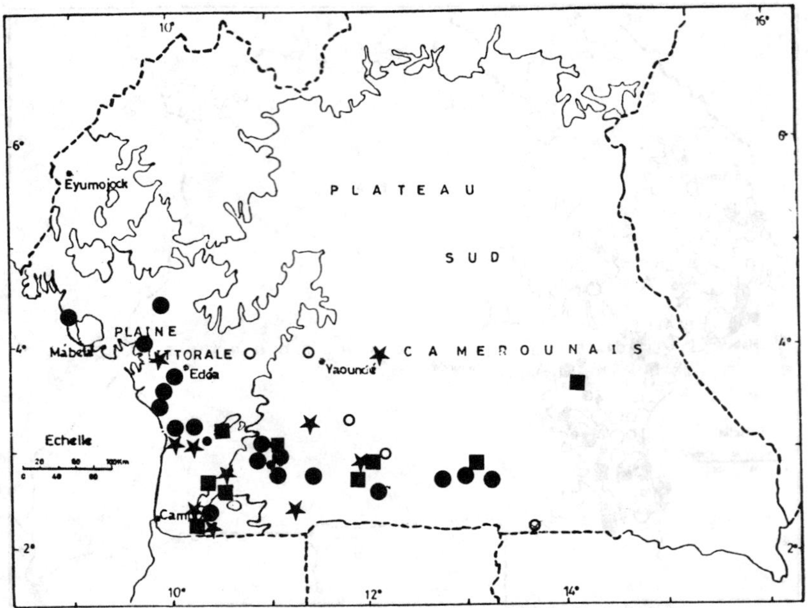

Figure 5. *Rinorea* communs à la plaine et à la forêt congolaise.
● *Rinorea longicuspis*, • *Rinorea ebolowensis*, ○ *Rinorea keayii*,
■ *Rinorea* sp. 7, ★ *Rinorea* sp. 14.

présente dans toute la plaine littorale. D'autres espèces relativement bien répandues de la plaine sont: *Rinorea gabunensis* Engl., *Rinorea parviflora* Chipp., *Rinorea* sp. 4.

Certaines espèces à large distribution dans la plaine ont des aires disjointes. Exemples *Rinorea exappendiculata* Engl., *Rinorea afzelii* Engl., *Rinorea talbotii* (Bak.f.) De Wild. Ces espèces présentes au sud de la plaine littorale disparaissent autour d'Edea et réapparaissent dans les forêts du nord de la plaine. La grande fréquence et l'abondance de ces espèces dans la plaine tranche avec leur absence absolue sur le plateau marquant une nette différence entre les deux milieux.

Une autre caractéristique fondamentale de la flore des *Rinorea* de la plaine littorale est sa richesse en taxons "archaïque". Le sous-genre *Etubulosae* et la section *Ardisianthus* totalisent 10 espèces et en raison de leur nombreux caractères primitifs doivent être considérés comme archaïques. Or parmi les 10 taxons se rencontrant tous dans la plaine 6 sont exclusifs de cette zone alors qu'aucun *Rinorea* archaïque n'est endémique du plateau. Ces particularités, jointes à la biodiversité maximale des *Rinorea* en forêt atlantique confirme le caractère de véritable musée vivant de cette dernière.

Les espèces communes à la plaine et au plateau
Certaines espèces rencontrées dans la plaine littorale sont aussi des taxons de forêt congolaise ou des espèces ubiquistes.

Espèces de plaine et de forêt congolaise (Figure 5)
L'aire de certaines espèces de la plaine s'étend aussi sur le plateau dans le secteur de

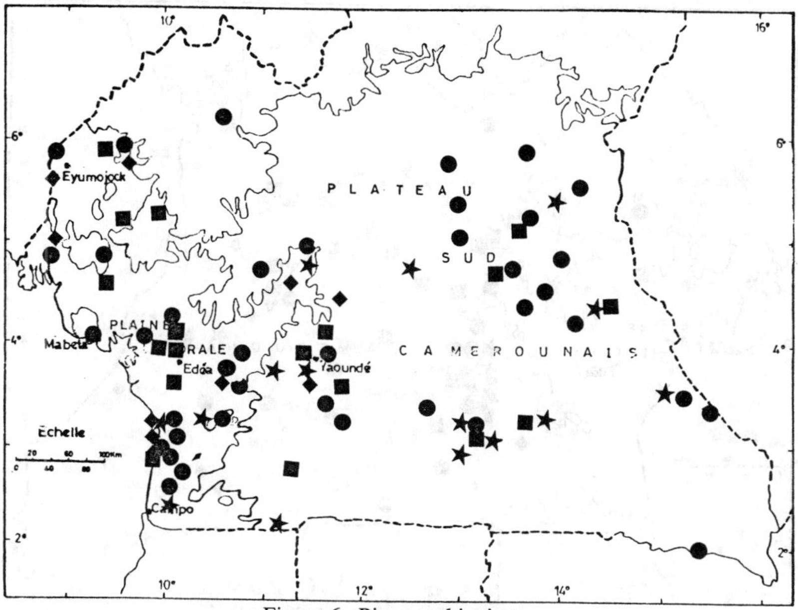

Figure 6. *Rinorea* ubiquistes.
● *Rinorea oblongifolia*, ■ *Rinorea dentata*,
★ *Rinorea subsessilis*, ◆ *Rinorea subintegrifolia*.

la forêt congolaise, sans atteindre la forêt sémi-décidue: *Rinorea ebolowensis* M. Brandt, *R. longicuspis* Engl., *R. keayii* Brenan, *Rinorea* sp. 14.

Espèces ubiquistes (Figure 6)
Plusieurs espèces vivant dans la plaine sont en réalité des espèces ubiquistes, dont l'aire de distribution s'étend à toute la forêt dense humide. Ce type de distribution concerne les espèces suivantes: *Rinorea convallarioides* (Bak.f.) Eyles, *R. angustifolia* (Thon.) Baill., *R. subsessilis* Brandt, *R. subintegrifolia* (Beauv.) O. Ktze, *R. dentata* (Beauv.) O. Ktze, *R. oblongifolia* (C.H. Wright) Marquand ex Chipp. Ces espèces sont par ailleurs des taxons répandus dans tous les blocs forestiers africains.

Le plateau sud-camerounais
Largement plus étendu que la plaine littorale, le plateau sud-camerounais n'héberge pourtant que 22 espèces de *Rinorea*, soit 40% seulement des espèces camerounaises (Figure 7). La richesse en *Rinorea* est nettement plus faible que celle de la forêt littorale. Parmi les 22 espèces vivant sur le plateau, 12 sont des espèces ubiquistes déjà signalées en forêt littorale. Le plateau a donc 11 espèces qui lui sont propres. Suivant les contours de leurs aires sur le plateau, on distingue les *Rinorea* de la forêt congolaise et les *Rinorea* de la forêt semi-décidue.

Espèces de la forêt congolaise
Quelques *Rinorea* du plateau ont leur aire de distribution nettement centrée sur la forêt

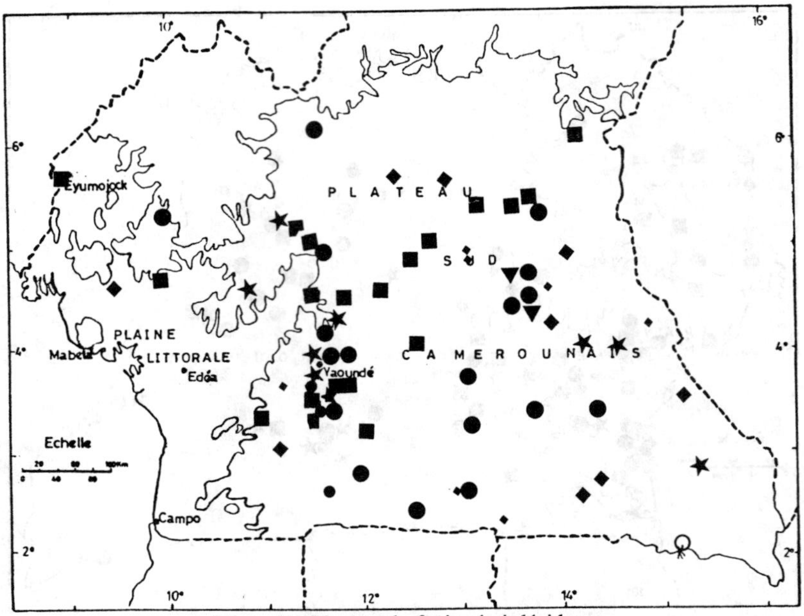

Figure 7. *Rinorea* de forêt sémi-décidue.
★ *Rinorea ilicifolia*, ● *Rinorea batesii*, • *Rinorea zenkeri*, ○ *Rinorea acutidens*, ◆ *Rinorea yaundensis*, ▾ *Rinorea* sp. 6, ■ *Rinorea* sp. 9.

congolaise, c'est le cas de *Rinorea* sp. 7, *Rinorea* sp. 8. Par endroits, l'aire de *Rinorea* sp. 7 déborde timidement sur les bords de la plaine (Lolodorf, Nyabesan) mais n'atteint pas la forêt littorale à *Lophira alata* et *Sacoglottis gabonensis*. Autres particularités de la forêt congolaise, certains taxons présents dans d'autres types forestiers y manifestent une fréquence plus grande et y forme des populations plus dense, c'est le cas de *Rinorea subsessilis*, *R. cerasifolia* M. Brandt, *R. keayii*.

Espèces de forêt sémi-caducifoliée
Certaines espèces du plateau se concentrent davantage en forêt sémi-caducifoliée, ce sont *Rinorea ilicifolia* (Welw. ex Oliv.) O. Ktze, *R. batesii* Chipp., *R. zenkeri* Engl., *R. acutidens* M. Brandt, *R. yaundensis*, *Rinorea* sp. 6, *Rinorea* sp. 9. Les aires de la plupart de ces espèces débordent souvent en forêt congolaise. Pour certaines espèces quelques populations très isolées ont été répertoriées en dehors du plateau: à Barombi Mbo, *Rinorea batesii* est observé en compagnie de *R. yaundensis*, cette dernière est signalé de Korup; à Eyumojok, dans la plaine de Manfé, *R.* sp. 9 n'est connu par une seule récolte.

On constate que au Cameroun, *Rinorea ilicifolia* est la seule espèce qui n'a jamais été signalée d'un autre type de forêt.

En définitive, si nous considérons la proportion des taxons occupant chaque zone biogéographique, nous obtenons les pourcentages suivants:

Endémiques de la plaine littorale	51%
Espèces communes à la plaine et au plateau .	25%
Soit pour la plaine	76%
Espèces propres au plateau	20%
Sans localité précise	4%

Variation de la biodiversité dans la plaine

La concentration des taxons archaïques, ainsi que la grande richesse spécifique de la plaine fait de cette zone un centre d'endémisme. A l'intérieur, on observe des variations importantes de la richesse spécifiques. Pour illustrer cette variation, nous comparons du Sud vers le Nord trois secteurs bien prospectés.

La région de Campo-Bipindi

C'est une zone de grande richesse spécifique. Elle héberge 33 *Rinorea* sur les 55 répertoriés au Cameroun. Parmi les 28 taxons littoraux du Cameroun, 22 s'y rencontrent. Tous les 10 taxons archaïques du Cameroun y sont représentés. Certains *Rinorea* ne sont connus que de cette région, ce sont: *R. umbricola, R. ovata, Rinorea* sp. 1, *Rinorea* sp. 2, *Rinorea* sp. 5. Cette zone correspond aussi à la limite Nord de certains taxons gabonais comme *R. albidiflora*, *R. soyauxii*.

La région d'Edea

Par rapport à Campo, c'est une région déjà très appauvrie. Elle héberge 16 *Rinorea* sur les 55 répertoriés. Parmi les 28 taxons littoraux, 12 dont 2 archaïques y sont représentés. La région d'Edea constitue la limite Nord des aires de certains taxons fréquents au Sud tels que *Rinorea* sp. 4, *Rinorea* sp. 14.

La région Mabeta Moliwe - Mokolo

Elle ne comprend plus que 13 *Rinorea* dont 8 littoraux, 2 archaïques.

Dans la plaine littorale, le nombre d'espèces diminue du Sud vers le Nord soit d'Est en Ouest. Cette diminution du nombre d'espèces d'Est en Ouest a déjà signalée par de nombreux auteurs pour les blocs forestiers africains.

Discussions et conclusions

Dans le sous-bois, les *Rinorea* se relaient d'un type forestier à un autre comme des bioindicateurs très sensibles aux variations du milieux. Les aires des espèces épousent les contours des différents faciès forestiers reconnus par d'autres méthodes. Au Cameroun, leur répartition reflètent particulièrement bien les différences entre les deux zones biogéographiques principales: la plaine littorale et le plateau sud-camerounais.

On constate aussi qu'il existe une correspondance remarquable entre les sous-zones de biodiversité des espèces de *Rinorea* et certaines zones de refuge reconnues au Cameroun notamment par Amiet (1987), Maley (1987) et plus récemment Sosef (1994). En particulier, la région de Campo correspond à la zone d'endémisme que Sosef désigne "South coastal Cameroon", équivalent de la zone de refuge que Maley place au sud de la Sanaga. Le nord-ouest de la côte, plus pauvre que le sud, indique bien que les deux zones n'ont pas eu la même évolution, ce qui est conforme aux hypothèses de Amiet et Maley qui s'accordent à penser que les forêts de part et d'autre de la Sanaga n'ont pas eu le même passé. D'après les *Rinorea*, la forêt au sud serait plus ancienne, c'est le refuge pour les *Rinorea*.

Cette correspondance est logique. En effet les *Rinorea* comme les Anoures de Amiet (1987) et les *Begonia* de Sosef (1994, et ce volume) n'ont aucun moyen de dispersion à grande distance. Les *Rinorea* ont des fruits déhiscents; les graines lourdes se dispersent à très faible distance du pied mère. Le transport zoochore est nul; les fruits ne sont pas consommés par les animaux.

Après une phase climatique aride les *Rinorea* comme les Anoures et les *Begonia* n'ont pas les moyens de suivre rapidement l'expansion des forêts. Seules les zones de forêts très anciennes pourront héberger les *Rinorea*. L'observation montre qu'aujourd'hui dans les zones de contact forêts-savanes, les *Rinorea* n'occupent pas les recrûs, mais se confinent en forêt plus âgée, notamment les bords humides de thalwegs près des cours d'eau.

Tableau 1. Noms des *Rinorea* spp. ined. correspondants aux numéros du texte.

Rinorea sp. 1	*Rinorea lobeensis*	*Rinorea* sp. 9	*R. villiersi*
Rinorea sp. 2	Nkongmeneck 413	*Rinorea* sp. 10	*R. thomasi*
Rinorea sp. 3	*R. amplexicaulis*	*Rinorea* sp. 11	*R. fausteana*
Rinorea sp. 4	*R. meziliana*	*Rinorea* sp. 12	*R. amietii*
Rinorea sp. 5	*R. kribiensis*	*Rinorea* sp. 13	*R. faurei*
Rinorea sp. 6	*R. dimakoensis*	*Rinorea* sp. 14	*R. bosi*
Rinorea sp. 7	*R. spongiocarpa*	*Rinorea* sp. 15	*R. soleensis*
Rinorea sp. 8	*R. vivieni*	*Rinorea* sp. 16	*R. dewildei*

Remerciements

Le Professeur J.L. Amiet a eu l'amabilité de lire et de critiquer les versions successives de ce travail, qui doit beaucoup à ses connaissances en biogéographie du Cameroun. Les dirigeants du programme ECOFIT de l'ORSTOM ont permis sa réalisation matérielle, notamment les prospections de terrain: qu'ils soient assuré de ma gratitude pour leur collaboration.

Bibliographie

Amiet, J.L., 1987. Aires disjointes et taxons vicariants chez les Anoures du Cameroun: implications paléoclimatiques. Alytes 6 (3-4): 99-115.

Letouzey, R., 1968. Etude phytogéographique du Cameroun. Encyclopédie Biologique 69: 1-508. Lechevalier, Paris.

Letouzey, R., 1985. Notice de la carte phytogéographique du Cameroun au 1:500000. Institut de la Carte Internationale de la Végétation, Toulouse. 240 p.

Maley, J., 1987. Fragmentation de la forêt dense humide africaine et extension des biotopes montagnards au quaternaire récent: nouvelles données polliniques et chronologiques. Implications paléoclimatiques et biogéographiques. Paleoecology of Africa 18: 307-334.

Morin, S., 1979. Geomorphologie - Atlas de la République Unie du Cameroun. Edit. Jeune Afrique, Paris: 13-15.

Sosef, M.S.M., 1994. Refuge Begonias: Taxonomy, phylogeny and historical biogeography of *Begonia* sect. *Loasibegonia* and sect. *Scutobegonia* in relation to glacial rain forest refuges in Africa. Thesis Wageningen Agric. Univ. and Studies in Begoniaceae V. Wageningen Agric. Univ. Papers 94-1: 1-306.

Changes in forest composition over the last 4000 years in the Ituri basin, Zaire

T.B. HART [1], J.A. HART [1], R. DECHAMPS [2], M. FOURNIER [3] & M. ATAHOLO [4]
[1] The Wildlife Conservation Society, International Programs, Bronx, N.Y. 10460, USA.
[2] Musée Royal de l'Afrique Central, B-3080, Tervuren, Belgium.
[3] ORSTOM, Centre de Bondy, 93143 Bondy Cedex, France.
[4] Centre de Formation et de Recherche en Conservation Forestière, Epulu, Zaire.

Abstract
Charcoal recovered from pits in mixed forest and forest dominated by *Gilbertiodendron dewevrei* in the Ituri watershed of NE Zaire provide a 4000 year record of vegetation change. 36 tree, shrub and liana species were identified from 1817 charcoal samples; 28 samples were radiocarbon-dated. Guineo-Congolian rain forest was present throughout the period. *Cynometra alexandri* is common in both the contemporary forest and the charcoal. Two notable changes in forest composition are the absence of *G. dewevrei* in the charcoal, and the disappearance from the contemporary flora of *Lebruniodendron leptanthum*, the most frequently identified species in the charcoal. Charcoal distribution and dates indicate that fires were widespread but probably mostly small. Increase in fire frequency in the last two millennia may reflect climate fluctuations and human activity.

Résumé
Le charbon de bois retrouvé dans des fosses creusées en forêt mélangée et en forêt à *Gilbertiodendron dewevrei*, dans le bassin de l'Ituri situé dans le nord-est du Zaïre, est témoin d'un changement de végétation sur une période de 4000 ans. On a pu identifier 36 espèces d'arbres, arbustes et lianes à partir de 1817 échantillons de charbon de bois; 28 d'entre eux ont été datés au carbone radioactif. La forêt ombrophile guinéo-congolaise était présente durant toute cette période. *Cynometra alexandri* y est commun tant dans la forêt actuelle que dans le charbon de bois. Les deux changements notables dans la composition de la forêt sont l'absence de *Gilbertiodendron dewevrei* dans le charbon de bois et la disparition de *Lebruniodendron leptanthum*, l'espèce qui a été le plus fréquemment identifiée dans le charbon de bois. La distribution du charbon de bois et sa datation indiquent que des feux étaient largement répandus mais probablement le plus souvent sous forme de petits foyers. Une augmentation de la fréquence des feux dans les deux derniers millénaires pourrait résulter de fluctuations climatiques et d'une activité humaine.

Key words
Fire, Forest history, Holocene, Tropical forests, Zaire.

Introduction

Our understanding of the history of the closed forests of northeastern Zaire is based in large part on interpretations of indirect evidence from modern faunal and floral distributions. Because of their high biological diversity these forests have been proposed as a refugia during Pleistocene and Holocene climate fluctuations (Grubb 1982, Hall & Moreau 1970, Hamilton 1982). This scenario of continuous forest cover has recently been challenged. An alternate hypothesis proposed to explain high primate diversity in the region is that of multiple immigrations into colonizing forest rather than persistence in ancient forest (Colyn et al. 1991). These alternative interpretations of forest history can not be evaluated without concrete data concerning past vegetation.

L.J.G. van der Maesen et al. (eds.), The Biodiversity of African Plants, 545-563.

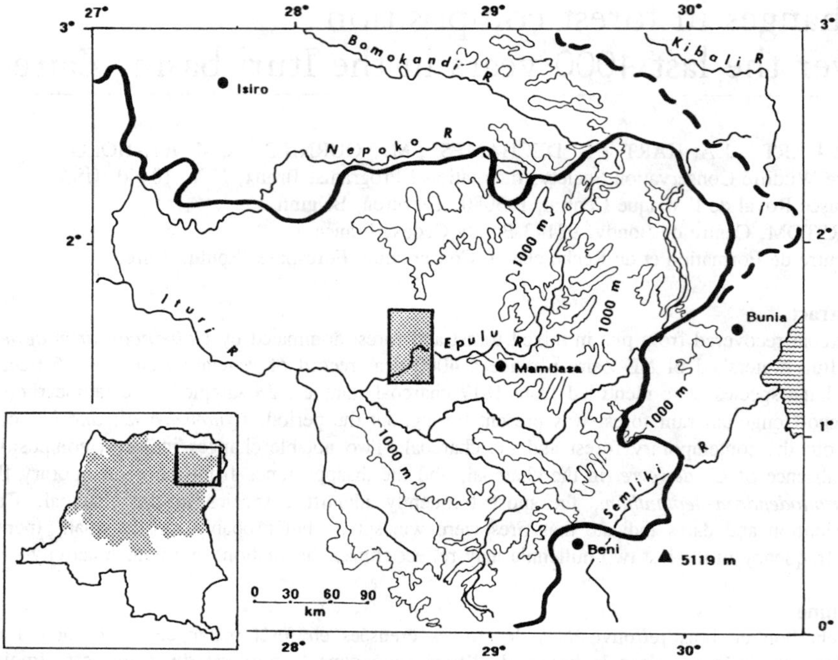

Figure 1. Lowland forest of the Ituri watershed, Zaire. Bold line indicates approximate border of continuous forest cover. Bold dashed line indicates approximate northern limit of forest/savanna mosaic. The isolines delimit the 1000 m contour. The stippled rectangle locates the Epulu region.

In this paper we present the first direct evidence of the Holocene vegetation history of the Ituri Forest of northeastern Zaire based on an analysis of charcoal samples found in subsurface soils. Our objectives are to present species determinations and radiocarbon dates for subsurface charcoal finds in the Epulu area of the central Ituri Forest and to compare these with the composition of contemporary vegetation formations in the region.

Charcoal discovered in the soils and lake sediments of tropical evergreen forest regions has been interpreted variously as evidence for past dry periods and/or periods of greater human activity (Burney et al. 1994, Fearnside 1990, Goldammer & Seibert 1990, Horn & Sanford 1992). Well-preserved charcoal presents an additional opportunity for the analysis of past vegetation. Not only can samples be dated by radiocarbon methods, but, based on wood structure, samples can be identified, often to species (Dechamps et al. 1988, Hopkins et al. 1993).

Site

The Ituri Forest (Figure 1) covers the upper watershed of the Ituri River between the equator and 03°N latitude and between 27°E and 30°E longitude. Elevation in the Ituri Forest ranges from about 600 m in the west to about 1500 m where the forest borders a mosaic of savanna and transition montane forest in the east. Soils in the central Ituri

Forest are derived from granitic Precambrian shield (Lavreau 1982, Sys 1972) and range from red and ochre sandy clays to loamy sands. Soils are generally homogeneous and lack distinct horizons beneath a shallow decomposition layer. Cation exchange capacity is low throughout (Hart 1985).

The Ituri Forest is currently composed of two principal types of Guineo-Congolian rain forest: mixed moist semi-evergreen forest and single-dominant evergreen forest (White 1983). Both forest types are dominated by Caesalpinioideae. The mixed forest has a high representation of *Julbernardia seretii* (De Wild.) Troupin and *Cynometra alexandri* C.H. Wright; the monodominant forest is dominated by *Gilbertiodendron dewevrei* (De Wild.) Léonard. Monodominant forest patches where *G. dewevrei* accounts for more than 70% of large trees are confined primarily to the south and west part of the Ituri region where they form a mosaic with mixed forest patches (Hart et al. 1989).

In addition to these major formations there are also small areas of specialized vegetation associated with specific environmental conditions. These include swamp forests that occur along streams in areas of impeded drainage and a xerophilous flora currently occurring in isolated patches and restricted to dry hilltops in a range of granitic inselbergs in the northern forest between 28° and 29°E longitude. This inselberg flora contains endemic species and a number of disjunct species linking it with floras of similar habitats in eastern and southern Africa and the Sudan (Bamps & Lisowski 1990, Lisowski 1992).

Field work for the study reported here was carried out in forest between elevations of 700 m and 850 m and within an area of 30 km of the Centre de Formation et de Recherche en Conservation Forestière (CEFRECOF), at Epulu (1°23' N, 28°35' E). The Epulu area is a transition zone between predominantly mixed forest to the north and a mosaic of monodominant forest to the south that includes stands of up to 10 km^2 in which *G. dewevrei* comprises more than 70% of the large trees.

Mean annual rainfall at Epulu was 1700 mm for the seven years, 1987 - 1993, with annual totals varying between 1307 and 2084 mm. Rainfall is seasonal with a single well-marked dry season from January to March. Dry seasons vary in their extent and severity. Total rainfall for these three months varied from 76 to 359 mm during the seven-year period (Figure 2). Forest fires are currently rare in the central Ituri Forest. We have recorded fires caused by lightning strikes that have burned one to three trees. Most fires in the forest occur during very dry periods, affect a small area, and can be traced to human activity.

Methods and analysis

A total 416 pits were inspected for buried charcoal at four sites in the Epulu area. Mixed forest sites were sampled on the Edoro Study Area (E). *Gilbertiodendron dewevrei* mosaic was sampled on the Lenda (L), Tofya (T) and Sambo (S) Study Areas (Figure 3). On the Edoro Study Area 380 pits, 2.5 m x 1 m x 1.75 m deep, originally dug for capture of okapi for a radio telemetry study, were used to sample charcoal. These pits had been dug in 44 clusters of 5 to 12 pits each in well drained sites over a 25 km^2 area. Pits in *G. dewevrei* mosaic were 2.5 m x 1 m x 0.8 m deep and were located in two clusters of 6 pits each across a transition from high to low *G. dewevrei* dominance at each site. At each pit the four walls were scraped and examined and all

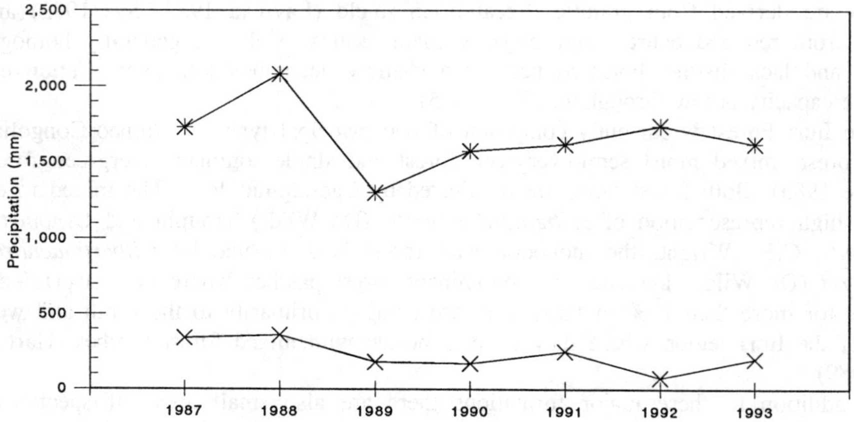

Figure 2. Total annual rainfall, Epulu, Zaire (1°23' N latitude, 28°35' E longitude), and rainfall during the dry season (January - March).

charcoal particles greater than 0.5 mm in diameter were removed. Multiple particles found within 10 cm of each other were lumped as a single collection unless otherwise clearly separable by stratum or appearance.

Charcoal collections were radiocarbon dated at Beta Analytic, Inc. in Miami, Florida and at ORSTOM laboratories in Bondy, France. Samples of charcoal were examined for rootlets then subjected to a series of acid, alkali and acid soakings to remove carbonates and humic acids. This was followed by benzene treatment and counting. A total of 28 charcoal samples were analyzed, 20 from the Edoro Study Area, the mixed forest site, and eight from the *G. dewevrei* mosaic sites.

Species determinations of charcoal collections were made at the Laboratoire d'Anatomie des Bois Tropicaux, Musée Royale de l'Afrique Centrale, Tervuren, Belgium on the basis of wood structure and comparison with the Museums's collection of wood samples from more than 3,000 African species (Dechamps 1975, Pinçon & Dechamps 1991). Wood structure in the charcoal was examined using a reflective light microscope after fracture of a fresh sample.

Different collections were generally used for species identification and radiocarbon dating. For six samples, however, subsamples were sent for radiocarbon dating after species determination.

Inventories of the contemporary flora were made at each site. Mixed forest trees were sampled at the Edoro Study Area by a 38 km x 5 m belt transect (n = 1800 stems ≥ 25 cm diameter at 1.3 m from the ground (dbh)), and on 48, 25 m x 25 m plots (n = 1288 stems ≥ 10 cm dbh) (Figure 3). Tree inventories in *G. dewevrei* mosaic included 45 km of belt transect at Lenda (n = 3318 stems ≥ 25 cm dbh) and 8, 25 m x 25 m plots each at Lenda, Sambo, and Tofya (n = 516 stems ≥ 10 cm dbh).

The CEFRECOF herbarium (established 1981, currently 2500 specimens) contains a collection of woody lianas, shrubs and trees of the Epulu region.

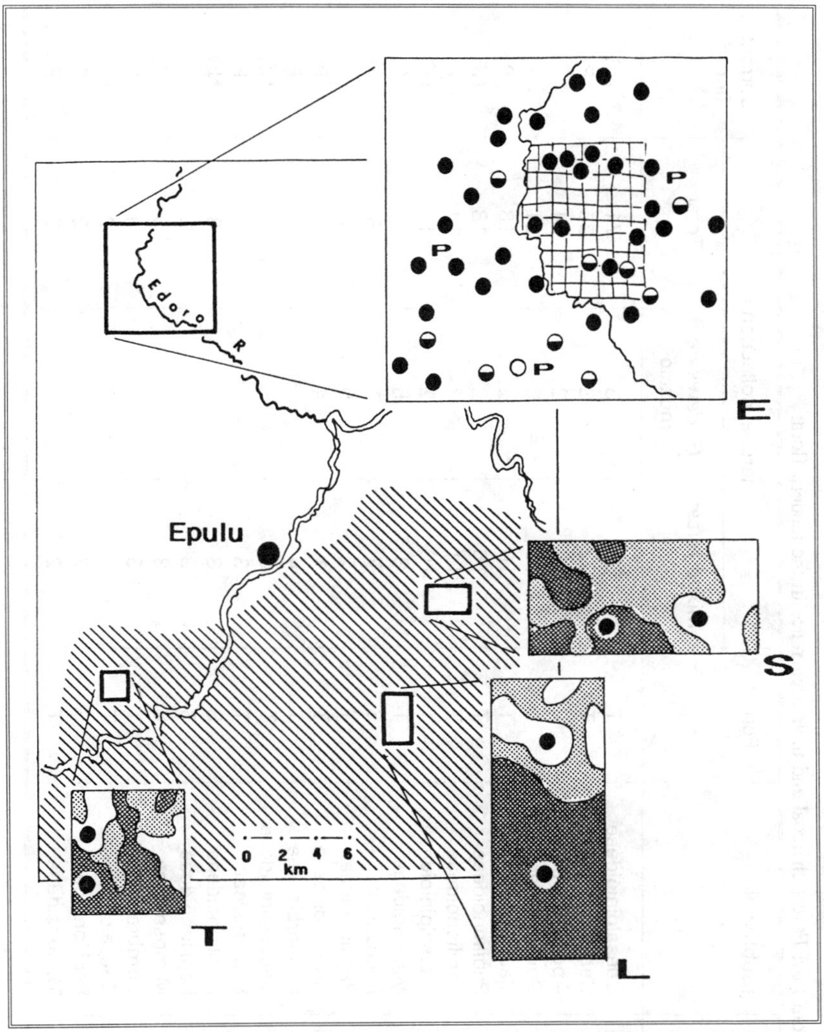

Figure 3. Study areas in the Epulu region. Diagonal hatching indicates area of mono-dominant *Gilbertiodendron dewevrei* and mixed forest mosaic. Insets locate study areas where floristic inventories were made and charcoal samples collected. Circles indicate pit groups sampled for charcoal; charcoal present in > 50% of pits (filled circles); charcoal present in > 50% of pits (filled circles); charcoal present in ≤ 50% of pits (half-filled circles); no charcoal recovered (open circles). Inventories of canopy trees on the Edoro Study area (E) included belt transects (irregular grid) and 16 plots of 25 m x 25 m at each of three sites (P). Percentage canopy composition by *G. dewevrei* on *G. dewevrei* mosaic study areas at Sambo (S), Lenda (L) and Tofya (T) is shown as > 65% (dark stippling); 20% - 65% (light stippling) and < 20% (no stippling). Plot inventories on *G. dewevrei* mosaic study sites were located in areas with > 65% canopy representation by *G. dewevrei*. Transect inventories included transition zones.

Table 1. Woody species identified from Ituri Forest charcoal and their occurrence in the modern flora.

Species	(Sub)Family	Form [a]	Charcoal collections [b]			Current flora [c]
			Mixed forest	*G. dewevrei* mosaic	Total	
Lebruniodendron leptanthum	Caesalpinioideae	T	68	3	71	NR
Cynometra alexandri	Caesalpinioideae	T	32	3	35	P
Landolphia spp.	Apocynaceae	L	22	2	24	
Agelaea spp.	Connaraceae	L	16	5	21	
Monopetalanthus microphyllus	Caesalpinioideae	T	16	2	18	P
Pterocarpus casteelsii	Papilionoideae	T	16	2	18	NR
Pterocarpus soyauxii	Papilionoideae	T	11	3	14	P
Brachystegia laurentii	Caesalpinioideae	T	10	2	12	NR
Albizia ferruginea	Mimosoideae	T	3	5	8	P
Diospyros sp.	Ebenaceae	S	7	1	8	
Pterocarpus sp.	Papilionoideae	T	2	5	7	
Lannea welwitschii	Anacardiaceae	T	1	4	5	P
Cleistanthus michelsonii	Euphorbiaceae	T	4		4	P
Cynometra hankei	Caesalpinioideae	T	3		3	NR
Eriocoelum microspermum	Sapindaceae	T	3		3	P
Baphia laurifolia	Papilionoideae	S	3		3	NR
Baphia sp.	Papilionoideae	S	3		3	
Samanea leptophylla	Mimosoideae	S	3		3	P
Combretum lokele	Combretaceae	T	3		3	NR
Oxystigma oxyphyllum	Caesalpinioideae	T		2	2	NR
Platysepalum violaceum	Papilionoideae	S	2		2	P
Albizia ealaensis	Mimosoideae	T	2		2	P

Table 1 (continued).

Anonidium mannii	Annonaceae	T	2		2	P
Julbernardia seretii	Caesalpinioideae	T	2		2	P
Enantia atrocyanescens	Annonaceae	S	1	1	2	NR
Popowia lucidula	Annonaceae	L	2		2	P
Ixora sp	Rubiaceae	S		1	1	
Milbraediodendron excelsum	Caesalpinioideae	T	1		1	P
Tessmannia anomala	Caesalpinioideae	T	1		1	P
Millettia mutans	Papilionoideae	S	1		1	NR
Cathormion altissimum	Mimosoideae	T	1		1	P
Piptadeniastrum africanum	Mimosoideae	T	1		1	P
Monodora louisii	Annonaceae	T		1	1	NR
Entandrophragma angolense	Meliaceae	T	1		1	P
Trichilia emetica	Meliaceae	T	1		1	NR
Tarchonanthus camphoratus	Compositae	S	1		1	NR
Hunteria congolana	Apocynaceae	S		1	1	P
Ongokea gore	Olacaeae	S		1	1	NR
Crataeva sp	Capparidaceae	S		1	1	
Jaundea pubescens	Connaraceae	L	1		1	P
Clitandra cymulosa	Apocynaceae	L		1	1	P
Cnestis urens	Connaraceae	L		1	1	P
Ventilago africana	Rhamnaceae	L	1		1	P
Strophanthus sp	Apocynaceae	L	1		1	

[a] Plant life form: T, tree. S, shrub, L, liana.
[b] Figures show number of sample pits where collections were identified. Mixed forest sites located on the Edoro Study Area (E). *G. dewevrei* mosaic sites located at Tofya (T), Lenda (L) and Sambo (S). See Figure 3.
[c] Occurrence in current flora: P, specimen present in CEFRECOF herbarium. NR, no specimen recorded.

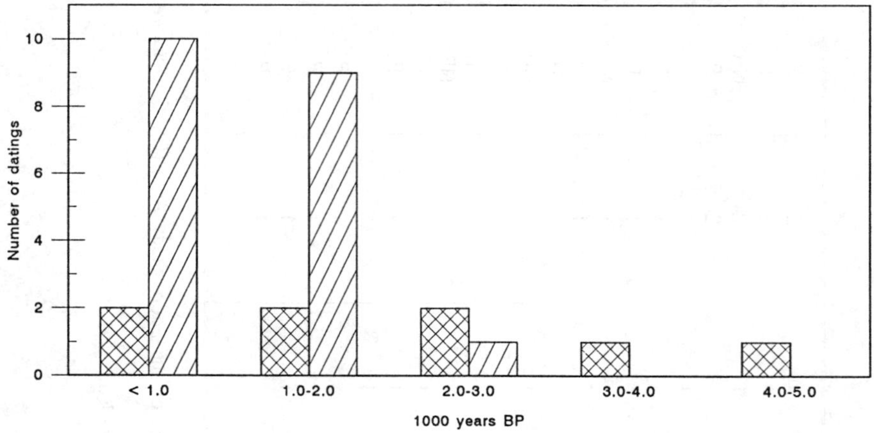

Figure 4. Radiocarbon dates of subsurface charcoal collections from the Epulu region, Ituri Forest Zaire. Collections from monodominant *Gilbertiodendron dewevrei* mosaic areas (S,L,T) are shown by checked bars. Collections from mixed forest areas (E) are shown by hatched bars.

Results

Charcoal deposits and dates

Subsurface charcoal is both widespread and frequent. Charcoal deposits or traces of charcoal were found in a total of 279 pits (67% of those examined). At the Edoro Study Area, 248 of the 380 pits had some charcoal. Charcoal was also found at all three monodominant mosaic sites in 31 of the 36 pits (Figure 3). Charcoal occurred in pits as distinct particles or diffuse deposits. In some pits only traces (< 0.5 mg) of charcoal were found. The maximum amount of charcoal recovered from a single level was 105 g.

Charcoal was found to a maximum depth of 80 cm. On the Edoro Study Area 58% of the collections were from 5 to 20 cm; an additional 31% were from 20 to 40 cm. In *G. dewevrei* mosaic average depth of charcoal finds was greater, 35% of the collections were from 0 to 25 cm and 41% from 25 to 50 cm.

The radiocarbon ages for charcoal varied from 130 $\pm$ 70 to 4190 $\pm$ 160 years before present (yr BP). Nineteen out of 20 dates from the Edoro mixed forest collections were less than 2000 yr BP. Ten dates were less than 1000 yr BP and five of these were less than 500 yr BP. In the *G. dewevrei* mosaic collections only two out of eight radiocarbon dates were less than 1000 yr BP and none were less than 500 yr BP. The oldest date was from *G. dewevrei* forest (Figure 4). Radiocarbon dates $\pm$ 1 standard deviation are given for each sample that was analyzed (Appendix 1).

Pottery shards were recovered from one pit on the Edoro Study Site. A date of 190 $\pm$ 90 yr BP was determined for charcoal taken from an adjacent pit.

Radiocarbon dates were made for multiple collections from the same pit at four sites on the Edoro Study Area. For three pits, radiocarbon dates overlapped or diverged by only a few decades and may have been formed, in each case, by a single burn event. In the fourth instance, radiocarbon dates for different collections diverged by 140 to

240 years. Charcoal in this instance could have resulted from separate fires or from different-aged wood burning simultaneously.

Dates for charcoal collections at the Edoro Study Area mixed forest sites are loosely grouped (Appendix 1) at approximately 200 yr BP (5 sites), 850 yr BP (3 sites), 1100 yr BP (2 sites), 1300 yr BP (4 sites) and 1450 yr BP (2 sites). Dates of *G. dewevrei* mosaic collections that fall within these periods include 850 $\pm$ 80 yr BP from Sambo and 700 $\pm$ 110 yr BP from Tofya. Tofya and Edoro also share two dates of 2180 yr BP. Two additional *G. dewevrei* mosaic dates of 1750 $\pm$ 100 yr BP and 1790 $\pm$ 60 yr BP overlap. These groupings suggest that cycles of localized fires recurred periodically over the Epulu area.

Species determinations of charcoal

A total of 1817 identifications were made from separate charcoal collections. These included 31 tree or shrub species and five liana species. In addition, three liana genera and five tree/shrub genera were identified for which the species could not be determined (Table 1). Less than 15% of the charcoal collections examined could not be determined to genus. Root material comprised almost half of the collections examined.

The tree species most frequently identified in charcoal samples were *Lebruniodendron leptanthum* (Harms) Léonard (71 pits) and *Cynometra alexandri* (35 pits). Other tree species that were identified from 10 or more pits include *Monopetalanthus microphyllus* Harms (18 pits), *Pterocarpus casteelsii* De Wild. (18 pits), *Pterocarpus soyauxii* Taub. (14 pits), and *Brachystegia laurentii* (De Wild.) Léonard (12 pits). Charcoal of all six of these species was collected from pits in both mixed forest on the Edoro Study Area and from *G. dewevrei* mosaic forest. Charcoal from liana species was also collected from both forest types. Charcoal of *Landolphia* spp. was recovered from 24 pits and charcoal of *Agelaea* spp. from 21 pits.

Fourteen of the 36 species identified in the charcoal have not been recorded in the contemporary flora of the central Ituri Forest, including three of the six most frequently recorded species, *L. leptanthum*, *P. casteelsii*, and *B. laurentii*. Only one species, *C. alexandri*, was both frequent in the charcoal flora, and widespread and abundant in the Epulu area today.

Two other charcoal identifications, *Tarchonanthus camphoratus* L. and *Trichilia emetica* Vahl are also exceptional. *T. emetica* is a small tree currently found in savanna woodlands north and east of the closed forest zone (De Wilde 1968); *T. camphoratus* is currently known only from rocky hillsides and deciduous woodlands in eastern and southern Africa (Pope 1992). Both species were identified on the Edoro Study Area from adjacent pits. The pit containing *T. camphoratus* was dated to 2180 $\pm$ 120 yrs BP. This pit and the pit with *T. emetica* charcoal also contained charcoal of *Cleistanthus michelsonii* Léonard and *Cynometra alexandri*, both species typical of moist semi-evergreen rain forest today.

Charcoal collections that were identified and then dated include *C. alexandri* (1090 $\pm$ 60 and 1340 $\pm$ 50 yr BP) and *Cathormion altissimum* (Oliv.) Gilbert & Boutique (1400 $\pm$ 40 yr BP), both elements of the contemporary flora, and *Lebruniodendron leptanthum* (250 $\pm$ 40 yr BP), *Millettia mutans* (620 $\pm$ 40 yr BP), and *Pterocarpus casteelsii* (990 $\pm$ 100 yr BP), none of which have been recorded in the Ituri Forest today.

Table 2. Common tree species of the contemporary flora of the Ituri Forest and their occurrence in the charcoal record [a].

Species	(Sub)Family	Percentage total stems				Presence in charcoal record [b]
		Mixed forest		*G. dewevrei* mosaic		
		Plot (n = 1288)	Transect (n = 1800)	Plots (n = 516)	Transects (n = 3318)	
Gilbertiodendron dewevrei	Caesalpinioideae	0.0	0.8	74.0	64.8	NR
Cynometra alexandri	Caesalpinioideae	14.9	27.9	0.4	2.1	P
Pancovia harmsiana	Sapindaceae	8.5	0.0	6.8	0.0	NR
Julbernardia seretii	Caesalpinioideae	4.4	7.0	1.2	8.8	P
Erythrophleum suaveolens	Caesalpinioideae	3.2	1.5	0.0	0.6	NR
Platysepalum violaceum	Papilionoideae	2.8	0.1	0.0	0.0	P
Cleistanthus michelsonii	Euphorbiaceae	2.7	5.1	1.3	0.9	P
Strombosia pustulata	Olacaceae	2.8	2.2	0.4	0.1	NR
Cola lateritia	Sterculiaceae	2.6	1.5	0.2	1.1	NR
Apodytes dimidiata	Icacinaceae	2.5	2.8	0.0	0.0	NR
Alstonia boonei	Apocynaceae	2.5	6.3	0.0	0.6	NR

[a] All species representing ≥ 2.5 % of the stems on inventories on mixed (E) or *G. dewevrei* mosaic (L,S,T) study areas are included. Stems ≥ 25 cm diameter at 1.3 m from the ground (dbh) censused on transects; stems ≥ 10 cm dbh censused on plots.
[b] Present (P), not recorded (NR).

Contemporary flora

Eleven species of trees each represented ≥ 2.5% of the total stems inventoried in mixed or *G. dewevrei* mosaic forests in the Epulu region today (Table 2). *Cynometra alexandri* and *Julbernardia seretii* are the most abundant large trees (≥ 25 cm dbh) in mixed forest on the Edoro Study Area. If stems ≥ 10 cm dbh are also included, *Pancovia harmsiana* Gilg. represents over 8% of the total stems. *Gilbertiodendron dewevrei* represented < 1% of the stems on transects on the Edoro Study Area. With the exception of *Apodytes dimidiata* Mey ex Bernh. and *Platysepalum violaceum* Welw. ex Bak., all common tree species are found in both forest types, although their percentage representation in *G. dewevrei* mosaic is reduced due to the dominance of the single species *G. dewevrei*. Of the 11 most abundant trees in the modern flora, seven, including *G. dewevrei*, were not recorded in the charcoal collections (Table 2).

Discussion

Fires and climate change

The first charcoal record of fire in the central Ituri region coincides with the beginning of a drying trend in central Africa. Evidence for this trend was found in Burundi as early as 4350 yr BP (Bonnefille 1987). Elsewhere in central Africa the trend may have started towards 3500 yr BP and was punctuated with abrupt climatic fluctuations (Maley 1990). Palynological and archaeological data from central Africa indicate a severe and widespread dry period between 2000 and 2500 yr BP (Schwartz et al. 1990, Maley 1992). This period coincides with a marked increase in the number of radiocarbon dates from the study area as well as an earlier reported radiocarbon date of 2290 ± 90 yr BP for charcoal from a pit near Epulu (Hart & Hart 1986). These dates may chronicle a period when successive and/or severe dry years increased the forest's vulnerability to fire and perhaps favoured human migration into the forest interior.

The greater frequency of fires in mixed forest during the last two millennia was likely a result of increasing human activity. The pattern of small fires scattered over the landscape is supported by clumped dates at single pits and suggests a scenario of escaped human fires into an environment that was only marginally and locally susceptible to burning. Forest fires in the Ituri, as we have known them over the last 20 yr, whether ignited by humans or lightning strikes, have generally burned only a few trees and affected less than one ha of forest. The fact that overlapping radiocarbon dates were found in widely separated pits suggests that small fires increased in frequency over large areas during periodic dry episodes.

Although recent fires were likely associated with human activity, these were probably wild-burning, not hearth fires. The charcoal was widely distributed and the species forming the charcoal represent a range of life forms including understorey shrubs, lianas and canopy trees. Primarily hard wood species were recorded, but charcoal of soft wood species was also represented.

Various natural phenomena, in combination with extreme dry seasons, may also make the forest prone to fire. Storm tracks in the form of corridors of treefalls are relatively common in the central Ituri Forest. The resulting broken and uprooted trees become potential tinder as shown in Malaysia where logging debris contributed to rain forest fires during an El Niño drought (Woods 1989). The prevalence of root material in the charcoal may have resulted from the carbonization of upturned root masses.

Change in vegetation type

The charcoal record provides no evidence for a major shift in vegetation type in the Central Ituri Forest over the last 4000 years. Two thirds of the species determined from the charcoal are still present in the contemporary forest. *Cynometra alexandri*, common today, was also present at least as long ago as 1300 yr BP. Lianas, abundant in the modern forest, were well represented and diverse in the charcoal flora. Tree species that are abundant in the charcoal record but have not been recorded in the contemporary flora of the Epulu region are, nevertheless, all species typical of Guineo-Congolian rain forests. Some elements of the charcoal flora are also indicative of single-dominant evergreen rain forest, e.g. *C. alexandri*, *B. laurentii*, and *J. seretii* (White 1983).

The continuous presence of moist semi-evergreen rain forest is confirmed by many of the rarer elements in the charcoal flora as well. The following are all trees or shrubs, identified from charcoal in only one or two pits, that today are found typically in Guineo-Congolian rain forest: *Platysepalum violaceum* Welw. ex Bak., *Albizia ealaensis* DeWild., *Anonidium mannii* (Oliv.) Engl. & Diels, *Julbernardia seretii*, *Enantia atrocyanescens* Robyns & Ghesq., *Oxystigma oxyphyllum* (Harms) J. Léonard, *Ongokea gore* (Hua) Pierre, *Monodora louisii* Boutique, *Piptadeniastrum africanum* (Hook. f.) Brenan, *Entandrophragma angolense* (Welw.) C.DC., *Milbraediodendron excelsum* Harms, *Tessmannia anomala* (Micheli) Harms, *Cathormion altissimum*, and *Hunteria congolana* Pichon.

Of the 1817 identifications made, there were only two that were of species clearly associated with drier habitats, *Tarchonanthus camphoratus* and *Trichilia emetica*. It seems unlikely that these indicate a replacement of forest by savanna vegetation in the Edoro region, particularly as identifications from charcoal collected in the same pits concern species typical of rain forest today.

Rather than large-scale forest conversion, it is likely that the dry period permitted expansion of various xerophilous species. Several rocky hills occur on the Edoro Study Area today and, although they currently support a forest vegetation, larger hills less than 20 km to the north are open and support elements of the specialized xerophilous flora described by Lisowski (1992). Vegetation change during the drier periods may have led to establishment of a mosaic of different forest types or a more fine-grained change in forest composition where, for a number of decades, drier elements were favoured in the normal recruitment process.

Changes in species composition

Of the three tree species that were abundant in the charcoal flora but are not known locally today, *L. leptanthum*, *P. casteelsii*, and *B. laurentii*, two have wide modern distributions in the equatorial forest zone. *B. laurentii* and *L. leptanthum* are known from the Biafrian forest of Cameroon (Letouzey 1968) as well as the forests of northern Kasai (Lebacq & Dechamps 1967); *B. laurentii* forms monodominant stands in forests surrounding Yangambi and *L. leptanthum* has also been collected from the vicinity of Yangambi. All three species have been described as components of closed evergreen forests of well-drained soils in the Tshuapa region of Zaire (Evrard, 1968).

The second most abundant species in the charcoal flora, *C. alexandri*, is widespread and locally abundant over a wide area of the Ituri Forest today. Other species that are common in the contemporary forest of the central Ituri and occur in the charcoal record are *Julbernardia seretii* and *Cleistanthus michelsonii* J. Léonard. *Julbernardia seretii* has a large modern distribution; *C. michelsonii* is restricted to the eastern forests of Zaire.

Gilbertiodendron dewevrei, which currently forms monodominant stands in the present-day central Ituri, was not identified in the charcoal flora. It is currently found at high densities in patches throughout the Guineo-Congolian forest zone west to Nigeria (Gérard 1960, Léonard 1952, Liben 1970). The apparently recent and rapid expansion of *G. dewevrei* in the Central Ituri is remarkable, especially since the species has a number of characteristics including shade tolerance, slow growth rates, and poorly dispersed seeds, that would appear to mitigate against rapid colonization (Hart et al. 1989). Factors permitting the expansion of *G. dewevrei* remain mysterious. It is not known whether the species has only recently arrived in the central Ituri or whether it formerly occurred at very low densities.

The loss of previously common species is dramatic. As little as two centuries ago, *L. leptanthum* was still present in the local flora. It is not known whether the loss of this or other formerly common species such as *P. casteelsii* and *B. laurentii* coincided with the expansion of *G. dewevrei*. Nor can the disappearance of these species be directly linked to climate change. Massive local die-offs of trees have been documented for temperate species in the historical and paleohistorical record (Allison et al. 1986, Peglar 1993). Similar impacts have not been documented for moist tropical forests. In the case of the central Ituri, the fires themselves may have played a role possibly in combination with other factors.

Dynamics of vegetation change

The charcoal record of vegetation change in the central Ituri Forest presents an enigma. There was a dramatic change in the species composition over the last four millennia but the major tree species lost and gained do not appear to differ significantly in many of their ecological attributes. Consequently, the appearance and disappearance of these principal forest components can not be simply explained by climate fluctuation alone.

Whatever the catalysts for change, the individual component species of the Ituri Forest appear to have reacted individually. The charcoal record documents local replacement of a number of mature forest species by other mature forest species. Drought, disease and fire, perhaps linked to human penetration of the forest, may all have played a role together or separately. These changes in forest composition suggest that as in temperate forests, tropical forest trees respond separately and not as communities to environmental perturbations (Bush et al. 1992, Davis 1981, Hunter et al. 1988, White 1993). Rather than being locked into co-evolved associations, the tree species comprising the closed semi-evergreen forests of the Ituri appear to react individualistically to environmental change and opportunity.

Conclusions

1 Subsurface charcoal is widespread in the central Ituri Forest. All 28 radiocarbon dates obtained were less than 4400 yr BP. Most dates are more recent than 2000 yr BP. Episodes of fires appear to have recurred in the

Epulu region periodically four or five times in the last two millennia. Rain forest fires are evidence of drier climatic pulses; however, the fires themselves may have been related to human activity in the forest.

2 There is no evidence of major change in vegetation type in the Ituri Forest over the period covered by the charcoal record. Charcoal at all periods is dominated by species typical of moist tropical forest. Two out of the 36 species identified are species clearly indicative of drier environments. These species could have co-occurred at low densities with moist forest species while their centre of concentration remained in dry microhabitats comparable to those occupied by a specialized xerophilous flora restricted to hill tops in the region today.

3 Although rain forest persisted throughout the period, its floristic composition changed radically. Over half of the most frequent elements of the contemporary flora have not been recorded in the charcoal sample and one third of the species identified in the charcoal have not been recorded in forests of the region today. The most notable changes are the complete absence in the charcoal record of *G. dewevrei*, a species which forms extensive monodominant stands throughout the southern and western Ituri Forest today and the complete absence from the modern flora of *L. leptanthum*, the most frequently recorded species in the charcoal, identified from 25% of the pits where charcoal was collected.

4 The patterns of floristic change over the last 4000 years do not indicate a directional climatic change over the period. Instead, the loss of some tree species and establishment of others suggest a process of opportunistic replacements of species that are ecological equivalents.

Acknowledgements

We thank J. Maley for discussions and J. Léonard and P. Bamps for assistance in the herbarium of the Jardin Botanique National de Belgique. J.-R. Makana and K. Muhongya aided in field work. An earlier version of the manuscript benefitted from critical reading by D. Burney and the corrections of R. Gereau. This work was supported by The Wildlife Conservation Society in partnership with the Institut Zaïrois pour la Conservation de la Nature.

References

Allison, T.D., R.E. Moeller & M.B. Davis, 1986. Pollen in laminated sediments provides evidence for a mid-Holocene forest pathogen outbreak. Ecology 67: 1101-1105.

Bamps, P. & S. Lisowski, 1990. A new species of *Encephalartos* (Zamiaceae) from northeastern Zaire. Mem. New York Bot. Gard. 57: 152-155.

Bonnefille, R., 1987. Evolution forestière et climatique au Burundi durant les quarante derniers milliers d'années. C. R. Acad. Sci. Paris 305 (sér. II): 1021-1026.

Burney, D.A., L.P. Burney & R.D.E. MacPhee, 1994. Holocene charcoal stratigraphy from Laguna Tortuguero, Puerto Rico, and the timing of human arrival on the island. Journal of Archaeological Science 21: (in press).

Bush, M.B., D.R. Piperno, P.A. Colinvaux, P.E. De Oliveira, L.A. Krissek, M.C. Miller & W.E. Rowe, 1992. A 14,300 yr paleoecological profile of a lowland tropical lake in Panama. Ecological Monographs 62: 251-275.

Colyn, M., A. Gauthier-Hion, & W. Verheyen, 1991. A re-appraisal of palaeoenvironmental history in Central Africa: evidence for a major fluvial refuge in the Zaire Basin. J. Biogeogr. 18: 403-407.

Davis, M.B., 1981. Quaternary history and the stability of forest communities. In: D.C. West, H.H. Shugart & D.B. Botkin (eds.), Forest Succession, Concepts and Application: 132-153, Springer-Verlag, New York.

Dechamps, R., 1975. Identification de Charbons de bois provenant de la Fosse de l'Age du Fer (K.70 A.31). Annexe VI. In: D. Cahen. Le site archéologique de la Kamoa (Région du Shaba-Rep. du Zaïre) De l'âge de la pière ancien à l'âge du fer: 390-395, Musée Royale de l'Afrique Centrale, Annales Sc. In, Sc. Hum. 84, Tervuren.

Dechamps, R., R. Lanfranchi, A. LeCocq & D. Schwartz, 1988. Reconstitution d'environnements quaternaires par l'étude de macrorestes végétaux (Pays Bateke, R.P. du Congo). Palaeogeography, Palaeoclimatology, Palaeoecology 66: 33-44.

De Wilde, J.J.F.E., 1968. A revision of the species of *Trichilia* P. Browne (Meliaceae) on the African continent. Mededelingen Landbouwhogeschool Wageningen 68-2: 1-207.

Evrard, C., 1968. Recherches ecologiques sur le peuplement forestier des sols hydromorphes de la cuvette centrale congolaise. Publ. de l'Institut National pour l'Etude Agronomique du Congo (I.N.E.A.C.), sér. sci. n° 110, Brussels.

Fearnside, P.M., 1990. Fire in the tropical rain forest of the Amazon basin. In: J.G. Goldammer (ed.). Fire in the Tropical Biota: Ecosystem Processes and Global Challenges: 106-116. Springer-Verlag, Berlin.

Gérard, P., 1960. Etude écologique de la forêt dense à *Gilbertiodendron dewevrei* dans la région de l'Uele. Institut National pour l'Etude Agronomique du Congo (I.N.E.A.C.), sér. sci. n° 87. Brussels.

Goldammer, J.G. & B. Seibert, 1990. The impact of droughts and forest fires on tropical lowland rain forest of East Kalimantan. In: J.G. Goldammer (ed.), Fire in the Tropical Biota: Ecosystem Processes and Global Challenges: 11-31, Springer-Verlag, Berlin.

Grubb, P., 1982. Refuges and dispersal in the speciation of African forest mammals. In: G.T. Prance (ed.). Biological diversification in the Tropics: 537-553. Columbia University Press, New York.

Hall, B.P. & R.E. Moreau, 1970. An atlas of speciation in African passerine birds. British Museum of Natural History, London.

Hamilton, A.C., 1982. Environmental history of East Africa: a study of the Quaternary. Academic Press.

Hart, T.B., 1985. The ecology of a single-species-dominant forest and of a mixed forest in Zaire, Africa. Ph.D. diss. Michigan State University, East Lansing.

Hart, T.B. & J.A. Hart, 1986. The ecological basis of hunter-gatherer subsistence in African rain forests: the Mbuti of eastern Zaire. Human Ecology 14 (1): 29-55.

Hart, T.B., J.A. Hart, & P.G. Murphy, 1989. Monodominant and species-rich forests of the humid tropics: causes for their co-occurrence. Am. Nat. 133: 613-633.

Hopkins, M.S., J. Ash, A.W. Graham, J. Head & R.K. Hewett, 1993. Charcoal evidence of the spatial extent of the *Eucalyptus* woodland expansions and rain forest contractions in North Queensland during the late Pleistocene. J. Biogeogr. 20: 357-372.

Horn, S.P. & R.L. Sanford, Jr., 1992. Holocene fires in Costa Rica. Biotropica 24 (3): 354-361.

Hunter, M.L., G.L. Jacobson & T. Webb, 1988. Paleoecology and the coarse-filter approach to maintaining biological diversity. Conservation Biology 2: 375-385.

Lavreau, J., 1982. Etude Géologique du Haut-Zaïre. Musée Royale de l'Afrique Centrale, Annales, série In-8, Sciences Géologiques 88, Tervuren, Belgium.

Lebacq, L. & R. Dechamps, 1967. Contribution à un Inventaire de Forêts du Nord-Kasai. Musée Royal de l'Afrique Centrale Annales, sér. In-8, Sciences Economiques no 5, Tervuren.

Léonard, J., 1952. Les divers types de forêts du Congo Belge. Lejeunia 16: 82-93.

Letouzey, R., 1968. Etude Phytogéographique du Cameroun. Encyclopédie Biologique LXIX. Editions Paul Lechevalier, Paris.

Liben, L., 1970. Répartition géographique de trois Caesalpiniaceae forestières d'Afrique équatoriale. Bull. Jard. Bot. Nat. Belg. 40: 295-298.

Lisowski, S., 1992. Note floristique de l'Ituri (Haut-Zaïre). Fragm. Flor. Geobot. 37 (1): 215-220.

Maley, J., 1990. L'histoire récente de la forêt dense humide africaine: essai sur le dynamisme de quelques formations forestières. In: R. Lanfranchi & D. Schwartz (eds.), Paysages Quaternaires de l'Afrique Centrale Atlantique: 383-389, ORSTOM, Paris.

Maley, J., 1992. Mise en évidence d'une péjoration climatique entre ca. 2500 et 2000 ans B.P. en afrique tropical humide. Bull. Soc. Géol. France 163: 363-365.

Peglar, S.M., 1993. The mid-Holocene *Ulmus* decline at Diss Mere, Norfolk, UK: a year-by-year pollen stratigraphy from annual laminations. The Holocene 3 (1): 1-13.

Pinçon, B. & R. Dechamps, 1991. Identification et interprétation de charbons de bois archéologiques. C. R. Acad. Sci. Paris 312: 1393-1399.

Pope, G.V. (ed.), 1992. Flora Zambesiaca, vol. 6. Flora Zambesiaca Managing Committee, London.

Schwartz, D., H. de Foresta, R. Dechamps & R. Lanfranchi, 1990. Découverte d'un premier site de l'âge du fer ancien (2 110 B.P.) dans le Mayombe congolais. Implications paléobotanique et pédologiques. C. R. Acad. Sci. Paris 310: 1293-1298.

Sys, C., 1972. Charactérisation morphologique et physico-chimique des profils types de l'Afrique Centrale. L'Institut National pour l'Etude Agronomique du Congo (I.N.E.A.C.), hors série, Brussels.

White, F., 1983. The Vegetation of Africa: A descriptive memoir to accompany the Unesco/-AETFAT/UNSO vegetation map of Africa. Unesco, Paris.

White, F., 1993. Refuge theory, ice-age aridity and the history of tropical biotas: an essay in plant geography. Fragm. Flor. Geobot. Suppl. 2 (2): 385-409.

Woods, P., 1989. Effects of logging, drought, and fire on structure and composition of tropical forests in Sabah, Malaysia. Biotropica 21 (4): 290 -298.

Appendix 1

Radiocarbon dates of subsurface charcoal from the Central Ituri Forest.

Years BP ± 1 S.D.	Laboratory No *	Site **	Identification
130 ± 70	B - 57211	Edoro	
190 ± 90	B - 57213	Edoro	
210 ± 50	B - 57210	Edoro	
240 ± 40	O - 1271	Edoro	
250 ± 40	O - 1210	Edoro	*Lebruniodendron leptanthum*
520 ± 40	O - 1145	Edoro [a]	
620 ± 40	O - 1147	Edoro [a]	*Millettia mutans*
700 ± 110	B - 59105	Tofya	
830 ± 60	B - 28200	Edoro	
850 ± 80	B - 57205	Sambo	
860 ± 40	O - 1273	Edoro	
900 ± 60	B - 57214	Edoro	
1010 ± 60	O - 1204	Edoro [b]	*Cynometra alexandri*
1090 ± 60	O - 1208	Edoro [b]	*Pterocarpus casteelsii*
1090 ± 60	O - 1202	Edoro [c]	
1180 ± 60	O - 1209	Edoro [b]	
1340 ± 50	O - 1203	Edoro [c]	
1330 ± 50	B - 28202	Edoro	*Cynometra alexandri*
1330 ± 40	O - 1212	Edoro [d]	
1400 ± 40	O - 1211	Edoro [d]	*Cathormion altissimum*
1430 ± 90	B - 28201	Edoro	
1750 ± 100	B - 57207	Lenda	
1790 ± 60	B - 59107	Tofya	
2180 ± 60	B - 57209	Lenda	
2180 ± 120	B - 57212	Edoro	
2470 ± 70	B - 59104	Tofya	
3020 ± 100	B - 59106	Tofya	
4190 ± 160	B - 57206	Sambo	

* Samples with an O prefix were analyzed at ORSTOM, Bondy, France. B-prefixed samples were analyzed at Beta Analytic, USA.

** See Figure 3 for site locations. Sites with the same superscript designation are collections from the same pit.

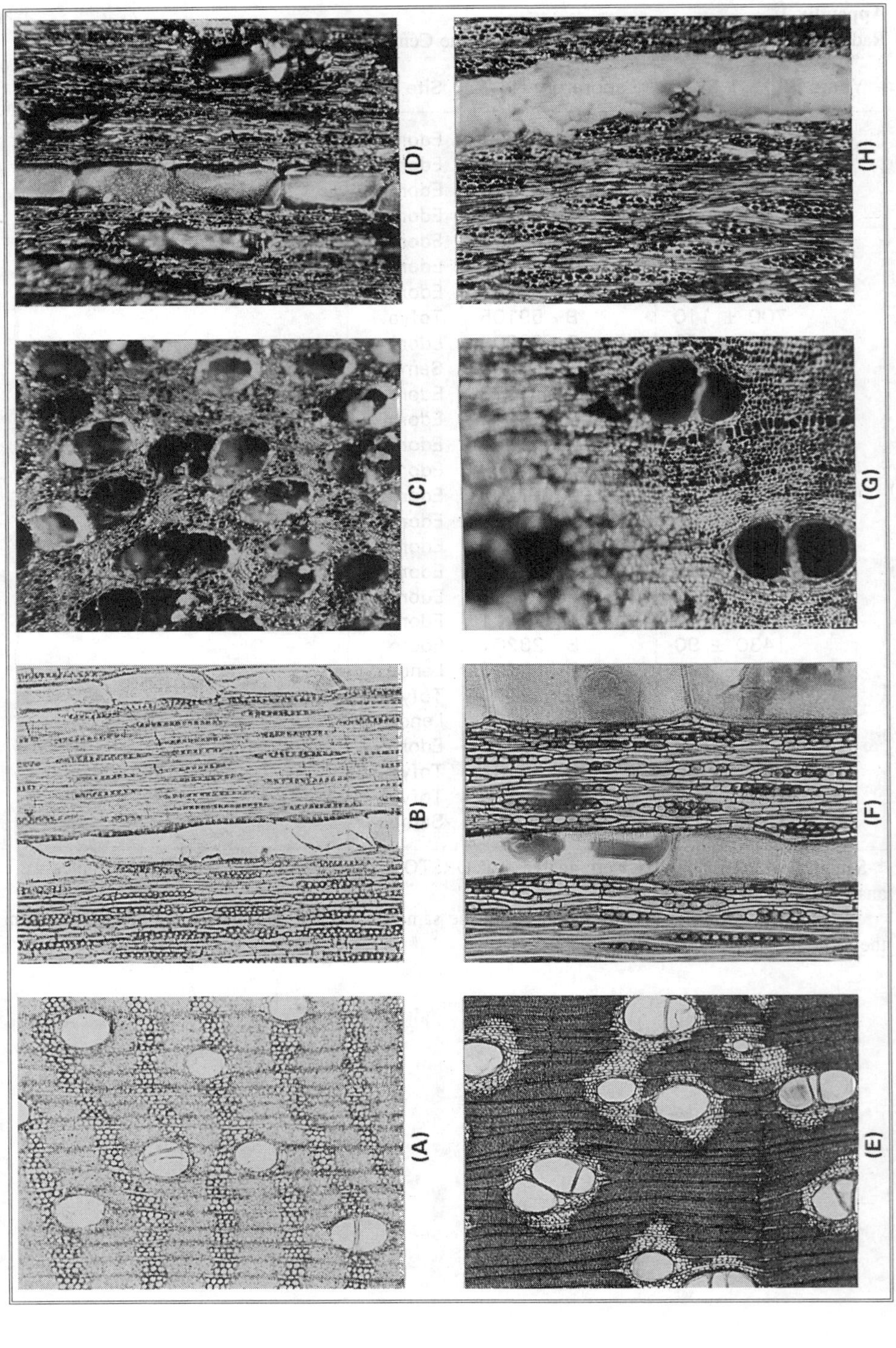
(A)
(B)
(C)
(D)
(E)
(F)
(G)
(H)

Appendix 2

Wood anatomy, or the microscopic study of wood structure, can serve to identify certain taxa from wood samples. This is done through comparison with wood that has been previously identified by other botanical characteristics (flowers, fruits, leaves, etc.). The same technique can be used for identification of charcoal and fossil wood. If both the samples to be identified and the reference samples are adequate and if the taxa concerned have low variability of wood structure characteristics, then identifications can be made to the species level.

The collection of wood samples at the Musée Royal de l'Afrique Centrale at Tervuren is one of the largest in the world, containing more than 52,000 specimens and representing more than 12,000 species and varieties of which approximately 3,500 are African. This was the reference collection used for identification of the Ituri Forest charcoal samples.

Two of the notable results of the study reported here were (1) the abundance in the charcoal of *Lebruniodendron leptanthum*, a species that is not known from the modern forest of the region and is not known to be abundant or gregarious anywhere today and (2) the absence of *Gilbertiodendron dewevrei* from the charcoal, a species that is locally gregarious and abundant in the Ituri region today. Critical and distinguishing wood characteristics of these two species include:

Lebruniodendron: Circummedullary parenchyma in relatively thick bands, thicker, for instance, than those of *Cynometra*. Finely storied rays. The only genus approaching this structure, but nevertheless distinguishable, is *Gilletiodendron* (also not present in the Ituri study area today). *Lebruniodendron* is monotypic.

Gilbertiodendron: Characteristic aliform circumvascular parenchyma not found in any of the charcoal samples. Structure somewhat similar to that of *Brachystegia* (rare in charcoal) but nevertheless distinguishable.

Also noteworthy was the identification from Ituri charcoal samples of *Tarchonanthus camphoratus* and *Trichilia emetica*, species typical of drier, seasonal environments. *Tarchonanthus* is of a rare group of Compositae with characteristic storied rays. *Trichilia emetica* is a savanna species with a large distribution and considerable anatomic and botanical variability, it nevertheless has a microscopic wood structure that distinguishes it from other *Trichilia*.

Figure 5 (left). Wood structure of modern wood from the Tervuren collection and charcoal from the Ituri forest.
Modern wood of *Lebruniodendron leptanthum*:
(A) transverse section x100, (B) tangential section x100.
Charcoal identified as *Lebruniodendron leptanthum*:
(C) transverse section x84, (D) tangential section x84.
Modern wood of *Gilbertiodendron dewevrei*:
(E) transverse section x40, (F) tangential section x100.
Charcoal identified as *Trichilia emetica*:
(G) transverse section x84, (H) tangential section x84.

Geography of African Rubiaceae with reference to glacial rain forest refuges

E. ROBBRECHT
Nationale Plantentuin van België, Domein van Bouchout, B-1860 Meise, Belgium.

Abstract
The Rubiaceae is among the most diversified and largest of the families in the African rain forest and consequently presents a suitable sample to test the theory of glacial forest refuges. Most forest Rubiaceae are ecologically selective and present a sufficiently low long distance dispersal potential to qualify. Generic diversity centres, patterns of endemism and disjunctions in lowland forest Rubiaceae generally corroborate the theory. It is also found that the refuges are characterized by generic endemism, a fact perhaps indicating that the ice ages had a catastrophic influence on the flora of tropical Africa. The data assembled (i) point towards modifications of the delimitation of accepted refuge areas, and (ii) indicate the existence of a few more refuge areas, e.g. in the Zaire Basin and in the Indian Ocean coastal belt.

Résumé
La famille des Rubiaceae est l'une des plus importantes de la forêt ombrophile africaine et constitue donc un exemple valable pour vérifier la théorie des refuges forestiers glaciaires. La plupart des Rubiaceae forestières ont une écologie suffisamment exigeante et une potentialité de dispersion à longue distance assez faible pour servir de témoin. Les centres de diversité générique, les modes d'endémisme et les disjonctions d'aires observées chez les Rubiaceae des forêts planitiaires corroborent cette théorie. Il s'avère également que les refuges sont caractérisés par un endémisme générique, ce qui pourrait indiquer que l'ère glaciaire a une influence catastrophique sur la flore de l'Afrique tropicale. Les données réunies (i) montrent des modifications dans la délimitation des aires de refuge acceptées et (ii) indiquent l'existence de quelques autres aires de refuge, notamment dans le bassin du Zaïre et dans la bande côtière de l'Océan Indien.

Key words
Africa, Chorological types, Disjunction, Endemism, Refuge theory, Rubiaceae.

1 Introduction

The Rubiaceae family (more than 630 genera and approximately 11,000 species) is the fourth largest of the flowering plants (Robbrecht 1988). On the subsaharan part of the African continent, Rubiaceae are represented by about 1,900 species in 165 genera (for comparison: 675 mostly endemic spp. in 83 genera on Madagascar and other islands in the western part of the Indian Ocean). Both the large size of the family as well as its concentration in the rain forest make the Rubiaceae an excellent sample to test existing hypotheses on the localisation of glacial rain forest refuges in Africa [Maley 1987, Sosef 1994, cf. also White's (1993b) criticism of the theory]. Many groups of African Rubiaceae are in need of revision at a continental scale to establish their distribution, but many others have been revised in the last decades, inter alia large taxa such as *Psychotria*, *Tricalysia* (now divided in two subgenera) and its segregate *Sericanthe*, or *Ixora* — the examples all in my institute which has a long tradition in the systematic study of African Rubiaceae (Robbrecht 1993a). There are now several hundreds of distribution maps of taxonomically consolidated taxa available. Hallé's (1966, 1970)

L.J.G. van der Maesen et al. (eds.), The Biodiversity of African Plants, 564-581.

Table 1. Lettering and numbering used for postulated refuge areas. (Mainly following Sosef (1994: Figure 12.1). His numbering and lettering between [ ]; areas not given there marked with *.)

	A: Upper Guinea
	A1: Mount Nimba and surroundings [1]
	A2: Cape Palmas area in Sierra Leone and Liberia, coastal [a]
	A3: Cape Three Points in Ghana, coastal [b]
	B: Lower Guinea (a chain of refugia from SW Cameroon to NW Angola)
*	B1: southwestern Nigeria
	B2: Mount Cameroon and surroundings [2]
	B3: the South Cameroon Plateau [3]
	B4: the Crystal Mountains [4]
*	B5: NE Gabon north of the Ivindo River
	B6: the Massif du Chaillu [5]
	B7: the Doudou Mountains [c]
	B8: Mayumbe (mouth of Congo river, i.e. SW Congo, Cabinda and W Zaire) [d]
*	B9: surroundings Golungo Alto, Angola
*	**C: western and central Congolia**
*	C1: Mai Ndombe Lake
*	C2: south Kwango
*	C3: interfluvium of lower Aruwimi and lower Lomami
*	C4: southern edge of Zaire Basin
	D: Kivu (eastern Zaire) [6]
*	**E: Indian Ocean coastal belt** (opposite Zanzibar)

partial local revision of the Rubiaceae is a welcome addition to these monographic data since it relates to Gabon, an area of particular importance to the topic of the present symposium.

The present paper reviews and discusses the available chorological data on Rubiaceae, which are important to localize former lowland rain forest refuge areas.

2 Methods

A survey of the generic distribution of the Rubiaceae (Robbrecht in prep.; summarized in Table 2) was used to compile distributional data with importance to the refuge-theory. Sosef's (1994) method (compiling outlines of areas of all species of a given group on a single map) was applied to endemic and disjunct lowland rain forest species of genera for which detailed distributional data are available. Reference to particular postulated refuge areas is made using a letter for the *main* areas and a number for the sub-areas inside them (Table 1 and Figure 1). A few references made to the classification of the Rubiaceae are according to my own system and the tribus abbreviations proposed there are used (Robbrecht 1988: 22, 1993b), e.g. Ise. for Isertieae.

3 The dispersal potential and ecology of rain forest Rubiaceae

A preliminary question should be examined. Is the dispersal potential of rain forest

Rubiaceae sufficiently limited to make them valuable indicators of former forest refuges? At first sight one would answer no. The majority of rubiaceous trees and shrubs have fleshy fruits (drupes or berry-like drupes) and are endozoochorous. They consequently mostly have good adaptations for long distance dispersal.

However, Rubiaceae possess some advanced morphological syndromes by which cross-pollination becomes obligatory, viz. secondary pollen presentation, in Rubiaceae often called "ixoroid pollen mechanism", a derived proterandrous syndrome, and heterodistyly, a well known breeding adaptation that occurs more often in the Rubiaceae than in any other family of the flowering plants. For a full discussion, see Robbrecht (1988: 120). In addition self-incompatibility may be expected to be the common situation, although experimental field pollination research is hardly done for tropical Rubiaceae. In the only existing study (Bawa & Beach 1983), however, it was found that 13 of 14 examined rubiaceous species from a plot of neotropical lowland wet forest are self-incompatible.

As a consequence, the breeding system of most Rubiaceae probably renders isolated cases of long-distance dispersal of single diaspores ineffective; the establishment of *a single* individual in a remote place will probably not result, for many taxa, in the dispersal of a species. One may thus accept that relict populations of such forest Rubiaceae will need a long time to gradually follow an extension of the forest in advantageous conditions after a period of forest reduction.

Refuge indicators should also be ecologically discriminating enough. This seems to be the case for many Rubiaceae, which are small shrubs or dwarfy monocauls typical for and even confined to the understorey of rain forest.

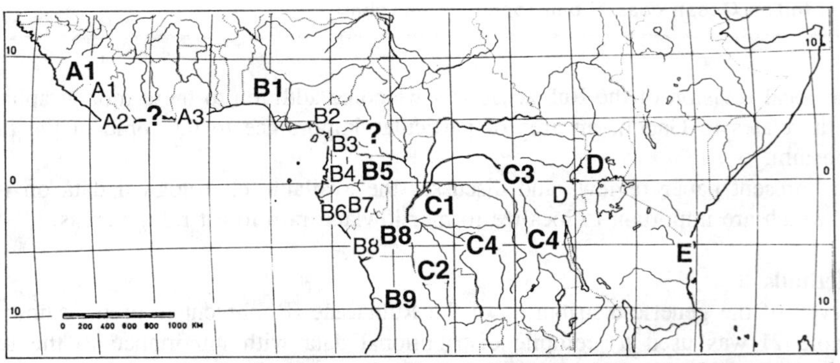

Figure 1. Areas of endemism supposed to indicate lowland rain forest refuge areas and sub-areas (normal letters: as summarized by Sosef 1994 or suggested by him; bold large: other areas from chorological data on Rubiaceae presented here); numbering and lettering explained in Table 1.

4 Evidence for former refuge areas of rain forest

4.1 Generic distribution patterns

It is evident that of the Rubiaceae, only the genera (in total 120) belonging to the main categories 1 to 3 and subcategory 81 of Table 2 are relevant to the subject dealt with in this paper.

4.1.1 Genera endemic of accepted refuge areas
In the Rubiaceae, several *genera* are endemic to the accepted main refuge areas (Table 2, category 34). This applies especially to the Lower Guinea Domain, which has ten endemic genera compared to two for Upper Guinea and only one (*Colletoecema*) for Congolia.

Ecpoma (Ise.) contains six endemic species which are all Gabonese or perigabonese. Distribution maps are not available, but N. Hallé's (1966) data are sufficiently detailed to state that *E. hiernianum* N. & F. Hallé and *E. gigantistipulum* (K. Schum.) N. Hallé have a somewhat wider distribution, while three other species are strictly endemic to refuge areas (B3: *E. apocynaceum* K. Schum.; B4: *E. geanthum* N. Hallé; B6: *E.* sp.); a fifth species is endemic on Sao Tomé. The monospecific genus *Coleactina* (Pav.) has only been recorded from two localities in the southern part of the Massif du Chaillu (Hallé 1966). *Petitiocodon* (Gar.-D.), a genus more closely related to *Tricalysia* than to *Didymosalpinx* in which the only species was accommodated formerly, is endemic to the Mount Cameroon area (B2) but not montane. The genus *Pentaloncha* (Pau.) has two species, each endemic to refuge area (B4 and B8 resp.).

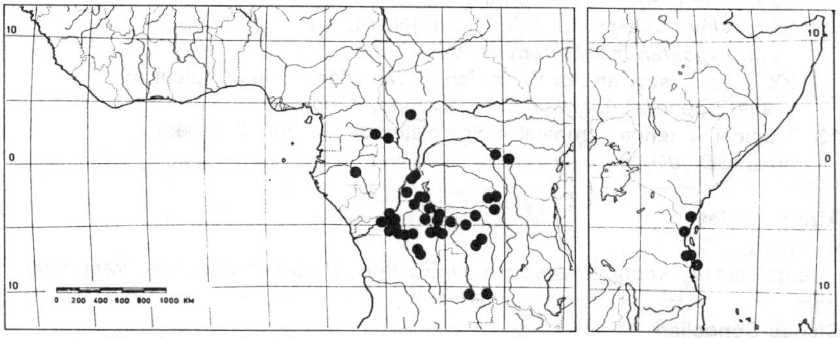

Figure 2. Distribution of *Colletoecema dewevrei* (left), the single species of the only rubiaceous genus which is endemic to the Congolia Domain; based on the specimens in BR. Distribution of *Cladoceras subcapitatum* (K. Schum. & K. Krause) Bremek. (right), the only species of a genus endemic to the rain forests of the east African coasts; according to specimens in Bremekamp (1940), Robbrecht & Bridson (1984) and Bridson & Verdcourt (1988).

The endemism of the monospecific genus *Cladoceras* (Pav.) (Figure 2), restricted to the coastal forests of S Kenya and Tanzania, is comparable and suggests the existence of a refuge area in the centre of the Zanzibar-Inhambane Region (Figure 1: E).

Colletoecema (? Psy.), the only genus endemic to Congolia, is monospecific; it has a ± wide distribution (Figure 2) not easily interpreted in terms of the refuge theory; dispersal from *several* refuge areas might be an explanation. The species' absence from the proper Zaire Basin (phytogeograpically distinguished as "secteur de la cuvette congolaise" by Léonard 1965) is striking. At any event, *Colletoecema* cannot be brought in relation with Congolia's generally accepted eastern refuge, the Kivu (D), which consequently differs from the western and central major refuges (A and B) by the absence of generic rubiaceous endemism.

Table 2. Generic distribution patterns in African Rubiaceae.

Summary of a survey of the distribution of the genera from the subsaharan (tropical and subtropical) part of the African continent, Madagascar and other islands in the western part of the Indian Ocean (eastern limit: the Mascarenes). Generic distribution patterns are classified against the framework of White's (1978, 1979, 1993a) phytogeographical division of Africa; however, "region" is used instead of White's periphrases for the higher phytochoria. Slight penetration in adjacent regions is not considered to qualify a genus as pluriregional.

Total number of genera in subsaharan Africa, Madagascar, the Comoro Islands, the Seychelles and the Mascarenes: 205.

1 Wides 39 genera (24 widely distributed in Africa), as follows **11 Pantropical**: 12 genera 111 largely present in Africa and/or Madagascar: 12 genera, e.g. *Ixora, Morinda, Psychotria* 112 poorly represented in Africa: *Cephalanthus, Galium, Guettarda* **12 Paleotropical**: 24 genera 121 largely present in Africa and/or Madagascar: 15 genera, e.g. *Gaertnera, Gardenia, Pavetta, Rothmannia* 122 poorly represented in Africa: 6 genera, e.g. *Argostemma, Mitragyna, Wendlandia* 123 Madagascar and/or W. Indian Ocean Islands plus tropical Asia: only 3 genera, *Antirhea, Cremocarpon, Timonius* **13 Tropical America - tropical Africa disjunctions**: only 3 genera: *Bertiera, Diodia, Sabicea*	**19%**
2 African wides 9 genera, e.g. *Heinsia, Keetia, Leptactina, Oxyanthus, Pseudomussaenda, Vangueria*	**4%**
3 Guineo-Congolian 52 genera, as follows **31 Guineo-Congolian wides**: 32 genera 311 also present in the eastern African coastal forest area: 11 genera, e.g. *Chazaliella, Parapentas, Rutidea* 312 not present in the eastern African coastal forest area: 21 genera, e.g. *Corynanthe, Pouchetia, Schumanniophyton* **32 present in two domains**: 5 genera 321 Upper & Lower Guinea: *Robynsia, Stelechantha* 322 Lower Guinea & Congolia: *Brenania, Oligocodon, Stephanococcus* **33 Guineo-Congolian disjunctions**: *Poecilocalyx, Preussiodora, Schizocolea* **34 Endemic of one domain**: 12 genera 341 Upper Guinea endemics: *Hutchinsonia, Monosalpinx* 342 Lower Guinea endemics: 10 genera, e.g. *Calochone, Commitheca, Ecpoma, Petitiocodon* 343 Congolia endemics: *Colletoecema* only	**26%**

4 Afromontane 8 genera, e.g. *Galiniera, Hedythyrsus, Temnocalyx*	**4%**
5 Soudanian, Somali-Masaian and Zambezian 24 genera, as follows **51 Pluriregional ("Soudano-Zambezian")**: 6 genera 511 continuous ("pericongolian arch"): *Crossopteryx, Feretia, Otomeria, Paraknoxia* 512 disjunct: *Batopedina, Lelya* **52 Endemic of one region**: 21 genera 521 Soudanian: *Thecorchus* only 522 Somali-Masaian: *Pentanopsis* only 523 Zambezian: 15 genera, e.g. *Amphiasma, Coddia, Conostomium, Fadogiella, Manostachya, Phellocalyx, Pygmaeothamnus*	**12%**
6 East African coast 10 genera, as follows **61 Zanzibar-Inhambane and Tongaland-Pondoland:** *Kraussia, Lagynias* **62 Zanzibar-Inhambane:** *Cladoceras, Lamprothamnus, Tennantia* **63 Tongaland-Pondoland:** *Burchellia, Eriosemopsis, Galopina, Hyperacanthus, Plectroniella*	**5%**
7 Southern Africa **71 Karroo-Namib**: *Crocyllis* **72 Cape**: *Carpacoce, Nenax*	**1%**
8 Afro-Madagascan 22 genera, as follows: **81 with wide distribution on the African continent:** 16 genera, e.g. *Anthospermum, Coffea, Polysphaeria, Tricalysia* **82 with limited distribution on the African continent:** 6 genera, *Alberta, Carphalea, Danais, Mitrasacmopsis, Phylohydrax, Triainolepis*	**11%**
9 Madagascar i.e. Madagascan and/or western Indian Ocean Islands (as specified in caption) endemics: 37 genera, as follows **91 Madagascar and the Mascarenes:** *Psathura* only **92 Madagascar:** 30 genera, e.g. *Canephora, Mantalania, Peponidium, Saldinia, Schismatoclada* **93 Mascarenes:** *Doricera, Fernelia, Myonima, Ramosmania* **94 Seychelles:** *Glionnetia, Paragenipa*	**18%**

4.1.2 Guineo-Congolian disjunctions

At the generic level, only three taxa show macro-disjunctions (as defined in 4.3.2). The monospecific genus *Preussiodora* (Gar.-G.; Robbrecht 1978b) is concentrated in the area B2 but also known from one unprecise record in Gabon (perhaps in B7), an outpost in Congo (in B8), and, more importantly, from a very distant outpost in the south of the Zaire Basin, viz. in a gallery forest of the Kisakwa River in Lower Shaba. *Poecilocalyx* (Pau.) was hitherto only recorded from Upper and Lower Guinea (two wides and two endemics); a fifth undescribed species occurs in the Kivu refugium [generically determined by N. Hallé, with *A. Léonard 3811* (BR)]. *Schizocolea* (Cou.) shows a similar macrodisjunction, which need to be documented in detail by distribution maps.

Most of the genera listed in categories 321 and 322 Table 2 are probably also essentially disjunct. In *Sericanthe* subg. *Macrocarpus*, a similar situation is met with: *S. adamii* (N. Hallé) Robbr. is endemic of the Nimba mountains (area A1); the second species *S. auriculata* (Keay) Robbr. is endemic of E Nigeria and Cameroon (micro-disjunct: areas B2 and B3).

Note: Generic disjunctions outside the Guineo-Congolian Region are also rare in African Rubiaceae. There are apparently only three cases. In *Knoxia*, the West-African species (formerly *Neobaumannia*) has a ± wide distribution from Ghana to Cameroon, while the Shaban species (formerly *Afroknoxia*) is endemic of the Manika and Biano Plateaus (cf. Puff & Robbrecht 1989). *Batopedina* (Robbrecht 1981b) has two species with a narrow distribution in Zaire-Shaba and one species occurring in Burkina Faso and Ghana. The genus *Otiophora* (Puff 1981) is essentially Zambezio-Madagascan, but *O. villicaulis* Mildbr. var. *villicaulis* is disjunct and occurs in the Zambezian Region and the area of the Cameroon mount. This taxon is submontane to montane, however.

4.2 Generic diversity centres

Several Rubiaceae genera for which detailed chorological data are available, generally corroborate the three major refuges accepted in West Africa, Lower Guinea and eastern Zaire. I deal with them in alphabetical order.

Aoranthe (? Ise.) (Somers 1988) is a small African genus of five species first accommodated under the artificial genus *Randia* and later erroneously associated with the Malesian *Porterandia*. Two species are Guineo-Congolian wides (but absent from Upper-Guinea), the three other ones are essentially confined to refuge areas [*A. castaneofulva* (S. Moore) Somers subendemic of B8; *A. penduliflora* (K. Schum.) Somers subendemic of E; *A. annulata* (K. Schum.) Somers restricted to B3, B4 and B6 though suggesting a larger extension of B4].

Corynanthe and *Pausinystalia* (Cop.; Stoffelen et al. in prep.) are a pair of sister genera restricted to the rain forest. These are huge forest trees with dry fruits and winged seeds; a condition by which they do not perfectly match the dispersal ability needed to reflect the localisation of former refugia (cf. paragraph 3). Still the species' distributions weakly correspond to refugia. Particularly noteworthy is a macro-disjunction shown by *P. lane-poolei* (Hutch.) Hutch. ex Lane-Poole (see 4.3.2.2).

Dictyandra (Pav.) is a small genus counting only two species (Robbrecht 1984); *D. arborescens* Hook.f. is a Guineo-Congolian wide. *D. congolana* Robbr. occurs from the surroundings of Kinshasa to the Nsah plateau in Congo, thus suggesting an extension of the Mayumbe refugium (B8) in eastern direction.

Hymenocoleus, a genus of rain forest floor herbs and monocaul dwarfs segregated from *Geophila*, is exclusively Guineo-Congolian. Not a single species is endemic to the accepted refugia, but many show micro-disjunctions, being restricted to two or

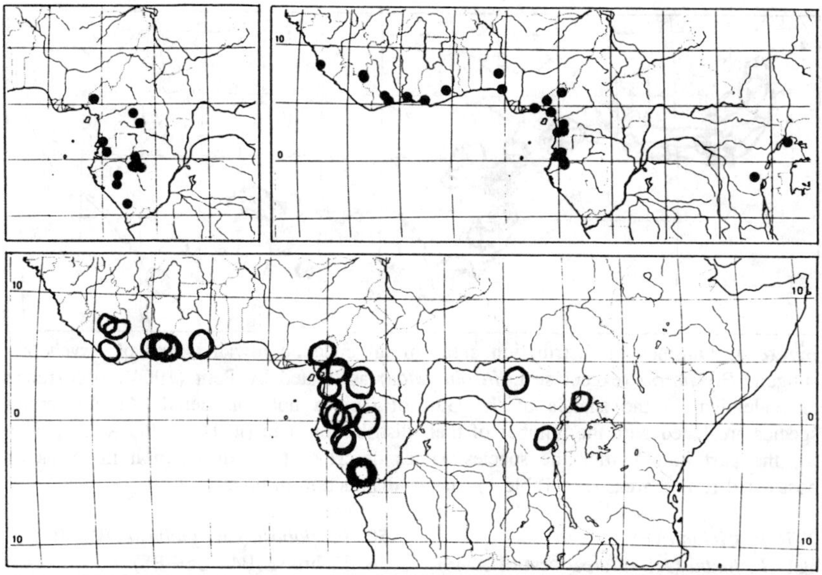

Figure 3. *Hymenocoleus*: above left distribution of *H. globulifer* Robbr. showing micro-disjunctions in Lower Guinea, and above right *H. rotundifolius* (A. Chev. ex Hepper) Robbr. showing a macro-disjunction between W and C Africa and the eastern part of the rain forest area; below outlines of areas of endemic species plus outlines of the two eastern outposts of *H. rotundifolius*. Species are listed with the number of their distribution map (in Distr. Pl. Afr. 12); the part in ( ) after the species names describes their distribution in terms of postulated refuge areas (see Table 1); macro-disjunctions in bold.

376 *H. nervopilosus* var. *nervopilosus* (A1 + appr. A2 + appr. A3)
377 *H. nervopilosus* var. *orientalis* (B3 + interval between B4 and B5)
378 *H. glaber* (interval between B2 and B3)
380 *H. axillaris* (**A1 + area east of B3 + B6 + B8**)
382 *H. rotundifolius* (**macro-disjunct:** largely distributed from Liberia to N Gabon **+ outposts in D and N Uganda**)
384 *H. multinervis* (A1 + interval between A2 and A3 + area east of A3)
386 *H. barbatus* (interval between B2 and B3 + B4 + B8 + N Zaire)
387 *H. globulifer* (B2 + area east of B3 + B4 + B5 + B8)

more local areas often congruent with postulated refugia (Figure 3). *H. glaber* Robbr., the most narrowly endemic species, is restricted to the Sanaga River interval between B2 and B3.

The pantropical *Ixora* (Pav.; De Block in prep.) is almost exclusively confined to lowland rain forest in Africa; only a few species occur in montane forest, and no one exist in the parts of Africa having a dry season. At the moment of the preparation of the present paper, distribution maps are not yet available for every species; the existing ones, however, show a pattern of endemism and disjunction as for the other genera discussed; some evidence is included in 4.3.

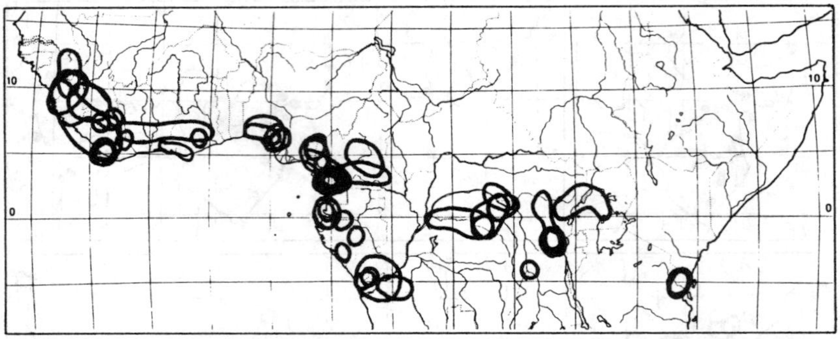

Figure 4. Outlines of distribution areas of all endemic lowland taxa of *Psychotria* subgen. *Psychotria* (except sect. *Involucratae*) as treated by Petit (1964); occurrence or endemism in the islands of the Gulf of Guinea not considered. All concerned species are listed with the number of their distribution map (in Distr. Pl. Afr. 4, 5 or 6); the part in ( ) after the species names describes their distribution in terms of postulated refuge areas (see Table 1); macro-disjunction in bold.

097 *P. triclada* (E)
101 *P. limba* (A1 + area west of A1)
103 *P. dorotheae* (**A1+B2**)
108 *P. articulata* (B1+B3)
109 *P. guineensis* (**A3+B2+B3+B4**)
125 *P. obovatifolia* (C3)
126 *P. walikalensis* (D)
141 *P. multinervis* (area west of A1 + A2) submontane in area west of A1
142 *P. subglabra* (between A1 and A2)
143 *P. klainei* (B4 but towards coast)
146 *P. reptans* (west of A1 + near A3)
147 *P. kitsonii* (sub-wide covering A1 and A2)
148 *P. rufipilis* (north of A2 and A3) submontane
149 *P. ebensis* (B3+B4)
150 *P. globosa* var. *globosa* (B1+B2)
156 *P. brieyi* (B4+B6+B8)
157 *P. globiceps* (B3 but extending east)
158 *P. avakubiensis* (**B3+B8+C3+D**)
160 *P. setistipula* (B8)
162 *P. liberica* (A1+A2)
163 *P. brevipuberula* (**C3+D**)
164 *P. rubropilosa* (B7)
167 *P. evrardiana* (C3, extending west)
170 *P. peteri* (E; submontane)
173 *P. lucens* var. *lucens* (B1 + B3 + area west of B2)
174 *P. lucens* var. *minor* (**C3+C4+D**)
175 *P. calceata* (B1+B3)
176 *P. lanceifolia* (B3)
177 *P. sadebeckiana* (B3)
178 *P. bukaensis* (B3)

The central African material of *Oxyanthus* (Gar.-G.) was recently revised (Sonké 1990). A few species are endemic to refuge areas (cf. 4.3.1) or show interesting disjunctions (cf. 4.3.2).

The large pantropical genus *Psychotria* (Psy.; Petit 1964) (maps and/or revision not available for subg. *Psychotria* sect. *Involucratae* and subg. *Tetramerae*) comprises a high number of Afromontane species. Many lowland species are endemic and their distribution either coincides well with postulated refuge areas or suggests additional refugia such as in the centre of the Zaire Basin (see Figure 4 and discussion). Remarkably, there are almost no endemic *Psychotria* species reported from Gabon. This is perhaps explained by the early date of the above mentioned revision, made before the intensive exploration of this country was started.

Sericanthe (Gar.-D.) is a small genus separated from *Tricalysia*; it comprises 17

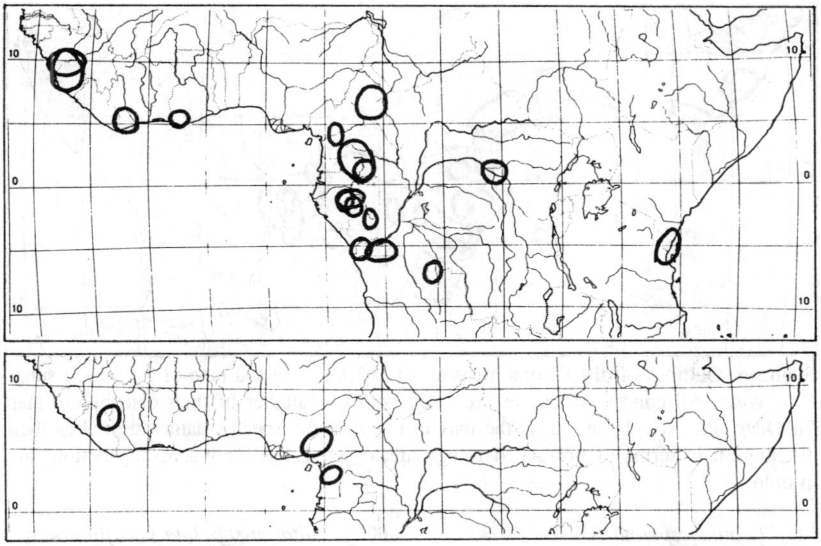

Figure 5. Outlines of distribution areas of all lowland taxa of *Sericanthe*. All concerned species are listed with the number of their distribution map (in Distr. Pl. Afr. 13 and 21); the part in ( ) after the species names describes their distribution in terms of postulated refuge areas (see Table 1).

subg. Sericanthe (map above)
415 *S. jacfelicis* (interval between B2 and B3 + area east of B6)
418 *S. odoratissima* var. *odoratissima* (E)
420 *S. pellegrinii* (B6+B8)
421 *S. petitii* (B6)
422 *S. raynaliorum* (far north of 2)
423 *S. roseoides* (B8)
425 *S. testui* var. *testui* (area east of B3/B4+B6+C2)
426 *S. testui* var. *pseudosalacia* (area east of B4)
427 *S. toupetou* (A2+A3)
428 *S. trilocularis* (two subsp.: area west of A1)
— *Sericanthe* sp.: C3
— *Sericanthe* sp.: C2

subg. Macrocarpus (map below)
709 *S. auriculata* (B2+B3)
710 *S. adamii* (A1)

species, half of which inhabit lowland rain forest; the other species are endemic montane or widespread Soudanian/Zambezian (Robbrecht 1978a, 1981a). Most of the lowland species are only recorded from refuge areas (Figure 5); other ones have "conflicting" distributions, most remarkable *S. jacfelicis* (N. Hallé) Robbr. which has a disjunct recorded occurrence, viz. in the Sanaga River interval between B2 and B3 and in an area far east of the Chaillu Massif.

Tricalysia has a very wide distribution from Senegal to Ethiopia and Somalia and to the southeastern part of the Republic of South Africa (Robbrecht 1979, 1982, 1983, 1987). Many taxa are adapted to the conditions of the Zambezian region, e.g. in becoming geofrutescent, or are montane. The majority of the species inhabits the lowland rain forest, however, and many of them have endemic or disjunct distributions which seem linked with the refuge theory, especially in Lower Guinea (Figures 6, 7).

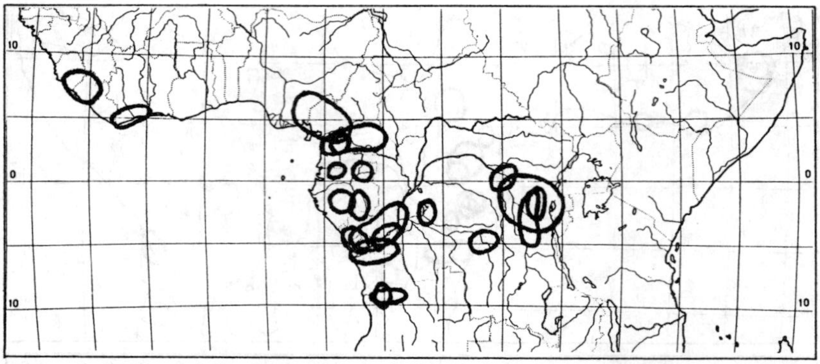

Figure 6. Outlines of distribution areas of all endemic lowland taxa of *Tricalysia* subg. *Empogona*. All concerned species are listed with the number of their distribution map (in Distr. Pl. Afr. 16 and 21); the part in ( ) after the species names describes their distribution in terms of postulated refuge areas (see Table 1); macro-disjunctions are in bold.

541 *T. aulacosperma* (D)
543 *T. gossweileri* (B3 and extensive area east of it +B4+B5+B8+B9)
546 *T. concolor* (B5)
555 *T. welwitschii* (B8+B9+ area in southern part of Zaire Basin)
558 *T. bequaertii* (**B6+B8+C3**)
559 *T. deightonii* (appr. A1+A2)
560 *T. filiformi-stipulata* var. *filiformistipulata* (**B2+B8+C1+D**)
561 *T. filiformi-stipulata* var. *epipsila* (C3 to D)
562 *T. talbotii* (B2 + interval between B2 and B3)
711 *T. breteleri* (area east of B6)

4.3 Species distribution

4.3.1 Species endemic of accepted refuge areas

Figure 3-7 show that many lowland taxa are endemic (or near-endemic, * below) to formerly postulated refuge areas, or areas suggested by evidence given in 4.2. The following list gives further examples from other genera:

- A2, Cape Palmas area: *Ixora aggregata* Hutch. (extending around A2)
- B2, Mount Cameroon and surroundings: *Calochone acuminata* Keay, *Petitiocodon parviflorum* (Keay) Robbr.
- B3, the South Cameroon Plateau: *Ecpoma apocynaceum* K. Schum. (extinct?); *Oxyanthus oliganthus* K. Schum.; *Pausinystalia brachythyrsum* (K. Schum.) W. Brandt (extinct?)
- B4, the Crystal Mountains: *Bertiera arctistipula* N. Hallé, *Bertiera lanx* N. Hallé, *Ecpoma geanthum* N. Hallé, *Pauridiantha venusta* N. Hallé, *Pentaloncha humilis* Hook. f., *Poecilocalyx crystallinus* N. Hallé, *Sherbournia ailarama* N. Hallé, *Virectaria salicoides* (C.H. Wright) Bremek.
- B5, hilly area N Ivindo River in NE Gabon: *Pauridiantha siderophila* N. Hallé, *Sabicea bigerrica* N. Hallé, *Sabicea caminata* N. Hallé, *Sherbournia kiliotricha* N. Hallé, *Tarenna calliblepharis* N. Hallé, *Virectaria belingana* N. Hallé

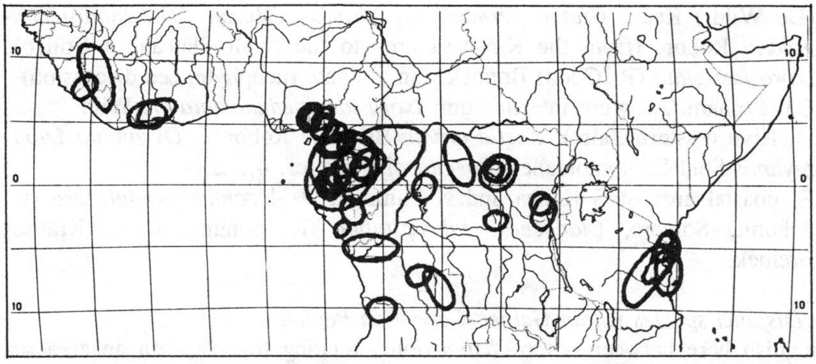

Figure 7. Outlines of distribution areas of all endemic and disjunct lowland taxa of *Tricalysia* subg. *Tricalysia* (a few outpost areas of wides added). All concerned species are listed with the number of their distribution map (in Distr. Pl. Afr. 23, 24, 31 and 32); the part in ( ) after the species names describes their distribution in terms of postulated refuge areas (see Table 1); macro-disjunctions are in bold.

760 *T. idiura* (B4)
762 *T. longituba* var. *velutina* (C3 + local area in S Zaire)
765 *T. elegans* (E)
767 *T. allocalyx* (E)
805 *T. elliottii* (outpost in B9) widely distributed in rain forest area
808 *T. trachycarpa* (very local in Zaire Basin)
809 *T. anomala* (**macro-disjunct:** sub-wide in Cameroon, Gabon and C.A.R. + in interval between A2 and A3+D+E; montane in D and E)
813 *T. ferorum* (B3)
1016 *T. bifida* (**B2+B5+B8+C1+D**)
1017 *T. lasiodelphys* subsp. *lasiodelphys* (interval between B2 and B3+B3)
1018 *T. lasiodelphys* subsp. *anomalura* (B5)
1019 *T. atherura* (B2+B3+B5)
1020 *T. pangolina* (B2+B4+B5+B6)
1021 *T. vadensis* (B2)
1028 *T. obanensis* subsp. *obanensis* (B2)
1029 *T. obanensis* subsp. *kwangoensis* (C2)
1032 *T. pedunculosa* var. *pedunculosa* (B3 to B4 and B5)
1033 *T. pedunculosa* var. *pilosula* (B4 + area south of B7)
1034 *T. soyauxii* (? sub-wide in S Cameroon and Gabon or micro-disjunct)
1035 *T. yangambiensis* (C3)
1038 *T. fangana* (area east B3+B4+B5)
1039 *T. potamogala* (area northeast of B4)
1040 *T. biafrana* (micro-disjunct: sub-wide from Liberia to north of Zaire Basin +B8+C2)
1041 *T. faranahensis* (A1 + area north of A1 + interval between A2 and A3)
1042 *T. micrantha* (B4)
1047 *T. obstetrix* (area east of B3 +B4+B5+B6) the record on the Ogooué River on map 1047 relates to erroneously identified material
1048 *T. werhamiana* (B2)
1049 *T. microphylla* (E)
1050 *T. parva* (appr. A1)
1052 *T. sylvae* (B2+B3+B4)
1053 *T. amplexicaulis* (B3)

- B6, the Massif du Chaillu: *Coleactina papalis* N. Hallé, (?) *Commitheca letestuana* N. Hallé, *Ecpoma* sp., (?) *Sherbournia myosura* N. Hallé
- B8, Mayumbe (mouth of the Congo river, i.e. SW Congo, Cabinda and W Zaire): * *Aoranthe castaneofulva* (S. Moore) Somers, *Calochone redingii*

(De Wild.) Keay, *Craterispermum inquisitorium* Wernh., *Dictyandra congolana* Robbr. (from the Kinshasa area to the Nsah Plateau in Congo), *Ixora fastigiata* (R. Good) Bremek. (an obligate rheophyte, see discussion)
- C3, Lomami/Aruwimi interfluvium: *Ixora longipedunculata* De Wild.
- D, Kivu (eastern Zaire): *Keetia ornata* Brids. & Robbr.**, *Oxyanthus letouzeyanus* Sonké (submontane), *Poecilocalyx* sp. (cf. 4.3.2.2)
- E, coastal area of S Kenya and N Tanzania: * *Aoranthe penduliflora* (K. Schum.) Somers, *Cladoceras subcapitatum* (K. Schum. & K. Krause) Bremek.

4.3.2 Disjunct species within Guineo-Congolian Region
I distinguish here between *macro-disjunctions*, relating to taxa with an area scattered over the main former refuge areas of lowland rain forest, and *micro-disjunctions*, mainly within each of the main areas.

4.3.2.1 Micro-disjunctions
Distribution of individual species confirms the need to recognize fragmented refugia. Many of them have already been cited above, e.g. *Sericanthe auriculata* (4.1.2) or are listed in Figure 3-7. A few other ones spotted are:

B2 to B3 + B6 + B8: *Ixora anemeinodesma* K. Schum.
B4 + B5: *Bertiera sphaerica* N. Hallé
B3 + B5 + C1: *Calycosiphonia macrochlamys* (K. Schum.) Robbr. (Distr. Pl. Afr. 22: map 758)
Gabon (one imprecise record surroundings Libreville) + B8: *Uncaria donisii* E. Petit
B8 + B9 + C1 + C2: *Corynanthe paniculata* Welw.

4.3.2.2 Macro-disjunctions
The following *specific* macro-disjunctions were noted:
- *Hymenocoleus rotundifolius* (A.Chev. ex Hepper) Robbr. (Distr. Pl. Afr. 12: map 382; copied here in Figure 3): widely distributed from Liberia to N Gabon; two eastern outposts: one in the centre of the Kivu refuge (D), one in N Uganda.
- *Oxyanthus brevicaulis* K.Krause (B2 + B6 + C3)
- *Oxyanthus setosus* Keay (A2 + B2 + B3)
- *Pausinystalia lane-poolei*: subsp. *lane-poolei* has a sub-wide distribution in the Upper Guinea Domain (A1 to A2 + A3); its subsp. *ituriensis* (comb. ined. in Stoffelen et al. in prep.) is endemic to the Kivu refugium (D) but

** Specimens from the mount Cameroon area received at Kew after the publication of *K. ornata* [Mapanja, *Tekwe 202, Thomas 9237* & *9239*] have similar fruit and may well be closely related but are montane (1400 - 1600 m), while the record of *K. ornata* in Kivu is from lowland forest (650 m). One may have here another interesting macro-disjunction, but the alliance of *Keetia* to which this (these) species belong(s) [i.e. *K. hispida* (Benth.) Bridson sensu lato, see Bridson (1988: 986)] remains unrevised. — D.M. Bridson & E. Robbrecht.

has an outpost east of the Chaillu Massif (B6).

- *Tricalysia anomala* (included in Figure 7). This species needs special mention; it occurs in the four main refuge areas and is absent in-between; in Upper Guinea it is restricted to the interval between A2 and A3; in Lower Guinea, it is sub-wide, occurring in southern Cameroon, Gabon and western C.A.R.; it further occurs in Kivu (D) and the eastern coastal forest zone (E). In the latter two, it is submontane or montane. The species consequently provides an example of possible shift from lowland to montane rain forest.

5 Discussion

A first remarkable finding of the present survey is the fact that endemism of postulated refuge areas is not only specific but also generic. It is plausible to explain the generic endemism established in 4.1.1 as *relict endemism*, as one may hardly expect that generic diversification took place after the last ice age. The present survey of endemism and disjunction in Rubiaceae indicates the ice ages as dramatic periods of decline of the African flora, perhaps also extinction, after which a number of genera never succeeded to occupy wide areas again.

The high generic and specific endemism of Lower Guinea is more than in the two other Domains explained by ice age events. White (1979: 31) questioned whether the Sangha interval between Lower Guinea and Congolia "reflects ecological conditions or has an historical explanation". Today it seems that the floristic differences between Lower Congolia and Guinea should mainly be explained by the ice age events.

The present contribution wanted to assemble the available rubiaceous evidence important to the context of the refuge theory in the first place. One should nevertheless bear in mind that vegetation history has been extremely complex, and most relations between actual plant distributions and the ice-age history may well be compound, as advocated by White (1993b). *Ixora aggregata*, confined to the Mayumbe refugium, is a simple example showing that endemism should not necessarily be linked to the refuge theory, since it is an obligate rheophyte which may well have accomplished its speciation after the last ice age.

From the theoretical viewpoint, it is important to document that the decline of rain forest taxa went hand in hand with the extension of savannah and woodland vegetation. In this respect one may postulate that the three generic macro-disjunctions recorded for Rubiaceae from outside the rain forest area may well be explained by ice-age events too. The actual continental African distribution of *Otiophora, Batopedina* and *Knoxia* is essentially Zambezian. The outposts north of the present rain forest area are perhaps explainable by the reverse of the phenomenon discussed here, i.e. extension and migration through the Zaire Basin during drier and cooler periods of the ice ages.

Tricalysia anomala (data given in 4.3.2.2) is another theoretically important case. This species combines a macro-disjunction over the four main refuge areas with a local shift to montane conditions. "Vegetation zones may have been altitudinally compressed as well as depressed at the time, with the consequence that there was less differentiation between lowland and highland forests" (Hamilton 1974: 32). Another consequence may have been subspeciation of certain taxa, as *T. anomala*, by adaptation to altitudinally different environments.

Generic diversity centres in the Rubiaceae corroborate the existence of three major refuge areas, one in West Africa, mainly Mount Nimba and surroundings, a second ± coastal from Nigeria through Cameroon, Equatorial Guinea and Gabon to the mouth of the Congo River, and a third in eastern Zaire (Kivu). The strongly narrow or disjunct distribution of many individual species corroborates the formerly proposed recognition of minor refuge areas within these three, but suggests the following modified circumscriptions [(1)-(5)]. Also, several areas of endemism exist outside the generally postulated former refugia [(6)].

1 The distribution of several Rubiaceae indicates that the refuge area of Mount Nimba (A1) perhaps extended rather far westward.

2 The small refugia A2 and A3 two coastal areas in Upper Guinea are not fully corroborated by rubiaceous distributions, since many endemic species occur also in the interval between them. They are perhaps only centres of one larger coastal refugium. Sosef (1994) already doubted the presence of a true refugium here.

3 Sosef's (op. cit.) hypothesis that the forest refugium of S Cameroon and Gabon consisted of a subcoastal chain of five small subareas is perhaps not fully correct. First, many species were found here to occur in B2 and B3 *and* in the Sanaga River interval between them; other species, e.g. *Hymenocoleus glaber* (Figure 3), are endemic to this interval. B2 and B3 may be the nuclei of a refugium extending as far as southeast Cameroon.

In northeast Gabon, the hilly area north of the Ivindo River may well have formed another inland forest refuge (B5), now indicated by the well-known endemism around the Bélinga iron mines. Sosef (1994: 119) equally tentatively suggested the Bélinga area as a refugium.

4 The Mayombe refuge (B8) may have extended eastwards along the Zaire River; distributions such as that of *Dictyandra congolana* indicate that it may have reached the Kinshasa area and even the Nsah plateau in Congo. Here, elevated areas coincide with the large Zaire River, which may have allowed the existence of fluvial refugia.

5 The Kivu refuge (D) at the eastern edge of the present rain forest area was perhaps larger than generally supposed or equally consisted of a chain of smaller refugia north and south of a more extensive area on the equator. To the north, Hamilton (1974: 31-32) suggested the existence of at least one lowland forest refuge in Uganda, the "Impenetrable-Kayonzo Forest" (Bwindi Forest).

The actually known records of many species endemic of the Kivu refuge area show a considerable disjunction between (a) the region of the upper Aruwimi River and (b) the lowland forests at the foothills of the Kahuzi mountain. This disjunction is probably not a chorological reality, since (a) and (b) correspond to the well-collected surroundings of Avakubi, Epulu, Mambasa and Walikale. The gap in between — where the Maiko National Park was recently created — is ill-known; the exploration of the mentioned Park will shed more light on the mentioned gap in the distribution records.

6 Endemism of and disjunctions observed in Rubiaceae point to the existence of two more sub-areas in the major refugium of Lower Guinea, viz. in Nigeria (B1) and in Angola (B9). In southwestern Nigeria, the Effon ridge (B1) may well meet the hilly requirements needed to explain a local refugium; endemism, however, was only met with in one of the genera studied in detail (Figure 4). Several endemics (*Tricalysia*

spp., Figure 7; *Corynanthe paniculata* Welw.), mainly of the main area B, have an outpost in the surroundings of Golungo Alto, Angola (B9); this is probably a true micro-disjunction (not due to undercollection) perhaps comparable with the situation at the southern edge of the Zaire basin discussed under (8) below.

Other areas of endemism may probably be considered to reflect the existence of a fourth and fifth *main* refuge area, next to the three classically accepted (central Upper Guinea, Lower Guinea and Kivu).

7 The eastern coastal area opposite Zanzibar (Figure 1: E) is a distinct area of endemism for lowland forest Rubiaceae. It corresponds to the Zanzibar-Inhambane Region's diversity centre on the eastern Usambara mountains and Shimba hills (White 1993b: 401) and was suggested as a possible refuge area by e.g. Hamilton & Taylor (1991; not seen, cited in Sosef 1994).

8 Several areas in the Zaire Basin (Figure 1: C1 to C4) show local endemism, most distinctly the ± triangular area between the lower parts of the Aruwimi and Lomami Rivers (i.e. at their confluence with the Zaire near Kisangani), but also the Mai Ndombe Lake area and several places at the actual southern edge of the Zaire Basin rain forest (all congruent with findings of Ndjele 1988). Contrary to the "classical" areas these ones seem to correspond to river-associated, not hilly refuge areas ("fluvial refugia"). Favourable edaphic conditions perhaps explain the survival of patches of dense evergreen forest in the southern part of the Zaire Basin during the cooler and drier periods of the ice ages, as they are still today characteristic of the transition between the Guineo-Congolian and Zambezian Region. In literature, the existence of ± fluvial, patchily distributed low forest remnants in the southern part of the Zaire Basin is repeatedly suggested (survey by and further corroboration from primate distribution in Colyn et al. 1991); these areas were not taken up in Sosef's (1994: Figure 12.1) survey, although he indicated one of the areas here included in C4 as a possible refuge (called by him the Sankuru area).

It consequently seems that plants have had two types of escape during the cooler periods of the ice age: either migration towards higher altitudes, or survival in edaphically favourable conditions ("fluvial" refugia in gallery forests).

Three final remarks are necessary. Firstly, the refuge theory may well shed a light on the enigma of some species only known from historical collections but confirmed by taxonomic monographs. These may well be extinct. I can refer to two cases in the Rubiaceae mentioned in 4.3.1.

Secondly, palaeobotanical research ideally should corroborate the conclusions reached here. It seems that only pollen is sufficiently available in fossil form to tackle this. Rubiaceous pollen surely exhibits an underexplored morphological diversity, such as in its endoapertures (cf. e.g. Huysmans et al. 1993). Yet I doubt that pollen grains will ever be diagnostic to distinguish between the more than 500 species of Rubiaceae inhabiting the African lowland rain forest.

Thirdly, all the conclusions above are based upon distributional data collected from herbarium specimens, most of them transposed on maps with a resolution of about one square degree. It is not easy to know whether present knowledge and the maps reflect the chorological reality. Both a better organisation of distribution records into databases as well as field work to gather additional distribution records (stated to be much needed; White 1993b: 405) may well allow a more reliable interpretation. Comparison with the known chorological data for common and wide species would e.g. allow to check whether the intervals repeatedly observed in cases of micro-disjunction are possibly the result of underexploration. Organisation of this kind of research has not been possible for the present paper, however.

Acknowledgements
I am grateful to Mrs D.M. Bridson (Kew) for her comments on my work document surveying the distribution of African Rubiaceae. Miss P. De Block kindly read the manuscript.

References (References to Distr. Pl. Afr. are given in the text)

Bawa, K.S. & J.H. Beach, 1983. Self-incompatibility systems in the Rubiaceae of a tropical lowland wet forest. Amer. J. Bot. 70: 1281-1288.

Bremekamp, C.E.B., 1940. *Cladoceras subcapitatum*. Hook. Ic. Pl., ser. 5, 5: t. 3411.

Bridson, D.M., 1986. The reinstatement of the African genus *Keetia* (Rubiaceae subfam. Cinchonoideae, tribe Vanguerieae). Kew Bull. 41: 965-994.

Bridson, D.M. & B. Verdcourt, 1988. Flora of tropical East Africa. Rubiaceae (part 2). Rotterdam, Balkema.

Colyn, M., A. Gauthier-Hion & W. Verheyen, 1991. A re-appraisal of palaeoenvironmental history in Central Africa: evidence for a major fluvial refuge in the Zaire basin. Journ. Biogeogr. 18: 403-407.

De Block, P., in prep. Revision of *Ixora* (Rubiaceae - Pavetteae) in Africa. Opera Bot. Belg.

Hallé, N., 1966. Famille des Rubiacées (1ère partie). Paris, Mus. Natn. Hist. Nat., Flore du Gabon 12: 278 pp.

Hallé, N., 1970. Famille des Rubiacées (2e partie). Paris, Mus. Natn. Hist. Nat., Flore du Gabon 17: 335 pp.

Hamilton, A.C., 1974. Distribution patterns of forest trees in Uganda and their historical significance. Vegetatio 29: 21-35.

Huysmans, S., E. Smets & E. Robbrecht, 1993. Endoaperture morphology in the Coptosapelteae (Rubiaceae-Cinchonoideae). In: D. Fürnkranz & H. Schantl (eds.), 11. Symp. Morph., Anat. Syst., Salzburg. Kurzfassungen: 66.

Léonard, J., 1965. Contribution à la subdivision de la région guinéo-congolaise d'après la répartition géographique d'Euphorbiacées d'Afrique tropicale. Webbia 19: 627-649.

Maley, J., 1987. Fragmentation de la forêt dense humide africaine et extension des biotopes montagnards au Quaternaire récent: nouvelles données polliniques et chronologiques. Implications paléoclimatiques et biogéographiques. In: J.A. Coetzee (ed.), Palaeoecology of Africa 18: 307-334. Rotterdam, Balkema.

Ndjele, M., 1988. Principales distributions obtenues par l'analyse factorielles des éléments phytogéographiques présumés endémiques dans la flore du Zaïre. Monogr. Syst. Bot. Missouri Bot. Gard. 25: 631-638.

Petit, E., 1964. Les espèces africaines du genre *Psychotria* L. (Rubiaceae) - I. Bull. Rijksplantentuin, Bruss. 34: 1-229.

Puff, C., 1981. Studies in *Otiophora* Zucc. 1. The tropical African and Madagascan taxa. Bull. Nat. Plantentuin Belg. 51: 119-141.

Puff, C. & E. Robbrecht, 1989. A survey of the Knoxieae (Rubiaceae - Antirheoideae). Bot. Jahrb. Syst. 110: 511-558.

Robbrecht, E., 1978a. *Sericanthe*, a new African genus of Rubiaceae (Coffeeae). Bull. Nat. Plantentuin Belg. 48: 3-78.

Robbrecht, E., 1978b. Some observations in *Preussiodora* Keay (African Rubiaceae, Gardenieae). Bull. Soc. Roy. Bot. Belg. 111: 3-9.

Robbrecht, E., 1979. The African genus *Tricalysia* A. Rich. (Rubiaceae - Coffeeae): 1. A revision of the species of subgenus *Empogona*. Bull. Nat. Plantentuin Belg. 49: 239-360.

Robbrecht, E., 1981a. Studies in tropical African Rubiaceae. 2. The position of *Tricalysia auriculata* and *Neorosea adamii*. Bull. Nat. Plantentuin Belg. 51: 167-173.

Robbrecht, E., 1981b. Studies in tropical African Rubiaceae. 4. The distribution of *Batopedina* and *Parapentas*. Bull. Nat. Plantentuin Belg. 51: 186-188.

Robbrecht, E., 1982. The African genus *Tricalysia* A. Rich. (Rubiaceae - Coffeeae) 2. *Ephedranthera*, a new section of subgenus *Tricalysia*. Bull. Nat. Plantentuin Belg. 52: 311-339.

Robbrecht, E., 1983. The African genus *Tricalysia* A. Rich. (Rubiaceae) 3. *Probletostemon* revived as a section of subgenus *Tricalysia*. Bull. Nat. Plantentuin Belg. 53: 299-320.

Robbrecht, E., 1984. The delimitation and taxonomic position of the tropical African genera *Leptactinia* and *Dictyandra* (Rubiaceae). Pl. Syst. Evol. 145: 105-118.

Robbrecht, E., 1987. The African genus *Tricalysia* A. Rich. (Rubiaceae). 4. A revision of the species of sectio *Tricalysia* and sectio *Rosea*. Bull. Nat. Plantentuin Belg. 57: 39-208.

Robbrecht, E., 1988. Tropical woody Rubiaceae. Characteristic features and progressions. Contributions to a new subfamilial classification. Opera Bot. Belg. 1: 271 pp.

Robbrecht, E., 1993a. Seventy years of systematics of tropical Rubiaceae at the National Botanic Garden. Bull. Nat. Plantentuin Belg. 62: 7-34.

Robbrecht, E., 1993b. Supplement to the 1988 outline of the classification of the Rubiaceae. Index to genera. Opera Bot. Belg. 6: 173-196.

Robbrecht, E., in prep. Generic distribution patterns in African Rubiaceae.

Robbrecht, E. & D.M. Bridson, 1984. The taxonomic position of the East African genus *Cladoceras* (Rubiaceae). Bull. Soc. Roy. Bot. Belg. 117: 247-251.

Somers, C., 1988. *Aoranthe* (Rubiaceae), a new genus to accommodate the African species of *Porterandia*. Bull. Nat. Plantentuin Belg. 58: 47-75.

Sonké, B., 1990. Systématique et biologie du genre *Oxyanthus* (Rubiaceae-Gardenieae) en Afrique centrale. 1: 88 pp.; 2: 130 pp. Thesis Université de Yaoundé.

Sosef, M.S.M., 1994. Refuge begonias. Taxonomy, phylogeny and historical biogeography of *Begonia* sect. *Loasibegonia* and sect. *Scutobegonia* in relation to glacial rain forest refuges in Africa. Wageningen Agric. Univ. Papers 94 (1): 306 pp.

Stoffelen, P., E. Robbrecht & E. Smets, in prep. A revision of *Corynanthe* and *Pausinystalia* (African Rubiaceae - Coptosapelteae).

White, F., 1978. The Afromontane Region. In: M.J.A. Werger (ed.). Biogeography and ecology of Southern Africa: 463-513. Junk, The Hague.

White, F., 1979. The Guineo-Congolian Region and its relationship to other phytochoria. Bull. Nat. Plantentuin Belg. 49: 11-55.

White, F., 1993a. The AETFAT chorological classification of Africa: history, methods and application. Bull. Nat. Plantentuin Belg. 62: 225-281.

White, F., 1993b. Refuge theory, ice-age aridity and the history of tropical biotas: an essay in plant-geography. Fragm. Flor. Geobot., suppl. 2 (2): 385-409.

Patterns of endemism in the woody flora of north-east and east Africa

J.C. LOVETT [1,2] & I. FRIIS [1]
[1] Botanical Museum, University of Copenhagen,
130 Gothersgade, DK-1123 Copenhagen, Denmark.
[2] Present Address: Department of Environmental Economics and Environmental Management, University of York, Heslington, York YO1 5DD, England, United Kingdom.

Abstract
Tree and shrub species of restricted distribution do not occur randomly throughout the vegetation types and geomorphological formations of eastern Africa, but rather, show distinct concentrations of high endemism. Four geographical areas of high endemism are considered, the Eastern Arc mountain forests, the Eastern Arc rain shadow, the dry transitional forests of north-east tropical Africa and the arid areas of north-east tropical Africa. Three types of endemics are recognised: phylogenetic relicts, geographical relicts, and those with closely related taxa in the same or adjacent regions. Species distribution patterns are concluded to be primarily due to tectonic events creating climatic changes, with areas rich in taxa of restricted distribution being climatically and geologically stable. In the stable areas, phylogenetic and geographical relicts occur in the same localities as taxa apparently derived by allopatric or sympatric speciation, suggesting that climatic stability is important both for survival of relictual taxa and for initial speciation. Stability may be the factor that creates the differentiation that is later interpreted as vicariance when the range of parental taxa is reduced. Geological heterogeneity leads to floristic heterogeneity, and may be important for local speciation.

Résumé
Les espèces arborescentes et arbustives à distribution restreinte ne se rencontrent pas au hasard parmi les types de végétation et les formations géomorphologiques de l'Afrique orientale, mais elles indiquent plutôt des concentrations distinctes de haut endémisme. Quatre zones géographiques de haut endémisme sont prises en considération, à savoir les forêts montagnardes de "l'Arc" oriental, les forêts claires à l'abri des pluies de "l'Arc" oriental, les forêts sèches de transition de l'Afrique tropicale du Nord-Est et les zones arides de l'Afrique tropicale du Nord-Est. On y reconnaît trois types d'endémiques: les relictes phylogénétiques, les relictes géographiques, ainsi que celles ayant des taxons très affins dans la même région ou dans les régions voisines. Il en ressort que les modes de distribution des espèces sont conditionnés primordialement par les mouvements tectoniques qui ont engendré des changements climatiques et que les zones riches en taxons à distribution restreinte sont climatiquement et géologiquement stables. Dans les zones stables, les relictes phylogénétiques et géographiques se rencontrent dans les mêmes endroits en tant que taxons apparemment dérivés par spéciation allopatrique ou sympatrique; cela laisse supposer que la stabilité climatique importe à la fois pour la survie des taxons relictuels et pour la spéciation initiale. La stabilité peut être le facteur qui crée la différenciation, interprétée plus tard comme vicariance lorsque l'aire de répartition des taxons parentaux est restreinte. L'hétérogénéité géologique entraîne une hétérogénéité floristique et peut s'avérer importante pour une spéciation localisée.

Key words
Africa, Climate, Endemism, Geology, Speciation.

L.J.G. van der Maesen et al. (eds.), The Biodiversity of African Plants, 582-601.

Introduction

Woody plant species of restricted distribution do not occur randomly throughout eastern Tropical Africa, but are concentrated in certain vegetation types or geographical areas. For example, the forests growing in areas of relatively high rainfall on the ancient crystalline Eastern Arc mountains and coast of eastern Tanzania and south-east Kenya have more than 25% of their species known only from that specific area of eastern Africa, with many species being found in only one or two localities (Hawthorne 1993, Lovett 1993a). This is in marked contrast to the moist forests of the Lake Nyasa rift and Kenyan highlands, which contain very few endemic species. In north-east tropical Africa, although the moist forests are largely composed of widespread plants, there is a concentration of taxonomically isolated restricted range species in the driest montane forests that are transitional to the arid Somalia-Masai vegetation (Friis 1983a, 1992). Many of these species also extend to the mountain slopes of the south-western part of the Arabian peninsula, which is floristically closely connected to the Somalia-Masai region of Africa (White & Léonard 1991). In the arid rain shadow west of the Eastern Arc mountains of Tanzania and south-east Kenya, there is another cluster of species of restricted distribution associated with low rainfall and *Acacia-Commiphora* woodland (Lovett 1988). A similar area of dry vegetation in the Ogaden-Borana region of Kenya, Somalia and Ethiopia, and in northern Somalia, contains an even larger number of restricted range species (Gilbert 1986a). Other arid areas in tropical Africa, such as the Sahel, contain few species of restricted distribution, although south-west Africa is noted for its rich and unusual flora.

The restricted range taxa represent widely different taxonomic ranks. Some are very distinct, with no obvious close affinities, for example the monotypic family Barbeyaceae, and are best regarded as phylogenetic relicts of partially extinct lines. Others are distinct taxa with readily recognised extant, but geographically distant, sister taxa. These can be regarded as geographical relicts (Cronk 1992). Others have their putative parents nearby and appear to have been derived by parapatric or even sympatric speciation. Sometimes their taxonomic status is a matter of opinion. The former two types of species have been referred to as palaeoendemics, and the latter as neoendemics, though there is no way of actually knowing when the species diversified (Cronk 1992). Both relictual endemics and parapatric or sympatrically derived endemics often occur in the same locality or general area. Interestingly, concentrations of restricted range species of birds which are considered to be part of a vicariance pattern have been observed to coincide with concentrations of older taxa with a relictual distribution (Fjeldså 1994a, 1994b).

There are a number of possible explanations for the present day patterns of species of restricted distribution. Current climate, past climate, geology and biotic factors may all have had a role. Much attention has been focused on Pleistocene climatic fluctuations as a possible mechanism for determining species ranges in Africa (Coetzee 1993, Hamilton 1976, 1982, Livingstone 1993) and as a mechanism for speciation through cyclical habitat contraction and expansion (Haffer 1974). Though, as noted by Richards (1973) it is important to seek explanations in existing environmental factors before looking to relatively little known past events. The question that we address here is to seek a common explanation for the patterns of concentration of both relictual and segregate restricted range species in both arid bush and forests in eastern and north eastern tropical Africa. We first provide examples from areas where many restricted range species are found, discuss their possible relationships and examine the following hypotheses:

1 The endemics are concentrated in biological refugia that existed intermittently as a result of Pleistocene climatic fluctuations.
2 The patterns are correlated with particular geological formations and tectonic events.
3 The patterns are associated with definite climatic regimes observable in the present-day climate.

Patterns of endemism

Forests, Eastern Arc

The Eastern Arc mountains of eastern Tanzania and south east Kenya are exceptionally rich in species of restricted distribution (Lovett 1993a). The endemic species are found throughout the elevational and moisture range of the mountains in many different families and all types of life forms. The endemics can be divided into three main groups: those with relatives outside continental Africa, those with relatives in Africa, but distant from the Eastern Arc mountains, and those with relatives on or adjacent to the Eastern Arc. The first group can be phylogenetic relicts, the second group are geographical relicts, whilst the third group could have arisen by parapatric or sympatric speciation. Authorities and families of endemic taxa mentioned in the text are listed in Table 1.

Eastern Arc endemic taxa of isolated taxonomic position or with putative nearest relative out of continental Africa: Three monotypic forest canopy tree genera, *Platypterocarpus*, *Cephalosphaera* and *Neohemsleya* appear to be phylogenetic relicts. *Platypterocarpus tanganyikensis* is only known from dry montane forest on the West Usambara mountains. It was reported to be fairly common, but has not been seen recently and may be extinct following clearance of large areas of forest in 1963 (Lovett & Stuart in press). It belongs in a monotypic genus in the subfamily Tripterygioideae which is otherwise known from Madagascar, eastern Asia, Mexico and Central and South America. It appears to have no clear close relatives, but shows some affinities with *Wimmeria* of Central South America (Dunkley & Brenan 1948). *Neohemsleya usambarensis* occurs in moist montane forest in the Shagayu and Baga forests of the West Usambara. It has also been seen on ridge-tops in the southern Nguru mountains. It represents a monotypic taxonomically isolated genus with affinities to *Sarcosperma* in India-Malesia (Pennington 1991). *Cephalosphaera usambarensis* occurs in submontane forest on the Shimba Hills, East and West Usambara, Southern Nguru, Uluguru and Udzungwa mountains. *Cephalosphaera* is a monotypic genus allied to the Madagascan *Brochoneura* which has three species (Kühn & Kubitzki 1993).

Eastern Arc endemic taxa with putative nearest relative distant, but on the African continent: There are many examples of Eastern Arc taxa with their nearest relatives in the forests of west and central Africa. The relationship has been recognised at different taxonomic levels, either with full species, or eastern subspecies of western species, or as disjunctions of the same species with some differences noted, for example *Mesogyne insignis* Engl. (Moraceae) which occurs on the Usambara, Uluguru and Udzungwa mountains, is also found on São Tomé where it is slightly different to the eastern

Table 1. Eastern Arc forest endemic taxa.

Eastern Arc endemic taxa of isolated taxonomic position or with putative nearest relative out of continental Africa:
Cephalosphaera usambarensis (Warb.) Warb. (Myristicaceae)
Neohemsleya usambarensis T.D. Penn. (Sapotaceae)
Platypterocarpus tanganyikensis Dunkley & Brenan (Celastraceae)
Eastern Arc endemic taxa with putative nearest relative distant, but on the African continent:
Alsodeiopsis schumannii (Engl.) Engl. (Icacinaceae)
Mammea usambarensis Verdc. (Guttiferae)
Octoknema orientalis Mildbr. (Octoknemataceae)
Polyceratocarpus scheffleri Engl. & Diels (Annonaceae)
Pterocarpus mildbraedii Harms subsp. *usambarensis* (Verdc.) Polhill (Leguminosae: Papilionoideae)
Sibangea pleioneura Radl.-Sm. (Euphorbiaceae)
Uvariopsis bisexualis Verdc. (Annonaceae)
Eastern Arc endemic taxa with nearest putative relative in the Eastern Arc or nearby mountains or lowlands:
Campylospermum scheffleri (Engl. et Gilg) Farron (Ochnaceae)
Croton dictyphlebodes Radl.-Sm. (Euphorbiaceae)
Garcinia semseii Verdc. (Guttiferae)
Hirtella megacarpa R.A. Graham (Chrysobalanaceae)
Schefflera lukwangulensis (Tennant) Bernardi (Araliaceae)
Trichilia lovettii Cheek (Meliaceae)
Uvariodendron oligocarpum Verdc. (Annonaceae)
U. pycnophyllum (Diels) R.E. Fr.
U. usambarense R.E. Fr.
Zenkerella capparidacea (Taub.) J. Léonard subsp. *capparidacea* (Leguminosae: Caesalpinioideae)
Z. capparidacea (Taub.) J. Léonard subsp. *grotei* (Harms) Temu
Z. egregia J. Léonard
Z. perplexa Temu

populations. Other examples of geographical relicts include: *Polyceratocarpus scheffleri* which occurs in submontane and montane forest on the Usambara, Uluguru and Udzungwa mountains. It is the only eastern representative of a tropical African genus of seven species. *Uvariopsis bisexualis* occurs in montane forest on the Udzungwa mountains. It is the only eastern African representative of a genus of about 11 species in Cameroon and Gabon. *Sibangea pleioneura* occurs in submontane forest on the Udzungwa mountains. It is an eastern representative of a tropical African genus of three species closely related to *Drypetes* (Radcliffe-Smith 1987). *Mammea usambarensis* occurs in montane forest on the West Usambara mountains. It is related to the lowland west and central African *M. africana* Sabine (Bamps et al. 1978), which is the only other continental African species and in which it was formerly placed (Bamps 1971). *Alsodeiopsis schumannii* occurs in submontane, montane and upper montane forest on the East and West Usambara, Southern Nguru, Uluguru and Udzungwa mountains, and is the only eastern representative of a tropical African genus of 11 species (Lucas 1968a). *Pterocarpus mildbraedii* subsp. *usambarensis* occurs in lowland forest on the East Usambara and Udzungwa mountains. It is the eastern subspecies of the west African *P. mildbraedii* subsp. *mildbraedii* (Polhill 1969). *Octoknema orientalis* occurs in submontane forest on Udzungwa and Mahenge Mountains. It is the only eastern representative of a tropical African genus of five species (Lucas 1968b).

Eastern Arc endemic taxa with putative nearest relative in the Eastern Arc or nearby mountains or lowlands: A number of species has the closest relatives in the Eastern Arc or on nearby mountains. Some belong to a small group related to each other that have apparently arisen by parapatric or sympatric differentiation, but geographically isolated from other related species. This raises the possibility of all the species having arisen by allopatric differentiation and subsequent loss of geographically distant closest relatives. For example three *Uvariodendron* species, all occurring on the Usambara mountains and all Eastern Arc endemics, appear to be related to each other (Verdcourt 1986): *Uvariodendron oligocarpum* from submontane forest on the West and East Usambara; *Uvariodendron pycnophyllum* from submontane forest on the East and West Usambara; and *Uvariodendron usambarense* from submontane forest on the East Usambara and Southern Nguru mountains. Other *Uvariodendron* species are in the eastern African coastal forests, the Kenyan highlands and west and central Africa. Another example is *Zenkerella capparidacea* subsp. *capparidacea* from submontane and montane forests on the Nguru and Uluguru mountains, the only other subspecies is subsp. *grotei* from the East and West Usambara (Temu 1990a). The species as a whole is restricted to the Eastern Arc mountains, as is *Zenkerella perplexa* from submontane forest on the Uluguru and Malundwe mountains, which appears to be a montane segregate of *Z. egregia* which occurs in riverine, lowland forest in the Uluguru, Nguru and East Usambara mountains (Temu 1990b). The only other *Zenkerella* is *Z. citrina* Taub. from west and central Africa.

Other species are montane segregates of more widespread lowland species, for example: *Hirtella megacarpa* from montane forest on the West Usambara and Udzungwa mountains appears to be a montane segregate of *H. zanzibarica* Oliv. (Graham 1960), to which it is sufficiently similar to be considered the same species (White 1976). *Campylospermum scheffleri* from upper montane, montane, and submontane forest on the Usambara and Uluguru mountains, is regarded as a higher altitude segregate of *C. sacleuxii* (van Tieghem) Farron, a coastal forest species restricted to eastern Tanzania related to the central African *C. elongatum* (Oliver) van Tieghem (Farron 1968). Some species are high forest forms of forest edge, woodland or riverine species. For example: *Garcinia semseii*, from lowland, montane and submontane forest on the Nguru and Uluguru, is a forest form of the widespread riparian and coastal forest *G. livingstonei* T. Anders. (Bamps et al. 1978). *Croton dictyphlebodes* is known only from a restricted area of montane and dry montane forest in the West Usambara mountains where it can be quite common. It is closely related to the widespread *C. sylvaticus* Hochst. complex (Radcliffe-Smith 1972).

Other species appear to have their closest relatives in adjacent areas, for example: *Schefflera lukwangulensis* from upper montane forest in the Uluguru, Southern Nguru, and Usambara is an Eastern Arc segregate of *S. umbellifera* (Sond.) Baill., a forest tree which occurs in montane forest from the southern Udzungwa to South Africa

Table 2. North-eastern tropical Africa dry forest endemic taxa.

North-eastern tropical African dry forest endemic taxa of isolated taxonomic position or with putative nearest relative out of continental Africa:
Barbeya oleoides Schweinf. (Barbeyaceae)
Buxus hildebrandtii Baill. (Buxaceae)
Cadia purpurea (Picc.) Ait. (Leguminosae: Papilionoideae)
Ceratonia oreothauma Hillc., G.P.Lewis & Verdc. (Leg.: Caesalpinioideae)
Dracaena ellenbeckiana Engl. (Dracaenaceae)
D. ombet Kotschy & Peyr.
Mimusops laurifolia (Forssk.) Friis (Sapotaceae)
Pistacia aethiopica Kokwaro (Anacardiaceae)
P. falcata Martelli
Renschia heterotypica (S. Moore) Vatke (Labiatae)
Sideroxylon mascatense (A.DC.) T.D. Penn. (Sapotaceae)
North-eastern tropical African dry forest endemic taxa with putative nearest relative distant, but on the African continent:
Sideroxylon oxyacantha Baill. (Sapotaceae)
North-eastern tropical African dry forest endemic taxa with nearest putative relative in the adjacent dry forests in nearby regions:
Vepris borenensis (M.G. Gilbert) Mziray (Rutaceae)

(Bamps 1974). *Trichilia lovettii* is a small montane forest tree or treelet on the Udzungwa mountains, related to *T. capitata* Klotzsch from low altitude thicket and woodland near streams in Zimbabwe, Malawi and Mozambique (Cheek 1989).

Dry forests of north-eastern tropical Africa

The dry forests of north-eastern tropical Africa under the Indian Ocean climatic system have a considerably smaller proportion of endemics apparently derived by parapatric or sympatric speciation than the other areas dealt with in this paper. These dry forests are transitional between the forests on the Ethiopian and Somalian plateaux and the bushland of the lowlands. The restricted range species occur both with widespread forest taxa such as *Juniperus procera* Hochst. ex Endl. (Cupressaceae), and with some of the species mentioned below under arid land endemics of the Horn of Africa, for example *Commiphora* and *Boswellia*. Wetter forests of north-eastern tropical Africa on the western side of the Ethiopian plateau are under the Atlantic climatic system and can be considered outliers of the main Guineo-Congolian region or are more or less strictly Afromontane (Friis 1992). The wetter forests also contain an number of restricted range species, for example, *Baphia abyssinica* Brummitt (Leguminosae: Papilionoideae) and *Rinorea friisii* M.G. Gilbert (Violaceae) in the lowlands and intermediate level, and at higher altitudes *Erythrina brucei* Schweinf. emend. J.B. Gillett (Leguminosae: Papilionoideae), *Millettia ferruginea* (Hochst.) Baker (Leguminosae: Papilionoideae), *Maytenus addat* (Loes.) Sebsebe (Celastraceae) and *Vepris daniellii* (Pich.-Serm.) Kokwaro (Rutaceae). These species have relatives in the Guineo-Congolian region or forests of eastern Africa. Authorities and families of endemic taxa mentioned in the text are listed in Table 2.

North-eastern tropical African dry forest endemic taxa of isolated taxonomic position or with putative nearest relative out of continental Africa: Barbeya oleoides is the only species in the single genus of the family Barbeyaceae, which is often placed in a monotypic order, Barbeyales, of uncertain affinity (Friis 1993). *B. oleoides* is a tree which occurs in dry *Juniperus* forest and open evergreen or semi-deciduous bushland on the eastern escarpment of Eritrea, Ethiopia, and in the mountains of north Somalia; it also occurs in a few similar habitats on the western escarpment of Yemen and south-west Saudi Arabia (Friis 1992). *Cadia purpurea* occurs largely in the same habitats and areas as *Barbeya oleoides*, but it extends into a few localities associated with the Great Rift in Kenya (Friis 1992). It is also associated with *Juniperus* forest, where it occurs in open areas or in evergreen scrub on slopes just outside the forest. It is the only species in continental Africa, the remaining species occur in Madagascar (van der Maesen 1970). *Buxus hildebrandtii* grows in association with the dry *Juniperus* forest on the east escarpment of Eritrea and Ethiopia, and in the mountains of north Somalia, where it occurs in open parts or in evergreen or semi-evergreen bushland (*Buxus-Acokanthera* scrub) outside the forest; it has also been recorded from Socotra. It is assumed to be rather isolated within the genus *Buxus*, with its closest relatives in America (Friis 1989). *Mimusops laurifolia* occurs associated with dry *Juniperus-Olea* forest, as large single trees in evergreen or semi-evergreen bushland on slopes outside the forest, in riverine forest or as isolated trees along semi-permanent or temporary streams on the eastern escarpment of Eritrea and Ethiopia, and in the mountains of north Somalia. It also occurs in a few similar habitats on the western escarpment of Yemen (Friis 1992). It does not appear to have any closely related species among the other African species of *Mimusops*. *Pistacia aethiopica* is a near-endemic, in other words it is relatively widespread compared to some of the other taxa discussed here. It occurs in *Juniperus-Olea* forest, *Juniperus-Olea-Barbeya* forest, and in various types of associated semi-evergreen and evergreen bushland on the east escarpment of Eritrea, Ethiopia, in the mountains of north Somalia and south to the Rift in Kenya and reaches the rain shadow behind the northern part of the Eastern Arc in Tanzania. (Friis 1992); its nearest relative seems to be *P. lentiscus* L. in the Mediterranean region. *Pistacia falcata* is a near-endemic which occurs in association with the dry *Juniperus-Olea* forest on the east escarpment of Eritrea, Ethiopia, and in the mountains of north Somalia, where it is found in evergreen or semi-evergreen bushland (*Buxus-Acokanthera* scrub) outside the forest. It also occurs as a pioneer species on recent lava-flows in the Ethiopian Rift, and is known from a few similar habitats on the western escarpment of Yemen (Friis 1992). Its nearest relatives seem to be the Asian species *P. chinensis* Bunge and the Middle East species *P. palaestina* Boiss. *Dracaena ellenbeckiana* is a near-endemic which occurs in association with the dry *Juniperus-Olea* forest on the south-east escarpment of Ethiopia, where it may be found in evergreen or semi-evergreen bushland (*Buxus-Acokanthera* scrub) outside the forest. It also occurs on rocky slopes in association with the Kenya Rift (Friis 1992). It is a very distinctive species. *Dracaena ombet* is a near-endemic which is here taken also to include the very similar *D. schizantha* Baker. The two taxa have a complementary distribution in association with the dry *Juniperus-Olea* forest on Jebel Elba in south east Egypt, the mountains of east Sudan, the east escarpment of Eritrea, Ethiopia, and in the mountains of north Somalia, where the taxa may also be found in evergreen or semi-evergreen bushland (*Buxus-Acokanthera* scrub) and similar habitats outside the forest. Established plants may persist on bare rocks (Friis 1992). Closely related species are *D. cinnabari* Balf. f. from Socotra and *D. serrulata* Baker from Saudi Arabia. It also somewhat resembles *D. draco* L. from the Canary Islands.

Renschia heterotypica is the only species of the genus *Renschia* (a woody member of the Labiatae, subfamily Ajugoideae) and is restricted to the Ahl Mountains in northern Somalia where for example it occurs at Meid, not far from the remnants of the dry *Juniperus-Olea* forests, though there is no record of it actually growing in the forest. Vollesen (1975) considers it most closely related to the much more widespread genus *Tinnea* (a genus predominantly of open land, but with some forest representatives such as *T. vesiculosa* Gürke from the Eastern Arc), but *Renschia* is supposed to represent an older line of evolution without any other close relatives. *Ceratonia oreothauma* is a restricted range species which occurs in the mountains of north Somalia and in similar habitats in Oman. The only other species in the genus is the cultigen *C. siliqua* L., thought to be of East Mediterranean origin (Irvin & Barneby 1981). *Sideroxylon mascatense* (syn.: *Monotheca buxifolia* (Falconer) A. DC.) is disjunct between Pakistan, Oman, the dry forests on the south-east escarpment of Ethiopia and the mountains of northern Somalia (Friis 1978, 1992). In spite of being now placed in *Sideroxylon* (Pennington 1991), its nearest relative in that genus is obscure.

North-eastern tropical African dry forest endemic taxa with putative nearest relative distant, but on the African continent: There are no clear-cut examples of geographical relicts, but *Sideroxylon oxyacantha* (syn.: *Spiniluma oxyacantha* (Baill.) Aubrév.) which occurs in dry *Juniperus* forest and in adjacent evergreen or semi-evergreen bushland on the east escarpment of Ethiopia and the mountains of northern Somalia (Friis 1978, 1992) is closely related to *S. discolor* Radl.-Sm. from Socotra. The following example could also be regarded as a geographical relict.

North-eastern tropical African dry forest endemic taxa with putative nearest relative in the adjacent dry forests or nearby regions: Vepris borenensis (syn.: *Teclea borenensis* M.G. Gilbert) occurs in the dry forest of southern Ethiopia and northern Kenya (Moyale and Mt. Marsabit; Friis 1992, Mziray 1992). According to Gilbert it is most closely related to *V. hanangensis* (Kokwaro) Mziray (syn.: *T. hanangensis* Kokwaro) which occurs in dry forest around Nairobi in Kenya and in northern and central Tanzania (Gilbert 1986b).

Arid land, the Eastern Arc rain shadow

West of the Eastern Arc mountains there is a concentration of restricted range species in the arid *Acacia-Commiphora* woodland associated with the Eastern Arc rain shadow. The area receives less than 500 mm of rain a year, in marked contrast with the moist forests only a few tens of kilometres distant. As with the Eastern Arc moist forests, restricted range relictual taxa whose relatives are geographically distant, occur with endemic taxa with more widespread relatives in adjacent areas. Endemics are found in a wide range of life forms, in herbs, shrubs, trees and parasites. Some species have been mentioned by Lovett (1988), though *Commiphora ruahensis* Mattick (Burseraceae) is now considered conspecific with the widespread *C. glandulosa* Schinz.

Some of the endemic species are confined to small areas, for example *Commiphora acuminata* Mattick, *Cordyla densiflora* Milne-Redhead (Leguminosae: Papilionoideae),

Erythrina greenwayi Verdc. (Leguminosae: Papilionoideae), *Euphorbia dumeticola* Bally & S. Carter (Euphorbiaceae), *E. proballyana* Leach, and *Ormocarpum flavum* J.B. Gillett (Leguminosae: Papilionoideae) are all restricted to the Ruaha Valley. Other arid land endemics occur throughout the Eastern Arc rain shadow, or extend into arid areas on the central Tanzanian plateau or south east Kenya. Authorities and families of endemic taxa mentioned in the text are listed in Table 3.

Table 3. Eastern Arc rain shadow endemic taxa.

Eastern Arc rain shadow endemic taxa of isolated taxonomic position or with putative nearest relative out of continental Africa:
Erythrina greenwayi Verdc. (Leguminosae: Papilionoideae)
Eastern Arc rain shadow endemic taxa with putative nearest relative distant, but on the African continent:
Carphalea glaucescens (Hiern) Verdc. ssp. *angustifolia* Verdc. (Rubiaceae)
Entandrophragma bussei Engl. (Meliaceae)
Millettia oblata Dunn subsp. *burtii* J.B. Gillett (Leguminosae: Papilionoideae)
Peterodendron ovatum (Sleumer) Sleumer (Flacourtiaceae)
Eastern Arc rain shadow endemic taxa with nearest putative relative in the adjacent dry forests in nearby regions:
Ormocarpum flavum J.B. Gillett (Leguminosae: Papilionoideae)
Pachystigma burtti Verdc. subsp. *hirtiflorum* Verdc. (Rubiaceae)

Eastern Arc rain shadow endemic taxa of isolated taxonomic position or with putative nearest relative out of continental Africa: There are no distinct phylogenetic relicts, but *Erythrina greenwayi* is thought to be very isolated in the genus without clear relatives (Verdcourt 1971).

Eastern Arc rain shadow endemic taxa with putative nearest relative distant, but on the African continent: Some of the arid land species are geographical relicts related to rain forest taxa. For example, the endemic *Peterodendron ovatum* is a shrub in a monotypic genus related to the west and central African *Poggea* (Sleumer 1938) which has 4 understorey species in rain forest, gallery forest, secondary forest and swamp forest (Sleumer 1974). A similar example, but with relatives much closer geographically is from the Itigi thicket, a curious vegetation type with many lianes west of the Eastern Arc rain shadow proper, where the restricted range *Millettia oblata* subsp. *burttii* occurs. The three other subspecies, *oblata*, *intermedia* J.B. Gillett and *teitensis* J.B. Gillett are found in rain forests on the Eastern Arc and at the northern end of Lake Nyasa (Gillett 1961).

Other endemic species have related taxa in distant arid areas, for example, *Entandrophragma bussei* is closely related to *E. spicatum* (C. DC.) Sprague which occurs in northern Namibia and southern Angola (Styles & White 1991). Although not woody, an interesting example of a similar distribution pattern is the relationship between two monotypic grass genera: *Chlorocalymma cryptacanthum* W.D. Clayton (Graminae) an Eastern Arc rain shadow endemic is related to the Angolan *Streptolophus* (Clayton 1970). Some species are disjunct to the north, for example: *Carphalea glaucescens* has an Eastern Arc rain shadow endemic subspecies, *angustifolia*, with the typical subspecies in north-east Africa, and *Commiphora swynnertonnii* B.D. Burtt (Burseraceae) has a disjunct distribution between the Eastern Arc rain shadow and northern Kenya.

Eastern Arc rain shadow endemic taxa with putative nearest relative in the adjacent dry forests in nearby regions: Ormocarpum flavum has affinities to the eastern, northern and south-west African *O. kirkii* S. Moore, which is closely related to the west and northern African *O. pubescens* (Hochst.) Cufod. (Gillett 1966). *Pachystigma burtti* subsp. *hirtiflorum* is only known from the Ruaha Valley, *P. burttii* subsp. *burtii* is restricted to dry thicket in central Tanzania, and the species as a whole is related to the Kenyan *P. kenyense* Verdc. (Verdcourt 1981).

Arid land, the Horn of Africa

The arid regions of the horn of Africa are particularly rich in species of restricted distribution. Geological heterogeneity, with limestones, sands and gypsums, is correlated with floristic heterogeneity (Gilbert 1986a). In addition, many species are disjunct between north-east and south-west Africa (reviewed in Werger 1978, with lists of taxa in Monod 1971, de Winter 1971). Authorities and families of endemic taxa mentioned in the text are listed in Table 4.

Horn of African arid land endemic taxa of isolated taxonomic position or with putative nearest relative out of continental Africa: Cordeauxia edulis, the Yeheb-nut represents a monotypic genus restricted to the eastern Ogaden and adjacent parts of central Somalia where it grows in semi-desert bushland and scrub on sandy soil (Polhill & Thulin 1989). *Cordeauxia* seems to be a fairly isolated genus, yet Brenan (1967) has stated that its closest relative is the monotypic *Stuhlmannia* (only species *S. moavi* Taub. restricted to lowland dry evergreen forest in north-east Tanzania). It is considered here because Brenan's opinion is not explicitly endorsed by Polhill & Vidal (1981). *Bottegoa* is an extremely distinctive monotypic genus with no clear relatives which occurs in bushland on basaltic and limestone rocks in the Ogaden of eastern Ethiopia, in Somalia and north-eastern Kenya. The monotypic genus *Puccionia*, with the subshrubby species *P. macradena*, occurs only on gypsum in northern Somalia. It is rather isolated and is placed in either the Capparaceae or the Cruciferae. The tree *Givotia gosai* occurs in central and southern Somalia, eastern Ethiopia and north-eastern Kenya in deciduous bushland on sandy soil; it has one related species in India and two on Madagascar.

There is a striking number of monotypic endemic (some shrubs, but mostly herbaceous) Amaranthaceous genera in the arid areas of the Horn of Africa: *Lopriorea* (with *L. ruspolii* (Lopr.) Schinz in bushland on limestone in central and southern Somalia and north-east Kenya), *Rosifax* (*R. sabuletorum* C.C. Townsend in sand in southern Somalia), *Eriostylos* (*E. stefanini* (Chiov.) C.C. Townsend in bushland of central and southern Somalia), and *Polyrhabda* (*P. atriplicifolia* C.C. Townsend on limestone in northern Somalia). A number of other endemic genera in the family have two or more species in the same general region or in adjacent dry areas: *Neocentema* (one endemic species, *N. robecchii* (Lopr.) Schinz, in the dry area of the Horn of Africa and *N. alternifolia* (Schinz) Schinz in Tanzania), *Pleuropterantha* (three endemic species in central and southern Somalia, *P. revoilii* Franch. being relatively widespread in the area and the other species, *P. undulatifolia* Chiov. and *P. thulinii*

Table 4. Northeastern Africa arid land endemic taxa.

Horn of African arid land endemic taxa of isolated taxonomic position or with putative nearest relative out of continental Africa:

Bottegoa insignis Chiov. (Sapindaceae)
Cordeauxia edulis Hemsley (Leguminosae: Caesalpinioideae)
Givotia gosai Radl.-Sm. (Euphorbiaceae)
Puccionia macradena Chiov. (Capparaceae)

Horn of African arid land endemic taxa with putative nearest relative distant, but on the African continent:

Atriplex nogalense Friis & M.G. Gilbert (Chenopodiaceae)
Dirachma somalensis D.A. Link (Dirachmaceae)
Gossypium trifurcatum Vollesen (Malvaceae)
Kissenia arabica Chiov. (Loasaceae)
Obetia sp. (= Burger 3368) (Urticaceae)
Tinnea somalensis Chiov. (Labiatae)
Turnera thomasii (Urb.) Story (Turneraceae)

Horn of African arid land endemic taxa with nearest putative relative within the Horn of African arid land or in arid areas closely nearby:

Boswellia brichettii (Chiov.) Chiov. (Burseraceae)
B. microphylla Chiov.; *B. neglecta* S. Moore
B. ogadensis Vollesen; *B. pirottae* Chiov.
B. rivae Engl.

Bourreria lyciacea Thulin (Boraginaceae)
B. orbicularis (Hutch. & E.A. Bruce) Thulin
B. teitensis (Gürke) Thulin

Commiphora confusa Vollesen (Burseraceae)
C. guidottii Chiov.; *C. gurreh* Engl.
C. longipedicellata Vollesen; *C. serrulata* Engl.
C. sphaerophylla Chiov.; *C. tenuis* Vollesen
C. truncata Engl.

Fadenia zygophylloides Aellen & C.C. Townsend (Chenopodiaceae)

Gossypium benadirense Mattei (Malvaceae)
G. brichettii (Ulbrich) Vollesen

Gyrocarpus angustifolius (Verdc.) Thulin (Hernandiaceae)
G. hababensis Chiov.

Lagenantha cycloptera (Stapf) M.G. Gilbert & Friis (Chenopodiaceae)
L. gillettii (Botsch.) M.G. Gilbert & Friis

Loewia glutinosa Urb. (Turneraceae)
L. tanænsis Urb.

Senna baccarinii (Chiov.) Lock (Leguminosae: Caesalpinioideae),
S. ellisiae (Brenan) Lock; *S. hookeriana* Batka
S. humifusa (Brenan) Lock; *S. ruspolii* (Chiov.) Lock
S. truncata (Brenan) Lock

Parkinsonia anacantha Brenan (Leguminosae: Caesalpinioideae)
P. raimondoi Brenan; *P. scioana* (Chiov.) Brenan

Xylocalyx carteriae Thulin (Scrophulariaceae)
X. hispidus S. Carter; *X. recurvus* S. Carter

C.C. Townsend occurring only on coastal sand), and the woody *Chionothrix* (two species, *C. latifolia* Rendle and *C. somalensis* (S. Moore) Hook.f., in the arid areas of the Horn).

Horn of African arid land endemic taxa with putative nearest relative distant, but on the African continent: A group of species which perhaps belong in the former category, have their closest relative on the island of Socotra which has a very similar geology and climate to northern Somalia. An interesting example is *Dirachma,* the only genus in the family Dirachmaceae, sometimes considered a subfamily in Geraniaceae, but perhaps better placed near Tiliaceae. It consists of two shrubby species: *D. somalensis* from deciduous bushland on sand over limestone in central Somalia, and *D. socotrana* Balf.f. on Socotra (Link 1991). Other species which clear relatives on the African mainland are: *Atriplex nogalense* occurs on gypsum plains in northern Somalia; it is a subshrub very close to the North African species *A. coriacea* Forssk. (Boulos et al. 1991). *Obetia sp.* (= Burger 3368) is a small shrub found in one locality in eastern Ethiopia, in deciduous bushland on limestone. It seems most closely related to *Obetia tenax* (N.E. Brown) Friis in dry regions of Zimbabwe and South Africa and *O. carretusiana* (Hiern) Rendle in Namibia and Angola; other less similar species occur in dry parts of East Africa, on Madagascar and the Mascarenes (Friis 1983b). *Gossypium trifurcatum* is only known from a few localities near Ras Hafun in north-east Somalia. According to Vollesen (1987) it is closely related to *G. anomalum* Wawra & Peyr. which is disjunctly distributed in Angola and Namibia (subsp. *anomalum*) and in a belt from Niger to Sudan and Eritrea (subsp. *senarense* (Wawra & Peyr.) Vollesen. *Tinnea somalensis* is one of c. 20 species of the genus *Tinnea* (woody members of the Labiatae, subfam. Ajugoideae). The genus is best represented in the Zambezian woodlands of south central Africa, but also occurs in the Sudanian woodlands and a few species elsewhere in open country; *T. somalensis* is restricted to the Borana Region of south Ethiopia and adjacent parts of south Somalia. According to Vollesen (1975) it seems to be rather distantly related to the other species of *Tinnea* in the region (*T. aethiopica* Hook.f.), and seems to most closely related to *T. barbata* Vollesen from dry habitats in Transvaal. The genus *Kissenia* a shrubby representative of the Loasaceae, has only two species: *K. arabica* in northern Somalia, eastern Ethiopia, Djibouti and South Yemen on limestone or sand, and *K. capensis* Endl. in Namibia and the Cape Province of South Africa. *Turnera thomasii* occurs in the Witu Desert of Lamu District in Kenya. A related species, *T. oculata* Story grows in the rain shadow of the Huambo Plateau in Angola (Story 1961). The third African *Turnera* species, which shows affinities to both the restricted range species, is the widespread *Turnera ulmifolia* L. from West Africa to the Seychelles and Madagascar.

Horn of African arid land endemic taxa with putative nearest relative within the Horn of African arid land or in arid areas closely nearby: Loewia glutinosa is restricted to arid areas in the horn of Africa. The closely related *L. tanænsis* is only known from northern Kenya, and may be conspecific (Thulin 1993b). The endemic or near endemic species of *Boswellia* which occur in the arid part of the Horn of Africa appear to be mutually rather closely related (Vollesen 1985). Some species are apparently very narrowly distributed. *B. ogadensis* is known only from limestone cliffs near Webi Schebele River at the border between Ethiopia and Somalia. The more widely distributed *B. microphylla* occurs in eastern Ethiopia, northern, central and southern Somalia and north-east Kenya, and the even more widespread *B. neglecta* is distributed in the dry part of northern Uganda, north and north-east Kenya, eastern Ethiopia and northern, central and southern Somalia. These species are also rather similar to the narrowly distributed *B. pirottae* which occurs in the deep river valleys (in rain shadow) in the Ethiopian highlands. Other, perhaps more distantly related, endemic species are of narrow distribution within the Horn, for example *B. rivae* which is confined to the limestone areas of the Ogaden and adjacent parts of Somalia, and *B. brichettii* which is only known from northern Somalia between Ras Haifun and Mait.

The genus *Commiphora* also provides numerous examples of local endemics in the Horn of Africa with putative relatives in the same region or in nearby arid areas. Many of these examples are taken from accounts of the family for Ethiopia by Vollesen (1989) and for the area covered by the Flora of Tropical East Africa by Gillett (1991). A few examples of local endemics with close relatives nearby (from Vollesen 1985) will be sufficient here in order to demonstrate the extraordinary degree of local speciation of this genus in the Horn of Africa: *C. tenuis* (of the section *Campestres* (Engl.) Vollesen) is fairly narrowly distributed on limestone in the Borana region of south Ethiopia and in the adjacent part of northeast Kenya. It is assumed to be closely related to *C. serrulata* (in the Borana region and the Ogaden of eastern Ethiopia and adjacent parts of Somalia) and to *C. gurreh* from Somalia. In sect. *Hildebrandtianae* there are two closely related species, *C. truncata* which is restricted to limestone areas in the Ogaden and adjacent parts of Somalia, and the fairly widely distributed *C. confusa* which occurs on a number of substrates and extends from the Borana region in south Ethiopia to the rain shadow behind the northern part of the Eastern Arc. In sect. *Arillopsidum*, *C. sphaerophylla* (syn.: *C. ellisiae* Vollesen) occurs on limestone in the Ogaden, according to Vollesen (1985) it is closely related to the very variable and widespread *C. erythraea* (Ehrenb.) Engl. which occurs from eastern Sudan and south-west Arabia to northern Tanzania. In the same section, *C. longipedicellata* which is restricted to south-east Ethiopia, north-east Kenya and central and southern Somalia, is closely related to *C. engleri* Guill. which is widespread in the dry areas of northern and central Tanzania. *C. guidottii* (syn.: *C. sessiliflora* Vollesen) from rocks rich in gypsum in south-east Ethiopia and Somalia is closely related to *C. ornifolia* (Balf. f.) J.B. Gillett from Socotra.

The genus *Gyrocarpus* (Hernandiaceae) has a number of species in tropical America, one pan-tropical species (*G. americanus* Jacq.) which in Africa has a subspecies in dry areas of north-east Africa and south-west Africa, and two closely related species in the arid areas of the Horn of Africa: *G. hababensis* (Somalia, eastern Ethiopia and north-east Kenya in bushland on sand) and *G. angustifolius* (central and southern Somalia, north-east Kenya in bushland on sand). Two narrowly distributed endemics belonging to the genus *Gossypium* are both closely related to the more widespread *G. somalense* (Gürke) J.B. Hutch., which occurs throughout the Ogaden and Borana region in Ethiopia, in north and south Somalia, and in the Sahel (Vollesen 1987). Of the two endemics, *G. brichettii* is restricted to central Somalia, and *G. benadirense* occurs in localities rich in gypsum in north and south Somalia and in the Daua Valley in northeast Kenya. *Xylocalyx* is a small genus of subshrubs with three species *X. recurvus*, *X. carteriae* and *X. hispidus* occurring in central Somalia, two other species occur in Socotra (Thulin 1987a). *Fadenia* and *Lagenantha* are two restricted range genera of subshrubs with succulent leaves, distinguished on the position of the wing on the fruiting perianth (Boulos et al. 1991). *Fadenia zygophylloides* occurs on saline soil or gypsum in central and southern Somalia, south-east Ethiopia and north-east Kenya, *Lagenantha cycloptera* occurs on gypsaceous rocks or soil in northern Somalia and Socotra, and *L. gillettii* on gypsaceous soil in southern Somalia and eastern Kenya.

The genus *Senna* has a number of more or less related restricted range species in the border region between eastern Ethiopia and central Somalia (Thulin 1993a): *S. baccarinii* (northern and central Somalia, eastern Ethiopia and adjacent part of north-east Kenya, in bushland), *S. ruspolii* (north and central Somalia, eastern Ethiopia and adjacent part of north-east Kenya, usually in bushland on sand), *S. humifusa* (southern Somalia, north-eastern Kenya, on sand), *S. truncata* (northern and central Somalia, eastern Ethiopia, bushland on sand) *S. ellisiae* (northern and central Somalia, eastern Ethiopia, in bushland on sand or limestone), *S. hookeriana* (northern Somalia, South Yemen, semi-desert bushland). *Parkinsonia* has three restricted range species in the arid areas of the Horn of Africa: *P. scioana* in dry bushland of east Ethiopia, south Somalia and the extreme north-east of Kenya; *P. anacantha* in subdesert scrub of north-east and east Kenya, and *P. raimondoi* from central and southern Somalia. There is one species in south Africa (*P. africana* Sond.), and 15 species in the drier parts of North and South America (Polhill & Vidal 1981). The genus *Bourreria* has five species in east Africa, three of which are restricted range species of the Horn of Africa and adjacent dry areas (Thulin 1987b): *B. orbicularis* is restricted to the Ogaden in east Ethiopia and adjacent parts of Somalia, *B. teitensis* is known from south Somalia, south-east Kenya and north-east Tanzania, and *B. lyciacea* is restricted to the arid areas of east Kenya and south Somalia. Two additional species occur in the coastal belt of Kenya, Tanzania and Mozambique, and on Mauritius and Madagascar. The remaining c. 50 species of *Bourreria* occur in central America.

Discussion

The examples cited here show that species of restricted distribution can be locally concentrated in a wide range of vegetation types, from arid woodland and bush to dry montane forest to moist submontane forest. Within the four areas of endemism there are four main patterns to explain. 1) The occurrence of phylogenetic relicts in all four areas; 2) The occurrence of geographical relicts in the Eastern Arc moist forests with affinities to western Africa; 3) The occurrence of geographical relicts in the arid areas with affinities to south-west Africa; and 4) the occurrence of restricted range species with close relatives nearby in the Eastern Arc, the Eastern Arc rain shadow and the Horn of Africa arid lands, though not in the Horn of Africa dry forests. In re-examining the hypotheses originally proposed as potential explanations of these patterns, and to seek a common explanation for restricted range species distribution in both arid bush and forests, the nature of narrow range endemics and widespread species needs to be contrasted. Species of restricted distribution are either too recently arisen to be widespread, have poor dispersal ability, or are adapted to surviving in specific environmental conditions and cannot compete outside their home range. Environmental stability is considered to be important for the survival of local isolates, which is a step towards speciation in metapopulations (Levin 1993), leading to an accumulation of species of restricted distribution. In contrast to narrow range endemics, widespread species will be adapted to environmental fluctuations, as by their very definition, they grow in a wide range of conditions (Stevens 1992). However, they may be at a disadvantage in areas of stability where they will be competing against plants adapted to a specific and predictable environment. A further consideration is distinguishing between vicariance events, when distribution patterns reflect geological and climatological changes, and long distance dispersal, when distribution patterns reflect dispersal routes.

Hypothesis 1. *The endemics are concentrated in biological refugia resulting from Pleistocene climatic fluctuations.* During the Pleistocene it has been suggested that there were alternating wet and dry periods (Bonnefille & Hamilton 1986, Hamilton 1982). During dry periods an arid ecosystem expansion could have connected arid areas in north-eastern and south-western Africa, with concomitant reduction in moist forest and a few areas of moist forest remaining as refugia. During wet periods an expansion of moist forests could have resulted in moisture-dependent species migrating out of dry period refuges enabling biotic exchange between forest areas, with arid rain shadow areas acting as refugia for arid-land species. Current climate and proposed Pleistocene refugia would probably be related, with present-day very humid areas being the proposed forest refugia and the drier arid areas being arid refugia. Pleistocene fluctuations could explain connections between the Eastern Arc and western Africa in wet periods, and north eastern and south western arid areas during dry periods. Extinction during fluctuations outside the refugia would then account for lack of phylogenetic or geographical relicts elsewhere. If distribution patterns are to be explained by vicariance due to climatic changes, then the main problem with this hypothesis is fitting both an east-west pan-African rain forest and a north-south arid corridor into present-day geomorphology. Wet and dry climatic extremes would need so great that it is unlikely that postulated refugia would have been sufficiently climatically stable to maintain their endemic flora. However, long-distance dispersal may have played a role if dispersal distances between wet or dry ecosystems where reduced.

Hypothesis 2. *The patterns are correlated with particular geological formations and tectonic events.* The Eastern Arc mountains are very old, being associated with rifts dating from the break-up of Gondwanaland with their most recent reactivation dating from the late Miocene (Griffiths 1993). Affinities between the Eastern Arc and moist forests of west and central Africa, especially with markedly disjunct taxa and taxa with poor dispersal abilities, suggest that the Eastern Arc forests are a relict of a former pan-African forest pre-dating Miocene uplift of the central African plateau (Axelrod & Raven 1978, Lovett 1993b). It is possible that affinities between the Eastern Arc and Madagascar date from separation of Madagascar from mainland Africa, but long-distance dispersal most likely played a role. The arid area west of the Eastern Arc will be as old as the mountains, being an orographic rain shadow from the Indian Ocean rainfall system. The arid corridor from north-eastern to south-western Africa was created by Pliocene uplift of the central African mountains delimiting eastern influence of the Atlantic rainfall system. The north-east tropical African dry montane forests grow on the Trap Lava series, an immense mass of lavas capping the Ethiopian plateau, which formation started in the Upper Cretaceous to Eocene and ended in the late Miocene (Mohr 1971), and on low limestone mountains in Somalia. Major tectonic movements in the Miocene and Pliocene uplifted the Ethiopian plateau, indicating the Miocene as the earliest age for the dry forests. The north-eastern arid area dates from the union of Africa with Eurasia in the mid-Miocene. In the Horn of Africa, different floras are found on limestones, sands and gypsum rocks and alluvia (Gilbert 1986a). Similarly the Ruaha valley in the Eastern Arc rain shadow is geologically complex with riverine sands, limestones, and crystalline rocks. Local edaphic variation could provide the conditions needed for speciation. This hypothesis overcomes the difficulties of the vicariance scenario in hypothesis 1 by placing the pan-African forest and arid corridor in different geomorphological settings with correspondingly different orographically determined climatic patterns. It also provides a mechanism for generating local isolates in the arid areas of endemism through geological heterogeneity.

Hypothesis 3. *The patterns are associated with definite climatic regimes observable in the present day climate.* Within the period covered by rainfall records, the African climate shows marked annual fluctuations in both wet and dry areas (Nicholson 1994). Long-term climatic stability is required both for survival of phylogenetic and geographical relicts, and, as indicated above, for local speciation. Although stability needs be in the order of millions of years to maintain phylogenetic relicts if their presence is due to vicariance rather than long distance dispersal, the four areas of high endemism are in present day areas of comparative climatic stability. The endemic rich Eastern Arc moist forests are under the direct influence of the Indian Ocean monsoon, which does not show the extreme climatic fluctuations of, for example, the Lake Nyasa rift (Owen et al. 1990), an area with few endemic species. Similarly the north-eastern African dry forests are under the Indian Ocean climate, in contrast to the moister forests of the Horn of Africa which are mostly composed of widespread species and are on the edge of the fluctuating Atlantic climatic system. The Eastern Arc rain shadow is dependant on the position of the Eastern Arc mountains, and so is climatically stable, as is the north-east African arid area. In contrast, the Sahelian region, with few endemics, has a climate which has fluctuated enormously in recent years (Nicholson 1994). Origin of endemics from more widespread taxa could be through differentiation in areas of environmental stability. Environmental changes outside centres of endemism could result in parental taxa changing the extent of their ranges, or becoming extinct and so leading to geographical or phylogenetic isolation of the restricted range taxon. Taxa which have apparently speciated through vicariance may have actually differentiated within the stability area before being geographically separated from their parents. This explains the coincident distributions of apparently relictual taxa and those seemingly derived by parapatric and sympatric speciation. Thus long term environmental stability allows differences in species richness and distribution patterns to develop: speciation does not need climatic fluctuations that lead to speciation by vicariance. Differences in distribution patterns of taxa growing in stable areas and areas where the climate fluctuates have been interpreted as "contact zones" and have been used as arguments for Pleistocene biological refugia (Mayr & O'Hara 1986), whereas in fact they represent primary speciation events (Endler 1977). So, this hypothesis provides an explanation for areas of low endemism without invoking massive extinctions, and provides a mechanism both for speciation and maintenance of restricted range species.

In summary, the patterns of endemism in the tropical north east and eastern African woody flora appear to be due primarily to climatic history determined by tectonic events which created areas of climatic stability. This combines hypothesis two and three. Environmental stability encourages speciation by allowing survival of local segregates. Allopatric differentiation may also play a role following vicariance or long-distance dispersal. Species apparently derived by parapatric speciation occur in the same localities as geographic and phylogenetic relicts, suggesting that climatic stability is also important for the creation of species, and may be the factor that creates initial differentiation that is later interpreted as vicariance. Geological heterogeneity is a factor in creating floristic heterogeneity, and may be important for local speciation. It is not necessary to invoke refuges resulting from Pleistocene climatic fluctuations to explain the present day distribution of restricted range taxa.

Acknowledgements

We are grateful to Roger Polhill, Jan Gillett, Kaj Vollesen and David Goyder of the Royal Botanic Gardens, Kew who helped with discussion on restricted range eastern African taxa. David Mabberley and Quentin Cronk commented on an earlier version of the manuscript. We thank the ornithologist Jon Fjeldså, Zoological Museum of Copenhagen, for stimulating discussions and comments. JCL was supported by the Danish Centre for Tropical Biodiversity during the preparation of the paper. Fieldwork in Tanzania was supported by the Missouri Botanical Garden, the National Geographic Society, the World Wide Fund for Nature, and Caltex Ltd. The Tanzania Commission for Science and Techonology kindly granted permission to conduct research in Tanzania. IF has been supported by the Carlsberg Foundation during large parts of the fieldwork in Ethiopia and Somalia which forms the background of this paper. Other valuable experience was obtained during work with the Flora of Ethiopia with M.G. Gilbert, Kaj Vollesen, Sebsebe Demissew, Ensermu Kelbessa and Mesfin Tadesse, under the supervision of Prof. O. Hedberg and Dr. Tewolde Berhan Gebre Exiabher; and with the Flora of Somalia under the supervision of Dr. Mats Thulin.

References

Axelrod, D.I. & P.H. Raven, 1978. Late Cretaceous and Tertiary vegetation history of Africa. In: M.J.A. Werger (ed.), Biogeography and Ecology of Southern Africa: 77-130, Junk, The Hague.

Bamps, P., 1971. Notes sur les Guttiferae d'Afrique tropicale. Bull. Jard. Bot. Nat. Belg. 41: 427-444.

Bamps, P., 1974. Contribution à l'étude des Araliacées africaines. Bull. Jard. Bot. Nat. Belg. 44: 101-139.

Bamps, P., N. Robson & B. Verdcourt, 1978. Guttiferae. In: R.M. Polhill (ed.), Flora of Tropical East Africa, Crown Agents, London.

Bonnefille, R. & A.C. Hamilton, 1986. Quaternary and late Tertiary History of Ethiopian vegetation. Symb. Bot. Ups. 26 (2): 48-63.

Boulos, L., I. Friis, & M.G. Gilbert, 1991. Notes on the Chenopodiaceae of Ethiopia, Somalia and Southern Arabia. Nordic J. Bot. 11: 309-316.

Brenan, J.P.M., 1967. Leguminosae subfamily Caesalpinioideae. In: E. Milne-Redhead & R.M. Polhill (eds.), Flora of Tropical East Africa, Crown Agents, London.

Cheek, M., 1989. A new *Trichilia* (Meliaceae) from Tanzania and its relationship with *Pseudobersama*. Kew Bull. 44: 457-463.

Clayton, W.D., 1970. Studies in the Graminae: XXII. A curious new genus from Tanzania. Kew Bull. 24: 461-463.
Coetzee, J.A., 1993. African flora since the terminal Jurassic. In: P. Goldblatt (ed.), Biological relationships between Africa and South America: 37-61, Yale, New Haven & London.
Cronk, Q.C.B., 1992. Relict floras of Atlantic islands: patterns assessed. Biological Journal of the Linnean Society 46: 91-103.
De Winter, B., 1971. Floristic relationships between the northern and southern arid areas in Africa. Mitt. Bot. Staatssamml. München 10: 424-437.
Dunkley, H.L. & J.P.M. Brenan, 1948. *Platypterocarpus*, a new tree genus of Celastraceae from Tanganyika. Kew Bull. 1948: 47-50.
Endler, J.A., 1977. Geographic variation, speciation and clines. Princeton University Press, Princeton.
Farron, C., 1968. Contribution à la taxonomie des Ourateae d'Afrique. Candollea 23: 177-228.
Fjeldså, J., 1994a. Geographical patterns for relict and young species of birds in Africa and South America and implications for conservation priorities. Biodiversity and Conservation 3: 207-226.
Fjeldså, J., 1994b. Geographical patterns of neoendemic and relict species of Andean forest birds: the significance of ecological stability areas. Mem. New York Bot. Gard. in press.
Friis, I., 1978. A reconsideration of the genera *Monotheca* and *Spiniluma* (Sapotaceae). Kew Bull. 33: 91-98.
Friis, I., 1983a. Phytogeography of the tropical northeast African mountains. Bothalia 14: 525-532.
Friis, I., 1983b. A synopsis of *Obetia* (Urticaceae). Kew Bull. 22: 221-228.
Friis, I., 1989. A synopsis of the Buxaceae in Africa South of the Sahara. Kew Bull. 44: 293-299.
Friis, I., 1992. Forests and Forest Trees of Northeast Tropical Africa. Kew Bulletin Additional Series 15: 1-396.
Friis, I., 1993. Barbeyaceae. In: K. Kubitzki (ed.), Families and genera of Flowering Plants, Vol. 2, Dicotyledones - Magnoliid, Hamameliid and Caryophyllid Families: 141-143, Springer Verlag, Berlin.
Gilbert, M.G., 1986a. The flora of south and south-east Ethiopia. Symb. Bot. Ups. 26 (2): 86-93.
Gilbert, M.G., 1986b. *Rhus neoglutinosua* sp. nov. (Anacardiaceae) and *Teclea borenensis* sp. nov. (Rutaceae) from Ethiopia. Nordic J. Bot. 6: 139-141.
Gillett, J.B., 1961. Notes on *Millettia* Wight & Arn. in East Africa. Kew Bull. 15: 19-40.
Gillett, J.B., 1966. The species of *Ormocarpum* Beauv. and *Arthrocarpum* Balf.f. (Leguminosae) in south-western Asia and Africa (excluding Madagascar). Kew Bull. 20: 323-355.
Gillett, J.B., 1991. Burseraceae. In: R.M. Polhill (ed.), Flora of Tropical East Africa, Crown Agents, London.
Graham, R.A., 1960. Rosaceae. In: C.E. Hubbard & E. Milne-Redhead (eds), Flora of Tropical East Africa, Crown Agents, London.
Griffiths, C.J., 1993. The geological evolution of East Africa. In: J.C. Lovett & S.K. Wasser (eds.), Biogeography and Ecology of the Rain Forests of Eastern Africa: 9-21, Cambridge University Press, Cambridge.
Haffer, J., 1974. Avian speciation in tropical South America. Publ. Nuttall Ornith. Club 14.
Hamilton, A.C., 1976. The significance of patterns of distribution shown by forest plants and animals in tropical Africa for the reconstruction of upper Pleistocene palaeo-environments: a review. Palaeoecol. Afr. 9: 63-97.
Hamilton, A.C., 1982. Environmental History of East Africa. A study of the Quaternary. Academic Press, London & New York.
Hawthorne, W.D., 1993. East African coastal forest botany. In: J.C. Lovett & S.K. Wasser (eds.), Biogeography and Ecology of the Rain Forests of Eastern Africa: 57-99, Cambridge University Press, Cambridge.
Irvin, H.S. & R.C. Barneby, 1981. Tribe 2. Cassieae. In: R.M. Polhill & P.H. Raven (eds.), Advances in Legume Systematics, Part 1: 97-106, Royal Botanic Gardens, Kew.

Kühn, U. & K. Kubitzki, 1993. Myristicaeae. In: K. Kubitzki, J.G. Rohwer & V. Bittrich (eds.), The Families and Genera of Vascular Plants Vol. II, Flowering Plants, Dicotyledons. Magnoliid, Hamamelid and Caryophyllid Families: 457-467, Springer-Verlag, Berlin.

Levin, D.A., 1993. Local speciation in plants: the rule not the exception. Systematic Botany 18 (2): 197-208.

Link, D.A., 1991. *Dirachma somalensis* D.A. Link sp. nov., a new species of a remarkable and highly endangered monogeneric family. Bull. Jard. Bot. Nat. Belg. 61: 3-13.

Livingstone, D.A., 1993. Evolution of African climate. In: P. Goldblatt (ed.), Biological Relationships between Africa and South America: 455-472, Yale University Press, New Haven & London.

Lovett, J.C., 1988. Endemism and affinities of the Tanzanian montane forest flora. In: P. Goldblatt & P.P. Lowry (eds.), Proceedings of the eleventh plenary meeting of the Association for the Taxonomic Study of Tropical Africa. Monographs in Systematic Botany from the Missouri Botanical Garden 25: 591-598.

Lovett, J.C., 1993a. Eastern Arc moist forest flora. In: Lovett, J.C. & S.K. Wasser (eds.), Biogeography and Ecology of the Rain Forests of Eastern Africa: 33-55, Cambridge University Press, Cambridge.

Lovett, J.C., 1993b. Climatic history of eastern Africa. In: J.C. Lovett & S.K. Wasser (eds.), Biogeography and Ecology of the Rain Forests of Eastern Africa: 23-29, Cambridge University Press, Cambridge.

Lovett, J.C. & S.N. Stuart, in press. Avifauna and vegetation of the Shume Juniperus forest in the West Usambara mountains, Tanzania. Scopus.

Lucas, G. Ll., 1968a. Icacinaceae. In: E. Milne-Redhead & R.M. Polhill (eds.), Flora of Tropical East Africa, Crown Agents, London.

Lucas, G. Ll., 1968b. Olacaceae. In: E. Milne-Redhead & R.M. Polhill (eds.), Flora of Tropical East Africa, Crown Agents, London.

Mayr, E. & R.J. O'Hara, 1986. The biogeographic evidence supporting the Pleistocene forest refuge hypothesis. Evolution 40 (1): 55-67.

Mohr, P.A., 1971. The geology of Ethiopia. Addis Ababa University, Addis Ababa.

Monod, T., 1971. Remarques sur les symmétries floristiques des zones sèches nord et sud en Afrique. Mitt. Bot. Staatssamml. München 10: 375-423.

Mziray, W., 1992. Taxonomic studies in *Toddalieae* Hook. f. (Rutaceae) in Africa. Symb. Bot. Upsal. 30 (1): 1-95.

Nicholson, S.E., 1994. Recent rainfall fluctuations in Africa and their relationship to past conditions over the continent. The Holocene 4 (2): 121-131.

Owen, R.B., R. Crossley, T.C. Johnson, D. Tweddle, I. Kornfielk, S. Davidson, D.H. Eccles & D.E. Engstrom, 1990. Major low levels of Lake Malawi and their implications for speciation rates in cichlid fishes. Proc. Roy. Soc. Lond. B 240: 519-553.

Pennington, T.D., 1991. The Genera of Sapotaceae. Royal Botanic Gardens, Kew & New York Botanical Garden.

Polhill, R.M., 1969. Notes on East African *Dalbergieae* Bronn (Leguminosae). Kew Bull. 23: 483-490.

Polhill, R.M. & M. Thulin, 1989. Subfamily Caesalpinioideae. In: S. Edwards & I. Hedberg, (eds.), Flora of Ethiopia, Vol. 3: 49-70, Asmara, Addis Ababa & Uppsala.

Polhill, R.M. & J.E. Vidal, 1981. Tribe 1. *Caesalpinieae*. In: R.M. Polhill & P.H. Raven (eds.), Advances in Legume Systematics, Part 1: 81-95, Royal Botanic Gardens, Kew.

Radcliffe-Smith, A., 1972. New and noteworthy Euphorbiaceae from Tropical Africa: I. Kew Bull. 27: 505-509.

Radcliffe-Smith, A., 1987. Euphorbiaceae (Part 1). In: R.M. Polhill (ed.), Flora of Tropical East Africa, Balkema, Rotterdam.

Richards, P.W., 1973. Africa, the "Odd Man Out". In: B.J. Meggers, E.S. Ayensu & W.D. Duckworth (eds.), Tropical Forest Ecosystems in Africa and South America: A comparative review: 21-26, Smithsonian, Washington.

Sleumer, H., 1938. *Peterodendron*, eine neue Gattung der Flacourtiaceen in Ostafrika. Notizbl. Bot. Gart. Mus. Berlin-Dahlem 13: 356-359.

Sleumer, H., 1974. Revision der Gattung *Poggea* Gürke (Flacourtiaceae). Bot. Jahrb. 94: 296-301.

Stevens, G.C., 1992. The elevational gradient in altitudinal range: an extension of Rapoport's latitudinal rule to altitude. The American Naturalist 140: 893-911.

Story, R., 1961. Turneraceae. Bothalia 7 (3): 493-496.

Styles, B.T. & F. White, 1991. Meliaceae. In: R.M. Polhill (ed.), Flora of Tropical East Africa, Balkema, Rotterdam.

Temu, R.P.C., 1990a. Taxonomy and distribution of *Zenkerella capparidacea* s. lat. (Leguminosae-Caesalpinioideae). Nordic J. Bot. 10: 425-430.

Temu, R.P.C., 1990b. A new species of *Zenkerella* (Leguminosae-Caesalpinioideae) from Tanzania. Nordic J. Bot. 10: 421-423.

Thulin, M., 1987a. A new species of *Xylocalyx* (Scrophulariaceae) from Somalia. Nordic J. Bot. 7: 267-269.

Thulin, M., 1987b. *Bourreria* (Boraginaceae) in tropical Africa. Nordic J. Bot. 7: 413-417.

Thulin, M., 1993a. Fabaceae (Leguminosae). In: M. Thulin (ed.), Flora of Somalia Volume 1, Pteridophyta; Gymnospermae; Angiospermae (Annonaceae-Fabaceae): 341-465, Royal Botanic Gardens, Kew.

Thulin, M., 1993b. Turneraceae. In: M. Thulin (ed.), Flora of Somalia Volume 1, Pteridophyta; Gymnospermae; Angiospermae (Annonaceae-Fabaceae): 34-36, Royal Botanic Gardens, Kew.

Van der Maesen, L.J.G., 1970. Primitiae Africanae VIII. A revision of the genus *Cadia* Forskål (Caes.) and some remarks regarding *Dicraeopetalum* Harms (Pap.) and *Platycelyphium* Harms (Pap.). Acta Bot. Neerl. 19 (2): 227-248.

Verdcourt, B., 1971. *Erythrina greenwayi* Verdc. Kew Bull. 25: 175-176.

Verdcourt, B., 1981. Notes on African Rubiaceae. Kew Bull. 36: 493-557.

Verdcourt, B., 1986. New taxa of East African Annonaceae. Kew Bull. 41: 287-297.

Vollesen, K., 1975. A taxonomic revision of the genera *Tinnea* and *Renschia* (Labiateae, Ajugoideae). Bot. Tidsskr. 70: 1-63.

Vollesen, K., 1985. Studies in Burseraceae of Northeastern Africa. Kew Bull. 40: 39-76.

Vollesen, K., 1987. The native species of *Gossypium* (Malvaceae) in Africa, Arabia and Pakistan. Kew Bull. 42: 337-349.

Vollesen, K., 1989. Burseraceae. In: S. Edwards & I. Hedberg (eds.), Flora of Ethiopia, Vol. 3: 442-478, Asmara, Addis Ababa & Uppsala.

Werger, M.J.A., 1978. The Karoo-Namib Region. In: M.J.A. Werger (ed.), Biogeography and Ecology of southern Africa: 231-299, Junk, The Hague.

White, F., 1976. Taxonomy, ecology and chorology of African Chrysobalanaceae (excluding *Acioa*). Bull. Jard. Bot. Nat. Belg. 46: 265-350.

White, F. & Léonard, J., 1991. Phytogeographical links between Africa and southwest Asia. Fl. & Veg. Mundi 9: 229-246.

Begonias and African rain forest refuges: general aspects and recent progress

M.S.M. SOSEF
Department of Plant Taxonomy, Herbarium Vadense, Wageningen Agricultural University, P.O. Box 8010, 6700 ED Wageningen, The Netherlands.

Abstract
In tropical Africa during the last glacial the climate was cooler and drier and the rain forest retreated. It remained present in several small forest refuge areas only. The species of two sections of *Begonia* are ideal to serve as bioindicators of the localities of former forest refuges. Their distribution patterns have been analyzed elsewhere but are discussed here in the light of new evidence and data. This is followed by a discussion on the validity of deducing localities of forest refuges by means of present day distributional data.

Résumé
En Afrique tropicale, durant la dernière période glaciaire, le climat était plus froid et plus sec et la forêt ombrophile se retira. Elle ne subsista que dans plusieurs petites superficies de refuges forestiers. Les espèces de deux sections du genre *Begonia* représentent des bioindicateurs idéaux pour situer les endroits de ces anciens refuges forestiers. Leurs modes de distributions ont été analysés ailleurs mais il en est discuté ici à la lumière de nouvelles preuves et données. S'en suit une discussion concernant la valeur que l'on peut accorder à la déduction des emplacements de refuges forestiers à partir des données actuelles de la distribution.

Key words
Begonia, Biodiversity, Biogeography, Forest refuges, Palaeoenvironments, Rain forest.

Introduction

It has often been thought that rain forest, being a climax vegetation, is a very stable ecosystem. Studying the climatological and vegetational history of tropical Africa, however, leads to the conclusion that most of the lowland rain forests are comparatively young, and have existed at their present localities for only 10,000 to 12,000 years, or even less. Recent biogeographical, palynological and climatological studies strongly support the rain forest refuge theory. This theory is based on the assumption that the oscillating climatic conditions during the late Quaternary had a profound impact on the area occupied by tropical rain forest. This geological period, starting at about 1 Myr BP, is characterized by 9 major and some 12 minor glacial periods (Hamilton 1982, Hamilton & Taylor 1991). In Africa most, but not all, of these were considerably drier than the interglacial ones. It has been shown recently (Dupont 1993) that indeed the extent of the rain forest did respond to the oscillating climatic conditions of the past few glacials.

At the glacial maximum of the most recent Ice Age, which lies around 18,000 yr BP, the estimated mean annual temperature in tropical Africa was around 4°C lower than today (Bonnefille et al. 1990). This strong decrease in temperature was also concluded from studies on the expansion and retreat of glaciers on the high African mountains. The precipitation was much less, while the dry season appears to have been

L.J.G. van der Maesen et al. (eds.), The Biodiversity of African Plants, 602-611.

more severe. Palaeoclimatologists have shown that cold water ocean currents along the West African coast were different from those today (Maley 1989). Cold water upwellings led to the development of stratiform cloud covers in certain coastal areas, decreasing precipitation even further (Maley 1987). Such major climatic changes must have had a profound impact on the area occupied by rain forest vegetation, and led to isolated patches of lowland rain forest: the refuges.

The montane rain forest probably occupied a larger area than today due to a descent of the montane vegetation zones. During the generally drier glacial the amount of precipitation is assumed to have primarily determined both lowland and montane rain forest distribution.

From 18,000 yr BP onward climatic conditions for lowland rain forest improved again. Forest species, both plants and animals, could expand their ranges, but it is assumed that not all of them were able to disperse out fast. The present-day rain forests surrounding the former refuges contain a species composition derived from that of the refuge areas. So, at present we may expect to find the highest diversity of rain forest species at locations of former refuges. Present-day distribution patterns of certain rain forest species may reveal those locations. Many authors have already applied this line of reasoning and postulated localities for several forest refuges (Aubréville 1962, Colyn et al. 1991, Diamond & Hamilton 1980, Grubb 1982, Kingdon 1980).

A second possibility to trace former refuge areas is through geological and palaeopalynological data. By dating a sedimentary core and identifying the pollen grains it contains one is able to reconstruct the vegetation history. Comparatively many such studies are available from the East African mountains (Livingstone 1982), but unfortunately very few have yet been performed in the lowlands of Central and West Africa, which seem to be the most interesting areas with respect to forest refuges. The most detailed studies in the latter area are those of Maley & Brenac (1987) and their co-workers, who traced a refuge area in western Cameroon, and of Elenga & Vincens (1990), who studied pollen cores from Congo. Figure 1 represents the most likely localities of rain forest refuges based on the combined work of biogeographers, geologists, and palynologists before I presented the results of my study on begonias (Sosef 1994a).

Nature conservation

In my opinion, the importance of tracing the localities of former rain forest refuges for activities concerning nature conservation has not been stressed strongly enough. As already mentioned above, these areas are centres of biodiversity of rain forest organisms, both plants and animals, as well as centres of endemism. Prime areas in the light of nature conservation, and hence knowledge on their exact location is highly desirable.

Several groups of organisms now seem to serve as bioindicators of former forest refuges. This is highly interesting as we may now predict high biodiversity, also for other extant groups of organisms, at these localities without having to complete a much more elaborate and intensive inventory of all of tropical Africa. We need to study only several indicator groups for this purpose.

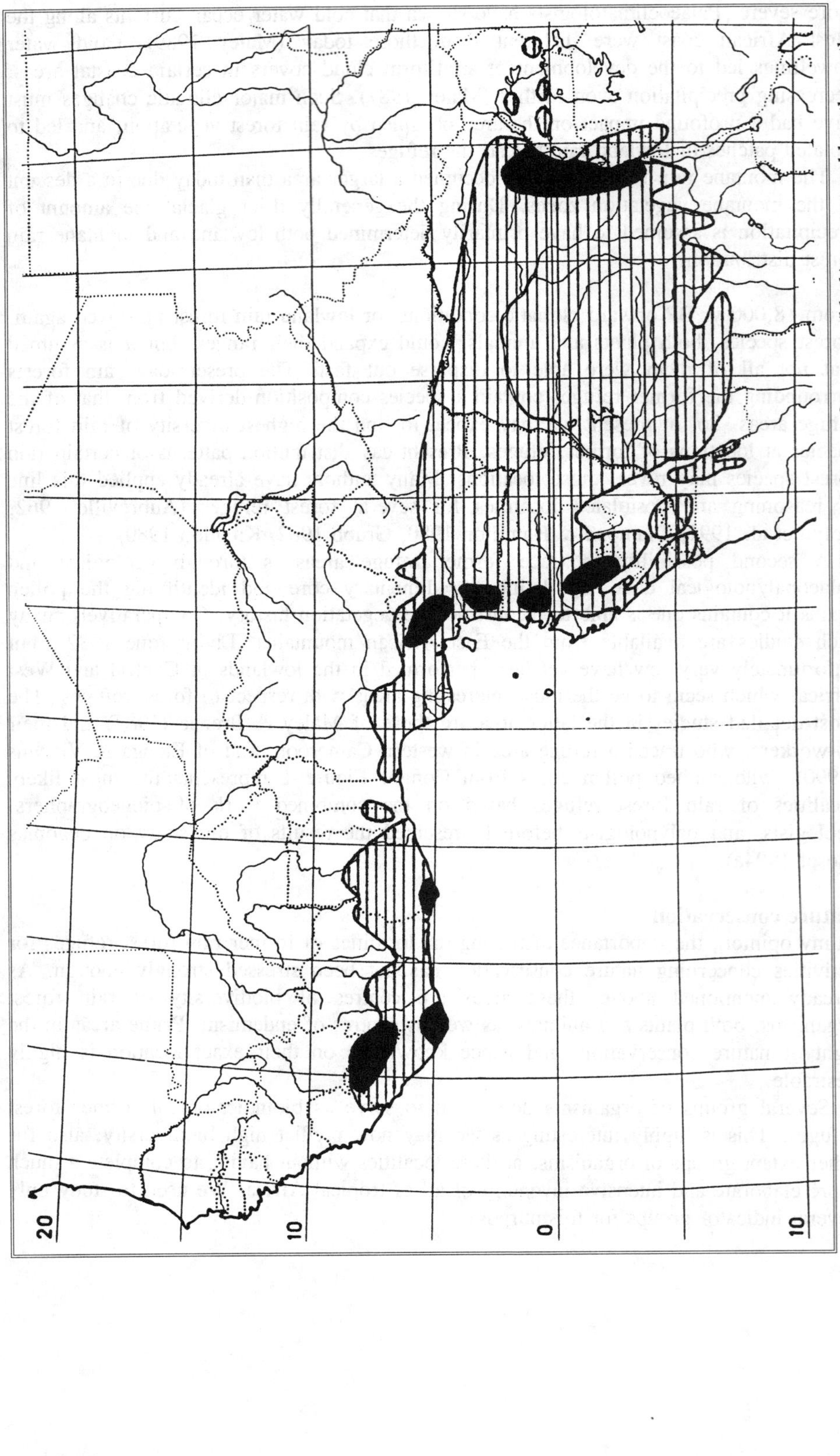

Figure 1. Present-day distribution of lowland rain forest (hatched, after White 1983) and forest refuge areas (black) as proposed by other authors.

Begonias as bioindicators

One group of bioindicator species for rain forest refuges is the group of 40 *Begonia* species encompassing the sections *Loasibegonia* and *Scutobegonia*. The pantropical genus *Begonia* consists of some 1000 species. Within Africa about 115 species can be recognized, which are divided amongst 11 sections. The closely related sections *Loasibegonia* and *Scutobegonia* are characterized by their terrestrial rhizomatous habit, their indehiscent fruits, and their male and female flowers with only 2 yellow or sometimes white perianth segments. The reasons why the species belonging to these sections serve as bioindicators are two-fold. First of all they show several biological features making it plausible that they will serve as such. These have been dealt with in detail before (Sosef 1992, 1994a, 1994b), but I will mention them briefly. The majority of the species is confined to heavily shaded moist conditions in primary rain forests, so they will have been able to survive through the last glacial only within the refuge areas. There are two reasons why they probably did not disperse far out of these areas again after expansion of the rain forest. Their seed dispersal ability is not sufficient in terms of long distance dispersal, and additionally many of the species appear to be self-incompatible. Finally, the species are usually narrow endemics, making a detailed study possible. The second reason why they can be looked upon as bioindicators of forest refuges is that their geographic distribution patterns coincide astonishingly well with the most recently postulated forest refuge areas of Figure 1, and as was shown by me earlier (Sosef 1994a & 1994b). The present paper discusses several additional important points regarding forest refuges which can be deduced via a thorough study of these begonias.

To eliminate the influence of chance events of sporadic dispersal, it is assumed that only the occurrence of a concentration of *Begonia* taxa (species and subspecies) belonging to these 2 sections points to a former forest refuge area. Figures 2 and 3 show the distribution and concentration of the taxa involved.

The Upper Guinea refuge areas

Three possible refuge areas have been indicated in this region (Figure 1): the Liberia refuge area, the Cape Palmas refuge area, and the Cape Three Points refuge area. The latter two showed comparatively high biodiversity figures for forest species (Aubréville 1962, Grubb 1982, Guillaumet 1967, Hamilton 1976), but were not accepted by Maley (1987). He based his deviating ideas on palaeoclimatic data concerning stratiform cloud covers causing a diminished precipitation in coastal areas and some contradictionary biogeographical evidence (see Maley 1989). However, also the distributional data of *Begonia* species support the presence of two coastal refuge areas (Figure 2 and Sosef 1994a) and I reinstated them again (Sosef 1994a). In my opinion the strong biogeographical evidence cannot be put aside. It may be that the climatic influence of the stratiform cloud cover was not as severe as assumed, or that its influence was felt only up to a few kilometres inland. Van Rompaey (1993, and this volume), in his thorough study of the forests of eastern Liberia and western Ivory Coast, also found strong evidence of a former refuge area near the coast of Cape Palmas. Actually, because of the recently presented new biogeographical evidence Maley also adapted his ideas recently (this volume) and accepted a refuge area near to Cape Palmas and another near to Cape Three Points.

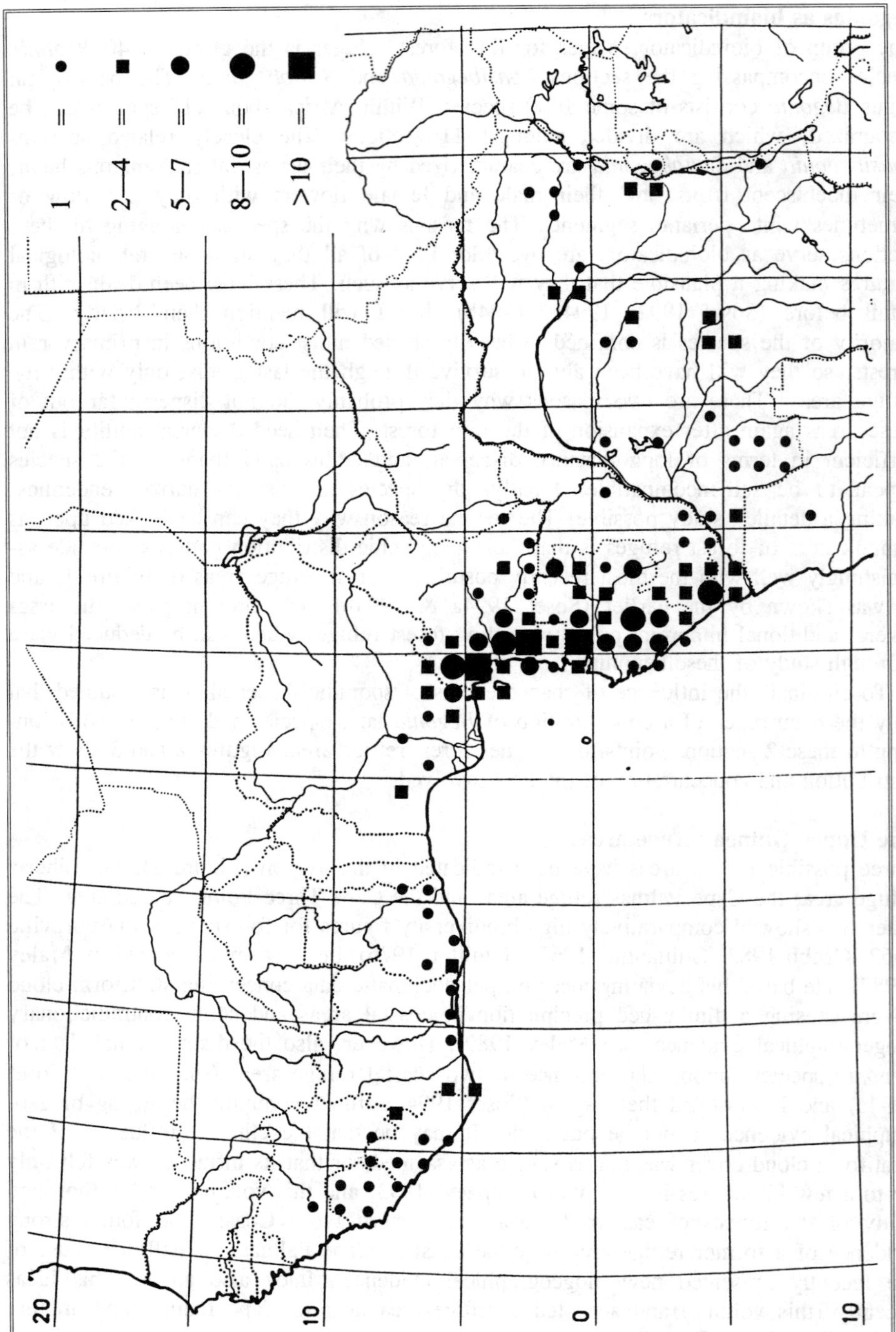

Figure 2. Density of taxa (species and subspecies) of *Begonia* sect. *Loasibegonia* and sect. *Scutobegonia*

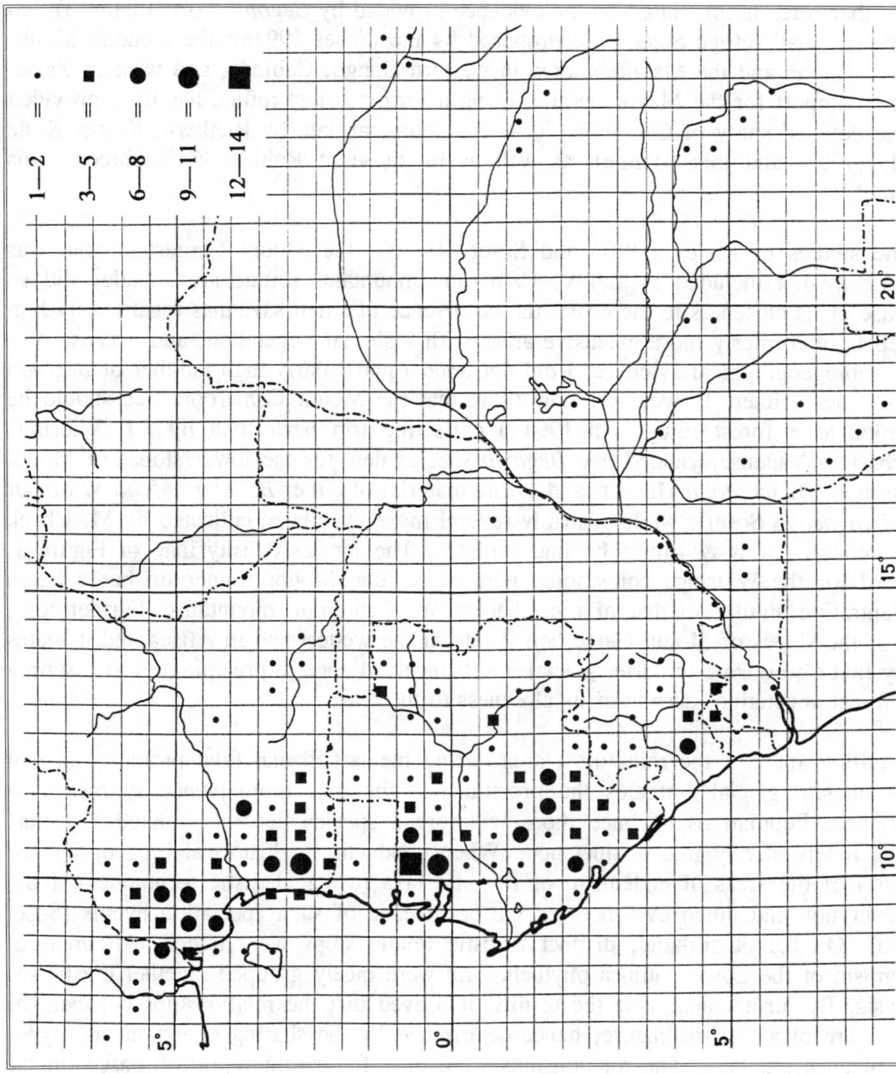

Figure 3. As Figure 2, but for the Lower Guinea region only.

The Lower Guinea refuge areas

Several studies have pointed to the presence of a single large forest refuge area in Cameroon, Equatorial Guinea, and Gabon, but all these were performed on a comparatively large scale. Maley (1987) combined the results of more detailed studies and suggested a series of four smaller refuges instead (see Figure 1 and Sosef 1994a: 116, for further references). Based on the evidence provided by *Begonia* taxa (Figure 3) two additional small refuge areas were proposed by me (Sosef 1994a): the Doudou Mountains in Gabon and the Mayombe area in western Congo, Cabinda, and western Zaire. Further support for the Mayombe area being a former forest refuge has been provided by the detailed study of Caesalpinioideae distributional data by Rietkerk, Ketner & de Wilde (1995, and this volume), as well as by those of Rubiaceae (Robbrecht, this volume).

In the studies of Maley (1987) and Sosef (1994a), the Mount Cameroon area was specifically not included in the W. Cameroon mountains refuge area. Maley did so because of its closeness to the coast, the occurrence of small savannas regarded by him as relicts of formerly more extensive areas with a similar vegetation, and because of a lack of biogeographical evidence. I did not trace a particularly high number of *Begonia* species there either. However, I now think that the Mount Cameroon area should be regarded as a forest refuge. In 1994 a collecting trip performed by J.J. Wieringa (Herbarium Vadense) yielded two *Begonia* species new for the lower slopes of Mount Cameroon (*B. scutifolia* Hook.f. and sterile material of either *B. laporteifolia* Warb. or *B. mbangaensis* Sosef), while probably several more have been collected by M. Cheek (pers. comm., Kew & Limbe botanic gardens). The species density map of Figure 3, adapted for the Wieringa collections, now shows that Mount Cameroon has a taxon concentration similar to that of the adjacent W. Cameroon mountains. Furthermore, part of the slopes of Mount Cameroon is one of the wettest site in Africa and it seems likely that also during the drier glacial it still received enough precipitation to sustain a rain forest vegetation, despite of its closeness to the sea.

The falling apart of the rain forest area during the last glacial (and probably also of more ancient glacials) creates the prerequisites for speciation events by means of vicariance. Populations of once more widespread species become isolated and may evolve independently into distinct taxa. When conducting a historical biogeographical analysis of the areas of endemism of *Begonia* taxa (using cladistic techniques), I did however not find much evidence for the occurrence of such speciation events (Sosef 1994a). On the other hand, distinct floristic relationships were traced. All areas of endemism of the Lower Guinea phytochorium were nicely grouped together during the analysis. By further analyzing the results, it proved that the relationships between the areas were of a floristic nature, hence determined by the sharing of extant and hypothetical ancestral taxa. The relationships were only for a minor portion based on the sharing of a common history of the areas involved.

What I díd find was evidence for a major floristic demarcation line across Equatorial Guinea (Sosef 1994a: 133). It deserves mentioning here that this demarcation coincides with a climatic demarcation. It marks the transition zone between areas under a mainly northern climatic influence, and ones with a mainly southernly influenced climate. The southern border of Cameroon corresponds largely with the most extreme northern influence limit of the Saint Helen anticyclone (Maley & Elenga 1993).

The Zaire refuge areas

The eastern Zaire region has been denoted as a large forest refuge area by virtually all scientists dealing with the subject. In fact, only few *Begonia* taxa occur in eastern Zaire (Figure 2). There is a concentration of taxa in a comparatively small area close to the border with Rwanda and Burundi. This area is much smaller than the refuge area proposed by others, and one may wonder what was its actual size. Could it be that, similar to the situation in the Lower Guinea region, the former studies were performed on too large a scale and hence denoted a too large refuge area? This question remains unanswered yet, but could be related to the problems addressed in the paragraph Discussion. The situation should get attention in future studies.

Several authors suggested the presence of a major refuge area in the Zaire River Basin (Colyn et al. 1991, Kingdon 1980). When analyzing the Figures 2 and 3, it struck me that within this Basin the *Begonia* taxa apparently occur mainly in the localities indicated by these authors. More evidence of forest remaining in the Zaire River Basin during the last glacial is provided by Robbrecht (this volume). Of course the presence of the few *Begonia* taxa indicates that some rain forest must have remained in this region, and in my 1994a study I stated that it might have been distributed patchily. Reconsidering my own data, and having knowledge of those of Robbrecht l.c., I think it is not unlikely, as also suggested by the latter author, that larger areas of rain forest, possibly located in comparatively narrow bands along rivers, did sustain during the last glacial.

Discussion

During personal discussions with several scientists (amongst others Van Rompaey, Maley, De Wilde, Wieringa), one subject seemed to arise particularly often. It sheds doubt on the validity of deducing the localities of former rain forest refuges by means of analyzing taxa distributions. The question is: "What are we looking at; what do the distribution patterns indicate?" Are the patterns of distribution, of biodiversity, we observe the result of historical events and circumstances, in particular during the last glacial period, or are they a reflection of the distribution of present-day ecological habitats? It is striking that most of the proposed forest refuge areas are located in hilly regions. Are such regions not always more diverse, hence occupied by a higher number of species, than more or less level and hence more uniform ones?

To me the answer seems obvious. Present-day distribution patterns are a reflection of a combination of both historical events and present-day ecological habitats. Yes, many of the begonias I studied do prefer hilly areas with small rocky creeks. But what is very important is that it can be made plausible that these species have very limited dispersal abilities. They will therefore, in my opinion, not have been able to disperse after the last glacial from refuge areas located elsewhere towards these hilly areas. This leads to the conclusion that such other refuge areas did not exist, ... or that the actual forest refuges were much larger than e.g. shown in Figure 1, and thus included the hilly sectors! Hence, the areas where these begonias live today are likely to have carried rain forest during the last glacial, but the actual forest refuge areas may have extended beyond the limits of the hilly areas.

So, it seems that enunciations concerning the surrounding plains as possible forest refuges must be based on distributional data of species preferring rain forest in such localities. Additionally, such species must of course also show very limited dispersal abilities. When their distribution patterns prove to be quite random throughout the plains, we must assume that the former refuges were much larger than suggested in the present paper. When, on the other hand their distributions are centred around the proposed small refuges in the hilly areas, it is plausible that during the last glacial the rain forest did not expand far beyond these. Species preferring level grounds then most likely survived directly around and possibly also locally within the hilly refuge areas.

After these conclusions I can of course only state that deductions based on distributional data of begonias, and likely also of other organisms, should be confirmed by still other evidence. Besides the need to study now distribution patterns of species preferring rain forest on plain areas and with restricted dispersal capacities, we most urgently need confirmations from palynological research. The results of an analysis of a pollen core are much more hard evidence of any former vegetation cover than distributional data can ever provide. Furthermore, palaeoclimatological research provides a valuable additional dataset.

References

Aubréville, A., 1962. Savanisation tropical et glaciations quaternaires. Adansonia 2: 16-84.

Bonnefille, R., J.C. Roeland & J. Guiot, 1990. Temperature and rainfall estimates for the past 40,000 years in equatorial Africa. Nature 346: 347-349.

Colyn, M., A. Gautier-Hion & W. Verheyen, 1991. A re-appraisal of palaeoenvironmental history in central Africa: evidence for a major fluvial refuge in the Zaire Basin. J. Biogeogr. 18: 403-407.

Diamond, A.W. & A.C. Hamilton, 1980. The distribution of forest passerine birds and Quaternary climatic change in tropical Africa. J. Zool., Lond. 191: 379-402.

Dupont, L., 1993. Vegetation zones in NW Africa during the Brunhes Chron reconstructed from marine palynological data. Quaternary Sci. Rev. 12: 189-202.

Elenga, H. & A. Vincens, 1990. Paléoenvironnements Quaternaires récents des Plateaux Batéké (Congo): étude palynologique des dépôts de la dépression du Bois de Bilanko. In: R. Lanfranchi & D. Schwartz (eds). Paysages Quaternaires de l'Afrique centrale Atlantique: 271-282. Didactiques, Travaux et Documents ORSTOM, Paris.

Grubb, P., 1982. Refuges and dispersal in the speciation of African forest mammals. In: G.T. Prance (ed.). Biological diversification in the tropics: 537-553. Columbia University Press, New York.

Guillaumet, J.L., 1967. Recherches sur la végétation et la flore de la région du Bas-Cavally (Côte d'Ivoire). Mémoires ORSTOM no 20. ORSTOM, Paris.

Hamilton, A.C., 1976. The significance of patterns of distribution shown by forest plants and animals in tropical Africa for the reconstruction of upper Pleistocene palaeoenvironments: a review. In: E.M. van Zinderen Bakker (ed.). Palaeoecology of Africa 9: 63-97. A.A. Balkema, Cape Town.

Hamilton, A.C., 1982. Environmental history of east Africa. A study of the Quaternary. Academic Press, London.

Hamilton, A.C., & D. Taylor, 1991. History of climate and forests in tropical Africa during the last 8 million years. In: N. Myers (ed.). Tropical forests and climate: 65-78. Kluwer Academic Publishers, Dordrecht. (= Climatic Change 19).

Kingdon, J.S., 1980. The role of visual signals and face patterns in African forest monkeys (guenons) of the genus *Cercopithecus*. Trans. Zool. Soc., London 35: 425-475.

Livingstone, D.A., 1982. Quaternary geography of Africa and the refuge theory. In: G.T. Prance (ed.). Biological diversification in the tropics: 523-536. Columbia University Press, New York.

Maley, J., 1987. Fragmentation de la forêt dense humide Africaine et extensions des biotopes montagnards au Quaternaire récent: nouvelles donnees polliniques et chronologiques. Implications paléoclimatiques et biogéographiques. In: J.A. Coetzee (ed.). Palaeoecology of Africa 18: 307-334. A.A. Balkema, Rotterdam.

Maley, J., 1989. Late Quaternary climatic changes in the African rain forest: forest refugia and the major role of sea surface temperature variations. In: M. Leinen & M. Sarnthein (eds). Paleoclimatology and paleometeorology: Modern and past patterns of global atmospheric transport. NATO Adv. Sci. Inst. Series, C: Math. & Phys. Sci. 282: 585-616. Kluwer Academic Publishers, Dordrecht.

Maley, J. & P. Brenac, 1987. Analyses polliniques préliminaires du Quaternaire récent de l'Ouest Cameroun: mise en évidence de refuges forestièrs et discussion des problèmes paléoclimatiques. In: G. Cambon, P. Richard & J.-P. Suc (eds). Palynology et milieux tropicaux. Mém. Trav. Ecole Prat. Htes Inst. Montpellier 17: 129-142.

Maley, J., & H. Elenga, 1993. Le rôle des nuages dans l'évolution des paléoenvironnements montagnards de l'Afrique tropicale. Vielle Climatique Satellitaire 46: 51-63.

Rietkerk, M., P. Ketner & J.J.F.E. de Wilde, 1995. *Caesalpinioideae* and the study of forest refuges in Gabon: Preliminary results. Bull. Mus. natl. Hist. nat., Paris, 4e sér., 17, section B, Adansonia, nos 1-2: 95-105.

Sosef, M.S.M., 1992. New species of *Begonia* in Africa and their relevance to the study of glacial rain forest refuges. In: J.J.F.E. de Wilde (ed.). Studies in *Begoniaceae* III. Wageningen Agric. Univ. Papers 91-4: 117-151 (also Belmontia n.s. 24, no 148).

Sosef, M.S.M., 1994a. Refuge begonias. Taxonomy, phylogeny and historical biogeography of *Begonia* sect. *Loasibegonia* and sect. *Scutobegonia* in relation to glacial rain forest refuges in Africa. Wageningen Agric. Univ. Papers 94-1: 1-306 (also Belmontia n.s. 26, no 151).

Sosef, M.S.M., 1994b. Glacial rain forest refuges in relation to speciation in *Begonia* sect. *Loasibegonia* and sect. *Scutobegonia*. In: J.H. Seyani & A.C. Chikuni (eds). Proceedings of the XIIIth plenary meeting of AETFAT, Zomba, Malawi, 2-11 April 1991, vol. 2: 1297-1302. National Herbarium and Botanic Gardens of Malawi, Zomba.

Van Rompaey, R.S.A.R., 1993. Forest gradients in West Africa. A spatial gradient analysis. PhD thesis, Wageningen Agricultural University. 142 p.

White, F., 1983. The vegetation of Africa. Natural Resources Research No 20. UNESCO, Paris.

Saprophytic flowering plants of Mount Cameroon

M. CHEEK [1] & N. NDAM [2]
[1] Royal Botanic Gardens, Kew, Surrey TW9 3AE, United Kingdom.
[2] Limbe Botanic Garden, Limbe, B.P. 437, S.W.P., Cameroon.

Abstract
Mount Cameroon probably has Africa's greatest concentration of saprophytic flowering plants: twelve species in seven or eight genera and four families. It is the only known site in Africa for *Oxygyne* and the type locality of two of the three described species of *Afrothismia* (Burmanniaceae) are at Mount Cameroon, one being strictly endemic to the mountain. This paper reviews all that is known of these plants on Mount Cameroon: geography, phenology and taxonomy. It also assesses these species as indicators of refugia: some do not seem useful, others, such as *Afrothismia* and *Oxygyne*, do.

Key words
Mount Cameroon, Pleistocene Refugia, Saprophyte, Thismieae.

Introduction

Saprophytic flowering plants are more strictly speaking mycotrophic plants. They are only indirectly saprophytic since they depend on their symbiotic fungi to provide metabolites. Holosaprophytes are the subject of this paper. They lack chlorophyll entirely and are wholly dependent on the fungi they parasitize. Many plant species are hemisaprophytic, having both green tissue for photosynthesis and mycorrhiza. Hemi-saprophytic and holosaprophytic species can occur in the same genera. For example, *Sebaea* (Gentianaceae) and *Burmannia* (Burmanniaceae). Maas et al. (1986) is the most detailed modern account of tropical saprophytes, albeit largely of neotropical species. There is no comparable work for Africa.

Saprophytes are mostly confined to moist evergreen forest. They seem to prefer areas of leaf-litter and often grow near streams in lowland areas, although this is less often so in montane forest. They can congregate in one area of forest: if one species is discovered it is always well-worth searching the immediate vicinity exhaustively for more. Maas et al. (1986) reports finding as many as four (five if the fern *Schizaea fluminensis* Miers ex Thurm. is included) saprophytic species in various families within one 100 m² quadrat in Costa Rica. As many as three species have been located recently in 100 m² in the eastern foothills of Mt. Cameroon, and Schlechter collected six species (including two new genera) in consecutive numbers from Moliwe in 1905. It is not known why they should be gregarious. They may all share the same fungus or such sites may have some other factor important for their establishment and survival.

Saprophytes are often only rarely collected for several reasons:

A They are usually inconspicuous and so overlooked.

B They are not visible above the ground unless they flower, and this may not occur every year. Horich (1982) pointed out that *Epipogium*, a saprophytic orchid, can remain strictly subterranean for dozens of years.

L.J.G. van der Maesen et al. (eds.), The Biodiversity of African Plants, 612-617.

C They are often rare. Very many of the species of Thismieae (Burmanniaceae) are only known from a single area. Some are only known from a single collection or believed extinct, such as *Thismia americana* Pfeiffer (Lucas & Synge 1978).

Saprophytes as indicators of refugia

Saprophytic species are not evenly distributed throughout the world's rain forests. For example, of the ten neotropical species of *Thismia*, five are strictly endemic to the Alto Macahé of Rio de Janeiro and most are described as "endangered and probably extinct" because they have not been collected for one hundred years or more. Saprophytes, as rain forest herbs producing minute seeds, seem to have limited opportunities for dispersal, apart from the orchid and some gentian species which unusually seem able to disperse their seed aerially from this most wind becalmed of habitats (Maas et al. 1986). Sensitive to desiccation, restricted to moist evergreen forest, with many local endemics, saprophytes, particularly the Thismieae, are surely candidates as indicators of Pleistocene forest refugia.

Just as the Alto Macahé with its concentration of saprophytes, particularly narrowly endemic Thismieae, is part of a neotropical refuge (Prance 1985), so is the Mount Cameroon area. The Gulf of Guinea area, from the Cross River in the SE corner of Nigeria, through Cameroon and into Gabon is postulated as one of two primary refuges in continental Africa (Hamilton 1976, Maley 1989). Mount Cameroon has twelve species of saprophyte in four families (Table 1). These species are mostly distributed in the foothill areas rather than on the main massif, probably because of the prevalence of recent lava flows there, inimical to species with very narrow ranges and low powers of dispersal. An undescribed species of Thismieae is exceptional in occurring on the main massif, but its site in the saddle between Fako and the much older, volcanically inactive Etinde, is unusual in being protected from lava flows.

There are four described species of Thismieae in Africa, in the genera *Oxygyne* Schltr. and *Afrothismia* Schltr. (Engler 1908). Three of these species are known from Mt Cameroon, and two of these are strictly endemic to Mount Cameroon. In addition, there is the fourth, undescribed species mentioned above, also strictly endemic to the mountain so far as is known. *Oxygyne triandra* Schltr., long regarded as monotypic, has its closest relatives in Japan and with the new species of Etinde. Generic limits have not yet been resolved here.

Afrothismia Schltr. was described from two species, collected with *O. triandra* from forest near Moliwe, namely *A. pachyantha* Schltr. and *A. winkleri* (Engl.) Schltr. (Schlechter 1908). *A. pachyantha* is only known from that single collection, whilst *A. winkleri* is known from Neu Tegel, near Limbe, also from Bipindi, SE Cameroon and from the Akwe forest reserve in Nigeria and the Budongo Forest Reserve in western Uganda. A third species, *A. insignis* Cowley has recently been described from a single collection in central Tanzania made in 1962. All occurrences apart from that from Nigeria tally well with proposed African refugia.

Endangered species

Oxygyne triandra and *Afrothismia pachyantha* are each known only from a single collection made in 1905, and have not been found since, despite searching in recent years

Table 1. Specimens of saprophytic flowering plants from Mount Cameroon.

Family, name	Locality	Month of flowering	Collector	Total range
Triuridaceae				
Sciaphila ledermannii Engl.	Onge	February	D. Thomas s.n.	SE Nig. & Cameroon
Orchidaceae				
Epipogium roseum (Don) Lindl.	Ambas bay	February	Mann 784	Eq. Africa to SE Asia
"	Moliwe	February	Dalziel 8205	"
"	Mimbia 1200m		Richards 9346	"
Auxopus macranthus Summerh.	Bakingili 250m	December	Cheek et al. 5860	Cam., Zaire, Uganda
A. kamerunensis Schltr.	S. Bakundu	January	Keay FHI 28570	Liberia to Cameroon
Gentianaceae				
Sebaea oligantha (Gilg) Schinz	Moliwe	September	Schlechter in obs.	Nigeria & Cameroon
"	Mabeta-Moliwe		Cheek & Ndam 3220	"
"	"	July & Nov.	Cheek s.n.	"
Burmanniaceae, Euburmannieae				
Burmannia hexaptera Schltr.	Mabeta-Moliwe	November	Cheek et al. 3979	"
"	Man O' War Bay	October	Schlechter 15785-86	"
B. congesta (Wright) Jonker	Mabeta-Moliwe	November	Cheek et al. 3980	Liberia to Angola
"	Moliwe	September	Schlechter 15787	"
B. densiflora Schltr.	Moliwe		Stammler	Cameroon
Burmanniaceae, Thismieae				
Afrothismia winkleri (Engl.) Schltr.	Buea, Neutegel	July	Winkler 225	SE Nig., Cam., Uganda
"	Moliwe	September	Schlechter 15788	"
A. pachyantha Schltr.	Moliwe	September	Schlechter 15789	Only this collection
Oxygyne triandra Schltr.	Moliwe c.100m	September	Schlechter 15790	"
O. (?) sp. nov.	Etinde, 1300m	October	Cheek et al. 3816	"

near the type locality of Moliwe in the eastern foothills. They must be considered endangered, possibly extinct. Expansion of farming activity following logging in the neighbourhood of Moliwe undoubtedly results in habitat destruction because saprophytes abhor turning of soil or disturbance of leaf litter and are prone to desiccation (Horich 1982). *Ex situ* cultivation of saprophytic orchids has proved difficult. Horich (1982), attempting to grow five European species, found that despite taking every care, all died within two years. However, greater success has been had with the Australian underground saprophytic orchid, *Rhizanthella gardneri* (Cribb, pers. comm.).

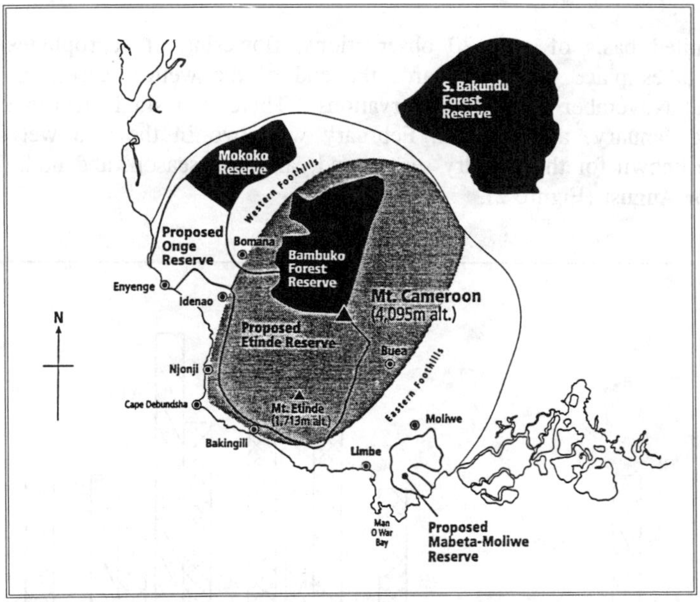

Figure 1 Collecting localities and protected forests on Mount Cameroon.

Mount Cameroon as a centre of diversity for African saprophytes

No detailed study, nor checklist exists for saprophytes in continental Africa. This makes a comparison of Mt. Cameroon's saprophyte complement with that of the whole continent difficult. Horich (1982) gives a world generic list for orchids. Of the three genera and five species he lists for Africa, two genera (*Auxopus* and *Epipogium*) and three species occur on Mt. Cameroon. The remaining two species are found in the genus *Schwartzkopffia* which is unusual, as a saprophyte, in preferring dry forest. Not listed by Horich, but also saprophytic in Africa though absent from Mt. Cameroon are three species of *Eulophia*, mostly known from Tanzania to Malawi, and *Didymoplexis africana* J.L. Stewart, only known from Natal. Amongst gentians one of the two known from Africa occur on Mt. Cameroon; amongst Triuridaceae one of the two African species of *Sciaphila* is found in the western foothills of the mountain, though *Seychellaria*, a rarity of East African coastal forest, the Seychelles and Madagascar, is absent. In the Burmanniaceae, all three (some authorities only recognize two) of the African saprophytic species of *Burmannia* occur and the Thismieae, already discussed in detail, are very well represented. Thanks to Jonker's monograph (1938), this family is well-known. Pyrolaceae, Monotropaceae and Corsiaceae are not present in Africa.

Three other species may be expected on Mount Cameroon. *Gymnosiphon longistylus* (Benth.) Hutch. (Burmanniaceae) is widespread, though rare, in West Africa, *Eulophia galeoloides* Kraenzl. occurs in Ghana, Nigeria, Zaire and East Africa and *Voyria primuloides* Baker (Gentianaceae) is known from adjoining Gabon and Nigeria.

All known specimens of saprophytic species from Mt. Cameroon are presented in Table 1. Figure 1 depicts collecting localities and protected forests on the mountain.

Phenology

On the limited basis of only 20 observations, flowering of saprophytes on Mount Cameroon takes place largely towards the end of the wetter season in September, October and November (total 11 observations). There is a single record each in December and January, and three in February with two in the mid wet season. No records are known for the late dry season and early wet season, and nothing has been recorded for August (Figure 2).

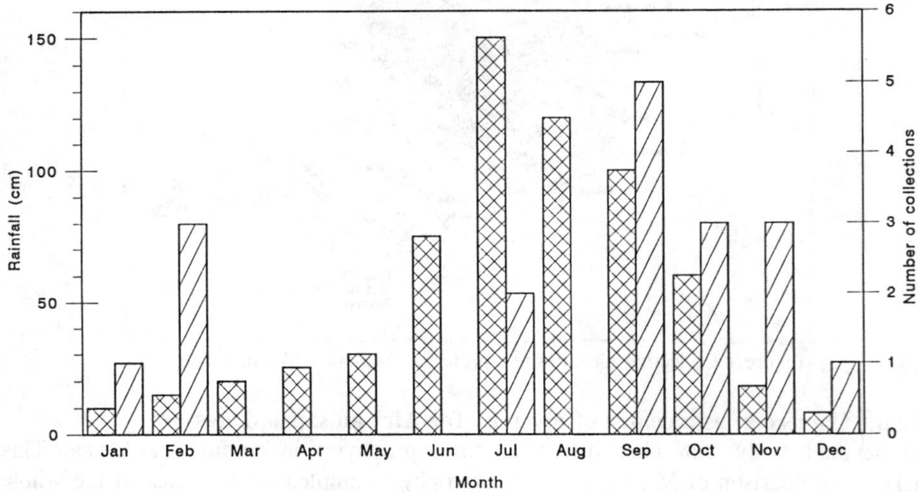

Figure 2. A comparison of monthly rainfall and flowering of saprophytes on Mount Cameroon. Flowering of saprophytes takes place largely towards the end of the wetter season.

Populations of even the most common species, *Sebaea oligantha* do not flower every year. At one site, abundant flowers were seen of this species in October 1991, none at all in the following May and July and only a single, inconspicuous inflorescence in November of 1992 with no evidence of fruits or withered flowers from the preceding month. Intervals of many years between flowering are well-known in saprophytes. Little seems to be known of what triggers flowering, though it is recorded that *Epipogium aphyllum* (W.F. Sch.) Sw. "appears in times of unusual frequency of summer lightning storms" (Horich 1982).

Further work

The reproductive biology of saprophytes, excepting some orchids such as *Rhizanthella*, is very poorly known. Nothing is known, though there has been much speculation, of the nature of floral visitors in the Burmanniaceae (Maas 1986). Seed dispersal, the occurrence of vegetative apomixis and population studies are other aspects in need of further work. A detailed examination of the correlation between specimen distribution of saprophytes and putative refugia in Africa would be of interest. All botanists are urged to scrutinize the shadier parts of rain forest more carefully in the future.

Acknowledgements

The authors are grateful for support from their respective institutes and for comments from E.J. Cowley (Burmanniaceae), Sarah Thomas, Joyce Stewart and especially Phil Cribb (Orchidaceae) and Mark Coode. Some of the fieldwork contributing to this paper was performed in the course of British ODA consultancy work. Other fieldwork was sponsored by Earthwatch Europe. We are grateful to both organizations for supporting botanical inventory work around Mount Cameroon.

References

Engler, A., 1908. Burmanniaceae. Die Pflanzenwelt Afrikas II: 400-403. Leipzig.

Hamilton, A., 1976. The significance of patterns of distribution. Palaeoecology of Africa 9: 63.

Horich, C.K., 1982. Orquídeas saprófitas. Orchideología 15 (2-3): 190-201

Jonker, F.P., 1938. A Monograph of the Burmanniaceae. Meded. Bot. Mus. Herb. Rijks Univ. Utrecht 51: 1-279

Lucas, G. & H. Synge, 1978. The IUCN Plant Red Data Book. 99-100. Morges, Switzerland.

Maas, P. & collaborators, 1986. Saprophytes Pro Parte. Flora Neotropica Monographs, 40, 41, 42. New York Botanical Garden, New York.

Maley, J., 1989. Late quaternary climatic changes in the African rain forest: forest refugia and the major role of sea surface temperature variations. In: Lewen, M. & M. Sarnthein (eds.), Palaeclimatology and Palaeometeorology: Modern and Past Patterns of Global Atmospheric Transport. Kluwer, London.

Prance, G.T., 1985. The changing forests. In: Prance, G.T. & T.L. Lovejoy (eds.), Amazonia. Pergamon Press, Oxford.

Caesalpinioideae and the study of forest refuges in central Africa

M. RIETKERK [1], P. KETNER [1] & J.J.F.E. DE WILDE [2]
[1] Dept. of Terrestrial Ecology and Nature Conservation. Wageningen Agricultural University, P.O. Box 8080, 6700 DD Wageningen, The Netherlands.
[2] Dept. of Plant Taxonomy, Wageningen Agricultural University, P.O. Box 8010, 6700 ED Wageningen, The Netherlands.

Abstract
The present study shows that distribution patterns of Caesalpinioideae species can be used in locating the position of pleistocene forest refuges in central Africa. Preliminary results indicate that previously proposed locations of forest refuges in central Africa are not correctly situated. Further, the presence of a recently proposed lowland glacial forest refuge area, located in the Mayombe region of Congo and Zaire, is confirmed by our data.

Résumé
La présente étude montre que les modes de distribution des espèces de Caesalpinioideae peuvent servir à localiser la position des refuges forestiers en Afrique centrale. Des résultats préliminaires indiquent que les localisations des refuges forestiers initialement proposés en Afrique centrale ne sont pas correctement situés. Aussi, un nouveau refuge forestier de basse altitude situé dans la région de Mayombe est proposé.

Key words
Biogeography, Caesalpinioideae, Climatic change, Distribution, Pleistocene forest refuges.

Introduction

During glacial times the vast expanse of tropical lowland rain forest had been repeatedly reduced to a few, relatively small and isolated patches (tropical forest refuges) due to unfavourable climatic conditions. Evidence for the geographic position of these tropical rain forest refuges can be deduced from data on biotic diversity combined with biogeographic, geological and (palaeo)palynological information. Maley (1987) published a scheme (Figure 1) representing postulated tropical lowland rain forest refuges in central Africa.

Caesalpinioideae play(ed) a dominant role in the ecology of lowland rain forest in central Africa. The present distribution of Caesalpinioideae provides additional data for delineating the location of glacial lowland forest refuges in central Africa. This is based on the assumption that some species "captured" in refuges during the last glacial were not or hardly capable of migrating out of these again when ecological conditions improved, due to their extreme low dispersability. In case this is true, modern distributions should show centres of frequency of occurrence as well as concentrations of species.

Material and methods

Step 1: A selection of 28 Caesalpinioideae taxa from the 158 species cited in the *Flore du Gabon* (Aubreville 1968) was made (Rietkerk et al. 1995). This selection included

L.J.G. van der Maesen et al. (eds.), The Biodiversity of African Plants, 618-623.

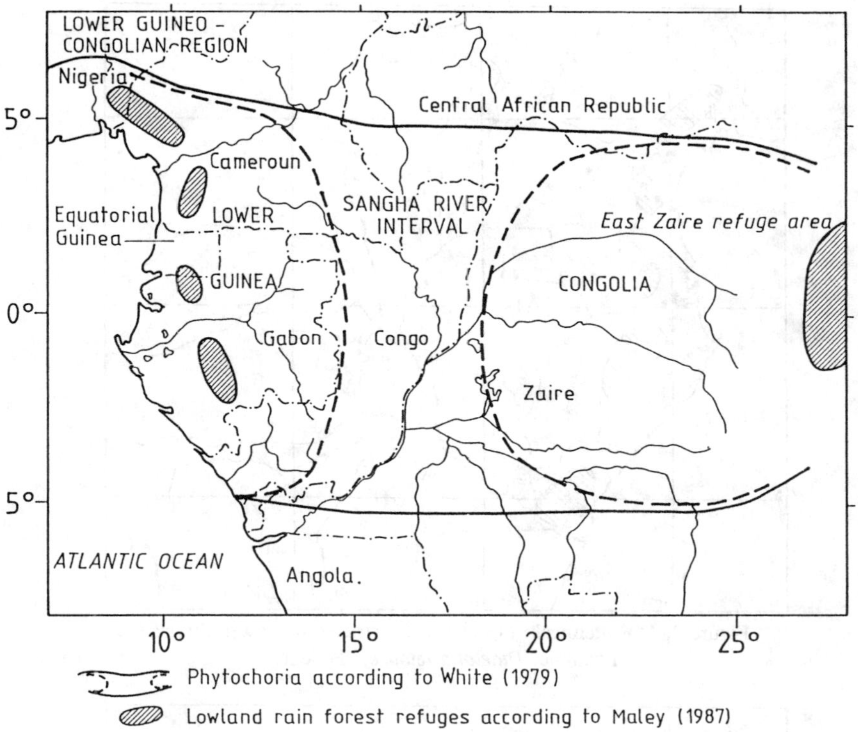

Figure 1. Location of lowland rain forest refuges in central Africa (Maley 1987) and phytochoria of the Lower Guineo-Congolian Region (White 1979)

species with different fruit and seed characteristics and is based on the underlying hypothesis that these differences could consequently lead to different dispersal capacities and ultimately to different distribution patterns. For each species a spot distribution map was made for central Africa, compiled from specimens present in Herbarium Vadense (WAG), Jardin Botanique National de Belgique (BR) and Muséum National d'Histoire Naturelle, Paris (P). The identification of the specimens is based on determinations, most often by specialists, annotated on the collections.

Step 2: The resulting distributions of the species were analyzed in a search for groups of species (biogeographical elements) showing discrete patterns.

Step 3: In order to characterize the thus found species groups and their patterns, these were compared with categorized groups of species and recognized biogeographical units within the forest flora of central Africa (White 1979) (see Figure 1). White's (1979) analysis of the distributions of 337 species included trees, shrubs and lianas belonging to 12 families.

Step 4: The last step was to compare the patterns found in the distributions of the studied Caesalpinioideae species with the position of the postulated refuges in the region.

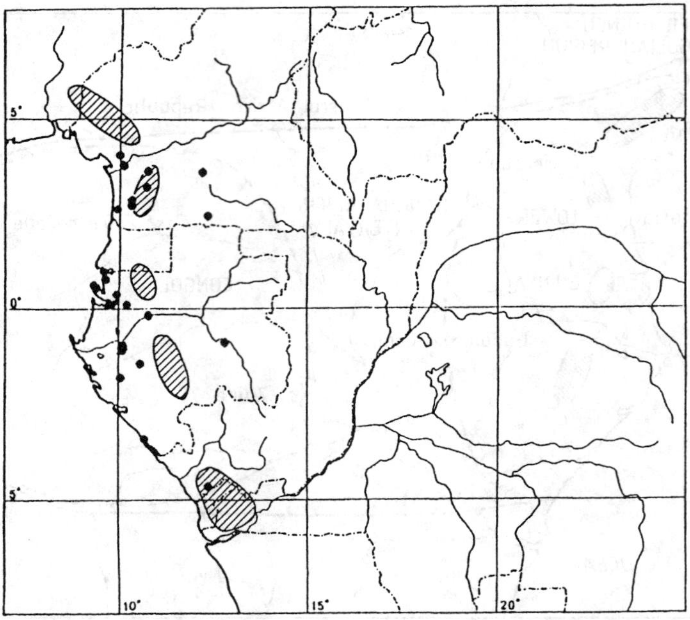

Figure 2a. "**Endemics**": (13 species) restricted to Lower Guinea. Example: *Didelotia letouzeyi* Pellegr.

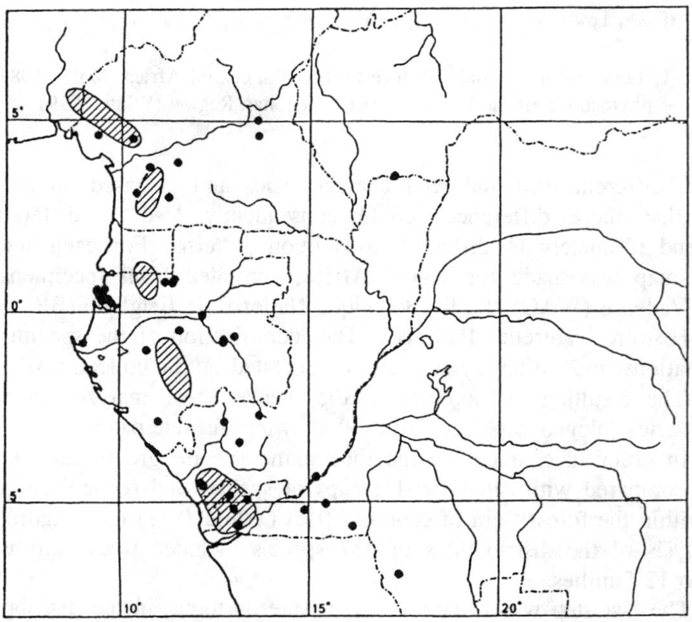

Figure 2b. "**Marginal intruders**": (5 species) characteristic of Lower Guinea, also in adjacent phytochoria. Example: *Hylodendron gabunense* Taub.

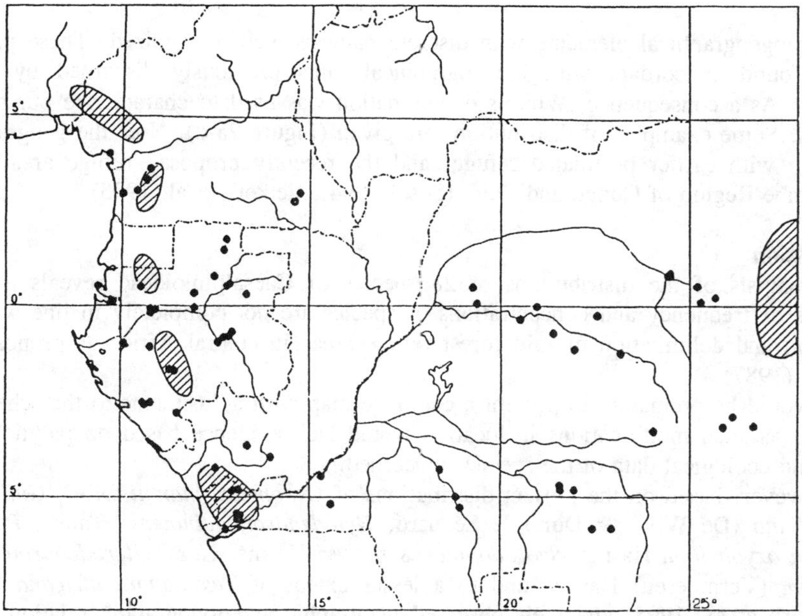

Figure 2c. "**Linking species**": (6 species) characteristic of the whole Lower Guineo-Congolian Region, limited in the Sangha River Interval. Example: *Scorodophloeus zenkeri* Harms.

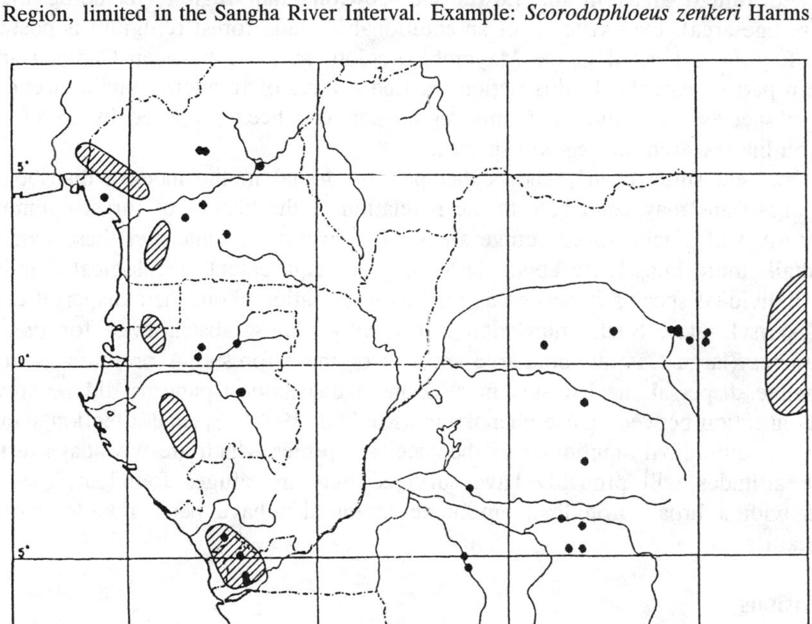

Figure 2d. "**Disjunct linking species**": (4 species) similar to "linking species", but not occurring in the Sangha River Interval. Example: *Gossweilerodendron balsamiferum* (Vermoesen) Harms.

Results

Four biogeographical elements with discrete patterns were recognized. These patterns were found concordant with biogeographical units previously discerned by White (1979). As a consequence, White's denomination was used to characterize our species groups. Some examples of distributions are given (Figure 2a-d). Note the geographical relation with earlier postulated refuges and the recently proposed refuge area in the Mayombe Region of Congo and Zaire (Sosef 1994, Rietkerk et al. 1995).

Discussion

The analysis of the distributions of 28 species of Caesalpinioideae reveals that the centres of frequency and concentrations of species are not completely in line with the location and delimitation of rain forest refuge areas in central Africa as proposed by Maley (1987).

It would be premature to present a complete map with adjustments to the scheme of Maley, as such modifications in location would lack evidence based on palynological and ecological data of the species concerned.

However, based on the present distributions of *Cynometra mannii* Oliv., *Guibourtia arnoldiana* (De Wild. & Dur.) J. Léonard, *Hylodendron gabunense* Taub., *Pterygopodium oxyphyllum* Harms, *Scorodophloeus zenkeri* Harms, *Gossweilerodendron balsamiferum* (Vermoesen) Harms, and to a lesser extent of *Tessmannia africana* Harms (Rietkerk et al. 1995) and their observed geographic and topographic relation to the postulated refuge areas in the Lower Guineo-Congolian region (including the East Zaire refuge area), the existence of an additional lowland forest refugium is postulated. This refugium is located in the Mayombe Region, in Zaire between Cabinda and the northern part of Angola. In this region we find centres of frequency and concentrations of these species. A refugium in this region has also been proposed by Sosef (1994) based on his research on *Begonia* species.

The present study clearly shows that patterns found in the modern distribution of forest organisms may contribute to the revelation of the history of forest communities (and as such of glacial forest refuge areas). But in order to interpret these patterns in any detail, more knowledge about the ecological requirements (ecological amplitudes) of the individual species is needed as well as information about their dispersal capacity and dispersal rates. Such knowledge is presently almost absent even for the major components (large Caesalpinioid tree species) of these forests. A preliminary analysis of possible dispersal mechanisms in relation to distribution patterns did not reveal a clear connection between these phenomena (Rietkerk 1993). Special attention should be paid to the altitudinal amplitudes of the species. Species which are nowadays restricted to low altitudes will probably have survived only in refuges found at low levels. Others with a broad altitudinal amplitude might also have been able to survive at higher altitudes.

Conclusions

* Centres of frequency and concentrations of Caesalpinioid species do not coincide with the postulated refuges.
* Earlier proposed locations of postulated refuges of lowland forest should be adjusted.

* Proposed new locations are situated to the west (and to the east) in the peripheral lower parts of the hitherto postulated refuges.
* The presence of an additional lowland forest refuge area located in the Mayombe Region is in line with findings based on distributions of *Begonia* species.

Acknowledgements
We are grateful to Prof.dr. P. Morat, Director of the Laboratoire de Phanérogamie in Paris (P) and to the staff of Jardin Botanique National de Belgique (BR) for their cooperation. Especially Dr. P. Bamps helped us with tracing collecting localities of specimens. The comments of Dr. J. Maley (Laboratoire de Palynologie, Montpellier), Dr. J.J. Bos and Dr. M.S.M. Sosef (Herbarium Vadense, Wageningen) on earlier versions of the manuscript were much appreciated. For stimulating discussions and Caesalpinioideae-expertise thanks are due to Dr. F.J. Breteler and Ir. J.J. Wieringa (Herbarium Vadense, Wageningen). The first author received a grant from the "Wageningen Universiteits Fonds" to visit the herbarium in Paris, which is gratefully acknowledged.

References

Aubréville, A., 1968. Légumineuses-Caesalpinioidées. Flore du Gabon 15. Muséum National d'Histoire Naturelle, Paris.

Maley, J., 1987. Fragmentation de la forêt dense humide africaine et extension des biotopes montagnards au Quaternaire recent: Nouvelle données polliniques et chronologiques. Implication paleoclimatiques et biogeographiques. Palaeoecology of Africa 18: 307-334.

Rietkerk, M., 1993. Caesalpinioideae diversiteit en verspreiding als indicator voor klimaatsverandering sedert het laat-kwartair in het sedimentaire bekken van Gabon. Unpubl. MSc thesis, Wageningen Agricultural University, Dept. of Nature Management & Dept. of Plant Taxonomy, Wageningen.

Rietkerk, M., P. Ketner & J.J.F.E. De Wilde, 1995. Caesalpinioideae and the study of forest refuges in Gabon: preliminary results. Bull. Mus. natl. Hist. nat., Paris, 4e sér., 17, section B, Adansonia, nos 1-2: 95-105.

Sosef, M.S.M., 1994. Refuge begonias. Taxonomy, phylogeny and historical biogeography of *Begonia* sect. *Loasibegonia* and sect. *Scutobegonia* in relation to glacial rain forest refuges in Africa. PhD thesis. Wageningen Agric. Univ. Papers 94-1: 1-306.

White, F., 1979. The Guineo-Congolian Region and its relationships to other phytochoria. Bull. Jard. Bot. Nat. Belg. 49: 11-55.

Rain forest refugia in Liberia

R.S.A.R. VAN ROMPAEY
Dept. of Plant Taxonomy, Wageningen Agricultural University,
P.O. Box 8010, 6700 ED Wageningen, The Netherlands.

Abstract
Inside Liberia, the lowland area is the most complete refugium for large Caesalpinioid tree species. The Nimba region constituted a partial refugium. A hypothesized partial refugium was also confirmed in southwest Ghana. Semi-deciduous forest was present outside these evergreen rain forest refugia. A map using only two classes (forest and non-forest) cannot represent this gradient. On paleo forest cover maps at least four classes should be distinguished: evergreen, semi-deciduous, montane and non-forest.

Key words
Liberia, Rain forest refugia.

Introduction: Liberia's forests

Liberia's rain forests received little attention from biogeographers until recently (Parren & de Graaf 1995, Van Rompaey 1993). Botanical explorers however discovered numerous endemic plant species in the country or near its border (Aubréville 1959, Guillaumet 1967, Rouw 1991, Voorhoeve 1965). The precise distribution and population density of these species have been studied only recently (Bongers et al. in prep.). In the present paper the actual state of knowledge on the distribution of a number of large Caesalpinioid tree species, endemic to Liberia or the Upper Guinea forest block, is summarized. Their distribution along a gradient from the coast inland to the Nimba mountains leads to a discussion on the location of Pleistocene rain forest refugia in the region.

Maley (in press) supposed that most of Liberia was continuously covered with rain forest throughout the Pleistocene. Apparently most rain forest species managed to survive during the drier and colder episodes of the last glacial-interglacial cycles, especially during the last glacial maximum at 18 ka BP. The continuity of favourable conditions for forest growth is related to the rainy, equatorial climate of Liberia. Its coastal zone is one of the rainiest regions of Africa, receiving up to 4 m rainfall per year and lacking a dry season in most years. These conditions are comparable to those in West Cameroon, Gabon and other supposed rain forest refuge areas (Sosef 1994).

On his glacial maximum forest cover map, Maley (in press) distinguishes three refugia for the Upper Guinea rain forest block, being the forest zone west of the Dahomey Interval, from Sierra Leone to Ghana. The largest one stretches along the coast from south Sierra Leone through Liberia to southwest Côte d'Ivoire and reaches some 100 to 150 km inland. Besides this major refugium, two smaller ones are identified: one near the foothills of the Nimba mountains in Liberia and one in the southwest of Ghana (actual Ankasa forest reserve). In this paper, the relative importance of these three Upper Guinean refugia is examined, by analyzing the distribution of large Caesalpinioid trees actually restricted to the wettest forest in Upper Guinea.

L.J.G. van der Maesen et al. (eds.), The Biodiversity of African Plants, 624-628.

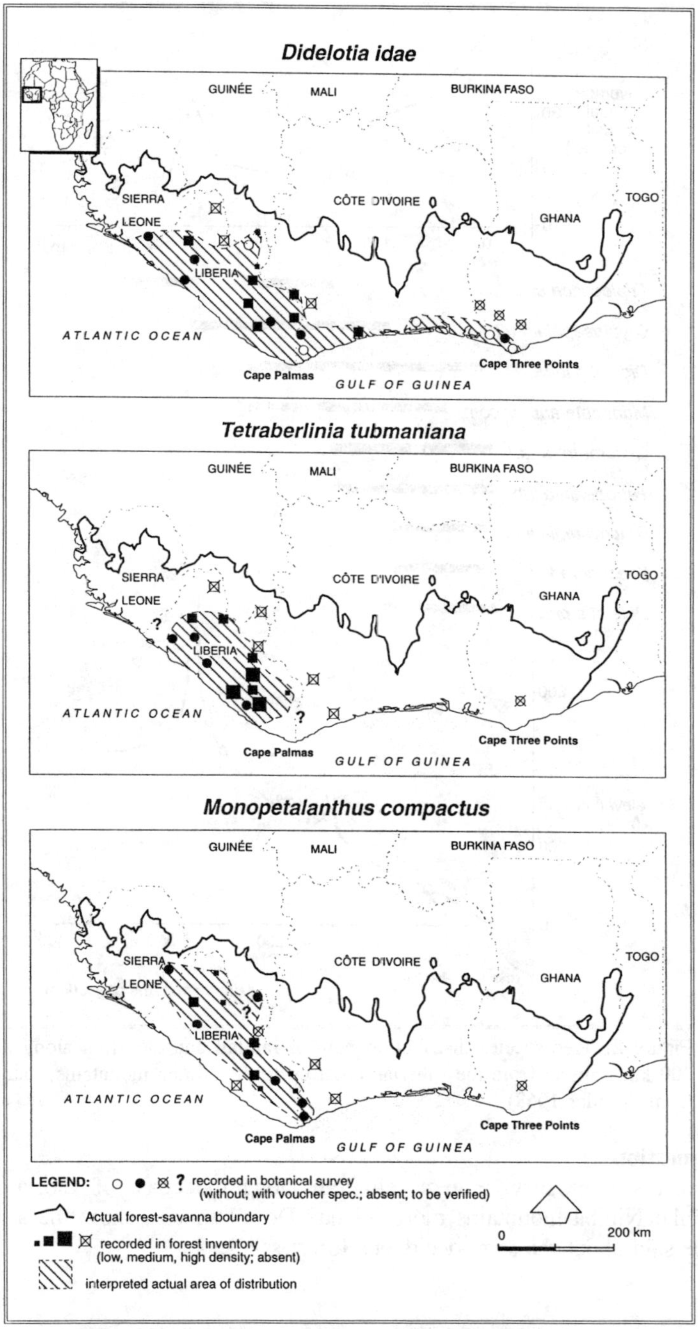

Figure 1. Observed and interpolated distribution of large Caesalpinioid tree species, endemic to the Upper Guinea forest belt.

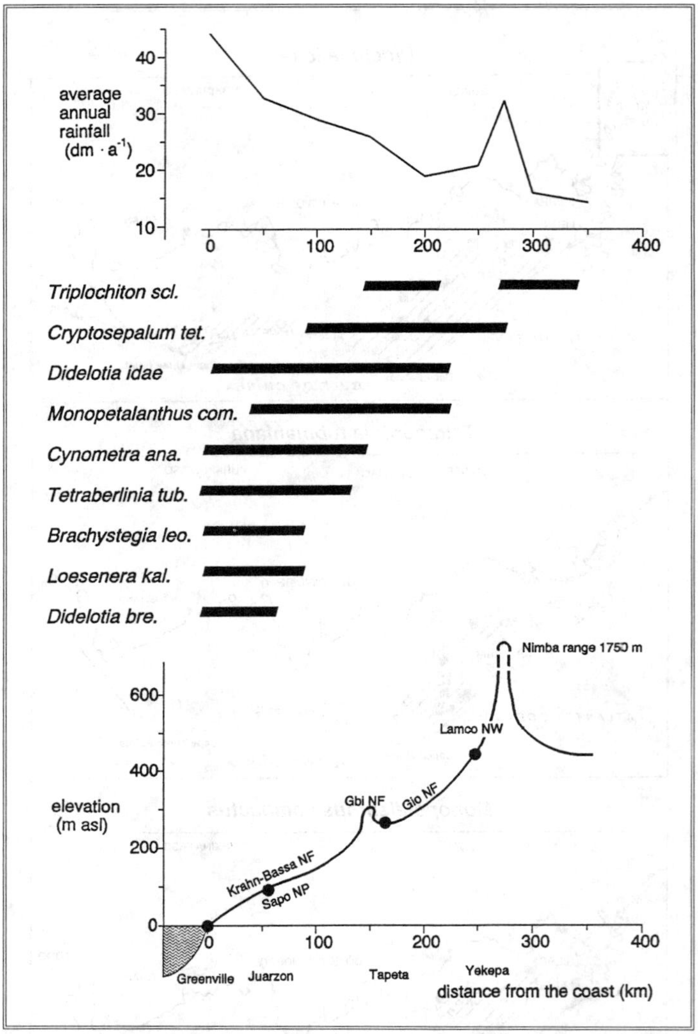

Figure 2. Tree species distribution, annual rainfall and elevation along a 300 km transect from the Liberian coast up to the Nimba mountains (data from Sachtler 1968).

Research questions

Did the forest species survive in a refugium along the coast of Liberia, or in the foothills of the Nimba mountains more inland? Do all species show the same pattern? What can be said about the semi-deciduous forest species?

Methods

I combined information from herbarium specimens (WAG, BR, P), botanical surveys (Guillaumet 1967, Hawthorne 1993, Voorhoeve 1965), forest inventories (Hawthorne

1993, Sachtler 1968, SODEFOR 1979) and permanent sample plots (Bertault 1986, Poker 1992, Van Rompaey 1993).

Herbarium specimens give verifiable information on presence. Botanical survey records give both presence as well as absence information, depending on sampling intensity and design (preferably systematically). Forest inventories provide, besides presence or absence, also reliable quantitative data on population density of the large trees over large areas (sometimes only the commercial ones). Permanent sample plots have the advantage that one has been forced to identify all trees (not only the flowering ones or the easily recognisable ones), and that they are marked and can be checked later. However, they only give information on a small area.

Results

All this information is drawn on distribution maps using different symbols. Also absence information is mapped. The limit of the area of distribution is interpolated between presence and absence points. Uncertainties are indicated by question marks (Figure 1). Along a transect from the Liberian coast up to the Nimba mountains the distribution of these species is summarized against the background of annual rainfall and elevation (Figure 2).

Conclusions

1 Coastal Liberia was the only refugium for the "wettest" Caesalpinioid tree species *Didelotia brevipaniculata, Tetraberlinia tubmaniana* and *Loesenera kalantha*. These species are not found near Mt Nimba, nor in Ghana. Thus, **lowland Liberia is the most complete refugium for large Caesalpinioid tree species**.

2 The "wet" species *Brachystegia leonensis* and *Monopetalanthus compactus* are actually found up to Nimba region, but it is not yet clear whether the Nimba populations are isolated from the populations from lowland Liberia. This confirms Maley's supposition that the Nimba foothills were forested during the ice age but only "intermediate" species hibernated there. **Nimba thus constituted a partial refugium**.

3 The high abundance of *Cryptosepalum tetraphyllum* and *Chidlowia sanguinea* (not in Figure 1) in the Nimba region may be an indication that they survived only in the Nimba region and not in lowland Liberia. However, the number of such "lower montane" species is very limited, as is the number of Nimba endemics (Adam 1983).

4 The "intermediate" species *Didelotia idae* and *Cynometra ananta* (not in Figure 1) show isolated populations **in southwest Ghana**, confirming the hypothesized **partial refugium** there.

5 Many "semi-deciduous forest" species (*Triplochiton scleroxylon, Mansonia altissima, Celtis* spp.) are actually only found outside these refugia. They probably hibernated outside (Lézine & Le Thomas 1993). So, **semi-deciduous forest was present outside these evergreen rain forest refugia.** A map using only two classes (forest and non-forest) cannot represent this gradient. **On paleo forest cover maps at least four classes should be distinguished: evergreen, semi-deciduous, montane and non-forest.**

References

Adam, J.G., 1983. Flore descriptive des Monts Nimba (Cote d'Ivoire, Guinée, Liberia). Editions du Centre National de la Recherche Scientifique.

Aubréville, A., 1959. La flore forestière de la Côte d'Ivoire. Publ. CTFT no 15 (3 tomes), Nogent-sur-Marne.

Bertault, J.-G., 1986. Etude de l'effet d'interventions sylvicoles sur la régénération naturelle au sein d'un périmètre expérimental d'aménagement en forêt dense humide de Côte d'Ivoire. Thèse de doctorat, Université de Nancy, Faculté des Sciences.

Bongers, F., M.P.E. Parren, L. Poorter & R.S.A.R. van Rompaey, in prep. Distribution of emergent forest tree species in West Africa: response curves to a climatic gradient. Compte Rendu Schnell Symposium 4-7 juillet 1993.

Guillaumet, J.L., 1967. Recherches sur la végétation et la flore de la région du Bas-Cavally (Côte d'Ivoire). Mémoires ORSTOM no. 20, ORSTOM, Paris.

Hawthorne, W., 1993. FROGGIE: Forest reserves of Ghana: graphical information exhibitor. Part 1 (Manual of the program), ODA/FIMP/Ghana Forestry Dept., Kumasi.

Lézine, A.M. & A. Le Thomas, 1993. Colloque Schnell international de phytogéographie tropicale, Paris: résumé.

Maley, J., in press. L'impact des phases arides sur la forêt dense humide africaine au cours de son histoire géologique. In: W. Weber, A. Vedder & H.S. Morland (eds.). African rain forest ecology and conservation. New York Wildlife Conserv. Soc. Publ., Yale Univ. Press.

Päivinen, R. & R. Witt, 1989. The methodology development project for tropical forest cover assessment in West Africa. Unpublished report UNEP/GRID, Geneva, Switserland.

Parren, M.P.E. & N.R. de Graaf, 1995. Forestry in West Africa: the quest for natural forest management: lessons to be learned and examples to be followed. Tropenbos series 13, The Tropenbos Foundation, Wageningen.

Poker, J., 1992. Struktur und Dynamik des Bestandesmosaiks tropischer Regenwälder - Entwicklung eines Modellansatzes zur Simulation natürlicher Mischbestände. Dissertation, Universität Hamburg.

Rouw, A. de, 1991. Rice, weeds and shifting cultivation in a tropical rain forest. A study of vegetation dynamics. Ph.D. thesis, Wageningen Agricultural University. 263 p.

Sachtler, M., 1968. General report on National Forest Inventory in Liberia. Technical report no. 1, German Forestry Mission to Liberia, Monrovia.

SODEFOR, 1979. Inventaire forestier national. Résultats de la région nord-ouest. SODEFOR, Abidjan.

Sosef, M.S.M., 1994. Refuge begonias. Taxonomy, phylogeny and historical biogeography of *Begonia* sect. *Loasibegonia* and sect. *Scutobegonia* in relation to glacial rain forest refuges in Africa. Ph.D. thesis. Wageningen Agric. Univ. Papers 94-1: 1-306.

Van Rompaey, R.S.A.R., 1993. Forest gradients in West Africa. A spatial gradient analysis. Ph.D. thesis. Wageningen Agricultural University. 142 p.

Voorhoeve, A.G., 1965. Liberian high forest trees. Ph.D. thesis, Wageningen Agricultural University. 416 p.

Floral biology and pollination

25 August 1994

Convenors: S. Liede, D. Lachaise, H. Baijnath

Floral structure and pollination in *Pelargonium*

M. STRUCK & J.J.A. VAN DER WALT
Department of Botany, University of Stellenbosch,
Private Bag X5018, Stellenbosch 7600, South Africa.

Abstract
The heterogeneity of *Pelargonium* in terms of structural blossom types and pollination syndromes indicates an independent and repeated evolution of convergent flower morphs in the genus and even in the sections. This suggests that pollinators played an important role as a selective force during the evolution of *Pelargonium*. Evolutionary lines in floral characters are surveyed and examples of striking convergencies in flower morphs of species placed in different sections are highlighted and discussed in the light of the current subdivision of the genus.

Résumé
La hétérogénéité de *Pelargonium* dans les sens des types structurels des fleurs et des syndromes de pollinisation indique l'évolution indépendante et répétée des morphes floraux convergents dans les genre et même dans les sections. Cela suggère que les pollinisateurs ont joué un rôle important comme force sélective pendant l'évolution de *Pelargonium*. Des lignes évolutionnaires des caractères floraux sont passées en revue et des exemples de convergences frappantes dans les morphes floraux d'espèces placées dans différentes sections sont soulignés et discutés tenant compte de la subdivision courante du genre.

Key words
Floral structure, *Pelargonium*, Pollination.

Introduction

Pelargonium L'Hérit. is the largest genus of the Geraniaceae. It comprises c. 250 species of which about 80% are endemic to the south-western part of South Africa (Van der Walt & Vorster 1983). The synapomorphic characters of the genus are (1) flower zygomorphic, (2) 7 fertile stamens (out of 10 androecial organs), and (3) a nectar tube in a modified receptacle. Pelargoniums, also known under the misleading vernacular names of "geraniums" or "malvas", immediately attracted the attention of the early travellers to the Cape. Numerous cultivars were derived which enjoy a tremendous popularity as garden and pot plants all over the world.

Perhaps the most striking feature of the genus *Pelargonium* is its wealth of flower morphs. This suggests that pollinator relationships played an important role as a selective force during its evolution. Hence, insight into the floral ecology of the genus should provide valuable information for a better understanding of its phylogenetic history.

In this paper we present the first results of a current project which aims to evaluate the ecological and taxonomic significance of floral characters in the genus *Pelargonium*. Floral characters are surveyed. Developmental lines in floral characters and examples of convergent flower morphs are highlighted and discussed in the light of the current subdivision of the genus.

L.J.G. van der Maesen et al. (eds.), The Biodiversity of African Plants, 631-638.

Current subdivision of *Pelargonium*

The latest taxonomic revision of *Pelargonium* was published by Knuth in 1912. He acknowledged 15 sections which were delimitated by morphological characters only. Later contributions yielded the recognition of new species, as well as the rectification of many synonyms, but also challenged Knuth's infrageneric system. According to current knowledge karyological features provide the key characters to a natural classification of the genus (e.g. Albers et al. 1992, Gibby et al. 1990, Van der Walt et al. 1990a). The current subdivision of *Pelargonium* comprises 16 sections:

1. *Pelargonium*
2. *Glaucophyllum*
3. *Campylia*
4. *Eumorpha*
5. *Ciconium*
6. *Dibrachya*
7. *Peristera*
8. *Isopetalum*
9. *Otidia*
10. *Polyactium*
11. *Cortusina*
12. *Reniformia*
13. *Ligularia* s.l.
14. *Jenkinsonia*
15. *Myrrhidium*
16. *Hoarea*

Developmental lines and convergencies in floral characters

The typical *Pelargonium* flower is pentamerous and zygomorphic. The nectar tube opens at the base of the posterior sepal. The tube is formed by passive excavation through the conspicuous elongation of the receptacle (Japp 1909). Nectar is secreted at the base of the tube. The posterior (upper) petals are usually marked with nectar guides. The 10 parts of the androecium are fused into a short column, the number of fertile stamens varies between 7 and 2. From this basic ground plan several modifications have evolved:

Perianth symmetry and shape

Perianth symmetry varies from almost actinomorphic to highly zygomorphic. An almost actinomorphic funnel-shaped perianth is found in *P. cucullatum* (L.) L'Hérit. (sect. *Pelargonium*). The small disc-shaped flowers of *P. incarnatum* Moench. (sect. *Campylia*) are almost perfectly actinomorphic. These flowers resemble those of the genus *Erodium* into which this species has sometimes been placed (Van der Walt et al. 1990c). Another example is *P. chelidonium* (Houtt.) DC. (sect. *Hoarea*) with flowers resembling those of *Oxalis* species. However, most flowers are zygomorphic (e.g. *P. reniforme* Curt., sect. *Reniformia*; *P. trifoliatum* (Eckl. & Zeyh.) E.M. Marais, sect. *Hoarea*). Examples of extremely zygomorphic flowers are found in *P. oreophilum* Schltr. (sect. *Ligularia* s.l.) and *P. punctatum* (Andr.) Willd. (sect. *Hoarea*).

Pelargonium flowers can mostly be classified as a primitive form of a flag blossom (*sensu* Faegri & van der Pijl 1979): the visual attraction is mediated mainly by the posterior petals and pollination is sternotribic, i.e. pollen is attached to the ventral side of flower visitors.

Two evolutionary lines can be distinguished: Firstly, the lateral anterior petals are turned by about 90° to form a landing platform alongside or above the protruding sexual organs, as seen in *P. antidysentericum* (Eckl. & Zeyh.) Costel (sect. *Jenkinsonia*) or in *P. multicaule* Jacq. (sect. *Myrrhidium*). In flowers of this type the middle anterior petal is often reduced in size or completely absent. Examples occur in several sections, including sectt. *Campylia* (*P. caespitosum* Turcz.), *Jenkinsonia* (*P. praemorsum* (Andr.) Dietr., *P. tetragonum* (L.f.) L'Hérit.), and *Myrrhidium* (all species). An odd modification of the latter flower type is found in *P. dolomiticum* Knuth ex Engl. and *P. griseum* Knuth of sect. *Ligularia* s.l.: in both species the claws of the posterior petals form false tubes. In a few species all three anterior petals are absent with the protruding stamens serving as a landing platform, including 5 species of sect. *Hoarea* (e.g., *P. dipetalum* L'Hérit.) and *P. quercetorum* Agnew (sect. *Ciconium*).

Secondly, a "papilionaceous" flag flower has evolved in sect. *Hoarea*. In the sister species *P. gracillimum* Fourc. and *P. luteolum* N.E. Brown the anterior petals are very broad to cover the reproductive organs, but, in the closely related *P. rapaceum* (L.) L'Hérit., the anterior petals form a typical "carina". In all these flowers the anterior petals need to be "tripped" to expose the reproductive parts. The unusual flowers of the closely related *P. grandicalcaratum* Knuth and *P. otaviense* Knuth (sect. *Ligularia* s.l.) show a similar architecture: the anterior petals also enclose the reproductive parts while the posterior petals remain almost straight. The orientation of these flowers and the size of the nectar tube resemble the ornithophilous flowers of *P. fulgidum* (L.) L'Hérit. (sect. *Ligularia* s.l.; see below), although they do not show a typical "bird colour".

In section *Otidia* and in a few species of other sections (*P. coronopifolium* Jacq., sect. *Campylia*; *P. hispidum* (L.f.) Willd. and *P. tomentosum* Jacq., sect. *Pelargonium*; *P. oreophilum*, sect. *Ligularia* s.l.) the opening of the nectar tube is concealed by the auricled bases of the posterior petals. Similarly, the entrance to the nectar tube is somewhat disguised in *P. triphyllum* Jacq. of sect. *Hoarea*.

Arrangement and number of fertile stamens

The androecium of *Pelargonium* consists of 10 organs. Seven fertile stamens are regarded to represent the primitive condition in *Pelargonium* (Albers & van der Walt 1984). A reduction in number of fertile stamens to 5 and more rarely to 6, 4, and 2 occurs throughout the genus. Furthermore, several modifications in the arrangement of stamens and presentation of anthers can be distinguished which are closely related to the blossom types and pollination syndromes (see below):

1 Stamens spreaded in a fan-like fashion in species with 7 stamens (e.g. *P. cucullatum*, sect. *Pelargonium*).
2 Stamens staggered in species with 7 to 4 stamens providing a usually longish "pollen carpet". The stamens might be deflected (e.g. *P. laxum* (Sweet) G. Don, sect. *Otidia*), slightly to sharply declined and reflexed into the centre plane of the perianth (e.g. *P. tetragonum*, sect. *Jenkinsonia*), or curved towards the posterior petals (e.g. *P. punctatum* Willd, sect. *Hoarea*).
3 Stamens of similar length collected into a ram-like column with anthers orientated in an outward-upward direction forming a "pollen pad" in species with 5 to 4 stamens (*P. caroli-henrici* Nordenstam, sect. *Hoarea*).
4 In a few species the 5 stamens are arranged in a ± circular fashion (e.g. *P. tricolor* Curt., sect. *Campylia*).

5 In a few species the number of fertile stamens is reduced to two straight fertile stamens protruding from the perianth (e.g. *P. punctatum* (Andr.) Willd., sect. *Hoarea*).

The staminodes are usually very short and do generally not seem to be of particular functional significance. However, in *P. fumariifolium* Knuth of sect. *Hoarea* some staminodes are long (Marais 1994). In *P. ovale* (Burm. f.) L'Hérit. (sect. *Campylia*) the posterior staminodes are markedly reflexed (Van der Walt & Van Zyl 1988). These modifications are obviously designed to fine-tune pollinator movements.

Nectar tube

The nectar tube contained in an elongated hypanthium is one of the synapomorphic characters for *Pelargonium*. The nectariferous tissue is always located at the base of the tube, but is absent in *P. incarnatum* (sect. *Campylia*) (Van der Walt et al. 1990c). The hypanthium is usually straight and slender, yet typically curved in *P. grenvilleae* (Andr.) Harv. and *P. incrassatum* (Andr.) Sims. (sect. *Hoarea*; Marais 1994). The nectar tube in *P. fulgidum* and, to a lesser degree, in *P. grandicalcaratum* and *P. otaviense* (all sect. *Ligularia* s.l.) is conspicuously enlarged.

The length of the nectar tube is a highly variable character on genus, section and species level. In several species the variation covers a span of 40-50 mm and even 60 mm (Marais 1994). Yet, the following size classes can be distinguished (M. Struck, unpublished data):

1 Nectar tube short, not more than 15 (-20) mm: e.g. *P. cucullatum* (sect. *Pelargonium*).
2 Nectar tube medium-sized, 20-40 mm: e.g. *P. laevigatum* (L.f.) Willd. (sect. *Glaucophyllum*).
3 Nectar tube long to very long, >40 mm: e.g. *P. carneum* Jacq. (sect. *Hoarea*); the longest nectar tube length (60-100 mm) has been recorded for *P. appendiculatum* (L.f.) Willd. of sect. *Hoarea* (Marais 1994).
4 A nectar tube reduced to a shallow cavity has been found in *P. cotyledonis* (L.) L'Hérit. (sect. *Isopetalum*) and *P. incarnatum* (sect. *Campylia*); nectariferous tissue is absent in the latter species.

The distribution in nectar tube sizes on section level can be summarized as follows:

1 Short or almost absent nectar tubes are restricted to sectt. *Campylia*, *Otidia*, *Isopetalum*, *Pelargonium*, and *Peristera*.
2 Medium-sized nectar tubes occur mostly in e.g. sect. *Cortusina*.
3 Long nectar tubes are common in sectt. *Ciconium* and *Reniformia*, but very long in the nocturnal sect. *Polyactium*.
4 Sectt. *Hoarea* and *Ligularia* s.l. show an extremely wide range of nectar tube sizes, varying from short to very long.

Floral behaviour

Pelargonium flowers are generally proterandrous, exhibiting distinct male and female phases. Though flowers may be self-compatible this rather effectively prevents self-pollination (though not necessarily geitonogamy) and clogging of the stigmas with selfed pollen (e.g. in *P. dolomiticum* Knuth ex Engl.; Zietsman 1993). In several species the sexual phases are even separated by a short lag, but imperfect dichogamy is also often encountered (M. Struck, unpublished data). A conversion to homogamy and (facultative) cleistogamy has been found in populations of the short-lived *P. mollicomum* Fourc. (sect. *Ligularia* s.l.; Schmitz 1993) and *P. candicans* Spreng. (sect. *Myrrhidium*; M. Struck, unpublished data): Anther dehiscence and stigma receptivity already occur in the bud stage. The reproductive parts are positioned within close contact, and self-pollination can result immediately. Presumably in response to weather conditions, the perianth may still open indicating that chasmogamy may still occur. Both species flower profusely and seed set has been observed to be high.

Pollination syndromes

Pollinators of pelargoniums include bees, long-proboscid flies, butterflies, hawkmoths and sun birds (Marloth 1908, McDonald & Van der Walt 1992, Scheltema & Van der Walt 1990, Struck 1990 and unpublished data, Van der Walt et al. 1990b, Vogel 1954, Zietsman 1993, F. Albers, S.D. Johnson and D.J. McDonald, pers. comm.). Judged from field observations and floral characters the following pollination syndromes can be distinguished:

1 Allophily - e.g. *P. cucullatum* (sect. *Pelargonium*): flowers with large to medium-sized, funnel-shaped perianth, short nectar tube, and nectar guides on the posterior petals visited by various classes of pollinators, including bees and long-proboscid hovering tabanid flies (Marloth 1908).
2 Melittophily (*sensu* Vogel 1954, including mio-melittophily (Whitehead et al. 1987), i.e. pollination by flies specialized in pollen and/or nectar feeding, e.g. Bombyliidae) - e.g. *P. tricolor* (sect. *Campylia*): flowers medium-sized with funnel-shaped perianth and dark embossed blotches on the posterior petals visited by bees and the small dipterean *Megapalpus capensis* Wiedemann. The latter has been identified as the major pollinator (McDonald & Van der Walt 1992).
3 Psychophily - e.g. *P. inquinans* (L.) L'Hérit. (sect. *Ciconium*): typical scarlet red tube flower with an expanded medium-sized, almost actinomorphic perianth, inserted stamens and a medium-sized nectar tube.
4 Rhinomyophily (Rebelo et al. 1985; Whitehead et al. 1987: pollination by diurnal long-proboscid hovering flies of the Tabanidae and Nemestrinidae) - e.g. *P. laevigatum* (sect. *Glaucophyllum*): tube flower with cream to pinkish perianth, reddish purple feather light nectar guides on posterior petals, short stamens and medium-sized nectar tube. Flower visitors are the tabanid *Philoliche gulosa* Wiedemann and *P. rostrata* Wiedemann (D.J. McDonald, pers. comm.).
5 Sphingophily - e.g. *Pelargonium lobatum* (Burm.f.) L'Hérit. (sect. *Polyactium*): typical night-scented tube flower which opens into a flat, expanded, circular, medium-sized perianth. The petals are dull yellow-green with large longish dark purple to black areas. The contrasting coloration of the petals results in a star-like appearance in dim light (Vogel 1954). The stamens are inserted and the nectar tube is long.
6 Ornithophily - *P. fulgidum* (sect. *Ligularia* s.l.): typical bird-attracting red

tube flower with the sexual organs slightly exserted and curved upwards, and a conspicuously inflated nectar tube. The perching sun bird *Nectarinia chalybea* has recently been identified as the pollinator (F. Albers, pers. comm.).

Discussion

A wide range of flower morphs and pollination syndromes are characteristic for the genus *Pelargonium*. Striking floral similarities between species assigned to different sections are common. Conversely, there are many examples of closely related species with different floral structures that are adapted to various classes of pollinators. The heterogeneity of the genus in terms of structural blossom types and pollination syndromes, therefore, indicates an independent and repeated evolution of convergent floral characters in the various sections of *Pelargonium*. Flower morphology in *Pelargonium* can, therefore, be very misleading as far as taxonomic affinities are concerned.

How can the floral heterogeneity be explained? In the absence of clear evidence we can only speculate about the present state. However, a few lines of reasoning - which at the same time outline potentially rewarding lines of research - shall briefly be discussed here.

There is evidence from studies in other plant groups that floral morphology and pigmentation are under simple genetic control (H. Stubbe 1966, cited after Von Denffer et al. 1978; unpubl. data cited in Wolf et al. 1993). In *Ipomopsis* species (Polemoniaceae) exhibiting distinct corolla shapes these differences become manifest in the final stages of development only, through subtle differences in cell elongation (see Wolf et al. 1993 for references). For *Pelargonium* sect. *Myrrhidium* might be a case in point: these species differ mainly in the length of the nectar tube and the size of the perianth which results in a transition between pollination syndromes.

The composition of pollinator assemblages varies within the distribution area of *Pelargonium*: e.g. butterflies are uncommon in the fynbos region of the south-western Cape whereas long-proboscid flies are an important component of the insect pollinator fauna in this area (Johnson 1992). Long-proboscid tabanid and nemestrinid flies are, however, rare in the arid parts of the western Cape (Struck 1994a, b). Moreover, flowering times of pelargoniums follow the general trend from early to later, in a northwest-southeast cline in the western Cape and a west-east cline in the south-western and southern Cape (Marais 1994). This may lead to the evolution of "pollination ecotypes" (Grant & Grant 1965) in the range margins of widespread species (Nilsson 1989, cited after Olesen 1992) and possibly further speciation.

Many *Pelargonium* species show a considerable plasticity in flower pigmentation and nectar tube size. Differences in floral characters of allopatric varieties and populations have been explained by habitat preferences of pollinators (see Olesen 1992 for references). Habitat preferences of pollinators might also account for differences in pollination success (Johnson & Bond 1991).

We conclude that intensive studies of the pollination biology and ecology of *Pelargonium* will not only give further insight into the evolution of this genus, but will also provide a valuable contribution towards the understanding of the structure and function of its native floristic region.

Acknowledgements
We wish to thank F. Albers (University of Münster), S.D. Johnson (University of Cape Town), E.M. Marais (University of Stellenbosch), and D.J. McDonald (National Botanical Institute, Kirstenbosch) for allowing us to use unpublished data. M. Gerbaulet (University of Cape Town) provided the French version of the abstract. The work was funded by the Foundation for Research Development, Pretoria, through a Student Support Programme grant.

References

Albers, F. & J.J.A. van der Walt, 1984. Untersuchungen zur Karyologie und Mikrosporogenese von *Pelargonium* sect. *Pelargonium* (Geraniaceae). Plant Systematics and Evolution 147: 177-188.

Albers, F., M. Gibby, & M. Austmann, 1992. A reappraisal of *Pelargonium* sect. *Ligularia* (Geraniaceae). Plant Systematics and Evolution 179: 257-276.

Faegri, K. & L. van der Pijl, 1979. The principles of pollination ecology. 3rd edition. Pergamon, Oxford.

Gibby, M., F. Albers & B. Prinsloo, 1990. Karyological studies in *Pelargonium* sectt. *Ciconium*, *Dibrachya*, and *Jenkinsonia* (Geraniaceae). Plant Systematics and Evolution 170: 151-159.

Grant, V. & K.A. Grant, 1965. Flower pollination in the *Phlox* family. Columbia University Press, New York.

Japp, G., 1909. Ueber die morphologische Wertigkeit des Nektariums der Blüten des *Pelargonium zonale*. Verhandlungen des naturforschenden Vereins in Brünn 47: 201-215.

Johnson, S.D., 1992. Plant animal relationships. In: Cowling, R.M. The Ecology of Fynbos. Nutrients, Fire and Diversity: 175-205, Oxford University Press, Cape Town.

Johnson, S.D. & W.J. Bond, 1992. Habitat dependent pollination success in a Cape orchid. Oecologia 91: 455-456.

Knuth, R., 1912. *Pelargonium*. In: A. Engler (ed.). Das Pflanzenreich 53: 316-545. Berlin.

Marais, E.M., 1994. Taxonomic studies in *Pelargonium* section *Hoarea*. Ph.D. thesis, University of Stellenbosch.

Marloth, 1908. Some observations on entomophilous flowers. South African Journal of Science 4: 110-113.

McDonald, D.J. & J.J.A. van der Walt, 1992. Observations on the pollination of *Pelargonium tricolor* section *Campylea* (Geraniaceae). South African Journal of Botany 58: 386-392.

Olesen, J.M., 1992. How do plants reproduce on their range margin? In: Thanos, C.A. (ed.). Plant-Animal Interactions in Mediterranean-type Ecosystems. Proceedings of the 6th International Conference on Mediterranean Climate Ecosystems: 117-122. Athens.

Rebelo, A.G., W.R. Siegfried & E.G.H. Oliver, 1985. Pollination syndromes of *Erica* species in the south-western Cape. South African Journal of Botany 51: 270-280.

Scheltema, A.G. & J.J.A. van der Walt, 1990. Taxonomic revision of *Pelargonium* section *Jenkinsonia* (Geraniaceae) in southern Africa. South African Journal of Botany 56: 285-302.

Schmitz, B., 1993. Zur Blütenbiologie der Gattung Pelargonium (Geraniaceae). Eine Bestandsaufnahme und eigene Beobachtungen. M.Sc. thesis, University of Münster.

Struck, M., 1990. Vergleichende phänologische und blütenbiologische Studien an Daueruntersuchungsflächen im ariden Winterregengebiet des südlichen Afrika. Ph.D. thesis, University of Hamburg.

Struck, M., 1994a. Flowers and their insect visitors in the arid winter rainfall region of southern Africa: Observations on permanent plots. Composition of the anthophilous insect fauna. Journal of Arid Environments 27.

Struck, M., 1994b. Flowers and their insect visitors in the arid winter rainfall region of southern Africa: Observations on permanent plots. Insect visitation behaviour. Journal of Arid Environments 27.

Van der Walt, J.J.A. & P.J. Vorster, 1983. Phytogeography of *Pelargonium*. Bothalia 14: 517-523.

Van der Walt, J.J.A. & L. van Zyl, 1988. A taxonomic revision of *Pelargonium* section *Campylia* (Geraniaceae). South African Journal of Botany 54: 145-171.

Van der Walt, J.J.A., F. Albers & M. Gibby, 1990a. Delimitation of *Pelargonium* section *Glaucophyllum* (Geraniaceae). Plant Systematics and Evolution 171: 15-26.

Van der Walt, J.J.A., D.J. McDonald & N. van Wyk, 1990b. A new species of *Pelargonium* with notes on its ecology and pollination biology. South African Journal of Botany 56: 467-470.

Van der Walt, J.J.A., H.J.T. Venter, R. Verhoeven & L.L. Dreyer, 1990c. The transfer of *Erodium incarnatum* to the genus *Pelargonium* (Geraniaceae). South African Journal of Botany 56: 560-564.

Vogel, S., 1954. Blütenbiologische Typen als Elemente zur Sippengliederung. In: W. Troll & H. von Guttenberg (eds.). Botanische Studien 1: 1-338, Fischer, Jena.

Von Denffer, D., F. Ehrendorfer, K. Mägdefrau, & H. Ziegler, 1978. Lehrbuch der Botanik. 31st. edition. Fischer, Stuttgart.

Whitehead, V.B., J.H. Giliomee & A.G. Rebelo, 1987. Insect pollination in the Cape flora. In: Rebelo, A.G. (ed.). A preliminary synthesis of pollination biology in the Cape flora. South African National Scientific Programmes Report 141: 52-82.

Zietsman, P.C., 1993. Pollination in *Pelargonium dolomiticum* Knuth (Geraniaceae). South African Journal of Botany 59: 259-264.

The one-to-one species specificity of the *Ficus*-Agaoninae mutualism: how casual?

J.Y. RASPLUS
INRA-CNRS Laboratoire: Populations, Génétique et Evolution,
Bat. 13, F-91198 Gif-sur-Yvette Cedex, France.

Abstract
The fig tree - fig wasp mutualism has long been considered to be strictly species-specific: each fig tree species is obligatorily pollinated by one fig wasp species, and each wasp species can only reproduce in one fig species. The first exceptions cited were thought to result from misidentifications of either figs or wasps, but as for most dogmas in biology, increasing knowledge of the mutualism has led to question the strict one-to-one rule. More and more examples of co-occurrence of at least two pollinators per fig species are provided from all over the world. I define below 6 categories of co-occurrence of more than one species of pollinators and 3 categories of pollination of different species of *Ficus* by the same species of pollinator. I give an accurate description of the state of our knowledge on the exceptions to the one-to-one rule in the different biogeographic regions and discuss the relatively high number of breakdowns to the rule, in South-Eastern Asia.

Résumé
Depuis des décennies, on considère la relation mutualiste entre les figuiers et leur pollinisateur comme strictement spécifique, c'est-à-dire que chaque espèce de figuier est obligatoirement pollinisée par une seule espèce d'hyménoptère et que chaque espèce de pollinisateur ne peut se reproduire que dans une seule espèce de figuier. Les premières exceptions mentionnées ont tout d'abord été considérées comme résultant d'erreurs d'identification de la plante ou du pollinisateur. Mais comme pour tous les dogmes en biologie, l'augmentation de nos connaissances nous a amené à remettre en question cette règle de stricte spécificité. Dans toutes les régions du monde on a mis en évidence des cas de cohabitation de plus d'un pollinisateur par espèce de figuier. Ci-dessous, je définis 6 types de coexistence de plusieurs pollinisateurs et 3 types de pollinisation de différentes espèces de figuiers par un même pollinisateur. Je donne aussi une description précise de l'état de nos connaissances sur les exceptions à cette règle dans les différentes régions biogéographiques et je discute le nombre relativement important de ces exceptions en Asie du Sud-Est.

Key words
Agaoninae, *Ficus*, Pollination, Species-specificity.

Introduction

The c. 750 species of *Ficus* (Moraceae) are consistently pollinated by agaonine chalcid wasps (Agaonidae). The fig tree - fig wasp mutualism has long been considered to be strictly species-specific: each fig tree species is obligatorily pollinated by one fig wasp species, and each wasp species can only reproduce in one fig species. Following Boucek (1988), we now consider all pollinators to represent one single subfamily, the Agaoninae. Until 1985 and the first report of co-occurrence of two pollinating species in one *Ficus* species (Michaloud et al. 1985), known exceptions of the one-to-one rule were thought to result from misidentifications of either figs or wasps. Wiebes (1963) discussed most of the cases known so far, and concluded that most of them (about 20) needed further investigation.

L.J.G. van der Maesen et al. (eds.), The Biodiversity of African Plants, 639-649.

Wiebes (1964) reported the "unexpected fact" that both *Ceratosolen arabicus* Mayr and *C. galili* Wiebes develop within the same receptacles of *Ficus sycomorus* L. Studies done by Galil & Eisikowitch (1968, 1969) and later on by Compton et al. (1991) showed that *Ceratosolen galili* acts as a "cuckoo" never using its pollen pockets to carry pollen. *C. galili* can passively pollinate a few flowers by carrying pollen on its body surface. The females enter the figs regardless of whether or not the pollinator *C. arabicus* is present, and oviposit into flowers without any act of pollination. During the seventies and early eighties, most cases were considered to be a coexistence of one pollinating fig wasp and lodgers which pollinate the fig only occasionally. Boucek et al. (1981) suggested this hypothesis for *Ficus thonningii* Blume which supports three different Agaoninae (*Elisabethiella stuckenbergi* (Grandi), *Alfonsiella brongersmai* Wiebes and *A. longiscapa* Joseph).

In 1985, Michaloud et al. reported the first record of two actively pollinating wasps in *Ficus ottoniifolia* (Miq.) Miq. The two pollinators show some habitat preference, *Courtella camerounensis* (Wiebes) is dominant in rain forests while *C. gabonensis* Wiebes prefers open habitats. They also report occurrence of two species of pollinators in *Ficus sur* Forskåll in Africa.

Since then and with the publications of several monographs (Berg & Wiebes 1992, Wiebes 1994, Wiebes & Compton 1990) the number of known cases of two (or more) pollinators in one species of *Ficus* has significantly increased. However, Wiebes (1994) recently questioned the cases of double pollination and argued that more research may demonstrate that each species (or other entities below the species level) of fig has its own species of pollinator wasp. The aim of this short review is to give an exact idea of the state of our knowledge on exceptions to the one-to-one rule: one *Ficus* - one pollinator. Species specificity breakdown is structurally two-faced either two or more species of wasps pollinate a single species of *Ficus*, or two fig species are pollinated by one and the same species of Agaoninae.

1 Co-occurrence of more than one species of pollinators

This kind of specificity breakdown poses no theoretical problems since it does not result in any gene flow between genetically different host plants. The questions are rather how two different gall-makers can exploit a unique resource and why some fig trees are currently pollinated by at least two Agaonine species. I describe below six different situations which summarize most of the different cases wherever two pollinators co-occur.

1.1 Taxonomic artefacts

Among the cases reported, a few are probably of taxonomic nature, and might be resolved if the relevant type material is available and inspected. Misidentifications of either the pollinator or the fig tree may result in apparent co-occurrence of two species of pollinators. The focus is here on a few cases demonstrating the effects of taxonomic uncertainties. *Eupristina saundersi* Grandi was reported from *Ficus religiosa* L. (Grandi 1916) which is pollinated by *Platyscapa quadraticeps* (Mayr) and from *Ficus microcarpa* L. (Boucek 1988), pollinated by *Eupristina* (*Parapristina*) *verticillata* Waterston. But likely both records are erroneous and the true host tree remains unknown. *Ceratosolen occultiventris* Mayr is reported from *Ficus panifica* Delile (a synonym of *Ficus sur*) in the Nile valley. Because the type cannot be located, the name of the species is considered a nomen oblitum (Wiebes 1989a). However, the host record is valid, and the actual identity of the fig wasps species remains obscure. We need to clarify such cases and establish whether it is another regional pollinator of *Ficus sur* (see part 1.4.) or an older synonym of one of its known pollinators that would cause name changes. Another example is the questionable Australian *Ceratosolen nigriscapus* (Girault) which could be, considering the host association, a younger synonym of *C. fuscipes* (Mayr). Type examination is highly needed to revise nomenclature and hence eliminate apparent cases of co-occurring pollinators.

1.2 Differences in biological characteristics of the different Agaoninae

A few cases of co-occurrence can be explained by differences in biological characteristics of the different species of fig wasps entering the syconium. One is the true pollinator, the other a "cuckoo" wasp which never pollinates. Thus, although provided with pollen loading adaptive devices, clear evidence of a previous pollinating activity, *Ceratosolen galili* has lost the pollination behaviour (i.e. biting of the stigmas, active unloading of pollen from the pockets...). This agaonine species is no doubt a cheater, but some other species are assumed to be also "cuckoo" wasps, e.g. the Oriental *Waterstoniella errata* (Wiebes) in syconia of *Ficus stupenda* Miquel pollinated by *W. masii* (Grandi), and some of the Agaoninae found in *Ficus thonningii* Blume syconia in Africa (Boucek et al. 1981). Advantages of being a cheater remain largely uncertain (Compton et al. 1991) and such a rare shift may reflect the local loss of a *Ficus* species and the subsequent colonization of another sympatric fig tree by the widowed partner.

1.3 Differential speciation in plant and wasps

The presence of co-occurring pollinators can also be explained by differential speciation in plants and wasps. Subspeciation in plants may lead to different species of pollinators In Africa, *Ficus* species of the subsections *Caulocarpae* and *Cyathistipulae* of *Galoglychia* frequently show such phenomena, (e.g. *F. ottoniifolia* (Figure 1) and *F. cyathistipula* Warb.). *F. cordata* Thunb. is another example. This may be considered as a by-product of some co-cladogenesis discrepancies.

In Australia *Ficus obliqua* var. *obliqua* Forst. and *F. obliqua* var. *petiolaris* (Benth.) are pollinated by two different species of *Pleistodontes, P. greenwoodi* (Greenwood) and *P. imperialis* Saunders respectively.

In the Oriental region two subspecies of *Ficus sinuata* Thunb., two varieties of *F. aurantiacea* Griff., *F. erecta* Thunb., *F. grossularioides* Burm., *F. heteropleura* Blanco, *F. obscura* Blanco, *F. ribes* Reinw., *F. sumatrana* Miq., *F. superba* Miq., *F. tinctoria* Miq. and *F. uniglandulosa* Wall., three varieties of *F. benjamina* L. and *F. virgata* Reinw., four varieties of *F. crassiramea* Miq. (Figure 2) and two forms of *F. subulata* Bl. are pollinated by different species of Agaoninae (Wiebes 1994). This peculiar pattern of the Oriental fauna may reflect a need of revision of the Australasian flora of *Ficus*, but recent revision of African *Ficus* (Berg 1989) resulted in a decreasing rather than an increasing number of valid species of *Ficus*. In most of these cases of co-occurrence, the two pollinators are allopatric and consequently they could also illustrate the next category: different pollinators in different parts of the *Ficus* range.

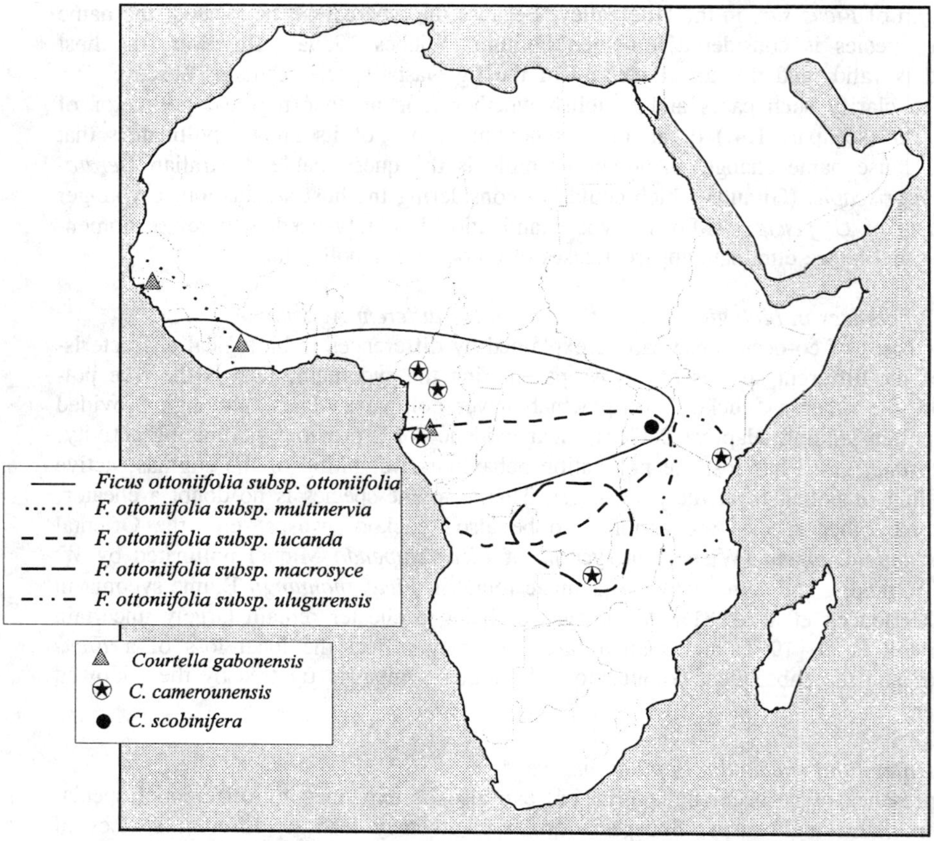

Figure 1. Distribution of the different subspecies of *Ficus ottoniifolia* in Africa, and their known pollinators.

I suggest that high insularity and natural fragmentation of the oriental flora may have favoured differential speciation phenomena in pollinators and *Ficus* because of differences in their population dynamics and in their aptitude of dispersion (seeds are transported by vertebrates, mainly bats or birds, pollinators can only colonize islands separated from the continent by at most 100 km).

1.4 Different pollinators in different parts of the geographic range of the Ficus

Species of *Ficus* with relatively large repartition can be pollinated by different species of Agaoninae in different parts of their distribution area. In Africa, *Ficus sur* Forsskål shows such pollination pattern: *Ceratosolen flabellatus* Grandi is its pollinator in West Africa together with *Ceratosolen silvestrianus* Grandi except in the Monts Nimba (Guinea). It also occurs in East Africa (Kenya, Uganda, Ethiopia) together with *Ceratosolen capensis* Grandi which is found from Ethiopia to South Africa where it is the only pollinator (Figure 3).

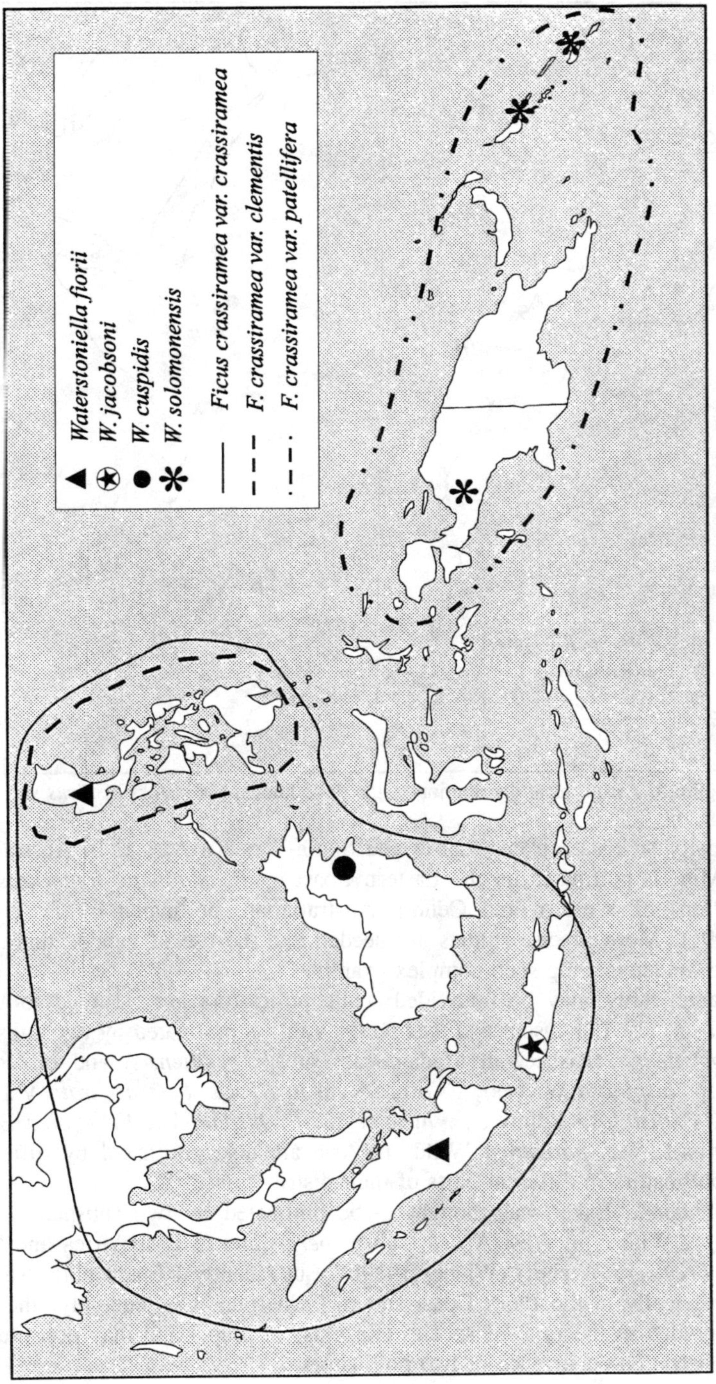

Figure 2. *Ficus crassiramea* and its pollinators in South-Eastern Asia.

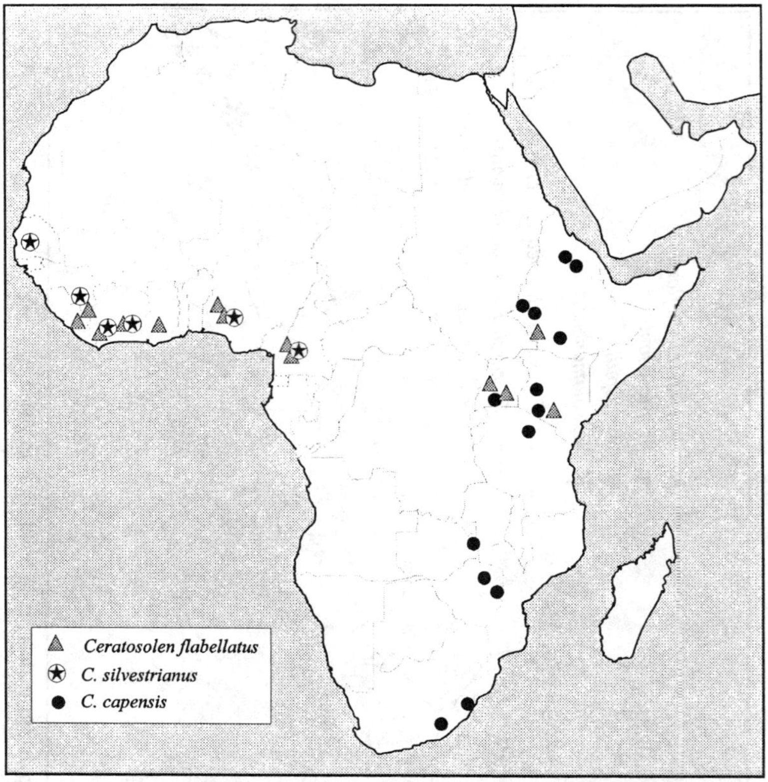

Figure 3. Afrotropical distribution of the three known pollinators of *Ficus sur*.

We crucially lack data from Central and South-Western Africa. In Australia, *Ficus platypoda* Miq. is pollinated on the Eastern Coast by *Pleistodontes imperialis* whereas the populations of Western and Central Australia are pollinated by *P. greenwoodi* (Wiebes 1991). More precise studies are needed and the use of genetic markers could be helpful in understanding such complex examples.

The botanic entity may be undivided while pollinators are assumed to belong to different subspecies. For instance, *Ficus polita* Vahl. is pollinated by two subspecies of *Courtella bekiliensis* (Risbec) in Madagascar (subsp. *bekiliensis*) and on continental Africa (subsp. *bispinosa* Wiebes); *F. ovata* Vahl in Africa, *F. exasperata* Vahl in Asia and Africa, *F. fistulosa* Reinw., *F. hirta* Vahl, *F. hispida* L., *F. septica* Burm., *F. tinctoria* Forst.f. var. *parasitica* Willd. in Asia are also pollinated by different subspecies of pollinators in different parts of their distribution.

Ficus trichopoda Baker was thought to be pollinated by two subspecies of *Elisabethiella bergi* Wiebes in West Africa (subsp. *bergi*) and in East Africa and Madagascar (subsp. *breviceps* Wiebes) (Wiebes 1989b). But recently I found both "subspecies" living sympatrically in the Ivory Coast and in Cameroon. Consequently, the two taxa must be upgraded to the rank of species (Rasplus, in prep.) and may constitute a new case of sympatric co-occurrence of two pollinators.

1.5 Two sympatric pollinators
Most of the remaining cases may illustrate actual co-occurrence of two active pollinators in adjacent areas, on the same tree, or even in the same syconium. Both *Ceratosolen flabellatus* and *C. silvestrianus* pollinate actively and effectively *Ficus sur* in the Ivory Coast (Kerdelhué 1994). Other African examples (Compton et al. 1994) may occur in figs of *F. abutilifolia* Miq., *F. artocarpoides* Warb., *F. natalensis* Hochst., *F. ottoniifolia* (Michaloud et al. 1985), *F. thonningii* (Boucek et al. 1981), *F. trichopoda* (see above). In Asia, *Blastophaga* (*Valisia*) *compacta* Wiebes and *B.* (*V.*) *inopinata* Grandi on *Ficus fulva* Reinw. ex Bl.; *Pleistodontes blandus* Wiebes and *P. mandibularis* Wiebes on *F. glandifera* Summerh., *Waterstoniella masii* (Grandi) and *W. errata* Wiebes on *F. stupenda* Miq. are other examples where two effective pollinators may co-occur (Wiebes 1994). In South America, the only reported case (Ramirez 1970) i.e. *Ficus tuerckheimii* Standley and its pollinators *Pegoscapus carlosi* (Ramirez) and *P. mariae* (Ramirez) also illustrates this category of species-specific breakdown.

Very little is known about the modality of double pollination but preference for different habitats could explain sympatric occurrence (Michaloud et al. 1985). We need a comprehensive study of such cases to understand the coexistence of two closely related species on the same resource.

1.6 The case of introduced trees
To be complete, I have to discuss a peculiar case of co-occurrence of two pollinators in the figs of one single species of *Ficus*. Some species of *Ficus* are introduced as ornamental trees in parks and gardens far from their natural distribution area. Such trees may cheat autochthonous pollinators normally associated with local *Ficus* species. The misled wasps may successfully reproduce, and pollinate the alien host plant; hybrid seeds may therefore be produced. This case is reported by Compton (1990) and Compton et al. (1994) in *Ficus lutea* Vahl which is normally pollinated by *Allotriozoon heterandromorphum* Grandi but can also be colonised by *Ceratosolen capensis* (the pollinator of *Ficus sur*) and *Elisabethiella stuckenbergi* Grandi (pollinator of *F. thonningii*), reproducing successfully into the unusual host.

2 Pollination of different species of *Ficus* by the same pollinator
Inversely one species of pollinator may pollinate two different species of *Ficus*. This may occur in different parts of the geographic range of the Agaoninae (allopatry) or in adjacent areas within the normal range of dispersion of the pollinator (sympatry). This case raises the question of how different species of *Ficus* can conserve genetic integrity in such conditions.

2.1 Different Ficus in different parts of the geographic range of the wasp
In Africa, species pollinating different *Ficus* in different parts of their distribution area are rare. *Elisabethiella socotrensis* (Mayr) known in the north from *Ficus vasta* Forssk. (Socotra, Ethiopia and Saudi Arabia) and in the south from *Ficus wakefieldii* Hutch. (Kenya and Zambia) is an example.

Such cases seem to be a little more frequent in the Oriental region. We can cite *Deilagaon annulatae* Wiebes and *D. chrysolepidis* Wiebes, *Ceratosolen constrictus*

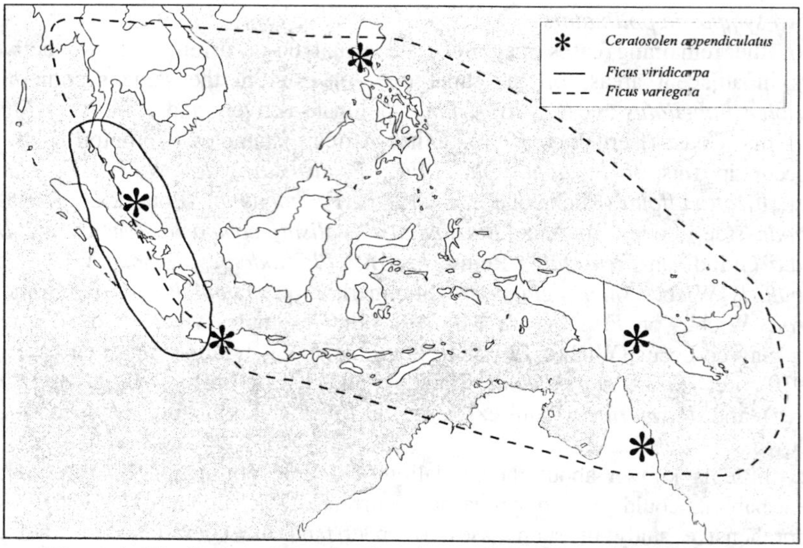

Figure 4. Distribution of the two *Ficus* hosts of *Ceratosolen appendiculatus* in Australasia.

constrictus (Mayr) and *C. appendiculatus* (Mayr) (Figure 4), *Pleistodontes greenwoodi* and *Eupristina* (*Parapristina*) *cyclostigma* Wiebes (Wiebes 1994) which pollinate different species of *Ficus* in different countries of the region.

2.2 Different Ficus in different adjacent habitats

Surprisingly, this category is well represented both in Africa and Australasia. In Africa three species belonging to the genus *Alfonsiella* are poorly specific (Berg & Wiebes 1992), they could develop in figs of different *Ficus*, but their exact biology (pollinator or cuckoo) is unknown. *A. michaloudi* Wiebes pollinates *Ficus craterostoma* Mildbr. & Burret and *F. lingua* De Wild. & Durand subsp. *lingua* in Gabon; *A. fimbriata* Waterston is recorded in West Africa from *Ficus natalensis* Hochst. subsp. *leprieurii* (Miq.) Berg and *F. kamerunensis* Mildbr. & Burret, and *A. longiscapa* Joseph in *F. thonningii* and *F. natalensis* Hochst. subsp. *natalensis* in Zambia. In such cases, both *Ficus* species are closely related taxonomically and morphologically. Four other African Agaoninae are not species-specific: *Ceratosolen arabicus* and *C. galili* which oviposit in *Ficus mucuso* Welw. ex Ficalho and *F. sycomorus*; *Agaon kiellandi* Wiebes occurs in three different *Ficus* in Gabon: *F. conraui* Warb., *F. cyathistipula* Warb. subsp. *pringsheimiana* (Braun & Schum.) Berg and *F. densestipulata* De Wild.; *Kradibia gestroi afrum* (Wiebes) develops in figs of *Ficus exasperata* Vahl and *F. asperifolia* Miq. in most of Africa and in *Ficus capreifolia* in Ethiopia and South Africa. Similar cases are known from Asia: *Blastophaga silvestriana* Grandi in Hong Kong, *Pleistodontes imperialis*, *Ceratosolen* (*Rothropus*) *humatus* Wiebes, *Ceratosolen* (*Rothropus*) *notus* (Baker) and *Kradibia commuta* Wiebes also pollinate different closely related species of *Ficus*. The last but probably the most widely known example, is *Blastophaga psenes* L. which pollinates both *Ficus carica* L. and *F. palmata* Forssk. in the

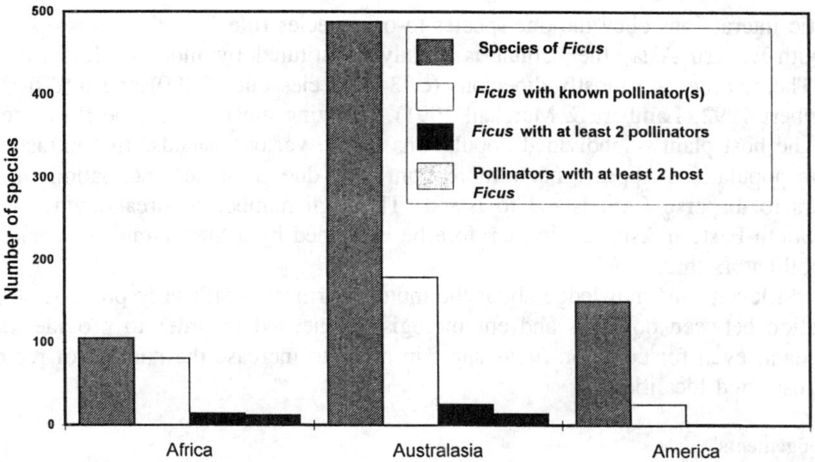

Figure 5. World number of species of *Ficus*, number of known pollinators, and number of exceptions to the one *Ficus*-one pollinator rule.

Oriental and Palearctic regions. The exact status of *F. palmata* is sometimes discussed, and the species could be considered as a form of *F. carica*.

2.3 Pollinators misled by introduced species of Ficus

When introduced in some area, exotic *Ficus* species may obtain its pollinator from an indigenous fig. Such records are scarce in literature, but Corner (1985) reports the case of a *Blastophaga* species entering the figs of introduced *Ficus obliqua* Miq. in Fiji, species normally pollinated by a *Pleistodontes* species. In Florida both *Ficus altissima* Blume and its pollinator (*Eupristina altissima* Balakrishnan, Joseph & Abdurahiman) are established. This pollinator also enters the figs of three other introduced *Ficus* species whose pollinators are absent (Patel et al. 1993). It seems that the closer the two fig hosts will be phylogenetically, the more likely the host may shift. Very few cases of co-occurrence of the same pollinator in two or more species of *Ficus* are due to such pollination mistakes.

Conclusions

This short survey of the one species - one species worldwide breakdown in the mutual relationships between Agaoninae and *Ficus* shows that these events are exceptional, but not as rare as usually thought. They represent about 13% of the described associations (Figure 5). In Africa, the Moraceae were recently revised (Berg 1989) and the pollinators of about 75% of the *Ficus* species are described (see Wiebes & Compton 1990 for a review). The cases of co-occurrence of at least two pollinators reach 17% of the known partners while c. 15% of the Agaoninae are found on more than two *Ficus*. In Australasia, these percentages drop to 13% and 7% respectively. Despite Wiebes' efforts, our knowledge in that part of the world is still incomplete, and we only know the pollinators of 35% of the c. 450 species of *Ficus*. In South America where little is known about both *Ficus* and fig-wasps, only one single exception is reported. Nevertheless, it should be stressed that the overwhelming majority, that is 80 percent, of the mutualistic interactions obey the one species-to-one species rule.

In South-Eastern Asia, the habitat is mainly constituted by more or less isolated islands. There, *Ficus* are mostly dioecious (c. 340 species out of 490) and bird-dispersed (Lambert 1992, Lambert & Marshall 1991), implying means that gene flow occurs between the host plant's subdivided populations on the various islands. In contrast, the pollinator populations appear much more confined, due to lesser inclination of the pollinators to disperse from island to island. The high number of breakdowns to the rule, in South-Eastern Asia, could therefore be explained by a higher rate of speciation for the pollinators than in *Ficus*.

At world level, our knowledge about the mutual partners is still very poor. A strong collaboration between botanists and entomologists is needed in order to provide more accurate data, even for common *Ficus* spp., in order to increase the number of records and of prospected localities.

Acknowledgements

I would like to thank Carole Kerdelhué and Daniel Lachaise for their help and valuable comments on the manuscript.

References

Berg, C.C., 1989. Moraceae. Flora of Tropical East Africa: 1-95.

Berg, C.C. & J.T. Wiebes, 1992. African fig trees and fig wasps. Verhandelingen der Koninklijke Nederlandsche (Nederlandse) Akademie van Wetenschappen 89: 1-298.

Boucek, Z., 1988. Australasian Chalcidoidea (Hymenoptera). CAB Int., Wellingford, U.K.: 1-832.

Boucek, Z., A. Watsham & J.T. Wiebes, 1981. The fig fauna of the receptacles of *Ficus thonningii* (Hymenoptera, Chalcidoidea). Tijdschrift voor Entomologie 124 (5): 149-233.

Compton, S.G., 1990. A collapse of host specificity in some African fig wasps. South African Journal of Science 86: 39-40.

Compton, S.G., K.C. Holton, V.K. Rashbrook, S. van Noort, S.L. Vincent & A.B. Ware, 1991. Studies on *Ceratosolen galili*, a non-pollinating Agaonid fig wasp. Biotropica 23: 188-194.

Compton, S.G., A.B. Ware & S. van Noort, 1994. African figs and fig wasps: the wasp's eye view of *Ficus* species. In: J.H. Seyani & A.C. Chikuni (eds), Proceedings of the XIIIth Plenary Meeting AETFAT, Malawi, 1: 451-459.

Corner, E.J.H., 1985. *Ficus* (Moraceae) and Hymenoptera (Chalcidoidea): figs and their pollinators. Biological Journal of the Linnean Society 25: 187-195.

Galil, J. & D. Eisikowitch, 1968. On the pollination ecology of *Ficus sycomorus* in East Africa. Ecology 49: 259-269.

Galil, J. & D. Eisikowitch, 1969. Further studies on the pollination ecology of *Ficus sycomorus* L. (Hym., Agaonidae). Tijdschrift voor Entomologie 112: 1-13.

Grandi, G., 1916. Gli Agaonini (Hymenoptera Chalcididae) raccolti nell'Africa occidentale dal Prof. F. Silvestri. Bollettino del Laboratorio di Zoologia Portici 10: 121-286.

Kerdelhué, C., 1994. Etude de la dynamique de pollinisation et de l'entomofaune de deux figuiers du sous-genre *Sycomorus* en milieu tropical. Mémoire de DEA Ecologie générale et productions végétales, Paris VI - Paris XI - INA-PG. 38 pp.

Lambert, F., 1992. Fig dimorphism in bird-dispersed gynodioecious *Ficus*. Biotropica 24: 214-216.

Lambert, F.R. & A.G. Marshall, 1991. Keystone characteristics of bird-dispersed *Ficus* in a Malaysian lowland rain forest. Journal of Ecology 79: 793-809.

Michaloud, G., S. Michaloud-Pelletier, J.T. Wiebes & C.C. Berg, 1985. The co-occurrence of two pollinating species of fig wasp and one species of fig. Proceedings of the Koninklijke Nederlandse Akademie van Wetenschappen (C) 88: 93-119.

Patel, A., M. Hossaert-McKey & D. McKey, 1993. Ficus - pollinator research in India: Past, present and future. Current Science 65 (3): 243-253.

Ramirez, B.W., 1970. Taxonomic and biological studies of neotropical fig wasps (Hymenoptera: Agaonidae). University of Kansas Science Bulletin 49: 1-44.

Ware, A.B., S.G. Compton, 1992. Breakdown of pollinator specificity in an African fig tree. Biotropica 24 (4): 544-549.

Wiebes, J.T., 1963. Taxonomy and host plant preference of Indo-Australian fig wasps of the genus *Ceratosolen* (Agaonidae). Tijdschrift voor Entomologie 106: 1-112.

Wiebes, J.T., 1964. Fig wasps from Israeli *Ficus sycomorus* and related East African species (Hymenoptera, Chalcidoidea: Agaonidae). Entomologische Berichten Amsterdam 24: 187-191.

Wiebes, J.T., 1989a. Agaonidae (Hymenoptera Chalcidoidea) and *Ficus* (Moraceae): fig wasp and their figs III. (*Elisabethiella*). Proceedings of the Koninklijke Nederlandse Akademie van Wetenschappen (C) 92: 117-136.

Wiebes, J.T., 1989b. Agaonidae (Hymenoptera Chalcidoidea) and *Ficus* (Moraceae): fig wasp and their figs IV. (African *Ceratosolen*). Proceedings of the Koninklijke Nederlandse Akademie van Wetenschappen (C) 92: 251-266.

Wiebes, J.T., 1991. Agaonidae (Hymenoptera Chalcidoidea) and *Ficus* (Moraceae): fig wasps and their figs VII (*Pleistodontes*). Proceedings of the Koninklijke Nederlandse Akademie van Wetenschappen 94: 137-152.

Wiebes, J.T., 1994. The Indo-Australian Agaoninae (pollinators of figs). Verhandelingen der Koninklijke Nederlandsche (Nederlandse) Akademie van Wetenschappen 92: 1-208.

Wiebes, J.T. & S.G. Compton, 1990. Agaonidae (Hymenoptera Chalcidoidea) and Ficus (Moraceae): fig wasps and their figs VI. (Africa concluded). Proceedings of the Koninklijke Nederlandse Akademie van Wetenschappen (C) 93: 203-222.

Sex expression in *Canthium mundianum* (Rubiaceae)

K. BALKWILL, J.R. SEBOLA & E.R. ROBINSON
C.E. Moss Herbarium, Department of Botany, University of the Witwatersrand,
Private Bag 3, WITS, 2050, South Africa.

Abstract

While investigating the validity of the segregation of *Canthium gilfillanii* (N.E. Br.) O.B. Miller from *Canthium mundianum* Cham. & Schlechtd., we noticed that flowers in a population were of different sizes and either female or hermaphrodite. Preliminary conclusions of the former study, suggest that *C. gilfillanii* and *C. mundianum* are indeed conspecific. The morphology of the style is specialised for pollen presentation in hermaphrodite flowers and for pollen reception in females. The anthers of hermaphrodite flowers contain copious viable pollen while those of female flowers are barren. The stigmatic area in both sexes is receptive, but enlarged in female flowers. All flowers on the same plant are of the same sex, throughout the flowering season and in successive seasons and the species is thus gynodioecious. Seed set was very successful in females, but very low in hermaphrodites, regardless of the amount of rainfall. Vegetative reproduction has led to the development of extensive clones.

Key words

Canthium, Clone, Fruit set, Gynodioecy, Sex expression.

Introduction

The validity of the segregation of *Canthium gilfillanii* (N.E. Br.) O.B. Miller from *Canthium mundianum* Cham. & Schlechtd., which is based mainly on the density of hairs on the leaves, has been questioned several times. We viewed this as an ideal study for the application of the technique of enzyme electrophoresis and this work is still underway. Preliminary results suggest that these species are conspecific, however. At Melville Koppies Nature Reserve, Johannesburg, South Africa, forms of *Canthium* exist that can be assigned to both *C. mundianum* and *C. gilfillanii* and we thus considered Melville to be an ideal study site due to its proximity to the University and ease of access. During the course of collecting material for isozyme studies, we noticed that flowers in the population were of different sizes and more detailed observations showed that they were indeed female or hermaphrodite. Unisexual flowers have only been recorded in *Canthium* section *Bullockia* before (Bridson 1987), so we thought sex expression in this species to be worthy of investigation.

Dioecy ensures that self-pollination cannot occur, but it places the female plants at more risk than the males, as the production of fruit and seed is far more expensive than the production of pollen. Thus, populations of dioecious species can potentially become dominated by male individuals, which can out-compete the females during the time that they bear fruits. Monoecy, in which unisexual flowers of both sexes are borne on the same plant, enhances, but does not ensure, outbreeding, but this condition reduces the possibility of male plants competing with female individuals of the same species. Gynodioecy is an intermediate condition, in which some individuals reproduce by cross-breeding only, and others by selfing or crossing. All individuals in the

L.J.G. van der Maesen et al. (eds.), The Biodiversity of African Plants, 650-655.

population potentially produce similar numbers of fruits and thus individuals of the species do not significantly compete with reproductive individuals of that species. Gynomonoecy enhances outbreeding, but could facilitate geitonogamy. Immelman (1984) has found that plants of *Kirkia wilmsii* Engl. produce flowers of one sex for a period of about a week and then swop to the other sex for the subsequent week. This promotes outbreeding, but without some plants bearing more reproductive costs than others. Another strategy that we think plants may adopt is to swop sexes on alternate years and thereby accumulate resources in the year that the plant is male and these resources can be tapped in the year that the plant is female. Van Wyk and Lowrey (1988) found that drought influenced flowering in *Eugenia* (which is androdioecious) and we thought it likely that rainfall may affect the level of seed set in flowers of different sexes of *Canthium*.

At Melville Koppies Nature Reserve, *Canthium mundianum* occurs in woodlands on scree slopes below the crests of quartzite ridges. There is considerable variation in size, shape and pubescence of leaves and in presence and absence of hairs in domatia. Large numbers of adjacent individuals are identical, but remarkably different from plants in other stands. Plants in morphologically identical stands are all of the same sex.

Methods

Between two and seven plants were selected in several neighbourhoods at Melville Koppies Nature Reserve. The neighbourhoods are distributed over approximately 600 m in a line from east to west. Plants were initially marked with permanent markers on masking tape, but these were replaced with plastic nursery tags. Fresh material was collected in the field and taken to the laboratory for scrutiny with a dissecting microscope. Pollen viability was tested with Alexander's stain (Alexander 1969) and stigmatic receptivity with Mayer's Haemalum and Napthyl Acetate in fast blue salt (Heslop-Harrison & Shivanna 1977).

Observations

Floral Structure and Function

Canthium mundianum bears two sizes of flowers; the smaller are female and the larger hermaphrodite. The anthers of the hermaphrodite flowers are larger and contrast with the corolla, making the flowers of this sex more conspicuous. It is thus possible that material with hermaphrodite flowers and fruiting material is collected more often than material with female flowers.

The morphology of the style is different in the flowers of the two kinds. The pollen presenter of the hermaphrodite flower is swollen and deeply longitudinally grooved (Figure 1), while the stigmatic surface is small and does not protrude over the upper rim of the pollen presenter. The pollen presenter of the female flower is under-developed, but the stigma is elaborated and protrudes markedly over the upper rim of the pollen presenter (Figure 2). The stigmatic area in both sexes is receptive. Thus the style of the hermaphrodite flower is specialised for pollen presentation and that of the female flower for pollen reception. These differences are easily observable with a 10x lens in the field and even on herbarium material with the aid of a dissecting microscope.

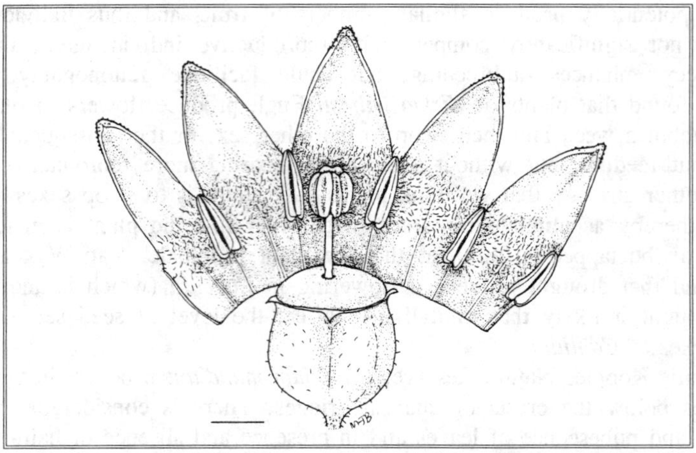

Figure 1. Hermaphrodite flower of *Canthium mundianum*, with corolla opened to show fertile stamens and markedly grooved pollen presenter, scale bar = 0,5 mm.

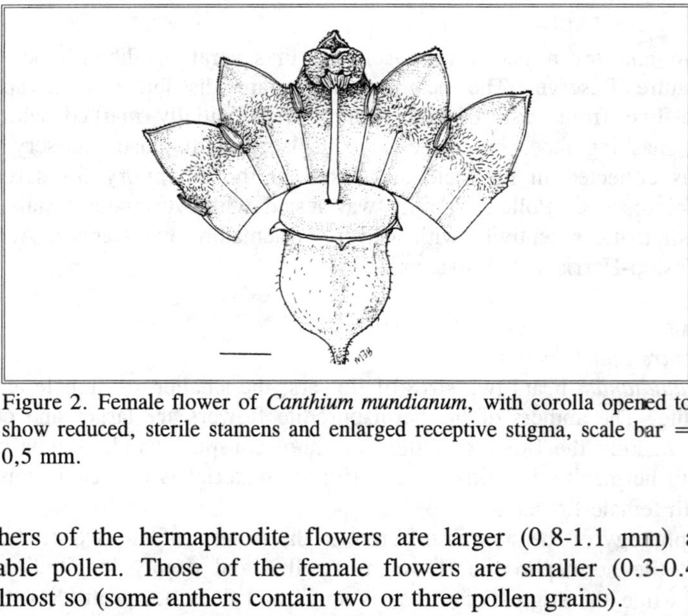

Figure 2. Female flower of *Canthium mundianum*, with corolla opened to show reduced, sterile stamens and enlarged receptive stigma, scale bar = 0,5 mm.

The anthers of the hermaphrodite flowers are larger (0.8-1.1 mm) and contain copious viable pollen. Those of the female flowers are smaller (0.3-0.4 mm) and barren or almost so (some anthers contain two or three pollen grains).

Viscous nectar is produced at the tip of the ovary, even after the corolla has fallen. The flowers are visited by many honey bees, which appear to be the pollinators.

Observations on sex expression

Observations of flowers on ten branches of a plant (approximately 70° apart and from near the top and the bottom of the crown) establish that all flowers simultaneously present on the same plant are of the same sex. The flowers are either hermaphrodite or female, so the species is morphologically gynodioecious, not gynomonoecious. Observations in the course of a month (the full flowering period) indicate that each plant remains to be of the same sex throughout the flowering season. Thus, *Canthium mundianum* does not display the temporal changes in sexual expression that have been observed in *Kirkia wilmsii*. Plants have maintained the same sex over the course of three seasons. *Canthium mundianum* thus is not alternating sexes from year to year and thus is not spreading the high reproductive costs borne by female individuals.

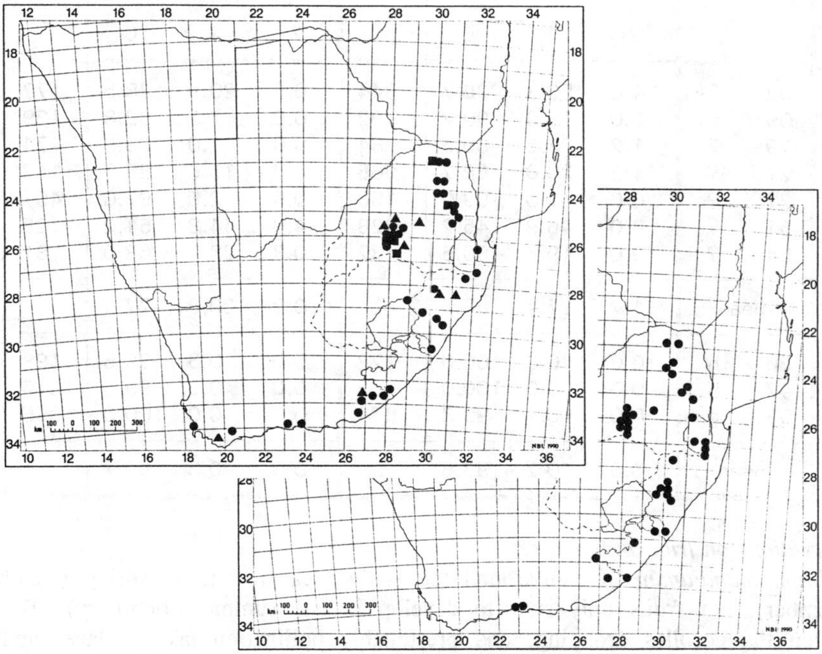

Figure 3. Distribution of *Canthium mundianum* in southern Africa. Left: based on flowering specimens, ▲ with female flowers in a ¼° square, ● with hermaphrodite flowers in a ¼° square, ■ with both female and hermaphrodite flowers in a ¼° square. Right: based on fruiting specimens.

Geographical distribution of sexes

Herbarium material at J and PRE was used to map the distribution of plants with female and hermaphrodite flowers and those with fruits (presumably female). Specimens with hermaphrodite flowers (Figure 3a) and with fruits (Figure 3b) show very similar distribution patterns, but specimens with female flowers (Figure 3a) have only been collected sporadically. These latter specimens stretch over almost the full southern African range of the species, although they are concentrated in the Pretoria-Witwatersrand area, which is the area in which many more specimens have been collected. This apparent pattern might be due to differences in collecting because the hermaphrodite flowers are more noticeable than the female flowers or alternatively, it may point to the sex ratios being different in other parts of the country.

Table 1. Records of percent of fruit and seed set in 1991 and 1993; n represents the total number of flowers that were produced on the inflorescences that were investigated, * indicates that plant 53 could not be relocated in 1993 and thus plant 30, another hermaphrodite was sampled; ♀ = female, h = hermaphrodite.

Plant	Sex	1991				1993			
		number of seeds per fruit			n	number of seeds per fruit			n
		2	1	0		2	1	0	
02	♀	14.0	59.3	26.7	344	0.5	20.9	78.5	177
09	♀	0.0	10.0	90.2	92	0.0	7.2	92.8	139
19	♀	1.2	27.6	69.5	84	0.0	31.0	68.9	74
21	♀	1.2	28.0	68.7	965	0.7	13.6	85.7	147
27	♀	0.0	1.0	99.0	190	0.5	2.5	97.0	407
51	♀	8.0	46.9	45.2	1628	4.0	44.8	51.2	125
54	♀	0.0	19.5	80.5	82	0.3	28.7	68.0	181
mean		1.2	27.4	68.5		0.8	20.9	77.4	
26	h	0.0	0.0	100.0	122	0.0	0.5	99.5	198
47	h	0.0	0.0	100.0	35	0.0	0.5	99.5	216
53	h	0.0	0.7	99.3	874	0.0	0.0	100.0	*356
mean		0.0	0.2	99.8		0.0	0.3	99.7	

Observations on fruit and seed set

Canthium mundianum in and around the study area flowers in spring (October to November), but fruits only start to develop in late summer (February). The time between the corollas dropping, and presumably pollination taking place, and fruit maturing, is two to three months. The fruits are fleshy, taste somewhat like dates and are probably dispersed by birds. Fruit set in female flowers is approximately 25%; that in hermaphrodite flowers is about 0.25% (Table 1), demonstrating that the hermaphrodite flowers are largely functionally male.

Total rainfall in the July 1991 to June 1992 season was 352 mm, while that in the July 1993 to June 1994 season was 808 mm. If the amount of rainfall affects fruit set in the hermaphrodite flowers, it would be expected that this would be observed in these two seasons, during which the rainfall had been so divergent. However, fruit set in these seasons was almost equal in the hermaphrodite flowers (Table 1).

Observations of vegetative reproduction

Plants of *Canthium mundianum* usually occur in dense stands. Individuals near the centre of stands are mainly adult and many smaller individuals occur at the edges of the stands. The area was protected from fire for several years, leading to the build-up of a considerable fuel load. An unplanned fire swept through the area in the winter of 1993, and killed the aerial parts of some of the more isolated individuals that had encroached into the grassland. In spring, many young shoots broke through the ground at the base of one of the damaged adults. Excavation revealed that these were root suckers, suggesting that the extensive stands of morphologically and sexually identical individuals are clones.

Future work

Future work must determine whether pollination is required for seed set or whether apomixis occurs in this species. Self-compatibility must be tested in hermaphrodite flowers, as well as whether hand pollination will affect seed set in these plants. Sex ratios in populations, especially those not in the Pretoria-Witwatersrand area should be investigated and it must be investigated whether or not hermaphrodite plants produce seeds in other areas.

Conclusions

Morphologically *Canthium mundianum* is a gynodioecious species, but functionally it is almost fully dioecious. Plants retain the same sex in successive seasons and rainfall does not affect the success of fruit set in the hermaphrodite flowers. Extensive vegetative reproduction occurs forming extensive clones. Unisexual flowers have only been recorded in *Canthium* subgenus *Bullockia* before; this record of unisexual flowers in subgenus *Canthium* is new and this record of gynodioecy is new for the genus.

Acknowledgements

We are very grateful to the University of the Witwatersrand (Herbarium Programme), the Foundation for Research Development and the Ford Foundation for financial support. We thank the Curator of the Johannesburg Botanical Garden for providing rainfall data and the Curator of the National Herbarium (PRE) for facilitating access to material of *Canthium*. Dr M.-J. Balkwill is thanked for providing the drawings of flowers.

References

Alexander, M.P., 1969. Differential staining of aborted and nonaborted pollen. Stain Technology 44: 117-122.

Bridson, D.M., 1987. Studies in African *Rubiaceae-Vangueriae*: a new circumscription of *Pyrostria* and a new subgenus, *Canthium* subgen. *Bullockia*. Kew Bull. 42: 611-639.

Heslop-Harrison, Y. & K.R. Shivanna, 1977. The receptive surface of the angiosperm stigma. Annals of Botany (London), n.s. 41: 1233-1258.

Immelman, K.L., 1984. Simaroubaceae, Flowering in *Kirkia wilmsii* Engl. Bothalia 15: 151-152

Van Wyk, A.E. & T.K. Lowrey, 1988. Studies on the reproductive biology of *Eugenia* L. (Myrtaceae) in southern Africa. Monogr. Syst. Bot. Missouri Bot. Gard. 24: 279-293.

Difficulties with cross-pollinating five cowpea lines: technique development

H.M. VENTER
Roodeplaat Grassland Institute, Private bag X05, Lynn East, 0039, South Africa.

Abstract
Five lines of cowpea, *Vigna unguiculata* (L.) Walp., selected for evaluation in a forage trial, proved difficult to cross-pollinate. Pot trials were executed in glasshouses at controlled temperatures and relative humidities. A pot trial was also done outside the glasshouses under field conditions. Six different pollen transfer techniques were tested under high relative humidity condition. Results showed that high relative humidity benefits pod production following cross-pollination using technique 1. Technique 6, using a small soft paint brush for pollen transfer at sunrise, proved the most successful. Controlled glasshouse conditions were more beneficial to successful cross-pollination than the variable field conditions.

Key words
Cowpea, Relative humidity, *Vigna unguiculata*.

Introduction

Vigna unguiculata (L.) Walp. is an important hay and fodder crop species in southern Africa (Rachie 1985). Breeding programmes have been established in a number of countries to improve and produce cowpea lines better adapted to specific conditions. Cowpea flowers exhibit self-pollination under most environmental conditions (Fery 1985), but limited outcrossing has been reported under certain temperature and relative humidity (RH) conditions. According to Rachie & Silvestre (1977) and Williams & Chambliss (1980) these conditions need to be identified. Cross-pollination does occur in more humid regions (Ojofmo 1970).

At the Roodeplaat Grassland Institute (25°35'S, 28°21'E) research is being done to evaluate, select and breed cowpea cultivars suitable for different environmental conditions and areas. Five promising lines are being used in this programme. Difficulties have been experienced with obtaining pods after cross-pollinating the lines. The aim of this study was to investigate and determine the best conditions and technique for obtaining successful cross-pollination.

Materials and Methods

The five cowpea lines studied originate from the 20 most promising lines selected from the 175 progeny types derived from a population, reported to have undergone natural outcrossing, at the Plant Protection Institute (Rietondale, Pretoria). This original population consisted of 30 cultivars originating from Botswana, Mozambique and the United States of America. The selected lines are identified by numbers given to them by the breeders and these five lines are: 9(128/4), 15(123/2), 18(122/4), 19(126) and 20(99).

1. Ten plants of each of the five lines were established in two glasshouses, 5 plants per glasshouse. Different RH and temperature conditions were maintained in the two glasshouses, a lower RH glasshouse (GH1: RH = 67.75% ± 9.9) with high day

L.J.G. van der Maesen et al. (eds.), The Biodiversity of African Plants, 656-660.

Table 1. Experimental design. The number of pods formed are given between brackets.

Technique	Low RH	High RH	Field
1	83 (6)	35 (7)	
2		44 (0)	
3		26 (0)	
4		14 (2)	
5		43 (9)	
6		51 (21)	93 (14)

(35° C) and low night (17° C) temperatures and a higher RH glasshouse (GH2: RH = 74.26% ± 7.4) with moderate day (27° C) temperature and moderate to high (22 - 25° C) night temperature. Hand cross-pollinations, using technique 1, were done and the total (results for lines pooled) pod production for each glasshouse was determined. Aborted pistils were collected from all the lines in the different glasshouses and studied using the ABF-technique (Martin 1959) to determine whether pollen germination did occur, the extent of pollen tube growth and if an incompatible reaction could be identified.

2. Six different pollination techniques, discussed below, were carried out in the higher humidity glasshouse (GH2). The total (results for lines pooled) pod production for each technique was compared. (1) Pollen was deposited using dehisced anthers. Emasculation and pollination were done simultaneously between 08:00 and 12:00. The position of the pollinated bud was sketched. (Technique originally used for the humidity/temperature study); (2) Pollen was deposited using a small soft paint brush. Emasculation and pollination were done simultaneously between 08:00 and 12:00. The pollinated bud was marked with white correction fluid; (3) All the flower parts except the pistil were removed and the stigmas pollinated simultaneously by means of the small paint brush between 08:00 and 12:00; (4) The buds were emasculated and pollinated with the small paint brush, between 08:00 and 12:00. The pollinated buds were marked with jewellers tags; (5) The buds were emasculated and pollinated, with the paint brush, between 08:00 and 12:00. All the old flowers and pods were removed. The position of the pollinated buds was sketched; (6) The buds were emasculated in the late afternoon and pollinated, using the small paint brush, early the next morning, at first light between 05:45 and 07:30. The buds were marked with jewellers tags.

3. Using pollination technique 6, the pod production of the plants established in a pot trial under field conditions was compared with that obtained for the plants in the higher humidity glasshouse.

4. The experimental design for this study is presented in tabular form (Table 1). Total number of flowers pollinated per treatment is indicated, no distinction is made between the lines.

Results and discussion

1. Different relative humidity and temperature treatment

A definite improvement in total pod production was achieved in the higher humidity glasshouse-GH2 (Figure 1). Because of problems experienced with temperature regulation in the glasshouses the average temperature obtained in the glasshouses did not differ much, varying from 26 to 28°C. The increase in pod production can thus be attributed to the difference in the humidity of the two glasshouses. When pod production in the higher humidity glasshouse (GH2) was compared with that obtained under field conditions (Figure 2), pod production was significantly better in the glasshouse. The temperature inside and outside the glasshouse did not differ much but the RH inside the glasshouse was much higher than the RH outside. It is of interest to note that, on average, there was only a 7% difference between the RH of the two glasshouses. This indicates that cross-fertilization of cowpeas may be very sensitive to changes in RH.

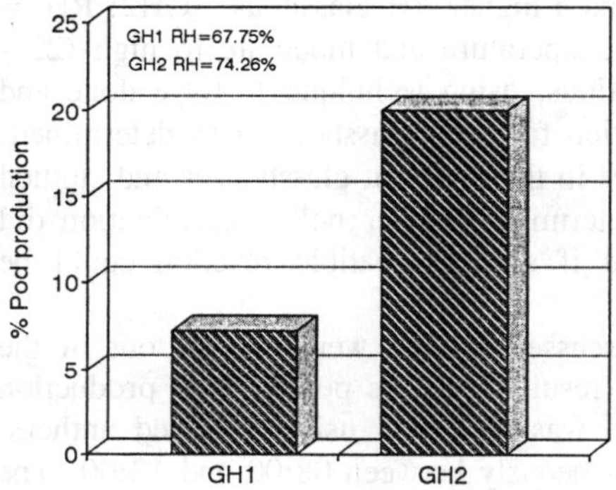

Figure 1. A comparison of pod production in the higher humidity (GH2) and lower humidity (GH1) glasshouses, using technique 1 as pollination technique.

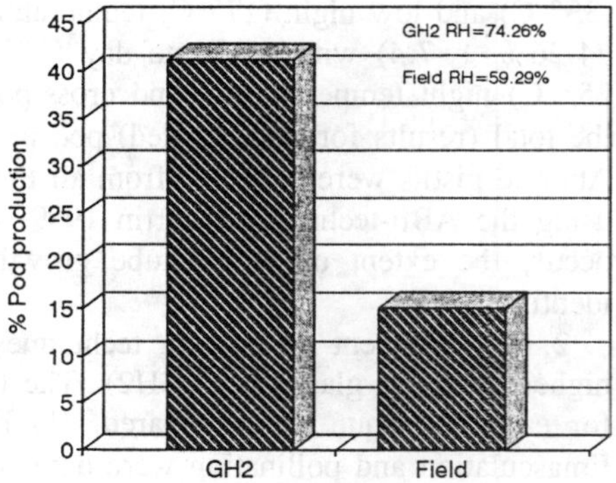

Figure 2. A comparison of pod production in the higher humidity glasshouse (GH2) and in the field, using technique 6 as pollination technique.

Ojofmo (1970) observed that in humid regions under field conditions 12% cross-pollination can occur in cowpeas but generally less than 5% cross-pollination does occur. After hand cross-pollinations an improvement of 13% was achieved under higher humidity conditions during this study. Generally, in breeding programmes, pod production of less than 8% is unacceptable while 48% pod production is preferable (V. Aggarwall, pers. comm.). Only 20% pod production could, however, be achieved at this stage of the study. This may be due to incompatibility between the lines. In the literature incompatibility has been reported within the African *Vigna* gene pool, and it may be determined by the direction in which the cross is made (Sakupwanya et al. 1978). It is also reported that some breeders find a 4% pod production, after cross-pollination, quite acceptable due to the occurrence of incompatibility between wild and domesticated *Vigna* lines.

In the study of aborted pistils, pollen germination had been more frequent on the stigmas of pistils collected from the higher humidity glasshouse-GH2 (Figure 3) indicating that a higher humidity environment is beneficial to pollen germination. However, in all the pistils, collected from both glasshouses, where pollen germination was observed the pollen tubes did not grow through the "saddle"-like receptive area of the stigma. At this stage it is not possible to attribute this observation to an incompatible reaction as none of the features (swollen tips, non-directional growth etc.) generally known to be associated with the incompatible arrest of pollen tube growth on the stigma, could be identified. Furthermore no indication of directional incompatibility could be identified, but the possibility that incompatibility may exist is not completely ruled out.

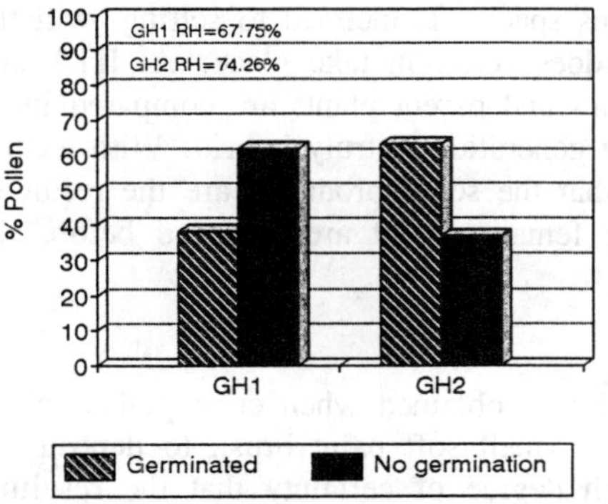

Figure 3. Pollen germination is more frequent on the stigmas of the aborted pistils collected from the high humidity glasshouse (GH2) than on those collected from the low humidity glasshouse (GH1)

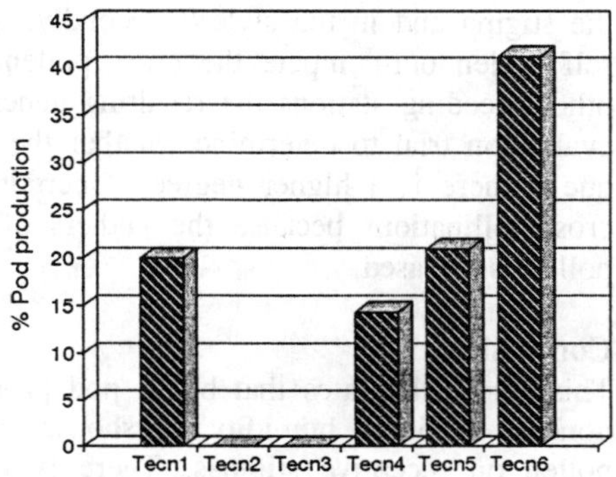

Figure 4. Under conditions of higher relative humidity the pod production of 5 different pollination techniques was compared with that obtained with technique 1. Total pod production was doubled using technique 6.

2. Pollination technique

As incompatibility could not be indicated as the main reason for the low pod production achieved in the higher humidity glasshouse, it was suggested that the pollination technique was inadequate. Of the six different techniques tried only 4 resulted in pod production (Figure 4). Techniques 1, 5 and 6 produced the most pods. There was very little difference in the pod production of techniques 1 and 5 but significantly better pod production was obtained with technique 6. The use of a small soft paint brush to deposit the pollen on the stigma is more accurate with less possibility of damage to the stigma. Identification and emasculation of swollen buds of which the wing and flag petals will open at sunrise the following morning, ensures that the pollen will be deposited, with a high degree of certainty, on receptive stigmas. Emasculation of the protandrous flower ensures that the stigmas will not be contaminated by their own pollen when they become receptive early in the morning. Pollen is usually released in the keel of the flower before the wing and flag petals open.

Technique 6 may have some advantages over the hooking technique generally used by cowpea breeders at the International Institute of Tropical Agriculture (IITA). The hooking technique consists of pushing the stigma of a young bud (younger and smaller than used for technique 6) through the hole in the keel. The protruding stigma is then hooked with a stigma covered in pollen that was collected from the male parent plant. The stigma covered with pollen prevents the hooked stigma from pulling back into the keel and germination of the top layer of male parent pollen can take place (V. Aggarwall, pers. comm.). However, if the hooked stigma is not receptive and does not become so before the other stigma dries and falls off, the hooked stigma will pull back into the keel and when it becomes receptive will be self-pollinated by the pollen released in the keel. Even if the hooked stigma was receptive at the time, when it pulls back into the keel it will also be covered with the pollen released in the keel. Two questions need to be answered. Firstly, does germination of the self pollen take place and what is the degree of competition occurring between the self and cross pollen on the stigma and in the style ? Secondly, as this species is inclined to selfing, will the self pollen out-compete the cross pollen or does certation take place? At IITA and other breeding stations the resulting generations and parent plants are compared in an evaluation trial to determine whether the new generation is truly hybrid. With technique 6 there is a higher degree of certainty that the seeds produced are the result of cross-pollination, because the anthers of the female parent are removed before the pollen is released.

Conclusion

This study illustrates that better pod production is obtained when cross-pollination is done in a higher humidity glasshouse using a small soft paint brush to deposit the pollen on receptive stigmas. There is a high degree of certainty that the resulting generation are hybrids when using this technique. However, pod production of 50% or more could not be achieved in this study. The fact that success was achieved very late in the season may have influenced pod production, but the possibility of a degree of incompatibility cannot be ruled out and requires further investigation.

References

Fery, R.L., 1985. The genetics of cowpeas: A review of the world literature. In: S.R. Singh & K.O. Rachie (eds.), Cowpea research, production and utilization: 25-62. John Wiley & Sons Ltd.

Martin, F.W., 1959. Staining and observing pollen tubes in the style by means of fluorescence. Stain techn. 34: 125-128.

Ojofmo, O.A., 1970. Pollination, fertilization and fruiting characteristics of cowpeas (*Vigna unguiculata*). Ghana J. Sci. 10: 33-37.

Rachie, K.O., 1985. Introduction. In: S.R. Singh & K.O. Rachie (eds.), Cowpea research, production & utilization, John Wiley & Sons Ltd.

Rachie, K.O. & P. Silvestre, 1977. Grain legumes. In: C.L.A. Leakey & J.B. Wills (eds.), Food crops of the lowland tropics: 41-73. Oxford University Press, London, UK.

Sakupwanya, S., R. Mitten & T. Mutangandura-Mhlanga, 1978. Studies on the African *Vigna* gene pool II: Studies with *Vigna unguiculata* var. *tenuis* and var. *stenophylla*. FAO/IBGR Plant Genetic Resources Newsletter 78/79: 5-9.

Williams, C.B. & O.L. Chambliss, 1980. Autocrossing in southern pea. Hort.Sci. 15: 179.

Uses of African plants

27 August 1994

Convenors: C.R. Peters, F.E.M. Cook

Introduction to the symposium Uses of African plants

C. KABUYE
East African Herbarium, National Museums of Kenya, Nairobi, Kenya.

"Uses of African plants" is a title that calls to mind a number of elements. The first thing we think about is food. At one end of the spectrum it includes the developed crops, their wild relatives and foods still gathered from the wild or at some stage of domestication, and at the other end includes plants with indirect uses for food such as honey plants and forage for livestock.

Although Africa is often shown to be a poor contributor to the world's granary, there are examples to show that the African had something to survive on before the arrival of maize. There are the cereals sorghum, millet and tef, but there is also the provision of genetic material to improve crops such as barley, and a wealth of under-utilized wild foods waiting to be developed.

Then comes medicine. Africa abounds in medicinal plants. Besides the fact that many people in Africa depend on traditional herbal medicines, governments are beginning to recognize their importance in health care while outsiders are cashing in to obtain plant material for developing various medicines.

Next to medicines are natural products such as resins, gums, dyes, pesticides, insecticides and cosmetics produced from plants. Besides the *Aloe* extract from which cosmetics are made, the recently established Natural Products Research Network for Eastern and Central Africa (NAPRECA) should reveal a number of plants exploited for natural products.

Shelter is another area where plants are used to a great extent, whether the buildings are temporary or permanent. Wooden poles and worked timber, together with thatching materials are the main examples of this category.

Use of bark for clothing and other materials for ornament is another aspect. Objects of cultural significance come in here. Connected to this are crafts, general utility items and implements including various carvings.

Almost the same kinds of plants cover the demands for fuel in the form of fuelwood and charcoal.

The larger part of the aforementioned uses is based on indigenous or local knowledge of what plants are best for a certain use.

At a time when we want to get maximum benefit out of the resources there is urgent need to collect and document such knowledge. This calls for more concerned efforts in ethnobotanical and botanical studies. I am sure this symposium will show some examples of what is going on in this line. The fact though is that in some fields the old people who have the knowledge are dying out without passing it on, thus the urgency in retrieving the information before it is too late, so that it can be used in development plans.

I should not forget the use of plants as ornamentals either in the landscape or for home gardens. The use made of plants to preserve the environment: human and animal

L.J.G. van der Maesen et al. (eds.), The Biodiversity of African Plants, 663-664.

habitats e.g. water catchment areas should be mentioned here.

Some of the uses, however, may not be sustainable. Uncontrolled harvesting may bring some species to near extinction. Whether this is for home use or for trade, solution need to be found in some conservation measurements. A balance between use and conservation to sustainable levels is required.

Reference is made here to the Convention on Biological Diversity which set in motion some responsible actions. In connection with the Convention: what are our responsibilities? I have already alluded to conservation and sustainable use of plant biodiversity. There is another responsibility we have as botanists and ethnobotanists to the local communities that we work with.

In rushing to get information from the local communities we need to protect the rights of the people to their knowledge and resources through well worked-out modalities including making them aware of their rights.

Some interesting aspects of the uses of African plants will be highlighted by the speakers of this symposium.

African wild plants with rootstocks reported to be eaten raw: the Monocotyledons, part III

C.R. PETERS
Department of Anthropology, Baldwin Hall, University of Georgia,
Athens, Georgia 30602, U.S.A.

Abstract
The Iridaceae includes a number of geophytes with apparently edible rootstocks. This is particularly true of the tropical species of *Lapeirousia*, the corms of four of which are reported to be eaten raw: *Lapeirousia bainesii*, found in semi-arid portions of the southern summer-rain climate region, locally common in sandy soil; *Lapeirousia coerulea*, primarily central and northern Namibia (summer-rain), seasonally damp locations, locally common; *Lapeirousia erythrantha*, primarily the northeastern portion of the southern summer-rain climate region with outliers in northern (summer-rain) Nigeria, sometimes found in large masses; *Lapeirousia schimperi*, primarily central portions of the southern summer-rain climate region with disjunct populations in eastern Africa, including west central Sudan, found in seasonally damp locations, locally common. *Ferraria glutinosa*: warning, corms said to be eaten raw, but also reported to be a hallucinogen and poisonous.

Résumé
Les Iridaceae comprennent un certain nombre de géophytes avec des rhizomes apparemment comestibles. C'est notamment le cas des espèces tropicales de *Lapeirousia*, dont les rhizomes de quatre d'entre elles sont signalés comme étant consommés crus: *Lapeirousia bainesii*, que l'on trouve dans les zones semi-arides de la région climatique méridionale à pluviosité estivale, commun par endroits sur sol sableux; *Lapeirousia coerulea*, principalement dans le centre et le nord de la Namibie (pluviosité estivale), dans des endroits périodiquement gorgés d'eau, commun par endroits; *Lapeirousia erythrantha*, principalement dans la partie nord-est de la région climatique méridionale à pluviosité estivale, avec une aire isolée dans le nord du Nigeria (pluviosité estivale), parfois trouvé en grands peuplements; *Lapeirousia schimperi*, principalement dans la partie centrale de la région climatique méridionale à pluviosité estivale, avec des populations isolées en Afrique orientale, y compris le centre-ouest du Soudan, dans des endroits périodiquement gorgés d'eau, commun par endroits. Attention au *Ferraria glutinosa*; ses rhizomes seraient consommés crus, mais on signale aussi qu'ils seraient hallucinogènes et vénéneux.

Key words
Africa, Edible wild rootstocks, Iridaceae, *Lapeirousia*.

Introduction

This report completes a major portion of the Iridaceae, begun at the 1988 congress of AETFAT in Hamburg (Peters 1990). The project emphasis is on constructing brief dossiers evaluating particular species, and on revealing the continent-wide ecological and chorological patterns among taxa with edible rootstocks that can be eaten raw. The focus is on rootstocks "eaten raw" because of: 1) their utility without the elaborate processing (including cooking) required for the more common poisonous species; 2) their potential significance for children; 3) their potential significance for the diet of ancient peoples, whose environmental resources prehistorians attempt to reconstruct as

L.J.G. van der Maesen et al. (eds.), The Biodiversity of African Plants, 665-677.

an aid in interpreting early patterns of land use; and 4) their possible role in the evolution (and perhaps future) of the human diet. The plants no longer play a prominent role in the rural adult diet, except in times of famine. Children are the commonly acknowledged main consumers of wild plant foods in recent history, but unfortunately they are not among the main informants usually consulted in ethnobotanical studies. Subsequently we must piece together the information we need from as many sources as possible. Initial identification of edible species was made through the published materials collated in Peters et al. (1992). Supplemental citations were appended as I became aware of them, especially through herbarium work at the Royal Botanic Gardens, Kew. Some personal observations were made in the field.

The phytogeographical regions referred to herein follow White (1983). The seasonality-of-precipitation climate regions follow the map in Peters (1990).

Results

IRIDACEAE

Lapeirousia Pourret

The genus *Lapeirousia* has presented a number of classificatory problems, overcome for the most part by Goldblatt's (1972, 1990) recent systematic treatments, invaluable for my studies. Many herbarium specimens, with his redeterminations, were indispensable. The following introductory notes on this genus are taken from his works unless otherwise noted.

The species of *Lapeirousia* are lily- or iris-like herbaceous geophytes, with star-shaped flowers; deciduous in the dry season. It seems that 25 species of the approximately 38 species in the genus are restricted to the southern African winter-rainfall region, with Namaqualand (especially the central west coast of South Africa) as their area of maximum species richness. This is unlike typical Cape Flora elements, whose speciation is greatest further south, in the southwest Cape proper. It is also common for members of the Cape Flora with species extending into the summer-rain climate region to follow the well-watered mountains of the eastern escarpment. For *Lapeirousia*, in contrast, the tropical species have a wide distribution across the semi-arid and subhumid drier parts of summer-rain south central Africa, with maximum species richness in the northern half of Namibia, and two Zambezian species with outliers in the northern African summer-rain climate region (Ethiopia and Nigeria). It is the tropical species that are of special interest here, and Goldblatt (1990: 438) wonders whether the unusual chromosome variability that he has discovered in these species might not be adaptively related to their "having radiated in semiarid habitats more severe than those that most genera of Iridaceae favour". The limitations to their geographic distributions and the unusual disjunctions of some of these (see below) are a particular puzzle to me. The means of dispersal for these plants are apparently unknown. The seeds are small (diameter ranges from 1.3-2.5 mm) but not particularly well-described (especially their number, much less their viability) in large part because the capsule is so seldom collected. I have the impression that few seeds are produced, but in good years the plants may flower more-or-less continuously while the ground is moist. A number of notes indicate that *Lapeirousia* species leaf and flower not long after the first soaking rains. For *L. avasmontana* Dinter (a Namibian endemic) Goldblatt (1990: 448) has also observed that "New branches continue to be produced from the cauline leaf axils as long as the ground remains moist, and a late rainfall can stimulate production of a second flush of flowering from new branches on the same flowering stem". Vegetative reproduction is also common in *Lapeirousia*, by means of tiny cormlets that are produced in lower leaf axils (Goldblatt, in litt.). Before learning of this I had expressed my doubts to Goldblatt about the possibility that *Lapeirousia* could have sustained substantial exploitation by humans including the corm in their diet, but he assures me this is plausible and notes that predation of the corms by baboons, porcupines and mole rats occurs in the wild on a huge scale.

The rootstock in *Lapeirousia* is a corm. This is the only part that is eaten. It is a specialized flat-based, bell-shaped corm, with more-or-less woody to coriaceous or fibrous tunics. This corm is the primary synapomorphy for the genus (variation between species is largely floral). A new layer of the tunic is produced each year, and although the corm itself is only about 1 cm in diameter it can appear much larger if the tunics of previous seasons persist. The corms are produced entirely by a bud near the base of the flowering stem and attached laterally to the flowering axis. As is usual in dry area geophytes, the new corm is nearly full size before flowering begins (Goldblatt, in litt.).

The corms of a number of the tropical species are reported to be eaten. Goldblatt cites several species (some of which reference to was provided by the present author in earlier correspondence), and they are included in the following. As one might expect, most of the reports are from the Namibian heartland of tropical *Lapeirousia*. Goldblatt especially notes that the early report by Dinter (1912) mentions *L. coerulea* Shinz, *L. odoratissima* Baker (as *L. juttae* Dinter), *L. littoralis* Baker (as *L. caudata* Schinz), and *L. schimperi* (Aschers. & Klatt) Milne-Redhead (as *L. uliginosa* Dinter). For others, Seydel 3419 (K) states *L. gracilis* Vaupel is edible, and Rodin (1985) adds *L. bainesii* Baker to this list. Not mentioned by Goldblatt, the de Winter and Leistner 5481 (K) voucher states "corms edible" for a specimen of *L. otaviensis* R. Foster. Outside of the Namibian heartland a few important references can be added. First, Simpathu 60 (K) reports that the corms of *L. erythrantha* (Klatt) Baker are edible. Second, Wehmeyer (1986) includes *L. sandersonii* Baker in his list of edible plants, and provides some information on the nutrient content of its corms. His analysis indicates that the corms are about one-half water and one-half carbohydrates with significant values for vitamin C and some minerals. Also previously unmentioned, Borle 43 (K) reports "bulb edible, fairly common" (Matopos, Zimbabwe), for a collection identified as *L. setifolia* Harms by Goldblatt from vouchers deposited at PRE and SRGH. Fourth, R.B. Drummond (pers. comm.) has recorded the corm of *L. masukuensis* Vaupel & Schlechter is eaten in Mozambique. With regard to generalizing from Wehmeyer's nutrient values, other species notes (see species dossiers below) suggest the carbohydrate portion is primarily starch, and Marshall (1976) provides a note on *L. odoratissima* that states the corm is juicy and eaten for its water content. As a final general point, with one very doubtful exception, *Lapeirousia* has not been reported to be poisonous or harmful when eaten (e.g. Watt & Breyer-Brandwijk 1962). Moreover, four species have been reported to be eaten raw. Given this apparently credible constellation of information, dossiers have been prepared on those four species as part of our project characterizing monocotyledonous species whose rootstocks can be eaten raw.

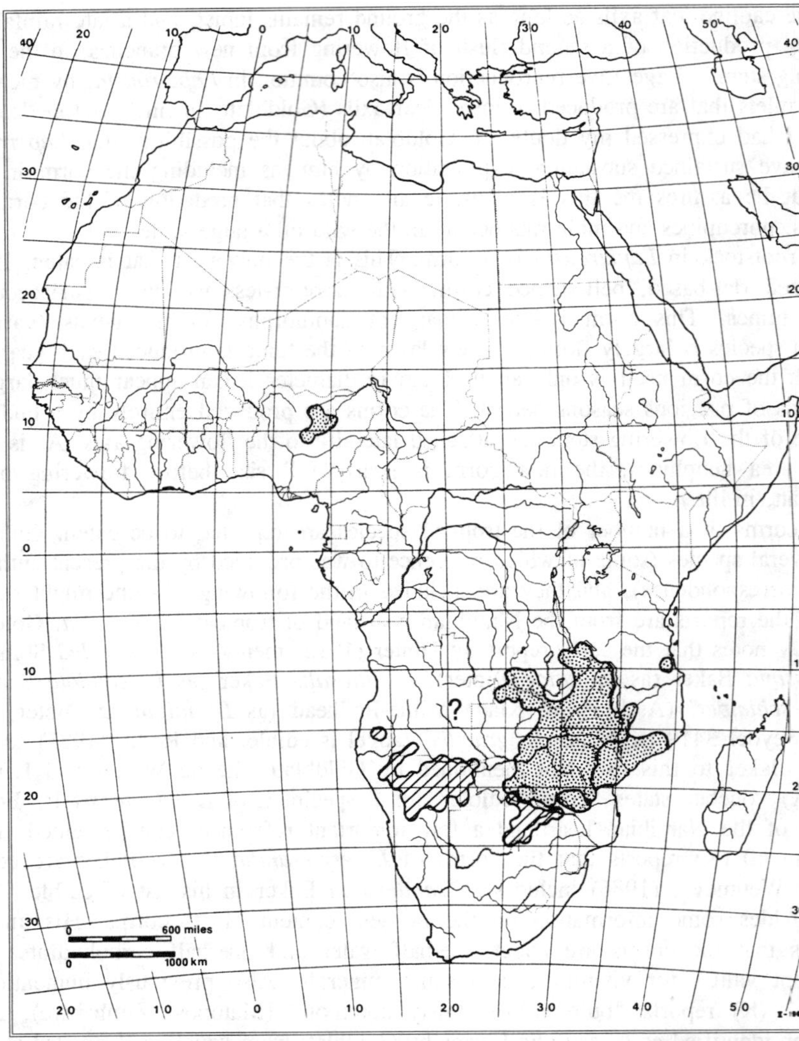

Figure 1. Geographical distribution of *Lapeirousia bainesii* (diagonal hatching) and *Lapeirousia erythrantha* (stippling and question mark). After Goldblatt, 1990.

Lapeirousia bainesii Baker
Medium to tall herbaceous geophytes, 30-60 (-90) cm in height (Goldblatt 1990, Maguire 1978); found in the northwestern Transvaal, eastern Botswana, and western Botswana through central Namibia northward into southern Angola (see Figure 1, based for the most part on Goldblatt 1990). The apparent geographic disjunction in the distribution of *L. bainesii* seen in Figure 1 is probably the result of the very limited collecting done in Botswana. Silberbauer (1965, 1972), however, records it as present there in the Central Kalahari Game Reserve, but provides the further note (1981) that

Lapeirousia is not common in the game reserve. One other geographic note deserves correction Peters et al. (1992) list a record of edibility for *L. bainesii* from the East African Herbarium Card Catalogue on Plant Uses. This is a mistake in data entry. The original entry lists only the genus name *Lapeirousia* with the note that the "tubers" are edible. Judging by Goldblatt's (1990) work this unnamed east African species is likely to be *L. schimperi* (see below). The white to pale pink flowers of *L. bainesii* are described as both sweetly (Pole Evans 4305, K) and strongly (Lugard 179, K) scented, open in the mid to late afternoon (lasting through the night into the following day), produce abundant nectar, and are thought to be pollinated by hawkmoths because of their pale colour, fragrance, long (25-35 mm) perianth tubes and night-flowering (Goldblatt 1990).

Corms bell-shaped with a flat base, c. 13-20 mm in diameter at the base (Goldblatt 1990), eaten raw or roasted after the firm fibrous tunic has been removed. Maguire (1978) and Silberbauer (1965, 1981) report that the corms are roasted before they are eaten. Maguire (1978: 356) further notes that "the roasted corms are sometimes pounded into a meal, which is then mixed with water (boiled or not) to make a porridge". Rodin (1985) provides another report that the corms are roasted (under the syn. *L. vaupeliana* Dinter), but he adds the fact that they are also eaten raw. Silberbauer (1965) lists the food value of *L. bainesii* as starch and the seasonal period of availability as January into June. He also notes that in the dry summer-rain climate of the Central Kalahari Game Reserve most of the rain usually falls in January and February or March. Silberbauer (1972) lists the season of availability as April into July, while Silberbauer (1981) lists it as March to July. For the Kalahari area some 300-400 km to the NW, Maguire (1978: 356) reports that "The corms are obtainable for about 5 months of the year—from January or February onwards". He also notes that the corms are locally "regarded as one of the more important foods", although he saw only a few plants, at one location. He adds, the dried corms can be stored, but if they are "greenish" or become damp during storage, they rot. With regard to their period of availability ending in the early winter dry season, he reports that the above-ground parts of the plant dry up in the winter and are easily blown away by the wind, after which the corms are untraceable, or found only with difficulty. Another point commented on by various collectors is that the corms are deep in the ground (e.g. Immelman 526, K; Rodin 9295, K), and this may explain why only one of the herbarium specimens that I examined had its corm. Maguire (1978) reports that the corm is usually 10-25 cm below the ground surface (the broken stems of the herbarium specimens suggest longer than 12 cm) and Goldblatt (1990) states that the corms may be as deep as 30 cm below the ground surface.

Lapeirousia bainesii inhabits the west-central interior (as far as c. 30° longitude east) of Africa's southern summer-rainfall climate region. The flowering season of January to April (rarely late November or December) (Goldblatt 1990) corresponds approximately to the middle to late rainy season. This species of the interior is found at moderate elevations, typically encountered on sandy soil in grassland to woodland vegetation (various vouchers). Goldblatt (1990) concludes that it favours hard flat sandy ground, and in the Kalahari, Maguire (1978) notes that it is apparently a plant of the flats and poorly defined shallow valleys that alternate with the broad, low E-W-tending dunes that cover much of the area.

A few vouchers note that *L. bainesii* can be locally common. North-central Namibia: for a Quickborn area collection, Bradfield 196 (K) notes it is common in sandy soil; on the farm Mex in the Gobabis District, Immelman 526 (K) reports that it is common in deep red sand in open bushveld. Central-eastern Botswana: for a Mahalapye-area collection, Yalala 356 (K) notes it presence as "locally high" on shallow gritty sandy loam. Northwest Transvaal: near the Dongola Reserve, Codd and Dyer 3780 (K) report it as locally common in mopane bushland on open red sandy flats at c. 600 m elevation; for "Dongola", Pole Evans 4305 (K) notes that it occurs over "large stretches". The plant's natural means of dispersal are apparently unknown.

Lapeirousia coerulea Schinz

Small to medium herbaceous geophytes (12-) 15-30 (-40) cm in height (Goldblatt 1990, Rodin 1985), found in central and northern Namibia just into northwestern-most Botswana (see Figure 2, based on Goldblatt 1990). Flowers blue to light purple with very short perianth tubes and scentless; the overall simple flower structure highlighted by Goldblatt's (1990) cladistic analysis leads him to conclude that *L. coerulea* appears to be close to the basal stock of the tropical (i.e. summer-rain) species. No notes on anthesis or pollination are available to me, but for the similar flowers of *L. avasmontana* Dinter, Goldblatt (ibid.) notes that no measurable nectar is produced and suggests (following S. Vogel) that both species depend on short-tongued pollen-foraging bees for pollen transfer.

Corms bell-shaped with a flat base, c. 12-16 mm in diameter at the base (Goldblatt 1990), eaten raw or roasted after the fibrous tunic has been removed. With one exception (Lee 1979) all the reports on edibility come from Namibia. Goldblatt (ibid.) cites the earliest of these as Dinter (1912), noting *L. coerulea* corms were a valued food eaten after roasting. Unmentioned, the 1935 notes by Schoenfelder 35 (K) confirm this and provide additional information, including that they were a staple food and that they "can be kept" (also see notes above, for *L. bainesii*). Only Rodin (1985) has reported that they are also eaten raw. Schoenfelder 35 (K) notes they are available in the later rainy season. Skarpe S-213 (K) reports a corm depth of 5-10 cm under the soil surface. My judgements for corm-bearing voucher specimens at the Kew is that corm depth is usually 3.5-10.0 cm below the ground surface, at most (cf. Schinz 13; Dinter 6957; Giess 11248A & 11292).

Lapeirousia coerulea inhabits the southwestern portion of Africa's southern summer-rain climatic region. The flowering season of (January) February to April reported in Goldblatt (1990) corresponds approximately to the main part of the rainy season: e.g. Rodin (1985) notes for central northernmost Namibia an annual average rainfall of 520 mm, scattered showers usually starting in October or November, with heavier rains falling in December through March. Goldblatt (1990: 445) characterizes *L. coerulea* as "Typically a species of damp, low-lying places....where it occurs in seeps, seasonal vleis, damp poorly drained grassland, and shallow soil in rock outcrops where water accumulates in the wet season". I would emphasize the pans, vleis, and shallow depressions in omuramba (dry river and stream courses), plus variable soils (including clay, loam, sand, shallow gravel), and periodic flooding (various vouchers).

A few of the Namibian vouchers record *L. coerulea* as locally common. For a gathering on the Neudam Exp. Farm (Windhoek), van Vuuren 1027 (K) notes it as

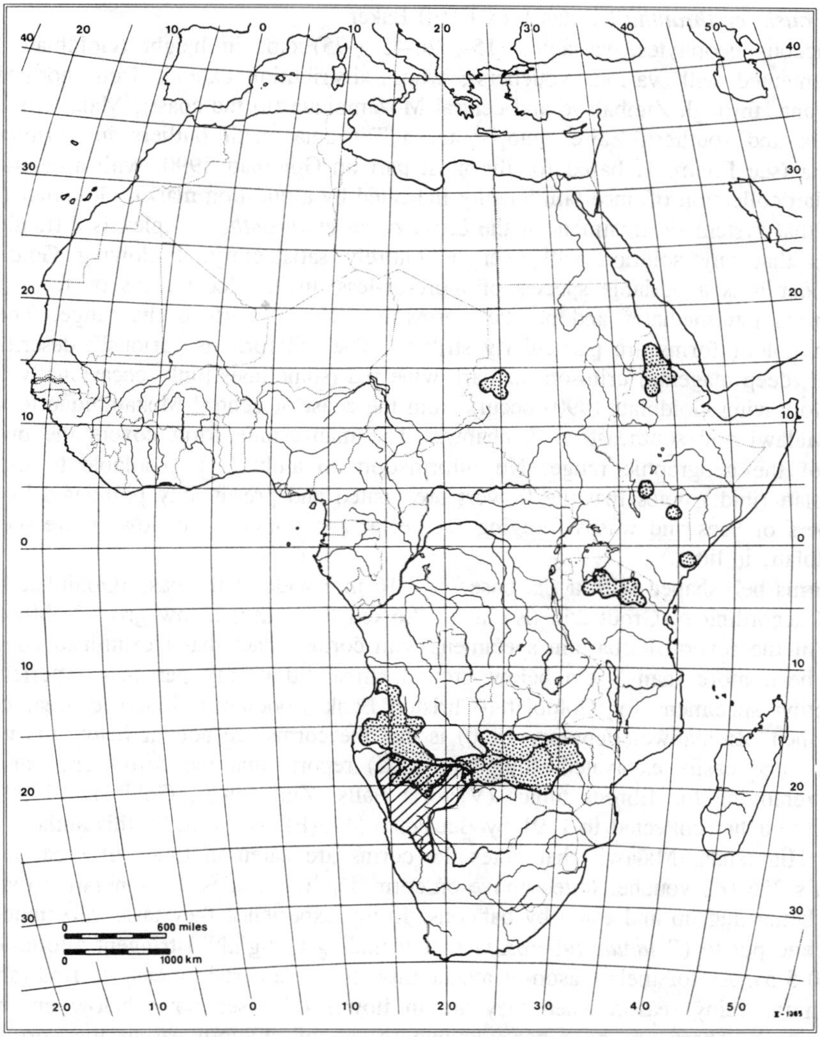

Figure 2. Geographical distribution of *Lapeirousia coerulea* (diagonal hatching) and *Lapeirousia schimperi* (stippling). After Goldblatt, 1990.

common. Wanntorp & Wanntorp 724 (K) report it as locally abundant on moist grassy ground by the road 10 km from Steinhausen (Gobabis District). Schoenfelder 35 (K) notes it is common on red loamy flats in Grootfontein District. Story 6121 (PRE) reports it as locally common at Tsumkwe in northeast Namibia. Goldblatt (1990) comments it is apparently rare in its distributional extension into nearby northwestern-most Botswana, which is consistent with Lee's (1979) notes wherein it has no special status in the diet of the hunting and gathering people of that area. The plant's natural means of dispersal and the reasons for its limited distribution are apparently unknown.

Lapeirousia erythrantha (Klotzsch ex Klatt) Baker
Herbaceous geophytes, variable, (15-) 20-45 (-75) cm, in height (Goldblatt 1990, supplemented with various vouchers); whose distribution extends from northeastern Botswana, through Zimbabwe and central Mozambique (to the coast), Malawi, parts of Zambia and southern Zaire, into southern Tanzania, with outliers in north-central Nigeria (see Figure 1, based for the most part on Goldblatt 1990, with a central-east Angolan collection of uncertain locality indicated by a question mark). Goldblatt (ibid.) notes that systematic treatment of the *Lapeirousia erythrantha* complex is difficult, and doubts that any solution will ever be entirely satisfactory. Following Goldblatt's treatment it is a variable species of more-or-less distinct local races or forms, with numerous intermediates and parallel forms in different parts of its range. The two flower colour forms are particularly striking. The red form is variously described as scarlet, deep magenta, crimson, or dark wine red (sometimes pink, occasionally white) and (following Goldblatt 1990) occurs from the coast of central Mozambique to southern Malawi and southeastern Zimbabwe. The mauve-blue form covers the broadest part of the geographic range. No information on anthesis is available to me, but Goldblatt (ibid.) notes that the flowers are visited and presumably pollinated by large numbers of bees and wasps foraging for nectar and pollen. The flowers are scentless (Goldblatt, in litt.).

Corms bell-shaped with a flat base, c. 8-16 mm wide at the base (Goldblatt 1990), found according to Grout 256 (K) about 7.6 cm (3 inches) below ground. Moreover, none of the several herbarium specimens with corms intact that I examined appear to have been more than 3 cm below ground. In addition, my personal experience in collecting specimens for Goldblatt (Christon Bank woodlands, Mazowe area, central watershed, Zimbabwe, February 1990) is that the corms are not far below the ground surface and easily excavated. Simpathu 60(K) reports that the corms are eaten raw after removing the fibrous tunics (Victoria Falls, Zimbabwe). Goldblatt (1990) also cites a voucher collected in 1891 by Buchanan [426(E), not seen by this author] in the Shire Highlands, Malawi, that notes the corms are eaten in times of great famine. Grout's 256 (K) voucher (cited above on corm depth) includes the annotation that they were being dug up and eaten by baboons. In my experience they taste very much like the white potato (*Solanum tuberosum* L.), including a slightly astringent aftertaste (pH = 5.0-5.5). As for their season of availability, they are certainly easy to find (and dig up) in the rainy season when they are in flower (also see notes below on relative abundance). There are no reports known to me of children eating the corms, but Seddon 3 (K) provides the following interesting notes suggesting that in some places they know these corms very well: "when the rootstock is dry the children pierce it, and threat through it a piece of string; then they hold the string taut and swing the root round and round as quickly as they can and in so doing they produce vibrations giving several musical notes; this process is one of the regular games after the rains have finished" (Mozambique, east shore of Lake Niassa; see related notes by Seddon on relative abundance, below).

Lapeirousia erythrantha inhabits a large block of the southern African summer-rain climate region where it is found at low to moderate elevations: near sea level on the east African coast but more broadly from c. 300-1600 m (highest reported elevation 2100 m, southwestern Tanzania, Richards 8781 (K)). This east central Zambezian area corresponds to what we could call the species" geographic heartland. The outliers on the north-central plateau of Nigeria are in the Sudanian belt of the northern African summer-rain climate region. In the southern summer-rain region the flowering season for *L. erythrantha* is usually December into March, with occasionally flowering recorded as early as September to November or as late as April or May (various vouchers), probably depending upon variation in when the summer rains begin and end. The months of heavy rain in the northern summer-rain area inhabited by the Nigerian outliers are usually May through September (Thornthwaite & Mather 1962), but the small sample of vouchers only confirms flowering in May, June and July. Goldblatt (1990) comments that *L. erythrantha* is the most common species of *Lapeirousia* in south central Africa. He concludes (p. 452) that it

> ... can frequently be found in rocky outcrops throughout its range, or less often in open grassland or woodland. Its frequency in thin soil in rocky sites suggests that the corms are protected in such habitats, while in deeper ground they are accessible to rodents, porcupines, and perhaps other animals. There seems no *a priori* reason for the apparent preference for rocky habitats, and plants grow well in cultivation in almost any type and depth of soil.

Judging by the annotations accompanying the herbarium specimens, I would emphasize that it occurs in a variety of non-arid habitats, with grassland to open woodland vegetation associations. Moreover, the sites where it is reported to occur in relative abundance would appear to be the more telling.

From eastern Botswana (Northern District, between Francistown and the Nata River on the road to Maun), West 2898 (K) reports it is locally abundant in mopane, a "pretty blue monocot making a great show". From western Zimbabwe (Nyamandhlovu District Pasture Station) Plowes 1540 (K) reports it as fairly common on basaltic red sandy loams. To the east in central Zimbabwe (Lomagundi District, Audley End Farm, Darwendale) Gordon s.n. (K, SRGH) reports its occurrence as masses in a vlei (January 1979), making a blue sea of colour after *Moraea carsonii* had done the same in December. From easternmost Zimbabwe in the upland grasslands of the Chimanimani Mountains at c. 1675 m elevation, it is reported to be common on damp patches of whitish sand (of quartzite origin): patches of c. 20 m across may be covered with these flowers (notes pieced together from Kew and SRGH specimens Goodier 183, Grosvenor 369 and Plowes 2855). For the red-flowered form seen in south-eastern Zimbabwe (Chibi District, Kopje near Madzivire Dip) Moll 484 (K) reports it is locally common in grassland. Also from the Masvingo area [road verge 5 miles from Fort Victoria (=Masvingo) on the main Sabi road] Goldsmith 10/69 (K) reports a beautiful bulbous plant on wet pan and sandy soil in dense clusters, flowers numerous, deep wine red. From the same collection stop Goldsmith 11/69 (K) reports large patches of the blue-flowered form on wet sandy granite soil. From Mozambique (Tete District, Angónia) Macuácua 1363 (K) reports it is very abundant at one locality on the outskirts of Ulongue. Also from Mozambique (Niassa Province, near Metangula, east shore of Lake Niassa, Msumba) Seddon 3 (K) provides the following notes: "common after the first heavy rains, growing on slopes or meadow land where the soil is moist, the flowers make a mass of colour and the hillside on which my house is built looked very much like a field of bluebells, covered as it was with these graceful flowers".

From southern Malawi (Matope, Blantyre) Jackson 1777 (K) reports that it is often gregarious in *Brachystegia* scrub woodland. From Zambia, Astle 5794 (K) reports that it is common in siltstone areas in the Luangwa Valley near the Kapanba River, and Mutimushi 668 (K) reports it is a common herb of hill woodland at Kitwe. From Zaire, Hirschberg 60 (K) reports that it is common in a "vacant stand" in Lubumbashi (as Elisabethville), and Quarré 3103 (K) reports that in the environs of Elisabethville it is (very) abundant in damp soil in the valley of the Lubumbashi. From Tanzania, Semsei 701 (K) reports it to be "very common, more common in native cultivation", at Lindi. From Nigeria (Bauchi Plateau) Lely P259 (K) reports that it occurs "in shallow pans of soil on flat rocks, in large masses". In addition to these herbarium notes, my own very limited collecting experience in Zimbabwe (referred to above in the notes on corms) was that this species was more frequent in footslope and upper valley floor locations than on the hill slope or rocky hilltops. That it occurred in discrete patches up to c.10 m in diameter, consisting of a dozen or so individual plants, the patches being a few hundred meters apart, suggests autochthonous seeding from previous flowering seasons, possibly originating from a single individual. The plant's means of dispersal and the reasons for its remarkable geographic disjunction are apparently unknown.

Lapeirousia schimperi (Aschers. & Klatt) Milne-Redhead
Medium to tall herbaceous geophytes, (14-)30-80(-122) cm in height (Goldblatt 1990, supplemented with various vouchers); found from southwestern Angola and north-central Namibia across northern Botswana, southern Zambia and most of Zimbabwe, with disjunct populations in northern Tanzania, eastern Kenya, southern Ethiopia, northernmost Ethiopia, and west-central Sudan (see Figure 2, based for the most part on Goldblatt 1990; also see Wickens 1976). The usually white flowers are described as scentless by Faulkner A390 (K) (Angola). This agrees with Goldblatt's experience in Namibia, but he notes some collections report the flowers have a sweet fragrance. He also notes the flowers open in late afternoon, lasting through the night but gradually wilting the next day, and suggests that they may be pollinated by hawkmoths (because of their pale colour, very long perianth tubes and night-flowering).

Corms bell-shaped with a flat base, c. 18-22 mm in diameter at the base (Goldblatt 1990). Rodin (1985: syn. *L. cyanescens* Baker) reports that they are eaten raw or roasted, presumably after the fibrous tunic has been removed. His notes are for Ovamboland in central northernmost Namibia. Approximately 500 km to the southeast, but still in northern Namibia, Marshall (1976: syn. *L. cyanescens*) reports that for the local hunting and gathering people these corms are a plant food that they especially enjoy. She also notes the children play a jump rope game singing a song called //Haru, which is the local name of this plant food. Other notes about the jump rope by Marshall (ibid.) indicate this play occurs in the season of heavy rains. Rodin's (1985) notes state that the corm is "deep". My examination of herbarium specimens indicates those successfully collected were c. 3-8 cm below ground (various vouchers). Cross 340 (K) reports the corm is c. 7.5 cm below ground surface in tall grass *Acacia* woodland near Bulawayo, Zimbabwe.

The unusual discontinuous distribution of *L. schimperi* across eastern Africa may not be simply an artifact of collection. Goldblatt (1990: 474) notes that it is "almost certainly absent from Zaire, Uganda, central and western Kenya and southern Tanzania". He also reports that it "exhibits relatively little variation for a species with such a wide distribution". Except for the eastern Kenyan localities it is essentially an inhabitant of the southern and northern summer-rain climate regions, occurring at elevations from near sea level in Kenya and Angola to c. 1500-1700(-2230) m in the mountains of north central Tanzania and especially west-central to southeastern Angola (various vouchers, including citations in Goldblatt 1990). The highest reported elevation is 2375 m in the montane grasslands of the Jebel Marra Massif in west Sudan (Wickens 1976). Generally *L. schimperi* flowers in the mid to late rainy season (various vouchers; also Goldblatt 1990), probably after the main rains. In addition, Wickens (1976: 31) reports a definite time-lag in the floral development of the herbaceous vegetation above 1800-2000 m elevation. He estimated that an increase in 300 m elevation produces a delay of approximately three weeks in floral maturity. Goldblatt (1990) concluded that *L. schimperi* grows in seasonally moist locations in otherwise largely arid country. These locations include vleis, river banks, the margins of seasonal marshes, the edges of pans, and cracks in granite domes, with variable soil types including peat and heavy clays to sand, some derived from lava and basalt (various vouchers).

A few vouchers note that this species can be locally common. From west-central Angola (Cuima area, Eleude Mission, c. 1460 m elevation) Faulkner A390 (K) reports that it is common around Eleude, growing in dry ground in the bush. Since he also notes that it can be up to 122 cm high (4 feet) one might suspect that this mountainous location receives quite a bit of moisture. From western Zimbabwe (Wankie, Shapi Camp, c. 1030 elevation) Rushworth 269 (K) reports that it relatively frequent in a sparse short-grassed mopane area on moist Kalahari Sand after recent rains. To the east from northern Zimbabwe (N of Sipolilo) Whellan 313 (K) reports it is locally frequent over a small area, and from central Zimbabwe, Davey 1 (K) reports a "very beautiful delicate starlike flower making quite a show for a short period" at one place on marshy ground in mopane on either side of the main Que Que to Gwelo road. From north-central Tanzania south of Lake Victoria (Shinyamga, Huruhuru-Mantini road, c. 1160 m elevation) Burtt 3511 (K) reports the white flowers occurring "in vast numbers among young grass" in the margin of a seasonal swamp. From Kenya (Kenya-Somalia border, consolidated sandhills) Gillespie 250 (K) reports it is frequent in open grass with wiry *Cyperus* on grey-white sand; but the plants are only 15-38 cm in height (Second Oxford Expedition to the Kiunga Archipelago). For the nearby Kenyan coast at Kiunga we have the 1910 report by Battiscombe 231 (K) that it is common in grassland. For the same area we have the 1961 report by Gillespie 176 (K) that in a grassy shady clearing on brown moist sandy soil it is infrequent to frequent on moister soil (Second Oxford Expedition to the Kiunga Archipelago). There is also one specimen annotation cited by Goldblatt (1990: 475) that contains a piece of information indicating relative abundance. This is a voucher by Giess (not seen by this author) from north-central Namibia (Grootfontein District) that states for the farm Kumkauas, "large colonies among tough grasses". One might conclude from these notes that in association with the main rains of the summer period there are localities probably well known to the indigenous people, that perhaps were once more numerous across south-central and eastern Africa, where this food plant is abundant enough to merit a special song in the light-hearted games children play in that season. It is unfortunate that its natural means of dispersal and the reasons for its markedly discontinuous distribution are apparently unknown.

Ferraria glutinosa (Bak.) Rendle: A Note of Warning.
These medium to tall geophytes are the only widespread and summer-flowering species in a small Karoo and Cape flora genus, most of whose species are found primarily in the southern African winter-rainfall region along the southwest coast (de Vos 1979). *F. glutinosa* in contrast is found from the Kalahari of the Cape Province northward through Botswana, via central Namibia up into west-central Angola, and through western/central Zimbabwe and central Zambia up to southeastern Zaire as far as just west of central Lake Tanganyika (ibid.). Usually not common (more often noted to be occasional or rare); apparently growing in seasonally damp locations, e.g., between rocks, dambo margins, open woodlands (various vouchers). The corms are 1-5 cm in diameter occurring in superposed clusters representing a number of seasons, and found not far below the ground surface.

Rodin (1985) reports the corms of *F. glutinosa* are eaten raw or roasted, especially by boys herding cattle, but I have come to question whether these corms are edible raw for the following reasons. Firstly, although the corms may be eaten after they are cooked [following Dinter 1912 (cited by Maguire; not seen by this author) and Rodin 1985], as Maguire (1978) points out it is notable that none of the ethnographic studies of Kalahari hunting and gathering peoples have reported *F. glutinosa* to be eaten. Only one, that I know of, includes any notes on *Ferraria*. This is Lee (1979), who reports *F. glutinosa* to be an hallucinogen. Finally, under the synonym *F. welwitschii* Bak., Watt & Breyer-Brandwijk (1962) cite a veterinary experiment that indicates the corm can be poisonous.

Discussion
The Iridaceae seems to be one of the more important families of wild monocots that provide edible rootstocks. There are, however, very few reports of species that can be eaten raw, and a number are known to be poisonous. Species of *Babiana* (Peters 1990) and *Lapeirousia* are notable exceptions. *Lapeirousia* corms are available after the first heavy rains, in the mid to late summer months, and could be a dietary resource (water, starch, some minerals, vitamin C) wherever they occur in relative abundance. Their habitat preference is for seasonally wet locations. Their potential dietary significance appears similar to *Babiana hypogea* Burchell.

An overview of reports, presented on a limited number of species (Peters 1990, 1994) suggests that three main chorological patterns characterize the African monocotyledons providing rootstocks that can be eaten raw. 1) Southern African endemics, e.g.,: Iridaceae, *Babiana hypogea* Burchell, *Moraea fugax* (Delaroche) Jacq. subsp. *fugax*; Orchidaceae, *Neobolusia tysonii* (Bolus) Schltr.; Tecophilaeaceae, *Walleria nutans* J. Kirk. 2) Tropical (east) African taxa with an apparent southern African heartland, e.g.: Orchidaceae, *Eulophia clavicornis* Lindley var. *nutans* (Sonder) A.V. Hall; Typhaceae, *Typha capensis* (Rohrb.) N.E. Brown; Cyperaceae, *Cyperus fulgens* C.B. Clarke. 3) Broadly African to Afro-Eurasian species, e.g.,: Gramineae, *Phragmites australis* (Cav.) Steudel; Cyperaceae, *Cyperus esculentus* L. The genus *Lapei-*

rousia has species that exhibit chorological patterns number one and number two, plus a fourth pattern with an east-central Zambezian heartland and outliers in north-central Nigeria. Whether there exists significant additional diversity in geographical pattern will become clearer when more families are brought into the analysis.

Acknowledgements
A special word of thanks to J. Cowley, R. Drummond, P. Goldblatt, and S. Mavi. The base map is slightly modified from that of the Distributiones Plantarum Africanarum, courtesy of P. Bamps, Jardin Botanique National de Belgique.

References

DeVos, M.P., 1979. The African genus *Ferraria*. J.S. African Bot. 45: 295-375.
Dinter, K., 1912, Die Vegetabilische Veldkost Deutsch-Südwest-Afrikas. Selbstverlag, Okahandja.
Goldblatt, P., 1972. A revision of the genera *Lapeirousia* Pourret and *Anomatheca* Ker in the winter rainfall region of South Africa. Contributions from the Bolus Herbarium 4: 1-111.
Goldblatt, P., 1990. Systematics of *Lapeirousia* (Iridaceae-Ixioideae) in tropical Africa. Ann. Missouri Bot. Gard. 77: 430-484.
Lee, R.B., 1979. The !Kung San. Cambridge University Press, Cambridge.
Maguire, B., 1978. The Food Plants of the !Khũ Bushman of North Eastern South West Africa. Master of Science thesis, University of the Witwatersrand, Johannesburg.
Marshall, L., 1976. The !Kung of Nyae Nyae. Harvard University Press, Cambridge, Massachusetts.
Peters, C.R., 1990. African wild plants with rootstocks reported to be eaten raw: the monocotyledons, part I. Mitt. Inst. Allg. Bot. Hamburg 23: 935-952.
Peters, C.R., 1994. African wild plants with rootstocks reported to be eaten raw: the monocotyledons, part II. In: J.H. Seyani & A.C. Chikuni. Proceedings XIIIth Plenary Meeting AETFAT, Malawi 1: 25-38.
Peters, C.R., E.M. O'Brien & R.B. Drummond, 1992. Edible Wild Plants of Subsaharan Africa. Royal Botanic Gardens, Kew.
Rodin, R.J., 1985. The Ethnobotany of the Kwanyama Ovambos. Monogr. Syst. Bot. Missouri Bot. Gard. 9: 1-163.
Silberbauer, G.B., 1965. Report to the Government of Bechuanaland on the Bushman Survey. Bechuanaland Government Printer, Gabeorone.
Silberbauer, G.B., 1972. The G/Wi bushman. In: M.G. Bicchieri (ed.), Hunters and Gatherers Today, pp. 271-326, Holt, Rinehart and Winston, Inc., New York.
Silberbauer, G.B., 1981. Hunter and Habitat in the Central Kalahari Desert. Cambridge University Press, Cambridge.
Thornthwaite, C.W. & J.R. Mather, 1962. Average Climate Water Balance Data of the Continents, Part 1, Africa. Publications in Climatology 15 (2). Laboratory of Climatology, Centerton, New Jersey.
Watt, J.M. & M.G. Breyer-Brandwijk, 1962. The Medicinal and Poisonous Plants of Southern and Eastern Africa. E. & S. Livingstone Ltd., Edinburgh.
Wehmeyer, A.S., 1986. Edible Wild Plants of Southern Africa. CSIR National Food Research Institute, Pretoria.
White, F., 1983. The Vegetation of Africa. UNESCO, Paris.
Wickens, G.E., 1976. The Flora of Jebel Marra. HMSO, London.

Utilization and conservation status of wild food plants in Kenya

P.M. MAUNDU
East African Herbarium, National Museums of Kenya, P.O. Box 40658, Nairobi, Kenya.

Abstract
Of the 7000 or so vascular plant species in Kenya, about 800 wild plant species are recorded to be used for food. The taxa are distributed over 105 families and over 300 genera. The paper summarises the results of four years of research and extension work carried out by the Indigenous Food Plants Programme (IFPP).

Résumé
Sur les quelque 7000 espèces de plantes vasculaires du Kenya, environ 800 espèces sauvages se sont révélées avoir une valeur alimentaire. 105 familles et plus de 300 genres sont concernés. Les résultats de quatre années de recherche et de travail subséquent, réalisés par le "Indigenous Food Plants Programme" (IFPP) sont présentés succinctement.

Key words
Conservation, Ethnobotany, Food plants, Kenya.

Introduction

One out of every 9 species of wild vascular plants is used as food by the various communities in Kenya. As can be seen in Table 1, these species represent about 105 of the approximately 222 families of vascular plants found in East Africa.

Table 1. Families of food species in Kenya.
1st column: Family (or subfamily).
2nd column: Food types: B - Beer, Fl - Flower, F - Fruit, G - Gum, R - Resin, S - Seed, So - Soup, St - Stem, T - Tuber/Root, Te - Tea, To - Tonic, V - Leafy vegetable.
3rd column: Number of edible spp.
4th column: Approximate total number of spp.
5th column: Percentage of edible spp.[a]

Spermatophyta, Angiospermae				
(102 of 197 (sub-)families)				
Dicotyledones				
Acanthaceae	V	19	246	8%
Aizoaceae	V	4	23	17%
Amaranthaceae	V	19	57	33%
Anacardiaceae	F	21	24	88%
Annonaceae	F	12	30	40%
Apocynaceae	So,F	14	54	26%
Aquifoliaceae	V	1	1	
Asclepiadaceae	St,T	30	180	17%
Balanitaceae	F	5	5	100%
Balanophoraceae	Te	1	2	
Basellaceae	V	1	1	
Bignoniaceae	B	1	13	8%
Bombacaceae	F,V	2	4	
Boraginaceae	F	13	62	21%
Burseraceae	F,R,Te,To	10	69	15%
Caesalpinioideae	F	14	56	25%
Campanulaceae	V	4	42	10%
Capparaceae	F,T,V	14	70	20%
Caryophylaceae	F	1	20	5%
Celastraceae	F,So	9	33	27%
Chenopodiaceae	V	2	19	11%
Combretaceae	F	3	39	8%
Compositae	V	18	417	4%

L.J.G. van der Maesen et al. (eds.), The Biodiversity of African Plants, 678-683.

Connaraceae	F	2	9	22%
Convolvulaceae	T,V	12	105	11%
Crassulaceae	Fl	2	39	5%
Cruciferae	V	2	58	
Cucurbitaceae	F,V	30	76	39%
Dichapetalaceae	F	1	4	
Ebenaceae	F,To	9	14	64%
Euphorbiaceae	F	29	251	12%
Flacourtiaceae	F	10	26	38%
Grossulariaceae	F	1	2	
Guttiferae[c]	F	4	20	20%
Hydnoraceae	Fl	2	2	
Icacinaceae	F	2	4	
Labiatae	F,Te	11	165	7%
Loganiaceae	F	14	25	56%
Loranthaceae	To	1	58	2%
Malpighiaceae	F	2	6	33%
Malvaceae	F,V	23	101	23%
Melastomataceae	V	3	24	13%
Meliaceae	F	3	26	12%
Menispermaceae	To	1	17	6%
Mimosoideae	G,Te	16	67	24%
Moraceae	F	17	50	34%
Moringaceae	F,V	2	5	40%
Myrsinaceae	F	1	10	10%
Myrtaceae	F	2	20[b]	
Nymphaeaceae	T	2	2	
Ochnaceae	F	2	12	17%
Olacaceae	F	2	2	
Oleaceae	F	3	16	19%
Onagraceae	V	3	11	27%
Opiliaceae	F	1	3	
Oxalidaceae	V	2	7	29%
Papilionoideae	S,T,V	41	554	7%
Passifloraceae	F	3	23[b]	13%
Pedaliaceae	V	7	13	54%
Piperaceae	F	1	9	
Polygalaceae	F,V	4	30	13%
Polygonaceae	V	7	26	27%
Portulacaceae	V	6	14	40%
Ranunculaceae	V	1	14	7%
Resedaceae	V	1	4	
Rhamnaceae	F	9	17	53%
Rhizophoraceae	F	1	10	10%
Rosaceae	F	12	29	41%
Rubiaceae	F	37	207	18%
Rutaceae	F,Te	3	39	8%
Salvadoraceae	F	3	4	
Santalaceae	Te	1	11	9%
Sapindaceae	F,So	8	39	21%
Sapotaceae	F	16	27	59%
Scrophulariaceae	To,V	5	123	4%
Simaroubaceae	F	1	3	
Solanaceae	V	10	65	15%
Sonneratiaceae	F	1	1	
Sterculiaceae	F,Fl,S	7	52	13%
Thymelaeaceae	F	1	15	7%
Tiliaceae	F	43	60	72%
Ulmaceae	F	1	8	13%
Urticaceae	V	4	28	14%
Verbenaceae	F	20	92	22%
Violaceae	F	1	16	6%
Vitaceae	F	16	64	43%
Zygophyllaceae	F,V	4	10	40%
Monocotyledones				
Agavaceae	Fl	1	30	
Alismataceae	T	1	2	
Aponogetonaceae	T	1	6	17%
Araceae	V	7	13	54%
Commelinaceae	V	14	38	37%
Cyperaceae	T	2	207	1%
Dioscoreaceae	T	4	9	4%
Gramineae	S	14	401	4%
Hypoxidaceae	T	2	8	25%
Liliaceae[d]	Sr,T	8	48	17%
Palmae	F	4	10	40%
Taccaceae	T	1	2	
Typhaceae	T	1	2	
Zingiberaceae	F	3	14	21%
Spermatophyta, Gymnospermae (1 of 4 families)				
Zamiaceae	S	1	5	
Pteridophyta (2 of 22 families)				
Schizaeaceae	To	1	5	
Selaginellaceae	To	1	2	
105 of 222 families		777	5109	15%

[a] Small families with less than 5 members in Kenya have been omitted from the % column. A better comparison would involve the number of genera.

[b] Number includes several exotic species.

[c] Includes Hypericaceae.

[d] Includes Aloeaceae.

Some families have contributed more edible wild food plant taxa for their size than others. Examples are Annonaceae, Rosaceae, Moraceae, Flacourtiaceae, Anacardiaceae, Commelinaceae and Amaranthaceae. Some families tend to contribute more to certain food types than others. Amaranthaceae for example are known for leafy vegetables; Asclepiadaceae and Convolvulaceae for tubers/roots*; Mimosoideae for edible gums; Rutaceae for spices; Anacardiaceae, Moraceae, Rhamnaceae and Boraginaceae for fruits. In general, fruits tend to be more common in the families dominated by trees, and leafy vegetables tend to be more common in the families with many herbs and shrubs.

Dependence on and use of these wild food species is greatly influenced by the presence or absence of the species, existing indigenous knowledge and the economic pursuits of the local communities. The use of wild food species is minimal in the high potential agricultural areas of Central Kenya partly due to lack of the species, loss of knowledge on indigenous plants and presence of modern agricultural crops. At the other extreme are the pastoral groups and hunter-gatherers in Kenya's Rift Valley province, whose daily wild plant food intake may contribute, during some seasons, up to 80% of the food taken. These communities have over the years acquired immense knowledge of the utilization of wild food species and have developed appropriate processing technologies for converting seemingly inedible to delicious food. One of the processes involves washing out bitter poisonous plant compounds by boiling for as much as 10 hours with frequent replacement of the water.

Utilization status

Up to about 800 species have been recorded as used in one way or another as food in the wild in Kenya. About 400 (50%) are fruit species, 200 (25%) provide vegetables, 100 (12.5%) tubers/roots, 30 (3.6%) edible gums/resins, 30 (3.6%) spices/flavourings; the remainder provide other types of food including soup plants, flowers, bark, chewing sticks and tonics.

Wild plant food plays a major role in supplementing other foods, especially in the rural areas. In some areas these foods are major sources of vitamins, and among some hunter-gatherers and pastoral communities they provide the main diet of the day at certain times of the year.

Wild food species form a considerable proportion of the higher plants we see around us in Kenya. Many of the common trees and shrubs in the immediate environment of communities in rural areas are used by the local people in one way or another as food. In some instances, the proportion of edible shrubs and trees may be 50% or more, especially among the hunter-gatherers and pastoral communities. However, in agricultural communities this may be lower than 5% (1 in 20), the determining factors being existing local knowledge and the economic pursuits of the people, and hence the nature of resources obtained from the environment.

The ratio for common edible herbaceous species is much lower than that for trees and shrubs, but in many communities it still ranges from just below 5% to about 20%.

---

* In this paper, the term *tubers/roots* refers to all of the underground food types contributed by the vascular plants.

Some communities however have a culture rich in utilizing herbaceous plants. Examples are the Giriama, the Luhya, the Ng'ikebootok and the Pokot, who use many herbaceous plants as leafy vegetables. Others have a culture rich in plant (especially herbal) tonics. This is common with the Maasai, the Turkana and the Pokot.

Of the nearly 800 edible species, some are widely used as food while others are only used locally by certain communities. Some are of "minor" significance and are usually termed "food for children". Others are important famine foods, or form the main meals in some communities at certain times of the year. Some food plants have the potential to become important crops if subjected to research and development.

Despite their importance, many food plants are diminishing while the survival of some very rare ones is at stake as humans encroach on their natural habitats. The use of these plants for food has also progressively declined over the years, with people tending more towards "modern feeding habits", and discarding their traditional foods. Fast-growing exotic trees are replacing indigenous tree species in most farmlands in agricultural areas. Among the reasons advanced for this trend are: low regard for indigenous plants and their foods, urbanization and market economies, monoculture, population pressure, and over-exploitation of plant resources. These have resulted in less variety in diets especially in the arid and semi-arid zones, malnutrition, loss of genetic diversity in traditional food species, loss of traditional knowledge about plants and a decline in cultural diversity.

Conservation status

Despite their value, wild food plants have not been spared by man in his destructive activities. There is an ongoing loss of wild food species at the farm and village level, and of traditional knowledge about edible species. As a result, there is decreased utilization of indigenous food species.

The degree of loss of species and knowledge at the farm level is different in different parts of the country, but mainly depends on the economic pursuits of the community. Loss of species, loss of knowledge, and decreased utilization is highest at farms and villages in the agricultural areas and least in the arid lands utilized by pastoral groups. In those areas with a high potential for agriculture, much of the original vegetation has been cleared. The original vegetation has been replaced by crop land, forests of exotic trees and infrastructure. Semi-arid lands, originally occupied by pastoralists, are being turned into agricultural land. More and more people are moving into these areas from the already overpopulated high-potential areas, clearing the natural vegetation indiscriminately. Arid pastoral lands have less human activity, the major one being livestock keeping. The population is thinly distributed. Apart from some cases of overstocking there are no other serious human factors that endanger plant species in this region. Wild plant food still plays a significant role in the diet of these people: much of their vitamin requirements is provided by wild plant foods.

The species listed in Table 2 are threatened in some regions although they may still be available in others. Conservation of these species in threatened areas is necessary to maintain genetic variability.

Domestication

More and more organisations are promoting the utilization, preservation and even

Table 2. Species threatened in some regions of Kenya.

Species	Family	Affected areas	Main cause*
Annona senegalensis	Annonaceae	Nyanza, Western Kenya	hd
Berchemia discolor	Rhamnaceae	Northern Kenya, Kitui	d/oex
Borassus aethiopum	Palmae	Coastal, Western Kenya	oex
Carissa edulis	Apocynaceae	Machakos	hd/oex
Ceropegia spp.	Asclepiadaceae	Masailand	og
Dioscorea bulbifera	Dioscoreaceae	Western Kenya	hd
Diospyros mespiliformis	Ebenaceae	S.E. Kenya, Coastal area	oex
Dobera glabra	Salvadoraceae	Northern Kenya	d/lrr
Elaeis guineensis	Palmae	Coastal Kenya	hd
Ficus spp.	Moraceae	Agricultural areas	hd/oex
Ipomoea lapathifolia	Convolvulaceae	Machakos/Makueni	hd
Lablab purpureus (wild)	Papilionoideae	Coastal, R. Valley	hd
Lannea schimperi	Anacardiaceae	Central highlands, West Kenya, Rift valley	hd
Manilkara mochisia	Sapotaceae	Countrywide	hd/oex
Mondia whiteii	Asclepiadaceae	Central, Western Kenya	hd
Moringa stenopetala	Moringaceae	Northern Kenya	d
Myrianthus holstii	Moraceae	Central highlands	hd
Parinari curatelifolia	Rosaceae	Embu	hd
Saba comorensis	Apocynaceae	Countrywide	hd
Sclerocarya birrea	Anacardiaceae	Homabay, Machakos	hd
Stathmostelma propingua	Asclepiadaceae	Kajiado	hd/og
Syzygium guineense	Myrtaceae	Countrywide	oex
Ximenia americana	Olacaceae	Machakos, Siaya,	oex
Zanha africana	Sapindaceae	Machakos, Makueni	hd
Zanthoxylum chalybeum	Rutaceae	Countrywide	hd/oex

* d - desertification, hd - habitat destruction, lrr - low regeneration rates, oex - over-exploitation, og - overgrazing.

planting of indigenous food species. Species with additional advantages such as timber, shade, etc. stand a better chance of being accepted for planting and conservation. A few species, especially leafy vegetables, have been domesticated. Examples are: *Amaranthus hybridus*, *A. dubius*, *A. lividus*, *Basella alba*, *Corchorus trilocularis*, *C. olitorius*, *Crotalaria brevidens*, *Gynandropsis gynandra*, *Sesamum angustifolium* and *Solanum nigrum*. These species may also be spared when weeding.

Domestication of trees and shrubs for fruit and other foods has been less successful. The few that have been planted are also valuable for other purposes such as shade, timber, etc. The following species have been seen in tree nurseries but it has been difficult to encourage local communities to plant them: *Acacia senegal, Adansonia digitata, Annona senegalensis, Balanites aegyptiaca, Carissa edulis, Cordia sinensis, Dialium orientale, Dobera glabra, Manilkara sansibarensis, Pappea capensis, Salvadora persica, Sclerocarya birrea, Strychnos spinosa, Syzygium guineense, Tamarindus indica, Vangueria infausta, V. madagascariensis, Ximenia americana* and *Ziziphus mauritiana*. Important food trees are also spared while clearing bush for cultivation.

Conclusion

Wild food plants play an important role in supplementing other foods as well as enhancing the quality of the environment. Knowledge acquired by local communities over a long time has enabled them to efficiently exploit these food resources. With diminishing local knowledge and the rising demand for more crop land, many of these species are becoming threatened. While there is need to promote the utilization of these wild species, there is also need to conserve them both by way of conserving existing wild habitats and local indigenous knowledge, and by incorporating the species into existing farming systems.

References

(some sources not referred to in the text)

Beentje, H., 1994. Kenya Trees, Shrubs and Lianes. National Museums of Kenya, Nairobi.

Blundell, M., 1987. Collins Guide to the Wild Flowers of East Africa. Collins, London, 464 pp.

Greenway, P.J., 1939. A Swahili-Botanical-English Dictionary of Plant Names. Dar-es-Salaam.

Kabuye, C.H.S., P.M. Maundu & G.W. Ngugi, in prep. An illustrated guide to some indigenous food plants of Kenya.

Kew Herbarium, 1994. List of Unpublished FTEA Families.

Strategies for enhancement of utilization potential of edible woody forest species of south-eastern Nigeria

J.C. OKAFOR [1], H.C. OKOLO [2] & M.A.N. EJIOFOR [3]
[1] Fame Agricultural Centre, 3, Kingsway road, P.O. Box 3856, Enugu, Nigeria.
[2] Project Development Institute (PRODA), P.O. Box 609, Enugu, Nigeria.
[3] Federal College of Agriculture, Umudike, P.M.B. 7008, Umuahia, Nigeria.

Abstract
In Nigeria, deforestation, which leads to tremendous loss of biodiversity, is well advanced and about 90% of its rain forest has already been cleared. The Nigerian environment is also beset with many other problems including a food crisis. To address these problems a research project on "Development and Utilization of some endangered edible forest species of South-eastern Nigeria" was sponsored by World Wildlife Fund, Washington D.C. The Project investigated chemical evaluation, formulation of new edible products and *ex-situ* conservation techniques of 16 study species and produced significant results in each of these aspects.

Résumé
Au Nigeria, la déforestation, qui entraîne une énorme perte de la biodiversité, est fort avancée et environ 90 % de la forêt ombrophile a déjà été défrichée. L'environnement du Nigeria est aussi confronté à de nombreux autres problèmes, notamment à la crise alimentaire. Pour aborder ces problèmes, un projet de recherche sur "Le développement et l'utilisation des espèces forestières comestibles dans le sud-est du Nigeria" a été subsidié par le "World Wildlife Fund" de Washington D.C. Ce projet a permis de procéder à une évaluation chimique, de détecter de nouveaux produits comestibles et d'établir des techniques de conservation *ex-situ* pour 16 espèces étudiées; des résultats significatifs ont été obtenus pour chacun de ces aspects de recherche.

Key words
Chemical analysis, Commerce, Foods, Forest species, Nigeria, Propagation.

Introduction

The South-eastern Zone of Nigeria lies between latitude 4°20' and 7°00 N and Longitude 5°25' and 9°35' E. It is bounded on the east by the Republic of Cameroon, on the south by the Atlantic ocean (on the Bight of Bonny), on the west by the River Niger and on the north by Kogi, Benue and Taraba States. The zone comprises seven (out of 30) states of Nigeria namely Enugu, Anambra, Imo, Abia, Akwa Ibom, Cross River and Rivers States. The climate and vegetation of the area are described in Agboola (1979), Keay (1959) and NEST (1991).

Much change has taken place in the natural vegetation in south-eastern Nigeria since the mid-1920's. It has been one long and depressing story of decline in extent and type. Even relict forests, in the form of sacred or Juju groves or "bad bush", which escaped disturbance until the late 1920's and early 1930's in Onitsha, Nnewi, Ukpor and many other places, have now either completely gone or have been badly mutilated and encroached upon. On a national level, deforestation is well advanced in Nigeria, where about 90% of its rain forest has already been cleared (WWF 1989). According

L.J.G. van der Maesen et al. (eds.), The Biodiversity of African Plants, 684-695.

to Myers (1989), Nigeria is one of the 12 countries whose deforestation rates account for over 80% of the global total.

Conservation through village farm forests represents a most effective strategy in which useful trees and shrubs and other useful plants are protected and/or planted around homesteads and villages in most towns and urban areas in the zone. Most of the trees are either protected as volunteers during forest clearing and village development or planted in various land use locations such as boundaries (Okafor & Fernandes 1987). Ultimately, a vegetation of complex structure and species richness is developed that usually envelops or surrounds the homesteads. Such village farm forests are contiguous between towns, more or less concealing the homesteads and even obliterating the boundaries between towns. Typical species include *Elaeis guineensis, Chrysophyllum albidum, Dacryodes edulis, Treculia africana, Garcinia kola, Pterocarpus* spp., *Irvingia gabonensis, Cola acuminata, C. nitida, Dennettia tripetala, Vernonia amygdalina, Pentaclethra macrophylla, Alstonia boonei, Milicia excelsa* and *Albizia ferruginea*.

Aims

This paper reports on the research project "Development and Utilization of fast disappearing and under-utilized edible woody forest species of South-eastern Nigeria" sponsored by Biodiversity Support Program, World Wildlife Fund, Washington D.C. The full report of this project is provided in Okafor et al. (1993).

The project aims were to enhance the potential of edible woody forest species, especially for food, industrial raw materials and environmental conservation in African countries. All the species selected for this study (Table 1) can be used in agroforestry systems such as alley farming, intercropping, improved planted fallow, barrier hedges and live fences (Okafor 1982, Okafor & Fernandes 1987). If people can be encouraged to cultivate local forest species, especially edible ones, in their farming systems and under social and urban forestry programmes (Okafor 1991), this could contribute significantly to the conservation of biodiversity.

The main objective was to develop adequate research and development efforts on indigenous African woody plants by:

* Providing basic data and information necessary for sustainable commercial exploitation of threatened species.
* Developing suitable propagation and field management techniques, as well as provision of planting materials.
* Involving local participation in conservation programmes and generating employment in the project area (South-eastern Nigeria).

Methods

The strategies and methods were multi-disciplinary with scientists from three separate institutions, namely Project Development Institute (PRODA), Enugu; Federal College of Agriculture, Umudike; and Fame Agricultural Centre, Enugu, collaborating, and local farmers also participated. Five major strategies were included in the research:

1. Selection of 19 fast disappearing woody species (see Table 1) which are potential sources of spices, fruits, nuts, seeds and leafy vegetables.
2. Nutritional and chemical evaluation of the edible parts of studied species.

Table 1. Studied species.

Botanical Name	Family	Vernacular Name	Part eaten	Traditional food type
Trees				
Chrysophyllum albidum	Sapotaceae	Udara/Star apple	Fruit pulp	Fruit
Dacryodes edulis	Burseraceae	Ube/Pear	Fruit pulp	Fruit
Dennettia tripetala	Annonaceae	Mmimi	Fruit	Spicy seeds
Garcinia kola	Guttiferae	Akilu/Bitter kola	Seeds	Seeds as kola
Irvingia gabonensis	Irvingiaceae	African mango	Fruit, Seeds	Fruit pulp; seed condiment
Monodora myristica	Annonaceae	Ehuru/Nutmeg	Seeds	Spice
Pentaclethra macrophylla	Mimosoideae	Ukpaka/Oil bean	Seeds	Fermented product
Pterocarpus mildbraedii	Papilionoideae	Oha	Leaves	Leafy vegetable
P. santalinoides	Papilionoideae	Nturukpa	Leaves	Leafy vegetable
P. soyauxii	Papilionoideae	Oha	Leaves	Leafy vegetable
Treculia africana	Moraceae	Ukwa/Breadfruit	Nuts	Main dish; roasted nuts
Xylopia aethiopica	Annonaceae	Uda	Seeds	Spice
Climbers				
Dioscoreophyllum cumminsii	Menispermaceae	Utobili/Serendipity berry	Fruit	Sweetener
Gnetum africanum	Gnetaceae	Okazi	Leaves	Leafy vegetable
G. buchholzianum	Gnetaceae	Okazi	Leaves	Leafy vegetable
Gongronema latifolium	Sapotaceae	Utazi	Leaves	Leafy vegetable
Piper guineense	Piperaceae	Uziza/Guinea pepper	Seeds	Spice
Plukenetia conophora	Euphorbiaceae	Ukpa/Conophor	Nuts	Nuts
Shrubs				
Vernonia amygdalina	Compositae	Onugbu/Bitter leaf	Leaves	Leafy vegetable

3 Investigation into their commercial potential through:
   a Extension of traditional utilization by developing improved processing, preservation, packaging and storage techniques.
   b Formulation and production of new food products including composite seasonings, beverages, enriched baby foods and convenience food products.
   c Recipe and culinary development involving local food items leading to their popularization among the Nigerian population and expatriates.
4 Development of nursery practices by collection of seeds and/or budwood of the species and studying germination and vegetative propagation techniques (bud grafting and stem cuttings).
5 Integrating the species into the traditional farming system, with field establishment trials using agroforestry techniques (e.g. alley cropping) in co-operation with 23 participating farmers.

Results

Selected Species

Table 1 lists the 19 species which were selected for the study. They were selected because previous studies (Okafor 1981, 1989, 1990a) had shown them to be of great potential but as yet remained undeveloped. These species are also threatened with extinction due to high population and deforestation pressures, as well as destructive harvests such as uprooting.

Chemical Analysis

In total 19 samples (15 species) were analyzed. The results obtained are similar to those of proximate analysis reported by Okafor (1981). Data obtained show that many of the samples investigated are valuable sources of minerals, protein, carbohydrates and fats (Table 2).

Two critical parameters: % ash and % fat/oil determined in the course of this project, need special comments. Ash: *Dacryodes edulis, Dennettia tripetala, Gnetum africanum, Gongronema latifolium, Plukenetia conophora, Piper guineense* and *Vernonia amygdalina* are good sources of minerals (Na, K, Ca, Mg and Fe). Fat/Oil: Tables 2 and 3 show the percentage oil for each species. The high fat/oil content (more than 40%) in *Dacryodes edulis, Monodora myristica, Pentaclethra macrophylla* and *Plukenetia conophora* seeds indicates they could be commercially exploited as sources of edible and industrial fats. *Irvingia gabonensis* also has potential in pharmaceutical preparations and confectionery, because of its relatively high melting point (44-45°C) (Okafor et al. 1993). Table 4 shows the oil characteristics for the four main oil seed species.

The acid values are advantageously low in *Irvingia gabonensis* (2.00) and the iodine value obtained for *Pentaclethra macrophylla* (178.65) and *Monodora myristica* (113.43) classify them as drying and semi-drying oils, respectively. The fat contents of three vegetables of dietary importance, namely *Gnetum africanum, Pterocarpus soyauxii* and *Vernonia amygdalina*, are significant (14.20%, 3.60% and 4.50%, respectively).

Table 2. Chemical analysis of fruits, nuts, seeds and leafy vegetables.

Scientific Name	Plant Part	Moist. (%)	Ash (%)	Oil (%)	Protein (%)	Na (ppm)	K (ppm)	Ca (ppm)	Mg (ppm)	Fe (ppm)	Zn (ppm)	Cu (ppm)
Chrysophyllum albidum	Pulp	73.56	1.22	6.94	5.43	1.3	2	2.70	4.25	1.69	0.17	0.05
Dacryodes edulis var. *edulis*	Pulp	58.94	3.39	66.87	3.63	15.0	74	3.92	3.50	1.33	0.30	Nil
D. edulis var. *parvicarpa*	Pulp	70.25	4.65	36.63	2.55	23.0	98	6.79	11.00	0.75	0.27	0.08
Dennettia tripetala	Carpel	70.86	4.92	1.92	15.41	21.0	78	4.17	6.50	1.15	0.29	Nil
Garcinia kola	Seeds	50.27	1.30	5.31	2.47	12.0	30	2.70	2.25	2.22	0.48	Nil
G. kola	Pulp	85.94	1.52	10.38	1.69	-	-	-	-	-	-	-
Gnetum africanum	Leaves	37.39	4.72	14.20	10.18	26.0	126	28.35	41.75	5.23	0.49	0.06
Gongronema latifolium	Leaves	71.14	10.94	18.77	25.66	58.0	336	20.75	56.00	8.17	0.90	0.12
Irvingia gabonensis	Seeds	4.53	2.65	66.26	10.62	8.8	70	11.26	20.50	1.87	0.37	0.12
I. gabonensis	Pulp	38.26	2.29	-	10.18	9.4	82	9.31	16.50	1.78	0.67	0.06
Monodora myristica	Seeds	10.41	2.12	41.15	9.25	8.2	62	3.04	13.50	1.42	0.32	Nil
Pentaclethra macrophylla	Seeds	4.34	2.50	58.15	26.71	8.4	66	4.51	10.75	1.69	0.51	0.05
Piper guineense	Leaves	78.58	15.56	9.30	18.54	35.0	320	7.00	147	6.12	2.05	0.15
P. guineense	Seeds	66.13	2.34	11.53	10.80	38.0	200	15.04	23.00	5.76	0.37	0.15
Plukenetia conophora	Seeds	49.80	6.09	61.04	18.27	21.0	109	14.63	40.25	1.24	0.68	0.10
Pterocarpus soyauxii	Leaves	74.69	2.83	3.60	31.84	1.8	11	19.37	35.50	3.79	0.84	0.21
Treculia africana	Seeds	42.92	2.12	15.65	14.00	8.3	50	3.75	7.50	1.60	0.50	0.08
Vernonia amygdalina	Leaves	79.67	10.13	4.50	23.24	46.0	304	40.00	58.00	4.72	0.81	0.10
Xylopia aethiopica	Carpel	14.52	4.59	26.85	9.86	17.0	140	16.30	24.75	4.08	0.65	0.10

Table 3. Mean chemical contents of fruits and seeds.

Scientific Name	Plant part	Dry Matter (%)	Oil (%)	Crude Fibre (%)	Ash (%)	Crude protein (%)	Total Carbon (%)	Vit C (mg/100g)	Vit A (mg/100g)	Amino Acid (mg/g)
Dacryodes edulis var. *edulis*	Pulp	12.6	71.3	0.50	2.40	13.13	10.33	209.0	0.65	9.81
D. edulis var. *parvicarpa*	Pulp	11.2	68.5	0.54	2.45	14.08	12.04	194.7	0.67	10.23
Dennettia tripetala	Carpels	73.6	1.5	4.35	2.68	14.88	76.30	98.7	0.60	11.78
Monodora myristica	Seeds	54.5	42.0	2.46	2.94	14.00	41.00	100.1	0.60	17.94
Piper guineense	Seeds	70.4	12.0	1.85	2.62	4.90	70.72	167.0	0.53	7.09
Plukenetia conophora	Seeds	48.7	23.1	2.33	6.11	17.85	42.10	116.3	0.70	10.22
Xylopia aethiopica	Pericarp	69.4	13.8	3.25	2.75	6.58	67.64	192.7	0.43	3.44
X. aethiopica	Seeds	60.3	26.0	2.92	2.40	12.25	54.81	161.9	0.54	6.01

All measurements except dry matter were calculated over dried sample.

Table 4. Characteristics of oils extracted from seeds containing more than 40% oil (dry weight).

Sample	Parts Tested	Oil %	Specific Gravity at 27°	Refractive index at 27°	Viscosity[1] at 27°	Saponification Value	Free Fatty Acid	Iodine Value
Dacryodes edulis var. *edulis*	Pulp	66.87	0.9420	1.4569	305	191	-	-
Irvingia gabonensis	Seeds/ kernels	66.26	0.9125[2]	1.4460[2]	210[3]	290	2.00	-
Monodora myristica	Seeds	41.15	0.9197	1.4715	709	201	14.89	113.43
Pentaclethra macrophylla	Seeds	58.15	0.9118	1.4668	849	224	2.35	178.65

[1] Viscosity: the efflux time of 50 ml in a Redwood № 1 Viscometer at ambient temperature unless otherwise stated.

[2] At 46°C.

[3] At 61°C, hot water bath.

Figure 1. Product development from seeds of *Treculia africana* (African breadfruit).

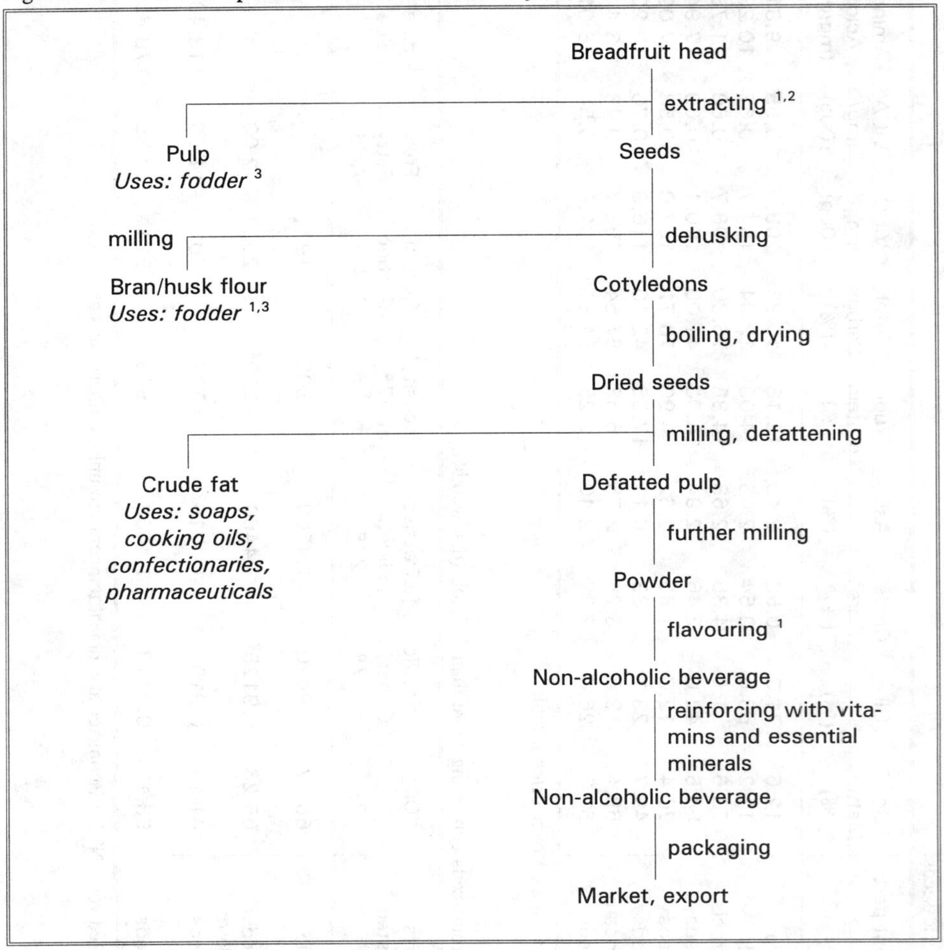

[1] Ejiofor et al. 1988, [2] Okafor 1981, [3] Okafor 1982.

Commercial development of plant products

In addition to the direct nutritional contribution of local trees and shrubs, there is significant potential for developing cottage industries based upon them. These industries, in turn, could further increase the use of these edible products through improved processing, preservation and food formulations.

A number of food products that could be produced commercially from the study species are described below. For more examples see Okafor & Lamb (1994).

Jams, Jellies and Fruit Juice: Suitable species include *Chrysophyllum albidum* and *Irvingia gabonensis* (Anazonwu-Bello 1981, Okafor et al. 1993).

Beverages: A non-alcoholic beverage which does not need milk and sugar can be prepared from powdered *Treculia africana* seeds (Ejiofor et al. 1988; see Figure 1).

Figure 2. Product Development from seeds of varieties of *Irvingia gabonensis*.

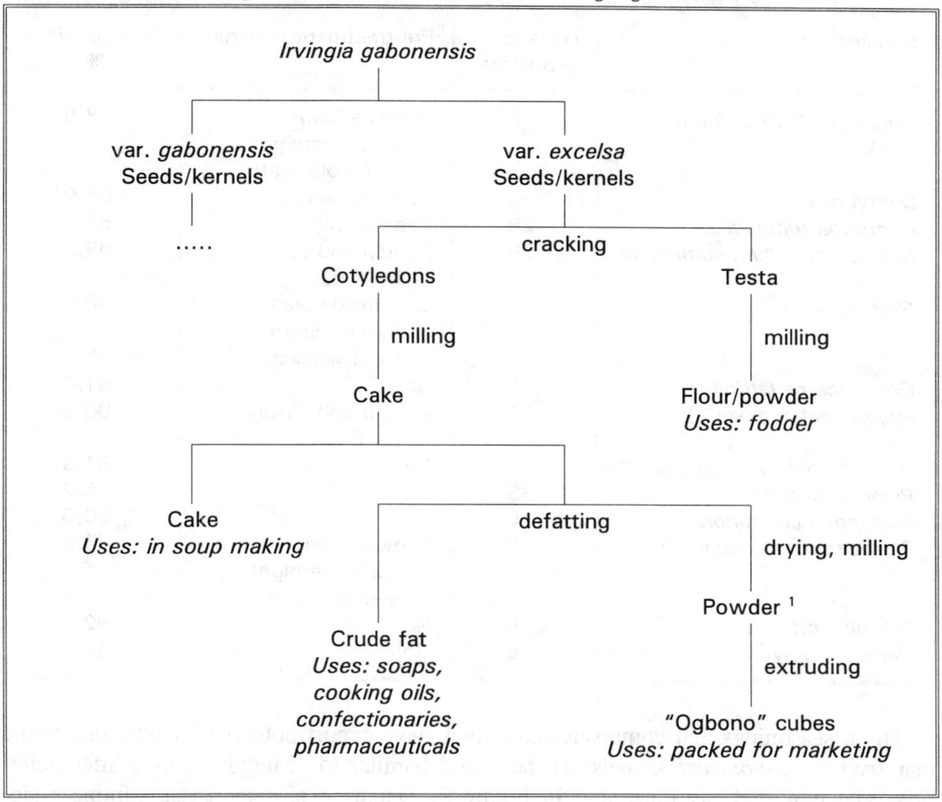

[1] Ejiofor et al 1987.

The use of dehusked, dried and ground seeds of *Garcinia kola* in flavouring soft drinks produced a pleasant drink, which could be regarded as a health drink considering the medicinal value of *G. kola* in the improvement of liver functions (Iwu et al. 1987). Bitters from seeds of *G. kola* are a potential export product.

Leafy vegetables: Leafy vegetables of *Vernonia amygdalina* and *Gongronema latifolium* should be explored as possible export crop.

Sweeteners: Sweeteners can be extracted from *Dioscoreophyllum cumminsii*.

Seasoning: Several local species including *Piper guineense, Xylopia* spp., *Monodora myristica* and *Dennettia tripetala* can be blended into seasoning (Ajayi 1986). Combinations of undefatted powders of seeds of these species can be used to improve the flavour and aroma of stews eaten with rice. They are also suitable for flavouring biscuits produced from flour of wheat and *Treculia africana*. Okafor et al. (1993) provide results of sensory evaluations of formulations of combinations of powders of these species in stews and schnapps, a locally manufactured alcoholic beverage. Defatted samples of fermented oil bean seeds (*Pentaclethra macrophylla*) improve the flavour and aroma of soups, pastries, biscuits, chin-chin and cakes.

Table 5: Optimum germination conditions for each species.

Species	Days to germinate	Pre-treatment required	Germination %
Chrysophyllum albidum	18	Remove testa and soak overnight with cold water	98.9
Dacryodes edulis	12	Remove endocarp	85.0
Dennettia tripetala	28	Depulp only	66.0
Dioscoreophyllum cumminsii	68	Depulp and sun-dry for 1 day	69.6
Garcinia kola	55	Bury inside plantain pseudo-stem for 3 weeks	80.0
Gongronema latifolium	6	No	81.6
Irvingia gabonensis	14	Depulp and air-dry for 2 days	80.0
Pentaclethra macrophylla	21	No	84.3
Piper guineense	33	No	16.2
Plukenetia conophora	15	No	80.0
Pterocarpus soyauxii	7	Remove pericarp and soak overnight in cold water.	86.0
Treculia africana	16	No	92.0
Vernonia amygdalina	8	No	89.0

These seasonings and composite seasonings have export potential. It was also found that *Irvingia gabonensis* kernels can be cubed (similar to "Maggi Cubes") after defatting, and exported, as illustrated in Figure 2. Defatting also increases sliminess significantly (Udolisa 1985). The flow-chart in Figure 2 also indicates other by-products and uses of this species.

Enriched Baby Foods: Powdered seeds of *Plukenetia conophora, Pentaclethra macrophylla* and *Treculia africana* which are rich in protein (Table 2) can be used to enrich baby foods based on cereals, such as maize (*Zea mays*).

Livestock Feeds: Kernels of *Dacryodes edulis* which contain 3.2% protein (Okafor & Okolo 1974), and fruit pulp and bran of *Treculia africana* which contain 9.4% and 5.7% protein, respectively, (Ejiofor et al. 1988) can be used in livestock feed formulations.

Thickeners: It was found that seeds of *Plukenetia conophora* (conophor), which are rich in protein, can be used to thicken soups in place of melon (*Citrullus colocynthis*), as well as to enrich yam pottage and other carbohydrate staple dishes, such as cocoyam, plantain and rice.

Nursery Practices

Table 5 shows the optimum germination conditions for each species and other germination results. Okafor et al. (1993) provide more detailed results on the germination of these species.

Table 6. Propagation by stem cuttings in various media in a non-mist propagator.

Species	Media					
	1	2	3	4	5	6
Gongronema latifolium (adult cutting)	55%	91%	91%	100%	45%	100%
Plukenetia conophora (adult cuttings)	0%	24%	0%	8%	0%	30%
P. conophora (juvenile cuttings)	0%	0%	0%	0%	0%	0%
Pterocarpus santalinoides (adult cuttings)	2.4%	2.4%	4.8%	2.4%	2.4%	0%
P. santalinoides (juvenile cuttings)	78%	78%	52%	78%	39%	70%
P. soyauxii (adult cuttings)	0%	0%	0%	0%	0%	0%
P. soyauxii (juvenile cuttings)	83%	83%	67%	50%	33%	33%
Treculia africana var. *inversa* (adult cuttings)	8%	-	-	-	-	-
Vernonia amygdalina (adult cuttings)	90%	80%	70%	60%	50%	40%
V. amygdalina (juvenile cuttings)	80%	80%	60%	90%	60%	70%

1: Topsoil, 2: Sawdust, 3: Riversand,
4: Topsoil, sawdust, riversand,
5: Topsoil, sawdust, riversand, manure,
6: Topsoil, sawdust, riversand, charcoal.

Bud grafting has been reported to reduce fruiting age from 10 years or more to 2-4 years and height of fruit set to 1-3 metres instead of 8 m or more in several fruit trees (Okafor 1978, 1980, 1981, Okafor & Lamb 1994, etc.). Budding trials were carried out for *Treculia africana, Irvingia gabonensis, Chrysophyllum albidum*, and *Pentaclethra macrophylla*. Budding of *T. africana* was most successful using Forkert technique (inverted V) with about 80% take. About 60% success was recorded in *I. gabonensis* but establishment in the field after transplanting was often unsuccessful. *In-situ* budding techniques in the field (rather than in the nursery) as described by Okafor (1990b), are therefore preferred. Only 10.2% and 10.7% success was achieved in *C. albidum* and *P. macrophylla*, respectively.

These percentages are much lower than were previously reported by Okafor (1981). It is speculated that the combined effect of poor quality budwood and unfavourable season may have been responsible for these poor results. Highest success was obtained, however, in *P. macrophylla* during the rains (in May).

Table 6 shows results of trials to propagate stem cuttings with various media and ages of cuttings for 6 species. The results show that charcoal, sawdust and riversand can be useful rooting media.

Involvement of local farmers

The local farmers were involved in establishing hedgerows with leguminous shrubs such as *Cajanus cajan* (pigeonpea) and *Pterocarpus santalinoides* for soil enrichment. They also planted out study species in their home gardens and farms under intercropping and alley farming systems.

The participating farmers in this project voluntarily formed a biodiversity conservation co-operative society, because of the importance and their corresponding interest in the project, as well as the support, both in training and supply of cash and other inputs which they readily received during the project life. Such local participation augurs well for conservation awareness and action.

Lessons from the project

The major lessons learned from this project are as follows:

1. Local participation is greatly enhanced if the objectives of the project are geared to the needs and priorities of the local people, and if the techniques adopted are simple and successful.
2. The results attest to the great potential of the study species for industrial exploitation and enhancement of food supplies.
3. Employment opportunities can be easily harnessed through development of propagation techniques by local farmers and children of school age.
4. Plant propagation techniques, developed by the project, are useful for *ex-situ* conservation of forest and savanna species, in home gardens and distant farms, to counteract unsustainable exploitation of wild resources due to demands of the population.
5. Involvement of students in institutions of higher learning can increase the scope of the investigations and spread conservation awareness to future African leaders.
6. Dissemination of project results through organised workshops and seminars is essential for adoption and follow-up action by policy makers, researchers, developers, hoteliers and farmers.
7. Finally, multidisciplinary teams should be used to implement research projects on conservation, development and utilization of biodiversity in Africa.

References

Agboola, S.A., 1979. An Agricultural Atlas of Nigeria. Oxford University Press, Oxford.

Ajayi, G.A., 1986. Production of composite seasoning from local species. Higher National Diploma Research Project. Dept. of Food Technology, Idah, Benue State, Nigeria.

Anazonwu-Bello, J.N., 1981. Indigenous foods and nutritional adequacy. Symposium on development of indigenous technology. Ministry of Science and Technology, Enugu, Nigeria.

Ejiofor M.A.N., S.N. Onwubuke & J.C. Okafor, 1987. Developing improved methods of processing and utilization of kernels of *Irvingia gabonensis* (var. *gabonensis* and var. *excelsa*). Int. Tree Crops 4 (4): 283-290.

Ejiofor, M.A.N., O.R. Obiajulu & J.C. Okafor, 1988. Diversifying utilities of African breadfruit (*Treculia africana* Decne. subsp. *africana*) as food and feed. Int. Tree Crops 5 (3): 125-134.

Iwu, M.M., O.A. Igboko, U.A. Onwuchekwa & C.O. Okonji, 1987. Evaluation of the anti-hepatotoxic activity of the biflavonoids of *Garcinia kola* seeds. Ethnopharmacol. 21: 127-138.

Keay, R.W.J., 1959. An Outline of Nigerian Vegetation. Federal Department of Forestry Research, Federal Ministry of Information, Lagos.

Myers, N., 1989. Deforestation Rates in Tropical Forests and their Climatic Implications. Friends of the Earth, London.

NEST, 1991. Nigeria's Threatened Environment. A National Profile. Nigerian Environmental Study Action Team (NEST), Ibadan.

Okafor, J.C., 1978. Development of forest tree crops for food supplies in Nigeria. Forest Ecol. Management 1: 235-247.

Okafor, J.C., 1980. Edible indigenous woody plants in the rural economy of the Nigerian forest zone. Forest Ecol. Management 3.

Okafor, J.C., 1981. Woody plants of nutritional importance in traditional farming systems of the Nigerian humid tropics. Ph.D. thesis. University of Ibadan, Ibadan, Nigeria.

Okafor, J.C., 1982. Promising trees for agroforestry in Southern Nigeria. United Nations University Publication, Tokyo, Japan, pp 103-107.

Okafor, J.C., 1989. Agroforestry aspects. Appendix 2 in: Caldecott et al. 1987. Cross River National Park. Oban Division: Plan for developing the park and its support zone. WWF, Godalming, UK.

Okafor, J.C., 1990a. Agroforestry development in the Boshi-Okwangwo Division of the Cross River National Park. WWF, UK, Godalming.

Okafor, J.C., 1990b. Development and selection of commercially viable cultivars from forest species for fruit. Proc. of the 12th Plenary Meeting of AETFAT. Hamburg.

Okafor, J.C., 1991. Improving edible species of forest products. Unasylva 165 (42): 17-23.

Okafor, J.C., M.A.N. Ejiofor & H.C. Okolo, 1993. Final Technical Report on the Project - Development and utilization of fast disappearing and under-utilized edible woody forest species of Southeastern Nigeria. Grant No. 7536 Biodiversity Supp. Prog. WWF, Washington D.C.

Okafor, J.C. & E.C.M. Fernandes, 1987. Compound farms of Southeastern Nigeria. A predominant agroforestry home garden system with crops and small livestock. Agroforestry System 5: 153-163.

Okafor, J.C. & A. Lamb, 1994. Fruit trees: diversity and conservation strategies. In: R.R.B. Leakey & A.C. Newton (eds.), Tropical trees: The Potential for Domestication and the Rebuilding of Forest Resources, pp 34-41, NMSO, London.

Okafor, J.C. & H.C. Okolo, 1974. Potentialities of some indigenous fruit trees of Nigeria. Keynote Paper. Seminar on National Conservation Strategy for Nigeria, Kano. 26-27, Feb. 1985.

Udolisa, M.O., 1985. *Irvingia gabonensis* meals: Composition and some functional properties. Unpubl. Higher Diploma thesis. Dept. of Food Technology, IMT, Enugu.

WWF, 1989. Cross River National Park, Oban Division: Plan for developing the Park and its support zone. WWF, Godalming.

The economic importance and research needs for *Uapaca* in Malawi

J.H. SEYANI
National Herbarium and Botanic Gardens of Malawi, P.O. Box 528, Zomba, Malawi.

Abstract

The five species of *Uapaca* (Euphorbiaceae) that occur in Malawi, namely *U. kirkiana, U. lissopyrena, U. nitida, U. robynsii* and *U. sansibarica*, are of great economic importance to many local people. The ripe fruits, mainly those of *U. kirkiana* and *U. nitida*, are a source of fresh fruits which are rich in carbohydrates, energy, fibre, minerals, proteins and vitamins. These nutritious fruits are also made into cakes, refreshing drinks, and wines. The fruits are sold throughout Malawi, and in Zomba some are sold to the Mulunguzi Winery. *Uapaca* fruits also form part of the main diet of many animals, including monkeys, squirrels and birds. The roots of most *Uapaca* species are associated with the growth of a great variety of edible mushrooms which are picked, eaten or sold during the rainy season. A herbal remedy for indigestion is prepared from the roots of *U. kirkiana*. Furthermore, the wood of mature trees is used for fuelwood, charcoal, building poles, and timber.

Ecologically, *Uapaca* species form one of the dominant, or co-dominant, vegetation types in Malawi and they play a vital role in the protection of watershed areas against excessive run-off. It is recommended that more multidisciplinary research work involving taxonomists, silviculturists, mycologists, and breeders should be undertaken in order to improve the productivity of *Uapaca* species in Africa and Madagascar.

Résumé

Les cinq espèces d'*Uapaca* (Euphorbiaceae) que l'on rencontre au Malawi, à savoir *U. kirkiana, U. lissopyrena, U. nitida, U. robynsii* et *U. sansibarica*, sont d'une grande importance économique pour de nombreuses collectivités locales. Les fruits mûrs, principalement ceux de *U. kirkiana* et *U. nitida*, sont consommés comme fruits frais; ils sont riches en hydrates de carbone, en énergie, en fibres, en minéraux, en protéines et en vitamines. Ces fruits nutritifs entrent aussi dans la fabrication de cakes, de boissons rafraîchissantes et de vins. Les fruits sont vendus un peu partout au Malawi et à Zomba, certains sont vendus à l'établissement vinicole de Mulunguzi. Les fruits d'*Uapaca* constituent également une partie de l'alimentation principale de nombreux animaux, notamment des singes, des écureuils et des oiseaux. Associés aux racines de la plupart des espèces d'*Uapaca*, une grande diversité de champignons s'y développent et sont récoltés, mangés ou vendus durant la saison des pluies. Un médicament contre l'indigestion est préparé à partir des racines d'*U. kirkiana*. De plus, le bois des arbres arrivés à maturité est utilisé comme bois à brûler, comme charbon de bois, comme perches dans la construction et comme bois d'oeuvre.

Au point de vue écologique, les espèces d'*Uapaca* représentent l'un des constituants des types de végétation dominants ou codominants au Malawi et ont un rôle vital dans la protection des bassins versants pour lutter contre un écoulement excessif. Il est souhaitable que soit entreprise une recherche davantage multidisciplinaire, concernant des taxonomistes, des sylviculteurs, des mycologues et des pépiniéristes, dans le but d'accroître la productivité des espèces d'*Uapaca* en Afrique et à Madagascar.

Key words

Economic botany, Malawi, *Uapaca*.

L.J.G. van der Maesen et al. (eds.), The Biodiversity of African Plants, 696-703.

Introduction

The indigenous flora continues to play a key role in the livelihoods of people in developing countries today. Ethnobotanical knowledge that has been passed on from generation to generation either within families or ethnic groups also continues to find valuable uses in present-day society. The socio-economic benefits arising from such resources are tremendous.

In Malawi, the flora offers many valuable consumptive and non-consumptive uses (Maghembe & Seyani 1991, Seyani 1990, Williamson 1975). Wild edible plants include cereals (c. 4 spp.), tubers (c. 21 spp.), fruits (c. 80 spp.), nuts (c. 6 spp.) and vegetables (c. 51 spp.). Other plant species provide fuelwood, timber, building poles, fibre, medicinals, dyes, wood for handcrafts, etc. Non-consumptive uses include the attractive wild flowers that adorn the various landscapes of Malawi. These play a vital role in beautifying the countryside, including places of attraction to tourists. Moreover, some plants, as components of climax ecosystems, provide vital ecological services such as water and soil conservation, water purification, increased precipitation, etc.

The aim of this paper is to document the economic importance of *Uapaca* species in Malawi.

Uapaca Baill.

Small to medium trees, 4-15 (-30) m high. Leaves alternate, simple, leathery; ovate to obovate, elliptic to oblong-elliptic, 5.2-21.5 x 2.5-10.0 cm; usually crowded near the ends of branchlets. Flowers dioecious, precocious: male flower in a dense head, up to c. 10 mm in diameter; female flowers solitary, ovary large, up to 10 mm in diameter. Fruits sub-globose to spherical, fleshy, yellow to brown, 1.5-3.0 cm in diameter when ripe. Seeds white, up to three per fruit, 1.5-2 cm long and 1.0-1.3 cm thick.

The genus *Uapaca* is a member of the family Euphorbiaceae (tribe Antidesmeae in subtribe Uapacinae). It consists of 60 species which are distributed in Tropical Africa and Madagascar (Radcliffe-Smith 1993). In Malawi, only five species of the genus are present, namely *U. kirkiana* Muell. Arg., *U. lissopyrena* Radcl.-Sm., *U. nitida* Muell. Arg., *U. robynsii* De Wild., and *U. sansibarica* Pax (Figure 1).

A study of the herbarium collection of *Uapaca* species in the National Herbarium of Malawi (MAL), in Zomba, shows that flowering in Malawi starts from January to May with a peak in February and March. Fruiting occurs from March to October, and fruits ripen from October to February.

Uapaca species are widely distributed in Malawi, although *U. lissopyrena* and *U. robynsii* have only been collected from the Dzalanyama Forest Reserve and the Nyika Plateau, respectively (Figure 1). *Uapaca* species are a major component of the *Brachystegia-Isoberlinia-Combretum* woodlands or occur as pure dominant stands, as in the cases of the widely distributed *U. kirkiana, U. nitida* and *U. sansibarica*. *Uapaca* species grow mainly in medium elevation areas and flourish on drier poor soils, especially on leached quartz sand or shallow immature soils. In some cases, *Uapaca* grow on fertile and deep soils. The various species of *Uapaca* are known by a number of common names: e.g. *U. kirkiana* (wild loquat: English; msuku, mtoto: Chewa, Yao; mompangwe: Lambya; musukuti, musukutimbininga: Sukwa); *U. nitida* (kasokolowe, msokolowe: Chewa; msukumpini: Nkhonde; msalinja, msanyinja: Tumbuka; msechera, mselechera: Yao); *U. sansibarica* (kasokolowe: Chewa; mtoto: Yao).

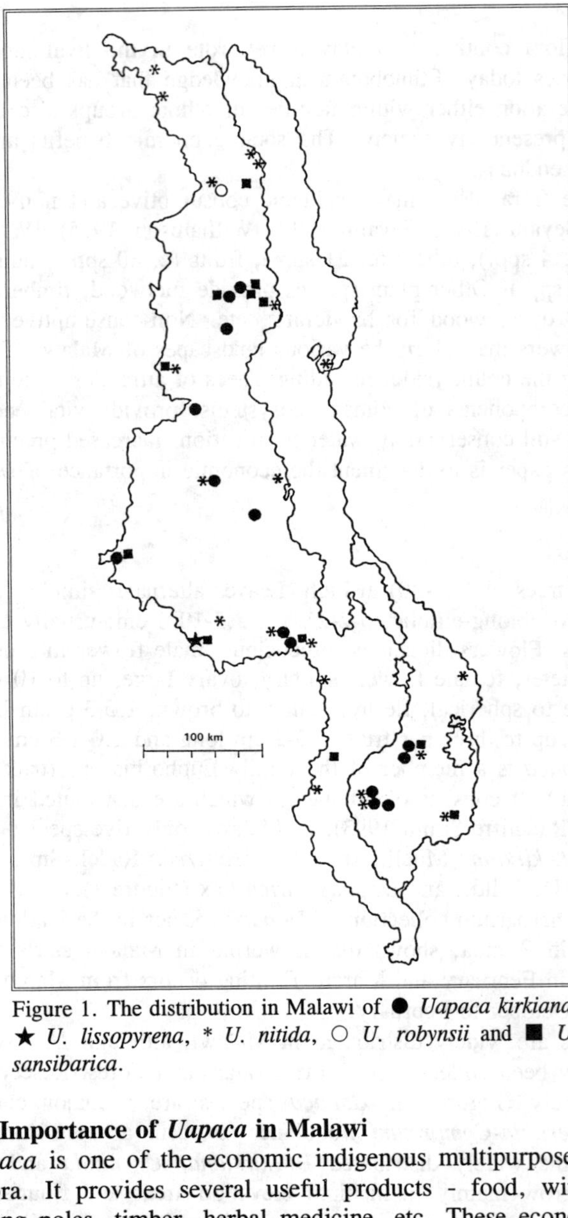

Figure 1. The distribution in Malawi of ● *Uapaca kirkiana*, ★ *U. lissopyrena*, * *U. nitida*, ○ *U. robynsii* and ■ *U. sansibarica*.

The Economic Importance of *Uapaca* in Malawi

The genus *Uapaca* is one of the economic indigenous multipurpose woody plants of the Malawi flora. It provides several useful products - food, wine, fuelwood and charcoal, building poles, timber, herbal medicine, etc. These economic products are described below.

Food

The fruits of all *Uapaca* species are edible (Blunt 1990, Coates Palgrave 1977, Maghembe & Seyani 1991, Williamson 1975), but those of *U. kirkiana* are much preferred because of their larger size and a sweet pleasant taste. Before Malawi achieved food self-sufficiency, the fruits of *U. kirkiana* formed an important part of the diet of some people and were used as a famine food since they ripened during the rainy season when other food was in short supply. In Barotseland, Zambia, the importance of this species as a food tree led to its declaration as a protected tree (Coates Palgrave & Coates Palgrave 1956). The ripe fruits are also a favourite food for many animals, including the savanna baboon (*Papio cynocephalus*); samango/blue monkey (*Cercopithecus mitis*); vervet/savanna monkey (*Cercopithecus aethiops*); greater or thick-tailed galago (*Galago crassicaudatus*); lesser galago/bush-baby (*Galago senegalensis*); squirrels; birds; etc.

In addition to eating the ripe fruits directly, cakes and a refreshing drink are prepared from the fruits. In making masuku (*Uapaca*) cakes, the ripe fruits are squeezed and water is added to the resulting pulp, followed by flour and sometimes eggs. The mixture is flattened into cakes that are later fried in cooking oil. A refreshing drink is prepared by pounding the de-seeded fruits with water and the extract is used to make a gruel. The latter is served as refreshing milk: "phala la masuku" (*Uapaca* porridge).

Because of their importance as a food source, the fruits are widely sold in the rural areas by the roadsides as well as in local markets. Some are ferried to urban markets where they are sold and are highly sought after by urban residents.

Nutritionally, *Uapaca* fruits are a good source of carbohydrates, energy, fibre, minerals, proteins and vitamins (Wehmeyer 1986). The carbohydrate and protein contents of *Uapaca* are higher than that of *Sclerocarya birrea* (A. Rich.) Hochst. The *Uapaca* fruits are therefore very important in supplementing the food and nutritional requirements of humans, especially where or when most of the popular domestic fruits such as mangoes, oranges and avocados are out of season or scarce.

A number of researchers have indicated that most fruits are important sources of carbohydrates, vitamins, Ca, Mg, and K (Peters & Maguire 1981, Peters & O'Brien 1984, Wehmeyer 1966). It has been shown that the composition of wild fruits does not appear to differ much from that of the better-known domestic fruits, and that their vitamin C content is in some cases even higher than that of domestic fruits. Wild edible fruits, therefore, may be very important in meeting the food and nutritional needs of people, especially in developing countries. Food scientists and nutritionists should therefore promote the consumption of many wild edible fruits that are found amongst our floras for good health and balanced diets.

Wine

The fruits of *Uapaca*, particularly *U. kirkiana*, are used to make wine. In Zomba, Malawi, the Mulunguzi Winery is making wine mainly from *U. kirkiana* fruits. This is the only wine factory in Malawi and perhaps one of the few in Africa that is based solely on indigenous fruits. Mulunguzi wine is widely sold in Malawi and some is exported. The abundant availability and ready supply of *Uapaca* fruits during the rainy season makes this venture a viable investment. It also shows that indigenous plants have economic potential for industrial and commercial development. This wine industry has suddenly transformed the *Uapaca* fruits into a cash-earning business for a number of households. In Nsomba Village, near Zomba, the local inhabitants earn considerable amounts of money from the sale of these fruits. They collect them in baskets on the slopes of the Zomba Mountain Forest Reserve, whose vegetation is dominated by *Uapaca*. This activity has now added economic value to the *Uapaca* trees and the monetary value of the fruits has led to the apparent ownership of the trees in the forest reserve by some households. Such economic interests are indirectly protecting the survival of the *Uapaca* trees from illegal cutting for fuelwood, charcoal, poles and other uses.

Fuelwood and Charcoal
The trees of *Uapaca* make very good firewood as the wood has a high calorific value. The wood also makes very good charcoal. These two uses have decimated *Uapaca* trees from most public land and as such most of the *Uapaca* trees today are found in protected areas; especially forest reserves, game reserves, national parks, graveyards, shrines, etc. Tree poachers are still cutting *Uapaca* trees from the various protected areas and if such illegal practices are not controlled *Uapaca* trees will become threatened in Malawi.

Building poles
Uapaca saplings are straight and this characteristic makes them attractive building materials for various local construction works. Saplings of *U. kirkiana, U. nitida* and *U. sansibarica* make good poles for construction of huts, houses, roofing, fences, etc.

Timber
The mature wood of *Uapaca*, particulary *U. kirkiana*, yields good timber which is fairly durable. It works easily and polishes well. The wood is of pleasing appearance and it almost resembles that of the American birch but of deeper reddish to red-brown colour (Coates Palgrave 1977, Williamson 1975). It is of medium weight, fine-grained and compact. The timber is also fairly resistant to termite attacks and boring insects. These attributes make *Uapaca* timber very popular to its users who are mainly rural artisans that manufacture various furniture types from it.

Herbal medicine
In Malawi, most edible plants are rarely used as herbal medicine. But an infusion of the roots of *U. kirkiana* is used as a remedy for indigestion (Coates Palgrave 1977), while an extract from the bark and leaves of *U. sansibarica* is used against stomach-ache (Clements 218: 1933, MAL).

Other uses
Uapaca trees form dominant and co-dominant vegetation types in Malawi. The plants protect watersheds, especially in montane and hilly areas and provide ideal habitats for various animal and plant species. For the latter, a variety of edible mushroom species grow where *Uapaca* species are dominant, suggesting therefore the existence of some mycorrhizal association. These mushrooms grow during the rainy season (November - April). They are an inexpensive source of nutrients and proteins, and are sold along roadsides and in markets, where the demand always outstrips the supply.

The research and development needs of *Uapaca* species

Little or no silvicultural, selection or breeding work on *Uapaca* species has been done in Malawi. But the food, nutritional, and medicinal values of the species make it a potential candidate for detailed improvement. The genetic variability of the species is poorly known and its responses to management are untested. Much unknown potential of this useful plant is being lost due to the degradation of the genetic base through deforestation and habitat loss. The development of *Uapaca* should be carried out through selection, silvicultural research, mycorrhizal research, and investigation of its pollination biology.

Selection and breeding

The present economic values of *Uapaca* species have been selected or bred by nature and man has made no attempts to improve the results of natural selection. There is an urgent need to make a country-wide survey of the species, aimed at selecting sweet and seedless provenances, bigger fruits, higher yielding trees, better resistance to bacterial, fungal or viral diseases, and insect attacks. The presence of five well-adapted *Uapaca* species in Malawi (Figure 1) offers great opportunity for breeding desirable characters. For instance *U. nitida* and *U. sansibarica* have very sweet but small fruits. The sweet character could be bred into *U. kirkiana* which has a bigger fruit. Another trait worthy of breeding is early maturity, as these species take a long time to mature. Detailed survey in the wild could identify some provenances possessing this desirable trait.

It would also be beneficial if such work extended to the other 55 species of *Uapaca* found elsewhere. The *Uapaca* species in Malawi may not possess all the characters that would breed a superior variety or hybrid for high quality fruit, fuelwood or medicine.

Silvicultural research

The silviculture of *Uapaca* species in Malawi has not been investigated. If we have to maximize the production potential of these species, there have to be some silvicultural trials for the identification of ideal climatic conditions, soils, altitude, temperature, fertilizer, spacing, pest control, etc. Such work should also look into the viability of seeds, their germination, and the propagation methods of the various species.

Mycorrhizal investigations

The *Uapaca* or *Brachystegia-Combretum-Uapaca*, or *Brachystegia-Isoberlinia-Uapaca* woodlands are habitats of many delicious and nutritious edible mushrooms during the rainy season (December-April). These mushrooms are picked in great amounts and are a source of income for many people who sell them along roadsides and in the local markets.

The occurrence of these mushrooms in the *Uapaca* woodlands is supposed to be associated with the *Uapaca* tree itself. A fungus-root association is suspected. This association has not been fully investigated and there is a need to initiate research on this important subject. If and when this mycorrhizal association is defined, then it will be possible to establish small *Uapaca*-integrated woodlots for the production of mushrooms, fruits, fuelwood, poles and medicine.

Pollination Biology
Uapaca species are dioecious but the mode of pollen transfer from the male to female plant is not known. As pollination is essential in the production of the fruits as well as viable seeds, knowledge of the pollination biology of *Uapaca* species would be very valuable. It has been reported that a heavy crop of *Uapaca* fruits is produced every two years (Coates Palgrave & Coates Palgrave 1956). Nobody has established the factors responsible for this phenomenon. Pollination, physiology and environmental factors could help explain such flowering and fruiting patterns.

General summary and conclusions

Malawi is blessed with a rich flora which contains a great variety of edible wild plants. *Uapaca* is one of the most valuable multipurpose plant groups which is widely distributed in Malawi. The species provide nutritious edible fruits, wine, fuelwood, building poles, timber and medicine. The species also form major vegetation types in Malawi woodlands, that act as watersheds for many rivers, and provide a habitat for many animals and plant species. A great variety of delicious and nutritious mushrooms is associated with *Uapaca* trees. The mushrooms are both a source of food, nutrients and income for many people.

Despite these valuable attributes, the potential of *Uapaca* species have not been harnessed to the full. The present products that humans are harvesting from these plants are the results of natural selection. There is therefore an urgent need for research to formulate a domestication and improvement programme for *Uapaca* species that should aim at selecting superior provenances, breeding for desirable traits, carrying out detailed silvicultural trials, studying mycorrhizal associations, and undertaking field studies on pollination biology. It is likely that such an approach can result in a number of improved *Uapaca* species. The end result will be improved cash earnings for the local people and further protection of *Uapaca* plants in the wild.

In conclusion, it is recommended that a team of taxonomists, silviculturists, mycologists and breeders should be formed which would work towards the improvement of *Uapaca* species in Malawi. Such a project should then be extended as a research network within the entire distributional range of *Uapaca*. The project should be funded by international organisations, if the objectives are to be achieved.

Acknowledgements
I would like to thank Prof. L.J.G. van der Maesen, General Secretary of AETFAT, for organising the XIVth AETFAT Congress and for funding my attendance of the Congress. Miss C. Mangasasa typed the paper, while Messrs. S. Kananji, E. Kathumba and I. Patel extracted *Uapaca* herbarium distribution data; to all of them I am grateful.

References
Blunt, S.J.L., 1990. Malawi's Changing Treescape. Dissertation for Diploma in Ecology. University of Kent. pp.140.
Coates Palgrave, K., 1977. Trees of Southern Africa. C. Struik Publishers, Cape Town, South Africa.
Coates Palgrave, O.H. & K. Coates Palgrave, 1956. Trees of Central Africa. National Publications Trust, Rhodesia & Nyasaland. University Press, Glasgow.

Maghembe, J.A. & J.H. Seyani, 1991. Multipurpose trees used by Smallholder Farmers in Malawi: Results of an Ethnobotanical Survey. AFRENA Report No. 42: pp. 30.

Peters, C.R. & B. Maguire, 1981. Wild Plant Foods of the Makapansgat Area: A Modern Ecosystems Analogue for *Australopithecus africanus* adaptations. J. Human Evol. 10: 565-583.

Peters, C.R. & E.M. O'Brien, 1984. On Hominoid Diet Before Fire. Curr. Anthropol. 25: 358-360.

Radcliffe-Smith, A., 1993. Notes on African Euphorbiaceae. XXI: *Uapaca*. Kew Bull. 48 (3): 611-617.

Seyani, J.H., 1990. The Potential Value of the Malawi Flora in Development. Paper presented at the National Conference on "Science and Technology for Development", Blantyre, Malawi.

Seyani, J.H., 1991. *Uapaca kirkiana* - Underutilized Multipurpose Tree Species in Malawi worth Some Development. International Commonwealth Science Council Workshop on "The Biodiversity of Traditional and Underutilized Crops", Valletta, Malta.

Wehmeyer, A.S., 1966. The Nutrient Composition of Some Edible Wild Fruits found in the Transvaal. S. Afr. Med. Journ. 40: 1102-1104.

Wehmeyer, A.S., 1986. The Edible Wild Plants of Southern Africa. National Food Research Institute, Pretoria. pp. 52.

Williamson, J., 1975. Useful Plants of Malawi. University of Malawi, Zomba.

Diversity of medicinal plants in West African habitats

N.H.A. COLE
Botany Department, Fourah Bay College, University of Sierra Leone, Freetown, Sierra Leone.

Abstract
This paper examines the habitats of 439 species of wild and cultivated plants in West Africa which are used in traditional medicine. About 30% of the medicinal plants in West Africa (138 species) are introduced into cultivation for other economic reasons. Most medicinal plants still grow in the wild, with 95 species in tropical closed forests (medicinal plant diversity H = 1.66), 80 species in secondary forests and regrowth (H = 1.34) and 82 species in savanna woodlands (H = 1.19). Sahel savanna arid lands and coastal sandy beach habitats are just as diverse with H = 1.55 but with only 31 medicinal species (7.1% of total). Wetland ecosystems are the least diverse (H = 0.90) and only a few medicinal plants are found in mangrove swamps and inland freshwater swamps (3.0%). The most important medicinal plants are understorey trees and shrubs in tropical closed forests (57.9%), not the timber trees; in tropical forest regrowth, the most important are scrambling shrubs and farmbush forbs (62.5%); trees and shrubs are important in savanna woodlands (75.6%) and also in the Sahel savanna.

Résumé
Les habitats de 439 espèces de plantes sauvages et cultivées de l'Afrique occidentale, utilisées en médecine traditionnelle, sont passés en revue. Environ 30% des plantes médicinales de l'Afrique occidentale (138 espèces) sont introduites en culture pour d'autres raisons économiques. La plupart des plantes médicinales croissent toujours à l'état sauvage, 95 espèces se rencontrent en forêt dense tropicale (H = 1,66), 80 espèces en forêt secondaire et recrû (H = 1,34) et 82 espèces en savane boisée (H = 1,19). Les savanes arides du Sahel et les formations littorales sur sable ont à peu près la même diversité (H = 1,55) mais ne comptent que 31 espèces médicinales (7,1% du total). Les écosystèmes des milieux humides sont les moins diversifiés (H = 0,9) et seules quelques plantes médicinales sont observées dans les marais des mangroves et ceux d'eau douce de l'intérieur (3,0%). Les plantes médicinales les plus importantes sont les arbres et arbustes du sous-bois des forêts denses tropicales (57,9%), à l'exclusion des grandes essences forestières; dans le recrû des forêts tropicales, ce sont les arbustes disséminés et les herbes des broussailles postculturales (62,5%); les arbres et les arbustes sont importants dans les savanes boisées (75,6%) et également dans la savane sahélienne.

Key words
Biodiversity, Ecological distribution, Habitats, Medicinal plants, West Africa.

Introduction

Habitat loss is of serious concern in the tropics as it threatens plant and animal species (genetic and biological resources) even before their potential economic use can be fully investigated. Nevertheless tropical forests and savanna woodlands are being deforested at an alarming rate.

Ecological study of medicinal plants is all the more urgent now due to the need for basic data and scientific information for use in management plans for in-situ conservation and sustainable exploitation of the genetic resources. Ecological study of habitat conditions of wild indigenous traditional drug plants will facilitate domestication and

L.J.G. van der Maesen et al. (eds.), The Biodiversity of African Plants, 704-713.

cultivation, not only in the geographic region of the species, but also in other countries with similar climate, soils and other growing conditions. Also, with the lack of foreign exchange to cope with the rising cost of drugs and other imports, and the low priority given to social services (health, education and rural development) under the Structural Adjustment Programme (SAP), many African countries are now taking measures to establish drug and pharmaceutical industries based on the traditional use of local medicinal plants. African Governments are now encouraging modern medical experts (doctors, pharmacists) to cooperate with local medicine-men in order to investigate traditional drug plants. Investigating drug plants in African countries today has to be an integrated research programme involving multidisciplinary team work by an analytical chemist, a taxonomic botanist and a pharmacist.

This paper aims to examine the habitats in West Africa where traditional drug plants occur in order to assess which are the most important habitats with regard to diversity of medicinal plants. This data will also be of value in future conservation plans.

Methods

To investigate the occurrence, distribution and habitat conditions of medicinal plants in West Africa, a list of such plants was compiled from various sources, e.g. the Fourah Bay College Herbarium and the Botanic Reserve (Cole 1990), the National Herbarium at Njala University College, the "Flora of West Tropical Africa" (Keay & Hepper 1954-1972), "The Useful Plants of West Tropical Africa" (Burkill 1985, Dalziel 1937) and Oliver-Bever's (1986) "Medicinal Plants in Tropical West Africa". About 90% of the medicinal plants listed in this study, are cited by Oliver-Bever (1986) with their active phytochemical compounds. The rest, listed from local herbaria and other sources, is commonly used in traditional medicine by various societies in West African countries. These form the basis for this comparative ecological description of the different plant communities in West Africa as resource bases for drug plants.

Each species of drug plant was assigned to the plant community where it occurs in the wild, based on detailed field knowledge of the species composition of the various plant communities in West Africa (Cole 1970, 1990). The physiognomic status of each of the species was also reviewed. The relative richness of the various West African habitats (or physiognomic fractions of these), in terms of medicinal plants, were then compared. Relative diversity was indicated in two ways: one by species richness, expressed as the total number of species of medicinal plants in each ecosystem; and the other by making use of information statistics:

$$H = - \sum_{i=1}^{S} p_i \ln p_i$$

where S is the total number of medicinal species and p_i is the proportion of species of drug plants in each of the plant communities comprising a major ecosystem. Shannon's H is one of the commonly used indices for species diversity based on the abundance of individuals in the sample (Shannon & Weaver 1963), and on equitability, i.e. the evenness of the distribution of individuals within the species population (Lloyd & Gelhardi 1964). In this case, based on a census of medicinal plants, the proportion of species (p_i) in the different plant communities within an ecosystem is used to calculate

Table 1. Habitats with medicinal plants in West African Plant Communities.

Major ecosystems	Plant communities under Keay's & Cole's physiognomic classification
Tropical Closed Forest	Tropical evergreen rain forests Tropical moist semi-deciduous forests
Tropical Swamp and Wetland	Inland valley swamp forests Gallery and fringing forests Mangrove swamp forests Riverine swamp grasslands
Secondary Forest Regrowth	Secondary forests Forest regrowth and thickets Farmbush weeds
Savanna Woodland	Guinean savanna woodlands Sudanian savanna woodlands Tall cane grass savanna
Sahel Savanna and Coastal Sandy Beach	Sahel steppe savanna Coastal park savanna Coastal strand Coastal thickets Lateritic pan grassland Sub-montane shrubs and grasses Montane grasslands
Cultivated Plants	

the medicinal plant diversity indices for a comparison between six selected ecosystems in West Africa.

This paper utilises the physiognomic classification of plant communities in Africa (Keay 1956), as elaborated by Cole (1970) into 19 plant communities commonly found in West Africa. During the assignment of medicinal plant species into these plant communities, it was found expedient to group the latter into six major ecosystems for a more meaningful analysis and comparison. Table 1 refers to the major groups of ecosystem used in this analysis, and the specific plant communities of Keay (1956) and Cole (1970) included in each group. It is noted that the plant communities within each group have similar physiognomy except for the heterogenous group - Sahel savanna and Coastal sandy beaches, where a similarity exists in physiological responses to aridity or to temperature.

Results

The results of the survey of drug plant distribution in West African plant communities are presented in Tables 2 and 3, and also in Appendix 1. Table 2 shows the total number of species and the percentages of medicinal plants recorded in each major ecosystem as defined in Table 1. Table 3 gives the distribution of medicinal plant species (numbers and percent of the group total) in terms of physiognomy or habitat types for each ecosystem group. The index of habitat diversity, H, of medicinal species in each ecosystem is also indicated. Medicinal plant species occurring in each life form or habitat group within each major ecosystem are listed in Appendix 1.

Table 2. Distribution of species of medicinal plants within ecosystem groups in West Africa, as shown by the number of species and their percentages.

Ecosystem groups	No. of species	% of species
Tropical Closed Forests	95	21.6
Tropical Swamps & Wetlands	13	3.0
Secondary Forest Regrowth	80	18.2
Savanna Woodlands	82	18.7
Sahel Savanna & Coastal Sandy Beaches	31	7.1
Cultivated Plants with medicinal properties	138	31.4
Total	439	100.0

Based on the number of medicinal plants species found in each major ecosystem (Table 2), tropical closed forests are the most diverse with 95 species. One must exclude the group of 138 cultivated plant species with medicinal properties as not forming a natural system, but one being manipulated for optimum exploitation by man. The least diverse ecosystems in terms of numbers of medicinal plant species are the Sahel savanna and coastal sandy beach (31 species) and the wetlands of mangrove swamps and inland freshwater swamps with only 13 species.

The medicinal plant diversity as calculated using the H diversity index is shown in Table 3. The closed tropical forest ecosystem with $H = 1.66$ and the Sahel savanna and coastal beach ecosystem with $H = 1.55$, are the most diverse. The latter is diverse in terms of available niches for medicinal plants rather than species number. The least diverse habitat for drug plant species is the wetlands ecosystem, i.e mangrove swamps and inland freshwater swamps, with $H = 0.90$ for the medicinal plant diversity.

More than half of the 95 drug plants in tropical closed forests occur as under-storey trees (C stratum with 28.4%) and shrubs (D stratum with 29.5% of the total), as shown in Table 3. Only 22.1% of drug plants are timber trees in the A or B stratum. In the case of secondary forest regrowth, most of the 80 drug plants found are farm-bush forbs and scrambling shrubs (62.5%).

In the savanna ecosystems, Table 3 shows that more than three-quarters of the 82 drug plants in the Guinea-Sudan savanna are trees or shrubs. The Sahel savanna and coastal sandy beach has 61.3% of the drug plants as Sahel shrubs (35.5%) or woody forbs (25.8%), and thorny Sahel trees account for 19.4%, as indicated in Table 3. Very insignificant numbers of drug plants were recorded in the tropical swamps and wetlands: 53.8% of these 13 drug plants are marshy shrubs and aquatics, and 38.5% are freshwater swamp forest trees.

Table 3. Medicinal plant diversity "H" and distribution of medicinal plant species in habitats or physiognomic groups in six major ecosystems in West Africa.

Ecosystem groups	No. of species	% of species
Tropical closed forests, H = 1.66		
A stratum timber trees	11	11.6
B stratum timber trees	10	10.5
C stratum understorey trees	27	28.4
D stratum understorey trees	28	29.5
E stratum ground herbs	8	8.4
Woody lianes	11	11.6
Total	95	
Tropical swamps and wetlands, H = 0.90		
swamp forest trees	5	38.5
mangrove swamp tree	1	7.7
marshy shrubs and aquatics	7	53.8
Total	13	
Secondary forest and regrowth, H = 1.34		
fast-growing trees	12	15.0
shrubby thicket forming scramblers	22	27.5
farmbush forbs	28	35.0
weedy herbs	18	22.5
Total	80	
Savanna woodlands, H = 1.19		
savanna trees	43	52.4
small trees and shrubs	19	23.2
woody forbs	8	9.8
savanna herbs	12	14.6
Total	82	
Sahel savanna and coastal sandy beach, H = 1.55		
Sahel trees	6	19.4
Sahel shrubs	11	35.5
coastal geophyte	1	3.2
woody forbs	8	25.8
scramblers	3	9.7
sand-binders	2	6.5
Total	31	
Cultivated plants with medicinal properties, H = 1.07		
food and cash crops	70	50.7
ornamental	50	36.2
industrial	11	8.0
medicinal	7	5.1
Total	138	

Appendix 1 lists 138 cultivated, exotic plants grown for economic reasons in West African countries. The majority are grown for non-medicinal purposes, but, incidentally, are also used medicinally. Most of these drug plants are food and cash crops (50.7%) and ornamentals (36.2%). Eight percent are used for industrial purposes and only a few cultivated plants are grown specifically for their medicinal value (5.1%).

Discussion

This study shows that of the 439 medicinal plant species growing in West Africa, only 31.4% of these have their gene pools protected by cultivation. The vast majority of West African drug plants, about 70% of the total, are still growing in the wild. These are threatened by deforestation, and their gene pools will be lost to the scientific and entrepreneurial community sooner or later. Furthermore, about 9.6% of the drug plants inventoried here have not been chemically analyzed, pharmaceutically tested or clinically screened for efficacy. Those cited in Oliver-Bever (1986) however, do have some information concerning this. The seriousness of the problem of loss of species diversity and genetic material form tropical closed forest and savanna woodland in West Africa can be put in its proper perspective when one considers that numerous other plant species are believed to have drug qualities and are being used by medicine-men in West Africa today! Since the earliest records by Dalziel (1937) which are now being revised by Burkill (1985, 1994), new records of plants used in traditional medicine are being uncovered regularly. These plant species, not yet scientifically explored for their drug and other qualities are fast disappearing from West Africa in the same way as other economic plant species are subjected to the loss of closed forest and savanna woodland habitats (Cole & Macfoy 1992).

This analysis shows that tropical closed forest is the most diverse habitat for West African drug plant species. The majority of these are found as C stratum understorey trees, shrubs and saplings. These C and D strata forest species are often destroyed during the harvesting of large timber trees in the A and B strata; but most seriously, they are the tree species cut down for firewood within protected forests under government licences. Most of the forested areas in West Africa are now covered in secondary forest regrowth of varying stages of maturity towards tropical closed forest with a five strata structure. It would be advisable for government licences to harvest firewood to be limited to the younger secondary forests where the fast-growing forest trees are dominant. Alternatively, firewood plantations with faster-growing exotic acacias can be developed to ease domestic energy demands on indigenous tree species, through agroforestry practices.

The problem of deforestation, loss of species diversity and of the gene pools of secondary forest regrowth plants is exacerbated by inadequate fallow periods for soil recuperation in abandoned farmlands after shifting cultivation, due to increased population pressure on arable land in West Africa. When the fallow period lasts for 10-15 years instead of the current 3-5 years in West Africa, then the secondary forest successional stages from weeds and farmbush species to thicket and then to fast-growing forest tree species are fully realized. This study shows that secondary forests and its regrowth are less diverse in medicinal plants than its climax - mature tropical closed forests.

Moreover, farmbush forbs and weeds form 10.5% of the drug plant species in West

Africa, indicating the potential usefulness of this resource. Since inadequate fallow periods with loss of soil fertility leads to a deflected succession of derived savanna (Cole 1970), this will eventually erode the gene pools for the farmbush species, the scrambling shrubby species forming thickets, and the fast-growing secondary forest tree species in West Africa, with loss of the drug plants and other economic species.

The medicinal plant diversity for savanna woodlands in West Africa is in the same range as that for secondary forests and regrowth, although deforestation is more intense in the savanna. The situation is most critical for the typical savanna tree and shrub species which constitute 75.6% of the drug plants in that community.

The Sahel savanna with its ecologically related coastal sandy beach, has a similar medicinal plant diversity index to tropical closed forest. The process of desertification, hover, is a limiting factor in the Sahel, as desert sand dunes creep over farmlands, roads, and human settlements. As population pressure from man and cattle increases during good spells of rainfall on the sparse Sahel vegetation, the arid and semi-arid land become denuded of its trees and shrubs up to 30-50 km around the large urban population centres. Thus, through desertification, many drug plants in the Sahel are threatened, such as: the typical low spiny trees *Balanites aegyptiaca* and *Acacia nilotica* and the bushy shrubs *Lawsonia inermis* and *Senna italica*.

Between half to three-quarters of known drug plant species found in the six major ecosystems in East Africa are in danger of being lost to science, through deforestation in the moist closed forests and savanna woodlands, and by desertification in the Sahel savanna. Therefore, this review recommends more intensive effort at *ex-situ* conservation of medicinal plants in botanical gardens, orchards, and as components in agroforestry and alley farming practices in West Africa.

References

Burkill, H.M., 1985. Useful Plants of West Tropical Africa, 2nd ed., vol. 1 - Families A-D. Royal Botanic Gardens, Kew, UK.

Burkill, H.M., 1994. Useful Plants of West Tropical Africa, 2nd ed. vol. 2 - Families E-I. Royal Botanic Gardens, Kew, UK.

Cole, N.H. Ayodele, 1970. The vegetation of Sierra Leone. Njala University College Press, University of Sierra Leone, Freetown.

Cole, N.H. Ayodele, 1990. Fourah Bay College Botanic Reserve: an ecological treasure for conservation in Sierra Leone. Tropical Africa Botanic Gardens Bulletin 2: 25-28.

Cole, N.H. Ayodele & C.A. Macfoy, 1992. Underexploited tropical economic plants in West Africa. Journal of the Faculty of Pure and Applied Science 1: 40-56.

Dalziel, J.M., 1937. Useful plants of West Tropical Africa, 1st ed. Crown Agents, London, UK.

Keay, R.W.J., 1956. African vegetation meeting at Yangambi. Nature 178: 1273-1274.

Keay, R.W.J. & F.N. Hepper, 1954-1972. Flora of West Tropical Africa. 2nd ed., vols I-III. Crown Agents, London.

Lloyd, M. & R.J. Ghelardi, 1964. A table for calculating the equitability component of species diversity. J. Anim. Ecol. 33: 217-225.

Oliver-Bever, B., 1986. Medicinal plants in tropical West Africa. Cambridge University Press, UK.

Shannon, C. & W. Weaver, 1963. The mathematical theory of communications. Univ. of Illinois Press, Urbana, USA.

Appendix 1
Medicinal plants in relation to West African Plant Communities
(* indicates species not listed in Oliver-Bever 1986).

Tropical closed forests
A stratum timber trees (11 species):
Antiaris africana, Bombax buonopozense, Bombax brevicuspe, Ceiba pentandra, Erythrophleum guineense, E. ivorense, Hunteria eburnea, Khaya ivorensis, Mansonia altissima, Pausinystalia johimbe, Terminalia ivorensis.*
B stratum timber trees (10 species):
Chlorophora excelsa, Corynanthe pachyceras, Enantia chlorantha, Erythrina mildbraedii, Garcinia afzelii, G. kola*, Lonchocarpus sericeus, Pachypodanthium staudtii, Pentaclethra macrophylla, Tetrapleura tetraptera.*
C stratum understorey trees (27 species):
Alstonia boonei, Anthocleista nobilis, Blighia sapida, Dialium guineense, Elaeophorbia grandifolia*, Enantia polycarpa, Erythrina vogelii, Ficus umbellata, Funtumia africana, F. elastica, Holarrhena floribunda, Hunteria elliotii, H. umbellata, Isolona campanulata, Mareya micrantha*, Morinda lucida, Phoenix reclinata*, Picralima nitida, Polyalthia oliveri, Rauvolfia caffra, Sterculia setigera, Soyauxia floribunda, Tabernaemontana crassa, T. pachysiphon, Zanthoxylum gilletii, Z. rubescens, Z. leprieuri.*
D stratum understorey shrubs (28 species):
Alchornea floribunda, Allophylus africanus, Caloncoba echinata, C. glauca, C. welwitschii, Dichapetalum toxicarium, Diospyros canaliculata, Harrisonia abyssinica, Lonchocarpus cyanescens, Massularia acuminata, Newbouldia laevis, Olax latifolia, Pauridiantha viridiflora, Penianthus zenkeri, Quassia africana, Quisqualis indica, Rauvolfia mannii, R. vomitoria, Strophanthus gracilis, S. gratus, S. hispidus, Strychnos afzelii, S. nux-vomica*, Synsepalum dulcificum, Tabernaemontana glandulosa, Vernonia conferta, Voacanga africana, V. thouarsii.*
E stratum herbaceous ground flora (8 species):
Acanthospermum hispidum, Adenia cissampeloides, Aframomum granum-paradisi, A. latifolium*, A. melegueta*, Costus afer, Cryptolepis sanguinolenta, Torenia thouarsii.*
Woody lianes (11 species):
Adenia lobata, Alafia multiflora, Chasmanthera dependens, Cnestis ferruginea, Combretum racemosum, Dioscoreophyllum cumminsii, Jateorhiza macrantha, Parquetina nigrescens, Stephania dinklagei, Strychnos camptoneura, S. usambarensis.

Tropical swamps and wetlands
Swamp forest trees (5 species):
Diospyros mespiliformis, Lonchocarpus sericeus, Mitragyna stipulosa, Nauclea diderrichii, N. pobeguinii.
Mangrove swamp trees (1 species):
Rhizophora racemosa.
Marshy shrubs and aquatics (7 species):
Aeschynomene indica, Eclipta prostrata, Hygrophila auriculata, Malouetia heudelotii, Nymphaea lotus, Polygonum senegalense, Spilanthes uliginosa.

Secondary forest regrowth
Fast-growing secondary forest trees (12 species):
Anthocleista vogelii, Anthostema senegalense, Canthium subcordatum*, Carapa procera, Craterispermum laurinum*, Dichrostachys glomerata, Maesopsis eminii, Mallotus oppositifolius, Millettia aboensis, Morinda geminata*, Musanga cecropioides*, Porterandia cladantha.*

Scrambling shrubs forming thicket (22 species):

Abrus precatorius, Alchornea cordifolia, Baissea leonensis, Cissampelos owariensis, Combretum rhodanthum, Embelia schimperi, Epinetrum cordifolium, Euphorbia lateriflora, Kolobopetalum auriculatum, Lantana camara, Millettia barteri, Morinda longiflora, Parquetina nigrescens, Pergularia daemia, Pleioceras barteri, Plumbago zeylanica, Ritchiea longipedicellata, Sarcostemma viminale, Secamone afzelii*, Strophanthus sarmentosus, Uncaria africana var. africana, U. africana var. angolensis, Uvaria chamae.*

Farmbush forbs (28 species):

Achyranthes aspera, Brucea guineensis, Cardiospermum halicacabum, Chenopodium ambrosioides, Coccinia grandis, Crotalaria retusa, Datura innoxia, Gongronema latifolium, Hyptis suaveolens, Ipomoea mauritiana, Justicia insularis, Leucaena leucocephala, Momordica charantia, Mostuea hirsuta, Mucuna pruriens, Ocimum gratissimum, Pueraria phaseoloides, Scoparia dulcis, Securinega virosa, Senna alata, S. occidentalis, S. podocarpa, S. tora, Smeathmannia pubescens, Solanum nigrum, S. torvum, Tephrosia vogelii, Waltheria indica.*

Weedy herbs (18 species):

Ageratum conyzoides, Alternanthera repens, Bidens pilosa, Boerhavia diffusa, Borreria verticillata, Centella asiatica, Chamaecrista absus, Desmodium gangeticum, Euphorbia hirta, E. prostrata, E. thymifolia, Gynandropsis gynandra, Phyllanthus niruri, Schwenckia americana, Sida acuta, S. cordifolia, Spigelia anthelmia, Vernonia cinerea.

Savanna woodlands (Guinea/Sudan)

Savanna trees (43 species):

Acacia nilotica, Adansonia digitata, Afrormosia laxiflora, Andira inermis, Annona senegalensis, Anogeissus leiocarpa, Anthocleista djalonensis, A. procera, Antidesma venosum, Bridelia ferruginea, Vitellaria (= Butyrospermum) paradoxa, Cassia sieberiana*, Crataeva religiosa, Crossopteryx febrifuga, Croton macrostachyus, Ekebergia senegalensis, Entada africana, Erythrina excelsa, E. sigmoidea, Grewia bicolor, Hymenocardia acida, Khaya senegalensis, Kigelia africana, Lonchocarpus laxiflorus, Lophira procera*, Mitragyna inermis, Nauclea latifolia, Phyllanthus discoideus, Piliostigma reticulatum, Pseudocedrela kotschyi, Sclerocarya birrea, Securidaca longepedunculata, Swartzia madagascariensis, Tamarindus indica*, Terminalia avicennioides, T. glaucescens, T. laxiflora, T. macroptera, Trichilia roka, Uapaca togoensis*, Vernonia colorata, Voacanga thouarsii, Vitex doniana.*

Savanna woodland, small trees and shrubs (19 species):

Afraegle paniculata, Callichilia barteri, C. monopodialis, Carissa edulis, Combretum sericeum, Feretia apodanthera, Grewia carpinifolia, G. lasiodiscus, Gymnema sylvestre, Leptactina densiflora, L. senegambica, Mundulea sericea, Nauclea latifolia, Paullinia pinnata, Pergularia chamae*, Phytolacca dodecandra, Strychnos spinosa, Teclea sudanica, Vernonia colorata.*

Savanna woody forbs (8 species):

Adenium obesum, Centaurea perrottetii, Lasiosiphon kraussianus, Opilia celtidifolia, Pachycarpus lineolatus, Tephrosia purpurea, Vernonia guineensis, V. nigritiana.

Savanna herbs (12 species):

Canscora decussata, Cissampelos mucronata, Cissus quadrangularis, Dioscorea dumetorum, Drosera indica, Oldenlandia affinis, Polygonum salicifolium, Urginea altissima, U. indica, Vernonia pauciflora, Xysmalobium heudelotianum, X. membraniferum.*

Sahel savanna and coastal sandy beaches

Sahel trees (6 species):

Acacia senegal, Balanites aegyptiaca, Capparis decidua, Commiphora africana, C. molmol*, Salvadora persica.*

Sahel shrubs (11 species):

Astragalus gummifer, Combretum micranthum, Diospyros tricolor, Guiera senegalensis,*

*Lawsonia inermis, Leptadenia hastata, L. pyrotechnica, Senna alexandrina**, *S. angustifolia, S. italica**, *Sophora occidentalis*.

Woody forbs and herbs (8 species):

Anagallis arvensis, Arnebia hispidissima, Atractylis aristata, Calotropis procera, Lupinus tassilicus, Tinospora bakis, Verbena officinalis, Withania somnifera.

Scramblers (3 species):

*Cocculus pendulus, Tiliacora warneckii**, *Zanthoxylum zanthoxyloides*.

Sand binders (2 species):

Dioclea reflexa, Ipomoea pes-caprae.

Coastal geophytes (1 species):

Hymenocallis littoralis.

Cultivated plants with drug properties

Food and cash crops with drug properties (70 species):

*Aframomum melegueta**, *Allium cepa* (onion), *A. sativum* (garlic), *Anacardium occidentale, Ananas comosus, Annona muricata, A. reticulata, A. senegalensis, A. squamosa, Arachis hypogaea* (peanuts), *Brassica oleracea, Cajanus cajan* (pigeon pea), *Camellia thea, Canavalia ensiformis, Capsicum annuum* (red pepper), *Capsicum frutescens* (cayenne pepper), *Carica papaya* (pawpaw), *Cinnamomum zeylanicum, Citrus aurantium, C. limonum, C. nobilis, Cocos nucifera, Coffea arabica, C. liberica**, *C. stenophylla, Cola acuminata, C. nitida, Corchorus olitorius, Cucurbita maxima, C. pepo, C. domestica, Cymbopogon citratus, Cyperus rotundus, Dioscorea dumetorum, Elaeis guineensis, Emilia sonchifolia**, *Glycine soja**, *Lepidium sativum, Mangifera indica, Manihot esculenta, Mentha piperita**, *M. spicata**, *M. viridis**, *Momordica charantia, Morus alba, Musa paradisiaca, M. sapientum, Myristica fragrans, Ocimum canum, O. gratissimum, Oryza sativa, Pachyrhizus erasus**, *Passiflora edulis, P. foetida, Phaseolus vulgaris, Physostigma venenosum, Piper guineense, P. nigrum, Pisum sativum, Psidium guajava, Saccharum officinale**, *Sesamum indicum, Syzygium cumini, S. guineense var. guineense, Tamarindus indica**, *Theobroma cacao**, *Triticum aestivum, Vernonia amygdalina, Zea mays**, *Zingiber officinale*.

Ornamental plants with drug properties (50 species):

*Acacia farnesiana, Acalypha wilkesiana, Afrormosia laxiflora, Albizia lebbeck, Allamanda cathartica, Aloe vera, Alpinia speciosa, Argemone mexicana, Asclepias curassavica, Azadirachta indica, Bauhinia purpurea, B. tomentosa, B. variegata, Bryophyllum pinnatum, Caesalpinia bonduc, C. pulcherrima, Cassia fistula, Catharanthus roseus, Chrysanthemum cinerariifolium**, *Cryptostegia grandiflora, Dalbergia sissoo, Datura candida, D. metel, D. stramonium, Derris microphylla, Haematoxylum campechianum, Hibiscus rosa-sinensis, H. vitifolius, Hymenocallis littoralis, Ipomoea purpurea, Jasminum sambac, Jatropha curcas, J. gossypiifolia, J. podagrica, Lagerstroemia speciosa, Medicago sativa, Melia azedarach, Mesua ferrea, Mimusops elengi, Nerium oleander, Nymphaea lotus, Plumeria rubra, Rhoeo spathacea, Samanea saman, Sapindus trifoliatus, Tecoma stans, Thevetia peruviana, Trigonella foenum-graecum, Vanilla planifolia**, *Vitex agnus-castus*.

Industry-based plants with drug properties (11 species):

*Agave sisalana**, *Aloe ferox**, *A. vera, Bixa orellana, Corchorus olitorius, Eucalyptus globulus, Gossypium barbadense, G. herbaceum, G. hirsutum, Haematoxylum campechianum, Santalum album*.

Cultivated medicinal plants (7 species):

*Argemone mexicana, Cannabis sativa, Cinchona calisaya, C. succirrubra, Citrus aurantifolium**, *Moringa oleifera, Ricinus communis*.

Misuses and abuses in self-medication with medicinal plants: the case of *Erythrophleum* in Ghana

D.K. ABBIW
Botany Department, University of Ghana, Legon, Ghana.

Abstract
A fairly large proportion of the people in both the rural areas and urban centres in developing countries still rely on plants for their everyday health needs. Usually the patient first consults a herbalist or spiritual healer who diagnoses the ailment and prescribes a remedy. However, instances of self-medication also occur, occasionally resulting in complications and sometimes death, either because the wrong plant was collected to start with or an overdose was taken. There are several causes leading to self-medication with medicinal plants, including instances of abuse and misuse. Recommendations are made to minimise or to prevent further recurrence of this practice, specifically the mistaken use of *Erythrophleum suaveolens* (the poisonous Ordeal Tree) for *Khaya senegalensis*.

Résumé
Une bonne partie des gens, tant dans les zones rurales que dans les centres urbains des pays en voie de développement, ont toujours recours aux plantes pour leurs problèmes de santé quotidiens. En général, le malade consulte d'abord un herboriste ou un guérisseur, qui donne un diagnostic de la maladie et prescrit le remède. Cependant, il existe aussi des cas d'automédication, pouvant parfois entraîner des complications et même des décès; ceci peut être dû soit à l'utilisation d'une plante mal identifiée soit à la consommation d'une dose excessive. Il existe différentes motivations qui conduisent à l'automédication à l'aide de plantes médicinales, y compris les abus ou les mauvais usages qu'elle peut entraîner. Il est recommandé de réduire au minimum ou de prévenir tout recours à cette pratique, notamment l'utilisation erronée d'*Erythrophleum suaveolens* (poison d'épreuve) à la place de *Khaya senegalensis*.

Key words
Medicinal plants, Mistaken self-medication, Secret knowledge.

Introduction

Humans have used plants as medicine for a very long time, and still continue to depend to the present day on plants, their extracts and plant-derived drugs for various ailments. In the rural parts of many developing countries, like Ghana, the proportion of people who rely on plants for their everyday medication is correspondingly higher than it is in the urban centres. The reasons for this dependence on plant medicine among rural dwellers are five-fold:

1 An estimated 80% of the population in developing countries live in rural areas, where there is one medical officer to 70,000 persons, as compared to the urban centres where the ratio is 1 to 4,000.
2 There are few hospitals or health posts in developing countries and they are not easily accessible to those who live in remote areas.

L.J.G. van der Maesen et al. (eds.), The Biodiversity of African Plants, 714-718.

Table 1. Common medicinal plants well-known to rural people in Ghana.

Species	Part used and preparation	Disease/Ailment
Azadirachta indica A. Juss.	Leaf decoction in hot bath	Fever
Phyllanthus niruri Thunb.	Decoction of whole plant	Fever
Alternanthera pungens H.B. & K.	Plant maceration in enemas	Purgative
Ananas comosus Merill	Peelings, decoction	Jaundice
Euphorbia hirta L.	Whole plant, added to food	Galactogogue
Secamone afzelii K. Schum.	Whole plant/latex	Chest pain/boils
Carica papaya L.	Chewed seeds	Vermifuge
Portulaca oleracea L.	Ground leaf paste applied	Boils
Boerhavia sp.	Plant, decoction	Amenorrhoea
Penianthus zenkeri Diels	Roots chewed or used to prepare bitters	Aphrodisiac
Sphenocentrum jollyanum Pierre	Chewed root	Aphrodisiac, Indigestion
Mangifera indica L.	Bark infusion	Diarrhoea
Senna alata (L.) Roxb.	Leaf decoction	Laxative
(= *Cassia alata*)	Crushed young leaves	Ringworm
Abrus precatorius L.	Chewed leaves	Cough
Manihot esculenta Crantz	Chewed tuber applied	Stop bleeding
Tamarindus indica L.	Chewed bark	Hiccups
Waltheria indica L.	Chewed root	Toothache
Xylopia aethiopica A. Rich.	Fruit paste to temple	Headache

3 Western or imported drugs are expensive and many of the poor rural dwellers just cannot afford to buy them.
4 Medicinal plants are more easily available and comparatively inexpensive in rural areas. They are often available at no cost at all.
5 The custodians of this knowledge and the experienced practitioners live mainly in rural areas.

Thus the first option when a person in rural areas is sick, is either to administer some sort of plant medication himself or, at best, to consult a herbalist. However, if a "supernatural" origin be suspected, a priest-healer is sought (Fink 1989).

Many common plants and the diseases they are normally used to cure are no secret among rural communities. This knowledge is freely exchanged or offered as need be. Table 1 gives some examples of these plants and the diseases they are used to cure.

Traditional healers and secrecy

The traditional medical practitioner, whether he is termed a medicine-man, native-doctor, herbalist, fetish-man, witch-doctor or juju-man, usually acquires this talent by patrimony or by initiative training, improving upon it with practice.

How the curative properties of herbs become associated with specific ailments, in the first place, is a subject of controversy. Elders and medicine-men have put up several possibilities: trial-and-error, observing the feeding habits of animals, observing plant behaviour, sympathetic magic, and understanding the language of herbs. However, revelation by ancestral or celestial spirits, deities, and dwarfs is, by far, believed to be the widest source of information. The revelation can occur either in a dream, a trance, a vision, or during spirit possession. It is also traditionally believed that witchcraft, devilry and the occult are sources providing insights to the secrets of medicinal plants.

Once the knowledge of the healing property of a particular plant is acquired, it remains the "copyright" of the medicine-man, a property he jealously guards. Consequently, a lot of secrecy surrounds traditional medicinal practice. The secrecy covers not only the plant or plant part used, but also its preparation, dosages, concentrations and ingredients added (if any). There are two main reasons for this secrecy. One is to maintain the reputation of the healer. For instance, Wickens (1987) observes in support of this that the active ingredients of many traditional folk medicines are often deliberately obscured by the presence of nonactive materials in order to conceal the identity of the former and preserve the reputation of the medicine-man. Secondly, the secrecy protects the copyright from infringement.

While these are advantages to the healer in concealing knowledge of plant remedies, there are equally serious disadvantages for the general public. Where the traditional healer is illiterate (as it is with many herbalists in Ghana), humankind, the world and science lose the wealth of knowledge accumulated over the years with the death of such a healer. This is a serious set-back. Also, since some people still do not appreciate the usefulness of protecting and conserving plants, some knowledge of their curative properties could perhaps help to change this attitude.

On the other hand, secrecy among herbalists helps to prevent the general misuse or abuse of plant medicine, in self-medication, and by quacks, false or unqualified herbalists and drug peddlers. Two of the well-known possible risks in buying from these peddlers are: 1) being sold the wrong plant through the ignorance of both parties; and 2) outright fraudulent practices, like peddling the seeds of *Adenanthera pavonina* L. (Bead Tree) as multivitamins because the seeds resemble this allopathic drug in colour, size and shape.

Self-medication

Self-medication may be defined as prescribing and administering a medication without consulting a qualified practitioner. In plant medicine it could also involve the collection of the plant.

With some of the common diseases and disorders like asthma, diabetes, earache, headache, constipation, indigestion, fever, ringworms, itch, sores, bleeding from minor cuts, etc., self-medication with plant medicine is one of the normal methods used as remedy by elders (parents, grandparents, knowledgeable individuals, etc.), besides the practitioners themselves. However, even with these elders, serious diseases like convulsions, tetanus, leprosy, elephantiasis, guinea worm, gastro-intestinal disorders, genito-urinary troubles, impotency, sterility, etc., are always referred to practising herbalists or to a specialist in that particular disease. There are even instances where qualified traditional medical practitioners refer patients to other practitioners.

Improper self-medication with medicinal plants arises when individuals are forced by circumstances like poverty, or are deceived by equally unqualified drug peddlers, or both. With the correct method of preparation, concentrations and dosages unknown, coupled with the possibility of collecting the wrong plant to start with, self-medication is not only risky but could be fatal. Of the four possible unknowns, the last one (collecting the wrong plant) has so far proved to be the most dangerous of all.

To a casual observer some plants do look alike, or are the same, though they may be taxonomically "miles apart", that is in completely different plant families with entirely different pharmacological properties.

The case of *Khaya* and *Erythrophleum*

Two such trees sometimes confused are *Khaya senegalensis* (Desv.) A. Juss. (Dry-Zone Mahogany; Meliaceae), and *Erythrophleum suaveolens* (G. et P.) Brenan (the Ordeal Tree; Caesalpinioideae). The confusion is compounded by the fact that both have been planted as avenue trees at the University of Ghana, Legon; at some places adjacent to each other.

Khaya is commonly debarked and sold in medicine markets. The decoction serves as blood tonic, febrifuge and virility tonic, among many of its attributes. *Erythrophleum* is poisonous to both man and beast. There are seven active principles including erythrophleguine which acts as a local anaesthetic and cardiac inhibitor. It decreases respiration, and in overdoses there are symptoms of depression of the circulation, difficulty in breathing, vomiting and convulsion, the latter resulting from direct action on the medulla oblongata (Abbiw 1990); and finally death. The remaining six poisons are cassaidine, nor-cassaidine, cassaine, cassaimin, erythrophlamine and homophleine.*

Statistics on fatalities in self-medication of medicinal plants generally are not available. However, in Legon alone four such cases resulting from confusing *Erythrophleum* with *Khaya* have been recorded in the last decade: one recorded incident each in Achimota and Akweteman; and Irvine (1961) reports an incident of *Erythrophleum* poisoning of two men in Saltpond. Incidentally, all the recorded instances were in urban centres, and those at the University of Ghana, Legon, happened within walking distance of the hospital.

---

* **Trial by ordeal**

The bark of *Erythrophleum*, like the seeds of *Physostigma venenosum* Balfour (the Ordeal or Calabar Bean), was used in former times in ordeal trials for serious offenses like alleged witchcraft or adultery. The belief then was that those who survived drinking a mixture made from it were innocent, but the guilty died. However, if he or she confessed, then an antidote or an emetic was given. By this system of trial, the investigation, sentence and punishment were carried out simultaneously.

Only decoctions prepared from the bark of young trees were used in these trials, because the bark of older trees contain such concentrations of these poisons that the victim died a few minutes after swallowing the decoction. Trial by ordeal was not only practised in traditional Africa. Gluckman (1963) writes, 'When David Livingstone came across the ordeal in the Zambesi, he told his African followers about the water-test formerly in use in Scotland: the supposed witch, being bound hand and foot, was thrown into a pond; if she floated, she was considered guilty, taken out and burned; but if she sank and drowned, she was pronounced innocent'. The author continues, 'Livingstone, with his usual dry humour, added: 'The wisdom of my ancestors excited as much wonder in their minds as their custom did in mine'.

Recommendations

1 A task force comprising the Ministries of Health and Information has to be set up to educate the general public, particularly those in the income group that tend to practise self-medication with medicinal plants, on the possible dangers involved.

2 It could be argued that one of the main reasons for trying plant medicine in self-medication is not the absence of hospitals or health posts, but the hospital fees and the cost of imported drugs. The Ministry of Health can go a step further by giving free medical care to the poor.

3 Since unqualified herbalists and drug peddlers are a major part of the problem, a system could be worked out:
   a. to weed out quacks, and
   b. require qualified peddlers and practitioners to label their drugs and provide some authentication assuring correct botanical identity and efficacy.

4 As a matter of urgency, the differences between *Khaya senegalensis* (Dry-Zone Mahogany) and *Erythrophleum suaveolens* (the Ordeal Tree) have to be made clear to all concerned to avoid possible confusion of one tree for the other in future.

5 The practice of planting the two species together as avenue trees as at the University of Ghana, Legon, and at the Achimota Secondary School, has to be avoided, because this increases the possibility of one tree being confused for the other.

References

Abbiw, D.K., 1990. Useful Plants of Ghana. Intermediate Technology Publications, London & Royal Botanic Gardens, Kew.

Fink, H.E., 1989. Religion, Disease and Healing in Ghana. A Case Study of Traditional Dormaa Medicine. Trister Wissenschaft, Germany.

Gluckman, M., 1963. Custom and Conflict in Africa. Basil Blackwell, Oxford.

Irvine, F.R., 1961. Woody Plants of Ghana. Oxford University Press, London.

Wickens, G.E., 1987. Ecosystem Data Base and Geographic Distribution of Molluscidal Plants. In: K.E. Mott (ed.). Plant Molluscides. Wiley & Sons, New York.

An alternative to *Eucalyptus*: *Terminalia sericea*

M.G. BINGHAM
Zambezi Livestock & Lands Ltd., P.O. Box 35355, Lusaka, Zambia.

Abstract
Terminalia sericea produces tough, durable and termite-resistant poles. The species is widespread on sandy soils in the medium to low rainfall areas of southern and south-central Africa. Where it produces straight poles, it is the preferred timber for building. Its variability in habit, and its apparent ability to produce hybrids with related species, indicate good prospects for genetic improvement. Although not as fast growing as *Eucalyptus*, its resistance to termites, both during growth and as timber, together with its tolerance of marginal sandy soils which have no agricultural potential, indicate a potential for development as a commercial timber.

Key words
Terminalia sericea, Timber, Zambesia.

Uses

Terminalia sericea DC. (Combretaceae) is the preferred timber for poles used in the construction of houses on the Kalahari Sands of south-western Zambia. The required length for most purposes is 2.5-3 m. After removal of the bark, poles are soaked in water for some weeks to remove the sap. This treatment is said to increase the durability. With the depletion of woodlands around the urban centres of Western Province, a trade in poles has developed to supply the urban markets.

The tree has many other uses in Zambia's Western Province (Barotseland). The timber is used to manufacture ox yokes. The bast fibre is used for ropes, the bark provides a yellow dye, and the leaves are used as a mordant in dyeing. Root extracts are used for treating venereal diseases and diarrhoea. Bees are attracted to the flowers.

Numerous records of the use of species of *Terminalia* for timber are to be found in the literature (Burkill 1985, Coates Palgrave 1977). Although the quality of the timber varies greatly from one species to another, many are recorded as being resistant to termites.

Distribution

The species has a Zambesian distribution, occurring in semi-arid to medium rainfall areas of southern and south-central Africa: Angola, Zaire (Katanga), Tanzania, Zambia, Malawi, Mozambique, Zimbabwe, Swaziland, South Africa, Botswana and Namibia.

Ecology

Terminalia sericea is particularly common on sandy soils in the low to medium rainfall areas of the Zambesian region. In miombo woodland it is much suppressed, occurring mainly as a pioneer at the dambo margins, where it frequently forms pure stands. In fire-degraded woodland it is more competitive than the miombo dominants, which it

L.J.G. van der Maesen et al. (eds.), The Biodiversity of African Plants, 719-721.

Figure 1. A bundle of 4 m poles of *Terminalia sericea*.

tends to replace. In towns and villages, where other miombo species may be at a disadvantage, *T. sericea* can grow to canopy size, indicating that it could do well on deep, well-drained soils if given protection from more competitive species.

The species is variable in habit and intermediate forms commonly occur wherever there are related species in the vicinity. There appears, therefore, to be scope for genetic improvement. *T. trichopoda* Diels, which is almost as widespread as *T. sericea*, but more local in occurrence, is a taller tree, with a longer bole. The escarpment species *T. sambesiaca* Engl. & Diels has an even better form. It is uncommon and restricted to river banks and rocky outcrops.

The relatively slow growth of *T. sericea* trees is compensated for by its ability to thrive on marginal soils which are of little or no value to agriculture. The young trees are a favourite browse of elephant and antelope, which may be a disadvantage in certain areas.

Habit, form and reproduction

Initially the tree is low growing, with a layered crown. The annual elongation of the stem results in successive layers of the canopy. The resulting stem is usually crooked, although the form varies with locality. After lopping a more or less straight pole is produced.

Terminalia spp. usually fruit prolifically every year, but most fruits appear to be sterile. Good germination occurs in some years. There is no evidence of suckering, which is prevalent in *Combretum*.

Commercial prospects

Silvicultural work is needed to develop the potential of the commercial timber potential of the *Terminalia* species in the Zambesian region.

References

Burkill, H.M., 1985. The Useful Plants of West Tropical Africa. Vol. 2. Families E-I. Royal Botanical Gardens, Kew.

Coates Palgrave, K., 1977. Trees of Southern Africa. G. Struik Publishers, Cape Town.

Les plantes antidiarrhéiques chez les Sundi de la sous-préfecture de Kindamba (Congo)

M. BITSINDOU & J. LEJOLY
Laboratoire de Botanique Systématique et de Phytosociologie, Université Libre de Bruxelles, Av. Paul Héger 28, CP 169, 1050 Bruxelles, Belgique.

Abstract
The Sundi of Kindamba (Congo) use 19 plant species to treat diarrhoea. In the Sundi traditional phytotherapy of diarrhoea, 18 recipes have been recorded from 10 mothers and 2 healers in Kindamba. Those recipes include 19 species often used separately. The species most often used are: *Garcinia huillensis* (Oliv.) Gilg (cited 8 times), *Mangifera indica* L. (8), *Psidium guajava* L. (6), *Alchornea cordifolia* (Schum. & Thonn.) Mull. Arg. (6), *Hymenocardia acida* Tul. (4) and *Trema orientalis* (L.) Blume (4). Ten species have already been cited in Congo and other African countries as antidiarrhoeic but the role of 9 species is indicated for the first time.

Key words
Diarrhoea, Congo, Sundi, Traditional medicine.

Introduction

La sous-préfecture de Kindamba est située entre 2°45' et 4°00' de longitude sud, 13°53' et 14°45' de latitude ouest. Elle est incluse dans la région naturelle de la Léfini, région dominée essentiellement par des formations herbeuses. Les formations forestières y sont peu importantes et irrégulièrement distribuées dans le fond des vallées, sous forme de galeries forestières plus ou moins dégradées, sur le sommet des collines ou sur les pentes les plus abruptes (Descoings 1975). Les Sundi, une des ethnies du groupe Kongo, forment la majorité de la population de cette sous-préfecture. En dehors du travail de Bouquet (1969) qui donne des indications thérapeutiques de quelques espèces en médecine traditionnelle chez les Sundi, aucun autre travail n'a été publié jusqu'à présent en particulier sur les plantes antidiarrhéiques. Ce travail est une contribution originale à la connaissance des plantes utilisées en médecine traditionnelle dans le traitement des diarrhées chez les Sundi.

Méthodes

La récolte des données sur le terrain a été effectuée durant le premier semestre de l'année 1992. Les données proviennent des enquêtes menées auprès des mères et pères de familles. Pour faciliter l'identification et la récolte des plantes, nous partions ensemble avec chaque informateur sur le terrain et l'interrogions sur les plantes qu'il utilise contre la diarrhée. Pour chaque plante indiquée, nous avons relevé le nom vernaculaire, l'organe utilisé, les modes de préparation et d'administration des remèdes.

Un herbier de référence a été constitué et déposé au Centre d'Etudes sur les Ressources Végétales (CERVE) à Brazzaville et à l'herbarium du laboratoire de Botanique Systématique et de Phytosociologie de l'Université Libre de Bruxelles (BRLU).

Une recherche bibliographique à partir de la banque de données Pharmel (banque de données de médecine traditionnelle et de pharmacopée), nous a permis d'identifier les

L.J.G. van der Maesen et al. (eds.), The Biodiversity of African Plants, 722-726.

plantes employées chez les Sundi et qui sont aussi connues comme plantes antidiarrhéiques au Congo et ailleurs en Afrique.

Résultats

La liste des recettes inventoriées est présentée ci-dessous. Pour chaque espèce médicinale, sont repris le nom scientifique, la famille, le nom vernaculaire, les modes de préparation et d'administration ainsi que le nombre d'informateurs qui ont indiqué la recette. Nous avons inventorié au total 19 espèces intervenant dans 18 recettes antidiarrhéiques. Ces espèces sont groupées en 19 genres appartenant à 11 familles dont les mieux représentées sont celles des Euphorbiaceae qui comptent 4 espèces, des Guttiferae, des Myrtaceae et des Zingiberaceae avec 2 espèces chacune.

Parmi les 19 espèces inventoriées, 6 d'entre elles sont indiquées au moins par 4 informateurs. C'est le cas de *Garcinia huillensis* (8 informateurs), *Mangifera indica* (8), *Psidium guajava* (6), *Alchornea cordifolia* (6), *Hymenocardia acida* (4) et *Trema orientalis* (4).

Les recettes traditionnelles employées dans le traitement des diarrhées sont:

Alchornea cordifolia (Schum. & Thonn.) Mull. Arg. (Euphorbiaceae), mubundjila.
Préparer une décoction aqueuse des racines. On prend un verre de décocté per os, le reste du décocté est employé en lavement.

Anisophyllea quangensis Engl. ex Henriq. (Anisophylleaceae), mbalango.
Préparer une décoction des racines et prendre le décocté aqueux per os contre la diarrhée.

Annona senegalensis Pers. (Annonaceae), mulolo.
Mâcher 2 à 3 jeunes feuilles et avaler le suc pour calmer la diarrhée.

Bridelia ferruginea Benth. (Euphorbiaceae), kikolokoto.
Mâcher 2 à 3 jeunes feuilles et avaler le suc pour soigner la diarrhée.

Caloncoba welwitschii (Oliv.) Gilg (Flacourtiaceae), n'tela.
Le suc des jeunes feuilles triturées est employé per os contre la diarrhée.

Combretum racemosum P. Beauv. (Combretaceae).
Préparer une décoction aqueuse des racines. Prendre un verre de décocté per os. Le reste de décocté est employé en lavement anal.

Desmodium velutinum (Willd.) DC. (Leguminosae-Papilionoideae), lunamanama.
Le suc des feuilles mâchées ensemble avec les graines d'*Aframomum melegueta* et la noix de kola est pris per os contre la diarrhée.

Euphorbia thymifolia L. (Euphorbiaceae).
Mâcher les feuilles avec la noix de palme et avaler le suc pour soigner la diarrhée.

Garcinia huillensis Welw. ex Oliv. (Guttiferae), mufingu.
Préparer une décoction des écorces de tronc et des racines; le décocté obtenu est administré per os pour soigner la diarrhée.

Hymenocardia acida Tul. (Hymenocardiaceae), muhété.
Le suc des jeunes feuilles mâchées soigne la diarrhée.

Mangifera indica L. (Anacardiaceae), mumanga.
La décoction des écorces de tronc est employée en bain de siège contre la diarrhée.

Millettia drastica Welw. ex Bak. (Leguminosae-Caesalpinioideae).
La décoction des jeunes feuilles est employée per os contre la diarrhée.

Psidium guajava L. (Myrtaceae), goyave.

L'eau de macération des feuilles ou le suc des jeunes feuilles triturées est employée en boisson pour calmer les diarrhées à raison d'un verre de macéré par jour.

Psorospermum febrifugum Spach (Guttiferae), kisosoka.

Mâcher 1 à 2 jeunes feuilles et avaler le suc pour calmer la diarrhée.

Renealmia polypus Gagnep. (Zingiberaceae).

Piler les feuilles fraîches puis filtrer. Le filtrat obtenu est administré per os contre la diarrhée amibienne.

Sapium cornutum Pax (Euphorbiaceae), ntihiti.

La décoction des jeunes feuilles est employée per os contre la diarrhée.

Syzygium guineense subsp. *macrocarpa* (Engl.) F. White (Myrtaceae), mukizu.

La décoction aqueuse des racines est employée comme antidiarrhéique. On administre per os un verre de décocté.

Trema orientalis (L.) Blume (Ulmaceae), ntiakala.

Préparer une décoction aqueuse des feuilles; le décocté est pris per os à raison d'un verre par jour.

Discussion

Parmi les 19 espèces recensées dans le traitement des diarrhées, 8 sont citées pour la première fois un usage antidiarrhéique au Congo (Tableau 1). Il s'agit de: *Anysophyllea quangensis, Caloncoba welwitschii, Euphorbia thymifolia, Millettia drastica, Renealmia polypus, Sapium cornutum, Syzygium guineense* subsp. *macrocarpa* et *Trema orientalis*. L'interprétation des usages de ces plantes devra être nuancée. Pour les espèces à aire de distribution limitée comme *Anisophyllea quangensis,* il est probable que l'information puisse correspondre à une nouvelle thérapeutique. Pour les espèces à large distribution, *Syzygium guineense* subsp. *macrocarpa*, *Trema orientalis, Sapium cornutum, Caloncoba welwitschii, Euphorbia thymifolia* et *Millettia drastica*, l'information reste peu fiable car ces espèces largement distribuées dans la zone guinéo-congolaise et soudanienne n'ont jamais fait l'objet d'indications dans les pays où les enquêtes ethnobotaniques ont été très poussées.

Au Congo, Bouquet (1969), Koechlin (1951), Descoings (1963) et Adjanohoun et al. (1988) ont signalé 5 espèces utilisées comme plantes antidiarrhéiques parmi celles employées chez les Sundi. Il s'agit de: *Combretum racemosum, Garcinia huillensis, Hymenocardia acida, Mangifera indica* et *Psidium guajava*; en dehors du Congo, 3 autres espèces sont citées ailleurs en Afrique. C'est le cas d'*Alchornea cordifolia* (Zaïre, R.C.A.), *Annona senegalensis* (Zaïre) et *Psorospermum febrifugum* (Zaïre). Il est fort probable que ces espèces possèdent un effet antidiarrhéique. *Mangifera indica* et *Psidium guajava* sont aussi employées comme antidiarrhéique en Guyane française (Grenand et al. 1987) et dans les Caraïbes (Séminaire Tramil 4 1989). Les propriétés antidiarrhéiques de *Psidium guajava* et *Mangifera indica* sont bien connues en pharmacologie. En effet, elles contiennent des tannins qui ont des propriétés antidiarrhéiques (Kerharo & Adam 1974). A défaut de pouvoir vérifier le bien fondé des remèdes traditionnels par un essai pharmacologique, le fait que des populations différentes et très distantes utilisent la même plante pour le même usage peut être révélateur de l'efficacité de la plante. On suppose que l'information est d'autant plus répandue que le résultat thérapeutique obtenu avec l'utilisation de la plante est meilleur.

Tableau 1. Plantes employées pour soigner les diarrhées chez les Sundi et citées au Congo et ailleurs pour les mêmes usages.

Nom scientifique	Nb	Organes utilisés	Pays dans lequel l'espèce est citée
Plantes employées par les Sundi, non citées au Congo et ailleurs comme antidiarrhéiques			
Aframomum melegueta	1	Gr	-
Anisophyllea quangensis	1	Ra	-
Caloncoba welwitschii	2	Fe	-
Euphorbia thymifolia	1	Fe, Tg	-
Millettia drastica	1	Fe	-
Renealmia polypus	2	Fe	-
Sapium cornutum	1	Fe	-
Syzygium guineense subsp. *macrocarpa*	1	Ec	-
Trema orientalis	4	Fe	-
Plantes employées par les Sundi et citées comme antidiarrhéiques au Congo et dans d'autres pays d'Afrique			
Alchornea cordifolia (1, 2, 3)	6	Fe	RCA, Congo, Zaire
Annona senegalensis (4, 5)	2	Fe	Zaïre, Congo
Bridelia ferruginea (6)	1	Fe	Togo
Combretum racemosum (4, 5)	3	Fe	Zaïre, Congo
Desmodium velutinum (6)	1	Fe	Togo
Hymenocardia acida (5, 4)	4	Ec, Fe	Congo, Zaïre
Garcinia huillensis (7)	8	Ra, Ec	Congo
Mangifera indica (8, 9, 10, 11)	8	Ec, Fe	RCA, Zaïre, Congo,...
Psidium guajava (12, 9, 10, 8)	6	Fe	Congo, RCA, Zaïre,...
Psorospermum febrifugum (4)	4	Fe	Zaïre

Légendes: Ec: écorce; Fe: feuille; Gr: graine; Nb: nombre d'informateurs ayant indiqué la recette dans la sous-préfecture de Kindamba; Ra: racine; Tg: tige.
1: Vergiat 1970; 2: Sillans 1953; 3: Wome 1985; 4: Lubini 1990; 5: Koechlin 1951; 6: Adjanohoun et al. 1986; 7: Adjanohoun et al. 1988; 8: Motte 1980; 9: Dhetchuvi & Lejoly 1990; 10: Bokdam & Droogers 1975; 11: Nyakabwa & Dibalula 1990; 12: Descoings 1963.

Conclusion

Les espèces suivantes: *Alchornea cordifolia*, *Mangifera indica*, *Psidium guajava*, *Garcinia huillensis* ont un même usage thérapeutique répandu au Congo et dans d'autres pays africains. Elles sont relativement bien connues au point de vue phytochimique et pharmacologique. Elles peuvent être recommandées pour une utilisation rationnelle en médecine traditionnelle. Les espèces nouvellement inventoriées constituent notre contribution à la flore médicinale du Congo. Des enquêtes complémentaires et des études phytochimiques et pharmacologiques doivent être effectuées pour évaluer leur potentiel médicinal.

Remerciements
Nous remercions le Centre d'Etudes sur les Ressources Végétales (CERVE) et la Faculté des Sciences de Brazzaville pour l'accueil qui nous a été reservé, ainsi que pour l'aide matérielle qu'ils nous ont apportés pour réaliser cette étude.

Bibliographie

Adjanohoun, E., A. Ahyi, L. Aké Assi, K. Akpagana, P. Chibon, A. Elhadji, J. Eyme, M. Garba, J.-N. Gassita, M. Gbeassor, E. Goudote, S. Guinko, K.-K. Hodouto, P. Houngnon, A. Keita, Y. Keoula, W.P. Klugaocloo, I. Lo, K.M. Siamevi & K.K. Taffame, 1986. Contribution aux études ethnobotaniques et floristiques au Togo. ACCT, Paris, 501 p.

Adjanohoun, E., A. Ahyi, L. Aké Assi, J. Baniakina, P. Chibon, G. Cusset, V. Doulou, A. Enzanza, J. Eyme, E. Goudote, A. Keita, C. Mbemba, J. Mollet, J.M. Moutsamboté, J.B. Mpati & P. Sita, 1988. Contribution aux études ethnobotaniques et floristiques au Congo. ACCT, Paris, 605 p.

Bokdam, J. & A.F. Droogers, 1975. Contribution à l'étude ethnobotanique des Wagenia de Kisangani, Zaïre. Mededelingen Landbouwhogeschool Wageningen 75-19: 1-74.

Bouquet, A., 1969. Féticheurs et médecines traditionnelles du Congo (Brazzaville). Orstom, Paris, 282 p.

Descoings, B., 1963. Essai d'inventaire préliminaire des plantes médicinales d'Afrique équatoriale. Bull. Inst. Rech. Scient. Congo: 7-24.

Descoings, B., 1975. Les grandes régions naturelles du Congo. Candollea 30: 91-120.

Dhetchuvi, M.-M. & J. Lejoly, 1990. Contribution à la connaissance des plantes médicinales du Nord-Est du Zaïre. 12e Congrès de l'AETFAT, Mitt. Inst. Allg. Bot., Hamburg, 23: 991-1006.

Grenand, P., C. Moretti & H. Jacquemin, 1987. Pharmacopée traditionnelle en Guyane, Créole, Palikur, Wayapi. Ed. Orstom, Paris, 564 p.

Kerharo, J. & J.G. Adam, 1974. Plantes médicinales et toxiques du Sénégal. Ed. Vigot & Frères, 1011 p.

Koechlin, J., 1951. Sur quelques usages de plantes spontanées de la région de Brazzaville. Bull. Inst. Etud. Centrafr. Nov. Sér. 2: 103-124.

Lubini, A., 1990. Les plantes utilisées en médecine traditionnelle par les Yansi de l'entre Kwilu-Kamtsha (Zaïre). Comptes Rendus de la 12e réunion de l'AETFAT, Mitt. Inst. Allg. Bot., Hamburg, 23b: 1007-1020.

Motte, F., 1980. Les plantes chez les pygmées Aka et les Monzombo de la Lobaye (Centrafrique). Soc. Et. Ling. et Anthrop. de France, Etudes pygmées V, 573 p.

Nyakabwa, M. & M. Dibalula, 1990. Plantes médicinales cultivées dans la zone de Kabondo à Kisangani (Zaïre). Afr. Stud. Monogr. 11(2): 87-99.

Séminaire Tramil 4, 1989. Vers une Pharmacopée Caraïbe. Recherche scientifique et usage populaire des plantes médicinales dans la Caraïbe. Robineau éd., Enda-Caribe, UNAH, 474 p.

Sillans, M.R., 1953. Plantes médicinales d'Afrique centrale. Extrait des Annales pharmaceutiques françaises, 11: 364-456.

Vergiat, A-M., 1970. Plantes magiques et médicinales des féticheurs de l'Oubangui (Région de Bangui). Journ. Agric. Trop. Bot. Appl., 17: 295-339.

Wome, B., 1985. Recherches ethnopharmacognosiques sur les plantes médicinales utilisées en médecine traditionnelle à Kisangani (Haut-Zaïre). Thèse Doct. Univ. Libre Bruxelles, 561 p.

Medicinal uses of plants from Guinea-Bissau

M.A. DINIZ [1], O. SILVA [2], M.A. PAULO [2] & E.T. GOMES [2]
[1] Centro de Botânica, Instituto de Investigação Científica Tropical, Rua da Junqueira 86, 1300 Lisboa, Portugal.
[2] Lab. de Farmacognosia, Faculdade de Farmácia da Universidade de Lisboa, Av. Forças Armadas, 1600 Lisboa, Portugal.

Abstract
Plants used by traditional Fulani healers to treat infectious diseases were revealed by an ethnopharmacological inquiry conducted in the Contúboel Sector of Guinea-Bissau. Data were collected on thirty species, including local plant names, parts of the plant used, plus formula, preparation and methods of administering various remedies. The plant extracts were screened for antibiotic activity. The screening results confirmed that most of the tested plant extracts exhibited antibiotic activity, especially *Terminalia macroptera* (Combretaceae) and *Cryptolepis sanguinolenta* (Asclepiadaceae) both of which presented interesting profiles of antibacterial activity, mainly against some of the more important etiologic agents of gastro-enteritis.

Résumé
Une enquête a été conduite dans quelques petites villages du Secteur de Contúboel en Guiné-Bissau sur des plantes utilisées dans la médecine traditionnelle, pour soigner des maladies infectieuses. Une trentaine d'espèces a été identifiée et des renseignements ont été recueillis chez des guérisseurs de l'ethnie Peul sur les recettes utilisées par eux. D'après ces données, des extraits de drogues ont été soumis à des essais de l'activité antibactérienne. Les résultats obtenus, surtout ceux qui concernent l'activité de *Terminalia macroptera* et de *Cryptolepis sanguinolenta* sur des organismes responsables pour des diarrhées d'origine infectieuse, sont très encourageants et sont susceptibles d'encourager des travaux en vue de la préparation de médicaments locaux.

Key words
Cryptolepis sanguinolenta, Guinea-Bissau, Medicinal plants, *Terminalia macroptera*.

Introduction

In February 1991 an ethnopharmacological project was initiated in Contúboel Sector, with local support from the Agricultural Research Department (DEPA) (Gomes & Diniz 1993). Data were collected on local remedies used to treat several infectious diseases and on some other drugs sold in the Bissau market of Bandim (dried roots and leaves of *Terminalia macroptera* Guill. & Perr. and dried roots of *Cryptolepis sanguinolenta* Schltr.). In all, about thirty plant species were identified. Ethanol extracts of the associated drugs were tested on several bacterial species and most of them showed some antimicrobial activity. Further studies on the chemical composition and antibiotic activities of two of the more frequently used species are very promising.

Geographical position

Guinea-Bissau, formerly (before 1974) Portuguese Guinea, is a west African coastal nation (area: 36,000 km^2), bounded by Senegal on the north and the Republic of Guinea (Conakry) on the east and southeast. The country is largely flat and low-lying ($<$ 40 meters); the highest elevation is in the south-eastern Boé region, on an outlier

L.J.G. van der Maesen et al. (eds.), The Biodiversity of African Plants, 727-731.

of Fouta Djallon plateau. The Contúboel Sector, in the northeastern region, is far from Bissau, the capital (180 km), and near to the border of Senegal.

Climate
The climate is warm and damp. There are two six-month seasons: the wet and warmer one starts in May and lasts to the end of October; the dry and cooler lasts from November until May. Rainfall decreases from the south-east to the north.

Flora
Due to human interference, climate, diverse soil and hydric regimes, a great diversity of vegetation types is found in Guinea-Bissau. Knowledge of some previous studies (Carvalho et al. 1956, Espírito Santo 1949, White 1983) and our own field work confirm that the vegetation of the Contúboel Sector shows features of transition between the Sahel-Sudanian regional zone and the Guinea-Congolian regional zone. The characteristic species of the main vegetation types of the mosaic in the Contúboel Sector are as follows:

* Dry woodlands: characterised by *Daniellia oliveri* (Rolfe) Hutch. & Dalz., *Khaya senegalensis* (Desr.) Adr. Juss., *Pterocarpus erinaceus* Poir., *Piliostigma thonningii* (Schum.) Milne-Redhead, *Combretum* spp., *Terminalia macroptera* Guill. & Perr., *Bombax costatum* Pellegr. & Vuillet, *Dichrostachys cinerea* (L.) Wight & Arn. and *Erythrina senegalensis* DC.
* Woody grasslands with trees or shrubs: mainly represented by andropogonoid grasses with scattered trees and shrubs of *Dichrostachys cinerea, Terminalia macroptera, Guiera senegalensis* Lam., *Combretum* spp., *Parkia biglobosa* (Jacq.) R.Br. ex Benth. and *Bombax costatum.*
* Gallery forests: chiefly represented by *Mitragyna ciliata* Aubrév. & Pellegr., *Morinda geminata* DC, *Sarcocephalus latifolius* (Smith) Bruce, *Holarrhena floribunda* (G. Don) Dur. & Schinz., *Parinari excelsa* Sabine and *Syzygium guineense* Jacq.
* Palm forest: characterised mainly by the occurrence of *Elaeis guineensis* Jacq. and *Anthocleista nobilis* G. Don.

Contúboel traditional medicine
Methods
Access to medical care, even in the main local hospitals we visited, was very limited and, at the time, there were no western pharmaceutical products available either in the hospitals or in the health care centres of Contúboel Sector.

Our contact with five traditional healers was organized by DEPA (Bissau and Contúboel) with the acceptance of the healers. Four male Fulani healers (Muslim religion) and one female healer of the Mandyako tribal group were interviewed on the main traditional remedies used in anti-infection therapy. The bark, root, leaf and plant aerial parts used in the remedies were collected by them for our analysis.

Results
Thirty species were identified (Table 1), from about twenty-two traditional recipes. Voucher specimens are deposited in the herbarium of the Centro de Botânica (LISC),

at the Instituto de Investigação Cièntifica Tropical (IICT), Lisbon, Portugal. Most of the traditional drugs made from the species are used to treat swellings, rheumatic disorders, wounds, purulent sores and veneral diseases. Data on the traditional recipes are published in Gomes & Diniz (1993). Two of the species are most promising and some additional information on them is provided below:

* *Cryptolepis sanguinolenta* - Dried root doses and bottled water macerates are sold in Bissau's Bandim market for the treatment of hepatitis and jaundice. Traditional recipe: small root fragment macerates are bottled with water, under sunlight and overnight, for 24 hours. A small cup is taken every morning during the treatment (till disappearance of jaundice symptoms). *Cryptolepis sanguinolenta* root extract is active against all but one of the tested bacteria. Both the ethanolic extract and the main indolic alkaloid cryptolepine show a very interesting profile of antibiotic activity against *Campylobacter* and *Vibrio cholerae* infective strains (Paulo et al. 1994).

* *Terminalia macroptera* - Decorticated dried root doses, bottled water macerates, and infusions are sold in Bissau's Bandim market. Traditional recipe: small pieces of dried and decorticated roots are boiled in water (for one or two hours), salt and cotton filtrate are added. The treatment of venereal disease is made by the oral administration of a glass of this extract, till the symptoms are cured. *Terminalia macroptera* root extract and the major water-soluble compounds (flavone derivatives) show antimicrobial activity against *Shigella dysenteriae*, *Vibrio cholerae*, *Vibrio parahaemoliticus* and *Staphylococcus aureus* (O. Silva, personal communication 1993).

Conclusion

Our present knowledge about the drugs used as anti-infective agents in Guinea Bissau (Contúboel) traditional medicine indicates that, after agricultural research projects on cultivation of the main useful species, development programs to enhance the preparation of local pharmaceutical formulae may be important to the progress of African countries with limited western health care resources.

Acknowledgements

JNICT, Portugal, for financial support to Project PMCT/C/SAU/749/90; Eng. C. Schwarz da Silva and Eng. I. Miranda of the DEPA-Bissau; to all the Directors and Technicians of DEPA-Contúboel. With our respect for the culture of the Traditional Healers and the rural populations of Contúboel Sector.

References

Carvalho J.A. de & S.J.S.F.P. Nunes, 1956. Contribuição para o Estudo do Problema Florestal da Guiné-Bissau. Estudos, Ensaios e Documentos, Junta Invest. Ultramar (Lisboa) 30: 194.

Espírito Santo, J.V. do, 1949. Contribuição para o conhecimento fitogeográfico da Guiné Portuguesa. Boletim Cultural da Guiné Portuguesa (Bissau) vol. IV. 13: 95-129.

Gomes, E.T. & M.A. Diniz, 1993. Plantas usadas na medicina tradicional na região de Contúboel Comun. IICT, Sér. Ciênc. agrárias 13: 153-165.

Paulo, A., M. Pimentel, S. Viegas, I. Pires, A. Duarte, J. Cabrita & E.T. Gomes, 1994. *Cryptolepis sanguinolenta* activity against diarrhoeal bacteria. J. of Ethnopharmacology, in press.

White, F., 1983. The Vegetation of Africa. UNESCO, Paris.

Table 1. Scientific names, Fulani names, plant parts and medicinal uses of twenty-one species.

Family/Species	Fulani name	Plant part	Uses
Annonaceae			
Uvaria chamae P. Beauv.	Sambafim-ô	Roots	Rheumatic diseases (Rd)
Apocynaceae			
Holarrhena floribunda (G. Don) Dur. & Schinz	Tchoraquidje	Bark	Anthelmintic
Asclepiadaceae			
Calotropis procera Ait.	Pama	Roots	Rheumatic diseases
Cryptolepis sanguinolenta Schltr.	Cuntésse	Roots	Jaundice, Hepatitis
Leptadenia hastata Decne.	Safarodje	Whole plant	Venereal diseases
Bombacaceae			
Adansonia digitata L.	Bóè	Roots	Urinary diseases
Cochlospermaceae			
Cochlospermum tinctorium A. Rich.	Djánderè	Rhizome	Jaundice, Hepatitis
Combretaceae			
Guiera senegalensis Lam.	Gélode	Whole plant	Vd, Tuberculosis
Terminalia macroptera Guill. & Perr.	Bóde	Roots	Vd, Hepatitis
Convolvulaceae			
Merremia kentrocaulos (Steud. ex. C.B. Clarke) Rendle var. *kentrocaulos*	Djon-musudjulo	Rhizome	Rheumatic diseases
Euphorbiaceae			
Hymenocardia acida Tul.	Coron-condé	Leaves	Wounds
Leg.-Caesalpinioideae			
Cassia sieberiana DC.	Sambasintchandje	Roots	Rheumatic diseases
Chamaecrista nigricans Vahl ex DC.	Bara-bubel	Whole plant	Wounds
Entada africana Guill. & Perr.	Padapar	Stem bark, roots	Wounds
Senna podocarpa Guill. & Perr.	Corotalindin	Leaves, roots	Laxative, Vd

Table 1. (Continued).

Family/Species	Fulani name	Plant part	Uses
Leg.- Mimosoideae			
Dichrostachys cinerea Wight & Arn.	Búrlè	Stem bark	Rheumatic diseases
Parkia biglobosa (Jacq.) Benth.	Neré	Stem bark	Anti-inflamatory
Prosopis africana (Guill. & Perr.) Taub.	Tchelem-tchelenadje	Stem bark	Anti-inflamatory
Leg.- Papilionoideae			
Erythrina senegalensis DC.	Botchochei	Roots	Rheumatic diseases
Erythrina sigmoidea Hua	Dolimbai	Roots	Vd, Diuretic
Piliostigma thonningii (Schum.)Milne-Redhead	Barquedje	Leaves, Stem bark	Wounds
Loganiaceae			
Strychnos spinosa Lam.	Saraculedje	Roots	Rheumatic diseases
Olacaceae			
Ximenia americana L.	Tcheme	Stem bark	Analgesic
Palmae			
Elaeis guineensis Jacq.	Tem-em-eih	Roots	Veneral diseases (VD)
Polygalaceae			
Securidaca longepedunculata Fresen.	Djúrô	Roots	Wounds
Rubiaceae			
Gardenia ternifolia Schumach. & Thonn. subsp. *jovis tonantis* (Welw.) Verdc. var. *goetzei* (Stapf & Hutch.)	Dignale	Roots	Analgesic
Morinda geminata DC.	Bacuré	Leaves, Roots	Rd, Hepatitis
Pavetta oblongifolia (Hiern) Bremek.	Cuncandjor	Whole plant	Urinary infections
Sarcocephalus latifolius (Smith) Bruce	Náudò-putcho	Roots	Rd, Venereal diseases
Verbenaceae			
Lippia chevalieri Moldenke	Ussum-coloma	Roots	Paludism, Antipyretic

Rwenzori Mountain Forest non-timber products and people

E.K.Z. KAKUDIDI, H. ORYEM-ORIGA, R. BUKENYA-ZIRABA & A.B. KATENDE
Makerere University, Botany Department, P.O. Box 7062, Kampala, Uganda.

Abstract
An inventory of non-timber products harvested from the Rwenzori Mountain Forest National Park in Bundibugyo District, Uganda was carried out over a sampled area of 4,250 m^2. 78 out of 118 plant species represented by 300 specimens were recorded as having some use-value to the Bwamba people. The species were used for various purposes including medicine (22), construction (12), craft-work (13), food (7), firewood (16) and other miscellaneous minor uses (30).

Key words
Baamba, Bakonjo, Bundibugyo, Ethnobotany, Inventory, Rwenzori.

Introduction
In 1941 the Rwenzori Mountain Forest was gazetted to protect an important water catchment area (Leggat & Beaton 1961) and since 1991 it has been a National Park. For generations, the forest has been a lifeline for the surrounding districts. It is on the border between Uganda and Zaire and covers parts of Bundibugyo, Kabarole and Kasese Districts; 0°6'-0°46' N, 29°47'-30°11' E. The study site is located on Bupompoli ridge in northern Rwenzori Mountain in Bwamba county, Bundibugyo district; 0°43'-0°45' N, 30°03'-30°01' E. The land rises to about 3,650 m, Karangora being the highest point in the county.

People
The people of Bwamba county, Bundibugyo District belong to two main ethnic groups: the Baamba and Bakonjo together with a small remnant population of Batwa pygmies. The Baamba (about 74% of Bwamba's population) occupy lowlands at the foot of the Rwenzori. The Bakonjo (about 23%) live on the lower slopes of the foothills of Rwenzori Mountains.

Bwamba county, by virtue of its inaccessibility, has remained seriously under-developed and government services are largely ineffective (Howard 1991). The area was made accessible by road in 1938 (Winter 1954), but transport to the area is still problematic. Thus the Bwamba people use footpaths across the forest to connect with neighbouring districts. Howard (1991) indicated that 8% of Baamba and 70% of Bakonjo use the forest as a transit route to Zaire or Kabarole.

Methods
This study was carried out between May and November 1991, along a major footpath passing over Bupompoli ridge from Bwamba to Kabarole District. Utility trails radiating from the main footpath were treated as transects. Transects of different lengths were sampled, to a total area of 4,250 m^2.

Specimens of all the plant species encountered along the transects were collected.

L.J.G. van der Maesen et al. (eds.), The Biodiversity of African Plants, 732-740.

Information about their uses was obtained from two field assistants (local to the area) during specimen collection. The 300 specimens were taken to the camp where people in the vicinity supplied further information on the uses. A visit to the medicinal practitioners could have related more but this was beyond the scope of this study. Voucher specimens were identified and deposited in the Makerere University Herbarium (MHU).

Results and discussion

The study revealed that 78 of the 118 species identified and documented were found to be useful to the Bwamba people. This level of utilization (66.1%) falls within the range of 48.6% to 78.7% that was found by Prance et al. (1987) among certain Amazonian communities. Table 1 summarises the various uses of these 78 species. Table 2 shows the relationship between plant use and habit.

Medicinal plants

Most of the medicinal plants treat or cure single medical conditions except for *Chenopodium ambrosioides*, *Maesa lanceolata*, *Momordica foetida* and *Plectranthus laxiflorus*, further details are in Table 3. Trees are not so much sought for medicine (only 4 out of 24 species) in this study contrasted to that of Butono (1988) who recorded 38 medicinal trees from the lowland Semuliki Forest Reserve.

Construction

Twelve species from the forest are used in construction, joinery, building poles and rafters. Howard (1991) reported that Bakonjo and Baamba collect 17% and 5% of building materials from forest reserves, respectively. He estimated that demands exceed the sustainable supply obtainable from the public lands by a factor of four in Baamba areas, and by a factor of seven in Bakonjo areas.

Timber exploitation from the Rwenzori Mountains has been limited by difficult terrain and inadequate technology. However, *Arundinaria alpina* stems/culms are used extensively and provide building poles for houses, granaries, roof rafters and beams whilst the leafless sheaths provide roofing material (Figure 1). Only dry old stems are harvested as poles/rafters since they are lighter to transport than fresh ones. They provide about 52% of construction products to the Bakonjo (Howard 1991).

Macaranga kilimandscharica, *Canthium oligocarpum*, *Dombeya elliottii* and *Xymalos monospora* are often used as supporting poles in houses. Others, which are often shaped into planks, include *Cordia millenii*, *Strombosia scheffleri* and *Cyathea manniana*. *Cordia millenii* is preferred because it is particularly resistant to attacks by termites.

Craftwork

Crafted household items include baskets from *Sida veronicaefolia* and *Smilax anceps*, and combs, axe handles, mortars and pestles from *Connarus longistipitatus*, *Canthium oligocarpum*, *Melchiora schliebenii* and *Rapanea melanophloeos*. The Bwamba are also known for their very beautiful handbags made from the vegetative parts of *Urera hypselodendron* and seeds of *Ensete ventricosum* (Figure 2). They are sold locally in big towns outside Bwamba.

Figure 1. Stems of *Arundinaria alpina* being transported from the forest.

Figure 2. A handbag made from the vegetative parts of *Urera hypselodendron* and seeds of *Ensete edule*.

Food

Food plants, edible fruits and vegetables make up 12.8% (10/78) of the useful plant species and 8.4% (10/118) of all the plant species recorded. Most food harvested from the forest, especially the fruits, is consumed within it, when harvesting other forest products. The Bwamba people are able to grow most of the food they need and harvesting food from the forest poses no threat.

Firewood

The vast majority of plant species from the forest fall into this category. Trees are preferred since they provide more durable wood than shrubs. Howard (1991) reported that the Bakonjo appear to be less self-reliant, as far as wood products are concerned, than the Baamba. The former collect 41% of the firewood from the forest reserves compared to the latter who collect 16%. Already 73% of the Baamba and 50% of the Bakonjo say that there is insufficient firewood in their home areas (Howard 1991). Inevitably there will be a progressive increase in time spent collecting firewood further and further away from their homes. Howard (1986) also estimated that over 52% of the forest cover which occurred at intermediate altitudes between the Semuliki and Rwenzori Mountain Forest Reserves has been lost in the last 33 years. Agricultural encroachment together with unregulated use of forest resources, and in particular harvesting for firewood, is already causing serious damage.

Miscellaneous uses and non-useful plants

Thirty plant species were found to be used for minor purposes within the Bwamba community (see Table 4). A relatively large number of plant species (41/118 from 21 families) were not known to have any apparent use to the Bakonjo and Baamba communities.

Acknowledgements

We are grateful to Dr A. Hamilton for encouragement and securing of funds from WWF to enable us to carry out the research. We are also grateful for the assistance of Mr A. Muhindo and Mr A. Kahigwa during the field work; Ms O. Maganyi for the herbarium work, the people of Kazimba village and Bupompoli for providing information. Makerere University provided various facilities.

References

Butono, F., 1988. The use of forest products other than wood with particular reference to traditional medicine. In: P.C. Howard (ed.). Conservation and Development of Natural Resources in Bundibugyo District, Uganda. WWF Project 3235 Res. Publ. WWF-International, Gland.

Howard, P.C., 1986. Agricultural encroachment in the Semuliki Forest Reserve, Western Uganda. WWF Project 3235. Unpublished report 8 pp.

Howard, P.C., 1991. Nature Conservation in Uganda's Tropical Forest Reserves. IUCN, Gland, Switzerland & Cambridge, U.K.

Leggat, G.J. & A.H.A. Beaton, 1961. Working Plan for Rwenzori Central Forest Reserves. First revision for the period 1st July 1961 to 30th June 1971. Uganda Forest Department, Entebbe.

Prance, G.T., N. Balee, B.M. Boom & R.L. Carneiro, 1987. Quantitative ethnobotany and the case conservation in Amazonia. Conservation Biology 1 (4): 296-310.

Winter, E.H., 1954. Bwamba Economy: the development of a primitive subsistence economy in Uganda. East African Institute of Social Research, Kampala.

Table 1. Habit and uses of 78 useful plant species.
Habit: Herb (H), Climber (C), Tree (T), Shrub (S). Uses: Medicinal (Md), Construction (Ct), Craftwork (Cr), Food (Fd), Firewood (Fr), Miscellaneous (Ms). Useful (+), not useful (—).

Family, species	Habit	Md	Ct	Cr	Fd	Fr	Ms
Acanthaceae							
Brachystephanus coeruleus S. Moore	H	—	—	—	—	+	—
Mimulopsis runssorica Lindau	S	—	—	—	—	+	+
Thunbergia mildbraediana Lebrun & Touss.	H	—	—	—	—	—	+
Amaranthaceae							
Cyathula cylindrica Moq.	H	—	—	—	—	—	+
Aquifoliaceae							
Ilex mitis (L.) Radlk.	T	—	—	—	—	+	—
Asclepiadaceae							
Mondia whiteii (Hook. f.) Skeels	C	+	—	—	—	—	—
Aspleniaceae							
Asplenium abyssinicum Fee	H	—	—	—	—	—	+
A. aethiopicum (Burm. f.) Becherer	H	—	—	—	—	—	+
A. elliottii C.H. Wright, *A. erectum* Willd.	H	—	—	—	—	—	+
A. inaequilaterale Willd., *A. linckii* Kuhn	H	—	—	—	—	—	+
Balsaminaceae							
Impatiens sp.	H	—	—	—	+	—	—
I. runssorensis Warb.	H	—	—	—	+	—	—
Basellaceae							
Basella alba L.	H	—	—	—	—	—	+
Boraginaceae							
Cordia millenii Bak.	T	—	+	+	—	—	—
Chenopodiaceae							
Chenopodium ambrosioides L.	H	+	—	—	—	—	—
C. procerum Moq.	H	+	—	—	—	—	—
Compositae							
Crassocephalum sp. 1	H	+	—	—	—	—	—
Crassocephalum sp. 2	H	+	—	—	—	—	—
Dichrocephala integrifolia O. Kuntze	H	+	—	—	—	—	—
Conyza floribunda H.B.K.	H	—	—	—	—	—	+
Helichrysum sp.	H	—	—	—	—	—	+
Vernonia adolfi-friederici Muschler	S	—	—	—	—	+	+
V. auriculifera Hiern	T	—	—	—	—	+	+
V. cinerea (L.) Less	H	—	—	—	—	+	+
V. syringifolia O. Hoffm.	S	—	—	—	—	+	+
Connaraceae							
Connarus longistipitatus Gilg.	S	—	—	+	—	—	—
Convolvulaceae							
Ipomoea sp.	C	—	+	—	—	+	—
Cucurbitaceae							
Coccinia mildbraedii Harms	H	+	—	—	—	—	—
Lagenaria sphaerica (Sond.) Naud.	C	+	—	—	—	—	—
Momordica foetida Schumach.	C	+	—	—	—	—	—
M. pterocarpa A. Rich.	C	+	—	—	—	—	—
Peponia vogelii (Hook. f.) Engl.	H	+	—	—	—	—	—

Table 1. (Continued).

Family, species	Habit	Md	Ct	Cr	Fd	Fr	Ms
Cyatheaceae							
Cyathea manniana Hook.	T	–	+	–	–	–	–
Cyperaceae							
Cyperus derreilema Steudel	H	–	–	+	–	–	–
Dennistaedtiaceae							
Pteridium aquilinum (L.) Kuhn	H	–	–	–	–	–	+
Euphorbiaceae							
Acalypha psilostachya Hochst.	S	+	–	–	–	–	+
Macaranga kilimandscharica Pax	T	–	+	–	–	+	–
Phyllanthus amarus Schumach. & Thonn.	H	–	–	–	–	–	+
Gramineae							
Arundinaria alpina K. Schum.	S	–	+	–	–	+	+
Panicum trichocladum K. Schum.	H	–	–	–	–	–	+
Labiatae							
Plectranthus laxiflorus Benth.	H	+	–	–	–	–	–
Leguminosae-Papilionoideae							
Desmodium repandum (Vahl) DC.	H	+	–	–	–	–	–
Eriosema sp.	S	–	–	–	–	–	+
Liliaceae							
Anthericum sp.	H	–	–	–	–	–	+
Dracaena afromontana Mildbr.	T	+	–	–	–	–	+
Malvaceae							
Sida veronicaefolia Lam.	S	–	–	+	–	–	–
Monimiaceae							
Xymalos monospora (Harv.) Warb.	T	–	+	+	–	+	–
Moraceae							
Ficus natalensis Hochst.	T	–	–	+	–	–	–
Musaceae							
Ensete ventricosum (Welw.)Cheesman	S	+	–	+	–	–	–
Myrsinaceae							
Embelia schimperi Vatke	S	+	–	–	+	–	–
Maesa lanceolata Forssk.	T	+	–	–	–	+	+
Rapanea melanophloeos (L.) Mez.	T	+	–	+	–	–	–
Olacaceae							
Strombosia scheffleri Engl.	T	–	+	–	–	–	–
Passifloraceae							
Adenia sp.	C	–	+	–	–	–	–
Passiflora edulis Sims	C	–	–	–	+	–	–
Piperaceae							
Piper capense L.	S	+	–	–	–	–	–
Plantaginaceae							
Plantago palmata Hook.f.	H	+	–	–	–	–	–
Polygonaceae							
Polygonum setosulum A. Rich.	H	+	–	–	–	–	–
Polypodiaceae							
Athyrium scandicinum (Willd.) Presl	H	–	–	–	–	–	+

Table 1. (Continued).

Family, species	Habit	Md	Ct	Cr	Fd	Fr	Ms
Rosaceae							
Alchemilla kiwuensis Engl.	H	+	−	−	−	−	−
Prunus africana (Hook.f.) Kalkm.	T	−	−	+	−	−	−
Rubiaceae							
Canthium oligocarpum Hiern	T	−	+	+	−	−	−
Galiniera saxifraga (Hochst.) Bridson	S	−	+	−	−	+	−
Psychotria mahonii C.H. Wright	T	−	−	−	−	+	−
Sapotaceae							
Aningeria adolfi-friederici (Engl.) Robyns & Gilbert	T	+	−	−	−	−	−
Smilacaceae							
Smilax anceps Willd.	C	−	−	+	−	−	−
Solanaceae							
Capsicum annuum L.	H	−	−	−	+	−	−
Discopodium penninervium Hochst.	T	−	−	−	−	−	+
Solanum aculeatissimum Jacq.	S	−	−	−	−	−	+
S. nigrum L.	H	−	−	−	+	−	−
Sterculiaceae							
Dombeya elliottii K. Schum.	T	−	+	−	−	+	−
Theaceae							
Melchiora schliebenii (Melchior) Kobuski	T	−	+	+	−	+	−
Thelypteridaceae							
Thelypteris dentata (Forssk.) E. St.John	H	−	−	−	−	−	+
Urticaceae							
Droquetia iners (Forssk.) Schweinf.	S	−	−	−	+	−	−
Urera hypselodendron (A. Rich.) Wedd.	C	−	−	+	−	−	−
Verbenaceae							
Clerodendrum c.f. *johnstonii* Oliv.	H	−	−	−	−	−	+
Total number of taxa for each use category		24	12	13	7	16	30

Table 2. Relationship between habit and plant use category.

Habit	Percentage contribution of habit type to each use category						
	Md	Ct	Cr	Fd	Fr	Ms	All uses
Trees	13.6	53.8	58.3	–	50.00	13.3	24.4 (19/78)
Shrubs	18.2	15.4	16.7	28.6	31.25	23.4	16.7 (13/78)
Climbers	18.2	7.7	16.7	14.3	6.25	–	11.6 (09/78)
Herbs	50.0	23.1	8.3	57.1	12.50	63.3	47.7 (37/78)

Md: Medicinal, Ct: Construction, Cr: Craftwork, Fd: Food, Fr: Firewood, Ms: Miscellaneous.

Table 3. Medicinal Plants of Bwamba.
Habit: Herb (H), Climber (C), Tree (T), Shrub (S). Uses: Fever (A), Stomach-ache (B), Whooping cough (C), Stimulating sexual potency in men (D), Hypertension (E), Boils (F), Worms (G), Muscular pains (H), Lameness in children (I), Eye problems (J), Diarrhoea (K), Evil spirits (L). Useful (+), not useful (—).

Family, species	Habit	A	B	C	D	E	F	G	H	I	J	K	L
Asclepiadaceae													
Mondia whiteii	C	—	—	—	+	—	—	—	—	—	—	—	—
Chenopodiaceae													
Chenopodium ambrosioides	H	—	+	—	—	—	—	—	+	—	—	—	—
C. procerum	H	—	+	—	—	—	—	—	—	—	—	—	—
Compositae													
Crassocephalum sp. 1	H	—	—	—	—	—	—	—	—	—	—	+	—
Crassocephalum sp. 2	H	—	—	—	—	—	—	—	—	—	—	+	—
Dichrocephala integrifolia	H	—	—	—	—	—	—	—	—	—	—	—	+
Cucurbitaceae													
Coccinia mildbraedii	H	—	+	—	—	—	—	—	+	—	—	—	—
Lagenaria sphaerica	C	—	+	—	—	—	—	—	—	—	—	+	—
Momordica foetida	C	+	+	—	—	—	—	—	—	—	—	—	—
M. pterocarpa	C	—	+	—	—	—	—	—	—	—	—	—	—
Peponia vogelii	H	—	+	—	—	—	—	—	—	—	—	—	—
Euphorbiaceae													
Acalypha psilostachya	S	—	+	—	—	—	—	—	—	—	—	—	—
Labiatae													
Plectranthus laxiflorus	H	—	—	—	—	—	+	—	—	—	—	—	+
Leguminosae-Papilionoideae													
Desmodium repandum	H	—	—	—	—	—	—	—	—	—	—	+	—
Liliaceae													
Dracaena afromontana	T	—	—	—	+	—	—	—	—	—	—	—	—
Musaceae													
Ensete ventricosum	H	—	—	—	—	—	—	—	—	+	—	—	—
Myrsinaceae													
Embelia schimperi	S	—	—	—	—	—	—	—	+	—	—	—	—
Maesa lanceolata	T	—	+	—	—	—	—	—	+	—	—	—	—
Rapanea melanophloeos	T	—	—	—	—	—	—	—	+	—	—	—	—
Piperaceae													
Piper capense	S	—	—	—	—	—	+	—	—	—	—	—	—
Plantaginaceae													
Plantago palmata	H	—	—	—	—	—	—	—	—	—	—	—	+
Polygonaceae													
Polygonum setosulum	H	—	—	+	—	—	+	—	—	—	—	—	—
Rosaceae													
Alchemilla kiwuensis	H	—	—	—	—	—	—	—	—	—	+	—	—
Sapotaceae													
Aningeria adolfi-friederici	T	—	—	—	—	—	—	—	—	—	—	—	+

Table 4. Plant species with miscellaneous uses.
Habit: Shrub (S), Herb (H), Tree (T), Shrine Construction (Sc), Granary cover (Gc), Protection of crops (P), Fodder (F), Toilet tissue (Tt), Carrying luggage (Cl), Trapping birds (Tb), Drinking straws & smoking pipes (Ds). Useful (+), not useful (—).

Family, species	Habit	Sc	Gc	P	F	Tt	Cl	Tb	Ds
Acanthaceae									
Mimulopsis runssorica	S	—	—	+	+	—	—	—	—
Thunbergia mildbraediana	H	—	—	—	+	—	—	—	—
Amaranthaceae									
Cyathula cylindrica	H	—	—	—	+	—	—	—	—
Aspleniaceae									
Asplenium abyssinicum	H	+	—	—	—	—	—	—	—
A. aethiopicum, A. elliottii	H	+	—	—	—	—	—	—	—
A. erectum, A. inaequilaterale	H	+	—	—	—	—	—	—	—
A. linckii	H	+	—	—	—	—	—	—	—
Basellaceae									
Basella alba	H	—	—	—	+	—	+	—	—
Compositae									
Conyza floribunda	H	—	—	—	—	+	—	—	—
Helichrysum sp.	H	—	—	—	—	—	+	—	—
Vernonia adolfi-friederici	S	—	—	—	—	+	—	—	—
V. auriculifera	T	—	—	—	—	+	—	—	—
V. cinerea	H	—	—	—	—	+	—	—	—
V. syringifolia	S	—	—	—	—	+	—	—	—
Dennistaedtiaceae									
Pteridium aquilinum	H	—	+	—	—	—	—	—	—
Euphorbiaceae									
Acalypha psilostachya	S	—	—	—	+	—	—	—	—
Phyllanthus amarus	H	—	—	—	+	—	—	—	—
Gramineae									
Arundinaria alpina	S	—	—	—	—	—	—	+	—
Panicum trichocladum	H	—	—	—	+	—	—	—	—
Leguminosae-Papilionoideae									
Eriosema sp.	S	—	—	—	—	—	—	+	—
Liliaceae									
Anthericum sp.	H	+	—	+	—	—	—	—	—
Dracaena afromontana	T	+	—	—	—	—	—	—	—
Myrsinaceae									
Maesa lanceolata	T	—	—	—	—	—	—	+	—
Polypodiaceae									
Athyrium scandicinum	H	—	+	—	—	—	—	—	—
Solanaceae									
Discopodium penninervium	T	—	—	—	—	—	—	+	—
Solanum aculeatissimum	S	—	—	+	—	—	—	—	—
Thelypteridaceae									
Thelypteris dentata	H	+	—	—	—	—	—	—	—
Verbenaceae									
Clerodendrum cf. *johnstonii*	S	—	—	—	+	—	—	—	—

Plantes médicinales utilisées par les Pygmées (Mbute) pour combattre le paludisme dans la zone de Mambasa, Ituri, Zaire

M. MAGILU, M. MBUYI & M.B. NDJELE
Faculté des Sciences, Université de Kisangani, B.P. 2012, Kisangani, Zaïre.

Résumé
Les enquêtes ethnobotaniques ont été menées auprès des Pygmées (Mbute) habitant les environs de Mambasa et Nduye en Ituri, Haut-Zaïre. Elles ont permis de recenser 24 espèces différentes qui interviennent dans 22 recettes utilisées pour le traitement de la malaria. Comparés à certaines ethnies qui habitent le Haut-Zaïre (Buta, Kisangani, Opala, Yangambi), des organes appartenant à 17 espèces ont été cités pour le même usage. D'après la Banque de données PHARMEL sur le paludisme, 5 de nos plantes sont également citées dans les rapports publiés dans d'autres pays africains. Certaines plantes recensées exclusivement chez les Pygmées, se sont montrées très efficaces dans le traitement de la malaria, surtout *Picralima nitida* (Apocynaceae).

Abstract
Ethnobotanic investigations have been made among the Pygmies (Mbute), the local population near Mambasa and Nduye in Ituri, Upper-Zaïre. We listed 24 different plant species used to treat malaria. Certain Bantu-tribes, also living in the Upper-Zaïre (Kisangani, Buta, Yangambi, Opala), use 17 similar plant species to treat the disease. According to the PHARMEL data on malaria, 5 of our plants are also mentioned in a report published in other African countries. Some species only used by Pygmies are very effective in the treatment of malaria: in particular *Picralima nitida* (Apocynaceae).

Key words
Ituri, Malaria, Mbute, Medicinal plants, Pygmies, Zaïre.

Introduction

Dans le Haut-Zaïre, les moustiques constituent le groupe d'insectes le plus nuisible. Ils peuvent être considérés parmi les pires ennemis naturels des habitants de cette région, à la fois par leurs attaques gênantes sur les personnes et surtout par leur rôle extraordinaire comme vecteurs de maladies. Le paludisme est la plus répandue de toutes les maladies transmises. Ses conséquences sur la santé publique sont actuellement fort inquiétantes.

Pour le moment, malheureusement, il n'existe à Kisangani, aucun programme de lutte contre les moustiques. Au niveau individuel, les insecticides chimiques demeurent les moyens de lutte les plus répandus et les plus efficaces. Cependant, leur prix élevé ainsi que leur disponibilité forcément limitée, découragent l'expansion de la lutte chimique à Kisangani. En outre, certains moustiques deviennent de moins en moins sensibles aux insecticides chimiques. Ce qui est encore plus grave, il y en a qui présentent déjà une résistance à certains antibiotiques.

Devant ce désastre qui est devenu la préoccupation de tous les scientifiques, il nous a paru utile d'envisager d'autres moyens de lutte contre la malaria. Nous avons pensé que la pharmacopée traditionnelle est l'une des voies la mieux indiquée parce que c'est

L.J.G. van der Maesen et al. (eds.), The Biodiversity of African Plants, 741-746.

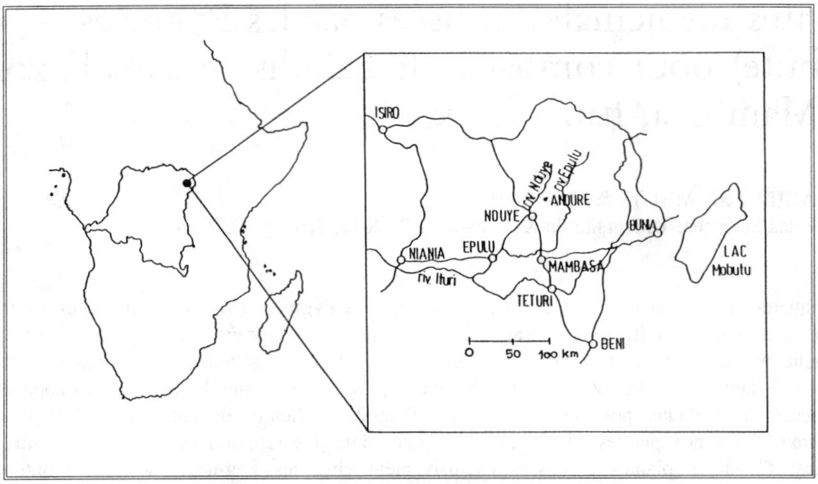

Figure 1. Localisation de la zone prospectée.

un axe important de la recherche pharmaceutique consistant justement à la mise au point des médicaments à partir des plantes. Dans ce contexte, nous avons orienté notre curiosité vers les Pygmées (Mbute) qui sont un peuple encore primitif habitant la forêt dense de l'Ituri, cadre idéal à ce genre de recherche, car éloigné de tout centre médical moderne. Nous pensons enfin que, l'intégration de la pharmacopée traditionnelle à la chimiothérapie peut nous permettre de concevoir pour l'avenir, une meilleure stratégie afin d'équilibrer les coûts des soins médicaux pour nos populations.

Méthodes d'enquêtes ethnobotaniques

Un de nous, accompagné d'un interprète, a effectué le déplacement dans la forêt de l'Ituri. Les environs de Mambasa et Nduye ont été particulièrement bien explorés (Figure 1).

D'après le questionnaire qui leur était soumis, les Pygmées interrogés ont livré les informations sur le nom des plantes, l'organe utilisé, les modes de préparation et d'administration ainsi que la posologie. La récolte des plantes a été effectuée sur le terrain en compagnie de l'informateur et du traducteur. Pour obtenir les renseignements, on a contacté au total 30 personnes différentes constituant ainsi l'effectif informateur.

Résultats

L'inventaire floristique des plantes antipaludiques utilisées par les pygmées apparaît dans les recettes citées ci-dessous. Au total, 24 espèces de plantes différentes interviennent dans 22 recettes pour le traitement de la malaria:

1. Faire un décocté aqueux dans un verre avec 5 g d'écorces de racines raclées de *Rauvolfia vomitoria*. Prendre par voie orale matin, midi et soir (Inf. 11).
2. Préparer une macération des feuilles pilées de *Rauvolfia vomitoria* remplissant le verre jusqu'au 1/5 dans l'eau. Après filtration, prendre par voie orale matin et soir (Inf. 2).

3. Faire un décocté de 2 bouteilles d'eau avec 15 à 20 feuilles de *Morinda lucida*. Prendre per-os 2 verres, matin et soir, pendant 3 jours ou en purgation (Inf. 12).
4. Les feuilles, rameaux florifères et fructifères de *Chenopodium ambrosioides* pilés, sont utilisés pour masser le corps et introduits dans les narines (Inf. 25, 27).
5. La graine dépochée de *Picralima nitida* est directement mangée, matin, midi et soir pendant 3 jours (Inf. 1, 2). Pendant la crise de malaria, il suffit de manger la graine et 30 minutes après, on est debout.
6. Préparer un décocté aqueux égal à un verre de râpures de 2 morceaux (1 dm x 0,5 dm) d'écorces du tronc de *Picralima nitida*. Prendre chaud ou froid par voie orale (Inf. 3, 6, 8, 9, 16, 21).
7. Placer les raclures de l'écorce de racine de *Tabernaemontana eglandulosa* dans un cornet d'une feuille quelconque. Par pressage de jus, instiller 2 gouttes dans chaque oeil ou narine dès le début de la malaria (Inf. 3, 8, 11).
8. Faire un décocté aqueux d'un litre des écorces des racines pilées de *Pentadiplandra brazzeana*. Après filtration, prendre en purgation toute la solution dans le cas d'un adulte malade, et la moitié chez l'enfant. A répéter après 2 jours (Inf. 12, 26).
9. Préparer une décoction dans un litre d'eau d'organes des plantes suivantes: écorces des racines de *Pentadiplandra brazzeana*, jus d'*Ananas comosus* mûr et de *Citrus limon*, 2 cuillerées à café des graines broyées de *Piper guineense*. Prendre per-os 1 verre 2 fois par jour pendant 1 semaine (Inf. 3).
10. Préparer un décocté aqueux avec 10 feuilles de *Morinda morindoides* dans un verre. Prendre par voie orale après filtration (Inf. 13).
11. Placer des raclures de l'écorce de la tige de *Baissea laxiflora* dans un cornet d'une feuille quelconque. Par pressage de jus, instiller 1 goutte dans chaque oeil ou narine (Inf. 3, 8, 10).
12. Placer les fleurs écrasées de *Adenostemma perrottetti* dans un cornet de feuille quelconque. Par pressage de jus, instiller 1 goutte dans chaque oeil. Ensuite, appliquer sur le front, une quantité de feuilles et d'inflorescences froissées (Inf. 13, 14, 15, 25).
13. Préparer une décoction aqueuse d'un mélange d'organes des plantes suivante: râpures du tronc de *Zanthoxylum gilletii* (= *Fagara macrophylla*), feuilles de *Psidium guajava* et de *Piper guineense*, écorces du tronc de *Petersianthus macrocarpus*. Prendre un bain de vapeur par inhalation en se couvrant avec une couverture (Inf. 2).
14. Faire un décocté de 1 litre d'eau avec 1 kg d'écorces du tronc de *Canarium schweinfurthii*. Prendre en purgation après filtration (Inf. 9, 12).
15. Préparer une décoction aqueuse d'un mélange des écorces des plantes suivantes: *Canarium schweinfurthii*, *Zanthoxylum gilletii*, *Petersianthus macrocarpus*, *Spathodea campanulata* et *Bridelia atroviridis*. Faire l'inhalation sous forme de bain de vapeur 3 fois tous les 2 jours (Inf. 7, 12).
16. Les jeunes feuilles froissées de *Vernonia amygdalina*, sont introduites par voie rectale à la manière d'un suppositoire (Inf. 17, 25).
17. Faire un décocté de 1 litre d'eau avec 500 g de racines pilées de *Piper guineense*. Après filtration, prendre la solution en purgation (Inf. 1).

18. Faire une macération de 5 bourgeons foliaires de *Vernonia amygdalina* avec 2 gobelets d'eau. Après filtration, prendre per-os un gobelet matin et soir pendant 3 jours (Inf. 25).
19. Boire 1/2 verre du latex de *Alstonia congensis* ou de *Alstonia boonei* avant sa coagulation (Inf. 4, 5).
20. Faire pendant une heure, une macération aqueuse d'un gobelet avec des raclures de 2 morceaux (1 x 0,5 dm) d'écorce du tronc de *Alstonia congensis* ou de *A. boonei*. Prendre par voie orale matin et soir pendant 3 jours (Inf. 1, 2, 4, 5, 7).
21. Faire une décoction d'une 1/2 bouteille d'eau en mélangeant les écorces des troncs des plantes suivantes: *Alstonia congensis* ou *Alstonia boonei*, *Zanthoxylum gilletii*, *Mitragyna stipulosa* et *Bridelia atroviridis*. Prendre per-os 2 verres matin et soir (Inf. 18, 19, 20, 22, 23, 28, 29).
22. Faire un décocté aqueux de 2 litres des râpures de 3 morceaux (3 dm x 1 dm) des écorces du tronc de *Autranella congolensis*. Après filtration, prendre en purgation 2 fois par semaine (Inf. 12, 21, 24, 30).

(Inf = Informateurs: 1. Chef Apuebo, 2. Lukaubo, 3. Chef Pascal, 4. Sakasho, 5. Fayizi, 6. Maria, 7. Modeste, 8. Mbakwe, 9. Ngwale, 10. Imulani, 11. Ramazani, 12. Ony, 13. Itwali, 14. Akense, 15. Leso, 16. Aposa, 17. Shamunionge, 18. Bamose, 19. Angalin, 20. Masudi, 21. Komba, 22. Ombakoy, 23. Abula, 24. Mapupu, 25. Lebro, 26. Rasa, 27. Wenie, 28. Eugène, 29. Papa Oney, 30. Papa Huseni).

Discussion

Parmi les 24 plantes recensées (Tableau 1), 7 sont exclusivement utilisées par les Pygmées; il s'agit des plantes suivantes: *Alchornea floribunda*, *Autranella congolensis*, *Baissea laxiflora*, *Bridelia atroviridis*, *Picralima nitida*, *Zanthoxylum gilletii* et *Petersianthus macrocarpus*.

Dix sept plantes sont employées pour le même usage dans la région du Haut-Zaïre (Kisangani, Buta, Yangambi, Opala) par les autres ethnies bantoues; il s'agit des plantes suivantes: *Alstonia boonei*, *Alstonia congensis*, *Canarium schweinfurthii*, *Chenopodium ambrosioides*, *Morinda morindoides*, *Vernonia amygdalina* (Toirambe 1989); *Tabernaemontana eglandulosa* (Batoko 1989); *Adenostemma perrottetti* (Ataholo 1988); *Ananas comosus*, *Citrus limon*, *Mitragyna stipulosa*, *Psidium guajava* (Wome 1985); *Rauvolfia vomitoria* (Lomba 1990, Toirambe 1989); *Spathodea campanulata* (Batoko 1989, Toirambe 1989); *Pentadiplandra brazzeana* (Toirambe 1989, Wome 1985); *Piper guineense* (Ataholo 1988, Toirambe 1989); *Morinda lucida* (Ataholo 1988, Batoko 1989).

Enfin, 5 plantes, utilisées à la fois par les Pygmées et les autres ethnies bantoues du Haut-Zaïre, sont également citées pour le même usage dans les rapports publiés dans d'autres pays d'Afrique (selon la banque des données PHARMEL); il s'agit des plantes suivantes: *Alstonia boonei* (Nigéria, Sénégal, Togo); *Rauvolfia vomitoria* (Nigéria, Togo); *Vernonia amygdalina* (Ouganda, Rwanda); *Morinda lucida* (Congo, Nigéria, Togo); *Morinda morindoides* (Ouganda).

Le *Picralima nitida* est cité comme antimalarique au Zaïre (Lejoly et al. 1992). Cette espèce a été étudiée par Clinquart (1927) et Michiels (1929); à l'époque, le principal alcaloïde contenu dans les graines, l'akuammine s'est révélé inactif sur les *Plasmodium*. Les organes les plus utilisés dans la préparation sont les écorces des

Tableau 1. Plantes médicinales recensées chez les Pygmées pour le traitement du paludisme.

Plantes médicinales	1	2	3	4	5	6
Apocynaceae						
Alstonia boonei De Wild.	—	+	—	—	—	NST
Alstonia congensis Engl.	—	+	—	—	—	—
Baissea laxiflora Stapf	—	—	—	—	—	—
Picralima nitida (Stapf) Th. & H.Dur.	—	—	—	—	—	Z
Rauvolfia vomitoria Afzel.	—	+	—	+	—	NTZ
Tabernaemontana eglandulosa Stapf	—	—	—	—	+	—
Bignoniaceae						
Spathodea campanulata P.Beauv.	—	+	—	—	+	—
Bromeliaceae						
Ananas comosus Merr.	+	—	—	—	—	—
Burseraceae						
Canarium schweinfurthii Engl.	—	+	—	—	—	—
Chenopodiaceae						
Chenopodium ambrosioides L.	—	+	—	—	—	Z
Compositae						
Adenostemma perrottetti DC.	—	—	+	—	—	—
Vernonia amygdalina Del.	—	+	—	—	—	OR
Euphorbiaceae						
Alchornea floribunda Müll.Arg.	—	—	—	—	—	—
Bridelia atroviridis Müll.Arg.	—	—	—	—	—	—
Lecythidaceae						
Petersianthus macrocarpus (P.Beauv.) Liben	—	—	—	—	—	—
Myrtaceae						
Psidium guajava L.	+	—	—	—	—	—
Pentadiplandraceae						
Pentadiplandra brazzeana Baill.	+	+	—	—	+	Z
Piperaceae						
Piper guineense Schum. & Thonn.	—	+	+	—	—	—
Rubiaceae						
Morinda lucida Benth.	—	—	+	—	+	CNT
Morinda morindoides (Bak.) Milne R.	—	+	—	—	—	O
Mitragyna stipulosa (DC.) O.Ktze	+	—	—	—	—	—
Rutaceae						
Citrus limon (L.) Burm.	+	—	—	—	—	—
Zanthoxylum gilletii (De Wild.) Waterman	—	—	—	—	—	—
Sapotaceae						
Autranella congolensis (De Wild.) A.Chev.	—	—	—	—	—	—

1 = Kisangani, Wome 1985. + = signalé antipaludique; — = non signalé antérieurement.
2 = Kisangani, Toirambe 1989.
3 = Buta, Atahola 1988.
4 = Yangambi, Lomba 1990.
5 = Opala, Batoko 1989.
6 = selon la banque des données PHARMEL aussi cité pour ces autres pays africains.
C = Congo; N = Nigéria; O = Ouganda; R = Rwanda; S = Sénégal; T = Togo; Z = Zaïre.
(Voir Adjanohoun et al. 1986, 1988, 1991)

troncs et des racines, et les feuilles. Le mode le plus important de préparation des remèdes est la décoction qui est prescrite pour la plupart des cas en prise orale ou en purgation. La macération et l'instillation sont également utilisées. Certaines recettes médicamenteuses sont obtenues en associant plusieurs plantes différentes.

Il est certain que les Pygmées ont cité comme antipaludiques, des plantes utilisées également pour soigner la fièvre parce qu'ils n'ont pas fait un examen de la "goutte épaisse" pour savoir si réellement la personne qui fait la fièvre, souffre de la malaria et présente donc des parasites du genre *Plasmodium* dans son sang.

Conclusion

A l'issue de cette étude préliminaire sur les plantes antipaludiques utilisées par les Pygmées de la Zone de Mambasa, la vertu curative de certaines plantes est confirmée notamment par la convergence d'utilisation dans d'autres zones du Haut-Zaïre et d'Afrique. Il nous paraît important de relever que parmi les plantes utilisées exclusivement par les Pygmées, certaines se sont montrées efficaces dans le traitement de la malaria chez les étudiants, professeurs, épouses et enfants de l'Université de Kisangani: c'est le cas de *Picralima nitida*.

Bibliographie

Adjanohoun, E., A. Ahyi, L. Aké Assi, K. Akpagana, P. Chibon, A. Elhadji, J. Eyme, M. Garba, J.-N. Gassita, M. Gbeassor, E. Goudote, S. Guinko, K.-K. Hodouto, P. Houngnon, A. Keita, Y. Keoula, W.P. Klugaocloo, I. Lo, K.M. Siamevi & K.K. Taffame, 1986. Contribution aux études ethnobotaniques et floristiques au Togo. ACCT, Paris, 671 p.

Adjanohoun, E., A. Ahyi, L. Aké Assi, J. Baniakina, P. Chibon, G. Cusset, V. Doulou, A. Enzanza, J. Eyme, E. Goudote, A. Keita, C. Mbemba, J. Mollet, J.M. Moutsamboté, J.B. Mpati & P. Sita, 1988. Contribution aux études ethnobotaniques et floristiques au Congo. ACCT, Paris, 605 p.

Adjanohoun, E., M.R.A. Ahiyi, L. Aké Assi, K. Dramane, J.A. Elewude, S.O. Fadoju, Z.O. Gbile, E. Goudote, C.L.A. Johnson, A. Keita, O. Morakinyo, J.A.O. Ojewole, A.O. Olatunji & E.A. Sofowora, 1991. Contribution to ethnobotanical and floristic studies in western Nigeria. CSTR/OUA, 420 p.

Ataholo, M.A., 1988. Plantes médicinales de Buta. Mémoire de Licence, Faculté des Sciences, Univ. Kisangani, 104 p., inédit.

Batoko, M.A., 1989. Plantes médicinales utilisées par les Bambole. Mémoire de Licence, Faculté des Sciences, Univ. Kisangani, 98 p., inédit.

Clinquart, E., 1927. Sur les alcaloïdes d'une graine congolaise, Picralima klaineana. Bull. Acad. Roy. Méd. Belge, 5e sér., 6: 492-504; Journ. Pharm. Belge 9, 187 p.

Lejoly, J., M.J. Polygenis-Bigendako & F. Maes, 1992. In: Janssens, P.G., M. Kivits & J. Vuylsteke (eds), Médecine et hygiène en Afrique Centrale de 1885 à nos jours. Fondation Roi Baudouin, Bruxelles et Masson, Paris.

Lomba, B., 1990. Plantes médicinales de la région de Yangambi. Mémoire de Licence, Faculté des Sciences, Univ. Kisangani, 166 p., inédit.

Michiels, M., 1929. Les graines de *Picralima klaineana* et leur toxicité. Bull. Acad. R. Méd. Belge, Séance 23 février: 21-25.

Toirambe, K., 1989. Plantes antipaludiques de Kisangani. Mémoire de Licence, Faculté des Sciences, Univ. Kisangani, 115 p., inédit.

Wome, B., 1985. Recherches ethnopharmacognosiques sur les plantes médicinales utilisées en médecine traditionnelle à Kisangani (Haut-Zaïre). Thèse de Doct. Univ. Libre de Bruxelles, 561 p.

Medicinal plants of the Maracheople of Kenya

P.S. MASINDE
East African Herbarium, National Museums of Kenya, P.O. Box 45166, Nairobi, Kenya

Abstract
A study carried out among the Marachi, of Western Province, Kenya, resulted in the recording of about 140 medicinal plant species from 42 (sub-)families. Taxa with over 5 species were: Compositae (20), Euphorbiaceae (11), Papilionoideae (9), Labiatae (8), Acanthaceae (6) and Caesalpinioideae (6). Common ailments treated were gastro-intestinal diseases like stomach-ache, skin problems like wounds and fungal infections, colds and coughs, headaches, fevers, sexually transmitted diseases, and respiratory diseases. Disease and treatment are perceived as the physical, psychological, and spiritual aspects of one's well-being, therefore treatment often involves rituals in addition to the administration of medicine. Most herbal medicines are harvested from the wild, and herbal practitioners are quite knowledgeable on the flora of the area.

Key words
Ethnobotany, Kenya, Marachi, Medicinal plants.

Introduction

Folk knowledge has been orally transmitted from one generation to another. Most of this knowledge is with the elderly people. Modernisation and the inevitable rapid changes in lives of the people in terms of education, customs and traditions, rural-urban migration, and environmental degradation, have disrupted folk knowledge transmission. Unlike other folk knowledge, knowledge about herbs is generally kept secret and confidential (Kokwaro 1976).

Among the Marachi, health and disease are perceived as the physical, psychological and spiritual dimensions of one's well-being. Therefore rituals are often performed in addition to the administration of medicine, in the treatment of certain diseases, for it is believed to be necessary to make the treatment effective.

Most herbal medicines are harvested from the wild. At the local level, herbal practitioners are quite knowledgeable on the plants of the area in terms of their identification, habitats, distribution and frequency. In Marachi, herbal medicine is the only traditional practice that utilises a wide range of plant species. This is true, to such an extent, that when local people see plant collectors or taxonomists in the field, they can only think of them as herbalists in disguise.

Study area and methods

The Marachi, a sub-group of the much larger Luhya tribe who are Bantu, live in Marachi Location, Busia District, in Western Province of Kenya. Most of the people are peasant farmers but the younger educated people have moved to urban areas to find work and only return to their villages occasionally.

This information is partly derived from the author's experience as a resident in Marachi for many years and from a study on the medicinal plants of the area in 1993 and 1994. The study was carried out by interviewing persons in the area. The first approach involved asking questions from a set questionnaire that had been memorised

L.J.G. van der Maesen et al. (eds.), The Biodiversity of African Plants, 747-753.

by the author. This was carried out orally without too much writing on the spot in order to make the process fast and relaxed so as to avoid impatience and suspicion from interviewees. The interviewees were visited at least twice, or more if possible, in order to exhaust the information they had. This is kept orally and cannot all be remembered at any one time. After an interview session the interviewee would be persuaded to accompany the author for a walk through the village bushes so that plants mentioned could be identified and collected. More importantly, seeing plants in the field prompted the recollection of otherwise forgotten information. Information obtained during any visit was compiled and a copy made for the interviewees to convince him/her that the process was not a way to steal their herbal secrets but to record them. Many were pleased to have copies of the information even though many are illiterate. Table 1 is a summary of information about the interviewees. The three age groups represent the young (15-35 years), middle age (36-49 years), and elders (50+ years).

Voucher specimens were collected, pressed, identified and deposited at the East African Herbarium, Nairobi.

Table 1. Age groups and gender of interviewees.

Age group (years)	Number and gender of interviewees		Total
	Male	Female	
15-35	3	4	7
36-49	2	3	5
50+	3	5	8
Total	8	12	20

Results

Herbal medicines are prepared and administered in various ways (Table 2; cf Kokwaro 1976). Mixtures of many plant species are often used in preparing medicines.

Common ailments of Marachi that are often treated with herbal medicine include the following: gastro-intestinal diseases like diarrhoea and stomach-ache, wounds, ulcers, cuts and burns, sexually transmitted diseases, malaria, colds and coughs, influenza, fevers, headaches, eye and ear infections, dental problems, skin problems like rashes, measles, boils, muscle and joint sprains, bone fractures, fungal infections, and gynaecological problems. There are several herbal medicines to deal with these ailments (Table 3).

The following are among the (sub)families important for medicinal plants with the number of species in brackets: Compositae (20), Euphorbiaceae (11), Papilionoideae (9), Labiatae (8), Acanthaceae (6), Caesalpinioideae (6), Vitaceae (5), Rubiaceae (4), Verbenaceae (4), Tiliaceae (3), and Anacardiaceae (3).

Discussion

There are three categories of herbal practitioners in Marachi. First, there are the lay people who happen to know some herbs because they have inherited the knowledge

Table 2. Methods of herbal medicine preparation and application.

Method	Medicine type	Application
Boiling & Steaming	Decoctions & Concoctions	Oral or vapours inhaled
Pounding	Paste, Pulp, Infusion, Ointment (ghee added)	Oral; applied to eyes, ears, wounds or used in massage
Chewing	Juice	Juice swallowed or retained in mouth for some time, then spat out
Drying & Grinding	Powder, Ointment, Infusion	Oral, external, or used in massage
Roasting & Burning	Warm leaf, Ash	Leaf for massage; ash licked, taken in water or food, or rubbed into cuts made in skin with a sharp blade

from their relations or were shown by someone else. In this category can be included nearly all Marachi people because traditionally one is supposed to know at least herbal medicines for first aid and for common problems like stomach-ache, colds, coughs, cuts, and so on. Secondly there are those who tend to specialise in treating particular diseases for example infant diseases, labour problems, venereal diseases, and so on. They are part-time practitioners and do not normally charge a fee but are mostly given tokens, especially when a patient recovers. Thirdly, there are professional full-time practitioners who are very knowledgeable in plants but also practice divination, foretelling and black magic. They are feared because they can use herbal medicine either to treat or to cause trouble. They use incantations and rituals in addition to the herbal medicine and charge consultation fees. Professional practitioners have medicines against any imaginable thing. For example, they employ medicines to make one have good luck and success, win an election, avert a misfortune that is coming, become rich, and so on.

The elders were most knowledgeable because of the many years of accumulated experience. Among the young and middle age, they often know the use of a plant but not its name. However, there are a few exceptions whereby a young person comes from a family with practising herbalists and has taken interest in knowing the plants.

In Marachi, the continuity of herbal medicine practice depends on sustainable use of wild plants. There are several practices that are geared toward plant conservation, many of which have been maintained as norms or taboos so that they are adhered to. Usually it is the herbs that are fast growing and abundant that are uprooted and the whole plant used as medicine. When roots are used as medicine, only a few of them are carefully dug out and then the soil is returned. The next time one has to dig roots from the same plant, one has to take them from a side that has not been recently dug. When farmers are clearing bushes for cultivation, they are careful to note any important medicinal plants, and will usually not clear such plants if known to be rare. They leave them to grow with their crop. When using the bark, they strip only a small piece for they do not want to kill the plant. Habitat destruction and poor harvesting are highly discouraged through taboos, to avoid local extirpation of some herbal medicines or rarity. Local extirpation and rarity of plants forces practitioners to travel long distances in bushes while searching for medicines. It is only in those cases where a medicinal plant grows far away, or is frequently used, or rare that it will be put in cultivation. Such medicinal plants are cultivated in homes like ornamentals or inter-planted with crops. Professional herbalists are always eager to learn about new plants in the hope that they could be new medicines.

There is a general belief that a disease should not be declared incurable by doctors in modern hospitals until herbal medicine has also been used for its treatment. Therefore, a question often asked of those who cannot be cured by modern medicine is "Have you tried our traditional herbal medicine?"

Herbal medicine is cheap and often times quite effective in the treatment of some diseases. Research on traditional herbal medicine will ensure that medicinal folk knowledge is not lost completely, will highlight the importance of many plant species, and aid in the conservation of the flora and culture of the people and at the same time supplement modern health services.

Acknowledgements

Sincere gratitude is extended to Longino Masinde, Peter Oyiero, and Immaculate Astieno, for providing a lot of information; to Dr. E.B. Knox for assistance with databasing the information; to Global Environment Facility (Nairobi) for providing part of the funds for the field work, and to the many interviewees for their cooperation.

Reference

Kokwaro, J.O., 1976. Medicinal Plants of East Africa. East African Literature Bureau. Nairobi.

Table 3. Selected Medicinal Plants of the Marachi tribe of West Kenya: family, species, Marachi name, plant part used/disease/comment.

Family	Species	Marachi name	Plant part used/disease/comment
Gastro-intestinal diseases			
Anacardiaceae	*Lannea edulis* (Sond.) Engl. var. *edulis*	Masungubale	Root decoction for diarrhoea
Apocynaceae	*Carissa edulis* (Forssk.) Vahl	Ochuoka	Root decoction for stomach problems
Caesalpinioideae	*Senna bicapsularis* (L.) Roxb.	Ingolingoli	Root added to meat of animals that die from disease to neutralise poisons before its consumption
	Senna didymobotrya (Fres.) Irwin & Barneby	Libinubinu	Leaf decoction cleanses stomach
Celastraceae	*Maytenus senegalensis* (Lam.) Exell	Induli	Root concoction for stomach problems
Compositae	*Berkheya spekeana* Oliv.	Lirakalu lisatsa	Root concoction for stomach-ache
	Tithonia diversifolia (Hemsl.) A. Gray	Liuwa lilulu	Leaf decoction for stomach-ache
	Vernonia hymenolepis A. Rich. Gray	Mululusia-musatsa	Root concoction for stomach-ache
Ebenaceae	*Euclea divinorum* Hiern	Mukato	Root decoction for constipation
Euphorbiaceae	*Acalypha villicaulis* A. Rich.	Not known	Root concoction for diarrhoea
	Bridelia micrantha (Hochst.) Baill.	Lupobo lusatsa	Root concoction for diarrhoea, and blood-stained stool
Menispermaceae	*Cissampelos mucronata* A. Rich.	Mulandira	Root decoction for stomach-ache
Mimosoideae	*Mimosa pudica* L.	Kukhuirukha	Leaf concoction for stomach-ache
Papilionoideae	*Abrus canescens* Welw. ex Bak.	Not known	Leaf infusion for stomach-ache
	Eriosema psoraleoides (Lam.) G. Don	Omuyenula	Root and leaf decoction for stomach-ache
	Indigofera circinella Bak. f.	Tsifunyi	Root and leaf decoction for intestinal worms
Pittosporaceae	*Pittosporum viridiflorum* Sims	Not known	Root concoction for diarrhoea
Simaroubaceae	*Harrisonia abyssinica* Oliv.	Mupedo	Root decoction for stomach problems
Tiliaceae	*Triumfetta rhomboidea* Jacq.	Lutsukutsiro	Leaf concoction for blood-stained stool
Vitaceae	*Rhoicissus revoilli* Planch.	Omulandira	Root decoction for stomach-ache
Wounds, ulcers, cuts, burns and skin problems			
Acanthacaeae	*Thunbergia alata* Sims	Msala kwe sikwata	Leaf chewed and solution spat out, for mouth ulcers
Caesalpinioideae	*Senna didymobotrya* (Fres.) Irwin & Barneby	Libinubinu	Leaf ointment for skin fungal infections

Table 3. (Continued).

Commelinaceae	*Murdannia simplex* (Vahl) Brenan	Not known
	Jelly from flowers rubbed on sites with athletes foot infections	
Compositae	*Ageratum conyzoides* L.	Nasirundu
	Leaf crushed and sqeezed to cuts and wounds to stop bleeding and for cure	
	Aspilia mossambicensis (Oliv.) Wild.	Litsomisi
	Leaf crushed and solution sqeezed onto cuts and wounds	
	Microglossa densiflora Hook.f.	Namukuu
	Powder from dried leaf or ash put on wounds and cuts	
Cucurbitaceae	*Momordica foetida* Schum.	Litunda
	Root concoction for boils	
Oxalidaceae	*Biophytum petersianum* Klotzsch	Kukhuirukha khatiti
	Dried leaf powder ointment for scabies	
Polygonaceae	*Oxygonum sinuatum* (Meisn.) Dammer	Sihenge
	Leaf ointment for boils	
Sterculiaceae	*Waltheria indica* L.	Not known
	Powder from dried leaf put on wounds and cuts	
Vitaceae	*Ampelocissus africana* (Lour.)Merr.	Libombwe likhasi
	Root decoction for a rash	
Sexually transmitted diseases		
Anacardiaceae	*Rhus vulgaris* Meikle	Owayo
	Root decoction for gonorrhoea	
Caesalpinioideae	*Senna didymobotrya* (Fres.) Irwin & Barneby	Libinubinu
	Root concoction for gonorrhoea and syphilis	
Euphorbiaceae	*Flueggea virosa* (Willd.) Vlogt.	Not known
	Leaf concoction for gonorrhoea	
Rubiaceae	*Gardenia ternifolia* Schum. & Thonn.	Esiuna
	Root concoction for gonorrhoea and syphilis	
Verbenaceae	*Lantana camara* L.	Obengele
	Root decoction for gonorrhoea	
	Ocimum basilicum L.	Obwali butiti
	Root decoction for gonorrhoea	
Colds, coughs, influenza, fever, and headache		
Anacardiaceae	*Rhus vulgaris* Meikle	Owayo
	Leaf chewed for irritated throat, and coughs	
Commelinaceae	*Cyanotis arachnoidea* C.B.Cl.	Linyolonyolo
	Leaves chewed for headache	
Compositae	*Schkuhria pinnata* (Lam.) O. Ktze	Khalulu
	Decoction from whole plant for fevers, infuenza, and malaria	
Euphorbiaceae	*Croton macrostachyus* Hochst. ex Del.	Omustwistwi
	Root and leaf decoction bathed for fever	
	Croton megalocarpus Hutch.	Not known
	Ash from bark licked for irritated throat, and colds and coughs	
Iridaceae	*Gladiolus newii* Bak.	Isamembe
	Ash from tuber licked for colds, coughs and sore throat.	
Labiatae	*Tetradenia riparia* (Hochst.) Codd	Ogita/Orokoma
	Leaf decoction for fevers, influenza and malaria	

Table 3. (Continued).

Melastomataceae	*Dissotis canescens* (Graham) Hook. f. Leaf infusion for colds and coughs	Nalunyulunyu
Papilionoideae	*Abrus precatorius* L. Leaves chewed for colds and coughs	Ombulu
Rutaceae	*Toddalia asiatica* (L.) Lam. Root chewed for sore throat and influenza	Naletukwachi
Other ailments or uses		
Acanthaceae	*Thunbergia alata* Sims Leaf concoction ointment for massage	Msala kwe sikwata
Anacardiaceae	*Lannea edulis* (Sond.) Engl. Root decoction given to mothers in labour to ease expulsion of placenta	Masungubale
Apocynaceae	*Carissa edulis* (Forssk.) Vahl Fruits eaten as an appetiser	Ochuoka
Asclepiadaceae	*Asclepias kaessneri* N.E. Br. Leaf pound and squeezed to get droplets for ailing ears	Lubondabondi
	Mondia ecornuta (N.E. Br.) Bullock Bark of root chewed as appetiser and as aphrodisiac	Mugombera
Caesalpinioideae	*Senna didymobotrya* (Fres.) Irwin & Barneby Root decoction stops too much blood flow during menstruation	Libinubinu
Compositae	*Bidens pilosa* L. Root infusion for convulsions	Inguyesi
Ebenaceae	*Euclea divinorum* Hiern Roots chewed for paining gums and teeth	Mukato
Euphorbiaceae	*Ricinus communis* L. Fruit solution used to induce vomiting	Liroborobo
Labiatae	*Tetradenia riparia* (Hochst.) Codd Leaf chewed for teeth and gum ailments	Ogita/Orokoma
Liliaceae	*Sansevieria trifasciata* Prain Leaf paste rubbed into scarification cuts as antidote for snake bites	Mukhuju
Moraceae	*Ficus natalensis* Hochst. Smoke from sticks directed into eyes when ailing	Muruba
Rubiaceae	*Gardenia ternifolia* Schum. & Thonn. Smoke from sticks directed into eyes when ailing	Esiuna
Sapindaceae	*Cardiospermum halicacabum* L. Leaf concoction for tuberculosis	Not known
Umbelliferae	*Centella asiatica* (L.) Urb. Leaf paste placed on snake bite wounds as antidote	Not known
	Diplolophium africanum Turcz. Leaf ash rubbed in scarification cuts to stop persons from having bad dreams and hallucinations	Omunya

Végétation et plantes alimentaires de la région de la Sangha (Nord-Congo)

J.M. MOUTSAMBOTÉ
Lab. de Botanique, Centre d'Etudes sur les Ressources Végétales, B.P. 1249, Brazzaville, Congo.

Résumé
Il existe au Congo d'importantes forêts denses humides renfermant, bien entendu, des plantes comestibles appartenant à la flore spontanée. Par la diversité floristique, la variation des sols, 5 types de végétation ont été identifiés dans la région; forêt mélangée, forêt riveraine, forêt marécageuse, forêt monospécifique à Limbali (*Gilbertiodendron dewevrei*), forêt secondaire. Nous disposons actuellement d'un inventaire floristique des plantes alimentaires utilisées par des hommes, fréquentes sur les marchés. La majorité des plantes légumières est vendue toute l'année, alors que les fruits et certains légumes dépendent de la phénologie des espèces.

Abstract
The vast rain forests extending over northern Congo of course include edible plants belonging to the spontaneous flora. Based on the variety of plant species, soils and soil water conditions, 5 vegetation types were identified in the region: mixed species forest, riverine forest, swamp forest, monospecific forest of Limbali (*Gilbertiodendron dewevrei*) and secondary forest. The inventory presently includes edible plants used by man, regularly sold on the market. Most species used as vegetable are sold throughout the year, whereas availability of fruits and a few vegetables depends on the phenology of the species.

Key words
Congo, Edible plants, Sangha area, Vegetation types.

Géographie
Le Congo est localisé en Afrique centrale. Il est traversé par l'Equateur entre les latitudes 3°34'N et 5°00'S, et les longitudes 11°11'E et 18°35'E. Il est limité à l'est par le Zaïre, au nord par le Cameroun et la République centrafricaine, à l'ouest par le Gabon, au sud par le Cabinda (Angola) et le Zaïre et au sud-ouest par l'Océan Atlantique. La superficie du Congo est de 342.000 km^2. Son climat est généralement équatorial. Le taux de précipitations est de 1.200 à 2.000 mm. La température moyenne annuelle atteint 27°C à travers tout le pays. La population humaine est d'environ 2,2 millions en 1990. La géomorphologie de la région de la Sangha est constituée de plateaux traversés par des vallées (300-600 m d'altitude). Les sols sur les plateaux sont des oxisols ferriques et dans les vallées humides, ce sont des oxisols ferriques hydromorphes.

Types de végétation
La région de la Sangha est une région généralement forestière. On y distingue cinq types de forêts: 1. Forêt hétérogène ou mélangée; 2. Forêt riveraine; 3. Forêt marécageuse; 4. Forêt monospécifique à Limbali (*Gilbertiodendron dewevrei*, Caesalpinioideae), appelée Bemba par les Baka, Aka et Bangombé; 5. Forêt secondaire couverte parfois par *Musanga cecropioides* ou autres espèces végétales secondaires.

L.J.G. van der Maesen et al. (eds.), The Biodiversity of African Plants, 754-756.

Site et méthodes d'étude

Le travail présenté ici a été réalisé dans la forêt du Parc National Nouabalé-Ndoki (district de Ouesso) et du Mont Nabemba-Garabizam (district de Souanké). Il existe dans la forêt marécageuse des clairières ou salines appelées baï en langue Bangombé. Ces salines sont couvertes d'espèces herbacées. Nous avons effectué des missions durant les années 1989, 1990 et 1992. L'enquête ethnobotanique a été menée auprès des Pygmées Baka, Bangombé et des Bantou. Nous disposons actuellement d'un inventaire de 60 espèces de plantes alimentaires utilisées par les populations locales.

Période de récolte

La phénologie influence la période de récolte des plantes comestibles dans la forêt suivant le climat. Dans la région de la Sangha, la période de récolte se situe entre les mois de juin et novembre. La production des plantes comestibles devient importante en juin, juillet et août. Durant cette saison de pluies, les populations récoltent des plantes à usage domestique et commercial vendues sur les marchés locaux. Les tiges et les feuilles comestibles sont récoltées durant toute l'année.

Fruits, feuilles et autres parties de plantes

Fruits

Les fruits les plus fréquemment consommés sont: *Trichoscypha* spp. (Anacardiaceae), *Anonidium mannii* (Annonaceae), *Clitandra cymulosa*, *Landolphia* spp. (Apocynaceae), *Allanblackia floribunda*, *Mammea africana* (Guttiferae), *Irvingia gabonensis*, *Klainedoxa gabonensis* (Irvingiaceae), *Myrianthus arboreus*, *Treculia africana* (Moraceae), *Chytranthus atroviolaceus* (Sapindaceae), *Gambeya* spp., *Lecomtedoxa biraudii*, *Synsepalum dulcificum* (Sapotaceae) et *Aframomum* spp. (Zingiberaceae).

Les graines de *Panda oleosa* (Pandaceae) et *Klainedoxa gabonensis* (Irvingiaceae) peuvent être mangées crues ou grillées. Les graines de *Tetracarpidium conophorum* (Euphorbiaceae) sont consommées bouillies, mais celles de *Coula edulis* (Olacaceae) sont consommées crues ou bouillies. Les graines d'*Irvingia gabonensis* (Irvingiaceae) et d'*Afrostyrax lepidophyllus* (Huaceae) séchées sont préparées avec de la viande, du poisson ou des légumes comme assaisonnement. L'huile extraite des graines de *Poga oleosa* (Rhizophoraceae) et de *Baillonella toxisperma* (Sapotaceae) est aussi utilisée dans la cuisson. La consommation des fruits diminue ou est presque inexistante pendant la saison sèche.

Feuilles

Parmi les feuilles comestibles nous citons: *Erythrococca chevalieri* (Euphorbiaceae), *Gnetum africanum*, *G. buchholzianum* (Gnetaceae) et *Afrostyrax lepidophyllus* (Huaceae). Ces feuilles sont les plus appréciées et utilisées dans la cuisson du poisson ou de la viande. Les jeunes feuilles de certaines Dioscoreaceae (*Dioscorea alata*) et de *Megaphrynium macrostachyum* (Marantaceae) sont parfois consommées.

Autres parties de plantes (tiges, tubercules)

Les tubercules de certains *Dioscorea* spp. (Dioscoreaceae) et les jeunes tiges de *Dioscorea mangenotiana*, de *D. praehensilis* et de *Lacosperma secundiflorum* (Palmae) sont importants dans l'alimentation des populations de la région de la Sangha.

Discussion et conclusion

L'inventaire des plantes effectué dans la forêt Nouabalé-Ndoki renferme 500 espèces parmi lesquelles 60 plantes alimentaires sont consommées par les populations locales et vendues parfois sur les marchés locaux ou urbains. Les (sub)familles des espèces les plus consommées sont: Anacardiaceae, Apocynaceae, Burseraceae, Caesalpinioideae, Guttiferae, Dioscoreaceae, Gnetaceae, Huaceae, Irvingiaceae, Moraceae, Pandaceae, Sapindaceae, Sapotaceae, Sterculiaceae et Zingiberaceae.

Après la pêche et la chasse, les plantes constituent le seul recours alimentaire des populations de la région qui ont développé une relation intime avec le milieu dont dépend leur survie. Malgré les variations saisonnières, la crise alimentaire ne s'observe pas chez les populations forestières. La nourriture disponible les épargne de toutes les adversités climatiques.

Bibliographie

Britta, A.-O., 1990. Dietary use of wild plant resources in rural Swaziland. Dans: Comptes rendus de la 12e réunion plénière de l'AETFAT: 895-910.

Gauttier-Beguin, D., 1989. Inventaire des plantes de cueillette à utilisation alimentaire dans le sud du V-Baoulé, Côte d'Ivoire. Dans: Comptes rendus de la 12e réunion plénière de l'AETFAT: 929-934.

Hladik, C.M. & A. Hladik, 1989. Les produits de la forêt dense. Dans: Se nourrir en forêt tropicale. UNESCO/MAB, Paris: 14-16.

Makita Madzou, J.P., 1985. Etude morphologique et phytogéographique des fruits comestibles de la flore spontanée du Congo. Université d'Orléans, Thèse 3e cycle: 159 pages.

Moutsamboté, J.M., 1994. Vegetation and list plant species identified in the Nouabalé-Ndoki forest, Congo. Dans: Floral lists from five study sites of apes in the African tropical forests. Tropics 3(3/4): 277-293.

Pagezy, H., 1991. Alimentation et saisonnalité dans la région du lac Tumba. Dans: Se nourrir en forêt équatoriale. UNESCO/MAB, Paris: 37-42.

Etude des effets des plantes molluscicides de la flore naturelle du Burundi sur *Biomphalaria pfeifferi*, hôte intermédiaire de la bilharziose

P. NDABANEZE [1], D. ENGELS [2] & P.C. KAVAMAHANGA [3]
[1] Université du Burundi, B.P. 2940, Bujumbura, Burundi.
[2] Ministère de la Santé Publique, Projet L.M.T.C., B.P. 337, Bujumbura, Burundi.
[3] Université du Burundi, B.P. 5142, Bujumbura, Burundi.

Résumé
L'action des espèces suivantes a été étudiée sur *Biomphalaria pfeifferi*, hôte intermédiaire de la bilharziose: *Agave sisalana*, *Anacardium occidentale*, *Annona senegalensis*, *Balanites aegyptiaca*, *Canna indica*, *Chenopodium ambrosioides*, *Phytolacca dodecandra*, *Polygonum senegalense*, *Sesbania sesban*, *Simarouba vesicolor* et *Solanum lycopersicum*. Ces espèces se sont révélées efficaces à divers degrés. *Phytolacca dodecandra*, *Balanites aegyptiaca*, *Agave sisalana* ont montré une toxicité élevée envers *Biomphalaria pfeifferi*. Elles exercent une action létale à faible dose (10 mg de poudre/l).

Abstract
Various plant extracts were tested for toxicity to the snail *Biomphalaria pfeifferi*, the intermediate host of Schistosoma, the causal agent of Bilharzia. *Agave sisalana*, *Anacardium occidentale*, *Annona senegalensis*, *Balanites aegyptiaca*, *Canna indica*, *Chenopodium ambrosioides*, *Phytolacca dodecandra*, *Polygonum senegalense*, *Sesbania sesban*, *Simarouba vesicolor* and *Solanum lycopersicum* are effective to different degrees. *Phytolacca dodecandra*, *Balanites aegyptiaca* and *Agave sisalana* have a toxic effect with the lowest dose (10 mg of powder/l).

Key words
Bilharzia, *Biomphalaria pfeifferi*, Plant extracts.

Introduction

Découverte en 1852 dans les pays tropicaux, la bilharziose fut signalée au Burundi pour la première fois en 1921. Elle est aujourd'hui répandue dans les régions de plaine situées entre 700 m et 1300 m, ainsi que dans les marécages et lacs du plateau central. Il s'agit d'une parasitose causée par un trématode à sexes séparés, *Schistosoma mansoni*, dont le stade de développement passe par un mollusque d'eau douce, *Biomphalaria pfeifferi*, hôte intermédiaire le plus important de la maladie (Mandahal 1959).

La lutte contre la bilharziose fait appel à plusieurs méthodes, notamment le traitement de masse et les soins de santé primaires. La lutte contre l'hôte intermédiaire est un support à la chimiothérapie et fait intervenir les molluscicides chimiques. Ces derniers constituent un polluant dangereux pour l'environnement d'une part et leurs prix croissent de façon permanente d'autre part (0,2 $ U.S./habitant/an en zone endémique au Burundi et 1-4 $ U.S. en Egypte). Le rôle de cette recherche est de mettre au point un support à la chimiothérapie à moyen terme par l'utilisation de moyens naturels et accessibles aux populations avoisinantes des zones infectées.

L.J.G. van der Maesen et al. (eds.), The Biodiversity of African Plants, 757-760.

Matériel et méthodes

L'intérêt s'est d'abord porté sur les plantes existant au Burundi (Troupin 1978, Troupin et al. 1983, 1985) et connues dans la littérature pour leur pouvoir molluscicide. Les espèces étudiées sont données ci-après en précisant les organes utilisés, tels que cités dans la littérature (Adjanohoun et al. 1988):

- *Agave sisalana* Perrine (plante entière);
- *Anacardium occidentale* L. (écorces et tiges);
- *Annona senegalensis* Pers. (tiges et écorces de tige);
- *Balanites aegyptiaca* L. (fruits);
- *Canna indica* L. (tiges);
- *Chenopodium ambrosioides* L. (feuilles);
- *Phytolacca dodecandra* L'Hérit. (fruits);
- *Polygonum senegalense* Meissn. (plante entière);
- *Sesbania sesban* L. (feuilles);
- *Simarouba vesicolor* St. Hil. (écorces de tige);
- *Solanum lycopersicum* L. (feuilles).

Le produit utilisé dans les analyses est la poudre des organes des plantes cités ci-dessus, après séchage et pilage.

Les échantillons de mollusques ont été récoltés dans des cours d'eau, dans des canaux de drainage et dans les étangs piscicoles. Les échantillons visibles ont été ramassés à l'aide d'une pince et conservés dans des bocaux en verre contenant un peu d'eau et des algues. D'autres qui étaient cachés sous les feuilles de plantes flottantes ont été récoltés à l'aide d'une passoire; ils ont ensuite été ramassés à l'aide d'une pince et conservés dans les mêmes conditions que les précédents. Les échantillons de mollusques récoltés ont été cultivés dans des aquariums. Ces derniers étaient d'abord remplis d'eau avant de recevoir les mollusques. Les échantillons expérimentés étaient nourris une fois tous les deux jours de salade fraîche et de petits poissons moulus. Après une semaine de culture, les mollusques étaient prêts à l'expérimentation.

Après la récolte, chaque organe de plante utilisé ou parfois la plante entière a été séché à l'étuve (40 à 60°) et pilé dans un mortier. Les poudres obtenues ont été conservées une semaine dans des bocaux en verre avant l'expérimentation. L'analyse au laboratoire a porté sur l'évaluation de la toxicité des poudres des différentes espèces sur *Biomphalaria pfeifferi*. Plusieurs doses de poudre exprimées en mg par litre d'eau ont été testées par plante jusqu'à obtenir la dose létale aiguë. Pour chaque dose, nous avons disposé de 40 mollusques: la moitié était placée dans l'aquarium témoin et l'autre dans l'aquarium expérimental. Le dénombrement des mollusques morts se faisait 24 heures après le début de l'expérimentation afin de tester la toxicité aiguë.

Résultats et discussion

Le Tableau 1 résume les résultats obtenus avec les différentes plantes testées et aux différentes doses. A partir de ce tableau, nous observons que la dose létale de départ n'est pas identique pour les différentes plantes étudiées.

Elle est de 10 mg/l chez *Phytolacca dodecandra* (1), *Balanites aegyptiaca* (3) et *Agave sisalana* (6) et correspond à des taux de mortalité respectivement de 21.6%, 15% et 25%. Pour *Sesbania sesban* (2), *Annona senegalensis* (5) et *Solanum lycopersicum* (11), la toxicité commence à se manifester à la dose de 100 mg/l correspondant

Tableau 1. Taux de mortalité de *Biomphalaria pfeifferi* en fonction de la dose.

Dose	1	2	3	4	5	6	7	8	9	10	11
10	21,6		15,0			25,0					
20	38,3		26,6			30,0					
30	50,0		30,0			45,0					
40	63,3		43,3			53,3					
50	68,3		50,0			66,6					
60	81,6		75,0			75,0					
70	86,6		85,0			85,0					
80	91,6		96,6			90,0					
90	98,3		100,0			93,3					
100	100,0	31,6			15,0	100,0					33,3
200		43,3		21,6	18,3						41,6
400		61,6		35,0	28,3		21,6				51,6
600		75,0		46,6	30,0		26,6				63,3
800		83,3		68,6	38,3		41,6				76,6
1000		86,6		80,0	46,6		48,3	31,6	21,6	30,0	85,0
1200		96,6		88,3	63,3		60,0	35,0	38,3	40,0	93,3
1400		100,0		93,3	68,3		73,3	45,0	58,3	58,3	100,0
1600				100,0	78,3		85,0	48,3	63,3	68,3	
1800					85,0		93,3	58,3	75,0	73,3	
2000					93,3		100,0	68,3	78,3	78,3	
2200					100,0			71,3	88,3	88,3	
2400								81,3	93,3	93,3	
2600								88,6	100,0	100,0	
2800								93,3			
3000								100,0			

1: *Phytolacca dodecandra*; 2: *Sesbania sesban*; 3: *Balanites aegyptiaca*; 4: *Anacardium occidentale*; 5: *Annona senegalensis*; 6: *Agave sisalana*; 7: *Chenopodium ambrosioides*; 8: *Polygonum senegalense*; 9: *Simarouba vesicolor*; 10: *Canna indica*; 11: *Solanum lycopersicum*.
Les doses testées varient entre 10 mg/l et 3000 mg/l.

à des taux de mortalité respectivement de 31,6%, 15%, 33%. Les poudres d'*Anacardium occidentale* (4) et *Chenopodium ambrosioides* (7) deviennent toxiques à partir respectivement de 200 mg/l et de 400 mg/l avec un taux de mortalité identique de 21,6%. Enfin, les poudres de *Polygonum senegalense* (8), *Simarouba vesicolor* (9) et *Canna indica* (10) agissent à partir de la dose élevée de 1000 mg/l correspondant à des taux de mortalité respectifs de 31,6%, 21,6%, 30%. Les taux de mortalité de 100% sont obtenus à des doses diverses par les différentes plantes:

- 100 mg/l chez *Balanites aegyptiaca*, *Phytolacca dodecandra* et *Agave sisalana*;
- 1400 mg/l chez *Sesbania sesban* et *Solanum lycopersicum*;
- 1600 mg/l chez *Anacardium occidentale*;
- 2000 mg/l chez *Chenopodium ambrosioides*;
- 2200 mg/l chez *Annona senegalensis*;
- 2600 mg/l chez *Simarouba vesicolor* et *Canna indica*;
- 3000 mg/l chez *Polygonum senegalense*.

Nous pouvons ainsi classer les différentes plantes étudiées en groupes suivants:

- Les plantes molluscicides à toxicité élevée: *Balanites aegyptiaca*, *Phytolacca dodecandra* et *Agave sisalana*.
- Les plantes molluscicides à toxicité moyenne: *Sesbania sesban*, *Solanum lycopersicum*, *Anacardium occidentale*, *Chenopodium ambrosioides* et *Annona senegalensis*.
- Les plantes molluscicides à toxicité faible: *Simarouba vesicolor*, *Canna indica*, *Polygonum senegalense*.

Conclusion

Les résultats obtenus confirment que toutes les plantes étudiées sont efficaces pour la lutte contre *Biomphalaria pfeifferi*, bien qu'elles agissent à des degrés divers.

Au stade actuel de notre étude, les plantes les plus intéressantes sont: *Phytolacca dodecandra*, *Balanites aegyptiaca* et *Agave sisalana*, dont la dose qui présente la toxicité aiguë se situe à 100 mg/l.

Dans le cas du Burundi, *Balanites aegyptiaca* et *Agave sisalana* sont les mieux indiquées car leur écologie correspond aux zones infestées par la bilharziose. Les possibilités d'adaptation de *Phytolacca dodecandra* peuvent être envisagées.

Bibliographie

Adjanohoun, E., A. Ahyi, L. Aké Assi, J. Baniakina, P. Chibon, G. Cusset, V. Doulou, A. Enzanza, J. Eyme, E. Goudote, A. Keita, C. Mbemba, J. Mollet, J.M. Moutsamboté, J.B. Mpati & P. Sita, 1988. Contribution aux études ethnobotaniques et floristiques en République Populaire du Congo. ACCT, Paris, 605 p.

Kenneth, E.M., 1978. Plant molluscicides. In: Wiley & Sons, L.T.D. (eds.) Special Program for Research and Training in Tropical Diseases. UNDP/WHO.

Mandahal, B., 1959. Les hôtes intermédiaires de Schistosoma, Biomphalaria, Bulinus africanus. OMS. Série de monographie n° 37: 1-45.

Troupin, G., 1978. Flore du Rwanda. Spermatophytes. Vol. I. Musée Royal de l'Afrique Centrale, Tervuren, Belgique.

Troupin, G. et al., 1983. Flore du Rwanda. Spermatophytes. Vol. II. Agence de Coopération Culturelle et Technique, Paris.

Troupin, G. et al., 1985. Flore du Rwanda. Spermatophytes. Vol. III. Agence de Coopération Culturelle et Technique, Paris.

L'homme et l'igname au Congo

J.S. N'KOUNKOU
Laboratoire de Botanique Systématique, Université libre de Bruxelles,
C.P. 169, 28 Avenue Paul Héger, B-1050 Bruxelles, Belgique.

Résumé
Au Congo, 17 taxons d'ignames sont recensés comme plantes utiles à l'homme. Deux d'entre-eux ne sont pas spontanés mais cultivés. Parmi les 15 spontanés, 8 sont mis en culture localement. La relation entre l'homme et l'igname est donnée pour chaque espèce; les organes et leurs caractéristiques alimentaires sont précisés. L'origine de certaines espèces est examinée.

Abstract
In Congo, 17 yam species have been registered as useful plants for man. Two taxa are cultivated, the other are spontaneous. Among the 15 spontaneous species, 8 are locally cultivated. The relationship between man and yam is given for each species. The nutrition characteristics are precised for plant organs. The origin of some species is examined.

Key words
Congo, *Dioscorea*, Man, Use, Vernacular names, Yams.

Introduction

Les ignames sont des plantes particulièrement intéressantes du point de vue de leurs utilisations; 17 espèces sont actuellement recensées comme plantes utiles à l'homme ou comestibles. En vue de l'analyse ethnobotanique des ignames (*Dioscorea*) du Congo, la place de l'igname dans la civilisation ou culture congolaise, la relation entre l'homme et la plante, et l'usage des espèces d'ignames ont été examinés. Pour chaque igname, la synthèse des organes et leurs caractéristiques alimentaires sur base d'enquêtes sont données.

Matériel et méthodes

Les enquêtes ethnobotaniques ont été menées de 1990 à 1991 dans le Congo central et méridional qui compte 7 districts phytogéographiques: Kouilou, Mayombe, Chaillu, Niarif, Cataractes, Léfini et Plateaux Batéké. Nos enquêtes ont été abordées selon 3 axes:

* Établir un inventaire des plantes indigènes ou exotiques cultivées et consommées par la population locale.
* Noter les noms vernaculaires.
* Connaître l'usage et la vertu de la plante auprès de la population.

Pour mener à bien ces enquêtes, nous avons interrogé les chefs de village et surtout les guides que ces derniers nous présentaient, dont la moyenne d'âge se situe entre 20 et 60 ans, dans les différentes contrées visitées. Les informateurs étaient composés de femmes et d'hommes rencontrés dans les différentes localités prospectées. Nous avons complété nos données par les notes d'herbiers d'autres récolteurs au Congo et par les références bibliographiques (Hecketsweiler et al. 1991, etc.).

Suivant le rapport séculaire entre l'homme et la plante, nous montrerons la place de

L.J.G. van der Maesen et al. (eds.), The Biodiversity of African Plants, 761-767.

la plante dans la civilisation, notamment dans des domaines tels que son utilisation alimentaire et médicinale, la mentalité végétative, les rites et origines des espèces du genre *Dioscorea*, afin de développer la conception de Portères (1961); nous utiliserons l'appellation igname à la place de *Dioscorea*, puisqu'il sera surtout question de l'alimentation.

Que ce soit pour des raisons vivrières, culturelles ou médicales, les ignames ont toujours joué un rôle important dans la société depuis des temps immémoriaux. Pendant la traite des esclaves, l'igname est restée l'aliment de consistance des Noirs transportés aux Amériques et certaines espèces d'origine africaine ont migré de l'ancien monde au nouveau monde, parmi lesquelles, *D. cayenensis* dont l'origine africaine est de nos jours prouvée par divers auteurs (Degras 1986, Miège 1952, Zoundjihekpon et al. 1990). En effet, Lamarck (1789) en découvrant cette espèce récoltée à Cayenne par Stoupy, a donné comme épithète à celle-ci, le nom de la ville où l'espèce a été cultivée par les noirs venus de l'ancien monde. Actuellement, l'homme les a largement diffusées dans toute l'Afrique tropicale pour des besoins alimentaires, portant parfois à confusion leur vraie origine et en modifiant leur phénotype (Degras 1986).

Résultats

Les ignames comestibles

Sur les 17 taxons comestibles au Congo, 2 espèces ne sont pas spontanées mais cultivées: *Dioscorea alata* d'origine asiatique avec 3 cultivars (cv. Mbungu-menga, cv. Mugenda et cv. Ngumvu), et *D. cayenensis* d'origine africaine avec 1 cultivar (cv. Akô). Parmi les 15 taxons spontanés, 8 sont localement mis en cultures: *D. bulbifera* var. *anthropophagorum, D. burkilliana, D. dumetorum, D. hirtiflora* subsp. *hirtiflora, D. liebrechtsiana, D. mangenotiana, D. praehensilis* et *D. semperflorens*; les 7 autres dont 5 espèces et 2 variétés font l'objet de cueillette et ne sont jamais cultivées: *D. baya* var. *baya, D. baya* var. *kimpundi, D. minutiflora, D. preussii, D. pynaertii, D. sagittifolia* var. *sagittifolia*, et *D. sagittifolia* var. *lecardii*. Ces 17 taxons peuvent être classés en 3 groupes au point de vue de leur utilisation par l'homme.

Ignames exotiques cultivées

Ce groupe est constitué de deux ignames introduites en cultures et non spontanées:

* *D. alata* L.: en Afrique centrale, comme dans son aire d'origine Indo-Malaise (Baquar 1992, Burkill 1960), l'espèce ne se rencontre pas en dehors des cultures. Trois cultivars ont été décrits: *D. alata* cv. Mbungu-menga, cv. Mugenda et cv. Ngumvu. Ces cultivars sont comestibles pour leurs feuilles, leurs bulbilles (*D. alata* cv. Mbungu-menga et cv. Mugenda) et leurs tubercules. Les feuilles sont utilisées comme légume. Le tubercule est utilisé par les Laari, les Kaamba, les Nsundi, les Ngangulu, les Téké et les Vili au Congo. Noms vernaculaires: cv. Mbungu-menga: Mbungu-menga (en laari); Gâtsuélé (en téké); Kikila (en beembe); kipanchi menge (en vili). cv. Mugenda: Mugenda (en laari); Mukondo (en kaamba); Mululu (en nsundi). cv. Ngumvu: Ngumvu (en laari); Lembié (en téké); Massambrela (en vili).

* *D. cayenensis* Lam. cv. Akô: l'origine de l'espèce demeure problématique; d'aucuns disent qu'elle provient de Cayenne, Porto Rico (Lamarck 1789), par contre d'autres affirment aujourd'hui qu'elle est bien d'origine africaine et qu'elle a été introduite aux Amériques lors de la traite négrière (Miège 1952). Pour notre travail,

nous ne nous plongeons pas dans cette polémique; l'espèce, bien que cultivée, étant présente en Afrique centrale dans certaines contrées reculées et dans les confins de la dition (Plateaux Batékés) peu explorés, nous croyons devoir affirmer qu'elle est bien d'origine africaine. Le cultivar est utilisé pour son tubercule et ses feuilles comestibles et il est surtout répandu dans le district des Plateaux Batéké (Congo). Il est consommé par les populations des grandes villes dans lesquelles sa commercialisation est faite par les Téké et les Ngangulu. Nom vernaculaire: Akô (en téké et en ngangulu).

Ignames indigènes domestiquées
Nous appelons ignames indigènes domestiquées, les ignames spontanées qui sont localement mises en culture; nous distinguons 8 taxons:

* *D. bulbifera* L.: à l'origine, l'espèce spontanée est toxique, mais une variété est mise en culture et utilisée pour ses grosses bulbilles à parenchyme médullaire beige jaunâtre, gluant, non âcre, à périderme terne parsemé de petits points bruns ou blancs: *D. bulbifera* L. var. *anthropophagorum* (A. Chev.) Prain & Burkill. Cependant, les bulbilles de l'autre variété, la var. *bulbifera*, toxique à la récolte sont parfois consommées par les habitants du district des Cataractes après ébullition, mais sa comestibilité est souvent restrictive. Noms vernaculaires: - var. *bulbifera*: Banga (en tiè); Gamba soko, Gamba massoko (en laari et en nsundi); Ivuba dumbala (en ndasa); Makambi, Niegulingwe (en yombe); Vuba (en mbaamba); Ngambela (en beembe); Sola-nkiti (en laali). - var. *anthropophagorum:* Soko, Massoko (en laari).

* *D. dumetorum* (Kunth) Pax: à l'origine l'espèce spontanée est toxique mais des formes cultivées sont comestibles pour leur tubercule en fascicule de (4-8) tubercules, à parenchyme médullaire jaune ou beige, à périderme marqué des rides transversales hérissées de radicelles éparses. L'espèce est consommée par les Laari, les Laali et les Beembe. Noms vernaculaires: Itsama (en ndasa); Kekama (en yaa); Kichunia, Tichunia (en laari); Kishama, Wami (en laali); Ngambala (en beembe).

* *D. burkilliana* Miège: c'est une plante spontanée, comestible pour son tubercule en forme d'amas d'oeufs digités à aspect mamelonné, à parenchyme médullaire blanc ou jaune. Les feuilles sont aussi utilisées comme légumes. L'espèce est consommée par les Laari et les Punu. Noms vernaculaires: Nkoloba (en laari); Divandi (en punu).

* *D. hirtiflora* Benth. subsp. *hirtiflora*: la sous-espèce est parfois domestiquée pour son tubercule aérien allongé-cylindrique, comestible. Les bulbilles sont consommées par les Laari. Nom vernaculaire: Nkeso (en laari).

* *D. liebrechtsiana* De Wild.: l'espèce est domestiquée pour ses jeunes pousses et parties jeunes des tiges utilisées comme asperges et pour son tubercule allongé à parenchyme médullaire blanc jaunâtre, à racines épineuses nombreuses sous un plateau ligneux. Les jeunes pousses et les parties jeunes des tiges sont commercialisées tout le long du chemin de fer Congo-Océan, notamment dans les gares des districts des Cataractes et du Mayombe. La plante est consommée par les Laari et les Yombe. Nom vernaculaire: Ntsinia (en laari).

* *D. mangenotiana* Miège: l'espèce est domestiquée pour ses jeunes pousses vert à rouge vineux utilisées comme asperges. Elle est consommée par les Kongo, les Laari et les Nsundi. Noms vernaculaires: Kimpundia (en laari); Tchimpundia (en nsundi)

* *D. praehensilis* Benth.: l'espèce est comestible pour son tubercule allongé-cylindrique, à parenchyme médullaire blanc, parfois violet blanc; les feuilles sont parfois

utilisées comme légumes. Elle est consommée par les Kongo, les Bakota, les Koyo, les Laari, les Mbaamba, les Mongo et les Nsundi. Noms vernaculaires: Esia (en mbaamba); Kobo (en Kôyô); Munkanga, Minkanga (en laari); Tchimpundia kuyu (en nsundi).

* *D. semperflorens* Uline: l'espèce est domestiquée et comestible pour ses bulbilles globuleuses ou ellipsoïdales; le tubercule est aussi comestible mais difficile à extirper. Les bulbilles sont consommées par les Laari et les Nsundi. Noms vernaculaires: Mfouessela (en laari); Mpeso (en sundi).

Ignames spontanées

Cinq espèces d'ignames dont 2 avec 2 variétés font l'objet de cueillette et ne sont jamais domestiquées:

* *D. baya* De Wild. var. *kimpundi* De Wild.: la variété est comestible pour ses feuilles utilisées comme asperges dans le Mayombe et dans les Cataractes. Les asperges sont consommées par les Bakongo, les Nsundi et les Yombe. Signalons également pour l'autre variété, la var. *baya* les feuilles seraient aussi consommées par les Nsundi. Nom vernaculaire: Muholoto (en nsundi).

* *D. minutiflora* Engl.: la plante est utilisée pour son tubercule digité renflé au bout; généralement on mange la partie renflée (boule), la partie supérieure étant filamenteuse. L'espèce est consommée par les Laari, les Punu et les Yombe. Noms vernaculaires: Bokoto (en laari); Lekakani (en yombe); Basadi (en punu).

* *D. preussii* Pax: l'espèce est comestible pour sa tige et les feuilles très recherchées en forêt par les Nzabi, utilisées comme légumes dans le Niari. Dans les autres districts, nous n'avons pas noté son usage. Le tubercule est vermiforme, toxique, difficile à extirper. Noms vernaculaires: Midoumbala, Mirombo (en nzabi et en punu); Libuenzo, Mbianzo (en téké).

* *D. pynaertii* De Wild.: la plante est utilisée pour son tubercule en bloc arrondi ou ellipsoïdal (tubercule de 1ère saison de récolte) ou digité cylindrique (tubercule de seconde saison de récolte). Elle est recherchée par les pygmées Baka des Plateaux Batéké; elle est consommée aussi par les Laari, les Nsundi et les Ngangulu. Noms vernaculaires: Nkoubi (en laari); Matadi (en nsundi); Ngambélé (en ngangulu).

* *D. sagittifolia* Pax var. *lecardii* (De Wild.) N'kounkou et var. *sagittifolia*: les deux variétés sont comestibles pour leurs tubercules allongés. L'igname est consommée par les Laali, les Laari et les Yombe.

Synthèse des organes utilisés et caractéristiques alimentaires des taxons

Cette partie est essentiellement une synthèse de l'usage de tous les taxons décrits au Congo (N'kounkou et al. 1993); elle montre d'une part les organes utilisés et d'autre part la comestibilité avec ou sans restriction et la toxicité si possible. L'inventaire et les utilisations alimentaires de ces différents taxons sont consignés dans le tableau 1.

Les rites et le pouvoir curatif de l'igname

Sur les 17 espèces d'ignames décrites au Congo (N'kounkou et al. 1993), 4 espèces se trouvent au centre de coutumes rituelles congolaises:

Tableau 1. Synthèse des organes et caractéristiques alimentaires des Ignames du Congo.

Ignames	Organes	1	2	3	4
D. alata, cv. Mbungu-menga, cv.	tige feuillée	+	—	—	L
Mugenda, cv. Ngumvu	tubercule	+	—	—	B
D. baya var. *baya*	feuille	+	—	—	L
	tubercule	+	—	—	B
D. baya var. *kimpundi*	jeune pousse	+	—	—	L
	tubercule	+	—	—	B
D. bulbifera var. *bulbifera*	feuilles	±	+	+	L
	bulbilles	±	+	+	B
D. bulbifera var. *anthropophagorum*	feuilles	±	+	+	L
	bulbilles	—	—	—	B
D. burkilliana	feuilles	+	—	—	L
	tubercule	+	—	—	B
D. cayenensis cv. Akô	feuilles	+	—	—	L
	tubercule	+	—	—	B
D. dumetorum	tubercule	+	±	±	B
D. hirtiflora	bulbilles	+	—	—	B
	tubercule	+	—	—	B
D. liebrechtsiana	jeune pousse	+	—	—	A
	tubercule	+	—	—	B, G
D. mangenotiana	jeune pousse	+	—	—	A
D. minutiflora	tubercule	+	—	—	B
D. praehensilis	feuilles	+	—	—	L
	tubercule	+	—	—	B
D. preussii	tige	+	—	—	L
	tubercule	—	+	+	L
D. pynaertii	tubercule	+	—	—	B, G
D. sagittifolia var. *sagittifolia*	tubercule	+	—	—	B
D. sagittifolia var. *lecardii*	tubercule	+	—	—	B
D. schimperiana	bulbilles	±	—	—	B
	tubercule	?	+	—	—
D. semperflorens	bulbilles	+	—	—	B
	tubercule	+	—	—	B
D. smilacifolia	tubercule	+	+	?	B

1 = comestibilité sans restriction
2 = comestibilité avec restriction
3 = toxicité totale
4 = Mode de consommation: A = asperge, B = bouillir, G = griller, L = légume.

1. *D. dumetorum*: utilisée quand on constate un cas d'ensorcellement au sein d'une famille humaine. Elle est employée dans les préparations magiques destinées pour éloigner les mauvais esprits (le diable); elle est plantée sur les tombes pour chasser les revenants lorsque la présence du défunt est soupçonnée (Bouquet 1969).

2. *D. mangenotiana*: chez les pygmées Baka des Plateaux Batéké, certains termes ethnobotaniques relatifs aux stades de développement du tubercule de *D. mangenotiana* (igname bà), sont utilisés dans le discours rituel pour distinguer deux tranches d'âge dans la généalogie jengi (Joiris 1991).

3. *D. bulbifera*: souvent cultivée en bordure des champs pour décourager et éloigner les herbivores (chèvre, mouton, etc.). Les bulbilles bouillies sont données aux enfants de moins de 5 ans pour éviter de tituber.

4. *D. preussii:* cette igname a une vertu de coupe-faim chez les nzabi du Niari au Congo. Dans le district administratif de Divénié cohabitent deux ethnies autochtones, les nzabi et les punu; les nzabi utilisent les tiges de *D. preussii* comme légumes; à l'opposé, les punu n'en mangent jamais. Lors des fêtes régionales ou de nouvel-an, quand un nzabi ne désire pas partager le repas avec un punu, il suffit qu'il mette dans un plat de viande boucanée par exemple les légumes de tige de *D. preussii*; le punu n'aura plus d'appétit.

Outre ces cérémonies rituelles, les ignames sont utilisées dans la pharmacopée traditionnelle et ont un pouvoir curatif prouvé souvent empiriquement. Quatre espèces sont recensées au cours de nos enquêtes comme plantes médicinales:

D. bulbifera: le jus obtenu en coupant une tige sert contre les ophtalmies purulentes et la lèpre; il est aussi employé contre les morsures de serpents.

D. praehensilis: la tige coupée en morceaux sert à préparer une tisane contre les maux de ventre et pour les rhumatisants.

D. preussii: les feuilles radoucies par le feu sont appliquées contre les ulcères.

D. smilacifolia: le suc chauffé avec l'eau est employé en pansement contre l'orchite.

Synthèse sur l'origine de quelques espèces du genre *Dioscorea*

En vue de faire la synthèse sur l'origine des espèces étudiées, nous nous sommes basés sur les travaux de Burkill (1960), Miège (1952) et Baquar (1992). Il est certain que les cultivars de quelques espèces proviennent de Birmanie et de Thaïlande, notamment ceux de *D. alata* et *D. dumetorum*. Les espèces de la section *Enantiophyllum* présentes au Congo (N'kounkou et al. 1993), à l'exception de *D. alata* d'origine asiatique, trouvent leur centre d'origine dans différentes régions d'Afrique. Les variétés de l'espèce *D. bulbifera* sont d'origine asiatique; quant à l'espèce *D. dumetorum*, elle est à la fois d'origine africaine et asiatique. En effet, Burkill (1960) signale l'ancêtre de cette espèce en Asie, en l'occurrence *D. hispida*, et fait remarquer que celui-ci avait pour origine l'Afrique et a migré de l'Afrique en Asie pendant le Miocène. Baquar dans son récent travail, confirme ces deux centres d'origine de l'espèce: l'Afrique centrale et l'Asie du Centre.

Discussion et conclusion

Dans cette étude, nous avons montré la place de l'igname dans les habitudes alimentaires de l'homme.

Sur les 17 espèces utiles à l'homme au Congo, 16 sont comestibles; pour chacune d'entre elles, nous avons précisé la comestibilité sans restriction, la comestibilité avec restriction, la toxicité et le mode de consommation.

Il ressort de cette étude que les espèces d'ignames sont pour la plupart comestibles, l'appréciation de la toxicité varie d'une contrée à une autre et parfois il devient difficile de le dire avec objectivité, d'où l'intérêt et le souhait d'entreprendre dans le futur l'analyse biochimique pour mieux connaître et mieux définir les substances toxiques.

Remerciements
Nous remercions cordialement le Professeur L.J.G. van der Maesen et tout le Secrétariat de l'AETFAT, pour la subvention à notre participation au XIVème Congrès de notre association.

Bibliographie

Baquar, S.R., 1992. Economic Importance of Yams (*Dioscorea* spp.) and their natural production with special reference to West Africa. Botany 2000: East and central Africa. Napreca Monograph 2: 19-23.

Bouquet, A., 1969. Les féticheurs et médecines traditionnelles du Congo (Brazzaville). Mém. O.R.S.T.O.M. 36: 106-107.

Burkill, I.H., 1960. The organography and the evolution of *Dioscoreaceae*, the family of the yams. Journ. Linn. Soc. Bot. 56, 367: 319-412.

Degras, L., 1986. L'igname, plante à tubercule tropicale. Techn. Agric. Prod. Trop. 36. Agence de Coopération Culturelle et Technique. Ed. G.P. Maisonneuve & Laroze, Paris: 409 p.

Hecketsweiler, P., C. Doumenge & J. Mokoko Ikonga, 1991. Le Parc national d'Odzala, Union mondiale pour la nature, CCE, Cambridge: 334 p.

Joiris, P.V., 1991. Ethnoécologie d'une igname sauvage et rituel associé chez les chasseurs-cueilleurs Baka (Sud Cameroun). Résumé Comm. Symp. Alim. Forêt Trop., Unesco, Paris: 80.

Lamarck, J.B.A.P., 1789. Ignames de Cayenne. Encyclopédie méthodique, Botanique, Paris, 3: 233.

Miège, J., 1952. Contribution à l'étude systématique des *Dioscorea* Ouest africains. Thèse de doctorat ès Sciences Naturelles. Univ. Paris: 266 p.

N'kounkou, J.S., J. Lejoly & D. Geerinck, 1993. Les *Dioscoreaceae* du Congo, Fragm. Flor. Geobot. Suppl. 2 (1): 139-182.

Portères, R., 1961. L'ethnobotanique, Place-Objet-Méthode-Philosophie. Journ. Agric. Trop. Bot. Appl. 8, 4-5: 102-109.

Zoundjihekpon, J., Sadi Essad & Bakary Touré, 1990. Dénombrement chromosomique des dix groupes variétaux du complexe *Dioscorea cayenensis-rotundata*. Cytologia 55: 115-120.

Medicinal plants used during ante- and post-natal periods and in early infant care in Busoga

E.N.K. OGWAL
Botany Department, Makerere University, P.O. Box 7062, Kampala, Uganda.

Abstract
This study investigates the medicinal plants - *Bidens pilosa, Cissampelos mucronata, Euphorbia prostrata, Ficus sp., Gynandropsis gynandra, Hoslundia opposita, Kalanchoe crenata, Senecio discifolius, Sida cuneifolia* - used to treat mothers and their infants in Busoga region, Uganda for purposes including prevention of miscarriage, induction of uterine contractions, expulsion of retained placenta, disinfectant of umbilical cord wounds and clearing stomach contents of newborn.

Key words
Infants, Medicinal plant, Mothers, Uganda.

Introduction

About 70-80% of the Ugandan population still rely on traditional healers for their day-to-day health care. The success of plant medicines is sometimes startling. Although great use is made of modern maternity services, most pregnant women still flock to the traditional midwife for protective medicine to cover the period of pregnancy and early infant care. This appears very important to both educated and uneducated.

According to Cunningham (1990) extensive use is made of medicinal plants in developing countries, either because people cannot afford Western pharmaceuticals or because traditional medicines are more acceptable. The World Health Organisation (WHO 1991) stated that of the people in developing countries about 80% still rely on traditional plant derived medicines, mainly due to their lower price. Cartwright (1979) reported that Chinese doctors admit that drugs used in Western medicines act more quickly and are more specific than herbal ones, but they claim that herbal remedies have fewer unwanted side effects and restore balance to the whole body.

Gelfand (1964) observed that the children of medicine-men and midwives much prefer the attractions of big cities to following in their parents foot-steps. Whereas, previously, indigenous knowledge was handed down from father/mother to son/daughter or to a close relative. This knowledge is now gradually being lost upon the death of the medicine-men and midwives. Another set back here in Africa has been the lack of records of reputed medicinal plants. Oliver-Bever (1986) reported great difficulty in finding reliable documentation, even on some of the best known traditional herbal remedies. Only in a few African countries like Kenya (Kokwaro 1979), West Africa (Oliver-Bever 1986) and recently Uganda (Adjanohoun et al. 1993) have efforts been made to record the surviving folklore. Lack of detailed descriptions has often made it impossible for researchers to identify these plants; the only source of reference being the local name of the herb. Local names are misleading and ambiguous because, not infrequently, they are used for more than one herb.

Renewed interest has now been shown in documenting the use of medicinal plants and scientific studies are beginning to explain some of the curative phenomena (Lambo

L.J.G. van der Maesen et al. (eds.), The Biodiversity of African Plants, 768-770.

1986). There has also been growing awareness by governments and scientific communities of the importance of medicinal plants in health care systems in many developing countries, which further encourages this research.

Methods

Locally reputed midwives were approached and interviewed in 6 villages in Busoga region. Interviews were conducted in the field, and the plants discussed were collected and pressed. Vernacular names were recorded. The voucher specimens were identified, illustrated and then preserved and stored in the Herbarium (MUH).

Midwives interviewed included: Tabitha Tibakawa and Zewulesi both of Kymagwa village; Mrs Kalikwani - Kayunga; Mrs Phillipa Isota - Nakabugu; Florence Nalubega - Kaliro; Beth Babyerabira - Iganga and Mrs Kiyuba of Buluba village. Their age ranged from 30-65 years.

Results

Ficus sp., L. (Moraceae)
Vernacular name: Omugaire
Uses: The red adventitious roots are either chewed or pounded and the extract drunk. This induces uterine contractions and is also used to expel retained placentas.

Euphorbia prostrata Ait., (Euphorbiaceae)
Vernacular name: Kasandosando
Uses: Unpounded shoots mixed with sesame, are boiled and eaten by pregnant women. This reduces the tendency to miscarriage due to rises of uterine temperatures.

Gynandropsis gynandra (L.) Briq., (Capparidaceae)
Vernacular name: Eyobyo
Uses: Roots mixed with those of *Bidens pilosa* L., induce stronger uterine contractions during delivery.

Bidens pilosa L., (Compositae)
Vernacular name: Obukaala
Uses: As explained above, used with *Gynandropsis gynandra*.

Cissampelos mucronata A. Rich., (Menispermaceae)
Vernacular name: Akavamagombe
Uses: Roots are used to induce uterine contractions during delivery. They are also baked and then chewed to expel retained placenta. The leaves, when crushed, are used to clean dirt from the infant's tongue.

Senecio discifolius Oliv., (Compositae)
Vernacular name: Katakyamusulo
Uses: After birth, the leaves are boiled and given to the infant. The extract clears the bowels of unnecessary contents.

Kalanchoe crenata (Andr.) Haw., (Crassulaceae)
Vernacular name: Ekiyondo
Uses: Shoots are squeezed around the umbilical cord wound. The extract acts as a disinfectant.

Hoslundia opposita Vahl., (Labiatae)
Vernacular name: Kamunye
Uses: Extract is drunk by mothers. It reduces abnormal uterine temperatures and clears any effects of syphilis which sometimes cause the skin of the infant to peel off before birth.

Sida cuneifolia Roxb., (Malvaceae)
Vernacular name: Akeyeyo
Uses: Leaves are eaten mixed with sesame; this prevents miscarriages.

Discussion

Further investigations of plants used by mothers and their infants are necessary in a country like Uganda where the majority of deliveries in rural and semi-rural areas are conducted by traditional midwives and attendants. This could help to improve Uganda's health care delivery base and could also be of special benefit to rural people by making them aware of the need to use plants sustainably.

Acknowledgements
Many thanks go the AETFAT Secretariat and to the Swiss Tropical Institute for the sponsorship which enabled me to attend the Wageningen congress. I am also indebted to the NUFU Programme of the Faculty of Science, Makerere University which provided the funds used for these preliminary investigations.

References

Adjanohoun, J.E. et al., 1993. Traditional Medicine and Pharmacopoeia. Contribution to Ethnobotanical and Floristic Studies in Uganda. O.A.U. Scientific Technical Research Commission.
Cartwright, L., 1979. Health from Herbs. Hemisphere 23 (4): 231-237.
Cunningham, A.B., 1990. People and medicines. The exploitation and conservation of traditional Zulu medicinal plants. Proceedings of the twelfth plenary meeting of AETFAT, Hamburg. Band 236: 979-990.
Gelfand, M., 1964. Medicine and Custom in Africa. E. & S. Livingston Ltd., Edinburgh and London.
Harley, G.W., 1941. Native African Medicine. Cambridge. Massachusetts, Harvard Univ. Press: 178.
Kokwaro, J.O., 1976. Medicinal plants of East Africa. The East African Literature Bureau. Nairobi, 1981. A review of research on plants for fertility regulation in Africa. Korean Journal of pharmacognosy 12: 149-152.
Lambo, T.A., 1986. In: Oliver-Bever, B., Medicinal plants in Tropical West Africa. Cambridge Univ. Press.
WHO, 1991. Biotechnology and Development. Monitor No. 9: 4.

Miscellaneous

Confusion de deux taxons spécifiques ou subspécifiques au sein du genre *Borassus* en Afrique de l'Ouest

L. AKÉ ASSI [1] & S. GUINKO [2]
[1] Université d'Abidjan, Centre National de Floristique, 22 B.P. 582, Abidjan 22, Côte d'Ivore.
[2] Faculté des Sciences et Techniques, Université de Ouagadougou,
B.P. 7021, Ouagadougou 03, Burkina Faso.

Résumé
Le genre *Borassus* L., paléotropical, était connu comme n'ayant, en Afrique intertropicale, qu'un seul représentant spécifique: *Borassus aethiopum* Mart. Or, il ressort des observations que nous avons faites, au cours d'une dizaine d'années, dans plusieurs peuplements de rôniers, dans différents pays d'Afrique, qu'il existe deux taxons distincts. Ceux-ci se distinguent par des caractères apparents, comme par exemple des épines, présentes ou non, sur les pétioles, la coloration, verte ou jaune, des fruits à leur maturation. L'un est *Borassus aethiopum*; *Borassus* sp., pourrait-il être *Borassus flabellifer*? Dans tous les cas, une confusion des deux taxons ne doit plus être faite.

Key words
Borassus, Confusion, East Africa, Palmae, Rônier.

Introduction

D'après nos connaissances actuelles, le genre *Borassus* L. (Palmae-Borassoideae), paléotropical, renferme 9 taxons, spécifiques ou subspécifiques, répartis de l'Afrique intertropicale à la Nouvelle Guinée, de la façon suivante: Afrique intertropicale, 2: *Borassus aethiopum* Mart., *B. deleb* Becc.; Madagascar, 2: *B. madagascariensis* Boj., *B. sambiranensis* Jumelle et Perrier; Inde, 2: *B. dichotomus* White, *B. flabellifer* L.; Malaisie, 1: *B. machadonis* Ridley; Iles de la Sonde, 1: *B. sundaica* Becc.; Nouvelle Guinée, 1: *B. heineana* Becc. Cette liste n'inclut pas des synonymes nouveaux.

Ce genre était connu comme n'ayant, en Afrique occidentale, qu'un seul représentant spécifique: *Borassus aethiopum*, considéré par certains comme synonyme de *B. flabellifer* var. *aethiopum* (Mart.) Warb. Or, ce ressort des observations que nous avons faites, au cours d'une dizaine d'années, dans plusieurs peuplements de rôniers et dans différents pays de la sous-région africaine (notamment au Bénin, Burkina Faso, en Côte d'Ivoire, au Mali, Niger et Nigéria), qu'il existe deux taxons qui se distinguent par des caractères apparents comme, par exemple, la présence ou non d'épines sur les bords des pétioles, la coloration verte ou jaune des fruits à leur maturation.

Caractéristiques des deux taxons observés

Borassus aethiopum Mart.

Münch. Gel. Anzeig. 639 (1838), 46 (1839); Russell, F.W.T.A. ed. 2, 3: 168 (1968). Ce rônier forme de beaux peuplements en zone de savane guinéenne et en région soudano-sahélienne, mais sans être exclusivement septentrional. Il descend jusqu'aux lisières de la forêt dense humide. Plus au Sud, dans la zone littorale, il réapparaît, en abondance, dans les savanes côtières (en Côte d'Ivoire notamment).

L.J.G. van der Maesen et al. (eds.), The Biodiversity of African Plants, 773-779.

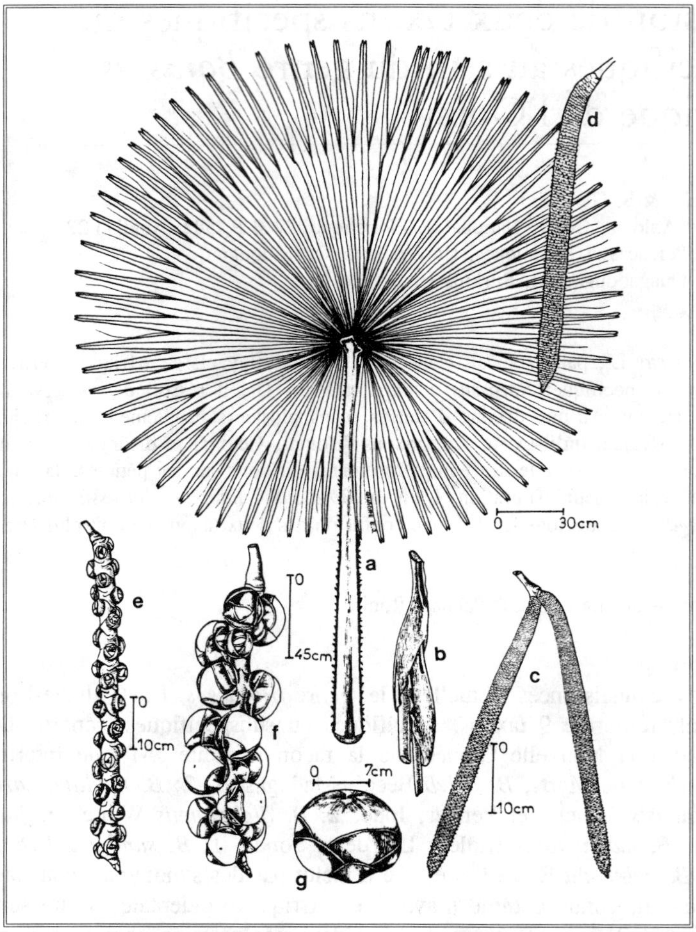

Figure 1. *Borassus aethiopum* Mart. a) feuille; b) portion d'inflor. mâle; c-d) spadices mâles; e) inflorescence femelle; f) infrutescence; g) fruit.

La plante vit, indifféremment, dans les dépressions inondées périodiquement, sur les terrains marécageux, au bord des cours d'eau, ou sur terrain sec sableux, argileux ou pierreux. En région soudano-sahélienne, elle est particulièrement abondante le long des cours d'eau où elle forme de vastes peuplements monospécifiques (Pendjari, fleuve Niger).

C'est un grand palmier dioïque, à stipe dressé, droit, un peu renflé au-dessus du milieu de sa hauteur, lisse, gris, atteignant 25 m de hauteur et de 30-40 cm de diamètre à 2 m du sol, à partie supérieure couverte de bases foliaires persistantes. Chez les jeunes individus, d'une taille comprise entre 2 et 6 m, les pétioles des feuilles desséchées et leurs gaines ne tombent pas et restent appliqués au stipe depuis le sol. A ce stade la base du stipe est considérablement épaissie et peut atteindre jusqu'à 90 cm de diamètre.

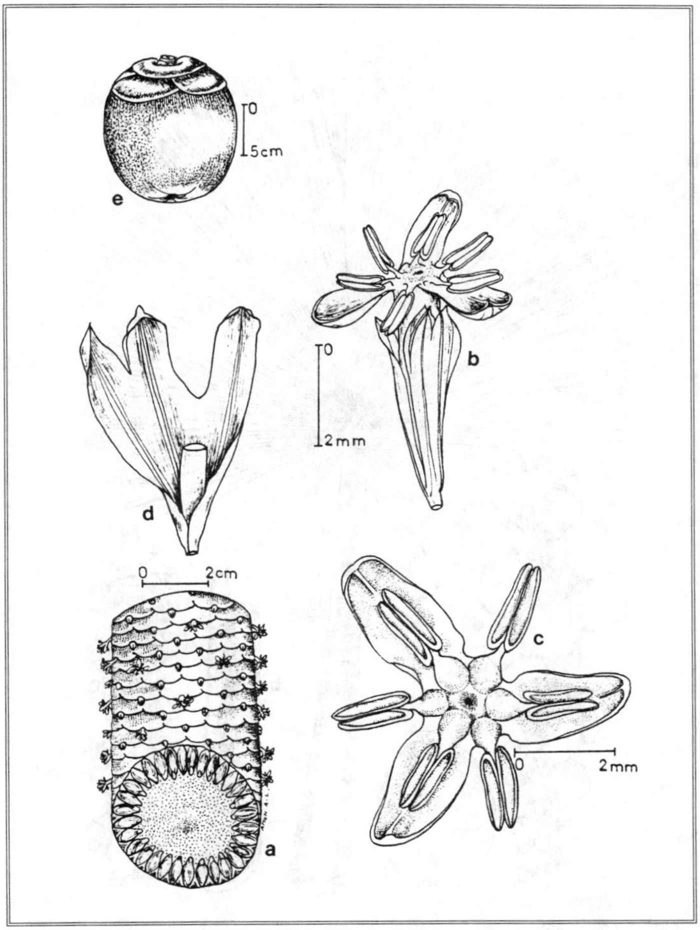

Figure 2. *Borassus aethiopum* Mart. a) portion de spadice mâle; b-c) fleurs mâles; d) tépales externes de fleur mâle; e) fruit.

Les feuilles, en bouquet terminal, sont grandes; le limbe, vert sombre, flabellé, de 2-2,5 m de longueur et de 3-3,5 m de largeur, est découpé en segments souvent bifides au sommet; les pétioles, de 1,5-3 m de longueur, sont garnis, sur les bords, d'épines irrégulières noirâtres; une ligule deltoïde, à sommet en gouttière, termine le pétiole à la base dissymétrique du limbe.

Les inflorescences mâles, axillaires, issues d'un axe principal mesurant environ 1,7 m de longueur, ramifié dès la base, sont d'abord enveloppées dans une spathe qui se déchire lors de sa croissance; les rameaux, apparaissant à l'aisselle de grandes bractées verdâtres lâchement imbriquées, portent de 1-3 spadices; ceux-ci cylindriques, compacts, grisâtres ou brunâtres, mesurent de 30-40 cm de longueur et de 3,5-4 cm de diamètre. Les fleurs mâles, petites, nombreuses, sont serrées les unes contre les autres, insérées à l'aisselle de bractéoles scarieuses imbriquées; les tépales externes, au nombre

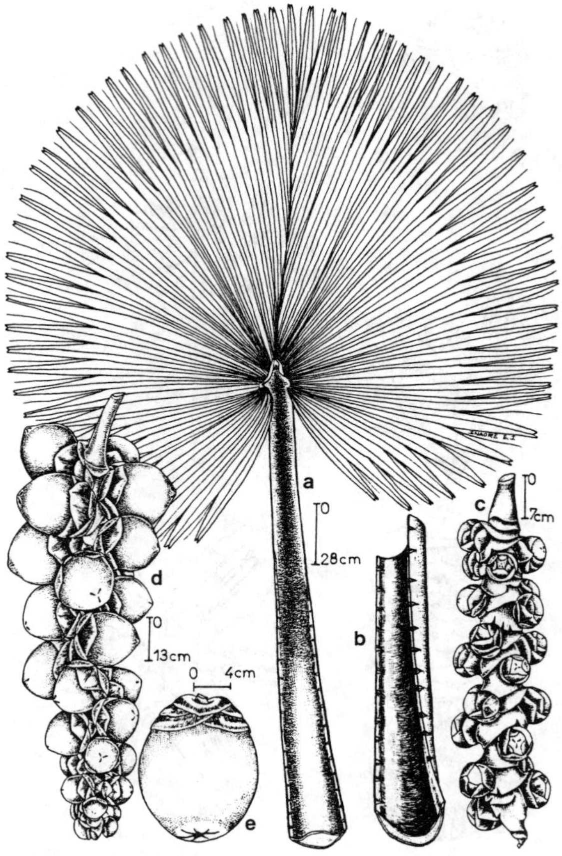

Figure 3. *Borassus* sp. (*B. flabellifer* L.?) a) feuille; b) portion du pétiole; c) inflorescence femelle; d) infrutescence; e) fruit.

de 3, sont libres; les 3 tépales internes sont soudés entre eux, à la base; les 6 etamines, à grandes anthères, ont un filet très court. Les inflorescences femelles, simples, comportent des fleurs relativement grandes, peu nombreuses; les tépales, concolores, charnus, réniformes, les externes étant plus grands que les internes; les staminodes, au nombre de 6-9, sont insérés à la base des tépales internes; l'ovaire, à 3 loges, est terminé par des stigmates sessiles.

Les infrutescences robustes atteignent jusqu'à 1,8 m de longueur. Les fruits, des drupes coriaces, lisses, subglobuleuses-ovoïdes, de 8-15 cm de hauteur et de 6-12 cm de diamètre, sont jaunes ou oranges, odoriférantes, à maturité; la base des fruits est entourée, des tépales externes persistants accrescents formant une cupule. Les 3 graines, à endocarpe très dur, sont entourées d'une pulpe filandreuse orange; l'albumen, blanc, est corné.

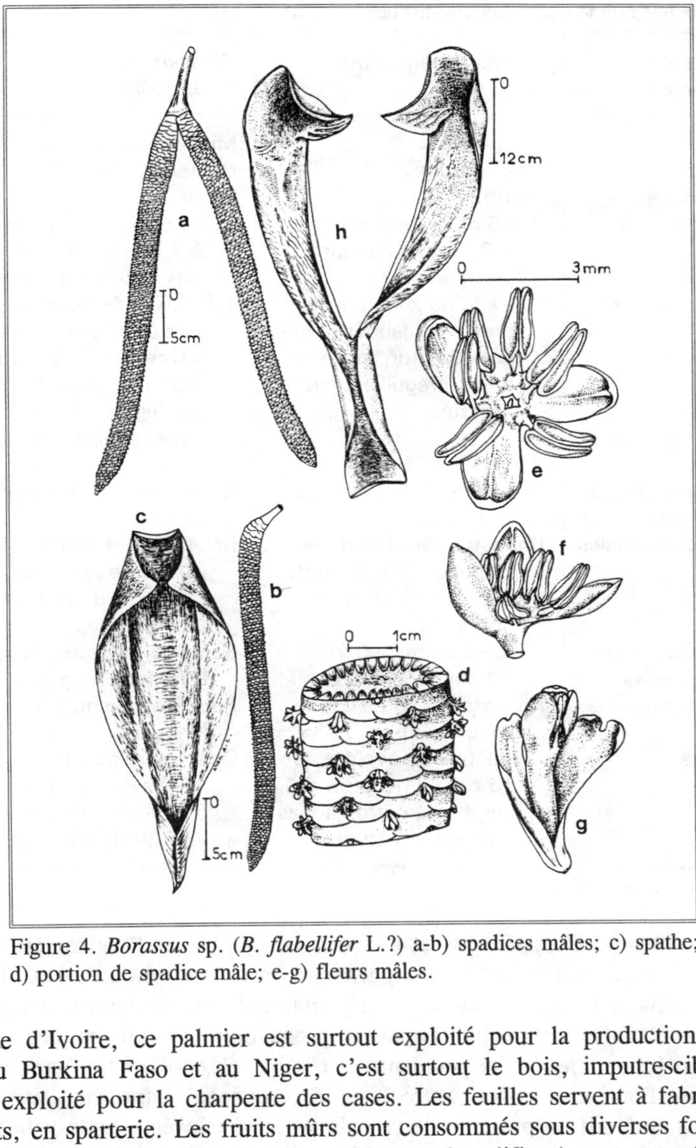

Figure 4. *Borassus* sp. (*B. flabellifer* L.?) a-b) spadices mâles; c) spathe; d) portion de spadice mâle; e-g) fleurs mâles.

En Côte d'Ivoire, ce palmier est surtout exploité pour la production du vin de palme. Au Burkina Faso et au Niger, c'est surtout le bois, imputrescible, qui est fortement exploité pour la charpente des cases. Les feuilles servent à fabriquer différents objets, en sparterie. Les fruits mûrs sont consommés sous diverses formes. Dans toute l'Afrique de l'Ouest, les plantules, charnues (amylifères) sont mangées, cuites à l'eau.

Borassus sp. (*Borassus flabellifer* L.?)
Cette plante a longtemps été confondue avec le taxon commun largement répandu (*Borassus aethiopum*). On la trouve dans toute l'Afrique de l'Ouest, aussi bien dans les formations naturelles qu'aux alentours des agglomérations.

Tableau 1. Résumé des caractères distinctifs des deux taxons.

Organes caractéristiques	*Borassus aethiopum*	*Borassus* sp. (*B. flabellifer*?)
Port	Robuste.	Mince.
Stipe	Gris sombre.	Gris clair.
Feuillage	Touffu.	Léger.
Limbe	2-2,5 m de longueur, 3-3,5 m de largeur, vert sombre.	1,5-2,5 m de longueur, 2-2,5 m de largeur, vert clair à glaucescent.
Pétioles	1,5-3 m de longueur, glabres, vert jaunâtre, marbrés de brun foncé, à bords irrégulièrement épineux.	1,5-2,5 m de longueur, vert clair, pruineux à glabrescents, à bords inermes, comportant des légères fentes noirâtres régulières décoratives.
Inflorescences mâles	Ramifiées, 1,5-1,7 m de longueur.	Ramifiées, 1,4-1,5 m de longueur.
Spadices mâles	Robustes, 30-40 cm de longueur, 3,5-4 cm de diamètre, à sommet obtus.	Relativement minces, 30-40 cm de longueur, 2,5-3 cm de diamètre, à sommet apiculé.
Inflorescences femelles	Simples, robustes, 70-80 cm de longueur.	Simples, robustes, jusqu'à 60 cm de longueur.
Infrutescences	Robustes, jusqu'à 1,8 m de longueur.	Robustes, jusqu'à 1 m de longueur.
Fruits	Subglobuleux-ovoïdes, 8-15 cm de hauteur, 6-12 cm de diamètre, jaunes, oranges à maturité.	Oblongs ellipsoïdes, 8-15 cm de hauteur, 8-12 cm de diamètre, verts, vert grisâtre à maturité.

Dans les villes de Ouagadougou et de Bobo-Dioulasso, au Burkina Faso, ce rônier est souvent planté en tant qu'arbre d'alignement. Dans la province de la Comoé, particulièrement dans la région de Banfora, on y trouve de magnifiques peuplements issus de plantation, abondamment exploités pour la production du vin de palme.

En Côte d'Ivoire, des petits peuplements ont été observés entre Yamoussoukro et Bouaflé, entre Bouaflé et Daloa, entre Bouaflé et Zuénoula ainsi que dans la ville de Bouaflé. Plus au Nord, la présence de ce palmier a été relevée dans la vallée du M'bé, entre Bouaké et Katiola.

Ce rônier diffère de l'espèce habituelle par les caractères suivants qui peuvent être ainsi résumés: port régulier, stipe gris clair. Feuillage clair, léger. Limbe de 1,5-2 m de longueur et de 2-2,5 m de largeur, vert clair, glaucescent; pétioles de 1,5-2,5 m de longueur, vert clair, pruineux à glabrescents, à bords inermes, comportant, perpendiculairement, des légères fentes noirâtres régulières et décoratives. Inflorescences mâles comprenant des spadices de 30-40 cm de longueur et de 2,5-3 cm de diamètre, à sommet apiculé. Inflorescences femelles atteignant 60 cm de longueur. Infrutescences mesurant jusqu'à 1 m de longueur. Fruits oblongs-ellipsoïdes, de 8-15 cm de hauteur et de 8-12 cm de diamètre, verts ou vert grisâtre, odoriférants, à maturité. Graines entourées d'une pulpe filandreuse orange.

Les usages de palmier sont identiques à ceux de *Borassus aethiopum*: matériaux de construction, sparterie, production de vin de palme, consommation des fruits et des plantules.

Conclusion

Deux sortes de rônier, facilement reconnaissables surtout par le feuillage, les pétioles et la coloration des fruits mûrs, sont présents en Afrique de l'Ouest. L'un, *Borassus aethiopum*, présente un feuillage touffu, un pétiole à bords nettement épineux et des fruits qui, à maturité, sont jaunes ou oranges. L'autre, *Borassus* sp., se caractérise par un feuillage léger, un pétiole à bords lisses (non épineux) et des fruits qui, à maturité, demeurent verts ou deviennent vert grisâtre.

Borassus sp. serait-il le taxon reconnu comme *Borassus flabellifer*, répandu dans les régions sèches d'Asie tropicale et surtout en Inde où il est particulièrement planté (Purseglove 1976)? Nous ne pouvons pas être affirmatifs car nous ne disposons pas d'une diagnose détaillée sur *Borassus flabellifer* nous permettant de faire la comparaison des caractères.

Cependant, contrairement aux pensées des botanistes et forestiers de l'Afrique de l'Ouest, le taxon *Borassus* sp. n'est pas synonyme de *Borassus aethiopum*. C'est une espèce ou une sous-espèce à part entière possédant ses propres caractéristiques. L'on ne doit plus confondre les deux taxons.

Note du rédacteur

Conformément a la suggestion des auteurs, le genre *Borassus* devrait être révisé, aussi selon l'avis de Dransfield et Beentje (pers. comm.).

Bibliographie

Busson, F., 1965. Plantes alimentaires de l'Ouest africain. Etude botanique, biologique et chimique. Ministère de la Coopération. Editions Leconte, Marseille: 498-504.

Corner, E.J.H., 1966. The Natural history of Palms. University of California Press: 393 pages.

Dransfield, J., 1988. The Palms of Africa and their relationships. Monogr. Syst. Bot. Missouri Bot. Gard. 25: 95-103.

Kerharo, J. & J.G. Adam, 1974. La Pharmacopée sénégalaise traditionnelle. Plantes médicinales et toxiques. Editions Vigot frères: 203-204.

McCurrach, J.C., 1960. Palms of the World. Harper & Brothers, New York: 290 pages.

Purseglove, J.W., 1976. Tropical Crops, Monocotyledons. Longman Group Ltd.: 420-421.

Russell, T.A., 1968. Flora of West Tropical Africa, ed. 2, 3: 159-169.

Sulyok, M., 1967. Pálmák. Mezögazdasági kiadó, Budapest: 143 pages.

Vanden Berghen, C., 1988. Flore illustrée de Sénégal, tome 9: 77-79.

Vuattoux, R., 1968. Le peuplement du rônier (Borassus aethiopum) d'une savane de Côte d'Ivoire. Ann. Univ. Abidjan, sér. E, tome 1: 1-138.

Rheophytes of Ghana

G. AMEKA [1], J. ADOMAKO [1], K.A.A. DE GRAFT-JOHNSON [2], M. CHEEK [3] & M. SWAINE [4]
[1] Department of Botany, University of Ghana, Legon.
[2] Institute of Aquatic Biology, Accra, Ghana.
[3] Royal Botanic Gardens, Kew, United Kingdom.
[4] Department of Plant and Soil Science, University of Aberdeen, Aberdeen, United Kingdom.

Abstract
A survey of rheophytes of Ghana, begun in early 1994, has been set up between the Botany Departments of the University of Ghana and Aberdeen, the Institute of Aquatic Biology, Accra and the Royal Botanic Gardens, Kew. The purpose of the survey is to assess whether rheophytes can be used as indicators of pollution from, for example, effluents from human settlements and farming activities. An identification guide to Ghanaian rheophytes is to be produced. This paper reviews progress made with the survey and the different classes of rheophytes in Ghana.

Key words
Biodiversity, Forest reserve, Rheophyte, River quality.

Introduction

Rheophytes are plant species of the beds and banks of streams and rivers. A study of the rheophyte flora of Ghana is important for three main reasons.

1 Compared with the rheophytes of S.E. Asia, enumerated by Van Steenis (1981), those of Africa have hardly been studied. Little is known about which species are rheophytic or where they occur.
2 The plants are useful indicators of river health.
3 The biodiversity of these plants is threatened by increased use of the land surrounding the rivers.

Reserved forest land in Ghana (20% of the forest zone) protects river catchments and headwaters as well as providing many direct benefits from timber production and other forest produce. Outside forest reserves rivers are sediment-rich and polluted to varying degrees, with an apparent loss of rheophyte diversity. The project aims to quantify the differences between protected and unprotected headwaters in terms of rheophyte biodiversity, by a survey of contrasting river systems. The results of the survey will allow comparisons between logged and unlogged forest reserves, forested and deforested areas and between high and low rainfall types. The results will also indicate the potential value of rheophytes as indicators of river quality, by correlating floristics with direct measures of water quality and river habitat. Collection, identification and curation of specimens will provide for a comprehensive account of the relatively poorly known rheophyte flora of the Ghanaian forest zone. If possible an illustrated guide to the rheophytes of West Africa will be produced. A duration of three years is envisaged for this Project.

Materials and methods

The area in and around Kakum Forest Reserve, in the forest zone of south central Ghana was selected as the first forest to be investigated for rheophytes because of its

L.J.G. van der Maesen et al. (eds.), The Biodiversity of African Plants, 780-782.

accessibility from the University of Ghana. It contains representatives of both forested and deforested catchments and is an area of a moderately wet type of forest, containing the headwaters of the Afia, Kakum and Obuo rivers. Surveys were started in April, at the end of the dry season (the wet seasons are from April to July, then September to October). At this time water levels and flow are low and many rheophytes are in their reproductive stage. Lengths of river 300 m long were sampled as follows. Water samples were taken at the downstream, then the upstream ends to avoid contamination of river water by disturbance of the substrate. Samples were taken from mid-stream in areas where the current was most swift. Temperature and pH were tested on the spot. Sulphide, phosphate, ammonium, nitrate, aluminium and phenols were measured the same day with a Hydrocheck photometer (WPA, Cambridge, UK). Subsamples were preserved by adding drops of hydrochloric acid so that potassium, calcium, sodium and iron could be evaluated later. It is intended in future to monitor levels of dissolved salts and chloride.

Estimates of tree cover, substrate type, conservation status and logging history were recorded. Rheophytes were surveyed on both banks as well as the riverbed, and herbarium specimens of each taxon recorded were made in sets of three or more. These vouchers are to be deposited at GC, K and WAG. Identifications are in progress at K. Illustrations for the "Guide to Rheophytes" will be made from herbarium specimens and using photographic slides taken in the course of the work. Apart from our own observations, the Flora of West Tropical Africa (Keay & Hepper 1954-72) will be used to produce a checklist of rheophytes for Ghana.

Results and discussion

Chemistry

Thirteen sites have each been sampled so far, six in April (end of dry season), seven in July (wet season). As yet, there has been insufficient replication to make a full analysis of chemical data worthwhile, but it has been noticeable that concentrations of solutes has been much lower in July than in April. A dilution factor is to be expected when river flows are higher during the wet season.

Botany

About 30 rheophytic species have been encountered. This represents a considerable increase over the numbers listed by Van Steenis (1981: 71) who gave figures of 21 species from Tropical Africa, including 19 endemic species, but excluding Podostemaceae. Rheophyte diversity is low in tropical Africa compared with S.E. Asia, and S. America. A provisional ecological classification is as follows:

- A Constantly submersed rheophytes. These are submerged even in the dry season, though they may produce flowers above the water surface, e.g. *Crinum natans* Bak. (Amaryllidaceae).
- B Periodically submersed rheophytes. These become exposed in the dry season, producing flowers and fruit before the water level rises, e.g. *Tristicha trifaria* (Bory) Spreng. and other Podostemaceae.
- C Emergent rheophytes. Species which are rooted in shallow water, but with leaves and flowers above the water. In the dry season they are completely exposed e.g., *Polygonum salicifolium* Brouss. ex Willd. (Polygonaceae) and

Cyperus fertilis Boeck. (Cyperaceae). Both these examples are facultative rheophytes and may also occur in other wet places, e.g. marshes.

D River margin annuals. These species grow on the banks of sediment exposed in the dry seasons, e.g. *Pentodon pentandrus* (Schum. & Thonn.) Vatke (Rubiaceae). These are also facultative river plants, and some workers would exclude these from the rheophytic category since they are not permanently exposed to flowing water. It is apocryphal that from this category arose rain forest weed species.

That we have not yet recorded any woody species of this biological group from Ghana is surprising. Van Steenis (1981) found that about half the species in his world census of rheophytes were woody. He admitted that African rheophytes were very poorly known. Woody rheophytes are known from Africa, e.g. *Coffea congoensis* Froehner (Rubiaceae) from the Congo basin.

Three species of Podostemaceae, *Saxicolella amicorum* J.B. Hall, *Inversodicraea bowlingii* J.B. Hall and *Polypleurum submersum* J.B. Hall, all endemic to Ghana (Hall 1971), are still only known from their type localities and new sites have not been discovered in the course of our survey, though a fourth species, *Tristicha trifaria* is widespread and frequently encountered. Why this should be is unclear.

In the course of our work we have noted two environmental variables (independent of pollution) which can have a profound effect on the rheophyte complement of a river. Firstly substrate: Podostemaceae, a family of obligate rheophytes seems to be restricted to streams where there are rock outcrops. Secondly shade: in areas of heavy shade, rheophyte diversity can be greatly reduced.

Acknowledgements

The authors are grateful for support from their respective institutes, and for funding to the Darwin Initiative scheme of the Department of the Environment, U.K.

References

Hall, J.B., 1971. New Podostemaceae from Ghana with notes on related species. Kew Bull. 26: 125-136.

Keay, R.W.J. & F.N. Hepper, 1954-72. Flora of West Tropical Africa. Second Edition. Crown Agents, London.

Van Steenis, C.G.G.J., 1981. Rheophytes of the World. Sijthoff & Noordhoff, The Netherlands.

Progress of the Cape Verde Flora

M.A. DINIZ [1], I. GOMES [2], J. PAIVA [3] & I. MOREIRA [1]
[1] Centro de Botânica, IICT, Rua da Junqueira n° 86, 1300 Lisboa, Portugal.
[2] Instituto Nacional de Investigação e Desenvolvimento Agrário, C.P. 84, Praia, Cabo Verde.
[3] Departamento de Botânica, Universidade de Coimbra, 3049 Coimbra Codex, Portugal.

Abstract
The Cape Verde Flora project is carried out since 1988. The vascular plant families (including the cultivated ones) in the Archipelago include 115 of Spermatophyta, and the Pteridophyta. The study of 66 families is completed (53 are in press) and that of 36 is continuing. Only 13 families, and the pteridophyta, have not yet been distributed to authors. The publication of each family will be in a separate fascicle.

Résumé
Le projet de la Flore de Cap-Vert a été programmé depuis 1988. Les familles de plantes vasculaires de l'archipel (en incluant les cultivées) sont au nombre de 115 pour les Spermatophytes, plus les Ptéridophytes. L'étude de 66 familles est terminée (53 sont sous presse) et l'étude de 36 autres est en cours. Seulement 13 familles, et les pteridophyta, n'ont pas encore été attribuées. La publication est programmée en fascicules à raison d'un par famille.

Key words
Cape Verde, Flora.

The Cape Verde Flora project was agreed upon in 1987 by the National Institute of Agrarian Research and Development (Instituto Nacional de Investigaçao e Desenvolvimento Agrário - INIDA) of Cape Verde and the Tropical Research Scientific Institute of Portugal (Instituto de Investigação Científica Tropical - IICT). The project was started in 1988, sponsored by Portuguese institutions of research financing. Nine botanists of the Botany Centre of IICT are involved, one of the Botany Institute of Coimbra University and two of INIDA.

The herbarium material was requested on loan mainly from CECV, COI, LISC, LISJC, LISU, P and, in special cases, from Lobin's herbarium (BONN) and O. Besides the valuable material received from these herbaria with prominent collections such as those of A. Chevalier and Grandvaux Barbosa, the specimens collected by Cardoso de Matos and by the Capeverdian botanist Samuel Gomes, identified or confirmed by the IICT Botany Centre proved very valuable. Part of the information of plant distribution and ecological characteristics is included in the descriptions of the maps of the Agro-ecological Zones and Vegetation (Diniz & Matos 1986, 1987, 1988a, b, 1993, 1994a, b). In the period of 1988 to 1994, six field expeditions were carried out by botanists of INIDA and IICTG, involved in the Flora project. The material collected is conserved in Cape Verde (CECV) and in Lisbon (LISC) herbaria.

The editorial board of the Flora is composed of six members, belonging to the INIDA (2), IICT (3) and Coimbra University (1). The Flora will be published in separate fascicles, containing one family each. The families are numbered according to Cronquist (1981, 1988). Each fascicle includes:

* diagnostic descriptions of the family, genera, species and infraspecific taxa, if present;
* dichotomous keys for genera, species and infra-specific taxa;
* citation of the correct name and synonymy relevant for the area with bibliographic references;
* distribution and diversity inside Cape Verde, with citation of one sample by each island with mention of the herbaria where the sample is kept;
* general geographical distribution;
* habitats and ecological information;
* common names;
* plant utilisation;
* iconographic tables, at least one for each genus.

The front cover of each fascicle is illustrated with a drawing of the endemic species *Echium vulcanorum* A. Chev. The number of Spermatophyta plant families is 115, of which 66 have been studied (53 are now in press) and 36 are distributed to prospective authors. The Pteridophyta and the 13 following families are not yet attributed: Agavaceae, Amaryllidaceae, Palmae, Brassicaceae (Cruciferae), Cupressaceae, Dioscoreaceae, Juncaceae, Liliaceae, Pinaceae, Rosaceae, Sapotaceae, Scrophulariaceae, Zingiberaceae.

Complementary to the Flora project, an identification manual of weeds is in preparation, the ecological study of the vegetation of Santiago Island is underway, and a study of the distribution of the endemic species of some families is being considered. Some information, related to these subjects, has been published (Basto 1987, 1988, Basto & Diniz 1993, Diniz et al. 1993, Gomes 1994).

References

Basto, M.F.P., 1987. Aditamentos à flora vascular de Cabo Verde. Garcia de Orta, Sér. Bot. 9: 1-15.

Basto, M.F.P., 1988. Plantas vasculares endémicas do arquipélago de Cabo Verde. Garcia de Orta, Sér. Bot. 10: 11-15.

Basto, M.F.P. & M.A. Diniz, 1993. Aditamentos à flora vascular do arquipélago de Cabo Verde - II. Garcia de Orta, Sér. Bot. 11: 75-82.

Cronquist, A., 1981. An Integrated System of Classification of Flowering Plants. The New York Botanical Garden, New York.

Cronquist, A., 1988. The Evolution and Classification of Flowering Plants, 2nd ed. The New York Botanical Garden, New York.

Diniz, A.C. & G.C. Matos, 1986, 1987, 1988a, 1988b, 1993, 1994a, 1994b. Carta de Zonagem Agro-ecológica e da Vegetação de Cabo Verde. I: Ilha de Santiago; II: Ilha do Fogo; III: Ilha de Maio; IV: Ilha da Boavista; V: Ilha do Sal; VI: Ilha de S. Vicente; VII: Ilha de Santa Luzia. Garcia de Orta, Sér. Bot., Lisboa, 8: 39-82; 9: 35-70; 10: 19-48; 10: 49-72; 11: 9-29; 12: 69-100; 12: 101-120.

Diniz, M.A., G.C. Matos & I. Moreira, 1993. Infestantes das culturas agrícolas de Cabo Verde. Comunicações Instituto de Investigação Científica Tropical, Sér. Ciênc. Agrárias 8: 17-21.

Gomes, S., 1994. Nomes vernáculos e vulgares de plantas de Cabo Verde. Garcia de Orta, Sér. Bot., 12: 127-150.

Biodiversity in the flora of Egypt

M.N. EL HADIDI & H.A. HOSNI
The Herbarium, Faculty of Science, Cairo University, Giza 12613, Egypt.

Abstract
The flora of Egypt consists of about 2,080 species of seed plants and vascular cryptogams. Of the 1083 species from the Mediterranean coasts, 255 species (incl. 18 endemics) are only known from this area. The Nile-land and the oases contain a total of 800 species; 134 of these species (incl. 10 endemics) are only known from this territory. Of the 1400 species from the deserts, the Sinai and the Elba mountains, 411 species (incl. 40 endemics) are known only from these territories.

Key words
Egypt, Flora, Floristic categories.

Introduction

Egypt is the meeting point of floristic elements belonging to four phytogeographical regions: the African Sudano-Zambesian; the Asiatic Irano-Turanian; the Afro-Asiatic Saharo-Sindian and the Euro-Afro-Asiatic Mediterranean.

Egypt is perhaps the most arid country in North Africa; desert ecosystems cover almost all of Egypt and extend south to the Sudan. Such hot ecosystems are distinguished into either arid or hyperarid provinces. Hyperarid provinces are characterized by a hot or very hot summer (mean temperature of 30° C in the hottest month) and a winter rainfall averaging 30 mm per year. Arid provinces are characterized by a hot summer (mean temperature of 20-30° C in the hottest month) and a winter rainfall ranging between 20 and 100 mm per year.

The natural vegetation in Egypt exhibits a boxed T-shaped pattern, and may be separated into the Mediterranean coast, the Nile land, and the deserts and mountains (Figure 1). The flora of Egypt consists of about 2,080 species of seed plants and vascular cryptogams.

The present investigation was based on fieldwork, as well as study and revision of the collections of Cairo University Herbarium (CAI). The phytogeographical treatment of the present study is in accordance with Wickens (1976).

The Mediterranean

The Mediterranean coast receives an annual rainfall of 20-100 mm. A total of 1083 species is recorded here. Of the 255 species known only from this territory (Table 1), 18 are believed to be endemic: *Silene biappendiculata* Rohrb., *Astragalus camelorum* W. Barbey, *Ebenus armitagei* Schweinf. & Taub., *Vicia sinaica* Boulos, *Euphorbia punctata* Del., *Helianthemum crassifolium* Pers. subsp. *sphaerocalyx* (Gaub. & Janchen) Maire, *Lycium aschersonii* Dammer, *L. schweinfurthii* Dammer, *Centaurea alexandrina* Del., *Sonchus macrocarpus* Boulos J.C. Jeffrey, *Biarum bovei* Blume, *Bellevalia salah-eidii* Täckh. & Boulos, *B. romana* (L.) Rchb., *Muscari salah-eidii* (Täckh. & Boulos) H. Hosni, *M. albiflora* (Täckh. & Boulos) H. Hosni, *Allium mareoticum* Bornm. & Gauba, *A. blomfieldianum* Aschers. & Schweinf. and *Pancratium arabicum* Sickenb.

L.J.G. van der Maesen et al. (eds.), The Biodiversity of African Plants, 785-787.

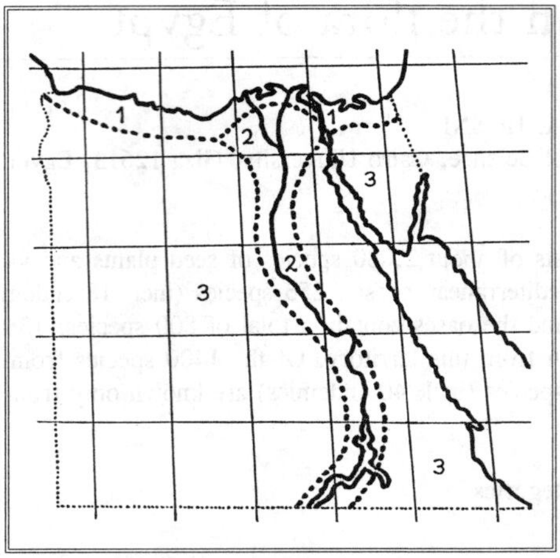

Figure 1. Egypt consists of the Mediterranean coast (1), the Nile land (2) and the deserts (3).

We established that 144 species are monoregional taxa; including Mediterranean (120), Saharo-Sindian (14) and Irano-Turanian (10) elements. Another 80 species are Mediterranean taxa with extensions to other regions (bi- or tri-regional taxa).

Nile land and the oases

In total 800 meso-, helo- and hydrophytic species are recorded from the Nile land and the oases, among which 134 species are confined to this territory (Table 1). Twenty species are of common occurrence in both territories and it is believed that these species were introduced to the farmland of the oases with the cultivation of certain crops, e.g. rice (El Hadidi 1993).

Tropical taxa dominate the weeds confined to the Nile land including taxa belonging to Sudano-Zambesian, Saharo-Sindian and Palaeotropical elements (54 species). Mediterranean taxa (mono-, bi- and tri-regional) are represented by 32 species.

A few cosmopolitan taxa are confined to these territories: *Amaranthus spinosus* L. *Cardiospermum halicacabum* L., *Cyperus alternifolius* L. subsp. *flabelliformis* (Rottb.) Kük, *Cyperus imbricatus* Willd. ex Link, *Pycreus mundtii* Nees and *P. polystachyos* (Rottb.) P. Beauv.

Endemics of the Nile land include: *Glinus runkewitzii* Täckh. & Boulos, *Sinapis allionii* Jacq., *Trigonella media* Del., *Tephrosia kassasi* Boulos, *Convolvulus austro-aegyptiacus* Abdallah & Saad, *Campanula dimorphantha* Schweinf., *Senecio belbeysius* Del. and *Bromus javorkae* Penzer. Those of the oases are: *Ducrosia ismaelis* Aschers. and *Pimpinella schweinfurthii* Aschers.

The deserts (arid lands)

The deserts of Egypt, that is the Libyan, Nubian, Isthmic, and Galala Desert; Red Sea coastal land; and the mountains of Sinai, Elba and Uweinat (El Hadidi 1980), can be grouped under the hyperarid province. The deserts, Sinai and Elba mountains contain 1400 species; 411 of these species are known only from these territories (Table 1).

Table 1. Floristic categories of the species confined to the Mediterranean coast (1), the Nile land and the oases (2), and the deserts (3).

Floristic categories	1	2	3
Endemics	18	10	40
Mediterranean	120	7	17
Medit. & Saharo-Sindian	14	5	
Medit. & Irano-Turanian	43	4	26
Medit. & Irano-Turanian & European	13	10	10
Medit. & European	10	6	
Irano-Turanian	10		46
Irano-Turanian & Saharo-Sindian			16
Saharo-Sindian	14	11	66
Saharo-Sindian & Sudano-Zambesian			8
Sudano-Zambesian		5	87
Cosmopolitan		6	
Palaeotropical		38	59
Pantropical		11	
Others	13	21	36
Total	255	134	411

At least 40 species are endemics: 10 in the Isthmic and Galala deserts, 3 in the Elba mountains and 27 in the Sinai. Sudano-Zambesian and palaeotropic taxa (87 and 59 species respectively) dominate the flora of the deserts. Saharo-Sindian taxa (66 species) are distributed at almost equal number in all deserts. A similar pattern is shown by Mediterranean taxa with extensions to Irano-Turanian territories (El Hadidi & Fayed 1994/1995).

References

El Hadidi, M.N., 1980. An outline of the planned flora of Egypt. Taeckholmia add. ser. 1: 1-12.

El Hadidi, M.N., 1993. Natural vegetation. In: G.M. Craig (ed.), The Agriculture of Egypt: 39-62. Oxford University Press.

El Hadidi, M.N. & A.A. Fayed, 1994/1995. Materials for excursion flora of Egypt. Taeckholmia 15: 1-2.

Wickens, G.E., 1976. The flora of Jebel Marra (Sudan Republic) and its geographical affinities. Kew Bull. add. ser. V. HMSO, London.

Some aspects of the biodiversity of the weed flora in the farmlands of Egypt

M.N. EL HADIDI, A.I. HOSNY & N. EL HUSSEINI
The Herbarium, Faculty of Science, Cairo University, Giza 12613, Egypt.

Abstract
Some 470 species (about 22.5% of the total number of the flowering plants represented in the flora of Egypt) constitute the weed assemblages in Egypt. They exhibit great diversity concerning their origin, behaviour, regional adaptation, etc. In total 185 species can be considered weeds of the farmlands of Egypt (Nile-land, Mediterranean coastal land and the oases). They comprise 110 winter, 49 summer and 26 all-year weeds. Among the winter weeds, 96 species are Mediterranean taxa (mono-, bi-, and pluriregional). Most summer weeds (40 species) and about half of the all-year weeds (12 species) are cosmopolitan or semi-cosmopolitan taxa.

Key words
Agriculture, Egypt, Weeds.

Introduction

The present investigations were based on intensive field work, as well as study and revision of the collections of Cairo University Herbarium (CAI), from the farmland territories 1-8 in Figure 1. The phytogeographical treatment is according to Wickens (1976).

During the early days of the Holocene (12,000-10,000 years BP), it is believed that the natural vegetation of Egypt included approximately 1,200 flowering plant species (El Hadidi 1993), which amount to half of the species known at present. With the adoption of agriculture and the introduction of domesticates during the Neolithic-Predynastic period (6,000-4,500 years BP), an estimated 170 species were introduced and naturalized. These constitute with other native species the early weed assemblages.

Following the introduction of other crop plants during Pharaonic Egypt (4,500-2,000 years BP), the number of adventive species is believed to have reached between 225 and 255 species. During the last two millennia, the numbers of weeds recorded in the Flora of Egypt increased to 470 species.

Statistical analyses of the present-day weed assemblages clearly show that the number of species belonging to Mediterranean (M), Sudano-Zambesian (S-Z) or Saharo-Sindian (Sa-Si) elements has remained about the same since Graeco-Roman times. The actual increase in the number of species during the last two millennia is related to the introduction of more widely distributed taxa which found their way to Egypt as associates of other plants of economic value. The present-day weed assemblages include 69 cosmopolitan (Cosm), 87 semi-cosmopolitan, 34 pantropical (Pan) and 22 palaeotropical (Pal) species. Other species may belong to European (Eur) or Irano-Turanian (I-T) elements.

It is obvious that the weed assemblages associated with an ancient crop such as wheat or barley, which are known to have been in cultivation some five millennia ago, must be different from those of maize, which was introduced into Egypt about 350

L.J.G. van der Maesen et al. (eds.), The Biodiversity of African Plants, 788-794.

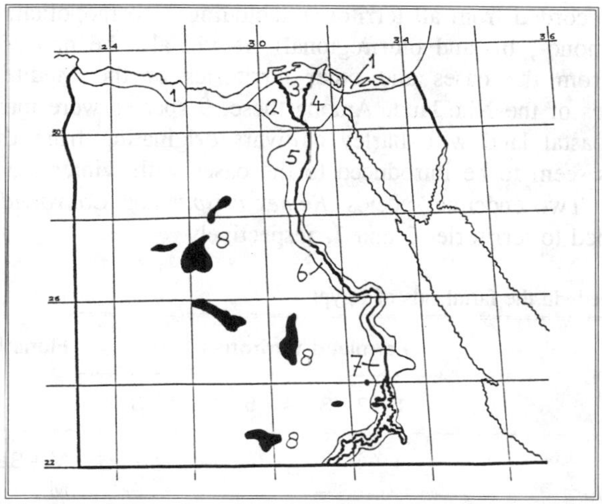

Figure 1. The 8 investigated farmland territories: 1. Mediterranean; 2. W. Delta; 3. Central Delta; 4. E. Delta; 5. N. Nile valley and Faiyum; 6. Central Nile valley; 7. S. Nile valley; 8. Oases.

years ago. The weed assemblages of wheat (*Triticum* spp.) are rather complex, comprising between 87 and 121 species, belonging to several mono-, di-, or pluri-regional elements. The weed assemblage of maize (*Zea mays*) does not exceed 40 species and includes pan- and palaeotropical species which were not previously known in Egypt.

Some 470 species constitute the current weed assemblages in Egypt (El Hadidi & Fayed 1994/1995). These can be grouped as follows:

* weeds of farmlands (agrestal weeds), comprising 218 species of which 185 species were recorded from the studied areas (Sinai excluded);
* wasteland and roadside weeds, comprising 86 species;
* canal bank weeds and water plants, comprising about 122 species;
* reclaimed desert weeds, comprising 42 species which are desert annuals that flourish during the early stages of reclamation of the desert outskirts of the farmland.

Weeds of farmlands

Field plots of the farmlands have weed assemblages which are fairly specific for each crop. Winter weeds abound in the cooler months of the year and are associated with winter crops. Summer weeds abound in the warmer months of the year and are associated with summer crops. Some weeds are biologically active throughout the year (all-year weeds) and are usually associated with perennial crops such as fruit crops and palms.

Winter weeds

Table 1 gives the chorology of 110 winter weeds in the investigated farmland territories. It may be noticed that Mediterranean taxa dominate the winter weeds. Thirty-two species are recorded from all territories including 7 cosmopolitan and 24 Mediterranean species (mono-, bi- and pluriregional). It will also be noticed that 23 species were recorded from the oases and other territories of the Mediterranean and the northern territories of the Nile land. Among these, 9 species were introduced from the Mediterranean coastal land with barley cultivars originating from this area. On the other, 14 species seem to be introduced to the oases with winter crops from the Nile Delta or Faiyum. Two endemic species, *Rumex crispus* and *Convolvulus austro-aegyptiacus*, are confined to territories 3 and 7, respectively.

Table 1. Winter weeds in the farmlands of Egypt.

Species	**Farmland territories**								**Floristic category**
	1	2	3	4	5	6	7	8	
Aegilops bicornis	+	+	–	–	–	–	–	–	M + Sa-Si
Amberboa tubuliflora	+	+	–	–	–	–	–	–	M
Ammi majus	+	+	+	+	+	+	+	+	M + I-T
Anagallis arvensis	+	+	+	+	+	+	+	+	Cosm
Apium graveolens	+	+	+	+	+	–	–	–	M + I-T
Asphodelus tenuifolius	+	+	–	–	–	–	–	+	M
Astragalus annularis	+	+	–	–	–	–	–	–	M + Sa-Si
Astragalus boeticus	+	+	–	+	–	–	–	–	M
Avena fatua	+	+	+	+	+	+	+	+	Cosm
Avena sterilis	+	+	+	+	–	–	–	–	M
Beta vulgaris	+	+	+	+	+	+	+	+	M + I-T + Eur
Brachiaria eruciformis	+	+	+	+	+	+	+	+	M + I-T + Eur
Brassica juncea	+	+	–	–	–	–	–	–	M
Brassica nigra	+	+	+	+	+	+	+	+	Cosm
Brassica tournefortii	+	+	+	+	+	+	+	+	M + I-T
Bromus diandrus	+	+	–	+	–	–	–	–	M + I-T + S-Z
Bromus fasciculata	+	+	–	–	–	–	–	+	M + I-T
Bromus madritensis	+	+	–	+	–	–	–	–	M + I-T
Bromus rubens	+	+	–	–	–	–	–	+	M + I-T
Bupleurum semicompositum	+	+	–	–	–	–	–	–	M
Calendula arvensis	+	+	+	+	+	+	+	+	M + I-T
Capsella bursa-pastoris	+	+	+	+	+	–	–	–	Cosm
Carthamus lanatus	+	+	–	–	–	–	–	–	M + Eur
Chenopodium ficifolium	+	+	+	+	–	–	–	–	Cosm
Chenopodium murale	+	+	+	+	+	+	+	+	Cosm
Chenopodium opulifolium	+	+	–	+	–	–	–	–	M + I-T + Eur
Cichorium pumilum	+	+	+	+	+	+	+	+	M + I-T
Convolvulus austro-aegyptiacus	–	–	–	–	–	–	+	–	Pal
Cutandia dichotoma	+	+	–	–	–	–	–	–	M + I-T + Sa-Si
Cutandia memphitica	+	+	–	–	–	–	–	+	M + I-T
Emex spinosa	+	+	+	+	+	+	+	+	M
Enarthrocarpus lyratus	+	+	–	–	+	–	–	+	M
Enarthrocarpus stragulatus	+	+	–	–	–	–	–	–	M + Sa-Si
Erodium malacoidis	+	+	–	–	–	–	–	+	M + I-T
Erodium pulverulentum	+	+	–	–	–	–	–	–	M + Sa-Si

Euphorbia peplus	+	+	+	+	+	+	+	+	Cosm
Fumaria densiflora	+	+	+	+	−	−	−	+	M + I-T + Eur
Fumaria judaica	+	+	−	−	−	−	−	−	M
Fumaria parviflora	+	+	+	+	+	−	−	−	M + I-T + Eur
Geranium dissectum	+	+	−	−	−	−	−	+	M + Eur
Hippocrepis arcolata	+	+	−	−	−	−	−	−	M + Sa-Si
Hordeum leporinum	+	+	−	+	−	−	−	−	M + I-T
Hypecoum aegyptiacum	+	+	−	−	−	−	−	−	M + Sa-Si
Ifloga spicata	+	+	−	−	−	−	−	−	M + Sa-Si
Lamium amplexicaule	+	+	+	+	+	−	−	−	Cosm
Lathyrus aphaca	+	+	+	+	+	−	−	+	M + I-T + Eur
Lathyrus hirsutus	+	+	+	+	−	−	−	−	M + I-T + Eur
Leopoldia comosa	+	+	−	−	−	−	−	−	M + I-T
Lepidium sativum	+	+	+	+	+	+	+	+	M
Lobularia arabica	+	+	−	−	−	−	−	−	M + Sa-Si
Lolium perenne	+	+	+	+	+	−	−	−	Cosm
Lolium rigidum	+	+	+	+	+	+	+	+	M + I-T + Eur
Lolium temulentum	+	+	+	+	+	+	+	+	M + I-T + Eur
Lotus corniculatus	+	+	+	+	+	−	−	+	M + I-T + Eur
Malva parviflora	+	+	+	+	+	+	+	+	M
Malva silvestris	+	−	−	−	−	−	−	−	M + I-T
Matricaria recutita	+	+	+	+	+	+	+	+	M + I-T + Eur
Matthiola longipetala	+	+	−	−	−	−	−	−	M + I-T
Medicago intertexta	+	+	+	+	+	−	−	−	M
Medicago littoralis	+	+	−	+	−	−	−	−	M
Medicago polymorpha	+	+	+	+	+	+	+	+	Cosm
Melilotus alba	+	−	−	−	+	−	−	−	M + I-T + Eur
Melilotus indica	+	+	+	+	+	+	+	+	Pal
Melilotus mesanensis	+	+	+	+	+	−	−	+	M + I-T
Melilotus sulcata	+	+	+	+	+	−	−	−	M
Mercurialis annua	+	+	−	−	−	−	−	−	M + Eur
Oxalis corniculata	+	+	+	+	+	+	+	+	Cosm
Papaver hybridum	+	+	−	−	−	−	−	−	M + I-T + Eur
Papaver rhoeas	+	+	−	−	−	−	−	−	M
Parapholis incurva	+	+	−	−	−	−	−	+	M + I-T + Eur
Parapholis marginata	+	−	−	+	−	−	−	+	M + I-T + Eur
Phalaris canariensis	+	+	−	−	−	−	−	−	M
Phalaris minor	+	+	+	+	−	−	−	+	M + I-T
Plantago cylindrica	+	+	−	−	−	−	−	−	M + Sa-Si
Plantago lagopus	+	+	+	+	+	+	+	+	M + I-T
Poa annua	+	+	+	+	+	+	+	+	M + I-T + Eur
Polygonum aviculare	+	+	−	−	−	−	−	−	Pan
Polypogon semiverticillata	+	+	+	+	+	+	+	+	M + I-T
Pseudorlaya pumila	+	+	−	−	−	−	−	−	M + Sa-Si
Raphanus raphanistrum	+	+	+	+	+	+	−	+	M + Eur
Rumex crispus	−	−	+	−	−	−	−	−	M + I-T + Eur
Rumex pulcher	+	+	−	−	−	−	−	+	M + I-T + Eur
Scorpiurus muricatus	+	+	+	+	−	−	−	+	M
Senecio desfontainei	+	+	+	+	+	+	+	+	M + I-T + Eur
Senecio vulgaris	+	+	+	+	+	−	−	+	M + I-T + Eur
Silene apetala	+	+	−	+	−	−	−	−	M

Silene behen	+	+	+	+	–	–	–	–	M
Silene colorata	+	+	–	–	–	–	–	–	M
Silene conoidea	+	+	–	+	–	–	–	–	M + I-T
Silene nocturna	+	+	+	+	–	–	–	+	M
Silene rubella	+	+	+	+	+	+	–	–	M + I-T
Sinapis allionii	+	+	–	+	–	–	–	–	M + I-T + Eur
Sisymbrium irio	+	+	+	+	+	+	–	–	M + I-T + Eur
Sisymbrium orientalis	+	+	–	–	–	–	–	–	M
Sphenopus divaricatus	+	+	–	–	–	–	–	–	M + I-T
Stellaria media	+	+	+	+	+	–	–	+	M + I-T
Stellaria pallida	+	+	+	+	+	–	–	+	M + Eur
Thesium humile	+	+	–	+	–	–	–	–	M
Torilis nodosa	+	+	+	+	+	–	–	+	M + I-T + Eur
Torilis radiata	+	+	–	–	–	–	–	–	M
Trifolium resupinatum	+	+	+	+	+	+	+	+	M + I-T + Eur
Trigonella hamosa	+	+	+	+	+	+	+	+	M + Sa-Si + S-Z
Urtica urens	+	+	+	+	+	+	+	+	M + I-T + Eur
Vaccaria hispanica	+	+	–	+	–	–	–	–	M + I-T + Eur
Vicia cineria	+	+	–	+	+	–	–	–	M + I-T
Vicia lutea	+	+	+	+	+	–	–	+	M + I-T + Eur
Vicia monantha	+	+	+	+	+	+	+	+	M + I-T
Vicia narbonensis	+	+	–	–	–	–	–	+	M + I-T + Eur
Vicia sativa	+	+	+	+	+	+	+	+	M + I-T + Eur
Vulpia fasciculata	+	+	–	–	–	–	–	–	M

Summer weeds

Table 2 gives the chorology of 49 summer weeds in the investigated farmland territories. It will be noticed that cosmopolitan, palaeo- and pantropical taxa dominate the summer weeds.

Twenty-two species were recorded from all territories including 7 cosmopolitan species, 7 palaeo-tropical species and 7 pantropical species.

Eight species are rice weeds recorded in the Nile Delta and the oases. The occurrence of these weeds in the oases is related to its introduction with rice cultivars from the Nile Delta and Faiyum.

Five species including the palaeotropical *Oldenlandia capensis* and *Laggera aurita*, the Saharo-Sindian *Convolvulus fatmensis* and *Echium rauwolfii* as well as the Sudano-Zambesian *Vigna membranacea* are summer weeds known only from Qena and Aswan Governorates (Territory 7).

Table 2. Summer weeds in the farmlands of Egypt.

Species	Farmland territories								Floristic category
	1	2	3	4	5	6	7	8	
Amaranthus graecizans	+	+	+	+	+	+	+	+	M + I-T
Amaranthus hybridus	+	+	+	+	+	+	+	–	Cosm
Amaranthus lividus	+	+	+	+	+	+	+	–	M + I-T
Amaranthus viridis	+	+	+	+	+	+	+	+	Cosm

Ammania auriculata	+	+	+	+	+	+	—	+	Pan
Ammania bacciflora	+	+	+	+	+	—	—	+	Pal
Antirrhinum orontium	+	+	+	+	—	—	—	—	M + I-T + Eur
Bergia capensis	—	—	—	—	+	—	—	+	Pal
Bromus unioloides	+	+	+	+	—	—	—	+	Cosm
Chenopodium album	+	+	+	+	+	+	+	+	Cosm
Convolvulus fatmensis	—	—	—	—	—	—	+	+	Sa-Si
Corchorus olitorius	—	+	+	+	+	+	+	+	Pan
Corchorus trilocularis	—	+	—	—	+	—	—	+	Pal
Cyperus rotundus	+	+	+	+	+	+	+	+	Pan
Dactyloctenium aegyptium	+	+	+	+	+	+	+	+	Pal
Dichanthium annulatum	+	+	+	+	—	+	+	+	Pal
Digitaria ciliaris	+	+	+	+	+	+	+	+	Pal
Digitaria sanguinalis	+	+	+	+	+	+	+	+	Pal
Dinebra retroflexa	+	+	+	+	—	+	+	+	Pan
Echinochloa colonum	+	+	+	+	+	+	+	+	Pan
Echinochloa crus-galli	+	+	+	+	+	—	—	+	Pan
Echium rauwolfii	—	—	—	—	—	—	+	—	Sa-Si
Eleocharis palustris	+	+	—	—	—	—	—	+	Pan
Eleusine indica	+	+	+	+	+	+	+	+	Pal
Eragrostis cilianensis	+	+	+	+	+	+	+	+	Cosm
Eragrostis pilosa	+	+	+	+	+	+	+	+	Pal
Euphorbia arguta	—	—	—	—	+	+	+	+	M
Euphorbia geniculata	+	+	+	+	+	+	+	—	Pan
Euphorbia hirta	—	—	—	+	—	+	+	—	Pal
Euphorbia hypericifolia	—	—	+	+	+	+	+	—	I-T + Sa-Si + S-Z
Galium tricornutum	+	+	—	—	—	—	—	—	M + I-T
Gynandropsis gynandra	+	+	+	+	+	+	+	—	Pal
Hibiscus trionum	+	—	+	+	+	+	+	+	Pal
Lactuca saligna	+	+	—	—	—	—	—	—	M + I-T + Eur
Laggera aurita	—	—	—	—	—	—	+	—	Pal
Lythrum hyssopifolium	+	—	—	—	+	—	—	+	Pal
Oldenlandia capensis	—	—	—	—	—	—	+	—	Pal
Panicum repens	+	+	+	+	+	+	+	—	Cosm
Paspalum distichum	—	—	+	+	+	+	+	—	Pal
Physalis angulata	—	+	—	—	+	—	+	—	Pan
Portulaca oleracea	+	+	+	+	+	+	+	+	Cosm
Setaria pumila	+	+	+	+	+	+	+	+	Cosm
Setaria verticillata	+	+	+	+	+	—	—	+	Cosm
Setaria viridis	+	+	+	+	+	+	+	+	Cosm
Sida alba	+	+	+	+	+	+	+	+	Pan
Sonchus asper	+	+	+	+	+	+	+	+	Cosm
Tagetes minuta	—	—	+	+	—	—	—	—	Pan
Trianthema portulacastrum	—	+	—	—	—	—	—	—	Pan
Vigna membranacea	—	—	—	—	—	—	+	—	S-Z

All-year weeds

Table 3 gives the chorology of 26 all-year weeds including 10 Mediterranean taxa. Twelve species are of diverse origin including 4 Mediterranean, 3 cosmopolitan and 3 palaeotropical species are recorded from all territories.

Seven mostly Mediterranean species are known from the first and second territories, as well as 4 Sudano-Zambesian and palaeotropical species are known only from the 7th and 8th territories.

Table 3. All-year weeds in the farmlands of Egypt.

Species	Farmland territories								Floristic category
	1	2	3	4	5	6	7	8	
Alternanthera sessilis	–	+	+	+	+	+	–	–	Pan
Boerhavia diandra	–	–	–	–	–	–	+	+	Pal
Convolvulus arvensis	+	+	+	+	+	+	+	+	Pal
Conyza aegyptiaca	–	–	–	–	–	–	+	–	S-Z
Cynanchum acutum	+	+	+	+	+	+	+	+	M + I-T + Eur
Cynodon dactylon	+	+	+	+	+	+	+	+	Pan
Cyperus difformis	–	+	+	+	+	+	–	–	Pan
Eclipta alba	–	+	+	+	+	+	–	–	Pan
Epilobium hirsutum	+	+	+	+	+	+	+	+	Pal
Euphorbia terracina	+	+	–	–	–	–	–	–	M
Launaea tenuiloba	+	+	–	–	–	–	–	–	M
Launaea nudicaulis	+	+	–	+	+	+	+	–	Pal
Lotus creticus	+	+	–	–	–	–	–	–	M
Oryzopis miliacea	+	+	–	–	–	–	–	–	M
Parietaria judaica	+	+	–	–	–	–	–	–	M
Plantago major	+	+	+	+	+	+	+	+	Cosm
Polygonum equisetiforme	+	+	+	+	+	+	+	+	M + I-T
Reichardia tingitana	+	+	+	+	+	+	+	+	M + Sa-Si
Rhynchosia minima	–	–	–	–	–	–	+	–	S-Z
Solanum elaeagnifolium	+	–	–	–	–	–	–	–	Pan
Solanum nigrum	+	+	+	+	+	+	+	+	Cosm
Sonchus oleraceus	+	+	+	+	+	+	+	+	Cosm
Sorghum virgatum	–	+	+	+	+	+	+	+	S-Z
Sorghum sudanense	–	–	–	–	–	–	+	–	S-Z
Trigonella maritima	+	+	–	–	–	–	–	–	M + Sa-Si
Urospermum picroides	+	+	+	+	+	+	+	+	M + I-T

Reference

El Hadidi, M.N., 1993. A historical flora of Egypt, a preliminary survey. Pp. 144-155 in: W.V. Davies & R. Walker (eds.). Biological anthropology and the study of ancient Egypt. British Museum Press.

El Hadidi, M.N. & A.A. Fayed, 1994/1995. Materials for excursion flora of Egypt. Taeckholmia 15: i-vii & 1-2.

Wickens, G.A., 1976. The Flora of Jebel Marra (Sudan Republic) and its geographical affinities. Kew Bulletin Additional Series V. HMSO, London.

Orchidaceae du district phytogéographique du Haut-Katanga (Zaïre): biodiversité et endémisme

D. GEERINCK [1], F. MALAISSE [2] & M. SCHAIJES [3]
[1] Jardin botanique national de Belgique et Université libre de Bruxelles, Rue Charles Pas 4, B-1160 Bruxelles, Belgique.
[2] Laboratoire d'Écologie, Faculté universitaire des Sciences agronomiques, Passage des Déportés 2, B-5030 Gembloux, Belgique.
[3] Rue de la Motte 8A, B-1390 Grez-Doiceau, Belgique.

Résumé
En prolongation de l'information assemblée dans la révision des Orchidaceae de la Flore d'Afrique centrale (Geerinck 1984, Geerinck et al. 1992) et de la publication récente relative aux orchidaceae du Burundi (Arbonnier & Geerinck 1994), l'article discute en premier lieu de la diversité spécifique et infraspécifique pour le district du Haut-Katanga au Zaïre. Des 587 taxons de la Flore d'Afrique centrale, 313 ont été reconnus pour le Haut-Katanga (53 %), dont 121 présentent une distribution limitée au territoire phytogéographique zambézien. Six aréotypes majeurs sont reconnus pour ces derniers dont certains à distribution très limitée. Deux exemples de spéciation sont analysés, à savoir celui du genre *Kryptostoma* (Summerh.) Geerinck et celui de la section Multipartitae Kraenzlin du genre *Habenaria* Willd.

Abstract
Starting from the information gathered in the revision of the Orchidaceae for the Flore d'Afrique centrale (Geerinck 1984, Geerinck et al. 1992) as well as from the recent paper dealing with the Orchids of Burundi (Arbonnier & Geerinck 1994), the present contribution firstly discusses diversity both at specific and infraspecific levels for the Upper Katanga district of Zaire. Amongst the 587 taxa of the Flore d'Afrique centrale, some 313 were recognized as present in Upper Katanga (53%), of which 121 present a distribution restricted to the Zambezian phytogeographical regional centre of endemism. Six main patterns of distribution are recognized, some with very restricted area. Two examples of speciation are analyzed, namely for the genus *Kryptostoma* (Summerh.) Geerinck and for the section Multipartitae Kraenzlin of the genus *Habenaria* Willd.

Key words
Haut-Katanga, Orchidaceae, Phytogeography, Speciation, Zaire.

Abréviations
Ang = Angola, Bot = Botswana, Bur = Burundi, HK = Haut-Katanga,
Mal = Malawi, Moz = Mozambique, Tan = Tanzanie, Tra = Transvaal,
Zam = Zambie, Zim = Zimbabwe.

Introduction

Dans le cadre de la Flore d'Afrique centrale (Zaïre, Rwanda, Burundi) 587 taxons d'Orchidaceae furent reconnus, dont 313 (53%) s'observent dans le district phytogéographique du Haut-Katanga (sud-est du Zaïre). Parmi ces derniers, 121 taxons ont une répartition limitée au territoire zambézien; ce sont essentiellement (92,5%) des taxons terrestres, le reste étant des épiphytes. Quelques biotopes ont privilégié une spéciation endémique plus ou moins importante. Ces taxons endémiques se regroupent selon six aréotypes majeurs qui peuvent être hiérarchisés comme suit.

L.J.G. van der Maesen et al. (eds.), The Biodiversity of African Plants, 795-801.

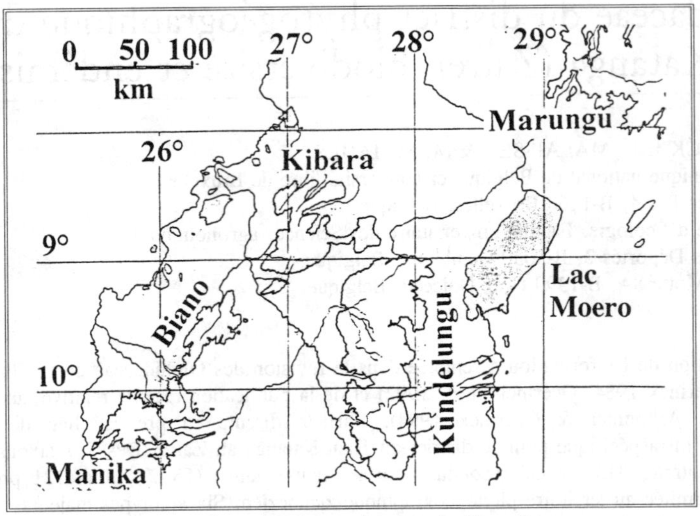

Figure 1. Carte des hauts plateaux du Haut-Katanga.

1. Endémisme des hauts plateaux du Haut-Katanga* (15 taxons)

Il s'agit de taxons dont la distribution est limitée aux zones altitudinales comprises entre 1400 et 2400 m, comprenant pratiquement d'est en ouest et du nord au sud, les massifs montagneux appelés Marungu, Kundelungu, Kibara, Biano et Manika (Figure 1), auxquels il faut adjoindre logiquement celui de Mwinilunga en Zambie. Ce territoire correspond à l'élément floral dilungien (Malaisse 1969). L'endémisme important du haut plateau de Mwinilunga, au niveau des Orchidaceae, mériterait d'être étudié par ailleurs. En effet non compris ceux inféodés uniquement à ce dernier massif, il existe actuellement 15 taxons ne croissant que sur l'un ou plusieurs de ces hauts plateaux:

Cynorkis summerhayesiana Geerinck [Marungu],
Disa danielae Geerinck [Manika],
D. roeperocharoides Kraenzlin, [Kundelungu, Biano, Mwinilunga],
Eulophia coutreziana Geerinck [Biano],
E. sylviae Geerinck [Manika],
Habenaria calvilabris Summerh. var. *crassocalcarata* Geerinck [Manika],
H. libeniana Geerinck [Kundelungu],
H. lisowskiana Geerinck [Marungu],
H. malaisseana Geerinck [Biano],
H. schaijesii Geerinck [Manika],
H. tenuifolia Summerh. var. *bianocensis* Geerinck [Biano],
H. vermeuleniana Geerinck J. Schaijes [Manika],
Kryptostoma geerinckiana Schaijes [Manika],
Platycoryne isoetifolia Cribb [Kundelungu, Manika, Mwinilunga] et
Polystachya billietiana Geerinck [Manika, Biano], apparemment seul épiphyte.

La plupart des taxons énumérés ci-dessus ont été reconnu récemment. Si la Manika (8 taxons) et les Biano (4 taxons) ont été relativement bien prospectés, il n'en est pas de même des Kibara (0 taxon) dont l'exploration méthodique reste à entreprendre.

*Il ne faut pas confondre l'entité phytogéographique du Haut-Katanga avec la division administrative du Shaba autrefois appelée Katanga.

2. Aire des moyens plateaux du Haut-Katanga (12 taxons)

Il s'agit des zones d'altitude moins élevée (1050-1400 m) dont les hauts plateaux émergent. Diverses formations végétales y sont rencontrées dont la plus important à caractère endémique marqué, est la forêt claire (6 espèces sur 13) qui couvrent par ailleurs 85% du territoire; d'autres biotopes intéressants sont notamment les affleurements rocheux et les dembos. Cette distribution correspond à l'élément floral altishabien. Les taxons endémiques sont:

Cynorkis clarae Geerinck [forêts claires],
Eulophia fernandeziana Geerinck [forêts claires à savanes arbustives],
E. lambinoniana Geerinck [forêt claires à affleurements rocheux],
E. letouzeyana Geerinck [savanes steppiques],
E. saintenoyana Geerinck [forêts claires],
E. schaijesii Geerinck [savanes steppiques],
E. vleminckxiana Geerinck [forêts claires],
Habenaria falciloba Summerh. [forêts claires],
H. kolweziensis Geerinck J. Schaijes [transition forêts claires-affleurements rocheux],
H. malacophylla Reichenb.f. var. *shabaensis* Geerinck [affleurements rocheux],
H. robbrechtiana Geerinck [forêts claires] et
H. vandenbergheniana Geerinck [forêts claires].

Les galeries forestières ne présentent que rarement des taxons endémiques; les espèces qui y sont observées, ont généralement une large distribution équatoriale (espèces de liaison et plurirégionales).

3. Aire des moyens plateaux du Haut-Katanga et de Zambie (13 taxons)

Il s'agit de distributions comparables à celles du groupe 2 mais s'étendant à la Zambie. Cet élément floral shabo-zambien a encore été appelé katango-rhodésien. Ces taxons sont:

Bulbophyllum injoloense De Wild. ssp. *pseudoxypterum* (J.J. Vermeulen) J.J.Vermeulen [forêts claires],
Disa aconitoides Sonder var. *dichroa* (Summerh.) Geerinck [forêts claires],
Eulophia angolensis (Reichenb.f.) Summerh. var. *mweruensis* (Cribb) Geerinck [endroits humides],
E. caricifolia (Reichenb.f.) Summerh. var. *antennisepala* (Reichenb.f.) Geerinck [prairies humides],
E. chilangensis Summerh. [forêts claires],
E. monotropis Schlechter [forêts claires],
Habenaria cataphysema Reichenb.f. [savanes],
H. edgari Summerh. [savanes],
H. pilosa Schlechter var. *grandiflora* Geerinck [forêts claires, savanes steppiques, affleurements rocheux],
H. pilosa Schlechter var. *kabompoensis* (Williamson) Geerinck [forêts claires],
H. splendentior Summerh. [forêts claires],
Platycoryne macroceras Summerh. [endroits marécageux] et
Polystachya epiphytica De Wild. [savanes arbustives].

4. Aire nord-orientale du territoire zambézien, du Burundi au Mozambique (25 taxons)

Les espèces citées ici ont une extension correspondant à la partie orientale du "wetter miombo type" de la carte AETFAT (White 1981, 1983). Ce territoire comprend le sud du Burundi (Mosso), le sud-ouest de la Tanzanie, le Haut-Katanga, la Zambie, le Malawi et le nord du Mozambique (Figure 2). Les taxons présents au Shaba et relevant de cet élément floral sont:

Centrostigma clavatum Summerh. (Tan-HK-Zam-Mal)
Disa cryptantha Summerh. (Bur-Tan-HK-Zam)
D. engleriana Kraenzlin (Tan-HK-Zam-Mal)
D. erubescens Rendle var. *carsonii* (N.E. Br.) Geerinck (Bur-Tan-HK-Zam-Mal)
Eulophia bouharmontii Geerinck (Bur-HK)

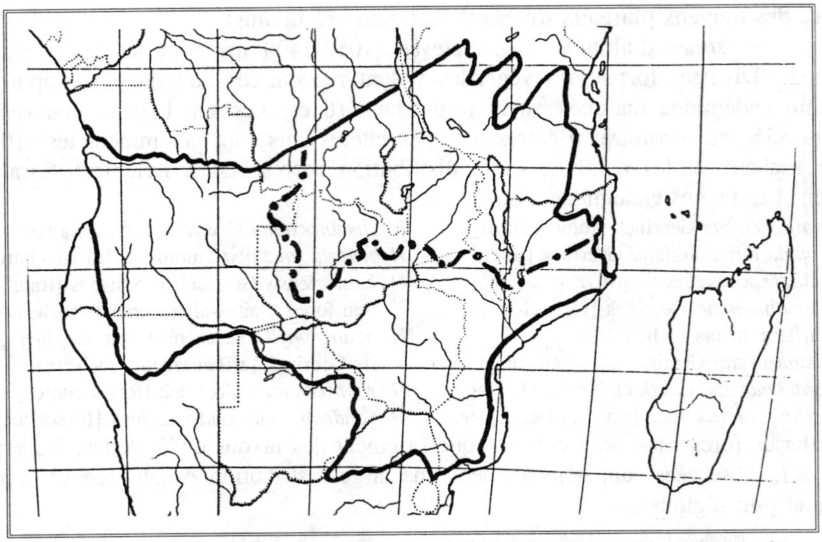

Figure 2. Carte du territoire phytogéographique zambézien (ligne pleine); l'extension du territoire nord-oriental est précisée (ligne pointillée).

Eulophia eylesii Summerh. var. *auquieriana* Geerinck (Bur-HK)
E. hirschbergii Summerh. (Bur-HK-Zam)
E. venulosa Reichenb.f. (Tan-HK
Habenaria galactantha Reichenb.f. (Bur-HK-Zam)
H. gonatosiphon Summerh. (Tan-HK-Zam)
H. kassneriana Kraenzlin (Tan-HK-Zam)
H. kyimbilae Schlechter (Tan-HK-Mal)
H. mirabilis Rolfe (Tan-HK-Zam)
H. parvifolia Summerh. (Tan-HK-Zam-Mal)
H. perpulchra Kraenzlin (Tan-HK-Zam)
H. retinervis Summerh. (Tan-HK-Zam-Mal)
H. rhopalostigma Rolfe ex Kraenzlin (Bur-Tan-HK-Zam)
H. tetraceras Summerh. (Tan-HK-Zam)
Holothrix papillosa Summerh. (Tan-HK-Mal)
Kryptostoma goetzeana (Kraenzlin) Geerinck (Bur-Tan-HK-Zam-Mal)
Nervilia stolziana Schlechter (Zam-Tan-Mal-Moz)
Polystachya brassii Summerh. (HK-Zam-Mal)
P. villosa Rolfe (Tan-HK-Zam-Mal)
Roeperocharis wentzeliana Kraenzlin (Tan-HK-Zam-Mal) et
Satyrium morrumbalense DeWild. (HK-Moz).

5. Aire où dominent [dans la partie nord du territoire zambézien] les forêts claires de type miombo humide (17 taxons)

Cette aire correspond au "wetter miombo type" (White 1981, 1983). Elle se différentie de l'aérotype précédent par le fait qu'elle comprend également une grande partie de l'Angola et notamment tout le nord-est de l'Angola qui correspond grosso modo au secteur Lunda (Symoens & Ohoto 1973). Il convient toutefois de se souvenir que ce dernier territoire a été peu prospecté. Une étude exhaustive des Orchidaceae de l'Angola reste à réaliser. Nous parlerons d'un élément floral nord-zambézien. Nous rattachons à cet aérotype les taxons suivants:

Brachycorythis angolensis (Schlechter) Schlechter (Tan-HK-Zam-Ang)
B. schlechteri Geerinck (Tan-HK-Ang)
Disa verdickii De Wild. (HK-Zam-Ang)
Disa aequiloba Summerh. (Tan-HK-Zam-Ang)
Disperis katangensis Summerh. (Bur-Tan-HK-Zam-Ang)

Eulophia abyssinica Reichenb.f. var. *euantha* (Schlechter) Geerinck (Tan-HK-Zam-Mal-Ang)
E. arenicola Schlechter (HK-Zam-Mal-Moz-Ang)
E. malangana (Reichenb.f.) Summerh. (Tan-HK-Zam-Mal-Ang)
E. rugulosa Summerh. (Tan-HK-Ang)
E. tricristata Schlechter (Tan-HK-Zam-Ang)
Habenaria macroplectron Schlechter (HK-Zam-Ang)
H. mechowii Reichenb.f. (Tan-HK-Zam-Mal-Ang)
H. tenuifolia Summerh. var. *tenuifolia* (Tan-HK-Zam-Mal-Ang)
H. uncicalcar Summerh. (Tan-HK-Zam-Mal-Ang)
H. verdickii (De Wild.) Schlechter var. *verdickii* (HK-Zam-Mal-Ang)
Polystachya zambesiaca Rolfe var. *malilaensis* (Schlechter) Geerinck (Tan-HK-Zam-Mal-Ang) et
Satyrium riparium Reichenb.f (Tan-HK-Zam-Mal-Ang).

6. Distribution (omni-)zambézienne (18 taxons)

Ce territoire correspond au centre régional d'endémisme zambézien. Il inclut notamment l'aire du "drier miombo type", qui était exclue de l'aérotype précédent, en s'étendant au Botswana, au Zimbabwe et au nord du Transvaal. Les Orchidaceae appartenant à l'élément floral zambézien sont:

Centrostigma occultans (Reichenb.f;) Schlechter (Tan-HK-Zam-Mal-Ang-Zim-Tra)
Disa ornithantha Schlechter (Tan-HK-Zam-Mal-Ang-Zim)
D. zombica N.E. Br. (Bur-Tan-HK-Zam-Mal-Moz-Zim)
Eulophia callichroma Reichenb.f. (Tan-HK-Zam-Mal-Moz-Zim)
E. coeloglossa Schlechter (Bur-Tan-HK-Zam-Mal-Ang-Zim)
E. holubii Rolfe (HK-Ang-Bot-Zim)
E. katangensis De Wild. (Tan-HK-Zam-Mal-Zim)
E. longisepala Rendle (Bur-Tan-HK-Zam-Mal-Moz-Zim-Tra)
E. obscura Cribb (HK-Zam-Mal-Zim)
E. tanganyikensis Rolfe (Bur-Tan-HK-Zam-Mal-Zim)
Habenaria anaphysema Reichenb.f.(Tan-HK-Zam-Ang-Zim)
H. calvilabris Summerh. var. *calvilabris* (Bur-HK-Zam-Zim)
H. disparilis Summerh. (HK-Zam-Mal-Zim)
H. magnirostris Summerh. (Tan-HK-Zam-Zim)
H. mira Summerh. (HK-Zim)
H. nicholsonii Rolfe var. *debeerstiana* (Kraenzlin) Geerinck (Bur-HK-Zam-Mal-Zim)
H. trachypetala Kraenzlin (Tan-HK-Zam-Mal-Zim)
Platycoryne protearum (Reichenb.f.) Rolfe (Tan-HK-Zam-Mal-Ang-Zim) et
Polystachya zambesiaca Rolfe var. *zambesiaca* (Tan-HK-Zam-Mal-Moz-Zim-Tra).

Les six aréotypes majeurs passés en revue ci-dessus totalisent cent espèces soit 82,5% des 121 taxons observés uniquement en territoire zambézien. Les 192 espèces restantes observées au Haut-Katanga présentent des distributions débordant le territoire phytogéographique zambézien.

Spéciation

Le genre *Kryptostoma* (Summerh.) Geerinck [= *Habenaria* Willd. sect. Kryptostoma Summerh.] fut établi en 1982. Ce taxon très particulier ne contient que trois espèces dont *K. geerinckiana* Schaijes, une endémique du plateau de la Manika, qui pourrait dériver de *K. goetzeana* (Kraenzlin) Geerinck à distribution plus étendue [Bur-Tan-HK-Zam-Mal], notamment par une réduction des dimensions de l'éperon du labelle.

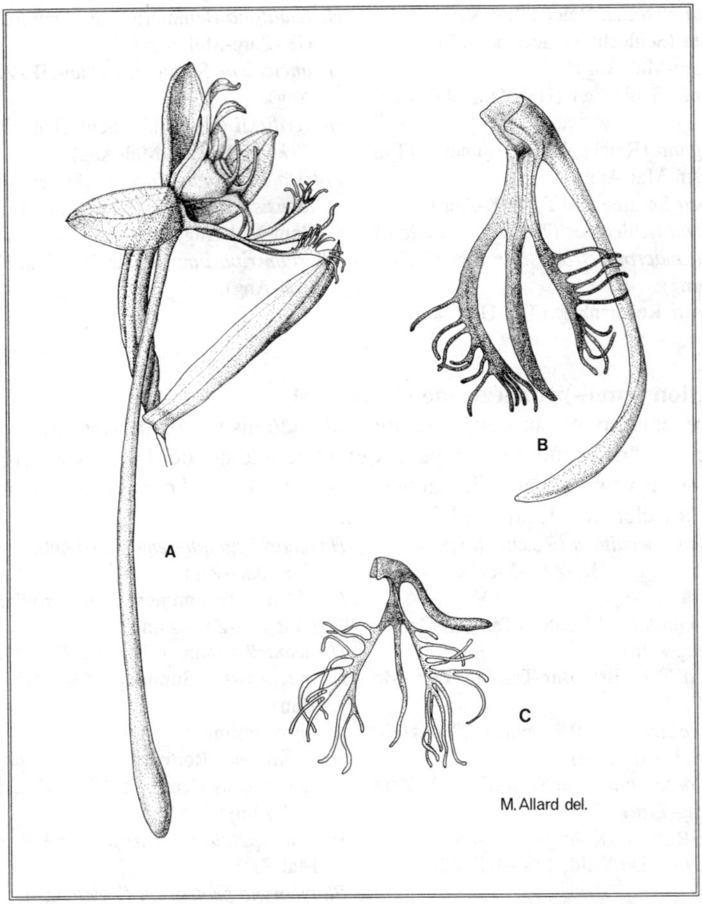

Figure 3. Fleurs d'*Habenaria* spp. (x 0.8): A. *H. schaijesii* Geerinck; B. *H. splendentior* Summerh.; C. *H. praestans* Rendle. D'après Flore d'Afrique centrale, Orchid. 1: 97 (1982).

Un autre cas de différenciation spécifique assez intéressant s'observe au sein de la section Multipartitae Kraenzlin du genre *Habenaria* Willd. Cette section comprend onze espèces distribuées en Afrique centrale et orientale; il s'agit surtout d'espèces montagnardes. *H. praestans* Rendle est la plus étendue (Uganda, Rwanda, Burundi, Tanzanie, Zaïre, Zambie, Malawi, Zimbabwe) et présente un labelle à segments latéraux laciniés sur toute leur longueur; *H. splendentior* Summerh., limité aux forêts claires du Haut-Katanga et du nord-ouest de la Zambie, a son labelle lacinié sur une longueur moindre et enfin *H. schaijesii* Geerinck, endémique du plateau de la Manika, n'est lacinié que vers le sommet des segments latéraux. Cette réduction relative de la partie laciniée de l'épichile est en corrélation avec une diminution générale du nombre et de la longueur des franges ainsi qu'un allongement sensible de la taille de l'éperon (Tableau 1 et Figure 3).

Tableau 1. Comparaison de dimensions des fleurs de trois *Habenaria*.
1: *H. praestans*; 2: *H. splendentior*; 3: *H. schaijesii*. Voir Figure 3.

Espèces	1	2	3
Longueur de l'éperon (cm)	1,5-2,5	6-7,5	8-10,5
Longueur de la partie non laciniée de l'épichile (mm)	7-10	22-30	17-20
Longueur de la partie laciniée de l'épichile (mm)	20-25	20-25	8-12
Rapport de la longueur de la partie non laciniée à la longueur totale de l'épichile (%)	70-73	45-48	32-37,5

Bibliographie

Arbonnier, M. & D. Geerinck, 1993. Contribution à l'étude des Orchidaceae du Burundi (1e part.). Belg. Journ. Bot. 126: 253-261.

Geerinck, D., 1984. Flore d'Afrique centrale (Zaïre-Rwanda-Burundi). Orchidaceae (1e part.): 1-296. Jardin botanique national de Belgique, Meise.

Geerinck, D., J.J. Vermeulen & B. Pettersson, collab. M. Schaijes, 1992. Flore d'Afrique centrale. Orchidaceae (2e part.): 297-780. Jardin botanique national de Belgique, Meise.

Malaisse, F., 1969. Note à propos de la flore des Kundelungu (Haut-Katanga, Rép. dém. Congo). Trav. Serv. Sylv. Pisc. Univ. off. Congo (Lubumbashi) 2: 1-10.

Symoens, J-J. & E. Ohoto, 1973. Les éléments phytogéographiques de la flore macro-phytique aquatique et semi-aquatique du Haut-Katanga. Verh. int. verein. theor. angew. Limnol. 18 (part 3): 1385 -1394.

White, F., 1981. Vegetation map of Africa (scale 1:5.000,000), UNESCO/AETFAT/UNSO. Oxford Univ. Press.

White, F., 1983. The vegetation of Africa. A descriptive memoir to accompany the UNESCO /AETFAT/UNSO vegetation map of Africa: 356 p. Paris.

Flora of Ethiopia and Eritrea

I. HEDBERG
Department of Systematic Botany, Uppsala University, Villavägen 6, S-752 36 Uppsala, Sweden.

Abstract
After several years of planning and fund-hunting the Flora of Ethiopia project was launched in 1980, funded by SAREC (Swedish Agency for Research Cooperation with Developing Countries) with the aim of producing a Flora of Ethiopia, training of Ethiopian students to Ph.D. in Systematic Botany, and strengthening the National Herbarium in Addis Ababa and its library. Four Ethiopian botanists have now received their Ph.D. in Uppsala and a considerable number of collections has been added in the National Herbarium. Vol. 3 of the Flora was published in 1989 and thanks to the combined efforts of many botanists accounts for most of the remaining families for Flora of Ethiopia are now available in a more or less finalized version.

Key words
Ethiopia, Flora.

The Flora of Ethiopia project, funded by SAREC (Swedish Agency for Research Cooperation with Developing Countries) was launched in 1980 after several years of planning and fund-hunting. A presentation of the project together with a summary of the history of botanical exploration of Ethiopia was published by O. Hedberg (1983) and short progress reports were given by I. Hedberg (1986, 1990). Hence only a more general overview will be given here.

To start planning for a new Flora during a period when molecular biology and biotechnology began their march over the world might seem rather optimistic, not least because those subjects were bound to get the lion's share of funding. The long period between the beginning of fund-hunting in late 1967 and the launching of the project in July 1980 demonstrates the difficulties in making donor agencies realize the need also for traditional basic research in botany, especially in the tropics. SAREC must therefore be given full credit for its long term support, thereby making it possible to produce a workable instrument for sustainable use of the plant resources of Ethiopia as well as for other branches of pure and applied sciences, such as ecology, plant breeding, medicine, and pest control.

As foreseen the work has not been without difficulties, but those have so far been caused mainly by factors outside the project. The move of the head office from Addis to Asmara University after a few years not only caused administrative difficulties but also burdened the project with unforeseen extra expenses, for instance because of the necessity to buy adequate equipment for the last mentioned university. Periodical political instability has fortunately not caused major problems but has naturally influenced the possibilities for undisturbed work at the National Herbarium.

The recent political development, resulting in Eritrea becoming an independent country, has also caused a change of the name to "Flora of Ethiopia and Eritrea".

According to plan, Vol. 3 (Pittosporaceae to Araliaceae) would be the first to be published since it contains inter alia the very important legume family. Since by the end

L.J.G. van der Maesen et al. (eds.), The Biodiversity of African Plants, 802-804.

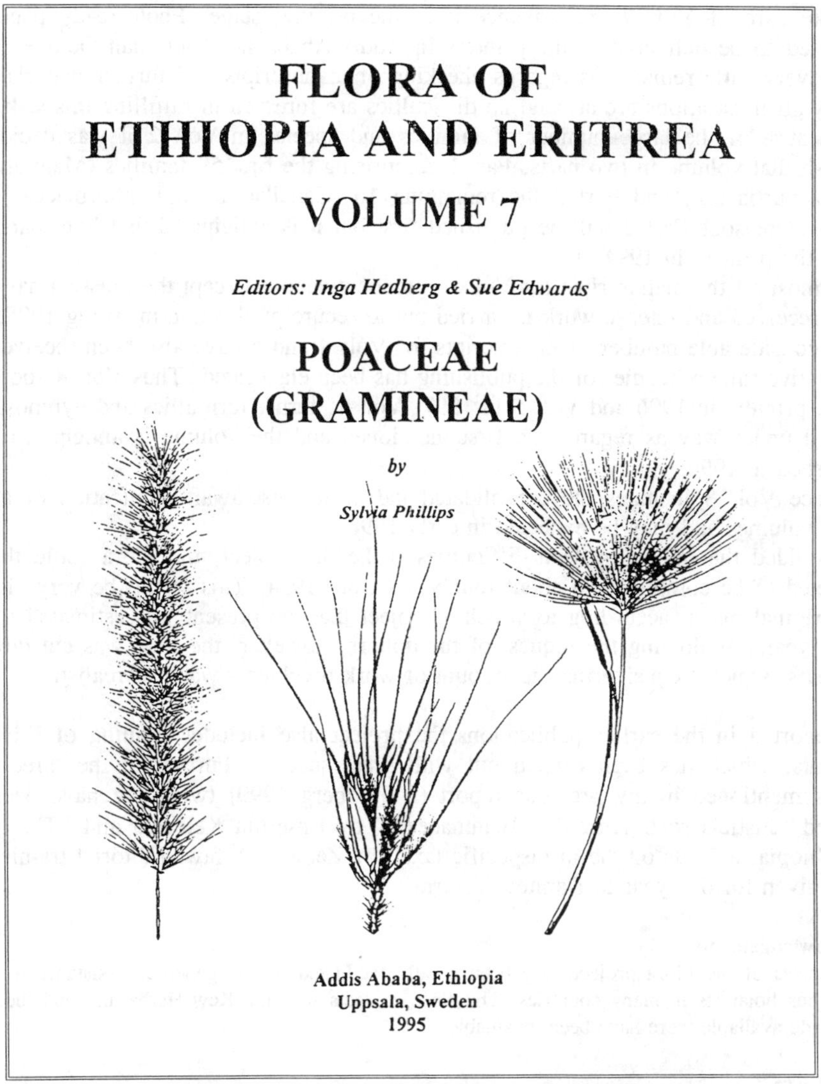

Figure 1. Flora of Ethiopia and Eritrea, Volume 7, Poaceae (Gramineae).

of 1993 only that volume had been published an Editorial Board meeting was organized in Addis Ababa in April 1994. On that occasion the situation concerning all the remaining volumes was reviewed.

The reports given by the editors on the number of manuscripts received made it evident that special arrangements had to be taken in order to speed up the publication, since all manuscripts both for Vol. 2 (Magnoliaceae to Euphorbiaceae) and Vol. 7 (Poaceae) were at hand. It was thus decided to aim at a more or less simultaneous publication of those volumes. In order to attain this goal the Uppsala office was asked to take care of Vol. 7 and advance it to the printing stage. Photo-ready pages are expected to be delivered to the printers in Addis Ababa not later than October 1994. Since very little remains as regards checking of manuscripts and further material, and nearly all illustrations are at hand no difficulties are foreseen in fulfilling this task.

Because of the large number of families and species in Vol. 2 it was decided to publish that volume in two parts, Part 1 comprising the first 59 families (Magnoliaceae to Flacourtiaceae) and Part 2 the remaining 16 (Canellaceae to Euphorbiaceae). For practical reasons Part 2 will be published first but it is anticipated that both parts will go to the printers in 1994.

Almost all the manuscripts for Vol. 6 (monocotyledons except the grasses) have also been received and intense work is carried out to secure publication in spring 1995.

A considerable number of manuscripts for Vols 4 and 5 have also been received and a tentative time schedule for the publishing has been elaborated. Thus Vol. 4 should go to the printers in 1996 and Vol. 5 in 1997. Vol. 1, ferns, fern allies and gymnosperms is well under way as regards the first mentioned and the volume is anticipated to be published in 1996.

Since Vol. 8 includes the consolidated index, it must await publication of all the other volumes and may be expected in early 1998.

Provided that no unforeseen difficulties strike the project, the Flora could thus be expected to be completed in about four years from 1994. This would be very close to the original plans, according to which a project like the present was estimated to take c. 20 years. Following the request of the donors, however, the time was cut down to 15 years, which - considering the amount of work involved - was less realistic.

As reported in the earlier publications the project also includes training of Ethiopian students, which has been carried out with great success. Further to the three Ph.D theses mentioned in my previous report (I. Hedberg 1990) two more have been published "Justicia sect. Ansellia (Acanthaceae)" by Ensermu Kelbessa and "The barley of Ethiopia: a focus on the infraspecific taxa" by Zemede Asfaw. Editorial training has been given for one year to Damtew Teferra.

Acknowledgements

All aspects of the Flora project have been greatly facilitated by the generous assistance given by numerous botanists in many countries. The close contacts with the Kew Herbarium and the facilities made available there have been invaluable.

References

Hedberg, O., 1983. Ethiopian flora project. In: D.J.B. Killick (ed.), Proceedings of the 10th AETFAT congress, Pretoria, 1982. Bothalia 14 (3 & 4): 571-574.

Hedberg, I., 1986. The Flora of Ethiopia: A progress report. In: I. Hedberg (ed.), Research on the Ethiopian Flora. Symbolae Botanicae Upsalienses 26(2): 17-18.

Hedberg, I., 1990. The Ethiopian Flora Project. In: H-D. Ihlenfeldt (ed.), Proceedings of the Twelfth Plenary Meeting of AETFAT. Mitt. Inst. Allg. Bot. Hamburg 23b: 745-746.

Floristics of Transkei sandstone forests (South Africa)

T.V. JACOBS
Department of Botany, University of Transkei,
P. Bag X1, UNITRA, Umtata, Transkei, South Africa.

Abstract
Early botanists exploring the South African flora have been highlighted. It is hypothesized that the sandstone regions of Transkei acted as edaphic islands, trapping the c. 60 now (near-)endemic species listed here, relicts of the Cretaceous Flora, linking the Cape flora with disjunct Transkei.

Key words
Cape flora links, Floristic exploration, Pioneer botanists, Transkei flora.

Introduction

Territories dealt with in the Flora of Southern Africa include the Republic of South Africa, Southwest Africa/Namibia, Swaziland and Lesotho, at latitudes between 17° and 35° South and longitudes between 12° and 37° East. Over 24,000 species of plants have been recorded from this region. Southern Africa has the most species-rich temperate flora in the world, making up about 9% of all known flowering plants of the world (Gibbs Russell et al. 1987). Apart from richness of species, South African flora is unique because of its high degree of endemism and the disjunct distribution of many of its species (Taylor 1978). As a phytogeographical unit, the Cape flora by itself is so distinct that it merited recognition as a floristic kingdom of its own (Good 1974).

Pioneer botanists interested in South African flora

The richness of the South African vegetation attracted the attention of explorers and botanists from the 16th century onwards. The first plant from South Africa to be described and illustrated appears to have been *Protea nerifolia*, in a work by Clusius published in 1605 (Werger 1978). Some of the early botanists who took interest in South African flora were Peter Jonas Bergius (1730-1790), Carl Friedrich Drege (1791-1867), Carl Gustaf Ekeberg (1716-1784), William Henry Harvey (1811-1866), Sir Joseph Dalton Hooker (1817-1911), Carl Linnaeus (1707-1778), Francis Mason (1741-1805), Andreas Franz Wilhelm Schimper (1856-1901), Thomas Robertson Sim (1858-1938), Anders Sparrman (1748-1820), Carl Petter Thunberg (1743-1828) and Carl Ludwig Philip Zeyher (1799-1858).

Early botanical exploration of Transkei

The first botanical expedition to visit Transkei was that of the Drege brothers, Johann and Carl, in 1832. They collected nearly 800 plants. The genus *Dregea* (Asclepiadaceae) commemorates Johann Frans Drege, one of the most successful 19th century explorers, who is supposed to have collected over 200,000 specimens (Jackson 1990).

Among the collectors who settled in Transkei, mention should be made of Leopold Richard Baur (1825-1889), who is remembered in *Erica baurii* H. Bol. (Ericaceae), as

L.J.G. van der Maesen et al. (eds.), The Biodiversity of African Plants, 805-808.

well as *Herschelia baurii* (H. Bol.) Kraenzlin (Orchidaceae) and *Rhodohypoxis baurii* (Bak.) Nel (Hypoxidaceae); James Bowker (1822-1900), a government official who is commemorated in *Bauhinia bowkeri* Harv. (Caesalpinioideae); Henry George Flanagan (1661-1719), who collected nearly 2,500 specimens, is commemorated in a number of species like *Gladiolus flanaganii* Bak. (Iridaceae), *Protea flanaganii* Phillips (Proteaceae), *Selago flanaganii* Rolfe (Selaginaceae), *Raphionacme flanaganii* Schltr. (Periplocaceae) etc.; his wife is commemorated in *Impatiens flanaganae* Hemsl., which she discovered at Port St. Johns (Gunn & Codd 1981); Marianne Harriet Mason (1845-1932), whose valuable paintings of Transkei plants were presented to Kew, is remembered in *Indigofera masonae* N.E. Br. (Papilionoideae), *Watsonia masonae* L. Bol. and *Crocosmia masonorum* (L. Bol.) N.E. Br. (Iridaceae); Thomas Robertson Sim (1858-1938), who published two papers on the forest flora of Transkei, is commemorated in *Exoecaria simii* (Kuntze) Pax (Myrtaceae) and in *Rhus simii* Schonl. (Anacardiaceae); William Tyson (1851-1920), after whom the genus *Tysonia* is named. Some of the species commemorating him are *Stachys tysonii* Skan (Labiatae), *Selago tysonii* Rolfe (Selaginaceae) and *Xysmalobium tysonianum* (Schltr.) N.E. Br. (Asclepiadaceae).

Analysis of the Transkei flora

Angiosperms and gymnosperms account for 160 families, 939 genera and 3331 species in Transkei. The largest family, Compositae, has 99 genera and 427 species, followed by Gramineae and Papilionoideae. The ratio of genus to species in Transkei is 1:3.55. For southern Africa the ratio is 1:7.7 (Johnson 1985). Bond & Goldblatt (1984) opine that a low genus/species ration indicates a flora that did not develop in isolation.

Link with the Cape flora

The richness of the Transkei flora is partly due to the migration of sub-tropical elements southwards. The evidence for this lies in the fact that the greater percentage of the 399 tree species recorded is found in the northern and south-eastern (Pondo-land) section of Transkei. Furthermore, Transkei also contains a temperate element which extends along the coast on sandstone outcrops. The following Cape genera are found in Transkei: *Erica* (15 spp.), *Leptocarpus* (1 sp.), *Leucadendron* (5 spp.), *Metalasia* (1 sp.), *Muraltia* (4 spp.), *Restio* (1 sp.), *Thamnochortus* (1 sp.) and *Watsonia* (3 spp.) (Johnson 1985).

Transkei endemics

Transkei has a rich endemic or near-endemic flora of around 60 species in 34 families: Amaryllidaceae: *Nerine gibsonii* Douglas, *N. masonorum* L. Bol.; Anacardiaceae: *Rhus acocksii* Moffett; Balsaminaceae: *Impatiens flanaganae* Hemsl.; Begoniaceae: *Begonia dregei* Otto & Dietr., *B. homonyma* Steud.; Bignoniaceae: *Podranea ricasoliana* (Tanf.) Sprague; Bruniaceae: *Raspalia trigyna* (Schltr.) Duemmer; Caesalpinioideae: *Bauhinia bowkeri* Harv., *Umtiza listeriana* Sim; Celastraceae: *Catha abbottii* Van Wyk & Prins, *Maytenus abbottii* Van Wyk, *M. bachmannii* (Loes.) Marais, *M. oleosa* Van Wyk & Archer, *Pseudosalacia streyi* Codd, *Putterlickia retrospinosa* Van Wyk & Mostert; Compositae: *Helichrysum pannosum* DC., *H. populifolium* DC., *Tarchonanthus trilobus* DC.; Crassulaceae: *Crassula streyi* Toelken, *C. planifolia* Schonl.; Ericaceae: *Erica baurii* H. Bol.; Euphorbiaceae: *Phyllanthus cedrelifolius* Verdoorn;

Flacourtiaceae: *Pseudoscolopia polyantha* Gilg; Geraniaceae: *Geranium subglabrum* Hilliard & Burtt; Iridaceae: *Aristea platycaulis* Bak., *Crocosmia masonorum* (L. Bol.) N.E. Br.; Labiatae: *Plectranthus ernstii* Codd, *P. hilliardiae* Codd, *P. oertendahlii* Th.Fr.Jr., *P. oribiensis* Codd; Lauraceae: *Cryptocarya latifolia* Sond., *C. wyliei* Stapf., *Dahlgrenodendron natalense* J.H. Ross; Liliaceae: *Aloe reynoldsii* Letty, *Kniphofia drepanophylla* Bak., *Lachenalia convallarioides* Bak.; Lythraceae: *Rhynchocalyx lawsonioides* Oliv.; Meliaceae: *Turraea pulchella* (Harms) Penn.; Melastomataceae: *Memecylon bachmannii* Engl.; Moraceae: *Ficus bizanae* Hutch. & Burtt- Davy; Myrtaceae: *Eugenia erythrophylla* Strey, *E. simii* Duemmer, *E. umtamvuensis* Van Wyk; *Syzygium pondoense* Engl.; Orchidaceae: *Corymborchis corymbis* Thouars, *Disa galpinii* Rolfe, *D. montana* Sond., *D. scullyi* H. Bol.; Palmae: *Jubaeopsis caffra* Becc.; Papilionoideae: *Tephrosia pondensis* (Codd) Schrire; Polygalaceae: *Polygala esterae* Chod.; Rhamnaceae: *Colubrina nicholsonii* Van Wyk & Schrire, *Phylica tysoni* Pillans; Rubiaceae: *Tricalysia africana* (Sim) Robbr.; Sapotaceae: *Manilkara nicholsonii* Van Wyk; Thymelaeacae: *Struthiola pondoensis* Gilg ex C.H. Wr.; Tiliaceae: *Grewia pondoensis* Burret; Violacae: *Rinorea domatiosa* Van Wyk, Stangeriaceae: *Stangeria eriopus* (Kunze) Baillon; Zamiaceae: *Encephalartos altensteinii* Lehm.

Hypothesis

The Transkei flora exhibits sub-tropical as well as temperate elements, which are commonly referred to as afromontane elements found on sandstone outcrops, along the escarpment and the coast. Nine Cape genera are represented among the Transkei flora, at a disjunction of nearly 1,000 km. Furthermore, Transkei flora contains around sixty endemic or near-endemic species. They show all the features of palaeoendemics. They can be considered relicts derived from the Cretaceous/Tertiary flora which prevailed in this part of Gondwanaland. The hypothesis is that the sandstone regions act as edaphic islands on which some species are trapped and conserved. It is also hypothesized that a direct sandstone link between Transkei and the Cape flora was provided by the Falkland Plateau which abutted the African plate before the break-up of Gondwanaland (Van Wyk 1992).

Gondwanaland - an elucidation

The presence of tillites (glacial deposits) in the continents of the southern hemisphere and the distribution of flora elements such as the seed fern, *Glossopteris*, restricted to these regions only, prompted the Austrian geologist, Eduard Suess to suggest the concept of a supercontinent, Gondwanaland, which included South America, Africa, peninsular India, Australia and Antarctica. The rock strata which contain the evidence for this are called the Karoo system in South Africa, the Gondwana system in India and the Santa Catharina system in South America.

Geologic regimes that led to the break-up of Gondwanaland

During the Precambrian the region of the Gondwana supercontinent was formed of numerous primitive cratons welding together along orogenic lines, with emission of juvenile basalt from below the crust. This annealing was supposed to have taken place at least five times, the last having been around 100 million years ago (Clifford 1970, Haughton 1963).

The second phase took place during the late Palaeozoic to mid-Mesozoic. This was marked by broad subsiding basins in which were laid layers of fresh water sandstones and fossil floras and faunas. In South Africa it is called the Karoo system with its counterparts in India and South America. Fossil record of this region indicates a progressive change of climate through polar glacial to warm desert. It is believed that Gondwanaland drifted across the South Pole to subtropical regions (King 1958).

The third phase occurred in a dramatic fashion. Gondwanaland was in turmoil. Huge emissions of basalt poured over the surface from every quarter of the supercontinent. According to Du Toit (1954), the volume of basalt involved in the South African region alone may have been over 200,000 cubic kilometres. This was followed by the fragmentation of Gondwana into the present southern continents, followed by the centrifugal drift of these fragments into their present positions (King 1978).

Conclusion

Following the disruption of Gondwanaland, when the Cape floristic region crossed latitudes from temperate to tropical regions, drastic climatic changes led to the extinction of many taxa that were trapped on sandstone substrates. A few of these managed to survive in isolated refuges. Hence it is suggested that the endemic flora of Transkei contains relict elements of the more tropical Cretaceous/Tertiary sandstone flora. A link between the Transkei flora and the Cape flora was probably provided by the Falkland Plateau which abutted the African plate before the disruption of Gondwanaland.

References

Bond, P. & P. Goldblatt, 1984. Plants of the Cape flora. S. Afr. Journal of Botany, Suppl. Vol. 13. NBI, Pretoria.

Clifford, T.N., 1970. The structural framework of Africa. In: T.N. Clifford & I.G. Gass (eds.), African magmatism and tectonics. Oliver & Boyd, Edinburgh.

Du Toit, A.L., 1954. Geology of South Africa. 3rd ed. Oliver & Boyd, Edinburgh.

Gibbs Russell, G.E. et al., 1987. List of species of southern African plants. 4th ed. Longmans, London.

Good, R., 1974. The Geography of the Flowering Plants, 4th ed. Longmans, London.

Gunn, M. & L.E. Codd, 1981. Botanical Exploration of southern Africa. Balkema, Cape Town.

Haugton, S.H., 1963. The stratigraphic history of Africa South of Sahara. Oliver & Boyd, London.

Jackson, W.P.U., 1990. Origins and meanings of names of South African plant genera. UCT, Cape Town.

Johnson, C.T., 1985. Survey of the wild flowers of Transkei. University of Transkei, Umtata.

King, L.C., 1958. Basic paleogeography of Gondwanaland during the late Palaeozoic and Mesozoic Eras. Geological Society of London Quarterly Journal 114: 44-70.

King, L.C., 1978. The geomorphology of central and southern Africa. In: M.J.A. Werger (ed.), Biogeography and ecology of southern Africa. W. Junk, The Hague.

Taylor, H.C., 1978. Capensis. In: M.J.A. Werger (ed.), Biogeography and ecology of southern Africa. W. Junk, The Hague.

Van Wyk, A.E., 1992. The sandstone regions of Natal and Pondoland: remarkable centres of endemism. In: K.H. Cooper & J.P.J. Swart (eds), Transkei forest survey. Transkei Forestry Council, Durban.

Werger, M.J.A., 1987. Biogeography and ecology of Southern Africa. W. Junk, The Hague.

Seed and seedling morphology of *Entada*, a study of Zambian species

S. LUNGU & A. CULHAM
Department of Botany, School of Plant Sciences, University of Reading, Whiteknights, P.O. Box 221, Reading RG6 2AS, England.

Abstract
A study of the seeds of *Entada* Adans. revealed three distinct species and one species group. In contrast, the seedlings of the genus could be identified to species by first leaf morphology. The two sections of *Entada* growing in Zambia were shown to have distinct patterns of germination: sect. *Entada* has epigeal germination while sect. *Neoentada* Brenan has hypogeal germination. Means of identification of these species are described and the taxonomic implications discussed.

Résumé
L'étude des semences de l'*Entada* Adans. a permis d'identifier trois espèces distinctes et un groupe d'espèces. Au contraire, les plantules du genre pouvaient être associées aux espèces sur la base de la morphologie de la première feuille. Il fut démonstré que les deux sect. d'*Entada* existant en Zambie présentaient des traits distincts de germination: la sect. *Entada* possedant une germination épigée tandis que la sect. *Neoentada* Brenan présentait une germination hypogée. Les méthodes d'identification de ces espèces sont décrites et les implications taxonomiques discutées.

Key words
Entada, Identification, Seed morphology, Seedling morphology.

Introduction

Seed morphology is considered useful in the identification and classification of many species (Corner 1976, Sporne 1974). Seed characters are known to be linked to microhabitat requirements for germination but are in general considered to be highly conservative characters. Seedlings have mostly been classified into two main groups epigeal and hypogeal, based on the position of the cotyledons and whether or not the testa and fruit wall are shed (De Vogel 1980). Differences in seedling morphology have sometimes been a basis for re-arrangement of classifications (De Vogel 1980). It had been assumed, sometimes tacitly, that, in natural genera, the seedlings must be similar.

This study aims to produce keys to identify predominantly Zambian *Entada* species that can be used by a wide range of people. The construction of keys for species at the seed and seedling stage will facilitate identification at this early stage.

Materials and method

Seeds of Zambian species were mostly collected during 1992 (Lungu 1993), further species were obtained from other sources (Table 1). They were observed under low magnification with the light microscope and photographed. For observation under the scanning electron microscope, the seeds were stuck to stubs with silver paint. They were left for 10 minutes to dry, then coated to 30 nm with gold, and were scanned on a JOEL JSM 35R Scanning Electron Microscope (SEM).

L.J.G. van der Maesen et al. (eds.), The Biodiversity of African Plants, 809-814.

Table 1. Seed characteristics of various *Entada* species used in this study.

Species, Collectors number, Date, Site	Weight	Length	Width	Thickness	Sample size (n)
	(mean (g, mm) ± σn-1)				
Entada abyssinica Steud. ex A. Rich.					
SL 004 (RNG), 21.08.92, Kateshi junction, Kasama	0.25 ±0.03	11.9 ±0.51	8.9 ±0.43	3.2 ±0.12	10
SL 005 (RNG), 25.08.92, Mbala road, 7 km from Kasama	0.32 ±0.04	12.2 ±0.51	9.7 ±0.34	3.3 ±0.11	10
SL 008 (RNG), 05.09.92, Phillipo P. School, Mporokoso	0.21 ±0.03	10.3 ±0.42	7.9 ±0.29	3.3 ±0.18	10
SL 012 (RNG), 15.10.92, Kalambo River, Mbala	0.33 ±0.04	13.1 ±0.40	9.2 ±0.33	3.6 ±0.14	10
E. bacillaris F. White					
SL 021 (RNG), 17.10.92, Kendulo farm, Mbala					
E. camerunensis J. Villiers					
SL 022 (RNG), 17.10.92, Kapata village, Mpulungu	0.76 ±0.04	20.2 ±0.98	11.6 ±0.55	4.2 ±0.18	10
E. chrysostachys (Benth.) Drake					
SL 010 (RNG), 03.10.92, V. Chibale, C. Sandwe Petauke	0.53 ±0.04	17.3 ±0.48	11.8 ±0.24	3.9 ±0.28	10
SL 031 (RNG)		14.1	10.5	3.7	1
E. dolichorrhachis Brenan					
SL 036 (RNG), 15.04.94, Misamfu, Kasama	0.10 ±0.00	8.8 ±0.21	6.6 ±1.16	2.6 ±0.10	3
E. gigas (L.) Fawc. & Rendle					
SL 028 (RNG), 04.12.92, Makeleti River, Mwinilunga	15.20 ±1.21	43.7 ±2.48	41.5 ±1.65	18.8 ±2.05	8
E. leptostachya Harms					
G&H 24820 (UPS), 16.06.83, Garba Harre, Somalia	0.25 ±0.07	11.7 ±1.77	10.1 ±0.78	3.6 ±0.14	2
E. mossambicensis Torre					
Torre 4750A (BM), 03.11.42, Nampula, Mozambique		10.4	9.2	2.3	1
E. rheedii Spreng.					
Hyland s.n. (RNG), 1993, Queensland, Australia	24.20 ±4.77	48.8 ±3.52	42.3 ±1.35	20.2 ±2.03	7
Knight s.n. (RNG)	24.00 ±1.45	49.0 ±1.41	41.0 ±1.41	21.8 ±0.28	2
E. spinescens Brenan					
Burtt 3611 (BM), 24.04.32, Mpwapwa, Dodoma, Tanzania		10.9	6.9	3.5	1

For germination tests, one set of seeds was soaked in water at room temperature for 24 hours and a second set of seeds was put in hot (95°C) water which gradually cooled while the soaking continued for 24 hours. They were then transferred to plastic boxes with moist tissue paper. They were observed on a daily basis to check the germination. *Entada gigas* and *E. rheedii* did not germinate in either case, therefore were scarified and acid-treated for 10 minutes. The seeds were placed in hot water and left to cool overnight, but for these gigantic seeds, soaking was done for two weeks. Except for *E. gigas* (3 seeds) and *E. rheedii* (3 seeds), 20 seeds were soaked in each experiment and observed for germination. When a rootlet appeared, the seeds were planted at 5 mm depth in pots filled with John Innes No. 1 compost with extra sand and their emergence and subsequent growth observed.

Results

Seed length, width and thickness measurements are shown in Table 1 and light micrographs in Figure 1. Most seeds have an oval shape with a length to width ratio of 1.3:1 with the exception of SL 028 which had a ratio of 1:1. A narrow shape is observed in SL 022 with L/W ratio of 1.7:1. Germination results are shown in Table 2.

Table 2. Germination and seedling morphology of *Entada* seeds soaked in cold and hot water.

Species, collectors number, germination type, first leaf arrangement	Cold water		Hot water	
	Days to emergence	Final germ. %	Days to emergence	Final germ. %
Entada abyssinica				
SL 004, hypogeal, alternate	—	0	5	65
SL 005, hypogeal, alternate	7	25	10	80
SL 008, hypogeal, alternate	7	10	—	0
SL 012, hypogeal, alternate	9	10	—	0
E. camerunensis				
SL 022, epigeal, opposite	12	10	10	85
E. chrysostachys				
SL 010, epigeal, opposite	7	15	8	75
E. gigas				
SL 028, not available	—	0	—	0

Key to seeds of *Entada*

Seeds somewhat compressed, elliptic, oval, subcircular or kidney-shaped in outline, about 10-50 x 8-50 x 2.7-22 mm, hard, brown or dark brown, with a small scar (hilum) on edge, with or without areole.

1 a. Seed small, < 20.3 x 11.7 x 4.5 mm 2
b. Seed large, > 40 x 40 x 18 mm 9
2 a. Areole open (even if only 0.5 mm) near the hilum 3
b. Areole completely closed making a continuous ring 6

3 a. Areole heart-shaped . . . . *E. bacillaris*
b. Areole elliptic 4
4 a. Seed small, < 10 x 8 mm areole ends far apart c. 2 mm . *E. dolichorrhachis*
b. Seed large, > 11 x 9 mm areole ends close c. 1 mm 5
5 a. Seed small, c. 11 x 10 mm, areole ends far apart, coat with mucilage when fresh . . . . *E. leptostachya*
b. Seed large, c. 20 x 11 mm, areole ends almost touching, coat without mucilage . . . . *E. camerunensis*
6 a. Areole elliptic . . . . *E. abyssinica*
b. Areole tapered at least on one end 7
7 a. Areole tapered at both ends . . . . *E. chrysostachys*
b. Areole tapered at one end only 8
8 a. Areole tapered towards hilum only . . . . *E. spinescens*
b. Areole tapered towards distal end only . . . . *E. mossambicensis*
9 a. Seed thickness > 19 mm, L/W ratio < 0.9 . . . . *E. rheedii*
b. Seed thickness ≤ 19 mm, L/W ratio > 1.0 . . . . *E. gigas*

Seed descriptions

E. abyssinica Steud. ex. A. Rich.: Seeds brown; rarely pale brown; elliptic or oval 10-13.5 x 7.5-10.5 x 3-3.5 mm; areole an open elliptic C-shape.

E. bacillaris F. White: Seeds brown; circular or oval; 14.5 x 12.7 x 3.4 mm (only seed available); areole heart-shaped, fading out near the hilum.

E. camerunensis J. Villiers: Seeds brown; elliptic rarely oval, unusually thick with thick smooth edges; 18.8-21.8 x 11.2-12.7 x 3.9-4.5 mm; areole open with the opening very close to the hilum.

E. chrysostachys (Benth.) Drake: Seeds brown; elliptic or oval with thin edges; 14-17.7 x 10.5-12.8 x 3.3-4.0 mm; areole open, elliptic, tapering towards both distal and hilar ends.

E. dolichorrhachis Brenan: Seeds brown; circular or oval; 8.6-9 x 5.5-7.8 x 2.5-2.7 mm; areole C-shaped with sharp ends far from the hilum.

E. gigas (L.) Fawcett & Rendle: Seeds brown; circular or oval; longer than wide; 40-47.8 x (38-)40-43 x 17-19.3 mm; areole absent. The seed tapers off towards the edges to form a thin border; much heavier compared to areolate *Entada* species.

E. leptostachya Harms: Seeds brown; with signs of mucilage on coat; elliptic or oval; 10.4-12.9 x 9.5-10.6 x 3.5-3.7 mm; areole open, elliptic, tapered towards both distal and hilar ends.

E. mossambicensis Torre: Seeds dark brown; rounded or oval; hilum end broader than the distal end; c. 10.4 x 9.2 x 2.3 mm (only seed available); areole closed and broader near the hilum than the distal end.

E. rheedii Spreng. (= *E. phaseoloides* (L.) Merr.; *E. pursaetha* DC.): Seeds brown; circular or oval; shorter than wide; 40.8-44.7 x 44.3-53.3 x 17.1-23 mm; areole absent; broad edges and of even thickness; very heavy compared to seeds of areolate *Entada* species.

E. spinescens Brenan: Seeds dark brown outside the areole and medium-brown within; oval, tapered towards the hilum; distal ends angled; 10.9 x 6.9 x 3.5 mm (only seed available); areole closed and narrow, in the centre of the seed.

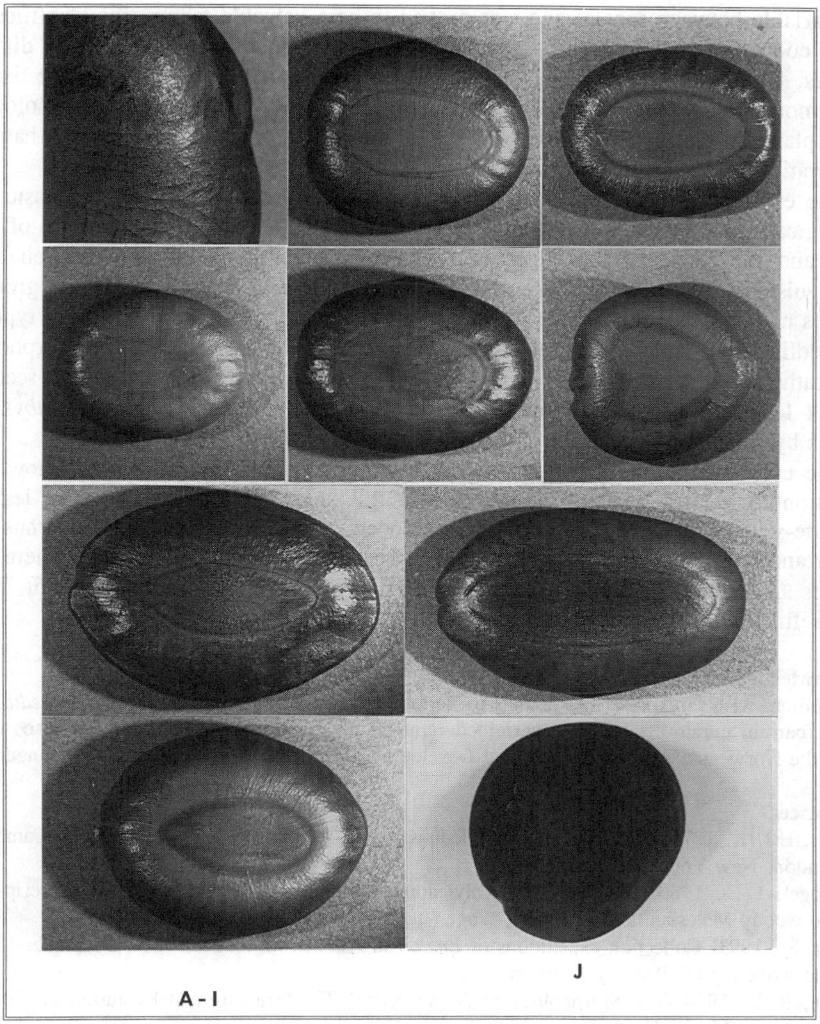

Figure 1. Seeds of *Entada*:
Entada gigas; *E. abyssinica* (SL 005); *E. abyssinica* (SL 004);
E. abyssinica (SL 008); *E. abyssinica* (SL 012); *E. bacillaris* (SL 021);
E. chrysostachys (SL 010); *E. camerunensis* (SL 022);
E. chrysostachys (SL 031); *E. gigas*.
All photos x2.5, except *E. gigas*, bottom right: x0.7.

Discussion

The seed morphology of *Entada* falls into two major categories; small and areolate, or large and exareolate. These categories distinguish the sections *Entada* and *Neoentada* respectively.

The shape, size and position of the areole was seen to be constant within a species but variable between species and was proved to be a valuable feature for identification. Field collections of *Entada* (SL 010 and SL 031) which seemed to be of different species, based on morphology in the dry season, appeared conspecific on the basis of seed morphology. This was later proven on the basis of vegetative morphology of these plants in cultivation. In this case, seed morphology was more valuable than field observations of the adult plant when leafless.

The exareolate species sampled so far can also be distinguished on the basic morphological features. The length versus width (90° to length on flat surface of seed) ratio and the shape of the edges of the seed (tapered or blunt) proved reliable in distinguishing *E. gigas* from *E. rheedii*. More extensive sampling of this group of species may enhance the fairly low number of characters available in this seed type.

Seedling morphology improves the species resolving power of seed morphology. The rather similar seeds of *E. camerunensis* and *E. chrysostachys* produce seedlings with 4-11 or 14-17 pairs of pinnae on the first leaf pair respectively. *E. abyssinica* differs by having hypogeal germination and an alternate leaf arrangement.

The combination of seed and seedling morphology is demonstrated to allow identification to species level of all the taxa in this study. It predicted that, at least for areolate-seeded taxa the key could be extended to encompass the whole genus. The importance of identifying species from their seeds and seedlings is evident where these species are to be grown on a large scale for e.g. agroforestry. Some species of *Entada* are useful for agroforestry purposes.

Acknowledgements
The authors wish to express their thanks to Bernie Hyland for providing seeds of *E. rheedii* and to the herbarium curators who have permitted removal of seeds from specimens. We also wish to thank the Norwegian Aid for International Development (NORAD) for support to Mr. Lungu.

References

Corner, E.J.H., 1976. The seeds of Dicotyledons Vol. 1. Cambridge University Press. Cambridge, London, New York, Melbourne.

De Vogel, E.F., 1980. Seedlings of Dicotyledons: Structure, Development, Type, Description of 150 woody Malesian taxa. PUDOC, Wageningen.

Lungu, S., 1993. Collection expeditions of *Entada* in Zambia 4th August - 31st December 1992: A field work report. Reading University.

Sporne, R.K., 1974. The Morphology of Angiosperms: The Structure and Evolution of Flowering Plants. Hutchinson Univ. Library. London.

Développements récents de l'Herbier de Dakar

B. SAMBOU [1], A.T. BÂ [1,2], D. DIONE [2], A. GOUDIABY [1],
J.E. MADSEN [1,3] & A.S. TRAORÉ [2]
[1] Institut des Sciences de l'Environnement, Faculté des Sciences et Techniques, Université Cheikh Anta Diop, Dakar, Sénégal.
[2] Département de Biologie Végétale, Faculté des Sciences et Techniques, Université Cheikh Anta Diop, Dakar, Sénégal.
[3] Department of Systematic Botany, Institute of Biology, Aarhus University, Nordlandsvej 68, 8240 Risskov, Denmark.

Résumé
Créé en 1960, l'herbier de Dakar renferme environ 8000 accessions. Comme pour d'autres herbiers d'Afrique, ses activités ont été fortement limitées pendant cette dernière décennie au point que certains spécimens soient attaqués par les insectes et que la structure se retrouve dans un état d'isolement notoire. Grâce à l'appui d'un projet de coopération avec l'Université de Aarhus (Danemark), cet herbier est en voie de réfection. Les principales innovations depuis un an ont porté sur la formation, les infrastructures et équipements et l'introduction de nouveaux spécimens.

Abstract
The DAKAR Herbarium created in 1960 holds approximately 8000 plant specimens. Like many other African herbaria, activities have been greatly limited during the last decade due to lack of funding which again has lead to scientific isolation. As part of a collaboration with the University of Aarhus (Denmark) the herbarium is now being renovated and it will soon be fully operative. The main innovations during the last year included training of researchers, implementation of infrastructure and equipment, and addition of new specimens.

Key words
Herbarium, Renovation, Senegal.

Introduction

La vitesse de destruction des formations végétales de la zone tropicale est devenue une des principales inquiétudes des institutions nationales et internationales soucieuses de la conservation de la biodiversité. Ces inquiétudes sont d'autant plus fondées que certaines espèces de cette zone ne sont pas encore bien connues et que les mesures de protection préconisées jusqu'ici sont restées peu ou pas efficaces. Face à ces difficultés, les herbiers apparaissent de plus en plus comme des structures devant jouer un rôle clé dans la conservation et la connaissance de la flore de cette zone.

En Afrique, 22% des pays ne disposent pas encore d'herbiers et 78% des collections se retrouvent dans 7 pays seulement. Il s'agit de l'Afrique du Sud, du Kenya, de l'Egypte, du Zimbabwé, de l'Algérie, du Zaïre et du Nigéria (Holmgren et al. 1990). Bien que 90% environ des espèces végétales du monde se développent en zone tropicale (Fosberg & Sachet 1965), le nombre de spécimens pour l'ensemble des herbiers du continent africain est inférieur à celui de certains herbiers d'Europe pris individuellement. Pour des raisons historiques (colonisation), les spécimens conservés dans les herbiers des pays d'Afrique sont généralement récents et les types relativement peu nombreux. Les plus importants spécimens de la zone tropicale sont conservés

L.J.G. van der Maesen et al. (eds.), The Biodiversity of African Plants, 815-818.

dans certains herbiers d'Europe et d'Amérique (Campbell & Hammond 1989).

Avec 118.000 échantillons, le Sénégal occupe, au plan du nombre de spécimens conservés dans les herbiers, la première place en Afrique occidentale et la neuvième sur le continent. L'herbier de l'Institut Français d'Afrique Noire (IFAN) devenu par la suite Institut Fondamentale d'Afrique Noire est le plus important en Afrique de l'Ouest (110.000 échantillons). Il a été mis en place en 1942 à Dakar pour abriter les collections de l'ensemble des pays francophones d'Afrique Noire. Toutefois, le chiffre avancé dans la huitième édition de l'Index Herbariorum concernant le nombre de types pour les herbiers du Sénégal (7000 types) est certainement une erreur.

L'objet de ce document est d'informer la communauté scientifique des développements récents de l'herbier de Dakar.

Historique de l'herbier de Dakar

Créé en 1960 par le Professeur J. Miège, l'herbier de Dakar (8000 feuilles) est le deuxième herbier du Sénégal et fait partie de l'Université de Dakar.

Parmi les auteurs qui ont déposé des collections dans cet herbier, on peut citer J. Berhaut qui y a laissé environ 2000 échantillons dont la plupart ont été citées dans la Flore Illustrée du Sénégal. F. Doumbia qui a étudié les forêts de Basse Casamance au Sud de Ziguinchor (Doumbia 1966) a laissé beaucoup d'échantillons incomplètes suite à une mort subite par accident. L'herbier renferme aussi des spécimens de A. Raynal (Raynal 1963) et des collections de Cyperaceae de M. Bodard (Bodard 1963). O.E. Kane (ancien technicien de l'herbier) y a également déposé un nombre d'échantillons appréciable. Nous avons retrouvé de même quelques collections de J. Miège. Il est à noter que l'herbier de Dakar ne dispose que de quelques spécimens types.

Pendant ces dix dernières années, les activités de l'herbier avaient été fortement limitées au point d'entraîner l'attaque de certains échantillons par les insectes. L'inexistence quasi totale de collaboration avec d'autres institutions spécialisées dans le domaine de la botanique a entraîné la structure dans un isolement notoire.

Les développements récents de l'herbier

A partir de 1993, l'herbier de Dakar a bénéficié de l'appui d'un projet de coopération inter-universitaire avec l'Université de Aarhus. Entre autres objectifs, ce projet a inscrit dans son programme la réfection des locaux de l'herbier, la formation de chercheurs en aménagement et en gestion des herbiers, le réarrangement de la collection, l'augmentation du nombre de spécimens et l'enrichissement de l'herbier, la relance de la collaboration.

Infrastructures et équipements

Les premiers travaux de rénovation de l'herbier ont porté sur la réfection du bâtiment. Ainsi, la salle de collection et deux autres pièces transformées en bureaux ont été rebadigeonnées. La climatisation, condition pour une bonne conservation du matériel végétal en climat tropical a été remise en marche. Pour un meilleur éclairage, des lampes néon ont été installées dans la salle de collection et dans les bureaux. Deux grands séchoirs électriques ont été installés dans une pièce qui fait office de salle de séchage. Ces premiers travaux de réfection ont d'ores et déjà rendu les conditions de travail acceptables dans cet herbier.

Activités

L'organisation d'un cours en aménagement et gestion des herbiers dispensé par le Directeur de l'herbier de l'Université de Aarhus a permis de former huit personnes dont quatre chercheurs. Le technicien de l'herbier de Dakar a effectué un stage d'un mois à l'herbier de Aarhus. Une équipe de quatre personnes s'occupe du montage des échantillons. Le matériel de montage et de conservation des échantillons est disponible sur le marché local.

L'une des premières tâches du groupe de chercheurs dans l'herbier a été la vérification de l'état de toutes les collections. Ces collections qui avaient été simplement rassemblées en fonction des collecteurs ont été regroupées en familles, elles-même classées par ordre alphabétique.

Les chercheurs effectuent de façon régulière (en moyenne une fois par mois) des missions de collecte d'échantillons à travers les différentes zones écogéographiques du pays et plus particulièrement dans la partie sud-est. Cette partie représente l'une des zones les plus diversifiées du pays au plan floristique.

Par ailleurs, certains échantillons vivants comme les géophytes sont conservés dans le jardin botanique.

En moins de deux ans d'activité, plus de 3000 nouvelles échantillons ont été introduites dans l'herbier. Une base de données déjà fonctionnelle facilitera au terme de la première phase de rénovation (dans 6 mois) la gestion de l'herbier et l'utilisation de l'information par les chercheurs. Cette base de données a été mise à jour avec les travaux de Berhaut (1971-1988), Hutchinson & Dalziel (1927-1972) et Lebrun (1973).

Collaborations

L'herbier de Dakar souhaiterait collaborer avec d'autres institutions spécialisées dans le domaine de la botanique. Cette collaboration pourrait revêtir diverses formes telles que l'accueil de chercheurs, l'échange de matériel végétal et de documents. En particulier, l'herbier de Dakar envisage de développer les échanges de collections avec d'autres herbiers. Il pourrait accueillir des visiteurs ou des chercheurs à partir du mois de janvier 1995.

Conclusion et perspectives

Au rythme de la disparition des espèces végétales dans la zone tropicale, les responsables des herbiers des pays tropicaux devraient se focaliser en priorité et de façon urgente sur la collecte et la conservation des espèces mal connues ou menacées. Plus d'efforts devraient être déployés par la communauté scientifique pour créer et/ou supporter les herbiers des pays tropicaux en dépit de certains problèmes d'entretien de telles institutions dans cette zone.

Avec les réfections entreprises depuis bientôt deux ans, la redynamisation des activités internes et des relations avec d'autres institutions, l'herbier de Dakar, qui a l'ambition de devenir un herbier moderne, a retrouvé des conditions de travail acceptables. En dépit du caractère peu spacieux de son local, l'herbier souhaite développer une collaboration active avec d'autres institutions. L'accueil de chercheurs et l'échange de spécimens de la sous-région avec d'autres herbiers représentent des priorités. Des moyens ont d'ailleurs été dégagés pour l'envoi d'échantillons. En plus de sa vocation principale de conservation du patrimoine végétal, cet herbier, organe d'une structure universitaire, se veut un instrument didactique et un cadre de recherche. Toutefois, la réalisation de ces ambitions dépend aussi des autres institutions collaboratrices.

Remerciements
Nos remerciements s'adressent d'abord à l'agence de coopération danoise (DANIDA) dont le financement a permis de réfectionner l'herbier de Dakar. Nous remercions également le Prof. A. Nongonierma, M.A. Sanokho et Dr. H. Balslev pour les informations qu'ils ont bien voulu nous livrer et les observations faites sur notre manuscrit.

Bibliographie

Berhaut, J., 1971-1988. Flore illustrée du Sénégal. Clairafrique, Dakar.

Bodard, M., 1963. Première contribution à la révision du genre Bulbostylis (Cypéracées) en Afrique. Anal. Fac. Sci. Univ. Dakar ser. vég. 9(2): 51-80.

Campbell, D.G. & D.H. Hammond, 1989. Floristic inventory of tropical country: The status of plant systematics, collections, and vegetation, plus recommendations for the future. The New York Botanical Garden.

Doumbia, F., 1966. Etude des forêts de Basse Casamance au Sud de Ziguinchor. Anal. Fac. Sci. ser. vég. 3(19): 61-100.

Fosberg, R.F. & M.-H. Sachet, 1965. Manual for tropical herbaria. Reg. Veg. 39.

Holmgren, P.K., H.N. Holmgren & C.L. Barnett, 1990. Index Herbariorum - Part 1: The herbaria of the World. Ed. 8. New York Botanical Garden: 290.

Hutchinson, J. & J.M. Dalziel, 1927-1936. Flora of West Tropical Africa. 2nd ed. revised by R.W.J. Keay and ed. by F.N. Hepper (1954-1972). Crown Agents, London.

Lebrun, J.P., 1973. Enumération des plantes vasculaires du Sénégal. Etude Botanique N° 2. Institut d'Elevage et de Medecine Vétérinaire des Pays Tropicaux, Maison-Alfort, Val de Marne.

Raynal, A., 1963. Flore et végétation des environs de Kayar (Sénégal): de la côte au lac Tamna. Anal Fac. Sci. ser. vég. 9(2): 121-231.

Tree-climbing on a free rope

J.J. WIERINGA
Department of Plant Taxonomy, Wageningen Agricultural University,
P.O. Box 8010, NL 6700 ED Wageningen, the Netherlands.

Abstract
A quick, relatively easy method to climb trees is described. In order to obtain sufficient and complete voucher specimens of tall forest trees, the free-rope climbing method with mountain climbing gear is appropriate. The method was demonstrated during the XIVth AETFAT Congress in a poplar tree behind the Wageningen International Congress Centre.

Key words
Collection methods, Free-rope climbing, Giant forest trees.

Collections of large forest trees

Of all the plant-forms growing in the tropical rain forest, strangely enough, the biology of large forest trees is relatively poorly known. Although these trees are the most conspicuous life forms present in those forests, and by itself easy to find, the number of specimens of most of these species is usually very low in herbarium collections, and a high percentage consists of poor or sterile material. The reason, of course, is their size; most botanists are not able to collect parts that are located in the canopy.

With clipper-poles one reaches up to 12 metres. Small trees (and lianas growing in them) can be collected by cutting down the tree. As most botanists are not equipped with a chain-saw, any tree with a DBH larger than 40 cm is impracticable to cut down. Many people dislike this rather destructive method anyway. It is evident that large trees may be obtained at sites were logging companies are felling trees. This, however, is often also rather unsatisfactory. The main reasons are: foresters usually cut a limited number of better-known species. Rare species are not wanted by the trade, unless these are removed for road construction. The foresters log most trees during the dry season which - in Equatorial Africa - is usually about the poorest period to collect flowering material; one is completely dependent on the location where logging is going on, and trees of botanical interest nearby cannot be sampled.

The only potential alternative of collecting good material from large trees is by climbing them. The trunk may be climbed using climbing irons or steel-spiked boots. However, this method has some disadvantages. If the aim is to obtain material from really large trees, climbing a trunk becomes more difficult if the diameter of the tree surpasses 50 cm DBH. If you are climbing a trunk, you often cannot get further than the first branch. From this point some leaved branches may be reached with clipper-poles, but since most trees are flowering only on the outer edge of their crown (often some 15 metres away from the first branch), flowers are hard to reach. By climbing a slim trunk next to the large tree you have less problems with the diameter, and probably will be able to get past the first branches, but flowers are still difficult to get. A disadvantage of this method is also the damage it causes. If the targeted tree has to be further monitored, the initial climb will have caused some damage, that may affect growth rates and possibly other features.

L.J.G. van der Maesen et al. (eds.), The Biodiversity of African Plants, 819-821.

Since I am working on a group of huge Caesalpinioideae trees, with diameters up to 2 m and heights up to 60 m, it became necessary to have a good technique to reach the flowering material I so desperately needed. I worked in Gabon and Cameroon, and here a good and economic option proved to be the free-rope climbing method. One can be fully equipped for less than $1000, and is very flexible as regards travel: the equipment fits in any vehicle.

Material and methods

Materials needed are: a cross- or handbow, arrows, 100 m of 0.5 mm fishing line (>160 N), a fishing mill, 80 m nylon rope (c. 3 mm), 70-80 m of climbing rope as used in speleology (mountain climbing rope is too elastic), a harness, c. 10 snap hooks, 2 shunts, a "figure eight" or "descendeur", some "Prusik-cords" of which at least one is 2 m long and another 4 m, a band of at least 16 m long, and some smaller ones, some lead plates and adhesive tape.

With the crossbow an arrow, connected to the 0.5 mm fishing line, is shot over the desired branch or crown. The arrow should be made heavier by tying or taping some lead to its tip. I used a 150 lbs (70 kg) crossbow and experience proves that 75 g arrows are most useful: a heavier arrow will not be shot away properly, a lighter arrow always gives more problems pulling the fishing line over the branches. If the arrow does not return to the ground by itself, short pulls followed by a swift release of the fishing line usually help. On the ground the arrow is disconnected and replaced by the 3 mm nylon rope. The fishing line is wound back on the mill, and the nylon rope pulled into the tree and back again. The use of this smaller rope is essential: fishing wire is not strong enough to pull a heavy climbing rope. With the nylon rope the climbing rope can then be hoisted into the tree. To prevent obstruction of the large rope by a forked branch, it is recommended to make such a knot that the nylon rope is connected to the end of the straight climbing rope. This can be realized by employing a Prusik-knot. To make the transition smoother, the nylon rope can be taped to the end of the climbing rope. Once this (permanent) connection is made, the nylon cord can be cut off a metre from the knot, after the climbing cord has past, facilitating the connection for a next climb. With the climbing rope in place, one end has to be secured to a tree or a car. For this purpose the 16 m circular band can be laid around a trunk and be connected to the rope.

Now the climb can start. Put on the harness, connect one shunt via the 2 m circular (so 1 m long double) Prusik cord to the harness. Pull the rope till it becomes fixed and connect the shunt. After the line has been tested by pulling to assure it can hold your own weight, connect the second shunt to the cord and connect the 4 m circular Prusik cord to both feet. An extra security Prusik connection can be made between harness and rope. By standing in the feet-cord the harness shunt can be hauled up, and when hanging from this shunt, you can pull in your legs and haul up the second shunt. This way you climb a few metres. When you are a few metres above the ground, a helper should try to hang on the rope as well to check if the supporting branch is strong enough. If it can hold two men, you can proceed. To be able to pull up clipper-poles or other equipment, it is advisable to connect the nylon cord to the harness to hoist the required items when needed.

While going up, often other interesting things will be seen, such as smaller flowering trees, lianas, and epiphytes. It is advisable to collect these on the way up rather than while going down. If these are out of reach, the helper on the ground can usually swing the rope for better approach. When the canopy is reached and the specimens are taken (with or without clipper-poles), connect the descendeur or "eight", disconnect the shunt and slide down to the ground. On the way back some branches which were thrown down, but got stuck halfway, may be picked up.

Results

With the free-rope climbing method a quantity of first-rate plant material has been collected during 1992-1994, including the first flowering specimens ever collected of some species, and even material that proved to belong to hitherto undescribed species. These novelties not only comprise tall trees, but also some of the lianas, epiphytes and hemiparasites collected in the canopy. Moreover, observations on e.g. anthesis, pollinators and other insects are made possible, and the method enables photography in-situ.

If you are not afraid of heights, it is a method very worthwhile to try. I suggest: give it a (cross-bow) shot!

Index to scientific names

Index of species, genera, and family names of plants, animals and other life forms. Some syntaxonomic units (vegetation associations) are also included.

Index to authors